Microelectronic Failure Analysis

Desk Reference
Fourth Edition

Technical Editors
Richard J. Ross
Christian Boit

Executive Editor
Donald Staab

Electronic Device Failure Analysis Society
An Affiliate Society of ASM International

**The Materials
Information Society**

Published by
ASM International®
Materials Park, Ohio 44073-0002

Library of Congress Catalog Card Number: 99-075422
ISBN: 0-87170-638-5
SAN: 204-7586

ASM International®
Materials Park, OH 44073-0002

Printed in the United States of America

PHOTONICS TO ATOMICS

The cover page depicts state-of-the-art fault isolation techniques, such as Scanning Probe Microscopy, used in the failure analysis of an EPROM that failed standby Idd. The photo emission image [Top Left] shows an n-channel transistor in saturation due to a gate short associated with the same long poly row decoder. The gate short is in an n-well area tied to the source side of a p-channel transistor. Since the gate short is to Vdd (n-well), the n-channel is saturated and cannot pull the poly line low. The Scanning Thermal Microscopy (STHM) image [Middle Left] shows the thermal energy liberated by the gate short in reference to the topography [Bottom Left]. The Electric Force Microscopy (EFM) image is represented in two parts:

A. The associated topography to the EFM image is shown [Top Right].
B. The associated node of the saturated transistor is stuck-high as shown by the arrow [Bottom Right].

Acknowledgement: We are grateful to Jim Colvin, Wafer Scale Integration and Yale Strausser, Digital Instruments, for sharing the photos and the text. We also thank Preston Scott of Texas Instruments for designing it. Photonics is a tradename of *Photonics Spectra* magazine.

Foreword

Failure analysis of electronic systems began around the mid-1960's. Component technology was less complex at that time as compared to the present. Usually, when a failure analysis program is instituted, the majority of failures first encountered are less complicated and more obvious. It was realized that the cause for failure was not known when the usual "remove and replace" philosophy left many unanswered questions. Demands for greater system performance and reliability supported the need to identify the cause of reduced system performance and shortened operational life. Initially analysis techniques were very limited and most had to be developed out of necessity. Then, as now, analysis techniques lag behind the analyst's needs. It seems the analytical technology is 1–2 generations behind the new component technologies. In other instances the needed instrumental resolutions or sensitivities are not what we would like to have[1-3].

An individual component's performance and reliability capability is a given at the completion of its' manufacturing process. This capability can be described as a multi-dimensional envelope that includes voltage, temperature, environmental and physical stresses with respect to time. Likewise, the application system requirements can be described as a multi-dimensional envelope. When the component's capability envelope resides within the application-system-requirements-envelope you should have desired system performance and reliability. The key to success is in verifying the components-capability-envelope and accurately determining the application-systems-requirements-envelope. In the end, if all the multitude of controls, testing, screening, inspection, analyses, etc. do not provide successful performance, the practical solution is root cause identification through failure analysis. Failure analysis can identify what action is required to correct the problem.

Failure analysis provides many challenges. Some can appear insurmountable at times. This must not be a deterrent. Failure analysis covers all technical disciplines. Important attributes of a failure analyst are: Patience, Persistence, Care and Quality of workmanship, and a Can-do-attitude. The failure characteristics must dictate the course of the analysis. Care must be taken to avoid masking or destroying crucial data. Jumping to conclusions or making assumptions will usually produce inaccurate analysis. Know your limitations and seek support from knowledgeable persons when needed. One must avoid getting too involved with the details of the analysis such that the obvious is overlooked. Once an incorrect assessment or interpretation is made, it is very difficult to realize it. Also, can the analysis findings stand up to the test of proof by contradiction?

Component complexity and reliability demands continue to grow and the key factor for all products is cost. Failure analysis is an important facet in supporting these factors through the identification of the root cause of failure. Through the implementation of timely and accurate corrective action we can meet the needs of the future.

To reach the goals of future analysis, it is important to share your experiences, new or improved analytical techniques and new failure mechanisms with the technical community. No organization can afford not to share and exchange their information. We can only be successful by working and sharing our information together. It is through this process that the information collected here in this Seminar Reference, could be made available to you. It is my hope that this information will be used as a foundation to develop new and improved analytical techniques and that you will share your work with our community. The purpose of ISTFA is to provide a forum for the interchange and dissemination of relevant information.

Jim Beall

---

[1] Evening Panel Discussion–Failure Analysis for the 90's and beyond (Back to the future!), ISTFA '92, p 443.
[2] Evening Panel Discussion–Failure Analysts for the 90's and beyond (Back to the future II), ISTFA '93, p 431.
[3] The Semiconductor failure analysis roadmap, ISTFA '95, p 1.

Preface

As the art and science of electronics move forward into the twenty-first century, the art and science of Electronic Device Failure Analysis must also advance to insure the timely introduction of new materials, processes, technologies, and products into the marketplace. The goals of this new edition of the Electronics Failure Analysis Desk Reference are to provide an overview to newcomers to the field and to provide a reference tool for the experienced analyst. The vast majority of the material in this edition has been written and/or updated within the past twelve to eighteen months, and the format has been revised to better reflect the flow of the Failure Analysis process. Sections have been added on device physics, circuit operation, test, and theory (as well as practice) of operation of many analytical tools and techniques. The editors have solicited tutorial and practical material from highly regarded experts in their respective fields, and we thank all of our authors and reviewers for their willingness and cooperation in preparing this work.

No project of this magnitude can be completed without the support and hard work of many people. We wish specifically to acknowledge the hard work of Ms. Bonnie Sanders of ASM International in preparing and publishing this book; the Board of Directors of the Electronic Device Failure Analysis Society (EDFAS) for their guidance and support; and our Executive Editor, Mr. Donald Staab, for his perseverance and forbearance as we strove to close the divergence of gathering new material and meeting publication schedules. We also wish to acknowledge our employers and families for their understanding and support of the time and effort involved in an undertaking of this scope.

Richard J. Ross
Christian Boit

Contents

Introduction/Overview

Test and Fail Verification

Decapsulation and Package Analysis

Fault Localization

Electrical Techniques

Electron/Ion Beam-Based Techniques

Thermal Techniques

Photonic Techniques

Deprocessing and Sample Preparation

Physical/Chemical Defect Characterization

Discrete/Passive Component Analysis

Failure Modes and Failure Mechanisms

Advanced Techniques

Laboratory Operations and Management

Afterword

Appendixes

Introduction/Overview

Failure Analysis Overview

Seshu V. Pabbisetty
Texas Instruments, Semiconductor Group
Stafford, Texas

ABSTRACT

This overview will cover the entire range of failure analysis activity in the semiconductor industry with emphasis on some of the joint efforts, technology directions, current and future strategies, and philosophies for failure analysis organizations. The impact of technology and the tools of the future will be examined along with the key challenges that must be overcome in order to prepare for the ever increasing demands of failure analysis in the new millennium.

INTRODUCTION

The term failure analysis (FA) has taken upon a myriad of meanings, depending upon the context. Ultimately, it entails the analysis of Integrated Circuit (IC) devices which have failed in qualification testing, engineering evaluations and customer applications.[1] In a wafer fab context, failure analysis may bring to mind physically or electrically isolating failures to determine process issues in current designs. To a designer, it may infer debug activities in new designs. Regardless of the perspective, however, failure analysis is an activity which involves many people operating strategically throughout a corporation.

The failure analysis flow is basically the same in all operations. There are several key subprocesses which comprise the complete failure analysis process. These are: Electrical Characterization and Verification, Failsite Isolation, Deprocessing, Defect Characterization, and Corrective Action. Forecasting the challenges that lie ahead in failure analysis necessitates careful examination of each of these subprocesses, as well as the technologies which drive the changes.

NEW MILLENNIUM FA CHALLENGES: GENERAL CONCEPTS

Failure analysis has long been considered a highly valued operation in the rapidly developing semiconductor industry for its role in accelerating the pace of improvements in performance and fabrication while reducing the cost of integrated circuits. Failures can occur at all stages of IC development, qualification, fabrication, and application; thus, conclusive failure analysis results are critical to effective problem solving in all of these phases. Despite the ongoing development of new tools and techniques, trends in the semiconductor industry increasingly challenge the ability to locate and identify defects in dense circuitry.

Semiconductor technology roadmaps point to a number of trends that will have a significant impact on FA capabilities. Transistor size, interconnect wiring dimensions, and power supply voltages have continued to decrease. Conversely, chip speed, transistor count, wiring density, the number of interconnect layers, and the number of input/outputs is on the increase. The biannual National Technology Roadmap for Semiconductors[2] (published by the Semiconductor Industry Association (SIA) provides a detailed picture of industry needs. Table I summarizes the key failure analysis drivers extracted from the SIA roadmap.[3] The road map was intended to provide a general outline of the anticipated capabilities that will be required through the year 2007 in the area of Complementary Metal-Oxide Semiconductor(CMOS) Ultra-Large Scale Integration (ULSI) failure analysis. The intended beneficiaries of this roadmap are the companies that supply FA capabilities, semiconductor FA operations and their management, and strategic planning organizations within the semiconductor companies such as design, test, and process development.

As table I reveals, there are many effects of IC technology enhancements that must be overcome from a failure analysis standpoint. Integrated circuits with smaller features, higher operating speeds, and lower supply voltages will naturally be sensitive to more subtle defects. Defects which were not prominent enough to affect the operation of larger geometries will have a significant impact as IC enhancements continue. As the subtlety of the failures increases, the ability to find them decreases. Failure analysis is sometimes rightly compared to looking for a needle in a haystack, and in this case, the haystack is getting larger, while the needle is getting smaller.[4] Reflecting on the trends that have been established, it is apparent that device technology developments will continue to push FA capabilities to their limit. Different packaging methods have also imposed limitations on the analyst's ability to physically locate defects. Packaging technologies like the "flip-chip," which sandwich active circuitry between the silicon die and the plastic or ceramic carrier are becoming more and more

Intro to manufacture date	1995	1998	2001	2004	2007
Minimum feature size, microns	0.35	0.25	.18	.12	0.1
Memory size (DRAM), MB	64M	256M	1G	4G	16G
Gates per chip	800K	2M	5M	10M	20M
Interconnect levels	4 to 5	5	5 to 6	6	6 to 7
Max power - high performance proc., W	15	30	40	40-120	40-200
Max power - portable proc., W	4	4	4	4	4
Min supply voltage, V	2.2	2.2	1.5	1.5	1.5
Number of I/O pins	750	1500	2000	3500	5000
Speed of processor - off chip, MHz	100	175	250	350	500
Speed of processor - on chip, MHz	200	350	500	700	1000
Chip size, microprocessor, mm sq	400	600	800	1000	1750
Chip size, DRAM, mm sq	200	370	500	750	1000
Package Technology, processor (leading)	Flip-chip	Flip-chip	Flip-chip	Flip-chip	Flip-chip
Package Technology, memory (leading)	DIP	Cube?/Others	Cube?/Others	Cube?/Others	Cube?/Others
Materials - Gate ox thickness Tox-eff, nm	7-12	4-6	4-5	4-5	<4
Materials - Metallization	Al	Al, Cu	Al, Cu	Al, Cu	Al, Cu
Materials - ILD	Oxide, Air, Polymide, Dielectric				
Design Methodology - Database	Flat	Hierarch	Hierarch	Hierarch	Hierarch
Design Methodology - Model	Behavioral	Behavioral	Symbolic	Symbolic	Megablock

The row labels are grouped as: **PRIMARY ROADMAP DRIVERS** (from Minimum feature size through Chip size, DRAM) and **ADDITIONAL DRIVERS** (from Package Technology through Design Methodology - Model).

Table I. Key failure analysis drivers extracted from the SIA roadmap[3]

prevalent. A number of traditional laboratory techniques are fast becoming obsolete with these new packages. Thus, the key challenges for future failure analysis capabilities will be in developing failsite isolation techniques for dense flip-chip parts, subtle fail mechanisms, and gigahertz operating speeds.[4]

Other challenges and limitations imposed on the FA community by way of these ever changing device properties are: i) Rise time limitations and stray impedance effects, ii) Fixturing costs and interconnection challenges, iii) The need for new tooling/re-mounting/re-bonding methods, iv) Masking and inaccessibility (i.e. signal nodes), v) Fail-site isolation, vi) De-processing, vii) Communication, and viii) Navigation. The increasing trend of global and inter-corporation product strategies augments the information challenge. Additionally, the ever shortening time-to-market goals call for a paradigm shift in the whole FA process. These goals include an increasing demand for better quality and reliability (lower failure rates) and better and faster resolution of failures leading to effective corrective actions and containment.

One such paradigm shift was seen in the formation of the Product Analysis Forum (PAF), under the auspices of the SEMATECH Quality Council.[3,4] This forum provides a platform for semiconductor failure analysis professionals of SEMATECH member companies to drive solutions to common issues affecting the successful completion of analysis and satisfaction by the end customer. Its strategies and operating guidelines are illustrated in Fig. 1. The Forum has recently been involved in a call for research and development of evolutionary and revolutionary approaches to

turn scientific concepts into fully-engineered and supported laboratory tools for routine use.

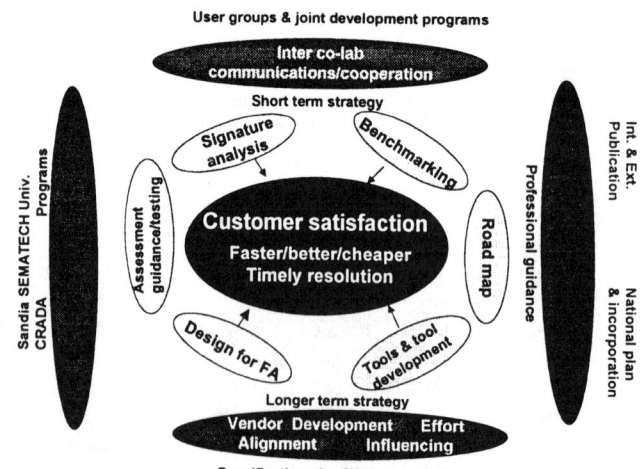

Fig. 1. SEMATECH PAF Strategies

Recent years have also seen an increased level of focus on FA at symposia and professional society meetings. The failure analysis content has been steadily rising at SPIE (Society of Photo-Optical Instrumentation Engineers) symposia and Institute of Electronic and Electrical Engineers (IEEE) meetings such as the International Reliability Physics Symposium (IRPS). A session at ITC (International Test Conference) was dedicated to FA activities in

failure analysts for a broad discussion of topics, such as trends and likely challenges, the key technological breakthroughs required, and cross-functional/leveraged concepts.[3-8]

FA - PROCESS FLOW

To thoroughly understand the challenges that lie ahead for FA, the subprocesses must be examined. The FA process can be seen as a linear progression of five different stages, as illustrated in Fig. 2. The FA road map is contingent upon the successful adaptation of each of these stages.

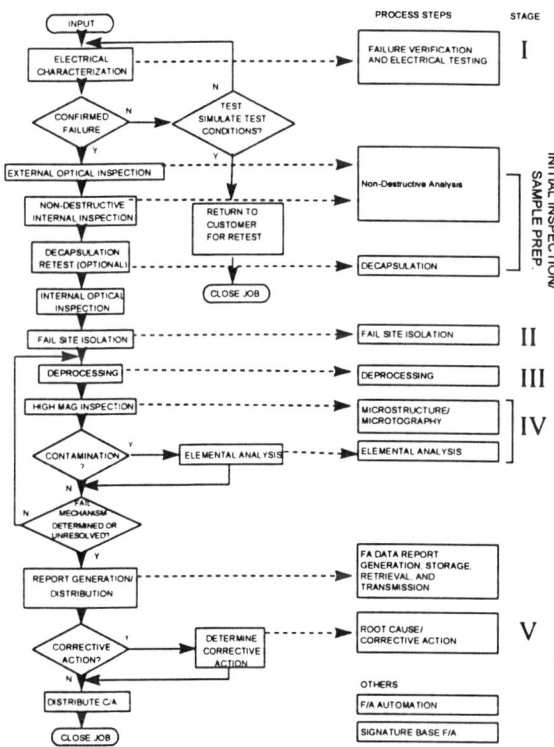

Fig. 2. A Typical Failure Analysis Flow Chart With Key Processes Identified

Failure Verification and Electrical Validity

The first stage in the FA process is to assess the validity of the failure. The goals in this process are not only to provide verification of the failure, but also to provide fail site isolation where possible. The initial electrical characterization usually occurs at the production ATE (Automatic Test Equipment), and these results can be verified using a laboratory bench tester. For packaged devices, the ATE will continue to be the method of choice for electrical stimulation and verification. This issue will be complicated, however, with direct flip chip designs, which will necessitate more creative solutions in the form of boundary scan designs and other techniques for failure identification. Fig. 3 illustrates the trends in electrical testing and characterization.

With increased levels of usage of flip-chip logic devices, fail site isolation from electrical testing is becoming more and more of a key issue. Design solutions, such as IBM's Level Sensitive Scan

Design (LSSD) incorporated into its C-4 technology will prove to be necessary in order to sufficiently isolate failures in the flip chip regime.[9] This approach makes it possible to diagnose and isolate faults using only the data gathered at the I/O pins of the device. This is accomplished by designing strings of latches throughout the chip to allow internal blocks of logic to be controlled and observed. Using a design for test (DFT) methodology, they can provide isolation to within a few nets on their devices. TI has also been active in developing scan architectures. These include TI's Modular Port Scan Design (MSPD) architecture, which allows modular access to designs with many scan domains.[10] More recently, TI has made use of a software based diagnostic tool called FastScan™ in IC fault diagnosis.[11,12] This tool performs diagnosis by simulating selected faults to determine the set of faults which most closely match the actual failure.[11]

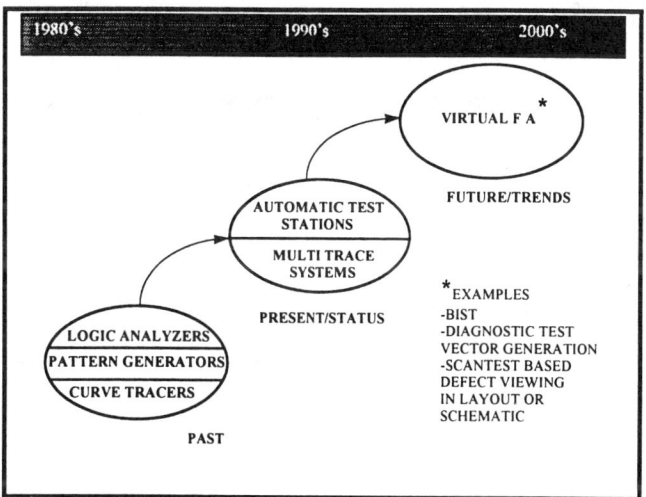

Fig. 3. Electrical Testing and Characterization

Intermediate Steps - Non-destructive Analysis (External) and Decapsulation

Before the localization and characterization of faults, there are two intermediate process steps that must be completed. These

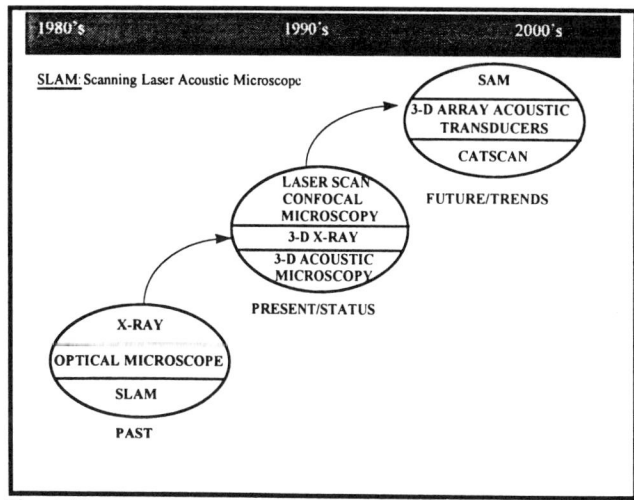

Fig. 4. Non-Destructive Analysis Techniques

steps are non-destructive (external) package analysis and decapsulation. The greatest concern in this area is, of course, the Ball Grid Array (BGA) package, with specific concerns for those with internal flip chip connections.[13] Sophisticated X-ray tools and scanning acoustic microscopy (SAM) have both proved to be viable options in this context. SAM workstations with high frequency (> 150 MHz) transducer have been able to penetrate these dense packages to provide excellent images of the package infrastructure. Fig. 4 illustrates trends in non-destructive analysis techniques.

Another key to successful fault isolation is the development of successful chip access techniques, in particular - decapsulation. The jet etch method has been an adequate option for many years and several more current designs, such as wire bonded BGA technologies, have been successfully decapsulated using this tool.[14] Fig. 5 shows the trends in device decapsulation technology.

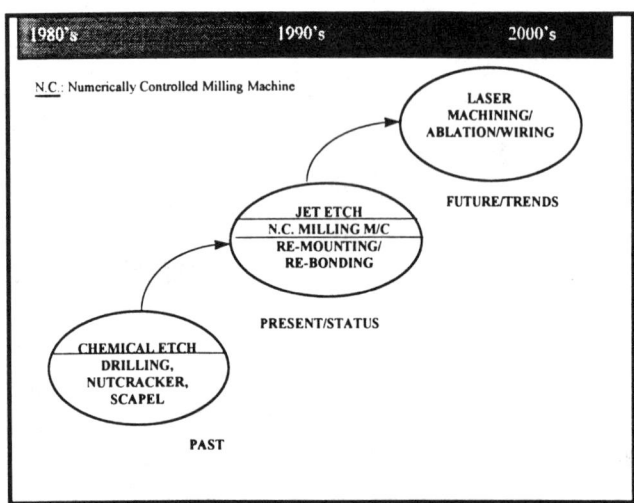

Fig. 5. Chip Access Techniques

More complex designs, such as the flip chip bonded BGA, have created many challenges for chip access. Past stop gap solutions have involved some amount of repackaging in order to complete analysis in some areas. Many current methodologies involve the removal of the backside heat sink in order to access backside silicon. The backside can then be thinned and polished for signal acqisition through the silicon or further silicon thickness reduction can occur via focused ion beam milling or chemically enhanced laser removal.

Failsite Isolation

Determining the fail mechanism of a device with electrical failures requires the isolation of the exact physical location of the failsite. Fault localization methods that can detect common failure mechanisms fall into two major categories. The first one is software-based diagnostics, which involves the use of simulation and tester data. The second one is hardware based diagnostics, which are based on secondary physical effects, such as light (photons), heat, or electron/ion beam radiation. Fig. 6 shows the

diagnostic options for fail site isolation in relation to the design and accessibility of each device.

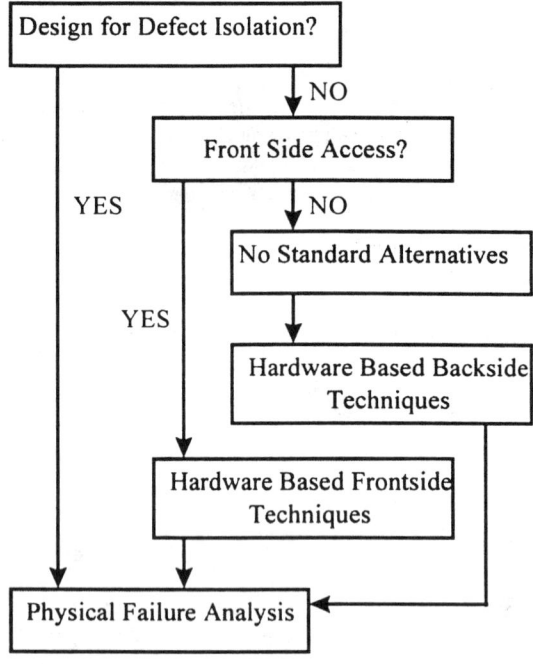

Fig. 6. Diagnostic Options for Failure Analysis

As shown in fig. 6, software based scan diagnostics are generally the first choice for fault isolation, especially with complex devices.[3] These techniques acquire probable fault locations by using a variety of fault simulation techniques and chip design data. The actual diagnosis can occur in two different ways. One approach uses pre-calculated fault dictionaries and the other approach implements post-test fault simulation techniques. Pre-calculated fault dictionaries are typically built during development by simulating faults at all known circuit nodes and noting the expected outputs for each type of fault. The software based technique involves post-test fault simulation in which only the outputs of suspected circuits are analyzed, greatly reducing the computation time. Designers can enhance the outcome of these techniques by structuring the device for maximum observability at the outputs.

In the absence of adequate built-in software diagnostics or otherwise, hardware based diagnostics see extensive use in the diagnosis and isolation of the three main electrical anomalies induced by physical defects: current, voltage, and frequency/timing. For current leakage failures, the failsites are usually associated with the the emission of photons or the generation of heat. For localization of photon (light) emitting defects,[15] photon emission microscopy (PEM) is perhaps the most extensively applied current leakage detection technique. In contrast, heat generating failsites are localized by thermal detection techniques such as liquid crystal thermography.[16,17] These techniques depend on local temperature rises induced by power dissipation within the circuitry.

The introduction of flip-chip technology challenges the effectiveness of the above methods of fail site isolation due to the lack of observability and access to the front side of these chips. The analyst must then resort to other methods, such as backside based fail site isolation (see Fig. 6). In regard to PEM, there is a convincing amount of evidence to support backside photoemission acquisition as a viable technique.[18,19] Since the emitted photons have energies ranging from the infrared (IR) to the visible regions, IR PEM can be used for analysis from the backside of the die. Heat detection from the backside has also proven to be a viable alternative.

Newer techniques in leakage detection include Fluorescent Microthermal Imaging (FMI).[20,21] Fig. 7 summarizes the trends in fail site leakage detection.

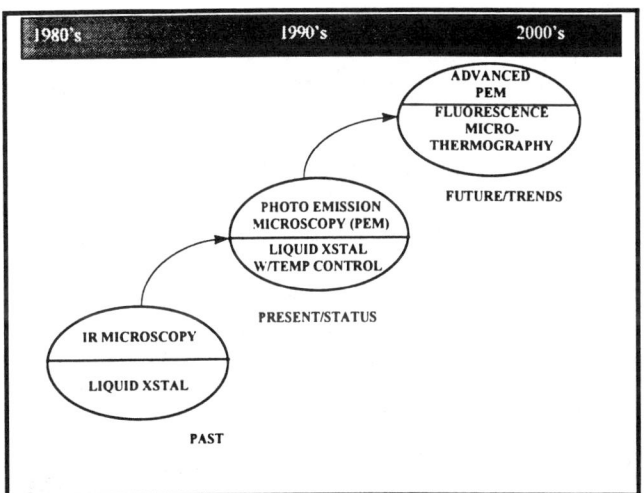

Fig. 7. Leakage Site Detection Techniques

The move toward the flip chip design has, in some cases, advanced the detection capabilities of the analyst rather than diminish them. Such is the case with PEM, where, problems with metallization coverage have decreased the effectiveness of PEM, even when frontside acess is available. Metallization coverage has increased to a point where many photoemissions simply cannot escape the device or are significantly reflected. Similarly for FMI, heat emission can be more accurately located through thinned silicon as heat loss is more uniform than through multi-level metallization.

Other fail site isolation methods involve detecting voltage and frequency anomalies within the circuitry, as shown in Fig. 8. These techniques use the response of the IC circuitry to external stimuli, such as electrons[22], ions, or photons.[23] Two such techniques are global and call for minimal knowledge of the internal circuitry of the device. These are Charge-Induced Voltage Alteration (CIVA)[22] and Light-Induced Voltage Alteration (LIVA).[24] They provide a fast, simple method for locating opens within conductors or other structures such as vias. Both of these techniques employ the same electrical approach, but use different beams; CIVA uses an electron beam while LIVA uses a photon (laser) beam. LIVA[24] and OBIRCH[25] (Optical Beam Induced Resistance Change) techniques have been shown to be applicable from the backside using IR (Infrared) lasers as light sources. OBIC (Optical Beam Induced Current) analysis can be performed on the backside of the die to detect latch-up defects and other junction

related defects. OBIC analysis has also been demonstrated in acquiring waveforms.[26]

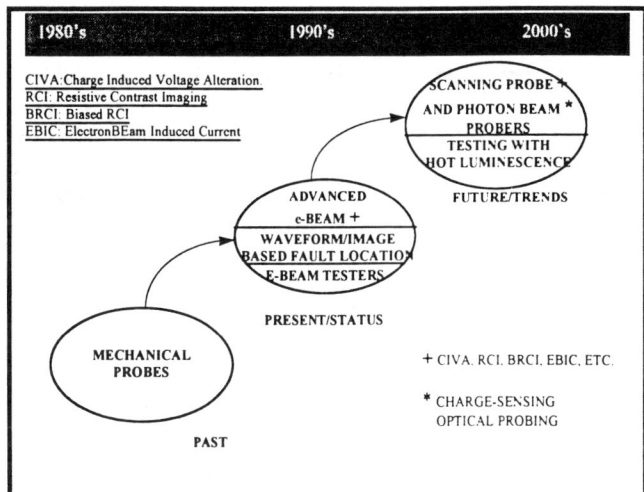

Fig. 8. Fail Site Isolation - Voltage/Frequency Techniques

Furthermore, these techniques are actually more suitably practiced from the backside, since there are fewer obstructions to keep the laser from penetrating into the junctions of the device. Thus, assuming that the backside thinning issues can be overcome, global voltage and frequency techniques are likely to continue to be valuable for fail site isolation.

When failures cannot be adequately isolated with global techniques, a more detailed approach may be necessary. In this case, either mechanical or electron beam probing can be employed to pinpoint the failures in DC and AC signals, respectively.[27,28]

Mechanical probing has seen a decrease in its usefulness because of the density of present ICs, the loading effects, and the difficulties in making electrical contacts to sub-micron lines. In contrast, electron beam probing equipment has seen an explosion of interest from design to FA.

Although improvements in electron beam testing equipment continue to provide better voltage, spatial, and timing resolution, they are facing fundamental limitations imposed by increasing interconnect complexity and flip-chip packaging. However, newer developments in the probing domain include the use of electro-optic techniques[29], backside laser probing[30], and scanning probe microscopy[8,31,32]. Backside laser probing presents a solution for devices which are inaccessible from the front side due to multilevel wiring or package mechanics by offering a method of acquiring signal measurements through bulk silicon.[30] One such laser probing system uses an interferometer to detect refractive index variations in the free carrier concentrations induced by electric fields in the n-channel and p-channel transistors[33] Other developments in circuit probing include the utilization of photon emissions as a method of probing individual gates.[34] In this case, the hot electron luminescence that occurs with logic state switching is detected and measured to reveal the propagation of signals through the gates. Table II presents the options in probing technology.

Method/Resolution	Spatial (um)	Temporal (ps)	Voltage, absolute $\left(\dfrac{mV}{\sqrt{Hz}} \right)$	Conditions
Electron Beam	0.5	150	2.5	metal/ (passivated)
Electro-optic	10	50	2	metal
Backside	1.5	50	Charge Sensitive	active semiconduct. region, backside polished
Scanning Probe	0.25	100	100	metal/ (passivated)

Table II. Typical performance capabilities for contactless probing technologies.[35]

Scanning probe techniques are emerging as the most promising probing tool for the future. SPM techniques have been used in the measurement of internal voltages and currents on ICs. Charge force microscopy (CFM) has been demonstrated to measure voltages on IC's with submillivolt sensitivity and has been instrumental in the development of an atomic force microscope (AFM) based voltage contrast technique for measuring waveforms up to 20 GHz. Magnetic force microscopy (MFM) uses a magnetized tip to detect magnetic force gradients; thus, MFM can be used in providing high sensitivity current contrast images of integrated circuits. The combination of CFM and MFM will allow the simultaneous measurements of internal voltages and currents.[8]

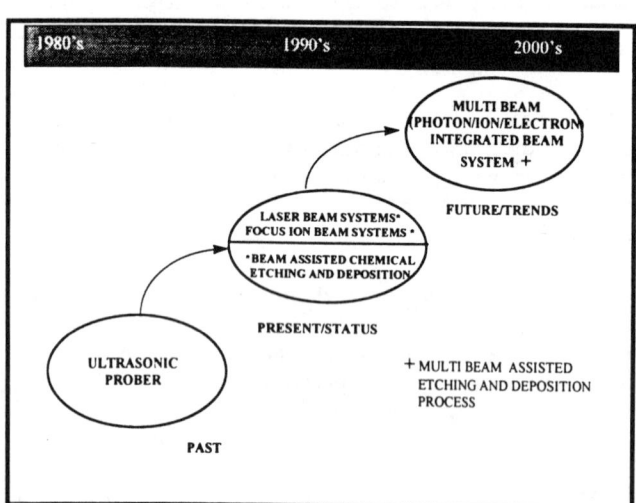

Fig. 9. Circuit Isolation and Microsurgery

These probing tools are particularly important in the design debug activity. Once again, flip chip devices (especially array bonded flip chips) present challenges to these types of probing methodologies. Therefore, in order to effectively use these techniques, the circuit must first be accessed and isolated. Traditionally, circuit isolation was done with laser cutters and ultrasonic mills. More recently, ion beam systems have been used extensively in circuit isolation and in milling for access.[36] Focused Ion Beam (FIB) vendors have successfully demonstrated the access of buried conductors through the use of ion beam milled trenches.[24] As a complimentary technique, chemically enhanced laser etching has also been introduced for the fast, easy removal of bulk silicon.[37] Vendors have also presented the option of electron beam induced etching as a means of exposing nodes which are buried.[28]

Fig. 9 shows the trends of circuit isolation and microsurgery. As indicated in Fig. 9, fully integrated multiple beam systems combining photon beams and charged particle beams could be the trends of the future.

Deprocessing

Once the fail site has been identified, one must perform the actual deprocessing of the die to gain access to the defect. This involves the physical process of uncovered the physical defect which has caused the electrical anomaly. These processes fall into two general categories: cross sectioning and delayering. Fig. 10 shows the evolution of deprocessing technologies. In cases where the defect location is accurately known, this may involve cross sectioning. Traditionally, cross sections have occurred through cleaving and encapsulated sectioning. More recent cross sectioning procedures have involved FIB milling. With recent advances in FIB technology (i.e. beam spot size), the FIB is expected to continue as the cross sectioning tool of choice.

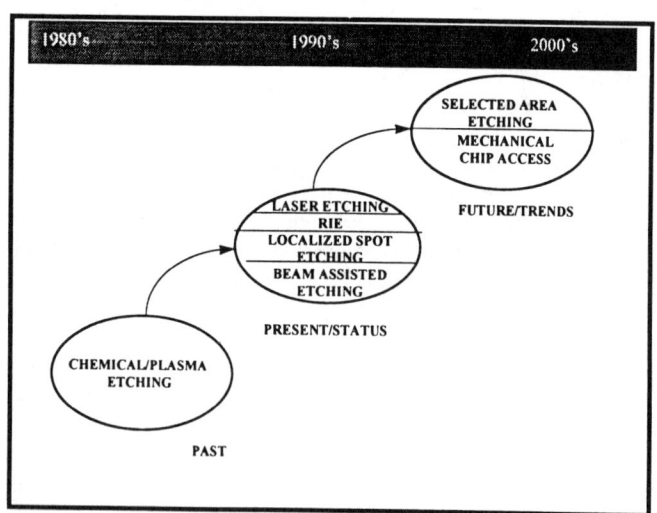

Fig. 10. Device Deprocessing Techniques

In contrast, delayering involves a layer by layer removal of the thin films which make up the device. This typically entails a host of complementary techniques including wet chemical etching and plasma or reactive ion etching. Dielectrics are commonly removed by reactive ion etching while wet chemicals remain the method of choice for metal layers, especially when there is no danger of etching through the vias. The methodology of material removal by ion bombardment may soon be in question as device miniaturization causes layers to become so small that selectivity is compromised. Enhancements to the RIE, such as magnetically enhanced RIE have been found to reduce charging damage. When global de-layering is not necessary, localized spot etching may become a technique to be considered in order to maintain the integrity of the rest of the die while exposing a small area. In the same way, the localized removal of heat sink material may be logical for devices that operate at higher power and must maintain operability. As new materials are added to the regime, the deprocessing techniques must constantly be developed. The integration of materials such as copper and newer dielectrics into devices may challenge the traditional deprocessing techniques, thus calling for new technique development.

Physical/ Chemical Characterization

The next stage of the FA process is the physical and chemical characterization of the failure. This is usually the final major step in a root-cause analysis effort. This stage involves inspecting the isolated areas of the die to find the responsible defect. The tools used for observation have evolved from optical microscopes to electron microscopy (see Fig. 11). Smaller defects and features have pushed optical microscopy beyond its resolution limit (around .2-.3 um using visible light). Scanning Optical Microscopy (SOM) and the use of IR lasers have extended the usefulness of conventional microscopy. SOM provides improved image resolution, while IR lasers permit observation through bulk silicon, as silicon is transparent to IR light.

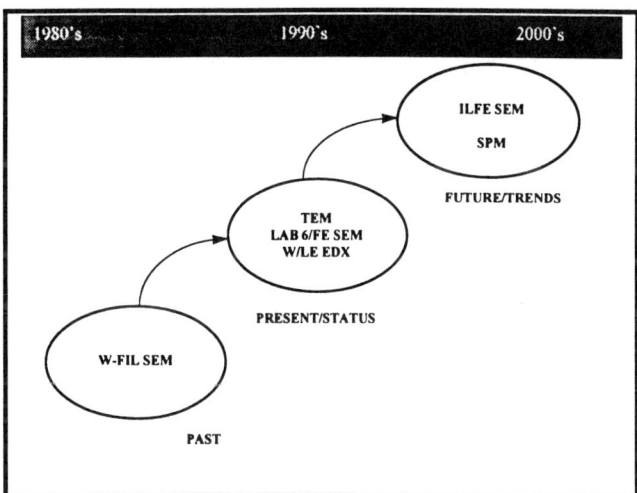

Fig. 11. Micro Structure and Micro Topography Instrumentation Trends

Various modes of the SPM (Scanning Probe Microscope) are showing remarkable potential for the future.[31] Using specially mounted sharp tips, SPM's will enable the observation of atomic scale topography and cross sections. One derivative of SPM technology is the AFM (Atomic Force Microscope). This high resolution tool will also facilitate the resolution of non-visual defects. Other modes such as Scanning Capacitance Microscopy (SCM) and Scanning Resistance Microscopy (SRM) will further enable the detection of process defects devoid of clear physical evidence. The SCM may be particularly useful in detecting implant defects through 2-D diffusion profiling. Many high resolution fail site isolation techniques are sure to evolve from the

SPM, much as e-beam probing, Electron Beam Induced Current (EBIC) and CIVA have evolved from the SEM.

Root Cause/ Corrective Action

The last stage of the FA process is, of course, the root cause and corrective action evaluation. This step ends the loop of the process that began in the design-development-manufacturing flow, where the failure was initially found. The fulfillment of this step eliminates the defect/problem repetition cycle. Proper documentation during this step communicates the final results and provides a permanent record of the findings.[38, 39] Fig. 12 outlines the trends in FA data management.

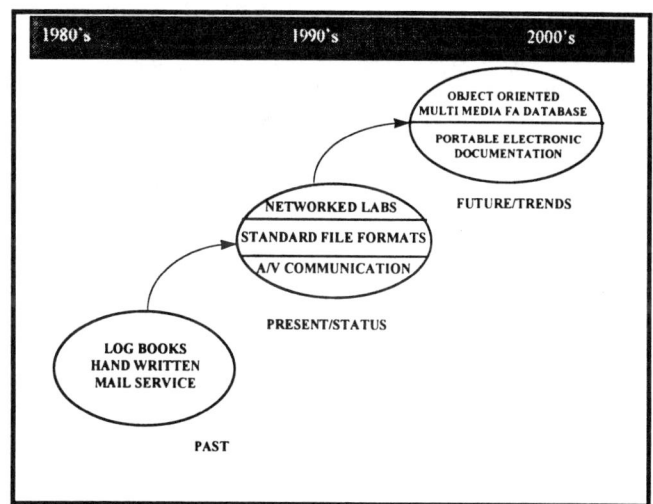

Fig. 12. Failure Analysis Data Management

The advent of the internet has greatly affected this communication trend, as well as the sharing of information and knowledge among laboratories and customers. Through the internet, the customer and analyst can jointly operate the FA tools and arrive at real-time solutions.

DIAGNOSTIC ACTIVITY IN THE FUTURE

The diverse diagnostic activities that make up the SC quality environment are becoming more and more focused on bringing new devices to high, stable yields quickly. The increasing correlation of yield with field quality should give added incentive for rapid ramping to high yields. This trend toward more up-front FA should reduce the pressures of customer return failure analysis. There is also an increasing trend to adopt Signature Analysis and thus shift the resources from post-mortem to preventative FA.[40]

There are many steps which can be taken during the planing stages to diminish ramp times and ensure greater fail site detection. Design for Testing (DFT) is important in both testability and application debug activity. Enhanced package modeling and FMEA (Failure Mode Effects Analysis) are also vital in diminishing complications in the qualification and application stages. The various stages of design debug, including traditional debug activity, developing working silicon, resolving application issues, and design/process interactions will be expected to occur in very short periods of time. Some have suggested plans which centralize all debug activities in one location at the engineering test floor, using the ATE for electrical stimulation.[41]

NECESSITY OF STANDARDS

Historically, failure analysis procedures and results had no broadly accepted standards for documentation. This created a discontinuity between organizations and even worse, between organizations and customers. Therefore, it is necessary to implement standards in order to improve the effectiveness of any process. Standards define common expectations between suppliers and users. They also enable the efficient flow of information within organizations and among inter-corporate groups. With this in mind, an FA subcommittee (JEDEC 14.6) was formulated under the auspices of the Electronic Industries Association (EIA)/Joint Electron Device Engineering Council (JEDEC) organization. Because the only product of a failure analysis operation is information, it was considered important to generate a set of standards and guidelines that would facilitate an effective communication of FA information across its broad spectrum of recipients. [42] Further joint efforts include activities such as the development of signature analysis guidelines.

CONCLUSIONS

In order to address the challenges that lie ahead for FA, there is much work to be done. Historically, much of the development in FA was left to the individual FA laboratory. This has led to a very reactive process in failure analysis development, as failure analysis labs were challenged to find solutions on-the-go to problems created by changes in designs, processes, and tests. However, with the continual advancements in semiconductor technology and faster design-to-market trends, a more proactive approach is imperative to even maintain current capabilities, let alone advance them.

These proactive measures involve more open lines of communication between FA groups and between different areas inside individual companies. In order to get ready for upcoming changes, product preparedness planning should be initiated to provide a link between FA technology development and the changing needs of the industry. Inter-company collaborations can be initiated to provide a forum for exchanges of ideas and drive solutions to common problems. Pre-silicon activities such as FMEA are important in predicting possible problems and identifying the tools and techniques which need to be developed.

In addition to initiating inter- and intra- corporate collaborative efforts, these new challenges have led to the participation of national laboratories and universities and with the vendors themselves. Due to the inherent tactical mode of operation, development is often sacrificed for current needs of FA laboratories. A consortium of shared ideas and shared costs that moves appropriate development projects out of industry FA operations and into the hands of suppliers and universities could

aid in solving these problems. Continuous dialogue with the supplier community could also aid in ensuring that the technological capabilities are in place when they are needed. With these kinds of proactive measures, the FA community can be certain of its ability to meet the challenges of the new millennium.

ACKNOWLEDGMENTS

We would also like to thank the SEMATECH Product Analysis Forum (PAF), the International Symposium for Testing and Failure Analysis (ISTFA) committee members, and the TI world wide FA managers for their advice and technical input.

REFERENCES

[1] S. Pabbisetty and L. Wagner, "Failure Analysis: An Overview," Texas Instruments Technical Journal, pp. 9-22 October-December 1997.

[2] Semiconductor Industry Association, The National Technology Roadmap for Semiconductors, 1994.

[3] S. M. Kudva et. al., "The Semiconductor Failure Analysis Roadmap," 21st International Symposium for Testing and Failure Analysis Conference Proceedings, ASM International, pp. 1-5, November 1995.

[4] D. Vallett, "Finding Fault With Deep-Submicron IC's," IEEE Spectrum, pp. 39-50, Oct. '97.

[5] S. V. Pabbisetty (moderator), "Evening Panel Discussion - Failure Analysis for the 90's and Beyond (Back to the Future!)," 18th International Symposium for Testing and Failure Analysis Conference Proceedings, ASM International, pp. 443-448, November 1992.

[6] S. V. Pabbisetty, "Advanced V/ULSI Device Technology Failure Analysis for Micro 2000: Status and Trends," Eigth Electronic Materials and Processing Conference Proceedings, ASM International, August 1993.

[7] S. V. Pabbisetty (moderator), "Evening Panel Discussion - Failure Analysts for the 90's and Beyond (Back to the Future II)," 19th International Symposium for Testing and Failure Analysis Conference Proceedings, ASM International, pp. 431-436, November 1993.

[8] E. Anderson, J. M. Soden, C. L. Henderson, "Future Technology Challenges for Failure Analysis," 21st International Symposium for Testing and Failure Analysis Conference Proceedings, ASM International, pp. 27-32, November 1995.

[9] E. B. Eichelberger and T.W. Williams, "A Logic Design Structure for LSI Testing," Proc. 14th Design Automation Conference., pp. 462-468, June 1977.

[10] S. Banerji and Martin Daniels, "Testability Design: A Tool for VLSI Systems," Texas Instruments Technical Journal, vol. 5, no. 4, pp. 138-144, July-Aug. 1988.

[11] J. Platt, K. Butler, G. Hetherington, G. Lorig, S. Pabbisetty, S. Venkataraman, "Fault Diagnosis on the TMS 320C80 (MVP) using FastScan,"

[12] M. Gala et al., "Design for Defect Isolation," Texas Instruments Technical Journal, 4Q '97 (to be published).

[13] D. Davis,et al., "Flip Chip Failure Analysis," Texas Instruments Technical Journal., 4Q '97 (to be published).

[14] R. Chowdury et al., "A Study of Chemical Decapsulation Techniques for New Generation Plastic Molded IC Packages," 21st International Symposium for Testing and Failure Analysis Conference Proceedings, ASM International, pp. 281-286, November 1995.

[15] J. Koltzer, et al., "Quantitative Emission Microscopy," J. Appl. Physics, 71 (11), pp. R23-R41, June 1992.

[16] D. J. Channin, "Liquid Crystal Technique for Observing Integrated Circuit Operation," IEEE Trans. On Electron Devices, Vol ED-21, pp. 650-652, Oct 1974.

[17] S. Khandekar and S. Wills, "Liquid Crystal Microscopy," Microelectronic Failure Analysis: Desk Reference, 3rd Edition, ASM International, 1993.

[18] N. M. Wu et. al., "Failure Analysis from the Backside of the Die", 22nd International Symposium for Testing and Failure Analysis Conference Proceedings, ASM International, p. 393, November 1996.

[19] D. L. Barton et. al., "Infrared Light Emission from Semiconductor Devices", 22nd International Symposium for Testing and Failure Analysis Conference Proceedings, ASM International, p. 9, November 1996.

[20] P. Tangyunyong et. al., "Localizing Heat-Generating Defects Using Fluorescent Microthermal Imaging and Light Emission Microscopy", 22nd International Symposium for Testing and Failure Analysis Conference Proceedings, ASM International, p. 55, November 1996.

[21] J-Y Glacet and S. Berne, "A User Friendly System for Fluorescent Microthermal Imaging and Light Emission Microscopy", 22nd International Symposium for Testing and Failure Analysis Conference Proceedings, ASM International, p. 63, November 1996.

[22] E. Cole and J. Soden, "Scanning Electron Microscopy Techniques for IC Failure Analysis," Microelectronic Failure Analysis: Desk Reference 3rd Edition, ASM International, pp. 163-175, 1993.

[23] F. J. Henley, "An Automated Laser Prober to Determine VLSI Internal Node Logic States, "IEEE/International Test Conference, pp. 536-542, 1984.

[24] K. Nikawa et. al., "Various Contrasts Identifiable from the Backside of a Chip by 1.3 micron Laser Beam Scanning and Current Change Imaging", 22nd International Symposium for Testing and Failure Analysis Conference Proceedings, ASM International, p. 387, November 1996.

[25] P. F. Ullman et. al., "A New Robust Backside Flip-Chip Probing Methodology", 22nd International Symposium for Testing and Failure Analysis Conference Proceedings, ASM International, p. 381, November 1996.

[26] J. Quincke et al., "Circuit Analysis in ICs Using the Scanning Laser Microscope," Proceedings of SPIE - The International Society for Optical Engineering," SPIE, September, 1988, pp. 211-216.

[27] S. V. Pabbisetty, et. al., "Electron Beam Testing and its Applications to VLSI Technology", SPIE Conference on Characterization of Very High Speed Semiconductor Devices and Integrated Circuits: Critical Review of Technology, SPIE, March 1987.

[28] J. Frosien, et. al., "State of the Art E-Beam Probing", 5th European Conference on Electron and Beam Testing of Electronic Devices, August 1995.

[29] G. Baur and G. Solkner, "Laser Diode Based Electro- Optic Measurement System with High Voltage Resolution," Rev. of Sc. Instr. 64,4, pp. 1081-1084, 1993.

[30] H. Heinrich et al., "Picosecond Backside Optical Detection of Internal Signals in Flip-Chip Mounted Silicon VLSI Circuits, " Microelectronic Engineering 16, pp. 313-324, 1992.

[31] T. Hochwitz, "Implementation, Characterization, and Applications of a Scanning Kelvin Probe Force Microscope," A Thesis Submitted to the Faculty of Thayer School of Engineering: Dartmouth College.

[32] H. K. Wickramasinghe, D. Rugar, Preface to IBM Journal of Research and Development: Proximity Probing Issue, Vol. 39, No. 6, November 1995.

[33] H. K. Heinrich, et. al., "Noninvasive sheet charge density probe for integrated silicon devices", Applied Physics Letters, 48 (16), p. 1066, April 1986.

[34] J. A. Kash and J.C. Tsang, "Dynamic Internal Testing of CMOS Circuits Using Hot Luminscence," IEEE Electron Device Letetrs, Vol. 18, No 7,pp. 330-332, July 1997.

[35] G. Sölkner, et. al., "Advanced Diagnosis Techniques for sub-μ m Integrated Circuits", private communication.

[36] S. V. Pabbisetty, et al.., "Focused Ion Beam Technology: Status and Trends," IEEE 4th International Symposium on the Physical & Failure Analysis of Integrated Circuits, November 1993.

[37] D.J. Ehrlich, et. al., "A review of laser-micromechanical processing", J. Vac. Sci. Technology B. Oct.-Dec. 1983.

[38] L. Bellay, et al., "Computers in Failure Analysis," 16th International Symposium for Testing and Failure Analysis Conference Proceedings, ASM International, pp. 89-95, November 1990.

[39] Cheryl Miller et al., "Customer Return Analysis: Creating a Better Customer Return System," Texas Instruments Technical Journal, 4Q '97, to be published.

[40] S. Lakshminarayan et al., "Signature Analysis for IC Failures," Texas Instruments Technical Journal, 4Q '97, to be published.

[41] David Patrick et al., "Test Floor Failure Isolation Techniques," Texas Instruments Technical Journal, 4Q '97, to be published.

[42] A. Kostic, "The Role of Standards Organizations in Improving Failure Analysis," presented at IBM Failure Analysis Workshop Technical Exchange '95, East Fishkill, NY.

CMOS Device Physics

Theodore A. Dellin
Sandia National Laboratories
Albuquerque, New Mexico

1. INTRODUCTION

At the heart of the silicon revolution is the transistor. Today, the dominant transistor technology is Metal Oxide Semiconductor (MOS). The MOS technology has enabled the ability to continuously improve the speed and functionality of integrated circuits at acceptable power levels. As the paper is being written transistors with 0.18μ features are being introduced into volume production enabling integrated circuits to contain over ten million transistors and operate at speeds over 500MHz.

The purpose of this chapter is to provide a summary of the physics behind the operation of the Silicon (Si) MOS transistor. Simplifying assumptions will be introduced to allow focusing on the most important effects. The major equations will be presented, but not derived. For more detailed information the reader is directed to the Suggested References at the end of this Chapter

Sections 2 to 5 provide the foundation for understanding the MOS transistor. In Section 2 the fundamental properties of semiconductors will be developed. It will be seen that there are both positive and negative mobile charge carriers and that their concentrations can be radically changed by doping the silicon crystal with other elements. Section 3 considers how to "read" energy band diagrams in order to visualize how devices operate. Section 4 will focus on the effects that occur at the junction between semiconductors that are doped differently. It will be seen that a pn junction has a built-in voltage and that it acts as a diode when external voltages are applied. Section 5 considers the MOS capacitor and how by applying a gate voltage an inversion layer of mobile charges can be created at the semiconductor's surface.

Having all the basics out of the way, Section 6 describes the operation of the MOS transistor in the linear, saturation and subthreshold operating regions. Section 7 considers the benefits that derive from scaling down the dimensions and voltages of MOS transistors. Section 8 concludes with a brief description of how the long channel transistor behavior is modified when dimensions are scaled down to submicron dimensions.

2. SEMICONDUCTOR PROPERTIES

An electric current is produced when there is net motion of mobile charges in a given direction. This section discusses the type of mobile charges in Si and how they are affected by temperature and by introducing other elements into Si. The two mechanisms that lead to a net current, the application of electric fields and concentration variation, are also considered.

2.1 Electrons and Holes

In an isolated atom the electrons are constrained to a set of discrete, separated energy states. When these atoms are brought together in a solid these states broaden out into allowed energy ranges, called "bands". Between the bands are ranges of energies in which electrons will not be present, called "band gaps."

There are lots of electrons in a silicon crystal. However, most of the electrons surrounding a Si atom are tightly bound to the atom and are not mobile. Only the outermost, most loosely bound, most energetic of the Si atom's electrons have the potential to be mobile charge carriers. The highest energy band that is filled at a temperature of absolute zero is called the valence band. The next highest energy band, which is empty at absolute zero, is called the conduction band.

Figure 1(A) is a "band diagram" of a semiconductor, like Si. In a band diagram the vertical direction represents increasing electron energy and the horizontal direction represents position in the crystal. In all the cases considered in Figure 1 there is no variation in the energies with distance so all the lines are horizontal. When we deal later with diodes and transistors we will see sloped lines reflecting the fact that electron energy will change with position.

Figure 1(A) is for pure Si without any impurities. The top of the allowed valence band states has an energy E_V. Then there is a band gap with no allowed energy states. Next we come to the bottom of the allowed conduction band energy states which start at an energy, E_C. The band gap energy,

$$E_G = E_C - E_V \qquad (2.1.1)$$

is 1.12eV for Silicon at room temperature (the band gap decreases slightly as the temperature is increased.)

To have an electron current requires two things. First, there must obviously be mobile carriers in a band since current is the motion of charges. Second, the band must not be completely full. A current requires a net motion of charges in one direction. In a completely

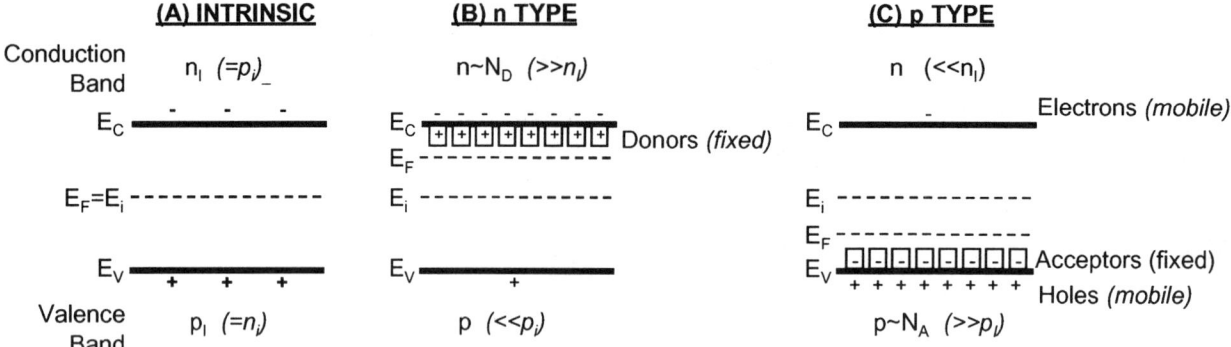

Figure 1. Band diagram (electron energy vertical, position horizontal) for (A) Intrinsic, (B) Extrinsic n type, and (C) Extrinsic p type semiconductors. E_V is the top edge of the valence band, E_C is the bottom edge of the conduction band, E_i is the Fermi Level in the intrinsic material and E_F is the Fermi Level.

full band, net motion is not possible. (The electrons are moving equally in all allowed directions.)

Applying these two criteria it can be seen that current flow is not possible at absolute zero in a semiconductor. The completely full valence band cannot conduct a current because it has no unoccupied energy states. The totally empty conduction band cannot conduct a current because it has no electrons.

Things become more interesting at temperatures above absolute zero. As the temperature is elevated the thermal energy of the electrons increase. There is now the possibility that electrons can jump from the valence band to the conduction band. Every electron that is excited into an energy state in the conduction band leaves behind an empty energy state in the valence band.

Current flow is now possible in *both* the conduction and valence bands. The conduction band now contains some electrons and has plenty of unoccupied energy states.

The current in the valence band results from the collective motion of the electrons in the valence into the few empty electron states. Fortunately, there is a simple way to account for currents in the valence band. The unoccupied energy states can be treated as a mobile particle, the "hole." The hole's charge is equal and opposite to the electron's charge. Due to their opposite charges, holes and electrons will move in opposite directions in an electric field.

There is one important asymmetry between electrons and holes. The holes do not move as easily through the Si crystal as do electrons. We will consider this effect in a later section when we discuss mobility.

2.2 Probability of an Electron State Being Occupied

The number of electrons at an energy E in the conduction band, n(E), is determined by

$$n(E) = N(E)f_n(E) \qquad (2.2.1)$$

where N(E) is the number of energy states at E and $f_n(E)$ is the probability that a given energy state will be occupied.

The probability is given by the Fermi-Dirac distribution. However, for our purposes, it will be sufficient to use a mathematically simpler approximation of this distribution, called the Boltzmann distribution. The probability of an electron state in the conduction band being *filled* is

$$f_n(E) = \exp\left[-\left(E - E_F\right)\big/kT\right] \text{ for } E > E_F \qquad (2.2.2)$$

where E_F is called the Fermi Energy or Fermi Level, k is the Boltzmann constant (1.38E-23 J/K) (E-23 is 10, not e, raised to the -23rd power) and T is the absolute temperature. E_F is the energy at which the probability = ½ and will be discussed in more detail below.

The probability of a hole (i.e., an *empty* electron energy state) in the valence band is

$$f_p(E) = \exp\left[-\left(E_F - E\right)\big/kT\right] \text{ for } E < E_F \qquad (2.2.3)$$

Note that the probability of an electron state being occupied falls off exponentially with energy. At 300K (room temperature) kT is about 0.026 eV. Thus the probability will fall 1/e every 0.026 eV or a factor of 10 every 0.06 eV.

There is a slight increase in the number of energy states as the energy increases in the conduction band. However, this effect is swamped out by the exponential decrease in the probability of the state being occupied, $f_n(E)$. Therefore, the number of electrons sharply decreases as one moves up in energy in the conduction band. Similarly, the number of holes decreases sharply as one decreases the energy below the top of the valence band.

The number of electrons in the conduction, n, and the number of holes in the valence band, p, is given by

14

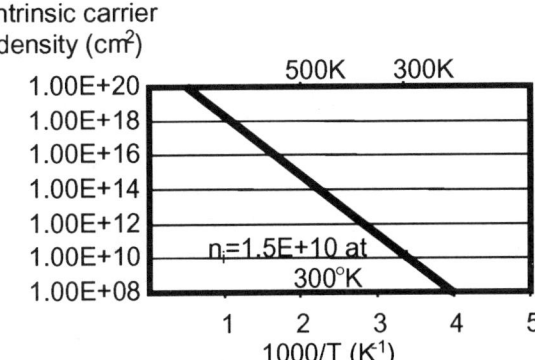

Figure 2. Intrinsic carrier density for Si as a function of inverse absolute temperature (after Sze).

$$E_{Fi} = E_i = \frac{E_V + E_C}{2} + \frac{kT}{2} \ln\left(\frac{N_V}{N_C}\right) \qquad (2.3.2)$$

E_i will be used is to indicate the Fermi level for the intrinsic semiconductor. E_i is essentially at the middle of the band gap as indicated in Figure 1(A).

2.4 Extrinsic semiconductors

An extrinsic semiconductor is one in which impurities are intentionally substituted for some of the Si atoms in the crystal lattice. The process of deliberately adding these impurities is called "doping." The object is to dramatically increase the number of either electrons or holes over their intrinsic values.

Si is located in Column IV of the periodic table. It has 4 valence electrons per atom which are used to form covalent bonds with its neighbors in the crystal lattice. To create an excess of electrons a Si atom is replaced with an element from Column V with 5 valence electrons (like P or As). Only 4 of the 5 valence electrons of the new atom are needed for the covalent bonding. The extra 5th electron is loosely bound and introduces a new allowed electron energy level in the bandgap that is located just below the conduction band edge. These are called "donor" elements because it takes very little energy to ionize these atoms and donate their 5th valence electron to the conduction band. At practical operating temperatures essentially all of the donor atoms are ionized. The material containing donor atoms is called "n type" (n for negative) because of the excess of electrons.

Figure 1 (B) shows the band diagram of an n type material. The doping levels used, N_D, range from 1E14 to 1E17 cm^3. These numbers are many orders of magnitude greater than the intrinsic electron density. Thus, to a good approximation, we can consider the electron concentration in an n type material to be equal to the density of donor atoms.

$$n_n - N_D \quad \text{(n type material)} \qquad (2.4.1)$$

At equilibrium the np product given in equation (2.2.3) does not depend on the doping level. Thus the concentration of holes is given by

$$p_n = n_i^2 / n_n = n_i^2 / N_A \quad \text{(n type material)} \qquad (2.4.2)$$

The effect on introducing donor atoms into the Si lattice is to drastically alter the concentration of electrons and holes. Relative to the intrinsic material the electron concentration is increased by orders of magnitude while the hole concentration is similarly decreased. The np product remains constant. For example, if N_D=1E16 cm^3 then n=1E16 cm^3 and p–2E4 cm^3.

In an n type material, the electrons are referred to as the "majority" carrier and the holes are the "minority" carrier.

Knowing the n_n and p_n densities for the n type material we can calculate the Fermi energy

$$n = N_C \exp[-(E_C - E_F)/kT] \qquad (2.2.4)$$

$$p = N_V \exp[-(E_F - E_V)/kT] \qquad (2.2.5)$$

where N_c is the effective number of energy states in the conduction band (~3E19/cm^3 at room temperature) and N_v is the effective density of states in the valence band (~1E19/cm^3).

One interesting result comes from multiplying Eq 2.2.4 and Eq 2.2.5 and using definition of E_G in Eq 2.2.1 gives

$$np = n_i^2 = N_C N_V \exp[-E_G/kT] \qquad (2.2.6)$$

Figure 2 shows n_i for Si as a function of temperature. At equilibrium, the product of the number of electrons and holes depends on the bandgap energy and on the absolute temperature, but not on the Fermi energy. At 300K, the intrinsic carrier concentration $n_i \sim 1.5E10$ cm^3. This relationship holds for both the intrinsic and extrinsic semiconductors discussed in the next sections.

2.3 Intrinsic Semiconductors

A Si crystal without any impurities is said to be intrinsic. We will use the subscript i to denote when variables apply to the intrinsic material.

In an intrinsic semiconductor every electron excited into the conduction band leaves behind a hole

$$n_i = p_i \qquad (2.3.1)$$

As mentioned above, at room temperature n_i is 1.5E10 cm^3.

The fact that the electron and hole concentrations are equal can be used to determine the position of the Fermi level in an intrinsic semiconductor. Setting Eq (2.2.3) equal to Eq (2.2.4) and solving for E_F gives

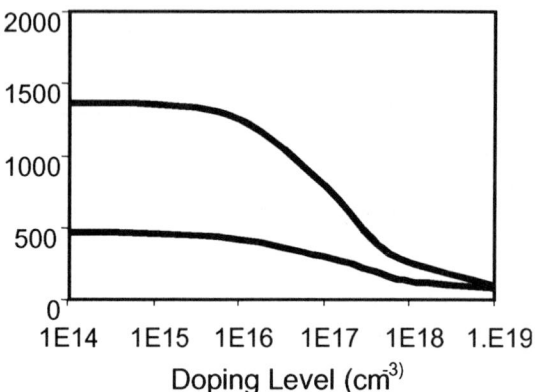

Figure 3. The bulk mobility of electrons (upper curve) and holes (lower curve) in Si as a function of doping level. (After Sze). The vertical axis is mobility in (cm²/Vs).

$$E_{Fn} = E_i + kT \ln\left(\frac{N_D}{n_i}\right) \quad \text{(for n type material)} \quad (2.4.3)$$

With the addition of donor atoms the Fermi energy moves upward from the intrinsic level, E_i, towards the valence band edge. The displacement from E_i is a logarithmic function of the donor doping density.

In a similar manner we can create p type material in which the dominant mobile species are holes by adding acceptor atoms from Column III of the periodic table. These atoms, like B, have only 3 valence electrons. They introduce *empty* electron states just above the valence band edge. It is very easy for valence band electrons to occupy these new states and leave behind holes. Again at practical operating temperatures essentially all of the acceptor atoms are negatively ionized. The p type equations are

$$p_p \approx N_A \quad \text{(for p type)} \quad (2.4.4)$$

$$n_p = n_i^2 / N_A \quad \text{(p type)} \quad (2.4.5)$$

$$E_{Fp} = E_i - kT \ln\left(\frac{N_A}{n_i}\right) \quad \text{(p type)} \quad (2.4.6)$$

In p type material holes are the majority carrier, electrons are the minority carrier. The Fermi level moves towards the valence band edge as acceptor atoms are added.

2.4.1 Charge Compensation Extrinsic semiconductors are charge neutral. For example, in n type material the negative charge of the extra electrons is compensated by the positive charge of the ionized donor ions. Similarly, in p type material the extra positive holes are compensated for by the negative charge of ionized acceptor ions.

2.5 Degenerate Semiconductors

At very high doping levels (>1E20 cm⁻³) the physics changes. The interaction between dopant atoms now needs to be considered. As a result of this interaction the energy levels introduced by the dopants broadens out into the bands. Furthermore, there are so many majority carriers that the Fermi level can move into the conduction band (n type) or valence band (p type). These materials are called degenerate semiconductors and the symbols n++ and p++ are frequently used to represent them. A very important application is the use of degenerately doped polysilicon as the gate material for transistors.

2.6 Mechanism for Producing Currents

The electrons and holes are in constant motion. Random motion with as many electrons going to the right as to the left will not produce an electric current. To produce a current we need to have a net motion of electrons and/or holes along a given direction. There are two physical processes that can produce new motion: drift and diffusion. Drift, the net motion of electrons in an electric field, is the major conduction mechanism in the on state of MOS transistors. Diffusion, the net motion caused by a concentration gradient, is responsible for the strong voltage dependence of the current in diodes and for the subthreshold leakage in MOS transistors.

2.6.1 Drift. If an electric field is present a net current will flow. At low fields the electric field, E, the electron and hole drift currents are given by

$$\mathbf{J}_{n,drift} = qn\mu_n \mathbf{E} \quad (2.6.1)$$

$$\mathbf{J}_{p,drift} = qp\mu_p \mathbf{E} \quad (2.6.2)$$

where n (p) is the electron (hole) concentration, q is 1.6E-19 coulomb, **E** is the electric field (V/cm), μ_n is the electron mobility (cm²/V), and μ_p is the hole mobility (cm²/Vs). The electron mobility (1500 cm²/Vs) at low doping concentrations is about 3 times greater than the hole mobility (450 cm²/Vs). Figure 3 shows the mobility as a function of doping in the bulk of the Silicon.

The mobility is determined by mechanisms that scatter charges. In the bulk there are two dominant mechanisms. At low temperatures the mobility is determined by scattering off of dopant atoms. This impurity scattering mobility decreases with increasing doping density. At higher temperatures the mobility is determined by scattering from the thermal vibration of atoms in the crystal (phonon scattering). Phonon scattering causes the mobility to decrease with temperature, but does not depend on doping density. For MOS transistors that conduct along the Si/SiO₂ surface the mobility can be lower due to scattering caused by charges at the interface and due to surface roughness.

Over the operating ranges of interest the mobility of electrons and holes decreases with temperature. This results in a decrease in the maximum operating frequency with temperature and can lead to speed failures at high temperatures.

The linear relationship between drift current and electric field in Eq (2.6.1) and (2.6.2) only holds at low electric fields. At fields above 1E4 V/cm the velocity starts to saturate. The drift velocity of electrons saturates at about 1E7 cm/s at electric fields above 1E5 V/cm and the drift velocity of holes saturates at about 8E6 cm/s. While at low fields the drift velocity of electrons is three times that of holes, at saturation the drift velocities are nearly equal.

2.6.2 Diffusion. The second transport mechanism we want to consider is diffusion. Diffusion results when there is a concentration gradient (i.e., when there is more of one species at a point than there is in an adjacent point). Thermal motion causes a net motion from regions of high concentration to regions of low concentration. A classic example is opening a perfume bottle in one corner of a room. Initially the aromatic compounds are confined to the bottle, but as time goes on the scent can be detected on the other side of the room.

The diffusion currents of electrons and holes are given by

$$\mathbf{J}_{n,diff} = qD_n \frac{dn}{dx} \qquad (2.6.3)$$

$$\mathbf{J}_{p,diff} = -qD_p \frac{dp}{dx} \qquad (2.6.4)$$

where D_n and D_p are the electron and hole diffusion coefficients (cm^2/s).

The mobility and diffusion coefficients are related by the Einstein equations

$$D_n = \frac{kT}{q} \mu_n \qquad (2.6.5)$$

$$D_p = \frac{kT}{q} \mu_p \qquad (2.6.6)$$

3. UNDERSTANDING BAND DIAGRAMS

Band diagrams are used to picture what is going on inside electronic devices. It is important to be able to "read" band diagrams. In this section we'll "read" the band diagram of the pn junction at zero bias shown in Figure 4(A) (how this particular band diagram comes to be will be discussed in the next section).

First, notice that the bottom edge of the conduction band, E_c, is always parallel to the top of the valence band edge, E_v. This is because the separation of the bands (bandgap) is constant.

Electron energy band diagrams are also electric potential (voltage) diagrams since, by definition,

$$\psi = -E/q \qquad (3.1)$$

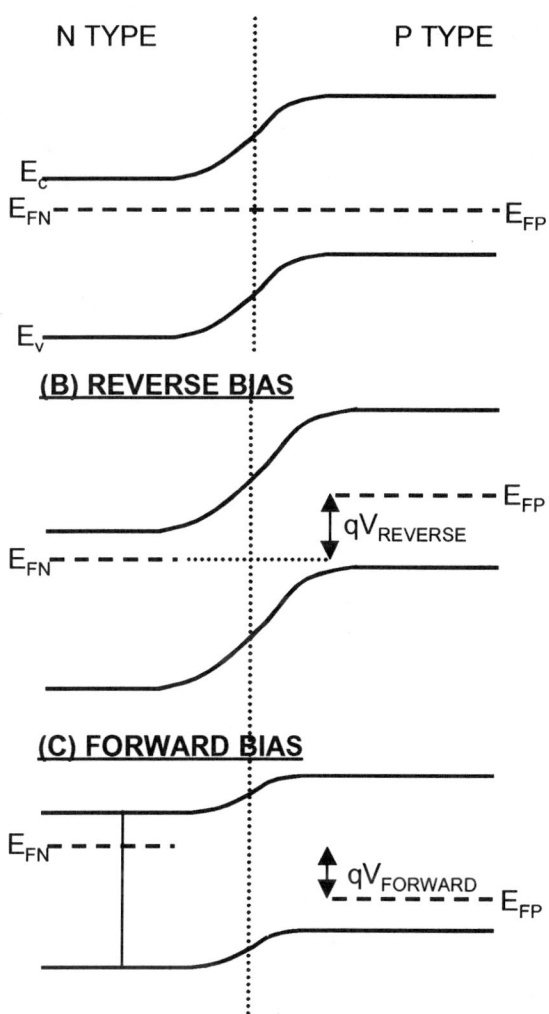

Figure 4. The band diagram of a pn junction under (A) Zero external bias, (B) Reverse external bias (p biased negative), and (C) Forward bias (p biased positive).

where ψ is the potential (V), E is the electron energy (eV), and the minus sign results since the electron's charge is negative, -1.6E-19 coulombs. Looking at Figure 4(A) it is seen that the electron energy (e.g., at the conduction or valence band edges) changes with position as we move from n to p. The potential energy also changes. When the energy increases the potential decreases.

The electric field $\mathbf{E}$ is related to the potential energy by

$$\mathbf{E} = -\frac{d\psi}{dx} \qquad (3.2)$$

The electric field, $\mathbf{E}$, is a vector having both magnitude and direction. Positively charged holes in the valence band will tend to drift in the direction of $\mathbf{E}$ while negatively charged conduction band electrons move in the opposite direction.

What does the band diagram in Figure 4(A) tell us about the electric field? Away from the junction the band energies are flat. The potential does not vary with position and hence the electric field is zero. However, near the junction the bands are bent and there is thus a variation in the potential. This varying potential produces an electric field as given by Eq 3.2. As we go from n to p the energy increases, the potential decreases and the resulting electric field vector points from n to p. Within this boundary region negative electrons will experience a force towards the bulk of the n type material and positive valence band holes will experience a force towards the p type bulk.

The band bending results from electric charges as given by Poisson's equation

$$\frac{d^2\psi(x)}{dx^2} = -\frac{\rho(x)}{\varepsilon} \tag{3.3}$$

where ρ is the charge density. This equation says that whenever there is a change in the slope of the potential versus position line, there will be an electric charge. Again, what can we infer from Fig. 4(A)? If we look carefully we can see that from the bulk of the n region towards the interface the slope of the energy line increases and thus the slope of the potential decreases. A decreasing slope of the potential indicates that there is a region of positive charge in the n type material at the interface. From the interface to the bulk of the p type material the slope of the energy line decreases, thus the slope of the potential line increases and thus there is a negative charge.

Finally, we notice that the Fermi level is constant from n to p type material. It can be shown that when the Fermi level is constant there is no net current flowing. (There may be equal and opposite currents flow, just no net current).

4. THE PN JUNCTION

In the previous section we saw that we use doping to create extrinsic materials that had either electrons or holes as the majority carrier. When we have an n type material in intimate contact with a p type material (e.g., by doping adjacent regions on a Si wafer with different atoms) we create a junction. Understanding the physical processes at a pn junction – drift, diffusion, built-in potential, depletion region – is necessary to understand how transistors work. In addition, the pn junction is a useful electrical device in its own right, a diode.

In this section we will treat the simple case that the junction is an abrupt boundary between two uniformly doped p and n materials. Also the currents will be low enough that the mobile (not fixed) charge in the depletion region (discussed below) can be neglected

In the pn junction there are four current components: electron diffusion, electron drift, hole diffusion and hole drift. The electrical properties of the pn junction result from the strong influence of electrical fields on diffusion currents, but not on drift currents.

To see what happens in a pn junction let's imagine that we initially have two physically separated pieces of n and p type

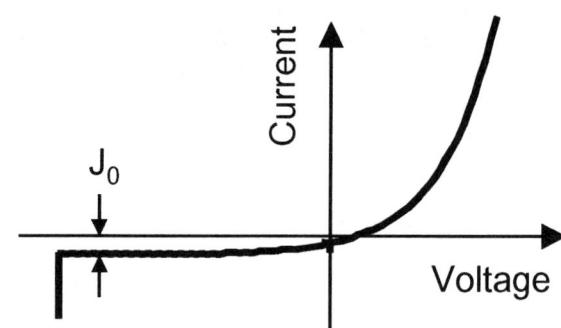

Figure 5. The current voltage curve of a pn diode. Positive voltage is forward bias (p positively biased).

material and then we bring them into intimate contact. Furthermore, let's initially just focus on what happens to the electrons in the conduction band. When isolated, the conduction band of the n type material has many more electrons than the conduction band of the p type material. When the materials are joined the electrons from the n type material at the boundary of the junction diffuse into the p type material. This leaves behind uncompensated positively charged donor atoms in the p type material.

These charges cause a change in the potential with position as shown by the change in the conduction band edge in Figure 4(A) . This band bending creates an energy barrier that reduces the number of electrons that can diffuse from n to p. At the same time the electric field induces a drift current of the minority carrier electrons in the p to the n material. This drift current is very small because of the small number of minority electrons in the n material. The direction of the drift current is opposite to the direction of the diffusion current. Equilibrium is reached when the potential barrier reduces the diffusion current to a level where it is equal and opposite to the drift current.

A similar process occurs with the holes. Holes diffuse from the p type material into the n type material. As holes leave the p type material a negative charge develops due to uncompensated acceptor ions. This sets up an electric field that opposes diffusion and also induces a drift current in the opposite direction. Equilibrium results when the diffusion current is reduced to a value that is equal and opposite to the drift current.

4.1 Built-in Potential

Even with zero bias there is a built-in potential at the junction due to the positive and negative charges.

$$\psi_{bi} = -(E_{Fn} - E_{Fp})/q \tag{4.1.1}$$

where the n type material is more positive than the p type.

When we apply a bias a net current will flow and the Fermi levels in the bulk of the n and p type material will be offset by the applied voltage, V. This modifies the potential across the junction

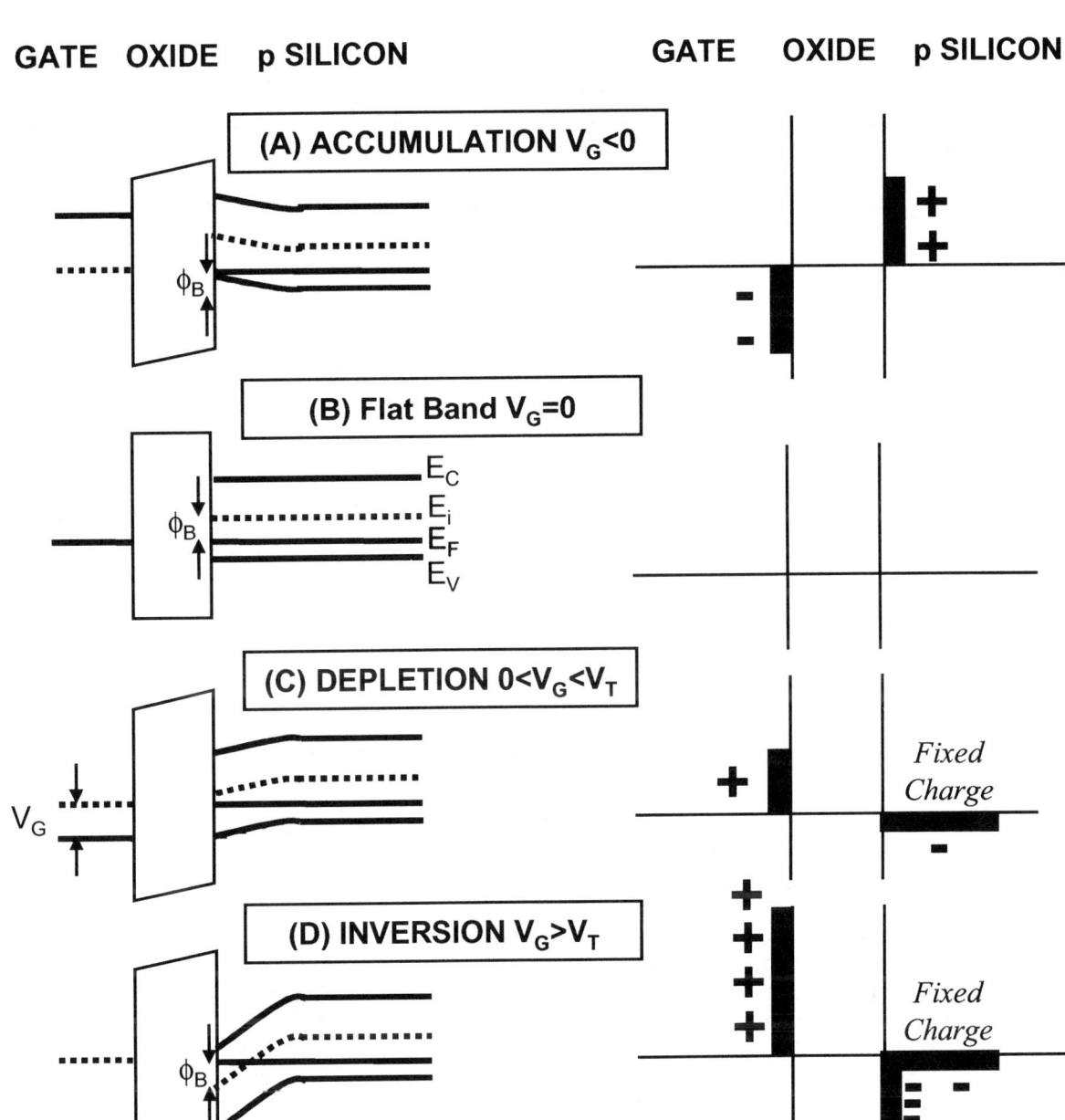

Figure 6. The band diagram (left side) and the charge distribution (right side) across a Metal-Oxide-p type Semiconductor capacitor for (A) Accumulation, (B) Flat Band, (C) Depletion and (D) Inversion.

$$\psi_{JUNCTION} = \psi_{bi} - V \qquad (4.1.2)$$

V is applied voltage defined as positive when the p material is biased positive and the n material negative ("forward bias"). The ability to change the junction potential leads to current rectification as described below.

4.2 Depletion Region

A depletion region is formed at the junction. The region is depleted of mobile carriers leaving behind uncompensated charged donor and acceptor ions. The magnitude of positive fixed charge on the n side is equal to the magnitude of the negative fixed charge on the p side.

The width of the depletion region is

$$W_d = \sqrt{\frac{2\varepsilon_{Si}(N_A + N_D)(\psi_{bi} - V)}{q N_A N_D}} \qquad (4.2.1)$$

It is possible (as in the source and drains of MOS transistors) to have one side of the junction much more heavily doped than the other side. If the n region is much more heavily doped than the p region (N_D>>N_A) then

$$W_d = \sqrt{\frac{2\varepsilon_{Si}(\psi_{bi} - V)}{qN_A}}$$

(4.2.2)

Observe that in this special case the width of the junction is dependent only on the lightly doped material. Furthermore, the depletion width goes as the inverse square root of the doping concentration and approximately as the square root of the voltage (when V is large compared to the built-in voltage.)

4.3 Forward and Reverse Bias

The pn junction is a diode – it conducts large currents when forward biased (p region positive) and only a small leakage current when reversed biased (n region negative). The equation for the current with the simplifying assumptions mentioned above is

$$\mathbf{J} = \mathbf{J}_0 \left[\exp(qV/kT) - 1 \right]$$

(4.3.1)

where positive voltage is defined when the p region is positive and $\mathbf{J}_0$ is the small leakage current that occurs during reverse bias. The current voltage curve for the diode is shown in Figure 5 including the breakdown that occurs at large reverse biases.

The rectifying action of the pn junction can be explained as follows. Application of an external bias results in a displacement of the Fermi levels in the bulk of the n and p materials as shown in Figures 4(B) and 4(C). This results in a change in the curvature of the bands. Reverse bias tends to increase the energy differences between the bands on the p and n sides. It also increases the width of the depletion region. On the other hand, forward bias reduces the energy barriers and the width of the depletion regions.

The electron and hole diffusion current components are very sensitive to the energy barriers, and thus to the applied voltages. In Section 2.2 it was shown that the probability of finding an electron decreases exponentially with increasing energy. As conduction band electrons from the n material try to diffuse to the p material they encounter an energy barrier (the energy of the conduction band edge rises from n to p). Only those electrons with energy greater than the barrier height can go from n to p. As the barrier height is raised there is an exponentially decreasing number of electrons that can diffuse. Similarly, if we lower the barrier height there is an exponentially increasing number of electrons that can surmount the barrier. Thus the electron diffusion is exponentially dependent on the barrier height. Since the barrier height is directly related to the applied voltage, the diffusion current is also exponentially dependent on the applied voltage. The holes face a similar barrier in diffusing from p to n and the hole diffusion current is also exponentially dependent on the applied voltage.

In contrast, the drift currents do not, to first order, depend on the barrier heights. The diffusion current components result from the motion of minority carrier electrons from the p material and from minority carrier holes from the n material. These minority carriers are injected into, and swept across, the depletion region. These diffusion currents are limited by the availability of minority carriers and not by the field in the junction. The applied fields do not affect the availability of minority carriers and thus do not affect the diffusion current components.

The rectifying action of the pn diode can now be explained. At 0 external bias there is no electron current or hole current since the energy barrier equilibrates at a value that reduces the diffusion current components to a level that is equal and opposite the drift current components. For reverse bias the energy barriers for both holes and electrons are raised and the diffusion currents are exponentially reduced to zero. All that's left are the small drift current components that comprise the negative reverse current. However, for forward bias the barriers are lowered and the diffusion current components grow exponentially.

5. THE MOS CAPACITOR

The next thing we need to investigate is the MOS (Metal Oxide Semiconductor) Capacitor. The capacitor consists of a top metal gate, an insulator, and an extrinsic semiconductor. There are two electrical connections to the gate and to the bottom semiconductor substrate. The substrate is grounded.

A number of simplifying assumptions will be introduced. Initially, we'll assume that there is no work function difference, no oxide charges and no interface states. We'll also assume that the semiconductor substrate is uniformly doped. The discussion below will be for a p substrate; n substrates behave in a similar manner.

Figure 6 shows the band diagram (left side) and charge distributions (right side) of a p type MOS capacitor for four different biasing conditions. In all of the cases there is no steady state current flow because the oxide is an insulator and because of the large barrier to injection of charges from the semiconductor. Since the currents are zero, the Fermi level in the semiconductor will be flat. The energy bands near the interface, however, will be bent by applied voltages. Viewing the changes in the position of the constant Fermi level relative to the curving band edges will allow us to picture the changes in the concentrations of electrons and holes at the surface.

First consider, Figure 6(B), the "Flat Band" condition which, with the assumptions we have made, occurs at $V_G = 0$. As the name applies the bands are flat as shown. Since the potential does not vary with position, there is no net charge in the Si. The Fermi level in the metal aligns with the Fermi level in the Silicon when $V_G = 0$. In the Silicon we show, starting at the bottom, the valence band edge (E_C), the Fermi level (E_F), the Fermi level of the intrinsic semiconductor (Ei) and the conduction band edge (E_C). Since this is a p type material the Fermi level is near the valence band edge and holes are the majority carrier.

In (A) we have the "Accumulation" condition. A negative gate voltage attracts the majority hole carriers to the silicon oxide

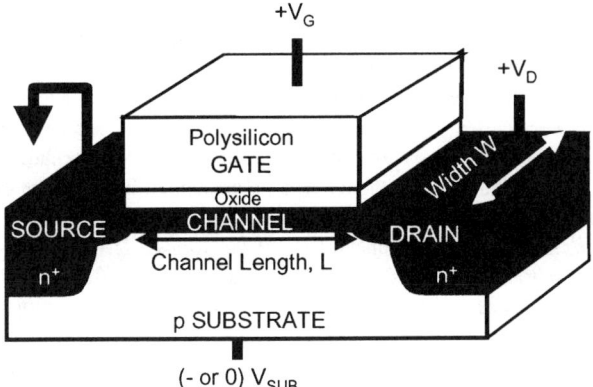

Figure 7. Schematic diagram of an n channel MOS transistor.

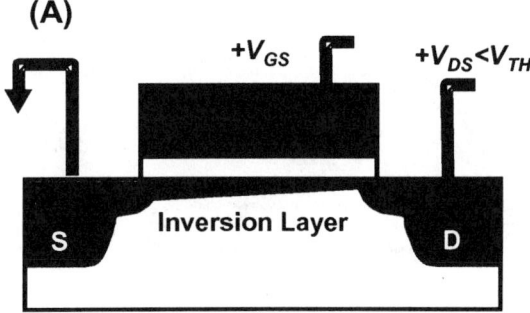

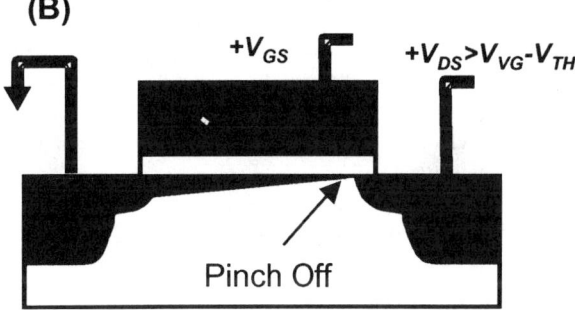

Figure 8. The inversion layer in an n channel transistor. In the linear region (A) the inversion layer extends from the source to the drain. In (B) the inversion layer is pinched off before it reaches the drain.

interface. The bands are bent upward. At the surface the Fermi level is even closer to the valence band edge consistent with the increased hole concentration at the surface. Due to the large number of majority carrier holes in the p substrate the device behaves like an ordinary parallel plate capacitor. (Accumulation is a biasing condition that is not used in MOS transistors.)

In (C) we show the "Depletion" condition. We apply a positive voltage to the gate that is less than the threshold voltage (to be defined below). The positive gate voltage repels the holes away from the interface. The result is a region of negative charge (called the depletion layer) resulting from uncovering the fixed acceptor ions. The bands are bent downward as we approach the interface. Note that the Fermi level moves away from the valence band edge, which is consistent with a reduction in hole concentration.

In Figure 6 (D) we apply an even greater voltage producing the "Inversion" condition. The bands are bent even more. The Fermi level at the surface is now so close to the conduction band edge that there are a large number of electrons in the conduction band. A very thin layer at the surface has been "inverted" from p to n type by band bending. The voltage where the inversion layer starts to form is called the threshold voltage. The depletion layer stops growing once inversion starts.

5.1 Work Function Differences

The flat band voltage is generally not at 0V for two reasons. First, the Fermi levels in the metal gate and in the semiconductor substrate do not, in general line up. The difference in Fermi levels is called the work function difference. This work function difference produces a built-in potential across the capacitor just as the work function differences produce a built-in potential across the pn junction. This work function difference shifts the flat band voltage.

The second reason that the flat band voltage shifts is that there can be extra charges in the oxide or at the semiconductor surface. As a result of processing (e.g., Na contamination) or radiation effects or from operation, charges can be present in the oxide. These charges also shift the flat band voltage.

5.2 Threshold Voltage

The threshold voltage is used to indicate the voltage at which the inversion layer starts to form. There are several possible definitions of V_{TH}. One popular version is illustrated in Figures 6(B) and 6(D). In (B) we see that the p substrate's Fermi level is *below* the intrinsic Fermi level by an amount we will call $q\phi_b$. The threshold is defined as the voltage that causes the Fermi level to be bent so that it is *above* the intrinsic level by the same amount, as shown in Figure (D). In essence threshold is defined when the minority carrier (electron) concentration at the surface in inversion is equal to the majority carrier (hole) concentration in the bulk.

Using this definition the threshold is given by

$$V_{TH} = V_{FB} + 2\phi_b + \frac{t_{OX}\sqrt{2\varepsilon_{sI}qN_A(2\phi_B)}}{\varepsilon_{OX}} \qquad (5.2.1)$$

where

$$\phi_B = \frac{kT}{q}\ln\left(\frac{N_A}{n_i}\right) \qquad (5.2.2)$$

As the voltage is changed from V_{FB} towards V_{TH} the depletion region grows so that the total depletion charge in the silicon equals the charge on the metal electrode. However, once the inversion layer is formed the depletion region essentially stops growing. The additional charges on the metal gate are balanced by a growth in the

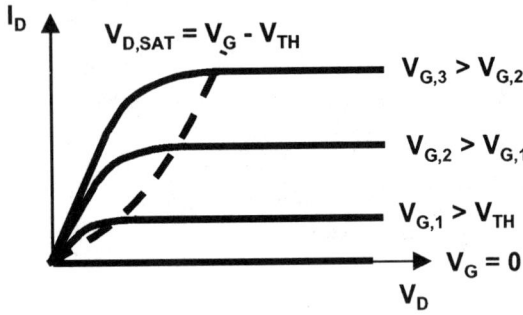

Figure 9. Schematic diagram of the drain current (solid lines) versus drain voltage with the gate voltage as a parameter for an n channel MOS transistor. The dashed line represents the curve of the drain voltage at which the drain current saturates.

inversion layer charge. The maximum width (thickness) of the depletion region is given by

$$W_{D,MAX} = \sqrt{\frac{4\varepsilon_{Si}kT\ln(N_A/n_i)}{q^2 N_A}} \qquad (5.2.3)$$

6. THE LONG CHANNEL MOS TRANSISTOR

A diagram of an n channel MOS transistor is shown in Figure 7. (We can also build p channel transistors.) The transistor is built on a p type silicon substrate. The source and drain are made up of n type silicon. The region along the surface of the silicon between the source and drain is called the channel. It has a length, L, and a width, W and thickness W_D. Above the channel is a thin oxide layer of thickness, t_{ox}. Above the oxide is the gate, which is typically made of degenerately doped polysilicon. The transistor is called n type because when the surface is inverted it will become n type and the conduction will be due to the drift of electrons.

The transistor is a four terminal electrical device. We'll consider the source to be grounded. Furthermore, in the following description we'll also consider the substrate to be grounded. When the gate is at 0V or the drain to source voltage is 0V the device is "off" and there is no current between source and drain. When the gate voltage is above the threshold voltage and the drain voltage is positive then the device is "on" and a current flows from drain to the source. Electrons move from the source to the drain, but since current is defined as the direction of positive charges, the current direction is opposite of the electron direction.

6.1 Long Channel Behavior

The transistor action results from the effect of the gate voltage and the drain voltage. In its most general form this is a multidimensional problem that needs to be analyzed on a 3-d numerical simulator problem. However, if the channel length is "long" and some simplifying assumptions are made an analytical solution is possible. (It's not the channel length, per se, but the

channel length in proportion to the gate oxide thickness and the depth of the source and drain that makes the difference between long and short channel transistors.) We will treat this long channel transistor first and later consider the impact of short channels.

The drain current versus drain voltage curves are shown in Figure 9 with the gate voltage as a parameter. When $V_G=0$ there is no current at any drain voltage. When the gate voltage exceeds the threshold voltage, V_{TH}, an inversion layer forms. The drain current now initially increases linearly with drain voltage (linear region) then assumes a parabolic shape and eventually saturates (saturation region.)

6.2 Linear Region

In the linear region the inversion layer will exist all the way across the channel from the edge of the source depletion region to the edge of the drain depletion region as indicated in Figure 8 (A). Furthermore, in this gate voltage region the perturbation of the inversion layer by the drain voltage is small. The channel of the MOS transistor behaves like a resistor, with the resistance determined by the inversion layer. The formula for the current in the inversion region is

$$I_{D,LIN} = \frac{\mu_{n,eff} W \varepsilon_{OX}}{L t_{OX}}(V_G - V_{TH})V_D$$

$$\text{for } V_D << (V_G - V_{TH}) \qquad (6.2.1)$$

where $\mu_{n,eff}$ is the effective mobility of electrons in the inversion channel.

6.3 Saturation Region

As the drain voltage is increased, the drain current vs. drain bias curve begins to saturate. The effect of the drain voltage on the inversion layer near the drain end of the channel cannot be ignored. The drain voltage reduces the potential at the surface along the channel. This, in turn, reduces the magnitude of the inversion layer which depends exponentially on the potential. At low voltages the effect is to narrow the inversion layer from source to drain as indicated schematically in Figure 8(A).

Eventually, as the drain voltage is increased the width of the inversion layer decreases and eventually vanishes (is "pinched off") near the drain end. Pinch off occurs when

$$V_G - V_D \le V_{TH} \qquad (6.3.1)$$

The drain voltage at which saturation occurs, $V_{D<SAT}$ is given by

$$V_{D,SAT} = V_G - V_{TH} \qquad (6.3.2)$$

The dashed line in Figure 9 shows the curve of $V_{D<SAT}$.

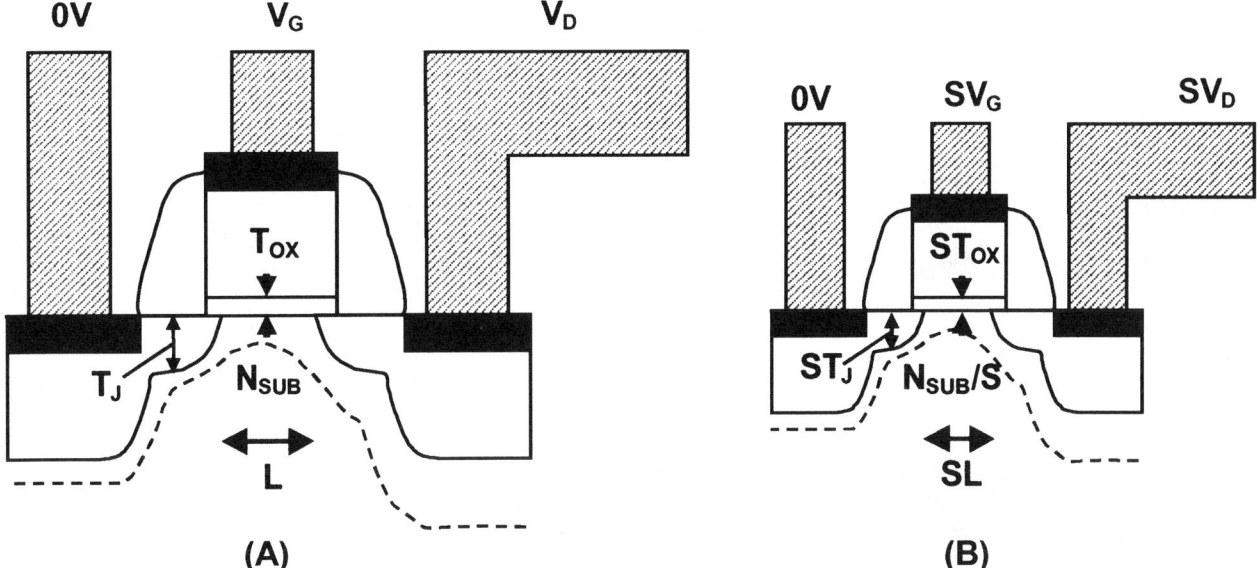

Figure 10. Illustration of ideal constant field scaling. The starting transistor is shown in (A). The scaled transistor is shown in (B). The scaling factor, S, is less than one (New generations of IC technologies have an S~0.7).

For drain voltages above $V_{D,SAT}$ the extra voltage is dropped across the depletion region, not along the inversion layer. Thus there is no increase in the drain current. The saturation value of the drain current is given by

$$I_{D,SAT} = \frac{\mu_{n,eff} W \varepsilon_{OX}}{2 L t_{OX}} (V_G - V_{TH})^2$$

$$\text{for } V_D \geq (V_G - V_{TH}) \qquad (6.3.3)$$

Notice that the saturation current depends on the square of the difference between the gate and threshold voltages.

6.4 Subthreshold Region

Finally, we need to consider subthreshold leakage. As the value of the gate voltage becomes less than the threshold, the carriers in the channel do not instantly go to the zero. Rather there is a region of "weak inversion" in which the carriers decrease approximately exponentially with decreasing voltage. This is the subthreshold region and the subthreshold current

$$I_{D,SUB} \propto \exp\left(\frac{qV_G}{kT}\right) \left[1 - \exp\left(\frac{-qV_D}{kT}\right)\right]$$

$$\text{for } V_G < V_{TH} \qquad (6.4.1)$$

The subthreshold current is exponentially dependent on the gate voltage and saturates at small drain voltages.

An important figure of merit used is the subthreshold slope which is defined as the gate voltage change required to reduce the drain current by a factor of 10. This slope cannot be less than 60mV/decade at room temperature. The slope varies linearly with absolute temperature.

6.5 p Channel Devices

It is also possible to construct p channel devices. In this case the substrate is n type and the source and drains are p++. The inversion layer consists of holes. The hole motion and the hole current are from source to drain. Negative voltages are required to form the hole inversion layer.

The effective mobility of holes is less than electrons. As a result, to get the same drive current a p channel device must have a larger width than the corresponding n channel device.

7. IDEAL SCALING

The density, speed and power of a MOS transistor can all be simultaneously improved by scaling down the voltages and feature sizes. Figure 10 shows an example of constant field scaling. The objective in this scaling scenario is to keep the value of the electric fields constant while scaling. Constant field scaling is good for reliability issues such as gate oxide breakdown times which decreases exponentially as the gate field increases. The dimensions and voltages are all reduced by a factor S (S is typically ~0.7 per technology generation). To reduce the depletion widths the substrate doping has to be increased.

Since the dimensions are reduced by a factor of S, the area of silicon required for a single transistor is reduced by S^2. This means that there can be $1/S^2$ more transistors in a given area. Next, the maximum frequency is increased by $1/S$. Finally, the power dissipated per transistor (including the effect of the higher frequency) is reduced by S^2. The power per chip of a fixed area stays constant (the power per transistor is reduced by S^2 but the number of transistors per area is increased by $1/S^2$.

8. CHALLENGES WITH DEEP SUBMICRON TRANSISTORS

There are a number of challenges when transistors are scaled into the deep submicron regime. An in-depth discussion is beyond the scope of this chapter. A few of these effects will be briefly mentioned in this section.

A major challenge is controlling the off state ($V_G=0V$) leakage of transistors. As the power supply voltage is scaled down the threshold voltage must be scaled down to maintain large saturation currents (Eq 6.3.3). However, the subthreshold slope does not scale. The threshold voltage divided by the subthreshold slope determines the number of decades that the current at $V_G=0V$ will be reduced from the saturation current. As V_{TH} is reduced the number of decades of current reduction at 0 volts decreases. This leads to an increase in the off state current per transistor increases. The off state current per chip grows at an even faster rate since, concurrent with reducing V_{TH}, the number of transistors per chip are being increased.

Short channel effects start to occur in submicron transistors. The drain and source depletion regions can become a significant fraction of the drain to source separation. This has the effect of reducing the channel length. This results in a lowering of threshold voltages in short channel devices. If the channel becomes so short that the source and drain depletion regions can "touch" then additional leakage can occur do to "punch through."

Furthermore, as transistor technologies have been scaled down, the electric fields in the channel have tended to increase. As mentioned in Section 2.6.1 the mobility is no longer proportional to the applied field at high fields. In the limiting case of velocity saturation the saturation current depends on (V_G-V_{TH}) instead of (V_G-V_{TH})^2.

The final effect we want to consider is parasitic resistances. In the above work it has been assumed that the resistances in the circuit interconnects and in the source and drain regions were negligible. Unfortunately, with scaling down comes a reduction in the cross sectional areas of the interconnect, source and drain. Since the resistivity of the materials does not change with scaling, the resistance of these elements increases. When these resistances become comparable to the resistance of the MOS channel the benefits of scaling on the performance of the integrated circuit are diminished. The need to reduce the resistance of interconnections is the motivation for the replacement of Al on-chip wiring with lower resistivity Cu.

For more details about submicron issues the reader is referred to the book by Taur and Ning in the Suggested References

ACKNOWLEDGEMENTS

This work was supported by the Electronics Quality/Reliability Center at Sandia National Laboratories (www. sandia.gov/eqrc).

Sandia is a multiprogram laboratory operated by Sandia Corporation, a Lockheed Martin Company, for the United States Department of Energy under Contract DE-AC04-94AL85000.

SUGGESTED REFERENCES

A good introductory text:
Streetman, Ben G., *Solid State Electronic Devices, 5th Ed*, New Jersey: Prentice Hall, 1999.

An up-to-date reference with in-depth treatment of the subtleties and tradeoffs in deep submicron transistors
Taur, Yuan and Ning, Tak H., *Fundamentals of Modern VLSI Devices*, New York: Cambridge University Press, 1998.

The classic reference:
Sze, S.M., *Physics of Semiconductor Devices, 2nd. Ed.*, New York:John Wiley, 1981

Reliability for the Failure Analyst

D. Burgess

PURPOSE

Failure analysts are key people in achieving the reliability the electronics world has enjoyed for the last few years. They have done this by providing the understanding required to eliminate failure causes. The work of the failure analyst is involved with the fixing of problems encountered when attempting to manufacture a product properly. The analyst is also involved when failures are encountered in the field. Figure 1 shows the typical "bathtub" curve describing product failure rate with time.

Despite this intimate role we have played, many analysts have little time to develop fluency in the mathematics of the quality and reliability they have helped create.

The purpose of this discussion will be to define and give an understanding of the terms of quality and reliability with respect to failure analysis.

At the expense of being rigidly correct, the discussion here will simplify quality and reliability statistics as they are commonly used. Just as a feel for the statistics of cards helps a poker player, a feel for reliability can help a failure analyst make wiser choices.

Quality is freedom from initial defects when the product is delivered. Reliability is how long the defect-free status is maintained.

TYPICAL TERMS

- AOQL
- AQL
- Confidence interval
- LPTD
- MTTF
- IFR
- Fit
- Normal
- Log normal
- Chi-squared
- Poisson
- Weibull
- MIL-STD-883
- MIL-STD-105
- Infant mortality
- Wearout failures
- Device-hours
- Acceleration factor

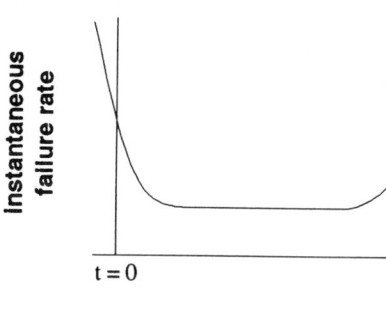

Fig. 1 Example of a typical "bathtub" failure rate curve. At the extreme left, infant mortality and early life failures constitute a decreasing failure rate. In the center, random or useful rate failures are constant and of low magnitude. At the right side of the curve, wearout failures have begun to occur. The product design should be obsolete long before wearout failures begin to occur.

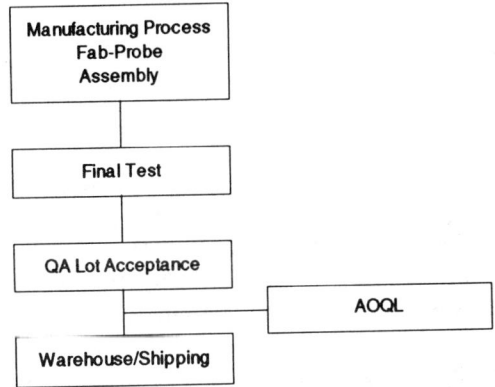

Fig. 2 Process flow block diagram, showing position of AOQL measurement.

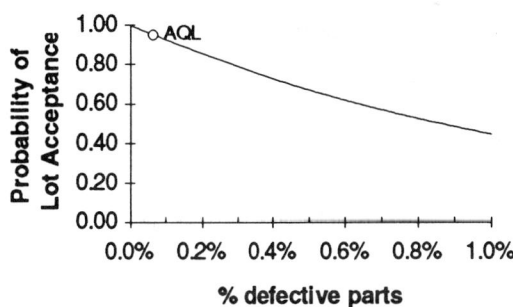

Fig. 3 Illustration of AQL.

25

QUALITY

At the manufacturer, measurement is done by QA sample randomly selected final test. Sample sizes depend on a sampling plan. An established and commonly used sampling plan is MIL STD 105, which specifies sample size as a function of desired AQL level and the lot size.

HOW IS AOQL MEASURED?

AOQL is a demonstrable fact, while AQL and LTPD are artifacts of a particular sampling plan which define only the user's and supplier's risks (a total of 100%) that product containing a given proportion defective will be accepted for delivery.

QA sampling is done at 0.25 AQL level. Sample sizes are determined by long established tables (MIL STD 105) according to lot size.

How can 0.25% AQL demonstrate 100 ppm? (0.25% = 0.25 / 100 = 2500 ppm)

WHAT DOES IT MEAN THAT THIS LOT PASSED 0.25 AQL SAMPLE?

If an accept on zero plan is used, we know n acceptable parts were chosen at random, what is the defect percent? If the probability that a part is defective is p, the probability of the lot being accepted is:

Prob. passing $= (1-p)^n$
$0.1 = (1 - p_{90})^n$
$0.4 = (1 - p_{60})^n$

From the plot, only 10% of the samples will come from lots with defects $p > p_{90}$. Solve directly or estimate values by the approximation $(1 - x/n)^n \cong e^{-x}$. Since $e^{-23} \cong 0.1$, the 90% confidence estimate is $p_{90} \cong 23/n$. [p_{90} has been named the LTPD (lot tolerant percent defective.)] Similarly, $e^{-0.92} \cong 0.4$ implies that $p_{60} \cong 0.92/n$. The actual 60% or 90% confidence estimate is less than p_{60} or p_{90}.

Estimative AOQ requires data from many lots—say over a month's time.

$L = \#$ lots sampled

$LR = \#$ lots rejected

$LR/L = \%$ lots rejected

$S =$ Total # samples per lot

$N =$ Total # samples

$D =$ Total # defective devices

for $D = 0$

$LR/L = 0$

60% confidence $AOQL \leq AOQL_{60} = 0.92/N$

For $D > 0$

This method uses a multiplier derived from the χ^2 distribution. The arithmetic is simple.

$$AOQL = \frac{D \cdot CF}{N}$$

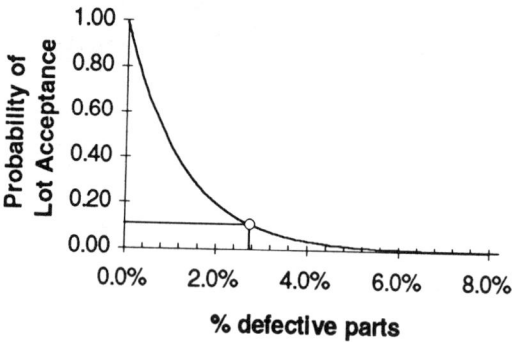

Operating Characteristic

Fig. 4 Probability of lot acceptance as a function of percent defective parts.

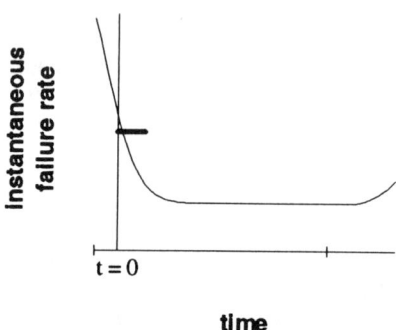

Fig. 5 Probability of instantaneous failure as a function of time.

where D is the number of defects, N is the total sample size, and CF is the confidence factor multiplier from the table below:

D	CF	
	60% confidence	90% confidence
1	2.02	3.89
2	1.55	2.66
3	1.39	2.23
4	1.31	2.00
5	1.26	1.85

EARLY FAILURE RATE (EFR) (ACCELERATED)

Accelerated-stress testing is legitimately used to estimate, and in some aspect and cases to control, both early and long-term reliability, not only the early. It is the only way to increase confidence in future reliability before the product is delivered and used.

Because of small number of failures, failure rate is assumed constant during a specified, but somewhat arbitrary, time. Typical are (a) 48 hours, 125 °C, 7V and (b) 168 hours, 125 °C, 7V.

EFR may be given as ppm value for the specified stress-time. In this case all calculations are exactly the same as for quality measurement. We are simply defining a test that extends over time.

For example, suppose 1000 devices per month are selected from inventory with no failures detected, $D = 0$:

$$EFR\,(60\%) = \frac{0.92}{1000} \times 10^6 = 9200 \text{ ppm}$$

The calculation from a year of tests at 1000 devices/month for $D = 0$ and $D = 1$:

$D = 0$ $\qquad\qquad\qquad EFR\,(60\%) = \dfrac{0.92}{12,000} \times 10^6 = 720 \text{ ppm}$

$D = 1$ $\qquad\qquad\qquad EFR\,(90\%) = \dfrac{1 \times 2.02}{12,000} \times 10^6 = 1667 \text{ ppm}$

EFR expressed as failure rate at normal use conditions (55 °C, 5V)

$$EFR = \frac{EFR \cdot 10^{-6}}{(HRS)(AF_{TEMP})(AF_{VOLT})} \text{ in failures/hour}$$

Acceleration factors AF_{TEMP}, AF_{VOLT} will be discussed later.

Use of cumulative lot-test data to estimate quality and/or reliability is acceptable for a stable design and manufacturing process; it is not for varying ones. Cumulative data do not influence the estimate of quality/reliability for a given lot of product.

INTRINSIC FAILURE RATE (IFR)

Again, because of small number of failures, failure rate is assumed constant during a specified time. Typical is 1000 hours, 125 °C, 7V. IFR is expected as a failure rate at normal use conditions.

$$IFR = \frac{CF \cdot (\# \text{ Fails})}{(HRS)(AF_{TEMP})(AF_{VOLT})} \times 10^9$$

DISCUSSION POINTS

1. Failure rate is shown for times $t < 0$, because the device is defined during the manufacturing process. Oxides are grown, contacts are made, etc. Defectives are removed at burn-in screen, other environment or electrical screen or at final test. These defects are yield loss.

2. Ratios EFR/IFR greater than 1 suggest the need for more/better production screening.

3. Failures occurring at final test, and during EFR and IFR are probably defect-generated. Reduction of value depends on decreasing defects.

4. If a semiconductor company's data shows AOQL < 100 ppm, observation of several failures by a customer is an unlikely event unless reject criteria, test conditions, or unique handling and processing conditions are involved. The failure analyst should not overlook lot history.

5. One "fit" is one failure in 10^9 device-hours (or unit-hours). If semiconductor company reliability monitor data show EFR $\cong$ IFR < 100 FITS, the observation of several fails by the customer suggest more than just statistical variation. Check lot history, special handling, testing, and fail criteria.

6. If field failure rate >> estimated failure rate, many reasons are more likely than statistical variation:

   a. *AF* factors. Using one accelerator has shown above, can be accurate only when the predominant failure mechanism is known and is properly modeled. The resulting number is useful if all conditions are met.

   b. Process variation has made other mechanisms significant.

   c. Field conditions include temperature cycles, humidity, and hot carriers not included in mode.

   d. Just as acceleration factor must be matched to failure mechanism, production screens are also mechanism specific. Screen will not be effective if defect types change.

REFERENCES

1. *Toshiba Semiconductor Reliability Handbook*

2. D.L. Burgess and O.D. Trapp, *Failure and Yield Analysis Handbook*, Technology Associates, 51 Hillbrook Drive, Portola Valley, California 94205. 415-941-8272.

3. J.M Juran and F.M. Gryna, *Juran's Quality Handbook*, McGraw-Hill, 1988.

4. T.P. Omdahl, *Reliability, Availability, and Maintainability (RAM) Dictionary*, ASCQ Press, 1988.

5. G.E.P. Box, W.G. Hunter, and S. Hunter, *Statistics for Experimenters*, Wiley, 1978.

6. D.J. Klinger, Y. Nakada, and M.A. Menendez, *AT & T Reliability Manual*, Van Nostrand-Reinhold, 1990.

Reliability Physics and Assurance

S. Khandekar and L. Gutai

Abstract:

The failure analysis laboratory in any semiconductor manufacturing company gets about 35-45% of work load through the reliability engineering. These failures are generated during the various environmental, wafer level reliability and other qualification tests done for product/process qualification or product monitoring. For a failure analyst, it is important to understand all the reliability tests performed as well as the reliability assurance program which generate these failures. Knowing the history of failure, the analysis becomes a much easier task. This tutorial presents discussion of 'what is quality v/s reliability, the role of reliability towards design, process reliability issues such as Electromigration (EM), Gate Oxide Integrity (GOI), Time Dependent Dielectric Breakdown (TDDB) and environmental reliability testing. Relationships between failure mechanisms and environmental testing will be established.

Introduction

The two most important implications of the continuous scaling trend in the semiconductor technology for designers, manufacturers and users are the (sometimes mutually exclusive) requirements for higher performance and better reliability on both the device level and circuit level. On the device level, the feature size is decreasing while the performance is increasing. On the circuit level, the circuit size, functionality and complexity are increasing. At the same time, quality and reliability requirements imposed on both the devices and the circuits are becoming more stringent.

Moore's Law

Dr. Gordon Moore, one of the founders of Intel Corporation stated several years ago, that "Computing power doubles every 18 months on an equivalent size chip". This is known in the industry as Moore's law. Moore's law has proven remarkably successful in characterizing the growth of the semiconductor industry for the past three decades. In the future, a discontinuity in basic semiconductor materials will be necessary for the industry to continue with the Moore's law. As the circuit complexity is increasing, circuit reliability has to increase at an even faster rate to follow Moore's law.

Saturation of Semiconductor Growth

In light of Moore's law, it is worthwhile to look at where the semiconductor industry is heading:

	1995	~2030
Semiconductor as % of Electronics	17%	5%
Electronics as % of GWP	4%	8%
Semiconductors as % of GWP	0.7%	3%
CMOS Technology	0.35μ	0.05μ
World Semiconductor Sales	$140B	$12,000B
Annual Growth Rate	16%	8%, same as GWP

Reliability Philosophy

For a typical semiconductor manufacturing company, the reliability program consists of build in reliability (BIR), design-in reliability (DIR), product qualification and product monitoring to ensure long-term reliability.

Key Components of the program are:

1. Reliability Assessment
2. Reliability Planning
3. Product Qualification
4. Product Monitoring
5. Failure Analysis

Reliability departments should be active right from the start of the design of a device.

Reliability Challenge

IC Reliability Requirements	↑	Scaling on Process	↓

According to the SRC roadmap, IC reliability needs to improve by three orders of magnitude by the year 2000 and scaling forces degradation in wear out mechanisms of roughly two to four orders of magnitudes, major emphasis in reliability physics research in the sub-half micron silicon era is to make certain that the wear out lifetimes are adequate for the 10 to 20 years lifetime required of a product.

If 1% of chips are allowed to fail in 10 years and each chip has 10^8 transistors, then only one failure is allowable in 10^{11} transistor-years. The age of the universe is only 10^{10} years, hence reliability by testing is no longer adequate.

The challenge is to increase the understanding of failure physics (oxide, metallization, hot electron) build-in reliability (process control, wafer-level monitor) and design-in reliability (DIR, reliability simulation, CAD).

To achieve this, help is needed from:
- New interconnect materials, e.g. barrier & under/over, copper
- layering
- Tungsten plugs, hot Al refill
- Newer gate oxide material, e.g. nitrous oxide
- Lower power supply
- Better drain engineering
- Better understanding of AC limits

Industry Reliability Goals
(From SRC Technology Roadmap)

1992	100 FIT
1992-1995	10 FIT
1995-1998	1 FIT
1998-2001	0.1 FIT

FIT and MTBF

FIT Rate (Failure In Time):
Calculated as Number of device failure per billion device hours

e.g. Life test on 10,000 devices for 1,000 hours
Total device hours = 10,000 x 1,000 = 10E + 6 hours
If there was one failure, the FIT rate will be;
(1/10E+6) x 10E+9 = 100 FITs

MTBF (Mean Time Between Failure)
    = 1/FIT i.e. for 100 FIT,
    the MTBF = 1/100x10E+9 = 10E+7 hours

Reliability v/s Quality

It is often misunderstood that both quality and reliability mean the same thing. Quality is a measure of products' compliance to its' specifications as it is shipped where as reliability measures the products' long term compliance to the specifications even after the product has gone through vigorous operation.

Quality thus is a measure of how well the product conforms to specifications.

Reliability is simply Quality over time.

What is Reliability?

The probability that an item will perform a specified function under specified conditions for a specified period of time. Reliability of electronic components generally, integrated circuits specifically, has been increasing since their creation. Today's expectation is, almost literally, that an integrated circuit will last forever.

In today's world quality is no longer expected, it is assumed. The important thing for the manufacturer is to establish a good reliability program and comply with it. Long term reliability is what the customers are now expecting.

The three classes of reliability failures for a semiconductor component are:

1. Failure due to catastrophic event (ESD, EOS, Latch-up)
2. Failure due to point defects (particles, substrate defects, conductor "stringers", metal voids)
3. Failure due to inherent process weakness (such as HCI, EM, SV, PMIC/contamination/corrosion, gate oxide wear-out)

Reliability Expectations

In a keynote address at the 1985 EOS/ESD symposium, Les Avery, the speaker noted on the then transition to sub -- $2\mu m$ technologies

"The continuous thrust to produce denser, faster integrated circuits has a profound impact on their ESD sensitivity. The thin oxides, shallow junctions, and small feature sizes result in significantly lower voltage and energy handling capabilities. Designers are already having to compromise between circuit speed and ESD capability..."

We have come a long way since 1985 and the issues regarding the thinner oxides, shallower junctions and small feature sizes are much more complex than what Les was commenting on.

Build In Reliability

Build in reliability (BIR) is a methodology that matches technology capability with design requirements to insure an acceptable customer perceived failure rate over the useful lifetime of an integrated circuit. A closed loop system is required to concurrently develop, understand and modify both technology capabilities and design requirements to achieve the ultimate goal.

True cross functional/cooperative/concurrent engineering efforts are needed to effectively build in reliability for the design, development, reliability, and manufacturing disciplines.

Improved reliability simulation tools allow for efficient optimization of design and process features

Deep Sub-micron processes reliability prediction

Scaling influences different reliability mechanisms in different ways. For example, with scaling, electromigration becomes more of a concern, while HCI (with the scaling down of the power supply voltages) might be less of a concern. However, in both cases, the "reliability margin" is becoming much smaller than it used to be. For the 1.0μ (and even for the 0.5μ) processes the conservative lifetime requirements were easily met in the past by *appropriately chosen design rules* (applied to *each* circuit element).

This practice provided an unnecessarily large reliability margin since in an actual circuit (paraphrasing the popular saying) "you stress some of the components all of the time, you stress all of the components some of the time but you don't stress all of the components all of the time". In addition, circuits operate mostly at ac rather than dc conditions and both Electromigration and Hot Carrier Injection lifetimes are orders of magnitude larger at ac conditions than at dc conditions. All these factors contributed to the high reliability (long lifetime) of the actual circuits. This explains the virtual lack of field returns for EM, HCI or Gate Oxide Integrity failures in the recent past.

Going below the 0.5μ feature size can radically change this picture mainly because, in general, improving device performance decreases device reliability. For example, the shrinking Al interconnect can no longer satisfy the speed, current, and heat requirements. Thinner oxide eliminates the Fowler—Nordheim (F-N) tunneling but increases leakage and poses new reliability problems. New solutions and new materials are needed that might cause reliability problems that have not been investigated in depth yet.

In the following, the some of the major wear-out mechanisms (EM, HCI, TDDB and GOI) will be discussed separately. According to the literature published in this area, the differences between reliability concerns for the 0.35μ and for the 0.25μ processes are quantitative rather than qualitative in general, therefore we will discuss them together.

Electromigration (EM)

Electromigration is the transport of ions through a conductor resulting from the passage of direct current. Metal ions (Al) in potential wells experience two forces, F_P and F_E. In Al metal films, $F_P > F_E$. Therefore positive ions migrate downstream while vacancies move upstream.

With decreasing metal line width, both the resistance and current density are increasing. The increased current density leads to reduced EM performance. With more layers of metal, the Joule heating effect becomes more relevant leading to further EM

degradation. Both the ideal scaling (S) and the quasi-ideal scaling ($S^{1/2}$) leads into three problems: i) Increased interconnect capacitance, ii) Inreased resistance and iii) Increased current densities.

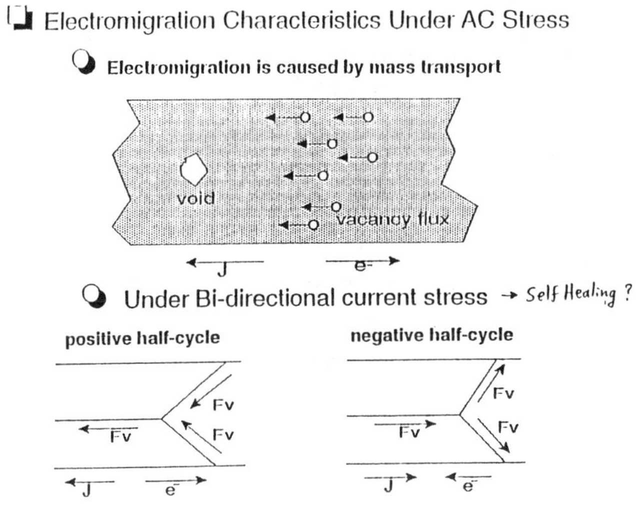

Electromigration Characteristics Under AC Stress

Electromigration is caused by mass transport

Under Bi-directional current stress → Self Healing ?

Figure 1. EM characteristics through a conductor

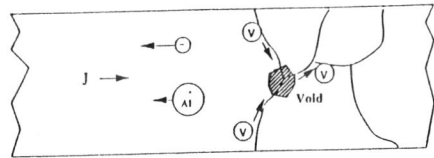

- Electromigration involves diffusion mechanism
- Diffusion is via grain boundaries
- Electromigration induced damage is caused by flux divergence

Figure no. 2. Schematics of diffusion via grain boundries

Concurrent and global optimization of the interconnect system is therefore necessary. In order to decrease the interconnect degradation, several alternatives have been proposed: i) Change the Al system to copper metallization; ii) Use additional metal layers with non-scaled metal lines for longe range global interconnect lines; iii) Use low permittivity dielectric to reduce capacitance, iv) Insert repeaters in the long interconnect lines, and v) Operate circuits at low temperatures (77 K) [Oh'95]. The scaling increases the current density by about an order of magnitude for each scaling step and therefore decreases the EM reliability. The conclusion of the paper cited above is that the Al metallization is not acceptable below 0.5 μm. Apart from the low temperature operation (might be applicable in special circumstances) the only realistic alternative is the Cu metallization.

31

The three major drawbacks of the Al metallization that causes circuit performance degradation and reliability problems (fairly high resistance, EM degradation and stress-induced voids) could easily be eliminated by the copper technology [Geppert'98]. Difficulties with copper fabrication and the fear of contamination have delaied its introduction.

All that changed recently when IBM and Motorola have (independently) announced a six metal layer copper system manufactured by the so called "dual-damascene" process: "copper is electrolytically plated into trenches etched in the wafer's surface and then chemically and mechanically polished" [Geppert'98]. The 40-45% resistivity decrease leads to dramatic performance increase in delay time and cross talk.

The biggest gain, however, comes in the reliability. Researchers from IBM found that the electromigration performance of copper lines is two orders of magnitude better than that of aluminum lines, and no stress voids were observable.

In summary, it appears that copper is the coice of metallization for the future. The difficulties with this new technology, however, do not make its introduction economical above the 0.18μ technology. We can conclude that both for the 0.35μ and 0.25μ designs, the AlCu system will be used with all its performance/reliability hinderances.

While designers push for higher current densities and circuits, with larger number of interconnects, the primary responsibility of the reliability engineering is to find the limits of the actual system and use reliability simulations to optimize the reliability performance of the product.

Based on the EM testing, it becomes easier to come up with the current density design rules which will guarantee long term reliability. For e. g.

Examples of current density design rules:

(Layer: Metal 1, 2, 3, 4, 5, Via 1, 2, 3, 4, Contact)

Structure	Current Density
Metal 1	0.50 mA/μm
Metal 2	0.50 mA/μm
Metal 3	0.50 mA/μm
Metal 4	0.50 mA/μm
Metal 5	0.80 mA/μm
Via 1	0.22 mA/via
Via 2	0.22 mA/via
Via 3	0.22 mA/via
Via 4	0.22 mA/via
Contact	0.40 mA/via

Stress Migration (SM)

Stress migration is the phenomenon in which voids are formed in metal lines due to stress relaxation processes (diffusional creep, dislocation glide, etc.). Al lines are highly tensile at operating temperatures. Narrower the lines, higher the stress. Higher the stress, greater the SM process. SM leads to two types of voids, Wedge and Slit type. Slit voids have lower free energy of formation and they lead to opens. Recent research has shown that wedge voids could transform to slit voids.

Interconnect Scaling

With each new generation of technology both performance and density have increased. R-C delays associated with wiring interconnect are playing a more significant role in determining the ultimate performance of a given technology. Increasing die sizes and circuit complexity are driving the need for more levels of metal. Both reliability and performance limits of today's metallization systems are pushed by the need for performance.

Hot Carrier Injection (HCI)

HCI is the phenomenon in which overly excited hot carriers (either electrons or holes are injected into the gate oxide causing damage to the oxide/Si interface or trapped charges in the gate oxide. Carriers become overly excited (hence, "hot") because they gain enough energy to cause impact ionization before they loose it by phonon scattering. Created electrons and holes (near the max. electric field region) gain more energy to cause impact ionization, etc. The reliability consequence is a degraded transistor in V_T, g_m I_{dsat}, $I_{dlinear}$, etc.

Hot Electron Induced Punch-Through Degradation
(P. Fang, etal, 1992 IRPS)

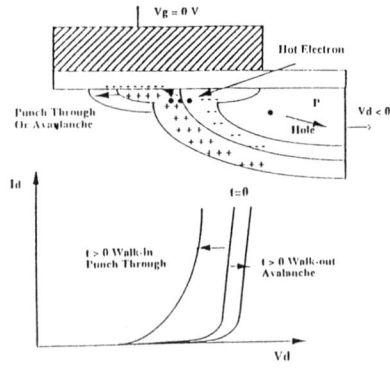

Figure 3. HCI induced punch through

HCI also causes substrate currents which, if severe enough, can induce bipolar latch-up in circuits. Interface generation is the main degradation mechanism for n-channel MOSFETs having normal quality oxides. Mechanisms of interface trap generation are not well understood. They do involve hot electrons, hot holes, and hydrogen.

Simplest model:

Hot electrons bombard the interface against a retarding E field. Breaks Si-H bonds and causes a trap.

"Things are getting worst before getting better". The DC HCI lifetime of the 0.5μ technology even with LDD drain construction is much lower (0.2 - 0.5 years) than that of the 1.0 μm technology. Further scaling, however, has increased the HCI lifetime. With gate oxides of 40 angstroms or less and drain voltages in the 1.5 V - 1.8 V range, better than 10 years of dc HCI lifetimes have been reported [Hwang'95, Rodder'95]. In the following, the intermediate (0.35μ to 0.25μ) case will be analyzed.

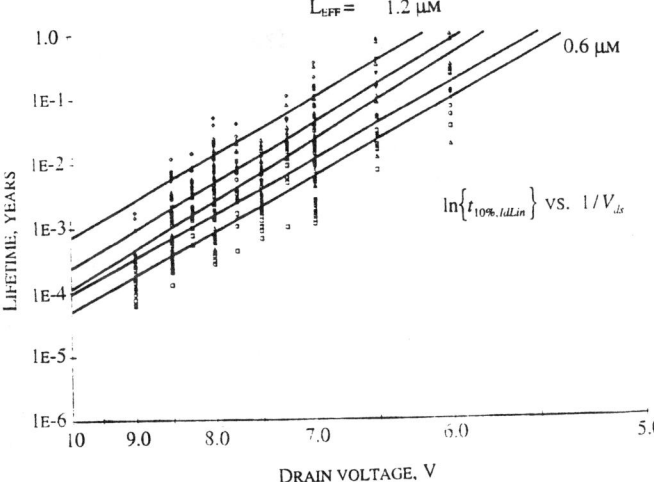

Figure 4. I_{dlin} lifetime v/s drain voltage with Leff as parameter

Even with reducing the supply voltage from 5V to 3.3V when scaling down LDD MOSFETs below 0.5μ, the hot carrier damage still remains a key limitation to reliability. In the optimization of conventional LDD transistors, a trade-off has to be made between electrical performance (drive current degradation) and reliability (HCI). The usual solutions (LATID, Inverse-T and GOLD) come with the price of more complex processing requirements and channeling or shadowing effects.

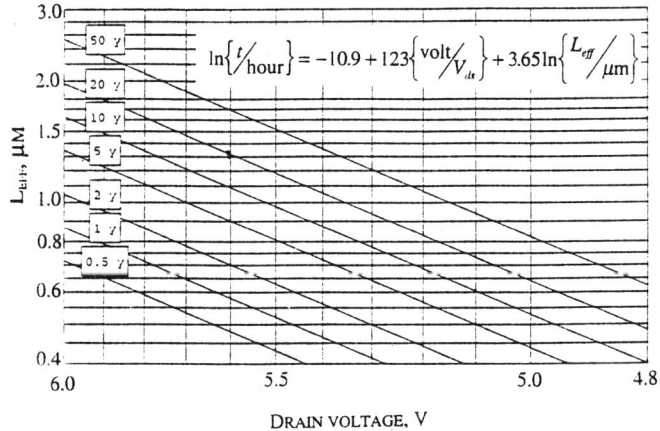

Figure 5. Hot carrier injection design rules for *n*-MOS transistors

The recently suggested FOND (Fully Overlapped Nitride-etch defined Device) technology can increase the HCI lifetime by two orders of magnitude for the 0.35μ design compared to the traditional LDD technology [Bellens'95] by controlling the gate-drain overlap independently of the implant conditions. The comparison of the initial device characteristics of devices fabricated by the LDD and by the FOND technology show an increased HCI resistance that can be "…explained by a smaller generation of damage and a lower sensitivity of the overlapped device to this damage." Detailed modeling and experimental work show the improvement in most degradation parameters. The ΔId/Id dc HCI lifetime of a 0.25μ FOND device exceeds 10 years at 3.3V drain voltage while the lifetime of a similar LDD device is 1 month.

Plasma etching poses further problems on the sub-micron MOS HCI lifetimes [Li'95]. The oxide charging damage is due to plasma induced Fowler-Nordheim current through the gate oxide, while edge damage is due to direct plasma exposure during the poly-Si over-etch period. In the 0.35μ to 0.50μ processes, the effects of these two different damage types are different on N-MOS and P-MOS transistors. Transconductance degradation of P-MOS transistors is accelerated probably by both types of plasma damage, while N-MOS devices are affected only by edge damage. The type of etching process has a significant effect on the lifetime mainly due to the differences in etcher damage between the different etch processes.

While HCI degradation mechanisms in the 0.6μ dimensions and above affected the N-MOS transistors to a much higher degree than the P-MOS transistors, the sub- (and deep sub-) micron P-MOS transistors have an increased HCI susceptibility.

The degradation of short channel P-MOSFET devices (<0.35μ), at any bias and temperature condition, results in a reduction in drive current due to donor type interface generation and positive charge formation [Rosa'97]. In advanced sub-micron technologies hot hole injection will control the channel hot carrier damage of short 'Leff' while the long channel P-MOSFET still suffers from electron trapping. It is expected that in future advanced sub-micron technologies, the negative bias temperature instability and the channel hot carrier effects will have major roles in predicting device lifetimes. At "high" temperatures (where most devices operate) the P-MOS degradation may play a similar or greater role than the n-MOS degradation in reducing the circuit performance.

Time Dependent Dielectric Breakdown (TDDB)

TDDB is the phenomenon of oxide breakdown (or excessive leakage) at high electric fields after a certain time. Oxide breakdown is generally believed to be caused by the positive charge buildup in oxide near the Si/SiO₂ interface. The source of the positive charges is from impact ionization deep in the oxide. Barrier lowering of the energy band leads to further electron injection which in turns leads to more hole trapping and hence a run away process.

The high current injected (I) at localized spots of "+ charge trapping" produces I^2R heating sufficient to melt the SiO_2. Prior to catastrophic oxide rupture, oxides are often found to be leaky. Weak spots in the gate oxide contamination causing oxide thinning impurities, mobile ions, metallic, etc. process damage high stress points.

Why Thinner is Better:

> MOSFET gate -- more current/speed, less short-channel effects
> DRAM capacitor -- same capacitance in smaller area
> Nonvolatile memory -- lower programming voltage

Sources for Oxide Defects:

> Particulates
> Gas Impurities
> Wafer Defects
> Impurities in Wafers (Metals, Mobile Ions)

Trends in Reliable Dielectrics:

No revolutionary alternative to SiO_2 gate. Defect reduction, e.g. CVD/thermal oxide, will allow 100Å oxide for 5.5V, 64Å for 3.6V. Back-end process impact on TDDB and stability will be a major issue. Control of H, N, CL, F can improve trapping. TDDB model will improve testing, burn-in, product reliability prediction.

Gate Oxide Integrity

Gate oxide scaling is a key element in increasing transistor performance. As gate oxides enter the sub-70Å region, the leakage mechanism changes from the Fowler-Nordheim tunneling to a direct tunneling that leads to the gate leakage rather than the gate oxide reliability as the limiting factor for operating voltages [Sweeny'96, p.3.1]. Recent theoretical work indicates that the gate current can increase significantly above the well known F.-N. tunneling current in direct tunneling mode.

It is well known that the normal passage of gate current creates damage and there are physical reasons to expect less gate oxide damage from direct tunneling. Comprehensive experimental data is still missing, and only extensive research in this field will show the actual reliability implications of the gate oxide scaling below 70Å. [Chatterjee'95]. Another serious reliability concern is the role of hydrogen in oxide defect generation [Cartier'97]. Since atomic hydrogen is very reactive in the SiO_2 network, it can passivate/depassivate interface states, create anomalous positive charge, can influence the generation of electron traps and recombination centers and induce leakage currents. In summary, without further reliability studies of the thin gate oxides, it would be premature to assume that they will cause no concerns for the lifetimes of the sub-micron CMOS devices.

Another important consideration is related to the mixed voltage I/O circuits. In mixed-voltage I/O circuits, during an accelerated product stress a new gate oxide breakdown mechanism has been found [Furukawa'97]. "Although no gate oxide breakdowns were expected from F.-N. stress, gate oxide fails were observed only in short channel *p*-FETs of the mixed voltage I/O circuits." The device was a dual gate CMOS with Ti-silicide diffusion, 0.25μ nominal channel length and 70Å gate oxide. In the experiment, the gate oxide breakdown was accelerated by the channel hot electrons and the acceleration increased with decreasing channel length.

Plasma Induced Damage
Antenna

Plasma induced damage is a serious concern for advanced silicon VLSI manufacturing. It is a problem that will become unavoidable as the industry continues to scale the thickness of the gate oxide. The charging damage need to be minimized. The plasma induced damage takes place when a wafer is exposed to plasma during the fab processing. Radiation by VUV photons, which are supplied in huge quantity in a processing plasma can cause damage to the gate oxide.

It is now established that plasma induced damage is due to high field tunneling (Fowler-Nordheim). The electron tunnel current can be either direction depending on the plasma system as well as the position on the wafer at any moment in time. The existence of an antenna – a conductor which exposed directly to the plasma and connected to the gate – can collect current from the plasma and channel it to the gate to worsen the damage.

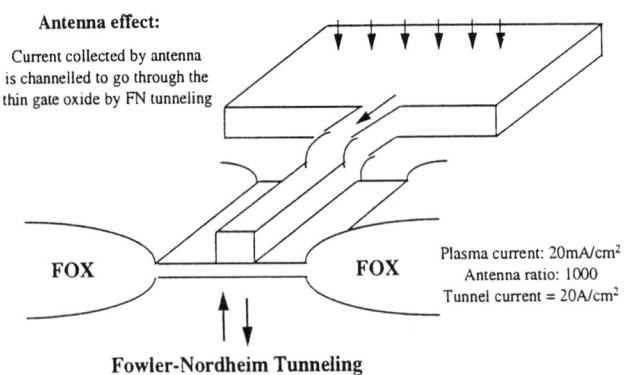

Figure 6. Antenna effect in a MOS device

One of the way to avoid this damage is to use the proper lengths for these conductors (antennas). During the design of a circuit, it is very important to follow the proper antenna ratios by working with the manufacturing fab. Below are some of the typical antenna design rules used in the industry.

When there are weak points in an oxide, the equivalent antenna ratio can be infinit. Even a very low density plasma is able to supply enough current to break them as long as the plasma can support the voltage.

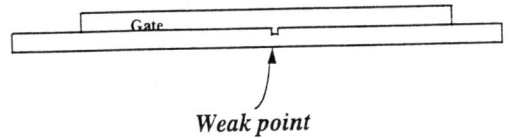

Weak point

Figure 7. Typical antenna induced damage caused to gate oxide

Antenna Design Rules:

(Layer: Poly, Metal 1, 2, 3 or 4)

Structure	Antenna Ratio
Maximum drawn ratio of field Poly perimeter area to the active Poly gate area connected directly to it.	100
Maximum drawn ratio of Metal1 perimeter area to the active Poly gate area connected directly to it.	400
Maximum drawn ratio of Metal2 perimeter area to the active Poly gate area connected directly to it.	400
Maximum drawn ratio of Metal3 perimeter area to the active Poly gate area connected directly to it.	400
Maximum drawn ratio of Metal4 perimeter area to the active Poly gate area connected directly to it.	400
Maximum drawn ratio of CO area to the active Poly gate area connected directly to it.	50
Maximum drawn ratio of Via1 area to the active Poly gate area connected directly to it.	50
Maximum drawn ratio of Via2 area to the active Poly gate area connected directly to it.	50
Maximum drawn ratio of Via3 area to the active Poly gate area connected directly to it.	50
Maximum drawn ratio of Via4 area to the active Poly gate area connected directly to it.	50

Accelerated Testing

Required because probability of failures under use conditions is very low. For example, for 100 FIT the median time to failure is 10^5 - 10^{11} hours.

Acceleration of certain conditions around the device in test is achieved to get meaningful information. The conditions accelerated are as follows:

- Temperature
- Voltage
- Current
- Humidity

Sample Size Selection

The samples for the tests are based on statistical sampling plan. MIL-S-19500 & MIL-M-38510, LTPD Sampling Plan Table is used to decide the sample size. The LTPD (Lot Tolerant Percent Defective) numbers such as 3% or 5% are mutually determined between customers and subcontractors. Various companies determine the sample size based on the customer requirements as well as their experience and viability to effective testing.

Product Qualification Testing (Plastic Packages)

* Die Related	* Package Related
Operating Life	THB
(hot and cold)	Temperature Cycle
High Temp Storage	Thermal Shock
Latch-up	Die Shear and Bond Pull
ESD	Solderability
Die X-section for process	Mark Permanency
evaluation	Physical Dimensions
EFR	Lead Integrity
	Autoclave
	Package DPA
	IR Pre-conditioning
	XFR Analysis
	Solder Dip

Environmental Tests

As explained earlier, reliability is Quality over time, it is important to assess it over time under various operating conditions. Since time can not be accelerated, to assure that the devices have been tested for long term reliability, other conditions such as voltage, temperature and humidity are varied and statistical techniques used to predict the components' reliability by testing a smaller sample size under elevated temperature, voltage and humidity.

As mentioned earlier in the semiconductor industry, there are several such tests which are used to evaluate the products' life times. The tests are as follows:

Test: High Temperature Operation Life (HTOL) Test

Condition: This test is done with a dynamic bias to the device to exercise as many nodes as possible, without generating excess heat.

Duration: 1000 hours (Qual point 500 hours) at 150°C
2000 hours (Qual point 1000 hours) at125°C

Purpose: The test simulates the working of the device in normal operation. The test is performed to accelerate wear-out mechanisms.

Failure Mechanisms Targeted:
Pinholes in gate oxide, dielectric, fabrication defects such as metal shorts, un-etched layer defects.

Caution: The die temperature is to be limited to <165°C, lower than the Tg of mold compound.

Sample Size: 3% LTPD, 129 (1) or 76 (0)

Figure 8. A hole in the interlevel dielectric found during HTOL

Test: Cold Temperature Operating Life (CTOL) Test

Condition: This test is done with a dynamic bias to the device to exercise as many nodes as possible.

Duration: 1000 hours (Qual pint 500 hours) at -20°C.

Purpose: The test is performed to check the hot electron susceptibility of the device. For geometries below 1.0μ, this test is very important.

Test: Temperature Humidity Bias Test (THB)

Two Types: Highly Accelerated Stress Test (HAST), conditions are 130°C and 85% Relative Humidity 85/85 test (done under the conditions of 85°C and 85% Relative Humidity).

Condition: This test is done with a static bias to the device pins, alternating with GND and Vcc to adjacent pins, such that the device does not draw more than 30~40 mA of current.

Duration: 200 hours (Qual pint 100 hours) for HAST test.

Purpose: This test simulates the static bias condition of the device in a system where the humidity is high.

Failure Mechanisms Targeted: Bond Pad corrosion is the primary mechanism. Internal corrosions been observed too.

Sample Size: 3% LTPD, 129 (1) or 76 (0).

Caution: Biasing configuration issued such that the device does not draw more that 30 ~ 40 mA of current thus not heating the device too much.

Test: High Temperature Storage (HTS)

Condition: This test is done with no bias. It is storage test.

Duration: 168 hours at 150°C.

Purpose: To evaluate the effect of ionic contamination which may surface after this test.

Sample Size: 3% LTPD, 129 (1) or 76 (0).

Test: Temperature Cycle testing, air to air (unbiased)

Condition: This test is done in a chamber capable of attaining the extreme temperatures of -65°C and +150°C.

Duration: 500 cycles (with a read point at 200 cycles).

Purpose: To accelerate mechanical failure of chip or package.

Evaluate thermal interfaces:
Variables Controlling Contact Thermal Resistance are:
   1. Surface Finish
   2. Material Properties
   3. Assembly
This test accelerates the conditions where the device may undergo sudden temperature changes

Failure Mechanisms Targeted: Delamination of the die off the die pad, lifting of the ball bonds, metal shifting, thin layer cracking and die cracking.

Sample Size: 5% LTPD, 77 (1) or 45 (0).

Figure 9. Die cracking observed after temperature cycling

Other Tests needed to assure reliability

Test: Latch-Up

Condition: The devices are tested at room temperature with five conditions:
1. Push current through a pin while holding all inputs to low.
2. Push current through a pin while holding all inputs to high.
3. Pull current from a pin while holding all inputs to low.
4. Pull current from a pin while holding all inputs to high.
5. Overvoltage test.

Purpose: Latch-Up testing is performed to check the susceptibility of CMOS devices to latch-up.

Failure Mechanisms targeted: Latch-up will cause gate oxide, poly and even metal lines to get thermally damaged.

Test: Electro-Static Discharge (ESD) testing

Condition: This test is done with the following conditions:
1. I/O pins are zapped positive with respect to GND one at a time with all other pins floating.
2. I/O pins are zapped negative with respect to GND one at a time with all other pins floating.
3. I/O pins are zapped with positive respect to Vcc one at a time with all other pins floating.
4. I/O pins are zapped negative with respect to Vcc one at a time with all other pins floating.
5. I/O pins are zapped ± one at a time to all other I/O.
6. Each unique power supply is zapped ± to each other one at a time. All other pins are left floating.

Purpose: The test simulates the handling of the devices during manufacturing, testing and assembling the devices at the customer.

Failure Mechanisms Targeted: Thermal damage to gate oxide. Gate oxide rupture, ploy meltdown, contact spiking, Si punch through.

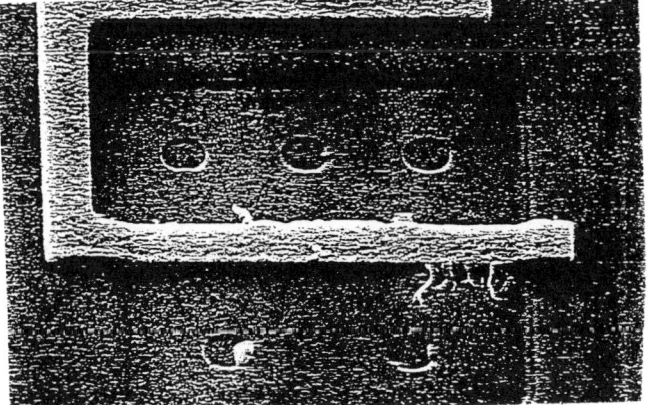

Figure 10. Poly filaments meltdown due to ESD induced damage

Test: Die X-section for Process Evaluation

Procedure: Two devices are inspected for passivation pinholes, then are cross-sectioned to look for step coverage's, layer thicknesses.

Purpose: To flag any process anomalies which can be fed back to the foundries.

Test: Package Destructive Physical Analysis (DPA)

Procedure: Five devices are inspected for the quality of assembly. The devices are cross-sectioned to check for manufacturability. Die attach coverage, bonding, mold compound etc. are inspected during this analysis.

Purpose: To flag any assembly anomalies which can be fed back to the assemblies.

Test: Die Shear and Bond Pull

Procedure: Thirteen devices are used for this test. Die shear is performed on all 13 and a total of 42 ball bonds are pulled.

Purpose: The die shear test is performed to check the quality and reliability of die attach used to attach the die to the die pedal. The bond pull test is performed to check the strengths of the ball bonds and their adhesion to the bond pads.

Test: Solderability Test

Procedure: Devices are dipped in a tank of molten solder after a steam aging process and later inspected with an optical microscope.

Purpose: To check for the solderability of the devices even after they stay (age) in a warehouse for a length of time. Customers should be able to solder the devices even after a few years of storage.

Sample Size: Eight Devices.

Test: Mark Permanency Test

Procedure: Effect of four different chemicals, on device marking is studied with this test.

Purpose: The test is conducted to study the reliability of the device marking. The chemicals used for the test are the ones commonly used during the printed circuit board assembly.

Sample Size: Four Devices.

Test: Physical Dimensions

Procedure: Measure all the dimensions of the package.

Purpose: The test is conducted to make sure that all our packages have dimensions within specifications.

Sample Size: Five Devices.

Test: Lead Integrity

Procedure: Bend lead five times with the lead fatigue tester.

Purpose: The test is conducted to make sure that the leads will stand the stresses of bending which is often performed during the assembly of PCBs.

Sample Size: Five Devices.

Test: X-Ray Fluorescence (XRF) Testing

Procedure: XRF equipment measures the thickness of solder on the leads as well as determines the composition of the solder.

Purpose: This test is conducted to determine the thickness and composition of solder plating in the device leads. It is important that those things are within specifications.

Sample Size: Two devices.

Test: Autoclave or Pressure Cooker Test (PCT) Unbiased

Procedure: Devices are subjected to this test in a pressure cooker chamber of 168 hours.

Conditions: The device temperature is 121°C, 100% Relative Humidity and at an atmospheric pressure of 2.0.

Purpose: This test is conducted to simulate warehouse storage conditions in high temperature and humidity.

Failure Mechanisms Targeted: Corrosion of metallization on bond pads as well as internal circuitry, delamination in passivation are the primary failure mechanisms.

Samples Size: 5% LTPD, 77 (1), 45 (0).

Test: IR reflow simulation (only or surface mount devices)

Procedure: The surface mount devices prior to the package related tests are subjected to moisture ingress or bake followed by five cycles of heat (IR) in convection oven where the temperature rises from room temperature rapidly to 240°C. The devices pass through it five times.

Purpose: This test is conducted to simulate the actual soldering.

Failure Mechanisms Targeted: Delamination of package to die and die to die attach interfaces. Popcorning of packages.

What Are Wafer Level Reliability (WLR) Tests?

WLR tests consist of test methods at wafer level which allows for the rapid assessment of reliability. WLR has been a growing concept in the semiconductor industry over the last decade. WLR typically involves electrical testing performed on specially designed test structures, and is performed in approximately several minutes v/s the long term testing performed on the packaged devices. This type of testing gives reliability information on specific failure mechanisms of the process.

WLR differs from traditional reliability testing since traditional reliability testing process takes two to three months to generate the data which the WLR can do it in several hours. The traditional reliability however tests the device package reliability also. It is important that the failure analysis is performed to find the cause and information fed back to the factory for correction.

WLR allows rapid feedback to factory for continuous improvement. The common uses of WLR are for debugging and optimizing of new process for reliability and robustness. Process qualification.

Semiconductor Failure Mechanism Activation Energies based on Jedec publication

Failure Mechanisms	Min. -- Max.
Surface/Oxide	0.75 -- 1.4
Hot Carriers	-0.1 -- 0.6
Electromigration (single layer Al)	0.48 -- 0.75
Electromigration (multi-layer Al)	0.49 -- 0.71
Stress Voiding	0.80 -- 1.67
Corrosion (Chlorine)	0.53 -- 0.95
Corrosion (Phosphorus)	0.30 -- 0.80
Package - wire intermetallics	0.85 -- 1.00

REFERENCES

1. **Sweeney**, "Reliability Challenges and Trends in Deep Sub-Half Micron Technology," 96 IRPS Tutorials, pp.3.1 – 33.

2. **Chatterjee** et al, "Trends for Deep Sub-Micron VLSI and Their Implications for Reliability," 95 IRPS, p.1.

3. **Hwang** et al, "Performance and Reliability Optimization of Ultra Short Channel CMOS Devices for Giga-bit DRAM Applications," IEDM'95, pp.435-438.

4. **Rodder** et al, "A Scaled 1.8 V, 0.18 μm Gate Length CMOS Technology: Device

5. Design and Reliability Considerations," IEDM'95, pp.415-418.

6. **Williams** et al, "Scaling Trend for Device Performance and Reliability in Channel-Engineered n-MOSFETs," IEEE Transactions on Electron Devices Vol. 45, No. 1, pp. 254-260, 1998.

7. **Yu** et al., "CMOS Transistor Reliability and Performance Impacted by Gate Microstructure," IRW'97, pp. 139-

8. **Fiegna**, et al., "Scaling the MOS Transistor Below 0.1 μm: Methodology, Device Structures, and Technology Requirements," IEEE Transaction on Electron Devices, Vol. 41, No. 6, June 1994, pp. 941-951.

9. -Y. **Oh** et al., "2001 Needs for Multi-Level Interconnect Technology," Circuits & Devices, January, pp. 16-21, 1995.

10. **Scarpulla** et al., "Reliability of Metal Interconnect After a High-Current Pulse," IEEE Transaction on Electron Device Letters, Vol. 17, No. 7, July 1996, pp. 322-324.

11. **Graas** et al., "Correlation Between Initial Via Resistance and Reliability Performance," IRPS'97, pp. 44-48.

12. **Geppert**, "Technology 1988, Analysis and Forecast," IEEE Spectrum, January, 1998, p.23.

13. **Bellens** et al, "Analysis and Optimization of the Hot-Carrier Degradation

14. Performance of 0.35 μm Fully Overlapped LDD Devices," IRPS'95, pp. 254-259.

15. Xiao-Yu **Li** et al, "Degraded CMOS Hot Carrier Lifetime - Role of Plasma Etching Induced Charging Damage and Edge Damage," IRPS'95, p.260.

16. LA **Rosa** et. al, "NBTI - Channel Hot Carrier Effects in PMOSFETs in Advanced

17. CMOS Technologies," IRPS, 1997, p.282.

18. **Cartier**, "Determination of Physical Parameters and Reliability of Ultra Thin Oxides," Tutorial A, IIRW, 1997, p.32.

19. **Furukawa** et al, "Accelerated Gate-Oxide Breakdown in Mixed-Voltage I/O Circuits," IRPS, 1997, p.169.

39

Test and Fail Verification

IC Testing: Background, Directions and Opportunities for Failure Analysis

Anne Gattiker
IBM Austin Research Lab

Phil Nigh
IBM Microelectronics

Thomas Vogels
Carnegie Mellon University

INTRODUCTION

Increasing complexity, speed and pincount and decreasing time-to-market pose many challenges for integrated circuit manufacturers. Integrated circuit testing is one of those challenges growing in importance and in overall contribution to the cost of a manufactured die. IC speeds that exceed automated test equipment (ATE) capabilities, embedded analog circuit macros and sheer number of transistors to be tested are examples of new realities that are making test more difficult and more challenging. However, as IC failure mechanisms and ICs themselves become more complex, testing is becoming not only more difficult, but also more important. Specifically, the increasing difficulty of physical failure analysis (PFA) coupled with the time-to-market-driven need for rapid yield-learning is creating an increasingly important role for test not only in separating good dies from bad, but also in giving feedback to the manufacturing process about imperfections occurring during fabrication.

This article presents an overview of microprocessor and ASIC integrated circuit testing, with emphasis on the dual roles of test in separating good product from bad and in diagnosing and characterizing manufacturing imperfections. The paper begins with a description of key industry trends impacting testing. Next, a basic overview of logic and memory test is given, where technical issues that may cause methodology changes are emphasized. The overview is followed by a description of a sample of test-based diagnosis techniques and a discussion of future opportunities for exploiting test-based techniques to discover the physical mechanisms underlying IC failure.

TEST-RELATED TRENDS

The majority of this paper is aimed at providing an overview of the test methods commonly used in the industry today. It is important, however, not only to understand where the industry is today, but also where it is heading. In this section, we highlight some key trends that will impact how ICs will be tested in the future.

There are some major challenges facing the IC industry related to the manufacturing test of future devices [1]. These key challenges include:

- **On-chip frequencies will exceed 1GHz in the near future.** The cost of testing these ICs at their full speed may become prohibitive.
- **Chip densities are expected to reach 40M logic circuits within the next two years.** Without test methodology changes, test times could exceed one minute and the size of test vectors could exceed 20 Gbytes.
- Manufacturers are moving toward **integrating many different types of circuits on a single die**--including putting analog and DRAM circuits on previously digital ICs. Test methods that have conventionally been used for various circuit types will change when these same circuits are embedded on a device with many other circuits.
- **Some testing methods**--particularly I_{DDQ} testing and high-voltage screening--**are becoming less effective** (or are projected to become less effective) due to device and voltage scaling. In addition, defects causing more subtle electrical fail types (e.g., timing-related failures) are projected to become more common in future technologies.
- **Rising test costs** may force new testing methods to become adopted. These rising costs are caused by the trends listed above, their impact on the capital cost of automatic test equipment (ATE) and the increased test times.

Despite these trends toward greater complexity, semiconductor market forces are trying to drive down testing costs and are allowing less time between product introductions. Meanwhile, the manufacturing cost per transistor is decreasing, but the cost to test a transistor remains at best the constant. Hence, market forces are directly in conflict with technology trends--either improved methods must be adopted or market expectations must change.

There are some interesting opportunities for testing to help meet the time-to-market requirement. An emerging opportunity is using test results and analysis techniques to understand process characteristics [2-6]. Yield ramping and reliability defect reduction would both be significantly improved if we could learn defect Paretos through testing and data analysis only--without the requirement of having to perform physical failure analysis. We discuss this opportunity in more detail at the end of this paper.

TESTING BASICS

In this section, we provide a brief description of the most common tests applied in the industry today. At least two levels of testing are normally applied to products before they are shipped: a test at the wafer level (often called "wafer probe test") and another test after the die is packaged. Some products may see additional test steps due to additional requirements related to temperature coverage, memory test redundancy verification or reliability screening. Figure 1 shows a typical test flow in which the IC undergoes a wafer level test, pre-

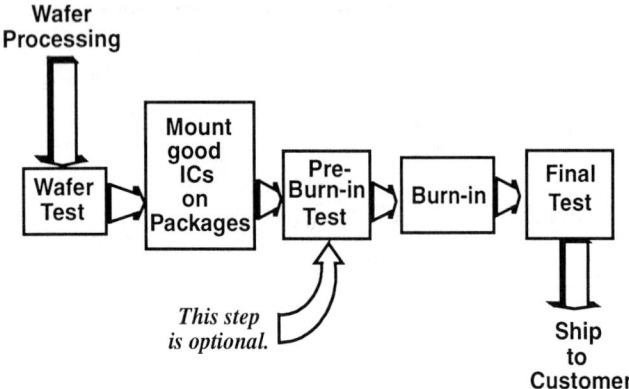

Figure 1. Typical test flow.

burn-in package test, burn-in and post-burn-in final test. Often extra "data collection" or "characterization" tests are applied for a sample of parts. The tests described in this section may be applied at either wafer or package test or both.

Types of tests

Contact / power-up tests. Normally the first test applied is aimed at ensuring that there is reliable contact between the tester and the device. Each device I/O should be tested for a low resistance connection to the ATE.

I/O parametric tests. These tests ensure that the parametric characteristics of the device's I/Os are within specification. Some of these specifications include guaranteed noise margins, drive capabilities and input leakage currents. For example, each I/O is functionally tested at its minimum voltage level to ensure it meets specification.

I_{DDQ} testing. Detecting defects by measuring the quiescent current of the power supply (typically V_{DD}) is called I_{DDQ} testing. This test method is most effective for fully complementary static CMOS circuits where there are no static paths of current.

There are many articles describing the effectiveness of I_{DDQ} testing [7-24]. Most defects cause abnormally high I_{DDQ}. Most defects are activated by some, but not all, circuit states so multiple I_{DDQ} measurements must be performed. Performing 4 to 30 I_{DDQ} measurements is most common in the industry (although many more measurements are performed for some products).

A benefit of I_{DDQ} testing is that it can detect reliability defects, i.e., defects that do not initially cause a circuit failure, but which will do so in the future. Some reliability failures pass all other tests, but have abnormally high I_{DDQ} during initial testing. Detecting these defective ICs at wafer or package test is a tremendous advantage.

As technology continues to advance, however, the effectiveness of conventional I_{DDQ} testing is getting worse since the normal background leakage currents are going up [25]. Figure 2 shows typical I_{DDQ} levels for a range of technologies [1]. Note that I_{DDQ} levels have been increasing by a factor of 10 with each technology generation.

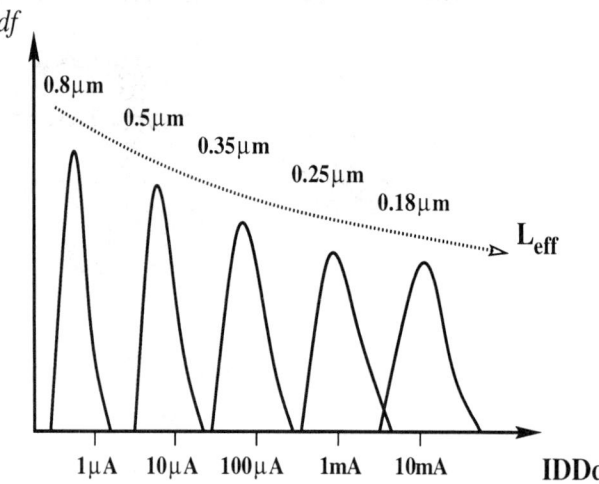

Figure 2. I_{DDQ} levels for good chips across technology generations. [1]

I/O performance testing. These tests are aimed at ensuring that the IC's inputs and outputs operate at the required speeds. Typical I/O timing specifications include the setup and hold times between signal inputs and clock lines.

Logical tests. The most widely known test method is where logical data is applied to the chip and the response is compared with expected values. The purpose of this testing is to verify that the boolean operation of the device is correct. Typically the logical input and output values are stored in the ATE. The speed of the device may or may not be measured during logical testing. There are two basic forms of logical testing: functional and structural testing. They are described in more detail later in this paper.

Speed binning tests. In addition to determining if a chip has a defect or not, it is often required to determine the speed at which the device fully operates. For example, the same microprocessor design may operate at 450MHz, 500MHz or 550MHz depending on the precise processing conditions seen by the IC. Manufacturers are eager to support speed binning since faster parts can be sold at a much higher price.

Characterization tests. In addition to determining the passing and failing devices and the device speed, manufacturers often apply additional tests to learn as much as possible about the semiconductor processing conditions. Such tests may include additional timing measurements, leakage current measurements, saturation current measurements and maximum/minimum voltage measurements. Typically manufacturing test stops with the first test failure to minimize test times. During characterization testing, however, all tests may be applied even for failing parts to gather information about all tests. Examples of such information are failing bits during logical tests and memory bitmaps, both of which can be used for fault localization.

Implementing characterization tests that can provide information about manufacturing defects is an emerging opportunity that is discussed later in this paper.

Functional and Structural Testing Definitions

The basic idea of functional testing is to test the device in the same way that it will be used in the final application. For example, an adder circuit is tested adding a sample of numbers. Typically, functional test patterns are manually created by designers whose goal it is to validate the correctness of the design. A subset of these functional verification patterns are then reused for manufacturing test of the ICs. The test vectors for functional testing are usually applied at the full speed of the device

The goal of structural testing, on the other hand, is to ensure that the logical circuit structure is defect-free. Test patterns are generated automatically. Automatic test pattern generation (ATPG) uses a gate-level representation of the circuit and a representation for defective circuits called a fault model. The most commonly used fault model (called the stuck-at fault model) represents defects as causing logical gate inputs or outputs to be stuck at either logical 0 or logical 1 levels. Other fault models (e.g., I_{DDQ}, transition, delay, bridging) have been proposed, but are used much less frequently [26]. Figure 3 illustrates several fault models.

While test generation based on the stuck-fault model has so far resulted in tests with adequate quality, new technologies may bring new fail mechanisms that require new fault models and thus an increased emphasis on developing improved fault models.

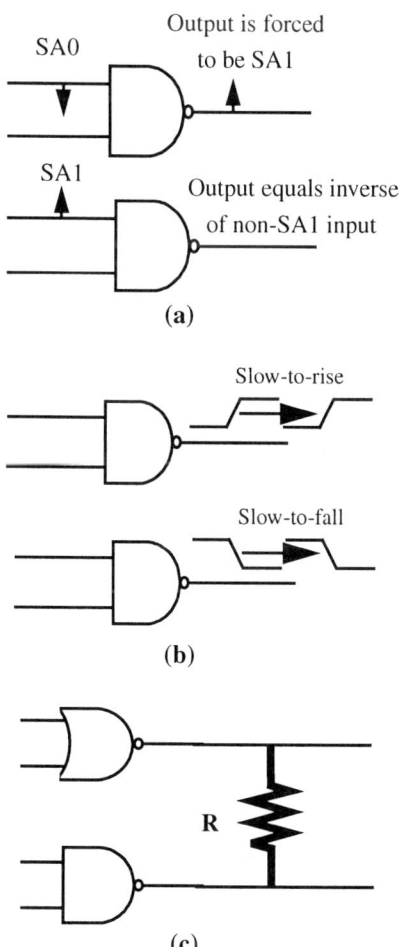

Figure 3. Fault models, stuck-at (a); transition fault (b) and bridging fault (c).

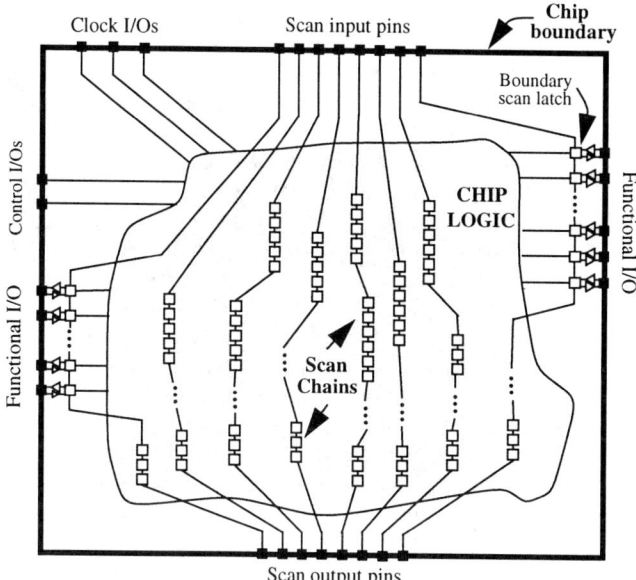

Figure 4. Scan design.

Scan Design

A key enabler of structural testing is scan design [27]. As illustrated in Figure 4, this is a design methodology in which all registers are connected into serial chains. The method tests the circuit by scanning inputs into the scan chains, evaluating the inputs in the combinational logic between the registers, latching the responses into the scan chains and serially shifting the responses out. Since the state of the circuits (i.e. the register contents) can directly be controlled and observed, the test generation problem is effectively reduced from a sequential one to a combinational one.

Designs for which sequential tests can be generated are limited to a size in the range of 50,000-100,000 logic gates. The design size limit is higher for combinational logic test generation where ATPG can be applied for devices larger than 4M logic gates (this number is increasing over time). Thus scan design effectively enables automated test pattern generation for all logic ICs. Without scan design, ATPG would be limited to small macros only.

Note that launch-capture delay testing is also possible using scan design. To implement it, inputs are shifted into the chain in such a way that they are seen only at an intermediate node in each scan latch--their values are hidden from the combinational logic driven by the latches. Then a fast sequence of special clocks shows the stored inputs to the combinational logic between the chains (i.e., it launches the values) and then quickly latches the results into the driven scan chain registers (i.e., it captures the responses) [28-30]. This method takes advantage of the excellent controllability and observability of scan design and therefore has the potential to provide very thorough delay test coverage.

To balance its many advantages, scan design does come with a price of area overhead and additional circuit delay. The improved test coverage, lower cost of test generation, better diagnostic capabilities, possibility of adding built in self-test (described below), and reduced tester requirements are benefits that help offset this cost. The diagnostic advantages of scan design are described later in this paper.

Functional vs. Structural Testing Trade-offs

Choosing functional versus structural testing requires considering a variety of trade-offs. Even long-time advocates of functional test such as Intel are revisiting their test approaches [31]. Some of the trade-offs are described below.

First, an advantage of functional testing is the possibility of "at speed" testing, i.e., operating the chip during test at system speeds. Functional testing can be used by the designer to verify that the most used or the most critical operations meet design specifications. A goal of functional testing is to mimic the ultimate system environment as closely as possible. For example, functional tests attempts to run the device at the final frequency of the system (or a little faster to provide a guardband). This "at-speed" testing also allows the already-mentioned speed binning--the process of classifying the chips according to the range of frequencies at which they will operate.

A disadvantage of functional testing is that functional testers are needed. Due to the required speeds, timing accuracy and power supply requirements, as well as the number of I/O pins, these testers are increasingly expensive, yet quickly outdated: commitment to functional testing entails huge capital investments.

The other reason why functional testing is becoming too expensive is that it relies on manual test generation. Often the designer is needed for the intimate understanding of his or her part of the design for the test generation. But the designer will also have to have a thorough understanding of the whole architecture of the chip in order to know how to get the necessary information to and from the part of the design. ATPG is critical to satisfy short product development cycles and to reduce the engineering effort in developing test patterns. Only structural, not functional, testing enables ATPG.

Since functional testing requires no changes to the design for testing, there is no area overhead involved. On the other hand, for structural testing additional design specifications are introduced to facilitate testing and making the test generation process automated. The area overhead incurred has to be compared with the lower test generation costs.

A major advantage of structural test is reduced ATE requirements. For example, scan design can be implemented at the design's I/Os (called boundary scan) to enable the entire chip to be tested with only a few I/O pins [32]. This type of "reduced pin-count testing" can decrease ATE cost both by requiring fewer tester pins and enabling easier ATE reuse for many different designs.

Similarly, verifying the speed of the circuit by performing latch-to-latch delay tests rather than applying the test vectors at the full-speed of the IC [28,29] can greatly reduce ATE cost and enable ATE reuse.

Built-in Self-Test (BIST)

An increasing number of logic designs are implementing built-in self-test (BIST). BIST dramatically reduces the ATE requirements and has the opportunity to reduce product development times, provide higher speed test capability and enable the devices to be retested at the system level. As design reuse becomes more common, BIST is an attractive opportunity to simplify the testing problem. The most common BIST implementation is called STUMPS [33] where the random pattern generators and signature analyzes are added to full scan designs.

Memory Testing

Historically, stand-alone memory ICs have been tested by applying a sequence of regular patterns to ensure that each of the cells can store and provide data. Various sequences of patterns are applied to ensure that the memory does not have any address or area sensitivities. Common test pattern types include marching 1s and 0s, gallup and neighborhood patterns [34-36]. These patterns are developed to verify address decoders (unique address generation), memory cell data retention, and absence of cross coupling between adjacent memory cells. Most memory architectures require an analog read-out of the memory cells which is achieved by an amplifier that may have unique test requirements. Normally, memory test patterns are applied at the full rated speed of the device.

DRAMs typically have more subtle sensitivities compared to SRAMs, so a wider variety of patterns and conditions may be applied to DRAMs. In addition, DRAMs typically require a larger effort to develop tests that cover all possible defective circuit behaviors.

For manufacturing test, memory ICs are tested in parallel (6 to 32 devices at a time) to reduce test costs. Normally SRAMs and DRAMs require an additional processing step of being exercised at high voltage and high temperature (burn-in). DRAMs (and some SRAM) normally contain "redundancy," or spare cells that can be used to replace defective cells. Bitmapping is widely applied to facilitate yield, test and reliability defect learning. Bitmapping is collecting all information about working/failing cells for the entire memory. Each of these steps are more difficult when the memory is implemented not as a stand-alone chip, but as a macro embedded within a logic chip.

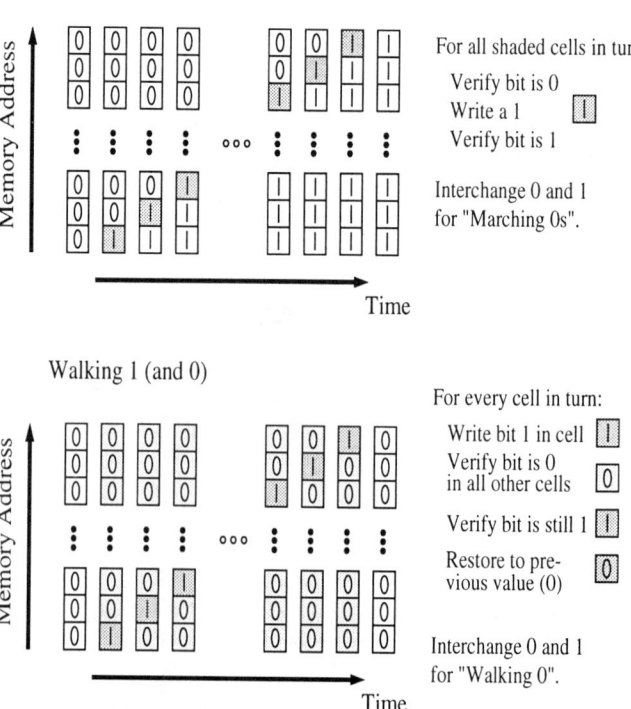

Figure 5. Memory test patterns

Embedded Memory Testing

Embedded memory as registers or register arrays has long been part of designs. However, the size of these register arrays now along with use of SRAM memory (e. g., on-chip caches for CPUs) has increased. Future designs will include embedded DRAM and embedded flash memories.

It is fairly straight-forward and cost effective to use an ATE to test stand-alone memory ICs. For memories embedded on logic chips, built-in self-test (BIST) is the preferred alternative. Because of the regular nature of memories, on-chip BIST capability is feasible. The BIST engine generates the addresses and data-in signals and verifies the data-out signals. Test patterns similar to the ones applied for stand-alone memories can be applied. Another capability that the BIST engine must support is wait periods that are needed between writing and reading data to verify data retention times. The BIST engine itself must be fully tested before memory testing can start.

Thorough testing and characterization of the embedded DRAM test is particularly important since DRAM requires unique process steps (e.g., capacitors in the memory cells) compared with digital logic circuits. During technology bring-up and yield ramping, the embedded DRAM may be the only vehicle to verify these additional steps. Yield-ramping requires an efficient approach to collect by bit fail maps using the BIST engine.

Analog circuit testing

Analog macros (e.g., phased-locked loops, digital-to-analog converters, analog-to-digital converters, sense amplifiers) are typically tested significantly different from digital circuits. Instead of checking boolean responses, analog test methods are normally functional tests of parametric behavior such as voltage measurements, linearity and frequency measurements, and jitter.

Because of the difference in test methodologies compared with logic circuits, ATE for analog circuits and the test development process is significantly different. ATE must have highly precise and accurate parametric and frequency instruments that are available to a number of pins. These instruments are much more expensive compared with simpler, digital ATE. Issues like power supply noise and precise parametric measurements are much more significant for analog circuits than for digital circuits. Because of the unique functional test requirements, normally test development for mixed-signal products is not automated. Thus, test development effort is much larger and development times are longer for mixed-signal products. Also, capabilities such as automated fault diagnostics is normally not possible.

System-on-a-chip (SOC) Testing

There has been much attention paid to the industrial trend toward systems-on-a-chip (SOC). The systems-on-a-chip concept suggests that functions that had previously been on independent dies are placed on a single die. SOC devices may include many circuit types including digital, memory and analog. Potentially, SOC devices may also contain multiple process technologies--e.g., CMOS and silicon germanium on the same IC.

SOC devices pose unique testing challenges. As unique circuit types (called embedded macros) are brought together on a single chip, it may not be possible to apply the same test methods that had been used to test the stand-alone chip. For example, the input and outputs (I/Os) of the macro may not be accessible from the chip I/Os. Inaccessibility of macro I/Os is a particular problem for testing analog circuitry.

The test economics for SOC devices will be challenged. As multiple macros are placed on a single IC, the cost to build the single die remains approximately constant (assuming the same die size and technology scaling where defect densities improve so that yield remains constant for a given die size). The test cost, however, may go up significantly if each embedded macro must be tested serially using conventional techniques. The testing-related costs may also go up if multiple, expensive testers are required for each individual macro. There are other technical issues associated with SOC devices such as embedded DRAM bitmapping, redundancy reconfiguration and mixed signal noise, which do not have industry-wide, accepted solutions at this time.

Automatic test equipment

Automatic test equipment (ATE) can be a large part of the cost of testing ICs. Typical testing costs are $3K to $8K per pin for ATE which results in total capital costs of $700K to $5M per tester. The cost drivers of ATE are the number of pins, the maximum frequency, pattern buffer storage and timing flexibility options.

Minimizing the cost of ATE is a high priority of semiconductor suppliers. This is done by optimizing manufacturing efficiency (e.g., parallel test) and reducing the requirements of the ATE cost drivers previously listed. For example, some test methodologies enable ICs to be tested using a reduced pincount interface. In general, structural test methods provide a better opportunity to exploit reduced ATE requirements than functional testing. Wider use of methodologies that significantly reduce test costs is expected to be a key trend in the next few years.

Reliability Defect Stressing

Burn-in is a step aimed at inducing early life failures to become hard failures by applying high temperature and high voltage to ICs. Some defects such as gate oxide shorts and high-resistance metal shorts may not disturb circuit functionality initially, but may later cause the IC to fail. Normally burn-in is an extra processing step where packaged ICs are placed in burn-in ovens for a significant length of time (e.g., 48 hours).

Testing methods can be applied at wafer or package test that can stress or detect these type of defects also. As mentioned earlier, I_{DDQ} testing may detect defects that do not cause a logical or performance abnormality, but which cause extra quiescent current. High-voltage stressing may be applied for a short duration at wafer test [37]. For some products, burn-in may be replaced by high-voltage stressing and I_{DDQ} testing.

DIAGNOSTIC TECHNIQUES

A key capability that is critical for failure analysis is being able to determine the physical location of defects. This capability is becoming increasingly important as advanced technologies are challenging conventionally applied failure analysis techniques [38].

The most commonly applied automated diagnostic technique is fault simulation-based software diagnostics. Commercial diagnostic tools are available from Mentor Graphics (Fastscan), Synopsys/Sunrise (Sherlock) and IBM (TestBench) [39]. Normally these tools are built off ATPG software using the same gate-level circuit representation and fault models as mentioned earlier. Given failure data collected at the tester, the list of possible faults can be trimmed by tracing back from the failing output or latch and identifying all possible fault sources (scan design dramatically simplifies this step). Then fault simulation can be applied for all possible faults until the faulty circuit response from the tester is matched.

There are some significant limitations with existing diagnostic software. The key limitation is that most of these software tools use the stuck-at fault model to represent defective circuit behavior. Since a significant percentage of defect types do not behave exactly like a stuck-at fault, diagnostic software may not be able to match the defective circuit response. Similar to ATPG fault models -- other models have been proposed, but are not widely used. Examples of the application of other fault models for diagnostic purposes include the transition fault model [40], the bridging fault model [41-45], a resistive short model [46] combinations of these [45-48].

Another diagnostic approach is a "full fault dictionary" where the failing vectors are stored for each fault. A state-of-the-art design may have millions of faults and millions of test vectors which makes the classical implementation of a full fault dictionary impractical for most designs.

Note that most examples of automated fault diagnostics are based on structural testing. For functional testing, fault diagnostics is normally a combination of ad-hoc methods and manual characterization on the ATE [49-52]. If the design supports full or partial scan, these scannable elements provide observation points that can provide some localization information.

DIAGNOSTIC OPPORTUNITIES

Because of the emphasis on reducing product development times and the increased difficulty of performing physical failure analysis, new approaches to physical failure analysis and defect learning should be pursued. One of the most promising new areas for physical defect analysis is to develop a methodology for understanding the relationship between defects and their electrical response -- also called "signature analysis." As described in [4], signature analysis proceeds by using historical data that links an IC's electrical response to a root cause failure mechanism (typically determined through PFA). The historical data is used to assign with a given confidence level the root cause mechanism underlying failure on a new faulty IC based on its response to electrical tests.

A key element of such a signature analysis is finding test results that can distinguish one type of defect mechanism from another. Possibilities include:

- I_{DDQ} vs. V_{DD} behavior [10,53]
- I_{DDQ} vs. temperature behavior [54]
- I_{DDQ} vs. time behavior [11]
- Comparison of I_{DDQ} on multiple test vectors [55-57]
- Timing vs. V_{DD} behavior
- Timing vs. temperature behavior
- I_{DD} transient test results [58-60]
- Power supply average current measurement [61]
- low & high voltage functionality [62]
- Low-temperature test results [51]
- Static test vs. timing test behavior [22]

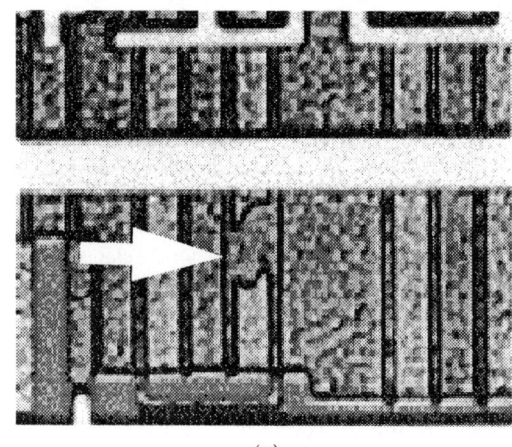

(a)

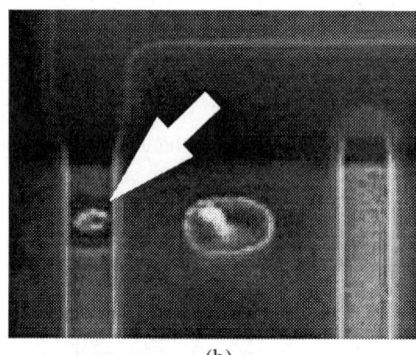

(b)

Figure 6. Photographs of two defects that behaved as "I_{DDQ}-only" failures. (a) Extra polysilicon that shorted two inputs to a NOR gate. (b) Gate oxide short on the input of an XOR gate.

Figure 6 shows two photographs of defects that passed all other tests (e.g., stuck-fault, functional and delay tests), but had abnormally high I_{DDQ} [48]. These defects were localized using both I_{DDQ} diagnostic software (using data collected on an ATE) and static photon emissions [47]. Although some of the characteristics measured for these chips were similar (e.g., pass/fail results, high I_{DDQ} for some patterns--but not others, passed post-burn-in tests), these defects will have significantly different electrical signatures for some of the previously listed tests (e.g., I_{DDQ} vs. V_{DD} and I_{DDQ} vs. temperature) due to the differing nature of the physical defect. The potential for signature analysis techniques is their capability for discriminating between defect types such as these.

The following figures show examples of characterizing dies using a few of the previously listed methods. First, Figure 7 shows the temperature sensitivity of the minimum latch-to-latch delays (using scan-based delay testing as described in [29]) for several dies. Two of dies fail delay testing and one is a control sample that passes all tests [63,64]. Note that for one of the failing ICs, the delay increases with higher temperature. For the other IC, however, the delay gets larger at low temperatures. These different dependencies on temperature suggest different underlying defect mechanisms.

48

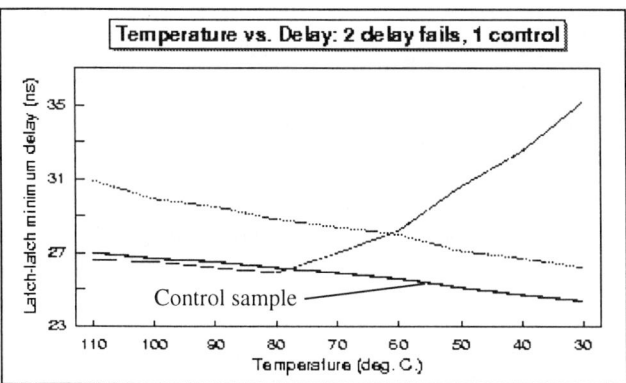

Figure 7. Delay versus temperature for two delay fails and a control sample that passes all production tests.

Figure 8 shows the results of a delay versus V_{DD} characterization test applied to five ICs that failed the delay test and a control sample [63,64]. Note the samples have different rates of delay increase as V_{DD} decreases. Also some defective ICs fail at low V_{DD} voltage independent of the appled timing [62].

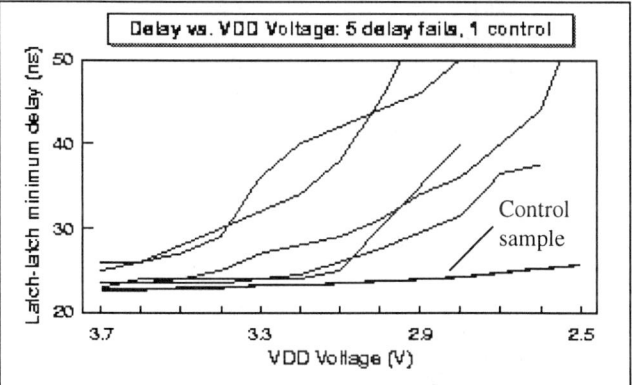

Figure 8. Delay versus VDD for 5 delay test failures and a control sample that passes all production tests.

Figure 9 shows I_{DDQ} versus V_{DD} characterization for three dies. Note that the shapes differ significantly. These differences clearly demonstrate that the current/voltage relationship is dependent on the underlying defect mechanism.

Figure 10 shows I_{DDQ} measurements on 195 test vectors in execution order. The I_{DDQ} appears to change over time. Such behavior should indicate an open that causes a wire or a floating transistor gate [11, 12].

Figure 11 shows "I_{DDQ} signatures" for two ICs. Such signatures compare I_{DDQ} on different test vectors, one to another [55]. In the figure the measurements are pictured in sorted order to show clearly steps between groups of measurements of similar magnitude. Figure 11a shows a signature that might be expected to result from a spot defect. For example, the signature is consistent with a resistive short connecting the output of a NOR gate to V_{DD}. The lowest I_{DDQ} level would correspond to circuit states where the defect is not activated.

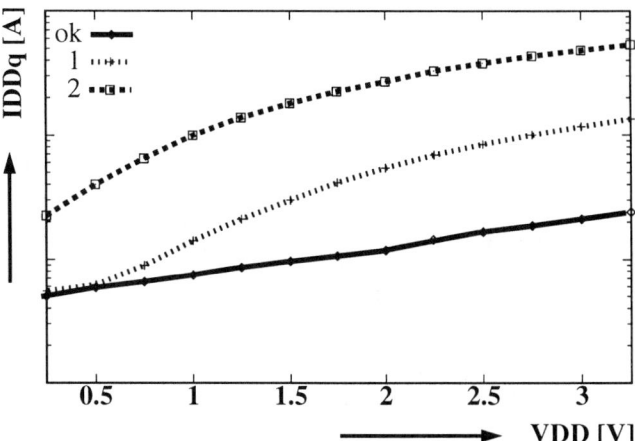

Figure 9. I_{DDQ} versus VDD characterization for three ICs. Note that the I_{DDQ} current is plotted on a log scale.

The middle level would correspond to one transistor in the pull-down path being turned on and the highest level would correspond to both transistors in the pull-down path being turned on. In contrast, Figure 11b shows a current signature that has, instead of a few levels, many unique levels of I_{DDQ}. Such a current signature may indicate multiple defects or a large area defect. Current signatures can also be compared from different test levels (e.g., before and after burn-in) to learn more about the underlying defects on a faulty die [56,57].

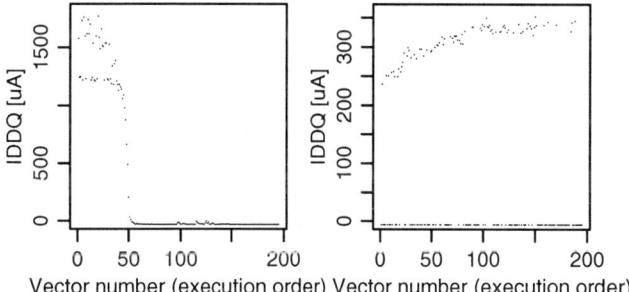

Figure 10. I_{DDQ} on 195 vectors in execution order showing time-dependent I_{DDQ} behavior on two ICs.

The above examples demonstrate the rich information about defect mechanisms available in test results. It is likely that no one type of characterization will point exclusively to a single root cause mechanism, but through experience, it may be possible to learn to use combinations of results on different tests to point to underlying defects.

Note that most of these tests would be applied only in characterization mode and only for defective parts. Despite requiring longer test times than production tests, they are orders of magnitude faster to apply than many physical FA techniques, are non-destructive and do not suffer from the many-metal-layer-imposed physical inaccessibility problem of many FA techniques.

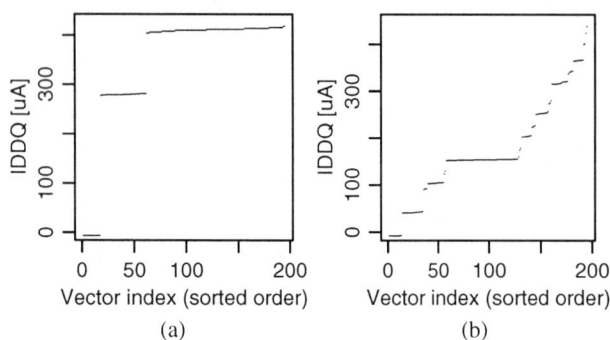

(a)   (b)

Figure 11. "I_{DDQ} signatures" comparing I_{DDQ} on different test vectors. Signature with few levels typical of spot defect (a); signature with many levels possibly indicating defects affecting multiple transistors (b).

CONCLUSIONS

This paper has given a basic overview of microprocessor and ASIC integrated circuit testing. IC testing is being strongly impacted by semiconductor industry trends toward greater complexity, speed and integration. These trends are forcing changes in test methodologies. The same trends are making PFA much more difficult now and perhaps impossible in some cases in the future, leading to a potential "brick wall" for technology development. At the same time, the increasingly short window between technology introductions makes it essential to find quick, effective alternatives. This need presents an opportunity for testing to play a new role beyond separating good chips from bad. The richness and relative ease of acquisition of test data makes it an excellent candidate as an alternative means of determining the physical mechanisms underlying IC failure. Opportunities exist, as shown by the examples in this paper; however, the test and FA communities must work together to develop and exploit them.

ACKNOWLEDGEMENTS

The authors wish to thank Wojciech Maly for providing the motivation for exploring many of these technical issues. We would also like to acknowledge Brian Kessler for providing information related to embedded memory testing.

REFERENCES

[1] The National Technology Roadmap for Semiconductors, 1997 Edition. Semiconductor Industry Association.

[2] W. Maly, "Testing-Based Failure Analysis: A Critical Component of the SIA Roadmap Vision," Int. Symp. on Testing and Failure Analysis, 1997, pp. 3-6.

[3] J. Soden, C. Hawkins and A. Righter, Identifying Defects in Deep-Submicron CMOS ICs, IEEE Spectrum, Vol. 33, No. 9, pp. 66-71, Sept. 1996.

[4] C. Henderson and J. Soden, "Signature Analysis for IC Diagnosis and Failure Analysis," Int. Test Conf., 1997, pp. 310-318

[5] K. Butler, K. Johnson, J. Platt, A. Kinra and J. Saxena, "Automated Diagnosis in Testing and Failure Analysis," IEEE Design & Test of Computers, Vol. 14 (No. 3), 1997, pp. 83 - 89

[6] Y. Kwon and D. Walker, "Yield Learning via Functional Test Data," IEEE Int. Test Conf., 1995, pp. 626-635.

[7] R. Wadsack, "Fault Modeling and Logic Simulation of CMOS and MOS Integrated Circuits," Bell System Technical Journal, Vol. 57, May-June 1978, pp. 1449-1474.

[8] C. Hawkins and J. Soden, "Electrical Characteristics and Testing Considerations for Gate Oxide Shorts in CMOS ICs," Int. Test Conf., Nov. 1985, pp. 544-555.

[9] C. Hawkins and J. Soden, "Reliability and Electrical Properties of Gate Oxide Shorts in CMOS ICs," Int. Test Conf., Sept. 1986, pp. 443-451.

[10] L. Horning, J. Soden, R. Fritzemeier and C. Hawkins, "Measurements of Quiescent Power Supply current for CMOS ICs in Production Testing,", Int. Test Conf., 1987, pp. 118-127.

[11] W. Maly, P. Nag and P. Nigh, "Testing Oriented Analysis of CMOS ICs with Opens," Int. Conf. on Computer-Aided Design, 1998, pp. 344-347.

[12] J. Soden, R. Treece, M. Taylor and C. Hawkins, "CMOS IC Stuck-open Fault Electrical Effects and Design Considerations," Int. Test Conf., Aug. 1989, pp. 423-430.

[13] F. Ferguson, M. Taylor and T. Larrabee, "Testing for Parametric Faults in Static CMOS Circuits," Int. Test Conf., 1990, pp. 436-443.

[14] T. Storey and W. Maly, "CMOS Bridging Fault Detection," Int. Test Conf., 1990, pp. 842-851.

[15] H. Hao and E. McCluskey, "'Resistive Shorts Within CMOS Gates," Int. Test Conf., 1991, pp. 292-301.

[16] R. Rodriquez-Montanes, J. Segura, V. Champac, J. Figueras, J. Rubio, "Current vs. Logic Testing of Gate Oxide Short, Floating Gate and Bridging Failures in CMOS," Int. Test Conf., 1991, pp. 510-519.

[17] C. Henderson, J. Soden and C. Hawkins, "The behavior and Testing Implications of CMOS IC Logic Gate Open Circuits," Int. Test Conf., 1991, pp. 302-310.

[18] P. Maxwell, R. Aitken, V. Johansen, I. Chiang, "The Effectiveness of I_{DDQ}, Functional and Scan Tests: How Many Fault Coverages Do We Need?" Int. Test Conf., 1992, pp. 168-177.

[19] J. Segura, V. Champac, R. Rodrigues-Montanes, J. Figueras and J. Rubio, "Quiescent Current Analysis and Experimentation of Defective CMOS Circuits," Journal of Electronic Testing: Theory and Applications (JETTA), Vol. 2, No. 4, Nov. 1992.

[20] H. Vierhaus, W. Meyer and U. Glaser, "CMOS Bridges and Resistive Transistor Faults: Iddq versus Delay Effects," Int. Test Conf. 1993, pp. 83-91.

[21] H. Hao and E. McCluskey, "Analysis of Gate Oxide Shorts in CMOS Circuits," IEEE Transactions of Computers, Vol. 42, No. 12, December 1993, pp. 1510-1516.

[22] C. Hawkins, J. Soden, A. Righter and F. Ferguson, "Defect Classes -- an Overdue Paradigm for CMOS IC Testing," Int. Test Conf., 1994, pp. 413-425.

[23] K. Wallquist, "On the Effectiveness of ISSq Testing in Reducing Early Failure Rate," Int. Test Conf., Oct. 1995, pp. 910-915.

[24] S. Ma, P. Franco and E. McCluskey, "An Experimental Chip to Evaluate Test Techniques: Experimental Results," Int. Test Conf., Oct. 1995, pp. 653-662.

[25] T. Williams, R. Dennard, R. Kapur, M. Mercer and W. Maly, "IDDq Test: Sensitivity Analysis of Scaling," Int. Test Conf., Nov. 1996, pp. 786-792.

[26] M. Abramovici, M. Breuer and A. Friedman, *Digital Systems Testing and Testable Design*, IEEE, 1998.

[27] E. Eichelberger and T. Williams, "A Logic Design Structure for LSI Testability," Design Automation Conf, 1978, pp. 165-178.

[28] B. Koenemann et al., "Delay Test: The Next Frontier for LSSD Test Systems," Int. Test Conf., Sept. 1992, pp. 578-587.

[29] O. Bula et al., "Gross Delay Defect Evaluation for a CMOS Logic Design System Product," *IBM J. Res. Develop.*, Vol. 34 (No. 2/3) 1990, pp. 325-338.

[30] P. Maxwell, R. Aitken, K. Kollitz and A. Brown, IDDq and AC Scan: The War Against Unmodeled Defects, Proceedings of the International Test Conference, pp. 250-258, Oct. 1996.

[31] S. Sengupta et al., "Defect-Based Test: A Key Enabler for Successful Migration to Structural Test," *Intel Technology Journal*, Q1, 1999, pp. 1-14.

[32] R. Basset et al., "Boundary-Scan Design Principles for Efficient LSSD ASIC Testing," *IBM J. Res. Develop.*, Vol. 34 (No. 2/3) 1990, pp. 339-354.

[33] P. Bardell and W. McAnney, "Self-Testing of Multichip Logic Modules," Int. Test Conf., Nov. 1982, pp. 283-288.

[34] R. Feugate and S. McIntyre, *Introduction to VLSI Testing*, Prentice Hall, 1988.

[35] A. Van de Goor, *Testing Semiconductor Memories*, John Wiley & Sons, 1991.

[36] A. Sharma, *Semiconductor Memories: Technology, Testing and Reliability*, IEEE, 1997.

[37] J. Chang and E. McCluskey, "SHOrt Voltage Elevation (SHOVE) Test for Weak CMOS ICs," VLSI Test Symposium, Oct. 1997, pp. 446-451.

[38] D. Vallett, "An Overview of CMOS VLSI FAilure Analysis and the Importance of Test and Diagnostics," Int. Test Conf. Lecture Series 2, 1996, pp. 1-7.

[39] Testbench Users Guide, IBM Corporation.

[40] J Waicukauski, E. Lindbloom, B. Rosen and V. Iyengar, "Transition Fault Simulation," IEEE Design & Test of Computers, April 1987, pp. 32-38.

[41] S. Chakravarty and M. Liu, "Algorithms for Current Monitorbased Diagnosis of Bridging and Leakage Faults," Design Automation Conference, 1992, pp. 353-356.

[42] S. Millman and J. Acken, "Diagnosing CMOS Bridging Faults with Stuck-at, IDDQ and Voting Model Fault Dictionaries," IEEE Custom Integrated Circuits Conference, 1994, pp. 17.2.1-4.

[43] T. Lee, W. Chuang, I. Hajj and W. Fuchs, "Circuit-level dictionaries of CMOS bridging faults," *Transactions on Computer-Aided Design of Integrated Circuits and Systems*, Vol. 14 (No. 5) 1995, pp. 596 -603

[44] B. Chess et al, "Diagnosis of realistic bridging faults with single stuck-at information," Int. Conf. on Computer-Aided Design, 1995, pp. 185 -192

[45] R. Aitken, "A Comparison of Defect Methods for Fault Localization with $I_{DD}Q$ Measurements," Int. Test Conf., Sept. 1992, pp. 778-787.

[46] D. Burns, "Locating High Resistance Shorts in CMOS Circuits by Analyzing Supply Current Measurements Vectors," ISTFA, Nov. 1989, pp. 231-237.

[47] P. Nigh, F. Motika and D. Forlenza, "Application and Analysis of IDDq Diagnostic Software," Int. Test Conf., Oct. 1997, pp. 319-327.

[48] P. Nigh, D. Vallett, A. Patel, J. Wright, F. Motika, D. Forlenza, R. Kurtulik and W. Chong, "Failure Analysis of Timing and IDDq-only Failures from the SEMATECH Test Methods Experiment," Proceedings of the International Test Conference, pp. 43-52, Oct. 1998.

[49] A. Carbine and D. Feltham, "Pentium(R) Pro processor design for test and debug," Int. Test Conf., 1997, pp. 294-303.

[50] Y. Hong and M. We, "The application of novel failure analysis techniques for advanced multi-layered CMOS devices," Int. Test Conference, 1997, pp. 304-309.

[51] W. Needham, C. Prunty and E. Yeoh, "High Volume Microprocessor test Escapes, An Analysis of Defects Our Tests are Missing," Int. Test Conf., 1998, pp. 25 -34.

[52] M. Levitt, et. al., "Testability, Debuggability and Manufacturability Features of the UltraSPARC-I Microprocessor," Int. Test Conf., Oct. 1995, pp. 157-166.

[53] R. Rodiquez-Montanes and J. Figueras, "Bridges in Sequential CMOS Circuits: Current-Voltage Signature," VLSI Test Symposium, 1997, pp. 68-73.

[54] J. Seo, S. Lee, S. Daniel and C. Yoon, "Temperature Dependence of Quiescent Currents as a Defect Prognosticator and Evaluation Tool," Int. Symp. for Testing and Failure Analysis, 1996, pp. 245-249.

[55] A. Gattiker and W. Maly, "Current Signatures: Application," Int. Test Conf., 1997, pp. 156-165

[56] A. Gattiker and W. Maly, "Toward Understanding IDDQ-Only Fails," Int. Test Conf., 1998, pp. 174-183.

[57] A. Gattiker, Ph.D. Dissertation, Carnegie Mellon Univ., May 1, 1998.

[58] J. Plusquellic, D. Chiarulli and S. Levitan, "Identification of Defective CMOS Devices Using Correlation and Regression Analysis of Frequency Domain Transient Signal Data," Int. Test Conf., Nov. 1998, pp. 40-49.

[59] J. Beasley, et. al, "IDD Pulse Response Testing of Analog and Digital CMOS Circuits, Int. Test Conf., Oct. 1993, pp. 626-634.

[60] M. Sachdev, P. Janssen and V. Zieren, "Defect Detection with Transient Current Testing and its Potential for Deep Sub-micron CMOS ICs," Int. Test Conf., Oct. 1998, pp. 204-213.

[61] B. Vinnakota, "Deep Submicron Defect Detection with Energy Consumption Ratio," Int. Conf. on Comp.-Aided Design, Nov. 1999.

[62] H. Hao and E. McCluskey, "Very-Low-Voltage Testing for Weak CMOS Logic ICs," Int. Test Conf., Oct. 1993, pp. 275-284.

[63] P. Nigh, W. Needham, K. Butler, P. Maxwell and R. Aitken, "An Experimental Study Comparing the Relative Effectiveness of Functional, Scan, IDDQ, and Delay-Fault Testing," VLSI Test Symposium, April-May 1997, pp. 459-463.

[64] P. Nigh, W. Needham, K. Butler, P. Maxwell, R. Aitken and W. Maly, "So What is the Optimal Mix? A Discussion of the SEMATECH Test Methods Experiment," Proceedings of the International Test Conference, pp. 1037-1038, Oct. 1997.

51

Decapsulation and Package Analysis

Chip Access Techniques

R. Raghunathan and D. Davis
Texas Instruments, Inc.
Stafford, Texas

INTRODUCTION

A defective junction may be one out of several million transistors on a chip and it is an enormous challenge to isolate the defect. Although leakage current in a semiconductor can be determined by straightforward electrical tests, pinpointing the exact site of leakage can be a very time consuming and arduous task. Fortunately, several types of chip level defects emit light or heat depending on the nature of the defects. Hence, by isolating the area of light or heat emissions, it is possible to locate the defective region on the chip. Emission microscopy [1] and liquid crystal techniques [2,3] have thus become an integral part of failure analysis. Several other techniques of fail site isolations like Electron Beam Induced Current (EBIC) [4,5] and Voltage Contrast Microscopy [6] are also extensively used by failure analysts to isolate the defect. Supplementing the fail site isolation techniques with physical deprocessing and inspection using SEM, it is possible to isolate the defect easily in most situations. However, for any of the analysis mentioned above, chip access is the first step. It is very critical to adopt good techniques during the chip access stage. If proven methods and logical operations are not followed, valuable specimen information may be lost. In this chapter we will focus on techniques used for accessing the chip during failure analysis for different types of devices and packages.

The evolution of high performance electronic devices has fostered the development of a wide variety of package types and materials. Chip access techniques are specific to the package types and the package materials used. In order to obtain a good understanding of the various chip access techniques it is important to understand the different package types and materials used and to know the trends in the packaging industry today. In the next section, we will briefly talk about various package types and materials prevalent in the industry today.

PACKAGING TYPES AND MATERIALS

Package Materials:

Essentially there are two different packaging materials used today: ceramic and plastic. Plastic packages are more prevalent in today's market for various reasons, the primary one being cost. In the early 1960's, plastic encapsulation emerged as an inexpensive and simple alternative to ceramics and metal casings and by the 1970's virtually all the packages made were plastic. Epoxy novolac was the first material used for encapsulation which was then replaced by the phenolics and silicones in the 1960s. However, these plastic packages had several reliability issues largely due to the poor quality of the encapsulation system. Over the decade numerous formulations of epoxies were developed which had lower curing shrinkages and contamination levels and thus became the dominant packaging material in the 1970's. A typical encapsulant used today is a complex mixture of cross-linkers, accelerators, flame retardants, fillers, coupling agents, mold release agents, flexibilizers in an epoxy matrix. There are hundreds of proprietary polymers in use for device encapsulation. They all fall into three major classes: Epoxies, Silicones and Urethanes.

A typical plastic package is shown in Fig. 1[7]. Polymer and metal are the flesh and bones of a molded plastic package. The plastic package which is usually non-hermetic in nature consists of a silicon chip, a metal support or lead frame, wires that electrically attach the chip circuits to a lead frame which incorporates external electrical connections and a plastic epoxy-encapsulating material to protect the chip and the wire interconnects. The lead frame is typically a copper (Cu) alloy, alloy 42 (42% Ni/58% Fe), or alloy 50 (50% Ni/50% Fe), and is plated with gold (Au) and silver (Ag) or palladium (Pd) over nickel (Ni) or nickel/cobalt. The silicon ship is mounted over the lead frame with an organic conductive formulation of epoxy. Wires, generally of gold but also of aluminum or copper, are bonded to the aluminum bonding pads on the chips and to the fingers of the lead frame. The assembly is then typically transfer molded in epoxy. Following the molding operations, the external leads are plated with a lead-tin alloy, cut away from the strip and formed into desired shapes.

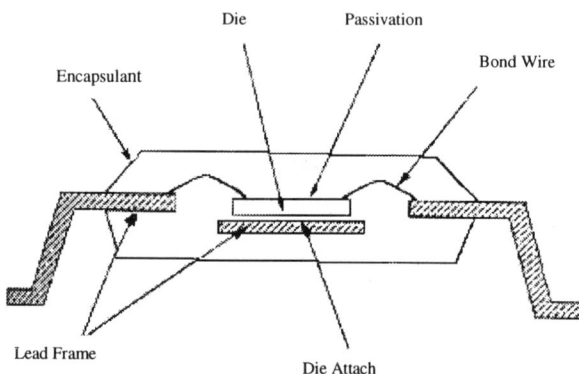

Figure 1   Cross section of a typical package type.

The evolving trend in the development of plastic packaging materials is to have higher glass transition temperature (T_g), lower stress and very low shrinkage of the encapsulants [8]. These new generation mold compounds are also very hard to etch. The new generation mold compounds and packaging are challenging the traditional methods of decapsulation of IC devices. As the package-to-die edge spacing decreases, the decapsulation of these packages becomes even more challenging. Later in this chapter, we will discuss various decapsulation techniques for the new generation mold compounds.

Plastic packages used extensively today, are low in cost but have low thermal conductivity. Ceramics on the other hand posses a combination of electrical, thermal and mechanical properties which are unmatched by any group of materials [9]. Ceramic substrates also provide the highest wiring density of all substrate technologies. Ceramic packages provide the most reliable performance of all package types and hence, are in use today where reliability is a prime concern. These include most defense and aerospace applications. Their reliability superiority over plastic is primarily due to three fundamental reasons:

(a) First, by their very nature, ceramics are hermetic. They do not absorb or retain moisture nor do they allow permeation of moisture. (b) Secondly, their dimensional stability during and after high temperature processing is exceptional. (c) Finally, the chemical inertness of most of the ceramics to water, acids, solvents and other chemicals is outstanding. This feature, that makes ceramics attractive as a packaging material, also makes it non-amenable to chemical deprocessing techniques. Hence, mechanical decapping techniques are usually employed for ceramic packages.

Package Types:

In this section of this chapter we will discuss the trends in the package industry today (shown in Fig. 2) and also describe the different package types. There are six main types of packages in the market today: Dual-in-line (DIP), Small Outline (SO), Chip Carrier (CC), Quad Flat Pack (QFP), Pin Grid Array (PGA) and Ball Grid Array (BGA). In addition to these, there are several derivatives in each of these categories. Also, there are some special-application packages (not discussed in this chapter), most notably among them is the Tape Automated Bonding (TAB) package. The trends in the packaging industry is shown in Fig. 2. As seen from Fig. 2, the trend in the packaging industry has moved from DIPs in 1970s to QFPs in 1980s to BGAs in 1990s. The package type envisioned for the next century is the Single-Level Integrated Module [10] commonly referred to as SLIM. The main motivating factors for innovations in the packaging industry is increased I/O pin density and decreased cost. In the following section we will briefly describe the different package types depicted in Fig. 2.

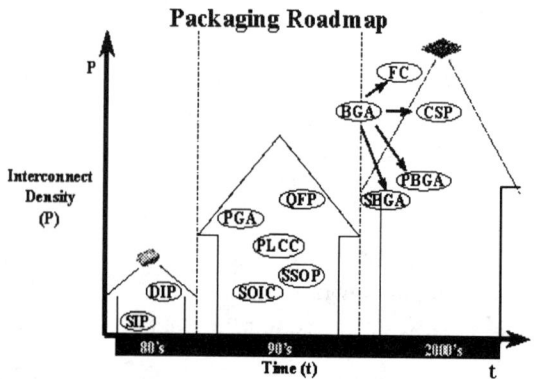

Figure 2   Tends in the packaging industry.

56

Dual In Line Packages (DIP):

One of the earliest packaging standards was the rectangular dual-in-line packages (DIP). DIP has I/O leads that extend from two opposite sides of the package and are bent downwards as shown in Fig. 3 [11]. DIP's were the mainstream of the microelectronics industry in 1980's, however its applications have reduced due to emerging technology such as Small-Outline Integrated Circuits (SOIC) and Pin-Grid Arrays (PGA). Plastic DIP's consist of a lead frame typically made of kovar. The lead frame is punched out of sheet metal in the desired shape, which typically includes a small platform or pad at the center with leads fanning out from this pad. The leads are formed so that they can bend down for insertion into the Printed Wiring Board (PWB). The lead frame is then plated with nickel as an underplate, or corrosion barrier, and tin or gold surface platting for wire bondability and/or solderability. The die is then mounted to the center pad of the lead frame and the I/O pins of the die are wire bonded to the appropriate leads. The entire assembly is then molded in plastic, forming the plastic body. In 1970's most plastic packages were made of Epoxy B or Novalac.

Figure 3 A Typical Plastic Dual-in-line Package (DIP).

Plastic DIP packages are available up to 68 leads. Above 68 leads, molding and lead frame problems limit the production of larger DIP's because of excessive costs. Plastic DIP's find their use in commercial applications, while military systems typically utilize ceramic (alumina) packages. In ceramic DIP packages, the die is sandwiched between two ceramic plate elements. The element on the bottom half of the sandwich holds the die while the section on the top half protects the die from mechanical stress during sealing operations. Ceramic packages are preferred for high quality electronic systems, where increased reliability takes precedence over increased cost.

Pin Grid Array:

Another type of through-hole pin package configuration is the Pin-Grid Array (PGA), which is typically used with VLSI chips having lead counts greater than 100. Schematic of a typical PGA is shown in Fig. 3.

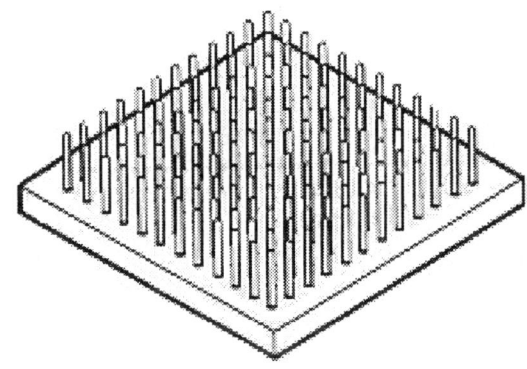

Fig. 3 Schematic of a Pin-Grid Array Package (PGA).

The PGA is often composed of a square, multilayered, ceramic chip carrier with the matrix of butt-brazed, through-hole pins onto its surface. The pins on the ceramic are placed on 100 mil centers over the complete component area. Since the leads in an array are over the entire area, the obtainable lead counts far exceed those for components with peripheral leads only. PGA's are available in cavity up and cavity down versions. The cavity up package has the chip mounting cavity on the substrate opposite the leads. With this configuration the entire lead side can be occupied with the pin grid, resulting in very high I/O density. The cavity down version has the chip cavity face on the same side as the pin grid arrays. This cavity is required for chips with high heat loads where heat exchangers are placed on top of the package.

Although the I/O density increased dramatically with the introduction of the pin grid array's, there were practical limitations in the through-hole assembly technique. The need for more leads resulted in finer pitch which in turn resulted in thinner pins. Thinner pins are more sensitive to handling and to insertion by machines into sockets and are generally more difficult to straighten. The trend has therefore lead to surface mount technologies like the Quad Flat Pack (QFP), Chip Carriers (CC) and the Ball Grid Array (BGA) packages.

Flatpack:

The emergence of surface or planar-mounted components has overcome some of the space problems associated with through-hole mounting and the high pin-count DIP packages. One of the earliest forms of surface mounted components is the flatpack. The flatpack is similar to the DIP with the exception that the leads are bent outwards to form a flat surface and are mounted on pads rather than inserted into holes on the board. The reason for a move in the direction of flatpacks was the need for an increase in the board density and a decrease in cost. Flatpacks are found in both dual-in-line called Dual Flat Package (DFP) and with leads which protrude on all four sides of the component package called Quad Flat Package (QFP) as shown in Fig. 4. Small Outline Integrated Circuits (SOIC) are similar to QFP's but are smaller in size and have lower interconnect densities.

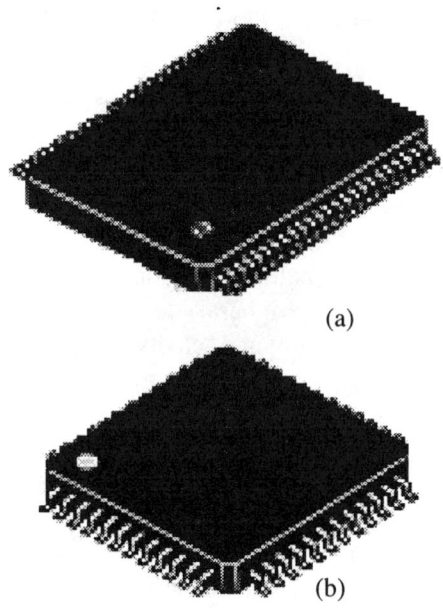

(a)

(b)

Figure 4 Schematic of a typical (a) Dual Flat Pack (DFP) (b) Quad Flat Pack (QFP).

Chip Carriers:

Another surface mount package type that was popular in the early 90's was the chip carriers. All ceramic chip carriers can be configured to be either leaded or leadless. The three main types of chip carriers were:

Ceramic and Plastic Leaded Chip Carriers and the Leadless Ceramic Chip Carriers (LCCC). Chip carriers offer higher packaging densities than their corresponding DIP packages. The Plastic Leaded Chip Carriers (PLCC) were developed in the early 1980's by Texas Instruments as a low cost alternative to leadless chip carriers. The leads on the plastic chip carrier can have the same configuration as the DIP's for insertion mounting. They can also be formed in a gull winged DIP configuration or a J lead configuration for surface mounting as shown in Fig. 5a and 5b respectively. J leads start off similar to DIP leads, but are bent up just below the package in the shape of a J.

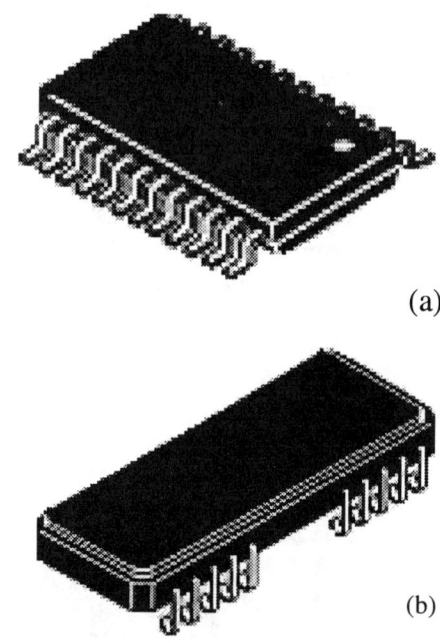

(a)

(b)

Figure 5 (a) Package with a Gull-Wing Lead (b) J leaded package.

Another popular configuration of the ceramic chip carrier which has the leads brazed to the top of the case, hence the name top brazed chip carrier. The top brazed chip carrier has leads that is formed into gull-winged packages for surface mounting. The schematic of the Leadless Ceramic Chip Carriers (LCCC) is shown in Fig. 6. The LCCC have castellations instead of leads. A semi-cylindrical column is cut out of the side of the chip carrier at every I/O and the insides of these columns are metallized. They are then plated, typically with nickel followed by gold. LCCC's with a typical pitch of 40 mils take up as little as 30% of the room required by DIP's that have the same lead count.

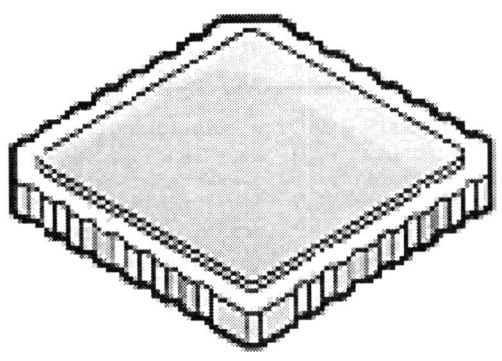

Figure 6   Schematic of a Leadless Ceramic Chip Carrier (LCCC).

Ball Grid Array (BGA):

Ball Grid Arrays (BGA's) are emerging as a successful package type due to efficient space utilization. Ball grid arrays are very similar in construction to the PGA except that the pins are replaced by solder balls. Fig. 7 shows schematic cross sections of the different types of BGA packages[10].

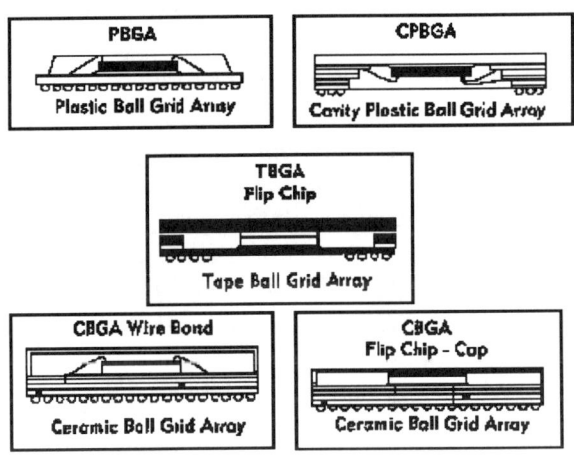

Figure 7 Schematic of different types of Ball Grid Array's (BGA).

Ball Grid Array (BGA) technology is emerging as a successful package option. There are many package configurations within the BGA family, such as the Flip Chip (FC), Chip Scale Package(CSP), Super Ball Grid Array (SBGA), etc. The BGA package configurations have significant performance advantages over some of the more traditional package types. Some of the performance advantages, mostly stemming from the inherent package construction, are increased switching speeds, increased interconnect capacity, and reduced signal cross talk. In addition, dimensionally the BGA package makes effective utilization of space, is commonly light weight, and has some of the smallest package profiles available. The ultimate goal and desire of any package designer is to minimize the package dimensions until they are very close to the actual size of the die. One of the most significant improvement is the packaging industry is the CSP where the package is hardly bigger than the die itself as shown in Fig. 8. Most BGA devices have lighter weight, higher density, low over-all dimensions and low power consumption which make them attractive for portable applications.

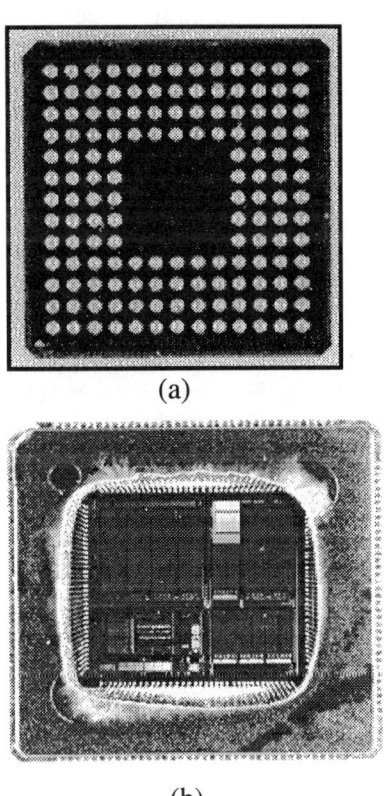

(a)

(b)

Figure 8 (a) Top of a CSP type BGA (b) CSP type of BGA with conventional wire bonding.

Currently the most popular technology amoung BGA type packages is the Flip Chip. In FC devices a direct electrical connection is achieved between the chip and the package through the solder bumps that connect bonding pads on the chip to corresponding pads on the substrate. In FCs the die is positioned face down to the substrate and hence his type of bonding is referred to as face-down bonding or the controlled-collapse soldering.

Unlike wire bonding technology, which allows only peripheral bonding, C4 bumps can be placed anywhere on the die. This leads to very low inductance power distribution to the chip. Some of the other advantages are higher I/O densities, and faster throughput times.

Achieving access to the die surface of a BGA device can prove to be very challenging. The package construction often hampers and some times prevent the access of the die's front or back surface. For example, the BGA device in Fig. 8 utilizes bond wires, polyimide materials and/or Teflon tapes, and solder (Pb:Sn) coated Cu traces to route the electrical signals from the die front surface to the bumps on the bottom side of the device. Often decapsulation these types of devices are extremely difficult because the acid or its fumes corrode the bumps or the tape, causing the device to lose continuity with the die's active circuitry. Placing a stiffener or metallic plate behind the device, when using automated decapsulation equipment, is often a necessity. Since the mechanical strength of the bumps on the package isn't very high compared to epoxy resin, the force of the pin used to hold the device in place on the etch machine has the potential to damage the bumps at the back surface of the device. Extended exposure of a device similar to the one depicted in Fig. 8 can corrode the bumps enough to separate them from the package. We will discuss the decapsulation techniques for BGA packages in later parts of this chapter.

Having described various package types and package materials used in the industry today, we will discuss the various chip access techniques used in failure analysis in the next section. Essentially there are two different approaches to decapping a unit: (a) mechanical decapping [12,13] and (b) chemical decapping techniques [12,14]. One could adopt either of the methods depending upon the nature of the analysis. Mechanical decapping techniques are usually destructive in nature and should be adopted only when electrical integrity is not required. If, on the other hand, the part is to remain functional electrically, chemical decapping techniques are adopted where a hole is made in the package locally above the chip by wet chemical etching processes.

MECHANICAL DECAPSULATION TECHNIQUES

Mechanical decapping is a quick and easy approach to chip access. In many situations, where the intention is only to inspect the chip surface, mechanical decapping is the ideal technique to use. Wet chemical processes have certain inherent risks; they can, if improperly handled, attack the metallization or damage the bond wires. Also, for small packages it becomes a challenge to use chemical etching techniques to access the chip without functionally damaging the part. This section describes a mechanical technique for decapsulation of devices. The technique has several unique applications as a failure analysis tool and can be performed more efficiently than traditional chemical or plasma techniques. The advantages, limitations and practical applications of mechanical decapping techniques are discussed.

Mechanical decapping techniques are primarily adopted for decapsulation of ceramic packages. Plastic encapsulated devices, with the exception of flip chip packages, cannot be decapped using mechanical decapping techniques without damaging the die. In flip chip packages, the die is positioned face down and hence decapping techniques for these devices is considerably different from that adopted for conventional packages. In the first half of this section we will focus our discussion on decapping techniques for non-flip chip devices and in the second half we will discuss various chip access techniques used for flip chip devices.

Mechanical Decapsulation of Ceramic Packages:

Ceramics are extremely resistant to chemicals and hence are not amenable to chemical decapping techniques. Hence, mechanical decapping techniques have to be adopted for ceramic packages. There are ceramic alternative packages to almost every plastic package configuration, the two main or most common types being the ceramic "side braised" and the "glass epoxy sandwich". The ceramic side braised package is essentially composed of a ceramic substrate or carrier, with a cavity carved or ablated into it and a lid attached using solder to the substrate material. The ceramic glass epoxy sandwich package is composed of a top and bottom ceramic lid, and a lead frame suspended in a glass epoxy medium.

De-Capsulation of "Side Braised" Packages

There are several standard methods of access for these package configurations. In this section we will discuss some common and successful methods of chip access for some popular ceramic package types. Chip access to a ceramic "Side Braised" package, and internal features, can be achieved by a number of methods, (1) shearing-off the lid, (2) de-soldering the lid, (3) or milling-off the lid.

Shearing Technique:

The materials required for shearing-off the lid of a ceramic 'side braised" package is as follows:

a. Vise
b. Needle-nosed pliers (ESD safe)
c. Single edge razor blade
d. Small metal or rubber hammer
e. Safety glasses
f. Cut resistant gloves

The procedure for shearing-off the lid is performed using the following steps (a) First, use the needle-nosed pliers to mount the package in the vice and apply enough pressure to hold the package in the vice. (b) Then, place the razor parallel to the package, just beneath the lid and the ceramic substrate, at one of the corner of the device as shown in Fig. 10. (c) Use a small hammer to tap lightly on the razor so as to detach the lid until approximately 1/4 of the lid area is lifted. (d) Finally, repeat this procedure at each of the remaining corners. Continue to tap the razor to the center of the lid or until the lid fully detaches.

Fig. 10. Picture depicting procedure for lid removal.

De-Soldering Technique:

The de-soldering method is the most cumbersome of the methods discussed in this section. The tools needed are the following

a. Soldering iron
b. Needle-nosed pliers
c. Safety glasses
d. Solder wick or solder vacuum pen (solder flux optional)

The procedure is preformed as follows: (a) Heat the soldering iron slightly above the melting point of the solder used to "side braise" the device. This temperature varies based on the specific composition of the solder used. (b) Then, place the solder wick in contact with the interface of the lid to the package. The solder will slowly move onto the solder wick leaving the edge of the lid detached. (c) Repeat this process until all the solder is removed. If the lid is not fully detached and almost all of the solder is removed, used the needle-nose pliers to lift the lid at an edge, then use the solder wick and iron to remove any underlying solder which may still be present.

Milling Technique:

The milling-off the lid method of exposing the die of a ceramic "side braised" package is a fast and convenient alternative to the aforementioned methods. The tools needed are the following:
(a) Vertical end mill
(b) Needle-nose pliers or tweezers
(c) Safety glasses.

The milling procedure is performed as follows: (a) First, mount the device with the needle-nose pliers to the base of the vertical end mill. (b) Then, turn on drill and place bit very close to the plane of the ceramic package and move along one edge of the lid until all of the solder at that edge is removed. (c) Proceed to mill away the solder interface on other edges of the lid and then remove lid with needle-nosed pliers and discard. The shearing-off and milling-off means of chip access can be combined to increase repeatability. This combination helps reduce the following difficulties experienced during each of the methods adopted (a) During shearing-off, the razors have a tendency to break at the corners of the lid during entry and (b) During milling-off, the bits tend to break due to overheating while working with large lids. A combination of the two techniques can be achieved by using the vertical mill to remove one corner of solder and lid and then using the razor to lift the remainder of the lid. The combination of methods is faster than the vertical mill method alone, much more repeatable and less destructive to the package. In addition, this approach extends the life of the razor and expensive mill bits.

De-Capsulation of "Glass Epoxy Sandwich" Packages

Glass epoxy ceramic sandwich packages are slightly more difficult to access than the "side braised" packages discussed earlier in the above paragraphs. There are two common methods of access: (a) the nutcracker method and (b) the shearing-off method. The key to accessing this package configuration is to take advantage of the glass transition temperature (T_g) properties of the glass epoxy medium in which the lead frame is embedded. The advantage is achieved through heating the package slightly above the T_g, thus reducing the adhesion of the glass to the ceramic lids while maintaining the connectivity of the lead frame and bonding wires.

The "NutCracker" Technique:

The nutcracker method has been very successful and removes a large amount of possibilities for operator error. The materials needed are as follows:

a. Fume hood (optional)
b. Needle-nosed pliers (ESD safe)
c. Hot plate (200 - 400 $^{\circ}$C)
d. "NutCracker" (device used in decapping ceramic units, see Figures 9a and 9b)
e. Heat resistant gloves
f. Safety glasses

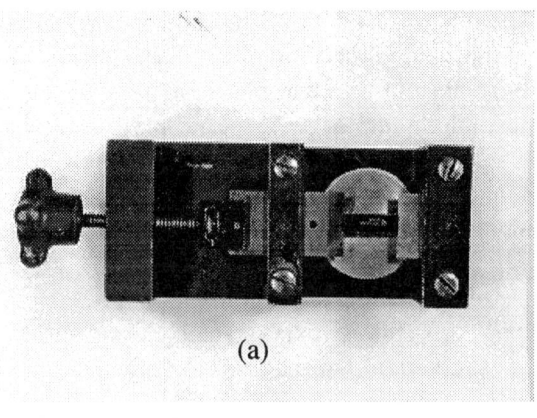

(a)

Fig. 11 (a) NUT CRACKER (top view) (b) NUT CRACKER (side view).

The fume hood may not be required if the work area is well ventilated and a lower plate temperature is used. The procedure is performed by pre-heating the hot plate to slightly above the T_g of the glass epoxy. A digital display/instrumentation and temperature controlled hot plate is recommended for the best results. The package is then placed onto a hot plate that has been pre-heated to approximately 300 - 400 $^{\circ}$C (Note: The plate should be explosion proof if used in a chemical fume hood). Do not place the device in direct contact with the hot plate, since the heat conducting properties of the leads may have a detrimental effect on the electrical signature of the device.

The glass epoxy will began to discolor slightly after a few minutes, indicating the T_g has been reached. This event can be detected by direct temperature measurement also. At this point, fumes are observed to come from the plastic (approximately 5 - 10 seconds later) and the unit is flipped over momentarily to obtain a uniform temperature. The fumes are an indication that the unit is ready for decapping and the unit will dissipate heat rapidly when the device is removed from the plate. The smoke that develops when heating the unit is toxic and should not be inhaled. Quickly remove the device and place the device on the nut cracker mechanism. Adjust the nutcracker stage height until the shearing fixture is in the same plane as the top of the ceramic lid. Now move the blade of the nutcracker shearing fixture until it is in direct contact with the ceramic lid, continue to turn in the blade until a cracking/shearing sound is heard indicating that the package has fractured. (Note: Cover the unit with gloved hand to deflect any air borne fragments). The fracture will most often occur at the chip/plastic interface and the two halves are easily separated. A typical fracture is shown in Fig. 12a and Fig. 12b.

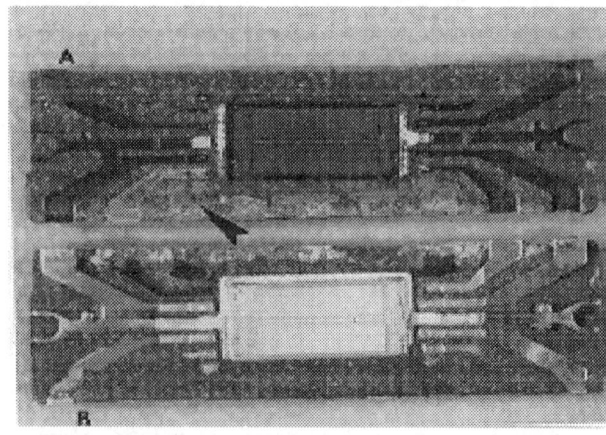

Figure 12 - Typical fracture (a) Imprint in mold compound (b) actual device.

If the fracture does not develop at this interface, the nut-cracker setting needs to be adjusted. After some practice, the fractures will consistently develop in the same area. When the fractures are unsatisfactory, the device is re-heated and the plastic is cracked away from the chip using an exacto knife, tweezers and/or wire cutters. If the chip does not come free, a decision should be made to abandon the process and use other traditional techniques to remove the die. There are several factors that may cause these improper fractures as follows:

a. Improper blade setting on nut-cracker
b. Improper temperature on hot plate
c. Type of plastic mold compound
e. Alpha shield overcoats (PIQ shields are the least successful)
e. Package size (large packages are the least successful)

Advantages and Limitations

Limitations of this technique develop when certain types of analyses are required. For instance, the technique obviously destroys electrical integrity of the package. This is undesirable if dynamic microprobing is needed. Additional problems arise when the method is performed on units that have seen extended high temperature environments (150 $^\circ$C operational life and temperature cycle). These units exhibit residue on the bar surface after decapping. The residue is suspected to be the release agent used in the mold compound. If detailed visual inspection is needed for these units, chemical cleaning will be required.

The advantages of mechanical decapping include a higher success rate in identifying contaminants that cause corrosion. Since the unit is completely dry decapped, there are no chemicals used that may distort or destroy these contaminants. Mechanical decapping is also very efficient and can be performed more quickly than traditional chemical etching and much more quickly than plasma techniques. Another advantage of this method is the unique ability to inspect the chip surface imprint in the top-side plastic. This imprint is indisputable evidence in cases where scratches or cracks are the cause of failure.

Shearing Technique:

The other method, shearing-off of the lid, is also performed very similarly. The materials needed are the same as indicated in the method described for shearing off the lid of a "side braised" ceramic package. In addition, the operator will need a hot plate. Pre-heat the hot plate to the T_g for the respective glass epoxy material. Then place the package on the hot plate and allow it to soak for several minutes. Once the T_g is reached, as indicated above, remove the package quickly from the hot plate and mount the device on the vice. Place the sharp edge of the razor against the point where the glass epoxy interfaces with the ceramic section (Fig. 10). Then proceed to tap lightly with the small hammer, the adhesive between the lid and glass will separate. Quickly continue until the entire lid is fully separated, as cooling of the device will make the separation more difficult. If necessary, re-heat the part to aid in the ease of separation of the lid.

CHEMICAL DECAPSULATION TECHNIQUES

Plastic packages are more prevalent in today's market compared to any other package type. As we discussed earlier, plastic encapsulated packages are not amenable to mechanical decasulation techniques. Mechanical decapsulation techniques are limited due to its destructive nature that results in loss of electrical integrity of the device. If electrical integrity of the device is to be maintained, it is necessary to adopt chemical deprocessing techniques. There are two different approaches to chemical deprocessing : (a) Wet chemical methods which use an acid etching process and, (b) Dry chemical methods which employ ashing with high frequency excited oxygen and gaseous freon. The major disadvantage of the dry chemical methods (both Reactive Ion Etching (RIE) and non-RIE) is their slow etch rates. These methods require hours to etch the encapsulant compared to a few minutes in the wet chemical method. Because of its poor processing capacity, dry chemical decapsulation is seldom used. Wet chemical decapsulation is a more common method and can be done using the following two techniques: (a) manual "drip and rinse" techniques or (b) automatic decapsulation (jet etch) systems.

Wet Chemical Deprocessing Techniques

New generation mold compounds and packaging are challenging the traditional methods of decapsulation of IC devices. As mentioned before polymer and metal are the flesh and bones of a molded plastic package. Plastic packages are non-hermetic but the polymer provides mechanical isolation of the die from the external environment. The evolving trend in the development of plastic packaging materials is to have higher glass transition temperature (T_g), lower stress and very low shrinkage of the encapsulants [8]. Unfortunately these new generation mold compounds are also very hard to etch. Choice of etchants should be made in such a manner that they dissolve only the plastic encapsulant and should not corrode any material composing the IC's, especially the bond pads. Nitric acid is widely used as a decapsulating etchant mainly due to its favorable ability to turn the surface of aluminum into a passive state. As the package-to-die edge spacing decreases, the decapsulation of these packages becomes even more challenging. These new generation plastic mold packages have rendered the traditional manual decapsulation methods ineffective. In this section we will describe the two different wet etching techniques and compare the advantages of automatic decapsulation equipment and techniques over the conventional manual "drip and rinse" techniques.

Both automatic and manual decapsulation techniques employ slot milling of the packages, over the die area, prior to acid exposure. This is necessary for the manual method and recommended for the jet etch method, to reduce the etch time and to prevent undercutting. The device-under-test (DUT) is milled as deep as possible (to within a few mils of bond wire loops) precisely over the die surface. Knowledge of die dimensions and plastic thickness is required for this step. A cross sectional X-ray of the device will provide this information [13]. From the X-ray image it is possible to determine the dimensions of the hole to be etched and the depth to be milled in the package. Milling can be done both manually or with computer aided numerical control. The advantage of computer aided control, is the high precision of the final milled depression. Computer aided milling also helps avoid time consuming measurements and plotting of the depression to be milled. However, where cost is an issue, manually milling is a cheaper alternative. Figure 13(a) shows a schematic diagram of a typical slot-milled plastic package. Once milled, the DUT is jet etched using an oversized template.

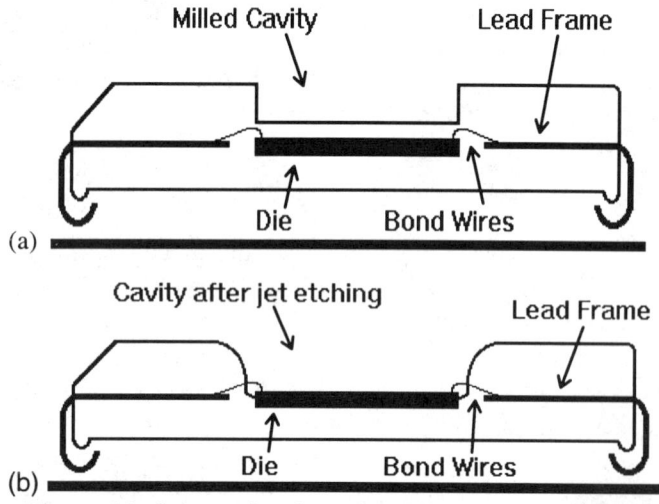

Figure 13: Schematic of (a) slot milled package and (b) DUT after decapsulation.

Manual "Drip and Rinse" Decapsulation Technique

In this technique, yellow or red fuming nitric acid is used to dissolve the mold compounds. The device is first milled mechanically using the techniques described above. It is important to keep in mind that the device should be devoid of any moisture so that there is no corrosion of the aluminum components. Hence, the chemical which is used for this purpose, must be moisture free or dehydrated before use. It is also advisable to drive out the moisture in the device by heating the device for about an hour at 125 °C – 150 °C. The unit is then placed on a hot chuck and acid is applied to the device using a disposable dropper to fill the milled 'slot'. Fuming nitric acid (~100 %), which is a solution of Nitrogen-dioxide (NO_2) in concentrated nitric acid (HNO_3), is commonly used as the etchant for plastic encapsulants. It is a very strong oxidizing agent and in addition to attacking the plastic it attacks copper (Cu) and silver (Ag) very vigorously. If damage to a Cu package or a silver-coated chip carrier island is to be avoided, HNO_3 acid should be mixed with 10% concentrated sulphuric acid. The specimen is kept hot on a hot plate set at 60 °C, to enhance the reaction of acids with the resins. Also it is advisable to pre-heat the acid to 60-70 °C before use, as cold nitric acid takes a very long time to etch. The nitric acid is then carefully dripped into the milled depression in the device using a pipette. After 10 seconds the part is removed from the hot chuck and sprayed with some dry acetone to remove reaction products and exposed filler. It is also very important to keep the exposure of the reactive acid very short in order to avoid accidental over-etching of the device. This process is repeated several times until the chip is sufficiently exposed. Finally, acid residues are removed from the surface using a glass beaker filled with acetone and an ultrasonic cleaner. This is followed by another ultrasonic clean in cold de-ionized water to remove all salt residues and a further ultrasonic clean in methanol to expel the water. It is required to standardize this procedure, using dummy units, in order to successfully use this technique.

Automatic Decapsulation Techniques

In this technique, a hot jet of acid is used to etch a desired void in the package, instead of manually dripping acid using pipettes. Jet etchers based on this principle are becoming increasingly popular and are replacing manual etch techniques. The main advantage of automatic decapsulating techniques using jet-etching is the precision of the etched void, which is rarely achievable by manual methods. The other advantage of using this technique is the gain in time, particularly when processing several parts of the similar construction. Also, in many cases, the expensive milling of the device is omitted and exposure of the chip by etching takes only 2-10 minutes. Another big advantage is the effective utilization of chemicals, as used acid flows back into the container and is re-circulated several times using this technique.

There are many available decapsulation systems in the market today, all of which use either a positive pressure, or a negative pressure spray method of removing the encapsulant. The positive pressure spray systems pump the etchant directly onto the DUT and the etchant then drains into a waste container without recycling. On the other hand, in the negative spray system, the device is placed top down and acid is sprayed by a nozzle from below using negative pressure. In this system the unreacted acid is then collected and recycled. The advantages and disadvantages of using either of these systems is listed in Table. 1.

	Advantages	**Disadvantages**
Positive Pressure Jet Etch System	• Fast etch rate • High temp. etch capability • Good repeatability • Programmable • H_2SO_4 usability • Fast heating of the acids.	• Poor etch controllability • High acid consumption • Incapable of etching a package of large die size-to-package size ratio.
Negative Pressure Jet Etch System	• Good etch control • Low acid consumption • Acid recycling • Multiple etch heads • Pause function • Capable of etching a package of large die size-to-package size ratio. • Etch thin packages.	• Longer etch time • No preset programs • Poor repeatability • Slow to heat acid • No high temperature etch capability.

Table. 1   The advantages and disadvantages of positive and negative pressure jet etcher.

In this section, we will discuss the jet etch system which employs a negative pressure injection and uses a fuming acid mixture (fuming nitric acid (HNO_3) : concentrated fuming sulfuric acid (H_2SO_4) = 9:1) at 65-70 °C. The HNO_3 is the primary etchant, the concentrated H_2SO_4 protects the copper lead frames and copper bonding wires from corrosion. Fig. 15 shows a block diagram of the operating principle of the negative pressure jet etch system [5]. After the DUT is de-encapsulated, it is dipped into acetone for 5 minutes to remove the excess acids from the device, otherwise it will start to corrode the metals in the DUT. A schematic diagram of a negative pressure jet etch system is shown in Fig. 14.

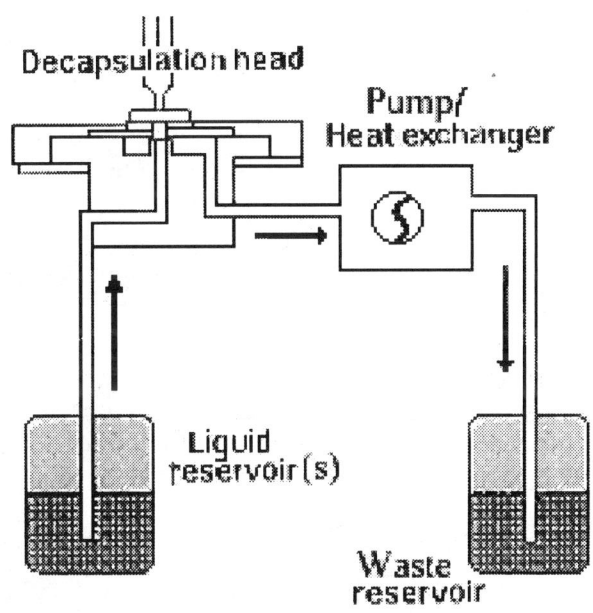

Fig. 14: Operating principle of a negative pressure jet etch system.

A brief description of the operating principle of the jet etcher is as follows: The device is placed top down on the decapsulation head, which is essentially a teflon block with a calotte-like depression on its surface. A nozzle juts into this from below, through which hot acid is sprayed against the surface of the part, etching a hole at the milled cavity. In the beginning, fuming nitric acid is taken from the nitric acid reservoir, with the aid of a geared pump, and sent through the heat exchanger. The fluid is circulated through the heat exchanger until it reaches a pre-set temperature. Even after the fluid reaches the set point temperature, the gear pump continues to circulate nitric acid to ensure that the entire system reaches the set temperature. Nitric acid is then sprayed at a reduced pressure to the sample for a predetermined time through the nozzle below the part. A timer is used to cut off the flow of the etchant to the device, depending upon package type and configuration. It is necessary to perform a few "dummy" runs in order to standardize times and ensure proper etching. The excess acid is collected and re-circulated for further etching of the device.

This technique is highly popular due to its high success rate and ease of equipment operation. Various jet-etching systems are now commercially available. In choosing an appropriate type one should pay particular attention to the corrosion resistance of the working parts and to package type and configuration of the device to be etched.

Dry Chemical Decapsulating Techniques

Yet another technique, not widely used, is the dry or plasma decapsulation technique. The reason this technique has not gained popularity is due to its lack of speed. A device that could be decapsulated in a few minutes using wet etching takes hours to do the same task using dry plasma etching. However, this technique allows a much more precise control over the etching process, compared to wet chemical etching processes. Although dry chemical techniques are not highly popular now, as integrated circuits become more complex, with smaller dimensions and newer materials, it is likely that plasma etching techniques will gain in popularity.

The two basic steps used for plasma decapsulation of the unit are Milling and Masking. Milling involves removing material from the area to be decapsulated before exposure to plasma, very similar to the techniques adopted for wet chemical etching. While, masking involves mechanically or electrically shielding specific areas of the device through the use of fixturing, thereby allowing only desired areas to be plasma etched. Fig. 15 schematically illustrates a parallel plate plasma system.

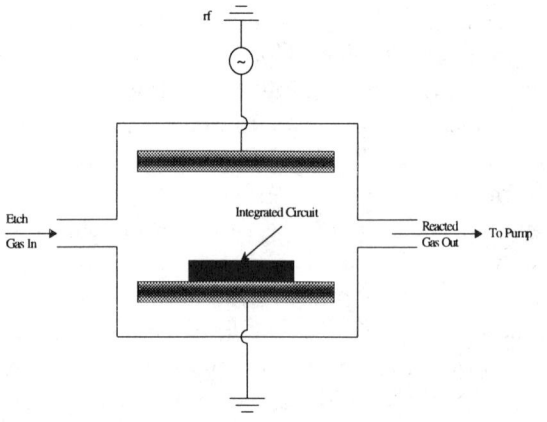

Fig. 15　Schematic diagram of a parallel plate plasma etcher.

The etch gasses flow between the parallel plate of the system, where they are ionized by RF energy. The ionized gas (plasma) chemically etches the packaging material from the device located on the lower electrode. Plasma etching can be done with both electrodes powered (RIE etching) or with just the top electrode powered (Non-RIE). The gaseous by-products are removed from the system using a vacuum pump. This RIE process is highly isotropic compared to the non-RIE process which is anisotropic like the wet chemical processes. We will not discuss this technique in detail in this chapter however, further information can be obtained from works of D. Thomasi et al [15].

CHIP ACCESS TECHNIQUES FOR FLIP CHIP DEVICES

We have focussed our discussion so far on chip access techniques for non-flip chip devices. For advanced electronic designs, mainstream packaging techniques are not cost effective, space efficient or electrically robust. Advanced designs often require higher interconnect densities and lower impedance to permit higher operational speeds. They also require better heat dissipation due to higher operational frequencies. As we mentioned before one solution to these challenges is flip chip technology. A schematic cross section of a of the flip chip configuration is shown in Fig. 16.

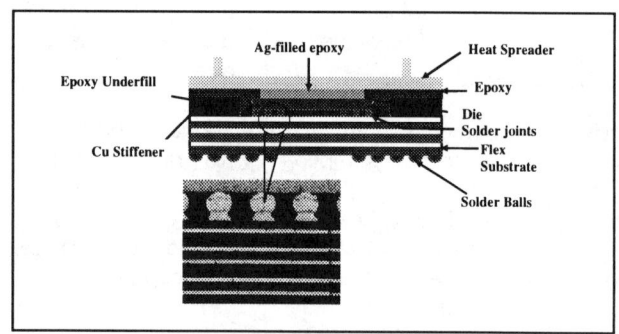

Figure 17　A schematic cross section of a of the flip chip configuration.

The flip-chip technology is based on the direct attach principle of the die to the substrate. The device in the flip chip configuration is electrically connected to the package (substrate) via bumps of Pb:Sn or other suitable electrically conductive materials. The absence of bond wires clearly enables packages to become more slim and compact, and also provides higher pin counts and higher-speeds [16].

In flip chip devices, the die is positioned face down which makes chip access for such devices considerably difficult. Traditional techniques of chemical and mechanical decapsulation are rendered ineffective for flip chip devices. Also, because the die is positioned face down it is not possible access the front surface of the die without compromising the electrical integrity of the device. New approaches have to be adopted and creative solutions have to be obtained.

One such solution is to perform electrical failure analysis (EFA) on the device from the backside of the die. With increasing complexity in device technology, such an approach may become necessary even for non-flip chip devices, as we will discuss in later sections of this paper. If, on the other hand, the intention is to inspect the different metal levels on the chip and electrical integrity need not be maintained, several mechanical decapping techniques may be adopted for flip chip devices.

MECHANICAL DECAPPING TECHNIQUES FOR FLIP CHIPS:

Mechanical decapping techniques can be adopted for die removal of both plastic and ceramic flip chip packages. The basic principle behind mechanical decapping of flip chip devices is as follows: Mechanical decapping of flip chip devices essentially consist of two steps, Heat Sink Removal and Die Removal. The die is usually attached to the heat sink on the backside through a lid attach material that softens typically at 80 $^{\circ}$C. Using this to our advantage, it is possible to heat the package sufficiently so as to soften the lid attach material and mechanically remove the lid. In order to then remove the die, a similar method is adopted. The die is heated to around 450 $^{\circ}$C, the temperature at which the underfill softens, and mechanically removed using the techniques described in detail below. It is necessary to keep in mind that these techniques are destructive and should not be used if the electrical integrity is to be maintained.

Heat Sink Removal Technique:

The materials required for mechanical removal of the die are as follows:
A. Fume hood (optional)
B. Hot plate (200 - 400 $^{\circ}$C)
C. Sharp razor blade
D. Small hammer
E. Vise
F. A pair of needle nose-pliers
G. A pair of thermal gloves
H. Exacto Knife

For removal of the heat sink preheat the hot chuck to 100$\pm$ 5 $^{\circ}$C. This temperature is sufficient for softening the lid attach material on the device. Then place flip chip package, heat sink down, on hot chuck for 3-4 minutes. Remove the unit and place it in a vise. Insert sharp single edge razor blade between heat sink and laminate as shown in Fig. 10. Tap lightly with small hammer to cut through adhesive.

Repeat this procedure on second corner. Return package to hot chuck if the unit has cooled down during this period. Insert razor between heat sink and laminate on remaining two corners and repeat the same procedure. Heat sink should fall off. If the heat does not fall off easily, return package to hot chuck for an additional 60 seconds and repeat the above procedure again.

Die Removal Techniques:

The next step is die removal. The procedure for die removal is as follows: First remove adhesive from back of die with acetone and scrape the blue underfill attach material, that extends out from the edges of the die surrounding it, with a sharp Exacto knife. Following this, set the temperature on the hot chuck to 450 $^{\circ}$C. Then place package, substrate side down, on hot chuck for a period of 3-4 minutes. When the die gets sufficiently hot you will hear a crackle sound and see smoke come out of the package. The fumes are an indication that the unit is ready for decapping and they will dissipate rapidly when the device is removed from the plate. At this stage, one can adopt either of the following two techniques for die removal:

Mechanical Lifting Technique:

On heating the die to 450°C the underfill sufficiently softens. Use a razor blade or an Exacto knife to scrape off any remaining underfill surrounding the die. After heating the die for 3-4 minutes, when fumes are seen coming out of the package, use the exacto knife to mechanically lift the die from the substrate. If the die does not lift easily, heat the package for another 1-2 minutes and the try lifting again. It is important to keep in mind that such high temperatures tend to reflow the solder bumps in flip chip packages and hence this technique cannot be used for the analysis of bump related defects.

Mechanical Twisting Technique:

The other technique that can be adopted for die removal is mechanical twisting. This technique can only be adopted for plastic packages because flexibility of the package is very essential in order to successfully perform this technique. Ceramic packages are usually very inflexible and hence not amenable to this technique.

The technique is performed as follows: After heating the package at 450 °C for 2-3 minutes, (smoke coming out of the package is a good indication that the unit is sufficiently heated) remove package with needle-nosed pliers and twist. Have a thermal glove placed below the unit in order to protect the die from impact. The laminate should buckle, deform and the die should come free. Fig. 16 shows an optical image of the laminate buckled, using this procedure, from which the die was successfully removed.

Fig. 16 Optical image of a laminate bucked during die removal using the twist technique.

Advantages and Limitations:

There are advantages and limitations to either of the two techniques. It is important to keep in mind that either of these techniques, destroy the electrical integrity of the package. This is undesirable if dynamic microprobing is needed. Mechanical twisting technique is quicker and simpler than the mechanical lifting technique. However, it requires a lot of practice and an experienced hand in order to have very high success rate. It is advised

that the analyst practice this technique on dummy units before performing it on actual devices. The advantage on the other hand of using the twisting technique is that it is less susceptible to damage compared to the mechanical lifting technique. Since we do not directly contact the die, there is less chance of damage to the passivation or the metallization on the die, unlike the mechanical lifting technique where we might scratch the die with the Exacto knife used. The analyst has to evaluate the advantages and limitation of both the techniques before adopting either of them.

BACKSIDE SAMPLE PREPARATION

In the previous section we discussed mechanical decapping techniques for flip chip packages. However, if the intention is to perform electrical failure analysis on the device, mechanical decapping techniques should not be used. Chip access with electrical integrity becomes considerably difficult for flip chip packages as the die is positioned face down. Even for the non-flip chip device families, new designs with higher I/O densities and complexities have evolved due to increasing demands on the performance of the chips. These designs usually include a higher level of metallization and higher density of metallization, with lead-on-chip packaging. These new technologies are challenging failure analysts using conventional techniques like emission microscopy for fail site isolation.

Conventionally, one could use photoemission techniques to detect emission from the failing site in a device. Most of the conventional package types require front side access of the device, which we have discussed in detail in the earlier chapters. However, with the increasing level of metallization in newer technologies, emissions from a defect deep within the lines of metallization never reach the surface. Light from the defects is usually completely reflected by the metal lines before it emerges from the surface. New approaches have to be sought if a failure analysts need to successfully isolate defect sites in a device. One such approach is backside analysis.

Backside analysis is based on the principle that silicon is transparent to light with wavelengths of 1070 nm. Fortunately for us, most of the defects emit light at frequencies ranging from 600 nm to 1200 nm depending upon the nature of the defect. Defects are isotropic in the emission of light, i.e. defects emit light equally in all direction. Although some portion of these emissions is filtered by transmission through silicon, we still have some emissions emerging from the backside. However, absorption of the emitted light in silicon drops off exponentially as a function of the thickness of silicon as given by the following equation:

$$I_T = (1 - R)I_0 \exp(-ax)$$

Where I_T is the intensity of the transmitted light, I_0 is the intensity of the incident light, R is the reflection index of silicon, a is the absorption coefficient and x is the thickness of the silicon. As seen from the equation above, the intensity of the transmitted light from the backside decreases exponentially with increase in the thickness of silicon. Hence, it is necessary to thin the backside of flip chip devices substantially in order to observe emissions at the failing site from the backside. In the next section, we will discuss backside access techniques that can be applied to access a variety of devices.

Backside Parallel Polishing:

There are a number of successful methods of backside sample preparations currently used. One of which is the global polishing method. This method uses varying grits of diamond films to reduce the thickness of the package and the die. The surface must also be polished to a mirror finish as surface roughness can have a detrimental effect on subsequent analysis [17,18]. For flip chip devices, global polishing techniques can be adopted after the heat sink removal. Since the die is already positioned face down, backside sample preparation of flip chip devices involves only grinding off of the die. For non flip chip packages, on the other hand, the analyst needs to grind the substrate on the backside before thinning the die. The technique however is the same regardless of the type of package used. In this discussion we will discuss the backside sample preparation techniques only for flip chip devices, as they are the emerging trends in the semiconductor industry.

The first step in the backside sample preparation technique for flip chip devices is heat sink removal. This technique was discussed in detail in earlier sections. After removal of the heat sink using the technique described earlier, the die is thinned using the following procedure:
1) First, mount the device on an allied polisher adapter plate using hot wax. To do this, heat the hot chuck to about 120 °C. Place the polisher adapter plate on the hot chuck, face up until it gains heat. Place enough wax on the adapter plate to suitably hold the package during polishing. When the wax melts, place the package face up on the adapter plate. Press the package down to ensure that the package is positioned evenly on the adapter plate and let the wax solidify. This is crucial to ensure uniform thinning of the die.
2) The second step involves the thinning and polishing of the silicon die. The initial thickness of the die must first be measured to get a baseline thickness. This is very important in order to monitor the die thickness constantly during the measurements. These measurements can be made using a dial gauge as shown in Fig. 17.

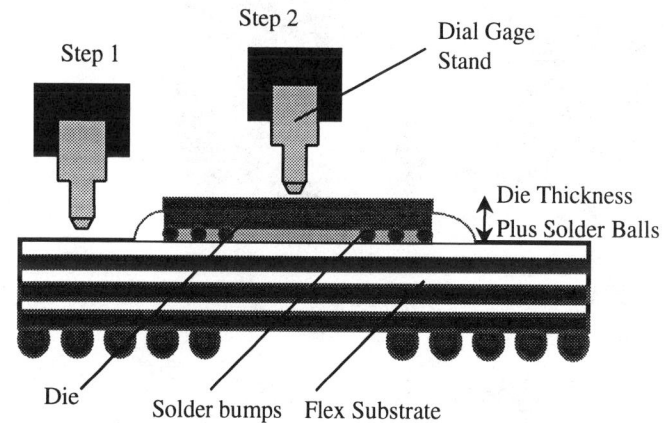

Figure 17. Schematic diagram of the die thickness monitoring using a dial gauge

Mount the adapter plate on the polisher and align it to make sure that it is initially planar with the polishing wheel. Since the silicon die thickness is usually known, it is possible to estimate the thickness of silicon to be removed in order successfully use backside emission analysis. This thickness is a function of doping and decreases with increasing level of doping in the substrate. Care must be taken during all polishing steps to take periodic measurements of die thickness.

Assuming that the intention is to reduce the thickness of the silicon by 400 μm, begin polishing using 180 grit Buehler special silicon carbide grinding paper. This paper is used to remove bulk of the material (300 out of 400 μm). The next step is to use 320 grit Buehler Special Silicon Carbide grinding paper to remove 50 more microns of silicon. Remove rest of the silicon using 3 micron, 1 micron, and 0.3 micron diamond polishing film. The last step in the polishing process is to polish the silicon on a cloth with 0.25 micron diamond paste.
3) The last step in the process is the adapter plate removal.

After the polishing is complete, the package must be removed from the adapter plate. To do this heat the hot chuck to 120 °C. Place the adapter plate on the hot chuck till the wax melts. The package should be easily removable. A razor blade can aid in this process. The remnants of wax on the adapter plate and the package can be removed using a Q-tip and acetone. A picture of the flip chip device after parallel polishing is shown in Fig 18.

Fig. 18 Picture of a typical flip chip package after backside parallel polishing.

The global polishing technique for non-flip chip packages are also very similar, except that there is an additional step of grinding the substrate on the backside of the die. Some package configurations are not amenable to the global approach described above, in these cases a more aggressive approach is necessary to access the back side successfully. The tools needed are a vertical end mill, x-ray equipment or package design details, diamond milling bits, a felt polish bit, 6 to 0.05 micron diamond paste, needle-nose pliers, and safety glasses. The procedure is performed in a three-step process as follows:

(a) First mount the package on the vertical end mill with the needle-nosed pliers, and mill off a window of the plastic package material large enough to expose the entire backside surface.

(b) Then chemically or mechanically remove the die attach metallic support and epoxy.

(c) Finally polish the backside of the package to the approximate desired thickness based on device technology [18] and complete the process by using the lowest grit size diamond paste available to achieve a mirror like finish. Figure 19 shows a completed QFP device from the backside.

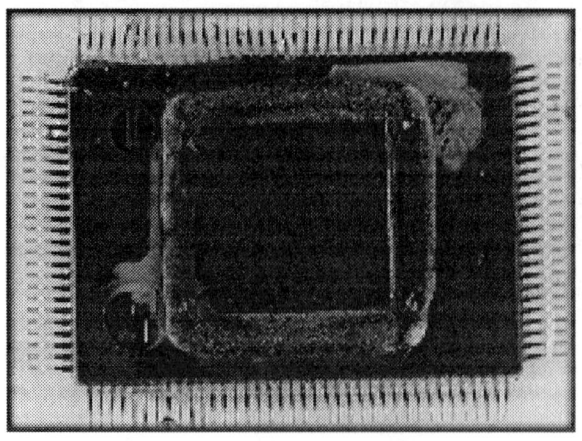

Figure 19 Picture showing a QFP package locally polished from the backside.

FUTURE TRENDS IN BACKSIDE SAMPLE PREPARATION TECHNIQUES

So far we have discussed both front and backside chip access techniques for a variety of devices. Some packages permit global access to the backside while others require local techniques. Cavity down pin grid arrays uses a substrate where access to the backside of the die is limited to a selective region of the package immediately behind the die. These packages are usually ceramic and very difficult to mill. When the pins are on the same side of the package as the backside of the die, a localized process is also required. Two local techniques are currently in use, both of which use a diamond bit. The bit is ultrasonically energized to improve lubrication of the diamond and to reduce heating of the die during grinding. Lasers also have been used to ablate the material from the backside of the die. Laser ablation has been tried on ceramic packages but has not been completely successful because of the high laser power required and the expense of such a laser. Future flip chip packages will have interconnects in the package over the backside of the die as well as heat spreaders. Stacked die processes, such as the memory cube, limit access only to the last die in the stack. To overcome these obstacles creative solutions to backside access must be developed. Both the diamond bit milling operations and the planar lapping of the package expose the die, but the surface is rough and the remaining silicon may be too thick for fail site isolation.

When access to the device has to be gained by local techniques, the surface must be mechanically polished to reduce roughness until it is optically smooth. This is necessary in order to reduce scattering of the incident light. Where access has been gained using the planar lapping, the polishing process is performed on lapping wheels with the final polish being done by lapping the surface with 0.05 micron diamond material. Mechanical polishing techniques, whether manual or automated, can scratch and crack the device. A laser chemical etch process is being developed to eliminate the need for the mechanical polish. The procedure is self-limiting in depth and only occurs at the depth where the laser is focused; it also can be scanned over the backside to remove material in a global mode [19]. Once the backside is polished and thinned to 100 microns or less, various fail site isolation techniques can be employed to image through the backside. Some optical techniques, such as photoemission microscopy (PEM), optical beam induced current (OBIC) and light induced voltage alteration (LIVA) can image through the backside of a die[20].

Techniques that cannot image through the backside, such as electron beam probing, are best performed if the location to be measured is brought to the surface. That is, there is no substrate material, conductive or non-conductive, between the contact point to be measured and the electron beam. To improve the signal received from these techniques, a deep trench needs to be cut in the backside. Currently, the most promising technique to make the trench is focused ion beam (FIB). The final cut is made to within 25 microns of the active junctions. Techniques also are under development that use laser processing to cut the final trench. These procedures are adequate for thinning the backside, but they do not provide direct access to the contact point of interest (e.g., a metal line). To gain access to the metal lines, holes must be milled from the thinned region to the metal. To prevent shorting of the substrate to the contact point to be measured, the milled hole is filled with a dielectric. A new hole is milled into the dielectric to access the metal line and this hole is then filled with metal. The signal can be seen at the backside from the exposed metal accessed through the hole. By bringing the metal lines out to the backside of the die, provisions can be made to reroute the interconnect signals in a similar manner as that performed on the topside. Once access has been gained to the backside of the device and the appropriate level of backside preparation has be accomplished, the analyst can begin fail site isolation.

CONCLUSION

Failure analysis is essential for any semiconductor manufacturing operation. Lack of failure analysis would lead to yield losses concerns during device fabrication and reliability. Chip access is an integral part of failure analysis. It is essential to adopt the proper chip access techniques for device failure analysis as it might otherwise result in loss of valuable information. As the chip access techniques used is dependent on the package types, we have briefly discussed in this chapter, some of the common package types available in the market today. Following this, we have discussed various chip access techniques that can be adopted for the different package types. Flip chip devices are unconventional in their design as the device is positioned face down. Analysis of flip chip devices are considerably different and hence is discussed in a separate section.

Armed with this information, an analyst can adopt any of the chip access techniques for sample preparation. It is recommended that the analyst have a good understanding of the package construction before adopting the techniques mentioned in this chapter.

ACKNOWLEDGEMENTS

We would like to thank Ron Parker for his invaluable help and critical review of the paper. We would also like to thank Robert York, Andy Vance, Charles Todd, Steve Nguyen and Preston Scott for their contributions to this paper. We would like to specially thank Scott Wills for his advice on failure analysis. We would like to thank Seshu Pabbisetty for giving us this opportunity to write this chapter.

References

1. J. Kolzer, C. Boit, A. Dallmann, G. Deboy, J. Otto, and D. Weinmann, "Quantitative Emission Microscopy", J. of Appl. Phys., 71 (11), June, 1992, pp. R23.
2. S. Khandekar and K. S. Wills, "Liquid Crystal Microscopy", Microelectronics Failure Analysis, Desk Reference, 3rd Edition, Ed. by T. W. Lee and S. V. Pabbisctty, 1993, pp141.
3. D. J. Burns, G. E. Jacobcic, and M. L. Wangler, "Liquid Crystal Display Techniques for Analyzing Microprocessors", Proceedings of Advanced Techniques in Failure Analysis, 1989, pp.27.

4. H. J. Leamy, "Charge Collection Scanning Electron Microscopy", J. Appl. Phys., 53(6), June, 1982, pp. R51.
5. S. M. Davidson and C. A. Dimitriadis, "Advances in the Electrical Assessment of Semiconductors using the Scanning Electron Microscope", J. of Microscopy, Vol. 118, Pt 3, March, 1980, pp. 275-290.
6. A. Gopinath, K. Gopinathan, and P. Thomas, "Voltage Contrast: A Review", Scanning Electron Microscopy, 1981, I, pp.375.
7. M. G. Pecht, L. T. Nguyen, "Plastic Packaging", Microelectronic Packaging Handbook, Part II, 1997, pp.395.
8. The National Technology Road map for Semiconductors, Semiconductor Industry Association (1994).
9. R. R. Tummala, P. Garrou, T. Gupta , N. Kuramoto, K. Niwa, Y. Shimada, and M. Terasawa, "Ceramic Packaging", Microelectronic Packaging Handbook, Part II, 1997, pp.285.
10. E. J. Rymaszewski, R.R. Tummala, and T. Watari,"Microelectronic Packaging – An Overview", Microelectronic Packaging Handbook, Part I, 1997, pp.285.
11. D. B. Harris, M. Pecht, and P. Lall, "Electronic Components", Handbook of Electronic Package Design, Ed. by M. Pecht, 1991, pp. 39.
12. T. W. Lee, "Mechanical and Chemical Decapsulation" Microelectronics Failure Analysis, Desk Reference, 3rd Edition, Ed. by T. W. Lee and S. V. Pabbisetty, 1993, pp 61.
13. "Integrated Circuit Failure Analysis – A Guide To Preparation Techniques", F. Beck, S. S. Wilson, Pub. Ed. by P. D. T. O'Connor, pp. 5.
14. R. Chowdhury, O. Adams, J. Bartlett, and C. Todd, "A Study Of Chemical Decapsulation Techniques For New Generation Plastic Molded IC Packages", 21st International Symposium for Testing and Failure Analysis (ISTFA), November, 1995, pp.281.
15. D. Thomasi, "Plasma Decapsulation Techniques", Microelectronics Failure Analysis, Desk Reference, 3rd Edition, Ed. by T. W. Lee and S. V. Pabbisetty, 1993, pp 61.
16. "Flip Chip Technologies", Ed. by J. H. Lau, 1995
17. N. M. Wu, et. al., "Failure Analysis from the Backside of the Die," 22nd International Symposium for Testing and Failure Analysis(ISTFA), November, 1996, p. 393.
18. T. Ishii et. al., "Functional Failure Analysis Technology from Backside of VLSI Chip," 20 th International Symposium for Testing and Failure Analysis (1994), p. 41.
19. D. J. Ehrlich, et. al., "A Review of Laser-Micromechanical Processing", J. Vac. Sci. Technology B. Oct.-Dec. 1983.
20. D. Davis et. al., "Flip Chip Failure Analysis," ", Texas Instruments Technical Journal, pp. 74-86, Nov. -Dec. 1997.

Acoustic Microscopy
of IC Packages

T.M. Moore and C.D. Hartfield

INTRODUCTION

Scanning acoustic microscopy (SAM) has been adopted by packaging researchers and IC failure analysis labs because it provides nondestructive imaging of moisture/thermal-induced damage, such as package cracks and delaminations. SAM is an important tool for aiding development of improved molded packages. Automatic polarity analysis of the echo pulse is used to assist in the identification of delaminated interfaces. Images of the three-dimensional internal structure of the package are produced that are often useful in recognizing package defects and in determining the mechanism for a package failure.[1]

The SAM used today in the IC industry (hereafter referred to as IC-SAM) is a hybrid instrument with characteristics of both the scanning acoustic microscope developed at Stanford in the early 1970s, and the C-scan which has been a part of the nondestructive test (NDT) industry since the 1950s.[2] The characteristics of each of these methods is briefly reviewed.

The term C-scan comes from early NDT nomenclature. The C-scan image is an image of a planar region at a constant depth within the sample. The C-scan image is formed by mechanically scanning a piezoelectric transducer above the specimen and electronically gating the signal in time. The broad-band C-scan transducer has a lens designed for sub-surface imaging. C-scan imaging has played a major role in the macroscopic imaging of sub-surface flaws in industrial components (rails, pipe, welds, etc.) with center frequencies in the range of 1-10MHz. C-scan inspection takes advantage of its ability to penetrate optically opaque solids and detect thin cracks.[3,4]

The SAM developed at Stanford and first demonstrated by Lemmons and Quate in 1973 employs a large numerical aperture (NA) lens in order to excite longitudinal, shear and surface waves in the sample.[6,7] Instead of the large water baths of C-scan, a tiny water droplet acoustically couples the transducer and sample. Image contrast is formed by the combined interference of longitudinal, shear and leaky surface waves. Narrow-band RF pulses with frequencies in the range of 100MHz-8GHz are used. The upper frequency limit, and spatial resolution limit, are determined by frequency-dependent attenuation in the couplant. Cryogenic fluids and high pressure gases have been used as the couplant for frequencies above 2GHz. Precision mechanical scanning is employed for sub-micron spatial resolution. Both the amplitude and phase of the reflected pulses are measured and used to produce images of the mechanical properties of the near-surface region.[5]

Development of the SAM at Stanford was strongly encouraged by biomedical researchers who anticipated improved contrast in images of tissue samples. Contrast in the SAM is determined by the large variation in elastic properties in these samples compared to a relatively small variation in dielectric properties, which determine optical contrast. Tissue samples normally required complicated chemical staining for optical microscopy. In the years following its discovery, the frequency of the Stanford SAM was continuously increased until the lateral resolution was comparable to that of optical microscopes.[8]

The application of reflection acoustic microscopy to the inspection of IC packages represents the convergence of the capabilities of focused C-scan and the Stanford SAM. The IC-SAM combines precision mechanical scanning for microscopic inspection of small samples, sophisticated RF signal analysis and display, and a broad-band transducer with a small NA lens for sub-surface imaging.[9,10] Center frequencies commonly used are in the range of 15-250MHz and are intermediate between the frequencies commonly used for C-scan and the Stanford SAM.

Scanning laser acoustic microscopy (SLAM) and x-ray radiography are other methods for nondestructive inspection of packaged ICs. Although these methods have overlapping capabilities with IC-SAM, the individual advantages of each technique make them complementary. The capabilities of these techniques will be briefly compared with those of pulse-echo acoustic microscopy.

Early SAM Applications in the IC Industry

SAMs operating at frequencies of 1GHz and higher became commercially available in 1983. The application of SAM in the semiconductor industry was initially for the high frequency inspection of the device layers near the surface of the die. Some SAMs were modified to take advantage of available broad-band NDT transducers in the intermediate frequency range of 30-100MHz for sub-surface studies such as die attach inspections. Die attach inspection through ceramic or metal packages had already been demonstrated using high resolution C-scan equipment. Prior to acoustic inspection, die attach inspection was performed primarily with x-ray radiography. Experience soon showed that reflected sound indicates the true percent area bonded while x-ray inspection reveals only large voids in the die attach material.[11-22]

Studies using pulse-echo acoustic microscopy for plastic package inspection began appearing around 1985. These early applications were performed mostly by Japanese IC manufacturers. These instruments incorporated precision microscope-type scanning and improved data analysis and presentation features. However, the echo signals in the studies typically were rectified and delamination detection was based solely on amplitude imaging. These early studies were instrumental in correlating the amount of damage detected in plastic packages after reflow soldering to the moisture content in the package.[23-28]

After 1988, the limitations of the detection of delamination by amplitude alone in plastic-packages ICs were recognized. Reports using SAM instruments dedicated to IC package inspection (IC-SAMs) began to appear.[1,29-32] Some of these instruments had the ability to detect phase inversion.

Acoustic microscopy became increasingly important as the IC packaging industry rapidly converted from the packaging of small dies in conventional through-hole dual in-line packages (DIPs) to the packaging of large high-functionality dies in space-efficient surface mount packages in the 1980's. This conversion called attention to a basic materials problem with the molded IC package. The molded package is made up of materials with widely varying coefficients of thermal expansion (CTE) (Figure 1) and is required to survive many large temperature excursions. The assembly of large-die surface mount molded packages can result in the development of moisture/thermal-induced stresses sufficient to exceed the mechanical strength of the materials and interfaces in the package. Studies were published which correlated electrical testing and destructive physical analysis with the results of acoustic inspection in order to understand the moisture sensitivity of modern surface mount packages during board mounting.[33-37]

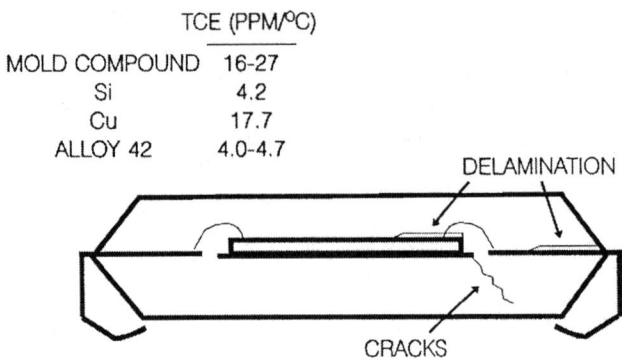

TCE (PPM/°C)

MOLD COMPOUND	16-27
Si	4.2
Cu	17.7
ALLOY 42	4.0-4.7

Figure 1: Coefficient of thermal expansion mismatches in plastic IC packages.

The internal stress (CTE) situation is further exacerbated by the fact that the plastic mold compound tends to absorb moisture from the air during shipping and storage. This is a problem when the device is soldered to the printed circuit board. Wave soldering of DIPs delivers a comparatively lesser thermal stress to the body of the package than that experienced by surface mount packages. When surface mount parts are reflowed, the entire package body is exposed to soldering temperatures. Absorbed moisture expands during the mounting operation. This greatly increases internal stresses and promotes delamination and package cracking. This "moisture sensitivity" is a problem primarily with surface mount package designs.

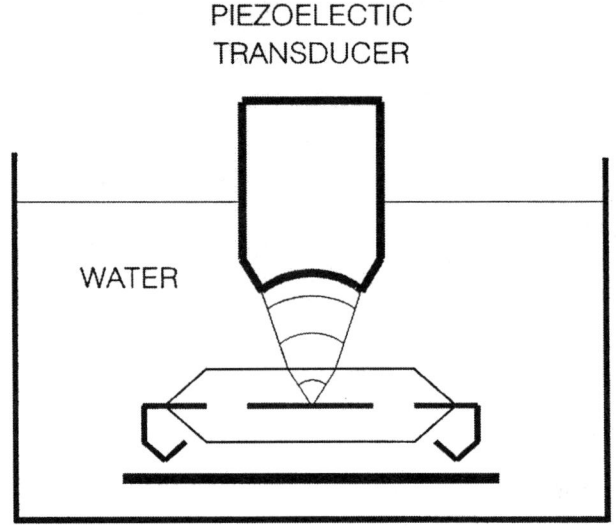

Figure 2: The inspection of IC packages with pulse-echo acoustic microscopy.

In the late 1980s, devices designated as being moisture sensitive began to be shipped in special dry bags. Limits were set for the maximum duration of exposure to air before assembly. These limits were based on a recommended moisture level threshold for the appearance of moisture/thermal-induced damage during mounting.[38] Dry packing did not significantly increase the manufacturing cost of the product. However, the possibility of mechanical damage and production delays associated with moisture control (such as in the baking of over-exposed devices) represented a risk for the assembly operation. IC manufacturers and mold compound producers worked to optimize package designs and mold compound characteristics in order to provide moisture insensitive packages.

Pulse-echo Acoustic Microscopy for IC Package Inspection

For IC package inspection, center frequencies in the range of 15-250MHz are currently used. The sample and transducer are acoustically coupled by a water bath. Broad-band acoustic pulses are focused to a point within the IC package (Figure 2). The pulse repetition rate is typically limited to 10KHz due to decay of the reverberations that occur between the transducer and sample. The transducer is precisely scanned in a plane parallel to the plane of the package for microscopic imaging. At internal interfaces, a fraction of the incident acoustic energy is reflected and detected by the same piezoelectric transducer and converted back to an electrical signal. The amount of incident acoustic energy that is reflected depends on many factors including the materials in contact at the interface, the mechanical properties of the interface, absorption, and the size and orientation of the interface. The echo signal is analyzed and characteristics of the signal are used to form images of internal structures and defects. Sophisticated signal analysis techniques are used to extract characteristics from the echo signal such as amplitude, phase and depth. Because sound is a matter wave, the technique is sensitive to cracks that are invisible to x-ray radiography.[1]

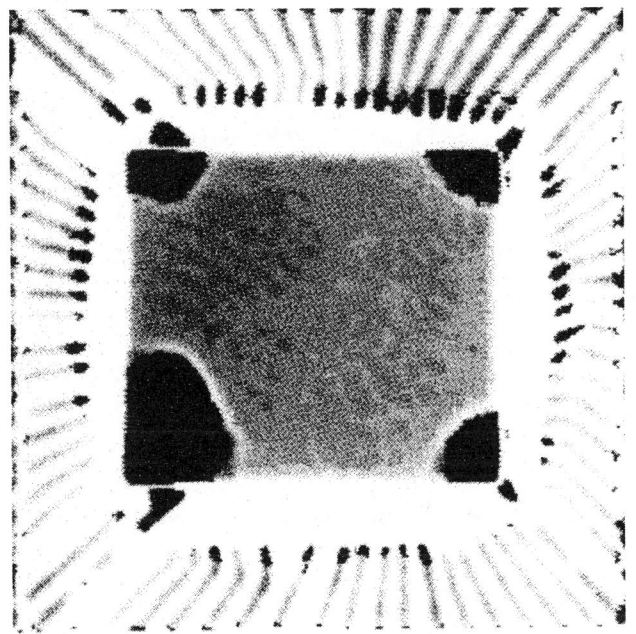

Figure 3: SAM image of the die area of a 68PLCC. The die is 3.6mm by 3.1mm. The reflected intensity image is presented as a gray scale image on a white background. The delaminated areas are indicated as black areas superimposed over the intensity image.

Signal Analysis for IC Package Inspection

Polarity analysis of acoustic echo signals provides information about delamination at internal interfaces in plastic-packaged ICs. Figure 3 shows a pulse-echo image (top view) of the die area of a 68-pin plastic leaded chip carrier (68PLCC). In Figure 3 the image of the reflected intensity of the primary sub-surface echo is displayed as a gray scale image on a white background. Those areas identified as delaminated by polarity analysis are indicated in black superimposed over the intensity image. Figure 4 shows typical examples of acoustic echo signals from an area with good adhesion, and a delaminated area, on the die of a 68PLCC such as the one shown in Figure 3. In each of the two echo signals in Figure 4, a reflection from the top surface of the package and a later sub-surface reflection from the die surface can be seen. Note the 180° inversion of the reflection at the delamination, and the deeper partially resolved reflections in the signal from the die surface area with good adhesion to the package.

For an explanation of the phase inversion phenomenon, consider the simplified example of plane wave reflection, at normal incidence, at an ideal interface (Figure 5).[39] The incident plane wave has the sinusoidal acoustic pressure amplitude P_I and reflected and transmitted pressures amplitudes P_R and P_T, respectively. As a result of the boundary conditions that the acoustic pressure and particle velocity in both materials must be equal at the interface, the frequency remains unchanged across the interface, and the reflected and transmitted pressure amplitudes can be described as functions of the acoustic impedances, Z_i, of the two materials (Eq. 1 and 2).

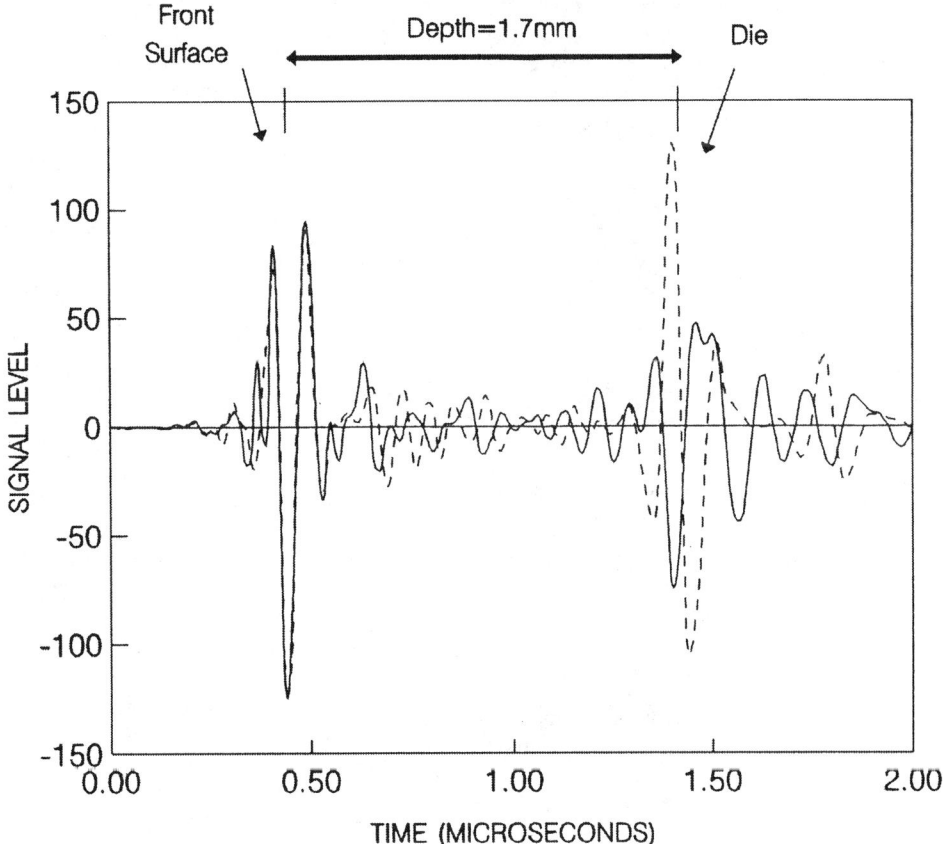

Figure 4: Typical acoustic echo signals (15MHz) from an area of good adhesion (solid) and a delaminated area (dashed) at the mold compound/die interface in a 68PLCC.

75

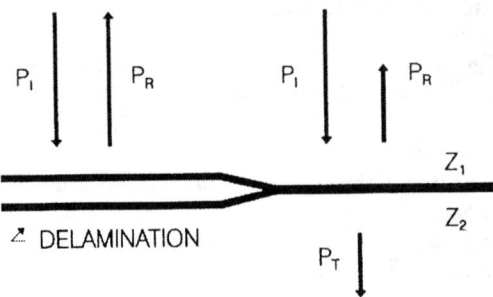

Figure 5: Reflection at normal incidence of a plane wave at a delamination (left) and at a bonded interface (right).

$$P_R = (Z_2 - Z_1)/(Z_2 + Z_1) \qquad [1]$$

$$P_T = 2 Z_2/(Z_2 + Z_1) \qquad [2]$$

The acoustic impedance is the ratio of the acoustic pressure to the particle velocity per unit area and is defined as the product of the density (ρ_i) and the speed of sound (v_i) in layer i.

$$Z_i = \rho_i v_i \qquad [3]$$

Equation 1 is plotted in Figure 6 for two values of Z_1. Curve 1 is calculated for Z_1 equal to the impedance of mold compound (plastic package), and Curve 2 for Z_1 equal to the impedance of Al_2O_3 (ceramic package). Note that each curve passes through the horizontal axis at the value of Z_2 for the appropriate package material. This indicates that there should be no reflection at an ideal interface between identical materials, as expected. In Figure 6, reflectivities less than zero refer to reflected pulses with inverted phase relative to the incident pulse.

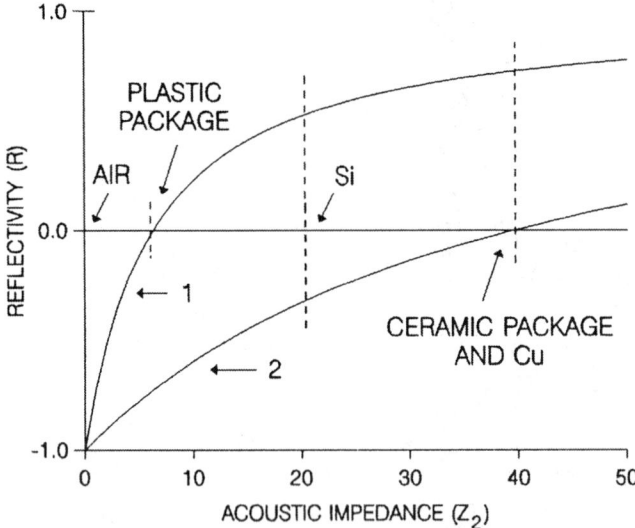

Figure 6: Ideal acoustic reflectivity (R) versus acoustic impedance of the second layer (Z_2) for plastic packages (Curve 1) and for ceramic packages (Curve 2). (Z units: $10^5 g/cm^2 sec$)

Curve 1 in Figure 6 shows that at bonded interfaces between the plastic mold compound and the die (MC/die), and between the mold compound and a Cu lead (MC/Cu), the transition is from lower to higher acoustic impedance. Therefore, P_R is positive at these interfaces

and there is no phase inversion. However, at a delamination or a package crack, which is represented by an interface between mold compound and air (MC/air), ideally 100% of the energy is reflected, and the phase of the reflected pulse is inverted relative to the incident pulse. This model does not include the effects of attenuation losses. Attenuation losses in plastic packages can often obscure the increase in the amplitude of signals reflected at delaminations, especially on the lead frame. Phase inversion is very important for reliable detection of delamination and cracks in plastic packages.

The simplified plane wave model (Eq.1) is useful in describing phase inversion of reflected acoustic pulses at delaminations and cracks. In practice, apparent phase shifts at similar interfaces due to multi-layer interference effects, frequency dependent attenuation and spatial resolution limitations can also be encountered. In addition, interfaces at a constant depth can produce reflections with a continuous variation in phase shift between bonded and delaminated areas that are not explained by the simplified model. These intermediate phase shifts may provide additional information on the condition of the interface.[48, 49] However, in spite of these practical limitations, the detection of phase inversion in reflected acoustic pulses has proven to be extremely useful in the detection of delaminations in plastic IC packages, and is a distinct advantage of pulse-echo inspection.

Curve 2 in Figure 6 describes the ideal reflectivities in ceramic (Al_2O_3) packages. In typical ceramic package applications, IC-SAM is used to inspect the die attach layer. Acoustic inspection offers an advantage over x-ray inspection of die attach quality in that acoustic images show the actual area of good adhesion while x-ray images indicate only voids in the die attach layer.

Table 1: Acoustic Parameters

Material	v, (m/sec)	ρ, (g/cc)	Z, $10^5 g/cm^2 sec$
Al_2O_3	10400	3.8	40
Cu	4400	8.9	39
Si	8430	2.4	20
Mold Comp.	~3500	1.8	6.3
Water	1480	1.0	1.5
Air	343	0.0012	0.00041

Table 1 shows the acoustic impedances of Al_2O_3 and Cu are very similar. The acoustic impedance of the ceramic package is so high that phase inversion detection is not applicable in ceramic package inspection. However, this is compensated for by the fact that the amplitude contrast in an image is typically very high. In a ceramic package with Cu leads, for example, the ceramic/Cu reflections are weak compared to a reflection from a crack or a disbonded lead. These defects provide almost 100% contrast. Inspection of eutectic die attach quality also typically shows dramatic contrast between bonds and disbonds. Polymeric die attach adhesives provide lower, but sufficient, contrast.

The speed of sound in materials is typically less than 13km/sec. This is roughly four orders of magnitude less than the speed of light. The time delay between returning echoes can be easily measured electronically and images with three-dimensional information can be displayed. This is a unique advantage of a pulse-echo acoustic technique and has been useful in determining the mechanism for package crack formation, for example. Figure 7 shows the acoustic time-of-flight image of a cavity-down ball grid array (BGA) package (die-side view). The contrast at the die surface indicates the variation

Figure 7: Time-of-flight image of a cavity-down BGA package. Darker areas are deeper in the package. A 3-D view is shown to assist visualization.

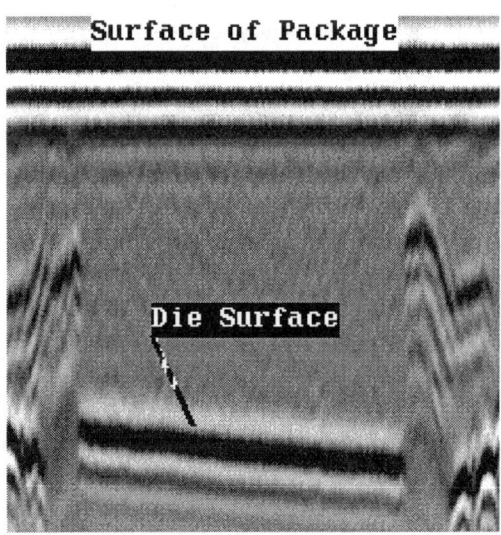

Figure 8: B-scan image showing die tilt, acquired with a 75MHz transducer focused at the die surface.

in depth of the die surface relative to package surface (darker means deeper). Figure 8 is a B-scan image of the same device. In the B-scan image, the depth information contained in the echo signals (A-scans) is plotted vertically for a single line scan across the device. The B-scan clearly shows the tilt of the die (caused by uneven die attach thickness and substrate warpage). The B-scan image simulates a nondestructive cross section through the package.

Plastic Package Inspection

Figure 9 shows two acoustic micrographs of the same 68PLCC at different times. The initial pulse-echo image of this device appears in Figure 9(a). This device was subsequently saturated with moisture (0.32 wt.%) during 168 hours of exposure to 85°C/85%RH (relative humidity), subjected to vapor phase reflow (VPR) mounting, and imaged again. This post-VPR image is shown in Figure 9(b). The micrographs in Figure 9 include both amplitude and phase information. The image of the amplitude of the sub-surface reflection is displayed

as a gray scale image on a white background. Delaminated interfaces are designated by total black superimposed over the amplitude image. The delaminated areas were identified by phase analysis of the reflected acoustic pulse.

In the initial pulse-echo image in Figure 9, the only significant delamination appears on the die pad periphery surrounding the die (Ag spot). After VPR, delamination has appeared on the entire die pad periphery, the corners of the die, on the leads (predominantly at the internal terminations), and at scattered locations on the lead tape. Studies have indicated that the primary reliability threat is wire bond degradation due to delamination at the die surface. It is likely that this delamination will spread from the shear stress maxima at the die corners toward the die center during subsequent temperature cycling. Typically, as the delamination spreads, stress-induced damage will occur at the die surface within the shrinking boundary of good adhesion. And in the delaminated corners, shear displacement between package and die will damage wire bonds. If delamination at the die surface is initiated during board mounting, it spreads during subsequent temperature cycling and leads to early device failure due to stress-induced damage to the device and wire bond degradation. Package cracks and delamination at the leads increase the risk of contamination-related failure of the device.[33,36]

Inspection of Packages Having High Velocity Substrates

Since the acoustic impedances of Cu and Al_2O_3 are so similar, Curve 2 in Figure 6 can be used to predict reflectivities for die attach inspections in ceramic packages, power heat sink packages or other high velocity substrates such as metal assemblies for microwave components. For example, Figure 10 is a sketch of a standard TO-220 molded power IC package. The package is characterized by its thick (1.2mm) Cu heat sink which is necessary to manage the heat produced by high power output transistors, for example. The die is attached directly to the inside surface of the heat sink with an experimental Pb/Sn die attach system. Figure 11 shows two 50MHz pulse-echo images of the die attach in the same device at different times. Figure 11(a) is an initial image, while Figure 11(b) was recorded after 200 cycles of temperature cycle reliability testing. The dark areas denote good adhesion and are areas where most of the energy in the acoustic pulse was transmitted into the die. The bright areas in the die attach indicate almost total reflection due to disbonding. Note the reduction

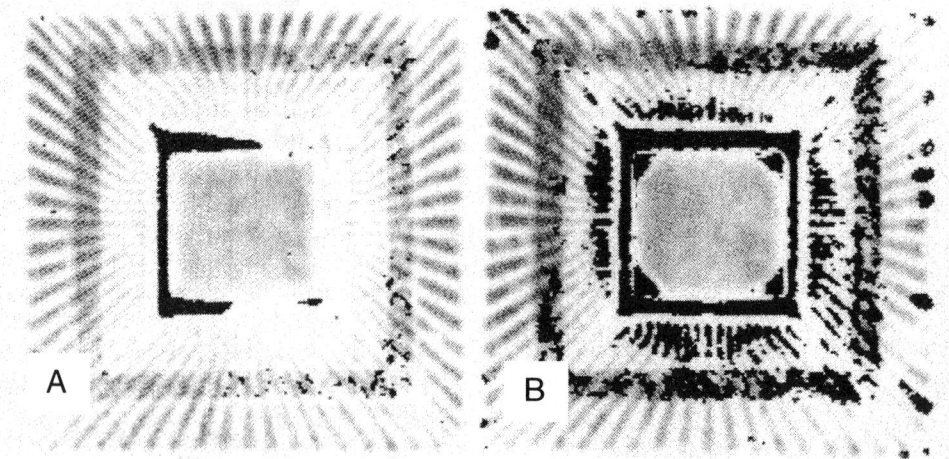

Figure 9: Delamination images of the same 68PLCC taken initially (a) and then after VPR exposure (b). The reflected intensity image is presented as a gray scale image on a white background. The delaminated areas are indicated as black areas superimposed over the intensity image. 68PLCCs are 24mm square.

of the area of good adhesion after temperature cycling. A real-time x-ray image was taken after temperature cycling and is shown in Figure 11(c). Due to x-ray attenuation in the thick Cu heat sink, the x-ray image required a significant amount of image processing to reveal contrast produced by the die attach layer. The die attach voids seen in the x-ray image taken after temperature cycling agree well with the features in the initial pulse-echo acoustic image. The acoustic image at 200 cycles indicates a reduction in the total area of adhesion that was not detectable by x-ray radiography because no significant increase in x-ray absorption was produced by the thin air gap in the delaminated areas.

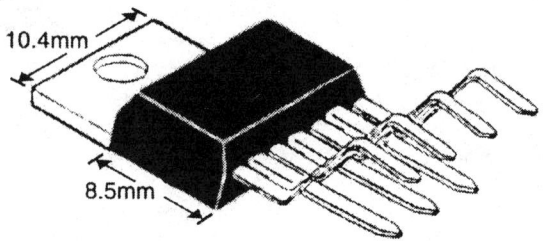

Figure 10: The TO-220 power IC package with Cu heat sink.

Resolution and Sensitivity of SAM Inspection

The spot size obtainable with a spherical lens is limited by diffraction effects. If the resolution is defined by the first zero in the Airy disk (Rayleigh criterion), the lateral resolution (d) is given by [40]:

$$d = 1.22 \, \lambda \, F / D \qquad [4]$$

Here, λ is the acoustic wavelength, F is the focal length, and D is the diameter of the lens. Perhaps a more representative resolution criterion with today's electronic signal processing technology is the detection of a significant local minimum in the signal between two points in the image (Sparrow criterion). Also, unlike the case with a telescope, the acoustic transducer lens both transmits and receives the signal pulse, so the point transfer function is squared at each point in the image. This

means that the constant in Eq. 4 may be as small as 1.02.

The value of F/D ranges from 2 to 4 for typical transducers used for subsurface inspection in IC packages. So, practically speaking, the best resolution obtainable is roughly two times the wavelength. At a center frequency of 75MHz the wavelength in water is approximately $20 \mu m$ and the expected lateral resolution is roughly $40-80 \mu m$. This does not account for frequency-dependent attenuation. Attenuation in a typical mold compound has been reported to be 40dB/cm at 15MHz and to increase rapidly with increasing frequency.[41] This attenuation acts as a low-pass filter and shifts the center of the pulse frequency distribution to a lower frequency. In a very highly attenuating mold compound (large irregular filler particles) the observed spatial resolution can be as large as $400 \mu m$ at a depth of 1.6mm (or 3.2mm round trip). Mold compound attenuation varies considerably from one formulation to the next. Both the penetration and resolution are noticeably degraded by temperature shock damage to the mold compound.

Depth (or axial) resolution is important for distinguishing reflections from closely spaced layers within the package. Depth resolution in the time domain is determined by pulse duration as well as frequency. The inherent decay time for the transducer, focusing properties of the lens and frequency dependent attenuation all contribute to pulse duration. A typical pulse duration for a broad band transducer is 2 periods at the pulse center frequency. This effect creates what has been termed the "dead zone" below an interface. For example, for a duration of two periods at the center frequency, reflections from one interface may interfere with the reflection from a deeper interface and make detection difficult, especially if the signal amplitude is diminished by losses at the first interface. Real-time frequency domain analysis techniques may become useful for reducing this effect.

The sensitivity of reflection acoustic imaging is superior to the lateral resolution of the technique. For example, $25 \mu m$ bond wires are often seen in a 20MHz IC-SAM image at a depth of 1.6mm when the spot size is significantly larger than $25 \mu m$. An object with lateral dimensions smaller than the spot size that is easily detected in reflection (compared to the noise background) may produce only a negligibly small contrast effect in transmission (compared to a large transmitted signal). However, the apparent size of a small reflector in

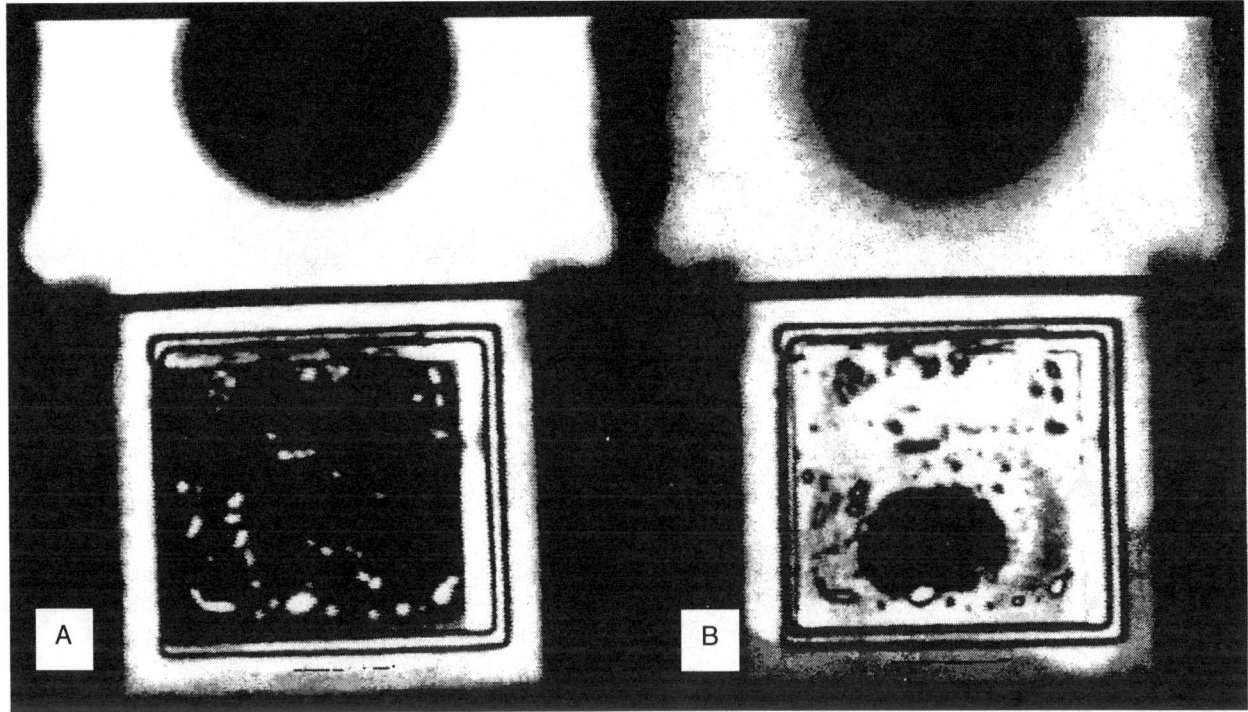

Figure 11: Die attach images in a TO-220 package before testing (a) and after 200 temperature cycles (b). Bright areas indicate a high reflected intensity. Dark areas in the die attach region indicate good adhesion. The heat sink is 7mm wide at the die attach region.

the image will be determined by the spot size.

Sound is a matter wave and depends on molecular vibrations for propagation. This is why acoustic inspection is much more sensitive to thin cracks than x-ray radiography. Theory predicts that crack openings greater than the particle displacement amplitude produced by the interrogating sound wave should be detectable. Experiments with steels, for example, indicate that air-filled cracks on the order of 10nm are detectable.[4,42]

Complementary Nondestructive Techniques

SAM is one of three complementary techniques used in the nondestructive inspection of packaged ICs. A transmission acoustic technique called scanning laser acoustic microscopy (SLAM), and x-ray radiography are also used for this application (Figure 13). These two techniques can each produce images of packaged ICs, and therefore have partially overlapping capabilities. However, unique capabilities of each technique make them complementary for the nondestructive inspection of IC packages.

Real-time x-ray inspection offers the best lateral resolution and is unsurpassed for nondestructive wire sweep inspection. Images are produced at TV rates and the sample is easily manipulated. Although commonly used for die attach inspection, x-ray imaging shows only the location of voids in the die attach layer, and not the total area of attachment. X-ray inspection often exhibits limited contrast for low mass thickness die attach layers on metal heat sinks (see Figure 12). Similarly, detection of package voids can be limited by lead frame shadowing.[32] X-ray is an excellent technique for detecting voids in flip chip bumps, but is not useful for detecting underfill voiding or delamination in flip chips.

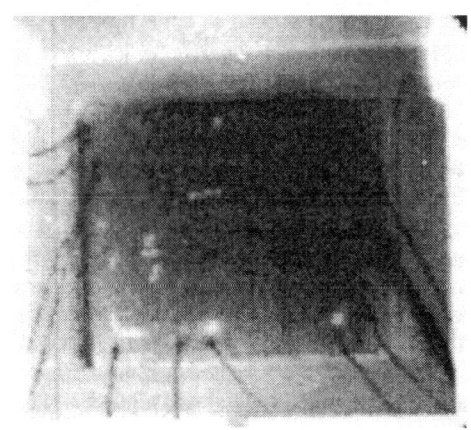

Figure 12: Real-time x-ray image of the device shown in Fig. 11 taken after temperature cycling.

X-ray laminography is a tomographic technique that produces an image of x-ray attenuation at a specific plane within the sample. X-ray laminography has seen limited use for component level inspection, but has been applied to the inspection of solder joint quality for surface mount process control.[43,44]

SLAM is a transmission acoustic technique developed in the early 1970's at Zenith and later transferred to Sonoscan in 1974.[45] SLAM incorporates concepts similar to those presented by Sokolov in 1936 for his proposed acoustic microscope.[46] In SLAM, a planar sound wave is transmitted through the sample onto a mirrored coverslip. The disturbances produced in the coverslip by the

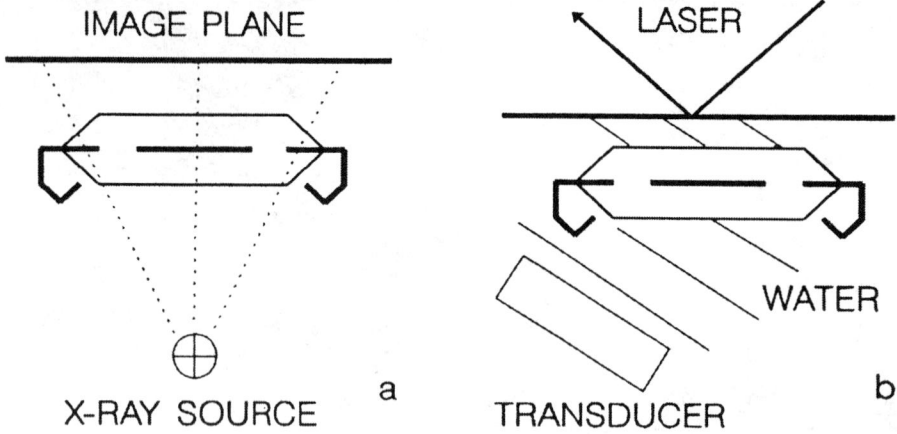

Figure 13: X-ray radiographic inspection (a) and scanning laser acoustic microscopy (SLAM) (b) of IC packages.

transmitted wave are imaged with a scanned laser beam. The technique produces either an acoustic shadow image or an interference pattern by comparison of the signal to a phase reference. SLAM offers the advantage of real time imaging which is more suitable for 100% inspection than a technique involving mechanical scanning. SLAM has been successfully applied to situations where real time imaging is critical, and in applications involving samples with irregular shapes and very thin multi-layer construction that cannot be imaged well by reflected sound due to the "dead zone" and reverberation effects described earlier. SLAM applications have included the inspection of tape automated bonding (TAB) packages, and thin film ceramic chip capacitors.[9] In practical application the spatial resolution of SAM images of IC packages is typically superior to that of SLAM images at the same frequency.[47] SLAM cannot be applied to die attach inspection in packages with air cavities, such as ceramic DIPs and pin grid arrays, unless the lid is removed. Similar information as provided by SLAM can be obtained from a SAM operating in through-transmission mode. In this mode, a separate transducer is placed on the other side of the package to receive sound transmitted through the package.

New Developments

The introduction of technologies such as ball grid array (BGA) and flip chip packages has driven improvements in acoustic inspection techniques. BGA packages often have laminated substrates composed of several layers. These packages can not always be reliably inspected by pulse-echo acoustic microscopy due to echo interference problems from the many thin layers in the package. Phase inversion detection, an important tool which assists delamination detection in molded surface mount packages, is not always useful for BGA packages due to very low acoustic impedances of some substrates. Additionally, the size and height of features has decreased, requiring increases in frequency to improve depth and spatial resolution. Alternative approaches such as through-transmission screening of BGAs and high frequency (>200MHz) pulse-echo inspection of flip chip bumps are addressing these new issues.

Initially, the Stanford SAM operated in through-transmission mode, but was later converted to pulse-echo for easier alignment and improved sample flexibility. Pulse-echo acoustic microscopy has the ability to temporally resolve echoes, allowing the identification of the specific plane where a defect occurs. Additionally, the pulse-echo mode allows the use of phase inversion to locate delaminations. In BGA packages, the substrates are often of lower acoustic impedance than the mold compound. Thus, phase inversion is invalid for delamination detection.

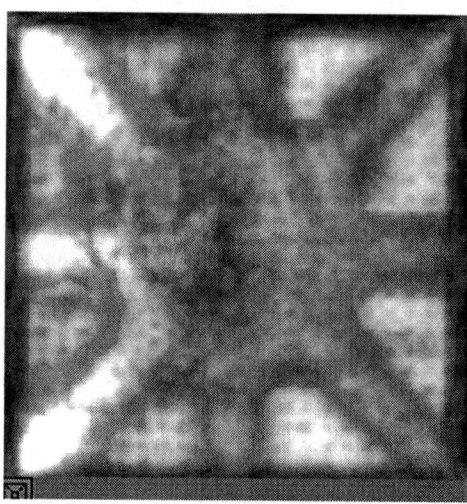

Figure 14: An attempted pulse-echo analysis of the die attach in an overmolded BGA package.

Also, the substrates in BGAs are composed of many thin layers, each of which produces echoes. These echoes overlap and often interfere with each other, making it difficult to locate the desired echo. Figure 14 shows a pulse-echo image produced during an attempted die attach analysis. The star-shaped pattern of the die pad is evident. Since delaminations appear bright in pulse-echo mode, this image would be incorrectly interpreted as having a large area of delamination between the copper die pad and die attach material. The through-transmission image in Figure 15 shows reliable data indicating only small areas of delaminations are present in the die attach area. However, this image alone does not reveal whether the delamination occurs between the die and die attach, die attach and solder mask, or solder mask and substrate. This is but one example of how real delaminations can be missed in pulse-echo inspection, and similarly bonded areas can appear delaminated, due to echo interferences. This is not necessarily a rare occurrence[49]. Unlike pulse-echo acoustic microscopy, echoes are not analyzed in through-transmission acoustic microscopy. Rather, the identification of delaminations is based upon the fact that air gaps (cracks, voids, delaminations) block sound transmission at MHz frequencies. Thus, a lack of transmission of sound through a package identifies the presence of a delamination. The application of through-transmission acoustic inspection to BGA packages both bypasses the echo interference problem and precludes the need for use of phase inversion to detect delaminations. This method has been applied successfully on virtually every type of BGA in production, including multi-shelf packages. However, due to a loss of resolution with this technique caused by scattering from the substrate fibers, it is possible that smaller delaminations, such as flip chip underfill delaminations, may be missed.

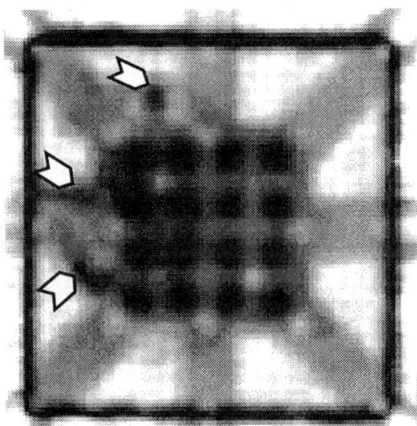

Figure 15: The corresponding through-transmission image for the package in Fig. 14. The 16 solder balls beneath the die are evident, as is the weave pattern in the substrate. Three small delaminations are shown in this image (arrows).

The development of flip chip technology has introduced the need for high frequency imaging of flip chip bump interconnects. Transducers have been developed that have improved lens design and deliver frequencies up to 250 MHz. These transducers allow excellent imaging of flip chip bumps and have detected defects such as non-wets of solder, missing bumps, excessive solder, and cracks. Figure 16 shows acoustic detection of defective bumps in a flip chip that had not been underfilled. X-ray inspection determined that the solder formed bridges on the substrate and did not make contact with the die surface. Although sometimes very large voids in the bumps can be detected with acoustic inspection, real-time x-ray offers superior resolution for void detection and is complementary for that purpose. To allow acoustic inspection of flip chip bumps, the backside of the die must be exposed. This reduces the absorption of the high frequency sound waves, which would degrade the resolution below that needed for bump inspection.

The continued need for higher speeds and increased circuit densities means package dimensions and I/O pitch will continue to shrink and continued improvements in resolution are required. The average flip chip bump today is approximately 100μm, and industry roadmaps forecast that by the year 2010 the bump size will be 20μm. A calculated return echo center frequency of 250MHz (assuming a 2 period pulse duration) will be required to temporally resolve echoes from the top and bottom of a bump this size (echoes are calculated to

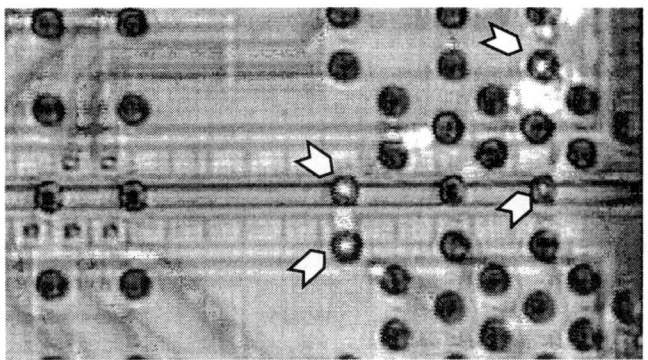

Figure 16: Acoustic inspection identifies bump defects in a flip chip. X-ray analysis revealed the solder did not make contact with the die surface at these locations.

be 8 ns apart). However, the frequency of the return echo will be adversely impacted by absorption of the signal in the coupling fluid and in the sample. Preferential absorption of the high frequency components of the broadband acoustic pulse lowers the center frequency and thus will determine the upper limit for transducer frequency. Attenuation of the signal, which is proportional to the frequency squared (in fluids), also imposes a practical upper limit upon frequency. Therefore, for a minimum acceptable 99.5% loss of return signal strength, the cutoff for the return echo frequency lies between 230 - 326MHz. Thus, a flip chip bump height of 20μm approaches the minimum size where improvements in axial resolution can still be obtained by increasing the frequency. Attenuation will remove echo frequencies above this range, regardless of the frequency of the initial incident signal.

For features smaller than 20μm, another method to temporally resolve echoes is required. With the emerging availability of streamlined algorithms, fast and inexpensive PCs, and dedicated digital signal processor chips (DSPs), it becomes feasible to apply frequency-domain signal analysis to routine inspection. In addition to enabling inspection of future packages having much smaller features than exist today, this type of analysis will also assist inspection of current packages where overlapping echoes from finely spaced layers is today a problem. For example, often there is no way to discriminate between the signal received from a bonded layer, or the signal received from a water-filled delamination. Also, at present bondline thicknesses beneath lids on flip chip packages can not be measured acoustically due to absorption of the high frequencies required for resolution by the thick copper lid. Development of real-time frequency-domain signal analysis may offer a non-destructive way to obtain these measurements.

The availability of high speed processors, inexpensive DRAM, and large capacity hard drives is also enabling analysis methods whereby after acquiring an image with one scan, the size and position of data gates can be re-set and through software analysis, the image is adjusted accordingly. This will enable re-examination of devices after they are no longer available (i.e. time zero data can be revisited after devices have been put through environmental testing), will allow engineers to verify the proper set-up was used for data acquisition, and will be a remarkable teaching tool to demonstrate to novice acoustic microscope users how positioning the data gates influences the resulting image. Within the next couple of years, this type of "virtual scanning" may become the routine method of choice.

SUMMARY

The SAM used today for IC inspection is a hybrid instrument with characteristics of both the Stanford SAM and the C-scan recorder.[9,10] For inspection in layered IC packages both amplitude and phase are measured and used to produce images of the internal structure and locations of interfacial delamination. Precise scanning and high frequencies are used for optimum lateral resolution. In addition, the depth of the primary sub-surface reflection is recorded for pseudo-three-dimensional representations of the package.

Acoustic microscopy is well suited to plastic package inspection. Although the plastic mold compound strongly attenuates sound and thereby limits resolution, the planar, layered design of IC packages is ideal for acoustic inspection. The typically featureless and planar front surface facilitates sub-surface imaging.

The acoustic microscope has been a key factor in the understanding of the moisture sensitivity problem in surface mount plastic IC packages. Polarity analysis of reflected acoustic pulses is extremely useful in the detection of delaminations in plastic IC packages, and is an important advantage of inspection with reflected sound. By quantifying limits on the amount of delamination that is acceptable from a reliability standpoint, SAM inspection can be incorporated into standard test methods and is preferred over destructive cross sectioning and electrical test methods.

The nondestructive nature of acoustic inspection means that in reliability evaluations involving a dense matrix of variables, fewer devices are required. Previously, a fraction of the remaining devices would have to be sacrificed for destructive analysis at each inspection interval. Also, the data of reliability evaluations using SAM are more valuable because the initiation and propagation of damage can be tracked on individual ICs throughout the test.

As packages have continued to evolve with time, adjustments to the application of acoustic microscopy and improvements in resolution have successfully overcome analysis challenges and ensured that SAM continues to be a non-destructive technique of significant importance for the development and evaluation of IC packages. Specifically, analysis of BGA packages in through-transmission mode bypasses two potential problems that exist for pulse-echo mode: echo interferences from the multi-layered substrates, and reduced acoustic impedances of the substrates that make phase inversion identification of delamination invalid. Transducers with center frequencies over 150MHz deliver the resolution necessary for flip chip bump inspection.

Due to frequency-dependant attenuation, eventually the feature size of interest will be smaller than SAM can resolve, regardless of how high a frequency is used. Frequency-domain analysis may overcome this limitation and is currently being developed. A one-dimensional or two-dimensional scanned array technology for acoustic imaging would have a dramatic impact on the effectiveness of SAM for real-time applications such as process control by eliminating the need for mechanical scanning.

REFERENCES

1. T. Moore, Proc. Int. Symp. Testing and Failure Analysis, 1989, pp. 61-67.
2. Y. Bar-Cohen and A.K. Mal, in Metals Handbook, 9th edn., ASM International, 17, 1989, pp. 231-277.
3. R.C. McMaster, Nondestructive Testing Handbook, Vol. 2, Ronald Press, 1959, p. 43.
4. J. Szilard, in Ultrasonic Testing, J. Szilard (ed.), John Wiley and Sons, 1982, pp. 1-23.
5. C.F. Quate, A. Atalar, H.K. Wickramasinghe, Proc. IEEE, 67, 1979, pp. 1092-1114.
6. R.A. Lemmons and C.F. Quate, App. Phys. Lett., 25, 1974, pp. 251-253.
7. R.A. Lemmons and C.F. Quate, Proc. 1973 IEEE Ultrasonic Symp., 1973, pp. 18-21.
8. C.F. Quate, IEEE Trans. Sonics and Untrason., SU-32 (2), 1985, pp. 132-135.
9. L.W. Kessler and S.R. Martell, Proc. Int. Soc. for Testing and Failure Anal., 1991, pp. 491-504.
10. B.T. Khuri-Yakub, in New Technology in Electronic Packaging, B.R. Livesay and M.D. Nagarkar, ASM International, 1990, pp- 311-315.
11. J.L. Rose and P.A. Meyer, Mater. Evaluation, 31 (6), 1973, p. 109.
12. G.J. Curtis, in Ultrasonic Testing, J. Szilard (ed.), John Wiley and Sons, 1982, pp. 495-555.
13. N.J. Burton and D.M. Thacker, Proc. Int. Soc. for Testing and Failure Analysis, 1985, pp. 187-192.
14. K. Shirai, K. Kobayashi, T. Noguchi and T. Goka, Proc. Int. Soc. for Testing and Failure Analysis, 1988, pp. 47-52.
15. M.J. Mirasole, Proc. Int. Soc. For Testing and Failure Analysis, 1988, pp. 77-88.
16. R.A. Lemmons and C.F. Quate, Appl. Phys. Lett., 25 (5), 1974, pp. 251-253.
17. R.G. Wilson, R.D. Weglein and D.M. Bonnell, Semiconductor Silicon 1977, 1977, pp. 431-435.
18. C.F. Quate, Semiconductor Silicon 1977, 1977, pp. 422-430.
19. A.J. Miller, Inst. Phys. Conf. Ser. No. 67: Section 8, 1983, pp. 393-398.
20. A.J. Miller, Acoust. Imaging, 12, 1982, pp. 67-78.
21. H.K. Wikramasinghe, J. Micros., 129, 1983, pp. 63-67.
22. H.R. Vetters, et al., Scanning Electron Microwscopy/III, 1985, pp. 981-989.
23. M. Sakimoto, et al., Proc. Int. Symp. for Testing and Failure Anal., 1985, pp. 173-177.
24. A. Kitayama, H. Tabata and H. Suziki, Proc. IMC, 1986, pp. 462-469.
25. S. Ito, et al., Proc. Elect. Comp. Conf., 1986, pp. 360-365.
26. S. Okikawa, et al., Proc. Int. Symp. for Testing and Failure Anal., 1987, pp. 75-81.
27. S. Kuroki and K. Oota, Proc. Elect. Comp. Conf., 1989, pp. 885-890.
28. A. Nishimura, S. Kawai and G. Murakami, Proc. Elect. Comp. Conf., 1989, pp. 524-530.
29. T.M. Moore, Texas Instruments Technical Report TR-088778, Feb. 1988.
30. L.W. Kessler and S.R. Martell, Proc. Int. Symp. for Testing and Failure Anal., 1990, pp. 491-504.
31. R. Birudavolu, Proc. Surface Mount 1989, pp. 751-766.
32. A. van der Wijk, K. van Doorselaer, Proc. Int. Symp for Testing and Failure Anal., 1989, pp. 69-74.
33. K. van Doorselaer and K. de Zeeuw, Proc. Elect. Comp. Conf., 1990, pp. B49-B53.
34. T. Moore, R. McKenna, S.J. Kelsall, Proc. Int. Symp. Testing and Failure Analysis, 1990, pp. 251-258.
35. T. Moore, R. McKenna, S.J. Kelsall, J. Surface Mount Tech., 4 (3), 1990, pp. 31-38.

36. T. Moore, R. McKenna, S.J. Kelsall, IEEE Int. Reliability Physics Symp., 1991, pp. 160-166.

37. K.R. Kinsman, J. Metals, 40 (6), 1988, pp. 8-13.

38. IPC-SM-786, "Impact of Moisture on Plastic Package Cracking," and IPC-Test Method 650-2.6.20, "Plastic Surface Mount Component Cracking," Institute for Interconnecting and Packaging Electronic Circuits (IPC), Lincolnwood, IL, 1991.

39. L.A. Kinsler, et al., Fundamentals of Acoustics, John Wiley and Sons, 1982, p. 125.

40. R.S. Gilmore, R.A. Hewes, L.J. Thomas, and J.D. Young, in Acoustical Imaging, 17, H. Shimizu, N. Chubachi, and J. Kushibiki (eds.), Plenum Press, 1989, pp. 97-109.

41. B.T. Khuri-Yakub, in New Technology in Electronic Packaging, B.R. Livesay and M.D. Nagarkar (eds.), ASM International, 1990, pp. 311-315.

42. J. Krautkramer and H. Krautkramer, Ultrasonic Testing of Materials, Springer-Verglag, 3rd edn., 1983, p. 28.

43. B. Baker, Electronic Manufacturing, Feb. 1989, pp. 20-22.

44. C. McBee, Circuits Manufacturing, Jan. 1989, pp-67-69.

45. L.W. Kessler, Proc. IEEE, 67, 1979, pp. 526-536.

46. S. Sokolov, USSR Patent No. 49, Aug. 31, 1936.

47. R.K. Mueller and R.L. Rylander, IEEE Spectrum, 1982, pp. 28-33.

48. T.M. Moore, Reliable Delamination Detection by Polarity Analysis of Reflected Acoustic Pulses, Proc. Int. Symp. For Testing and Failure Anal., 1991, pp. 49-54.

49. T.M. Moore and C.D. Hartfield, Proc. Characterization and Metrology for ULSI Technology, NIST, 1998.

Advanced Radiographic Techniques in Failure Analysis

Joe Colangelo
Texas Instruments, Inc.
Dallas, Texas

The use of x-ray energy for nondestructive evaluation has been common practice for years. The use of radiographic techniques reveals many defects, and internal features, in parts and materials.

In the past, the primary method for capturing an image was to place the sample on a sheet of film and expose it to x-ray energy for a predetermined time. The amount of x-ray energy that passes through the sample to reach the film is determined by the sample's density and atomic structure. The areas of higher absorption will allow less energy to pass, resulting in darker features on the film (assuming a positive imaging medium such as Polaroid film). This creates the familiar "shadowgraph."

The acquisition of an x-ray system which provides magnified images in real time (see Fig. 1) eliminates the trial-and-error method: expose sample, develop film, vent frustration, adjust position, adjust power, and repeat. Historically, this has been the typical method of x-ray evaluation.

A comprehensive failure analysis can only proceed in one direction since it involves destructive tests. Accurately capturing information pertaining to the failure in each phase of the analysis is critical because steps cannot be retraced. Clearly, the addition of this real-time capability provides a wealth of information at the front end of the analysis that was previously unattainable. X-ray analysis is as simple as mounting the sample on the five-axis manipulator and adjusting for the optimum viewing angle. The power level is continuously adjustable from 30 to 160 kV and up to 1 mA for penetration of a wide range of samples. Table 1 summarizes the performance features of a typical real time x-ray system:

Table 1   Performance Summary of the Feinfocus FXS 160.52

Magnification	Up to 200×
Resolution	4 to 6 µm
Maximum power	160 kV, 1 mA
Cycle time	5 minutes
Manipulation	Five axes, 0.01-mm resolution, joystick or CNC-controlled
Image processing	Pseudo color, frame integration, background subtraction, intensity profile/histogram, distance measurement, superimposed text
Data storage	Photograph, videotape
Customer interaction	During analysis

Instead of exposing film, the x-ray energy is converted to a video signal by an image intensifier detector tube. The x-ray tubehead is classified as a micro focus type, because the electron beam is focused to a very small spot on the target, approximately 5 microns in size. Because the x-rays are emitted in a conical pattern from a point source to the target, geometric magnification of the image in real time is achieved. This is done by projecting the image onto the image intensifier (see Fig. 2). By changing the positions of the image intensifier and sample with respect to the x-ray source, magnification of up to 200X is possible with very low distortion.

Fig. 1.   External appearance of the Feinfocus FXS 160.52 machine.

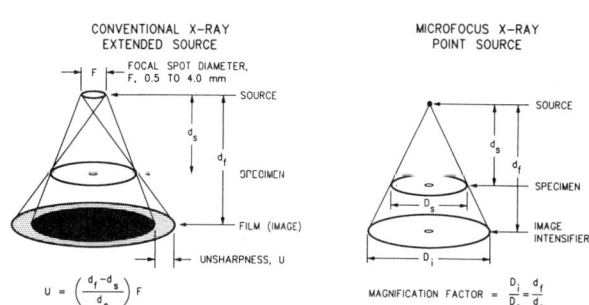

Fig. 2.   Illustration of x-ray magnifications and sharpness.

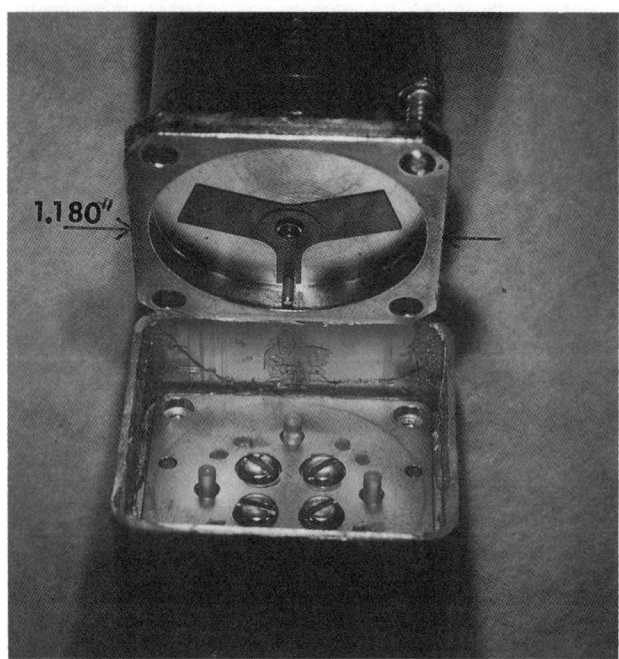

Fig. 3. RF shutter partially disassembled.

The purpose of this section is to demonstrate the capabilities of this system and to illustrate its direct impact on productivity. This will be done by describing actual case histories, although it is difficult to accurately present the benefits of real time imaging with photographs.

APPLICATION EXAMPLES

1. One program was experiencing repeated field failures on an RF shutter assembly. Figure 3 shows the unit cut open to expose the Y-shaped 0.007 inch-thick beryllium copper (BeCu) switch plate. This plate was driven by a solenoid, which moved the three clear plastic plungers in the lower section, when it was energized.

   The photos show, from a side view, the plate in the rest position (Fig. 4), and the failed unit with the coil energized (Fig. 5). Figure 6 shows the operation of a good unit for comparison. The plate has flexed under the pressure of the switch return springs and the switch contacts have changed position. The failure was due to a fatigue fracture in the small arm of the plate which rides on the locating pin. This allowed the plate to rotate out of position. The internal action of the switch was evaluated in real time, with no destructive analysis. The benefit of this became apparent when customer-owned material was returned from the field, and destructive analysis was not authorized. Subsequent failures were quickly identified.

2. In this example, an outside vendor was performing a hand-soldering operation of connectors and plated-through holes on a motherboard with inconsistent and unacceptable results. Because of the high thermal conductivity of the internal copper plane layers, an adequate temperature was not achieved for the solder to fill the barrels. The unsupported barrels would crack and become intermittent during environmental

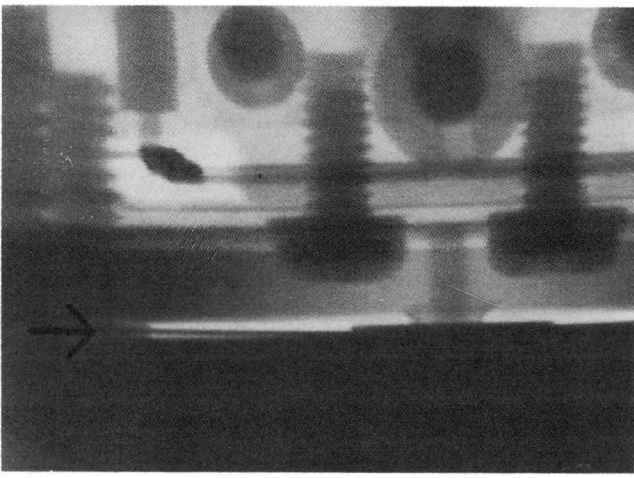

Fig. 4. X-ray image side view, with shutter de-energized.

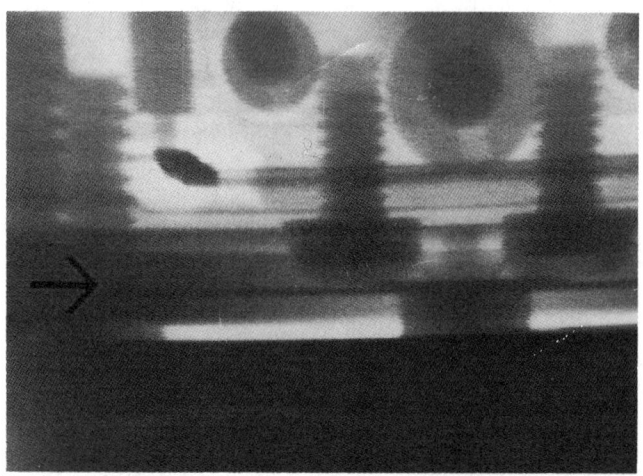

Fig. 5. Failed unit, energized.

stress screening. Conventional radiographic inspection would have been difficult and time consuming, even if a suitable large chamber was available. Using five-axis manipulation, the optimum viewing angle was quickly established.

Figure 7 is an overall view of the motherboard, with a 6 inch scale for reference. It was a 20 layer, 160 mil-thick polyimide PWB with 2 oz. (2.8 mil copper) plane layers. Figure 8 reveals a partially filled plated-through hole near one of the round connectors. Figure 9 is a close-up, and Fig. 10 is the same image processed to enhance the defect. The photograph also shows the annular ring of each layer, which gives an indication of layer registration. With a videotape of the x-ray results, the vendor initiated training sessions for the operators to eliminate the problem.

3. This example illustrates the importance of proper viewing angle to highlight failures with x-ray. Figure 11 is a conventional x-ray image using wet film (negative) processing. Many of these RF filter assemblies were failing in system burn-in. Conventional x-ray could not detect any anomalies; but,

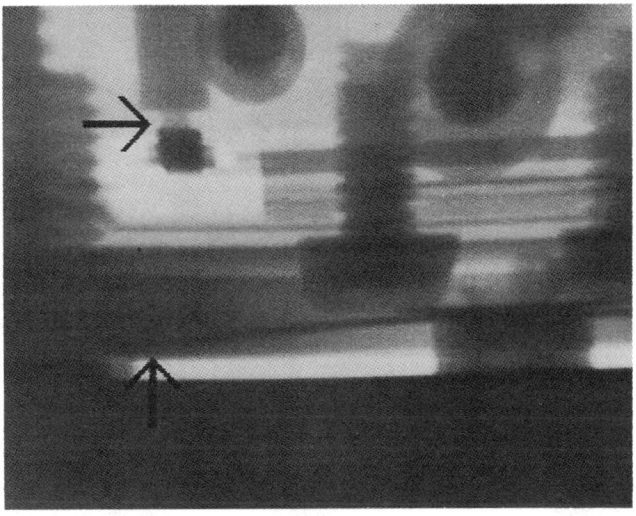

Fig. 6. Good unit, energized.

Fig. 7. Motherboard, 20 layer, polyimide.

Fig. 8. X-ray image of filled, plated-through holes.

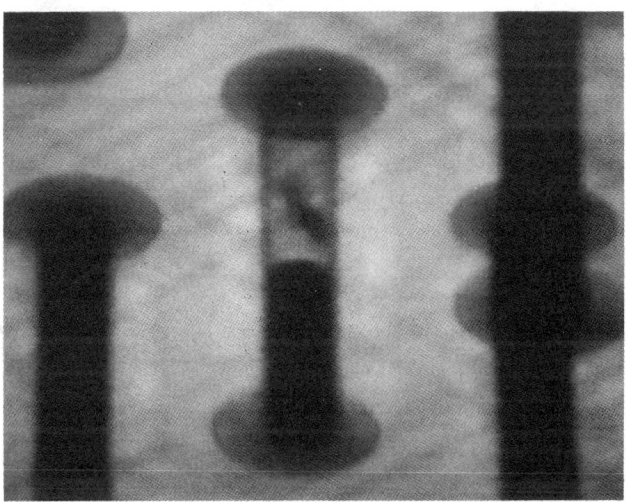

Fig. 9. Higher magnification of unacceptable solder fill.

when rotated while examining with real time x-ray (Fig. 12), the crack became evident when properly aligned with the x-ray source.

Figure 13 shows the crack from a different angle: looking down through the barrel of the SMA connector. Figure 14 is a scanning electron microscope image of the crack taken during the course of destructive analysis, to confirm the x-ray results. This problem was quickly resolved because the argument that the crack was caused by stresses induced during destructive analysis was no longer valid.

4. The all-ceramic hybrid in Fig. 15 failed electrically and also failed subsequent particle impact noise detection (PIND) testing. The construction of the package made lid removal nearly impossible without introducing contamination internally. X-ray examination (Fig. 16) quickly revealed a conductive epoxy moisture "getter" that had separated from the substrate and shorted the internal circuitry. In the processed image (Fig. 17), the getter and 1 mil gold wire bonds, a few of which are

fused, are clearly visible. Some minor voiding was also detectable in the eutectic under the large I.C.

5. Figure 18 shows a power hybrid which was evaluated for die attach integrity. Each of the four cells contained a power Schottky rectifier die that required a low thermal impedance to the 75 mil copper substrate. Using x-ray analysis, the process could be modified to reduce voiding to an acceptable level. Figures 19, 20, and 21 demonstrate some image processor functions that can assist in the inspection.

In Fig. 19, the intensity profile across the bottom of the screen represents the relative intensity of the area traversed by the horizontal cursor. Figure 20 shows a 7.09 mil diameter void which was measured using the cursors. Figure 21 shows the image processed to enhance the boundaries of the voided areas. The die size is 180 x 180 mils. This quick turn-around evaluation can reduce the time required for process optimization.

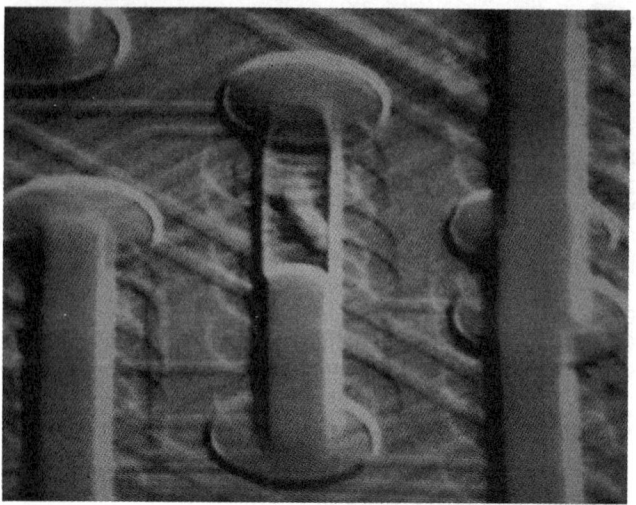

Fig. 10. Processed image revealing layer registration.

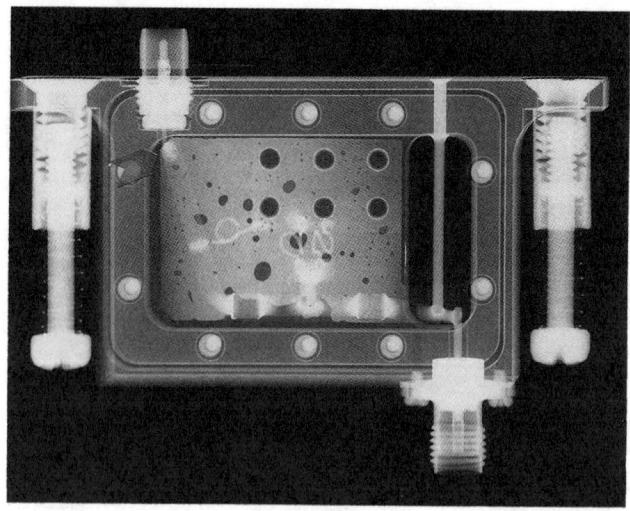

Fig. 11. "Wet-film" x-ray shadowgraph of filter module. (A) denotes problem solder joint.

Fig. 12. With manipulation in real time, crack becomes apparent.

Fig. 13. View of same joint by looking through SMA connector.

6. One critical (and mystifying) failure was resolved with a confidence level that would have been impossible with any other analysis technique, including standard x-ray. An encapsulated connector on a cable assembly failed during system burn-in (Fig. 22). The failure mechanism was the lack of solder where the wire terminated in the solder cup. Other connectors examined showed varying degrees of the same anomaly. The problem always occurred on an odd-numbered pin (i.e., on the same side as the marking ink). During assembly, each solder joint was visually inspected for a fillet prior to encapsulation, so it appeared that the solder may have reflowed either during or after the encapsulation process. The hypothesis was that the curing process for the marking ink on the outside of the connector involved a heat gun, and this uncontrolled process may have reflowed the solder. This did not seem likely because there was no evidence of heat damage on the connector body.

Using the real-time x-ray system, a test was set up to verify the hypothesis. A heat gun was placed 3 inches from the connector while it was examined. At approximately 90 seconds, the solder became molten and was completely wicked out of the odd-numbered solder cups by the silver-plated stranded wire (Fig. 23). The results were recorded on videotape and presented to the project. Other physical evidence was collected to support the x-ray data. Because of the convincing evidence, the problem was quickly resolved and corrective action was implemented.

7. Figure 24 demonstrates the usable magnification levels attainable in real time. This is a temperature-sensing device with a 0.4 mil Nichrome® wire sensing element. Decapsulation to expose this extremely fine wire was practically impossible. X-ray analysis revealed damage to some of the windings in the coil. This most likely occurred prior to the encapsulation process, while the exposed winding was susceptible to damage.

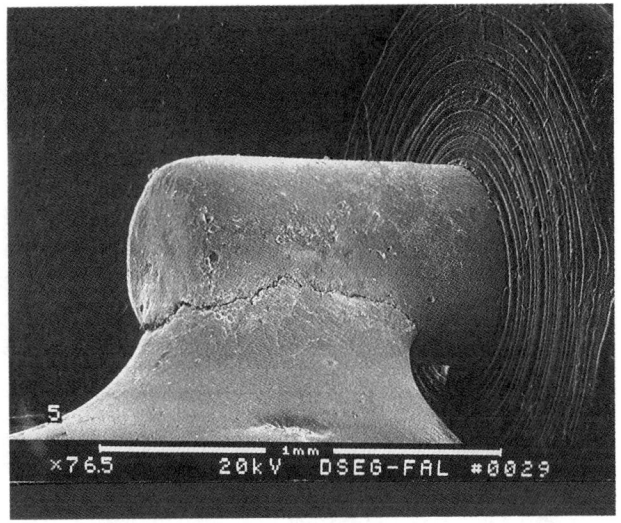

Fig. 14. SEM image of the crack confirmed x-ray results.

Fig. 15. All-ceramic hybrid assembly.

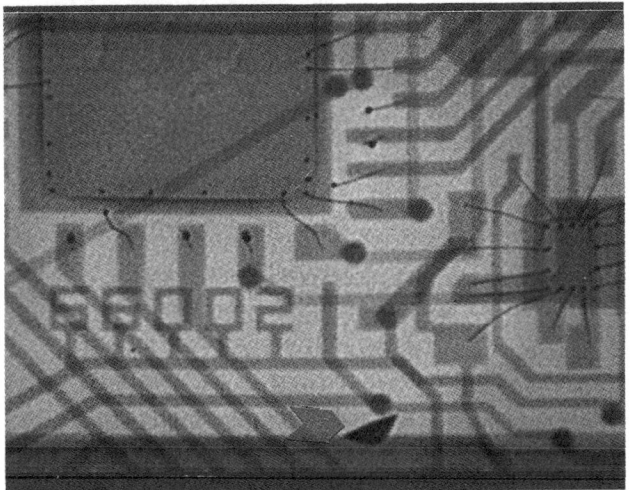

Fig. 16. X-ray image revealing particle, a conductive epoxy "getter."

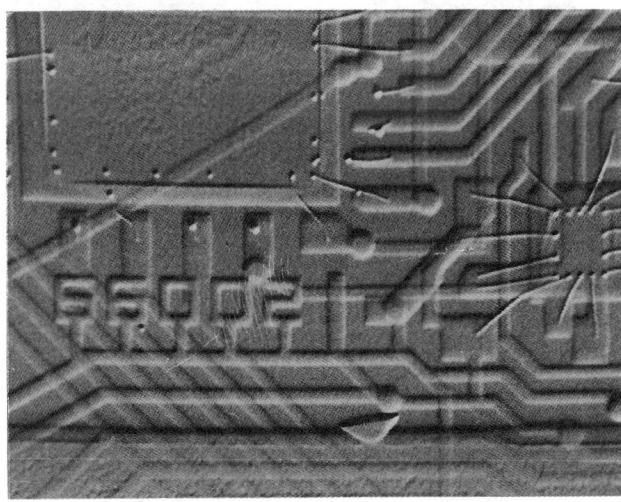

Fig. 17. Processed image, enhancing detail. Fused 1 mil bond wires are visible.

COMPONENT DEGRADATION

Using the x-ray analysis technique to examine solder joints on populated PWBs has been extremely successful. Voids in surface mount solder and unfilled plated-through holes are clearly visible. Because the inspection time is decreased to minutes, much less than conventional techniques, inspection of high risk production board samples is feasible. This system has also successfully been used to determine defects flagged with other inspection systems actually did not exist.

When examining active components, it becomes important to limit the total exposure so that the absorbed dose of radiation does not exceed amounts which might cause damage. Table II lists the maximum allowable dose ranges for various device technologies.

With most real-time x-ray systems, many factors determine the dose rate: accelerating voltage and current are continuously

Fig. 18. Quad power rectifier hybrid.

adjustable, distance to the source (inverse square relationship), duration of exposure, and the amount of inherent filtration in the packaging, if any. Preliminary tests were conducted to determine the actual dose rates under various conditions so that

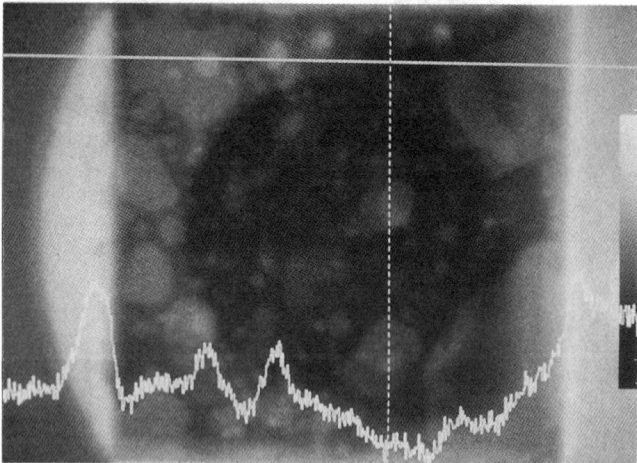

Fig. 19. Intensity profile of voiding along horizontal cursor.

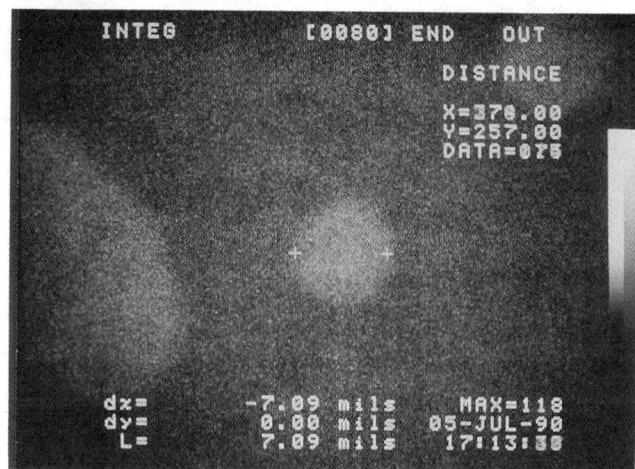

Fig. 20. Measurement of 7.09 mil diameter solder void.

Fig. 21. Processed image, enhancing detail. Die size is 180 x 180 mils.

Fig. 22. Connector with reference designator on the encapsulation material.

"live" boards can be examined with some confidence that reliability will not be affected.

The effectiveness of an x-ray filter material depends primarily upon thickness, density, atomic structure and accelerating voltage. The x-rays are produced by a stream of electrons striking a tungsten target. The energy is emitted in the form of a band of frequencies in the x-ray spectrum. Higher accelerating volt-

Table 2 Radiation Damage Threshold for Common Device Technologies

Generic Technology	Damage Threshold, RAD (Si)
ECL, TTL, GaAs	200 k to 10 meg
Linear, I²L	15 k to 1 meg
MNOS, PMOS	8 k to 100 k
CMOS, VMOS, NMOS	0.8 k to 10 k
Quartz crystals (natural)(a)	100 to 1 k

(a) Crystals will show part per million frequency shifts, although the effect is temporary and will anneal out.[3]

ages not only increase the radiation intensity but also cause a shift in the envelope towards the shorter wavelengths, as shown in Fig. 25. This short wavelength energy provides greater penetration.

The information necessary for image generation is contained in the x-rays of various intensity which pass through the sample. Most of the lower frequency, or longer wavelength, "soft x-rays," are absorbed by the sample and do not contribute to image quality. The purpose of filtering, therefore, is to block these soft x-rays so as to minimize the dose absorbed by the sample. At an accelerating voltage of 100 kV, a .030 inch-thick layer of aluminum absorbs approximately half of the x-ray energy. Hence, it is called the "first half value layer"(4). An equivalent absorber at this voltage is approximately 1.7 mils of copper. This is convenient when estimating the filtering characteristics of copper plane layers in a PWB.

Table 3 illustrates equivalence factors of metals at various voltages.

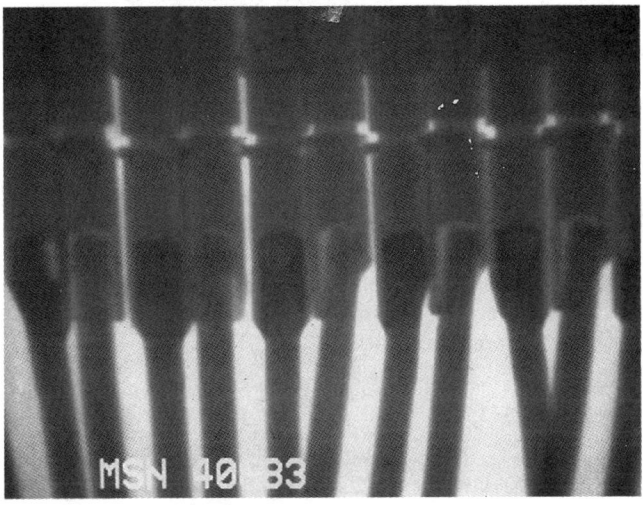

Fig. 23. Alternate cups without solder correspond to the side with the marking ink.

Fig. 24. Processed image of an encapsulated temperature sensing device with damaged 0.4-mil (10.2 μm) diameter Nichrome® wire.

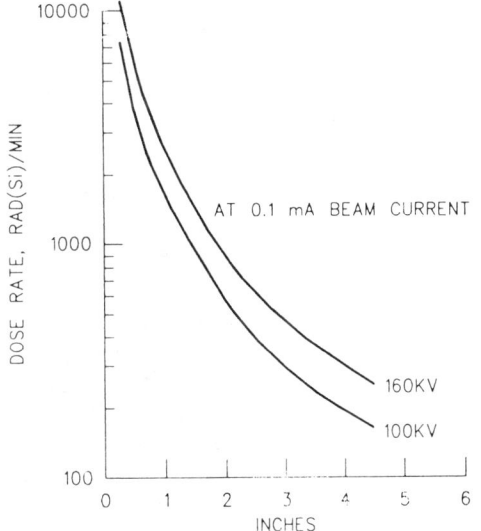

Fig. 25. Typical x-ray envelopes, as a function of accelerating voltage and beam current.

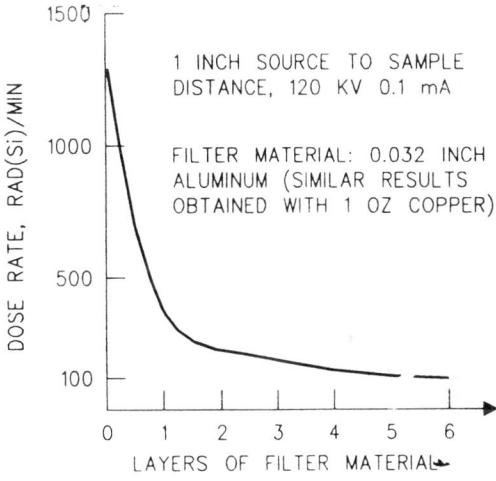

Fig. 26. Alternate cups without solder correspond to the side with the marking ink.

Table 3 Approximate equivalence factors of metals

Metal	Density, gm/cm^3	Atomic Number	50 kV	100 kV	150 kV	220 kV
Magnesium	1.74	12	0.6	0.6	0.05	0.08
Aluminum	2.70	13	1.0	1.0	0.12	0.18
2024 Al alloy	(2.99)b	(14)	1.4	1.2	0.13	0.14
Steel	(7.81)	(26)		12.0	1.0	1.0
18-8 Steel	(7.82)	(26)		12.0	1.0	1.0
Copper	8.96	29		18.0	1.6	1.4
Zinc	7.14	30			1.4	1.3
Brass(a)	(8.32)	(29)			1.4	1.3
Lead	11.4	82			14.0	12.0

Note: Aluminum is taken as the standard at 50 kV and 100 kV, and steel at the higher voltages.[5] **(a)** Tin or lead alloyed in the brass will increase these factors. **(b)** Numbers in parenthesis are estimates based on representative alloys.

It is apparent from these equivalence values that the x-ray attenuation properties of the materials aren't directly proportional to either density or atomic number. The relative effectiveness of these metals as x-ray filters will also vary as a function of accelerating voltage.

The testing performed on the Feinfocus FXS-160 involved measuring the total dose at different voltages, source to object distances, and with various layers of filtering. Thermoluminescent dosimeters (TLDs) were used to measure the ionizing total dose. They were exposed at various dose rates for a fixed amount of time. Since the effects of ionizing radiation are a function of the absorbing material, the dose rate must be converted to units of RADs (radiation absorbed dose) silicon for correlation to the limits in Table 2. A RAD is defined as "the absorbed dose of any ionizing radiation which is accompanied by the liberation of 0.01 joule of energy per kilogram of absorbing material"(4).

The data from some of these tests was plotted and is shown in Fig. 26. With this information, and by compensating for the filtration inherent in the packaging materials (or adding more), the dose rate in RADs silicon can be estimated. With the rate information and total dose ranges in Table 2, inspection time limits can be established which would ensure only insignificant levels of degradation.

SUMMARY

An advanced radiographic capability has been demonstrated for effective nondestructive evaluation of components and assemblies. Empirical data support the premise that examination of active components is feasible without sustaining permanent damage. Although many studies exist pertaining to the characteristics of x-rays and the effects of radiation on electronic components, many questions persist. We are currently attempting to quantify the true impact on components in the unique environment that exists with the Feinfocus system.

The data contained in this report are preliminary and the values presented should not be construed as absolute. Future emphasis will focus on the questions of: (1) the accuracy of the various measuring devices at the energy levels of interest, (2) the absorption characteristics of silicon as a function of accelerating voltage, and (3) assessing the true impact on semiconductors. This question arises because the data reflect absorption in bulk silicon. In most cases the damage occurs in the oxide layers, which represent a small fraction of the mass of the die, and may also possess different absorption characteristics than doped silicon.

ACKNOWLEDGEMENTS

The author would like to thank Frank Poblenz of the DSEG Radiation Effects Laboratory for his technical assistance with the dose rate measurements.

REFERENCES

1. *Fundamentals of Microfocus Radiography,* Application Note, Feinfocus USA Inc., Agoura Hills, CA, pp. 5-7.

2. MIL-HDBK-728/5, Radiographic Testing, 1985, p. 45.

3. A similar chart is available from IRT Corporation, San Diego, CA.

4. *Ionizing Radiation and Its Effects Upon Electronic Components,* Application Note, Nicolet Test Instruments Division, Madison, WI.

5. *Radiography in Modern Industry,* 2nd Edition, Eastman Kodak Company, 1957, p. 18.

Fault Localization

Electrical Techniques

Curve Tracer Data Interpretation
for Failure Analysis

D. Wilson

ABSTRACT

Determining the cause of failure on semiconductor devices requires a combination of electrical characterization and physical analysis. The curve tracer can provide the electrical characteristics, displaying voltage versus current. This paper will relate the information obtained on a curve tracer to physical effects on devices. Examples of degraded electrical characteristics and some basic curve tracer operating considerations will be discussed.

INTRODUCTION

Device acceptance testing generally is based on single point data for a given parameter. This assumes that the parameter response apart from this data point is characteristic of a typical device. When evaluating a suspect or failed device, this assumption should not be made. The curve tracer provides a display of current over a continuous voltage range. This can provide information on non-linearities, instabilities, channels, and soft breakdown characteristics that are not evident from simple point data measurements.

Relating these electrical characteristics to physical conditions will help to determine the course of testing to be taken during a failure analysis to determine the root cause.

SEMICONDUCTOR JUNCTION CHARACTERIZATION

An understanding of the information available from a curve tracer requires a fundamental understanding of semiconductor physics and basic device parametrics. There are a number of reference textbooks that address these topics [1-5]. This section contains information on semiconductor junction characterization and the type of curves that the failure analyst may encounter.

Forward-Bias

The typical junction should have a uniform smooth exponential current response with increasing voltage as shown in Figure 1. The curve for a silicon device becomes nearly asymptotic to the voltage axis in the 0.6-0.7 volt range.

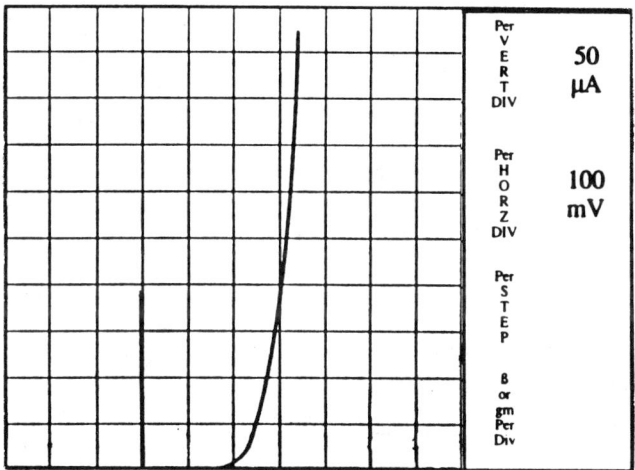

Fig. 1. Curve tracer I-V junction characteristic.

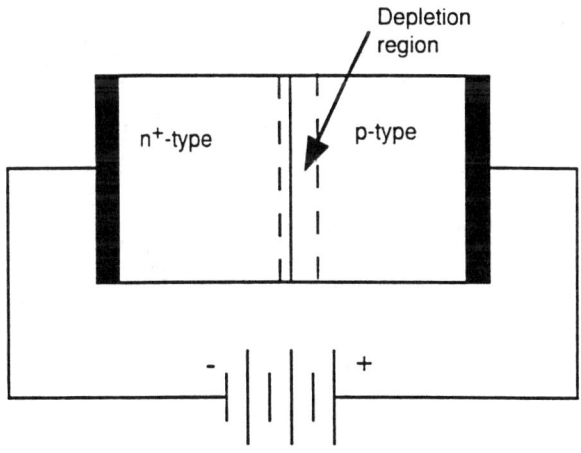

Fig. 2. Physical representation of semiconductor junction.

A physical representation of a p-n junction is shown in Figure 2. At the metallurgical junction between the n-type region and the p-type region there is a concentration gradient because of the large concentration of holes on the p-type side and the large concentration of electrons on the n-type side. There is a flux of electrons into the p-type side and a flux of holes into the n-type

95

side to eliminate this concentration gradient. This flow stops when an electric field builds up due to the remaining donor and acceptor ions. The region that the electrons and holes have flowed from is called the space-charge region or depletion region (depleted of carriers). The forward-bias curve on the curve tracer has the knee at the voltage level required to overcome this electric field (or built-in voltage). The knee will move to the right with higher dopant concentration and will vary for different semiconductor materials. The location of the physical area on the junction related to the depletion region is important for the failure analyst to understand. This region produces much of the voltage drop that can be affected by part damage.

The resistance of the junction after the knee is approximately 26/I, where resistance is in ohms and I (current) is in milliamps. For instance, at 1 milliamp the resistance would be 26 ohms and at 26 milliamps the resistance would decrease to 1 ohm. In addition to the junction resistance there is the ohmic body resistance and interconnect or contact resistance. For analyzing these series resistance effects, a high value of current should be used to reduce the resistance effect from the junction. Excessive current levels will lead to an increase in the effective series resistance and the 26/I approximation will no longer be valid.

Abnormal Characteristic

Abnormal characteristics include parallel or series resistance as represented in Figure 3.

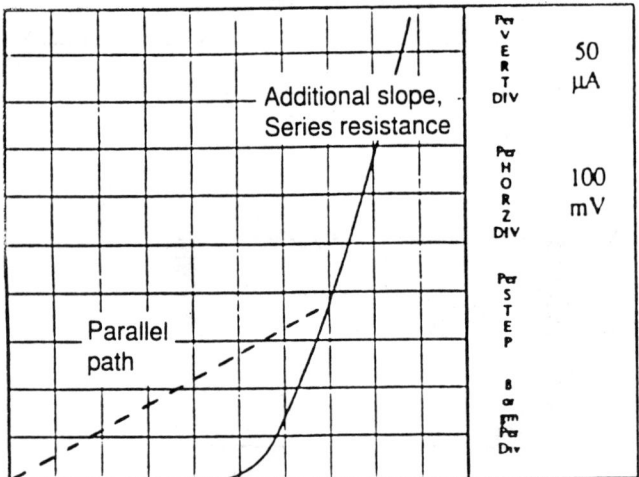

Fig. 3. Abnormal forward-bias junction characteristics; parallel resistance and series resistance.

Parallel resistance indicates that there is a path that shunts the depletion region. This could be at the metallurgical junction or as far away as the measuring points. Additional series resistance can be the result of problems at interfaces. The additional slope shown in Figure 3 represents a series resistance of approximately 500 ohms. A note of caution is appropriate at this point. When using a curve tracer to view a resistive slope, it is important to pay attention to the current and voltage scales. Figure 4 displays a 500 ohm resistor using three different current scales. The resistor can have the appearance of a short, a resistor, or an open simply by changing

the displayed current. This underlines the importance of specifying what "short" means in report verbiage. At one setting, it may appear to be excessive leakage; at another, an extremely low resistance, perhaps milliohms.

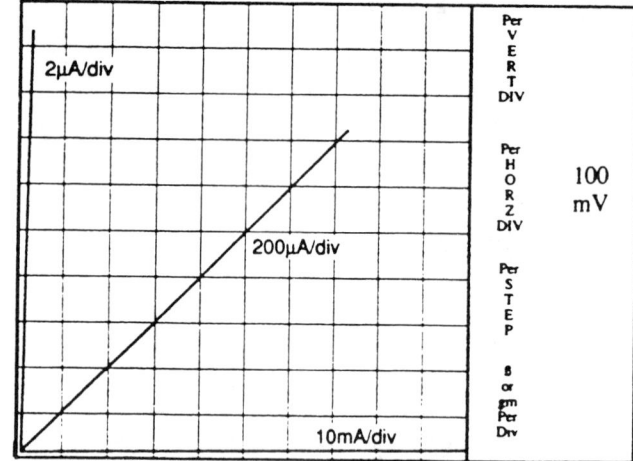

Fig. 4. Curve tracer display of a 500 ohm resistor with three different vertical current scales.

Reverse-Bias

Under reverse-bias conditions, the typical junction will have a constant low level of current (reverse saturation current) until the voltage reaches the point where breakdown occurs [6]. The two mechanisms responsible for junction breakdown are zener breakdown and avalanche breakdown.

Zener breakdown occurs in heavily doped semiconductor junctions due to the large electric field produced across the narrow depletion region. This field causes separation of valence electrons from their respective nuclei and a large current results at the breakdown voltage level.

Avalanche breakdown occurs in lower doped semiconductor junctions. The reverse-bias produces movement of holes and electrons across the depletion region. The acceleration of these particles increases with increased bias until the point is reached that they have enough energy to free valence electrons during collisions. These electrons produce additional free valence electrons in a process referred to as avalanche multiplication. Again, as in the case of zener breakdown, a large current results at the breakdown voltage level. Zener breakdown predominates for breakdown levels up to 5 volts. Between 5 and 8 volts both mechanisms are present, and above 8 volts avalanche breakdown predominates.

Due to the differences in breakdown mechanisms, the temperature coefficient of the two processes are opposite. The zener breakdown voltage decreases with temperature and the avalanche breakdown voltage increases with temperature. This can be a factor where devices are operated at high temperatures.

A typical avalanche reverse-bias breakdown curve is shown in Figure 5. This junction has a sharp breakdown just below 9 volts. Breakdown is normally specified at 10 microamps; this device measures approximately 8.8 volts at this point. This same junction is shown in Figure 6 with the current scale decreased by 3 orders of magnitude (the mode switch is set to leakage current).

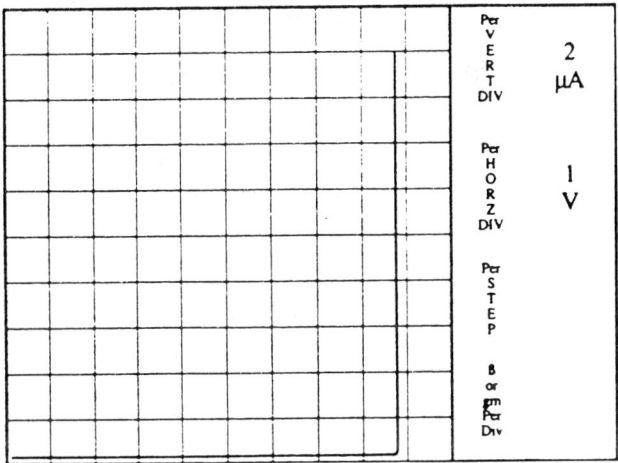

Fig.5. Reverse-bias avalanche breakdown characteristic.

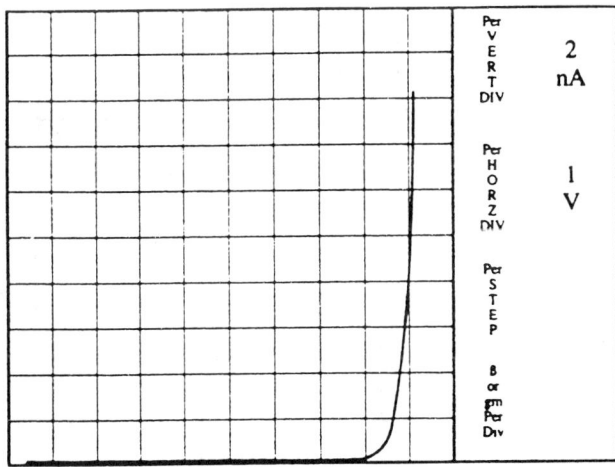

Fig. 6. Same junction breakdown as shown in Figure 5, displayed in leakage mode.

Comparing Figures 5 and 6 it is noted that at 10 nanoamps the voltage level is 9. This should be below 8.8 volts at this low current level indicating that the voltage displayed is higher than the voltage actually across the device. This higher level is due to the instrument configuration used for the leakage measurements (Tektronix 576 Curve Tracer). A resistor to ground is present when the mode switch is set to leakage. The effect in Figure 6 is a resistive slope of 25 megohms after the breakdown knee. The increase is 10 nanoamps times 25 megohms equals .25 volts. The equipment configuration is illustrated in Figure 7.

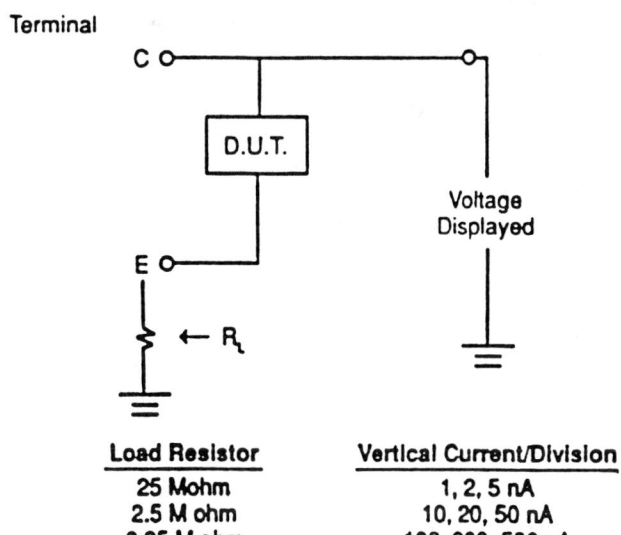

Fig. 7. Tektronix 576 Curve Tracer leakage current measurement implementation.

The resistor to ground varies depending on the current/division so the resistive slope will vary accordingly. It is possible to overstress and damage this resistor to ground so the leakage setting should be used with caution.

Reverse breakdown voltage measurement should not be performed on PIN diodes. The parameter specified for these devices is the leakage at a specific voltage rather than breakdown voltage at a given current. During avalanche breakdown, a junction with the intrinsic region is apparently damaged by injection of carriers in the reverse-bias direction. This may be aggravated by current concentration due to a negative resistance characteristic [7]. Reverse breakdown voltage measurement on PIN diodes can result in permanent damage to the breakdown level.

Abnormal Characteristics

A number of abnormal characteristics can occur for different reasons on a reverse-biased junction. These characteristics include:
1) Reduced breakdown voltage indication.
2) Soft breakdown knee.
3) Channeled curve.
4) Walkout of the breakdown voltage.
5) Unstable curve.

Conditions that produce these abnormal characteristics are electrical overstress, surface contamination, and manufacturing defects.

Electrical overstress is a common cause of degraded junctions. The overstress condition can be forward or reverse-bias and DC, AC, or transients. Forward-bias electrical overstress has been reported to cause a soft breakdown knee characteristic [8]. The phenomena is believed to be due to the formation of a positive charge sheet in the oxide at the oxide to silicon interface.

Degradation due to forward-bias electrical overstress is not as likely as degradation due to reverse-bias overstress because of the power limitations in the former.

Current is often limited by the interconnect wire size and the voltage is limited by intrinsic conduction of the diode. In many cases when forward-bias electrical overstress occurs the wire will be fused open but the junction characteristics will not be degraded.

Reverse-bias electrical overstress can cause both an apparent reduction in breakdown voltage levels and soft breakdown knees. The effect that occurs is related to the type and location of the damage. Active trap sites are generated and metal atoms migrate into the silicon. DC electrical stress causes softening of the knee as shown in Figure 8. This characteristic was produced by forcing 600 mA through a reverse-biased base-emitter junction. The avalanche breakdown voltage is still at 8.8 volts as it was prior to the stress. This indicates that damage sites have been produced near the junction and these sites contribute leakage current as they are incorporated into the depletion region. This is shown in Figure 9 in a physical representation of the junction.

Very fast transients, such as an electrostatic discharge (ESD), can cause reduction of the apparent breakdown voltage level. Figure 10 shows the reverse breakdown characteristic on a field effect transistor prior to a simulated ESD stress. Following this stress, the displayed breakdown voltage has been reduced as shown in Figure 11. There is slight softening of the knee and more resistance after the breakdown. If a high enough current were forced through the junction to follow this resistive path out to the original breakdown voltage level, avalanche breakdown would occur (Figure 12). The curve displayed in Figure 11 is not true avalanche breakdown since the electric field is not high enough to support multiplication. To have a high current flow below the level where the depletion region field can support avalanche multiplication, indicates that a source of electrons has been reached by the depletion region. This may be due to aluminum from the contact spiking into the silicon, producing high current when the voltage level is high enough for the depletion region to incorporate the aluminum (Figure 13).

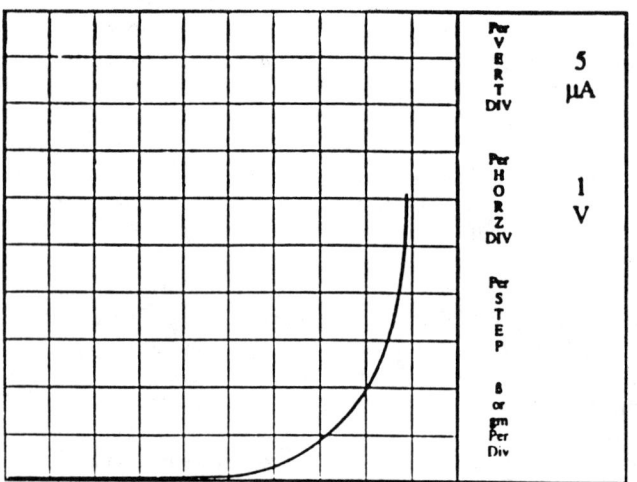

Fig. 8. Soft reverse-bias breakdown due to DC electrical overstress.

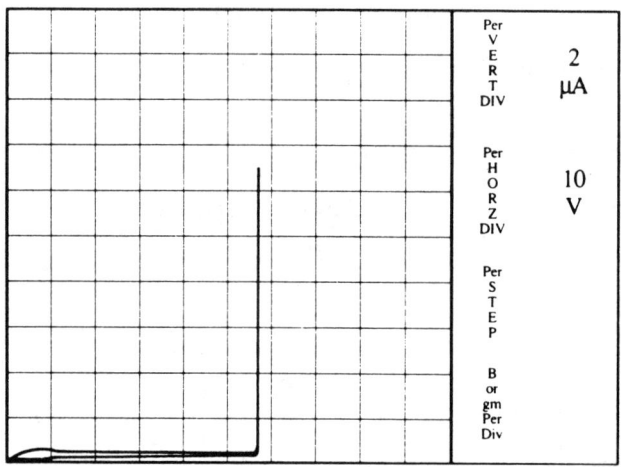

Fig. 10. Junction avalanche breakdown prior to ESD stress.

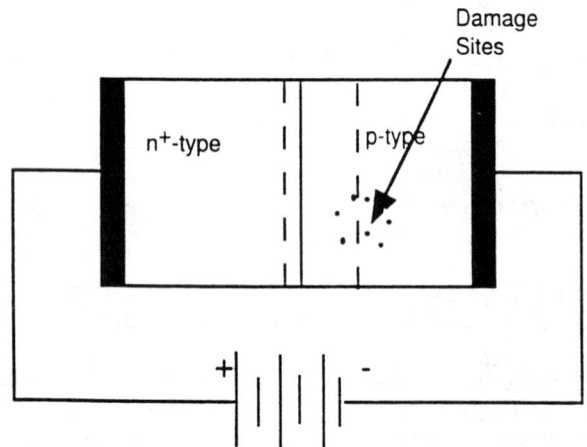

Fig.9. Physical representation of the junction shown in Figure 8.

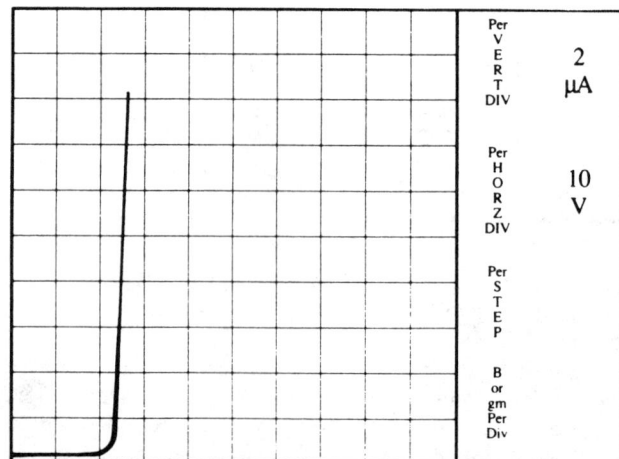

Fig. 11. Apparent breakdown voltage reduction due to ESD stress.

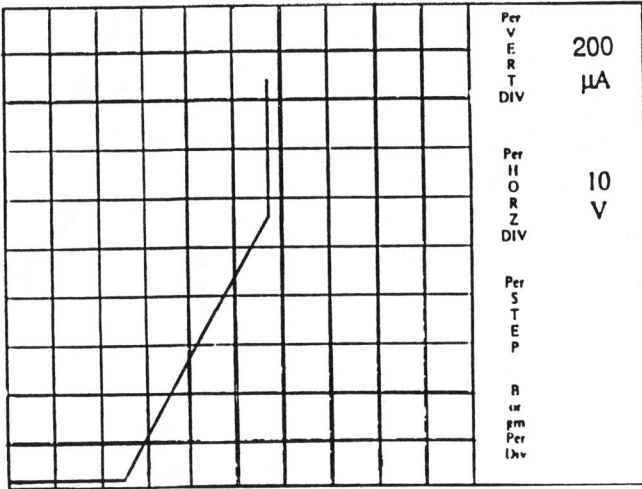

Fig. 12. Reverse-bias characteristic displayed at 200 microamps per division.

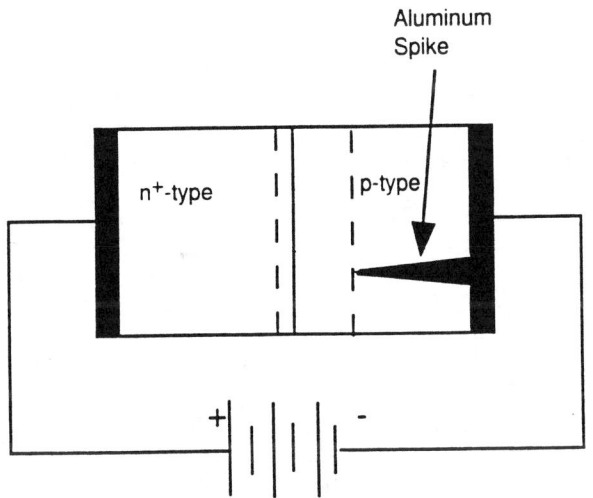

Fig. 13. Physical representation of aluminum spike into junction area.

The location for this type of damage may be shown by techniques such as liquid crystal, light emission, or Electron Beam Induced Current (EBIC) on a Scanning Electron Microscope (SEM). When using these techniques to identify the site, the current level should be kept below the avalanche breakdown level since after avalanche there will be current flow through other parts of the junction as well as the damage site.

Surface inversion due to ionic contamination of the oxide can cause reduction of the breakdown voltage and the formation of inversion channels [9,10]. Positive ions such as sodium are readily transported to the silicon-silicon dioxide interface and can invert the underlying silicon. The inversion layer will then modify the depletion region characteristics as indicated in the junction cross-section drawing in Figure 14.

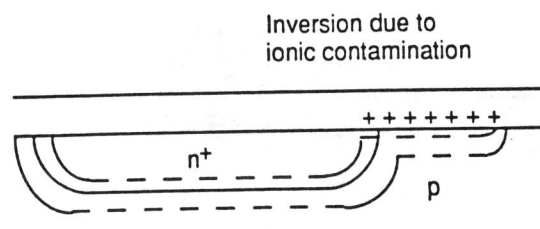

Fig. 14. Cross-section showing inversion layer created by ionic contamination.

If the inversion region incorporates a defect in the silicon, a curve as shown in Figure 15a will occur. If the junction needs to be reverse-biased to incorporate the defect, then the curve will appear as shown by in Figure 15b. In this case, the reverse-biased junction has little current flow until the reverse-bias is high enough to incorporate the carrier generation site. When the carrier generation site is incorporated, current flows through the inversion layer. The resistance associated with this inversion layer leads to the saturation characteristic (100 nA level).

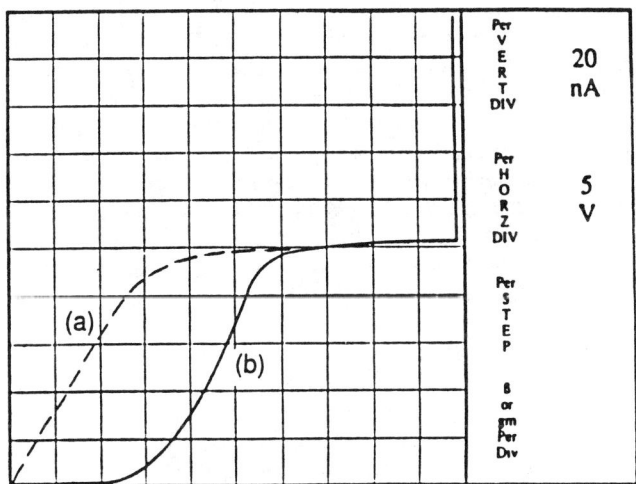

Fig. 15. Channel leakage characteristics that can be caused by ionic contamination.

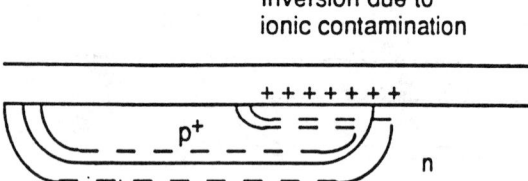

Inversion due to
ionic contamination

Fig. 16. Cross-section showing inversion layer in p$^+$ region created by ionic contamination.

An inversion layer on a p$^+$n junction produces a curve with similar appearance but does not require the incorporation of a carrier generation site. This physical representation is shown in Figure 16. The electrical characteristic is the same as shown in Figure 15b. The inversion layer is an n$^+$ region. The breakdown voltage of this induced p$^+$n$^+$ junction will be lower than the breakdown voltage of the p$^+$n junction. The knee at about 20 volts would be the induced junction breakdown with the normal avalanche breakdown at about 50 volts.

The same type of saturated channel characteristic has been seen due to electrical overstress. This has occurred on a junction field effect transistor (JFET) when there is electrical damage to either the source to gate junction or the drain to gate junction. The channel is seen when displaying the reverse-bias characteristic of the opposite junction. For example, with damage near the source contact, when the depletion region reaches this damage site current will be generated. This current has to travel through the pinched-off channel in the source/drain diffusion, to the contact to the drain. Analogous to the inversion layer resistance, the channel resistance produces the limited current characteristic.

Walkout of the breakdown voltage is commonly observed on devices due to positive charge in the oxide, either as a result of ionic contamination or following long periods of forward-bias stress [8]. This is depicted in Figure 17. The appearance of these curves is dependent upon the current limiting resistance on the curve tracer, and on the adjustments made to the collector supply as the voltage is increasing. Often a sawtooth appearance results. Walkout will also be observed as a thermal effect since breakdown voltage increases with junction temperature.

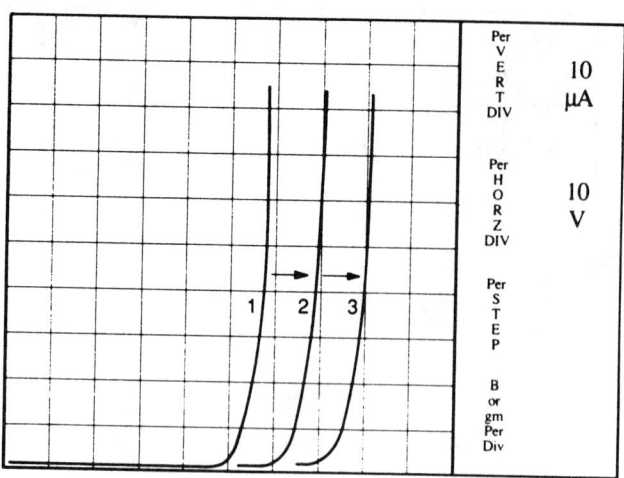

Fig. 17. Walkout of reverse-bias breakdown curve.

Manufacturing defects related to the silicon die can produce reduced breakdown voltage levels, soft breakdown knees or unstable curves. These characteristics will occur as a result of defects in the silicon within the depletion region.

An example of a defect that will produce an unstable curve is a cracked die. High current flow may occur before breakdown, or at breakdown the curve may jump between different characteristics. This is depicted in Figure 18.

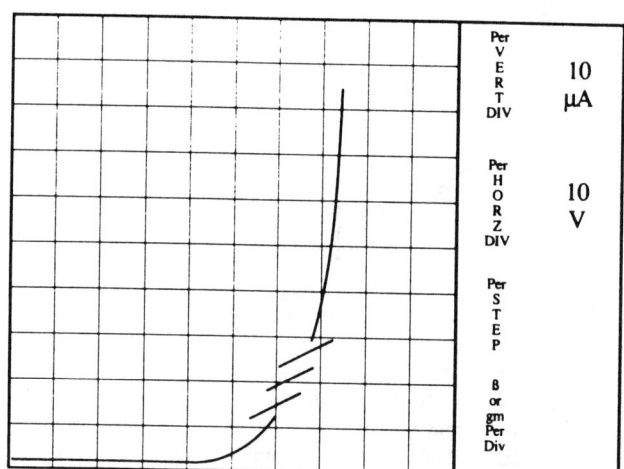

Fig. 18. Unstable curve due to cracked die.

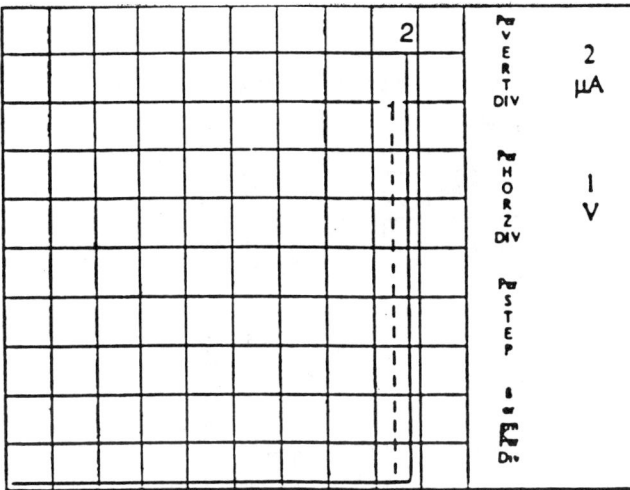

Fig.19. Bistable curves due to alternating breakdown sites.

Another characteristic that can readily be viewed on a curve tracer is a noise characteristic at voltage levels close to the breakdown point. In this situation current will flow at one voltage level and then the curve will jump to a second stable condition. The curves will alternate between these two characteristics in a bistable manner (Figure 19). This generally occurs at low current levels near the onset of avalanche breakdown. One explanation for the switching between the two curves is as follows. Current flows to one breakdown site and an appreciable voltage drop develops across the internal series resistance. Also there is an increase in the breakdown voltage level due to temperature rise. The field strength at the depletion region will decrease below that necessary to maintain avalanche, and so the current flow stops in this area and begins in another location that has a higher avalanche breakdown voltage(Figure 20). When the initial site cools down, breakdown at that location is reinitiated. Oscillation between these two locations then results. Different semiconductor noise sources include burst noise, zener noise, and microplasma noise [8,11].

TRANSISTOR CHARACTERIZATION

The information in the previous section on the characterization of semiconductor junctions is relevant for transistor base-emitter and base-collector junction characterization. Additional transistor parameters that are commonly measured on the curve tracer include:
1) Common-emitter current gain, beta or h_{FE}.
2) Collector-emitter characteristics, breakdown and leakage.
3) Base-emitter and base-collector saturation characteristics.

Common-emitter Current Gain

The low frequency common-emitter current gain, beta or h_{FE}, is defined as the ratio of the collector current to the base current (I_C/I_B). This can be measured directly on the curve tracer. The small-signal current gain, h_{fe}, is the change in the collector current divided by the change in base current. This is usually defined at a given frequency, and cannot normally be performed on the curve tracer. The current gain of a transistor is controlled by the manufacturing process with effects produced by the quality of silicon and silicon dioxide, the diffusion profile and geometry, and dopant concentration. The gain will vary with temperature, current level, and frequency due to the interaction of a variety of effects [12].

For the failure analyst, the low-current gain is typically the most important gain measurement [13]. This gain can be degraded by reverse-bias base-emitter junction current flow. The characteristics of this include:

1) Low gain at low collector currents(Figure 21).
2) Normal gain at higher collector current levels.
3) Increase of collector to emitter breakdown voltage in the low-current region.
4) Little or no damage to the base-emitter reverse-bias characteristic.
5) Nearly full recovery of the gain with high temperature bake.
6) Little or no degradation with a high temperature reverse-bias bake (as would be expected with ionic contamination).

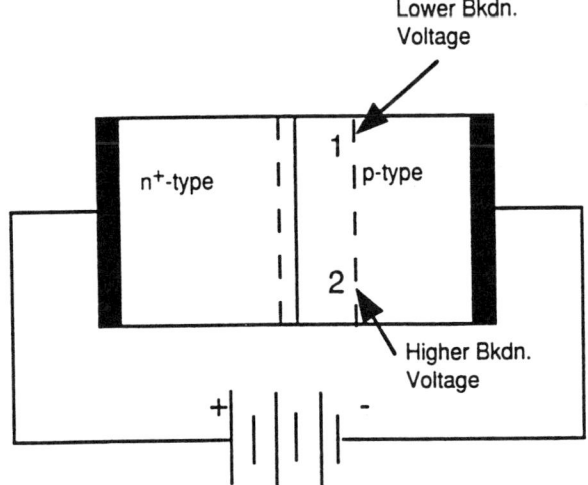

Fig. 20. Physical representation of the two alternating breakdown sites shown in Figure 19.

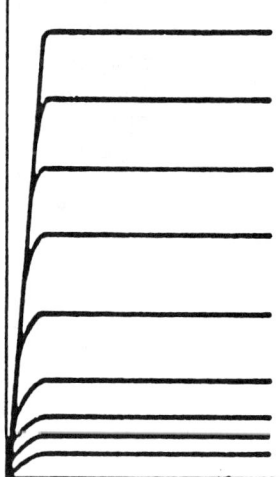

Fig. 21. Family of curves showing low gain at low collector current and normal gain at higher collector currents.

These six items provide the basis for assessing a device during a failure analysis. The low current gain is dominated by the recombination of electrons and holes in the emitter depletion layer. This recombination current is composed of both bulk and surface components. Electrical stress likely increases the surface recombination component and thus reduces the low current gain. The increase seen in the collector to emitter breakdown voltage is simply an effect of this gain degradation since BV_{CEO} is indirectly proportional to gain and the low current gain is the most severely degraded [12]. The recovery characteristic, item 5), is also consistent with surface effects.

Collector-emitter Characteristics

Measurement of breakdown voltage and leakage can be made in both polarities between the collector and emitter. These parameters are usually specified for collector positive on NPN transistors and collector negative on PNP transistors. The opposite polarity measurement has a breakdown voltage approximately equal to the base-emitter breakdown voltage. A common usage for this latter measurement is to verify continuity between the collector and emitter in the case when the base wire is open. A forward-bias overstress involving the base can open the base wire. This measurement will verify that the connections to the collector and emitter are still intact and also will show the condition of the base-emitter junction.

The breakdown voltage measurement between the collector and emitter is an important parameter for power transistors particularly in switching applications. Breakdown voltage can be measured with:
-The base open, BV_{CEO}.
-The base-emitter connected by a resistor, BV_{CER}.
-The base-emitter biased, BV_{CEX}.
-The base-emitter shorted, BV_{CES}.

The voltage versus current curves for these measurements are different with voltage increasing as the measurements go down the list. Caution must be exercised during the measurement of these parameters on the curve tracer due to the negative resistance characteristic encountered after current flow begins [14,15]. The most difficult to measure because of this instability is BV_{CEO}. Figure 22 depicts the negative resistance characteristics of BV_{CEO}. When power is left on for several seconds, the curve shifts as shown in Figure 23. The die heats up during the time the voltage level is increasing (right-hand trace) and this produces a higher current level during the decreasing voltage time period (left-hand trace). Monitoring the shift of the retrace voltage is a good indicator of the die temperature. As a rule BV_{CEO} should not be measured during a failure analysis unless there is a specific need. When the measurement is considered pertinent to the analysis, it is recommended that the device be tested using a system or circuit specifically designed to test this parameter.

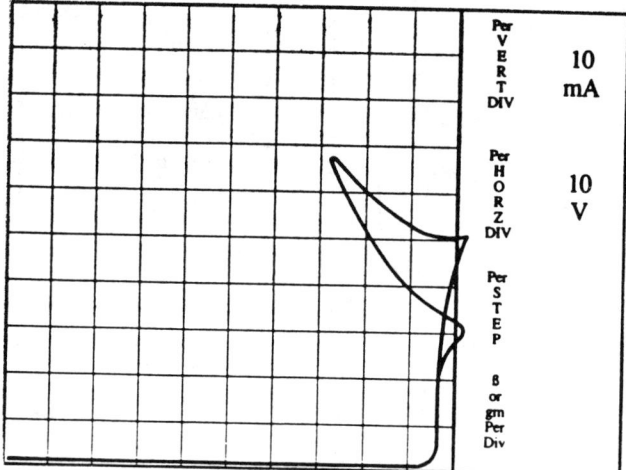

Fig. 22. Negative resistance characteristics of BV_{CEO}.

A note of caution is in order at this time. BV_{CEO}(sustained) is often specified at a relatively high current level and is to be measured under pulsed conditions. A common mistake is to perform this measurement with the base drive set to pulsed operation. This does not apply a pulse to the collector supply but rather forces it to be a DC supply. This is the worst case power dissipation situation and can easily result in the destruction of the device.

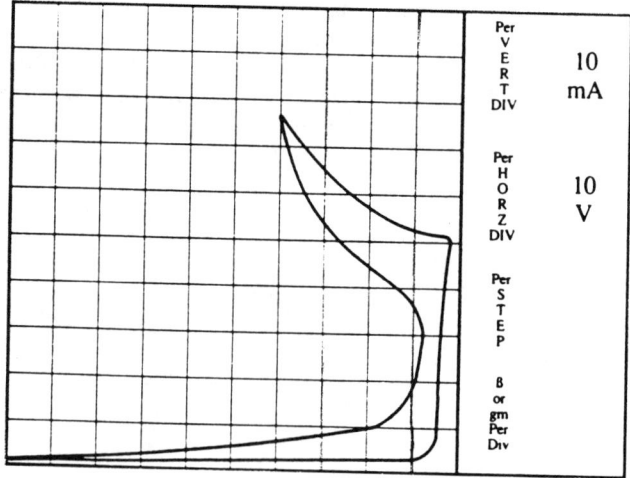

Fig. 23. Affect of heating on BV_{CEO} characteristic.

Measurement of leakage current between the collector and emitter can be made with the same connections as the breakdown voltage measurements. Two abnormal conditions encountered are a resistive short or a channel. The resistive short can be produced by a variety of overstress conditions [16]. The cause of this type of failure is normally due to second breakdown. First breakdown is avalanche breakdown as previously discussed. Second breakdown occurs when the temperature of the silicon becomes hot enough to thermally produce carriers in excess of the background concentration. At this point, large current flows and thermal runaway occurs [4]. The dopant in the emitter region diffuses through this very hot region and forms a pipe of opposite dopant material in the base region [14]. This is depicted in the transistor cross-section drawing in Figure 24. Electrically, the base-emitter breakdown may show little or no effect while the collector-emitter path becomes resistive.

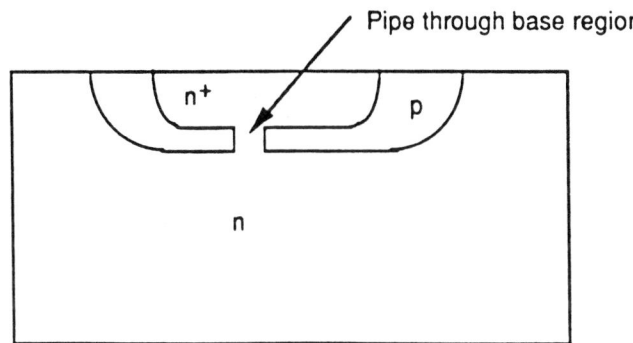

Fig. 24. Depiction of transistor cross-section showing n-type pipe through base region.

A channel characteristic can occur between the collector and emitter that has the appearance of a low-current gain curve. This effect can be due to surface inversion in the base or collector region (Figure 25). As with other inversion problems, a high temperature unbiased bake will cause the current level to decrease, and a high temperature biased bake will cause recurrence.

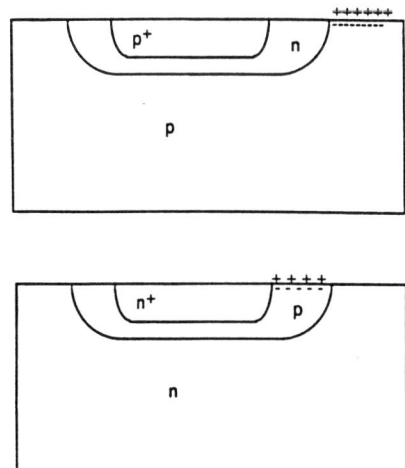

Fig. 25. Inversion of p-type collector region (top) or p-type base region (bottom).

Base-emitter and Collector-emitter Saturation Characteristics

Transistor base-emitter and collector-emitter saturation voltages are the minimum voltage levels that occur at collector saturation. Collector saturation is defined as the point where an increase in base current produces no significant increase in collector current. Saturation parameters usually provide a margin of safety to allow for beta degradation over operating life. This safety margin typically assumes a forced beta of 10, or for low gain power devices a forced beta of 5 is sometimes used. Base-emitter and collector-emitter saturation voltages are measured using the lower volts/cm or the expanded voltage feature to provide maximum sensitivity and readability. Saturation voltage margins can be measured by observing either the base-emitter or collector-emitter voltage with sufficient sensitivity, as the base current is gradually decreased from the forced beta level. When the base or collector voltage exceeds the maximum specified saturation voltage, this base current divided into the forced beta base current will provide the saturation safety margin. An abnormal saturation voltage characteristic indicates possible wrong die (no epitaxial layer), poor backside die eutectic solder wetting, bad bond, or excessive resistance in the lead crimp or weld terminals. A resistive collector-emitter saturation characteristic is illustrated in Figure 26.

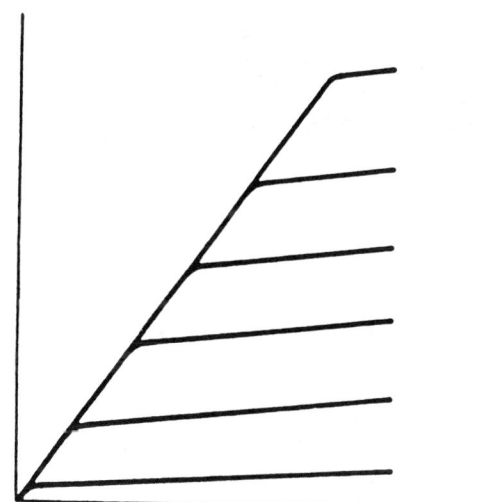

Fig. 26. Resistive slope on collector-emitter saturation characteristic.

Many junction characteristics have been described. The display of these characteristics on the curve tracer illustrates the power and versatility of this instrument. The I-V response curves provide an unequaled capability to the analyst for evaluation of parametric data. This characterization should be considered as a starting point for the failure analyst that can be expanded and built upon.

REFERENCES

1. J. Millman, C. Halkias, Electronic Devices and Circuits, Published by McGraw-Hill Book Company, Inc., 1967.

2. 2 . J. Lindmayer, C. Wrigley, Fundamentals of Semiconductor Devices, Published by D. Van Nostrand Company, Inc., 1965.

3. A. Grove, Physics and Technology of Semiconductor Devices, Published by John Wiley and Sons, Inc., 1967.

4. S. Ghandhi, Semiconductor Power Devices, Published by John Wiley and Sons, Inc., 1977.

5. M. Howes, D. Morgan, Reliability and Degradation - Semiconductor Devices and Circuits, Published by John Wiley and Sons, Inc., 1981.

6. "Motorola Zener Diode Manual," Published by Motorola, Inc. 1980.

7. H. Egawa, "Avalanche Characteristics and Failure Mechanisms of High Voltage Diodes," in IEEE Transactions on Electron Devices, Vol. ED-13, No. 11, 1966.

8. J. Schenck, "Burst Noise and Walkout in Degraded Silicon Devices," in Proceedings of 6th Annual Reliability Symposium, 1967, pp. 31-39.

9. W. Schroen, "Failure Analysis of Surface Inversion," in Proceedings of the Reliability Physics Symposium, 1973, pp. 117-123.

10. D. Fitzgerald and A. Grove, "Mechanisms of Channel Current Formation in Silicon P-N Junctions," in Proceedings of the Physics of Failure in Electronics, 1966, Volume 4, pp. 315-332.

11. C. Varker, "An Investigation of Microplasma Noise in Zener Diodes With the SEM," in Proceedings of the Reliability Physics Symposium, 1971, pp. 155-162.

12. J. Walston, J. Miller, Transistor Circuit Design, Published by McGraw-Hill Book Company, Inc., 1963.

13. M. Jensen, R. Milburn, "Diagnosis and Analysis of Emitter-Base Junction Overstress Damage," in Proceedings of the Electrical Overstress/Electrostatic Discharge Symposium, 1981, pp. 101-105.

14. W. Roehr, "Avoiding Second Breakdown," Motorola Semiconductor Products Inc. Application Note 415A

15. "How to Safely Check Sustaining Voltage on Power Transistors," Unitrode Corporation Design Note 5

16. J. T. May, "Limiting Phenomena in Power Transistors and the Interpretation of EOS Damage," in Microelectronics Failure Analysis Desk Reference, Published by ASM International 1990.

A Primer on Simple Device Problems and Curve Tracer Characteristics

Douglas McCormac
TRW Components International
Torrance, California

These curve trace drawings are to complement the previous article. Note that Figs. 1-9 and 12 are in quadrants 1-4 (A.C. mode, center zero)

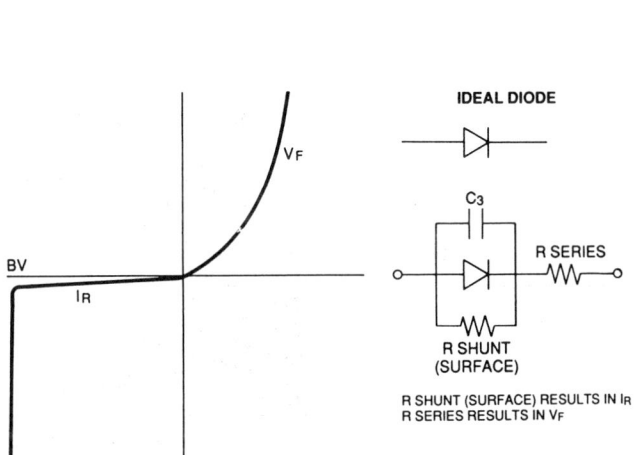

Fig. 1 Illustration of parasitic elements in real vs. ideal diodes, and effect on characteristic

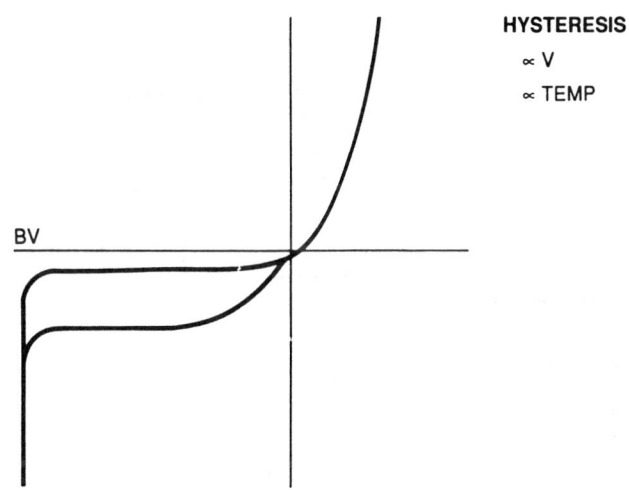

Fig. 2 Illustration of hysteresis, which can be proportional to voltage or temperature.

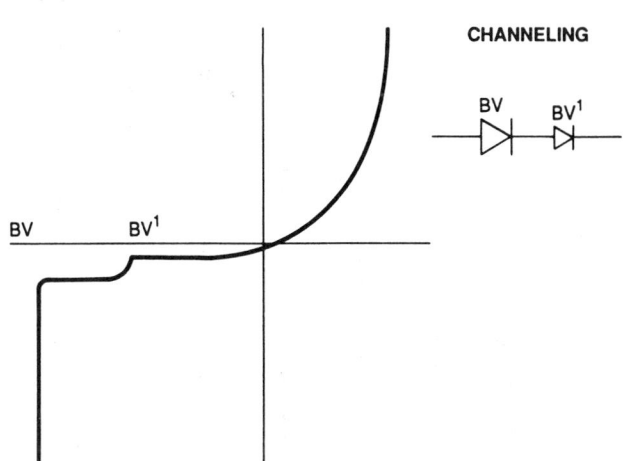

Fig. 3 During channeling, inversion of the surface results in abnormally high leakage.

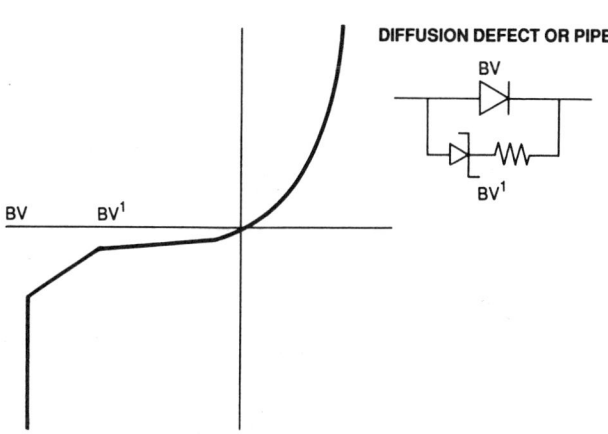

Fig. 4 A diffusion defect or pipe effectively parallels the function with a smaller device with a lower breakdown voltage.

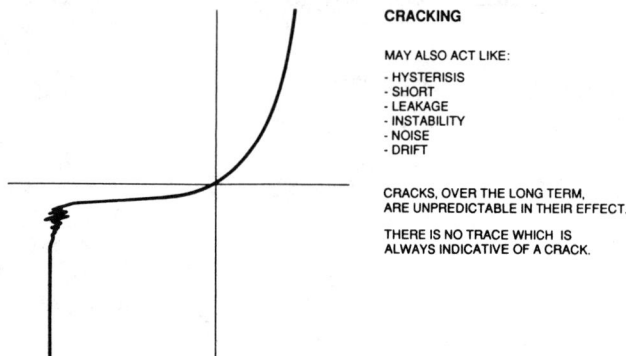

CRACKING

MAY ALSO ACT LIKE:

- HYSTERISIS
- SHORT
- LEAKAGE
- INSTABILITY
- NOISE
- DRIFT

CRACKS, OVER THE LONG TERM,
ARE UNPREDICTABLE IN THEIR EFFECT.

THERE IS NO TRACE WHICH IS
ALWAYS INDICATIVE OF A CRACK.

Fig. 5 A cracked die may imitate hysteresis, a short, leakage, instability, noise, or drift; while the above trace is typical, there is no trace which is always indicative of a cracked die.

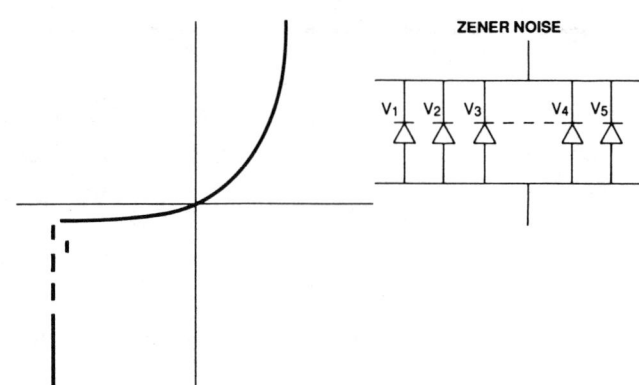

ZENER NOISE

Fig. 6 Zener noise can be represented as many small diodes in series, each with a slightly variable reverse breakdown voltage.

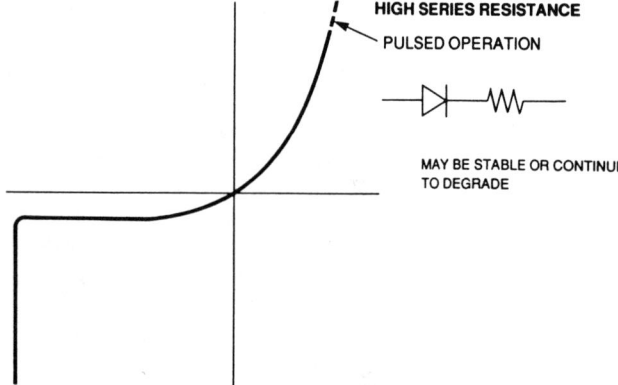

HIGH SERIES RESISTANCE

PULSED OPERATION

MAY BE STABLE OR CONTINUE
TO DEGRADE

Fig. 7 High series resistance can be stable or unstable

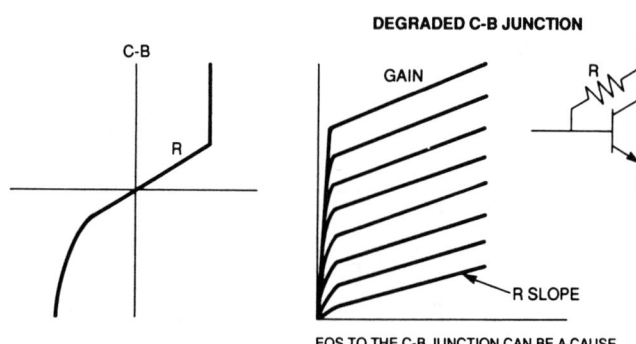

C-B

DEGRADED C-B JUNCTION

GAIN

R SLOPE

EOS TO THE C-B JUNCTION CAN BE A CAUSE

Fig. 8 A degraded C-B junction will offset the characteristic family by the amount of leakage

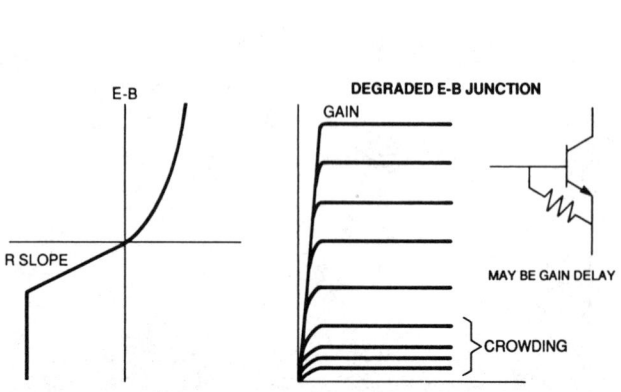

E-B

DEGRADED E-B JUNCTION

GAIN

R SLOPE

MAY BE GAIN DELAY

CROWDING

Fig. 9 A degraded E-B junction will decrease h_{FE}, result in poor carrier injection, current crowding, and high E-B reverse leakage.

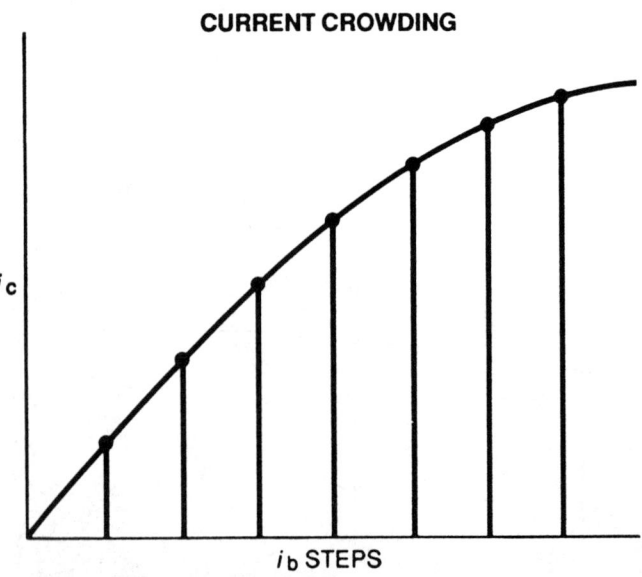

CURRENT CROWDING

i_c

i_b STEPS

Fig. 10 Detailed illustration of current crowding

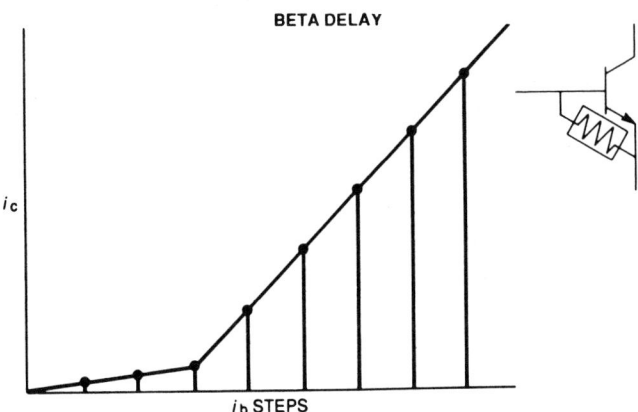

Fig. 11 Detailed illustration of beta or h_{FE} delay

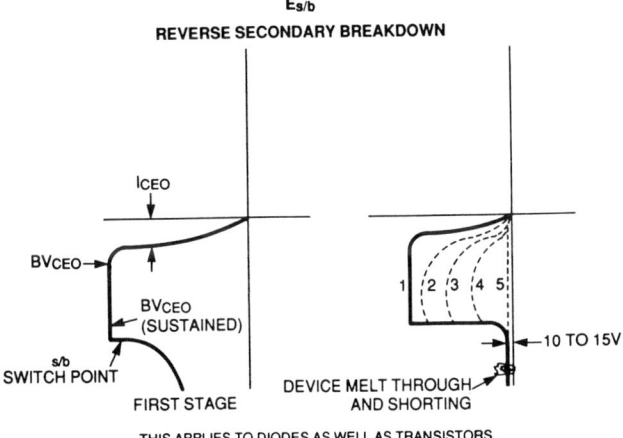

Fig. 12 Illustration of second breakdown progressing from stable BV_{CEO} (sus) condition

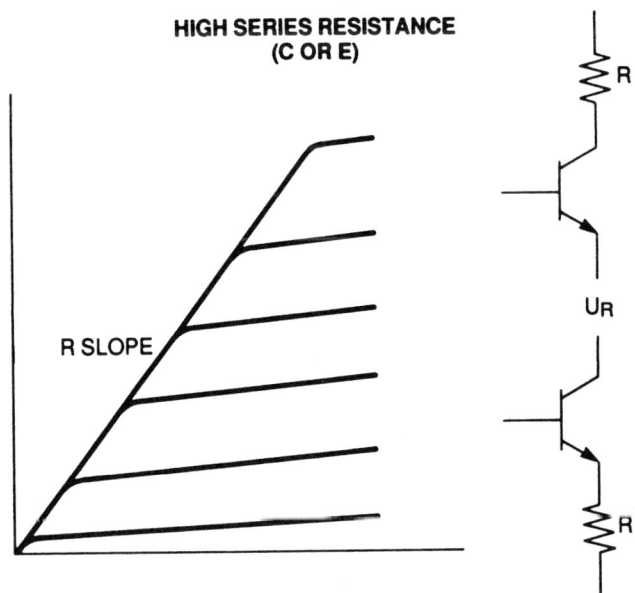

Fig. 13 Series resistance within a transistor collector or emitter can decrease the slope of the saturation curve.

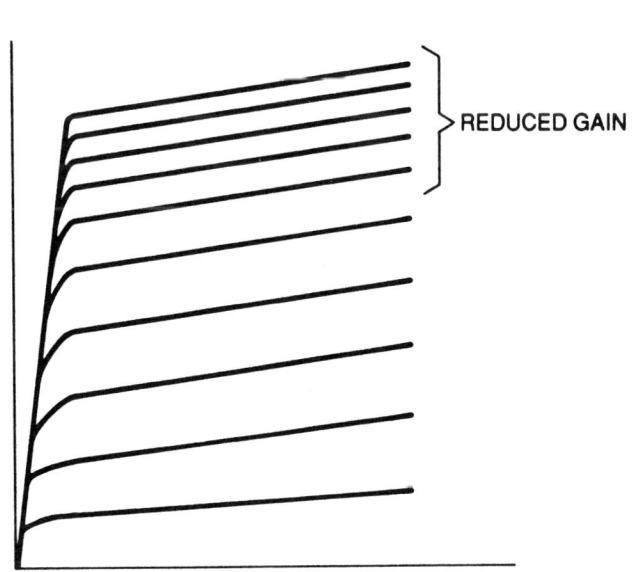

Fig. 14 Series resistance in the base region can compress the characteristic family at high current.

107

Electrical Failure Mode Characterization in CMOS ICs

Charles F. Hawkins and Jerry M. Soden
University of New Mexico and Sandia National Labs
Albuquerque, New Mexico

Introduction

Failure analysts must locate the defect site and electrically characterize the circuit failure mode. This requires reflexive knowledge of how CMOS elements respond to defects. This paper targets those for whom electronic depth may be lacking and who may have backgrounds in physics, chemical engineering, materials science, or biology. Transistor operation is reviewed and applied to electronic behavior in the presence of defects, such as bridges and opens. This paper cannot replace a course in electronics, but it puts that knowledge in perspective and provides direction.

Electronics spans a number of devices, their configurations, and properties. A challenge is to identify those electronic subjects essential for failure analysis. Transistor circuits with and without resistors are one object of our study. A theme is that failure analysts deal with relatively simple circuits, but deeply understand their operation. Logic circuits must be understood, but when a defective circuit gives the final electronic clues to a defect, they are typically analog. These electronic principles are developed and applied to an inexpensive CMOS failure analysis technique using power supply signature analysis.

MOSFET Operation and Three Bias States

Fig. 1 shows an n-channel transistor cross section with heavily n^+-doped drain and n^+-source regions and a p-well. When the gate voltage (V_G) is zero and the source and drain are grounded, only a few thermally generated free carriers exist in the p-well and the transistor is in the *off-state*. No drain current exists even when a voltage is across the drain-source.

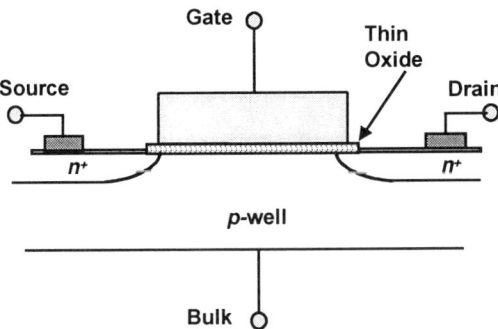

Fig. 1. n-channel transistor cross section.

When V_G is much larger than the drain (V_D) and source (V_S) voltages, then electrons (minority carriers) from the p-well are attracted to the thin oxide interface (hatched area, Fig. 2a). This second important bias state, called the *non-saturated state*, exists when a continuous electron inversion region connects the source and drain. If the drain voltage is positive, then the inverted free electrons drift in the electric field forming drain to source current. V_{GS} (gate to source voltage) has a minimum value necessary to sustain minority carrier inversion and that V_{GS} is defined as the gate threshold voltage V_T. The non-saturated and off-states exist in normal CMOS logic circuits when the clock pulse is off and all voltages have settled to their quiescent values.

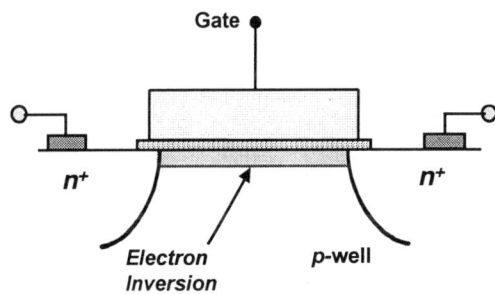

Fig. 2(a). Carrier inversion in non-saturated bias state.

The *saturated state* occurs when V_D is much larger than V_G and V_S. If the drain voltage in Fig. 2(a) is increased, then the difference between V_G and V_D diminishes in the oxide region near the drain. At some point, the local gate to drain voltage across the oxide drops below V_T ($V_{GD} < V_T$) and no inverted free carriers can exist near the drain. Since the n^+-doped drain has a positive voltage with respect to the p-well, a reversed bias pn junction exists between the drain and p-well. Fig. 2(b) shows this state and the fixed junction charges when $V_S = 0$, $V_{GS} > V_T$, and $V_D > V_G$.

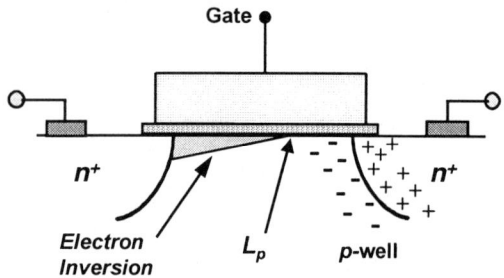

Fig. 2(b). Carrier inversion in saturated bias state.
($V_S = 0$, $V_{GS} > V_T$, and $V_D > V_G$)

The location where the *pn* junction touches the inversion layer is called the pinchoff point (L_P in Fig. 2b) and to the right of L_P is the pinchoff or depletion region. The existence of a pinchoff region puts the transistor in the saturated bias state. A high electric field exists between L_P and the drain with fixed positive charges in the drain and fixed negative charges in the substrate (*p*-well). When V_{DS} is increased, the junction widens and L_P moves to the left.

Importantly, the reverse biased drain-substrate *pn* junction does not prevent charge flow. Many inverted electrons in Fig. 2(b) are caught in the high electric depletion field and accelerated toward the drain. These electrons exit the drain terminal from the high impedance reverse biased *pn* junction and form I_D.

p-channel transistors use gate voltages that are negative with respect to the source to generate minority carrier (hole) inversion in an *n*-well. The negative gate voltage draws holes from the *n*-well to the thin oxide surface if the gate voltage amplitude is greater than the threshold voltage. The *n*- and *p*-channel transistors have opposite terminal voltage polarities for normal operation. The terminology and polarity conventions of both transistors are important and discussed later.

<u>Failure Analysis and Bias State</u>: Mobile electrons (or holes) that are accelerated across the saturated state depletion region cause impact ionization and photon emission that are readily seen in the drain region by photoemission microscopy [1]. Normal transistors are in the saturated state during most of their logic state transition (shown later) and emit photons. Normally transistors in the quiescent portion of the logic cycle are either in the nonsaturated or off-state so that no photons are emitted in the absence of a depletion region. However, bridge and open defects may hold one or more transistors in the saturated state where the presence of a light emitting transistor indicates proximity of a defect. The defect may be a bridge external to the transistor that sets up a saturated bias state or defects may lie in the transistor itself. Examples include gate shorts, soft *pn* junctions, or an open circuit on one of the terminals.

<u>Review</u>: MOSFETs have three bias or operating states: off, saturated, and non-saturated. CMOS digital ICs use all three states. The off- and non-saturated states exist when logic gates are in their quiescent period. These states set the high and low logic voltage levels through the power rails. The saturated state dominates when the transistors are switching. This state has analog voltage gain, a beneficial property that shortens switching time. The gate to drain voltage difference determines if a transistor is in the saturated or nonsaturated state. *p*- and *n*-channel transistors have similar operating mechanisms except the *n*-channel uses electrons as carriers and the *p*-channel uses holes.

MOSFET Terminal Characteristics

MOSFETs have four terminals: gate, drain, source, and substrate (or bulk) (Fig. 1) with terminal current and voltage relations well defined. These equations are a bit clumsy, but they must be learned (Table 1). Eq (1) relates I_D to V_{GS} when the transistor is in the saturated state. μ is the carrier mobility, ε is the dielectric constant of silicon, Tox is gate oxide thickness, W is gate width, L is gate (channel) length, and V_{Tn} is the *n*MOSFET threshold voltage. Eq (2) relates I_D to V_{GS} and V_{DS} for the nonsaturated state. (1) and (2) were developed for "long channel" transistors that may be approximated for L $\geq$ 0.5 μm. Models for short channel devices are more complex to account for field and charge interactions at the smaller dimensions. However, we still use (1) and (2) in modern failure analysis as the parameters are dominant and give good estimations without resort to computer calculations.

Table 1. *n*MOSFET Bias State Equations.

State	Equation	
Saturated:	$I_D = \dfrac{\mu\varepsilon}{2T_{ox}} \dfrac{W}{L} (V_{GS} - V_{Tn})^2$	(1)
Non-Saturated:	$I_D = \dfrac{\mu\varepsilon}{2T_{ox}} \dfrac{W}{L} [2(V_{GS} - V_{Tn})V_{DS} - V_{DS}^2]$	(2)
Off-State:	$I_D = 0$ for $V_{GS} < V_{Tn}$	(3)

(1) and (2) can be combined to model the terminal characteristics over a range of gate and drain voltages. Fig. 3 shows a typical I_D versus V_{DS} family of curves measured on a 0.35 um *n*-channel transistor. Each curve merges half of a parabola (nonsaturated region) with a flat portion (saturated region). Eq (2) applies to the parabola and Eq (1) to the flat line. A pinchoff point, indicated by the vertical bars, starts at the intersection of the two curves and exists for all of the saturated, flat line. We need the relation that defines the boundary of these two bias states, as without knowledge of the correct bias state, we cannot pick the proper model (Eq 1, 2, or 3).

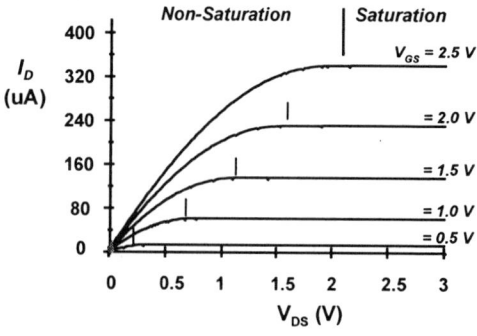

Fig. 3. I_D versus V_{DS} for a family of V_{GS} curves in an *n*MOSFET ($V_T = 0.4$ V).

Eq (2) is a parabola peaking at the boundary between the saturated and non-saturated regions. The bias condition at the boundary is found by differentiating I_D with respect to V_{DS} in Eq (2), setting the result to zero, and solving for V_{GS}. Eq (4) defines these voltage conditions that put the transistor at the boundary of the two states $V_T = V_{Tn}$.

$$V_{GS} = V_{DS} + V_T \qquad (4)$$

A common use of (4) is to define saturated and non-saturated states by

Saturated State: $\qquad V_{GS} < V_{DS} + V_T \qquad (4a)$

Non-Saturated State: $\qquad V_{GS} > V_{DS} + V_T \qquad (4b)$

Bias State Examples: Two transistors are shown in Fig. 4a,b with terminal voltages. An exercise is to use Eqs (3,4) and find the bias state.

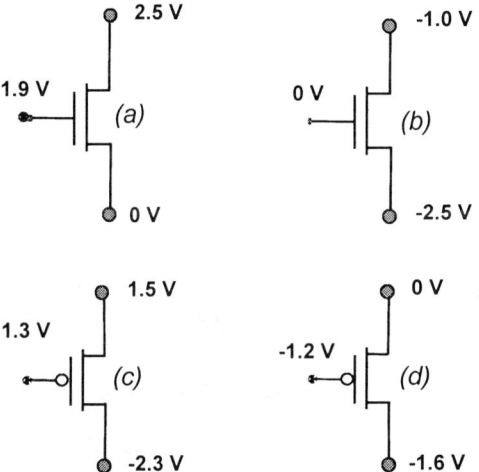

Fig. 4. (a-b) Transistor bias examples. ($V_T = 0.4$ V)

Fig. 4(a) is in the saturated state since $V_{GS} = 1.9 < 2.5 + 0.4$. Fig. 4(b) is in non-saturated state since $V_{GS} = (0 - (-2.5)) > (-1.0 - (-2.5)) + 0.4$.

pMOSFETs have similar equations, but the signal polarities are negative and I_D enters the drain terminal. The threshold voltage, V_{Tp}, is negative and hole mobility is typically less than half electron mobility. Table 2 shows these pMOSFET relations (Eq 7-9).

Table 2. pMOSFET Bias State Equations.		
State	Equation	
Saturated: $I_D = \dfrac{\mu \varepsilon}{2T_{ox}} \dfrac{W}{L} (V_{SG} + V_{Tp})^2$		(5)
Non-Saturated: $I_D = \dfrac{\mu \varepsilon}{2T_{ox}} \dfrac{W}{L} [2(V_{SG} + V_{Tp}) V_{SD} - V_{SD}^2]$		(6)
Off-State: $\quad I_D = 0$ for $V_{SG} < V_{Tp}$		(7)
Sat. State: $\quad V_{SG} < V_{SD} - V_{Tp}$		(8)
Non-Sat. State: $\quad V_{SG} > V_{SD} - V_{Tp}$		(9)

Bias State Examples: Two transistors are shown in Fig. 4c,d with terminal voltages. The source terminal is more positive than the drain terminal (upper terminal). An exercise is to use Eqs (5-7) and find the bias state.

Fig. 4(c) is in the off-state since $V_{GS} = |1.3 - 1.5| < V_T = |-0.4$ V|. Fig. 4(d) lies on the boundary of both saturated and non-saturated state, therefore Eq (5) or (6) may be used.

Analysis of a Complete Circuit

Transistors don't operate in isolation. They typically require a power supply (V_{DD}), ground (V_{SS}), an input signal, and a load. The load can be a resistor (Fig. 5) or another transistor. Eqs (1-3) allow analysis of node voltages and drain current (I_D).

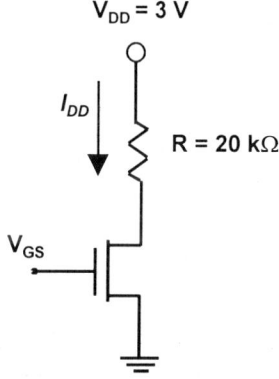

Fig. 5. nMOSFET 20 kΩ load circuit.

Example (Fig. 5): Calculate I_D, I_S, V_{DS}, power dissipated by circuit (P_{DD}), and by transistor (P_T) if $V_T = 0.4$ V, $V_{IN} = 1.0$ V, and the conduction constant

$$K = \frac{\mu \varepsilon}{2T_{ox}} \frac{W}{L} = 60 \ \mu A/V^2$$

Solution: We don't know which bias state exists so we will try the saturated state Eq (1) and see if the solution is consistent with Eq (4a).

$$I_D = K[V_{GS} - V_T]^2$$

$$= 60 \ \frac{\mu A}{V^2} [1 - 0.4]^2 = 21.6 \ \mu A$$

Kirchhoff's Voltage Law gives

$$V_{DD} = I_D R_D + V_{DS} \qquad (10)$$

$$V_{DS} = 3 - (21.6 \text{ uA}) (20 \text{ k}\Omega) = 2.568 \text{ V}$$

from Eq (4b) $\qquad V_{GS} < V_{DS} + V_T$

$$1.0 < 2.568 + 0.4$$

The transistor is in saturation so the first guess was correct and $I_S = I_D = 21.6 \ \mu A$. The power to the circuit is $P_{DD} = V_{DD}$

I_{DD} = (3 V) (21.6 μA) = 64.8 μW. Transistor power is $P_T = V_{DS}$
I_D = (2.568 V) (21.6 μA) = 55.5 μW.

What if the transistor was in the non-saturated state and we repeat the procedure assuming that the transistor is in saturation? Set V_{GS} = 2.5 V and use

$$I_D = K[V_{GS} - V_T]^2$$

$$= 60 \text{ μA } [2.5 - 0.4]^2 = 264.6 \text{ μA}$$
$$V^2$$

$$V_{DS} = 3 - (264.6 \text{ μA}) (20 \text{ kΩ}) = -2.29 \text{ V}$$

This is an error since $V_{GS} > V_{DS} + V_T$ violates our initial saturated state assumption. Also, a negative potential is not possible in this circuit. We then start again using the non-saturated Eq (2). Since Eq (2) has two unknowns (I_D, V_{DS}) we need another equation. Kirchhoff's Voltage Law states that the sum of the voltage drops in a loop in zero. Therefore

$$V_{DD} = I_D R_D + V_{DS}$$

and

$$I_D = [V_{DD} - V_{DS}]/R_D \tag{11}$$

Combine (11) with (2)

$$I_D = K[2(V_{GS} - V_T) V_{DS} - V_{DS}^2] \tag{2}$$

and solve the quadratic Eq for two values V_{DS} = 4.47 V, 0.559 V. The valid solution is V_{DS} = 0.559 V since 4.47 V violates our non-saturated state assumption and is also larger than V_{DD}. Eq (4b) is satisfied: 2.5 > 0.559 + 0.4 confirming the non-saturated state assumption. I_D is calculated by either (2) or (11) above as I_D = 122.1 μA.

This exercise shows how to use terminal equations (1-3) to calculate voltages and currents. If V_{GS} = 1.0 V, then we expect the saturated transistor to emit photons. No photons are emitted for V_{GS} = 2.5 V even though the drain current is much larger in the non-saturated state.

CMOS Inverter

The inverter is a small CMOS logic gate having one *n*- and one *p*-channel transistor (Fig. 6a). Fig. 6(b) illustrates the complementary action of the two transistors. When V_{IN} = 0 V, the *n*-channel is off and the *p*-channel with 5 V across the source to gate is driven hard into its non-saturated state. It is a low resistance switch so V_{OUT} = 5 V (logic high). As V_{IN} rises, the *n*-channel will turn on at V_{IN} = V_{Tn} going from off to the saturated bias state. The gate voltage is above threshold, but much less than the drain voltage. The *p*-channel is in its saturated state and drain current is drawn through both transistors (dotted line in Fig. 6b). As V_{IN} rises, the drive voltage to the *p*MOSFET gate (V_{SG}) drops and the *p*MOSFET current drive weakens. Near the midpoint, the *n*- and *p*-channel current drive strengths are about equal and maximum I_{DD} exists. When $|V_{DD} - V_{IN}| < |V_{Tp}|$, the *p*-channel is off, the *n*-MOSFET is driven hard into non-saturation, and I_{DD} and V_{OUT} are zero (logic low).

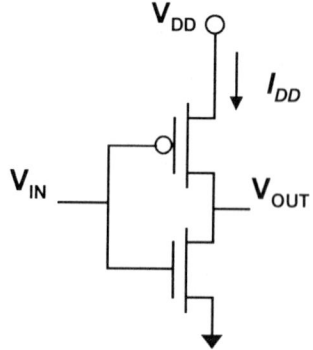

Fig. 6(a). Inverter circuit.

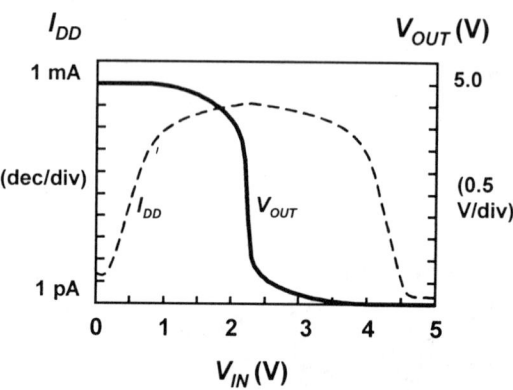

Fig. 6(b) Inverter voltage and current transfer curves.

A detailed analysis of the transfer curve using Eq s (1,2) shows that both transistors are in saturation for about the middle 50% of the transition (the straight-line segment). Each transistor is in saturation for about 75% of the transition. You can verify this by plotting Eq (4) overlaid on Fig. 6(b) for both *n*-channel (V_{Tn} = 0.8 V) and *p*-channel (V_{Tp} = - 0.8V) transistors. The analog voltage gain in the transition region can be estimated by measuring $\Delta V_{OUT} / \Delta V_{IN} \cong -8.0$. The negative gain means that as V_{IN} increases, V_{OUT} goes down, but the amplitude ratios are 8. Photon emission occurs during all of the transition since one or both transistors emit photons.

The power supply current (I_{DD}) and the voltage (V_{OUT}) shape should be committed to memory (Fig. 6). Defects alter the shape of I_{DD} or V_{OUT} versus V_{IN}. The power supply current I_{DD} is plotted as log(I_{DD}) and is near zero at the quiescent logic states. The drain current is junction leakage that is in femtoamps (fA) for small, single transistors. Deep submicron transistors have additional leakage mechanisms that complicate test and failure analysis [2].

Test Methods

The electronic response of a defective IC implies knowledge of the test stimulus that led to a failure. There are several test methods [3]

1. Functional testing has at least three meanings: (a) stimulation of the IC in a way that replicates how the customer will use the part, (b) applying the logic truth table to the circuit, (c) using the primary inputs pins to an IC in contrast to using a scan chain test port. Functional testing can not replicate customer use. For example, a 1K SRAM has 2^{1024} possible

112

states. If these patterns are run at 100 MHz then the test of full function will take about 10^{293} years.

2. Stuck-at-fault testing (SAF) uses a test abstraction that assumes all failures happen because one of the signal nodes is clamped to either the power or ground rail. SAF patterns apply voltages that represent Boolean logic patterns. A correct response for each pattern is stored in the tester and compared with the result measured on the IC.

3. At-Speed Testing is the third major voltage-based test method. The F_{MAX} test applies many functional patterns to the IC and finds the maximum functional clock rate. Another test generates delay fault patterns that target either logic gates or signal paths and measures the propagation delay through these elements.

4. I_{DDQ} testing measures the power supply current of a CMOS circuit during the quiescent period of the clock cycle. I_{DDQ} in a typical CMOS IC is in tens or hundreds of nanoamps. Deep submicron transistors elevate normal I_{DDQ} even more. Most defects in CMOS ICs elevate I_{DDQ} above its normal value.

5. Other test types include I/O pin current and voltage levels, set up and hold time measurements, measurement of power supply current during the transient period of the clock, and testing at low power supply voltages.

Failure analysts work with ICs that fail one or more of the tests described above. These assorted tests fall into three categories: (1) voltage-based tests of logic function at slow clock rate (scan stuck-at fault testing), (2) at-speed voltage-based testing, and (3) parametric tests (I_{DDQ}, pin I/O, etc.). Comparative test methods data show that I_{DDQ} is the most sensitive test method for CMOS IC defect detection, but a good test strategy uses current- and voltage-based testing [4,5].

Electronics Properties of CMOS Defects

CMOS IC defects have several types with many electronic patterns. Defects are classed as bridges, opens, or parametric delay types. Their properties and response to voltage- and current-based testing are described.

Bridging defects are unintentional connections between signal interconnect or power lines. Bridges can be tiny connections (Fig. 7) or may cover several interconnect lines. They also occur within the structure of transistors, such as with gate oxide shorts or soft *pn* junction breakdowns.

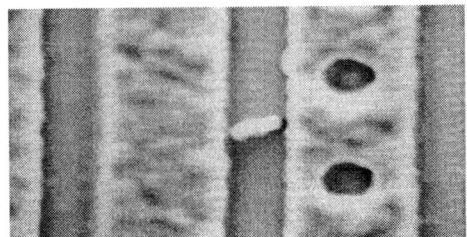

Fig. 7. SEM photo of bridge defect between two metal bus lines.

Fig. 8 shows a bridge defect between two logic gates. The effect of a bridge on functionality depends on bridge location and its resistance. If the resistance is large (≥ 5 kΩ), then little effect is seen on circuit functionality. A circuit will fail only when the nodes of the bridge are driven to opposite logic states and the resistance is sufficiently small. When failure occurs, one bridge node is correct and the other faulty. The current drive to one node overpowers the other.

Fig. 8. Bridge defect across output signal nodes of two inverters.

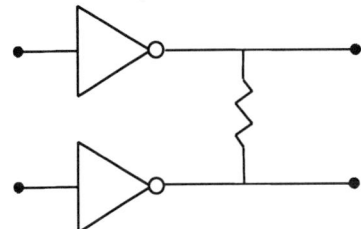

Fig. 8. Bridge defect across output signal nodes of two inverters.

The dominating property of bridge defects is called *critical resistance* [6]. Critical resistance (R_{crit}) is the minimum value above which the signal nodes are functionally correct. Imagine a 1 GΩ bridge between the output of two logic gates (Fig. 8). The effect on the signal node voltage levels is trivial and if this resistance was reduced to 1 MΩ or 10 kΩ, the effect is still small. We typically do not see circuit failures until most bridge defects get below 2 kΩ. This is surprisingly low, but verified with simulations and failure analysis.

Fig, 9 shows a buffered 2NAND gate in which variable gate to drain resistance's (dotted line) were simulated for the *n*-channel transistor. Fig. 10 shows the DC transfer curves for several gate-drain bridge resistances in an *n*-channel transistor showing that the circuit is functionally correct above about 1 kΩ ($R_{crit} \cong 1$ kΩ). The exact value of R_{crit} depends on the relative current driving power of the pull-up and pull-down transistors contending at each end of the bridge. When a stronger current drive transistor dominates a weaker one then R_{crit} is large. R_{crit} goes to zero when the pull-up and pull-down strengths are equal. A low value of R_{crit} means that voltage-based tests are weakened with respect to bridge defect detection.

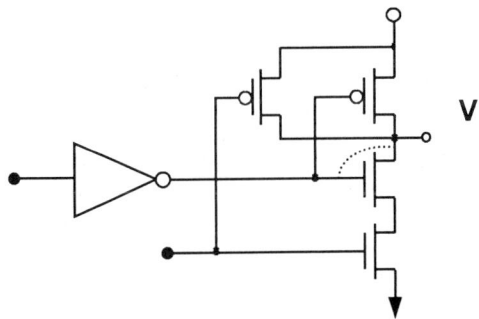

Fig. 9. Buffered 2NAND test circuit logic response to a gate-drain defect resistance (dotted line) [3].

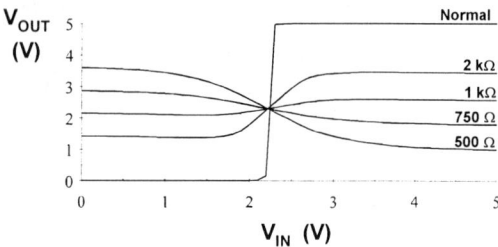

Fig. 10. Transfer curves of 2NAND when R_{def} across gate to drain is varied (Fig. 9) [3].

This R_{crit} effect explains why many quite dramatic bridge defects do not cause circuit failure. It is a reason why voltage-based tests such as, functional, stuck-at, or delay fault are weak in bridge defect detection. The I_{DDQ} test does however have sensitive detection capability for bridging defects. Fig. 11 shows mA level I_{DDQ} elevation for the same circuit as Fig. 9. I_{DDQ} values as low as 1 to 10 μA are detectable at test.

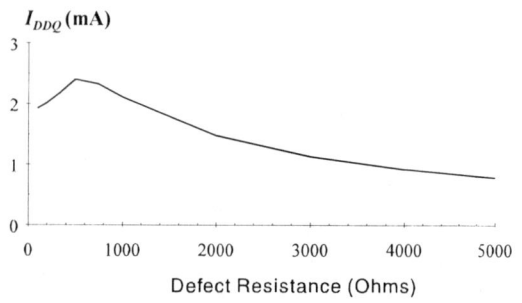

Fig. 11. I_{DDQ} response to gate-drain bridge defect (Fig. 9) [3].

Any bridge defect with at least one node tied to a signal line shows the critical resistance effect. Rail-to-rail bridges do not, but show a constant I_{DDQ} elevation for all test vectors.

Gate oxide shorts are defective electrical paths between the gate material and anything under the thin oxide [6-8]. A gate short is not a zero ohm structure. Gate shorts may have linear or nonlinear I-V properties. The inverter cross-section in Fig. 12 illustrates several forms of gate shorts. A connection in the n-channel transistor from n-doped polysilicon gate to n^+-drain or n^+-source creates a parasitic linear resistor (A,B). An n-doped gate to p-well connection creates a diode (C) that when powered up with positive logic on the gate forms a parasitic saturated bias state MOSFET whose gate and drain are connected. A connection between the source or drain of the p-channel to their gate also creates a diode (D,E). The n-doped gate to n-well connection creates a linear resistor that forms a base resistor of a parasitic pnp bipolar transistor when power is applied (F). Recent deep submicron pMOSFETs use a p-doped (boron) polysilicon gate. This will alter the expected electrical properties of gate shorts in pMOSFETs. We would expect that gate-drain and gate-source shorts to be Ohmic and gate-n-well shorts to form a parasitic p-channel transistor with gate tied to drain. Test detection of gate shorts is similar to linear bridges - voltage based tests are weak and I_{DDQ} is the only test that guarantees detection [6-8].

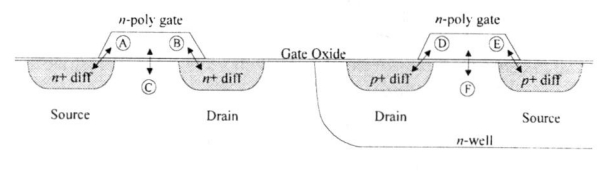

Fig. 12. Inverter cross section showing locations of different gate shorts.

Summary of Bridge Defect Electronics: A bridge defect with impedance above critical does not cause functional failure. It weakens node voltages, elevates I_{DDQ} and is often a reliability risk. The affect on propagation delay is negligible until the bridge resistance approaches the critical value. Bridge defects $< R_{crit}$ will show at least one signal node in error.

Open Defects have several forms and their electronic behavior is much more diverse than bridges. Defect location and physical dimensions are important open circuit defect variables. Six open defect behavior classes are [3]

1. Transistor with missing gate contact. Fig. 13(a) shows a test structure and its transfer function (Fig. 13(b)). The circuit is functional, but has a weak high voltage and an even weaker low voltage. I_{DDQ} is elevated in one logic state. The capacitive coupling between drain-gate and gate-source forms a capacitive voltage divider. When the drain voltage is sufficiently high, then the capacitor divider allows the gate voltage to exceed threshold and the transistor conducts. When the input voltage is high, the p-channel shuts off. The n-channel transistor stays on draining charge from the load capacitance until V_G is no longer above threshold. The output node is then a weak logic low level in a high impedance or floating state as shown by the horizontal lines to the right in Fig. 13(b).

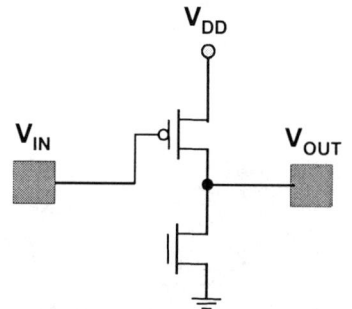

Fig. 13(a). An open defect (missing contact) at n-MOSFET gate [3].

114

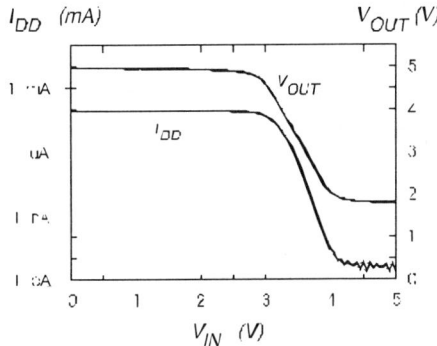

Fig. 13(b). Transfer curves for open gate contact [3].

2, 3. An interconnect break affecting both complementary transistors (Fig. 14). This open circuit creates a high impedance or floating node that acquires a steady state voltage. That voltage is dependent upon local topography, especially parasitic coupling capacitance. The isolated gate can float to either rail potential or any voltage in between allowing two possible behavior modes. If it floats to a rail, then that input node has a stuck-at behavior and I_{DDQ} is not elevated. If the floating node acquires an intermediate voltage greater than the p- and n-channel transistor thresholds, then both transistors are permanently on, I_{DDQ} is elevated, and the output is a weak stuck-at voltage. That form of open is detectable with a stuck-at fault or I_{DDQ} test.

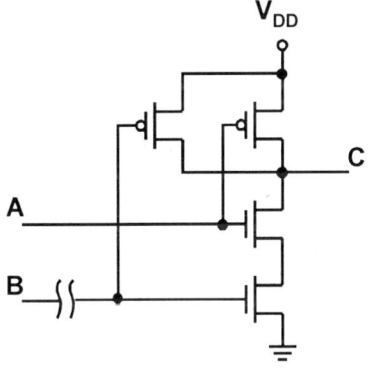

Fig. 14. An open defect to the logic gate input.

4. Open defect behavior in sequential circuits. Fig. 15 shows a master-slave flip-flop with an "X" denoting possible open defect sites. Analysis shows that any of these open defects will cause behavior consistent with the three classes described above. An open in one of the CMOS transmission gate (T-gate) transistor lines still allows functionality with the gate acting as a single pass transistor. The responses of these sequential open defects are either I_{DDQ} elevation only, functional fail only, or both.

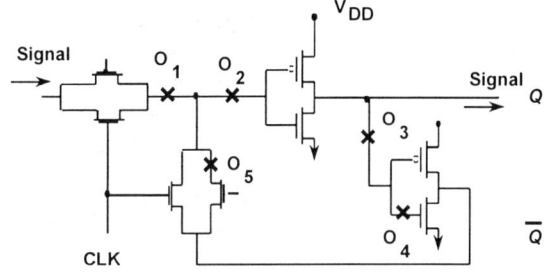

Fig. 15. Open defects in a sequential circuit.

5. CMOS Stuck-Open (Memory) Defect. The fifth open defect class is peculiar to CMOS ICs and occurs when an open circuit happens in the drain or source of a transistor [9]. This is also called the CMOS memory defect. Fig. 16 shows such a defect and its truth table response for a 2NAND gate. The first vector (AB = 00) sets the output to a logic high by turning on p-channel transistor A and turning off both n-channel transistors. The second vector (AB = 01) drives the output high through the good p-channel pull-up. The third test vector (AB = 10) attempts to turn on the defective p-channel pull-up, but no charge can pass since the drain has an open circuit defect. The output node goes into a high impedance state and might be indeterminate except that the previous logic state was a high. The load capacitance holds this high value and the third test vector is read correctly. The fourth vector pulls the output node low and the result is dramatic. No error occurred in the truth table for such a flagrant defect!

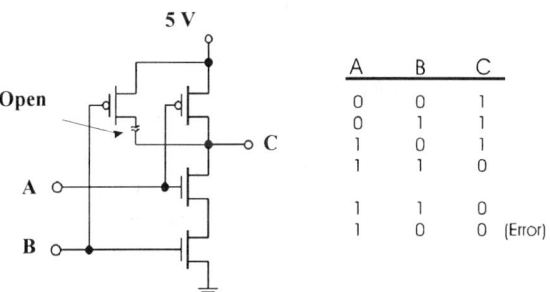

Fig. 16. Open memory defect in a 2NAND gate and its truth table.

The defect is detected when the sequence of vectors is changed. If AB = 11 is followed by AB = 10, then the AB = 10 vector is read at the output as a fail state zero (the previous logic state). Stuck-open defects occur and are a difficult failure analysis challenge. The IC sends conflicting information due to the memory response of that defective node. I_{DDQ} is often elevated due to design contentions caused by the unintended logic state or the node drifts with a 1-3 s time constant at room temperature and can turn on complementary pairs of load transistors [9].

6. Crack in an interconnect line (Fig. 17). This defect supports circuit functionality, especially for very narrow cracks [10]. A narrow crack allows electron tunneling across the barrier and ICs with this defect type can operate in the tens or hundreds of

MHz. They are called tunneling opens. These defects also occur in vias and contacts. Cold temperature lowers F_{MAX} since metal contraction widens the crack reducing the tunneling efficiency. Conversely, circuits with these defects run faster at hot temperatures. These unusual frequency-temperature properties are symptomatic of an interconnect, via, or contact crack.

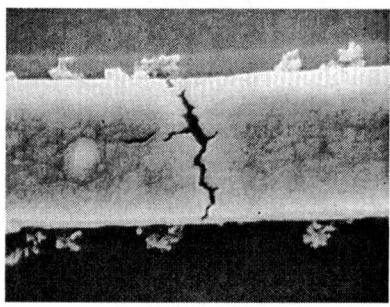

Fig. 17. Metal crack [10].

Summary of Open Defect Electronics: Six different behavior patterns were described that are dependent on the open defect location and size. A comprehensive test for open defects includes both voltage and current-based tests. Inattention to these details can frustrate failure analysis.

Parametric Delay Defects are the third and most difficult class of defects to detect. These include vias and contacts with resistance higher than normal. This defect often does not elevate I_{DDQ} and added signal timing delay can be in picoseconds. Other such defects can be abnormal V_T or W/L ratio. These defects pose reliability risk. More research is needed to develop tests that are sensitive to parametric delay defects.

Interconnect Defect Resistance Defects can alter resistance of the interconnect, however, the effects can be deceiving. Fig. 18 sketches two bus lines with a small bridge sliver connecting them. The size and connectivity of the defect will affect the critical resistance.

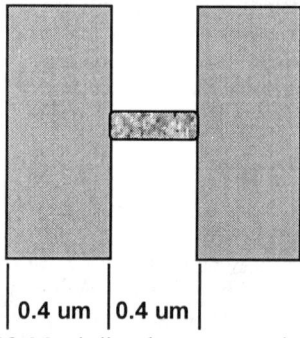

0.4 um | 0.4 um

Fig. 18. Metal sliver between two signal lines.

The resistance of the tiny sliver can be estimated. If we assume an aluminum sliver, then the resistivity $\rho = 3.0$ uΩ cm. If the dimensions of the sliver are: W = 0.2 μm, L = 0.4 μm, and thickness t = 0.4 μm, then the resistance is

$$R = \frac{\rho L}{W t} = 150 \text{ m}\Omega \qquad (10)$$

This value is well below all critical resistances. However, failure analysis shows that these types of defects have from tens to hundreds of Ω [11]. Possible reasons are that the bonding of the sliver to the bus lines may not be strong or the material may not be Al, but chrome or other foreign particulate having higher resistivity.

Another unusual metal property concerns what are called mousebits or voids. Fig. 19 sketches an idealized form of this defect. The question is the effect on line resistance when this void occurs.

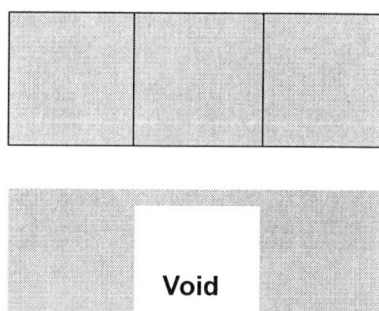

Void

Fig. 19. Metal mousebit (void).

This will be calculated using the sheet resistance ($R_\square$) of a metal line that is taken from (12) by setting W = L (one square)

$$R/t = \rho/t = R_\square = 75 \text{ m}\Omega /\square \qquad (13)$$

The stripe resistance is obtained by counting the squares (L/A) and inserting into (14)

$$R_{strip} = R_\square (L/A) \qquad (14)$$

If the void is 90%, then the number of squares in the stripe is 10 or $R_{strip} = 750$ mΩ. If the original line was 100 μm long and 0.4 μm wide, then the resistance increase was a negligible 18.75 Ω to 19.43 Ω (9%). Fig. 20 plots the increase in resistance of the stripe versus fraction of mousebit. The resistance increases dramatically beyond about 90% voiding. This observation is useful when visually looking at mousebits and predicting the impact on failure. These small resistance increases due to mousebits have a negligible affect on RC time constant of that line. Mousebits are difficult to detect, but pose an electromigration failure risk due to the increased current density in the stripe.

116

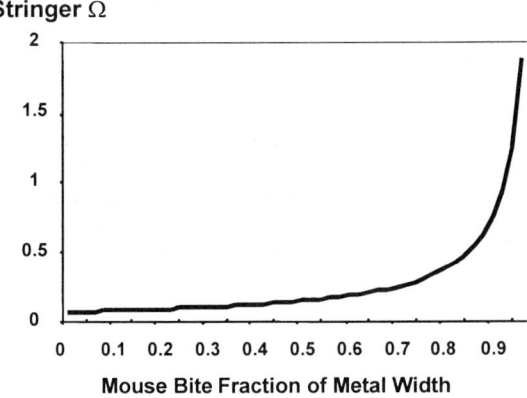

Fig. 20. Stringer effect on line resistance.

Signature Analysis with I_{DDQ} versus V_{DD}

CMOS IC designs with their low quiescent currents are a unique opportunity for failure analysis. Sandia National Labs has used I_{DDQ} versus V_{DD} signature analysis curves since the late 1970's to perform a rapid early assessment of failure mode. Fig. 21 shows the technique that sweeps the power supply pin voltage (V_{DD}) and monitors I_{DDQ}. The IC is first put in a logic state that draws abnormal I_{DDQ}. The input pins at logic 0 are tied to ground and the logic high pins are tied to V_{DD}. It is important that input high pins are at the same voltage as V_{DD} or there is risk that input protection circuits can be damaged. The measurement setup is simple and conducted essentially under DC conditions.

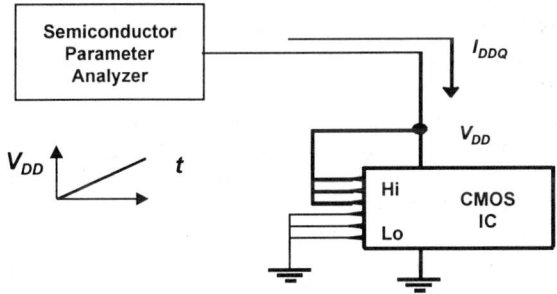

Fig. 21. Measurement setup for I_{DDQ} versus V_{DD} defect signatures.

I_{DDQ} versus V_{DD} signatures can rapidly indicate clues to the nature of the defect. The technique doesn't pin point defect location, but can often distinguish bridges from opens and power rail shorts from signal node shorts. A bridge defect with at least one end tied to a signal node will not draw current until V_{DD} rises above transistor threshold, V_T. In contrast, rail to rail bridge defects show conduction from the initiation of power. Figs 22 and 23 illustrate these two types of bridging defect signatures found in a microprocessor [12]. Fig. 22 shows a 1.9 MΩ rail to rail short whose I-V curve reflects conduction of the defect without contributions from driving transistors. This defect will also show a line with the same slope if V_{DD} is driven slightly negative.

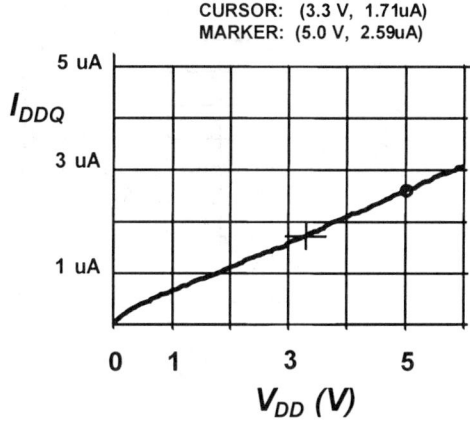

Fig. 22. A bridge defect between V_{DD} and V_{SS}.

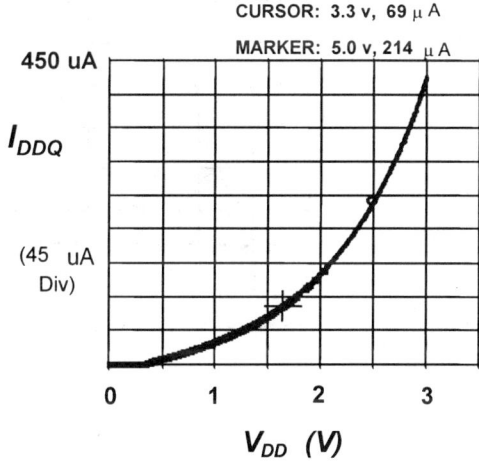

Fig. 23. Signature of n-channel gate oxide short.

Fig. 23 shows an n-channel transistor gate oxide short I-V signature [12]. Conduction doesn't begin until V_{DD} exceeds V_T. The signature now includes contributions from the defect, the defective transistor, and the driving transistor. Gate to substrate shorts in n-channel transistors typically show a pronounced parabolic curve [7,8]. The gate short forms a parasitic MOSFET transistor whose gate and drain are connected. This connection places the device in its saturated state whose I_D and V_{GS} response is approximate a square law (Eq 1).

Several classes of open defects were described and their signatures can take different forms. Often, an open defect signature shows a time instability when the defective IC is powered. Fig. 24 shows this type of behavior. Capacitive dividers may delay turn on of the transistor as shown at $\cong$ 1.5 V. There is need for more research analyzing open defect signatures.

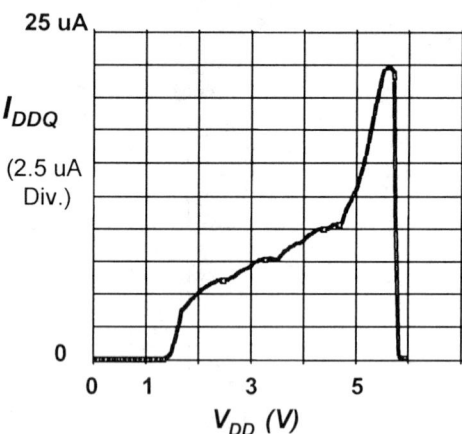

Fig. 24. An open defect response.

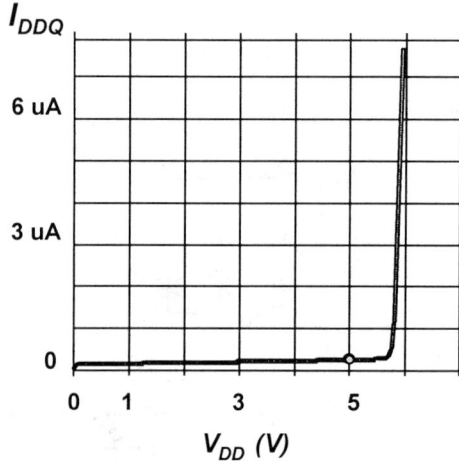

Fig. 25. Signature of *pn* junction early breakdown.

Soft breakdown of *pn* junctions also has a distinct signature (Fig. 25). This defect and most gate oxide shorts are strongly photon emitting.

The curvature of the signature can direct a course of action. Curved signatures usually have light emission occurring either at the defect or in the transistor driving the defect. Straight-line responses have not been found to emit light [1]. A transistor in its linear (non-saturated) state will not emit photons. MOSFETs in their saturated state emit considerable light in the drain depletion region as do parasitic bipolar transistors in linear bias. Wilson describes in this text a similar signature technique using curve tracer responses on I/O pins as opposed to the V_DD pin applications described here [13].

Summary

Failure analysts need skills in relating transistor currents and node voltages to the three bias states: off, saturated, and non-saturated. The CMOS inverter is the most fundamental logic gate and its current and voltage transfer curve properties must be understood in detail. The electronic behavior of opens, shorts, and parametric delay defects was described and it is essential for understanding the symptoms of a failing IC. A final application was given using I_{DD} versus V_{DD} signatures to identify types of defects. More detailed information on transistor theory and design may be found in electronic textbooks such as found in [14,15].

Acknowledgement

This work was performed at Sandia National Laboratories, a Lockheed Martin Company, for the U.S. Department of Energy under contract number DE-AC04-94AL85000.

References

1. C. F. Hawkins, J. M. Soden, E. I. Cole Jr., and E. S. Snyder, The Use of Light Emission in Failure Analysis of CMOS ICs, *Int. Symp. for Testing and Failure Analysis (ISTFA)*, Oct. 1990

2. A. Keshavarzi, K. Roy, and C. Hawkins, Intrinsic Leakage in Low Power Deep Submicron CMOS ICs, *Int. Test Conf.*, pp. 146-155, Nov. 1997

3. C. Hawkins, J. Soden, A. Righter, and J. Ferguson, Defect Classes – An Overdue Paradigm for CMOS IC Testing, *Int. Test Conf.*, pp. 413-424, Oct. 1994

4. P. Maxwell, R. Aitken, K. Kollitz, and A. Brown, I_{DDQ} and AC Scan: The War on Unmodelled Defects, *Int. Test Conf.*, pp. 250-258, Oct. 1996

5. P. Nigh, W. Needham, K. Butler, P. Maxwell, R. Aitken, and W. Maly, What is an Optimal Test Mix? A Discussion of the Sematech Methods Experiment, *Int. Test Conf.*, pp. 1037, Oct. 1997

6. R. Rodriguez-Montanes, J. Segura, V. Champac, J. Figueras, and A. Rubio, Current vs. Logic Testing of Gate Oxide Short, Floating Gate, and Bridging Failures in CMOS, *Int. Test Conf.*, pp. 510-519, Oct. 1991

7. C. F. Hawkins and J. M. Soden, Electrical Characteristics and Testing Considerations for Gate Oxide Shorts in CMOS ICs, *Int. Test Conf.*, pp. 544-555, Nov. 1985

8. J. Segura, C. DeBenito, A. Rubio, and C.F. Hawkins, A Detailed Analysis and Electrical Modeling of Gate Oxide Shorts in MOS Transistors, *J. of Electronic Testing: Theory and Applications (JETTA)*, pp. 229-239, (1996)

9. J. M. Soden, R. K. Treece, M. R. Taylor, and C. F. Hawkins, CMOS IC Stuck-Open Fault Electrical Effects and Design Considerations, *Int. Test Conf.*, pp. 423-430, August, 1989

10. C. L. Henderson, J. M. Soden, and C. F. Hawkins, The Behavior and Testing Implications of CMOS IC Open Circuits, *Int. Test Conf.*, pp. 302-310, Nashville, TN, Oct. 1991

11. P. Nigh, et al., Failure Analysis and IDDq-Only Failures from the SEMATECH Test Methods Experiment, *Int. Test Conf.*, pp. 43-52, Washington, D.C., Oct. 1998

12. T. Miller, J.M. Soden, and C.F. Hawkins, Diagnosis, Analysis, and Comparison of 80386EX I_{DDQ} and Functional Test Failures, I_{DDQ} *Workshop*, pp. 66-68, Oct. 1995

13. D. Wilson, Curve Tracer Data Interpretation for Failure Analysis, pp. 45-57, in *ISTFA Electronic Failure Analysis*. ed. S. Pabbisetty and R. Ross, ASM Int. Pub., 1998

14. D. Neamen, *Electronic Circuit Analysis and Design,* Richard D. Irwin Pub., Chicago (McGraw-Hill), 1996

15. R.C. Jaeger, *Microelectronic Design*, McGraw-Hill, 1997

Bipolar Device and Analog Circuit Characterization

Steve Frank
Texas Instruments, Incorporated
Dallas, Texas

INTRODUCTION

Characterization of microelectronic circuits for fault isolation and characterization of semiconductor devices for failure mechanism identification is an important part of a failure analyst's job. This paper covers the basics of bipolar device and analog circuit characterization with an emphasis on topics useful for failure analysis. The breadth of the subject matter does not allow for an in-depth treatment of every subject relating to bipolar device and analog circuit characterization. However, it is hoped that this paper gives a good introduction for the failure analyst new to the subject. The paper is divided into two sections. The first section gives a quick overview of characterization of bipolar semiconductor devices – namely PN junction diodes and bipolar junction transistors (BJTs). The second section discusses the characterization of analog circuits. These circuits are divided into five basic analog circuit blocks:

- Simple transistor circuits
- Current sources
- Voltage references
- Voltage regulators
- Op-amps

For each of these analog building blocks a simple description of it's operation will be given, the DC characterization of each is outlined, and for many, possible failure modes are discussed.

BIPOLAR DEVICE CHARACTERIZATION

In this section we briefly look at the characterization of bipolar semiconductor devices. It will by no means a comprehensive treatment of semiconductor characterization,

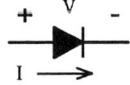

Figure 1. Schematic symbol of a diode showing the voltage and current conventions.

but focuses on the type of characterization commonly used in microelectronic failure analysis. With this in mind, the discussion will be limited exclusively to the DC characterization of bipolar devices.

PN Junction Diode

The PN junction diode is the simplest semiconductor device that is encountered by a failure analyst; it consists of just one PN junction. PN junctions are found everywhere on integrated circuits – in diodes, BJTs, MOS transistors, diffused resistors, and ESD protection circuits. Characterization techniques used for PN junctions can be applied to these other areas as well.

Theory of Operation. Figure 1 shows the schematic symbol for a diode. A PN junction diode has three regions of operation: forward bias, reverse bias, and reverse breakdown. The forward bias region is where $V > 0$ and the current through the diode increases exponentially. The reverse bias region is where $V < 0$ and there is a small (essentially zero) current flowing through the diode. The reverse breakdown region is where the reverse bias voltage has exceeded the

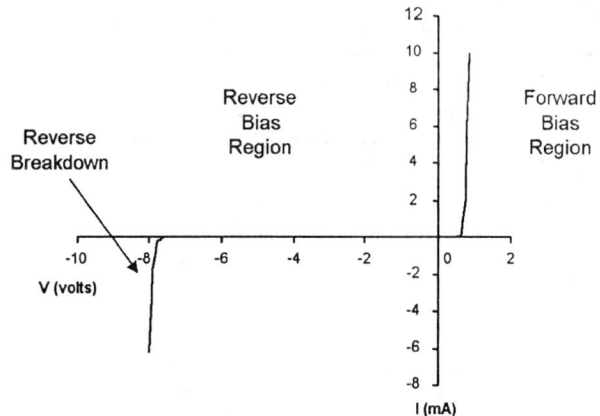

Figure 2. Measured I-V characteristics of a PN junction showing the three regions of operation.

breakdown voltage of the device and the current through the diode rises sharply due to avalanche multiplication. Figure 2

121

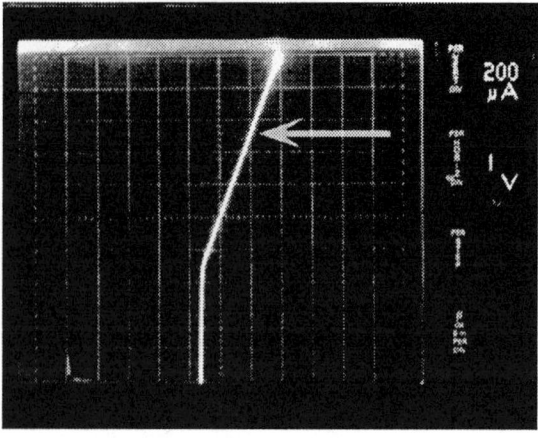

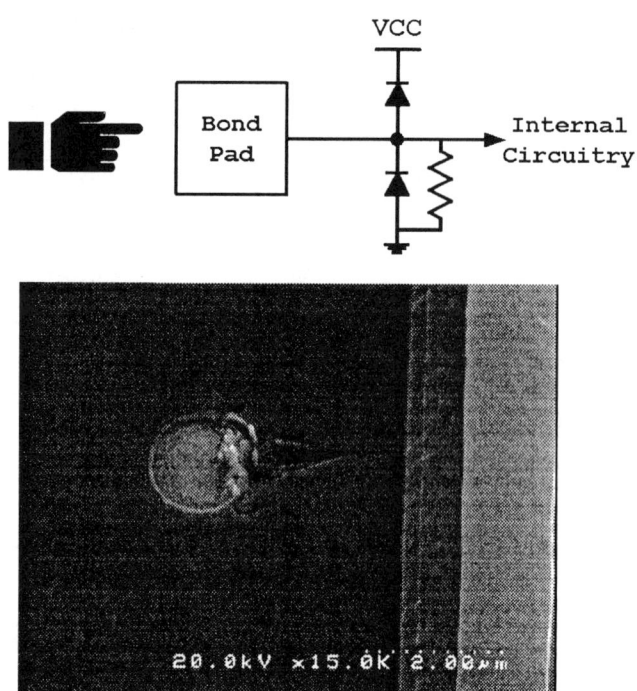

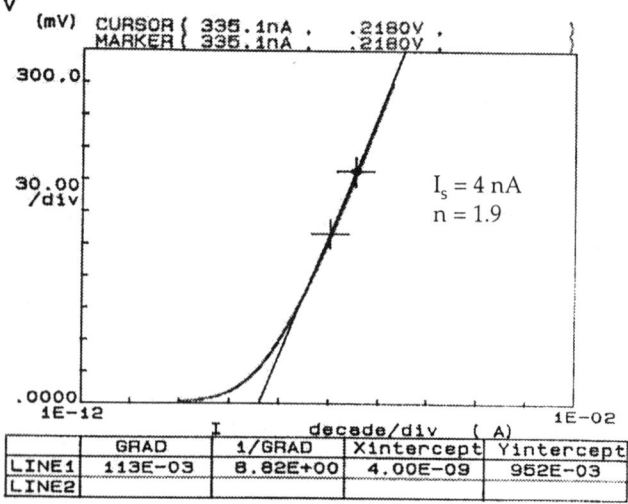

Figure 3. Standard curve trace of a damaged PN junction (this PN junction was part of the ESD protection circuitry.) The damage resulted in a parallel resistive path to ground. Physical analysis of the defect found it to be an ESD event resulting in contact punch through.

shows a measured I-V curve for a PN junction diode. The three regions of operation are marked.

Diode Characterization. The first characterization technique to be used on a PN junction should be a standard curve trace of the junction. A standard curve trace is made using a curve tracer or parameter analyzer and results in an I-V plot similar to that found in Figure 2. A curve trace is a quick and easy technique and it uncovers many types of defects that result in current leakage across the junction. The reverse breakdown voltage is also easily measured [1, 2]. Figure 3 shows an example of standard curve trace plot for a damaged junction. The I-V plot on the upper left portion of the figure shows that this PN junction is electrically behaving like a PN junction with a parallel resistive path to ground. After physical analysis was performed, the cause for the leakage was found. The lower right SEM photo shows that the damage to the junction was a result of ESD induced contact punch through.

In addition to the standard curve trace of the junction, the two diode parameters commonly measured are the saturation current and the ideality factor (n) [3, 4]. The current through a diode is related to the voltage across it by the following equation.

$$I = I_S \left[\exp\left(\frac{V}{nkT}\right) - 1 \right] \qquad (1.)$$

where I_S is the saturation current, n is the ideality constant (normally ranges from 1-2), V is the applied voltage, k is Boltzman's constant, and T is the temperature in Kelvins.

Saturation Current and Ideality Constant Measurements.

Figure 4. Measurement of the saturation current and ideality constant of a diode.

Measuring the saturation current and ideality constant is straightforward and can be accomplished using only one measurement. To do this, first approximate Eq. 1 by assuming that the diode is forward biased and $V \gg nkT$. With these assumptions the I-V relationship is simplified to

$$I = I_S \exp\left(\frac{V}{nkT}\right) \tag{2.}$$

which can be rearranged into a more useful form

$$V = nkT \ln[I] - nkT \ln[I_S]. \tag{3.}$$

Eq. 3 is in the form of a line $(y = mx + b)$ from which the saturation current and the ideality factor can be extracted. The procedure to do this is as follows:

Set up a parameter analyzer or similar system to measure the I-V characteristics of the diode.
- Plot V vs. $\ln(I)$.
- The saturation current, I_S, is the x-intercept of the resulting curve.
- Measure the slope, m, of the curve.
- The ideality constant is m/kT.

The ideality constant should range from 1-2. A measured ideality constant outside this range is indicative of an abnormal junction [5]. Figure 4 shows an example measurement of a discrete diode. The saturation current is the x-intercept of the curve and can be read directly from the graph. For this diode, the saturation current measured 4 nA. The ideality constant is calculated from the slope of the linear portion of the curve. For this diode, the slope can be recorded directly from the graph and measures 113 mV. The ideality constant is then

$$n = \frac{m \log_{10}(e)}{kT} = 1.9 \tag{4.}$$

($\log_{10}(e)$ is a conversion factor to convert between base 10 logarithm and natural logarithm).

Bipolar Junction Transistors

The next semiconductor device examined is the bipolar

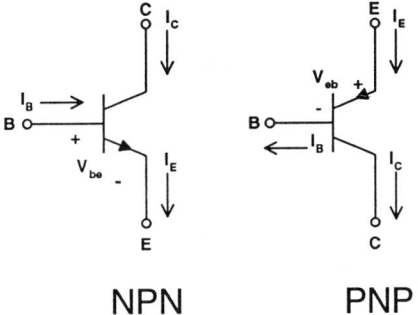

Figure 5. Schematic symbol for an NPN and PNP transistor showing the current conventions.

junction transistor. BJTs are three terminal devices that are composed of two back to back PN junctions. Figure 5 shows the schematic diagrams for an NPN and PNP transistor showing the current and voltage conventions.

Theory of Operation. In DC operation, a bipolar transistor has three regions of operation – the cutoff region, the active region and the saturation region. The cutoff region occurs when the emitter-base junction and the base-collector junctions are both reverse biased. In this region the transistor is turned off and no collector current flows. In the active region, the emitter-base junction is forward biased and the base-collector junction is reverse biased. In this region of operation, the transistor has gain and acts like a current controlled current source. A small base current controls a large collector current. The terminal equations for the transistor operating in the active region are

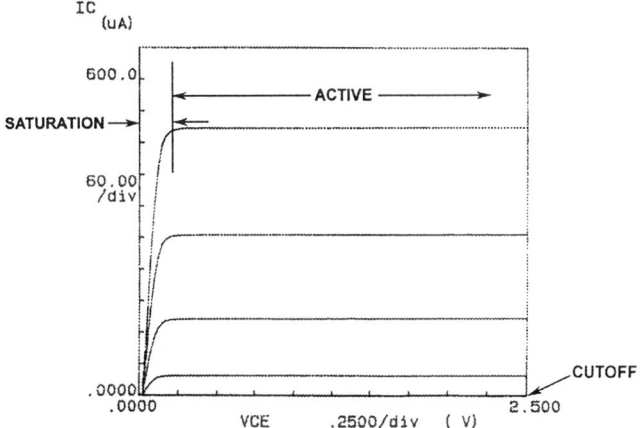

Figure 6. Characteristic curves of an NPN transistor showing the three regions of operation.

$$I_C = I_S \exp\left(\frac{V_{BE}}{kT}\right) \tag{5.}$$

$$I_B = \frac{I_S}{\beta} \exp\left(\frac{V_{BE}}{kT}\right) \tag{6.}$$

and

$$I_E = \frac{\beta+1}{\beta} \exp\left(\frac{V_{BE}}{kT}\right) \tag{7.}$$

where $\beta = I_C/I_B$ is the DC current gain of the transistor. Finally, in the saturation region both the emitter-base junction and the base-collector junction are forward biased. Here the collector current is dependent on the collector-emitter voltage. The regions of operation are illustrated in Figure 6, which is a plot of the characteristic curves for an NPN transistor.

Transistor Characterization. For failure analysis, there are three parameters that are useful to characterize. They are the characteristic I-V curves, the saturation current and ideality factor, and the DC current gain (β) [3, 4].

Transistor Characteristic Curves. Figure 6 shows typical characteristic curves for an NPN transistor. A curve tracer can be used to measure the characteristic curves of a transistor. Figure 6 was generated using a parametric

123

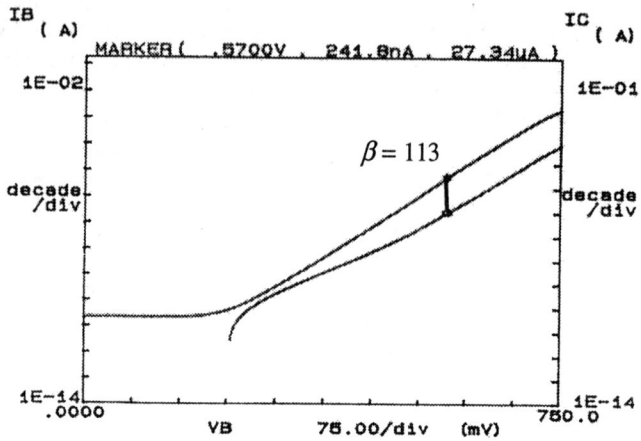

Figure 7. Gummel plot of an NPN transistor.

analyzer that was set up to source a series of fixed base currents to the device and measure the resulting collector current as a function of the collector-emitter voltage.

Saturation Current and Ideality Constant. The measurement and calculation procedure of these parameters is the same as for the diode; just replace I with I_C and V with V_{BE} in Eq. 1 and follow the diode procedure.

DC Current Gain (β). There are several ways to measure the DC current gain of a transistor:

- Use the characteristic curves. β is the ratio of the difference in collector currents to the difference in base currents for successive curves ($\beta = \Delta I_C / \Delta I_B$).

- Use a Gummel plot. Sweep the base voltage from 0 to 1V and plot I_C and I_B (log scale) vs. V_{BE} (linear scale). β is determined by the distance between the two curves. Figure 7 is an example of a Gummel plot. It shows the relationship between the base current and collector current for a discrete transistor. The measured β is 113.

- β can be measured by sweeping the base current of the transistor, measuring the resulting collector current and plotting I_C vs. I_B . β is the slope of resulting line.

ANALOG CIRCUIT CHARACTERIZATION

The remainder of this paper deals with characterization of

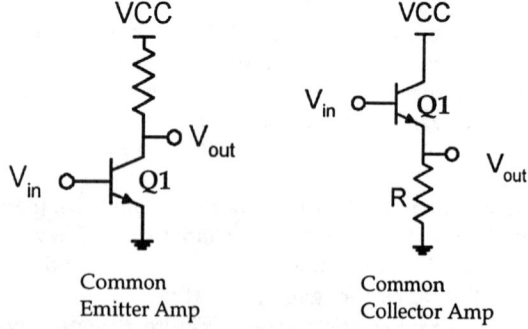

Figure 8. Schematic diagrams of a common collector and common emitter amplifier.

analog circuits commonly encountered in failure analysis. The scope again is limited to DC analysis of bipolar circuits.

Single Transistor Circuits.

The discussion of analog circuit characterization begins with the analysis of two simple single transistor circuits – the common collector and common emitter amplifiers. Figure 8 shows the schematics of the two circuits. For each, *Vin* is the input voltage and *Vout* is the output voltage.

Common Emitter (CE) Amp. For DC inputs, the common emitter amplifier acts like an inverter with three distinct regions of operation.

- *Region 1: Vin = 0V*
  - Q1 is cut-off.
  - IC1 = 0.
  - *Vout* = V_{CC} .

- *Region 2: Vin > 0V* and the base-collector junction is reverse biased.
  - Q1 is active.
  - $Vout = V_{CC} - I_S R exp\left(\dfrac{Vin}{kT}\right)$.

- *Region 3: Vin > 0V* and the base-collector junction is forward biased.
  - Q1 is saturated
  - *Vout* = V_{CEsat}

Figure 9 is a measured DC transfer function for a common emitter amplifier clearly showing the three regions of operation. Operation in region 2 is required for amplification.

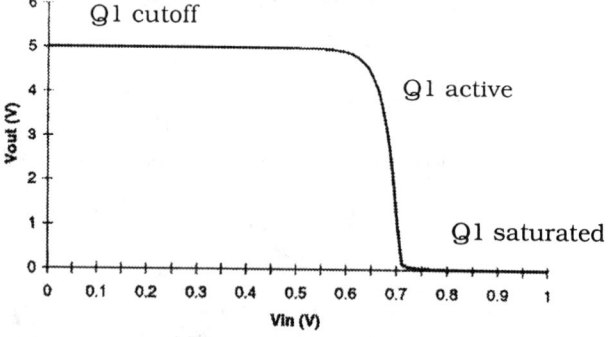

Figure 9. DC transfer function of a CE amplifier. The three regions of operation are marked.

Common Collector (CC) Amp. For DC inputs, a common collector amp acts like a voltage follower with two distinct regions of operation.

- *Region 1: Vin − Vout < 0.7V*
  - Q1 is turned off.
  - *Vout* = 0 .

124

- *Region 2: Vin −Vout > 0.7 V*
  - *Vout = Vin − 0.7V*

In region 2, the output of the CC amplifier is the input voltage minus a B-E voltage drop (typically 0.6 – 0.7 V).

Characterization for Failure Analysis. Characterization of analog circuits for failure analysis means "what type of characterization is needed to help isolate a failure". It has nothing to do with circuit characterization to determine if it complies with design specifications. With this in mind, there are two characterization strategies for the CC and CE amplifiers. First, a basic curve trace analysis of the transistor terminals with respect to ground and VCC is a good starting

	GOOD	B-E leakage	C-E Leakage	Diode from output to ground
VOH (VI = 0.2)	5.0 V	5.0	450 mV (**)	700 mV (**)
VOL (VI = 1.0)	40 mV	40 mV	40 mV	40 mV
IIL (VIL = 0.2)	0.0	2.0 mA (**)	0.0	0.0
IIH (VIH = 1.0)	23 mA	35 mA (**)	23 mA	23 mA
ICCQ (vin = 0.2)	0.0	0.0	4.5 mA (**)	4.3 mA (**)

Table 1 Response of a CE amplifier to simulated defects. The (**) indicate detectable failure modes for each test.

point. The I-V curves for each junction will detect leakage, shorts, and possibly opens in the circuit. Secondly, parametric testing of the circuit can be a method of characterizing CC and CE circuits. Parametric testing includes testing VOL, VOH for the outputs and IIH, IIL for the inputs. Table 1 details the response of a CE amp to several simulated defects. These defects are B-E leakage, C-E leakage, and the output clamped by a diode to ground. The table shows that not all tests are equally suited for all possible defect types. For detecting B-E leakage the input current tests are useful whereas for detecting C-E leakage, the output voltage, and IDDQ tests are useful (in this example IDDQ refers to the VCC current when Q1 is in region 1 of operation [cut-off].)

Current Sources.

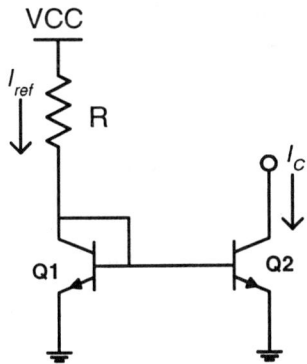

Figure 10. Simple bipolar current source.

Current sources are used for biasing and act as active loads. An active load would take the place of the resistor in

the CE amplifier. There are too many types of current source topologies to describe them all, so the following discussion will be limited to one simple current source. The basic functionality of various current sources are the same, they differ mainly in the implementation details.

Theory of Operation. Figure 10 shows a schematic of a simple bipolar current source. To simplify the analysis of this circuit, neglect the base current required for Q1 and Q2. The bases of the two transistors are tied together so that

$V_{BE1} = V_{BE2}$. Since $I_{REF} = I_{C1}$ and $I_C = I_S \exp\left(\dfrac{V_{BE}}{kT}\right)$ then

$$I_C = \frac{I_{S2}}{I_{S1}} I_{REF} \text{ [4].} \tag{8.}$$

I_C is the output of the current source that is used to bias other circuits. It is generated from I_{REF} and is scaled by the factor $\dfrac{I_{S2}}{I_{S1}}$.

Characterization for Failure Analysis. Characterization of current sources is straightforward. Usually it involves just measuring the bias current and comparing it to either a known good current source or to the designed value. Current sources that fail generally fall into two categories –they fail catastrophically or fail to have the correct bias current (parametric fails). Catastrophic failures may be due to EOS or gross misprocessing (open contacts, etc.) Parametric fails generally are caused by some type of nonideality (e.g. low β) or mismatch in the transistors that comprise the current source.

Voltage References.

A voltage reference is used to supply a stable voltage that is independent of temperature and power supply variations.

Bandgap Voltage Reference. Figure 11 is a block diagram of a bandgap voltage reference. The reference voltage is derived from the sum of two components – a base-emitter voltage and a voltage that is proportional to the thermal voltage. Since the base-emitter voltage of a bipolar transistor has a negative temperature coefficient (-2mV/°C) and the thermal voltage has a positive temperature coefficient (+ 0.0853 mV/°C), a constant is chosen which results in a zero temperature coefficient.

Theory of Operation. Figure 12 is a schematic of a Widlar bandgap voltage reference circuit. The node 'Vout' is the output of the voltage reference. Assuming the transistors have no base current the following current and voltage equations can be written.

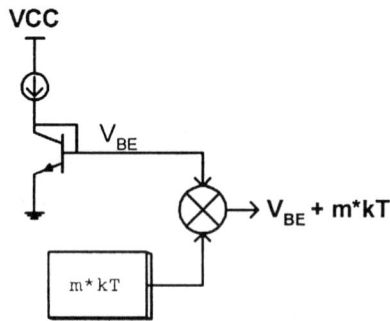

Figure 11. Block diagram of a bandgap reference voltage.

$$Vout = V_{BE3} + I_{C2}R_2 \tag{9.}$$

$$I_{C2} = \frac{V_{BE1} - V_{BE2}}{R3} = \frac{\Delta V_{BE}}{R3} \tag{10.}$$

$$\Delta V_{BE} = kT \ln\left(\frac{I_{C1}}{I_{C2}} \frac{I_{S2}}{I_{S1}}\right) \tag{11.}$$

combining Eqs. 9-11 yield

$$Vout = V_{BE} + \frac{R2}{R3} kT \ln\left(\frac{I_{C1}}{I_{C2}} \frac{I_{S2}}{I_{S1}}\right) \tag{12.}$$

if R1 and R2 are chosen so that $I_{C1}R1 = I_{C2}R2$ then $V_{BE3} = V_{BE2}$ and

$$Vout = V_{BE3} + \frac{R2}{R3} kT \ln\left(\frac{R2}{R3} \frac{I_{S2}}{I_{S1}}\right) \text{ [4, 6].} \tag{13.}$$

Characterization for Failure Analysis. From a failure analysis point of view, voltage references are not difficult circuits to characterize. You simply measure the output voltage. The only complication is that it may also be necessary to measure the output voltage over temperature. Failure modes associated with voltage references generally fall into two categories – no output voltage or an output voltage that is incorrect or doesn't track correctly over it's specified temperature range. Bandgap voltage references operate in a feedback configuration and will have two stable states – one stable state is at the correct output voltage and the other when the output is zero. The existence of a stable output state with *Vref* = 0 requires a voltage reference to have a start-up circuit to 'nudge' the output into the correct stable state. If a reference voltage reads zero volts, then characterize the circuit as follows:

- Perform a standard curve trace from VCC to ground to detect any leakage current that may be due to catastrophic damage such as electrical overstress .
- If no leakage current is measured, characterize the start-up circuitry to see if it is working properly.

Depending on the circuit architecture and layout, this may or may not be possible. If it is not possible to isolate the start-up circuitry, measure node voltages and compare them to a good unit.

Voltage references that have an incorrect output voltage or a reference voltage that doesn't track properly over

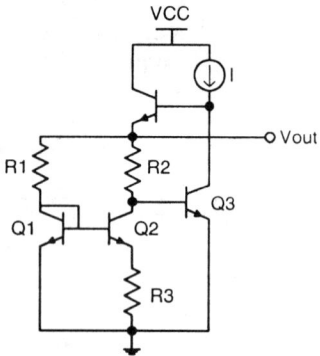

Figure 12. Schematic diagram for a bandgap reference voltage.

temperature require different characterization strategies.

- Voltage references are typically trimmed to the correct voltage at wafer probe. The trimming process can take a variety of forms, but one common method is to trim resistor values via fuses or zener diodes. An improperly blown fuse or zener diode may experience some unexpected behavior over time and temperature. The first step is to characterize the trims over temperature and see if they are stable.
- Frequently ΔV_{BE} is set up by a ratio of resistor values. If one of these resistor values is grossly off, say by an open contact, the reference voltage will be incorrect.
- A high input offset voltage in the summer circuitry will also affect the reference voltage. The characterization of input offset voltage is discussed in section on op-amps.

Voltage Regulators.

A voltage regulator outputs a known voltage that is stable over a wide range of output current [7].

Theory of Operation. Figure 13 is a block diagram of a series voltage regulator. Vo is the regulated output voltage. The sampling network, consisting of a voltage divider (R1 and R2) samples a portion of the output voltage and feeds it back to the error amplifier. The error amplifier is a differential amplifier (usually an op-amp) which compares this sampled voltage to a stable reference voltage and amplifies the difference of the two signals. If the gain of the error amp is high enough, the feedback loop will force the voltages at it's inputs to be equal (V+ = V-). This results in the output of the voltage regulator being

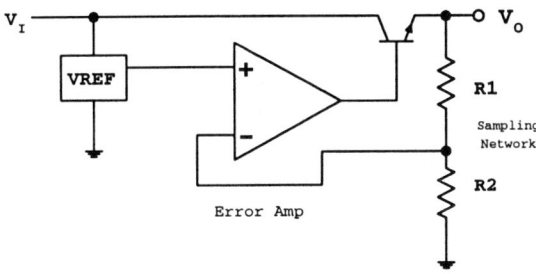

Figure 13. Block diagram of a series pass voltage regulator.

$$Vo = \left(1 + \frac{R1}{R2}\right)Vref .$$ (14.)

Characterization for Failure Analysis. When a voltage regulator requires failure analysis a common complaint is that the output voltage is incorrect or the unit is not regulating. There are other possible failmodes for features not shown on the block diagram, such a overcurrent limiting, but these will not be discussed. The first step in characterizing a voltage regulator is to do an output vs. input plot. From this, useful information is gathered that helps determine the direction of future characterization.

- *The regulator does not regulate.* This means that the output voltage is a function of the input voltage for all values of input voltage. If the regulator doesn't regulate, then the next step is to curve trace the regulator between the input and output pins. Many times the pass transistor is damaged or shorted and the curve trace will quickly find this type of damage.

- *The regulator regulates, but does so at an incorrect voltage.* This type of failure generally indicates that the feedback loop is operating properly, but there is an error in one of the components that make up the regulator. The contribution of each of the possible error sources in a voltage regulator is discussed in the following section.

Error Sources in a Voltage Regulator. There are several possible error sources found in voltage regulators.

1. The voltage reference could be output the incorrect voltage. The effect on the regulator's output due to this error is

$$\Delta V_o = \left(1 + \frac{R1}{R2}\right)\Delta Vref .$$ (15.)

2. There could be errors associated with the sampling network. The effect on the regulator's output due to this error alone is

$$\Delta V_O = \frac{R2}{R1}\left(\frac{\Delta R2}{R2} - \frac{\Delta R1}{R1}\right)Vref .$$ (16.)

3. Errors associated with the error amplifier will also have an affect on the regulator's output voltage.

$$V_O = \frac{Vref - V_{OS}}{\frac{1}{A_V} + \left(\frac{R2}{R1 + R2}\right)}$$ (17.)

where A_V is gain of the error amplifier and V_{OS} is the input offset voltage of the error amplifier.

Op-amps.

The final analog building block is the op-amp. Op-amps are either stand-alone integrated circuits or may be found internally on large analog designs.

Theory of Operation. An op-amp is a differential amplifier whose output voltage is given by

$$Vout = A_v(V_+ - V_-)$$ (18.)

where A_v is the gain of the amplifier and V_+ and V_- are the non-inverting and inverting inputs, respectively [4]. Although op-amps may have a variety of topologies, two and three stage op-amps are most common. Figure 14 is a block diagram showing a three-stage op-amp. The first stage is a differential amplifier that provides gain and often performs a differential to single ended conversion. Some op-amps do

Differential Amp   Gain Stage   Output Buffer

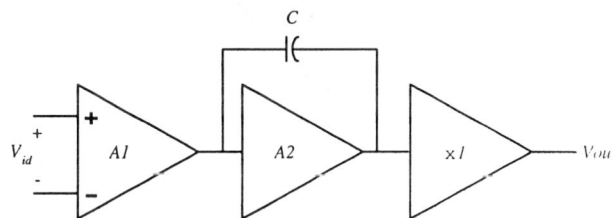

Figure 14. Block diagram of a three-stage op-amp.

have differential outputs. The second stage is a gain stage and usually has a compensation network to ensure stability. The third stage, if present, is a buffer that supplies the drive current needed to drive external loads.

An ideal op-amp has these qualities:
- Infinite gain.
- No input bias current.
- Zero output resistance.

Unfortunately, real op-amps are not ideal and these aspects of op-amp behavior need to be understood by the failure analyst.

Characterization for Failure Analysis. Real world op-amps depart from the ideal and have
- Finite gain (Av).
- Input bias current (IIB).

- Input offset current (IOS)
- Input offset voltage (VOS).
- Non-zero output resistance (Ro).

Failures of these and other parameters (e.g. output voltage swing, common mode rejection ratio) are likely to be encountered by a failure analyst who works on op-amps. The following sections outline the characterization techniques for parameter failures most likely to be encountered.

Input bias current (IIB). The input bias current for bipolar op-amps is the base current required for the input differential pair. For example, Figure 16 shows a simplified schematic of an input differential pair. The input bias current is the base current required to bias Q1 and Q2 to *IEE*/2 (plus any possible leakage in the ESD protection circuitry). To characterize IIB, use the following procedure:

- Do a standard curve trace on the input pins to see if there is any leakage current in the ESD structure. Finding leakage on a curve trace does not uniquely isolate the leakage to the ESD structure, but not finding leakage eliminates the ESD structure as being a potential failure site.

- If leakage current is measured during the standard curve trace, isolate the ESD structure from the op-amp circuitry. If the leakage goes away, then the failure is isolated to the ESD structure. If the leakage is still present, characterize the PN junctions of the input transistors to isolate the failure mechanism.

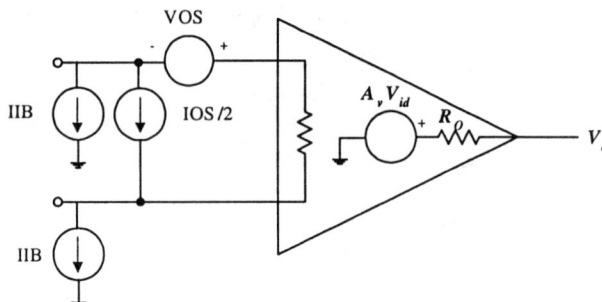

Figure 15. Block diagram of an op-amp showing the non-idealities.

- If a standard curve trace does not measure any leakage current, perform an IIB test to verify the failure. The test set-up will usually be found in the op-amp specifications. If the IIB failure is verified, characterize the input transistors (possibly for low *β*.)

Input offset voltage (VOS). The input offset voltage is the input differential voltage required to bring the output voltage of the op-amp to zero (assuming positive and negative power supplies). The input offset voltage of an op-amp can range in magnitude from tens of microvolts to tens of millivolts. There are several different ways to measure the input offset voltage. The first is to configure the op-amp as a non-inverting

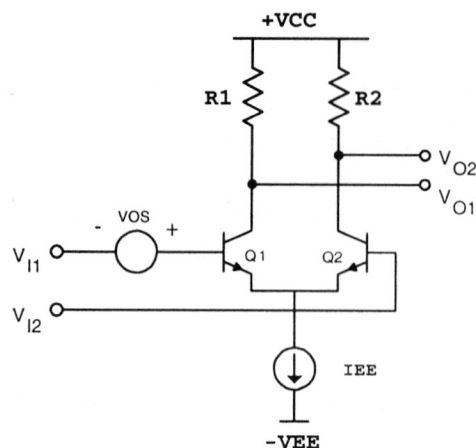

Figure 16. Schematic of a simple bipolar input differential pair amplifier.

amplifier whose inputs are grounded through resistors (see Figure 17). In this configuration

$$VOS = \frac{Vout}{\left(1 + \dfrac{R2}{R1}\right)}. \tag{19.}$$

The second method measures VOS directly in an open loop configuration. Figure 18 shows a setup diagram for this measurement. A low value resistor (~10 Ω) is connected across the input terminals of the op-amp. The inverting input is grounded and the non-inverting input is driven with a

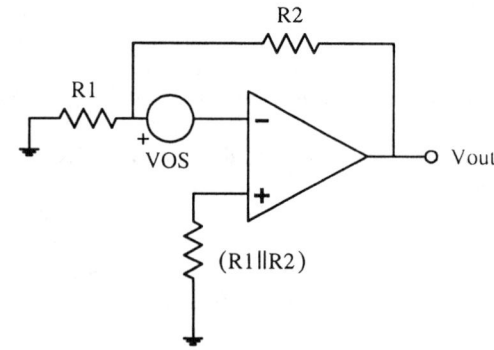

Figure 17. VOS measurement of an op-amp using a closed loop configuration.

current source.

In this set-up the input voltage is generated by the voltage drop across the resistor and is given by

$$V_{id} = IR \tag{20.}$$

Vo is plotted against *Vid* and the resulting curve will look similar to Figure 19. For this op-amp, the input offset voltage is 800 µV. This method is also useful in determining the DC gain of the op-amp – it's just the slope of the linear portion of the curve as plotted in Figure 18.

128

Factors that affect VOS. The factors that affect VOS are important to consider and will help an analyst isolate the cause of a VOS failure. Figure 16 is a schematic of a simple input differential pair. The input offset voltage is modeled as a voltage source in series with one of the inputs. Using a voltage loop consisting of VI1, VI2, Q1, and Q2 we have

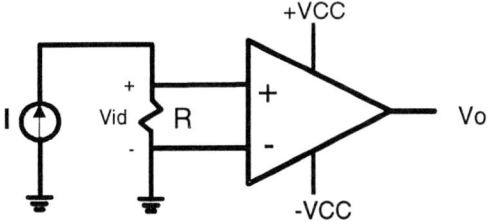

Figure 18. Measurement set-up for VOS using an open-loop configuration.

$$VOS = V_{BE1} - V_{BE2} \qquad (21.)$$

using Eq. 5 and solving for V_{BE} results in

$$VOS = kT \ln\left(\frac{I_{C1}}{I_{C2}} \frac{I_{S2}}{I_{S1}} \right). \qquad (22.)$$

Recall that *VOS* is the amount of differential input voltage required to bring the output of the op-amp to zero. For the input differential stage in Figure 16, the output is *VOD*, which is *VO1 – VO2*. To make this output voltage to equal

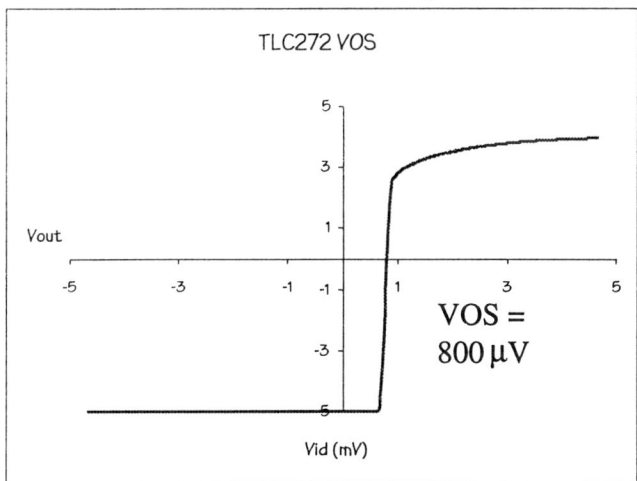

Figure 19. Measured VOS for a TLC272 op-amp using the open-loop method.

zero requires that $I_{C1}R1 = I_{C2}R2$. When this is satisfied, the offset voltage is

$$VOS = kT \ln\left(\frac{R2}{R1} \frac{I_{S2}}{I_{S1}} \right). \qquad (23.)$$

By defining new variables, Eq. 23 can be derived in a way that is particularly helpful for failure analysis. With

$$\Delta R = R1 - R2 \qquad (24.)$$

$$R = \frac{R1 + R2}{2} \qquad (25.)$$

$$\Delta I_S = I_{S1} + I_{S2} \qquad (26.)$$

$$I_S = \frac{I_{S1} + I_{S2}}{2} \qquad (27.)$$

the offset voltage is

$$VOS \approx kT \left(-\frac{\Delta R}{R} - \frac{\Delta I_S}{I_S} \right) [4]. \qquad (28.)$$

The input offset voltage is then proportional to the relative mismatch in the input transistors and the load resistors. And though this result was derived for this particular differential pair, in general the input offset voltage can be attributed mismatches in the components that make up the input differential pair.

CONCLUSION

This paper covered the basics of bipolar device and analog circuit characterization. Though the breadth of the subject matter did not allow for an in-depth treatment of every subject, it is hoped that it is a good introduction for the failure analyst who is new to the subject.

REFERENCES

1. J. Beall and D. Wilson, "Curve Tracer Applications and Hints for Failure Analysis" in *Microelectronics Failure Analysis Desk Reference*, 3rd ed., ASM International, 1993.

2. D. Appleman and F. Wong, "Computerized Analysis and Comparison of IC Curve Trace Data and Other Device Characteristics" in *ISTFA 1990*, pp. 271-277.

3. D. Schroder, *Semiconductor Material and Device Characterization*, John Wiley & Sons, 1990, pp. 147 – 193

4. P. Gray and R. Meyer, Analysis *and Design of Analog Integrated Circuits*, 3rd ed., John Wiley & Sons, 1993

5. B. A. McDonald, "Avalanche Degradation of h_{FE}", *IEEE Transactions on Electron Devices*, VOL. ED-17, No. 10, October 1970, pp. 871-878.

6. P. Brokaw, "A Simple Three-Terminal IC Bandgap Reference", *IEEE Journal of Solid-State Circuits*, VOL. SC-9, No. 6, December 1974, pp. 388-393.

7. R. Widlar, "New Developments in IC Voltage Regulators", *IEEE Journal of Solid-State Circuits*, VOL. SC-6, No. 1, February 1971, pp. 2-7.

Fault Localization

Electron/Ion Beam-Based Techniques

Beam-Based Localization Techniques
for IC Failure Analysis

Edward I. Cole Jr.
Infineon Technologies
München, Germany

INTRODUCTION

The scanning electron microscope (SEM) has become as standard a tool for IC failure analysis as the optical microscope. The SEM's advantages over light microscopy include greatly increased depth of field, much higher magnification, increased working distance, and improved imaging of surface topography. In addition, the interaction of the electron beam with the IC enables unique imaging and analytical capabilities. The strengths, limitations, and effects upon the device being analyzed must be understood to use the SEM effectively. SEM settings that are appropriate for one analytical technique may be unsuccessful, may give misleading results, or may damage the device being analyzed if used for a different technique.

More recently, the scanning optical microscope (SOM) has become a valuable tool for photon beam based analysis of ICs. Like SEM, the SOM can produce reflected light images of improved spatial resolution and depth of field compared to conventional optical microscopy. Additionally, by taking advantage of silicon's relative transparency to infrared wavelength's, the SOM has become one of the main vehicles for backside failure analysis tool development.

This workshop reviews conventional and new SEM techniques for IC analysis and new SOM analysis methods. All of these techniques can be performed on a standard SEM or SOM (using the proper laser wavelengths). The use of advanced electron beam test systems is also discussed. The workshop is designed to provide beneficial information to both novice and experienced failure analysts. Topics to be covered are (1) standard techniques: secondary electron imaging for surface topology, voltage contrast, capacitive coupling voltage contrast, backscattered electron imaging, electron beam induced current imaging, and x-ray microanalysis, (2) new SEM techniques: novel voltage contrast applications, resistive contrast imaging, and charge-induced voltage alteration (both high and low energy versions) and (3) new SOM techniques: light-induced voltage alteration, thermally-induced voltage alteration, and Seebeck Effect imaging. Each technique will be described in terms of the information yielded, the physics behind technique use, any special equipment and/or instrumentation required to implement the technique, the expertise required to implement the technique, possible damage to the IC as a result of using the technique, and examples of using the technique for failure analysis.

STANDARD SEM TECHNIQUES

Secondary Electron (SE) Imaging for Surface Morphology

Information Technique Yields: SE imaging generates a high resolution, large depth of field image (compared to optical microscopy) depicting the surface morphology of the sample examined. This is the most commonly used imaging mode of the scanning electron microscope.

Physics Behind Technique Use: SE image contrast is generated by differences in SE emission efficiency with topography. Figure 1 displays the electron beam interaction products from a passivated integrated circuit generated by a 10 keV primary electron beam. A primary electron beam incident on a solid will create many excited electrons in the target material. These excited electrons are scattered isotropically, some having enough energy to escape the solid at the surface. SEs are those electrons emitted from the surface with energy < 50 eV. A primary electron beam with energy > 100 eV will yield an SE energy distribution whose shape is determined by the work function, Fermi level, and other material parameters. (Below 100 eV numerous other factors determine SE emission.) The SEs originate only from the top 30 nm of passivated integrated circuits. Scattered electrons generated deeper than this do not have the energy required to escape the surface. Figure 2 displays primary electron beam scenarios with the surface at various angles relative to the electron beam. At steeper angles a larger portion of the primary beam interaction volume is generated nearer to the surface. This "edge effect" results in a larger number of SEs being generated and a "brighter" image. It is the change in SE emission efficiency with primary beam/sample angle that produces the SE image. Other considerations that can reduce or increase image quality are the relative detector location, electron beam energy and current, and the target material.

Difficulty to Implement: The only equipment necessary to acquire SE images is a scanning electron microscope. Whether or not any sample preparation is performed depends upon the sample type and information desired. Any packaged integrated circuit SE observation requires removal of enough of the package to expose the IC surface to be examined (metal lid removal, plastic grinding or etching, etc.). Sample charging by the electron beam is the most common problem encountered during SE imaging. A net charge will be produced at the integrated circuit surface if the number of electrons absorbed from the primary electron beam is not equal to the number exiting the sample. This generates a surface charge which increases in magnitude until equilibrium is reached. The charging effect is beneficial in other electron beam imaging techniques, but its effect in topography imaging is to reduce spatial resolution. To eliminate surface charging a thin conductive coating (~20 angstroms of C or Au/Pd) may be applied to the surface by vacuum evaporation. Careful selection of the primary electron beam energy will also reduce charging effects. For example, by grounding the substrate of an integrated circuit and using a high enough primary electron beam energy such that the surface is shorted to the substrate through the interaction volume, charging can be greatly reduced and an adequate image produced. This technique was used to produce the SE images in Figure 3. The surface passivation was removed in Figure 3.

Possible IC Damage: No direct "physical" damage occurs with electron beam testing. Alteration of the threshold voltage on MOS transistors is possible. This change in threshold voltage results from irradiation damage to the gate oxides of MOS transistors. The damage is generated by direct primary electron collisions and by x-rays generated through interactions in the surface layer(s). The primary electrons quickly alter the interface trap density and occupancy as well as the fixed charge levels in the gate oxide. If desired, low temperature annealing (100-150 ^{o}C) may be used to restore the threshold voltage to near its original value.

If surface layer removal is performed, there is the possibility of device damage during the removal process. Conductive coating to eliminate charging will prevent/hinder microelectronic operation and should be removed to restore device operation. Note that removal of any coatings may damage the IC under examination.

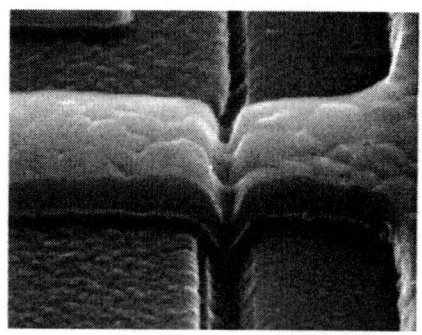

Figure 3. Example of SE morphology imaging.

Backscattered Electron (BSE) Imaging

Information Technique Yields: BSE imaging detects differences in elemental atomic number on and below the surface. It is primarily used to detect sharp atomic number gradients under the passivation, such as those caused by impurities and metal voids.

Physics Behind Technique Use: BSEs are electrons with energies E such that 50 eV < E< E*, where E* is the primary beam energy. Unlike SEs, the BSEs escape from much deeper in the surface, most coming from about 1/3 the maximum depth of the electron beam/device interaction volume (Figure 4). This greatly reduces any of the surface effects that hamper SE imaging. The major factor determining BSE contrast in images is the atomic number of the target, indicating that nuclear scattering is the principal electron interaction. Because of the complex elastic and inelastic scattering processes that create BSEs, no exact theory is possible, but two general statements about imaging expectations can be made. First, the spatial resolution is limited by the size of the interaction volume at the BSE source, therefore the spatial resolution will always be less than or equal to the SE image resolution. Second, experimentation with different beam energies can improve spatial resolution depending upon the BSE source and its depth. Figure 5 is a BSE image revealing electromigration voids in a metal conductor under a passivation layer.

Difficulty to Implement: The equipment necessary to perform BSE imaging is a scanning electron microscope and a BSE detector. The detector is normally a large area semiconductor, positioned directly over the sample. A digital imaging system is advantageous to image acquisition/manipulation, but not necessary.

Possible IC Damage: Same as with SE imaging.

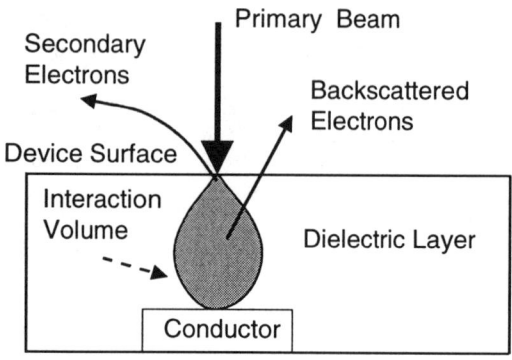

Figure 1. Primary electron beam interaction products.

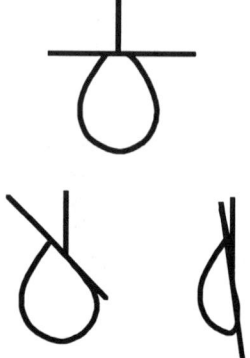

Figure 2. Electron beam interaction volume variation with angle.

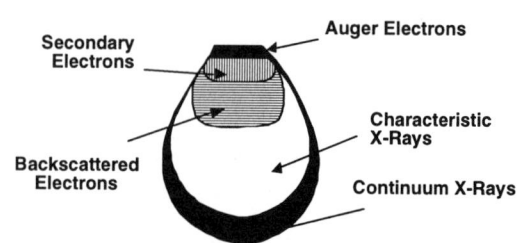

Figure 4. Distribution of electron beam interaction products.

Figure 5. Example of BSE imaging showing electromigration voids.

Voltage Contrast (VC) Imaging

Information Technique Yields: VC imaging creates an image in which the intensity is largely determined by the static voltages on an integrated circuit. By analyzing the variations in brightness, the logic levels of a digital integrated circuit can be determined and the voltage on internal circuit nodes can be measured.

Physics Behind Technique Use: VC imaging takes advantage of differing secondary electron (SE) emission efficiencies with applied bias on an integrated circuit. Figure 1 displays the electron beam interaction products between a passivated IC and a 10-keV primary electron beam. SEs are defined as those electrons emitted from the surface with energies of 50 eV or less. The physics of SE emission is complicated, involving SE generation, transport to the surface, and escape from the surface. The typical shape of the SE energy distribution is shown in Figure 6. This shape is largely independent of primary electron beam current and energy when using primary electron beam energies above 100 eV. Because of their low energies, SEs are generated only from about 30 nm or less from the surface. The ratio of the electron flux escaping the surface of an exposed device to the total electron flux injected into a device varies with the material examined and the primary electron beam energy used. For microelectronic materials at relatively low beam energies (< 2 keV) more electrons escape the surface than are injected. At higher energies and extremely low primary electron beam energies (< 100 eV) fewer electrons escape than are injected. Under both conditions the surface has a net charge. For the case of low beam energies a positive charge builds up on the surface. This charge will prevent lower energy SEs from escaping the surface and a positive equilibrium voltage (about 0.5 V with a 1.0 keV beam) will be established at which the escaping and incident electron flux are equal.

If the surface of the target is forced by an external power supply to a different voltage, then the SE image will have increased or reduced signal depending upon the difference between the forced voltage and the equilibrium voltage. The conductors of a depassivated integrated circuit may be forced in such a manner, as shown in the image of a CMOS static RAM cell in Figure 7. The lowest voltage regions (V_{SS}) are bright and the highest voltage regions (V_{DD}) are dark.

Passivated devices may also be observed using static VC. If the primary electron beam energy is increased to the point where more electrons are injected than escape, a negative charge develops on the surface. The voltage can equal that of the primary beam for the insulating glass, obscuring all image information. However, if

the primary beam energy is increased to the point where the primary electron interaction volume reaches the substrate of the integrated circuit, a charge leakage path through the substrate is generated. Because of hole-electron pair production in the interaction volume, this pear-shaped region is a conductor. This conducting volume will force the surface to the voltage of the buried conductors, thereby affecting the SE emission in a manner similar to the low beam energy situation described above. As a rule of thumb for typical IC materials (Si, Al, SiO_2) the following expression may be used as a rough estimate of primary electron beam penetration,

$$R = 0.022 \, E^{1.65} \qquad \text{Eq. 1}$$

where R is the primary electron beam penetration depth in microns and E is the primary electron beam energy in keV.

The techniques described here generate images which yield qualitative voltage information. Quantitative information may be obtained in two different ways. The first is the simple comparison of image intensity as a function of voltage. This method has an accuracy of about 0.5 V. The second, preferred method for voltage measurement is to employ an energy spectrometer on the emitted SEs. This technique has a voltage accuracy of 10 mV.

Difficulty to Implement: The equipment necessary to perform static VC are a scanning electron microscope, an electrical vacuum feed-through, and a voltage supply. By applying the physics described above, static VC images and voltage measurements are routinely acquired throughout the microelectronics industry. For device and state comparison, a digital image acquisition system is advantageous for image acquisition/manipulation, but not necessary. Factors which limit the voltage and spatial resolution include surface charging from incomplete beam penetration on passivated surfaces, surface contaminants, and local electric fields of structures from neighboring regions. These must be overcome/minimized to optimize performance.

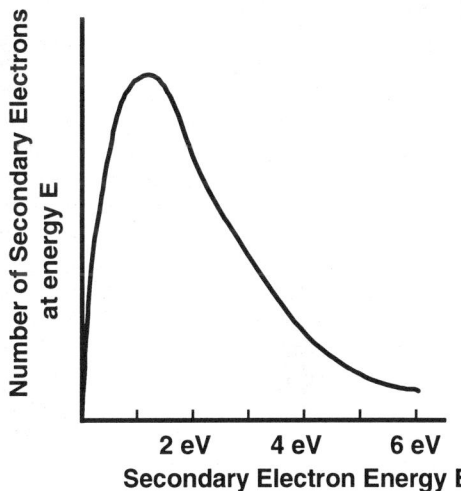

Figure 6. A typical secondary electron energy distribution.

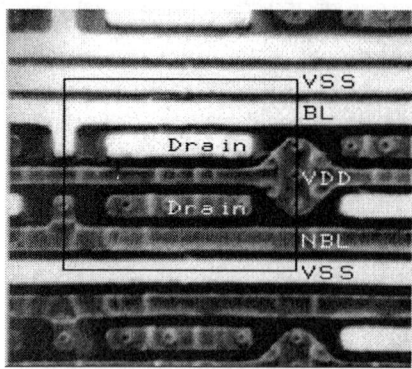

Figure 7. VC of a memory cell.

Possible IC Damage: No direct "physical" damage occurs with electron beam testing. Examination of devices at beam energies of 1 keV or less is non-destructive. The same considerations for MOS devices described for SE imaging apply here as well.

Capacitive Coupling Voltage Contrast (CCVC) Imaging

Information Technique Yields: The CCVC mechanism permits imaging and measurement of dynamic voltages beneath passivation layers. Through primary beam blanking, CCVC facilitates buried conductor as well as diffusion imaging and depth measurement with no applied bias. The major advantage of CCVC over static voltage contrast is the absence of any irradiation damage.

Physics Behind Technique Use: (NOTE: This description builds on the physics of secondary electron generation (SE) described in the static voltage contrast (VC) section. Please read this description before proceeding.)

CCVC imaging uses the passivation layer as a discharging capacitor to generate a dynamic image of changing subsurface voltages. CCVC is performed at fast scan rates to increase the time resolution of the dynamic signal. As described in the VC physics section, at low primary electron beam energies (< 2 keV) more electrons escape than are injected by the primary electron beam. A net positive charge will build up on the surface, preventing lower energy SE's from escaping the surface and decreases the SE image intensity. An equilibrium voltage is reached when the net charge accumulated on the device does not change with time. A bound surface charge will be produced at the Induced Conductive Surface Layer (ICSL) when structures below the maximum beam penetration depth change potential, and the material between them and the surface becomes polarized (Figure 8). The CCVC signal is the change in the number of SEs caused by this bound charge potential. As the primary electron beam scans across the surface the bound change is dissipated via differences in SE emission and equilibrium is reestablished. This differs from static VC imaging where a conducting path to the biased structure exists. The time for the bound charge to decay to the equilibrium potential depends on the passivation, the primary beam energy, the depth of the structure examined, and the primary electron flux. The smaller the incident electron flux, the longer the CCVC decay times. Unfortunately, the signal-to-noise ratio (SNR) decreases with reduction in incident electron flux. As with VC, image comparison and energy spectrometry may be used for voltage measurement. Figure 9

displays a CCVC image of a failing bit line (bright-negative transition) and a functional bit-not line (dark-positive transition). The time for the CCVC signal to decay to equilibrium may also be used to measure voltage.

An additional use for CCVC is the imaging of buried conductors with no applied bias. This is achieved by blanking the primary electron beam and allowing the passivation surface to "leak" to the stage potential, ground. This normally takes about 30 seconds. The beam is then scanned across the surface, which will go to a positive equilibrium voltage as described above. The time it takes the surface to reach the equilibrium voltage depends inversely on the depth of buried conductors under the passivation, permitting layer identification through different equilibrium times.

Difficulty to Implement: The equipment necessary to perform CCVC is a scanning electron microscope, an electrical vacuum feed-through, a voltage supply, and a switching mechanism to generate dynamic voltages. The scanning electron microscope must scan at or near video rates for adequate time resolution to observe CCVC. Because of the SNR-time resolution trade off, a frame grabbing digital image acquisition system is necessary to obtain adequate signal resolution via averaging. This same imaging system should be synchronized to a beam blanking unit to examine buried conductors with no applied bias. The factors that limit static VC imaging also must be dealt with when using CCVC.

Possible IC Damage: There is no damage associated with CCVC, other than package lid/plastic removal.

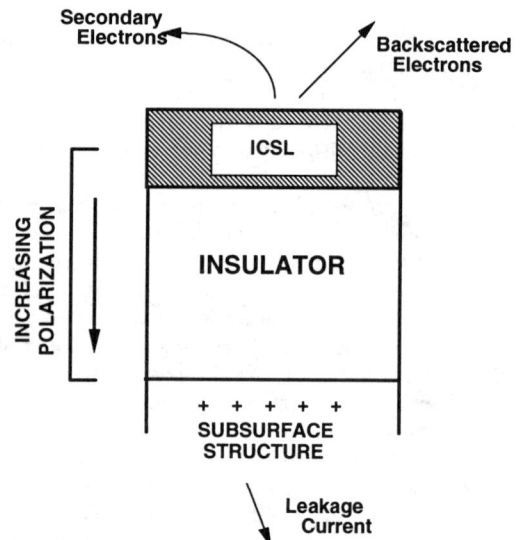

Figure 8. Physics of CCVC generation.

Advanced Electron Beam Test (EBT) Systems

Information Tool Yields: Advanced electron beam test (EBT) systems generate voltage waveforms for conductors on ICs under test. These waveforms can have voltage and timing resolutions of 10 mV and 10 ps respectively. The advanced EBT systems also permit comparison of control and failing devices through image subtraction as well as comparison of CAD layout data with SEM images.

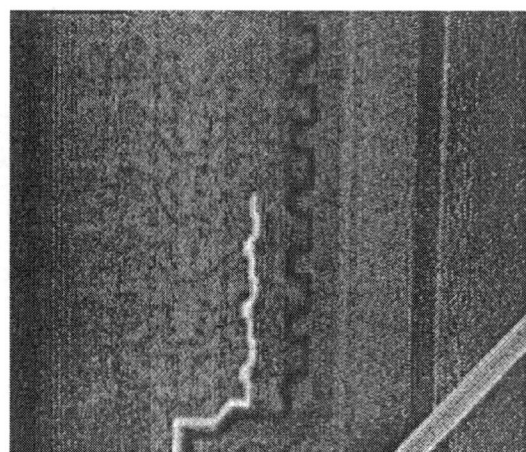

Figure 9. Example of CCVC imaging on an open conductor.

Physics Behind Tool Use: The data acquisition physics of advanced EBT systems is automated VC and CCVC. Time resolution is increased beyond that of standard SEM scan rates by employing a beam blanker. The beam blanker consists of electrostatic plates that can be biased at a fast rate to move the scanning or "spotted" electron beam away from and back to the IC under test. A VC image of the IC in a given state may be generated by triggering the beam blanking plates to permit electron beam interaction with the sample only when the IC is in the state of interest. By repeated cycling of the vectors applied to the IC, a VC image of a given logic state, showing the complete field of view, may be acquired. A digital image storage system is required for strobed VC image acquisition and display.

The waveforms generated by EBT systems employ a secondary electron energy spectrometer to obtain quantitative voltage and timing information. The energy spectrometer consists of a retarding grid in front of a secondary electron detector. The change in the number of secondary electrons reaching the detector with an applied retarding grid voltage may be used to determine the voltage at the surface of the IC. A waveform is generated using the beam blanking plates described above with the difference that the electron beam is not scanning when unblanked, but in spot mode over the conductor being measured. The waveforms in Figure 10 display single and 20 averages of three different conductors.

An additional feature of EBT systems is that CAD information on layout, schematic, and circuit modeling may be incorporated into the EBT system so that the CAD data may be used during IC analysis. The features of the CAD interface and its utilities depend on the vendor.

Difficulty to Implement: The equipment required to implement advanced electron beam testing of ICs is readily available from several vendors. The equipment cost ranges from ~$200K to ~$700K depending upon the vendor and options desired. Full utilization of EBT systems normally requires sophisticated stimulus to the IC under test, which adds to a system's cost. The compatibility of existing CAD databases with the EBT system is another factor that must be considered when evaluating EBT systems. One last consideration is that EBT systems normally operate only at low electron beam energies (< 1.25 keV). While this is ideal for most VC and CCVC observations, certain other SEM techniques cannot be used under these conditions.

Possible IC Damage: Because of the low primary electron beam energies used in EBT systems, there is no damage to ICs other than package lid/plastic removal.

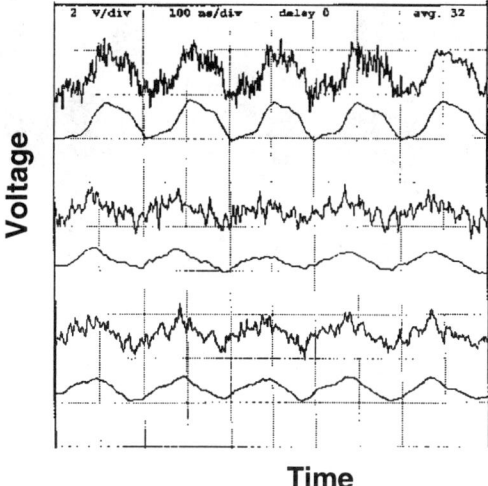

Figure 10. Example of an EBT waveform.

Electron Beam Induced Current (EBIC) Imaging

Information Technique Yields: EBIC imaging localizes regions of Fermi level transition. EBIC is primarily used to identify buried diffusions and Si defects.

Physics Behind Technique Use: When the primary electron beam is scanned across a sample, collisions in the target material form electron-hole pairs within the bulk of the sample. The relatively low ionization energies (less than 10 eV) of materials used in integrated circuit manufacturing allow a single 10 keV primary electron to produce an many as 500 to 1000 free electron-hole pairs. These pairs usually recombine randomly in the material; however, if production occurs in a space-charge (depletion) region, the charge carriers will be separated by the junction potential before recombination. The large number of pairs per primary electron generates an EBIC signal much larger than the incident beam current. The magnitude and direction of the induced current is used to generate an image localizing where junction potentials occur. In contrast to secondary electron and backscattered electron scanning electron modes, the EBIC signal detector is the device itself. By controlling the primary electron beam energy, the depth of the diffusions examined can be differentiated. Various device pins may be sampled to observe different EBIC signals.

Difficulty to Implement: The equipment necessary to perform EBIC imaging is a scanning electron microscope, current to voltage amplifier, and an electrical vacuum feed-through. The only sample preparation needed is lid/plastic package removal. The passivation may or may not be removed for EBIC analysis. No electrical driving equipment is necessary since the integrated circuit is driven only by the electron beam. Because of the small signal generated, a digital image acquisition system would be advantageous to image acquisition/manipulation, but not necessary. Electrical testing to determine proper node selection is also desirable. An example of an EBIC image is shown in Figure 11. Diffusions across the entire IC are visible in Figure 11.

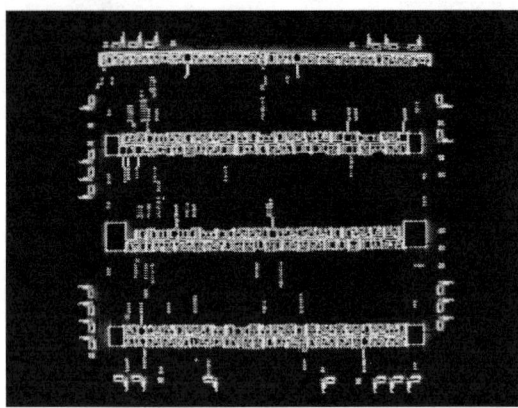

Figure 11. Example of EBIC imaging of an entire die.

Possible IC Damage: The same considerations for IC damage described for SE imaging apply to EBIC imaging.

X-Ray Microanalysis

Information Technique Yields: X-ray microanalysis permits identification and localization of elemental constituents of a sample, with resolution limits of 0.5 to 0.1 weight percent.

Physics Behind Technique Use: Two types of x-ray generation occur during primary electron beam interactions with target atoms. Bremsstrahlung or continuum x-rays are generated as the electrons are decelerated through interactions with the nuclear cores of atoms. These x-rays are generally considered the "noise" component of the total x-ray signal. The continuum x-rays have an intensity that varies as (Eo-E)/E, where Eo is the electron beam energy and E is the x-ray energy.

Characteristic x-rays are the second type of x-ray generation from electron beam interaction. When the electron beam displaces an inner core electron, an electron from a higher energy level will decay to replace the missing, inner electron (Figure 12). The transition from the higher energy level to the lower level will be accompanied by an energy release, whose magnitude is the difference between the two electron energy levels. A photon (x-ray) with this "characteristic" energy is one form of energy release. Because the inner core electrons of different atoms are separated by known, discrete energy levels, elemental identification is possible from observing the energies of the characteristic x-rays produced. Figure 13 summarizes some of the permitted energy level transitions for atomic electrons. The K, L, M, and N refer to different energy levels.

X-ray energy spectra of a sample may be acquired with the beam scanning over a sample area or localized to a particular point. A sample spectra is shown in Figure 14. These spectra identify the elements in that region. A scanned image or dot-map in which the contrast is modulated by detection of a certain x-ray energy displays the spatial distribution of that x-ray energy (element) source. A secondary electron image and dot-map for Au characteristic x-rays is shown in Figure 15.

Difficulty to Implement: The equipment necessary to perform x-ray microanalysis is an electron beam source, such as a scanning electron microscope or an electron-microprobe, and an x-ray detector. Two types of x-ray detectors are available. First, a wavelength dispersive x-ray (WDX) detector uses Bragg interference to generate an x-ray wavelength spectra. The WDX detector takes advantage of an interference crystal and the Bragg equation:

$$n\lambda = 2d\sin\Theta \qquad \text{Eq. 2}$$

where λ is the x-ray wave length, Θ is the x-ray angel of incidence, d is the interplane spacing of the reflection crystal, and n is a quantum number (1, 2, 3, ...). By changing the angle of incidence different wavelengths (energies) may be observed. (For photons, energy and wavelength differ by a constant.) The actual geometry for WDX detectors is quite complicated. More details are available in the reference section. The typical energy resolution of WDX detector is < 10 eV.

The second type of detector is an energy dispersive x-ray (EDX) detector. The EDX detector employs a semiconducting medium, usually lithium-drifted silicon, to convert x-ray energy into charge. The x-rays deposited into the semiconductor will create electron-hole pairs in the semiconductor. These hole pairs are collected by biasing the semiconductor and are converted into a voltage pulse. The larger the energy of the x-ray the greater the number of hole pairs and the larger the voltage pulse. The magnitude of the voltage pulse is then used to determine the incident x-ray energy. The typical energy resolution of EDX detectors is < 150 eV, but it is faster than WDX analysis.

Quantitative x-ray analysis is possible with both detection systems but requires implementing a complicated methodology which considers electron beam current, x-ray production efficiency, sample absorption of x-rays, and other factors. Information on quantitative analysis may be found in the references section.

Possible IC Damage: Same as for SE imaging.

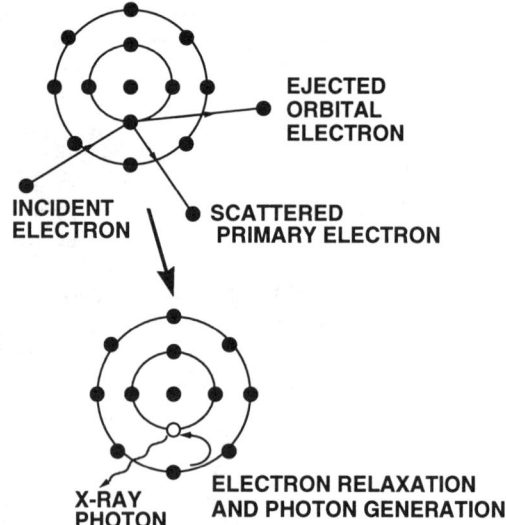

Figure 12. Characteristic x-ray generation.

138

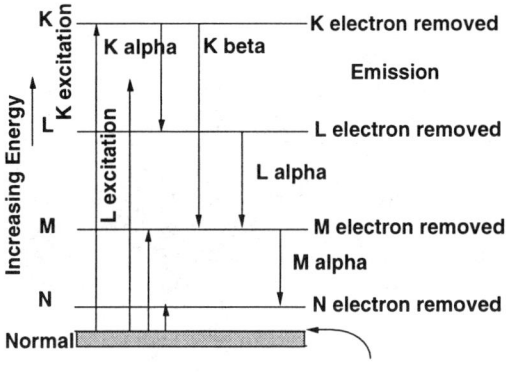

Figure 13. Atomic electron energy level transitions.

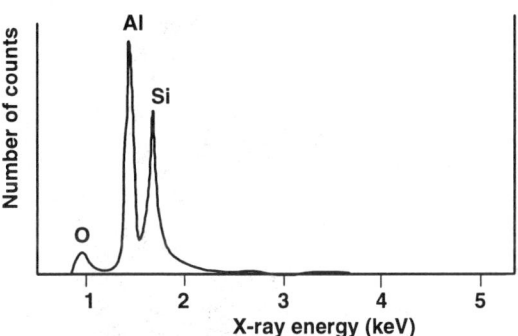

Figure 14. An example x-ray spectra.

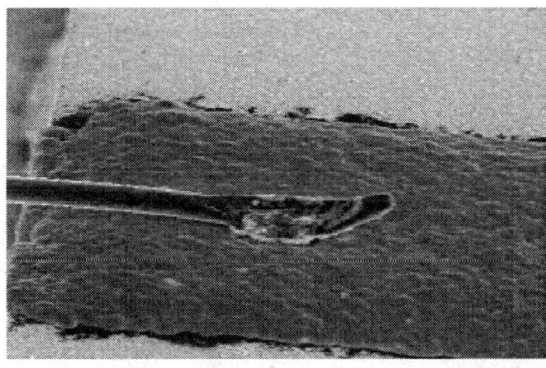

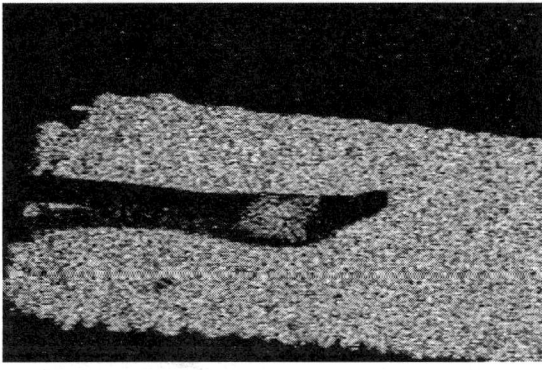

Figure 15. An example of a SE and a Au x-ray dot map with the same field of view.

ADVANCED SEM TECHNIQUES

Passive Voltage Contrast Imaging at Various Angles

Information Technique Yields: Conductor continuity and gate oxide ruptures can be located using the variation in VC with primary electron beam angle. This technique is used on depassivated ICs.

Physics Behind Technique Use: Passive VC imaging employs the physics described in the SE emission and VC sections along with the variation in secondary electron emission efficiency with tilt. The secondary electron emission efficiency increases with tilt angle as 1/cos(tilt angle). Low primary electron beam energies (~1 keV) are used for passive voltage contrast to avoid large amounts of deposited negative change and maximize the difference in image contrast between floating and grounded conductors. Grounded conductors will not change voltage with tilt because of a charge path to ground. Floating conductors will increase in voltage as a result of increased secondary electron emission, until an equilibrium voltage is achieved where the number of primary electrons injected equals the number of secondary electrons emitted. The change in the equilibrium voltage of floating conductors with tilt can be seen in the VC image to differentiate grounded and floating regions.

Figure 16 is an example of using passive voltage contrast to locate a faulty gate contact. The passivation and metal have been removed from the IC. The polysilicon contact highlighted by the arrow is bright, indicating a path to ground similar to most of the other contacts to the silicon substrate in Figure 16. The path to ground of the polysilicon contact is the result of a gate oxide rupture between the polysilicon and the silicon substrate. The dark contact to the right is electrically floating and charged positive because no ground path defect exists.

Difficulty to Implement: The equipment necessary to use passive voltage contrast is an SEM with low primary beam energy capability and the capacity to ground structures on an IC. Almost all SEMs have the capability to tilt the sample and ground connections in the vacuum chamber. Sample preparation and access to regions on the IC that need to be grounded are the major difficulties with this technique.

Possible IC Damage: Same as for CCVC imaging.

Figure 16. Example of passive voltage contrast with the grounded structures bright and the floating ones dark.. (From S. Bothra et al., "A New Failure Mechanism by Corrosion of Tungsten in a Tugsten Plug Process", Proceedings of the IRPS, 1998, pp. 150-156.)

139

CCVC Coupled with Quiescent Current (I_{DDQ}) Testing

Information Technique Yields: CCVC imaging coupled with I_{DDQ} testing permits localization of nodes generating anomalous I_{DDQ} values on a CMOS IC.

Physics Behind Technique Use: I_{DDQ} testing of CMOS ICs examines the quiescent current values (I_{DDQ}) after all switching transients. Large values of I_{DDQ} (above ~50 mA for most static CMOS) may indicate defects or a reliability risk. The I_{DDQ} testing technique is powerful because only one parameter needs to be examined for analysis but localization of the nodes producing high I_{DDQ} values can be difficult. The CCVC technique has been applied to localize the nodes generating high I_{DDQ} values. When performing I_{DDQ} testing the vectors that generate high I_{DDQ} are recorded. It is assumed that the same defect will produce the same or similar I_{DDQ} value when activated by different applied vectors. CCVC images are then acquired of the IC at the different vectors causing the elevated I_{DDQ} value. The CCVC images are then compared and processed to retain only the conductors that have the same potential shift. The left-hand side of Figure 17 is a single CCVC image with a high I_{DDQ} applied vector. The right-hand side of Figure 17 is the result of comparison with three other CCVC images with different vectors applied, but the same I_{DDQ} value. Only the information common to all four images is retained. The eventual outcome is that only the node(s) responsible for the high I_{DDQ} is present in the image.

Difficulty to Implement: The same hardware necessary for CCVC imaging is required. In addition, an IC tester capable of performing I_{DDQ} testing and vector recording as well as a digital image processing system for image comparison and processing is needed. No image alignment software is needed since all images for comparison will be from the same IC. The data acquisition time will be reduced if the entire IC die or area of interest can be put into the same field of view. If the IC must be moved the data acquisition time will increase accordingly.

Possible IC Damage: Same as with CCVC imaging.

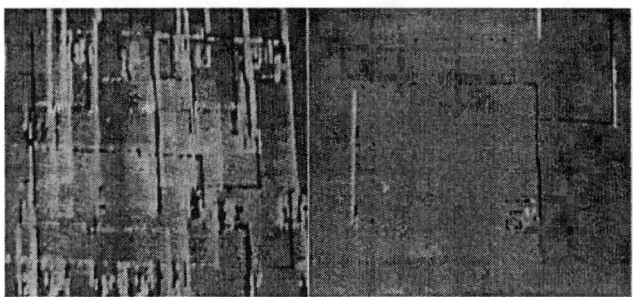

Figure 17. Example images of initial (left) and processed (right) images of CCVC coupled with I_{DDQ} testing. (From R. Bottini et al., "Failure Analysis of CMOS Devices with Anomalous IDD Currents", Proceedings of the ISTFA, 1991, pp. 381-388.)

Resistive Contrast Imaging (RCI)

Information Technique Yields: RCI generates a relative resistance map between two test nodes of a passivated integrated circuit. The map generated will display buried conductors on an integrated circuit and may be used to localize open conductors.

Physics Behind Technique Use: RCI obtains resistance information by using the integrated circuit as a complex current divider. Figure 1 displays the electron beam interaction products between a passivated integrated circuit and a 10-keV primary electron beam. To obtain RCI information the primary electron beam energy is increased until the tip of the interaction volume intersects the buried conductor of interest. A portion of the primary electron beam current will be injected into the conductor. Using an amplification configuration as shown in Figure 18, the currents induced by electron beam exposure will have a path out of the integrated circuit. The relative resistance between the electron beam position on the circuit and the test nodes determines the direction and amplitude of current flow. The current, on the order of nanoamps, is amplified and used to make a resistance map of the conductors. Usually the power and ground inputs are used as test nodes because of their global nature across the integrated circuit. However, other node combinations may be used if desirable/indicated. If a resistance change occurs along a conductor relative to the test node combination selected, such as an open conductor, the RCI image will display an abrupt contrast change at the open site. The RCI image in Figure 19 localizes an electromigration open along a clock line.

Difficulty to Implement: The equipment items necessary to implement RCI are: a scanning electron microscope, current to voltage amplifier, and an electrical vacuum feed-through. The only sample preparation needed is lid/plastic package removal. The passivation is not removed for RCI. By increasing the primary electron beam energy, multilevel conductors under metal may be observed. No electrical driving equipment is necessary since the integrated circuit is driven only by the electron beam. Because of the small signal generated, a digital image acquisition system would be advantageous to image acquisition/manipulation, but not necessary. Electrical testing to determine proper node selection is also desirable. Another induced current effect, Electron Beam Induced Current (EBIC) should be avoided. EBIC generates currents several orders of magnitude greater than RCI which can mask the RCI signal. EBIC signals are generated primarily from buried diffusions and can be avoided by using lower primary electron beam energies. With practice in selecting the proper primary electron beam energy and current RCI data may be acquired readily. Unfortunately, not all internal conductors with defects will be identified using RCI. Escape from detection occurs when the current paths from the defect-containing conductor to the IC pins are too convoluted and no difference in resistance occurs across the open site relative to the IC pins.

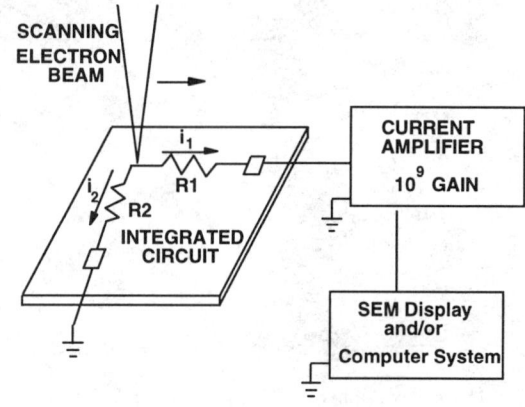

Figure 18. RCI imaging system.

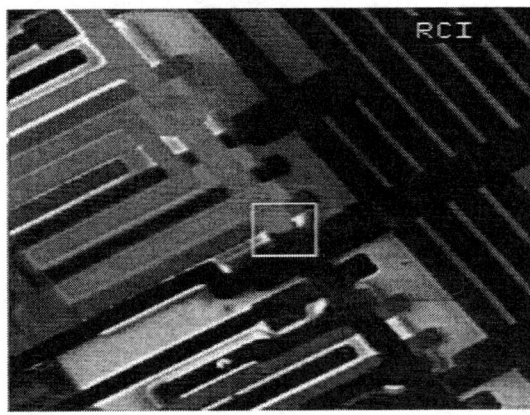

Figure 19. Example of RCI imaging locating an open conductor.

Possible IC Damage: No direct "physical" damage occurs with electron beam testing. A possible alteration of the threshold voltages on MOS transistors is possible. This change in threshold voltage results from irradiation damage to the gate oxides of MOS transistors. At the primary electron beam energies used for RCI, no direct primary electron/gate oxide interactions occur. However, some bremsstrahlung x-rays generated by interactions in the passivation layer will deposit energy in the gate oxide. This x-ray dose will alter the interface trap density and occupancy as well as the fixed charge levels in the gate oxide. Experiments on 3 micron, commercial grade, MOS transistors indicate that 40 images with 1 micron spatial resolution may be acquired before the threshold voltage shifts by 5%. If desired, low temperature annealing (100-150 C) may be used to restore the threshold voltage near or to its original value.

Charge-Induced Voltage Alteration (CIVA) Imaging

Information Technique Yields: CIVA was developed to localize open conductors on both passivated and depassivated CMOS ICs. CIVA facilitates localization of all open interconnections on an entire IC in a single, unprocessed image. CIVA has been applied to an analog bipolar technology with similar results.

Physics Behind Technique Use: CMOS ICs with open conductor lines may function at low to moderate (< 50 kHz) frequencies. The reason for this functionality is that significant quantum mechanical electron tunneling across the open can transport enough charge at low frequencies to maintain functionality. The maximum operating speed depends upon the nature of the open.

Even though the ICs may be functional with open conductors, charge injection into the floating portion of the conductor may cause significant loading that can overwhelm the open's tunneling capacity. CIVA takes advantage of this tunneling capacity to create an image of "loaded" areas. The CIVA image is generated by monitoring the voltage shifts in a constant current power supply as the electron beam is scanned over a biased integrated circuit. As electrons are injected into non-failing conductors the additional current, on the order of nanoamps, is readily absorbed and produces little change in the supply voltage. When charge is injected into an electrically floating conductor, the voltage of the negative conductor becomes more negative. This abrupt change in voltage on the floating conductor generates a temporary shift in the voltage demand of the of the constant current source supplying bias to the integrated circuit. The shift in power supply voltage can be either positive or negative depending on the proper state of the floating conductor. As the electron beam moves away from the floating conductor the previous equilibrium is quickly (~ 100 msec depending upon the bandwidth of the current source) reestablished. The shifts observed in the power supply voltage, even for opens that exhibit significant tunneling, are on the order of 100 mV with a 5 V supply voltage. These relatively large shifts produce images in which the contrast is dominated by the open conductors. Transistors with "weak" drive capacity have also been identified using CIVA.

Figure 20 displays the experimental setup used to generate CIVA images. Figure 21 shows an example of CIVA imaging on a passivated IC with an open conductor. The top left image in Figure 21 shows a CIVA image with no processing. The other three images show an overlay of the CIVA signal and secondary electron images at different magnifications. The highest magnification shows the open conductor at a polysilicon step.

Difficulty to Implement: The equipment necessary to perform CIVA imaging are a scanning electron microscope, a constant current source, and an electrical vacuum feed-through. The only sample preparation needed is lid/plastic package removal. A digital image acquisition would be advantageous to image acquisition/manipulation, but is not necessary. The passivation may or may not be removed for CIVA analysis. The IC must be biased into a non-contention state, but no complicated vector set is required. By increasing the primary electron beam energy, multilevel conductors under metal may be detected. Like RCI and BRCI, the selection of proper primary electron beam energy includes using energies that just reach the buried conductors and avoid EBIC signal generation. See the sections describing RCI and BRCI for a complete description of proper beam energy selection.

Possible IC Damage: Same as with RCI imaging.

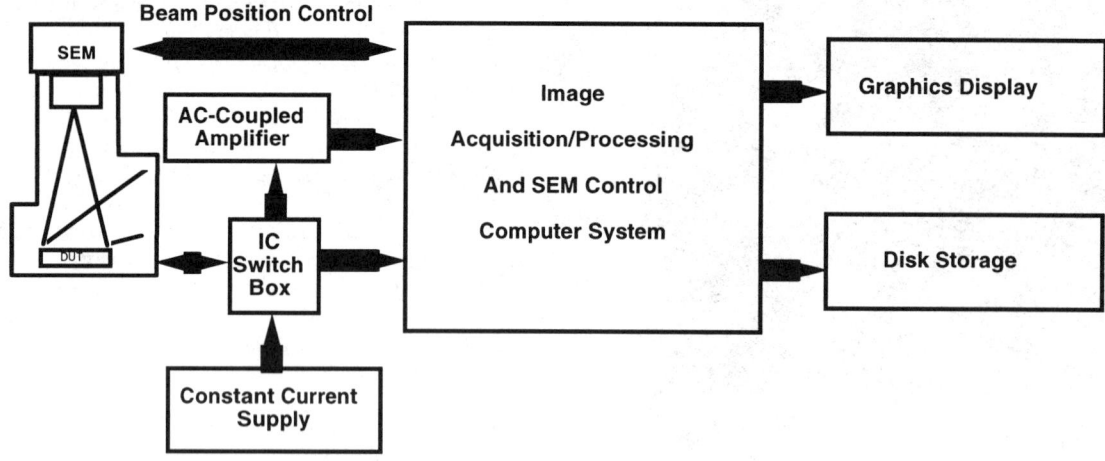

Figure 20. CIVA imaging system.

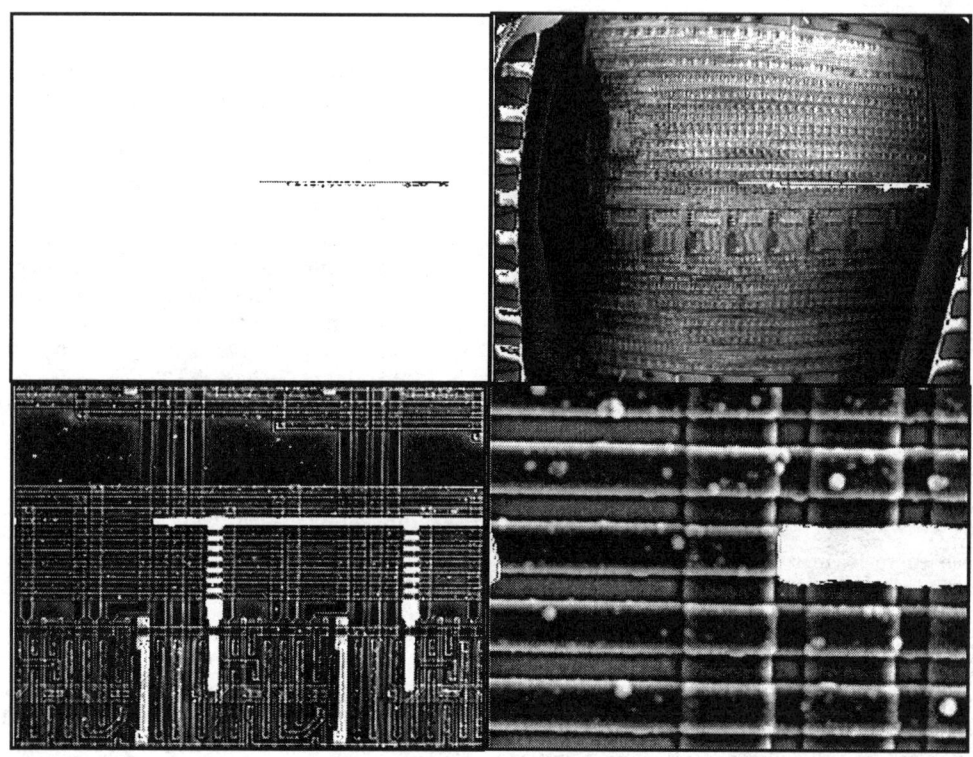

Figure 21. Image examples of CIVA at low magnification (upper left), overlaid with an SE image(upper right), and higher magnification CIVA/SE images localizing an open conductor (lower images).

Charge-Induced Voltage Alteration (CIVA) Using Low Primary Beam Energies

Information Tool Yields: CIVA at low primary electron beam energies (< 1.0 keV) localizes open metal conductors beneath passivation layers as with conventional CIVA. The major difference between the two approaches is that low beam energy CIVA does not introduce any irradiation damage to the IC under test. Additionally, low energy CIVA can be implemented on electron beam test systems which operate only at low primary

electron beam energies. Thus far, only single level metal CMOS ICs have been examined.

Physics Behind Tool Use: (NOTE: This description builds on the physics of CIVA and assumes the reader is familiar with conventional CIVA.)

Conventional CIVA localizes open conductors on an IC by powering the IC with a constant current source and monitoring the voltage fluctuations in the power supply with electron beam position. Passivated and multilevel interconnect ICs are examined by increasing the primary electron beam energy until electrons from the beam are injected into the buried conductors. This capability to

142

probe through buried layers is very powerful, but care must be exercised to avoid irradiation damage of the IC under examination. The requirement that the interaction volume of the primary electron beam interact directly with the buried conductor also limits the utilization of CIVA to SEMs with variable beam energy. Most electron beam testers have a fixed primary electron beam energy around 1.0 keV. CIVA at low primary beam energies (< 1.0 keV) circumvents these problems by producing CIVA images of buried IC conductors without direct interaction.

At low primary electron beam energies (< 1.0 keV) the surface passivation of an IC will reach a positive equilibrium voltage as with Capacitive Coupling Voltage Contrast (CCVC). While the time to reach the positive equilibrium voltage is directly proportional to the incident electron beam flux, the value of the equilibrium voltage has previously been thought to be independent of beam flux. Recent experimental work has shown that at very high primary electron beam currents (> 20 nA) the equilibrium surface voltage will become negative at low primary electron beam energies. During a low energy beam exposure, a surface initially at 0V is thought to go positive and then negative at high electron flux. The negative voltage is believed to be due to a depletion of the available secondary electrons at the surface during a high electron flux, low energy exposure. (This effect has **not** been previously documented in literature describing electron beam surface interactions. In fact, it is generally believed that the surface equilibrium voltage is independent of primary electron beam current.) This rapidly changing surface equilibrium voltage will polarize the passivation and produce a changing bound charge at the metal conductor-passivation interface, similar to the bound charge used to make a CCVC image. The changing bound charge will introduce a small voltage pulse on the buried conductor. As with conventional CIVA, this small voltage change is readily absorbed by non-failing conductors, but the voltage pulse can change the voltage of floating conductors and alter the power supply demands of the IC under test. By imaging the changing voltage demands of a constant current supply powering the IC under test, low power CIVA images are produced.

AC operation of the IC produces an improved low energy CIVA image by "depleting" the bound charge on the IC conductors. AC operation also increases the probability of localizing open conductors by increasing the chances that a given failing conductor will be placed in a logic state susceptible to CIVA observation. AC operation is facilitated by performing CIVA on commercial electron beam test systems which are normally configured for AC analysis of an IC.

Low energy CIVA images of open conductors are show in Figure 22. The images were acquired using a 300 eV primary electron beam energy.

Implementation: The equipment necessary to perform low energy CIVA is identical to that for conventional CIVA: a scanning electron microscope, a constant current source, and an electrical vacuum feed-through. In addition, electron beam testers can be used. The sample preparation needed is lid/plastic package removal. The IC must be biased in a non-contention state(s), but no complicated vector set is required. High electron beam currents are required for low energy CIVA analysis, so procedures to maximize the currents such as larger apertures and lower condenser lens settings are advisable.

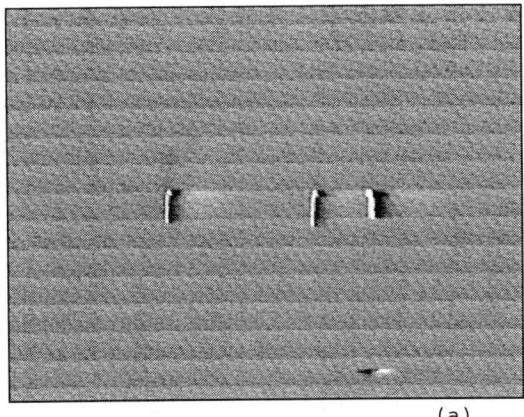

(a)

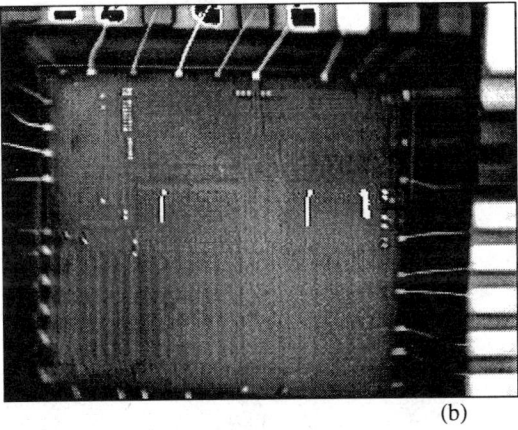

(b)

Figure 22. A low energy CIVA image (a) showing 3 open conductors. The CIVA image was acquired using a 300 eV primary electron beam. A combined secondary electron image (b) is shown for registration.

For standard IC technologies with single level metal interconnect low energy CIVA will be capable of imaging most structures. Higher interconnect layers between the passivation and the conductors of interest will absorb the bound charge of interest, preventing observation of the deeper conductor regions. Even if no metal layer obscures the bound charge path from the surface to a lower level conductor, the increased thickness of dielectric material will reduce the low energy CIVA effect just as the CCVC signal is attenuated by thicker dielectric layers. Under these conditions conventional CIVA may be advisable, but low energy CIVA should be applied first because of its almost totally non-destructive nature.

Possible IC Damage: No physical or irradiation damage occurs with low energy CIVA examination. A possible "damage" scenario for ICs is the creation of snap-back or latchup conditions in a biased IC, which can occur rarely in limited situations. This can be avoided by prudent selection of the current compliance of the biasing power supply.

143

OPTICAL BEAM BASED TECHNIQUES

Light-Induced Voltage Alteration (LIVA)

Information Tool Yields: LIVA is a Scanning Optical Microscopy (SOM) technique developed to localize diffusions associated with integrated circuit defects. The high selectivity of LIVA permits examination of the entire IC die in a single, unprocessed image. LIVA examination of an IC's surface from the front side is performed using a visible laser light source. By using an infrared light source LIVA has shown applicability to backside examination of ICs. The LIVA approach has also been used to identify the logic states of transistors with much greater sensitivity than previous optical techniques.

Physics Behind Tool Use: The effects of photon injection on ICs have been well documented in the literature. As photons interact with semiconductor materials they produce electron-hole pairs which can effect IC operation. The most widely know SOM technique utilizing electron-hole pair generation is Optical Beam Induced Current (OBIC). In OBIC non-random recombination of electron-hole pairs near diffusions generates a net current that can be used to image the diffusions. OBIC can also be used to identify biased transistor logic states by observing the magnitude of the "photo-current" generated by optical beam exposure.

LIVA, like OBIC, takes advantage of photon generated electron-hole pairs to yield IC functional and defect information. LIVA is an optical beam corollary to the electron beam technique, charge-induced voltage alteration (CIVA). In both techniques the device under examination is biased using a constant current power supply. LIVA images are produced by monitoring the voltage shifts in the power supply as the optical beam from the SOM is scanned across the device. Voltage shifts occur when the recombination current increases or decreases the power demands of the IC.

The LIVA measurement and imaging of voltage shifts rather than directly observing the photo-currents has two advantages. First, the IC will act as its own amplifier producing a much larger LIVA voltage signal than a photo-current signal. This is in part due to the difference in "scale" for voltage and current on ICs. For example, a CMOS IC biased at 5V had a photo-current increase of 100 nA (from 90 nA to 190 nA) when a transistor gate was exposed to the SOM beam. When the same conditions were repeated for the IC biased with a constant current supply set to yield 5V with no illumination (90 nA supplied), the voltage decreased by **2.4 volts** (from 5 V to 2.6 V). A second advantage is that IC voltages are much simpler to measure than IC currents. Current measurement is in series, with the measurement system becoming "part" of the IC. Because of this series measurement, most current amplification systems will have maximum current limits (typically 250 mA) that limit the operational range without modifications. There is also the added complication of sometimes needing to measure a relatively small photo-current against a large dc background current. Voltage measurements are made in parallel and are therefore "separate" from the IC with none of the current measurement limitations. Small changes in voltage are easily measured using an ac coupled amplifier immune to background dc voltages. This relative signal difference and simpler equipment setup make LIVA far more attractive than conventional photo-current methods.

Defects on ICs, such as diffusions connected to open conductors and electrostatic discharge damage, produce relatively large LIVA signals when compared to the LIVA signal of biased diffusions under similar conditions (10 to 1000 or more times greater). This relative difference is used to produce highly selective LIVA images for defect localization.

When no defects are present, or when they are "deactivated" by changing the state of the IC such that the diffusions associated with defects have little or no electric field, the LIVA signal from non-defective diffusions can be used to determine transistor logic state.

Backside LIVA examination is performed using an infrared light source. The source wavelength must be large enough to take advantage of silicon's greater transparency to infra-red light, but small enough to generate electron-hole pairs in the diffusion regions of the IC. An 1152 nm, 5 mW, HeNe laser have been used to identify LIVA defects from the backside of an IC die. This laser does not produce enough electron-hole pairs for backside logic mapping using LIVA. A 1064, 1.2 W Nd:YAG laser has been used to successfully identify CMOS transistor logic states from the backside of an IC die.

Figure 23 displays how a defect can be localized using LIVA from the backside of an IC. The IC is a radiation hardened version of the Intel 80C51 microcontroller. A 5 mW, 1152 nm laser was used to acquire the images. The LIVA image in Figure 23a displays a small signal in the lower right. Figure 23b is a reflected light image showing the same field of view. Note that the entire IC is being examined in this field of view. Figures 23c and 23d are higher magnification LIVA and reflected light images of the area identified in Figures 23a and 23b. Later analysis indicated that the LIVA signal is caused by an open metal-1 to silicon contact. Note that the open contact area is completely covered by a metal-2 power bus that would obscure any front side examination.

Figure 24 displays how logic state mapping can be performed using LIVA from the backside. A 1.2W, 1064 nm laser was used to acquire Figure 24. The field of view shows a portion of the SRAM embedded in the microcontroller examined in Figure 23. Figures 24a and 24b are same field of view LIVA and reflected light images respectively. The logically "off" p-channel transistors produce the dark contrast seen in Figure 24a. Logically "on" p-channel transistors produce a bright contrast. This difference in contrast permits reading of the SRAM's contents.

Difficulty to Implement: The equipment necessary to perform LIVA imaging are a scanning optical microscope (SOM), a constant current source, voltage amplifier (preferably AC coupled), and electrical connections to the IC under examination. Almost all commercial SOMs will have an auxiliary or OBIC input port suitable for the LIVA input. The IC must be biased in a non-contention state, but no complicated vector set is required. For front side LIVA examination any visible wavelength will generate enough electron-hole pairs for LIVA imaging. Backside LIVA examination must consider the wavelength restrictions mentioned above.

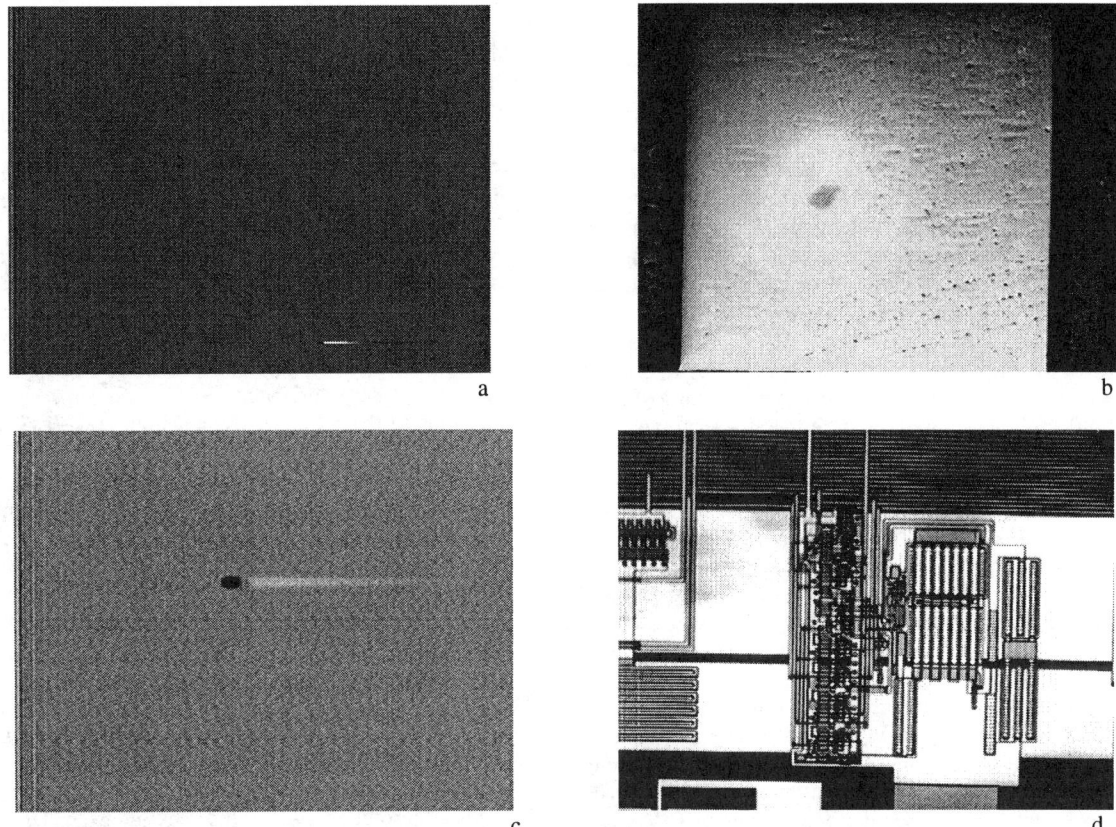

Figure 23. Backside LIVA examination of a microcontroller: LIVA image (a) indicating the area of an open metal-1 to silicon contact, backside reflected image (b) showing the same field of view as Figure 23a, higher magnification LIVA image (c) of the defect in Figure 23a, and a reflected light image (d) of the same field of view as Figure 23c. The defect site is completely covered by a metal-2 power bus from the IC's front side.

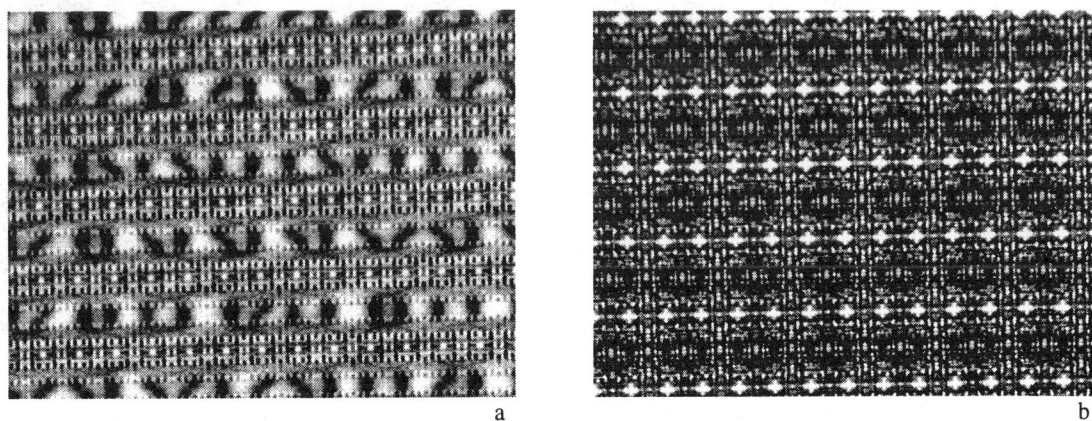

Figure 24. Backside LIVA image (a) showing the logic states of transistors in an SRAM embedded in a microcontroller and backside reflected light (b) image of the same field of view.

Possible IC Damage: No physical damage occurs with LIVA examination. Two possible "damage" scenarios for ICs are: (1) erasure of nonvolatile memories which are susceptible to certain light energies and (2) creating snap-back or latchup conditions in a biased IC. This can be avoided by prudent selection of the light source and the current compliance of the biasing power supply. Die thinning is not necessary for LIVA, but polishing the backside of the IC die to reduce light scattering greatly improves LIVA and reflected light image resolution. All of the examples shown above have had polished backsides.

145

Thermally-Induced Voltage Alteration (TIVA) and Seebeck Effect Imaging (SEI)

Information Tool Yields: TIVA and SEI are different variants of a Scanning Optical Microscopy (SOM) technique developed to localize opens (SEI) and short circuits (TIVA) on integrated circuits with localized heating. The high selectivity of both variants permit examination of the entire IC die in a single, unprocessed image. At the wavelengths used localized heating can be produced from the front or backside of the IC making defect localization possible from either side.

Physics Behind Tool Use: Both TIVA and SEI take advantage of the interactions of a localized heat source (a focused infrared laser) and two common IC defects (an open or a shorted interconnection). Both techniques use an infrared laser wavelength longer than 1.1 mm (the indirect bandgap in silicon) to produce localized heating on an IC without electron-hole pair production. In TIVA, the localized heating changes the resistance of a short site. If the short site is a conductor, the resistance will increase with temperature. If the short site is a semiconductor, the resistance will decrease with increasing temperature. In any event the resistance of the short site changes with temperature. As the resistance of the short site changes the power demand of the IC changes, assuming the short site has a voltage gradient across it. This effect of localized heating on IC short circuits and the subsequent IC power demand changes was first shown in the OBIRCH (optical beam induced resistance change) technique. TIVA displays an increase in detection sensitivity using the constant current biasing approach applied in CIVA and LIVA. (See the descriptions of CIVA and LIVA in this seminar for a description of the defect sensitivity advantages of constant current biasing over conventional constant voltage biasing.) Figure 25a displays an entire 1MB SRAM with a short site as indicated by the TIVA image. Note that the short site has been localized in a single image. Figure 25b is a reflected light image of the same field of view for registration. Figure 26 is a backside TIVA/reflected light image pair of a short site. The particle producing the short cannot be seen in Figure 26b because it is on top of the shorted metal-1 interconnections. The example in Figure 26 demonstrates how heat can travel through a metal layer to detect a defect where photons cannot.

Localized heating can also be used to locate open conductor sites from the front and backside of an IC. Open conductor sites are identified through the use of the Seebeck effect. Thermal gradients in conductors generate electrical potential gradients with typical values on the order of mV/K. This is known as thermoelectric power or the Seebeck Effect and refers to the work of Thomas Johann Seebeck (1770-1831). The most common application of thermoelectric power is the thermocouple, which uses the difference in thermoelectric voltages of two different metals to measure temperature. If an IC conductor is electrically intact and has no opens, the potential gradient produced by localized heating is readily compensated for by the transistor or power bus electrically driving the conductor and essentially no signal is produced. However, if the conductor is electrically isolated from a driving transistor or power bus, the Seebeck Effect will change the potential of the conductor. This change in conductor potential will change the bias condition of transistors whose gates are connected to the electrically open conductor, changing the transistors' saturation condition and power dissipation. An image of the changing IC power demands (via constant current biasing) displays the location of electrically floating conductors. For the laser and SOMs used to

date a maximum temperature gradient of about 30 °C has been achieved, with most work producing temperature changes of 10 °C or less. The resulting small changes in open conductor potential need the sensitive constant current biasing approach for defect detection. Figure 27a shows a backside SEI image of a FIB-opened conductor. Figure 27b is a reflected light image for registration. The open conductor can be seen as well as strong contrast at the metal-polysilicon interconnections. The enhanced contrast is believed to result from the thermopower difference in the two different materials (polysilicon and the metal conductor).

Difficulty to Implement: The equipment necessary to perform TIVA and SEI imaging are a scanning optical microscope (SOM) with the proper wavelength laser, a constant current source, voltage amplifier (preferably AC coupled), and electrical connections to the IC under examination. Like LIVA, almost all commercial SOMs will have an auxiliary or OBIC input port suitable for the TIVA/SEI input. The IC must be biased in a non-contention state, but no complicated vector set is required. Because the TIVA/SEI signal can be small depending on the details of the defect (especially for SEI), increased laser power will produce stronger signals. The use of a pulsed laser and a lock-in amplifier is presently underway to determine their applicability for increased signal strength.

Possible IC Damage: No physical damage occurs with TIVA/SEI examination. Die thinning is not necessary for backside analysis, but will increase the signal strength about 10X with each "halving" of the die thickness. Additionally, polishing the backside of the IC die to reduce light scattering greatly improves reflected light image resolution. All of the backside examples shown above have been polished, but not purposely thinned.

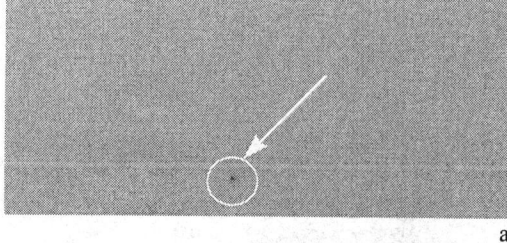

a

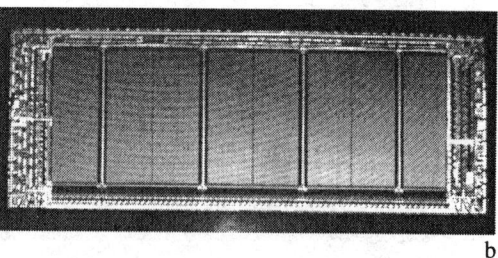

b

Figure 25. Entire die front side TIVA image of a 1MB SRAM showing (a) the site of a particle short. A reflected light image (b) is show of the same field of view for registration.

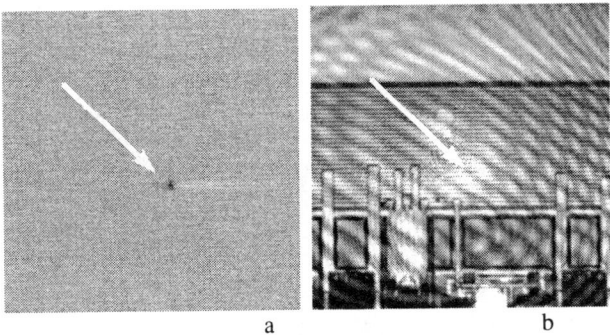

Figure 26. Backside TIVA image (a) localizing a short circuit site on a 1MB SRAM and backside reflected light (b) image of the same field of view. Note that the shorting particle on top of metal-1 is not visible from the backside in (b).

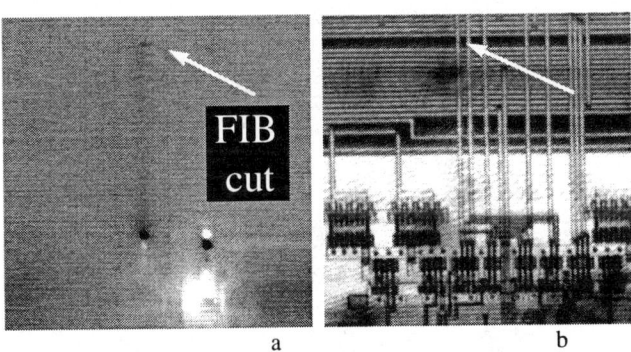

Figure 27. Backside SEI image (a) showing an open conductor resulting from an FIB cut and a reflected light image (b) for registration.

CONCLUSION

Because of the spatial resolution, depth of focus, ease of use, ready availability, and localizable interactions of a focused beam with microelectronic technologies beam-based analytical methods are powerful failure analysis techniques for IC failure analysis. For many of the photon based techniques an increase in defect sensitivity is desirable, especially for backside analysis of ICs with elevated power demands where the defect signal can be "lost" in the background signal. While averaging somewhat compensates for the weaker signals, further improvements are required to examine higher power ICs and to compensate for substrates with higher infrared absorption.

Two improvements to increase the defect signal strength in active photon systems is increased photon intensity and pulsed photon sources. While a more powerful laser source will increase the signal there are limits to the power that can be transmitted through the SOM optics without damaging the surface coatings. A pulsed laser and lock-in amplification of the active photon signals can also improve defect sensitivity. Although silicon has a high thermal conductivity, there are limits to how fast a laser pulse can be and still produce a useful change in temperature between pulses. Modeling at Sandia indicates that the time constant for silicon surface heating by a focused laser is in the range of 100 to 200 ns. The thermal response of the surface to a 1 MHz modulated laser should produce changes in temperature within 90% of the steady state heating conditions. Therefore pulsed laser examination of active thermal deposition should increase defect sensitivity. Similar work will be pursued for enhanced future LIVA analysis.

Near term improvements in the CIVA/LECIVA techniques will most likely come from equipment vendors supplying "IVA" options on existing equipment. For example, LECIVA can be a powerful addition to an existing electron beam/voltage contrast probe tool, incorporating the existing IC fixture infrastructure, voltage contrast analysis, and CAD database with open conductor localization. In addition to the equipment enhancements and automation, improving the detection sensitivity of CIVA and LECIVA for resistive, but not open, interconnections will be a topic of future development. This may require an incorporation of dynamic IC testing and analysis of functional and timing parameters for fault localization and diagnosis.

Because of their great effectiveness in failure analysis and future potential, beam-based photon and electron probing will continue to be mainstays in IC failure analysis for the foreseeable future.

ACKNOWLEDGMENTS

The author would like to thank Richard E. Anderson, Daniel L. Barton, Ann N. Campbell, Christopher L. Henderson, and Jerry M. Soden for their careful review of and valuable contributions to this seminar.

Sandia is a multiprogram laboratory operated by Sandia Corporation, a Lockheed Martin Company, for the United States Department of Energy under contract number DE-AC04-94AL85000.

REFERENCES FOR ALL SECTIONS

For SE, VC, EBIC, BSE, and X-ray:
L. Reimer, *Scanning Electron Microscopy*, Berlin, Springer-Verlag, 1985.

For VC, CCVC, EBIC, RCI, and BSE:
J.R. Beall, *Voltage Contrast Techniques and Procedures" and "Electron Beam-Induced Current Application Techniques*, Microelectronic Failure Analysis Desk Reference, Supplement Two, November 1991.

E.I. Cole Jr. et al., Advanced Scanning Electron Microscopy Methods and Applications to Integrated Circuit Failure Analysis, *Scanning Microscopy*, vol. 2, no. 1, 1988, p. 133-150.

For CCVC:
E.I. Cole Jr., et al., A Novel Method for Depth Profiling and Imaging of Semiconductors Devices Using Capacitive Coupling Voltage Contrast, *Journal of Applied Physics*, vol. 62 (12), 1987, p. 4909-4915.

For RCI:
E. I. Cole Jr., "Resistive Contrasting Applied to Multilevel Interconnection Failure Analysis", *Proceedings of IEEE VLSI Multilevel Interconnection Conference*, (Santa Clara, CA), June 1989, p. 176-182.

C.A. Smith et al., Resistive Contrast Imaging: A New SEM Mode for Failure Analysis, *IEEE Transactions on Electron Devices*, ED-33, No. 2, 1986, p. 282-285.

E. I. Cole Jr., "A New Technique for Imaging the Logic State of Passivated Conductors: Biased Resistive Contrast Imaging", *Proceedings of IEEE International Reliability Physics Symposium,* (New Orleans, LA), March 1990, p. 45-50.

For X-ray:
J.I. Goldstein et al., *Scanning Electron Microscopy and X-Ray Microanalysis*, New York, Plenum Press, 1984.

For Passive Voltage Contrast:
J. Colvin, "A New Technique to Rapidly Identify Gate Oxide Leakage in Field effect Semiconductors Using a Scanning Electron Microscope", *Proceedings of the International Symposium for Testing and Failure Analysis,* (Los Angeles, CA), October 1990, pp. 331-336.

For CCVC combined with IDDQ :
R. Bottini, et. al., "Failure Analysis of CMOS Devices with Anomalous IDD Currents", *Proceedings of the International Symposium for Testing and Failure Analysis,* (Los Angeles, CA), November 1991, p. 381-388.

For CIVA:
E.I. Cole Jr. and R.E. Anderson, "Rapid Localization of IC Open Conductors Using Charge-Induced Voltage Alteration (CIVA*)",* *Proceedings of IEEE International Reliability Physics Symposium,* (San Diego, CA), April 1992, p. 288-298.

For LECIVA*:*
E.I. Cole Jr. et al., "Low Electron Beam Energy CIVA Analysis of Passivated ICs.*", Proceeding of the ASM International Symposium for Testing and Failure Analysis*, (Los Angeles, CA), November 1994, p. 23-32..

For LIVA:
E. I. Cole Jr. et al., "Novel Failure Analysis Techniques Using Photon Probing With a Scanning Optical Microscope", *Proceedings of IEEE International Reliability Physics Symposium*, (Santa Clara, CA), April 1994, p. 388-398.

For OBIRCH and TIVA/SEI:
K. Nikawa and S. Inoue, "New Capabilities of OBIRCH Method for Fault Localization and Defect Detection", *Proc. of Sixth Asian Test Symposium,* 1997, pp.219-219.

E.I. Cole Jr. et al, "Backside Localization of Open and Shorted IC Interconnections", *Proceedings of IEEE International Reliability Physics Symposium*, (Reno, NV), April 1998, p. 129-136.

Electrical Fault Isolation
by Beam-Based Technology

Valluri R. Rao and Paul Winer
Intel Corporation
Santa Clara, California

INTRODUCTION

In identifying the root cause of an IC failure the first step involves the electrical Fault Isolation (FI) to the failing node. Physically a node is an electrical net in which every part of the net is at the same potential. Once the failing node is discovered, the precise X-Y coordinates of the defect location are found through a process known as Physical Failure Analysis (PFA). In the PFA step surface analysis tools (such as X-ray analysis, FIB, Auger, SIMS etc.) are used in conjunction with layer by layer stripback to find the physical defect location. For some classes of defects the X-Y coordinates of the defect location can be found directly in one step utilizing PhotoEMission (PEM) techniques. PEM techniques detect faint light emission that arise, due to hot electron effects even in indirect bandgap materials such as silicon. MOS transistor channels in saturation, P-N junctions in forward bias or reverse bias breakdown, latchup phenomena and gate oxide breakdowns due to pinholes all cause faint light emissions with unique signatures. The X-Y locations of these light emission sites can be quickly located with PEM techniques [1]. In this paper we primarily focus on the electrical Fault Isolation (FI) step of the process. Electrical Fault Isolation can be performed in a variety of ways. If the chip is designed using structured DFT methods such as full Scan Design (e.g. LSSD) [2], then it may be possible to isolate the failing node largely by the use of these testability features alone. Other testability features such as Built In Self Test (BIST) can be used isolate the fault to specific blocks in the chip. Other analysis techniques such as the two stage Fault Dictionary approach [3] can reduce the time required to perform fault isolation but becomes prohibitive for very large chips. IDDQ [4] can also be used to localize the failure in time (i.e. the clock phase for which the static current is higher than average), but, when used in combination with PEM or liquid crystal can be used to localize the defect location. There are however many situations in which none of these approaches are possible either due to inadequate DFT, multiple metallization layers (which obscures light emissions) or high background leakage (which makes IDDQ more difficult) etc. In these cases Fault Isolation must be performed by a combination of utilizing built in DFT features and test patterns to localize the problem as far as possible, followed by internal signal measurement and signal backtracing utilizing a variety of tools. Most of these tools rely on some form of charged particle or optical beam.

In this paper, the various tools and techniques that are available to us to aid in Fault Isolation, will be discussed. Generally the approach taken is that data measured from inside of the chip is compared with expected simulation data or with data taken from a golden or reference device to find the cause of the failure. In some situations it may be impossible to physically probe a node either because it is buried deep in the chip or it is covered by other metal lines. In these situations a Focused Ion Beam (FIB) mill can be used to locally generate probe points or cut away overlaying metal to make a node visible. In other situations particularly during the debug of a new product the probed data may indicate several possible reasons for a failure. The FIB mill can then be used to reroute signals and test the different theories so that the correct cause of failure may be ascertained. Most of these tools, use in some form or other, a charged particle beam (e.g. electron beam for contactless probing) or an optical beam. In the next section, the various techniques that are available for probing internal signals from ICs, will be discussed. In the last section some case studies will be presented. These case studies will illustrate how the cooperative use of some of the tools that are discussed in this paper can lead to rapid and effective fault isolation on complex VLSI chips.

SOME COMMON ISSUES

All beam based techniques rely on line of sight access to nodes (interconnects) in a chip. A typical integrated circuit has multiple levels of metal interconnects separated by inter-layer dielectrics. The final dielectric layer, on top of the upper interconnect level, protects the chip and is called the passivation layer. Most ICs also have a polyimide layer on top of the passivation. This layer reduces surface stresses, which could cause die cracking, during plastic packaging of the chip.

The typical layer build up of an IC is illustrated in Fig.1. Specimen preparation is required to decapsulate the device and also to etch the passivation layers to provide direct line of sight to the node to be probed. This is typically performed with plasma etching. More refined techniques such as Focused Ion Beam (FIB to be discussed

149

in more detail later) are used for locally removing passivation and to provide probing windows through obscuring metalization.

Also common to all analysis techniques is the problem of finding the required node and navigating to it. With modern complex IC's containing millions of transistors, finding the node is a major challenge. The only practical way to do this is by using X-Y coordinates from the design layout database of the chip to drive the precision positioning stages in the equipment. CAD tools also exist to map the circuit schematic nodes to the layout polygons and the netlist, as well as manually appended notations. A comprehensive CAD navigation package will seamlessly integrate all these functions. It is essential to have an established data processing flow to create all the required chip databases to ensure that the navigation works effectively.

Figure 1. A typical build up of an IC

Finally with techniques that require the chip to be operating or connected electrically to power supplies or a tester, the electrical interface must have adequate performance. In addition any special tests (e.g. short loops for stroboscopic waveform acquisition with E-beam probers) must also be available. Fig.2. illustrates the typical infrastructure required for effective use of these tools.

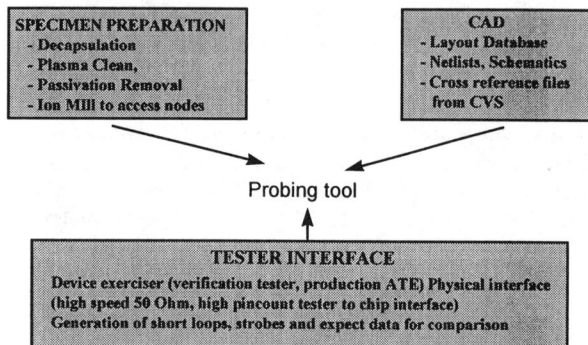

Figure 2. Typical Infrastructure for effective use of fault isolation tools

INTERNAL CHIP PROBING METHODS

A variety of techniques currently exist for probing internal signals from ICs. These include mechanical probes, Electron Beam probes, and Optical probes. The data that is measured by these different types of probes can be processed and displayed in a variety of ways to highlight various aspects of the failure. For example with an electron beam the secondary electron emission signal can be processed to give either Voltage Contrast images or waveforms. In addition if the leakage current from a reversed biased PN junction, which is being irradiated by the beam, is used to generate the

image, we get EBIC. Each of the probing techniques will now be discussed individually.

MECHANICAL PROBING

Mechanical probing is the oldest, simplest and least expensive method for probing internal chip signals. Although it is not a "Beam Based" technique it is included here for completeness particularly since it is still the best way of measuring precision analog signals and also asynchronous signals. The arrangement of a typical mechanical probing station is illustrated in Fig.3. A long working distance optical microscope gives an image of the chip and the lines being probed.

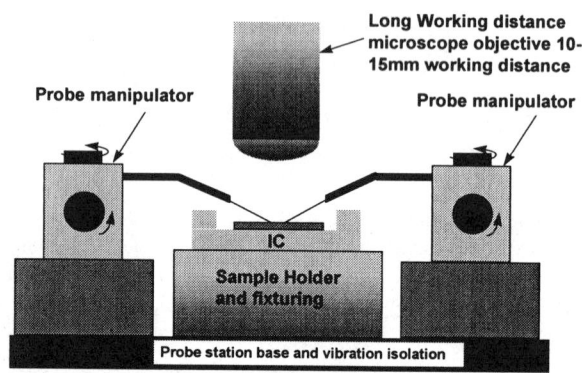

Figure 3. A typical mechanical probe station arrangement

In order to probe lines it is necessary to have line of sight access to the nodes so that the probe tip can physically touch the line. Laser ablation or Focused Ion Beam (FIB) or both can be used to locally mill away the polyimide and passivation to generate the node access. This is illustrated in Fig.4.

State of the art mechanical probes offer < 0.5 μm probe tips, capacitive loading of less than 50 fF and a bandwidth of greater than 1 GHz. Typically the probes with smaller capacitive loading and higher frequency response are "active" probes. These have small FET amplifiers built in very close to the probe tip. This is illustrated in Fig.5.

A typical problem encountered during mechanical probing is obtaining good physical contact between the probe tip and the metal line. A thin layer of native oxides on top the metal lines normally prevents good electrical contact unless the tip is slightly overdriven into to the line to punch through the oxide.

However one has to be careful not to damage the metal line during this procedure. Methods to obtain good contact with the metal line are shown in Fig.4 and Fig.5.

It is also possible to control the X-Y position of the probe relative to the chip so that precise navigation from a CAD layout is possible.

The mechanical probe is invaluable in making accurate analog measurements and also for capturing asynchronous signals. It's main drawbacks are spatial resolution and the associated difficulty of contacting very small lines with the probe tip and circuit loading due to the contact nature of the probe.

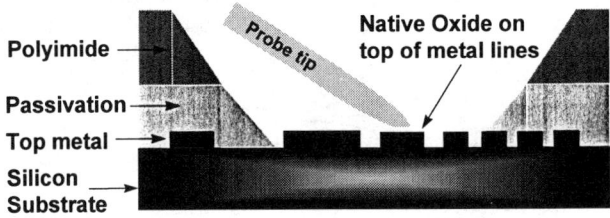

Figure 4. Local removal of polyimide and passivation using a laser or FIB for probing specific lines. The probe tip is over driven to punch through the native oxide that builds up on the metal lines

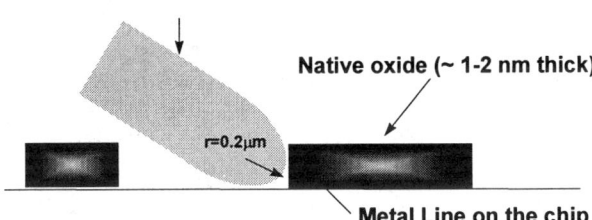

Figure 5. Using an active probe to probe a metal line. The probe is lodged into the edge making for a more effective contact to the line.

NON-INVASIVE E-BEAM PROBING

Electron Beam (E-Beam) probing utilizes a finely focused beam of electrons to replace the fine probe tip of a mechanical probe. An electron probe diameter of less than 100 nm is common. We will now discuss E-Beam probes is greater detail

When an electron beam bombards a sample, the primary electrons interact with the target atoms in a variety of ways to generate: a) Low energy secondary electrons with an energy in the range 0-50eV b) Auger electrons and characteristic X-Rays c) Photons, which, for example give rise to "Cathodo luminescence" d) Elastically reflected "Backscattered" electrons which have the same energy as the incident primary electrons. One feature of the secondary electrons is that their energy distribution dN(E)/dE is relatively independent of the primary *beam energy* and is only a strong function of the target material type and the angle of incidence of the primary beam with the target. The amplitude of the energy distribution is also proportional to the primary beam current. The ratio of the total secondary to the primary electron currents is commonly referred to as the *secondary emission coefficient* δ. The typical energy distribution of secondary electrons is given in Fig.6. In a Scanning Electron Microscope (SEM) the variation in δ, due to the change in incident angle of the primary E-Beam, as it scans over a surface leads to *topography contrast.* Contrast also arises in the SEM due to variations in sample electrical potential. This is called *Voltage Contrast.*

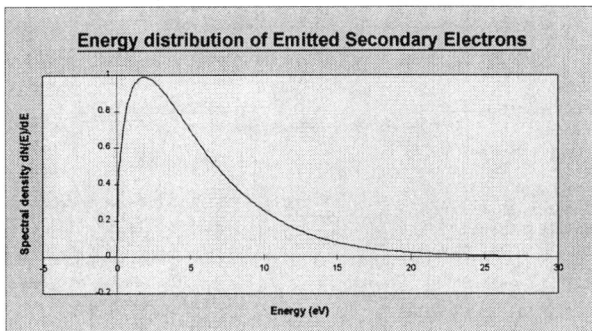

Figure 6. Typical secondary electron energy distribution

For a given primary electron beam and specimen conditions, the shape and magnitude of the secondary electron energy distribution is, to a first order, always the same. If the specimen potential is now made + (or -)V_S volts with respect to ground, every secondary electron attempting to escape from the sample, will be retarded (or accelerated) and will lose (or gain) an energy eV_S electron-volts. Hence the entire secondary electron energy spectrum shifts to the left (for +V_S) and to the right (for –V_S) along the energy axis (see Fig.7).

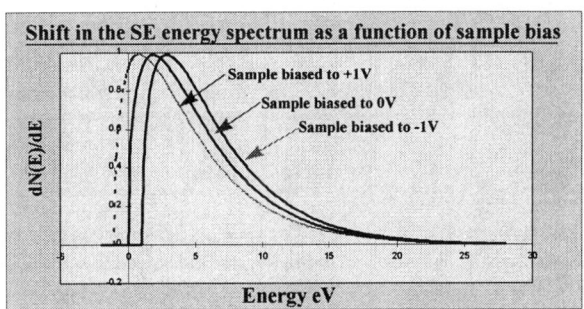

Figure 7. The phenomenon of Voltage Contrast illustrated by the shift in the secondary electron energy spectrum as a function of specimen bias

The average energy of a secondary electron is only about 2 eV. Hence a positive specimen bias greater than about $V_S = 2V$, will ensure that a significant portion of the energy spectrum has shifted to the left of the x-axis (illustrated by the dashed curve in Fig.7) implying that these electrons do not escape but are trapped by the surface. An alternative and qualitative way of looking at this is shown in Fig.8.

For any of the curves in Fig.7, the only current a detector can actually monitor, is the area under the curve for electron energy > 0.

A detector always monitors the *total* secondary electron current (i.e. the area under the energy distribution curves of Fig.7 to right of E=0). When the specimen biased positively, the detector will register a lower collected current than for the case when the specimen is biased negatively or at zero potential. Correspondingly, the positively biased specimen will appear darker in the SEM image than, when the specimen is negatively biased or grounded. This phenomenon is referred to as Voltage Contrast (VC) and forms the basis for all forms E-Beam probing. A typical Voltage Contrast image of a depassivated IC is illustrated in Fig.9.

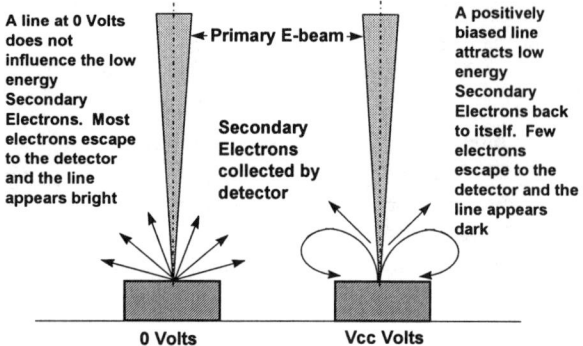

Figure 8. A qualitative explanation of voltage contrast

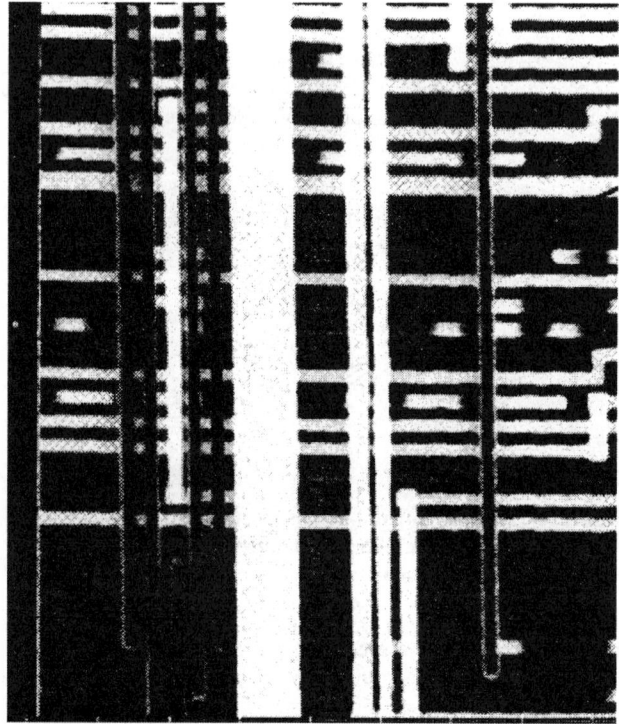

Figure 9. Voltage Contrast from an IC, depassivated to expose several interconnect layers. Bright lines are at ground while dark lines are at V$_{cc}$

It is clear from the above explanation that if the linear shift of the spectrum can be measured, the specimen voltage can be determined accurately. This is performed with the help of an *energy analyzer* (see Fig.10). It is also clear that the sample itself, acts as a high pass electron energy filter since it traps the electrons in the low energy portion of the spectrum. Before an energy analyzer can make an effective measurement of the shift in the energy spectrum, the local filtering action of the specimen must first be neutralized or made unimportant. One can do this either a) with the use a high electrostatic extraction field above the sample b) by only

concentrating on the high energy portion of the spectrum (i.e. greater than 5eV) where the high pass filtering action of the sample is unimportant. c) a combination of a) and b). Precise details of the energy analyzer and implementation will not be discussed here but can be found in [5]

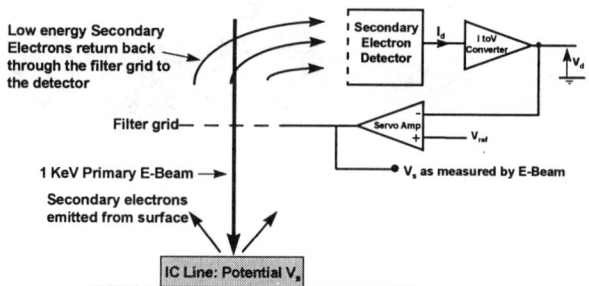

Figure 10. Operation of an energy analyzer for performing quantitative voltage measurements from ICs. The potential of the filter grid is caused to swing in unison with the potential on the line being probed by the electronic feedback loop. A condition for this to happen is constancy of the current that reaches the detector. The output of the servo amplifier, which drives the filter grid is therefore a direct readout of the voltage on the IC line.

Voltage Contrast therefore gives two very useful types of information when the specimen is a VLSI Integrated Circuit. Firstly the local high pass filtering action of the sample gives rise to a qualitative or Logic State contrast in which positively biased lines in the chip appear darker than grounded lines (Fig.9).

Secondly the use of an energy analyzer enables quantitative voltage measurements to be performed (Fig. 10). It should be mentioned that since voltage is measured by noting the energy shift of the secondary electron spectrum, between the measurement point in the chip and a reference electrode at a known voltage inside the energy analyzer, it is in principle possible to measure voltages from arbitrarily small lines. This is possible, even when the intended line is surrounded by other interfering lines, provided the *total energy* is measured and the primary beam *only* illuminates the intended line. The special requirement of measuring the total energy leads to *angle independent* detection and has been the subject of much research.

Time resolved measurements are performed with an E-Beam using a sampling (stroboscopic) technique (Fig.11). This is necessary because the secondary electron currents are very low (typically nA) and one runs into tradeoffs between frequency response, and signal to noise ratio. In the stroboscopic approach the only high speed component in the system is the sampling gate, which, in an E-Beam prober generates short electron pulses. The measurement electronics is low bandwidth and can be designed to be very low noise. There are however several drawbacks to this approach. Firstly, for the measurements, the chip must be exercising a precisely repeating test loop. In the case of a complex processor for example, there may need to be a significant amount of internal set up before a failure is made visible and this could lead to very long loop lengths leading to long acquisition times. A long acquisition time poses a challenge to keeping a 0.1μm diameter beam at the center of a sub-micron line. A second disadvantage of stroboscopic measurements is that asynchronous events that are intermittent in nature, cannot be measured.

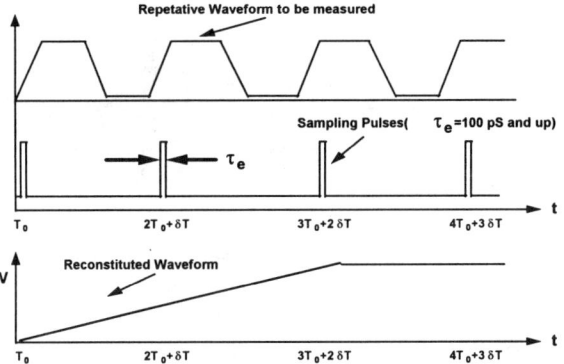

Figure 11. Illustrating Stroboscopic waveform acquisition. The burden of achieving high bandwidth in the measurement rests solely on the width of the sampling electron pulse. The pulse width (τ sec) and bandwidth (B Hz) are related approximately by $\tau B \sim 0.36$. τ is typically < 100pS

A major problem with E-beam probing in the past was the interface between the tester and the Device Under Test which must reside in the vacuum chamber of the SEM. This was particularly challenging in the case of high pin count devices since it was necessary to lead in many high frequency cables into the vacuum chamber with good reliability. In recently available commercial systems however, this has been elegantly addressed by placing the E-Beam column on a precision X-Y stage and keeping the DUT hardware in the vacuum chamber stationary. This arrangement permits a direct or "Hard Dock" between a test head and the device being probed thereby eliminating cables and enabling very high speed operation (>250 MHz) of high pin count chips inside the vacuum chamber. A different approach using flex cable technology, described by Argyrakis [6] has been used when high performance is required along with the flexibility of cables as when the DUT is mounted on a X-Y stage. Heating and cooling of the device is also essential particularly since there is no convective cooling in a vacuum chamber.

To aid in the physical location of nodes in a complex IC, CAD navigation tools have been developed [7], [8]. These tools run on workstations displaying netlists, schematics and layouts in separate windows. In most modern instruments the E-Beam prober itself is controlled from the same workstation.

To use these tools a cross-mapping database is first generated which associates schematic signal names with the corresponding polygons in the layout that physically makes up that signal in the chip. When the complete name of the signal is known and selected, the navigation tool will highlight this signal in the layout window. The X-Y coordinates of the particular location on the highlighted signal that the user selects, is then communicated to the X-Y stage in the E-Beam prober and the beam is accurately positioned over the selected location in the real chip.

The versatility of E-Beam probing lies in the fact that it is possible to capture Stroboscopic waveforms (voltage VS time – see Fig.12), Stroboscopic Voltage Contrast images and Stroboscopic Logic State Maps (Voltage Contrast image with time along one axis - see Fig.13).

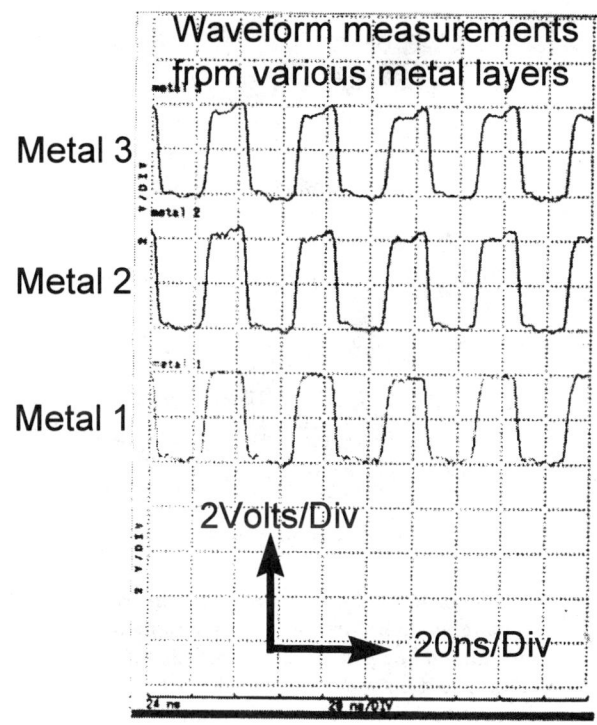

Figure 12. Typical E-beam waveform measurements from a 3 metal layer process.

In addition, the video signal from the SEM can be processed in a variety of ways to enhance the Voltage Contrast to speed up fault localization in specific situations. Examples include Dynamic Fault Imaging (DFI) [9] and Fault Contrast (FC) [10]. DFI is a technique in which Voltage Contrast images of "good" and "faulty" devices are acquired over a failing test sequence for every clock cycle and stored in computer memory. The images are subsequently digitally subtracted and the difference images are examined. For all the clock cycles at which the "good" and "faulty" images are identical the difference image will have no detail. At the first occurrence of a fault the failing node will be highlighted in the difference image. An advantage with DFI is that it provides a way to performing electrical fault isolation in complex ICs with little knowledge of how the chip operates. However with large chips and long test sequences the time to acquire the large number of stroboscopic images can become prohibitive.

Fault Contrast is a variation on DFI in which the failing signal is displayed in a slow scan real time image. The technique is primary useful when the failure mode is a voltage or a frequency marginality. In the case of a voltage marginality the power supply

153

to the chip is modulated to cause the chip to alternately pass and fail while the E-Beam in the SEM is scanned over the chip.

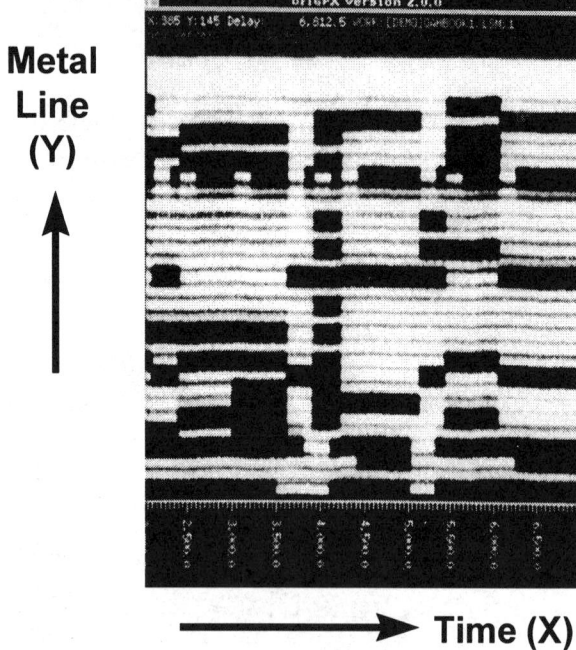

Metal Line (Y) ↑

Time (X) →

Figure 13. Illustrating a Logic State Map. This is a useful combination obtained by superimposing both imaging and timing information one picture. One dimension (Y) is used for spatial information and the other (X) for timing information. Because the spatial information in X is lost, this method is particularly useful for mapping out large parallel busses inside the chip.

The stroboscopically acquired Voltage Contrast video signal from the SEM for the passing and failing cases is subtracted on the fly on a pixel by pixel basis and the resulting difference signal is used to modulate the brightness of the SEM CRT. The faulty line appears highlighted on the SEM CRT and can be traced over the chip by panning the sample. Like with DFI the image acquisition times for long test sequences can become very long.

For waveform probing of lines the electrical bandwidth of the measurements can exceed 10 GHz and measurements from sub micron wide lines are possible. Because the E-Beam can be scanned it can be conveniently positioned on a small line while a SEM image of the line is visible at high magnification.

Finally we will say a few words about probing passivated devices. Although it is always desirable to probe unpassivated or depassivated devices, E-Beam probing gives us an opportunity to probe interconnects with passivation intact through a phenomenon called Capacitive Coupling Voltage Contrast (CCVC) [11]. The principle is illustrated in Fig 14.

The conductor can be thought of as a source of charge embedded in a dielectric medium, which is the passivation. In the absence of E-Beam bombardment the potential on the surface of the dielectric exactly follows the potential on the line but attenuated from the value on the line (from Gauss's law in electrostatics). When an E-

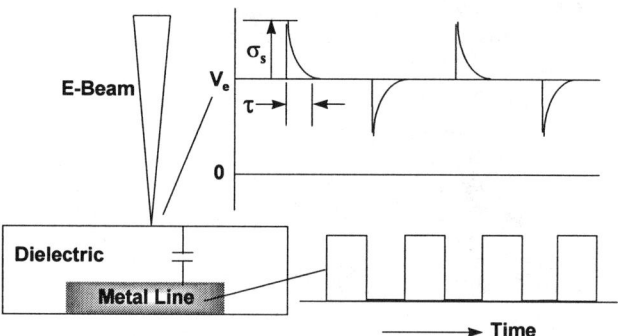

Figure 14. Capacitive Coupling Voltage Contrast enables probing through dielectrics when the removal of dielectric material is either impractical or would adversely effect the probing results.

Beam now bombards the surface, the necessary charge balance at the surface due to the secondary emission ($\delta > 1$ for beam voltages on the order of 1kV) causes the surface to charge up to a small positive voltage V_e (typically ~ +1 to +2 volts). This means that the surface is discharged (or charged) to V_e from wherever it was before the beam was switched on. For an AC signal therefore, the surface potential with continuous beam bombardment is as shown in Fig. 14

The sharp spike is due to the voltage switching on the metal line whereas the gradual charge or discharge towards V_e is due to relaxation caused by the beam. The surface potential has the same shape as one gets when a perfect square wave is passed through a high pass filter. This phenomenon is therefore known as Capacitive Coupling Voltage Contrast. If the frequency is high enough the waveform can be nicely recovered, as the droop, (caused by the surface attempting to reach a potential of V_e) is small. A voltage contrast image is also possible under these conditions. It is observed from Fig.14. that a number of factors can affect the quality of CCVC. These include beam current (this affects the surface relaxation time), beam voltage (which affects δ and surface relaxation) type of passivation and its thickness and the frequency of the signal being measured. All of these variables affect the value of τ. The larger the value of τ, the better is the measured CCVC signal.

INVASIVE ELECTRON BEAM PROBING

Voltage Contrast is essentially non-contact and non-loading. It is non-loading due to several reasons. Firstly, the beam currents used for E-Beam probing are less than 20nA. Secondly, the secondary emission coefficient δ of the metal lines in ICs are on the order of unity at the 1keV beam voltages utilized for probing. Hence, most of the primary beam current returns back from the surface in the form of a secondary electron current. The actual current that is injected into the interconnect being probed, is at most on the order of several nA which is orders of magnitude smaller than the hundreds of μa of current that the transistors switches. Finally the very low beam voltages utilized (1kV) cause negligible damage transistors.

The electron beam however can be used in an invasive manner also to give useful information about a failure. Electron Beam Induced Conductivity (EBIC) is a good example of an invasive E-beam probe. In this technique a high voltage E-Beam (10 keV or greater) is used to generate carriers in reversed biased P-N junctions in the chip. In a CMOS Integrated Circuit these regions are generally source-drain diffusions or any regions where there is a potential P-N junction. The high voltage E-Beam generates electron hole pairs in the depletion region of the P-N junction and these carriers are carried away to the contacts and are detected as an EBIC current. This current is amplified, processed and used to modulate the SEM CRT to generate an EBIC SEM image. The EBIC current is sensitive to any damage in the junction since this will alter the carrier recombination and cause a change in the EBIC current. This current modulation leads to EBIC contrast on the SEM CRT.

A variation on EBIC called CIVA [12] has been developed at Sandia labs to detect opens in passivated ICs. In this technique an E-Beam is used to bombard metal interconnects in the chip. If a particular metal conductor in the scanned field of view is connected to an MOS transistor gate *and* happens to be "open" (i.e. "floating") then the charge injection by the E-Beam will change the potential of the gate. During CIVA imaging the chip is powered from a constant current source and the changes in gate potential due to beam charging will lead to a change in the supply voltage to the chip due to the trans-impedance of an MOS transistor. The changes in the supply voltage as the beam scans the chip are used to modulate the SEM CRT to generate the CIVA image.

OPTICAL BEAM PROBING

There are a variety of ways in which a focused optical beam can be used to measure internal signals from an IC.

PHOTOEMISSION PROBING

In direct analogy with an E-Beam prober, which, generates *secondary electrons* when it impinges upon an IC line, a focused Ultra Violet light beam can be used to generate *photo-electrons* [13]. The energy of these photo electrons can then be analyzed with the use of an energy analyzer just like in an E-Beam prober to measure voltage waveforms. The photo emission probe still requires the chip to operate in a vacuum since we are still dealing with electron energy analysis for the measurement. In addition the spatial resolution of the probe will be inferior as compared to an E-Beam probe because of the limitations of optics, working distance etc. The principle advantage of a photo emission probe over the E-beam probe lies in its basically superior time resolution, which is possible if a mode locked pico-second laser is used as the light source and a stroboscopic measurement method is used. This delegates the use of this technique to specialized applications for making measurements from ultra fast devices.

ELECTRO-OPTIC PROBING

Electro-Optic (E-O) probes [14], [15] rely on the fact that the refractive index of a certain class of optical materials (such as Lithium Tantalate) is different along and perpendicular to the direction of an externally applied electric field. The electric field therefore induces a *birefringence* in the material. This birefringence is related to the applied electric field. For the Pockels effect, the change in refractive index is proportional to the applied electric field. Both of these effects are also extremely fast and extend to tens of GHz. If a suitable stroboscopic pulsed light source is used, pico-second time resolutions are possible [16]. A suitable stroboscopic light source is a mode locked laser which can generate pico-second optical pulses [16]. There are two types of E-O probes. In the external E-O probe a E-O crystal is brought into close proximity with the line being probed. A polarized light beam is made to impinge upon the line being probed after passing through the E-O medium. The polarization state of the reflected beam is measured since it is ultimately related to the voltage on the line. In the internal E-O probe [14] the natural E-O nature of GaAs is used to probe metal line in GaAs ICs by shining a laser beam through the backside of the IC. The substrate provides the interaction length between the electric field and the medium. Materials that are suitable for E-O probing are generally transparent in the infra red (wavelength ≥ 1 μm) part of the optical spectrum. A limitation of these techniques is the poorer spatial resolution. However as in the case of the photo emission probe the principle advantage of the E-O probe lies in the inherently high time resolution that one obtains when using mode locked lasers. The time resolution of the E-O probe is actually superior than the photo-emission probe since the latter is limited by the "transit time effect" which arises, because at very high frequencies, the low energy secondary electrons cannot escape quickly enough from the chip surface. Whereas the photo-emission probe is limited to time resolutions on the order of 10pS, the E-O probe has been demonstrated in measurements of less than 1pS. A second advantage of the E-O probe over the photo-emission probe or the E-beam probe is that the chip can operate in air rather than in a vacuum. This greatly simplifies interfacing and power management. Electro-optic probes again are used in very specialized applications involving measurements from very high speed devices. Also since the source of stroboscopic pulses is the laser, synchronization between the probe and the DUT is more complex.

OBIC

Both the photo-emission probe and electro-optic probe are *non-invasive* in that they do not affect the operation of the IC. Other optical probing techniques are *invasive* in that, some operational parameter of the chip is altered by the beam and this change is measured. Examples include OBIC [17] in which a focused laser beam is used to induce photo currents in the P-N junctions in the chip. The photo-current generation can be optimized for specific application by a correct choice of the laser wavelength. Light from a He-Ne laser (632.8 nm wavelength), or Argon Ion laser (514.5 & 488.0 nm wavelength) have been used. The resulting photo current is collected and after amplification and processing is used to modulate the brightness of the SEM CRT to generate the OBIC image. Since the amount of photo current generated is a sensitive function of junction defects, the latter appear as bright areas in an OBIC image. A typical OBIC setup utilizing a scanning laser microscope is illustrated in Fig.15.

The invasive techniques can provide valuable information about failures but their biggest drawback lies in the inaccessibility of diffusion areas in VLSI Integrated Circuits, which are covered with multiple layers of metal on the diffusion.

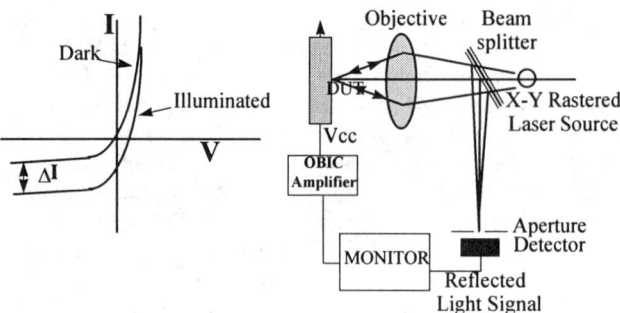

Figure 15. The I/V curve illustrates an increase in reverse saturation current when illuminated. A Schematic diagram of an OBIC setup in a Laser Scanning Microscope is shown to the right.

However these techniques become very powerful from the backside when Infra-Red lasers are used to reach diffusions through the substrate.

However both electron beam and optical probes can benefit greatly by the use of Focused Ion Beam (FIB) milling to selectively mill holes or to deposit metal lines to enable inaccessible nodes to be probed. FIB will be discussed in the next section.

FOCUSED ION BEAM MILLING AND DEPOSITION

A positively charged ion beam interacts with a target specimen to generate a variety of particles and radiation as follows: a) *Elastic* interactions between the ion beam and the target atoms lead to the generation of sputtered (mostly neutral) and backscattered ions. The sputtered neutral atoms have low energies in the region of 10-30 eV while the backscattered ions have energies centered around the primary beam energy. b) *Inelastic* interactions between the ion beam and the target electrons near the surface lead to the generation and emission of secondary electrons, photons and X-Rays. The sputtering of the target material by the FIB probe enables it to be used for micromachining with sub 0.1μm resolution [18]. In addition a FIB probe can be used to dissociate a metallo-organic gas that has been adsorbed on the target surface, to form, solid metal which sticks to the surface and volatile byproducts which are pumped out of the vacuum chamber. Two commonly used gases are Tungsten Hexacarbonyl (for tungsten depositions) and Trimethyl-methylcyclopentadienylplatinum (for Platinum depositions). With this technique well defined Tungsten or Platinum metal lines or pads can be deposited on an IC. During metal deposition there are two competing mechanisms that are occurring simultaneously. One is the dissociation and deposition of fresh metal and the second is the sputtering of the deposited metal. When the deposition rate exceeds the sputtering rate the net effect is that a metal trace is formed.

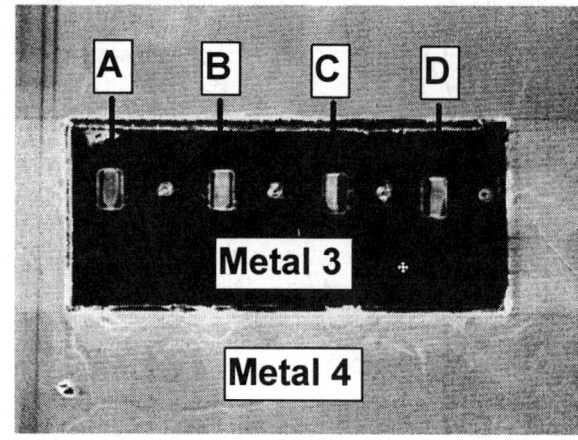

Figure 16. Probe holes A,B,C,D drilled into a metal 3 power bus to expose M2 lines for E-Beam probing. A portion of an overlaying Metal 4 had to be milled away to access the M3 line.

The ability of the FIB to both mill and deposit metal traces makes it invaluable during Fault Isolation.

The FIB can be used to open probe windows to nodes that would otherwise be inaccessible (see Fig.16). It can also be used to repair faulty circuitry on the silicon or to induce a failure on good silicon to test hypotheses of the cause of failure (see Fig.17).

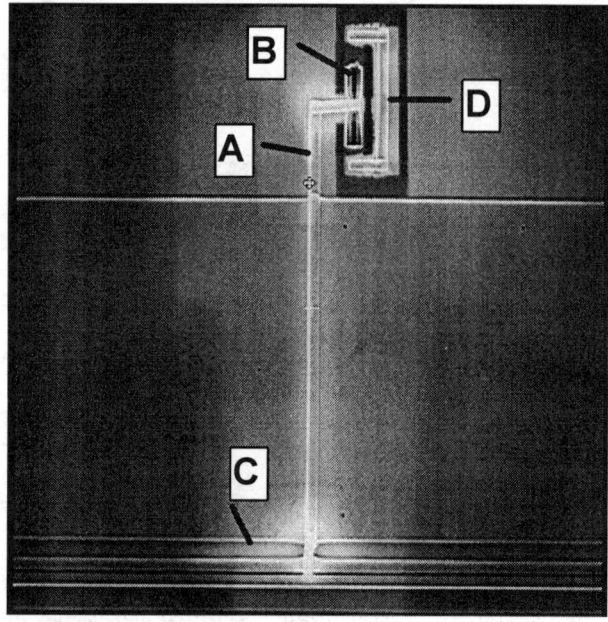

Figure 17. An FIB micrograph showing the use of an FIB to generate a complex on Silicon circuit edit. A signal A buried under a signal B is connected to a signal C using FIB metal. Signal B being in the way of signal A is first cut for access and then rerouted with FIB metal line D

The secondary electrons that are generated by the FIB probe can be used to generate high resolution images similarly to the SEM. This image is used to position the FIB probe on the point of interest.

REACTIVE ION ETCHING (RIE)

The Reactive Ion Etcher is primarily used in Fault Isolation to depassivate ICs so that clean metal is exposed for probing. It is primarily a tool for selectively removing dielectrics [6]. Whereas the FIB probe removes material (dielectric or metal) in a highly localized way, the RIE removes dielectric over the entire IC by a combination of momentum and chemical etching. Because the etch profiles generated with an RIE are anisotropic it can also be used to effectively uncover nodes on lower level metallization that are covered by dielectric (but not metal) without causing lift off of upper metal layers (see Fig.9). This greatly improves the node accessibility or observability of signals in the chip. Removal of dielectric over the entire chip will lower the interconnect capacitance and could cause speed related failures to recover. In such cases the FIB must be used to generate probe points.

CASE STUDIES

The case studies presented in this section illustrate the cooperative use of Voltage Contrast, FIB and Reactive Ion Etch techniques to electrically fault isolate failures on complex ICs.

Case study 1: The first case study (shown in Fig.18) is a classic race condition/speed path inside a microprocessor chip. The circuitry in question is shown in Fig.18a.

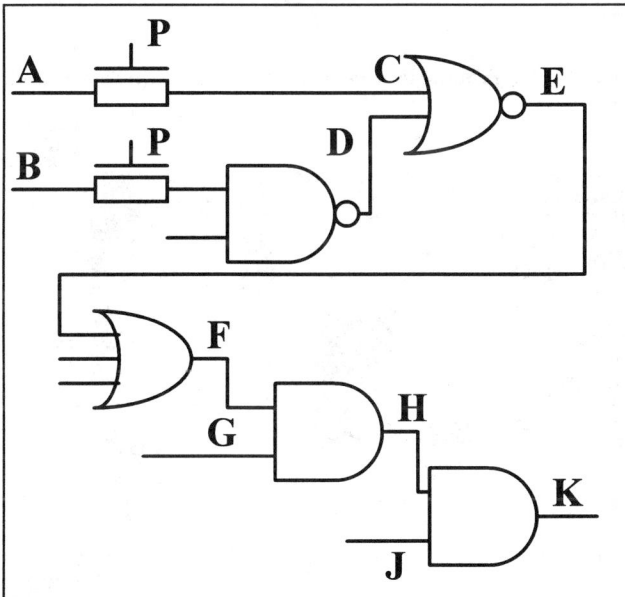

Figure 18a. Schematic of circuit under investigation in Case Study 1.

The problem was first logically isolated using internal DFT and test patterns to the suspected circuitry shown. The chip was then delidded and depassivated and internal waveforms were measured from various nodes using an E-Beam prober. The race condition between signals A and B causes a short pulse E which normally would not be a problem with the two phase non overlapping clocking scheme, since this pulse is not latched into subsequent circuitry. However in this instance a heavy capacitive load on the node H delayed this pulse sufficiently that it was erroneously

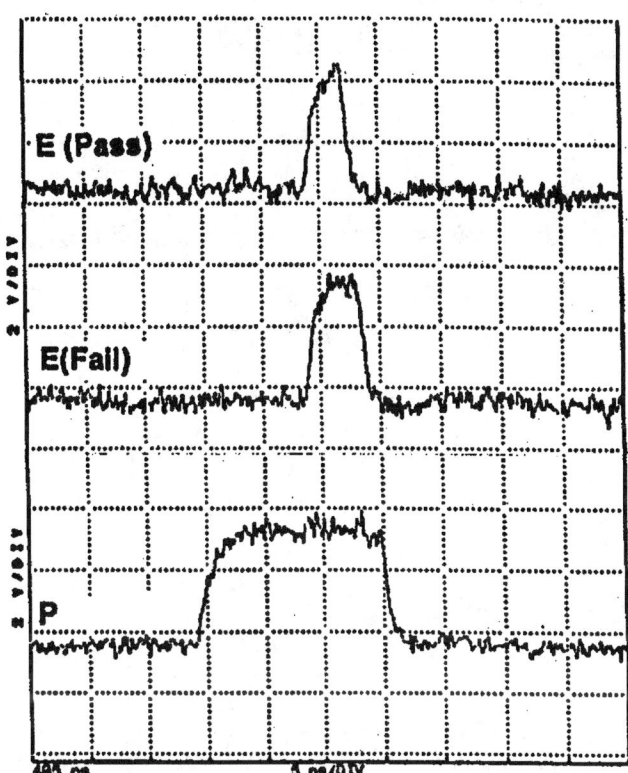

Figure 18b. Intermediate passing and failing waveforms measured by an E-beam prober at location E (Case Study 1).

transmitted through to the output (K). The passing and failing conditions were generated by varying the timing between signals A and B both of which were controllable.

Case Study 2: During the silicon debug of a product a combination of E-Beam probing, FIB milling and Reactive Ion Etching was used to isolate a race condition in the chip. The analysis was performed by docking an ATE to an E-Beam prober. After characterization on the tester the chip was depassivated using a Reactive Ion Etcher. The chip was recharacterized to ensure that the etching had not altered the failure mode. A tester schmoo plot was then measured while the chip operated inside the SEM and showed passing and failing regions. E-Beam probing was then performed while the chip was passing and failing. Suspected signals in the chip illustrated as A, B, C, D in Fig. 19a were probed. A glitch was observed on signal D in the failing condition and was suspected to be caused by a race between signals A and B (Fig. 19c and with a smaller time/div in Fig.19e). This glitch did not exist in the passing condition (Fig.19d). To verify this, the intermediate signal C was also measured. Unfortunately this signal was buried under a metal power bus. Therefore an FIB mill was used to open a probe hole (Fig.19b) through which the signal was acquired. Signal C was found to be the inverse of signal D (Fig.19f) which confirmed the hypothesis. The problem was fixed by circuit changes.

CONCLUSIONS AND ACKNOWLEDGMENTS

This paper has primarily reviewed the various beam based tools and techniques that are available to perform electrical fault isolation on on complex ICs. The use of these tools has been illustrated with a number of case studies, which illustrate the power of the cooperative use of these tools. Rick Livengood at Intel is acknowledged for providing valuable inputs to this paper.

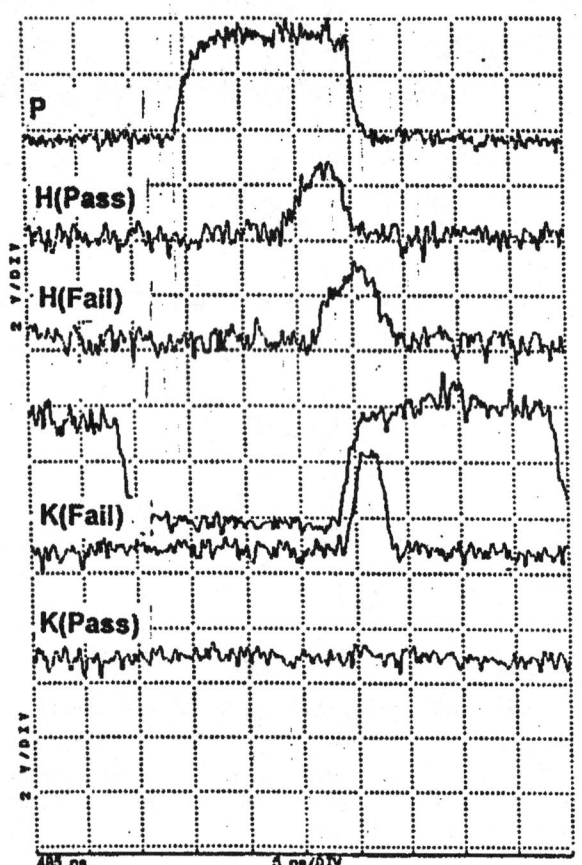

Figure 18.c. Additional representative waveforms described in Case Study 1.

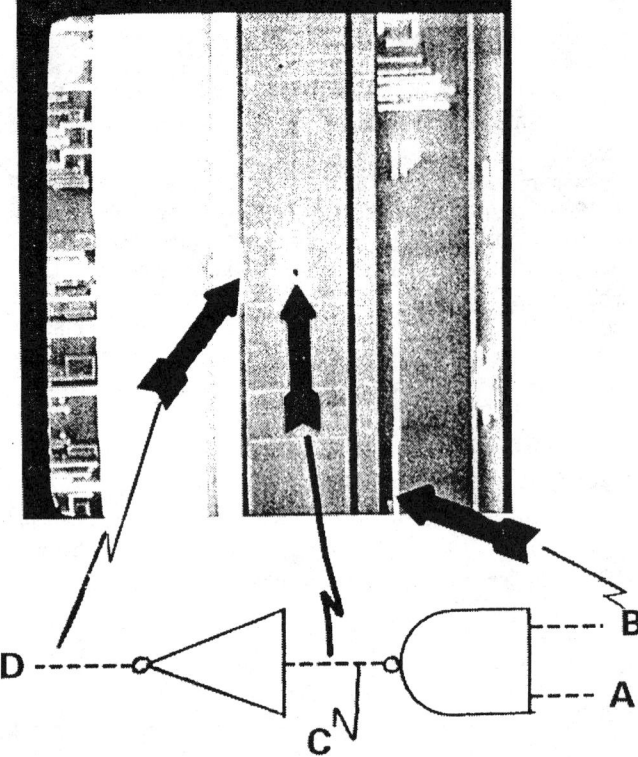

Figure. 19a. Probing locations and schematic representation for Case Study 2.

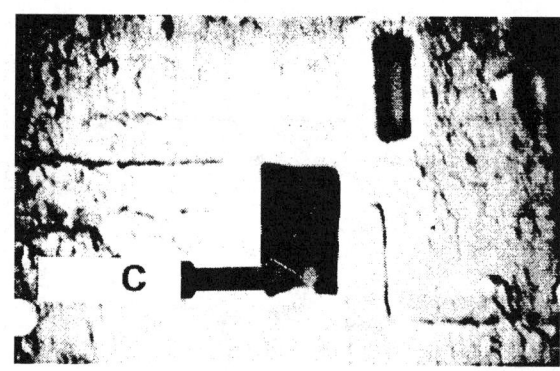

Figure 19.b. Detail of probing location C (Case Study 2).

REFERENCES

[1] See papers on Photo-Emission by Christian. Boit; G. Shade & S. Wills in this publication

[2] Thomas W. Williams, & Kenneth P. Parker, "Design for Testability - A Survey" *Proc. of the IEEE*, vol. 71, no. 1, Jan 1983, pp98-112

[3] Paul G. Ryan, Shishpal Rawat & W.Kerr Fuchs, "Automated Diagnosis of VLSI Failures" *Proc. IEEE VLSI Test Symposium*, 1991, pp187-192

[4] J.M. Soden, C.F. Hawkins, R.R. Fritzmeier & L.K. Horning, "Quiescent Power Supply Current Measurement for CMOS IC Defect Detection," *IEEE Trans. on. Indust.Electron,*.vol. 36, no. 2, pp211-218, May 1989

[5] E. Plies, *Microelectronic Engineering* **12** (1990) pp189-204, Elsevier Science Publishers B.V

[6] S. Argyrakis, "High Frequency/High Density High Vacuum Transmission Line Interface" *Connection Technology*, Feb 1991

[7] S. Concina et al, " Workstation driven E-Beam prober", *Pro.c International Test Conference*, 1987, pp 554-560

[8] A. Hu & H. Nijima "New approach to integrate LSI design databases with E-Beam tester" *Proc. of International Test Conference*, 1990, pp1040-1048

[9] T.C. May, G.L. Scott, E.S. Meieran, P. Winer & V.R.M. Rao "Dynamic Fault Imaging of VLSI random logic devices", *Proc. of the International Reliability Physics Symposium*, 1984, pp95-108

[10] A.R. Stivers & D.C. Ferguson, "Fault Contrast: A new voltage contrast VLSI diagnostic technique", *Proc. of the International Reliability Physics Symposium*, 1984, pp109-114.

[11] W.Reiners, "Fundamentals of electron beam testing via capacitive coupling voltage contrast", *Microelectronic Engineering* 12 (1990), Elsevier Science Publishers B.V

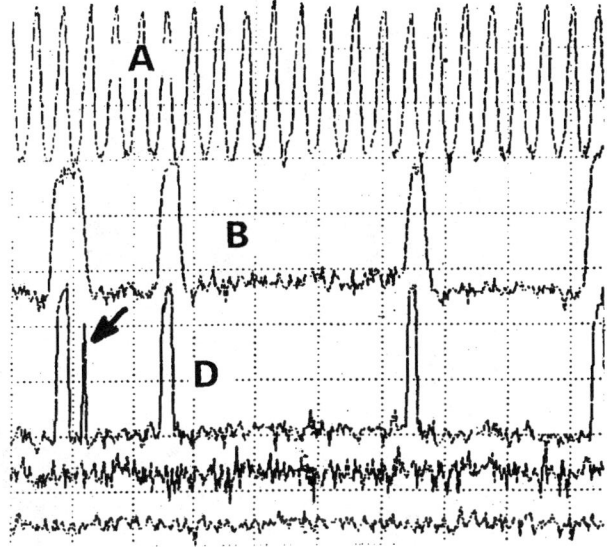

Figure. 19c. Probed waveforms for case study 2. Waveforms are 2 volts/division vertical and 50 nS/division horizontal. Note glitch on waveform D at arrow showing failure.

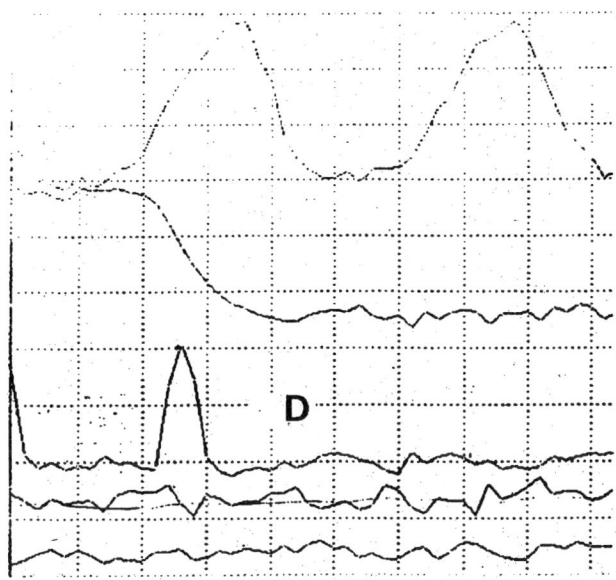

Figure. 19e. Expanded horizontal scale of 5nS/division showing glitch on signal D.

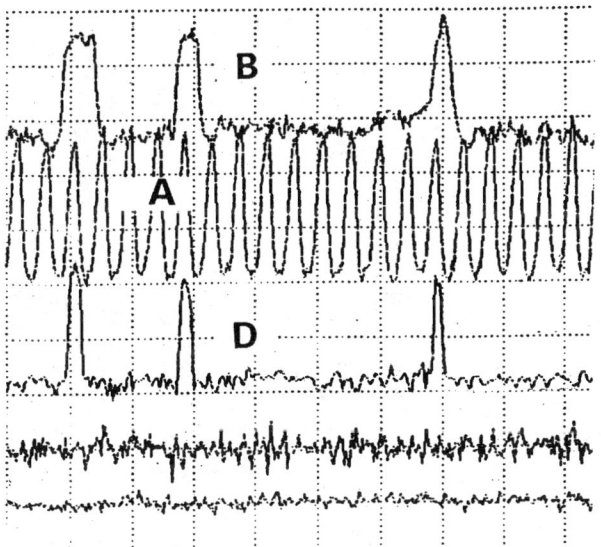

Figure. 19d. Probed waveforms for case study 2. Passing case.

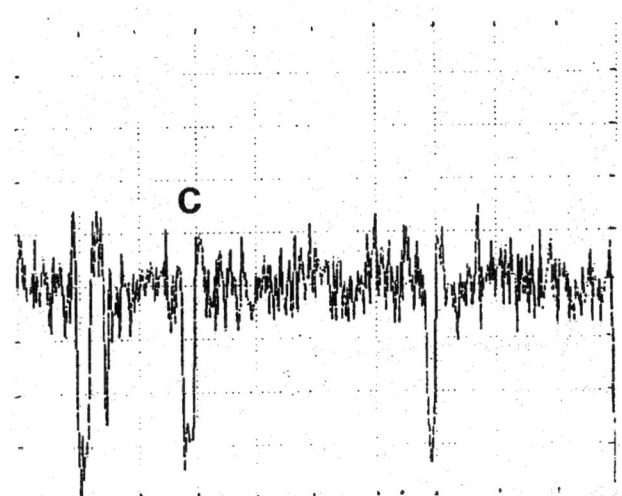

Figure. 19f. Glitch on signal C probed through ion milled hole. Horizontal scale is 50nS/division. Case Study 2

[12] E.I. Cole Jr & R.E. Anderson, "Rapid localization of IC open conductors using Charge Induced Voltage Alteration (CIVA)", *Proc. of the International Reliability Physics Symposium*, 1992, pp288-298.

[13] H.K. Seitz, A. Blacha, R. Clauberg, H. Beha, & J. Feder, "Contactless, high speed waveform measurements on Gallium Arsenide ICs" *IEEE design and test of computers*, Feb 1990, pp20-25

[14] B.H. Kolner & D.M. Bloom "Electrooptic sampling in GaAs integrated circuits", *IEEE journal of quantum electronics*, Vol. QE-22, NO.1, Jan 1986, pp79-93

[15] J.F. Whitaker et al "External electrooptic integrated circuit probing", *Microelectronic Engineering* 12 (1990), pp369-379, Elesvier Science Publishers B.V.

[16] J.A. Valdamnis, G. Mourou & C.W. Gabel" Subpicosecond electrical sampling," *SPIE Proc.*, vol.439, pp142-148, 1983

[17] K. S. Wills, T. Lewis, G. Billus & H. Hoang "Optical Beam Induced Current applications for failure analysis of VLSI devices", *Proceedings of ISTFA* 1991, pp 21-26

[18] R. Boylan, M. Ward, & D. Tuggle, "Failure analysis of micron technology VLSI using focused ion Beams" *Proceedings of ISTFA* 1989, pp249 – 255

[19] W.Baerg, V.R.M. Rao & R. Livengood, "Selective Removal of Dielectrics from Integrated Circuits For Electron Beam Probing" *Proc of the IEEE International Reliability Physics Symposium*, 1992, pp 320-326

Voltage Contrast in the FIB
as a Failure Analysis Tool

Ann N. Campbell and Jerry M. Soden
Sandia National Laboratories
Albuquerque, New Mexico

INTRODUCTION

The use of focused ion beam (FIB) systems for integrated circuit (IC) failure analysis has increased greatly in the past decade. FIB systems were initially used primarily for precision cross sectioning and modification of ICs and microelectronic devices in support of root cause failure analysis and design debugging [1,2]. While they continue to be used extensively for those applications, it has also been recognized that it is possible to take advantage of the charged beam in the FIB system to obtain electrical information from ICs [3]. In particular, voltage contrast (VC) information that is available in FIB images of ICs and microelectronic structures is being used increasingly for applications such as the localization of defects and sub-surface feature localization on planarized ICs. This chapter describes how to perform VC imaging in a FIB system, discusses the optimum operating conditions for obtaining good FIB voltage contrast, and provides several examples of the application of FIB VC imaging in microelectronics failure analysis.

BACKGROUND

Electron Beam Voltage Contrast

VC techniques based on scanning electron microscopes (SEMs) have been widely used for over 20 years to analyze the internal operation of ICs (4-6). These techniques take advantage of the interaction between a charged beam and local electric fields resulting from the presence of voltage (charge) on the surface. Local electric fields on the device in turn modulate the secondary electron image intensity. E-beam voltage contrast imaging is typically performed at low beam energy to minimize hydrocarbon contamination, radiation damage to active regions, and surface charging. The rate of electron emission is greater than that of electron arrival (primary beam current), so the sample surface acquires a net positive charge.

Voltage contrast information can be obtained both from unbiased samples (also called passive voltage contrast, or PVC, imaging), and from those which are powered [4-10]. Static voltage contrast (SVC) imaging is performed with the top dielectric layer (passivation) removed, and the image intensity is largely determined by the static voltages on the IC. Variations in image brightness are used to determine logic levels of digital ICs and the voltages on internal test nodes. Capacitive coupling voltage contrast (CCVC) imaging is a method for viewing dynamic voltages on conductors beneath passivation layers and can also reveal unbiased, subsurface conductors. The passivation layer acts as the dielectric of a capacitor to generate the CCVC image of changing subsurface voltages. The transient contrast due to switching voltages decays very rapidly. Signal averaging may be used to enhance CCVC images.

Ion Beam Voltage Contrast Imaging

FIB VC has been used increasingly during the past few years. A number of examples in the literature involve static, unbiased voltage contrast imaging in a FIB system to detect open conductor defects on deprocessed ICs [11,12] as well as on IC cross sections [13, 14,15]. The capability to electrically bias ICs in the FIB has become commercially available, permitting biased VC imaging of ICs. This capability enables detailed examination of an IC's electrical operation while in the FIB system [3].

Voltage contrast in FIB system images is similar to that produced by e-beam VC [3]. Commercial FIB systems use a gallium liquid metal ion source to produce a beam of positive (Ga^+) ions. When a sample is exposed to the ion beam, Ga^+ ions are implanted into the sample and secondary particles (electrons, ions, and neutrals) are emitted from the surface by sputtering. The sputter yield of secondary electrons is about 10 times greater than that of secondary ions (most of which are positively charged). This effect, combined with the implantation of positively charged Ga^+ ions, results in a net positive charge on insulating sample surfaces. Because both 1 keV electron and 25 - 50 keV ion beams produce a positive surface potential, ion beam VC effects in a FIB system are analogous to those observed during low energy e-beam VC in an SEM.

The secondary electrons emitted from a sample surface exposed to the Ga^+ ion beam in the FIB have an energy distribution from 0 to approximately 5 eV, with an average of about 2 eV for a 30 keV ion beam energy [16]. The electrons are accelerated toward the detector by the extraction field provided by both the microchannel plate

(MCP) detector and the electron floodscreen (described in the Experimental Approach section below). Simulations describing the modulation of the secondary electron trajectory as a function of voltage on the sample surface have been performed using the SIMION program [17]. For adjacent unpassivated conductor lines these simulations show that most of the secondary electrons sputtered from a 0 V line are detected and that most of the secondary electrons generated from a +5.0 V line are recaptured.

EXPERIMENTAL APPROACH

Imaging in the FIB System.

Imaging in the FIB system is accomplished by rastering the beam over the sample and collecting the secondary particles, either positive (ions) or negative (electrons) on a pixel by pixel basis. The image contrast at each pixel is proportional to the number of detected secondary particles. The image is a superposition of the "normal" or topographical image plus the modulation due to VC effects.

VC information in FIB images arises from modulation of the secondary electron emission caused by surface charging due to the ion beam and the presence of applied voltages on the sample. Local voltage differences on a sample surface can be detected in the FIB by using secondary electron mode imaging.

Secondary Particle Detection. Most commercial FIB systems use a microchannel plate (MCP) detector [18] mounted at the bottom of the FIB column directly above the sample surface for secondary particle detection. The MCP detector consists of an array of hundreds of hollow tubes, or channels, in a lead glass substrate that acts as an electron multiplier. An electron shower is generated inside a channel when a particle (secondary electron or ion) strikes the channel's interior surface. By applying a voltage between the top and bottom faces of the MCP detector, these electrons are accelerated along the channels, generating more electron showers when they strike the surface deeper into the channel. The total charge is collected at the output of the channels and sampled at specific time intervals. The MCP measured charge is indicative of the number of collected secondary particles from the region (pixel) struck by the ion beam.

The channels generate electron showers regardless of whether the impinging particle is an electron or ion. Therefore, the MCP detector can easily detect either secondary ions or secondary electrons (and vice-versa) by simply changing the bias on the detector. Typical MCP biases for imaging are +400 V for secondary electrons and -1400 V for positive secondary ions. The MCP detector is shown in Figure 1.

Charge Neutralization. The FIB system used for the experiments described in this chapter is a Micrion 9000D FIB system that operates with a maximum 30 keV primary ion beam energy. This FIB system has an electron floodgun that directs low energy electrons onto the sample surface to neutralize the positive charge buildup due to implanted Ga^+. The electron floodgun is normally active in secondary ion mode (SIM) and is effective in reducing sample charging and electrostatic discharge damage. The floodgun cannot be active during secondary electron mode imaging because the MCP detector would be swamped by electrons from the floodgun, preventing detection of secondary electrons. A mesh

screen between the MCP detector and the sample is used to steer low energy electrons down to the sample surface. The electron floodscreen bias voltage is negative when the FIB is enabled for secondary ion mode imaging and is positive (~80 V) during secondary electron mode operation. The electron floodgun and floodscreen are also shown in Figure 1.

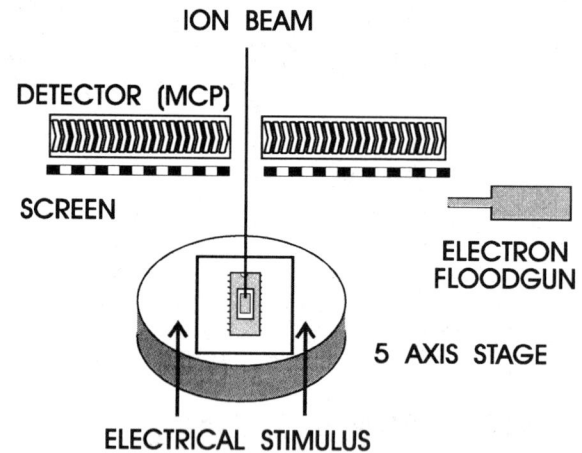

Fig. 1. The MCP detector, electron floodgun, electron floodscreen, and sample surface in the FIB system.

Implementing Voltage Contrast Imaging in the FIB System.

Electrical Biasing in the FIB System. Secondary electron imaging in the FIB system is used to acquire voltage contrast information. The implementation of unbiased SVC imaging does not require any system modification. An electron floodgun for charge neutralization is very helpful for performing CCVC in the FIB system. The biased SVC and CCVC techniques require that power be applied to the sample in the FIB chamber. Several options are commercially available for electrical biasing in FIB systems, ranging from electrical probes that can be added to a basic FIB system to fully integrated FIB systems designed for full electrical biasing and testing *in situ*. The FIB system used in this study has a modified stage that provides 120 pin electrical biasing capability. The device to be analyzed is inserted into a socket on a printed circuit board that is transported into the FIB process chamber through the system's load lock. Electrical connection to the PC board is made *in situ* via a clamshell style motor-driven clamping mechanism. The electrical socket and clamping mechanism are illustrated in Figure 2. Electrical stimulus is provided to the device by appropriate equipment, such as a parameter analyzer interfaced through a switchbox or a digital tester. In addition to the electrical biasing capability, the FIB system's stage has the capability to rotate a full 360° and can be tilted up to a 60° angle.

CCVC Imaging in the FIB System. Controlling the ion dose delivered to the sample is important for successful CCVC imaging of passivated biased and unbiased conductors. This is required because the local electric field effects that give rise to CCVC are quickly overcome by the charging effect of the ion beam. The ion dose is determined by the raster parameters, the number of rasters performed, and the ion beam current. The raster parameters that can be controlled are the dwell time at each pixel (i.e., the scan rate), the number of pixels in the raster, and the field of view (FOV). The number of rasters can be controlled by "grabbing" one image raster at a time rather than by continuous imaging. The ion beam current is determined by the size of the beam-limiting aperture. The FIB system used for these experiments has apertures and associated beam currents ranging from 25 μm/2.3 pA to 750 μm/7.9 nA.

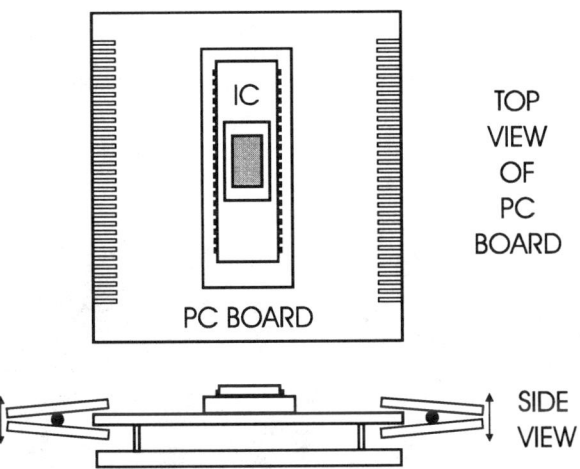

Fig. 2. The PC board, load lock transport frame, and clamshell clamping mechanism for electrical testing of ICs in the FIB system. The side view shows the clamps in the open position.

In general, VC imaging was performed with the lowest possible beam current. CCVC effects could be achieved only with the smallest possible ion dose, and the results reported here were obtained by "grabbing" single raster scans. Often the effects disappeared after several subsequent rasters. Once the contrast faded due to charging from the ion beam, the VC effect could be restored by turning on the electron floodgun (i.e., by selecting secondary ion mode imaging) for a short time to neutralize the positive surface charge and then imaging once again in the secondary electron mode.

By comparison with CCVC, SVC is relatively insensitive to the ion dose. Although SVC imaging can be performed with larger beam currents and ion doses, it is not advisable to do so since the sample is being sputtered away as it is imaged, and the surface damage increases with ion dose. In addition, it is known that FIB irradiation can alter transistor parameters such as threshold voltage and that the amount of change in the transistor parameters increases with ion dose [19].

FIB VOLTAGE CONTRAST IMAGING EXAMPLES

Unbiased Voltage Contrast Imaging.

SVC imaging effects arise during imaging with a Ga$^+$ ion beam in the FIB system when portions of a sample acquire a net positive charge relative to the surrounding areas. Passive SVC effects are readily apparent in many secondary electron mode images taken with a FIB system, including cross sections. As an example, consider the cross section of an IC shown in Figure 3. The dielectric materials acquire a positive charge due to implanted Ga$^+$, recapture secondary electrons, and appear dark as a result. The conductors (metal and polysilicon) normally have a bright contrast because they are conductive enough to neutralize the implanted Ga$^+$, and hence the secondary electron yield is much higher from these layers than from the surrounding insulating layers. The metal-1 (M1) layer in Fig. 3 illustrates this. However, if a portion of a conductor is electrically floating (as is the case for the M2 line and the polysilicon gate electrode in this figure), it will also charge positively and have dark image contrast.

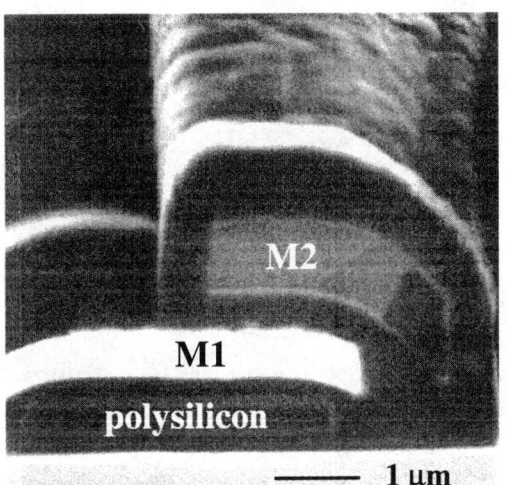

Fig. 3. Voltage contrast effects in a secondary electron mode image of a FIB cross section. The M2 and polysilicon are floating conductors.

Another example of this effect is observed in Figure 4, which shows an unpassivated (i.e., no dielectric material on top of the metal interconnects) metal via chain structure containing an open contact. This device is a test structure for Sandia's 0.5 μm CMP-planarized CMOS 6 technology. The interconnect material is Al-0.5% Cu with Ti interfacial layers and W studs for vias.

Figure 4a is a secondary electron mode image showing that the section of the chain at the upper right appears dark compared with the metal in the rest of the chain. This implies that the dark region is electrically disconnected from the rest of the test structure and has acquired a net positive charge from the Ga$^+$ ion beam. As a result, secondary electrons emitted by that region of the conductor are recaptured, resulting in a darker image contrast for that portion of the chain. By comparison, the electrically disconnected portion of the chain does not have a different contrast in a SIM image with the same FOV, as shown in Figure 4b. The electron floodgun is active during SIM imaging, neutralizing the effect of charge buildup due to Ga^+ implantation.

The contacts at the end of the conducting portion and the beginning of the isolated portion of the chain were cross sectioned with the FIB system. An FESEM was used to image the cross section, and the result is shown in Figure 4c. The contact via on the left hand side of the figure is properly filled with tungsten, but the via on the right (the contact to the electrically isolated section of the chain) is voided and does not contain W but rather (as shown by subsequent analysis) a corrosion product.

This example illustrates the power of passive SVC imaging in a FIB system. The FIB is used both to identify and to cross section the defective contact.

grab single frame images, neutralizing the surface each time or as needed by using the electron floodgun to restore the CCVC.

The passive CCVC effect can be used for navigation on planarized or partially planarized devices and is illustrated in Figure 5 for a 256K SRAM which is planarized except for the top level metal. This IC is implemented with a 2-level metal, 1-level polysilicon, 0.8 µm CMOS technology. Figure 5a is a secondary electron mode image that shows an area in the peripheral circuitry of the IC where the M1 lines (horizontal in the figure) are visible due to CCVC. The vertical lines in the figure are unplanarized M2. This image was the first raster scan acquired after neutralizing the surface with the floodgun. Figure 5b is the secondary electron mode image collected on the second scan without changing the MCP detector gain; the CCVC effect had decayed and only the topology due to the unplanarized metal-2 lines is visible in this image. Figures 5a and 5b were obtained with a 12 pA beam current and employed a 1 µs dwell time and 512 x 512 pixels in the single frame image of a 375 µm x 375 µm area.

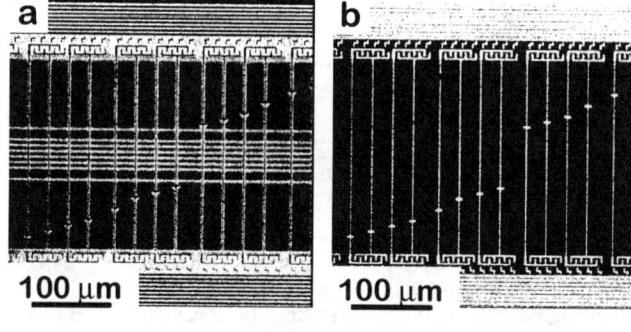

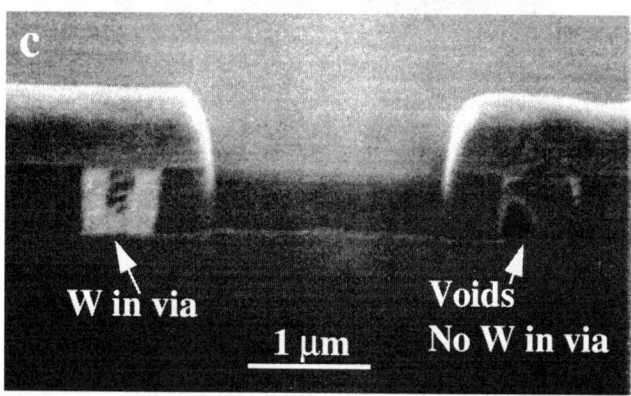

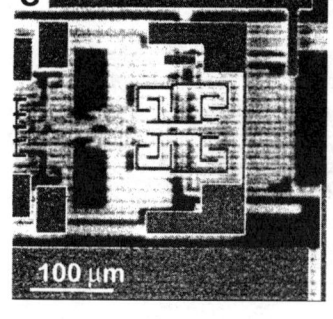
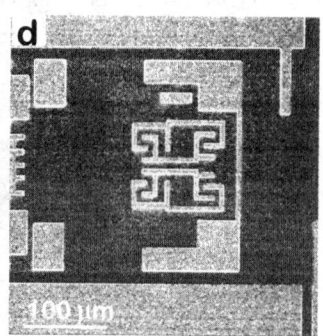

Fig. 4. (a) Secondary electron mode image showing extinction of contrast from the electrically disconnected portion of the via chain. (b) Secondary ion mode image of the same region, showing no difference in contrast. (c) FESEM image of FIB cross section through the defective contact.

Fig. 5. (a) SVC image showing planarized M1 lines (horizontal). 375 µm FOV. (b) SVC effect is no longer visible in next scan. (c) Highly detailed SVC image. (d) SVC effects not visible in subsequent scan.

CCVC effects which are observed on unbiased, passivated devices are of short duration, lasting only until exposure by the ion beam overwhelms the small localized differences in surface charge. The ideal approach is to navigate to the area of interest and then

When the beam current is further reduced to 2.3 pA, it becomes possible to obtain images with significantly greater detail and/or higher magnification. For example, the images shown in Fig. 5c and 5d were taken with a 2.3 pA beam current. The pixel dwell time was 16 µs, the field of view was 150 µm x 150 µm, and the number of pixels in the image was 256 x 256. The image in Fig. 5c is a combined CCVC and topology image, showing a great deal of detail about the planarized M1 level. The image in Fig. 5d was grabbed subsequently. Surface charging has obscured most of the CCVC from the M1 level. As for the previous example, the power of the technique illustrated in Fig. 5 is the ability to use the FIB both to locate and to modify or cross section a feature of interest (in this case, a subsurface conductor on a planarized device).

Biased Voltage Contrast Imaging.

All that is required to perform biased voltage contrast imaging in a conventional FIB system is to be able to apply power to the sample. As described in a previous section, our system was modified to include a 120-pin electrical biasing capability. While the MCP detector bias and electron floodscreen voltage can be varied to modify the secondary electron extraction field for certain applications, in general the typical operating floodscreen and detector biases were used.

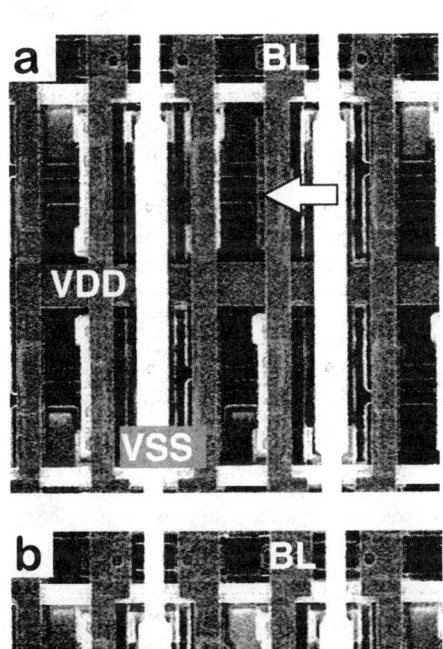

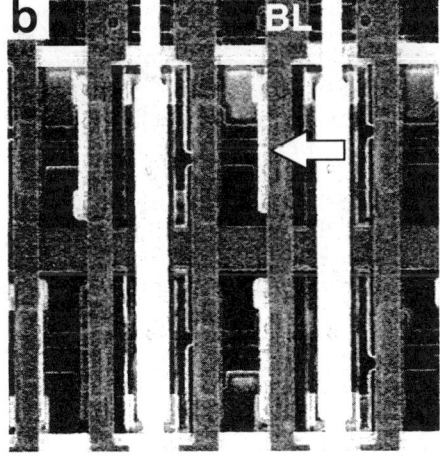

Fig. 6. SVC imaging of a 2 μm, 2-metal CMOS technology SRAM, 40 μm FOV. (a) Several SRAM cells indicating the cell components. The arrow points to the M1 drain strap. The indicated cell is written to the 1 state. (b) The same memory cell, now written to the 0 state.

Static Voltage Contrast Imaging. A 16K SRAM fabricated with a 2 level metal, 2 μm CMOS technology was examined with SVC imaging, and the results are shown in Figure 6. The IC was deprocessed to remove the glass passivation to permit SVC analysis. Power was applied to the IC. For this SRAM, V_{DD} = 5 V and V_{SS} = 0 V. Several cells in the memory array are visible in Figure 6a. When power is applied to the SRAM, each memory cell latches to a logic 0 or logic 1. The voltage on the M1 drain strap (indicated by the arrow in Figure 6a) indicates the logic state of each cell. If the drain contact adjacent to the bit line (BL) is at 0 V (bright contrast), the cell is in the 0 state. The indicated cell in Fig. 6a is in the 1 state. In Fig. 6b, this cell is shown in the 0 state. The fact that an SVC image is obtained of the voltage on the M1 drain strap indicates that M1 as well as M2 was exposed during deprocessing. A cross section of one of the tested devices confirmed this.

SVC was performed successfully with both small and moderately high ion beam currents and doses. The SVC image persists during continuous scanning, permitting the change from one logic state to another to be observed. While the SVC effect was not sensitive to the ion beam parameters, it was noted that prolonged exposure to the ion beam led to increased power supply current of the SRAMs and, in some instances, to failure of the ICs.

Rather than using RIE or a wet etch to remove the passivation from the entire chip, FIB milling with or without gas enhanced etching could be used for local, selective removal of the passivation on the M1 drain straps or other features of interest on the IC. This would permit SVC imaging of the voltages at those points without requiring that the entire IC be deprocessed.

Dynamic Capacitive Coupling Voltage Contrast Imaging. Moving to the next level of complexity, we used CCVC imaging to visualize the voltages on metal interconnects in a passivated 16K SRAM with the V_{SS} lines biased at 0 V and the bit lines at 5 V. Two images of the IC are shown in Figure 7. Figure 7a was imaged with a beam current of 12 pA and a 1 μs pixel dwell time. The V_{SS} lines have bright contrast and the V_{DD} lines have dark contrast. Clearly, much more detail is present in Fig. 7b which used a beam current of 2.3 pA and a pixel dwell time of 16 μs.

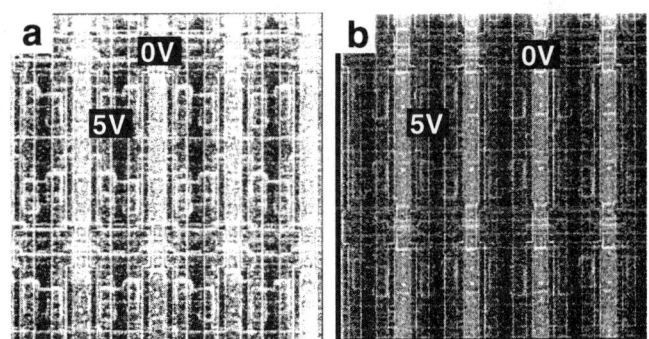

Fig. 7. CCVC images of the 2 μm SRAM, 40 μm FOV, showing the V_{SS} lines at 0 V (bright contrast). (a) 12 pA beam current. (b) 2.3 pA beam current. The use of the lower beam current permits a longer pixel dwell time, allowing higher resolution images to be acquired.

Combined Static and Capacitive Coupling Voltage Contrast Imaging. Finally, it is possible to obtain both SVC and CCVC information from an IC when the passivation is removed from a portion of the surface. An example of this is shown in Fig. 8 for the 16K SRAM. In this case, gas assisted etching (GAE) with XeF_2 was used to selectively remove the passivation covering one of the memory cells. Once the passivation was removed, SVC images can be obtained, showing the logic state of the cell. Simultaneously, CCVC information is available for the various signal lines in the vicinity (Fig. 8a). One limitation of this approach is evident in Fig. 8a. In order to perform GAE, the IC surface was first imaged to localize the cell of interest. The area that was imaged several times with a 400 pA beam appears as a dark square surrounding the region where the M1 drain straps and M2 bit lines were exposed (bright contrast in Fig. 8a and 8b). The dark region contains a greater amount of implanted Ga than the surrounding area. A CCVC signal could not be obtained from this dark region. An alternative method for localized removal of the passivation that will not produce a Ga stain is laser milling.

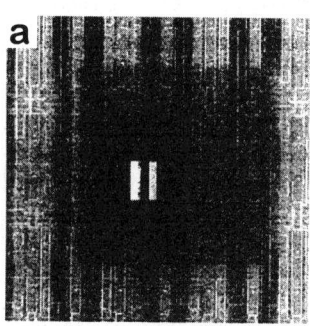 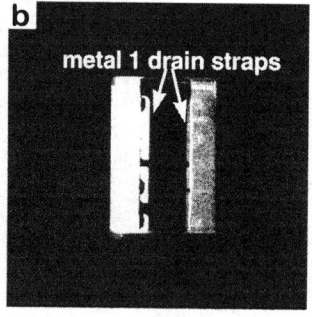

Fig. 8. (a) SVC image of bit lines and M1 drain straps in the 16K SRAM (bright area in center) in an area with passivation removed by GAE. CCVC information for surrounding signal lines is also present in the image. (b) Higher magnification image of exposed region of the die.

Electrical Effects of Voltage Contrast Imaging.

Our experiments with the 16K SRAM indicated that the power supply current, I_{DDQ}, was sensitive to ion irradiation *while the IC is biased*. Certain portions of the IC were more sensitive than others, and passivated 16k SRAMs were found to be more sensitive than those with the top passivation removed. Increases in I_{DDQ} from the nominal value of a few hundred nA to 10 - 20 μA were commonly observed.

It is also known that FIB irradiation can lead to significant parameter shifts of *unbiased*, isolated transistors [19]. The amount of damage varies with technology, and transistors with smaller gate areas tend to be more susceptible. This damage has also been shown to be bake recoverable.

Damage of a more permanent nature, such as electrostatic discharge (ESD) damage can also occur in the FIB system. ESD occurs when a large amount of charging occurs due to the ion beam, and the charge built up finds a path to ground through a sensitive

structure. The likelihood of ESD damage increases as the ion beam current increases; thus, in addition to being a necessity for CCVC imaging, the use of low beam current is advisable for ESD avoidance. Good sample grounding can help minimize the occurrence of ESD damage in the FIB system.

DISCUSSION AND CONCLUSIONS

As expected, many of the ion beam voltage contrast effects observed are analogous to those of electron beam (SEM) VC. Examples of SVC and dynamic CCVC imaging for unbiased and biased IC samples were presented. Two important differences were noted between electron and ion mode voltage contrast imaging. First, the implantation of Ga^+ into the sample leads to permanent physical change of the device as well as sputtering of the sample surface. We observed that this gallium implantation eventually impairs the ability to obtain voltage contrast information from the sample, leading to a reduced number of secondary electrons reaching the detector, even after neutralization of the surface with the electron floodgun. Second, we found that focused ion beam imaging of an IC (either passivated or with the passivation removed) *when the IC is powered* can lead to greatly increased power supply current. This effect is apparently due to sample charging.

We have presented several examples of the use of voltage contrast imaging in the FIB system for defect localization and IC analysis. These examples have demonstrated the ability to localize circuit defects and features of interest *in situ*. The required FIB modifications or cross sections can then be made and the modifications can be verified before removing the sample from the FIB chamber.

The use of ion beams for IC analysis poses certain challenges. Ion beam irradiation is potentially more damaging to microelectronic devices than the low energy (1 keV) electron beams typically used for e-beam probing and VC. The sample surface is gradually sputtered away during ion beam imaging, gallium is implanted into the sample, charging effects can be severe, and alteration of transistor parameters is possible. However, all these effects are minimized when SVC and CCVC are performed with the minimum possible ion dose to the sample. The use of very low ion doses is an absolute requirement for successful CCVC.

It is to be stressed *that FIB based biased VC is not a substitute for e-beam VC*, which is far less damaging to ICs. SVC and CCVC on unbiased IC samples are probably the most useful of the FIB VC techniques. However, the ability to bias the IC and perform VC imaging in the FIB chamber provides the ability to identify defects and verify circuit modifications *in situ* which adds a great deal of versatility to the tool.

ACKNOWLEDGMENTS

The authors thank Bruce Draper, Rich Flores, and Henry White of Sandia for providing test samples, and Tom Olson of Micrion Corp. for simulations of secondary electron paths. This work was performed at Sandia National Laboratories, a multiprogram laboratory operated by Sandia Corporation, a Lockheed Martin company, for the U. S. Department of Energy under contract number DE-AC04-94AL85000.

REFERENCES

1. D. Perrin and W. Seifert, Solid State Technology, October 1994, p. 95-96.

2. F. A. Stevie,, T. C. Shane, P. M. Kahora, R. Hull, D. Bahnck, V. C. Kannan and E. David, Surface and Interface Analysis, Vol. 23, 1995, p. 61- 68.

3. A. N. Campbell, J. M. Soden, R. G. Lee, and J. L. Rife, in Proceedings of the 21st Internat. Symp. for Testing and Failure Analysis (Santa Clara, CA), Nov. 6-10, 1995, p. 33-41.

4. E. I. Cole Jr., C R. Bagnell Jr., B.G. Davies, A.M. Neacsu, W.V. Oxford and R.H. Propst, Scanning Microscopy, Vol. 2, 1988, p. 133-150.

5. E. I. Cole Jr. and J.M. Soden, *Microelectronics Failure Analysis Desk Reference*, 3rd ed., ASM International, Materials Park, Ohio 1993, p. 163-179.

6. J. M. Soden and R.E. Anderson, Proc. IEEE, Vol. 81, 1993, p. 703-715.

7. S. Gorlich, K. D. Hermann, W. Reiners and E. Kubalek, Scanning Electron Microscopy, II, 1986, p. 447-464.

8. E. I. Cole Jr., C. R. Bagnell, F. A. DiBiance, D. Johnson, W. Oxford, C. Smith and R. H. Propst, Appl. Phys. Lett., Vol. 48, 1986, p. 599-600.

9. J. Colvin, in Proceedings of the 16th Internat. Symp. for Testing and Failure Analysis, (Los Angeles, CA), Oct. 29-Nov. 2, 1990, p. 331.

10. E. I. Cole Jr., C. R. Bagnell Jr., B. Davies, A. Neacsu, W. Oxford, S. Roy and R. H. Propst, J. Appl. Phys., Vol. 62, 1987, p. 4909-4915.

11. H. T. Lin, J. F. McDonald, J. C. Corelli, S. Balakrishnan and N. King, Thin Solid Films, Vol. 166, 1988, p. 121-130.

12. H. Ogawa, K. Tamuru, K. Matsuyama, M. Fukumoto and H. Iwasaki, in Extended Abstracts of the 22nd Conference on Solid State Devices and Materials, (Sendai, Japan), 1990, p. 405.

13. M. Abramo and R. Wasielewski, Semiconductor International, October 1997, p. 133-134.

14. X. Yang and Y. Song, in Proceedings of the 23rd Internat. Symp. for Testing and Failure Analysis, (Santa Clara, CA), Oct. 27-31, 1997, p. 115-119.

15. D. Luo and X. Song, in Proceedings of the 23rd Internat. Symp. for Testing and Failure Analysis, (Santa Clara, CA), Oct. 27-31, 1997, p. 339-343.

16. N. Benazeth, Nuclear Instruments and Methods, Vol. 194, 1982, p. 405-413.

17. D. A. Dahl, Idaho National Engineering Laboratory, Idaho Falls, ID, Document No. EGG-CS-7233, 1994.

18. J. L. Wiza, Nuclear Instruments and Methods, Vol. 162, 1979, p. 587-601.

19. A. N. Campbell, K. A. Peterson, D. A. Fleetwood, and J. M. Soden, in Proceedings of the 35th Internat. Rel. Phys. Symp., (Denver, CO), April 8-10, 1997, p. 72-81.

Fault Localization

Thermal Techniques

Liquid Crystal Microscopy

Shekhar Khandekar
Level One Communications, Inc.
Sacramento, California

Kendall Scott Wills
Texas Instruments, Semiconductor Group
Stafford, Texas

ABSTRACT

To perform successful failure analysis on a semiconductor device, precise knowledge of the exact fail site location is important. Several techniques are used by failure analysts to determine the failing location. Photo emission microscopy, voltage contrast and liquid crystal are among these techniques. Liquid crystal microscopy is the most economical and simplest to set up. The technique is based on the change in electro-optical properties of the liquid crystals as a function of voltage, current or temperature caused by the defect.

1. INTRODUCTION

The most critical problem concerning failure analysis of integrated circuits lies in determination of the precise location of electrical defects. Locating the defect site in current devices is complicated by the small geometries, 1 micron or less, of the transistors used to form the integrated circuit and the large number of transistors which are combined to generate the required logic function. Internal probing, once the main stay of failure analysis, has become more difficult. The ability to sharpen the needles to less than 0.5 μm and then place them repeatedly on a metal line without damage to the metal has not developed as fast as the increment in integration.

Several new tools have been developed to test the current generation of semiconductors. The new techniques are electron beam (E-beam) and laser probing (1), Emission Microscopy, electron beam induced current (EBIC) and optical beam induced current (OBIC). A complimentary technique to these techniques is liquid crystal microscopy (LCM). The change in the electro-optical properties of the crystal as a function of voltage, current or temperature is used to visualize the action of the failing circuit (2). Liquid crystal microscopy (LCM) is important, easy and economical as it can be set up for a small fraction of the cost as is required for E-Beam or Emission Microscopy.

2. LIQUID CRYSTALS

2.1. Principle Of The Technique

The technique works on the principle of liquid crystal's inheritant property of "visible transition from anisotropic state to isotropic state".

2.2. Types Of Liquid Crystals

Two types of liquid crystals are typically used in failure analysis: Cholesteric and Nematic. There are magnetostrictive and voltage sensitive liquid crystals available on the market. Some market research is required to find the correct product for one's needs. The demand for liquid crystals is low in comparison to other chemicals. The low demand has set up a rather volatile industry. One company, BDH, however, has continued to be a constant supplier of liquid crystals to the failure analysis.

Cholesteric LCs display color changes in response to localized temperature differences by monochromatically reflecting incident light. Some of the Cholesteric LCs are not sensitive enough to detect low power dissipation shunts.

Nematic LCs possess optical birefringence properties. The crystals are rotated by electric fields, changing the material's property of reflection of polarized light. Defects like polysilicon to diffusion shorts, typical of ESD failures, are easily isolated with high spatial resolution, down to 1 micron, when nematic liquid crystals are used.

2.3. Selection Of Liquid Crystal Type

Various liquid crystals are commercially available. For example, 4-cyano-4 'hexyl-biphenyl, commercially known as K-18 by BDH, is available from various sources. K-18 is popular LC since its state transition temperature (STT) is 29.9 °C which is very close to the room temperature. Picart *et al*. (3) described several of the LCs used for the LCM. Bahr *et al*. (4) reported the properties of the smectic 'C' phase, and smectic 'A' phase of compounds with high spontaneous polarization.

LC K-21 (BDH) has a STT of 42.83 °C. Other LCs are available with their STT ranging from −30 °C to 60 °C. Merck T74 and T75 display color bands as a function of temperature. LC for special application are commercially available with STT of 83.5 °C, 60 °C, as well as that of 107 °C.

Several vendors sell LC kits in which they label these LCs as LC1, LC2 etc.

Fig. 1 A typical setup for liquid crystal testing contains A) probe station with polarizers, B) temperature controller to adjust the temperature of the device under test to the transition temperature of the liquid crystal, C) some visual monitor - in this case a video camera.

171

3. APPLICATION

The liquid crystal microscopy is very useful in locating failures related to excessive leakage current. The technique has been effectively used on CMOS, MOS, Bipolar and BiCMOS technologies. Many types of current conducting phenomena can be detected by this technique. Care should be taken when selecting a LC to ensure the temperature range of the transition is compatible with the fail mechanism.

4. EQUIPMENT REQUIREMENTS

Failure analysis laboratories already possess almost all the equipment required for this technique, a power supply or some form of a stimulus (a tester or a pattern generator), a temperature controller and a microscope with two polarizers. The first polarizer is in the initial light path to the device under test. The second polarizer, analyzer, is in the reflected light path. The polarizer near the light source should be rotatable to align the line of polarization with the optic axis of the liquid crystal. As the liquid crystal changes state, the optical axis of polarization of the liquid crystal changes from in line with the polarizers, to a polarization out of phase, or in some cases, random with respect to the analyzer.

Areas of the test device which do not change state remain clear. The light passes through the polarizers unattenuated. Regions of the test device which do change state cause the polarization to shift. The analyzer then attenuates the light causing enhanced contrast between the two regions of the test die.

The power supply or the stimulus is used to power-up the device and to run the failed pattern. To increase the viewability of the failure, the power to the device is pulsed. The power pulse causes the fail-site to wink or flicker. The power pulse also helps control the temperature of the device. On devices where pulsing the power is not acceptable, the suspected failing circuit must be activated at a frequency which can be seen by the human eye, and which causes the temperature of the fail-site to increase sufficiently to cause the liquid crystal to change states.

The temperature controller is used to vary the temperature of the device around the STT. The closer to the STT without going over the more sensitive to state changes the liquid crystal becomes. One equipment supplier adds a temperature jogger to the temperature controller to allow the temperature to slowly fluctuate across the transition temperature. During the time the temperature is infinitely close to the transition temperature the ultimate sensitivity is reached. The ultimate sensitivity is only supported for a short time, typically 2 to 10 seconds. Pictures of the fail-site must be taken at this time. If the work is to be done on a wafer or a device which has lost its bond wires, a probe station can be set-up to do liquid crystal microscopy.

Other accessories required, other than the liquid crystal bottle are: small bottles for mixing the liquid crystal, syringe, a glass beaker, latex gloves, filter paper, a small soft brush, methyl alcohol (methanol) or ether. Acetone can also be used to dilute the LC. Acetone is required to clean the device after the LCM. Test the solvent with a small test sample before use. Some solvents react with the liquid crystal causing the crystal to gel.

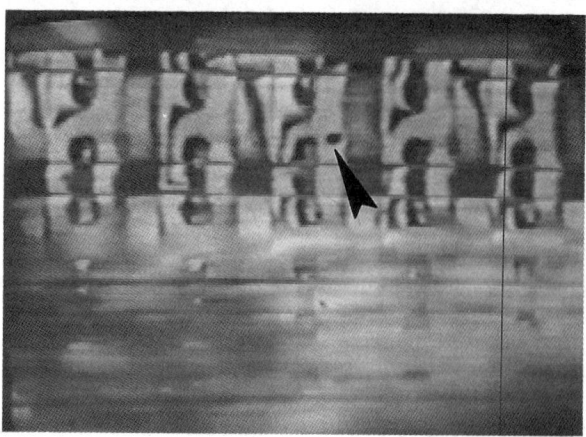

Fig. 2 A video image of liquid crystal testing where the defect is shown by the black spot at the arrow. Note no black spot appears on the structures which are identical next to the defective structure.

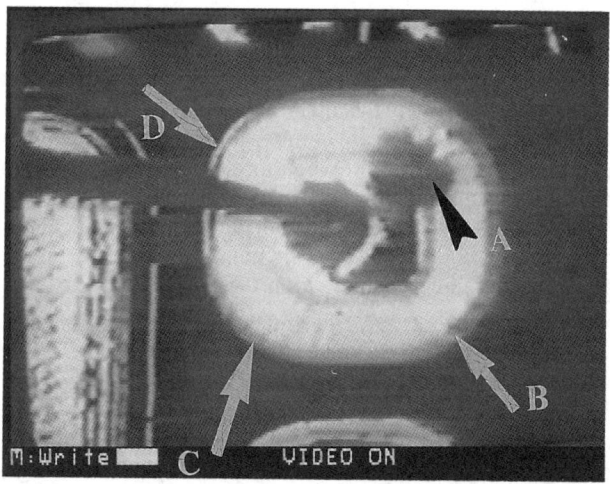

Fig. 3 A video image of liquid crystal testing of a diode. The black spot at arrow A is the location of the defect. Compare the black spot to the other corners of the diode, arrows B, C and D.

5. LIQUID CRYSTAL MICROSCOPY

5.1. Liquid Crystal Preparation

Take about 1 ml of liquid crystal with the syringe and pour it into the bottle for mixing. Take 10 ml of methanol and pour it into the bottle with liquid crystal thus giving you a 1:10 mixture of LC and methanol. This ratio works well for applying a uniform and thin coat of LC on the surface of the die. Pure LC is very viscous and does not spread uniformly on the surface of the die.

Some liquid crystals are supplied as a powder. In such cases place a small piece of crystal on the sample. Heating the sample above the transition temperature will cause the crystal to melt and flow over the sample.

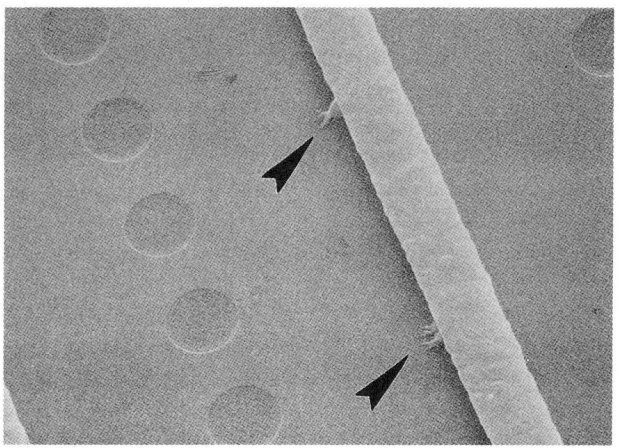

Fig. 4 A SEM image of poly filaments caused by electrical over stress. The defects are pointed to by the arrows.

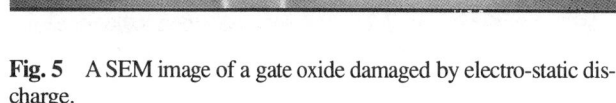

Fig. 5 A SEM image of a gate oxide damaged by electro-static discharge.

5.2. Preparing The Device For CM

The device to be analyzed is decapsulated by using an appropriate technique for the device which maintains electrical integrity. The device is then mounted on a socket under the microscope which has the polarizers. The device is connected to the test stimulus and readied for testing.

The LC is then applied to the die with the help of syringe. This is done by applying a few drops of diluted LC on the die surface and spreading the crystal with a small brush or filter paper. The brush or filter paper will eliminate any additional LC from the die. The die can be spun to get a uniform coating if the coating is applied before insertion into the electrical test fixtures.

The device is then powered up as per test conditions. The device is heated beyond the STT of the LC being used. At this time make sure that the polarizers are in place. Let the device cool down. As the device cools down to near the STT of the LC, a hot spot will be detected. Try to rotate the polarizer to get the best viewing. Remember, the temperature reading on the controller may be different than the temperature at the surface of the device.

The device is then deprocessed by selectively etching the various layers which form the integrated circuit to identify the defect.

6. SENSITIVITY OF LIQUID CRYSTALS

Good results are obtained at a power dissipation of 1 uW or better. A resolution of 1 micron square, and temperature sensitivity of 0.1 °C can be obtained. For smaller structures, the sensitivity is much lower. For buried layers, a loss in resolution has been experienced.

7. LIMITATIONS

Only current-conducting failure mechanisms can be detected. Failures which occur at temperatures other than those

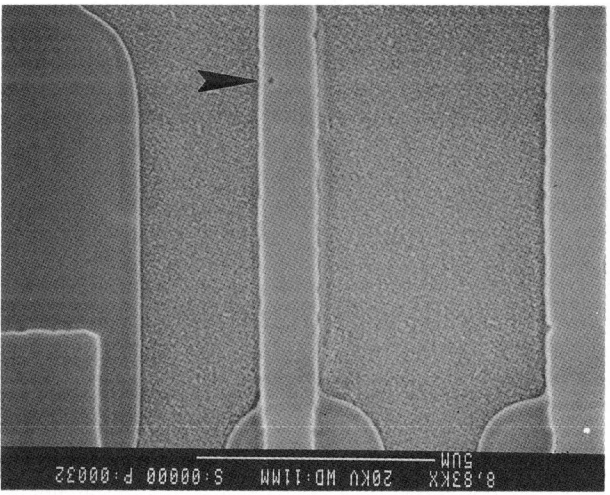

Fig. 6 A SEM image of a gate oxide pinhole. The pinhole is shown at the arrow.

close to the liquid crystal's state transition temperature will be difficult to detect with this technique.

Limited work has been performed on the magnetostrictive and voltage sensitive liquid crystals. The primary reason for the limited work is the rather low device operating frequency with which these crystals work, less than 1 MHz. The rather low operating frequency and the difficulty of obtaining a consistent supply of these liquid crystals has prevented them from becoming prominent in the failure analysis laboratory.

8. SAFETY

Although no evidence of any major effects on the human body have been documented, irritation has been observed on

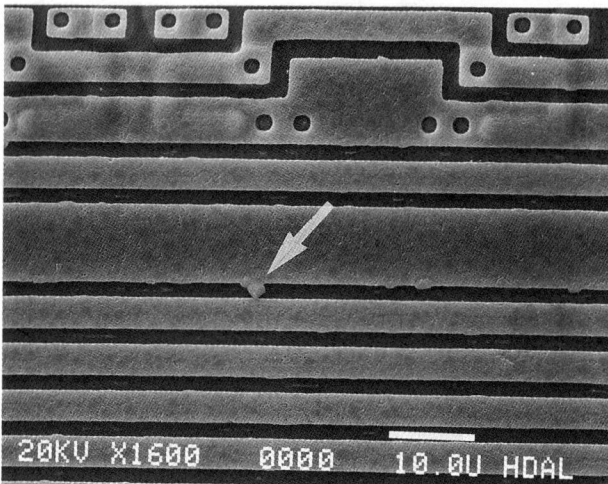

Fig. 7 A SEM image of a metal short.

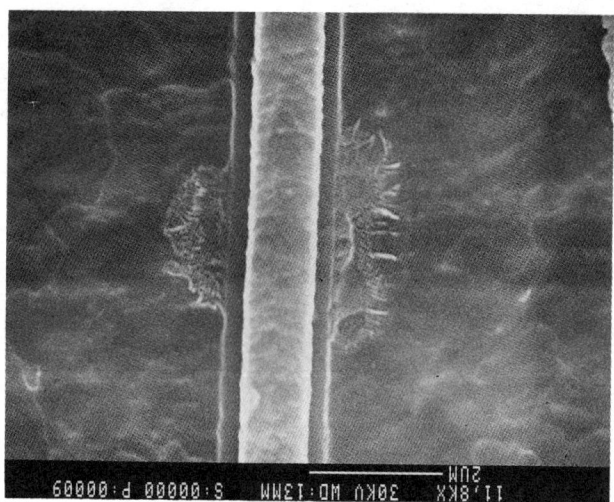

Fig. 8 A SEM image of the residue left from a contaminate. Residue shown at arrow.

the skin of laboratory rabbits. LC is not known to be toxic material. Nevertheless, care should be taken not to expose the body to liquid crystal. If accidental contact does occur, the LC should be washed off. The proper MSDS (Material Safety Data Sheet) should be obtained from the vendor of the LC before use. When using LC's at elevated temperatures some of them produce a foul odor. They need to be used where the area around the test die can be vented.

9. CONCLUSIONS

An inexpensive method of liquid crystal microscopy for fail- site isolation has been presented. This technique is fast and easy to set up but must be customized by the user per the requirements of each device to be analyzed. The technique is limited to defects which will generate enough heat to change the state of the liquid crystal being used. For defects which generate heat the power dissipation of the failure must be greater than 1 uW. Liquid crystals which are magnito constrictive and voltage sensitive are available with some research into potential suppliers.

10. EXAMPLES OF PROBLEMS FOUND WITH LIQUID CRYSTAL

See Fig. 1-8.

Acknowledgments

Thanks is extended to the various sources of liquid crystal listed in the following section. Without their help this paper would not be possible. Special thanks to Temptronic Corp. for providing video tape of equipment setup and applications.

Bibliography

1. Proceedings of the 1st European Conference on Electron and Optical Beam Testing of Integrated Circuits, Grenoble, France, 1987. Microelectronics Engineering, 7, 1987.

2. Bauman *et al.*, 5th International Conference on Reliability and Maintainability, Biarritz, France, 1986.

3. Picart *et al.*, "Visualization if VLSI Integrated Circuits by means of Ferro electric Liquid Crystals," SPIE Vol. 1080, Liquid Crystal Chemistry, Physics and Applications (1989), p 131-139.

4. Bahr Ch. *et al.*, Discussion Meeting on Physics and Chemistry of Unconv. Org. Material, Wiesbaden-Naurod, West Germany, 29 Apr-1 May 1987, p 925-927.

5. Kubalek *et al.*, Microelectronics Engineering, Volume 10, Numbers 1-4, May 1990.

6. Strnad R., "Sensing of Cold Spots on Integrated Circuits by Using Liquid Crystals," Sensors Expo Proceedings, 1987, p 369-374.

7. Doane *et al.*, "Field Controlled Light Scattering from Nematic Micro Droplets," *Applied Physics Letter*, Vol. 48, No. 4, 27 Jan. 1986, p 269-271.

8. Fleuren E. M., "A Very Sensitive, Simple Analysis Technique using Nematic Liquid Crystals," IRPS proceedings, 1983, p 148-149.

9. West G. J., "A Simple Technique for Analysis of ESD Failures of Dynamic RAMS using Liquid Crystals," IRPS Proceedings, 1982, p 185-187.

10. Crow *et al.*, "A New Liquid Crystal for Field-Effect Viewing of 5V Vcc CMOS Logic Families," IRPS Proceedings, 1982, p 179-184.

11. Goel *et al.*, "Liquid Crystal Technique as a Failure Analysis Tool," IRPS Proceedings, 1980, p 115.

12. Hiatt J., "A Method of Detecting Hot Spots on Semiconductor Using Liquid Crystals," IRPS Proceedings, 1981, p 130-133.

13. Burgess *et al.*, "Improved Sensitivity for Hot Spot Detection Using Liquid Crystals," IRPS Proceedings, 1984, p 119-121.

14. Taylor *et al.*, "Leakage Detection Techniques: A Comparative Study," ISTFA Proceedings, 1989, p 5-13.

15. Batchman *et al.*, "Liquid Crystals in Failure Analysis of Analog and Digital IC Chips," IRPS Proceedings, 1986, p 133-136.

16. Batchman T.E., "The Use of Liquid Crystal Materials for Fault Detection in VLSI Circuits".

17. Fleuren G., "Liquid Crystal Microthermography State of the Art".

174

Fluorescent Microthermal Imaging

D.L. Barton and P. Tangyunyong

INTRODUCTION

The need for a technique that would produce high spatial and thermal resolution images of microelectronic devices has been around for many years. In the early days of microelectronics, design rules and feature sizes were large enough that sub-micron spatial resolution was not needed. Infrared or IR thermal techniques were available that calculated the object's temperature from infrared emission. As will be shown in this tutorial, there is a fundamental spatial resolution limitation dependent on the wavelengths of light being used in the image formation process. As the integrated circuit feature sizes began to shrink toward the one micron level in the late 1980's, the limitations imposed on IR thermal systems became more pronounced. Something else was needed to overcome this limitation. Liquid crystals have been used with great success, but they lack the temperature measurement capabilities of other techniques. Liquid crystals provide a binary response, indicating if the hot area is above the crystal's transition temperature or not [1-3].

The fluorescent microthermal imaging technique (FMI) was developed to meet this need [4,5]. This technique offers better than 0.01 °C temperature resolution and is diffraction limited to 0.3 μm spatial resolution. While the temperature resolution is comparable to that available on IR systems, the spatial resolution is much better. The FMI technique provides better spatial resolution by using a temperature dependent fluorescent film that emits light at 612 nm instead of the 1.5 μm to 12 μm range used by IR techniques.

This tutorial starts with a review of blackbody radiation physics, the process by which all heated objects emit radiation to their surroundings, in order to understand the sources of information that are available to characterize an object's surface temperature. The processes used in infrared thermal imaging are then detailed to point out the limitations of the technique but also to contrast it with the FMI process. The FMI technique is then described in detail, starting with the fluorescent film physics and ending with a series of examples of past applications of FMI.

BLACKBODY RADIATION - BASIC PHYSICS [6]

It is well known that heated objects emit radiation. Most of us have observed a red hot heating element on a stove or have looked at a red hot poker that had been withdrawn from a fire. Items such as these appear hot to us not because we felt the heat, but because we could see the radiation emitted by the object. As objects are heated beyond the red hot temperature, approximately 700 °C, the amount of radiation being emitted in the visible range increases and the objects begin to turn orange, then yellow, and eventually to a bluish-white color. For example, the sun has a surface temperature of 5700 K and gives off a visible light with a peak around 510 nm. The North Star, by comparison, has a surface temperature of 8300 K and gives off radiation peaked at 350 nm. What is not readily apparent to us is that most of the light being emitted by hot objects is infrared. As natural evolution would have it, our eyes have developed so that their spectral response is well suited to the peak wavelengths being emitted by our sun. The spectral sensitivity of the human eye is shown in Fig. 1.

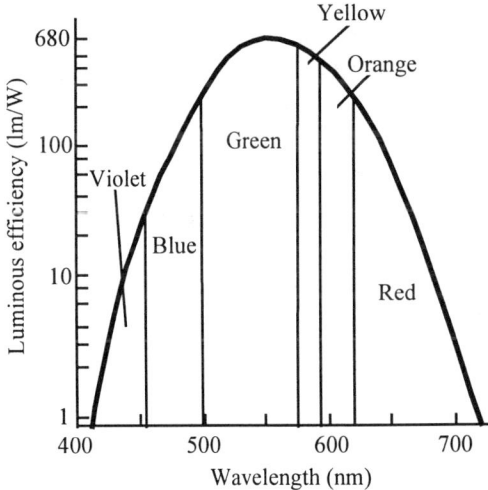

Fig. 1 - Spectral response of the human eye [7]

The connection between the peak radiation wavelength and temperature forms the basis for infrared (IR) temperature measurement. Physicists in the late 19th century knew the relationship between the temperature of a blackbody, a body that absorbs all of the radiation incident upon it, and the peak wavelength of the radiation being given off by that body, but could not describe that relationship mathematically. The most important observation was that all blackbodies at the same temperature emitted radiation with the same spectral distribution, regardless of their composition. The spectral distribution of the radiation emitted by a blackbody is known as the spectral radiancy, $R_T(\nu)$, where ν is the frequency of the radiation. The quantity $R_T(\nu)$ is defined so that $R_T(\nu)d\nu$ is the energy emitted per unit time in the frequency range ν to $\nu + d\nu$ from a unit area on a surface at temperature T. Example spectral radiances of blackbodies at temperatures ranging from 1000 K to 2000 K are shown in Fig. 2.

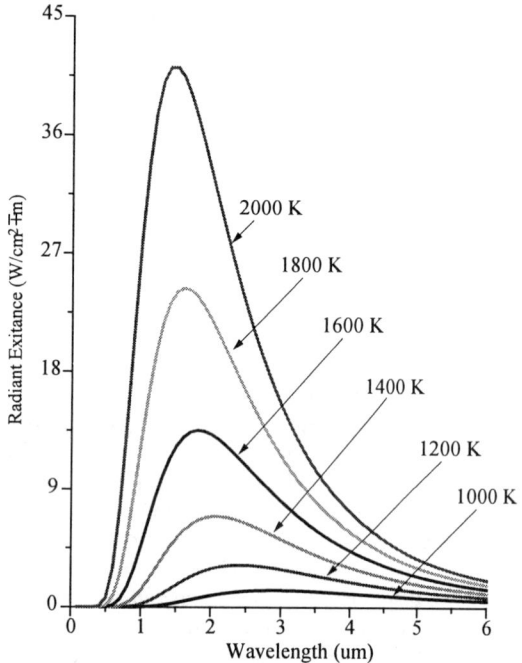

Fig. 2 - Spectral radiance of blackbodies at 1000 K to 2000 K

The data plotted in Fig. 2 has been plotted as a function of wavelength instead of frequency as is described above. The integral of the spectral radiancy $R_T(\nu)$ over all frequencies is known as the radiancy, R_T, or

$$R_T = \int_0^\infty R_T(\nu)d\nu.$$

In Fig. 2, it is evident that the radiancy increases with increasing temperature of the body. The relationship between the radiancy and temperature was first formulated in 1897 and is known as Stefan's Law,

$$R(T) = \sigma T^4.$$

Though Stefan's Law was defined empirically in its original form, the constant σ came to be known as the Stefan-Boltzmann constant with a value of

$$\sigma = 5.67 \times 10^{-8} \frac{W}{m^2 \cdot K^4}.$$

Also evident from Fig. 2 is the relationship between the wavelength of maximum spectral radiance and temperature. It is evident that the peak wavelength decreases with increasing temperature. The relationship between wavelength peak and temperature is known as Wein's displacement law, which can be written,

$$\lambda_{MAX} = \frac{2.898 \cdot 10^{-3}}{T}$$

where T is the temperature in degrees Kelvin. The result of the equation is λ_{MAX} in meters.

Classically, if we consider a small hole in a cavity as an approximation of a blackbody, (this happens to be a very good blackbody), we can easily find the number of modes present in the cavity and assign an energy to each of these modes which should describe blackbody radiation. If we realize that the spectral radiancy of a blackbody is directly proportional to that of a cavity,

$$\rho(\nu) \propto R(\nu)$$

and use the equapartition of energy theory where each mode of the cavity is assigned an energy $\bar{\varepsilon} = kT$, we end up with the Raleigh-Jeans formula for blackbody radiation

$$\rho_T(\nu)d\nu = \frac{8\pi\nu^2 kT}{c^2}d\nu.$$

This formula proved to be accurate at low frequencies, but failed miserably at high frequencies in the ultraviolet end of the spectrum. This theory was ultimately known as the ultraviolet catastrophe. The reason for the failure of this theory is simple. The equapartition theorem is only valid for a continuous distribution of energies. However, the energy of an electromagnetic wave is quantized in units of $h\nu$. The theory of energy quantization led Max Planck to the correct equation to describe the blackbody radiation spectrum

$$\rho_T(\nu)d\nu = \frac{8\pi\nu^2}{c^3} \cdot \frac{h\nu}{e^{\frac{h\nu}{kT}} - 1}d\nu.$$

In most applications, the object under examination will not be a perfect blackbody. As such, the radiancy will have to be corrected for the emissivity of the material. The emissivity, e, is defined through Stephan's Law which relates the radiancy to temperature,

$$R(T) = e \cdot \sigma T^4.$$

By definition, the emissivity of a material is the ratio of the energy radiated by a given object to the energy radiated by a blackbody at the same temperature. The emissivities of many materials have been measured and are generally available. Table 1 lists several common materials and their emissivity values.

Now that the theory of radiation from heated objects has been reviewed, we are ready to understand the emission spectrum of objects encountered in the microelectronics industry. Returning again to Fig. 2, we see that even for objects at 2000 K, there is only a small portion of the radiated energy in the visible portion of the spectrum. In fact, the majority of the information relating to the object's temperature is well into the infrared region of the electromagnetic spectrum. In the semiconductor industry, there are few devices that operate at such high temperatures. If we expand the scale shown in Fig. 2 to temperatures around room temperature, we begin to understand the difficulties in performing high spatial resolution thermal imaging on semiconductor devices. Fig. 3 shows blackbody radiation spectra for objects between 250 K and 350 K.

Of particular interest in this figure is the curve for 300 K. It is clear from this curve, that very little energy is emitted at wavelengths less than about 3 μm and most of the energy is emitted at wavelengths greater than 5 μm. In order to collect the infrared radiation information from objects near room temperature, a detector sensitive well into the infrared range is needed.

Table 1 - Emissivity of various materials [8,9]

Material	Emissivity
Ideal black body	1.0
Lampblack	.95
Asbestos paper	.95
White Lacquer	.95
Bronze paint	.8
Carbon, rough plate	.76
Oxidized steel	.7
Polished brass, oxidized copper	.60
Aluminum paint	.55
Oxidized monel metal	.43
Cast iron - polished	.25
Copper - polished	.15
Nickel - polished	.12
Aluminum - highly polished	.08
Platinum - highly polished	.05
Silver - highly polished	.02

INFRARED (IR) THERMOGRAPHY [8-12]

It is evident from Fig. 3 that to use the physics presented in the previous section on semiconductor devices we must analyze the radiation emitted by the sample in the 3 μm to 12 μm range. Most commercially available IR thermography systems use one of two types of detectors; indium antimonide or mercury cadmium telluride. Indium antimonide (InSb) detectors are sensitive in the wavelength range 1.5 μm to 5.5 μm. Mercury cadmium telluride (HgCdTe) detectors are sensitive over the range of 8 μm to 12 μm. Both detectors offer similar temperature sensitivities and ranges, but InSb operates at shorter wavelengths and should have somewhat better spatial resolution. Table 2 lists several manufacturers of IR thermography equipment and their performance characteristics.

The data in Table 2 shows that, while IR thermal imaging systems have excellent potential for temperature resolution, they suffer from a fundamental limitation on spatial resolution. In general, the spatial resolution of a microscope is limited by the wavelength of light. The relationship between resolution and wavelength as given by Lord Rayleigh's criteria can be written [13]

$$\text{Re solution} = \frac{0.61 \cdot \lambda}{\text{N.A.}}$$

where λ is the wavelength of light being imaged and N.A. is the numerical aperture of the microscope. This result although very simple, provides a good estimate of the resolving power of a microscope system. In Lord Rayleigh's own words, "*The rule is convenient on account of its simplicity and it is sufficiently accurate in view of the necessary uncertainty as to what is meant by resolution.*" [13]

This relation clearly shows that IR thermal systems that gather radiation in the 1.5 μm to 12 μm range will not be able to resolve sub-micron structures that are found on modern VLSI integrated circuit technologies. Before discussing how to circumvent this limitation, we should review the general theory of operation of IR thermal systems.

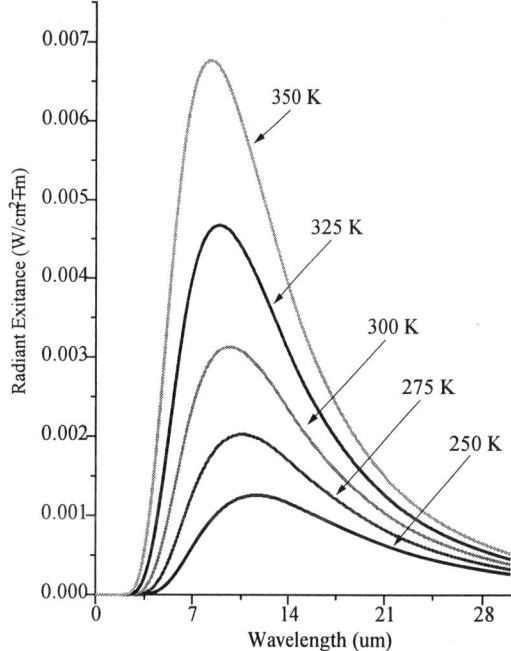

Fig. 3 - Spectral radiance of blackbodies at 250 K to 350 K

In general, there are several ways to measure the temperature of an object. The simplest method is measure the radiance and, if the emissivity of the material is known, directly compute the temperature of the sample. Systems of this type use a very simple photovoltaic type detector that is sensitive in the IR wavelengths. Cooled InSb photovoltaics are perhaps the most common. These detectors are generally sensitive in the 1.5 μm to 5.5 μm range.

To directly measure the temperature of a sample, the radiance is measured first. The radiance from a sample at a given temperature is related to the radiance that would be collected from a blackbody at the same temperature by the emissivity, or

$$R_T = e \cdot R_{TBB}$$

where R_T is the radiance from the sample, e is the sample's emissivity, and R_{TBB} is the radiance of a blackbody at the same temperature. Ideally, if the sample's emissivity is known, its temperature can be calculated if the relationship between the radiance collected by a given system from a blackbody and its temperature is known.

In order to increase the accuracy of the temperature measurement, the radiance that is reflected by the sample must be accounted for. Thus, the total radiance collected by a system from a sample would be

$$R_{Total} = R_T + (1-e)R_0$$

Table 2 - Thermography equipment for semiconductor and hybrid production applications [14]

Company/Models	Detector (std/opt)	Max. IR spectral range in µm (std/opt)	Temperature range (low/high °C)	Temperature resolution (°C/@°C)	Spatial[1] resolution
AGEMA IR Thermovision 782LWB	InSb/ HgCdTe	2-5.6 8-12	-20/1600	0.1/30	15 µm
Barnes Engineering Computherm RM-2A Microscope	InSb InSb	1.5-5.5 1.5-5.5	amb/600 amb/500	0.1/75 0.1/75	15 µm** 8 µm
Hughes Probeye Series 4000	InSb/ HgCdTe	2-5.6 8-12	-40/1500	0.1/22	2.2 mrad
Inframetrics Model 600	HgCdTe	3-12	-20/1500	0.1/30	30 µm
Mikron Instrument Thermo Tracer 6T61	HgCdTe/ InSb	8-13/ 1.5-5.5	-50/2000	0.1/70	100 µm
UTI CCT-9000	HgCdTe	8-14	-50/350	0.1/75	600 e/l
notes: std = standard opt = optional	amb = ambient e/l = elements per line	1. Spatial resolution using microscope objective	** Their latest system quotes 4 µm spatial resolution		

where R_0 is the radiance emitted by the ambient background. Combining the last two equations, the total radiance collected by the system is

$$R_{Total} = e \cdot R_{TBB} + (1 - e)R_0 \,.$$

If the emissivity is known and the ambient radiance can be accounted for, the temperature of the sample can be found.

Lastly, the spectral response of the system must be considered. In general, the system response will not be constant over the spectral range on interest. The introduction of an "effective" blackbody radiance representing a convolution of the blackbody radiation spectrum and the IR microscope/detector response must be known. This information has usually been characterized by the manufacturer and is incorporated into their blackbody radiance to temperature conversion algorithm.

If the emissivity of the sample is not known, a comparison between the radiance measured from the unknown at a known temperature and from a blackbody can be performed. While measurements of this type can be easily done on uniform macroscopic samples, making these measurements on VLSI integrated circuits is difficult, but not impossible.

More elaborate IR imaging systems have been developed which scan the sample and create a two-dimensional image of the surface radiance [9]. Non-scanning systems simply add up the radiance from the entire field of view, or a pre-selected portion of it, and convert that radiance to temperature. Scanning systems allow surface temperature measurements to be made on samples with multiple areas of different emissivity values in the field of view. While the theory of operation is as before, calibration of the system requires that the user measure the radiance of the sample at two known temperatures using a sample stage heating unit. This measurement allows for the creation of an emissivity map of the sample surface over the temperature range of interest.

FLUORESCENT MICROTHERMAL IMAGING [15 - 22]

History

The concept of using a film with a temperature dependent fluorescence quantum yield to generate high resolution thermal maps of integrated circuits was first published in 1982 by Paul Kolodner and J. Anthony Tyson at Bell Laboratories [4]. Their first work described a technique which could yield thermal images with a thermal resolution of 0.006 °C and a spatial resolution which was equipment limited to 15 µm. They noted that spatial resolutions of better than 1 µm were achievable with different optics and a better camera. This improvement in spatial resolution was demonstrated less than a year later in the second work published by the same authors [5]. The goal of this second work was to demonstrate the spatial resolution capability of the technique, which they measured to be 0.7 µm. Our subsequent research [15-17] has developed the understanding of the fundamental limitations of FMI and the operational procedures needed to insure maximum performance when applied to modern IC technologies. This research has also refined the hardware used for FMI to improve its usability.

This tutorial will lead readers through the theory behind the technique, describe the hardware and image processing requirements needed, and discuss some example applications. The preceding sections on blackbody radiation and IR thermography were included so that the readers can understand thermal imaging concepts, the nature of the signals that are present, and why there is a fundamental limitation on spatial resolution imposed by using imaging IR radiation. While IR thermography is invaluable to the microelectronics industry, its applications are limited when the areas of interest are sometimes an order of magnitude smaller then the available spatial resolution. Applications such as these are where FMI is unchallenged.

EuTTA Compound Specifics: [18-22]

During the late 1950's and early 1960's, there was a great deal of laser research. Some of this work dealt with the use of rare earth chelates as sources for use in liquid-based lasers. Rare earth chelates were identified as possible sources because of their well known fluorescence responses to UV or near-UV excitation sources. One of these compounds, EuTTA (europium thenoyltrifluoroacteonate) is the focal point for the fluorescent microthermal technique. The chemical structure of EuTTA is shown in Fig. 4. The availability of compounds such as EuTTA is the main reason that the fundamental limitation on spatial resolution encountered in IR thermal systems could be overcome.

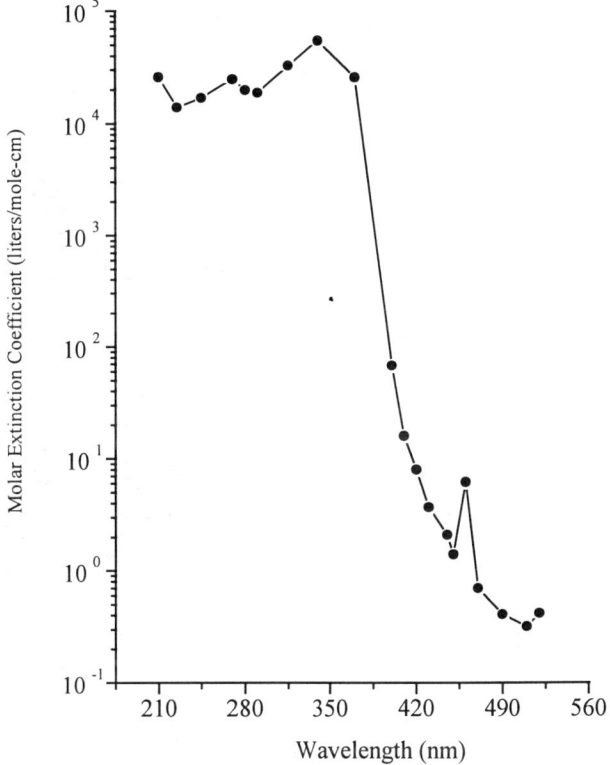

Fig. 4 - Chemical structure of EuTTA

EuTTA is not the only compound available for FMI. In fact, there are chelates of all of the rare earth elements which include La, Sm, Eu, Gd, Tb, Dy, Tm, Yb, and Lu. The europium system was ultimately selected as the most suitable because of its temperature characteristics, emission/absorption characteristics, availability, and other qualities. There are several other europium compounds which might be suitable for FMI. Other β–diketone chelates of europium are available such as europium benzoylacetonate, europium dibenzoylmethide, and europium hexfluoroacetonate in addition to EuTTA. EuTTA however, has the best fit for temperature dependent fluorescence quantum yield in the temperature range near room temperature.

To better understand the theory behind FMI, we need to discuss the process which gives the fluorescing film a temperature dependent fluorescence quantum yield. Fig. 5 shows the molar extinction coefficient (or loosely, the absorption spectra) versus wavelength for EuTTA in an ethanol solution. While FMI requires that EuTTA be suspended in a solid matrix, the data in Fig. 5 indicates the excitation wavelengths of interest. First, there is a broad absorption peak centered around 335 nm. This is where the TTA ligand absorbs energy. After about 360 nm, the amount of incident radiation that is absorbed falls off strongly. The two peaks at 460 nm and 525 nm are consistent with Eu^{3+} levels and are not of interest from an excitation viewpoint. What is of interest is the lack of absorption for wavelengths much above 500 nm. This allows for a strong separation between the excitation source and the fluorescence emission.

The ultraviolet radiation used to excite the EuTTA fluorescence does so through a series of intermolecular energy transfers as illustrated in Fig. 6. The TTA ligand absorbs the UV light then transfers the energy to the europium ion. While several fluorescence lines are excited, the transition from the Eu^{3+} $^5D_0$ energy level to the $^7F_2$ level, as is shown in Fig. 7, is the most efficient. This transition generates the bright fluorescence line at 612 nm which is used for FMI. Fig. 8 shows the emission spectrum for crystalline EuTTA at 25 °C.

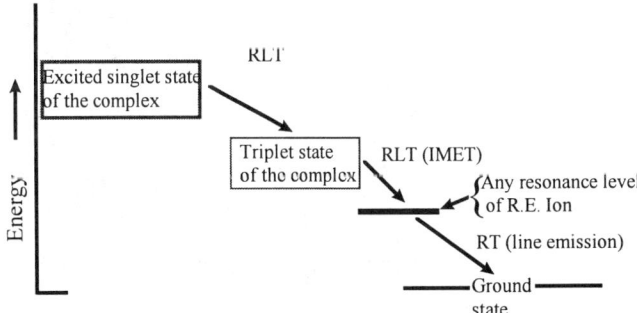

Fig. 5 - Absorption spectra for EuTTA (in solution) [18]

Fig. 6 - Intermolecular energy transfer processes. RLT denotes non-radiative transitions while RT is the radiative transition.

For thermal imaging applications, we need to know how the emission spectra of the compound changes with temperature. The temperature dependence of this europium chelate was considered a problem for liquid laser applications, but it is what FMI relies upon for image formation. Fig. 9 shows the measured absolute quantum yield versus temperature and Fig. 10 shows the decay time of the fluorescence yield versus temperature. Both of these plots were generated for EuTTA in an ether:isopentane:ethanol (5:5:2) solution. For the FMI application, a curve will need to be generated for each compound mixture that is used. These data have been included to illustrate the temperature dependence of EuTTA.

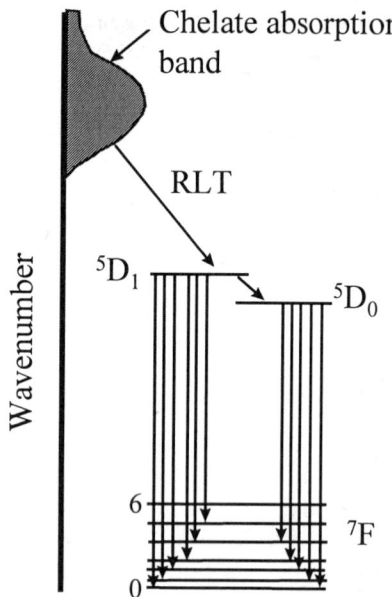

Fig. 7 - Europium chelate fluorescence

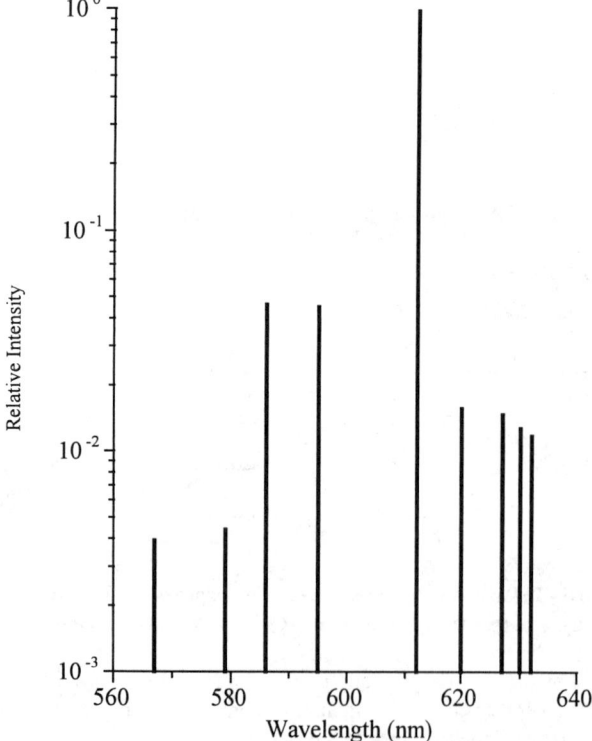

Fig. 8 - Emission spectra of EuTTA (Crystalline) at 25 °C [18]

The fit curve in Fig. 9 was plotted using the equation

$$Q(T) = 0.398 - 0.07 \cdot e^{0.031 \cdot T}$$

where Q(T) represents the quantum efficiency of the compound. The behavior of this compound over this broad temperature range is very predictable and provides a simple way to calculate the temperature of an object by imaging its change in quantum yield.

The information in Figs. 9 and 10 illustrates the intermolecular energy transfer between the TTA ligand and the europium ion. First, notice that the quantum efficiency falls off faster than the decay time with increasing temperature. This shows that we may have a low quantum efficiency even at low temperatures due to a loss of excitation energy of the Eu^{3+} ion. The change in quantum efficiency with temperature is an indicator of the quenching of the whole system, while the fluorescence decay time is an indication of quenching of the fluorescence in the europium ion itself.

The standard application technique for using EuTTA for FMI is to incorporate the chelate into a PMMA (polymethylmethacrylate) matrix. A typical starting point is a solution consisting of 1.2 wt% EuTTA, 1.8 wt% PMMA, and 97 wt% MEK (methylethylketone). The MEK is a very high vapor pressure solvent that evaporates rapidly leaving the EuTTA/PMMA mixture on the sample [4]. Typically this mixture is spun on the sample and allowed to cure in an oven at 125 °C for about 30 minutes. Ideally, the film should only be several optical absorption lengths thick. At an excitation wavelength of 365 nm, a 300 nm film is approximately 3.5 optical absorption lengths thick. The idea is to have the film thick enough that most of the UV light is absorbed, but thin enough that the thermal profile of the sample surface is not distorted. As we will find out in later sections, the image processing required to create a thermal image reduces the influence of film non-uniformity on image quality. As such, the film should be as uniform as possible, but great pains to achieve perfect uniformity of film composition and thickness are not necessary.

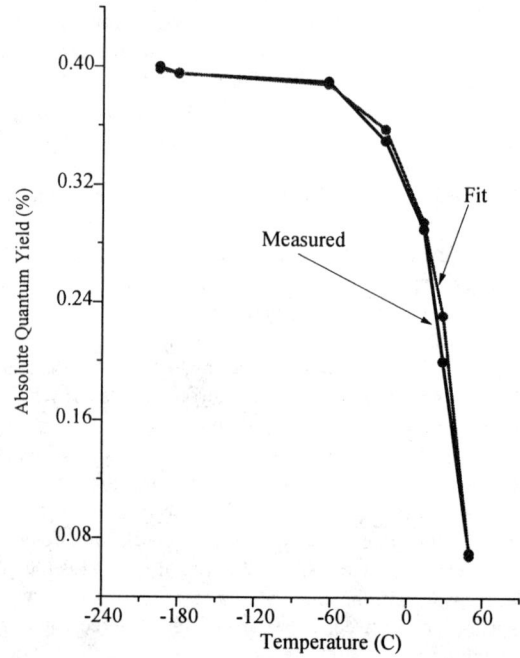

Fig. 9 - Absolute quantum yield for EuTTA [19]

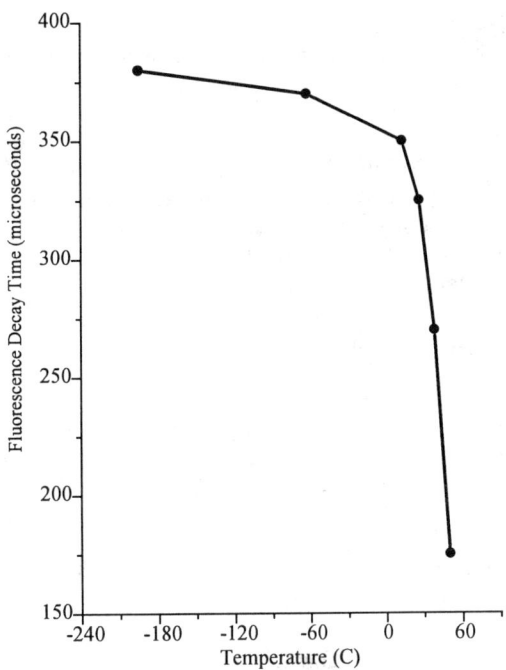

Fig. 10 - Fluorescence decay time for EuTTA [19]

The EuTTA/PMMA composition can be varied as needed for any specific application. Adjusting the EuTTA content will change the amount of fluorescent light emitted from the coated sample and changing the amount of MEK in the mixture will thin the solution out for applications where spinning the sample is not practical. For instance, integrated circuits in packages are sometimes difficult to mount on a photoresist spinner. The use of a thinned out mixture would allow a thin film to be deposited without spinning the IC. Usually, spinning a packaged IC will cause the mixture to accumulate around the ball bonds leaving a thick film in these areas. The thick film is often not a problem, unless the signal input structures are the areas of interest. For these applications, a thinner mixture or a higher spin rate would be in order.

The advantage of PMMA is that it can easily be removed once the thermal analysis is completed. Rinsing the sample in acetone will dissolve the film in several minutes. The use of other polymers such dPMMA, (perdeutero-poly-methylmethacrylate) will provide a stronger temperature dependence, but the additional cost of dPMMA is not justified. Other matrixes, such as Owens-Corning GR650 spin-on glass have been successfully used, but make film removal more difficult.

Regardless of the film type used, accurate absolute temperature measurements are possible but, because of the differences in the logarithmic slope of the quantum efficiency versus temperature curves for different materials, an accurate film calibration should be done for each type of pre-mixed solution. The calibration curve can be easily obtained by using a hot/cold stage, a calibrated thermocouple, and the camera to be used for FMI. Simply record the total emission observed by the camera in a given time period with the sample at a given temperature. Varying the hot/cold stage temperature over as large a range as possible will give the best results. Samples such as blank wafers would be good for this process since, during the measurements, they will be close to the hot

chuck temperature. The emission versus temperature data can easily be plotted and a logarithmic slope can be found. Unless the composition of the mixture changes drastically, this measured slope need only be done once, especially if only relative temperature measurements are needed.

For higher temperature measurements, another europium compound, perdeutero-(tris-6,6,7,7,8,8,8-heptafluoro-2,2-dimethyl-3,5-octandionato) europium (dEuFOD) may be used up to about 200 °C. dEuFOD has a much weaker temperature dependence both for fluorescence quantum yield and fluorescence lifetime than EuTTA [22].

Image Processing

Now that we have an understanding of the fluorescent film properties which allow us to use this technique to generate thermal images, we need to cover how that information is converted to temperature data. Returning to Fig. 9, we recall that the quantum efficiency versus temperature can be represented as an exponential. Fitting an exponential function to the data in Fig. 9 gave us,

$$Q(T) = 0.398 - 0.07 \cdot e^{0.031 \cdot T}$$

as the quantum efficiency versus temperature relationship. The light intensity at a given point, (x,y), on the image can be represented by

$$S(x,y) = I(x,y) \cdot \eta(x,y) \cdot r(x,y) \cdot Q(T(x,y))$$

where $I(x,y)$ is the illumination intensity, $\eta(x,y)$ is the optical collection efficiency, $r(x,y)$ is the sample reflectivity, and $Q(T(x,y))$ is the quantum efficiency. In order to remove all spatial artifacts included in the I, η, and r terms, and leave an image containing only thermal information, we can divide an image taken with the sample under bias, i.e. a hot image, by one without bias, i.e. a cold image. The result is a map of the ratio of quantum efficiencies between hot and cold images

$$S_R(x,y) = \frac{S_H(x,y)}{S_C(x,y)} = \frac{I(x,y) \cdot \eta(x,y) \cdot r(x,y) \cdot Q(T_H(x,y))}{I(x,y) \cdot \eta(x,y) \cdot r(x,y) \cdot Q(T_C(x,y))} = \frac{Q(T_H(x,y))}{Q(T_C(x,y))} \cdot$$

If we were working with a pure exponential, this ratio would be directly proportional to the difference in temperature, T_H-T_C. The problem with doing this directly is the leading constant (equal to 0.398 in this example) in the fit for Q(T) given above.

Once way around this problem is to create a carefully measured calibration curve for a given film mixture, using a given illumination intensity, on a given optical setup, etc. and use that curve as Q(T) above. This method allows accurate *absolute* temperature measurements to be made, but adds a great deal of difficulty to the FMI process. Problems with this process arise from small changes in equipment creating changes in light collection from a sample at a given temperature. For example, since the fluorescence intensity *decreases* as the film gets hotter and the film degrades (loses fluorescence intensity) under exposure to UV light, the sample will appear hotter after repeated imaging sequences. The degradation will require repeated removal and re-application of the film for accurate results.

Since in most applications, relative rather than absolute temperature measurements are needed, the image math can be simplified greatly with only a slight loss in accuracy. We need to stress that the slight accuracy change does not decrease the sensitivity, or the smallest change in temperature that can be resolved, of the technique.

To modify the process to allow for relative temperature changes, continue with the equation for the ratio of quantum efficiencies that we had before, take the natural logarithm, and plot the result for temperatures around room temperature. This gives us,

$$\ln(S_R(T)) = \ln\left[\frac{Q(T_H)}{Q(T_C)}\right] \propto \delta T \ .$$

This equation is plotted in Fig. 11 with the cold temperature, T_C, set at 28 °C.

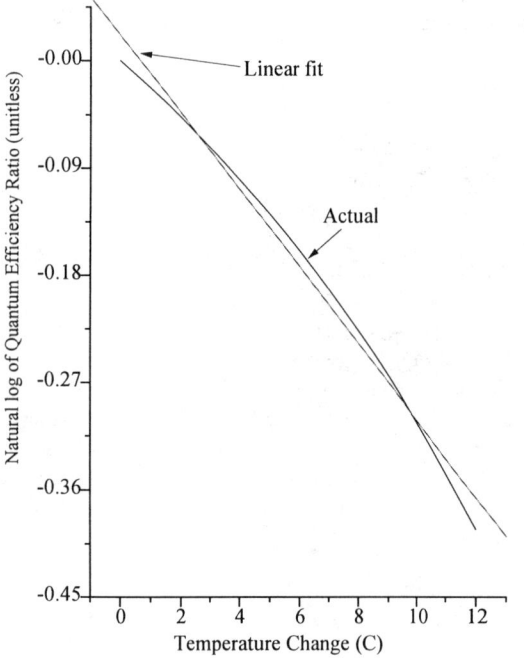

Fig. 11 - Plot of quantum efficiency ratio around room temperature

The linear fit in Fig. 11 can be represented by the equation,

$$y = 0.02132 - 0.03233 \cdot x.$$

The standard deviation in the slope is 0.0006909, or about two percent. The slope of the linear fit is what we need for temperature conversion. Once we have the slope, the rest is easy. Simply divide the natural log of the quantum efficiency relation by the slope. The result is the relative temperature change at a given pixel location.

$$\delta T \approx \alpha^{-1} \cdot \ln\left[\frac{Q(T_H)}{Q(T_C)}\right] = \alpha^{-1} \cdot \ln\left[\frac{S_H(x,y)}{S_C(x,y)}\right].$$

By simply taking the natural log of the ratio of the light intensities from a hot and cold image, we can divide the result by a constant and have a relative temperature measurement. This is the method presented in the literature [4,5] for use in FMI.

The examples that have been used here are based on the quantum efficiency versus temperature from EuTTA in a solvent solution, as described in the previous section. For EuTTA combined in a polymer or glass matrix, the fluorescence quantum efficiency will have a similar temperature dependence, as this is dictated by the intermolecular energy transfer process, but they will be different. Published values for the slope of the linear fit in Fig. 11 for EuTTA in a dPMMA matrix are approximately -0.047 /°C [5]. This number will be different for EuTTA in PMMA or GR650 glass, as well as for EuFOD in dPMMA. Whatever chemistry is chosen, a calibration curve as outlined above must be made.

As an example, if a camera with 16-bit gray scale resolution is used and assuming the published value of the logarithmic slope of -0.047 /°C, the best possible thermal resolution would be:

$$\delta T \approx \left(\frac{1}{0.047}\right) \cdot \ln\left(\frac{65535}{65534}\right) = 0.324 \times 10^{-3} C \ .$$

Comparing this with an 8-bit gray scale camera:

$$\delta T \approx \left(\frac{1}{0.047}\right) \cdot \ln\left(\frac{255}{254}\right) = 83.6 \times 10^{-3} C \ .$$

This result will be referred to in the next section where hardware requirements are discussed.

Standard techniques can be used to reduce noise in the resulting thermal image. Image averaging on both the hot and cold images can be done prior to processing. Various techniques can be used after processing to enhance the thermal artifacts.

System Hardware

There are essentially three main system components that are required for FMI. These components include a light source, a camera system, and an optical platform. This section discusses each of these areas and indicates differences between each of the possible choices.

Fluorescence Excitation Source. The light source used for fluorescence excitation has the most possible choices. The absorption spectra for EuTTA (Fig. 5) indicates that we are interested in ultraviolet sources in the approximate range of 210 nm to 365 nm. Sources with wavelengths much greater than about 400 nm will require higher intensities to excite the fluorescence and thus are not practical. UV sources always present a hazard to the human eye so the use of proper eye protection is mandatory.

Most of the FMI systems currently in operation use arc lamps as their excitation sources. The most common lamps for UV applications are mercury, xenon, and mercury/xenon types. Mercury bulbs have well known spectral peaks in the UV range and several that extend into the visible range. Xenon lamps have peaks in the upper end of the visible range, but have a broad continuum of output that extends usefully into the UV. Mercury/Xenon lamps combine the two spectra to create a broad range, general purpose light source.

Arc lamps have been manufactured for many years. As such, an old lamp may be in storage somewhere at your facility. While this old lamp may be good for demonstrating the FMI technique, modern arc lamps have benefited from years of research and are more stable sources. Stability of the light source, as evident from the last section, is one of the limiting factors in creating high temperature resolution images. If you decide to use an arc lamp as an excitation source, inquire about the stability of the light output as a key factor toward deciding which unit to purchase.

The other obvious choice for excitation source is a laser. Lasers differ from arc lamps in that all of their optical energy is in one narrow wavelength band rather than being spread over a continuum. Lasers are more efficient since all of the light output is within the wavelength range of interest. Table 3 lists some of the different laser systems and their respective lines. About half of the systems are listed as CW, or continuous wave, systems. These lasers have a continuous light emission. The remaining systems are pulsed, meaning they cannot sustain continuous emission and emit light in short pulses, usually with repetition rates up to several kilohertz. The choice between laser systems is usually cost. Many of the CW systems listed are very large laser systems, such as argon ion lasers, that have a weak line suitable for FMI. The problem is that you may have to purchase a laser with an output of several watts to yield several milliwatts for the UV line you need.

Pulsed systems, such as the dye lasers or solid state lasers, can be relatively inexpensive, but for FMI, two successive exposures (one for the cold image and one for the hot image) may contain different numbers of pulses from the laser. Although this may sound like a small problem, it adds another process that may reduce the temperature resolution of the system. Dye lasers do have the ability to be tuned over a broad range of wavelengths giving them an advantage when and if other fluorescent compounds are developed.

Table 3 - Laser systems suitable for FMI excitation sources

Wavelength (nm)	Laser System	Pulsed or CW
190-1000+	Dye	pulsed
220-390	Frequency doubled dye	CW
325	HeCd	CW
330-380	Neon	CW
333-364	Argon Ion	CW
333.6	Argon	CW
337	Nitrogen	pulsed
345-500	Ti:Sapphire	pulsed
351	XeFl	pulsed
351.1	Argon	CW
351 or 355	Nd:glass	pulsed
365	Nd:YAG	pulsed

Camera Systems. The next step for system design is to choose a camera system. Existing systems, without exception, use slow-scan CCD cameras. In this case, slow-scan refers to the frame rate at which data is read out of the CCD array. Television cameras adhere to the NTSC (or other) video standard where the CCD array in a CCD camera would be read at a rate of 30 frames per second. While this frame rate is good for television cameras, for quantitative analysis of the image content it is relatively poor. High quality TV cameras can approach 400 lines of information in about 500 fields with about 8-bits of dynamic range.

As we saw in the previous section, using a camera system with 8-bits of dynamic range would limit system sensitivity to roughly 0.4% change in quantum efficiency, or 1 part in 256. Slow-scan cameras are available with 12 to 16-bits of dynamic range and can thus image changes in intensity from 1 part in 4096 to 1 part in 65536 in an image of size ranging from 512 by 512 to more than 4096 by 4096 pixels. This translates into an order of magnitude gain in *possible* difference in temperature resolution.

Slow-scan cameras, since they do not adhere to TV standards, are designed to stare at a field of view and integrate for a variable length of time. For a situation where there is a very small amount of light being emitted, the camera can stare at the field of view for several minutes to several hours or until the detector becomes saturated. In contrast, when using TV cameras for low light situations, it is necessary to grab and add video frames together, which also adds noise. Image averaging can be used to help remove noise, but it doesn't boost signal.

Noise in collected images is a second factor that will limit temperature resolution. Slow-scan cameras are almost always either peltier or liquid nitrogen cooled. Cooling helps to eliminate thermal generation of electron-hole pairs which can fill up CCD charge wells with noise instead of images signal. Peltier, or liquid/peltier, cooled systems generally operate at -39 °C or so and, as a result, generate only several electrons per second of noise with a well capacity of several hundred thousand electrons. Liquid nitrogen cooled systems, by cooling to a much lower temperature, keep the noise down to several electrons for every ten to one hundred seconds. With these systems, the readout electronics also add a small, but predictable amount of noise to the image, typically several electrons or tens of electrons per pixel. The noise qualities of these cameras and their integration capabilities, along with spectral sensitivities to wavelengths longer than 1 µm has prompted manufacturers of light emission systems to use them in lieu of TV cameras.

In general, virtually any camera which can yield an image in digital format, either directly or by frame grabbing, can be used for FMI. While TV cameras will work, they may become the limiting factor for thermal resolution. In applications where the technique is needed, but funds to build a system are scarce, TV cameras may be a good solution. However, the use of slow-scan CCD cameras will insure that the camera is not the limiting factor in system performance.

Optical Platform. Lastly, the optical platform to house the excitation source and camera must be decided upon. In order to electrically bias the sample, probe stations provide an obvious choice for an FMI system. Most systems used for probing fine geometries have optics boxes that have TV camera ports. The most common optics assemblies are made by Bausch & Lomb and Mitutoyo. Since they are used during probing, the lenses on these systems are extra-long working distance, but have to sacrifice some optical quality.

Standard metallographic microscopes generally have superior optical systems, but suffer from short working distance objectives and limited facilities for electrical biasing of the sample. Most microscopes do have c-mount camera ports or other ways of attaching a camera.

Depending on the application, whether it is primarily packaged ICs or wafer level analysis, the optical platform that best suits the most frequent use of the system should be used.

Other System Components. Once the three main system components have been selected, the remaining considerations are relatively small. A computer with image processing capabilities that can handle the requirements outlined in the previous section is needed. The number of possibilities for this part of the system is too great to warrant much further discussion. It is sufficient to state that virtually any computer that you can get the image to will work without affecting performance, other than image processing speed, of course.

Traditional FMI systems input the excitation source through a UV grade fiber optic cable onto the sample at an oblique angle [4]. This does remove many headaches from the optical system, but tends to limit the amount of light that can be easily be sent to the sample, especially when using high magnification, shorter working distance lenses. Even on a probe station, a 50x lens will have a short enough working distance to complicate sample illumination. The use of a "through-the-lens" type of illumination removes the problems of sample illumination, but adds the problems encountered with non-UV transparent optics found in most microscopes. Generally, standard optical components offer transparency for light with wavelengths greater than about 370 nm. UV grade components are available because of the market for people doing fluorescence work, but the components, such as lenses, tend to be very limited in application and type, (e.g. magnification, working distance, numerical aperture, etc.).

In order to minimize the amount of time that UV light from the excitation source is focused on the fluorescent film, a shutter must be placed in the beam path between the UV source and the sample under examination. The shutter is usually synchronized with the camera shutter. As will be shown in the next sections, the UV light which excites the fluorescence also degrades the film which can generate noise in the thermal images. Although FMI can and has been done without this shutter, its inclusion is a simple way to control one of the dominant noise sources in an otherwise well optimized system.

The final system component needed is an interference filter to filter out all of the fluorescence except the dominant line at 612 nm as shown in Fig. 8. Most manufacturers of optics or filters will be able to put together a filter sandwich that can meet the requirements of FMI. A filter with a bandwidth of about 2 - 4 nm will be sufficient.

Example Systems. Now that we have discussed some of the system design considerations, it would be beneficial to describe some of the systems that have been assembled to do FMI.

First, in their original work, Kolodner and Tyson [4,5] describe what they felt would be a typical system for FMI. They did not provide details about the optical system, but they did use a probe station and mounted the slow-scan CCD camera with the interference filter in front, on the TV camera port. Their first CCD array was a 100 x 100 array but they later upgraded to a camera with a 384 x 576 array that was peltier cooled. They sent the light from a 100 watt Hg arc lamp through IR and blue-glass filters and a fiber optic cable and brought it to focus on the sample at an oblique angle. Image processing, remember that this is circa 1983, was done on a PDP 11/73 type computer. This system worked well enough that they could measure 0.01 °C thermal and 0.7 μm spatial resolution. The only systems that were known to be commercially

marketed to FMI were virtually identical to this system, but benefited from some computer improvements. A system configured in this manner is illustrated in Fig. 12.

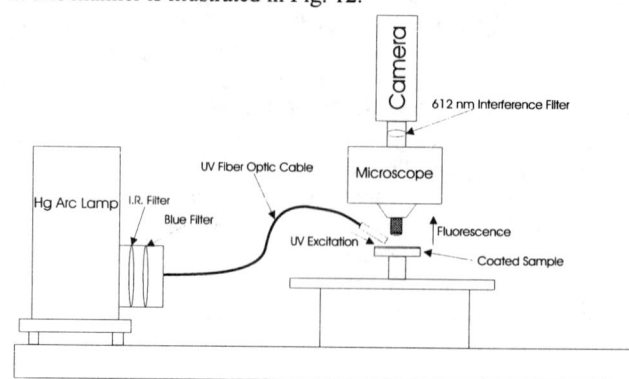

Fig. 12 - Implementation of FMI Using Hg Arc Lamp and Oblique Illumination

Another system developed in the early 1990's represented a good contrast to the system just described. The system was setup on a scanning laser microscope (even though the scanning laser feature was not utilized). The fluorescence excitation source was a 15 mW HeCd laser operating at a wavelength of 325 nm. The laser light passed through a shutter to a UV grade fiber optic cable and through some UV transparent (reflecting) objectives. This arrangement is illustrated in Fig. 13. Since the LSM optics were not UV transparent (for wavelengths less than 365 nm), a special filter block was manufactured to allow the UV light to be introduced in the optical path just above the objectives. The special block was designed to position the beamsplitter to face the side rather than the back (where the white light illumination source normally resides) of the microscope. While this system was used successfully for FMI, it did not provide for spatially uniform illumination and had problems with mode stability (small movements in the fiber optic cable caused noise in the thermal images).

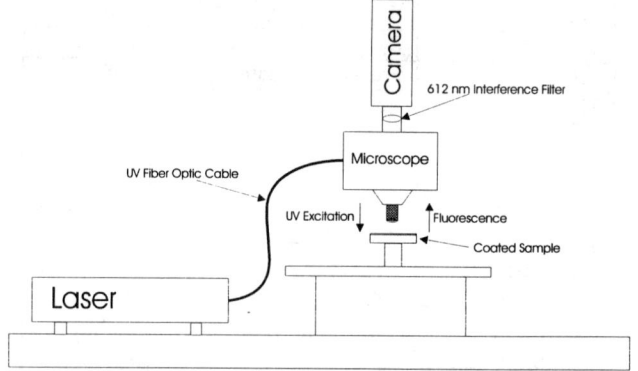

Fig. 13 - Implementation of FMI Using Laser and Coincident Illumination

The present iteration of FMI system design is shown in Fig. 14. This has proven to be the easiest to assemble and best performing system design. The basic premise is to use light from a Hg arc lamp in the wavelength range from 365 - 390 nm as the excitation source. This, at first glance, is a contradiction of the desirable system design criteria for excitation source selection. The absorption spectra shown in Fig. 5 indicates that light at wavelengths longer than 365 nm will be stimulate the fluorescence but with much less efficiency than at shorter wavelengths. While this is true, systems assembled using this method have no problems with a lack of fluorescence. The system still uses the scanning laser microscope as an optical platform, but now has a 200 W Hg arc lamp and shutter replacing the normal white light illumination source. The most important components in this application are the filters and beamsplitter. The excitation filter was a short wavelength pass (SP) filter with a 390 nm cutoff wavelength. The beamsplitter had to reflect the UV light passed by the excitation filter but allow the visible fluorescence to pass with minimal attenuation. The interference filter used was the same 612 nm wavelength filter used in earlier systems. This system has been successfully used with standard microscope objectives from 1.25 X to 150 X from several different manufacturers. All of the example FMI applications described a the end of this tutorial were done on this system.

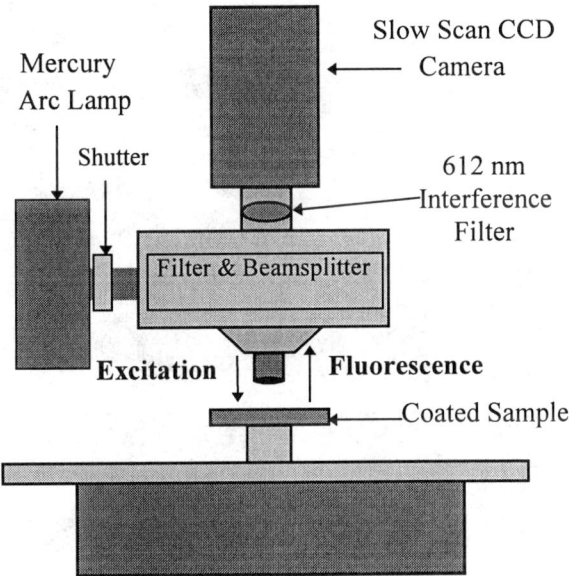

Fig. 14 - Implementation of FMI Using an Arc Lamp and Coincident Illumination

For all of the systems described above, the same slow scan CCD camera was used. A Photometrics slow-scan CCD camera was chosen that featured a liquid/peltier cooling system and was interfaced into a workstation. The cooling of the camera has been subsequently upgraded to liquid nitrogen to reduce the array's dark current primarily for light emission applications where long exposures are needed. The liquid/peltier cooling system is sufficient for FMI as typical exposure times are in the 0.1 - 10 second range and FMI is a large signal (the fluorescence is very bright) imaging application. The CCD array chosen was a 512 by 512 pixel array with charge wells averaging over 650,000 electrons. The deep wells enable the use of the full 16-bit dynamic range (available as an option on the camera electronics). The array was also thinned and is back illuminated to enhance the light collection efficiency. Larger arrays, such as the of 1024 by 1024 or 2048 by 2048 types are available, but they added significant cost and reduced the cell dimensions and thus the full well capacity.

Photon Shot Noise and Signal Averaging

In absence of all other noise sources, the signal-to-noise ratio in high photon flux imaging applications is limited by photon shot noise. The statistics describing photon shot noise follows a Poisson distribution. From elementary probability and statistics, a random variable, x, is said to have a Poisson distribution if the probability mass function is

$$p(x; \lambda) = \frac{e^{-\lambda} \lambda^x}{x!}$$

for some $\lambda > 0$. The variable, λ, is frequently a rate per unit time per unit area. In order to obtain any meaningful thermal information, it is crucial that we can separate the thermal signals from the photon shot noise. Fig. 15 and 16 are shown to illustrate this point. Fig. 15 shows a histogram of pixel intensity for a FMI thermal image of an **unbiased**, metal test structure coated with the standard FMI film. The image was taken at a magnification where the test structure did not fill the field of view, thus including a background area as well as a hot region as would be found in most applications. Since no current runs through the metal test structure, the histogram in Fig. 15 contains no thermal information which means that the peak observed is mostly due to photon shot noise. As expected, the histogram shows a relatively good fit to the Poisson distribution, except on the right-hand side of the curve. The slight deviation is the result of ultraviolet bleaching that will be subject of discussion in the next section. Fig. 16 shows a series of histograms of FMI thermal images for the same test structure biased at currents of 0 mA, 35 mA and 50 mA, respectively. In both the 35 mA and 50 mA histograms, an extra peak on the right hand side of the Poisson peak was observed. This peak contains thermal information associated with the heating of the metal test structure. As expected, this peak becomes more pronounced as the current in the test structure increases .

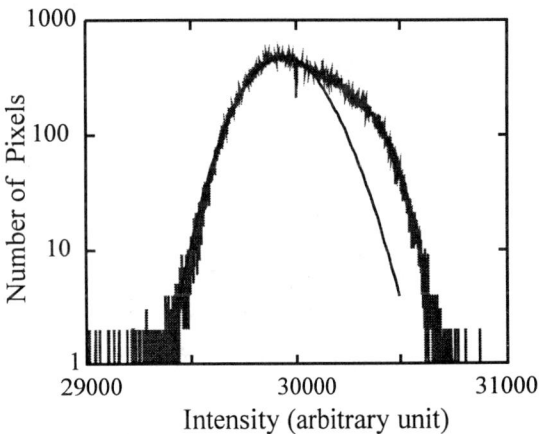

Fig. 15 - Histogram of a FMI image of an unbiased test structure coated with a base-mixture film showing a relatively good fit to a Poisson distribution except on the right-side of the curve.

Another interesting feature in Fig. 16 is the shift in the Poisson-peak position as the current increases. We attribute this shift to the increase in the background temperature surrounding the test structure. We can easily convert the position shift to the relative increase in temperature using equation (4), with the slope (α) for the base-mixture film being obtained experimentally (see the

185

section on the film dilution). For the histograms in Fig. 16, a background temperature increases of 0.6 °C and 0.9 °C are calculated from the 35 mA and 50 mA histograms, respectively. Interestingly, the relative increase in background temperature varies linearly with respect to the input power of the test structure (Fig. 17).

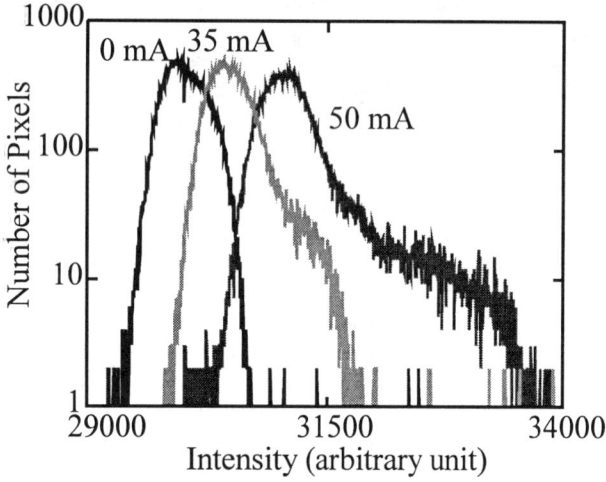

Fig. 16 -Histograms of FMI images for a test structure coated with a base-mixture film and biased with currents of 0 mA, 35 mA and 50 mA, respectively.

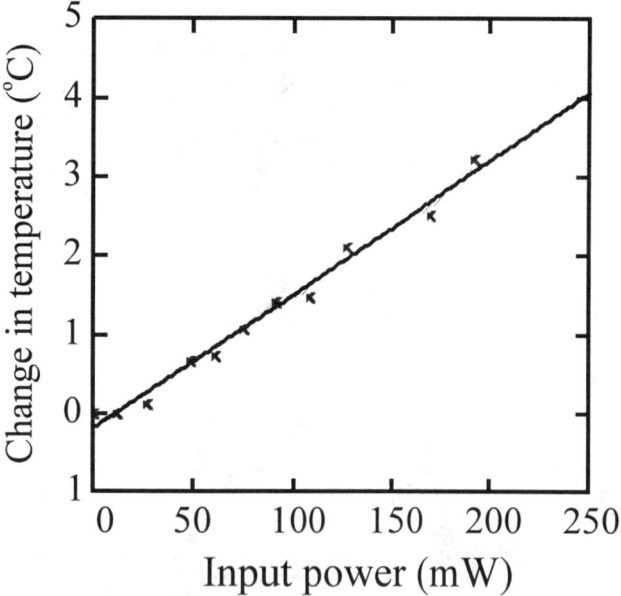

Fig. 17 - Change in background temperature calculated from the position shift in the Poisson peak as a function of input power to the test structure.

Since photon shot noise is due to the quantum nature of light, there is no way to eliminate it totally; however, it is possible to reduce its effects through signal averaging. Fig. 18 shows a comparison of two FMI images, one from a single cold/hot image pair and the other from an average of ten cold/hot image pairs. All the images were made with a current of 50 mA flowing through the metal test structure during the hot image exposure and a five second per frame exposure time on a UV-stabilized film. The image made

from a single cold/hot image pair shows a grainy texture away from the test structure while the averaged image has a very smooth appearance. The effect of signal averaging can also be illustrated through the image histograms. Fig. 19 shows histograms of a single cold/hot image pair and of an average of ten pairs. The Poisson peak from the averaged image shows a narrower width than the single image one, while there is virtually no change between the two thermal peaks. Clearly, signal averaging has improved the signal to noise of the FMI image by reducing the photon shot noise.

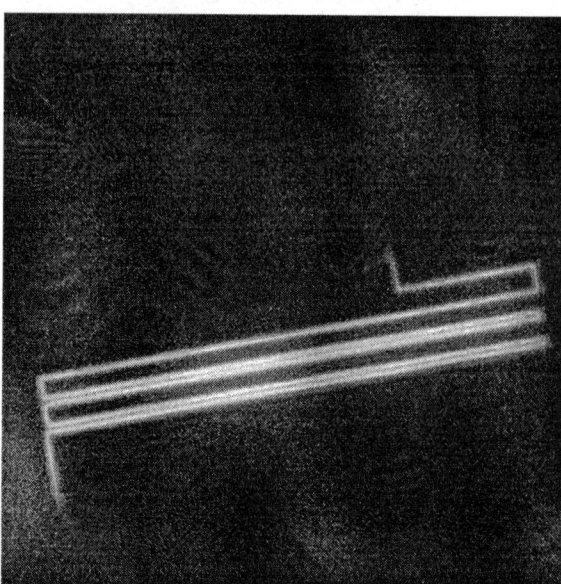

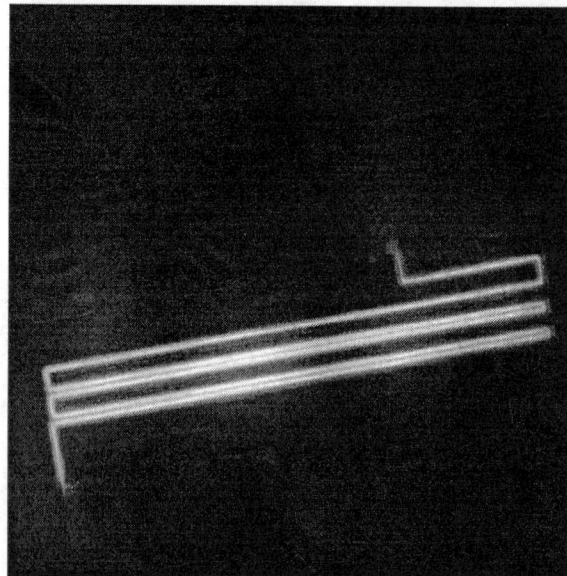

Fig. 18 - A comparison of thermal images made with single cold/hot image pair (left) and an average of ten pairs (right).

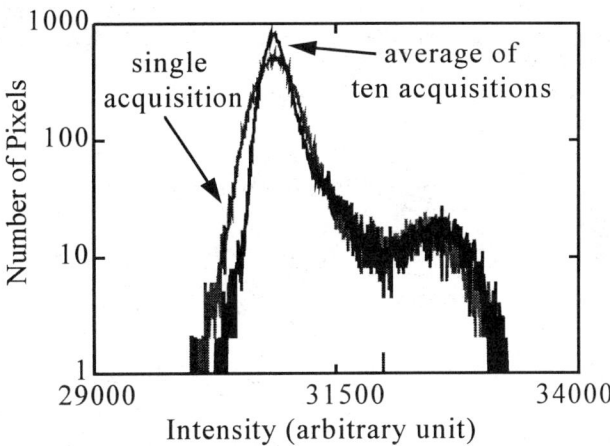

Fig. 19 - Histograms of FMI images of a test structure, showing an improved signal-to-noise ratio after signal averaging. Note that the Poisson peak obtained from an average of ten acquisitions has a narrower peak width than that obtained from single acquisitions while there are no changes in thermal signals.

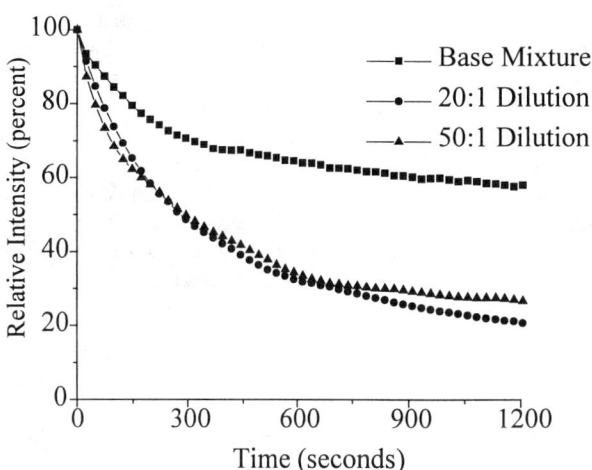

Fig. 20 - Changes in fluorescent yield versus UV exposure for three different mixture dilutions.

Ultraviolet Film Bleaching

Inorganic-based films such as EuTTA are known to gradually lose their ability to fluoresce under UV illumination leading to decreasing fluorescent intensity with increasing UV exposure time. This gradual loss in fluorescent intensity is known as bleaching. Since bleaching is unavoidable in EuTTA films, we must characterize its behavior to identify methods to minimize its effect. Fig. 20 shows the bleaching behavior of three EuTTA/PMMA films of various dilutions to continuous exposure to UV light. In all three films, fluorescent intensity decays rapidly initially and stabilizes after approximately 20 minutes of UV exposure. The diluted films, however, show a larger decrease in intensity before stabilization than the base mixture film.

In FMI, a thermal image is generated either by dividing the signals of a cold image by those of a subsequent hot image or vice versa. Ideally, the fluorescent signals of the hot and cold images should be identical in areas where no temperature changes occur. With bleaching, this ideal condition is not possible, resulting in the generation of unwanted non-thermal signals, the so-called spatial artifacts. To demonstrate the effect of UV bleaching on FMI image quality, a series of FMI images were taken after 20 seconds, 10 minutes and 20 minutes of UV exposure, respectively. In all three images, a metal test structure was biased with a current of 50 mA using a base-mixture film (Fig. 21). The spatial features are very pronounced in the top image (taken after 20 seconds of UV exposure) due to initial rapid decrease in fluorescent intensity and less noticeable in the bottom image (after 20 minutes of exposure) when the fluorescent intensity stabilizes.

Fig. 21 - Fluorescent microthermal images illustrating the reduction in spatial features after UV film stabilization (top to bottom: t = 20 sec, t = 600 sec, t = 1200 sec).

To better understand how bleaching creates spatial artifacts, we have taken a series of histograms of FMI images for the same test structure shown in Fig. 21 and (coated with a base-mixture film) taken after 20 seconds, 10 minutes and 20 minutes of UV exposure (Fig. 22). The histograms were generated with no current flowing through the structure. It is clear from Fig. 21 that bleaching causes a shift in the position of the Poisson peak and has the apparent effect of changing the background temperature as was previously described. The maximum shift occurs between the 20 second exposure and 10 minute exposure histograms, giving an apparent change in temperature of 0.60 °C. The *apparent* thermal signals resulting from bleaching manifest themselves as the spatial artifacts observed in Fig. 21.

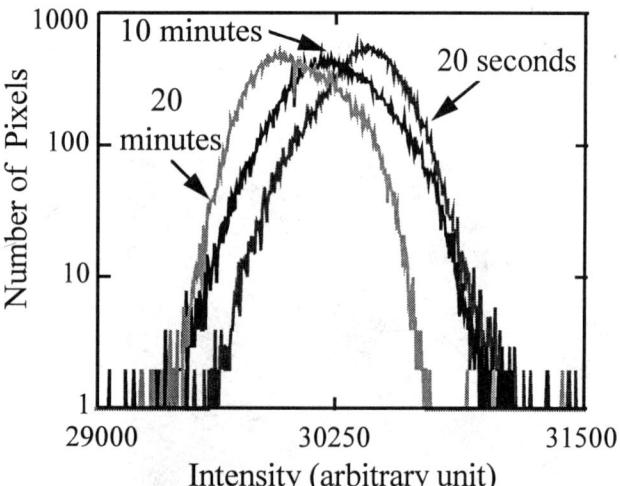

Fig. 22 - Histograms of FMI images of an unbiased test structure, showing the effect of UV bleaching in creating *apparent* thermal signals.

Effects of Film Dilution on Thermal Sensitivity

In many situations it is impractical to spin the EuTTA/PMMA mixture onto packaged integrated circuits. For these applications, a thinned mixture is typically applied without spinning. Until now, it has always been assumed that the dilution of the mixture has no effect on the logarithmic slope and hence the sensitivity of the film. In order to examine this effect, three film dilutions were characterized. The base mixture (1.8% PMMA, 1.2% EuTTA, and 97% MEK), a 20:1 dilution with MEK, and a 50:1 dilution with MEK were applied to several slices of scrap silicon. The change in fluorescent yield above room temperature was measured by heating the coated wafers with a hot stage. Fig. 23 shows the results of this experiment. Based on equation (4), the slopes (α) extracted from the curves in Fig. 23 were calculated to be 0.0245, 0.015 and 0.008 for base-mixture, 20:1 mixture and 50:1 mixture films, respectively. This data shows that α is not independent of film dilution as has usually been assumed.

Fig. 24 illustrates the effect of a change in logarithmic slope on thermal image quality. The image on the left was made with a current of 50 mA using a base mixture film after twenty minutes of UV exposure while the image on the right was made with a 20-minute bleached 20:1 film at the same 50 mA current. To ensure all images have the same signal-to-noise ratio (i.e., the same amount of photons are collected), the exposure time for the right image is increased by a factor of 10 over the left image. It is clear from Fig.

24 that the FMI image of the base-mixture film is far superior to that of the 20: 1 film with stronger thermal signals due to larger value of slope (α). In addition, the spatial artifacts are much less pronounced in the image made from the base-mixture film.

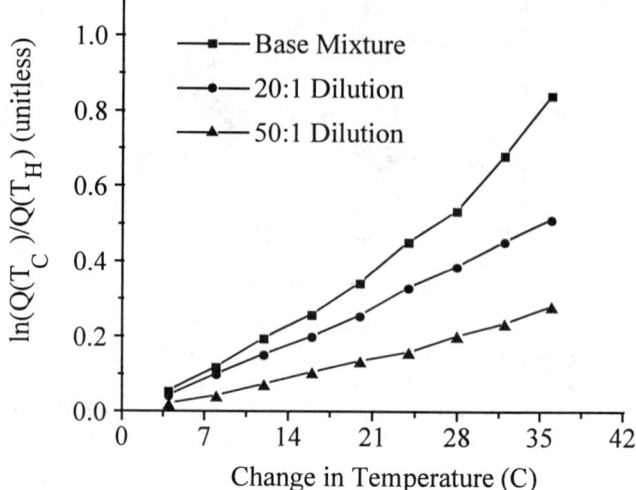

Fig. 23- Temperature dependence of the natural log of the quantum efficiency ratio for several different mixture dilutions.

Effects of Curing EuTTA/PMMA

Typically for FMI applications, the EuTTA/PMMA films are allowed to cure in an oven at 125 °C for about 30 minutes [4]. For the measurements in this paper, however, we cured the films by letting them dry in air for about 8 hours (although the applied film is useable after drying for only a few minutes in air). Fig. 25 shows the bleaching behavior of four different samples coated with base-mixture films after 8 hours of drying. The fluorescent intensity in all samples decays rapidly initially but stabilizes after 20 minutes of UV exposure. As a comparison, Fig. 26 shows the bleaching behavior of several base-mixture samples that had not been cured, i.e. the fresh samples. It is clear from Fig. 26 that the bleaching behavior of the fresh samples does not follow any set pattern and there is no stabilization in fluorescent intensity even after 20 minutes of UV exposure. As expected, we did not see any reduction in the spatial artifacts in FMI images using these uncured base-mixture films. In addition to the base-mixture films, the effect of curing on the bleaching behavior of 20:1-dilution and 50:1-dilution films was investigated. In both cases, no differences in the bleaching behavior between the fresh and cured samples were observed.

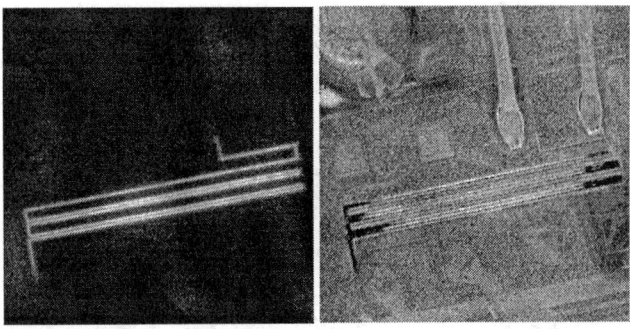

Fig. 24 - FMI Images of a test structure (biased at a current of 50 mA) using a base-mixture film (top) and 20:1 film (bottom). Note that the thermal signals are much stronger in the base-mixture film due to enhanced thermal sensitivity of the film as shown in Fig. 23.

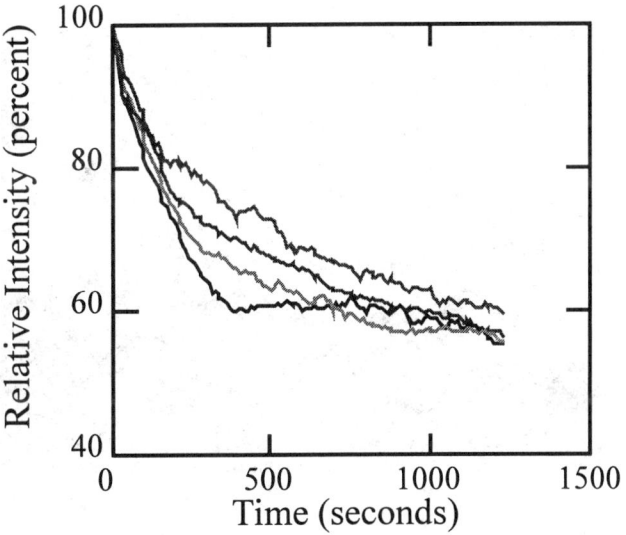

Fig. 25 - Changes in fluorescent yield versus UV exposure for four base-mixture films after drying for 8 hours in air.

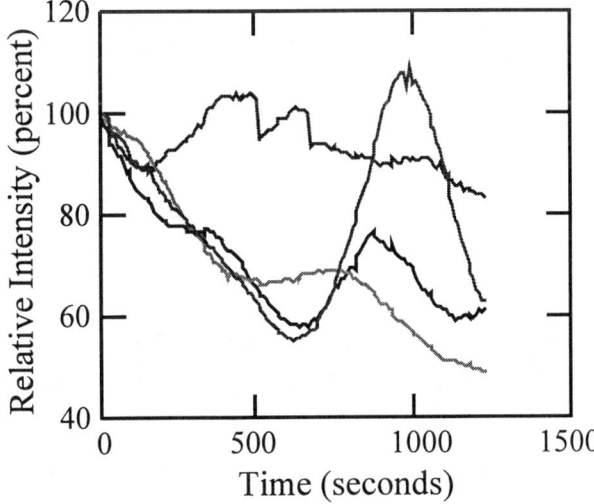

Fig. 26 - Changes in fluorescent yield versus UV exposure for four uncured base-mixture films.

Example Applications:

In this final section, examples of past applications of the FMI technique to IC technology problems will be presented. In some of the examples, the defect sites could be identified with both FMI and light emission. However, in some examples, no light emission was observed leaving FMI as the most efficient technique for locating the defect. We will show how FMI in conjunction with other failure analysis techniques was used to identify the root causes of several device failures.

Example 1. The example images shown in Fig. 27 illustrate the utility of FMI for analyzing ICs from low yield wafer lots. These images were taken on two SRAMs which were fabricated in a 0.5 μm, 3- level metal, CMOS process which had low resistance shorts from V_{DD} to V_{SS}. The parts were first examined using light emission equipment which failed to detect any emission. FMI was subsequently applied and was successful in locating the cause of the shorts in a matter of minutes. The FMI images showed that there was a metal-1 patterning problem through which metal to metal short paths were created between V_{DD} and V_{SS} with resistances from 20 to 30 Ω.

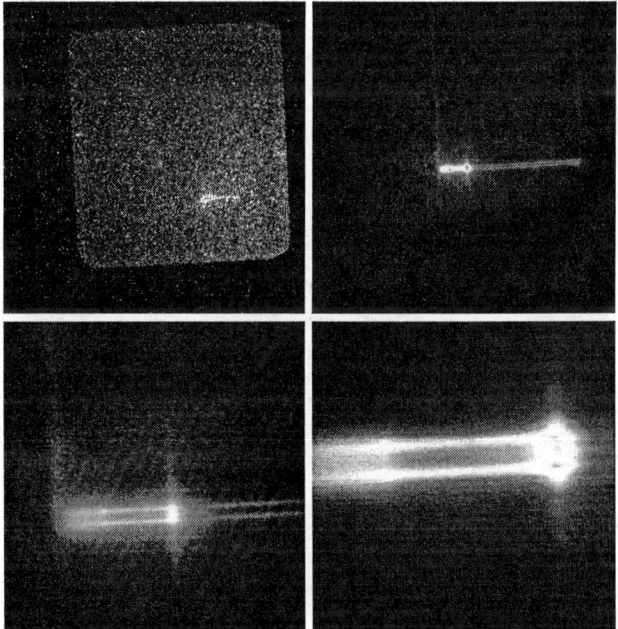

Fig. 27 - FMI images of a SRAM with a low resistance (20 Ω) short from V_{DD} to V_{SS}.

The images in Fig. 27 were taken at magnifications from 25X to 1000X to demonstrate how FMI can be used to first locate the defect at a low enough magnification to image the entire IC and subsequently determine the root cause of the problem by imaging the hot spots at successively higher magnifications. In each of the images, a current of 171 mA was supplied to the IC. This current was more than sufficient to localize the defect and was used to allow imaging of the full current path between V_{DD} and V_{SS}.

Example 2. The second example is a 1-megabit CMOS SRAM. It is a 128K x 8 SRAM with a nominal operating voltage of 5 V and a standby I_{DDQ} of < 1 μA. It has an effective channel length (L_{eff}) of 0.5 μm (0.6 μm drawn), gate oxide thickness (t_{ox}) of 16 nm, 3 layers of metal and 2 layers of polysilicon. It was designed to be I_{DDQ}-testable [24]. The SRAM cell design is a standard six-transistor cell. After dynamic burn-in, the SRAM failed the functional test and had an elevated I_{DDQ} of 3 mA. As shown in the I-V curve in Fig. 28, the IC has low I_{DDQ} below 1 V. Above 1 V, the I-V curve shows a linear characteristic, indicating that the high I_{DDQ} is a result of a resistive short. As expected, no light emission was detected on this IC (resistive defects often do not have photon generation mechanisms). With FMI, two "hot spots" were quickly located (Fig. 29). Using SEM imaging and FIB cross sectioning (Figs. 30 & 31), the root cause of failure was identified to be an embedded particle that produced an ohmic short between two adjacent metal lines. The size of the particle estimated from the SEM image is ~ 5 μm in diameter. The resistance of the short was estimated to be ~ 1200 Ω based on the linear region of Fig. 28. The FIB cross section showed that the particle was encapsulated by a passivation layer, indicating the particle was most likely deposited onto the IC right after the last metal etching process but before the passivation step. Energy dispersive x-ray (EDX) analysis of the particle in Fig. 32 showed the presence of iron, nickel and chromium, indicating it was a stainless steel particle.

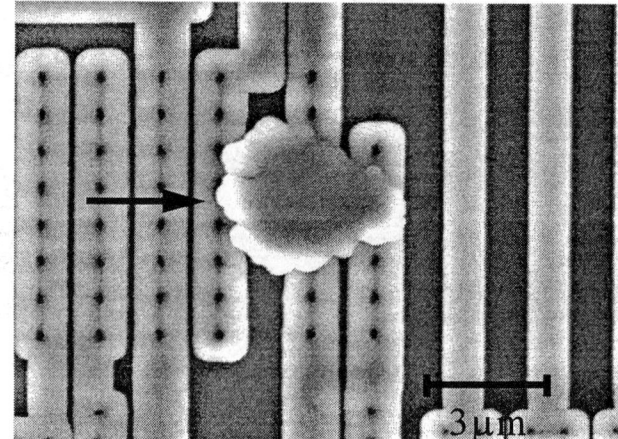

Fig. 30 - SEM image showing an embedded particle creating an ohmic short between two metal lines.

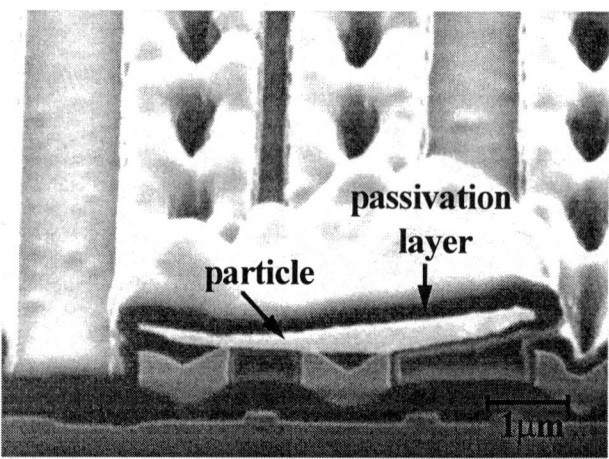

Fig. 31 - FIB cross section of the embedded particle shown in Fig. 30.

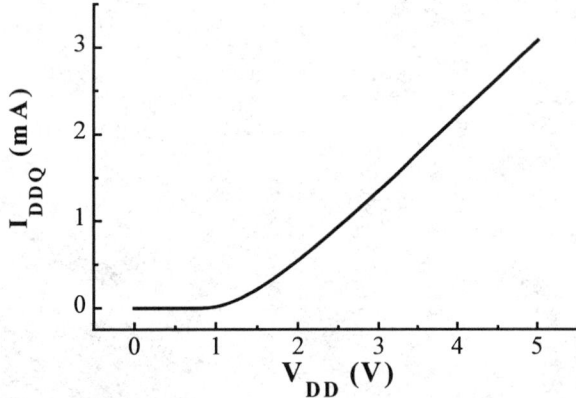

Fig. 28 - I_{DDQ} versus V_{DD} for the failed SRAM in example 2, showing a linear region above 1 V.

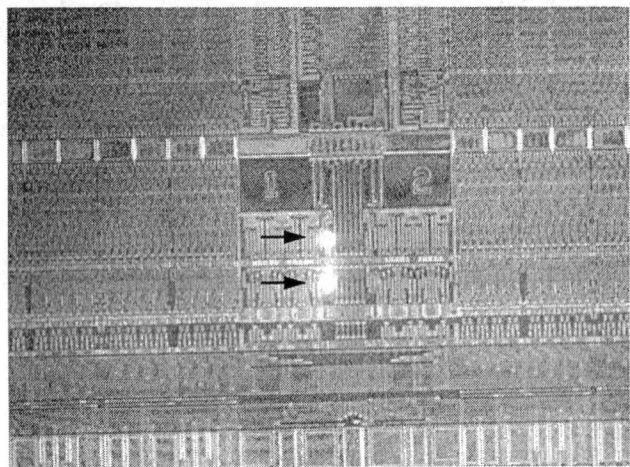

Fig. 29 - Overlay of FMI and reflected light images showing the locations of heat-generating defects of the failed SRAM in example 2.

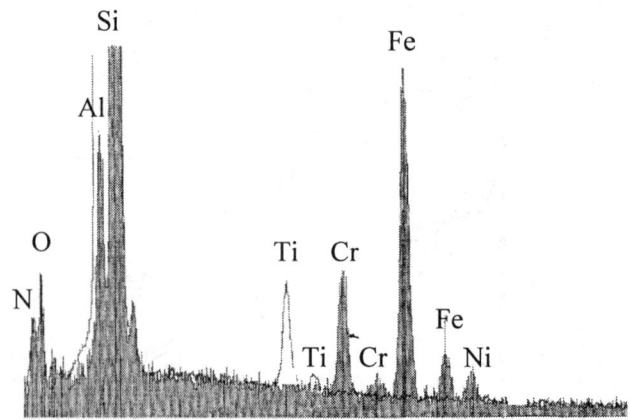

Fig. 32 - EDX spectra of the embedded particle shown in Fig. 30 and the adjacent metal line.

Example 3. The third example is another 1-megabit SRAM that has the same layout and functional operating characteristics as the SRAM described in example 2. This SRAM failed the functional test after an elevated voltage stress and had an elevated

I_{DDQ} of 5 mA at 5 V. The I-V curve of this IC is shown in Fig. 33 and it is different from that shown in Fig. 28 in two respects. First, I_{DDQ} increases greatly when V_{DD} is above 1.6 V (versus 1 V for Fig. 28). Second, the curve does not have a single linear region above 1.6 V. The I-V curve strongly suggests that the failure of this SRAM involves more than just a simple ohmic short and most likely involves other processes such as those that cause light emission. In fact, three light-emitting areas were located in this IC (Fig. 34(a)). A heat-generating area was also located with FMI. The locations of the light-emitting and heat-generating areas, however, did not coincide. Upon further SEM examination (Fig. 35) and visual inspection, the root cause of the failure was determined to be an embedded particle that was identified using FMI (Fig. 34(b)). The particle produced an ohmic short between two adjacent metal lines. By design, one of these metal lines was electrically connected to the gates of several transistors in series. The ohmic short caused these transistors to go into saturation, resulting in light emission.

Fig. 34(b) - overlay of FMI and reflected light images showing the location of a "hot spot". The black rectangular outline in 34(a) indicates the area where the FMI image in 34(b) was taken.

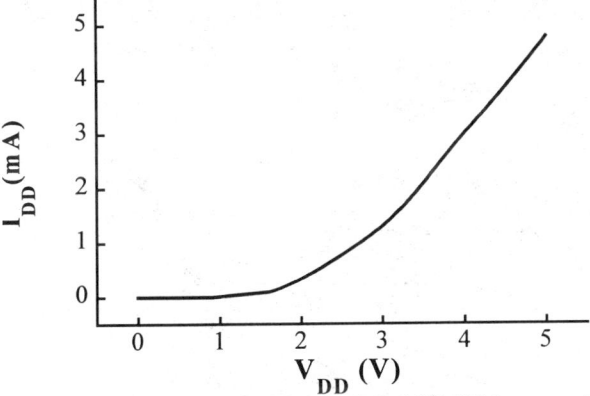

Fig. 33 - I_{DDQ} versus V_{DD} for the failed SRAM in example 3.

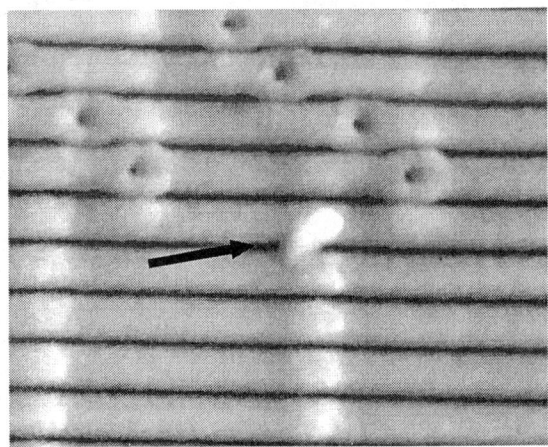

Fig. 35 - SEM image showing an embedded particle that produced an ohmic short between two adjacent metal lines in the failed SRAM in example 3.

Fig. 34(a), overlay of light-emission and reflected light images showing light-emitting regions of the failed SRAM in example 2.

Example 4. The fourth example is an SRAM similar to those described in examples 2 and 3. This SRAM failed the functional test after an elevated voltage stress and had an elevated I_{DDQ} of 18 mA at 5 V. The I-V curve of this IC (Fig. 36) has a linear region and passes through the origin. This IV characteristic indicates that there is a direct ohmic short between V_{DD} and V_{SS}. Using FMI, a "hot spot" (Fig. 37) was located in this IC. Interestingly, this "hot spot" also emitted light. Upon further visual inspection, the root cause was identified to be *pn* junction damage in one of the input protection diodes, producing a resistive short of 200 Ω (estimated from the I-V curve) between V_{DD} and V_{SS}.

191

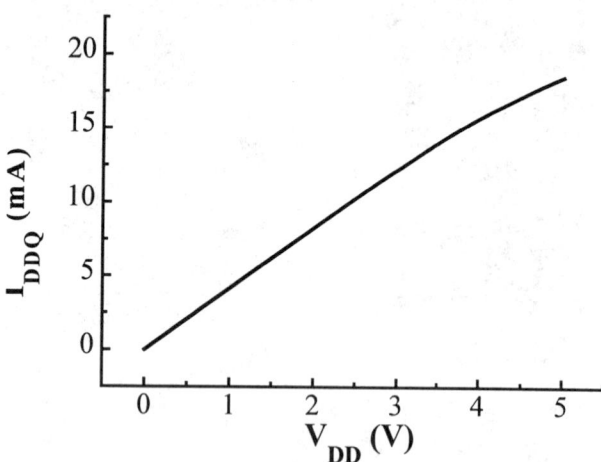

Fig. 36 - I_{DDQ} versus V_{DD} for the failed SRAM in example 4, showing a linear region that passes through the origin.

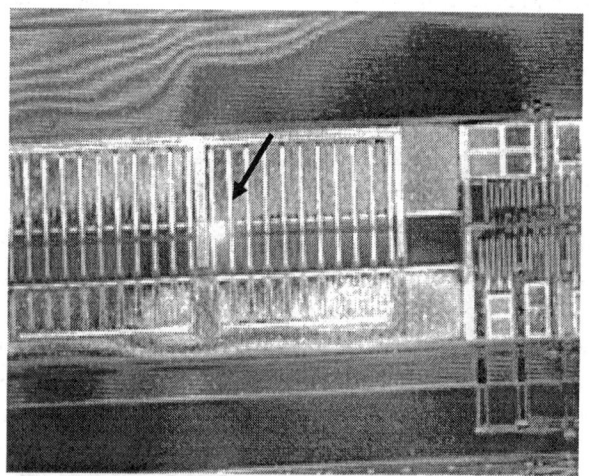

Fig. 37 - Overlay of FMI and reflected light images showing the location of a heat-generating defect of the failed SRAM in example 4.

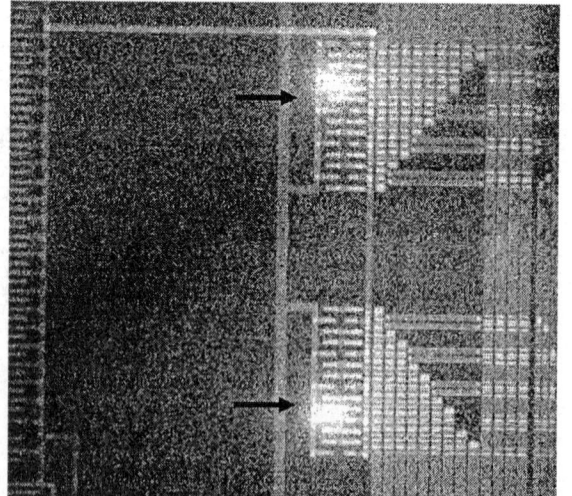

Fig. 38 - Overlay of FMI and reflected light images showing two heat-generating areas of the failed IC in example 5.

Example 5. The fifth example involves failure of an assembly test chip (ATC) designed for measurement of mechanical stress and thermal resistance. The ATC was fabricated using a 2 μm, twin-tub, CMOS technology with two levels of metal. After fabrication, a functional test was performed on this ATC and it was found that one of the ring oscillators failed functionally. Using FMI, two heat-generating areas were located in the output buffer region (Fig. 38). Upon visual inspection, mechanical damage was found near the heat-generating areas (Fig. 39). This mechanical damage produced an open circuit in the metal line that connected the gates of several transistors to a driving circuit. The floating gates, in turn, caused the transistors to go into saturation, resulting in heat generation.

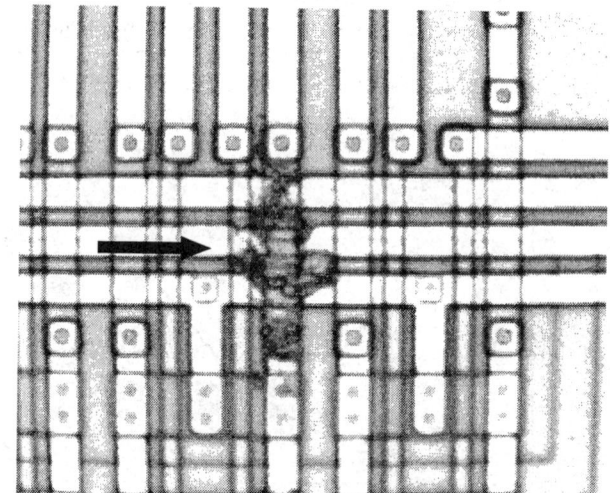

Fig. 39 - Optical image showing mechanical damage near the heat-generating areas in Fig. 38.

Example 6. The sixth example involves failure of a 256k-bit SRAM. The 64k x 8 SRAM was fabricated using a 0.5 μm N-well CMOS technology with three levels of metal. It has a nominal operating voltage of 3.3 V and a nominal standby I_{DDQ} of < 200 μA. After packaging, a functional test was performed on this SRAM and it failed functionally and had a high I_{DDQ} of > 100 mA at 3.3 V. The I-V curve of this IC showed a linear region, indicating an ohmic short. FMI was used to locate the heat generating area (Fig. 40). Temperature information derived from the FMI image showed that there was an average rise of ~ 2 °C above room temperature in the background surrounding the heat-generating defect. On the other hand, an average rise of ~ 3.5 °C above room temperature was observed within the heat-generating area. The temperature information indicates that the defect is not a localized heat source. Upon visual inspection, a patterning defect was observed at the heat-generating area (Fig. 41). FIB cross section (Fig. 42) shows that part of the oxide layer is missing in the heat-generating area, resulting in a direct electrical contact between several polysilicon lines and the silicon substrate. The silicon substrate has high thermal conductivity and this explains why there is a significant rise in the background temperature surrounding the defect.

Fig. 40 -Overlay of FMI and reflected light images showing a heat-generating area in the failed SRAM in example 6.

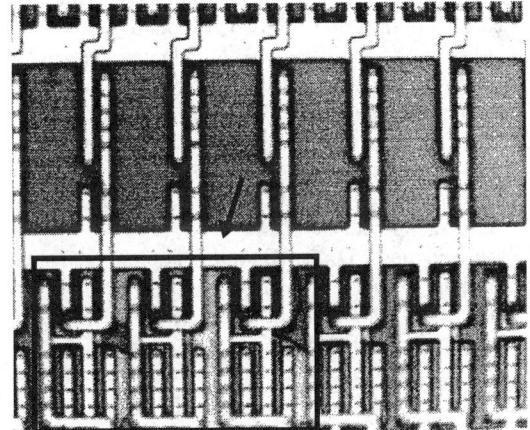

Fig. 41 - Optical image showing a patterning defect in the failed SRAM in example 6. The black rectangular outline indicates the boundary of the defect.

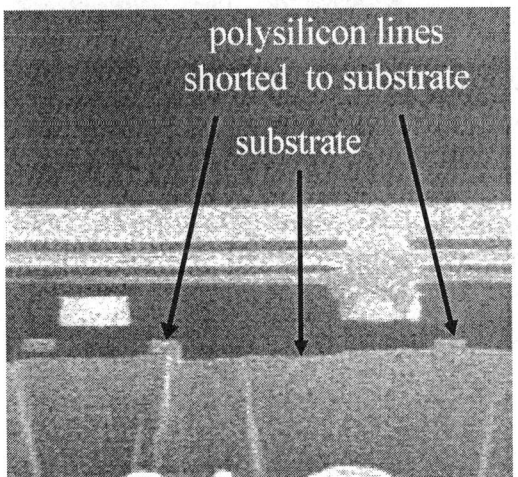

Fig. 42 - FIB cross section of the heat-generating area in the failed SRAM in example 6.

Example 7. The final example also involves the failure of a 256k-bit SRAM. The SRAM has the same layout and functional operating characteristics as described in example 6. This SRAM was part of a low-yield lot that failed functionally and had a high I_{DDQ} of > 100 mA at 3.3 V. The I-V curve of this device also showed a linear region that passed through the origin, indicating an ohmic short. No light emission was detected in this device. FMI was used to quickly locate the heat generating area (Fig. 43). The FMI image in Fig. 43 shows not only the location of the hot spot but also the current paths leading to it. Upon visual inspection, the root cause of this failure was determined to be a metal 1 patterning defect (Fig. 44) which produced an ohmic short between V_{DD} and V_{SS}.

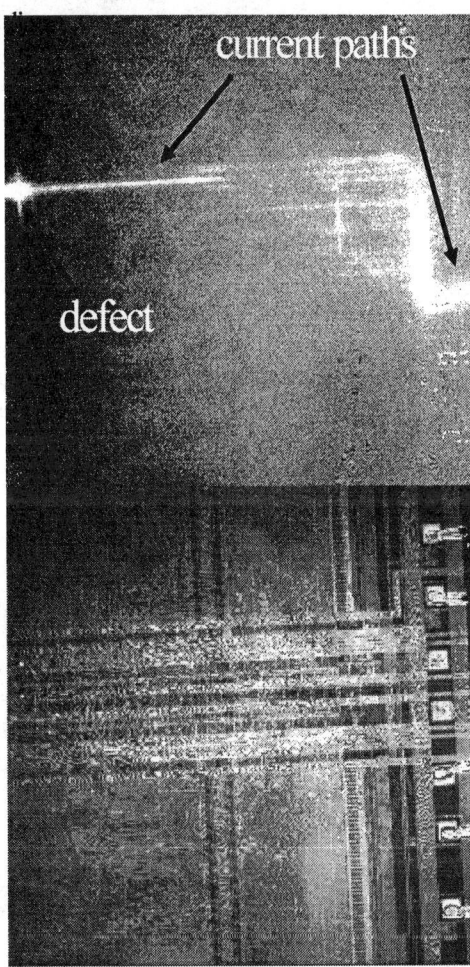

Fig. 43 - FMI image (top and corresponding bright field image (bottom) of a SRAM with a low resistance (20 Ω) short from V_{DD} to V_{SS}. The current in the left hand image has been increased to show path to V_{DD}.

CONCLUSION

In this tutorial, the subject of microthermal imaging was reviewed in detail. By introducing the physics that describes the spectral properties of heated bodies, the physical origins of the limitations of IR thermal techniques were studied. We found that the thermal resolution of IR systems can be quite good for IC applications, but the spatial resolution is quickly becoming an order of magnitude too large. Fluorescent microthermal imaging (FMI) was originally invented to overcome the spatial resolution limitations imposed by imaging IR radiation. In its present form, FMI relies on light emitted by a fluorescing compound at 612 nm to achieve a spatial resolution of approximately 0.3 µm.

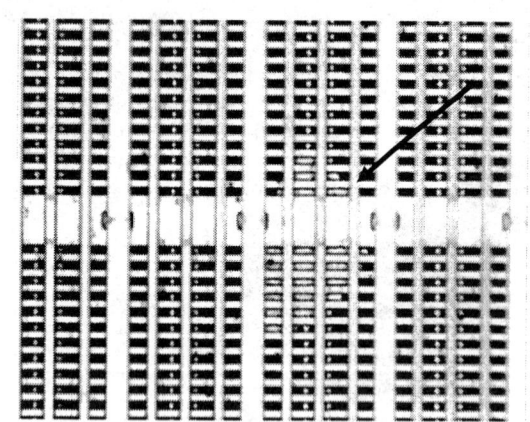

Fig. 44 - Optical image showing a metal 1 patterning defect in the failed SRAM in example 7.

The study of FMI was broken down into its basic elements so that each area could be detailed properly. The technology behind the temperature-dependent fluorescing film was presented, followed by the image processing techniques needed to turn changes in quantum efficiency into changes in temperature. Once the theory behind FMI was covered, system hardware considerations were addressed. The multitude of possibilities of excitation sources and imaging hardware were covered in general to give readers an understanding of the tradeoffs involved when deciding upon a particular platform. Finally, some examples of applications of FMI were presented.

In closing, FMI is a technique that has been around for almost a decade and has seen growing interest throughout the microelectronics industry. One of the goals of this tutorial is to enlighten those not yet familiar with this technology to encourage more people to try FMI for themselves and learn what it can do for them.

ACKNOWLEDGMENTS

The authors would like to thank C. L. Henderson, J. M. Soden, and W. M. Miller for reviewing this manuscript. This work was performed at Sandia National Laboratories and is supported by the U.S. Department of Energy under contract DE-AC04-94AL85000.

REFERENCES

1. J. Hiatt, "A Method of Detecting Hot Spots on Semiconductors Using Liquid Crystals" in Proc. IRPS, 1981, pp. 130-133.
2. G. D. Dixon, Cholesteric Liquid Crystals in Non-Destructive Testing, *Materials Evaluation*, June 1977, pp. 51-55.
3. A. Geol, A. Gray, "Liquid Crystal Technique as a Failure Analysis Tool", Proc. IRPS, 1980, p. 115.
4. P. Kolodner, J. A. Tyson, "Microscopic fluorescent imaging of surface temperature profiles with 0.01 °C resolution", Appl. Phys. Lett. 40, 782 (1982).
5. P. Kolodner, J. A. Tyson, "Remote thermal imaging with 0.7 mm spatial resolution using temperature dependent fluorescent thin films", Appl. Phys. Lett. 42, 117 (1983).
6. R. Eisberg, R. Resnick, *Quantum Physics*, John Wiley and Sons, 1974, ch. 1.
7. E. Yang, *Fundamentals of Semiconductor Devices*, McGraw-Hill, 1978, ch. 6.
8. Barnes Infrared Radiometric Microscope Model RM-2A Instruction Manual
9. Barnes Infrared Micro Imager Model RM-50 Instruction Manual
10. C. T. Elliott, D. Day, D. J. Wilson, "An Integrating Detector for Serial Scan Thermal Imaging", Infrared Physics, Vol. 22, pp. 31-42, 1982.
11. D. Pote, G. Thome, T. Guthrie, "An Overview of Infrared Thermal Imaging Techniques in the Reliability and Failure Analysis of Power Transistors", Proc. ISTFA, pp. 63-75, 1988.
12. G. J. Zissis, "Infrared Technology Fundamentals", Optical Engineering, Vol. 15, no. 6, pp. 484-497, 1976.
13. E. Hecht, A. Zajac, *Optics*, Addison Wesley, 1974, ch. 10.
14. P. Burgraaf, "IR Imaging: Microscopy and Thermography", Semiconductor International, pp. 58-65, July, 1986.
15. D. L. Barton, "Fluorescent microthermographic imaging," Proceedings of the 20th ISTFA, 1994, pp. 87-95.
16. D. L. Barton and P. Tangyunyong, "Fluorescent Microthermal Imaging - Theory and Methodology for Achieving High Thermal resolution Images", Microelectronic Engineering, volume 31, Numbers 1-4, February 1996, pp. 271 - 280. (Proceedings of the Fifth European Conference on Electron and Optical Beam Testing of Electronic Devices, August 27 - 30, 1995, Wuppertal, Germany)
17. P. Tangyunyong, D. L. Barton, "Photon Statistics, Film Preparation, and Characterization in Fluorescent Microthermographic Imaging", Proc. ISTFA, 1995, pp.79 - 86.
18. H. Winston, O. J. Marsh, C. K. Suzuki, C. L. Telk, "Fluorescence of Europium Thenoylfrifluoroacetonate. I. Evaluation of Laser Threshold Parameters", J. Chem. Phys., vol. 39, no. 2, pp. 267-270, July, 1963.
19. M. Bhaumik, "Quenching and Temperature Dependence of Fluorescence in Rare-Earth Chelates", J. Chem. Phys., Vol. 40, (3711), 1964.
20. G. Crosby, R. Whan, R. Alire, "Intramolecular Energy Transfer in Rare Earth Chelates. Role of the Triplet State", J. Chem. Phys., Vol. 34, (743), 1961.
21. E. Bowen, J. Sahu, "The Effect of Temperature on fluorescence of Solutions", J. Phys. Chem., Vol. 63 (4), 1959.
22. P. Kolodner, A. Katzir, N. Hartsough, "Noncontact surface temperature measurement during reactive-ion etching using fluorescent polymer films", *Appl. Phys. Lett.* 42 (8), 15 April 1983.
23. J. M. Soden and C. F. Hawkins, "I_{DDQ} Testing and Defect Classes - A Tutorial," Custom Integrated Circuits Conf., p. 633 (1995).

Fault Localization

Photonic Techniques

Photoemission Microscopy:
Parts I and II

Christian Boit
Infineon Semiconductors

Gary Shade
Ford Microelectronics Inc.

PREFACE:

The photoemission microscopy tutorial is provided in two parts. Part I focuses on the analyst who is reading this material for the first time., and Part II provides added detail for the experienced user or someone who needs specific technical information for purchasing equipment. Both parts are complete in themselves and redundancy has been reduced from the earlier versions of this tutorial.

PART I (by Gary Shade):

This section explains why photoemission occurs, how to classify emissions and how to relate these to potential defects. In addition, useful techniques are provided for first time users and those desiring to improve their success rate while avoiding mistakes.

A method of classifying photoemission sources is provided based on the detection capability and concern for each photoemission. Guidelines are provided on how to prepare for the analysis, what steps to take and how to resolve problems with the technique. Several examples are provided with photographs of the emission and the responsible defect.

PART II (by Christian Boit)

This section discusses the theory for each type of photoemission (both real concerns and artifacts) using device structure and bandgap theory. This information is then used to provide a method of classifying defects based on emission spectra rather than detection capability as in Part I. (Indeed this is the natural 'learning curve' users actually experience. Users start with classifying by detection and concern. Then with experience, this is replaced with classifying by the cause of the defect. To support this approach the components of the emission microscope are described with their impact on measuring emission spectra allowing the reader to make better equipment choices for implementing the technique.

This section also adds information on advanced opportunities such as imaging devices from the backside and how to 'dock' the microscope with a full functional tester (ATE). In addition, many more details are provided on how the emission microscope is constructed and how each part can be optimized for your individual needs. This is especially helpful for someone planning to purchase this type of equipment, or upgrade the technique for the next generation of devices.

Photoemission Microscopy— Basic Theory/Application

Gary Shade
Ford Microelectronics Inc.

Statement Of Need

Current designs in very large scale integrated (VLSI) circuits are too complex to determine logically what is failing. Even with design for testability (DFT) in mind during the initial design stages, computer algorithms only guess at the location of the failure. In most cases the logical approach only determines the circuit logic affected. The data register, for example, may contain 64 bits at 7 transistors per bit for a total of 448 transistors. Without photoemission microscopy a failure analyst would need to test all 448 transistors by needle probe. With photoemission microscopy the single latch which is defective can be located. The failure is isolated to 7 transistors at most.

The nature of the semiconductor industry is to have better quality with faster speed at a lower cost. This nature makes rapid design debug, quick failure analysis and immediate corrective action imperative. Point by point inspection of VLSI devices is no longer adequate. Even discrete components are affected by the push to bigger, faster and better. Discrete components in many ways can be more difficult to analyze than VLSI because there are fewer visual clues to help locate the problem.

The implication of what the semiconductor industry needs is simple. Failures must be located in a timely manner. All failure analysis must be accurate and detailed. Testing must be direct. As much information as possible must be gleaned from a single quick test.

Photoemission microscopy is a powerful and useful tool when applied properly to meet this need. Its uses impact semiconductor devices during the design [9, 12, 21] fabrication [13, 34] and test cycles [13, 21, 26] as well as in analyzing failures that occur in use [2, 3, 4, 5, etc.]. Despite continued increases in the complexity of ICs, (including 5 layer metal and sub-0.5 micron dimensions), this technique continues to have a major impact on leading edge integrated and discrete device analyses.

Complimentary Techniques

Photoemission microscopy provides direct testing of semiconductor devices. The following techniques compliment the photoemission technique by providing more detailed information about the electrical aspects of the failure or the thermal properties of the failure.

Stroboscopic Voltage Contrast: Voltage contrast uses a scanning electron microscope (SEM) to monitor the electrical activity of a semiconductor device. Its main advantage is that the electrical signature of the failure can be determined. Like photoemission microscopy, the failure is isolated to a single circuit. The disadvantages of voltage contrast are the inability to precisely locate the fail site and its inability to force internal device nodes. It is difficult to locate the failure to a single transistor. The voltage system requires a skilled operator.

Liquid Crystal: Liquid crystal is the least expensive technique to implement next to visual inspection. Two polarizers and a microscope are all the equipment that is needed. The technique is quick to setup. The device under test is placed under a microscope, liquid crystal is applied and the device is activated in such a mode the associated transistors are dissipating power. The disadvantage is the limited spatial resolution and the relatively high power required at the fail site to visualize the failure.

External Electrical Testing: Device failures are caught because the device failed to perform correctly at the external pins. To do external testing there is no deprocessing of the device. An electrical signature can be developed for the failure which may limit the possible fail sites on the device. External testing in many cases is incapable of isolating the failure to a single transistor.

Internal Mechanical Probe: Internal needle probe, like voltage contrast, can determine the electrical signature of the defect found. Unlike voltage contrast, internal needle probing can force nodes on the device. This gives the test engineer greater ability to isolate the problem to a single transistor. A problem arises with the size of structures available for needle probing. Structures as small as 0.5 micron can be probed with needles but the possibility of damage is great.

Optical Beam Induced Current - OBIC: Optical beam induced current works by inducing hole electron pairs in the silicon. If there is a defect or junction, the OBIC current varies from the current of the native silicon. The current variation is converted into contrast which can be seen through the use of a monitor. OBIC is quick and relates directly to junction damage. The primary disadvantage is the technique can not be performed when metal covers the junction.

Electron Beam Induced Current - EBIC: Electron beam induced current is somewhat like OBIC. Contrast is generated due to the interaction of the primary electron beam of a scanning electron microscope (SEM) and the defects in the silicon or semiconductor junction. The SEM monitor displays the location of the defect. SEMs have good position accuracy so the defect can be specified to within 0.5 micron. The EBIC technique only verifies a defect is present. No determination of what the defect is can be made from the EBIC signal. The major problem with EBIC is the amount of work required to ready the device for test. In most cases to perform EBIC testing, the device must be completely deprocessed.

Resistive Contrast Imaging - RCI: Resistive contrast imaging is similar to EBIC. The primary

electron beam is used to excite the defect. In this case, however, the defect can be in an oxide above the substrate or be a metal to metal short. The requirement is for the SEM primary beam current to reach the substrate by some unique path through the defect. Contrast is generated by the difference in current through the defect and the current from the surrounding area. No deprocessing is required to image metal shorts. Some deprocessing may be required to image other types of defects such as polysilicon filaments shorted to substrate. Typically this technique will not find opens.

The Photoemission Phenomenon

Visible and near-infrared light emission from semiconductors and related materials (near room temperature) has been noted and studied since the early 1950s [74, 75, 76]. The materials and the energy levels involved determine the wavelength and efficiency (ultimately the quantity) of the light emitted. This effect is dominated by theories involving quantum particle interactions and statistical distributions of particle energies. For this reason, the discussion of theory will be left to "Photoemission Microscopy: Part II" along with the references provided at the end of this tutorial.

The photoemission phenomenon relates to failure analysis of electronic devices formed from:
- optoelectronic materials (GaAs, InP, etc.)
- discrete silicon devices
- silicon integrated circuits
- passive devices (in some cases)
- insulators used in the above

The photon emission from any of these is typically the direct result of recombining electrons and holes. Thus the technique is a detector of defects that involve a leakage current.

Direct bandgap semiconductors, such as gallium arsenide (GaAs) and indium phosphide (InP) are efficient emitters as indicated by the bright light from Light emitting diodes (LEDs) and solid state lasers. In contrast, silicon and germanium have an indirect bandgap and are very inefficient. Insulators have a large bandgap and are inefficient at conducting and photoemitting. With the exception of the intentional light emitters, light emission is useful in solving device problems. For the light emitters, the device can be studied at bias levels below normal emission or in bias conditions that do not normally cause photoemission (e.g. reverse bias). Because the emissions are weak, a photoemission microscope is necessary that can collect and amplify the image for viewing.

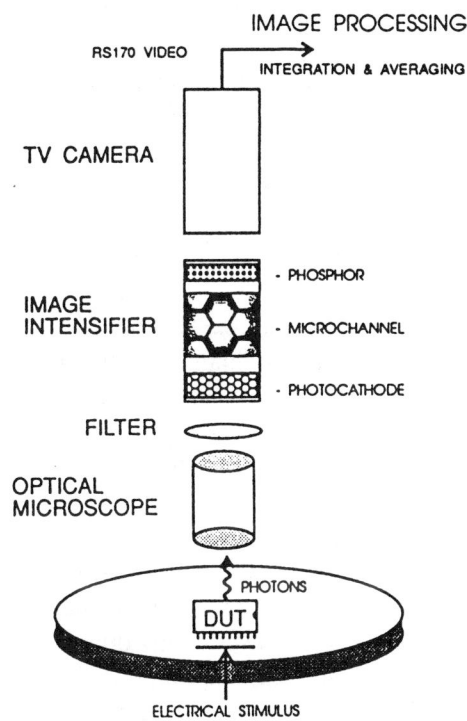

Fig. 1 Typical photoemission system with Image Intensifier.

Photoemission Equipment

The Basic system: The basic photoemission system may vary somewhat, but the basic fundamentals shown in Figure 1 will be the same. The system needs to be able to detect the weak emissions while also collecting the normal image of the device. When these two images are overlaid, the emission can be localized. The basic system has a stage to hold the device under test (DUT). The device must be stimulated electrically and thermally in such a manner that the defect site is active. Generally, the best results are obtained with a system which puts all the pins in a known state. Floating pins or uncontrolled pins can lead to erratic photoemission results.

An optical microscope is used to image the DUT onto a photocathode of a microchannel plate image intensifier. A set of filters may be placed in the beam path to the photocathode to obtain the emission spectra. The photocathode converts the photons to electrons which are amplified by the image intensifier. A phosphor plate on the end of the image intensifier converts the electrons back to photons generating an image which can be seen by a video camera (e.g. CCD). Because the photon count is low, the image from the video camera is averaged over many frames and manipulated by an image capture board to improve signal to noise ratio.

Classifying Emission by Detection and Concern

All too often the photoemission microscope is wrongly assumed to be an absolute defect detector. This misconception can be prevented by either of two disciplines. The first option is to apply a system of classifying defects; the second option is obtaining a detailed understanding and application of photoemission theory. The first option is described here as it is simpler and more generally applicable to the first time or less frequent user. (For the second option, see Part II of this subject by Christian Boit.)

A system of classifying defects is instrumental in identifying the actual causes of leakage current and how the emission microscope can be applied to detect them. The following list shows how the various causes of leakage current can be fit into four classes. In each case, the list makes it clear whether the defect is possible to detect and whether or not it should be a concern to the analyst.

Detected emissions

> **True concerns** (Emissions representing a weakness or defect condition)

- Junction leakage
- Contact spiking
- Hot electrons (saturated transistors)
- Junction avalanche
- Latch-up
- Oxide current leakage
- Polysilicon filaments
- Substrate damage

Detected emissions (continued)

    Artifacts (Emissions that are artifacts of a design or test condition)

- Floating gates

- Saturated bipolar transistors

- Saturated analog MOSFETs

- Forward biased diodes

Non-detected Emissions

    Non-Emitting (Leakage sites that do not have an emission in the visible spectrum.)

- Ohmic shorts

- Shorted metal interconnects

- Surface inversion

- Silicon conduction paths (e.g. diffused resistor)

- Sub-threshold conduction

    Emission masked (Emissions can not reach the detector.)

- Buried junctions

- Leakage sites under metal

Technique

Analytical procedure: The objectives of the photoemission technique are to locate potential fail sites, separate false from true failures, identify potential device weak points and to study device physics. To meet these objectives, the following sequence is highly recommended.

1. Perform functional test on failure

2. Perform IddQ testing

3. Reproduce failure on bench

4. Decapsulate or de-lid component

5. Visually inspect the die

6. Reconfirm bench test (preferably in the photoemission microscope)

7. Inspect component with photoemission microscope (acquire emission image)

8. Compare results to known good component

9. Isolate emission location at highest magnification.

10. Verify leakage site by direct technique

As with any good testing technique there is some preparation to do before the device can be tested. In the case of photoemission microscopy, the most important item is the DUT hardware. Any test sockets, cabling or test programs should be made ready prior to photoemission testing. In the high pin count VLSI devices, the test sockets are difficult to wire or must be custom built. Waiting until the last minute will not produce results in a timely manner. For unexpected probe or socketing requirements a versatile setup should be prepared that can adapt to as many needs as possible. A source for cooling and heating the device may be necessary if the failure occurs only at a given temperature. When testers are not adequate or unavailable, the circuit board with the suspect IC or device can be located in the acquisition chamber for the analysis. This has been successfully done for complete radios, modules, and computer boards [25].

The next most important item in photoemission testing is to compare the suspect unit to a known good unit. This verifies that the test conditions are correct and that any new emissions are abnormal. If a reference unit is not available, a similar area of the same device may be useful. If neither is available, the analysis should be diligent in correlating the observed emission to the final defect identified. The last item is to check the test program to see if the program puts all the pins in a known state. Remember that the photoemission microscope is a tool which locates potentially defects. Be prepared to follow-up with additional techniques to isolate root cause and relate the emission to the defect.

For the first time user of the emission microscope, a simple analysis is to take an IC and perform an input stress test on a device pin. This can be accomplished by connecting a power supply to any pin of an IC with the Vss or ground pin of the device connected to the power supply ground. This may be destructive, so a part from the scrap bin or a low cost part from a nearby IC supply source will work fine. Using a variable supply that can go from 0 to 10 volts is perfect; however a leakage tester such as a curve tracer or parametric analyzer is even better. The device should begin emitting whenever the supply exceeds the design range of the device. The simple case is to bias the pin negative relative to the Vss by a few volts. Once this has been successful, try biasing in 0.5 volt steps above the design value (i.e. 5-6 volts typical) with positive bias to the pin.

Optimizing for success: Lack of photoemission is not necessarily an indication that the device does not have a failure. The test procedure and the device setup must be checked before concluding that no photoemission should be seen. The following lists items which should be checked when no photoemission is seen.

What To Check When No Photoemission Is Seen:

1. Readjust the acquisition parameters (i.e. acquisition time) until the noise level is reached.

2. Increase the device voltage to maximum rating.

3. Increase the stress level on the failing pin.

4. Vary the clock frequency.

5. Vary the component temperature (i.e. for latch-up)

6. Check that component is still in the failing state

7. Try additional failures (different pin or device)

8. Review the possible leakage modes and consider the non-emitting types; are they possible?

9. Alter the voltage ramp rate or the ramp rate on various pins to determine if the failure has a relationship to how the pin is driven.

10. Try an alternate technique

When too many photoemission sites are seen, the device may have gone into latch-up or could be in an improper logic state.

What to Check When There Are Too Many Photoemission Sites:

1. Reduce the acquisition time or sensitivity

2. Reduce the supply voltage, and/or clock frequency

3. Reduce the stress to the failing pin

4. Try to use fewer functional vectors that will stress smaller portions of the component.

Success Rate: The photoemission technique has been shown to find 80% of the failures on CMOS when combined with liquid crystal techniques [41, 42]. The caveat is the device must be tested correctly and the location of the defect must be in a region of the device which can be activated by the test. In ESD structures, for example, the defect may be in a transistor which has a gate connected to ground. Without a fast high voltage spike to activate the defect, it may never be caught. The photoemitting site is not necessarily at the location of the defect. In some cases, the defect causing the failure will cause the entire die to photoemit. An example is a latch-up prone device which photoemits over most of the die even though only one area of the die is the cause of the problem. The most important advantage of photoemission testing is that the defect does not need to propagate to the external pins. In addition, even for complex circuits, there is a 50% chance of finding the defect on a CMOS IC on power up with no additional stimulus.

The photoemission technique can not identify root cause for defects by itself. The complimentary techniques can help isolate the failure to the level in the device. OBIC, EBIC and RCI help to compliment the photoemission system. Of course voltage contrast, internal probing and liquid crystal help as well.

Troubleshooting: With photoemission microscopy as with any test, there are compromises which must be made. For best results using photoemission microscopy, a full ATE tester should be used. Attaching power and ground alone can be used on simple devices or devices where the operating conditions are not very complex. On complex VLSI devices, a tester is a must to put all the pins at a know state. However, simpler tests may be possible if IddQ vectors are available and can be applied to the pins with a static test connection.

In some cases, such as inactive ESD structures, a defect may not be able to be activated except during an ESD event. In these cases, a simulated ESD stress should be applied or ESD testing done to failure. Then the defect can be analyzed.

When a defect is on a power bus pulling down the voltage, photoemission may occur at several sites in the die. In this case the apparent fail site (photoemission site) is not at the defect site. The inability of the photoemission microscope to locate the actual defect site causes a problem with analysis. Alternate techniques such as voltage contrast, internal probing and liquid crystal may be needed to complete the analysis by correlating the emission to a defect and to the electrical fault.

Dynamic random access memories (DRAMS) can have a charge pump. Unless a reference device that is working correctly is supplied, the photoemission from the charge pump could be misconstrued as a fail site. Large clock circuits also exhibit the same confusion. They can photoemit under heavy load. Only the use of a reference device will determine if the photoemission is causing the failure.

Timing related failures and defects which hide under metal and ohmic leakage paths can generally not be detected with the photoemission technique. (See Reference [31, 45] for some documented exceptions of ohmic leakage.) With the application of liquid crystal techniques, some of the hidden defects and ohmic leakage paths can be found, but timing defects can not be found without the help of voltage contrast or internal probing, unless the timing failure causes a short which heats the device or causes photoemission. For photoemitting timing failures, they may occur at small dwell times with respect to the total cycle time of the clocks. If this is the case, a gated photoemission camera would be useful. To date little use has been made of gated photoemission techniques [53].

Applications of the Photoemission Technique

Typical Applications Include:

- Substrate defects (crystal defects, stacking faults, gettering damage, mechanical damage)

- Specialized test structures
- Finished die (wafer and package level)
- Screen for "Latent" defects (finished die)
- Location of known defect
- In-process control
- Yield analysis
- Reliability studies
- Design faults
- Processing faults

To illustrate how a defect might affect a CMOS inverter circuit, different defect configurations are shown in the following figures. It is left to the reader to consider the effects of different bias states and resistance values of the defect. In Figure 2, the defect is connected to the output of the inverter. When the state of the inverter turns line B to ground, there is the possibility of photoemission from the n-channel transistor. Depending upon the nature of the defect, there is also a possibility the defect might photoemit. As long as the defect is a true resistor as shown in Figures 2-4, no photoemission will occur at the defect site. Only the photoemission at the n-channel transistor will appear to be the problem. If the defect is connected to ground, then the photoemission will occur when B is at V_{dd}. Then the p-channel transistor will photoemit.

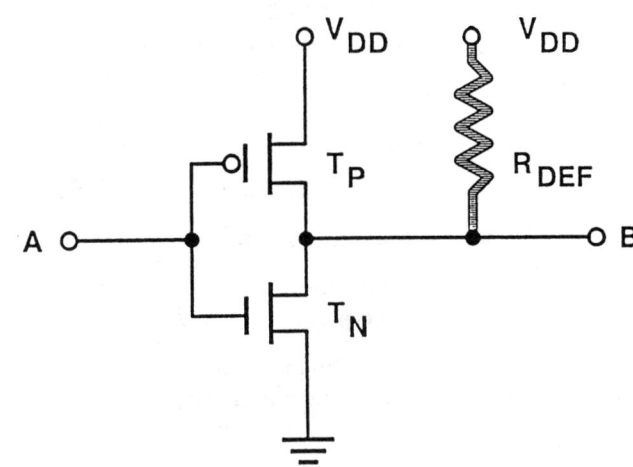

Fig. 2 Inverter with a bridging defect on the output.

Figure 3 shows a 2-NAND with a bridging defect on the output. When the state of the system tries to make line C go to ground, then transistor T_{NA} will photoemit. Transistor T_{NB} might photoemit, but the photoemission would be less. The greater photoemission in T_{NA} is due to the current being diverted into the substrate. Transistor T_{NB} will not see the same Id as T_{NA}.

Figure 4 is a more typical situation. A defect R_{DEF} controls the voltage on line B. Photoemission can occur at different points depending upon the value of R_{DEF}. For example if R_{DEF} can hold B at 3.5 V then T_{N1} will

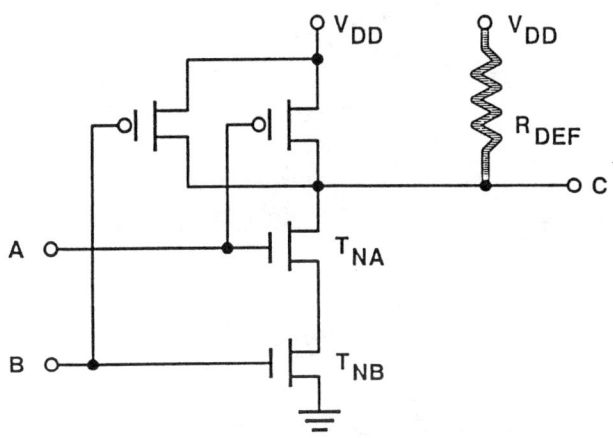

Fig. 3 Circuit diagram of a 2-NAND with a bridging defect on the output.

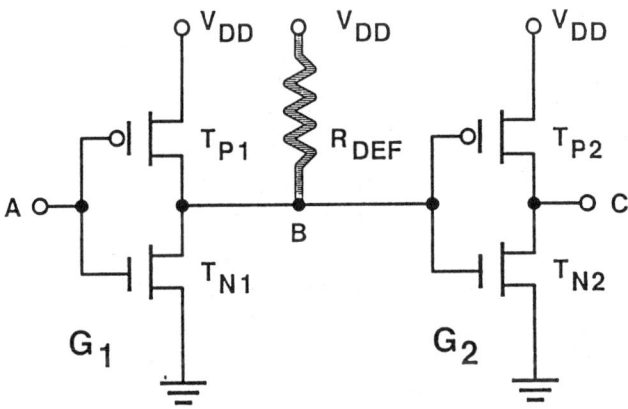

Fig. 4 Two inverters with a bridging defect (R_{DEF}) on the output of the first.

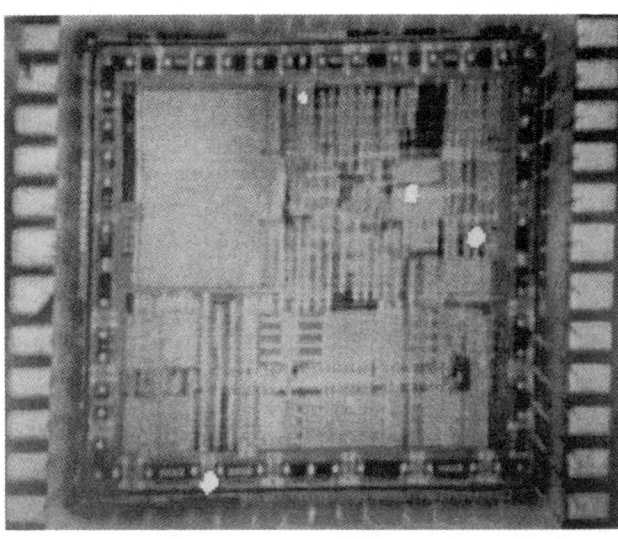

Fig. 5. Four light emitting gate shorts visible as 14K vectors are cycled repeatedly and light emission is integrated.

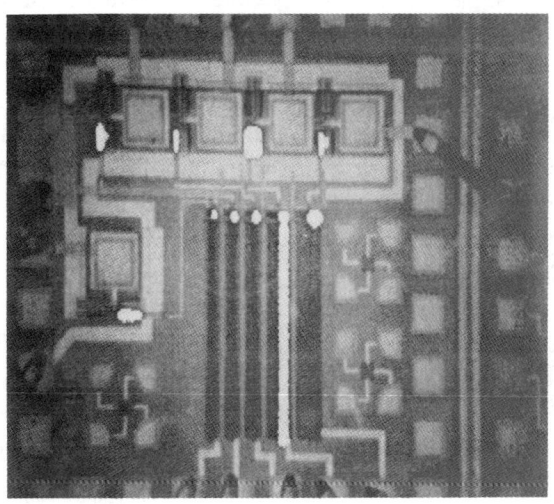

Fig. 6 Legal photoemission from a ring oscillator operating at 5V and 30 MHz.

photoemit due to excessive current to the defect. Transistors T_{N2} and T_{P2} can photoemit depending upon their parameters. As the voltage on B is raised, n-channel transistor T_{n2} would photoemit more than T_{P2}. If the voltage on B is lowered, the photoemission would be more prominent on T_{P2}. As described earlier, the photo-emission intensity of a transistor is dependent upon its parameters. Which transistor will be seen as

the photoemitting transistor will depend upon the particular transistor parameters in the circuit being tested.

In Figure 5 photoemission is seen on a microprocessor. The processor is being cycled through 14K vectors and the photoemission is being integrated over all test vectors.

Figure 6 shows a common acceptable emission in a ring oscillator operating at 5 V and 30 MHz.

205

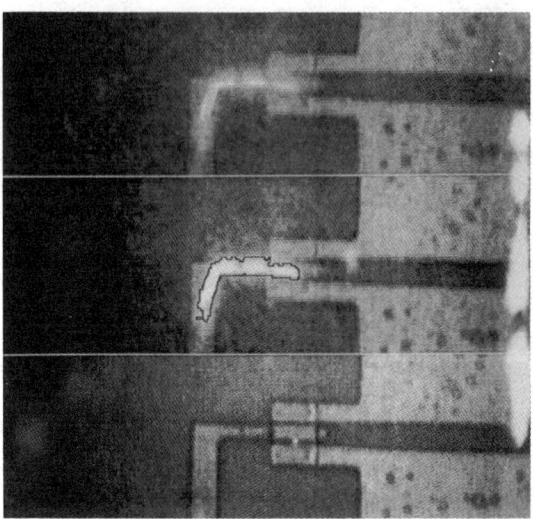

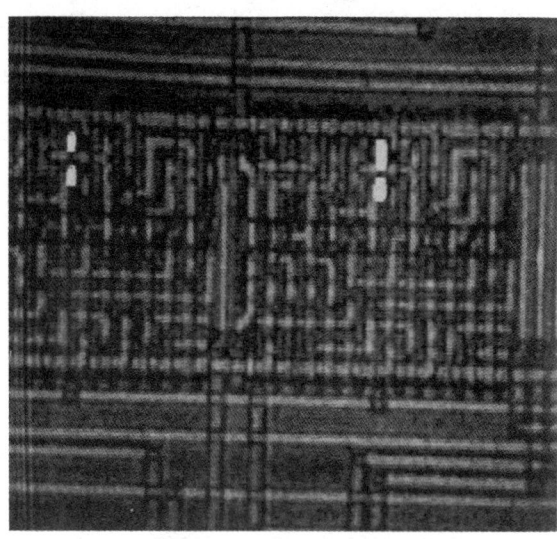

Fig. 7 Gate oxide rupture. The flash of visible light at the instant of gate rupture. The spot in the lower image shows light emission from the gate short.

Gate oxide rupture due to ESD can be seen in Figure 7. The top image is the start of the ESD stress pulse. The middle image is during maximum current conduction. The photoemission is due to current crowding in the polysilicon. The bottom image has a white dot at the eventual fail site.

The next image, Figure 8, shows a long chain of flip-flops which photoemit. The defect is in the flip-flop to the right of the right-hand-most photoemission. Figure 8 shows how one defect can cause many photoemission sites. Figure 9 shows a close up of the

Fig. 9 Light emission from the row of flip-flops at higher magnification. The n-channel transistors are the light emission sources.

sites. The transistors which are photoemitting are the n-channel devices.

ESD is a constant problem for VLSI devices. To determine the ESD sensitivity of a device, one test technique is the Human Body Model (HBM). The results of HBM testing can be variable and somewhat erratic. To locate the fail site, photoemission testing is very helpful. Figure 10 shows the result of some typical photoemission testing to determine the ESD fail site. The photoemission results are compared with the results of curve tracing. Notice that the curve tracer did not find any of the failing pins.

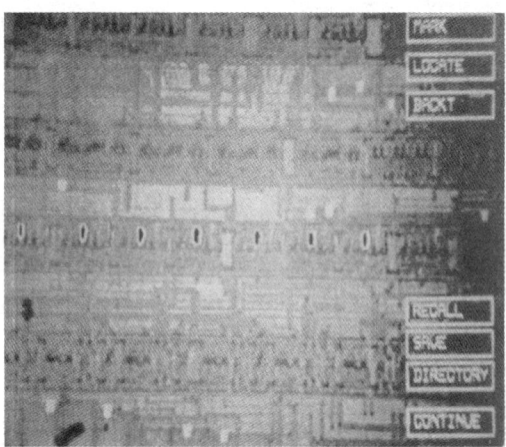

Fig. 8 Light emission from a row of flip-flops at power-up. The line of light emitting areas suddenly end on the right before the flip-flop chain does.

ESD STRESS TEST

CASE #	STRESS VOLTAGE	PHOTOEMISSION YES/NO	BY	PROBLEM CURVE TRACER	PROBLEM BY SEM
1	500V	YES	GG* GH*	NONE	GATE EDGE BLOWN CONTACTS
2	500V	YES	GH	NONE	BLOWN CONTACTS GATE EDGE
3	500V	YES	GH	NONE	BLOWN CONTACT
4	1000V	NO		NONE	BLOWN CONTACTS
5	1000V	YES	GH	NONE	BLOWN CONTACT
6	1000V	YES	GH	NONE	GATE EDGE

*GG — Gate grounded
*GH — Gate high

Fig. 10 Results of photoemission testing after ESD stress using the HBM ESD stress model.

Parametric testing, which is normally used as the indication of good ESD performance, also did not indicate a problem. Photoemission testing did locate the fail site most of the time. In one case, the fail site could not be found parametrically or by photoemission.

Contamination in various oxides can cause photoemission due to field enhancement. In Figure 11 the photoemission was caused by the contamination in the gate oxide. The contamination is the black spot at the arrow.

When devices are not powered up correctly, the device may go into an unknown state. Figure 12 shows photoemission from a device which latched due to floating (uncontrolled bias) pins. Complimentary techniques do not always point to the same location as the fail site. Figure 13 shows locations found by liquid crystal and photoemission. The locations do not match. In this case, the defect was at location B found by liquid crystal.

Fig. 11 Gate oxide contamination found using the photoemission techniques.

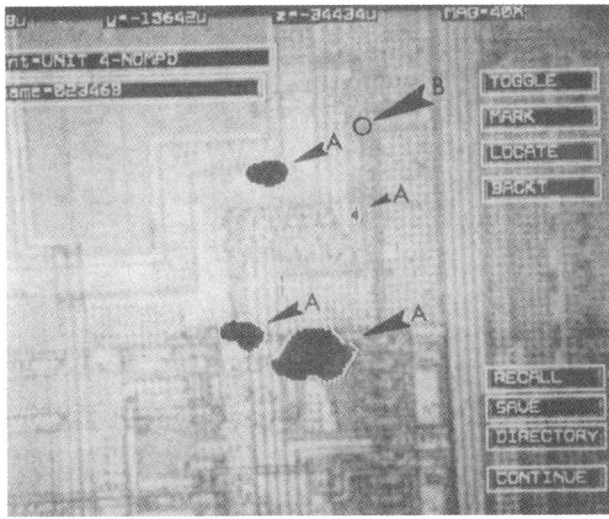

Fig. 13 Photoemission and liquid crystal testing did not locate the same defect site. In this case the defect is at the liquid crystal site.

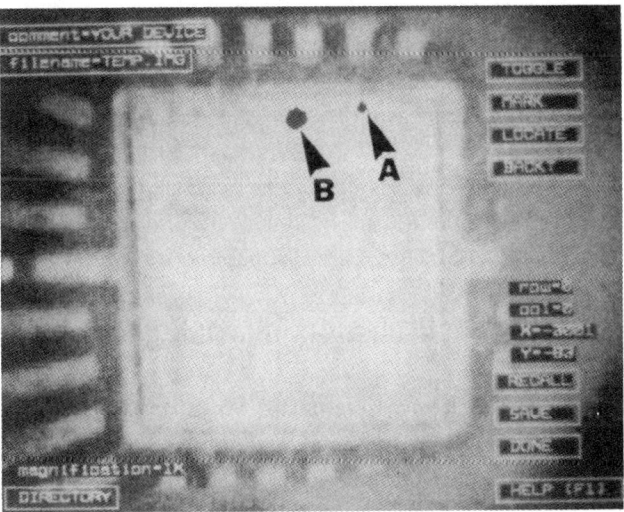

Fig. 12 Photoemission found at the clock circuit and an internal node. The photoemission at B is not valid. The photoemission was due to latch-up.

Fig. 14 Gate oxide pinhole found by photoemission microscopy.

EOS and ESD damage can both cause photoemission. Oxide rupture can cause photoemission as shown in Figure 14. In the image the photoemission was caused by oxide breakdown in a gate oxide pin hole. Extreme EOS can cause massive damage. In the examples of Figures 15 and 16, the EOS event caused silicon migration as can be seen by the silicon filaments from the gate polysilicon to the substrate

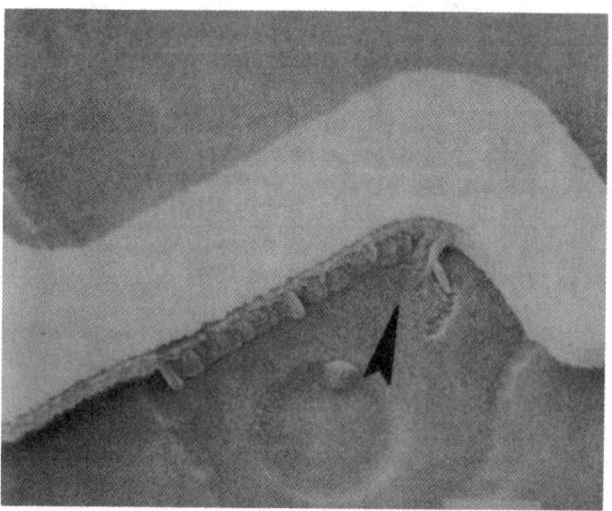

Fig. 15 Extreme EOS failure site found by photoemission.

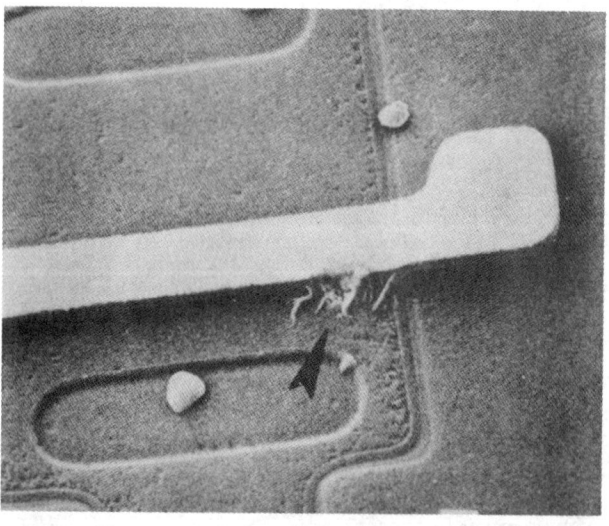

Fig. 16 Extreme EOS failure site found by photoemission.

Baselining / Correlation of Equipment: In this case, baselining refers to the standardized testing of the photoemission system. This baseline is needed to understand when the system is not functioning correctly.

It is also needed to insure the system is in a known state so that the measurements from the system are repeatable. Baselining is essential to permit the consistent testing of devices over a long period of time for the purposes of process control.

Why should a photoemission system be calibrated? Who cares if the absolute value of the data is different from day to day? The answer to these questions is simple. If failure analysis labs are to be able to compare data, the data must be consistent and accurate. A failure analyst must be assured when no photoemission is seen, that the sensitivity of the equipment has not changed. The results are believable. Suppliers of photoemission equipment also should be able to tell when the equipment is not performing correctly and what is wrong. Ultimately, the performance of the system must be trustworthy. So trustworthy, in fact, that less skilled operators can use the equipment.

Additional benefits can be gained as well. Baselining allows for a better understanding of the system, better characterization, and facilitates routine maintenance. Baselining increases measurement confidence and permits verifiable training. When upgrades are made to the system, baselining permits the user to verify the upgrade benefits.

The desired baselining procedure would provide on a routine basis, system characterization. The measurements would be traceable. The baselining procedure would provide system diagnostics. This list of requirements can be accomplished by checking the main camera sensitivity, by checking the linearity of the main camera and determining the minimum detection limit of the main camera. A test fixture with varying spectral frequencies can determine the spectral response of the main camera and the spatial response uniformity of the main camera. The best possible baselining procedure would require low skill level individuals, take a short time and be done at reasonable cost [14, 17].

Obtaining an Emission Microscope:

Purchasing a new tool usually requires locating a source for the technology and a source for the funds to buy it. There are only a few sources for the emission microscope at present and each has its own strengths and weaknesses. The following lists some of the major suppliers:

* Hypervision - 46560 Fremont Blvd. #414, Fremont, CA; Tel. 510-651-7768

* Quantum Focus Instruments (formerly EDO/Barnes) - 88 Long Hill Cross Roads, Shelton, Conn. Tel. 203-926-4203

* Hamamatsu Photonics - 790 Lucerne Dr. Sunnyvale, CA   Tel. 408-727-8447

* Alpha-Innotech - 14743 Catalina St. San Leandro, CA Tel. 800-795-5556

One popular technique for justifying this tool (obtaining funds) is to use time on a vendor's demo unit and analyze your own parts. This helps to establish how the tool will work on your specific technology. It is limited however as there may be a need for specific hardware from your lab that is hard to transport to the vendor.

Many of the current tools are modular so care should be taken to configure a system that will be flexible, but meet the specific needs of your analyses. Two major decisions are the type of detector (optimized for front-side or backside) and the sample mounting technique (wafer prober, sockets for packages, or a docking system that mates with your ATE equipment). In general, the vendors are skilled in helping to make these decisions. See also Part II of this subject for insight into how to optimize the equipment configuration.

Conclusions

Photoemission microscopy is a powerful and efficient tool when applied properly. Its uses impact semiconductor devices during the design, fabrication and test cycles as well as in analyzing failures that occur in use. Despite continued increases in the complexity of ICs, (including 5 layer metal and sub-0.5 micron dimensions), this technique continues to have a major impact on both leading edge and more classic discrete device analyses. Results can be obtained quickly in all but a few cases. When the device is powered up in a known state, the results are consistent and reliable.

Care must be taken in interpreting the cause of the photoemission. A photoemitting site can be generated by a defect which is in a different location on the die.

The photoemission is always in the circuit which failed, not necessarily at the site of the photoemission. With proper care and complimentary techniques, the weak, defective, or failed site can be located and proper corrective actions implemented.

Acknowledgments
The following provided materials for this tutorial:

Kendall Scott Wills
Texas Instruments, Inc.
Stafford, Texas
(ASM Tutorial Instructor)

Edward Isaac Cole Jr.
Sandia National Laboratories
Failure Analysis, Division 2142
Albuquerque, New Mexico 87185
(IRPS Tutorial Instructor)

REFERENCES (in chronological order)
1996 - 1952

1. D.L. Barton, P. Tangyunyong, J.M. Soden, and A. Y. Liang, "Infrared Light Emission from Semiconductor Devices," Proc. of the Int. Symp. for Test. and Fail. Analy., pp. 9-17, 1996.

2. G.F. Shade, "Two Unique Case Studies Performed with Photoemission Microscopy (PEM)," Proc. ISTFA '96, pp. 41-46, 1996.

3. G. F. Shade and K.S. Wills, "Photoemission microscopy," Microelectronic Failure Analysis Desk Reference, 3rd Ed., AMS/ISTFA, Nov., 1993.

4. K. Van Doorselaer, U. Swerts, L. Van Den Bempt, "Broadening the Use of Emission Microscopy," Proc. of the Int. Symp. for Test. and Fail. Analy., pp. 57-65, 1995.

5. J.S. Seo, S. S. Lee, C.S. Choe, S. Daniel, K.D.. Hong, C.K.Yoon, "Photoemission Spectrum Analysis - A Powerful Tool for Increased Root Cause Success," Proc. of the Int. Symp. for Test. and Fail. Analy., pp. 73-78, 1995.

6. L. Baker, R. Currence, F. Martin, W. Moore, W. Reyes, E. St. Peter, D. Golijanin, J. Patterson, "A Simplified Application of a Slow Scan CCD Astronomy Camera to Emission Microscopy and Fluorescent Microthermography," ISTFA, pp. 9-10, 1994.

7. J. Sweeney, M. Phillips, P. Thomas, and V. Soorholtz, "Emission microscopic identification of nonuniform submicron transistors with hot carrier characteristics, " Proc. of the ISTFA, 113-117, 1991.

8. A.N. Campbell, E.I. Cole, Jr., C.L. Henderson, and M.R. Taylor, "Case history: failure analysis of a CMOS SRAM with an intermittent open contact," Proc of the ISTFA., pp. 261-269, 1991.

9. K.S. Wills, D. DePaolis, and G. Billus, "Advanced photoemission technique for distinguishing latch-up from logic failures on CMOS devices," Proc. of the Int. Symp. for Test. and Fail. Analy., pp. 335-341, 1991.

10. J. Balog and M. Lin, "Failure analysis of degraded voltage regulator using light emission microscopy," Proc of the ISTFA, pp. 343-352, 1991.

11. R. Mann, and D. McElfresh, "Emission microscopy applied to optoelectronic emitter failure analysis," Proc. of the Int. Symp. for Test. and Fail. Analy., pp. 353-362, 1991.

12. S. Kiefer and M. Oyler, "Evaluation of the gate defects in GaAs MESFETs by emission microscopy," Proc. of the Int. Symp. for Test. and Fail. Analy., pp. 363-368, 1991.

13. K. Symonds, M. Bahrami, and P. Skerry, "Functional failure analysis using photoemission microscopy," Proc. of the Int. Symp. for Test. and Fail. Analy., pp. 369-375, 1991.

14. K.M. Baker, "An Economical Approach to Correlation and Calibration of Light Emission Microscopes," ISTFA-1991.

15. F. Magistrali, D. Sala, G. Salmini, F. Fantini, M. Giansante and M. Vanzi, Proc. IEEE Annual Int. Rel. Phys. Symposium IRPS-91, pp. 224-233, 1991.

16. K.S. Wills, et al, "Spectroscopic Photoemission Techniques to Understand Microprocessor Failures," Third International Symposium of the Physical and Failure Analysis of Integrated Circuits-Pan Pacific Singapore, 1991.

17. J. Kolzer, "Quantitative aspects of emission microscopy," Proc. CERT '91 (Components Engineering, Reliability and Test Conference), 1991, pp. 86-94.

18. Y Uraoka, H. Yoshikawa, N. Tsutsu, and S. Akiyama, "Evaluation of gate oxide reliability using luminescence method," Proc. IEEE Int. Conf. on Microelectronic Test Structures, Vol. 4, 1991, pp. 69-74.

19. Y. Uraoka, N. Tsutsu, Y. Nakata, and S. Akiyama, "Evaluation Technology of VLSI Reliability Using Hot Carrier Luminescence," IEEE Trans. on Semiconductor Manufacturing, Vol. 4, No. 3, Aug. 1991.

20. N.C. Das and B.M. Arora, "Luminescence Spectra of an N-channel Metal-Oxide Semiconductor Field Effect Transistor at Breakdown," Applied Physics Letters, vol. 56, No. 12, pp. 1152-3, (19 March 1990).

21. C.F. Hawkins, J.M. Soden, E.I. Cole and E. Snyder, "MOSFET Photon Emission for Analysis of CMOS ICs," ASM International Symposium for Testing and Failure Analysis, 1990.

22. C. Hawkins, J. Soden, E. Cole, and E. Snyder, "The use of light emission in failure analysis of CMOS ICs," Proc. of the Int. Symp. for Test. and Fail. Analy., pp. 55-67, 1990.

23. K. de Kort and P. Damink, "The spectroscopic signature of light emitted by integrated circuits," First ESREF Symp. (Italy), September, 1990.

24. J.T. May and G. Misakian, "Failure analysis of a turn-on degraded transistor using photoemission microscopy," Proc. of the Int. Symp. for Test. and Fail. Analy., pp. 69-71, 1990.

25. C. Khandekar, M. Hennis, P. Brownell, and D. Bethke, "Photoemission detection of vendor integrated circuit failures during system level functional testing," Proc. of the Int. Symp. for Test. and Fail. Analy., pp. 73-79, 1990.

26. Tsutsu, N., Y Uraoka, Y. Nakata, S. Akiyama, H. Esaki, "New Detection Method Of Hot Carrier Degradation Using Photon Spectrum Analysis Of Weak Luminescence On CMOS VLSI" IEEE ICMTS, vol. 3, no. 1, p. 143, 1990.

27. K. de Kort and P. Damink, "The Spectroscopic Signature of Light Emitted by Integrated Circuits," Proceedings of the Workshop on Emission Microscopy, 1990, pp. 45-52.

28. B. Doisneau, R. Michel and C. Plougonven ESREF 90 Proceeding, p. 101. 1990.

29. F.M. Roche, S.D. Bocus, B. Pistoulet, Analyse de Construction et de Defaillance, Atliere de la SEE, Cargese (1990).

30. H. Ishizuka, M. Tanaka, H. Konishi, and H. Ishida, "Advanced method of failure analysis using photon spectrum of emission microscopy" Proc. ISTFA '90. 1990, pp. 13-19.

31. G.F. Shade, "Physical mechanisms for light emission microscopy," Proc. ISTFA '90, 1990, pp. 121-128.

32. E. Zanoni, S. Bigliardi, R. Cappelletti, P. Lugli, F. Magistrali, M. Manfredi, A. Paccagnella, N. Testa, and C. Canali, "Light emission in AlGaAs/GaAs HEMT's and GaAs MESFET's induced by hot carriers," IEEE Electron Device Letters, Vol. 11, 1990, pp. 487-489.

33. M. Oyler, and S. Cohen, "Gate-drain breakdown mapping in gallium arsenide power MESFETs by emission microscopy," Proc. ESREF '90, 1990, pp. 85-92.

34. A. Dallman, and G. Deboy, "Characterization of trench-trench punch-through mechanisms by emission microscopy," Proc. ESREF '90, 1990, pp. 69-76.

35. D. Weinmann, C. Boit, and J. Kolzer, "Characterization of leakage currents by emission microscopy," Proc. ESREF '90, 1990, pp. 61-68.

36. C.L. Chiang, "Light emitting phenomenon in Si technology and its applications," Proc. of the VLSI Int. Educational Workshop and Symposium (VIEWS), 1990, VLSI Technology Inc.

37. C. Boit, J. Kolzer, H. Benzinger, A. Dallman, M. Herzog, and J. Quincke, "Discrimination of parasitic bipolar operating modes in ICs with emission microscopy, " Proc. IEEE/IRPS '90, 1990, pp. 81-85.

38. O. Sirch, J. Kolzer, and C. Boit, "Emission microscopy on integrated bipolar transistors," Proc. ESREF '90, 1990, pp. 93-100.

39. M. Hannelamm, and A. Amerasekera, "Photon emission as a tool for ESD failure localization and as a technique for studying ESD phenomena," Proc. ESREF '90, 1990, pp. 77-84.

40. Y. Hiruma, and E. Inuzuka, "Application of photon counting imaging to semiconductor failure analysis," Proc. ESREF '90, 1990, pp. 53-60.

41. K.S. Wills, S. Vaughan, Charvaka Duvvury, O. Adams, and J. Bartlett, "Photoemission Testing for EOS/ESD Failures in VLSI Devices: Advantages and Limitations," ASM International Symposium for Testing and Failure Analysis, p. 183, 1989.

42. T.S. Taylor, T. Dao, T.B. Haddock, and Lawrence C. Wagner, "Leakage Detection Techniques: A Comparitive Study," ASM International Symposium for Testing and Failure Analysis, pp. 5-13, 1989.

43. C. Canali, F. Corsi, M. Muschitiello, E. Zanoni, IEEE Trans. Elec. Dev., vol. 36, no. 5, (1989).

44. T. Aoki, and A. Yoshii, "Analysis of Latch-up-induced photoemission," Proc. IEEE/IEDM '89, 1989, pp. 281-284.

45. N. Khurana, "Second generation emission microscopy and its application," Proc. ISTFA '89, 1989, pp. 277-283.

46. C.G.C. de Kort, "Integrated circuit diagnostic tools: Underlaying physics and applications," Philips J. Res., Vol. 44, 1989, pp. 295-327.

47. S. C. Lim and E.G. Tan, "Detection of Junction Spiking and its Induced Latchup by Emission Microscopy," IEEE Annual Proc. of the Reliability Physics Symposium, pp. 119-125, 1988.

48. S.N. Chu, S. Nakahara, M.E. Twigg, L.A. Koszi, E. J. Flynn, A.K. Chin, B.P. Segner and W.D. Johnson Jr., J. Appl. Phys. vol. 63, n. 3, 1 Feb. 1988, pp. 611-623.

49. C. Canali, M. Giannini, A. Scorzoni, M. Vanzi, E. Zanoni, IEEE Journal of Solid State Circuits, vol. 23, no. 2, (1988).

50. M. Herzog, and F. Koch, "Hot-carrier light emission from silicon metal-oxide-semiconductor devices," Appl. Phys. Lett., Vol. 53, 1988, pp. 2620-2622.

51. N. Khurana and C. Chiang, "Dynamic imaging of current conduction in dielectric films by emission microscopy," Proc. of Int. Rel. Phys. Symp., pp. 72-76, April 1987.

52. A. Toriumi, M. Yosimi, M. Iwase, Y. Akiyama, and K. Taniguchi, "A study of photo emission from n-channel MOSFET's," IEEE Trans. on Electron Dev., Vol. ED-34, No. 7, pp. 1501-1508, July, 1987.

53. N. Khurana and C.L. Chang, "Analysis of Product Hot Electron Problems by Gated Emission Microscopy," IEEE Annual Proc. of the Reliability Physics Symp., pp. 189-194, 1986.

54. C.L. Chiang and N. Khurana, "Imaging and Detection of Current Conduction in Dielectric Films by Emission Microscopy," IEEE International Electron Devices Meeting, pp. 672-675, 1986.

55. A.R. Leblanc and W.W. Abadeer, "Behavior of SiO2 Under High Electric Field/Current Stress Conditions," IEEE Annual Proc. of the Reliability Physics Symp., pp. 230-234, 1986.

56. T. Tsuchiya and S. Nakajima, "Emission Mechanism and Bias-Dependent Emission Efficiency of Photons Induced by Drain Avalanche in Si MOSFET's," IEEE Transactions on Electron Devices, vol. ED-32, no. 2, Feb. 1985, pp. 405-412.

57. N. Das and W. Khokle, "Visible light emission from silicon MOSFETs," Solid-State Electronics, Vol. 28, No. 10, pp. 967-977, 1985.

58. S. Tam and C. Hu, "Hot-Electron-Induced Photon and Photocarrier Generation in Silicon MOSFET's, IEEE Transactions on Electron Devices, vol. ED-31, no. 9, Sept. 1984, pp. 1264-1273.

59. N. Khurana, "Pulsed infra-red microscopy for debugging latch-up on CMOS products," Proc. IEEE/IRPS, 1984, pp. 122-127.

60. T.C. Ong, K.W. Terrill, S. Tam and C. Hu, "Photon Generation in Forward-Biased Silicon p-n Junctions," IEEE Electron Device Letters, vol. EDL-4, No. 12, pp. 460-462, December, 1983.

61. T.N. Theis, J.R. Kirtley, D.J. DiMariea, and D.W. Dong, "Spectroscopic Studies of Electronic Conduction in SiO2," in Insulating Films on Semiconductors. J.F. Verweij and D.R. Wolters (ed.), North-Holland, pp. 134-140, 1983.

62. S. Tam. F. Hsu, P. Ko, C. Hu, and R. Muller, "Spatially resolved observation of visible-light emission from Si MOSFET's," IEEE Electron Dev. Letters,Vol. EDL-4, No. 10, pp. 386-388, October 1983.

63. David, J.R. Stictch, J.E. , and Stern, M.S., "Gate-Drain Avalanche Breakdown in GaAs Power MESFETs" IEEE Trans. Electron Devices, Vol. ED-29, pp. 1548-1552, Oct. 1982.

64. P. Solomon and N. Klein. "Electroluminescence at High Electric Fields in Silicon Dioxide," J. Appl. Phys., vol. 47, pp. 1023-26, March, 1976.

65. J.I. Pankove, Optical Processes in Semiconductors, Dover Publications, Inc., 1975.

66. W. Haecker, "Infrared radiation from breakdown plasmas in Si, GaSb, and Ge: Evidence for direct free hole radiation," Phys. State Sol., Vol. 25, 1974, pp. 301-310.

67. M. Lenzlinger and E.H. Snow, "Fowler-Nordheim Tunneling into Thermally Grown SiO2", Journal of Applied Physics, Vol. 40, p. 278, 1969.

68. H. Kressel, "A review of the effect of imperfections on the electrical breakdown in silicon p-n junctions," RCA Rev., vol. XXVIII, No. 1, p. 181, March 1967.

69. J. Shewchun and L.Y. Wei, "Mechanism for Reverse-biased Breakdown Radiation in p-n Junctions," Solid-State Electronics, vol. 8, 1964, pp. 485.

70. E. Kamieniecki, "Hot Carriers in Microplasmas and their Radiation in Germanium and Silicon," Physical Status Solidi, vol. 6, 1964, pp. 877.

71. T. Figelski and A. Torun, "On the Origin of Light Emitted from Reverse Biased p-n Junctions," Proceedings of the International Conference in the Physics of Semiconductors, 1962, pp. 853.

72. L.W. Davies, and A.R. Storm, Jr., "Recombination radiation from silicon under strong-field conditions," Phys. Rev., Vol. 121, 1961, 1961, pp. 381-387.

73. M. Kikuchi, "Visible light emission and microplasma phenomena in silicon p-n junctions," J. Phys. Soc. Japan, Vol. 15, p. 1822, October, 1960.

74. A.G. Chynoweth et al., "Photon emission from avalanche breakdown in silicon," Phys. Rev., vol. 102, no. 2, p. 369, 1956.

75. Newman, R., Dash, Hall, Burch, "Visible Light From Si p-n Junctions," Phys. Rev., Vol. 100, p. 700, 1955.

76. W. van Roosbroeck, and W. Shockley, "Photon-radiative Recombination of electrons and holes in germanium," Phys. Rev., Vol. 94, 1954, pp. 1558-1560.

Photoemission Microscopy—
Advanced/Theory of Operation

Christian Boit
Infineon Semiconductors

1) INTRODUCTION

1.1) Where to use PEM in Failure Analysis Process Flow

- Fail recorded in electrical test
- In depth electrical investigation of the fail
- **localization of the electrical fail**
- preparation and imaging of the physical root cause

1.2) Spotlight on PEM

PEM detects faint visible and near IR radiation emitted from silicon devices and circuits under numerous, not only failing, conditions. Failures may show up due to high local current densities (leakage, circuit failure) or an undesired operation of an active device. The spectral distribution of the light emission delivers information about the operation mode of semiconductor devices. As long as chip metallization and package technologies leave an optical path from the silicon surface to the detector, a chip in a locally decapsulated package can be investigated without further preparation. For these reasons PEM became the #1 technique in *localizing electrical failures* in microelectronics. It is quick (a complete IC can be checked for photoemission by only one image shot), versatile (setup is based on a probe station and/or can be docked to a tester), clean (no films etc. required), simple (no interaction with probes, no artefacts) and sensitive (leakage currents down to the pA range, depending on local failure current density). Emission spots are not necessarily failures; PEM failure localization may require a golden device for comparison. As the photoemission is generated by semiconductor effects, circuit failures in the metallization levels can only be detected by PEM if they create irregularities inside silicon. Direct access to metallization level failures delivers hot spot-thermography. Most of the PEM machines are equipped for the application of thermography as well. PEM is not a thermography of silicon - black body radiation is not in the range of the typically detected spectrum for temperatures well below 500K.

1.3) Alternative Methods

- Thermography (Liquid Crystal, Fluorescent film)
- Optical / Electron - Beam Induced Current OBIC, EBIC
  and derivatives LIVA, CIVA etc.
- E-Beam Testing (Stroboscopic Voltage Contrast)

2) BASICS

The two basic effects that contribute to the photoemission signal are most comprehensible by studying the two operation modes of a pn-junction in silicon. **A forward biased junction** reduces the pn-energy barrier to a level (Fig.2.1) that allows the majority carriers to diffuse into the alternate part as minority carriers (minority carrier injection), which is accompanied by strong electron-hole recombination activity.

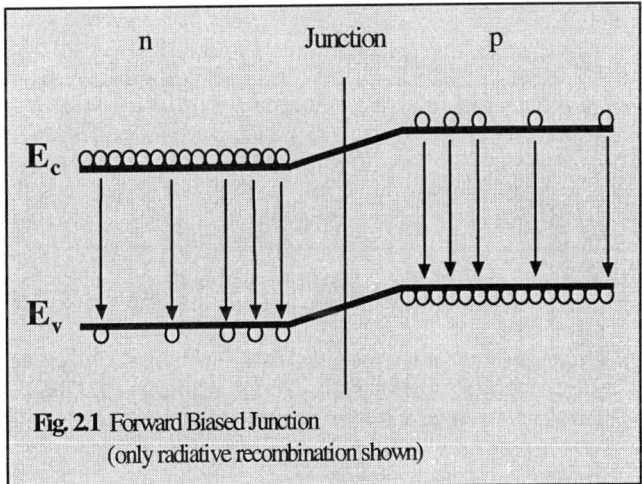

Fig. 2.1 Forward Biased Junction
(only radiative recombination shown)

As silicon is an indirect semiconductor, the bottom of the conduction band and the top of the valence band have different momentum coordinates (Fig.2.2).
A localized center in the bandgap has due to the uncertainty principle a blurred momentum coordinate.
This raises the probability for non-radiant recombination via centers several orders of magnitude above the one for **radiant band-band recombination** which requires a third partner, a phonon, for the momentum exchange.
In the forward bias case, the recombination activity via centers close to midgap level is enhanced proportional to the excess minority carrier concentration and the radiant recombination proportional to the excess minority carrier concentration, multiplied by the majority carrier concentration.

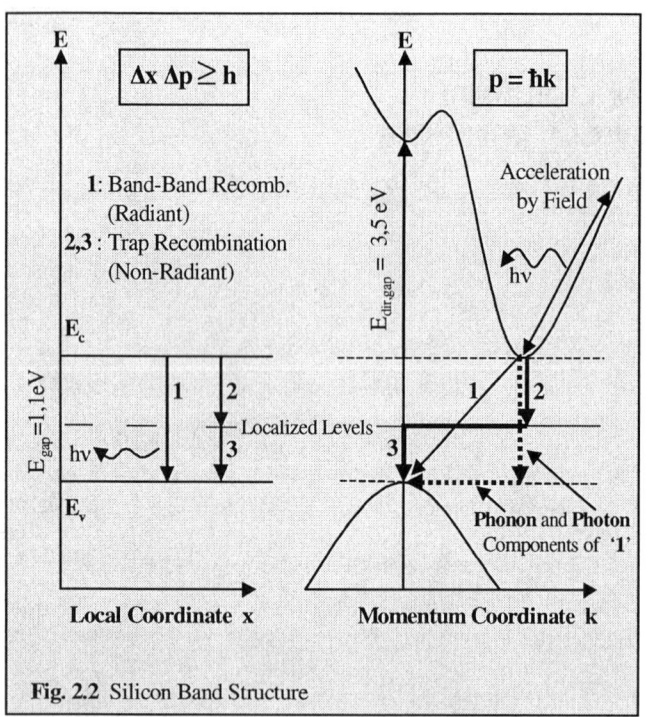

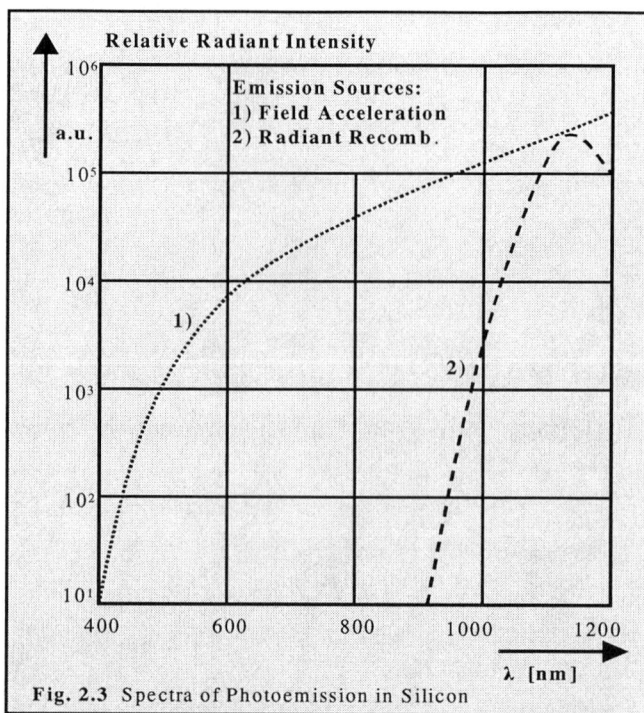

Fig. 2.2 Silicon Band Structure

Fig. 2.3 Spectra of Photoemission in Silicon

The recombination activity A decays with the distance from the junction x by:

$$A = A_0 \cdot \exp(-x/L) ,$$
$$\text{with } L = \sqrt{D \cdot \tau} \qquad (1)$$

τ: carrier lifetime
A: Recomb. Activity
A_0: A at junction
L: diffusion length
D: diffusion constant
x: distance from junc.

A_0 is the activity at the junction and L the diffusion length as the square root of the product of the diffusion constant D and the carrier lifetime τ. The radiation wavelength is about 1.1μm (1.1 eV band gap). The intensity I is very faint, even at high injection levels and the spectral distribution follows

$$I = I_0 \cdot \exp\left[-\left[\,|(E - E_G)|\,\right]/kT_e\right] \qquad (2)$$

E_G: Bandgap energy
I_0: Intensity at E_G
T_e: electron temp.
E: electron energy

with T_e being the electron temperature, E_G the bandgap energy and I_0 the intensity at E_G. In the recombination case only heat contributes to T_e and the Intensity decreases by orders of magnitude within a wavelength range of 100nm (Fig.2.3).

A reverse biased junction raises the pn-energy barrier and creates a large space charge region (SCR) which accelerates the few passing carriers. They lose their gained energy by **scattering** via phonons, crystal defects and charged coulomb centers which may be accompanied by recombination (Fig.2.4a). The energy quantum per impact varies and the impact probability decreases exponentially with increasing energy. Scattering via charged coulomb centers results in a light emission with a broad spectral distribution in the visible to near IR region (Fig.2.3). The intensity decreases for E>E_G again like **Eq 2.** with T_e defined by the energy the electrons gain in the electrical field. The emission intensity is usually below detection limit of PEM unless avalanche condition is approached, which is accompanied by unstable emission conditions and

breakdown device operation. Leakage currents are detectable by PEM before reaching avalanche condition (Fig.2.4b).

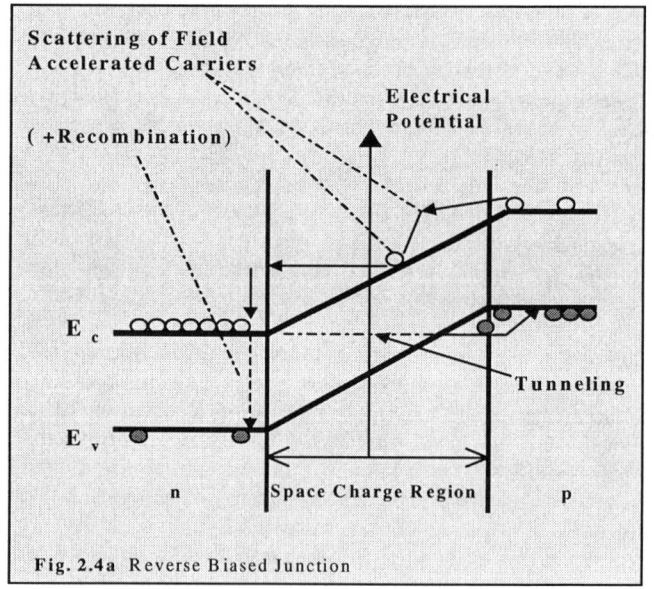

Fig. 2.4a Reverse Biased Junction

214

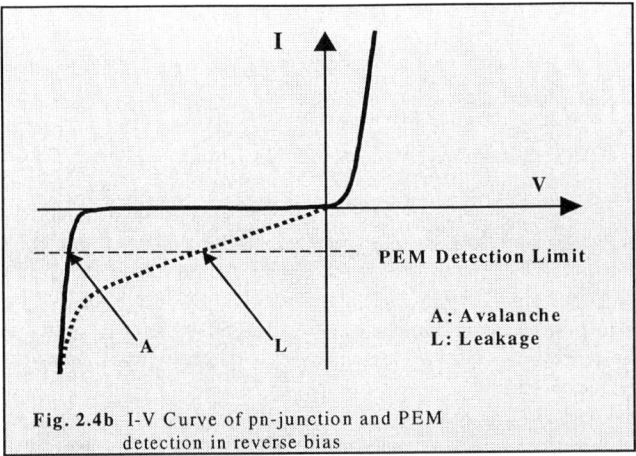

Fig. 2.4b I-V Curve of pn-junction and PEM detection in reverse bias

These two effects which correlate to the active properties of semiconductor devices are the source for PEM. 'Ohmic' currents in homogeneously distributed density within their wires or designed current paths do not show in PEM unless they are accompanied by one of the photoemission processes. PEM is no thermography. Black Body Radiation in the detected spectral range is only significant at temperatures well above 500K (Fig.4.7).

3) CLASSIFICATION OF EMISSION SOURCES

3.1) Photoemission Generated by Scattering of Field Accelerated Carriers (F-PE)

Most photoemission spots in microelectronic failure analysis find their origin in the effect that has been discussed in chapter 2 with the reverse biased junction, as described in Tab 1. The electrical field induced events include most leakage currents in silicon. Even ohmic leakage currents as they occur for example at gate oxide (GOX) defects can be detected if the current density is locally high enough to create F-PE. GOX-defects usually need a local voltage drop of 2V or higher for PEM-detection. Fowler-Nordheim Currents create F-PE due to scattering in silicon or polysilicon, respectively. Effective field variations of the GOX area, resulting from lateral polysilicon resistance or local GOX thinnings, are accompanied by FN-current variations that can be recorded by PEM.

3.2) Photoemission Generated by Radiant Electron-Hole Recombination (R-PE)

R-PE is subject to IC failure analysis mainly for latch up investigations. Forward bias effects occur here and there if wells are floating. It is very important for power devices and their switching properties.

F-PE	Space Charge Region	Reverse Biased Junction
		Silicon Leakage Currents
		Saturated MOS Transistors
		ESD Protection Breakdown
		Bipolar Transistors
	Locally High Current	GOX-Defects / -Leakage
	Fowler-Nordheim Current	GOX Leakage Current
R-PE	E-H-Recombination	Forward Biased Junction
		Bipolar Transistors
		Latch up

Table. 1 Classification of Photoemission Sources

3.3) Leakage Currents in Silicon

In chapter 2 has been stated avalanche condition is necessary to get an emission signal from a reverse biased junction - this holds only for a good junction. A leaky diode (Fig.2.4b) still builds up a considerable electrical field and mostly offers a current density well above PEM detection limit (exception: well-substrate diodes: if a current is leaking all over their large area, current density may never reach detection limit).

Leakage currents in silicon are the classic PEM application for failure localization. They produce emission spots when passing through or created in an SCR (F-PE). Contact spiking is one more kind of junction leakage. Highly localized junctions or capacitors allow detection even in pA range.

3.4) Saturated MOSFETs/ Hot Electrons

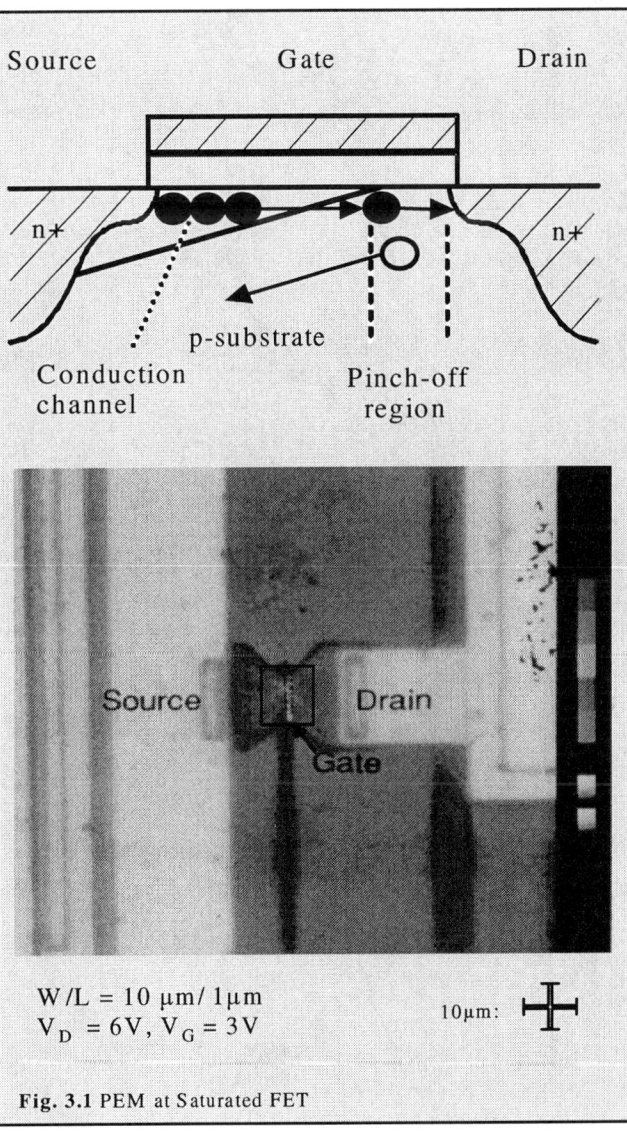

$W/L = 10 \ \mu m / 1 \mu m$
$V_D = 6V, \ V_G = 3V$

10μm:

Fig. 3.1 PEM at Saturated FET

The current through the inversion channel of a MOS transistor is not accompanied by light emission. Only in saturation, it passes an SCR from pinch-off to drain (Fig. 3.1). The light emission is proportional to the back bias/substrate current (Fig.3.2).

215

This means, the PEM signal is proportional to the carrier multiplication in the SCR. Some of the carriers generated by multiplication in the pinch off region penetrate the gate oxide and become hot carriers. Light emission in an FET can be used as a measure of hot electron damage. Evaluation of the emission spectra allows to determine the electron temperature of hot electrons. The maximum electrical field between the pinch-off point and the drain (effecting a substrate current maximum) occurs at gate voltages well below drain voltages which happens during switching of the FETs or analog operation. Inverters or flip flops deliver emission spots if they are not in a defined condition. Sometimes this effect is unwanted, i.e. with floating pins or incomplete power-down mode. Fig. 3.10 shows the spectral distribution to correlate with F-PE and a typical electron temperature of 1500 to 2500K.

3.5) Fowler-Nordheim (FN) Current

Leakage currents of intact gate oxides consist of a tunneling phase and an "ohmic" part of the path in which the carriers get accelerated by the electrical field (Fig.3.3). The emission mechanism is again F-PE, but the spectral distribution (Fig.3.10) is very different from the other sources: The emission intensity is almost equally distributed over the recorded spectral region with a relative maximum at 1.8eV. As discussed before, the slope of the intensity over photon energy expresses the energy distribution of the carriers which undergo the light emitting impacts. In this case, only carriers contribute to the current that pass the Si-SiO₂ barrier, lowered by the tunneling condition. FN-current is exponentially dependent on the electrical field.

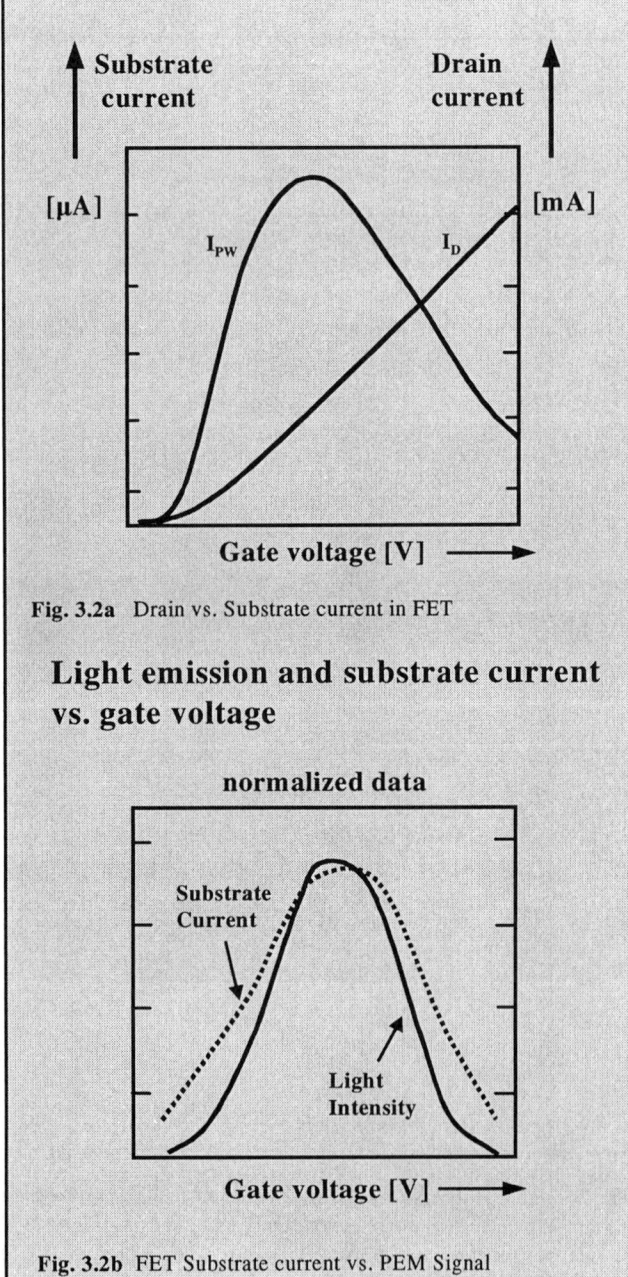

Fig. 3.2a Drain vs. Substrate current in FET

Light emission and substrate current vs. gate voltage

normalized data

Fig. 3.2b FET Substrate current vs. PEM Signal

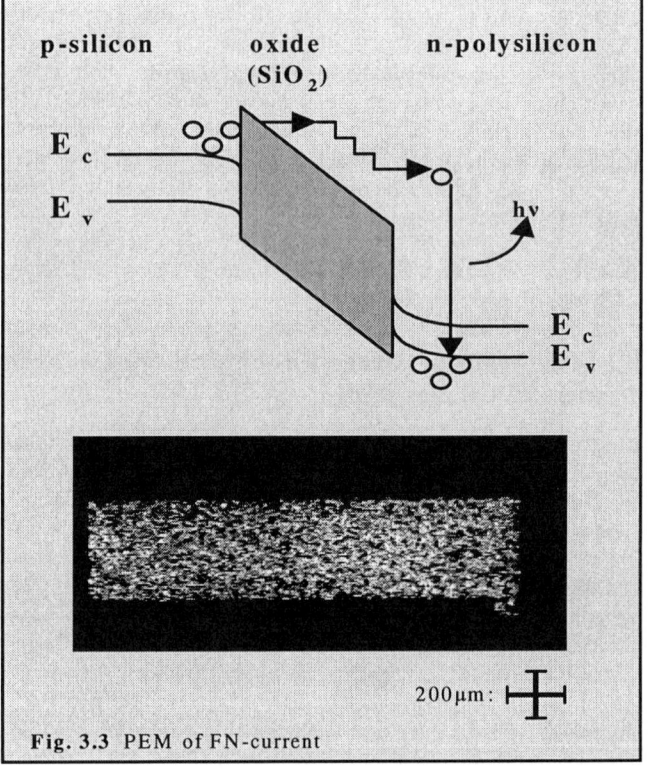

Fig. 3.3 PEM of FN-current

As the voltage is constant at the silicon and poly electrode, oxide thickness variations create locally changing FN-currents that can be identified by PEM. Quantitative evaluations of the thickness usually fail because the varying sites are of nm dimensions which is out of reach of PEM optical resolution (see chapter 5.4).

One more drawback for quantitative evaluation of PEM in the FN-current case is the relative weak and noisy signal. Whereas scattering in the other examples happens mostly within the electrical field, here the majority that contributes to the PEM signal is supposed to do so in silicon or poly area and less in the oxide. Outside the field, more carriers are assumed to thermalize i.e. non-radiant relaxation. For studying FN-currents, the most sensitive detector in the correlated spectral region is recommended.

216

3.6) Locally High Current Density / Thin Oxide Breakdown

GOX-defects belong to the most important failures which get localized by PEM. But the breakdown of thin oxides does not necessarily produce an SCR especially when poly and well are doped of the same type. The explanation is: the current density is high enough to create a voltage drop at the fail site which in turn implies F-PE in PEM spectral range (typically >2V). This makes GOX-breakdown usually observable in PEM (Fig.3.4). It also explains why not every breakdown can be found in PEM. Some emission spots are unstable or vanish after a while (local high current melts breakdown region, enlarges broken area and reduces current density) and some fails even heal electrically during operation (Joule heat blows off the poly piece and creates an open circuitry around breakdown area).

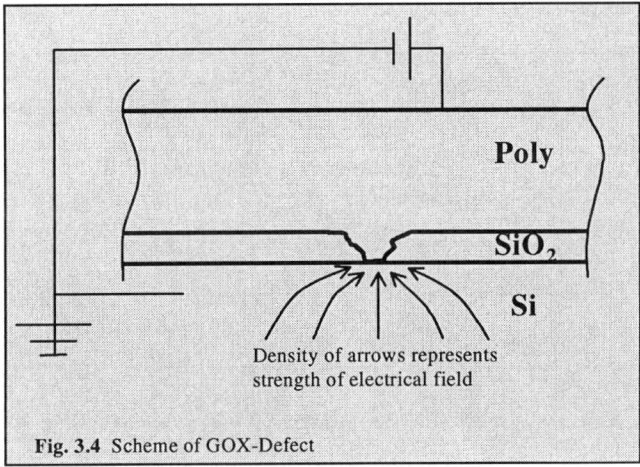

Fig. 3.4 Scheme of GOX-Defect

3.7) Diode

The principal behavior of a diode in PEM has already been discussed in chapter 2. Here the focus is on the characteristic emission images from diodes in ICs or related test structures. Fig.3.5 shows a diode built of diffusions which are covered by a MoSi₂-alloy.

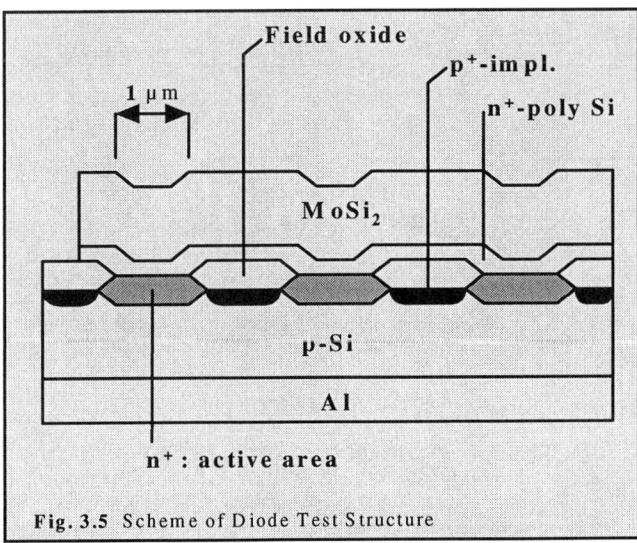

Fig. 3.5 Scheme of Diode Test Structure

At forward bias (Fig.3.6a), the silicide is blocking most of the light emission underneath, so only the recombination of the carriers diffused away from the diode area deliver a PEM signal. The diffusion length can be easily calculated from this emission image. Proof of the blocking character of the silicide is given in Fig.3.6b. The silicide has been removed by chemical delayering techniques, and the recombination radiation in the diode area shows the emission image. However, the signal decreases already inside the diode area with the distance the diode cells have from the location of the probe tip. The resistance of the polylayer after removal of the silicide is not low enough to turn on the complete diode area to the same level.

In the reverse biased case the image (Fig.3.6c) looks completely different: Generally, no photoemission is observed. Once the breakdown slope of the curve is reached, sharply localized spots are visible, even with the silicide layer on top. In real time mode of the camera, these light spots fluctuate strongly in intensity and location. Reverse current light emission is too faint for PEM unless avalanche multiplication raises the current density - and so the scattering probability - by orders of magnitude.

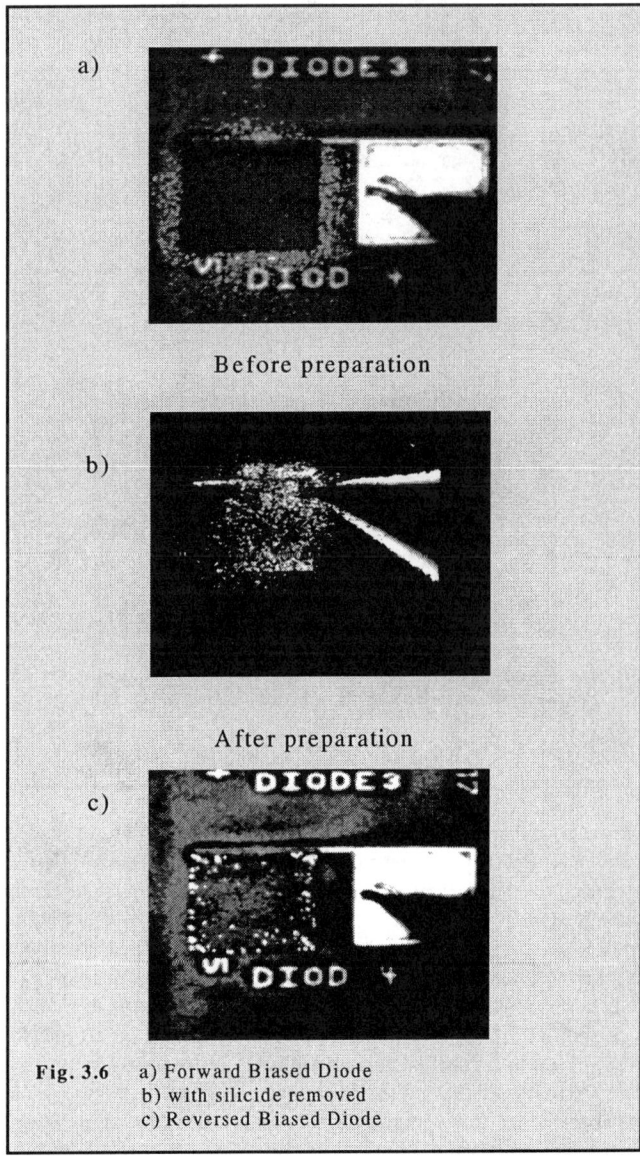

Fig. 3.6 a) Forward Biased Diode
b) with silicide removed
c) Reversed Biased Diode

Light spots appear at field maxima like corners of the SCR and are sharply localized. Fluctuations of avalanche multiplication are well known and multiplication maxima consist of so many carriers - they produce light flashes even visible with the bare eye and can easily be detected through the silicide although the signal is attenuated by orders of magnitude.

3.8) Bipolar Transistor

For better understanding, in Fig.3.7 a MOS transistor has been driven in its parasitic bipolar mode, with the well under the gate acting as base. This way, the structure is laterally orientated, whereas regular integrated bipolar transistors are vertically structured and the explained effect not so obvious. A bipolar transistor in active mode operates the emitter-basis junction forward biased and the basis-collector junction reverse biased. In saturated mode both junctions are forward biased. The emission images show a very sharply localized signal like F-PE in active, a broader signal like R-PE in saturated mode. (Fig.3.7). The reason is, in active mode the reverse biased base-collector diode produces a much steeper diffusion slope in the base and concentrates the current into the collector direction.

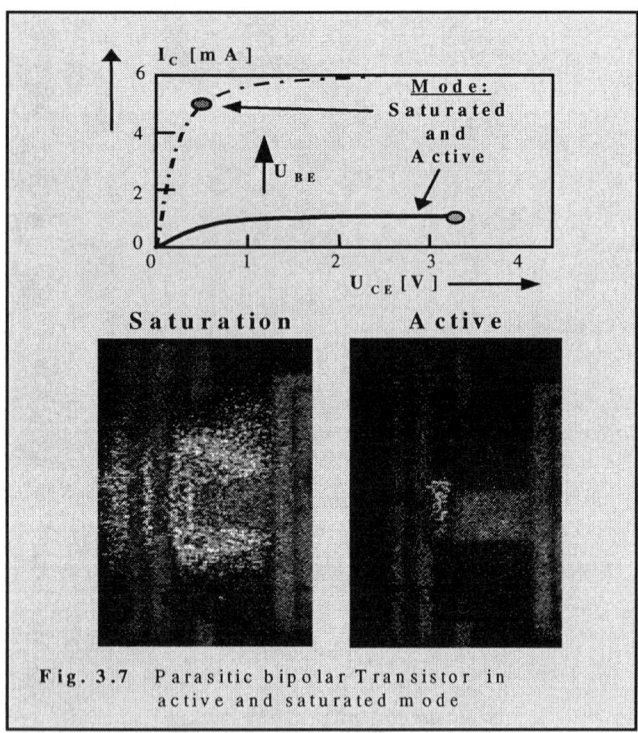

Fig. 3.7 Parasitic bipolar Transistor in active and saturated mode

On the other hand, the recombination probability in the thin base is very small. So, a typical recombination signal relies on the absence of SCRs because they separate carriers and keep them from recombination even if there is a high excess carrier concentration. Proof is given in the saturated mode which differs from active mode by the absence of the base-collector SCR. The lateral signal distribution is characterized rather by carrier diffusion than by field acceleration. From PEM point of view, an advantage of a transistor over a diode is the chance to elevate blocking currents by raising the base current and make them detectable by PEM without all the drawbacks like fluctuations, degradation or destruction as under avalanche condition.

3.9) Latch up

Latch up is a parasitic bipolar effect of two wells and their diffusions which define a four-layer device like a thyristor. During latch up the thyristor turns on in forward direction (from the viewpoint of the two diffusions) by strong minority carrier injection of the diffusions which floods the wells and the junction of the two wells. Turn on condition is a combined amplification factor >1 of the two folded bipolar parasitic transistors (Fig.3.8). This may be reached by a parasitic base current as in the example of Fig.3.9, where the minority carrier diffusion from the protection diode serves as additional injection or base current in the latching system and is turning it on. In other cases, a break over condition may be reached by field peaks combined with a tight design and avalanche produces latch up. The emission signal of a stable latch up is always R-PE because of the high excess carrier density of electrons and holes. At the turn on moment in the breakover case F-PE occurs due to avalanche multiplication.

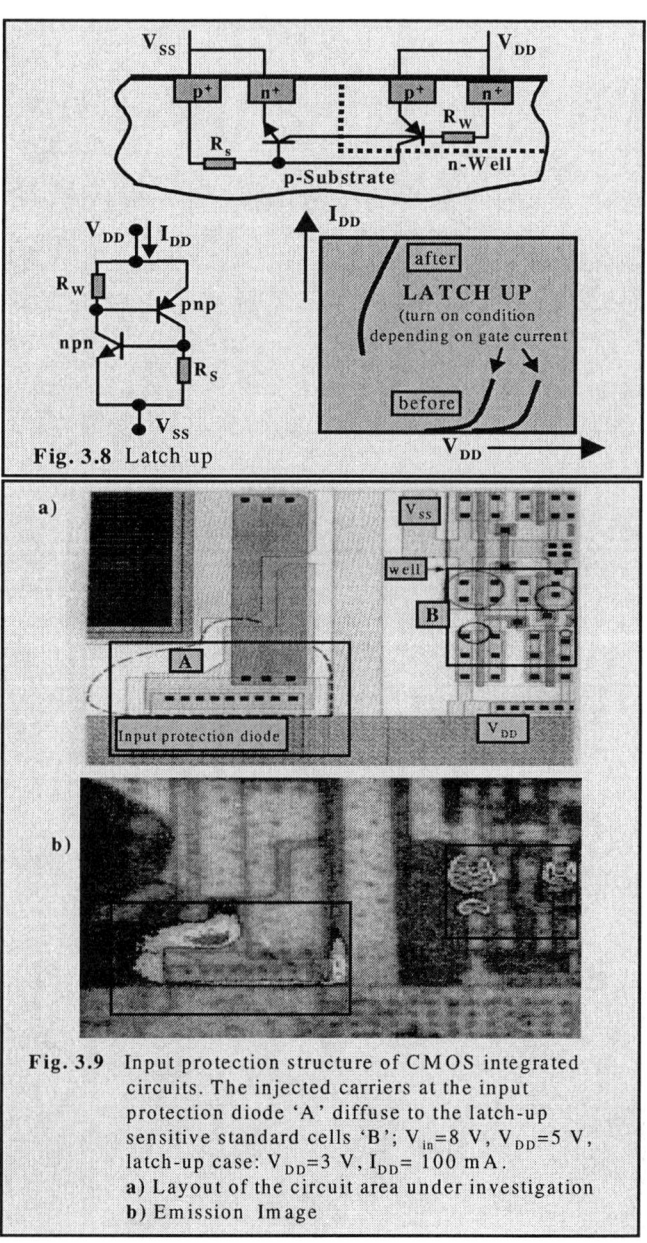

Fig. 3.8 Latch up

Fig. 3.9 Input protection structure of CMOS integrated circuits. The injected carriers at the input protection diode 'A' diffuse to the latch-up sensitive standard cells 'B'; $V_{in}=8$ V, $V_{DD}=5$ V, latch-up case: $V_{DD}=3$ V, $I_{DD}=100$ mA.
a) Layout of the circuit area under investigation
b) Emission Image

3.10) Emission Spectra

All the emission spectra of the chapters 2 to 3.9 are concentrated in Fig. 3.10. The curves of the separate effects have already been discussed previously. In this overview three basic types of spectra directly catch the eye:

1) The recombination radiation (R-PE) with a steep slope due to the **low electron temperature** representing only heat (especially obvious in the latch up case: latch up heats the device significantly - even the slope shows).

2) Radiant scattering of field accelerated carriers (F-PE) with the carriers still approximated to **obey Boltzmann** statistics, so the exponential slopes represent field-supported electron temperatures of 1500K and more.

3) Radiant scattering of field accelerated carriers (F-PE) of FN-currents through intact thin oxides with the carriers that contribute to the current **not obeying Boltzmann** statistics due to FN-tunneling and therefore not showing a significant slope in the spectra.(except for the relative peak at 1.8 eV).

All the quantitative evaluation of photoemission is based on the hypothesis that emission intensity is directly proportional to the carrier density supporting the emitting effect. Spectral resolution techniques of PEM are presented in chapter 4.2. There are certain limitations to quantitative evaluation. The detected light intensity depends on optical transmittance of layers on top of the emission sources. There have been many efforts to file fingerprints of the failure types for automatic recognition - none of these has become routine application because of the ever changing layers the light has to pass like filters - even within one technology.

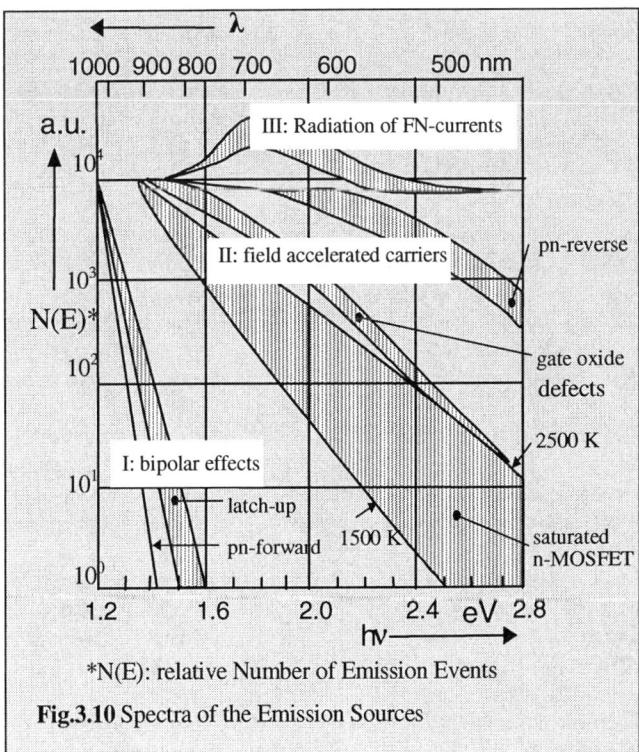

*N(E): relative Number of Emission Events

Fig.3.10 Spectra of the Emission Sources

3.11) Extracts for The Failure Analyst

1) A powered-down digital CMOS IC (i.e. DC-condition) should not produce PEM-signals - FETs should not switch and have small substrate current, pn-junctions are all in reverse bias. Every emission spot represents a failure as mentioned above - leakage, spiking, GOX-defect, FET in incorrect state, current density higher as designed, ESD protection damage, latch up etc. If certain areas show emission cumulations which may even alter each time the circuit is turned on again or even buses show up, the chip is probably not in a defined state > Recheck if Power-Down is realized correctly. AC-analysis of digital CMOS requires usually a good chip for comparison, to separate functional emission spots from failure indicators. In general, the emission spot is not always identical to the failure location. A flip flop may not switch correctly, and the result can be an undefined state of another FET which may emit light, and so on.

The key is to always verify the reason for the light emission in terms of device and circuitry operation.

Often there are more than one possible explanations. The analysis job is not done before they all have been investigated.

2) Analog CMOS and AC driven (switching) digital ICs contain FET devices operated at low gate voltages at which they emit light. That means, emission signals in analog circuits occur in good chips. In terms of failure analysis by photo emission, analog circuits can be more complicated than digital, because circuit elements like current or capacitance mirrors and the wide variety of operating modes of the devices may lead to a number of emission signals of devices which indicate a failure only in comparison with a good chip. Exact understanding of the electrical circuit is required to nail down the failing origin which itself may not show emission (i.e. wrong resistive path in current mirror unit etc.).GOX defects are relatively easy to detect with PEM in all CMOS technologies - the hard part is to highlight the defect in chemo-physical analysis afterwards.

3) Bipolar ICs show numerous emission spots during regular operation of many transistors. Leakage currents due to failures are usually higher than in CMOS and tend to give a stronger signal than the operational spots do.

The vertical structure of integrated bipolar transistors may cause hidden emission sources in the buried collector layer - the light can get re-absorbed in the upper silicon layers, especially the highly doped emitters.

4) Low supply voltages reduce the probability of light emission because fewer carriers will be accelerated to the required energy levels. The spectra displayed in Fig. 3.10 allow the expectation that the signal intensity will further increase into the IR regime, and spectra published in ref /13/ prove this expectation. Table 2 shows signal intensities of saturated FETs at substrate current maximum obtained with different detectors that have from left to right increasing sensitivity into the IR range (the count rates should not be compared quantitatively between the systems).

The result complies with the observation in chapter 3.6 that typically 2V or more are required to obtain a PEM signal with the detectors currently in use.

PEM Signal at Low IC Supply Voltage
- FET in Saturation (Substrate Current Maximum)

FET Supply [V] \ Detector [counts]	Phemos 200		IREM 1
	Si-CCD Background Subtr.	Image Intensifier GaAs⁺ Disc. = 90	HgCdTe-CCD
3	13761		
2,5	1398	7518	55960
2,2	-	964	
2,0	-	239	9380
1,9	-	-	
1,8	-	-	2780
1,5	-	-	460

Data: Wirth/Grützner, Infineon FA; Zaplatin, Infrared Labs

Table. 2: IC Voltage Supply and PEM Signal

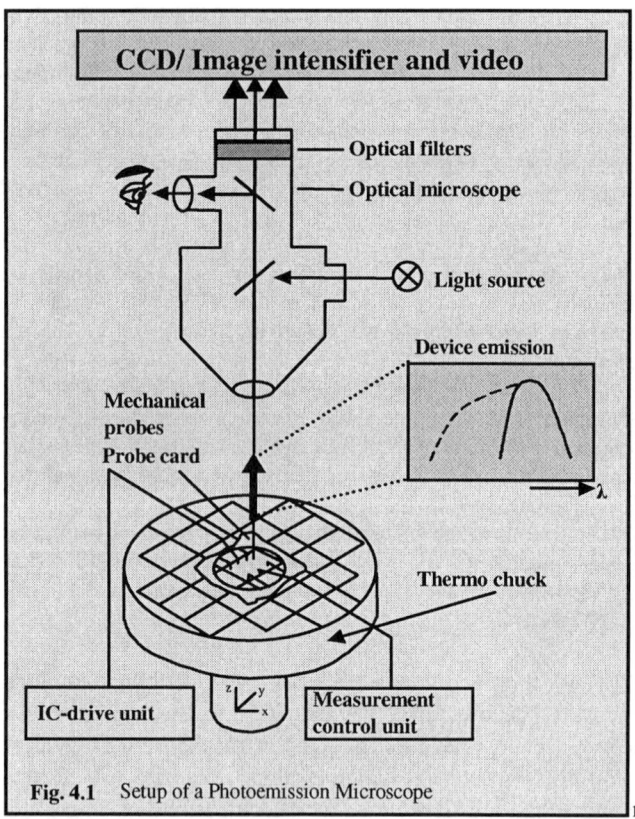

CCD/ Image intensifier and video

Optical filters

Optical microscope

Light source

Device emission

Mechanical probes
Probe card

Thermo chuck

IC-drive unit

Measurement control unit

Fig. 4.1 Setup of a Photoemission Microscope

A PEM tool (Fig.4.1) is usually based on a wafer prober with the product or structure under investigation being driven via electrical probes / probe cards, or a locally opened IC positioned on the stage. The microscope undergoes a few optimizations as is presented in chapter 4.2. The faint light must be detected very sensitively, either with an image intensifier or a cooled CCD (Fig.4.3). For localization, the resulting PEM image needs to be overlaid with a micrograph of the structure which requires some image processing.

4.2) Optical Path

The microscope, on a prober for adjustment reasons only, has a crucial function in PEM. For an image of the light reflected from the IC surface, the lamp is bright enough. The faint photoemission source needs the microscope as optical path to get transmitted to the detector. Any loss of intensity should be avoided. Most important is a high numerical aperture (NA) of the objectives with the high working distance (WD) necessary for manipulation still maintained. The NA is a measure for the light flux F transmitted with a magnification M from a light spot to a CCD pixel:

$$F= f(NA/M)^2 \qquad (3)$$

M : magnification
NA: numerical aperture
F : light flux

Table 3 shows numbers for typical objectives:

Magnification	NA	rel. Flux
0.8	0.025	0,13
0.8 Macro	0.40 (eff. 0.14)	321 (39)
5	0.14	1
20	0.42	0,6
100	0.50	0,03

Tab. 3 Light flux of PEM microscope objectives

The highest flux occurs at the higher magnifications. But 0.8 magnification is essential for PEM because it allows to image the complete chip with one image shot. Here, optimum sensitivity is required to check the chip for the presence and rough localization of photoemission sources. This can only be assured with a macro objective which is complicated to fit into a microscope optical path. The resulting compromise in handling is more than balanced by the quick investigation for PEM sources it gains. If spots are detected at 0.8x, the time spent to continue with higher magnifications is justified.

The relative flux as in Tab.3 represents ideal microscope objectives. In reality, some intensity gets lost and the effective NA of macro objectives is significantly lower. The resulting relative signal intensity for each objective may differ from the calculated numbers, even by some extent.

High transmission means removal of all dispensable lenses and mirrors, i.e. zooms, during the recording of the emission image. Absolute darkness should be provided, any noise reduces the sensitivity significantly. The spectral resolution of the signal is usually in failure analysis applications of minor interest, as most effects are understood today (chapter 3). If required, one approach is via optical filters.

There is a trade off between spectral resolution and sensitivity. In most cases filters of 40nm bandwidth still deliver signals that are quantitatively valuable. For correct evaluation, the filter transmission spectrum must be folded into the spectral sensitivity curve of the detector (Fig.4.2a). Still, a recording of a complete spectrum is a serial process which may be accompanied by degradation or destruction of the emission source.

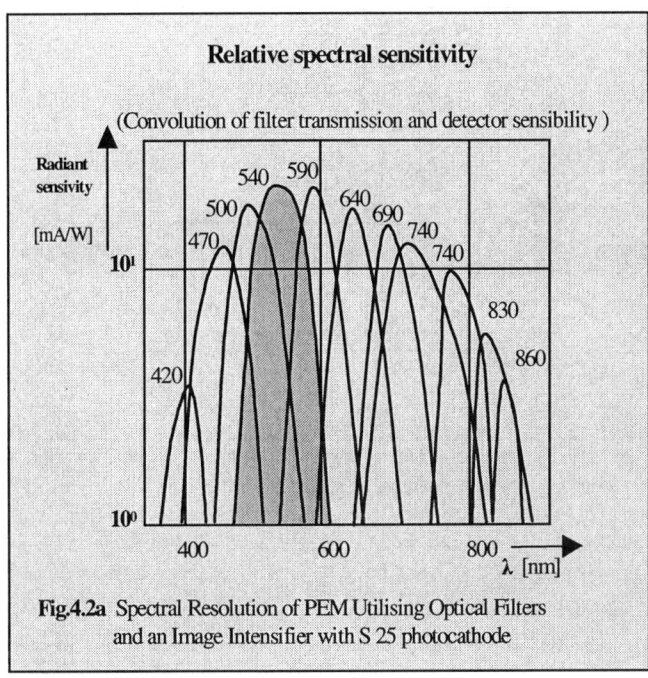

Relative spectral sensitivity

(Convolution of filter transmission and detector sensibility)

Fig.4.2a Spectral Resolution of PEM Utilising Optical Filters and an Image Intensifier with S 25 photocathode

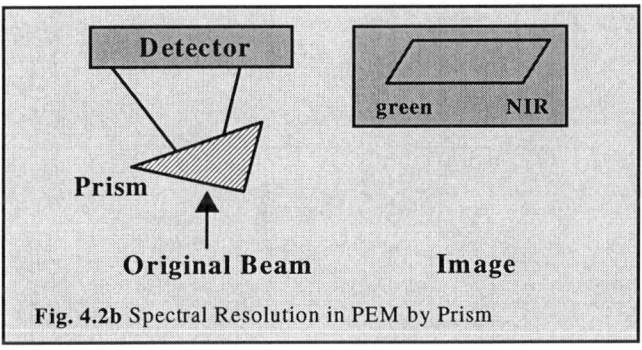

Fig. 4.2b Spectral Resolution in PEM by Prism

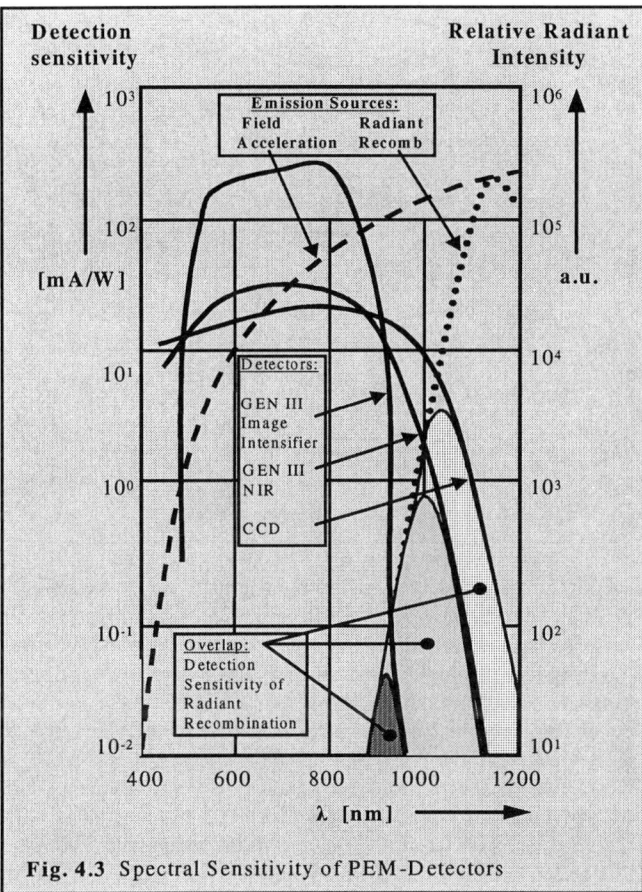

Fig. 4.3 Spectral Sensitivity of PEM-Detectors

Λ workaround for that problem is the use of a prism in the optical path (Fig.4.2b). In that case, the prism should only be in the optical path at low magnification, because for optimized spectral resolution the emission source should come closest to a point light source.

A useful approach to access basic spectral information like F-PE / R-PE discrimination is the application of bandpass filters which reduces the number of required filters significantly and conserves most of the signal intensity.

4.3) Detectors

The ideal detector is sensitive in the spectral range from 500 to 1200 nm wavelength, images the lateral resolution of the microscope, has MHz frequency range with stroboscopic gating option and operates at TV image frequency. An overview about spectral sensitivity is given in Fig.4.3 .

Cooled Si-CCDs have a broad spectral range but decrease by an order of magnitude in the recombination radiation region. Typical pixel fields of 1056x1056 allow imaging with microscope resolution. Integration times for emission images are in the order of seconds and prohibit live imaging. CCDs operate only DC with no stroboscopic gating available.

HgCdTe CCDs (and of some other compound semiconductors) extend their spectral sensitivity further into the IR regime. These detectors will play an important role in the future for PEM from chip backside (chapter 4.4) and low voltage supply of ICs (3.11.4).

They are not discussed here in detail because not many of those detectors are currently in PEM use. They are quite expensive today, require cooling far below room temperature for noise reasons and heat radiation (chapter 4.6) as an emission source needs a closer look.

Image intensifiers (II) are systems built of a photocathode which determines the spectral sensitivity, a microchannel plate (MCP) that multiplies the photoelectrons in the channels by some orders of magnitude and a phosphorus screen which is illuminated by the multiplied electrons (Fig. 4.4), which in turn illuminates a TV scan CCD. The photoelectrons in an II undergo an acceleration of several hundred volts. The MCP-channel density is much smaller than the CCD pixel density which limits the lateral resolution to ca. 1.5 µm. Live imaging is possible if the emission source is not too weak.

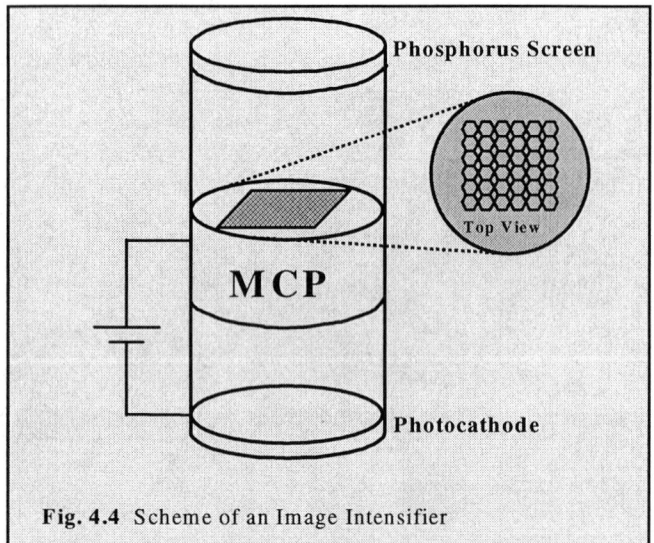

Fig. 4.4 Scheme of an Image Intensifier

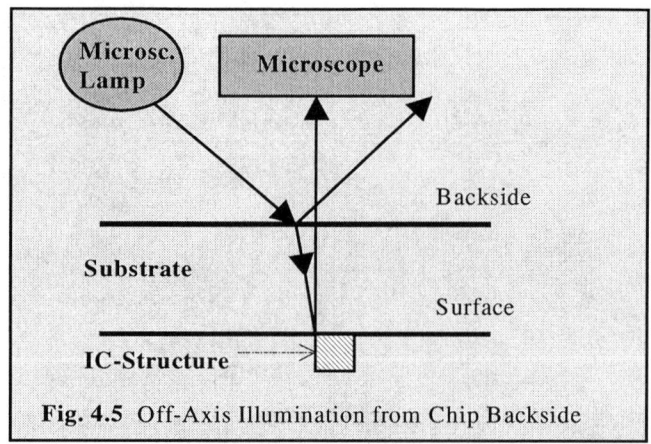

Fig. 4.5 Off-Axis Illumination from Chip Backside

An II offers a gating opportunity by externally driven switching of the high voltage. Rise times down to the nanosecond range can be achieved but high repetition rates may decrease the rise time due to the physical limits of the high voltage generator. Spectral sensitivities depend on the photocathode (names and construction depend on the vendor, but the properties are similar): GEN III is the most sensitive in the 500-800nm range and makes it best suited for GOX currents but insensitive to bandgap related phenomena. GEN III NIR extends its sensitivity into the 1000-1100nm range but is there less sensitive than a CCD, and in the range below 900nm is less sensitive than a regular GEN III. Other detectors with spectral ranges further into the infrared are not included here as they need liquid nitrogen cooling or have other properties which complicate their routine application in a failure analysis laboratory.

Noise is a different effect in IIs and CCDs. CCDs have a dark current which is the noise level after long integration time, whereas the noise of IIs raises with the signal and undergoes multiplication. In Fig.4.3 the sensitivity of CCD is compared with the IIs on a signal to noise ratio (SNR) by assuming a typical noise level at 5s integration time. CCD quantum efficiencies are higher but in PEM SNR determines the effective sensitivity.

4.4) Backside Inspection

Multi level metallization, packaging innovations or flip-chip technology are obstacles for PEM or other localization techniques, even direct probing from the top of the chip. Silicon is transparent above 1100nm spectral range due to its bandgap. This spectral part of photoemission is accessible also from the back side of the chip which makes PEM in the future an even more valuable tool. However, there are some challenges:

1) The reflection micrograph is hard to get, the microscope lamp is comparably dim in this range and the reflection from the backside surface is an enormous source of noise. Perfect polishing helps, Laser Scan Microscopes deliver acceptable results but are not easy to integrate into PEM (at least with high WD and macro objective) and an off-axis illumination is on the market that keeps reflection from the backside out of the optical path and results in a reflection image from the IC structures similar to a darkfield image (Fig.4.5)

2) The only routine detector applicable to backside investigations is the CCD due to the spectral properties required (Fig.4.6), which may need even more sensitivity (backside-illuminated, 512x512 pixels) with the disadvantage of reduced lateral resolution.

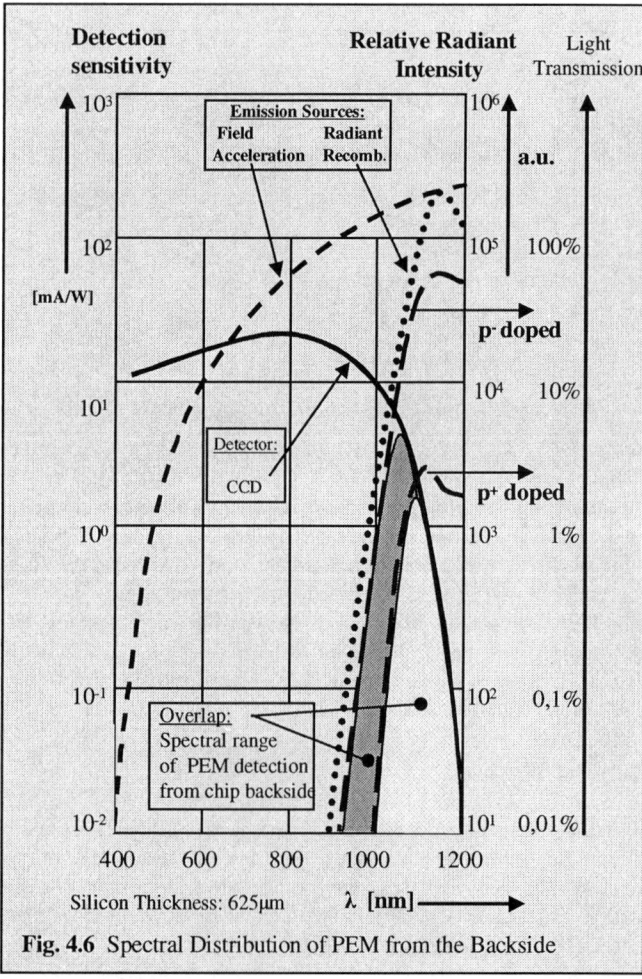

Fig. 4.6 Spectral Distribution of PEM from the Backside

3) Heavy doping of the substrate reduces transmission even above 1100nm seriously (bandgap-narrowing and absorption of free carriers). Such material needs to be thinned down to ca. 50µm to assure effective PEM. For every substrate doping level, the maximum bulk silicon thickness that still provides good backside

PEM results, should be determined. Thinning and polishing of the chip backside are challenges for the mechanical stability of the chip and can harm the electrical functionality of the device.

4) The lateral resolution of backside imaging is reduced because of infrared light, the above mentioned different CCD properties and problems of the optical path in silicon.

5) As the bulk silicon is acting like an optical filter that cuts off everything below 1000nm wavelength and the CCD like an optical filter that cuts off light above 1200nm, spectral analysis would not gain much information.

For this aspect, CCDs with an extended spectral range into the IR regime could open up not only new ways to gain a stronger signal for PEM from the chip backside but also a different set of spectral information to classify emission sources.

This topic is discussed in detail by 'Flip chip and backside techniques' in this book.

4.5) PEM Docked to IC-Tester

The localization power of PEM can be directly applied to the failing test pattern of the IC if PEM merges with the tester. Light tight coupling of the microscope to the test head and stroboscopic gating of the signal with high repetition rates can be essential for the results.

Light tight coupling to the test head requires meticulous construction efforts. For most of the common testers docking stations are commercially available.

Stroboscopic gating is optionally available with image intensifiers as described in chapter 4.3. The rise time / repetition rate trade off is usually no severe problem with a pretrigger-driven gating as long as the emission evidence of the desired test signal is just requested. Time resolution of the signal itself in nanosecond range or even better requires a more complicated detector system with a streak camera or optical switches. For all those applications jittering control is an issue.

The advantages of CCD detection can be gained if the failing test vector can be repeated in a short pattern so that a sampling condition of the failing emission spot is detectable by CCD integration mode.

4.6) Heat Radiation

Heat does not interfere much with PEM as long as the spectral range is not extended into the IR (3.11.4, 4.3, 4.4). Black Body radiation emits in the regularly detected spectral range only at temperatures significantly higher than specified for ICs (Fig.4.7). Rarely, metallization shorts may reach this temperature range for a short time and produce an emission spot, but evidence should be verified in this case. Even if heat contributed to the signal, it is hard to determine the temperature exactly. Standardization measurements would be necessary with the detector used, and still confidence would not exceed +/- 25 K. If temperature is of interest, thermography methods (chapter 5.1) should be pursued.

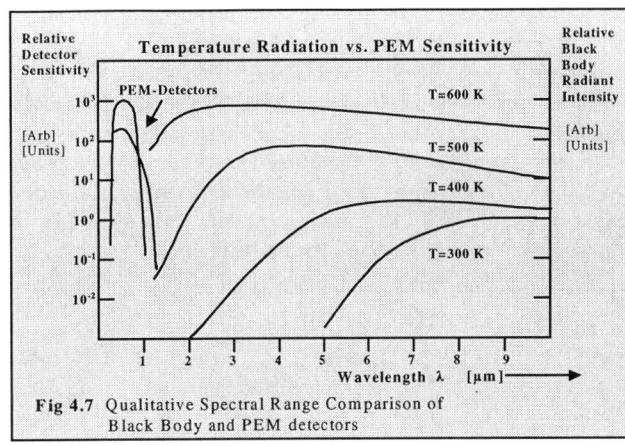

Fig 4.7 Qualitative Spectral Range Comparison of Black Body and PEM detectors

4.7) Versatility

One of the main advantages of PEM is the versatility the prober or tester remains in. There is almost no restriction to the use of micro probes. No vacuum, deprocessing or film deposition is necessary as long as light finds its way into the optical path of the microscope. A little space taken by the macro objective may reduce the number of micro probes to be positioned. In case of power devices or multilevel-metallization the emission source is sometimes covered by metal layers. Focused Ion Beam (FIB) with selective gas-assistance can help open up small areas for PEM or even E-Beam Testing, as is displayed in Fig.4.8, with maintaining electrical functionality.

Fig. 4.8 Local delayering of metallization and isolation layers with FIB. The passivation with support of XeF_2, the metal 2 with Cl_2 and again IMD with XeF_2

5) INTERFACE TO RELATED TECHNIQUES

5.1) Thermography

- Liquid Crystal (LCT)

LCT has been for years and may still be the most widely spread IC failure localization technique. With an electrical prober at hand, the method can be implemented at almost no cost. Following Fig.5.1, filters for polarizing the light of the microscope lamp, analyzing the reflected image signal before it reaches eyepiece or TV-camera and a precise hot chuck, is all the necessary hardware equipment. In other words, it is easy to equip an emission microscope with LCT. The thermo-sensitive liquid crystal is brushed on the IC surface as a film.The liquid crystal has a critical Temperature T_c -the value is depending on its composition-, at $T<T_c$ the polarization angle of the microscope light is turned. The hot chuck is driven to a temperature right below of T_c. Hot spots in the IC exceed T_c and appear dark in a bright environment or vice versa. The sensitivity is limited by the minimal temperature step of the chuck (10-100mK).

Mechanical Probe Station: Hot Spot Detection

Fig. 5.1 Setup for Liquid Crystal Thermography

Photoemission spots can be hot as well and show up in both techniques (PEM is more sensitive by orders of magnitude). LCT can add value to PEM because metallization failures that are blind spots for PEM can be localized with LCT. The lateral resolution depends on the skills of the engineer, how thin the film is deposited and how inventive a pulsing signal is created. LCT is hard to ban on a micrograph properly. The best achievable resolution is in the micron range, but it is hard to rely on.

Close to wire bonds the film thickness changes strongly. Thus, I/O- or ESD protection devices are hard to analyze.

LCT delivers a temperature step function. A temperature profile can only be gained by overlay of many LCD images at varying chuck temperatures. It is time consuming and complicated.

- Fluorescence Film (FMI / FMT)

The abbreviation is not fixed yet - **M**icro**T**hermography or **M**icrothermal **I**maging- which indicates that the technique is not yet well established. But it is powerful as the resolution and the sensitivity are only limited by the detector, i.e. improvable. These days both properties reach or exceed LCT already, temperature profiles are produced inherently and there is no dependence on hot chuck accuracy. A fluorescent film is spinned on the IC (sensitive temperature range depends on composition), the film is exposed to UV light and fluorescent at 612nm. The quantum efficiency of the

fluorescence is temperature dependent. So, a hot and a cold image must be overlaid, the two must be divided and the result is the thermographic profile. As Fig. 5.2 shows, two bright images divided produce the net signal which is by orders of magnitude smaller. Successful FMI requires CCD cameras of at least 8 bit sensitivity. Again, the setup for this method is based on an emission microscope (for example Fig.5.3) or an LSM (see next chapter).

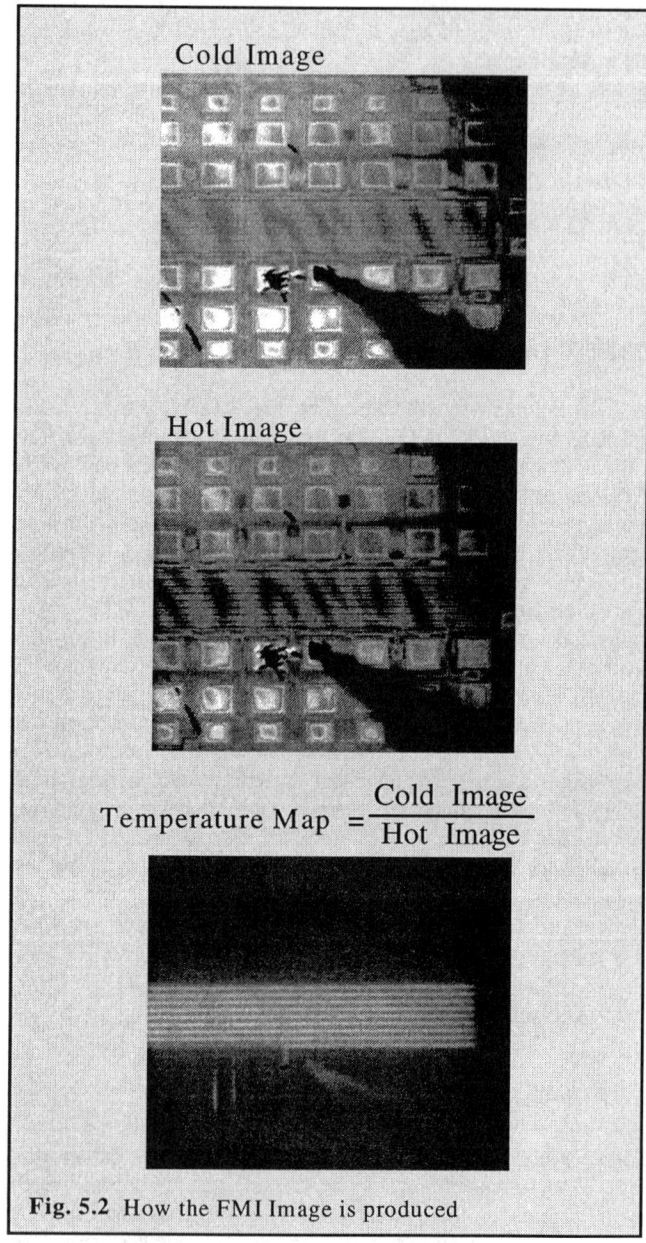

Cold Image

Hot Image

$$\text{Temperature Map} = \frac{\text{Cold Image}}{\text{Hot Image}}$$

Fig. 5.2 How the FMI Image is produced

Details of this method are discussed in a separate contribution to this desk reference. As final impression may serve Fig.5.3 which demonstrates the power of FMI with temperature profiles at high sensitivity of electromigration test structures (resolution 6 mK).

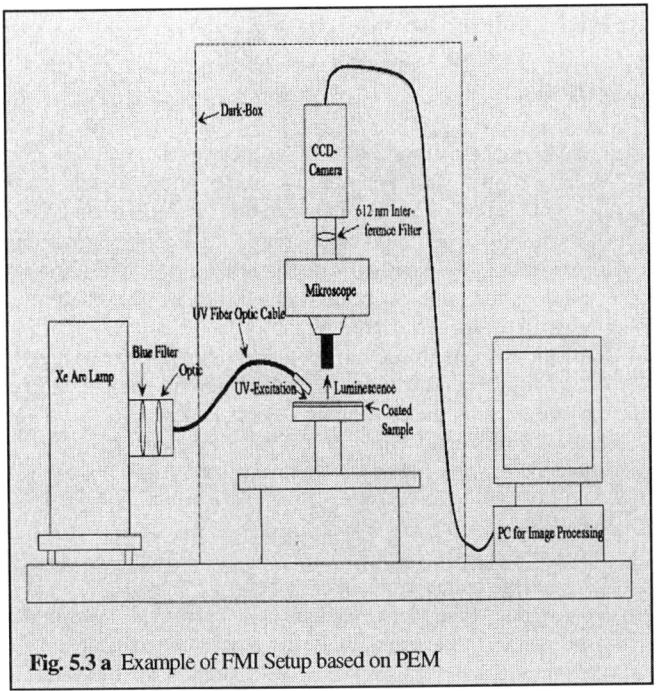

Fig. 5.3 a Example of FMI Setup based on PEM

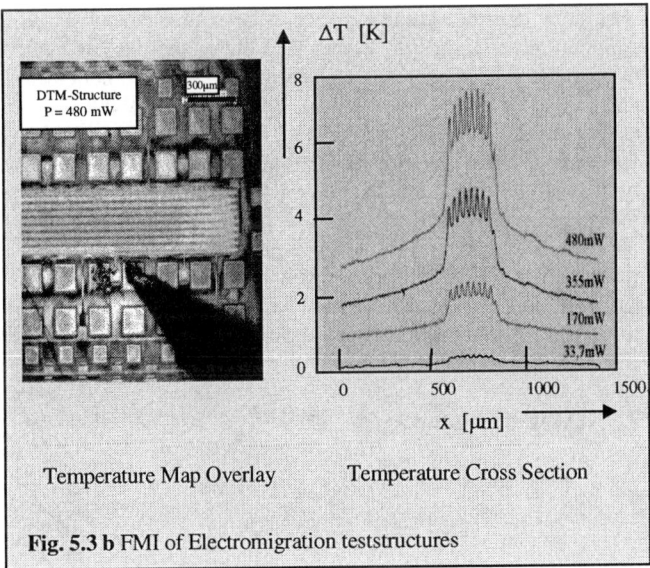

Temperature Map Overlay   Temperature Cross Section

Fig. 5.3 b FMI of Electromigration teststructures

5.2) Optical Beam Techniques

With the advantage of sub-0.5μm resolution of confocal microscopy, in a Laser Scan Microscope (LSM) an optical (laser-) beam probe creates electron-hole pairs in silicon. An electrical field in the neighborhood (a few diffusion lengths) separates the pairs and creates an optical beam induced current (OBIC) or, in a highly resistive circuit a photo voltage, like a solar cell (Fig.5.4).

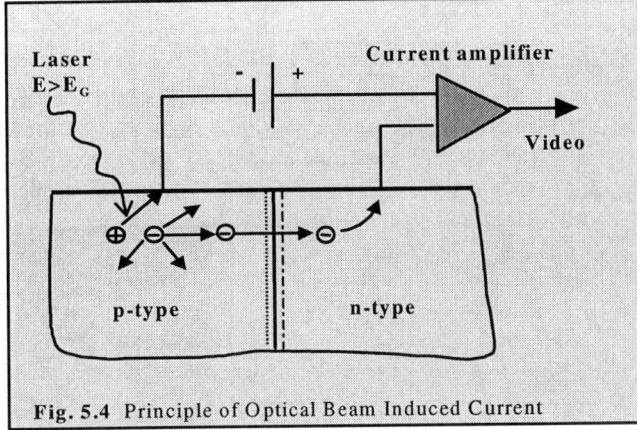

Fig. 5.4 Principle of Optical Beam Induced Current

These effects are very close to PEM. In fact, a much smaller electrical field is sufficient to produce an OBIC signal than is necessary for F-PE, because the carriers only need to be separated not accelerated. GOX-breakdowns can be localized much more accurate utilizing OBIC than PEM and with a much smaller breakdown current and there is no danger like broadening the breakdown location by PEM detection (see Fig.5.5 for comparison). This holds at least for test structures - in an IC it is often hard or impossible to extract the small photocurrents. OBIC can turn on latch up-sensitive systems by contributing a base current via minority carrier diffusion (see chapter 3.9). A setup for this purpose is presented in Fig.5.6. A latch up triggered at one scanning spot must be maintained even after switched off laser beam to prove thyristor behavior. Before the scanner moves on to the next spot to be checked for latch up sensitivity, the thyristor is switched off purposely. LSM can be equipped with PEM as well, but without the advantage of the macro objective (chapter 4.2).

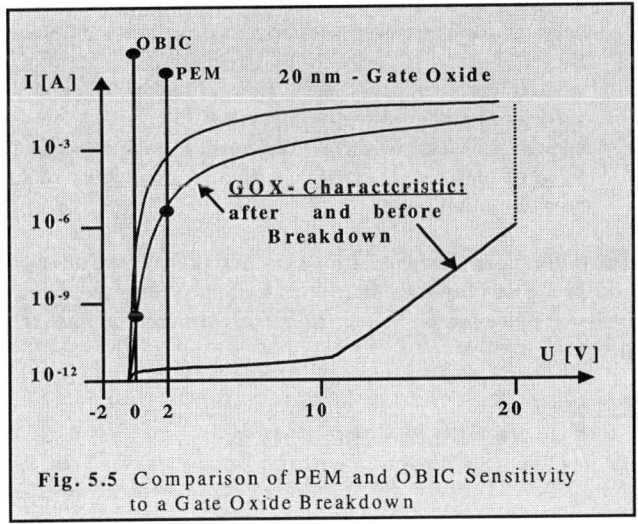

Fig. 5.5 Comparison of PEM and OBIC Sensitivity to a Gate Oxide Breakdown

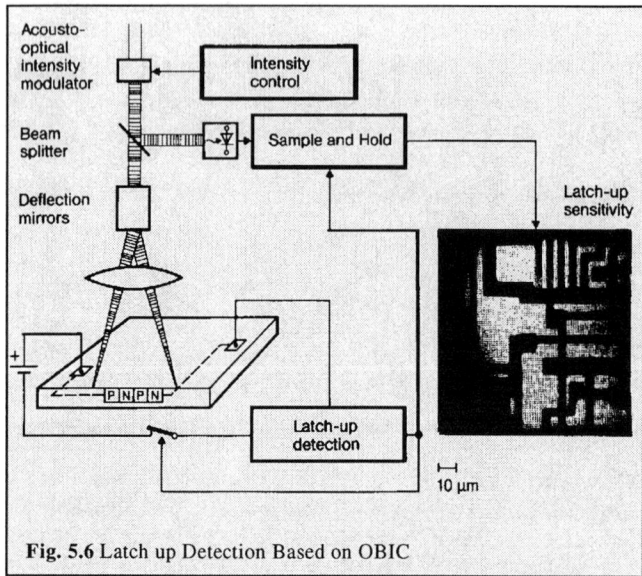

Fig. 5.6 Latch up Detection Based on OBIC

There are interesting further techniques based on LSM like LIVA and OBIRCH which can only be recited here but are completing the application range of PEM. Also FMI can be improved if based on LSM instead on PEM. The potential of LSM for failure localization seems not to be exhausted yet.

6) SUMMARY

Successful failure localization with PEM requires light emission spots generated in silicon. The key issue is a locally concentrated high current density that creates:

- **a local electric field to accelerate a sufficient number of carriers for scattering accompanied by light emission in the PEM-sensitive range (GOX-Breakdown),**
- **enough carriers in a space charge region to certify a sufficient number of scattering events** or
- **enough carriers to diffuse as minority carriers in the forward bias case to certify a sufficient number of radiant recombination events.**

The failure analyst should always try and understand a possible and probable reason in the circuit that can cause the detected emission phenomenon before further preparation because these processes are final.

For Starters:

Check out and verify PEM phenomena on test structures first, they are more flexible in terms of electrical parameters. Find out about thresholds for emission signals, forward bias / reverse bias effects, about their shapes on the screen etc., vary frequencies, temperatures and compare to alternate techniques.

This is helpful to get a feeling which parameter needs some more or some less stress to receive the desired localization result or how far from destruction you can operate your device until an emission signal can be recorded. With those experiments in mind, it is much easier on the failing chip to squeeze out the much more limited frame of electrical variability and get the highest possible success rate in localizing failures with PEM. There are not many failures in ICs that are inaccessible to PEM.

7) REFERENCES

7.1) History

Electroluminescence from active silicon devices has been observed in the early 1950's. Back then, reverse bias light emission has been rather an indication of obstructive operation (avalanche mode) than something useful to get information from. As a diagnostic tool it has been utilized in the early 60's for power devices. Forward-bias recombination radiation has been recorded with S1-photomultipliers because it indicates the degree by which the middle layer of pin-diodes or thyristor is flooded by an electron-hole plasma. Laterally resolved measurements could detect the carrier diffusion length that the forward bias resistance of those devices is depending on. Time-resolved measurements allowed exact calculations how, in the switching period from forward to reverse bias, this plasma gets removed from the basis and the SCR starts to penetrate the device - a dangerous moment for a power device because there is still high current to remove the carriers from the base and already the high blocking voltage on the terminals.

These methods have been meticulously developed through the seventies and even the eighties. Power devices offered the amount of light and the geometry to comply with detector sensitivities and the resolution of the optical paths available by them. In the eighties, integration density and the victory of MOS technologies led to light emission observations of some IC failure analysts by looking through the microscopes of their probe stations due to hot electrons and leakage currents. At the same time, night vision technology offered sensitive image intensifiers. Bringing together these two approaches, today's photoemission microscopy has been presented by N.Khurana and C.L. Chiang 1986. Until 1990, semiconductor industry checked out the range of application and the easiness of operation of the method with a still small number of those machines running. Most of the information how to understand the phenomena has been collected in that period of time. From then on, with the news spread and the tools optimized, volume application started with the full potential and range of use still not exhausted yet.

7.2) Acknowledgements

The author wants to acknowledge
- Dr. Edward I. Cole of Sandia National Laboratories, Albuquerque NM for contributing the figures on the correlation of FET substrate current and photoemission,
- Dr. Karolin Bernhard-Hoefer, Siemens Munich Germany for her multiple contributions, especially on PEM inspection from chip backside,
- Mrs. Wiltrud Golombek and Mr. Eric F. Dulkeith, Siemens Munich Germany for word processing and graphics.

7.3) Literature

This is only a selected list of publications about PEM. Many case histories and reports of PEM events are published in Proc. of the Int. Symp. for Test. and Fail. Analy. (ISTFA) from 1989 on.
Some principal aspects are discussed in more detail in: "Quantitative emission microscopy (J.Koelzer, C. Boit, A. Dallmann, G.Deboy, J. Otto, D. Weinmann) J. Appl. Phys. 71 (11), 1 June 1992, R23-R41.

Classification of Emission Sources

/01/ Mayer, D.C.; Ferro, R.J.; Leung, D.L.; Dooley, M.A.; Scarpulla, J.R.; "Application of photoemission microscopy and focused ion beam microsurgery to an investigation of latch-up", ISTFA '96, Proceedings of the 22nd International Symp. for Testing and Failure Analysis, p. 47-51

/02/ Inuzuka, E.; Suzuki, H.; "Emission microscopy in semiconductor failure", Conf. Proc. 10th anniversary, IMTC/94, Advanced Technologies in I&M, 1994 IEEE Instrumentation and Measurement Technology Conference

/03/ Golijanin, D., "Photoemission microscopy – A novel technique for failure analysis of VLSI silicon integrated circuits", X-Ray Optics and Microanalysis 1992, Proc. of the 13th Int. Congress, UMIST, Inst. Of Physics, 1992, p. 465-468

/04/ J.T. May and G. Misakian, "Failure analysis of a turn-on degraded transistor using photoemission microscopy", Proc. Of the Int. Symp. for Test. and Fail. Analy., pp. 69-71, 1990

/05/ G.F. Shade, "Physical mechanisms for light emission microscopy" Proc. ISTFA '90, 1990, pp. 121-128

/06/ A. Dallmann and G. Deboy, "Characterization of Trench-Trench Punch-Through Mechanisms by Emission Microscopy," Proc. ESREF '90, 1990, p. 69-76

/07/ D. Weinmann, C. Boit and J. Koelzer, "Characterization of leakage currents by emission microscopy,", Proc. ESREF '90, 1990, pp. 61-68

/08/ C. Boit, J. Koelzer, H. Benzinger, A. Dallmann, M. Herzog, and J. Quincke, "Discrimination of parasitic bipolar operating modes in ICs with emission microscopy," Proc. IEEE/IRPS '90, 1990, pp. 81-85

/09/ Y. Uraoka et al., "Evaluation technique of gate oxide reliability with electrical and optical measurements," in IEEE ICMTS, Vol. 2, No. 1, p. 97, 1989

/10/ Brahme, U.; Li, D.; "Analysis of device gate oxide problems by photoemission microscopy," Materials Letters (1989) vol. 9, No. 1, p. 10-13

/11/ T. Aoki and A. Yoshii, "Analysis of Latch-up induced photoemission," Proc. IEEE/IEDM '89, 1989, pp. 281-284

/12/ S.C. Lim and E.G. Tan, "Detection of Junction Spiking and its Induced Latch-up by Emission Microscopy," IEEE Annual Proc. of the Reliability Physics Symp., pp. 119-125, 1988

/13/ M. Herzog and F. Koch, "Hot carrier light emission from silicon metaloxide semiconductor devices," Appl. Phys. Lett., Vol. 53, 1988, pp. 2620-2622

/14/ Lim, S.C.; Tan, E.G.; "Detection of junction spiking ant its induced latch-up by emission microscopy", 26th Annual Proceedings, Reliability Physics 1988, 12-14 April 1988, Monterey, pp. 119-125

/15/ N. Khurana and C. Chiang, "Dynamic imaging of current conduction in dielectric films by emission microscopy," Proc. of Int. Rel. Phys. Symp., pp. 72-76, April 1987

/16/ A. Toriumi, M. Yosimi, M. Iwase, Y. Akiyama and K. Taniguchi, "A study of photo emission from n-channel MOSFETs," IEEE Trans. on Electron Dev., Vol. ED-34, No. 7, pp. 1501-1508, July 1987

/17/ N. Khurana and C.L. Chiang, "Analysis of Product Hot Electron Problems by Gated Emission Microscopy," IEEE Annual Proc. of the Reliability Physics Symp., pp. 189, 1986

/18/ C.L. Chiang and N. Khurana, "Imaging and Detection of Current Conduction in Dielectric Films by Emission Microscopy," IEEE International Electron Devices Meeting, pp. 672-675, 1986

/19/ T. Tsuchiya and S. Nakajima, "Emission Mechanism and Bias-Dependent Emission Efficiency of Photons Induced by Drain Avalanche in Si MOSFETs," IEEE Transactions on Electron Devices, Vol. ED-32, No. 2, Feb. 1985, pp. 405-412

IC-Functionality - PEM

/01/ Adams, T.; "Emission microscopes reveal IC defects", Test & Measurement World (1995) vol. 15, no. 2, p. 51-2

/02/ Finotello, A.; Gallesio, A.; Marchisio, L.; Riva, D.; "Emission microscopy: a powerful tool for IC's design validation and failure analysis", Proc. Of the 18th Int. Symp. f. Testing and Failure Analysis, p. 289-293, 1992

/03/ C. Khandekar, M. Hennis, P. Brownell and D. Bethke, "Photoemission detection of vendor integrated circuit failures during system level functional testing," Proc. of the Int. Symp. for Test. and Fail. Analy., pp. 73-79, 1990

/04/ C. Hawkins, J. Soden, E. Cole and E. Snyder, "The Use of Light Emission in Failure Analysis of CMOS IC's," ISTFA '90, pp. 55-67

/05/ R. Lemme, M. Gentsch, R. Kutzner, H. Wendt and H. Haudek, "Defect analysis of VLSI dynamic memories," Proc. ISTFA '89, 1989, pp. 9-14

PEM for ESD Failures

/01/ Salome, P.; Leroux, C. Chante, J.P.; Crevel, P.; Reimbold, G,; "Study of a 3D phenomenon during ESD stresses in deep submicron CMOS technologies using photon emission tool", 1997 IEEE International Reliability Physics Symposium Proceedings, p. 325-3

/02/ C. Boit, J. Koelzer, C. Stein, J. Otto, H. Benzinger and M. Kreitmaier, "Characterization of Field and Diffusion Currents in ICs with Emission Microscopy" Proc. CERT '90 (Component Engineering, Reliability and Test Conference), 1990, p. 110-114

/03/ M. Hannelamm and A. Amerasekera, "Photon emission as a tool for ESD failure locaization and as a technique for studying ESD phenomena," Proc. ESREF '90, pp. 77-84

/04/ K.S. Wills, S. Vaughan, Charvaka Duvvury, O. Adams and J. Bartlett, "Photoemission Testing for EOS/ESD Failures in VLSI Devices: Advantages and Limitations,"Proc. ASM Intern. Symposium for Testing and Failure Anal., p. 183, 1989

/05/ K.S. Wills, C. Duvvury and O. Adams, "Photoemission testing for ESD failures, advantage and limitations," Proc. of Electrical Overstress/Electrostatic Discharge Symp., pp. 53-61, September 1988

/06/ N. Khurana, T. Maloney and W. Yeh, "ESD on CHMOS devices, equivalent circuits, physical models and failure mechanisms," Proc. IEEE/IRPS '85, 1985, pp. 212-223

PEM Setup

/01/ Nothnagle, P.E.; Zinter, J.R; Ruben, P.L.; "Macro lens for emission microscopy", SPIE-Int. Soc. Opt. Eng: 1995, vol. 2537, p. 13-26

/02/ An Economical Approach to Correlation of Light Emission Microscopes, K.M. Baker, Proc. of the Int. Symp. for Test and Fail Analy. 1991

/03/ N. Khurana, "Second generation emission microscopy and its application," Proc. ISTFA '89, 1989, pp. 277-283

Spectral Analysis

/01/ M. Rasras; I. De Wolf; G. Groeseneken; H.E. Maes; S. Vanhaeverbeke; P. De Pauw; "A Simple Cost Effective, and Very Sensitive Alternative for Photon Emission Spectroscopy"; 23rd ISTFA 1997, p. 153-157

/02/ Tao, J.M; Chim. W.K.; Chan, D.S.H.; Phang, J.C.H.; Liu, Y.Y.; "A high sensitivity photon emission microscope system with continuous wavelength spectroscopic capability [semiconductor device failure analysis]", 1996 IEEE Intern. Reliability Physics Proceedings,34th Annual, p. 360-5

/03/ Golijanin, D.; ed. Kenway, P.B.; Duke, P.J.; Lorimer, g.W.; Mulvey, T.; Drummond, I.W.; Love, G.; Michette, A.G.; Stedman, M.; "Photoemission microscopy -a novel technique for failure analysis of VLSI silicon integrated circuits", X-Ray Optics and Microanalysis 1992, p. 465-8

/04/ Bruce, V.J.; "Energy resolved emission microscopy" IEEE 1993, p. 178-83

/05/ K.S. Wills, D. Depaolis and G. Billus, "Advanced photoemission technique for distinguishing latch-up from logic failures on CMOS devices," Proc. of the Int. Symp. for Test. and Fail Analy., pp 335-341, 1991

/06/ T. Wallinger; "Characterization of Device Structure by Spectral Analysis of Photoemission" , Proc. of the Int. Symp. for Test and Fail. Analy. 1991

/07/ K.S. Wills, et. al. "Spectroscopic Photoemission Techniques to Understand Microprocessor Failures," Third International Symposium of the Physical and Failure Analysis of Integrated Cicuits-Pan Pacific Singapore, 1991

/08/ N.C. Das and B.M Arora, "Luminescence Spectra of an N-channel Metal-Oxide Semiconductor Field Effect Transistor and Breakdown," Applied Physics Letters, Vol 56, No. 12, pp. 1152-1153, (19 March 1990)

/09/ K. de Kort and P. Damink, "The spectroscopic signature of light emitted by integrated circuits" Proc. ESREF, 1990

/10/ N. Tsutsu et al., "New detection method of hot-carrier degradation using photon spectrum analysis of weak luminescence on CMOS/VLSI," IEEE ICMTS, Vol 3, No. 1, p. 143, 1990

/11/ H. Ishizuka, M. Tanaka, H. Konishi and H. Ishida, "Advanced method of failure analysis using photon spectrum of emission microscopy. " Proc. ISTFA '90, 1990, pp. 13-19

/12/ A. Toriumi, M. Yoshimi, M. Iwase and K. Taniguchi, "Experimental Determination of Hot Carrier Energy Distribution and Minority Carrier Generation Mechanism Due to Hot-Carrier Effects," IEEE International Electron Devices Meeting, pp. 56-59, 1985

/13/ S.D. Borson, D.J. Di Maria, M.V. Fischetti, F.L. Pesavento, P.M. Solomon and D.W. Dong, "Direct measurement of the energy distribution of hot electrons in silicon dioxide," J. Appl. Phys. Vol. 58, 1985

/14/ T.N. Theis, J.R. Kirtley, D.J. Di Maria and D.W. Dong, "Spectroscopic Studies of Electronic Conduction of SiO2" in Insulating Films on Semiconductors, J.F. Verweij and D.R. Wolters (ed.), North-Holland, pp 134-140, 1983

PEM Inspection from Chip Backside

/01/ S.-S. Lee; J.-S. Seo; N.-S. Cho; S. Daniel; "Application of Backside Photo and Thermal Emission Microscopy Techniques to Advanced Memory Devices", Proc. ISTFA 1997, pp. 63-66

/02/ K. Naitoh; T. Ishii; J.I. Mitsuhashi; "Investigation of Multi-Level Metallization ULSIs by Light Emission from the Back-Side and Frontside of the Chip", Proc. ISTFA 97, pp.145-151

/03/ Wu, N.M.; Tang, K.; Ling, J.H.; "Back side emission microscopy for failure analysis", SPIE-Int. Soc. Opt. Eng: 1996, vol 2874, p. 238-47

/04/ Vallett, D.P.; "An overview of CMOS VLSI, failure analysis and the importance of test and diagnostics", Proceedings International Test Conference 1996, p. 930

/05/ Daniel L. Barton et al.; "Infrared Light Emission from Semiconductor Devices", Proc. ISTFA 96, pp. 9

PEM on IC Testers

/01/ J.A. Kash; J.C. Tsang; " Dynamic Internal Testing of CMOS Circuits Using Hot Luminescence"; IEEE Electron Device Letters, Vol. 18, N0. 7, 1997, pp. 330-335

/02/ Van Doorselaer, K.; Swerts, U.; Van Den Bempt, L.; "Broadening the use of emission microscopy", Proc. ISTFA '95, p. 57-65

Related Techniques

Liquid Crystal

/01/ S. Ferrier, "Thermal and Optical Enhancements to Liquid Crystal Hot Spot Detection Methods", Proc.of the 23rd Intern. Symp. for Test. and Fail. Analys. (ISTFA) 57-62, 1997

Thermography: FMT/FMI

/01/ C. Herzum, C. Boit, J. Koelzer, J. Otto, R. Weiland, "High Resolution Temperature Mapping of Microelectronic Structures Using Quantitative Fluorescence Microthermography", THERMINIC, 1996, Budapest, Hungary, Microelectronics Journal 29, 1998, pp. 163-170

/02/ D.L. Barton, "Fluorescent Microthermographic Imaging", Proc. of the 20th Intern. Symp. for Test. and Fail. Analys. (ISTFA) 87-95, 1994

/03/ Cole, E.I.; Jr.; Soden, J.M.; Rife, J.L.; Barton, D.L.; Henderson C.L.; "Novel failure analysis techniques using photon probing with a scanning optical microscope", IEEE 1994, p. 388-98

/04/ Hamilton, D.K.; Wilson, T.; "Infrared sub-band-gap photocurrent imaging in the scanning optical microscope of defects in semiconductor devices", Micron and Microscopica Acta (1987) Bd. 18, No. 2, p. 77-80

Optical Beam Techniques

/01/ J. Quincke, C. Boit, D. Fuehrer, "Electroluminescence measurements with temporal and spatial resolution for CMOS latch-up investigations", Proc. 2nd European Conference on EOBT, Oct. 1989, Duisburg, Germany, in Microelectronic Engineering, Vol. 12, 1-4, May 1990, p. 157-162

/02/ J. Quincke, E. Plies, J. Otto, "Circuit Analysis in ICs Using the Scanning Laser Microscope", SPIE Vol. 1028, Scanning Imaging (1988), p. 211-216

History

/01/ K. Penner, "Electroluminescence from silicon devices - a tool for device and material characterization," Journal de Physique, Coll. c4, Suppl. 9, Tome 49, 1988, pp. 797-800

/02/ C. Boit, "Quantitative evaluation of High Injection effects in forward biased pin-diodes by means of recombination radiation measurements" Ph.D. Thesis, Technical Univ. Berlin, Germany 1987

/03/ S. Tam and C. Hu, "Hot Electron-Induced Photon and Photocarrier Generation in Silicon MOSFETs, " IEEE Transactions on Electron Devices, Vol. ED-31, No. 9, Sept. 1984, pp. 1264-1273

/04/ S. Tam, F. Hsu, P. Ko, C. Hu and R. Mueller, "Spatially resolved observation of visible-light emission from Si MOSFETs," IEEE Electron Dev. Letters, Vol. EDL-4, No. 10, pp. 386-388, October 1983

/05/ F. Berz and J.A.G. Slatter, "Solid State Electronics Vol. 25, No.8, p. 693 (1982)

/06/ H. Schlangenotto, H. Maeder, W. Gerlach, phys. Stat. Sol (a) 21, p. 357, (1974)

/07/ F. Dannhäuser, J. Krausse: Solid State Electronics Vol. 16, (1972), pp. 861-873

/08/ T. Figelski and A. Torun, "On the Origin of Light Emitted from Reverse Biased p-n Junctions," Proceedings of the International Conference in the Physics of Semiconductors, 1962, pp. 853

/09/ L.W. Davies and A.R. Storm, Jr., "Recombination radiation from silicon under strong-field conditions," Phys. Rev. Vol. 121, 1961, pp. 381-387

/10/ A.G. Chynoweth et al., "Photon emission from avalanche breakdown in silicon," Phys. Rev., Vol 102, No. 2, p. 369, 1956

/11/ R. Newmann, Dash, Hall and Burch, "Visible Light From a Si p-n Junctions," Phys. Rev., Vol. 100, p. 700, 1955

/12/ W. van Roosbroeck and W. Shockley, "Photon-radiative Recombination of electrons and holes in germanium," Phys. Rev., Vol. 94, 1954, pp. 1558-1560

/13/ R.N. Hall , Phys. Rev., Vol. 83, p. 228, 1951 and Vol. 87, p. 387, 1952

Compound Semiconductors

/01/ Roesch, W.J.; "Light emission as an analysis tool for GaAs ICs", III-Vs Review, 1997, vol. 10, no. 1, p. 24-7

/02/ E. Zanoni, S. Bigliardi, R. Cappelletti, P. Lugli, F. Magistrali, M. Manfredi, A. Paccagnellla, N. Testa and C. Canali," Light emission in AlGaAs/GaAs HEMT's and GaAs MESFET's induced by hot carriers," IEEE Electron Device Letters, Vol. 11, 1990, pp. 487-489

Deprocessing and Sample Preparation

Wet Chemical
Deprocessing Techniques

S. Perungulam and Kendall Scott Wills

INTRODUCTION

The scaling down of device sizes to achieve higher speeds and densities has led to an increase in the complexity of the semiconductor processes. As a result, the failure analysis of these products is getting increasingly complex and challenging. Physical failure analysis of ultra large-scale integrated circuit (ULSI) chips is now a greater challenge as the margin for error is shrinking. In a typical Failure Analysis (FA) flow, the failure sites are isolated using one or more of the following techniques: electrical testing, non-destructive evaluation, fail site isolation techniques and visual inspection. As a final step in the FA process, the device is stripped back to the location of the defect to determine the failure mechanism and the root cause of failure. Deprocessing is usually the last step in FA as there is no going back once the device is deprocessed. Typical Integrated Circuit (IC) chips have several layers of metals and oxides on top of the transistors patterned on the Silicon substrate. These different layers can be deprocessed using either dry or wet etching techniques or a combination of both wet and dry etching.

SELECTIVITY AND ANISOTROPY

One of the most important properties of an etch recipe is its selectivity. Selectivity can be defined as the ratio of the rate of etching of the material being removed to the rate of etching of the material around it. For instance, if the purpose of a particular etch is to remove the metal layer, the metal to oxide selectivity must be high, i.e. the etch rate for the metal must be much higher than for the oxide. Selectivity of A to B is given by:

$$S_{AB} = V_A/V_B = \text{etch rate of A/etch rate of B}.$$

To ensure accurate pattern definition during the photolithography process and for uniform coverage of the conducting layers, the dielectric layers have to be planarized. This can be achieved by resist etchback, spin on glass (SOG), Chemical Mechanical Planarization/ Polishing (CMP) or a combination of these techniques. Planarization is also achieved by either a reflow of the dielectric (the reflow temperature can be reduced by the addition of dopants such as Boron and Phosphorous). With the increasing use of planarization processes in wafer fabrication, there is a large variation in the layer thicknesses, especially in the dielectrics. It is very important for the etch to be highly selective against underlying layers as a certain degree of overetch is essential for complete removal of the dielectric.

Another important parameter in the etching process is the degree of anisotropy. A purely anisotropic etch is unidirectional. An isotropic etch has the same rate of etching in all directions. For instance, ion beam milling has a degree of Anisotropy close to 1.
The degree of Anisotropy A is given by:
$$A = 1 - V_l/V_v$$
where V_l and V_v are the lateral and vertical etch rates.

Dry etching processes have varying degrees of anisotropy. Ion milling has $A \sim 1$ while RIE etching is more isotropic. Wet chemical etch processes often have $A \sim 0$.

Yet another parameter often used to compare different etches is the Etch Bias. Etch Bias is an indication of the extent of undercut of a masking material M as material A is etched (Fig.1). The etch bias is then defined as:
$$\text{Bias } B = l_m - l_a = (\text{undercut}) \times 2.$$
Anisotropy can now be defined as $A = 1 - B/2t$, where t is the thickness of the feature just etched to completion (i.e. $t = V_v \times$ time of etching).

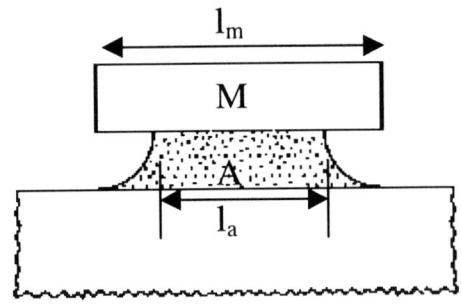

Figure 1: Schematic diagram showing the etch bias.

Dry versus wet etching

Dry etching techniques such as Plasma Etching and Reactive Ion Etching can have a high degree of anisotropy. An advantage of these techniques is that the etch rate can be controlled by regulating various parameters such as the gas flow rate, power and temperature. Another plus is that end-point detection can, in principle, be implemented. On the other hand, these dry etching methods irradiate the devices with high energy electrons, photons and ions, that can cause damage to the devices. Dry etching techniques use Trifluoro methane (CHF_3), Carbon Tetrafluoride (CF_4), Sulphur hexafluoride (SF_6), Boron trichloride (BCl_3) and Chlorine (Cl_2), among other gases. Several of these gases are corrosive and/or toxic. In addition, the selectivity for several dry etching processes is not high as compared to wet chemical techniques. In many cases, dry etch processes can leave debris after an etch. The debris can be reaction products, sputtered material being redeposited (gold is a typical example), polymers, material from the chamber, etc. This may necessitate a reduction in the power level, the use of special fixtures or the use of exotic cleaning procedures, both wet and dry.

Ion milling is a highly anisotropic technique and can replicate patterns with a high degree of accuracy. High etch rates can be achieved at the expense of selectivity and damage to underlying layers. Enhancing gases can be used to improve the etch rates and selectivity. Many of the problems with contamination are similar to those encountered with RIE.

Wet chemical etching processes typically have a higher degree of selectivity. There is no danger of damage to underlying layers as long as the process is selective. Debris left behind in most wet etch processes can be easily removed either using other solvents or by an air-gun. The equipment used for wet etches is inexpensive. There are negative aspects to wet etching. The use of acids and other chemicals can be a safety hazard. Wet etching is usually isotropic and in many cases, end point detection is difficult to implement.

Despite some of the disadvantages, wet etching is an essential tool in FA for deprocessing IC chips. The following observations illustrate this point:

- Several etch artifacts that are caused by dry etching processes can be avoided by wet chemical etching.
- As long as there is good control over etchant composition and etching times, the damage to underlying layers is minimal.
- Proper use of isotropic and highly selective etches result in effective removal of overlying layers during deprocessing.
- When deprocessing down to the substrate level, wet etching is the safe etch route as dry etch techniques can cause significant damage to the substrate.
- Wet etching techniques can achieve much higher etch rates as compared to the dry processes.
- Often, chemical reactions lead to the evolution of gases such as H_2 or CO_2. The evolved gas bubbles to the surface through the solution. This can be an effective endpoint detection tool for the etch process.

WET DEPROCESSING TECHNIQUES

SEPARATING THE DIE FROM THE PACKAGE

Removal of the plastic mold compounds: Plastic mold compounds can be dissolved in hot nitric or sulfuric acids or a mixture of the two. These mold compounds are thermosetting plastics with large cross-linked polymer chains. The optimal acid composition and temperature required to dissolve the mold compounds is a function of the composition of the compound as well as the extent of cross linking. Discussion of the different mold compounds is beyond the scope of this paper. Therefore, it is advisable to contact the manufacturer to obtain optimal conditions for removal of the mold compounds. The acids can be dispensed using the automated jet etch equipment to remove the mold compound while maintaining the electrical integrity of the device.

Removal of dies from flip chip packages: Plastic packages can be dissolved in hot nitric acid separating the die from the package. Thermal Coefficient of Expansion (TCE) mismatch between the package and the die can potentially alter failure analysis results by stressing the bump structures, causing die level failures. Swelling of the plastic during package removal can also cause die level failures.

For the ceramic packages, the heat sink is mechanically removed from the bottom of the package. Boiling the package in organic solvents such as n-methyl pyrrolidone (NMP), Ethylene Diamine Anhydrous (EDA) or Uresolve Plus* can dissolve the underfill epoxy material and separate the* die from the package.

---

* Uresolve Plus is a tradename of Dynaloy, Inc.

Removal of solder bumps and bond wires: Pb-Sn solder balls can be removed by placing the unit in hot fuming nitric acid. The acid needs to be refreshed every 5 minutes or so for effective bump removal. Typical time for removal of solder bumps is around 15 minutes.

Gold bond wires can be dissolved in aqua regia or in a 1:1 mixture of Hydrochloric and Yellow nitric acid. Aqua Regia is a mixture of nitric and hydrochloric acids. The ratio of these two acids in varies from 1:3 to 3:1. Aqua regia is used to dissolve several metals including the noble metals: Gold (Au), Platinum (Pt) and Palladium (Pd). Transition metals such as Niobium (Nb), Titanium (Ti) and Tantalum (Ta) do not dissolve in this reagent. Therefore, aqua regia has been a key ingredient in refining complex metallic ores. Another effective method of dissolution of gold wires is to use a mixture of Potassium Iodide and Iodine in DI water. Typical time for dissolution of the bond wires is 3 to 5 minutes. It is important to note that leaving the die in the etchant longer than necessary may lead to the aluminum at the pads being etched away.

Detailed descriptions of the mechanical and chemical decapsulation techniques can be found the the ASM Microelectronic Desk reference (1, 2).

DIE DEPROCESSING

Passivation and Dielectric removal

The passivation layer protects the die from mechanical damage, contamination, humidity, external stress, impurity diffusion, etc. The passivation can be composed of any of the following: Silicon dioxide, quartz, Silicon nitride, Silicon oxynitride or nitride over oxide.

Silicon Nitride: Silicon Nitride is often used as a passivating layer on ULSI chips. Normally, the passivation is removed using dry etch techniques. Hot Phosphoric acid can be used to remove the nitride chemically. The acid needs to be heated to 150°C. This etch is highly selective against oxide layers but attacks the Aluminum. Recently, Malberti and Ciappa (3) have reported a new wet etch recipe to selectively etch nitride layers. This recipe consists of 20 ml ethylene glycol, 60 ml glacial acetic acid, 12 ml 65% nitric acid and 4 gms Ammonium fluoride. The etchant has been found to work most effectively at 70°C with a stirring motion. The etching time for the passivating nitride layers is about fifteen minutes.

Intermediate oxide levels: Silicon oxide (SiOx) (stoichiometric Silicon dioxide- SiO_2), is used as a passivation layer, interlevel dielectric between metal layers, isolation oxide between polysilicon and metal and to isolate diffusion regions (as field oxide). The oxide can vary from undoped SiOx to doped phosphosilicate (PSG), Borosilicate (BSG) and Borophosphosilicate (BPSG) glasses. These doped oxides are used in planarization and as an effective block against the diffusion of Sodium ions. The presence of the dopant helps reduce the reflow temperatures of the oxide.

The limit on the Boron concentration is 4% in BSG and the limit on the Phosphorous concentration is around 8% in PSG. Doped oxides used for passivation or interlevel insulation contains 2 to 8% Phosphorous. The dopant prevents the diffusion of ionic impurities and helps reduce the reflow temperature. Higher (>7%) Phosphorous content results in the hygroscopic nature of the P_2O_5 becoming dominant, leading to the formation of Phosphoric acid and can lead to corrosion of the metal lines.

Wet etching of the oxide layers is tricky to handle as often, the 'whole' oxide layer is made up of several different oxide layers and the chemicals have a different etch rate for each oxide composition. The most effective etchant for the oxides is HF. The chemical reaction can be written as:

$$SiO_2 + 6HF \rightarrow H_2SiF_6 + 2H_2O$$

Normally, HF is buffered with Ammonium fluoride (NH_4F) for oxide etching. Common etch compositions are HF: $NH_4F :: 5:1$ or $10:1$. The time of etch is very critical as metal lines beneath the oxide, if exposed, can get corroded.

Another effective oxide etch is the etch developed by Shankoff, consisting of 67 ml ethylene glycol ($HOCH_2CH_2OH$), 5 ml 49% Hydrofluoric acid and 11 gms Ammonium fluoride in 40 ml distilled water (4). This etch is selective to aluminum but etches nitride layers at a slower rate than oxide layers.

Metal and barrier layer removal

Aluminum can be etched using either hot phosphoric acid or hot hydrochloric acid. Stirring or use of surfactants is recommended to prevent formation of a passivating layer of hydrogen gas bubbles on the metal surface. Phosphoric acid etches Aluminum at a uniform rate and is highly selective against oxide layers.

Barrier layers: A primary disadvantage of Aluminum based systems is the susceptibility to electromigration and impurity diffusion. Electromigration is basically a diffusion phenomenon under a driving force. The possible consequences of electromigration include open circuits, cracked protective coatings leading to corrosion and short circuits from metal protrusions. The high mass flux can lead to failures in metal lines and at contacts. Direct contact between W(used in the plugs) and Al can lead to embrittlement of the Al layers. Direct contact between the Al (in the metal contacts) and Si can lead to spiking in the Silicon.

Therefore, there is a need for diffusion barriers over metal layers and at contacts and interconnects. Other reasons for a barrier layer include improving adhesion properties and lowering contact resistivity. The barrier layer is usually a thin coat of a material through which the diffusion of impurities or metal atoms is low. The barrier itself must be stable in the presence of the surrounding materials at all processing temperatures. Other properties include good adhesion and low contact resistivity. Outlined below are the common barrier layers used in ULSI technology and the chemical methods for the removal of these barrier layers.

TiN/Ti: Titanium Nitride (TiN) is often used as an anti-reflection coating/ barrier layer in ULSI chips. A combination of ammonium hydroxide, hydrogen peroxide and water in even proportion can be used for the removal of TiN layers. The mixture can be heated to accelerate the dissolution of the TiN. Often, there is a thin coat of Titanium metal below the TiN. Titanium can be removed by using a mixture of nitric acid and Hydrofluoric acid in the ratio 100:1.

TiW: Titanium-Tungsten (TiW) is another common barrier layer used in ULSI. This falls under the class of stuffed barriers, wherein the grain boundaries of the barrier layer are "stuffed" with atoms that reduce the diffusion rates. TiW can be removed by hot H_2O_2. The etch rate is about 1000A/min at 75°C.

Other barrier layers include refractory metal silicides such as $MoSi_2$, $TaSi_2$, WSi_2, $TiSi_2$ and Platinum silicides ($PtSi$, $PtSi_2$, $PtSi_3$). Most of these silicides can be etched away in a mixture of HF, HNO_3 and NH_4F. The metal is oxidized and the reaction products are dissolved in HF.

Tungsten: Tungsten plugs are often used in interconnects between metal levels. A mixture of HF and HNO_3 can be used to etch tungsten (W)/Tungsten Silicide (WSi). Another effective etchant for W/WSi is a warm mixture of Hydrogen Peroxide(H_2O_2) and Ammonium Hydroxide(NH_4OH).

As mentioned earlier, noble metals such as Gold(Au), Platinum(Pt) and Palladium(Pd) can be dissolved in a mixture of nitric and hydrochloric acids, popularly known as Aqua Regia.

Polysilicon and gate oxide

Polysilicon can be removed by using a mixture of Hydrofluoric and Nitric acid buffered with acetic acid. Typical etch times are of the order of 5-15 seconds. The polysilicon is often coated with a barrier metal layer. To maintain a uniform etch process, the barrier metal must be removed before the polysilicon etch.

Gate Oxide pinhole decoration: Dilute Sodium hydroxide (NaOH) solution is sometimes used to decorate pinholes in the gate oxide. Another effective method is the use of Copper sulfate in solution with Nitric and Hydrofluoric acids to decorate pinholes. The gate oxide can be removed using buffered HF.

Substrate Silicon

The substrate can be etched using a mixture of nitric, hydrofluoric and acetic acids. The etching involves an oxidation-reduction (REDOX) reaction followed by the dissolution of the oxidized product.

The nitric acid acts as the oxidant and the oxidized product is dissolved in hydrofluoric acid. The chemical reaction is described as follows:

$$3Si + 4HNO_3 + 18HF \rightarrow 3H_2SiF_6 + 4NO + 8H_2O$$

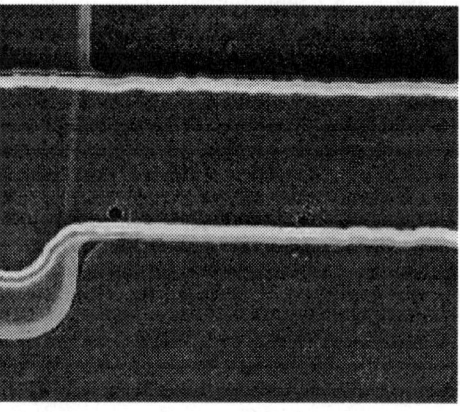

Figure 2: SEM image showing pinholes in the gate oxide caused by ESD.

The presence of acetic acid leads to better control over the etching. With proper control of etch temperature and composition, the etching can be made to proceed in an isotropic fashion. In the HF rich systems, the oxidation process is the rate-limiting step and the etching is anisotropic. The oxidation step is sensitive to substrate orientation, doping, defect concentration and structure.

In the nitric acid rich systems, the rate-limiting step is the dissolution of the oxidation product and the etching is isotropic. In the temperature range 30-50°C the reaction kinetics are diffusion controlled and the diffusion of the reaction products across the boundary layer is the rate controlling mechanism. A typical HNO_3 rich system is a 4:1:3 mixture of 79%HNO_3, 49%HF and CH_3COOH.

Superior smoothness can be achieved using an alkaline system to etch oriented Silicon. Alkaline etches are anisotropic and the etch rate depends on the Silicon orientation. The number of dangling bonds present on the silicon surface apparently dominates the reaction. A typical alkaline etch composition would be 45%KOH in H_2O. The chemical etch of Silicon is believed to proceed as follows:

$$Si + 2KOH + H_2O \rightarrow K_2SiO_3 + 2H_2$$

The liberation of H_2 gas can be seen as bubbles evolving from the reaction surface.

CRYSTALLOGRAPHIC DEFECTS

The substrate silicon may have inherent crystallographic defects such as dislocations and stacking faults. Implantation processes, diffusion processes, mechanical abrasion, thermal effects, chemical reactions, and presence of interstitials or substitutional elements in the silicon can cause these defects. During epitaxial growth, the substrate wafer acts a seed crystal. Any crystallographic defects present in the substrate grow into the epitaxial layer. Therefore, the defect concentration in the epitaxial layer is always greater than or equal to that in the substrate. A thermal oxidation process at temperatures around 1000°C can lead to the formation of Oxidation Induced Stacking Faults (OISF). OISF density is greatly influenced by impurity diffusion. Stacking faults are planar defects that are bounded by dislocations. These crystallographic defects may be the cause of failure of the die. A small deviation in parameters during epitaxial growth can cause dislocations and slip in silicon. Most of the naturally occurring dislocations in silicon are rendered inactive by hydrogen sintering. Nevertheless, the presence of dislocations in the active area of a transistor can significantly alter the characteristics of the device.

Defect staining

Any crystallographic defect, such as a dislocation, has a certain strain field associated with it. When the unit is dipped in the etchant, the strain field causes a change in the etch rate in the vicinity of the defect as opposed to bulk silicon. Therefore, these defects are preferentially decorated or 'stained'.

Broadly the etchants used for defect decoration in Silicon can be classified into two major categories: those based on the $HF-CrO_3$ system and those based on the $HF-HNO_3$ system (5). Outlined below are some of the common etch solutions used to decorate crystallographic defects.

Etchants based on the HF-CrO₃ system

Sirtl Etch: The Sirtl etch is used to decorate defects in the substrate. Von Erhard Sirtl and Anne-Marie Adler, describe in their 1961 paper (6), the decoration of substrate defects in the presence of chromic acid. The etch consists of Chromium Oxide (CrO_3) and Hydrofluoric acid (HF) in DI water. The original recipe had a composition of 46 gms CrO_3 in 100 gms of 40% HF. Stacking faults and dislocations can be identified by the formation of etch pits in the Silicon (Fig.3, Fig.4). Typical etching time is on the order of 10-20 seconds. The reaction is believed to proceed as follows (4):

$$4CrO_3 + 24H^+ + 12F^- + 3Si \rightarrow 4Cr^{+3} + SiF_4 + 12H_2O$$

Schimmel Etch: This etch consists of Chromium oxide and HF buffered with acetic acid. The defects are delineated in 5-15 minutes, with regions of lower resistivity etching quicker. All crystallographic defects and inclusions such as oxygen clusters in <100> Si can be stained out using this etch.

Secco Etch: This is a mixture of Hydrofluoric acid and Potassium dichromate ($K_2Cr_2O_7$) in DI water (7). This etch consists of two parts. Part A is composed of two parts HF and one part 0.15 molar dichromate. Part B is two parts 49%HF. The two parts are mixed together before the etch. The stacking faults and dislocations are etched out in 5 to 15 minutes. Typically, elliptical features are formed at <100> and <111> dislocations (Fig.5).

Yang Etch: This etch was developed by K.H.Yang and is described in detail in Yang's original paper (8). 150 gms of CrO_3 is dissolved in 1 liter water the volume doubled with D.I. water. All common defects can be delineated using this etch composition. Process related defects are stained out in 2-3 minutes and defects in the starting material are delineated in 10-15 minutes.

Etchants based on the HF-HNO₃ system

Dash Etch: This recipe was first developed for germanium by Dash (9) to delineate substrate and diffusion defects. It is a mixture of 1 part hydrofluoric acid and 3 parts nitric acid buffered with acetic acid. This etch is used to delineate junctions and substrate defects. The p-doped regions are preferentially stained using this particular etch composition. Typical etch time is around 10 seconds for junction delineation. It may take up to 1 minute for dislocation decoration. Variations of this composition have been given different names such as dislocation etch, silicon etch and white etch.

White Etch: This is an extremely aggressive etch. It consists of HF:HNO₃ ::1:4. The reaction with Silicon is exothermic (i.e. there is an evolution of heat) and red fumes are released as a result of the reaction. This etch is used to chemically reduce Si wafer thickness.

Sopori Etch: This etch was originally developed and reported in 1984 by B.L. Sopori (10). This etch consists of 36 parts of 49%HF, 20 parts of CH_3COOH and 1-2 parts of HNO₃. Most substrate defects are delineated in less than a minute.

A commonly used etchant for delineating substrate crystallographic defects, such as stacking faults, is the Wright (Jenkins) Etch. This etch combines the two major etchant systems and was developed by Margaret Wright Jenkins (11). This etch is a mixture of Hydrofluoric acid, nitric acid, cupric nitrate, chromium oxide and acetic acid. The presence of a defect leads to an increase in the stress at the interface. This high stress region is preferentially etched by the etchant (Fig.6, Fig.7, Fig.8). The cupric nitrate helps in enhancing the defect region. The result is the formation of pits on the silicon where there is a preferential etch. The shape of the pits depends on the defect orientation. Thus it is possible to differentiate between stacking faults in (111) Si and those in (100) Si.

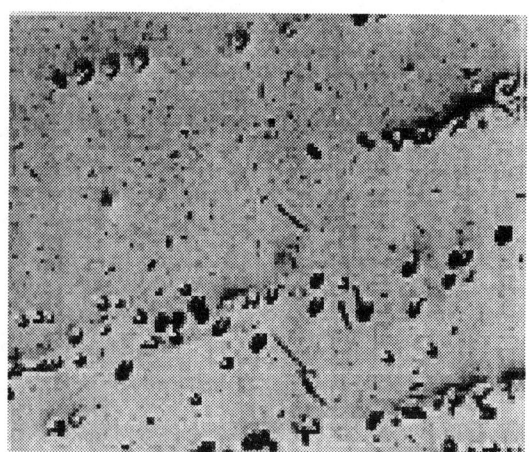

Figure 5: Circular and elliptical etch pits revealing dislocations on <100> Silicon. The etchant used was Secco Etch. (From VLSI Technology by S.M.Sze, McGraw-Hill, 1988(12))

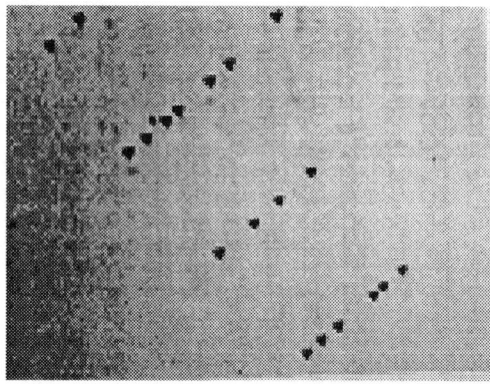

Figure 3: Dislocations in <111> Silicon, revealed as triangular etch pits stained out using Sirtl etch. (From VLSI Technology by S.M.Sze, McGraw-Hill, 1988)

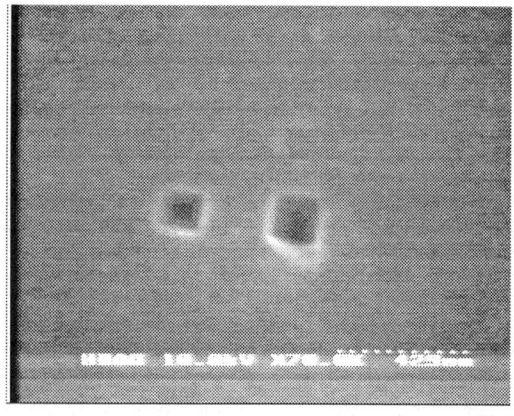

Figure 4: SEM picture showing the etch pits at stacking faults in <100> Silicon stained out using the Sirtl etch for 20 seconds.

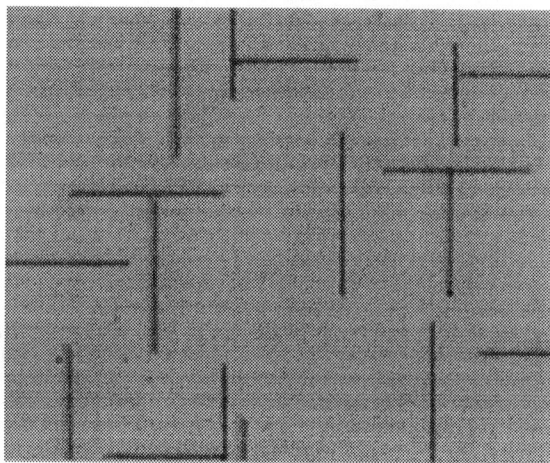

Figure 6: Optical image showing oxidation induced stacking faults in <100> Si. The faults were stained out using Wright etch for 30 seconds.

Other substrate-related etches

There are several non-crystallographic defects that can be caused during the wafer processing steps. These include abnormal diffusion depths or profiles, abnormal channel lengths, blocked implants, contact overetch, etc. All these defects can affect the performance of the device. In most cases, the use of the Wright etch for a few seconds can delineate the defects. To view the diffusion depths, profiles and to measure channel lengths, the units need to be cross sectioned and then stained with the etchant.

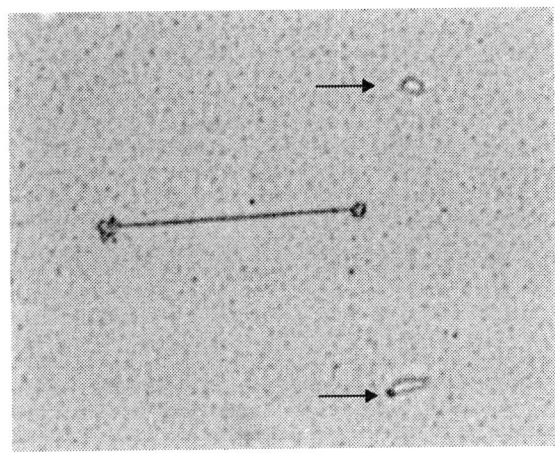

Figure 7: Optical image showing a stacking fault in <100> Si. Two dislocation loops, indicated by the arrows, have also been stained out. The etchant used was Wright etch for 30 seconds.

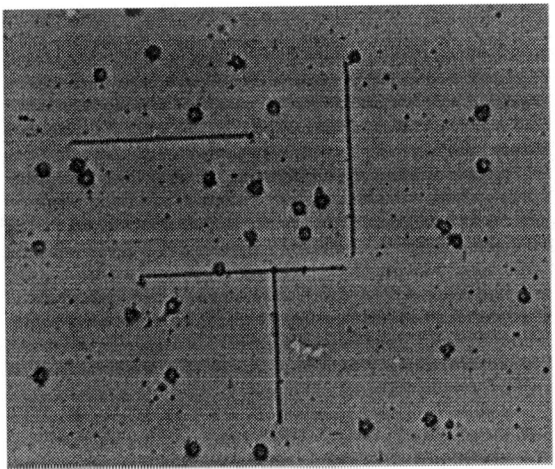

Figure 8: Stacking faults and Carbon precipitates in high Carbon <100> Si. This sample was stained using the Wright Etch for 2 minutes.

CROSS SECTION DELINEATION

Delineating metal lines: After the cross section is polished to a smooth finish using metallographic sample preparation techniques, the cross section needs to be preferentially stained to bring out the features of interest.

The Slow Aluminum etch can be used to delineate the metal (13). Any Si or Cu precipitates present in the Aluminum are also stained out after about a 90 second etch. This etch consists of Phosphoric acid, Nitric acid, Acetic Acid and DI water in a 6:1:1:3 ratio. The etch rate is about 50 nm per minute at room temperature.

Schottky diodes are used in bipolar circuits as collector clamps and as diodes in memory cells. The silicide height is critical and often there is a barrier layer of Ti compounds (Ti, TiW, TiN) to prevent reaction between the silicide and the metal during the high temperature processing steps. The barrier layers can be delineated from the silicides using the Silicide delineation etch (13): 6 gms Sodium Hydroxide(NaOH) and 20 gms Potassium Ferrocyanide ($K_4Fe(CN)_6.3H_2O$) in 50ml Distilled water. This mixture is heated to 90°C and the specimen is dipped in it for 10 seconds to enhance the barrier metal regions.

Enhancing polysilicon: To enhance the conducting filaments that may short poly lines or between poly lines and metallization, the interlayer oxide can be etched back thereby delineating the filament. The etch composition is as follows:
80 ml Glycerine, 400 ml Ammonium Fluoride(NH_4F) and 20 ml 49%HF.
The polysilicon itself can be stained to show the grain boundaries and surface topography. HNO_3, CH_3COOH and HF in the ratio 40:20:1 is a standard poly etch composition.

The use of a 10:1 Buffered Oxide etch for 5-10 seconds etches back the oxide layers and helps delineate the diffusion regions. Cross sectional samples can also be dipped in Potassium Hydroxide (KOH) for a few seconds to enhance the diffusion regions. Solutions based on Hydrofluoric acid are commonly used to stain out the junction regions. As mentioned earlier, the HF-HNO_3 systems etch based on a REDOX reaction in conjunction with differential etching. Based on the ratio of these two acids, P-doped or N-doped Silicon can be preferentially stained.

When the ratio of Nitric to Hydrofluoric is small, the P doped region is stained out. When the ratio of Nitric to Hydrofluoric acid is high, the N doped region gets preferentially stained. As the shelf life for this mixture is short, best results are obtained when the acids are mixed just before use.

Given below are some of the other common etchants used to stain out the substrate and diffusion regions.

SD 1 etch: This etch can be used to emphasize p-doped regions. The p-doped regions are smoother and well defined while the n-doped region is rough. This etchant is effective on all orientations of Silicon (14). The crystal defects are not particularly well highlighted. The etching time is on the order of a few seconds.

CP Etch: This etch is used to delineate the epitaxial regions and p-doped regions. This etch proceeds very quickly and the p-doped areas are strongly decorated while the n-doped regions are faintly demarcated. Epitaxial layers can be delineated but crystal defects do not stain out.

P/N etchant: A 97 to 3 mixture of HF to HNO$_3$ can be used to decorate the n-doped regions. The etch proceeds very quickly and to preserve the junction profile, the stain must be no longer than 10 seconds.

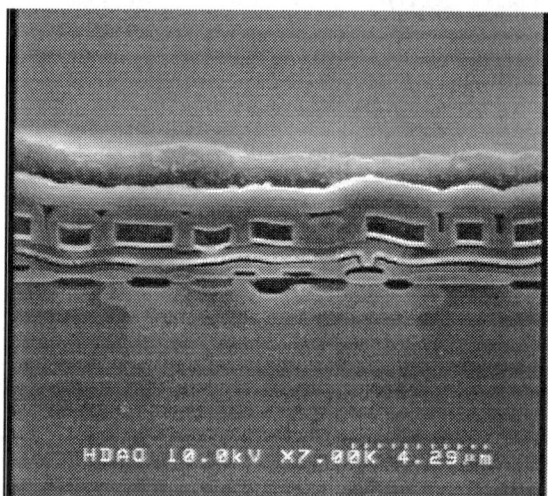

Figure 9: SEM image of the cross section of a 2 metal layer technology. The unit was dipped in Buffered HF for a few seconds followed by a dip in KOH. The oxide layers are etched back and the diffusion profiles are clearly delineated.

When a device is cross sectioned using the Focussed Ion Beam (FIB), the milling process leads to a local annealing of the sample. Cross section staining and delineation steps typically take longer time for the FIB samples as compared to a mechanically polished cross section.

Often, the precise location of the defect (i.e. the x,y coordinates as well as the depth level) can be identified before the device is deprocessed. Under these circumstances, layer by layer deprocessing is often a waste of time. Alternate methods such as mechanical polishing or the use of a strong etchant can be adopted to get down to the required level. Often, particle contaminants can be visually observed through the oxide layers. Signature analysis may reveal the presence of certain defects at a particular level. ESD events can often lead to blown contacts that can be identified at substrate level.

49% HF can be used to remove all the layers and strip devices down to substrate. The etch to substrate can take up to 15 minutes depending on the number of layers and the thickness of these layers. HF buffered with acetic acid can be used to strip the devices down to poly. The time for etching depends on the composition of the mixture (typically 1:20 to 1:100). This process can take anywhere from 1 to 4 hours. Another effective method is to mechanically polish to the level above the defect and then use chemical etching

to get to the defect quickly and without damage to the area of interest.

ETCHING OTHER SEMI-CONDUCTOR MATERIALS

Several materials have been studied and researched as possible alternatives for Silicon in the semiconductor industry. Of these, the III-V compounds are most promising. To date, there has not been much documented on the wet etching techniques for failure analysis of these materials. Gallium Arsenide (GaAs) can be etched uniformly using a mixture of an acid (HCl, H$_2$SO$_4$, H$_3$PO$_4$) and H$_2$O$_2$ in DI water. Indium Phosphide (InP) can be etched using warm HCl or a mixture of HCl and H$_3$PO$_4$. A dilute solution of Bromine (Br) in methanol is an aggressive etch for most III-V compounds.

CONCLUSION

Wet chemical etching is a very useful technique for Failure Analysis and offers an alternate method of deprocessing to RIE. Substrate damage and etching time can be greatly reduced using wet chemistry. Wet etchants are necessary for substrate staining and cross section delineation. The selectivity against underlying layers, reaction temperature and time of etching are key parameters that need to be ascertained before using any wet etch recipe. As the feature size shinks and the number of metal levels increases, the risk of metal undercutting and lift off becomes greater with wet etch processes. Several etch recipes have been documented in this paper. Attention needs to be paid more to the concepts behind these etch recipes and the exact time and composition are less significant. Etch times and compositions can vary over a broad spectrum. Ambient conditions such as humidity and storage can greatly affect the performance of an etchant. Process conditions also affect the etch rate significantly. Finally, it is advisable to be aware of the hazards some of these chemicals can pose to the analysts as well as the devices.

ACKNOWLEDGMENTS

The authors would like to acknowledge Ron Parker and other members of Houston Device Analysis Operations (HDAO) for their support and assistance.

REFERENCES

1. Tom Lee, Mechanical and Chemical decapsulation, Microelectronic Failure Analysis, Desk Reference, 3rd Edition, 1993, pp 61-74.

2.   R. Raghunathan and D.Davis, Chip Access Techniques, to appear in Microelectronic Failure Analysis, 4th Edition, Desk Reference.

3   P. Malberti and M. Ciappa, Selective wet etch of Silicon Nitride passivation layers, to appear in Proc. of ISTFA, 1998.

4.   T.A Shankoff, et al, Controlling the interfacial oxide layer of Ti-Al contacts with the CrO_3-H_3PO_4 etch, Journal of Electrochemical Society, March 1978, 125, 3, pp467-471.

5.   Chemical etch formulations and history, Thomas W. Lee, Microelectronic Failure Analysis, Desk Reference, 3rd Edition, p 111.

6.   E.Sirtl and A.Adler, Z. Metallkd., 52, 1961, p 529.

7.   Dislocation Etch for (100) planes in Silicon, F. Secco d'Aragona, Journal of Electrochemical Society, 119, July 1972, p 950.

8.   An etch for delineation of defects in silicon, K.H.Yang, Journal of Electrochemical Society, 131, 1984, p 1140.

9.   W.C.Dash, Journal of Applied Physics, 27, 1956, p1193.

10. B.L.Sopori, Journal of Electrochemical Society, Solid-State Science and Technology, Technical notes, March 1984, pp 667-672.

11. A new preferential etch for defects in silicon crystals, Margaret Wright Jenkins, Journal of Electrochemical Society, 124, 1977, p 757.

12. S.M.Sze, VLSI Technology, McGraw-Hill International Edition, 1988.

13. Chip and Device Sectioning Techniques, J.J Gajda, Microelectronic failure Analysis, Desk Reference, 3rd Edition, p 97.

14. Integrated Circuit Failure Analysis, A guide to preparation techniques, F. Beck, J.Wiley & Sons, 1997.

Additional references for wet etch techniques and etch formulations:

1.   Mike Jaques, "Tough analysis problems that have a solution", Proceedings of ATFA, 1978, pp124-136.

2.   Mike Jaques, "The chemistry of Failure Analysis", proc. of IRPS, 1979, pp. 197-208.

3.   W.E.Beadle et al, Quick Reference manual for Silicon Integrated Circuit Technology, Section 5, "Chemical Recipes", Wiley Interscience.

4.   Failure Mechanisms in Integrated Circuits, Dan Corum et al, Microelectronic failure Analysis, Desk Reference, 3rd Edition, p 277.

Plasma Delayering
of Integrated Circuits

A. Crockett and W. Vanderlin

Increased circuit densities, smaller feature sizes and ever increasing multilayer technologies have created many challenges for today's failure analysis engineer. Wet etch techniques have long been unable to do an acceptable job deprocessing and even many dry etch techniques fall short in etching the more difficult ICs. This paper gives a brief introduction to the basic concepts in plasma processing, discusses today's state-of-the-art techniques for dry processing in failure analysis and provides an outline of dry etch recipes for etching critical layers on new technologies. Also discussed are solutions to the major pit falls of dry etching such as RIE grass and keeping devices active.

PLASMA BASICS

What is a plasma?

Plasma is called, by many, the fourth state of matter. It is different from the other three, in that, it contains free disassociated electrons and ions in a balanced steady-state condition. This, by definition, is a plasma. A plasma contains in a disassociated state; free radicals, ions, electrons and unexcited molecules. The ratio of the ions to the rest of the molecules constitutes the "ion density" of the plasma.

For practical use, a plasma is simply a method for turning nonreactive molecules into electrically charged reactive molecules. Many safe inert gases when broken up yield extremely reactive by-products. Freons, for instance, are extremely long lived inert compounds; when broken up they yield relatively large concentrations of fluorine and chlorine (some of the most reactive compounds known). Since, this process is controlled directly from the application of an electric field, the reactivity and directionality of the process can be controlled by the applied power. This is what makes plasma processing useful.

How are plasmas made?

A plasma is made by introducing energy into matter. This is accomplished in many ways, such as: through heat, radiation and (as in our case) an electric field.

The plasmas used in Semiconductor processing are very specific in nature. Processing semiconductor devices requires a relatively low temperature plasma. To create this low temperature plasma, the plasma has to be created through the application of an electric field to conductive gases. Fortunately, gases are electrically conductive at easily achieved, moderately low pressures (on the order of 1 torr).

HISTORY OF PLASMA REACTORS

Plasma processing was introduced to the Semiconductor Industry in the 60s. The first systems were of the "Barrel" type and were typically used for stripping photoresist. Previous to this time, wet chemical solvents were used. These solvents were potentially carcinogenic and expensive to dispose of. Plasma processing, on the other hand, is far more effective in removing positive resist, uses orders of magnitude less chemicals, and is therefore far more kind to the environment.

Barrel Reactors

The first barrels systems were inductively coupled (see figure 1) and consist of a quartz bell jar with a coil wrapped around it, turned on its side. Since the quartz chambers etch in fluorinated gas, these systems are usually only used to stripping photoresist with Oxygen. Later versions of barrel systems were capacitively coupled and consist of a cylindrical aluminum chamber with an inner concentric cathode (figure 2). Since the anode in capacitively coupled barrels is the barrel wall itself, these barrels can be made of aluminum; and because the aluminum is inert to the fluorinated etch gas, these systems can be used for etching.

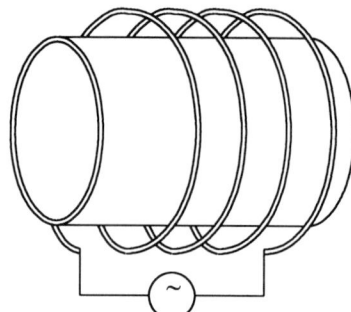

Fig. 1. Inductively Coupled Barrel

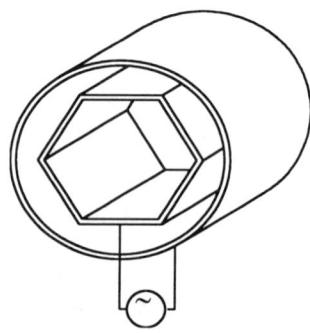

Fig. 2 Capacitively Coupled Barrel

Despite the fact this is an older technology and they etch isotropically (due to the chamber geometry), barrel systems are very versatile and are still widely used in the industry today.

Parallel Plate Reactors

Parallel Plate reactors (see figure 3) are by definition capacitively coupled, they are either bottom or top powered. Etching in a top powered reactor is referred to as "Plasma mode" etching or "PE mode" and etching a bottom powered reactor it is referred to as "Reactive Ion Etching". This is a bit of a misnomer, because both systems are generically "plasma etchers".

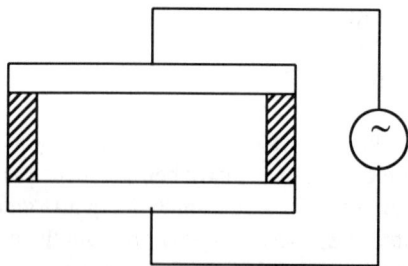

Fig. 3. Parallel Plate Reactor

The first widely used Parallel Plate Reactor was the "Reinburg" reactor (see figure 4), which was developed at Texas Instruments in 1972. There are many variants of this design, but they are all basically a bottom powered electrode system, large enough to accommodate twenty five, 100 mm wafers.

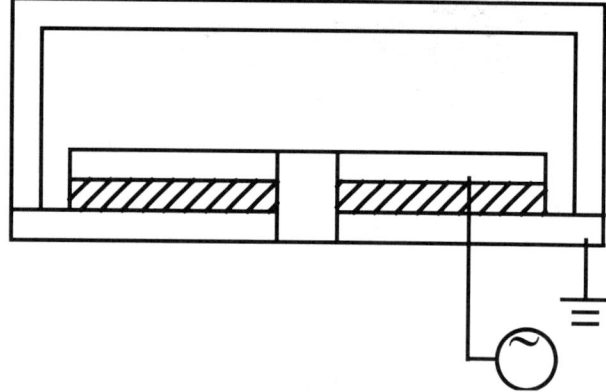

Fig. 4. Reinburg Reactor

The first fully automated, single wafer parallel plate system was introduced to semiconductor production lines in 1979, by Tegal corporation. Because of the superior process results that single wafer systems offer, almost all production etch systems are of this configuration.

Reactive Ion Etching

The name "Reactive Ion Etching" is misleading. It really should be called "Ion Assisted Etching". The percentage of ions in a plasma is very small, and if they were the only species participating, the overall etch rate would also be very small. The actual mechanism is a three step process:

1. Chemical adsorption of reactive molecules (free radicals) on the surface.
2. Impact of an ion on the surface (reactive or not).
3. Physical disassociation of many reaction by-products from the surface.

Reactive Ion Etch processes (see figure 5) were, until the late 80's, the only production plasma processes that produced an anisotropic etch. This phenomena is a result of the chamber geometry, plasma physics and operating pressures of the systems. In general, Reactive Ion Etching is defined by two things, a bottom powered reactor and an operating pressure below 100 mtorr. This is also misleading, because these are only two contributing factors in achieving an anisotropic etch.

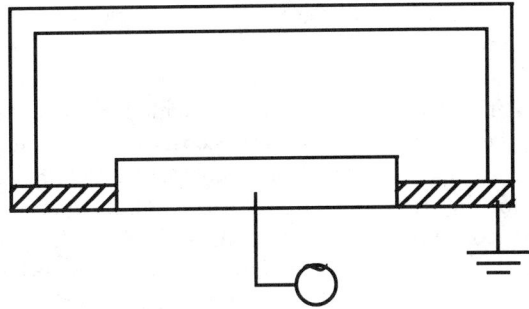

Fig. 5. Reactive Ion Etch Configuration

Hybrid Reactors (Triode, ECR and ICP)

Within the last few years, there have been many new types of reactors introduced into the industry. These are Triode, ECR (or downstream microwave) and ICP reactors. We call these "Hybrid" reactors, because they introduce power into the reactor through a secondary source. This is advantageous, because you can add power to the plasma without coupling it through the sample. Hybrid reactors, therefore, have much lower charge damage, sputtering and operating temperature; and they have higher etch rate and selectivity.

PLASMA ETCH CHEMISTRIES

The five most common materials that are etched in deprocessing ICs are: polyimide, silicon nitride, silicon dioxide, aluminum and polysilicon. Table I shows the most common etch chemistries for each of these materials. Later, is a more in depth discussion of these chemistries and why each are used for each material provided.

Material	Etch Gases	Reactive Species	By-product
Polyimide	Oxygen	monatomic oxygen and/or ozone	CO and H20
Silicon Nitride	Freon 14 (CF4) / 8% O2 or Sulfur Hexaflouride	free fluorine	SiF4 and N2
Silicon Dioxide	Freon 14 and Freon 23	CF3	SiF4 and CO
Aluminum	Boron Trichloride and Chlorine	free chlorine	AlCl3
Polysilicon	CF4 / 8% , SF6 or Chlorine	free fluorine or free chlorine	SiF4 or SiCl4

Table I. Common materials and etch chemistries.

BASIC DEPROCESSING RECIPES

Polyimide

There are many varieties of Polyimide on the market today. They have different curing properties, solids content, etc. However, they are all hydrocarbons, and most etch readily in Oxygen plasmas. This reaction is a simple oxidation of the organics as shown in the following equation:

$$C_xH_y \text{ (s)} + O_2 \text{ (g)} + \text{plasma} \text{ --------> } CO \text{ (g)} + H_2O \text{ (g)}$$

The main problem, in the removal of Polyimide, is insuring that the Imide does not over heat during the process. If this occurs the Imide can carbonize, leave a grass like residue, and be virtually impossible to get off. Reducing the ion bombardment solves this problem; this can be accomplished by running very low power, and/or "floating" the sample in the plasma, and/or using a hybrid reactor.

A good starting recipe for Polyimide* removal is given in Table II.

Parameter	Value	Comment
Pressure	250 mtorr	Relatively high pressure = low voltage
Power	100 watts	Relatively low power = low voltage
Oxygen Flow	50 sccm	Sufficient flow for most processes
Etch Rate	500 nm/min	

***Note: Addition of up to 20% CF4 to the oxygen process can increase the polymide etch rate up to 2 microns per minute. However, the CF4 may cause an increase in grass formation and decrease the selectivity to underlying materials.**

Table II. Polymide etch recipe.

Silicon Nitride

Silicon Nitride typically comprises the final passivation layer of an IC. It etches readily in plasmas that contain a lot of free fluorine (such as Sulfur Hexaflouride and CF4/O2 plasmas). These chemistries are isotropic by nature. However, in this case, this property is actually advantageous in removing the Nitride sidewalls surrounding top metal.

The equation for this reaction is:

$$Si_3N_4 \text{ (s)} + SF_6 + \text{plasma} \text{ -------> } SiF_4 \text{ (g)} + SF_2 \text{ (g)} + N_2 \text{ (g)}$$

A good starting recipe for Nitride is:

Parameter	Value	Comment
Pressure	250 mtorr	relatively high pressure = low voltage
Power	100 watts	relatively low power = low voltage
SF6 Flow	50 sccm	sufficient flow for fast rate
Etch Rate	200 nm/min	

Table III. Nitride etch recipe.

Silicon Dioxide

There are many types of Silicon Dioxide in use today. They all etch in the same chemistry; however the recipes and etch rates vary a little with the type. Typically, highly doped oxides etch faster and oxides with high carbon content etch dirtier. The chemical reaction for this process is given below:

$$SiO_2 (s) + CF_4 (g) + plasma ---------> SiF_4 (g) + CO (g)$$

Silicon Dioxide etching is intrinsically anisotropic due to the fact that the strong chemical bond between the Silicon and Oxygen requires ion bombardment to break.

A good starting recipe is:

Parameter	Value	Comment
Pressure	50 mtorr	low pressure to reduce polymerization
Power	100 watts	
CF4 Flow	50 sccm	
Etch Rate	100 nm/min	

Table IV. SiO$_2$ etch recipe.

Aluminum

Pure aluminum, by itself, etches readily in a Cl_2 plasma. However, all aluminum films are covered by a native oxide layer. Pure Cl_2 does not etch this oxide, so BCl_3 is added to increase the amount of sputtering and to scavenge the oxygen in the aluminum oxide layer. In most modern integrated circuits the aluminum is alloyed with small amounts (0.5% to 2%) of silicon and copper. Copper in particular is difficult to etch by RIE, so BCl_3 is also useful for increasing the amount of physical sputtering which removes the copper. Typically the aluminum itself is removed very rapidly, but complete removal of other metal residue requires a longer etch. A variety of wet chemical etchants can also be used to remove the metallization, but the chemicals will enter the vias and isotropically undercut the next lower metal line.

It is important that Aluminum etching be done in a separate reactor (or one that has been cleaned thoroughly) than the reactor used for Silicon compound etching. The reason for this is that the chemistries "poison" each other. Fluorine containing polymer residues from oxide etching will react with Aluminum to form Aluminum Fluoride on the metal surface, which is inert to Chlorine etch. Aluminum Chloride by-products, left behind from the Aluminum etch, form an Aluminum Fluoride powder when exposed to a fluorinated plasma; this powder subsequently falls and contaminates the sample surface. Another, important consideration in Aluminum etching is moisture contamination, for this reason, as well as safety considerations, a vacuum load-lock is highly recommended for this application.

Obviously, Aluminum etching is one of the most difficult processes. However, if done correctly very good etch results can be obtained. A good starting recipe is:

Parameter	Value	Comments
Pressure	180 mtorr	Best pressure for good uniformity
Power	100 watts	High Etch Rate, but necessary to remove native oxide
BCl3 Flow	30 sccm	
Cl2 Flow	10 sccm	Too much of this can cause undercutting
Etch Rate	500 nm/min	

Table V. Aluminum etch recipe.

Polysilicon

Polysilicon can be etched anisotropically in Chlorine gas, but an anisotropic etch usually is not necessary for this layer and the extra cost of incorporating Chlorine may be prohibitive. A very selective (to oxide) isotropic etch can be obtained with a low power SF$_6$ plasma. Selectivities of 100:1 can be obtained; a good starting recipe is:

Parameter	Value	Comments
Pressure	100 mtorr	
Power	30 watts	low power = high selectivity = low voltage
SF6 Flow	50 sccm	
Etch Rate	300 nm/min	

Table VI. Polysilicon etch recipe.

ANISOTROPIC DIELECTRIC REMOVAL VS. SEQUENTIAL DELAYERING

Depending on the desired information, DELAYERING of the integrated circuit is usually performed by either of two strategies: (a) anisotropic removal of all dielectric layers, or (b) sequential removal of all layers including conductors.

Anisotropic removal of dielectric layers ("Skeleton Etch")

In this procedure, all dielectric layers are removed anisotropically down to the silicon surface. Metal conductors will be left sitting on top of pedestals of dielectric material, see figure 6. An anisotropic etch must be used to prevent undercut of the metal lines, or else the stress in the metal will usually cause delamination. When the silicon dioxide etch approaches the polysilicon gate material, a $CF_4 + CHF_3$ gas mix is used to improve selectivity to silicon and decrease erosion of the polysilicon lines.

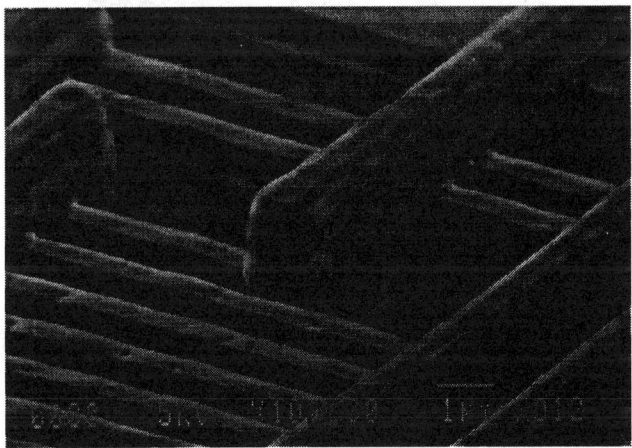

Fig. 6. "Skeleton" Etch

Sequential Removal

Anisotropic dielectric removal is useful, but in many situations defects or other features of interest may lie underneath conductors. In this case, metal and dielectric layers must be removed sequentially. Naively, one would expect that a simple reversal of the etch processes used to fabricate the circuit would delayer it. However, it is important to remember that when reactive ion etching is performed during fabrication there is photoresist on the circuit surface to cover up areas that should not be etched, and each etch step usually terminates on an "etch stop", i.e. a layer which has a very low etch rate for the etch. During sequential delayering, neither of these things is true.

Proper etch recipe selection is crucial to prevent inadvertent removal of layers not intended to be etched. It is

important that lower level metals layers are not exposed when the top level metal is etched, or the lower level metals will be removed prematurely.

Oxide Sidewalls

Because interlayer dieletrics are usually quite planar, in principle it is possible to maintain a planar surface during alternate delayering of metal lines and dielectric layers. However, the top most layer of dielectric, usually referred to as the passivation, is not usually planarized. Because this silicon dioxide or silicon nitride layer is conformal, anisotropic RIE removal tends to leave an "oxide sidewall" around the metal line. The sidewall may also contain the aluminum native oxide and a sidewall polymer created during IC fabrication, both of which are very resistant to plasma etching. The sidewall features are difficult to remove by RIE and tend to simply propagate downward during any anisotropic etch. Some success has been achieved at removing the oxide sidewalls using a dilute solution of buffered HF or PSG etch solutions.

Planar Delayering

In order to maintain planarity during sequential delayering of an IC, it is desirable to stop each dielectric etch when a level is reached that is even with the next metal layer to be etched. Thus when the passivation is etched, it should be etched level with the base of the metal line. Since there is no etch stop between the dielectric layers, this process requires a timed etch. The time will have to be determined by trial and error for any given integrated circuit process using several test pieces for process development. Failure to time the etch process properly will result in a metal line that is either sitting on top of an oxide pedestal, or contained within a trough. Once the metal line is removed, this pedestal or trough geometry will be propagated downward during the next oxide etch step because the oxide etch is anisotropic. If delayering is not done in a planar fashion, more and more topography will be created that will propagate downward, leaving a very irregular surface. If the objective is simply to remove all metal and dielectric layers so defects on the polysilicon can be viewed, it is probably easiest to dip the part in dilute HF (hydrofluoric acid). HF will isotropically etch all dielectric materials and will undercut the metal lines, thus removing all conductors except the polysilicon.

RIE GRASS

RIE grass is an undesirable artifact of etching which prevents a clean delayering of the integrated circuit and interferes with failure analysis. RIE grass occurs when an etch resistant material accumulates in small patches on the sample surface. These patches cause micro-etch-masking, which results in formation of cones. The basic mechanism of cone formation during plasma etching has been understood since the 1940's [3]. We have found that micro-etch-masking can be due to a complex interaction of sample materials and plasma conditions. It is important to note that once created, the cones tends to propagate downward during the etch, even if the original mask material has been sputtered off. Therefore analysis of the sputter resistant material must be performed early in the development of the cone structures.

Previous reports have identified the cause of RIE grass as sputter re-deposition of package materials, [2,4,5,6] sputter re-deposition of cathode materials, [6,7] and polymer formation. [7] Proposed solutions to prevent RIE grass included covering all gold on the package, [2,4,5,6,7] reducing the RF power, [5] pre-conditioning the chamber, [8] and adjusting the plasma gases and plasma potential.[7] However, a clear presentation of the fundamental sources and solutions to RIE grass has only been recently presented. These mechanisms have been confirmed by Auger analysis .

RIE grass has three basic causes, each of which results in a different physical appearance. The first type of grass is due to sputter re-deposition of package materials and gold leads. This can only be eliminated by masking these materials from exposure to the plasma. The second type of RIE grass is due to chemical attack of exposed aluminum metallization which results in micro-etch masking by aluminum or aluminum compounds. This can be avoided by eliminating O_2 from the gas supply whenever aluminum lines are exposed. The third type of grass is due to a plasma etch chemistry which results in polymer formation on the sample surface. This can be eliminated by reducing the gas pressure which increases plasma potential and therefore increases sputtering.

The creation of grass and other plasma effects is strongly dependent on the equipment used. Therefore our results are not necessarily applicable to all RIE systems. It is observed that as an RIE system is "conditioned" over time the etching results will drift somewhat from the results obtained in a brand-new system. This conditioning is presumed to be caused by deposition of etch residue on the walls of the reaction chamber. Consistent cleaning of the reaction chamber will bring the etching results to a steady state condition after a short time of use.

TYPES OF RIE GRASS

Sputtered Grass

When packaged parts are delayered, it is common that foreign materials on the package will sputter re-deposit onto

the die surface causing micro-etch masking and grass formation. By far the most serious problems are produced by gold which is commonly used in ceramic packages (a) in the die cavity for gold/silicon eutectic die attachment, (b) as a metallurgical coating on the lid seal ring, and (c) as gold bond wires. Examples of grass produced by gold re-deposition are shown in figure 7.

Fig. 7. "Sputtered" Grass

Gold is easily sputtered by RIE, and the re-deposition of gold onto an IC surface results in severe grass formation. As measured by scanning Auger microscopy, fluorine was not associated with the patches of re-deposited gold. Furthermore the XPS spectra of the gold 4f peaks at 84 eV and 87 eV did not have any binding energy shifts as expected for a gold compound. Therefore we believe that the gold is re-deposited as pure gold.

In order to achieve clean etching, it is essential that all gold on the package be protected from exposure to the plasma. Plastic covers can be used to protect the lid seal ring from the plasma, but gold bond wires and gold die attach in the cavity are difficult to cover because of their close proximity to the die. Photoresist or carbon dag painted on the bond wires and into the cavity will eliminate the gold grass, however the presence of extra organic material can alter the plasma chemistry and increase problems with a different type of grass, such as polymer grass.

Aluminum Grass

Like gold, aluminum can also sputter re-deposit onto the dielectric surfaces and cause grass. Unlike gold, aluminum is virtually always present within the chip circuitry itself, therefore it is not possible to mask it from the plasma as with gold. Fortunately, etch recipes are available which minimize re-deposition of the aluminum.

For most CMOS processes, the dielectric layers between the conductors are silicon dioxide, and silicon nitride is used only for the top passivation layer. Although a pure CF_4 plasma etches silicon nitride slowly, the addition of O_2 will increase the fluorine free radical concentration and increase the etch rate of silicon nitride.[10] Therefore it is common to use CF_4 + O_2 to remove silicon nitride passivation. However, it has been found that once aluminum metal is exposed, the CF_4 + O_2 plasma will attack aluminum, re-depositing the etch by-products onto the dielectric resulting in "aluminum grass," see figure 8. Over time the aluminum lines will become significantly eroded by this etch process, which is also undesirable. Therefore, once the silicon nitride passivation is removed, we change our etch recipe to pure CF_4. The pure CF_4 recipe does not attack aluminum. Because a simple change in chemistry eliminates this type of grass, it is reasonable to assume that the aluminum attack is chemical in nature, not physical as commonly believed. This recipe is also more anisotropic and thus less likely to undercut aluminum lines. Unfortunately, pure CF_4 is more likely to cause polymer grass, as discussed in the next section.

Fig. 8. "Aluminum" Grass

Some integrated circuit processes use silicon nitride as an interlayer dielectric. For those ICs it is necessary to use SF_6 (sulfur hexafluoride) gas to etch the interlayer dielectric, as exposure of aluminum to CF_4 + O_2 will cause aluminum erosion. Unfortunately, to maintain anisotropy the SF_6 process needs to be run at a lower operating pressure than CF_4, and the turbo pump life is adversely affected by sulfur deposits.

Polymer Grass

Under certain plasma conditions, carbon-fluorine polymers can form on an integrated circuit surface.[10] This polymer can micro-etch mask the sample surface and produce RIE grass. Polymer grass commonly has the appearance of a fat cylinder with the axis oriented normal to the sample surface. The top end of the cylinder is often slightly concave. This type of grass is sometimes referred to as "tube worms", see figure 9. Under slightly different etching conditions, the polymer grass appears as narrow cylinders.

Fig. 9. "Polymer" Grass

Scanning Auger microscopy of the polymer grass shows a high concentration of carbon and fluorine on the grass. XPS analysis of this layer shows evidence of chemical bonding between the carbon and fluorine. In addition to the usual carbon 1s peak near 285 eV, there is a second peak shifted to 289 eV. This chemical shift is typical for a carbon atom bonded to a single fluorine atom.[11]

The tendency for polymer grass to form is sensitive to both the chemistry of the plasma and materials on the sample surface. The tendency of a plasma to produce a polymer is usually described in terms of the fluorine-to-carbon ratio model,[10] where a low ratio of fluorine to carbon is more likely to result in polymer formation. The addition of O_2 to a CF_4 plasma increases the fluorine atomic concentration in the plasma and tends to reduce polymer formation. Thus polymer grass is not formed by the CF_4 +

O2 plasma used for etching silicon nitride. Addition of H2 or CHF3 to a pure CF4 plasma decreases the fluorine concentration in the plasma by forming HF, and this makes polymer deposition more likely. It has been also observed that other sources of carbon on the IC itself can increase the tendency of polymers to form.

The following factors tend to make polymer grass <u>worse</u>:

a. High chamber pressure with pure CF4, especially if the chamber pressure is above 200 mtorr.
b. The addition of CHF3, which is used to improve silicon dioxide to silicon etch rate selectivity.
c. The presence of carbon containing dielectric layers such as spin-on-glass (SOG) or tetraethyl orthosilicate (TEOS) glass.
d. The presence of Photoresist or carbon paint used to cover up gold on packaged parts.

The following factors tend to <u>eliminate</u> polymer grass:

a. Reducing chamber pressure, which increases plasma potential and therefore increases sputtering of the surface and removes the polymer micro-etch masks. Typically the silicon dioxide etch was performed at 70 Mtorr, which requires a turbo pumped chamber.
b. Addition of O2 to the CF4 chemistry. However, this has the disadvantage of attacking the aluminum and causing aluminum grass.

Our standard silicon dioxide etch works well for silicon dioxide, TEOS, or SOG. If polymer does begin to form, it is necessary to either reduce the chamber pressure or add small amounts of O2 to the gas mixture.

HYBRID REACTORS AND DEEP OXIDE ("SKELETON") ETCHING

Keeping devices active during deprocessing can be a major frustration for reliability engineers. Reactive ion etching (RIE) provides a rapid, controlled and acid-free method for delayering integrated circuits. However, RIE places a working device directly on a powered electrode, and this can produce surface contamination, charge damage, and waste many man-hours by destroying one-of-a-kind parts. New HYBRID reactor types solve these problems.[14]

Hybrid reactors are also invaluable in etching deeper than 3 levels of metal. RIE tends to errode the exposed Aluminum as the etch progresses; Hybrid reactors on the other hand are much more delicate in their etching and allows etching which exposed 5 or even 6 levels deep.

Of all the Hybrid reactors, Inductively coupled plasma systems are fast becoming the system of choice in front end production areas.[12, 13] They offer many advantages of RIE systems such as faster etch rates, cleaner and more selective etches, and much lower plasma damage due to the lower operating voltages at the sample surface.[13] These benefits are also important to the failure analysis engineer. Inductively coupled plasma systems typically contain two RF power sources, and each of these power sources introduces energy into the system in different ways. The ICP source introduces energy by the use of an RF induction coil upstream from the sample. This source uses less power and is needed to form a plasma potential at the sample and obtain an anisotropic etch. In this way, the inductively coupled plasma system can directly control the voltages at the sample surface without changing the other plasma parameters, this method achieves two results, fast etching and low device damage.

Deprocessing Results

The optimum operating conditions for RIE and ICP on a 200 mm wafer are given below in Table 1. Figures 10 and 11 show electron micrographs of three level metal devices that were deprocessed down to metal 1. Note that the ICP etched devices are etched more cleanly and with less erosion to the metal lines than the RIE etched devices. It is important to note that the ICP source does not directly add any power through the sample, it merely enhances the etch rate by increasing the number of reactive species in the reactor. The net effect is the process can be run at lower voltages and maintain a fast anisotropic etch.

Fig. 10. RIE 3 Level Etch

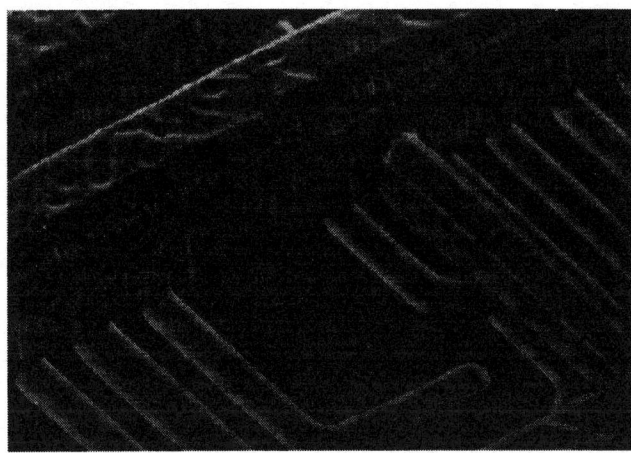

Fig. 11. ICP 3 Level Etch

Elimination of RIE Grass

All of the mechanisms for RIE grass previously discussed, also apply to hybrid reactors. ICP reactors have an added advantage, because they dramatically increase the operating range of the system. They can be operated at pressures far below those possible in a RIE system (about 1 mtorr). At this pressure no polymer grass can form. Also, they use a lot less power for the RIE bias, and thereby eliminate sputtered grass.

Electrical Damage

The ability to control the voltage at the sample surface is essential to the elimination of charge damage in integrated circuit devices. This is especially important when delayering completed devices to be used for electrical test. The consequence of the reduced voltage in the ICP system should be an improvement in device survival.. For parts etched down to M1, we found there was no threshold voltage shift for ether RIE or ICP etched devices. However, if the polysilicon gates are exposed by deprocessing the damage may be much greater.

Acknowledgements

The authors would like to acknowledge David Scott Kiefer for his contributions in writing this paper and allowing us to use excerpts from his earlier paper (reference 2).

References

1. J. Beall, "Plasma etching," in *Microelectronics Failure Analysis Desk Reference, 3rd Edition,* T. Lee and S. Pabisetty, editors, pp. 121-123, ASM, Materials Park, Ohio, 1993.

2. D.S. Kiefer, "Reactive ion etch recipes for failure analysis," in *Microelectronics Failure Analysis Desk Reference, 3rd Edition,* T. Lee and S. Pabisetty, editors, pp. 121-123, ASM, Materials Park, Ohio, 1993.

3. O. Auciello, "Historical overview of ion-induced morphological modification of surfaces," in *Ion Bombardment Modification of Surfaces, Fundamentals and Applications.* O. Auciello and R. Kelly, editors pp. 1-25, Elsevier, New York, 1984.

4. W. Baerg, V.R.M. Rao, and R. Livengood, "Selective removal of dielectrics from integrated circuits for electron beam probing, *"Proceedings of the 30th IEEE IRPS Symposium,* pp. 320-326, IEEE, Piscataway, New Jersey, 1992.

5. S. Prasad, G. Lindberg, H. Zhang, S. Jacobson, and R. Huerta, "Failure analysis of multilevel high density ASICs by integrated E-beam tester and the reactive ion etcher, *"Proceedings of the 20th International Symposium for Testing and Failure Analysis,* pp. 111-115, ASM, Materials Park, Ohio, 1994.

6. B.R. Peters, "Magnetically enhanced reactive ion etch for failure analysis, *"Proceedings of the 20th International Symposium for Testing and Failure Analysis,* pp. 379-384, ASM, Materials Park, Ohio, 1994.

7. M. Sanada, S. Suzuki, T. Numaziri, T. Omata, and N. Yoshida, "Fundamental evaluation of LSI's using anisotropic reactive ion etching, *"Proceedings of the 21th International Symposium for Testing and Failure Analysis,* pp. 87-92, ASM, Materials Park, Ohio, 1995.

8. M.T. Abramo, E.B. Roy, and S.M. LeCours, "Reactive ion etching for failure analysis application, *"Proceedings of the 30th IEEE IRPS Symposium,* pp. 315-319, IEEE, Piscataway, New Jersey, 1992.

9. L.I. Matiengo and S.O. Grim, "Interactions of some free phosphorus (III) compounds with gold vapor detected by means of x-ray photoelectron spectroscopy, "Analytical Chemistry, Vol. 46, pp. 2052-2055, November 1974.

10. S. Wolf and R.N. Tauber, *Silicon Processing for the VLSI era, Volume 1: Process Technology,* p. 550, Lattice Press, Sunset Beach, CA, 1986.

11. D. Briggs and M.P. Seah, *Practical Surface Analysis by Auger and X-ray Photoelectron Spectroscopy,* p. 363, John Wiley & Sons, New York, 1983.

12. E. Korczynski Solid State Technology, April, 63-73 (1996).

13. P. Singer, Semiconductor International, July, 52-57 (1992).

14. W.E. Vanderlinde, C.J. Von Benken, C. Davens, and A.R. Crockett, "Fast, clean low damage deprocessing using inductively coupled and RIE plasmas, "submitted for publication in *Proceedings of the 22th International Symposium for Testing and Failure Analysis,* ASM, Materials Park, Ohio, 1996.

15. W. E. Vanderlinde, C. J. Von Benken, and A. R. Crockett, "Rapid integrated circuit delayering without grass, "Proceedings of the 1996 SPIE Symposium on Microelectronics Manufacturing," pp. 260-271 (1996).

Metallographic Techniques for
Semiconductor Device Failure Analysis

Francis G. Trudeau and Thomas W. Joseph
IBM Microelectronics Division

HISTORICAL PERSPECTIVE

Semiconductor components are most likely to fail if subjected to harsh and abusive environments. However, it's the fail that occurs when those conditions are not present that warrant investigation, an investigation that ultimately will employ the use of metallographic techniques. Metallographic techniques are not new to the semiconductor industry. If fact, metallography has been used in the study of semiconductor fails for over three decades. The first fails analyzed were

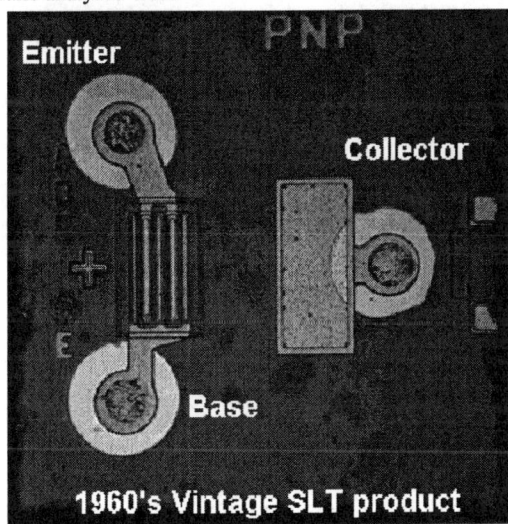

FIG 1. Optical photo of 1960's vintage single transistor chip.

discrete devices such as shown in figure 1. These single transistor chips were usually bevel sectioned and stained to highlight junctions and then inspected on optical microscopes, FIG 2. The beveled angle was used to expand the vertical geometry of the junctions to facilitate the measurement of their depths with an optical microscope. It was important to obtain accurate information on the beveled angle. This was accomplished with the use of a goniometer. Most cross sectioning was accomplished by encapsulating the devices in an epoxy compound, a technique that is still in use today, although not as frequently. The

encapsulated device would be metallographically prepared using a combination of silicon carbide and diamond abrasive polishing wheels. The

FIG 2. Beveled angle section highlighted to show placement of junctions.

final polishing would be accomplished on diamond impregnated polishing cloths mounted to polishing wheels. This technique worked fine for blind navigation into large structures. However, when a single artifact needed to be examined in cross section, the device had to be mounted on a stainless steel beveling block and cross sectioned using a frosted glass wheel. Aluminum slurry was applied to the rotating wheel to aid in both grinding unwanted materials away and as a lubricant to keep the device from chipping.

In the late 1960's and early 1970's, semiconductor technology matured as more devices found their way onto a single silicon chip. The introduction of the scanning electron microscope to the failure analysis lab increased imaging power many times over optical microscopy. There was no longer the need to optically expand the vertical geometry. Cross sections could be examined, as prepared at 90 degrees, allowing for geometries to be observed with a natural aspect ratio and allowing measurements to be taken directly from SEM micrographs. It also became increasingly difficult to navigate through epoxy to a defect at 90 degrees with an optical microscope. As a result, devices were being mounted to the vertical edge of beveled, stainless steel blocks and cross sectioned directly on the glass wheels.

As device geometries decreased and multiple layers of metallurgy were added, advancements in mounting techniques became necessary to maintain a level cross section face to accommodate the inspection of several features in a row, such as contacts and interconnects.

Fig 3. Photo of early 1970s vintage block.

More powerful SEMs became available in the late 1970's to help the failure analyst meet the demanding analysis challenges of 1 micron technology that was looming over the horizon in the next decade. New polishing media (Al2O3, SiC and diamond) on paper and mylar films were introduced as well as new slurries and polishing cloths. The surface of rotating glass wheels were no longer adequate for providing a suitable surface to cross section defects. 0.1 micron grit papers and mylar films and 0.05 micron diamond suspensions on nap less cloths paved the way towards analyzing sub micron defects. Innovative block and SEM mounting fixtures were designed and built to meet new sample handling challenges.

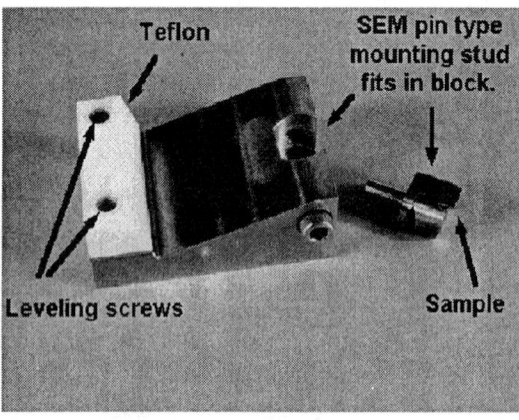

Fig 4. Photo of early 1980s vintage block.

As the 1980's gave way to the 1990's we saw the introduction of the focused ion beam as a promising new tool to take advantage of the metallographic techniques being applied on semiconductor devices to find root cause of fails. Metallographic delayering techniques are being used to expose hidden layers to facilitate navigation to devices that require TEM cross section preparation in the FIB. Fib voltage contrast of lower circuit levels require a similar intervention of metallographic techniques.

In the interest of enhancing performance by reducing circuit RC time constants, new materials are being used in the fabrication of semiconductor devices. This coupled with the drive to shrink dimensions to 0.1 microns will demand even more aggressive approaches to metallographic technique development in the future. The procedures that are described in the following sections are based on years of experience and are presented here to provide a foundation for those interested in meeting those challenges.

CHIP CROSS SECTIONING TECHNIQUES

Cross sectioning is a metallographic technique available to the semiconductor failure analyst that is used as a means, to pinpoint a location in a fabrication process where a defect occurs, verify design ground rules and to improve process yields. By selectively removing unwanted, bulk material, the analyst can obtain an alternate view of a structure under study, as well as defects embedded within these structures, that may not be observed in other ways. Valuable information, related to the position of artifacts within the vertical structure of a semiconductor device; can lead the analyst to the root cause of a failing component, alert production engineers to process variations responsible for reliability fails and lower yields and aid design engineers in understanding and predicting design models.

This selective removal of unwanted bulk material can be accomplished in several ways. It is the purpose of this section to provide a description of the variations available to the analyst so that an informed decision can be made in support of an analysis strategy that will assure success for the analyst. There are several choices available to accomplish any cross sectioning goal, including encapsulated cross sectioning, non-encapsulated cross sectioning, fracture cleaving and Focused Ion Beam milling as a method of cross sectioning. Each cross sectioning approach has it's benefits as well as complications. The differences between consumable materials as well as fixtures used to handle samples will be discussed as well as the variety of laboratory equipment that is required to support an analysis lab capable of providing full service cross sectioning capabilities.

ENCAPSULATED CROSS SECTIONING

In this first section a technique will be described that facilitates the easy mounting of samples for either angle or perpendicular sections. The technique is used on devices with or without full metallization and bonding pads and is not limited to discrete devices. There may be instances where it will be desirable to cross section through an entire packaged device to investigate wire bond integrity, materials analysis of ball bond interconnects or package materials. When polished sections of the highest quality are required on samples composed of several metals with different hardness, this procedure is used. Implementation of this procedure requires a special set of fixtures and tooling. Plastic molds are fabricated to produce 3X, 5X and 10X magnifications of the vertical structure under investigation as well as a perpendicular pre form. The angled pre forms were relied on in the past to expand the vertical structure because of the imaging limitations of optical microscopes. Today's labs are equipped with high powered scanning electron microscopes that eliminate this need to stretch the vertical dimension, and allows imaging of vertical structures, resulting in a 1:1 aspect ratio, that facilitates both qualitative and quantitative inspections. Figure 5 shows examples of mounting a die at a predetermined angle as well as 90 degrees by using these pre formed

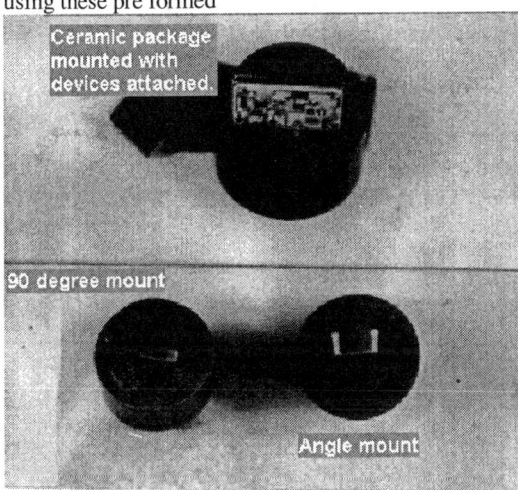

FIG 5. Examples of plastic mounts

plastic mounts. In situations where a precise angle must be maintained, the plastic mount can be placed in the polishing fixture, shown in figure 6, where the top of the mount can be polished away to define an angle that the device is aligned to during the mounting procedure.

SAMPLE PRE-PREPARATION

The materials required to prepare the sample for encapsulated cross sectioning include plastic molds , a suitable adhesive, a

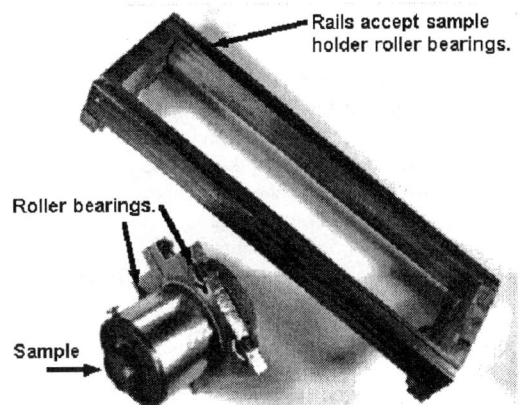

FIG 6. Polishing Fixture

low power stereo microscope (up to 40X), a scale to assure accuracy when mixing two part epoxies, tweezers, a goniometer if precise angle measurements are going to be required, and mating plastic thin walled cups that the plastic molds tightly fit into.

A silicon chip is bonded to the angled surface on the plastic mold with a suitable adhesive. This can either be double sided tape or Super-glue (polymethyl-methacrylate). If using Super-glue, a thin, uniform layer of the adhesive is spread on the surface to keep the sample at the desired angle. At this point in the preparation, a low powered stereo microscope is used to aid in the alignment of the device to the polished edge (flat part) of the mount. Care must be taken at this point to assure that the feature of interest be positioned above the polished surface (flat part) of the mount resulting in the final position of the polished plane passing through only the device and the epoxy. Any size device, up to the diameter of the plastic mold, can be mounted in this manner including packaged devices. When mounting packaged devices, it may be necessary to cut off pins and grind the backside of the package to provide a planar surface to help with adhesion to the plastic mold. Some semiconductor devices, when bonded to their lead frames, can be slightly tipped, resulting in an unknown angle after polishing is complete. If an accurate account of the polished angle is required measurements can be made with a optical goniometer with an accuracy of degrees and minutes. A goniometer uses a focused light beam that is reflected off of a beveled surface resulting in the splitting of the light beam. The part of the focused light beam that is reflected off of the unpolished surface is projected onto a referenced comparator. The goniometer is then rotated so that the light beam that is projected from the polished, beveled surface lines up with the initial referenced point. The angle is measured by the goniometer as the difference between these two referenced points.

After the sample is secured to the plastic mount, as previously described, the plastic mount is then placed into a slightly larger, thinned walled, plastic container. The epoxy is then poured into the container, completely covering the plastic mount and sample. Areldite has been found to be a most desirable epoxy. It has proven to be sufficiently hard in preventing rounding of edges during angle polishing as well as being resistant to many of the acids used in junction delineation etches. Areldite's low viscosity allows it to flow into small cavities, allowing it to adhere to the smallest surface features which helps reduce separation during polishing. Room temperature curing can take up to 10 hours, however, this can be reduced to 10 minutes with a curing temperature of 55 degrees C.

CROSS SECTIONING PROCEDURE

Rough Grinding

The first step in performing an encapsulated cross section is very important in determining a successful and accurate result. For an encapsulated cross section to be successful, the specimen must be kept perpendicular to the polished surface at all times during the rough grinding step, a task that is nearly impossible to accomplish by hand. Therefore, a suitable fixture was fabricated to tightly hold the mounted sample during this process. Figure 4 shows a mounted sample held securely in the polishing fixture. To maintain a perpendicular relationship between the sample and the polished surface, during the bulk removal process, a set of rails were designed to accommodate the fixture. The entire fixture is placed on these runners (rails), shown in the background of figure 4, and is allowed to move along the rails in a back and forth direction with the aid of roller bearings. Bulk material removal is accomplished by placing the fixture and rail on a coarse grit media and moving the sample along the rails in a back and forth direction. Adjustment of the sample toward or away from the grinding media is accomplished with the micrometer that is built into the fixture. Micrometer adjustments are made to control the amount of material to be removed, down to an accuracy of 0.5 mils. This feature is important when approaching the target area on the sample.

Prior to setting the mounted sample in the fixture, the top edges of the Areldite should be rounded off to prevent tearing of the grinding media. Preliminary bulk removal of Areldite is started on a 320-grit silicon carbide paper, lubricated with water. Grinding should continue until only a thin amount of Areldite remains over the sample. Care should be taken not to grind too close to the target area with this rough media, severe damage could result to the brittle silicon resulting in loss of time and effort, not to mention the area of interest. With a thin layer of Areldite still around the device, the media should be changed to a 440 grit and then a 600 grit paper. The edge of the semiconductor device should be exposed with the 600 grit paper. The progress of the bulk silicon removal should always be checked under the microscope. Before the feature of interest is reached, the 600 grit polishing should be terminated.

Intermediate Grinding & Fine Polishing

The polishing of the silicon will be accomplished on much finer grit diamond paste and Al2O3 on cloths. The plastic mount can be removed from the grinding fixture and the edges rounded off, a second time, on a rotating grinding wheel. This is necessary to prevent the sharp edge of the mount from grabbing and tearing the polishing cloth. The sample should be thoroughly washed before preceding with the following steps.

During intermediate and final polishing, diamond paste on PAN-K pellon paper was found most satisfactory for suitable preparation of silicon surfaces. The hardness of the diamond provides uniform abrasion producing a flat surface practically free of polishing damage. Clean, fast-cutting action is obtained while extreme long life of the media is assured. A medium grade of diamond is used (4um - 5um). The coarse grades tend to chip and gouge semiconductor materials. The low nap feature of the PAN-K paper helps keep the sample flat during this stage of the polishing as well as act as a soft cushion that hold the diamond particles security. Polishing with the 4um - 5um diamond paste is continued until a clear surface is obtained. A high wheel speed promotes clean and fast cutting. The area of interest, for large defects, is reached using this step. Subsequent steps will remove any scratches, but will be done so at a much slower pace.

The specimen is transferred to another polishing wheel that has been setup with 1um - 2um diamond paste on PAN-K paper to remove the 4um & 5um damage. Any remaining scratches are then removed on another wheel with 0.25um diamond slurry on "Struers MOL" cloth. A clean-up polish, using 0.1um Al2O3 powder on "Buehler Microcloth, should only be necessary for a few seconds on a rotating wheel. Both of these final polishing steps should be kept to a minimum to prevent rounding of the sample in the area of interest. The polished semiconductor should be scratch free to obtain reproducible sample delineation results.

A successful alternate method of eliminating surface damage is to use colloidal silica slurry for final fine polishing in place of the 0.5um diamond paste and the 0.1um Al2O3 slurry. This procedure has proven to be a reliable sample preparation technique for imaging AlCu via interconnect structures, uncovering mechanisms such as resistive interfacial films and micro cracks.

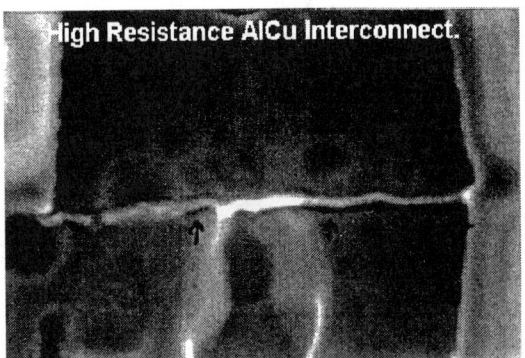

Fig 7. Resistive Interfacial Film in AlCu Interconnect.

When using colloidal silica slurry, it is very important that the polishing wheel be free of any silica particles that may have separated from the slurry prior to placing the sample on the wheel. This is accomplished by lightly touching the rotating wheel (low to medium speeds) with your fingers while colloidal silica is slowly dripped onto the low nap cloth. If any particles are detected, then the wheel needs to be cleaned by flushing with water while lightly scraping the polishing cloth with the edge of a stainless steel block.

Unlike the diamond paste and Al2O3 polishing sequences for encapsulated cross sectioning previously mentioned, the colloidal silica slurry procedure is more successful when a hands free set up can be accomplished. The only additional items needed to accomplish this are a ring stand, a 500ml separatory funnel with a stopcock plug, Tygon tubing and a hose cock clamp. The Tygon tubing is fitted over the lower end of the funnel containing the silica slurry allowing delivery of the slurry to the rotating wheel. The hose cock clamp controls the colloidal silica's rate of delivery to the rotating wheel. An automatic polishing arm holds the encapsulated sample and moves it across the rotating wheel in a periodic fashion. Less than 100 grams of weight are placed on the sample during this final polishing procedure. This hands free operation results in a procedure that is repeatable by removing the variations associated with holding the sample on the wheel such as angle and pressure on the sample as well as rotation of the sample on the wheel. Once a wheel has been set up for polishing with colloidal silica slurry it becomes dedicated for that purpose because it is not compatible with the diamond paste and Al2O3 slurries.

An analyst should always be aware of the type of material that is being prepared for micro-inspection. Advanced semiconductor processing incorporates many different materials, of different characteristics, that are interfaced together. There is no single final polishing approach that will guarantee the same high quality results desired for all the interface combinations you will encounter on today's technology. During an analysis you can encounter layered, composite or homogeneous structures consisting entirely of hard materials, hard and soft

materials or soft materials only. The only difference in the final polishing approach for this variation in material hardness is the use of diamond slurries in place of Al2O3 or silica slurries. When polishing hard materials, as well as hard materials interfaced with soft metals, the diamond slurries do the best job in maintaining a flat surface while providing a highly polished surface.

Diagram #1 is a graphical representation of approaching a defect consisting of hard materials only.

Metallographic Techniques

Polishing Hard Structures

GOAL: TO END UP WITH THE BEST MECHANICAL
POLISH FOR HIGHLIGHTING INTERFACES.

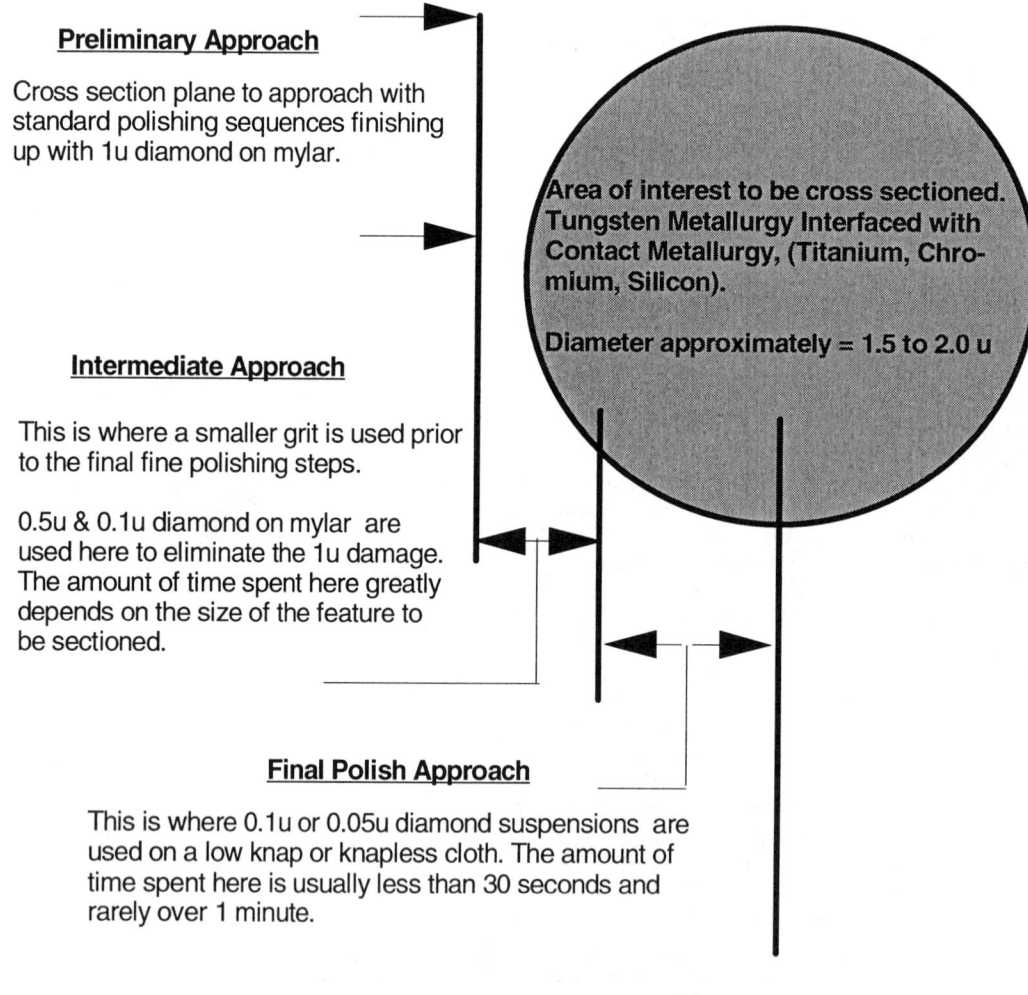

Preliminary Approach

Cross section plane to approach with
standard polishing sequences finishing
up with 1u diamond on mylar.

Area of interest to be cross sectioned.
Tungsten Metallurgy Interfaced with
Contact Metallurgy, (Titanium, Chromium, Silicon).

Diameter approximately = 1.5 to 2.0 u

Intermediate Approach

This is where a smaller grit is used prior
to the final fine polishing steps.

0.5u & 0.1u diamond on mylar are
used here to eliminate the 1u damage.
The amount of time spent here greatly
depends on the size of the feature to
be sectioned.

Final Polish Approach

This is where 0.1u or 0.05u diamond suspensions are
used on a low knap or knapless cloth. The amount of
time spent here is usually less than 30 seconds and
rarely over 1 minute.

DIAGRAM #1

Metallographic Techniques

Polishing Soft Structures

GOAL: TO END UP WITH THE BEST MECHANICAL
POLISH FOR HIGHLIGHTING INTERFACES.

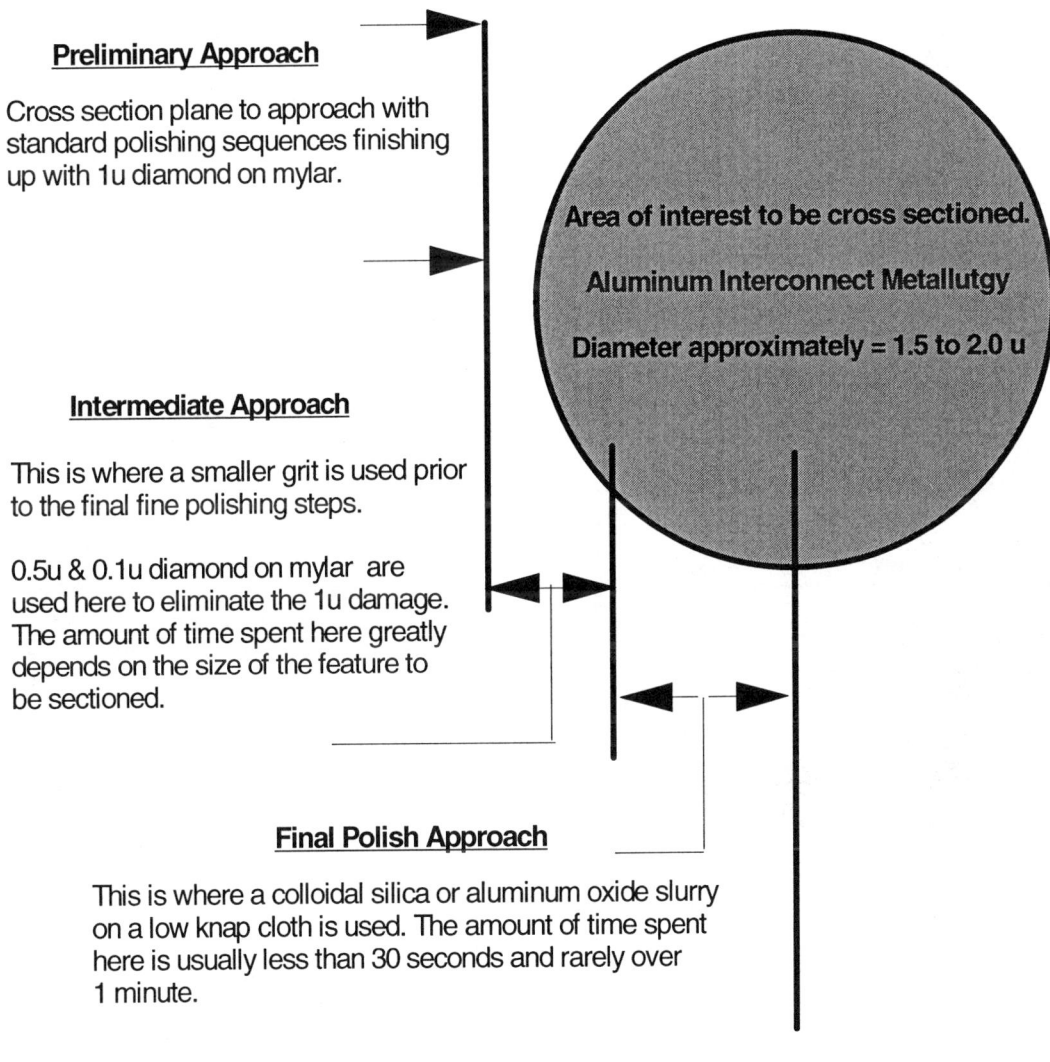

Preliminary Approach

Cross section plane to approach with
standard polishing sequences finishing
up with 1u diamond on mylar.

Area of interest to be cross sectioned.

Aluminum Interconnect Metallutgy

Diameter approximately = 1.5 to 2.0 u

Intermediate Approach

This is where a smaller grit is used prior
to the final fine polishing steps.

0.5u & 0.1u diamond on mylar are
used here to eliminate the 1u damage.
The amount of time spent here greatly
depends on the size of the feature to
be sectioned.

Final Polish Approach

This is where a colloidal silica or aluminum oxide slurry
on a low knap cloth is used. The amount of time spent
here is usually less than 30 seconds and rarely over
1 minute.

DIAGRAM #2

Diagram #2, shown on the next page, is a graphic that shows the approach that should be considered when polishing soft structures as well as soft structures interfaced to hard materials.

NON-ENCAPSULATED CROSS SECTIONING

Early non-encapsulated cross sectioning was accomplished on devices without metallization, where junction depths were measured from beveled sections. Devices were glued down to stainless steel blocks with known angles, polished on glass wheels and stained for junction control. When device geometries got smaller this technique was adapted to 90 degree sectioning. The devices were simply attached to the top of the stainless steel blocks with glycol phthalate wax and polished on a glass wheel. Although this early method did not produce metallographic quality surfaces, the benefits were numerous. The devices were no longer encased in a semi transparent media. This allowed optical inspection of the top surface of the device throughout the cross sectioning procedure resulting in a high degree of spatial precision on dense circuit patterns. Sample handling was much easier, allowing for additional analysis techniques to be applied if needed. It didn't take long to realize that the quality of the polished surface need to be improved. Aluminum metallization land patterns and interconnects were soft, and as a result, prone to distortion by the surface of the glass wheel as shown in figure #8.

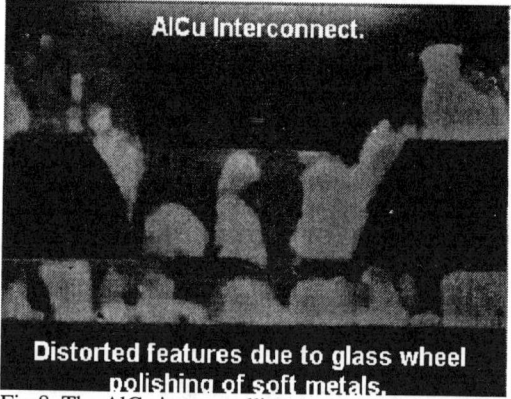

Fig 8. The AlCu intermetallics are smeared and not cut by the glass wheel polishing.

It became important to produce a damage free surface in order to study metallization and interconnect fails which became the dominant factor limiting VLSI device reliability in the 1980's. Fail modes such as electromigration, corrosion, stress induced deformation and resistive interfacial films have been successfully analyzed using these new polishing procedures developed for non-encapsulated cross sectioning.

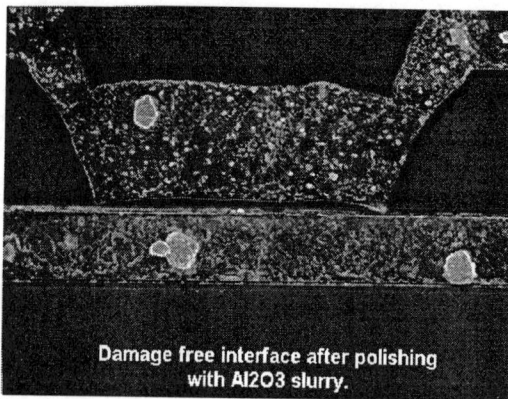

Fig 9. Final polish was with an Al2O3 slurry on a polishing cloth.

Cross Sectioning Procedure

This procedure for non-encapsulated cross sectioning will include sample pre-preparation, mounting techniques, sectioning soft metals, sectioning hard metals and navigational considerations.

Sample Preparation

In the previous sections, there wasn't much mention about the size of a sample and the location of a defect on that sample. With earlier technology, the samples were just broken, via scribe and cleave, to a manageable size, mounted and polished without too much concern with navigation to the defect. Defects occurring on today's technology demand considerable thought on how to prepare the sample prior to mounting it for cross sectioning. The defects, that are responsible for today's fails, may have many hours of analysis invested in them by the time a decision is made to perform a cross section. This demands that accurate navigational skills be applied to assure success in hitting your target to keep loss of time and resources to a minimum. The final size of the sample must also be taken into consideration, depending on the type of inspection tool that will be used. SEM stages and sample holders vary considerably. In-lens type stages can limit the size of a sample to just 3mm to 4mm high compared to conventional type SEM stages that can accept an entire cross section block.

It is important that the area of interest be recorded with photos. Navigational marks placed around the area of interest can make approaching the defect during cross sectioning much easier. Lasers have been used in the past for this purpose, with excellent success, but care should be exercised so not to be too robust in the marking. Too large, or deep of a mark could setup stresses in the brittle

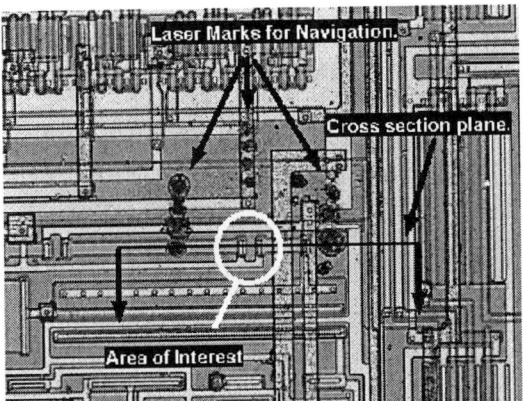

Fig 10. An example of marking your approach to the defect with a laser.

semiconductor material that could result in a fracture in the cross section in a most undesirable location. An alternate to laser marking a defect area is to use the FIB tool to do the job. The FIB tool is extremely accurate resulting in a clean, controllable cut that is easy to navigate to while cross sectioning.

Once you have the sample sized properly and the area of interest is well documented and marked for navigation you need to mount the sample onto the cross sectioning fixture. The fixture shown in figure 12 is another innovative design resulting in a common polishing and SEM fixture that minimizes sample handling. The sample is

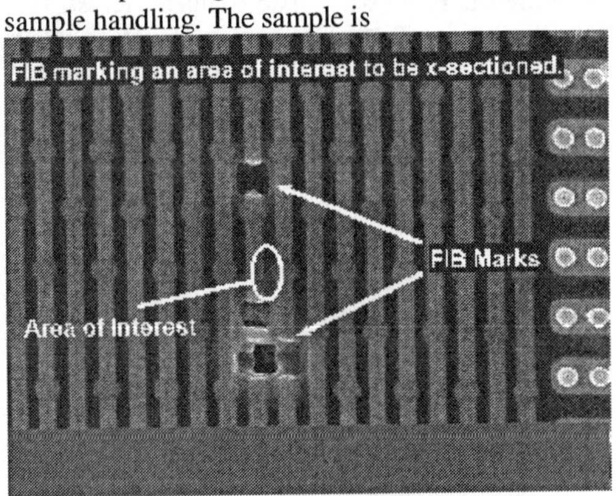

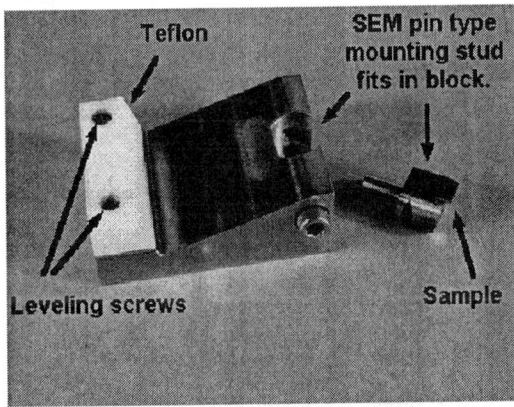

Fig 11. An example of FIB navigation marks.

mounted on the stainless steel insert with glycol phthalate wax, using a hot plate set to approximately 130 C. The sample should be aligned with the area of interest and protruding out past the edge of the insert approximately 1/8 inch. This alignment is easily accomplished with the use of a stereo microscope. It is very important to remove all excess glycol wax with acetone to keep the sample as clean as possible during the polishing. One method that works well in removing the fillet of wax behind the sample is to wet a Q-tip with acetone and gently rub the wax while rotating the Q-tip. The insert is then placed into the polishing block. After all polishing and surface etching is performed, the entire insert, with the sample, is placed into the SEM chamber for inspection. Additional polishing can be performed

after SEM analysis without the need of remounting the sample.

Leveling of the sample is important if you need to polish into a row of structures. This should be accomplished early on in the polishing and should be checked and corrected with each change that is made in the grit that is used to remove material. When viewed on an inverted microscope, the degree to which the polished edge is parallel to the target row of structures can be determined. Adjustments can be made with the leveling screws, located at the opposite end of the polishing block from the sample, to steer the polished edge into a parallel approach. The amount that the screws need to be turned will be learned with experience.

Fig 12. Block with leveling screws

Polishing is accomplished by using various grit Al-Oxide, Silicon Carbide or Diamond Polishing Pads followed by either Al2O3, Silicon oxide or diamond slurries. The heavy grit media is either a mylar or paper pad that is used for bulk material removal. Typical grits used are 30, 15, 6, 3, 1, 0.5 and 0.1 micron. The grit to be used depends on the distance to be polished and the desired finish. Heavier grits are used for rapid removal of material and finer grits are used to achieve a more polished surface.

Typically you would start sectioning with a 15 micron grit media. Use water as a lubricant. If the distance to the defect is far away (>6mm), start with a 30 micron grit media. If the distance to the defect is less than 4mm, then start out with a smaller grit media such as 5 microns. When you get to within 2 microns of the defect area you need to switch to a 1 micron grit media. Stay on the 1 micron grit until you are approximately at the edge of defect area.

Up to this point in the sectioning procedure it didn't make too much difference to what type media was used. Although the diamond grit on mylar cuts faster, lasts longer and results in a somewhat cleaner edge, it is more expensive. The Al2O3 and silicon carbide on paper is less expensive, but it will wear out faster resulting is a less consistent rate of removal. Care should be taken when changing to a fresh pad in the middle of an approach to your defect. The unanticipated increase in removal rate could result in sectioning through the defect.

The materials used for the final fine polishing of the section is determined by the material that is being polished. For materials that are soft, such as Aluminum interconnect material, you need to proceed with the section using a 0.5 micron diamond grit on mylar film followed by a 0.1 micron diamond grit on mylar with touch up being performed with Al2O3 slurry, such as Glanzox.

The 0.5 micron polishing should continue until you are approximately ¼ of the way into the defect for a defect greater than 1 micron in diameter. For defects that are smaller than 1 micron, you need to polish just up to the edge of the defect with the 0.5 micron diamond mylar film. The 0.5 micron damage is then removed by switching to a 0.1 micron diamond grit on mylar film. Wheel speeds are generally set to be less than 100 RPM resulting in polishing times ranging from a few seconds on upwards to a minute. Too much time spent on the polishing wheel here will result in rounding of the edge.

The final polish is completed using 0.1 and 0.05 micron colloidal silica or aluminum oxide slurry on a low nap, or nap less, polishing cloth. The amount of time spent here is usually less than 30 seconds with a wheel speed less than 10 RPM. The direction of polishing is usually performed from the back end of the polishing cloth towards the front end where the sample is mounted. This is opposite the direction used when using the mylar or paper backed media.

The final polish sequence for hard materials, such as tungsten metallurgy, involves the use of 0.1 and 0.05 micron diamond slurries in place of the aluminum oxide or colloidal silica slurries. The time spent on using the 0.05u diamond slurry is very short, usually several seconds, and rarely over 30 seconds. A low nap, or nap less cloth is used for this final polish and the wheel speed can be as low as 1 RPM, especially when applying the final polish on a very small feature. An example of this 0.05 final polish is shown in the following photo where both soft and hard materials were polished to the same plane.

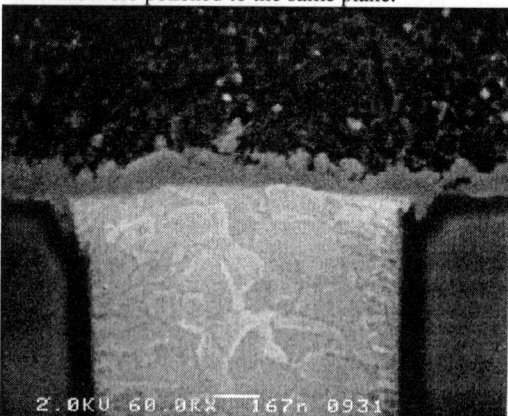

Fig 13. Final polish example of an interface consisting of hard and soft materials.

FRACTURE CLEAVING

Single crystal Silicon's brittle nature and propensity to fracture along crystal planes make fracture cleaving the quickest method of preparing a cross section. With practice, accuracy of about 500uM can be achieved with a simple carbide scribe. Commercial equipment exists which is accurate to less than 10uM. Fracture cleaving is most often used to measure blanket layer thickness or to examine the vertical structure of large arrays of features.

A disadvantage of this technique is the mechanical stress created during the fracture can damage structures of interest or create artifacts that obscure desired details. Soft materials which are present (e.g. Al, polyimide) do not fracture well and become distorted. Often a decoration etch (see section below) is needed to highlight the layers which are present.

A fracture cleave can be done on an in-process or completed device. Full wafers are easiest to start with but pieces as small as 5 mm on a side can be cleaved successfully. The simplest method of preparing a cleaved section is to nick the edge of the wafer (or die) with a carbide scribe so that the cleave line will follow the crystal planes through the area of interest to the opposite edge of the wafer. The point of the scribe is then placed under the wafer at the nicked spot and gentle pressure is applied on either side of the nick to fracture the wafer. A technique which improves the accuracy of cleaving uses a laser to create a trench or closely spaced holes across the wafer. Typically an Nd:YAG laser of 30 or more watts focused through a suitable microscope is used. Starting at the area of interest and following the crystal planes to the edge of the sample, successive laser shots are used to define the cleavage plane. The wafer is then held and gently flexed so that it will break along the laser scribed line.

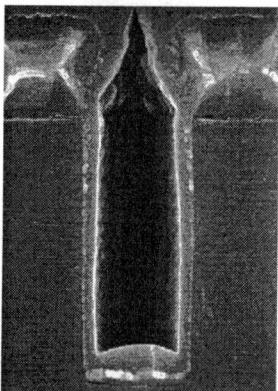

FIG. 14 SEM view of cleaved via. Al fracture is ductile but liner has brittle fracture and can be characterized.

Commercially available systems use a third method of cleaving which is the most accurate. It uses a scribe which is positioned under a microscope to score the edge of the wafer. Then, with the wafer under tension perpendicular to the desired section plane, a hammer impacts the edge of the wafer opposite to the scribed edge causing the wafer to fracture.

FOCUSED ION BEAM CROSS SECTIONS

Preparation of cross sections by focused ion beam (FIB) milling through a defect or layered stack of materials has several advantages. First, FIB is inherently clean in that it is performed in a vacuum and the only material present, other than the sample itself, is the beam material (typically Gallium). There is no opportunity for contamination from polishing media or rinsing, especially important where chemical analysis will be performed. Second, only a small area of the sample undergoes analysis. This leaves the remainder of the sample for other analyses, or in the case of a wafer, for further processing. Third, ion bombardment is gentle compared to mechanical polishing so that voids are not distorted, delaminating layers do not peel apart, and soft materials are less likely to smear.

Several aspects of FIB cross sectioning can also be a disadvantage. These include practical size limitations on the size of the area to be sectioned due to milling times. If many structures in a plane of section need to be examined, FIB may be time prohibitive. For a true 90 degree section, a piece needs to be cut out and mechanically polished close to the area of interest before performing FIB. Likewise for good signal collection in materials analysis (EDS, Auger, etc.) a true 90 degree section is required. Materials with dissimilar milling rates or voids may cause artifacts by allowing the beam to mill faster or "smear" material (called the waterfall effect) down the face of the section.

The details of focused ion beam operation and use are treated thoroughly in the chapter dedicated to this subject. The discussion that follows will specifically address some of the important aspects of FIB as it pertains to cross sectioning.

Sample pre-preparation for FIB cross sectioning is similar to that of unencapsulated mechanical sectioning. The area of interest should be documented and noted with respect to easily identifiable features or marked with a laser. This is especially important for FIB as the image of the sample seen in the instrument is only of the (near) surface. Structures below the surface will not be visible and must be able to be located with respect to a surface feature, such as a laser mark or device structure.

Since material is removed from the surface of the sample during image formation in the FIB, care must be taken when defects are at or near the surface. Often it is prudent to coat the area of interest with several hundred nanometers of oxide or metal film to prevent erosion of the entire surface (and the area of interest). Both metals and oxides can be deposited in-situ with the FIB or in a dedicated deposition system prior to sample loading in the FIB. With the sample loaded in the FIB and

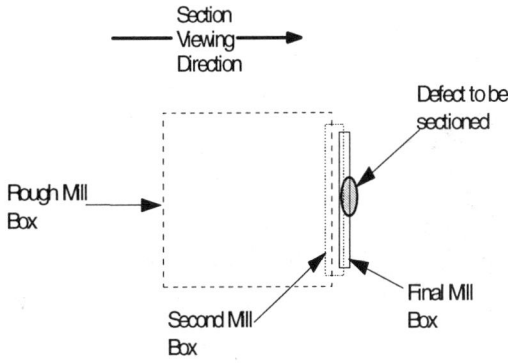

FIG 15. Plan-view schematic showing placement of milling boxes for FIB cross section

the region of interest located, a milling box is set so that one edge is parallel to the desired plane of section and a micron or so from this plane. This box must be long enough in the direction perpendicular to the plane of interest so that the sample can be tilted and the wall of the box through the area of interest viewed in the FIB or in an SEM. A typical viewing angle of 70 degrees requires a box with a 3:1 length to depth ratio. To save milling time, the box can be made shallower at the side away from the desired plane of section and taper down towards the plane of section.

Once the first box has been milled (analogous to rough polishing), successively smaller slices can be taken from the edge of the box nearest to the region of interest until the desired plane is achieved. The actual plane of section can be determined by either directly viewing the face of the section or by using the reference marks on the sample surface.

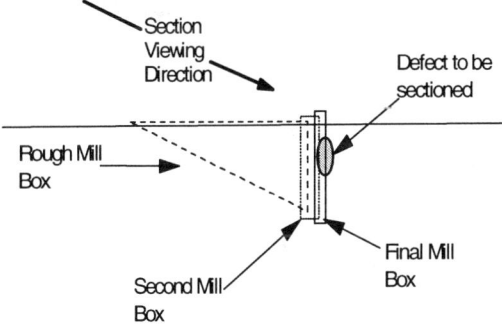

Fig. 16. Cross section schematic showing placement of milling boxes and viewing angle for FIB cross section.

When the desired plane is achieved, analysis can proceed by performing chemical delineation, SEM imaging, etc.

In the case where a true 90 degree section is to be prepared by FIB, an unencapsulated section must first be prepared. Mechanical polishing is stopped

263

approximately 10 to 20um from the desired plane of section. The sample is then loaded into the FIB, and using suitable navigation aids, a box is milled from the polished edge back through the plane of interest. Successively smaller slices may be taken from the back of the box to place the plane of section exactly in the desired area.

Accuracy of cross sectioning in the FIB depends on several variables but can be in the vicinity of 200nm. Important aspects of

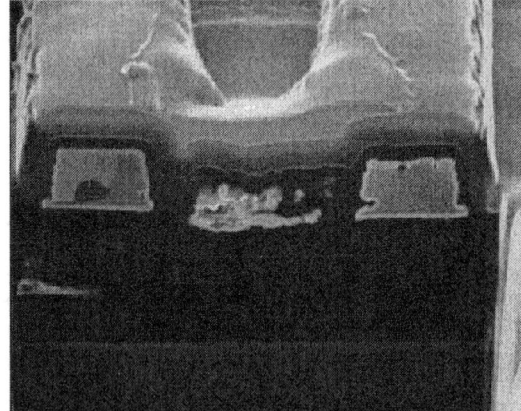

Fig 17. FIB cross section through foreign material embedded between metal lines. Tungsten layer was deposited over defect area before sectioning.

instrument capability are beam size, stage stability and image drift. Operator skill plays a large role as experience with a particular type of sample, materials, etc. can help in selecting the operational parameters for a particular analysis. Placement of navigation aids and the ability of the FIB operator to determine exactly where the plane of interest lies will also affect the outcome.

DELINEATION OF CROSS SECTIONS

Often a delineation or highlighting technique must be used on a cross section to make layers distinguishable. Highlighting techniques include chemical etching (wet or dry) and ion milling. These techniques are especially important in secondary electron imaging of cross sections as little signal contrast is generated on an extremely smooth polished surface.

It should be noted that delineation can cause artifacts and that in general, the lighter the delineation, the better. Etching times will vary based on the sample and the desired effect but are often in the range of seconds to a minute. It is helpful to know the materials present (layers, etc.) and what layer is to be highlighted at the expense of another layer when choosing a delineation technique. A reference sample can be prepared along with a defective sample to help determine what etch is to be used and if any artifacts are created.

In the case of an unknown sample or material (such as a foreign particle in a process stack), choice of an etch can help identify it.

Fig. 18. FIB cross section in two planes through contact giving three dimensional view.

This procedure may be an iterative approach including repolishing, re-etching with a different etch, and SEM examination to determine if the unknown has reacted.

Wet chemical etches for cross sections fall generally into the categories of oxide etches, metal etches, and Silicon etches. Many etches will attack other materials but this general categorization is helpful in choosing an etch. Several commonly used etches will be covered but it must be remembered that there are always tradeoffs when choosing an etch; see the chapter on etch formulations for a broader view and more detailed descriptions.

Wet etches are typically applied either by immersing the sample in a beaker of solution or by applying a droplet of etch to the sample with an eyedropper. The sample should be rinsed thoroughly in deionized water after etching and dried completely before further analysis is performed.

Hydrofluoric acid based etches are typically used for highlighting oxide layers. HF is often buffered (e.g. BOE) with ammonium fluoride to give controllable etch rates. Glycerin may be added to BOE to inhibit etching of metals while still etching oxides.

Metal etches must be chosen for the particular material to be highlighted. Aluminum can be decorated with dilute Hydrochloric or Phosphoric acid. Tungsten and Titanium can be decorated with Huang A ($H2O/H2O2/NH4OH$) etch. Interfaces between metal layers require an etch formulation that will distinguish the interface from the layers themselves. For example, a TiN barrier between Al layers can be highlighted using a dilute HCl etch.

Silicon etches are commonly used to show junctions but may also be used to highlight polysilicon layers. These include HF-HNO3 based (e.g. Dash) etches and CrO3-HF based (e.g. Sirtl, Wright) etches. These etches are very aggressive towards metals and oxides which must often be sacrificed to show junctions.

Dry chemical etches can be used in either the isotropic form (plasma) or anisotropic form (RIE) to decorate cross sections. They provide an often cleaner and more controllable method for decoration than wet etches.

CF4 based etches have fairly good selectivity with polysilicon and Silicon being attacked the fastest, silicon nitride being attacked less, and silicon dioxide and metals (with the exception of silicides) being attacked least. O2 may be mixed with CF4 to vary etch rates and selectivity. Dry etches are generally a good first pass etch if unknown materials are present as they are the least invasive and usually allow for repolishing and re-etching without sacrificing much material.

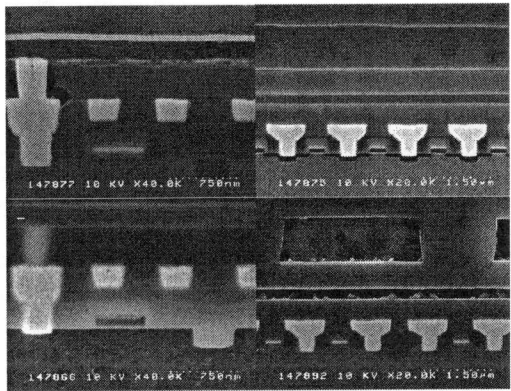

FIG. 19. SEM images of similar die structures after different highlighting etches. Clockwise from top left: 7:1 BOE, Dash etch, dilute HCl etch, CF4 plasma etch.

SF6 can be used as a very selective etch for polysilicon and Si while CHF3 can be used as an oxide etch. Various combinations of gases can be developed for specific applications. Tool parameters (power, pressure, etc.) can also be varied to achieve different effects.

Ion milling of cross sections can be performed to clean the specimen surface or to give slight relief to aid in SEM examination. In contrast to the focused ion beam, the ion mill uses a broad beam, typically several mm in diameter, at an oblique angle to the surface. The most common ion used is Argon and some instruments allow the sample to be rotated during milling to provide more uniform material removal.

Polishing into an area of interest and ion milling for 30 seconds or one minute can provide enough topography to make secondary electron imaging possible while preserving the sample for further polishing or etching. This brief milling will also help to eliminate smearing of soft materials that may have occurred during polishing.

Another application of the ion mill is to allow sectioning into a defect without introducing polishing slurry. For example, in the case of a corroded or voided metal line, the final polish will be stopped so that the feature is still encapsulated in several hundred nM of insulator. Subsequent ion milling will then break into the area of interest without the introduction of foreign material. Also, the physical structure of delicate material such as voids or interfacial separations will not be changed. Chemical analysis can then be performed (EDS, Auger, etc.) to determine what materials are present.

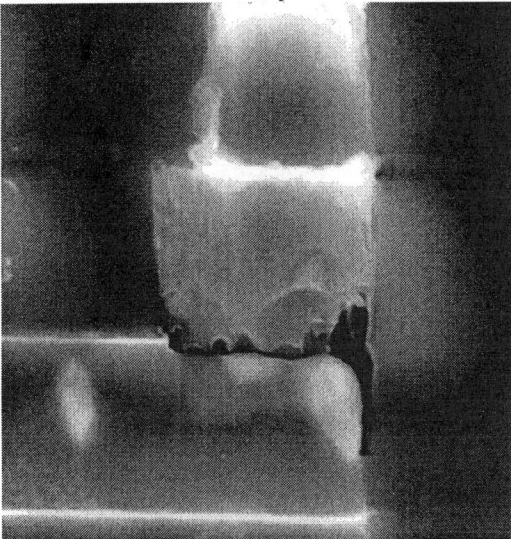

Fig. 20. Ion mill cross section showing interfacial separation in contact.

It should be noted that ion milling for long periods of time will create undesirable topography and make observation difficult. This is due primarily to differential milling rates between different materials which results in roughening of the surface as longer milling times are used.

CONCLUSION

Cross sectioning is a powerful approach to understanding integrated circuit structures and determining the cause of many of the defects encountered by the failure analyst. Vertical and layered structures can be characterized and compared to expectations. Defects can be observed in relation to the process stack, giving unambiguous identification of the process step where they originated. Several different cross sectioning techniques have been described which the analyst can choose from based on the type of sample, what data is required, and available equipment. With practice and patience, these techniques can be successfully implemented in any failure analysis laboratory.

REFERENCES

1. Dansky, A. H., 1881. "Bipolar Circuit Design for a 5000-Circuit VLSI Gate Array," IBM Journal of Research and Development: 25, 116.

2. Danyew, R. R. and Hammond, B. R., 1989, "A Micro Finishing Technique For Semiconductor Failure Analysis". International Symposium for Testing and Failure Analysis Proceedings, pp. 161-165.

3. Favaron, J., 1990. "Plasma Delineation of Silicon Chip Cross Sections". International Symposium for Testing and Failure Analysis Proceedings, pp. 117-120.

4. Gajda, J. J., 1993. "Chip and Device Sectioning Techniques", Microelectronic Failure Analysis Desk Reference, 3RD Edition, pp. 97-110.

5. Gajda, J. J., 1974. Techniques in Failure Analysis. Proceedings of International Reliability Physics Symposium, pp. 30-37.

Physical/Chemical Defect Characterization

Optical Microscopy

Lawrence C. Wagner
Texas Instruments, Inc.
Dallas, Texas

INTRODUCTION

Optical microscopy remains a mainstay of the failure analyst. However, a shift in usage can readily be noted towards more SEM (scanning electron microscope) utilization relative to optical microscopy. This is in large part due to the resolution and depth of field limitations of the optical microscope, which become much more significant with smaller geometry devices and increased number of layers in devices respectively. It is also a reflection of the increased ease of use of the scanning electron microscope (SEM) and enhancements to resolution achieved with field emission sources. However, optical microscopy continues to have one significant advantage over SEM in that defects can be viewed through transparent dielectrics. Other advantages are ease of use and straightforward interpretation. While brightfield/darkfield applications are the dominant application of the optical microscope in failure analysis, other modes such as interference contrast find useful niche applications. For example, interference contrast is particularly useful in the documentation of crystalline defects after deprocessing to silicon and decoration. Fluorescence can prove useful in the evaluation of contamination, for example observing the extent of an organic contaminant. Polarized light microscopy finds applications in such areas as enhancement of liquid crystal transitions.

DISCUSSION AND GENERAL PRINCIPLES

While a detailed understanding of the optical principles behind the microscope are not essential for the failure analyst, a general understanding of some of the principles are helpful. Some of these basic principles will be discussed below.

Resolution

The concept of resolution is very significant for semiconductor applications because of the small feature sizes. Resolution is often defined as the ability to resolve or define and or space between small objects. It can often be better understood in terms of its negative results. As resolution becomes worse, an edge or space becomes less defined and more blurred. This can be well understood through an evaluation of light sources. A physical limit to resolution is defined by the wavelength of the incident radiation. Although the discussion below largely ignores this fundamental limit, it remains an important element in microscopy. This type of fundamental limit, for example, dictates that the ultimate resolution of IR optical systems is not as good as a visible light microscope. It also dictates that SEM has potentially and in fact has better resolution because the wavelength of the electrons is shorter than the wavelength for visible light.

Light sources

The source of the light is one of the key factors, which impacts resolution. A near-point source can result in improved resolution as illustrated in Figure 1. (This example actually can arise in X-radiography.)

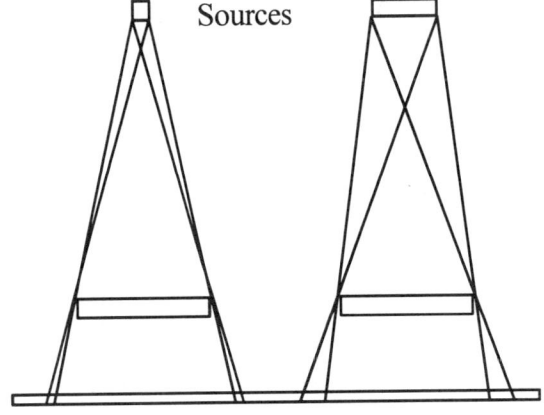

Sources

Fig. 1. This simple example illustrates how light source size impacts resolution. A smaller light source produces improved resolution.

To understand how the light source impacts resolution, consider the case shown of an edge of the sample illuminated from opposite sides of the light source and examine the exposure or image on the plane below. With a small light source, the gray area, which is the image of the edge, is small or exhibits good resolution. With a larger source, the gray area is broad, creating and edge that is more blurred. The introduction of apertures and lens complicate the situation but the same general principles apply: A more point-like source will result in improved image resolutions

Apertures

The use of apertures is a concept common to many forms of microscopy. Optical microscopes generally include apertures to control the light entering the optical system. The apertures in the optical microscope perform much the same function as apertures in the SEM. Smaller apertures close off parts of the incident radiation source. This generally results in an increase in resolution but with a corresponding loss in depth of field. This trade off of resolution for depth of field is common to many forms of microscopy. The impact of apertures can be understood in terms of the simple example shown in figure 2.

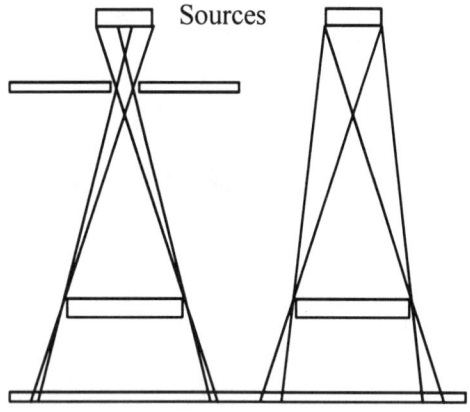

Sources

Fig. 2. This simple example illustrates how apertures impact resolution.

Here a large light source is used. In the case without an aperture, a poorly defined edge is imaged on the lower plane. In the second case, the aperture is reduces the light illuminating the sample and the imaged becomes fainter but the area of the light source, which actually illuminates the edge is reduces. This is equivalent to having a smaller light source. In a limiting case, we can achieve resolutions approaching the physical or diffraction limits but not have a very useful image. With SEM's, this tradeoff is very apparent, as the spot size is reduced, the resolution is improved but the image becomes very grainy.

Confocal microscopy provides a method for obtaining the maximum physically possible resolution from optical microscopy but with a minimum depth of field. Microscopes, in general, have a diffraction-limited resolution, which is on the order of magnitude of the wavelength of the incident radiation. The confocal microscope relies on a very small aperture, a pinhole to enhance resolution. Resolution is enhanced in both optical and electron microscopy through the use of smaller apertures. This is done by trading off transmission for resolution, which impacts signal-to-noise ratio as well as depth of field. New computer controlled scanning laser confocal microscope systems allow merging of images from different focal planes into a composite with maximum resolution of a confocal microscope without the depth of field limitations. The limitation in transmission are overcome by the use of lasers to improve signal to noise rations and reduce image acquisition times. In addition to providing optical microscope enhancement, the confocal laser microscope provides an ideal platform for OBIC (Optical Beam Induce Current) analysis. Spinning disc confocal microscopes use a large number of pinholes to create a merge image with enhance resolution.

Infrared microscopy finds a range of applications dependent on the infrared transparency of silicon and other semiconductors. This has become particularly important as flip-chip becomes a more common and the number of layers of metallization continues to increase. The IR microscope finds application in detection of subsurface defects such as cracks and die-attach anomalies. It has also been used to view devices from the backside. This application has been effective for detection of metallization corrosion, electrical overstress damage and bond formation.

Numerical Aperture

Numerical aperture is a measure of the light gathering capability of an optical system such as a microscope. It is commonly used as an indication of the quality of the lens in a microscope. It is dependent on the refractive indices of the lens and transmission medium as well as the range of collected light (typically measured as the half angle of the cone light entering the optical system). This concept is particularly important when one considers the failure analyst's requirement for long working distance lens for probing and observing devices in packages. Generally, longer working distance lenses will have lower resolution with increased depth of field.

APPLICATIONS

Brightfield/Darkfield

Brightfield imaging is the optical microscope application most commonly employed with semiconductors. The incident light beam is directed normal or perpendicular to the beam and the reflected light, which is collected by the lens system, is that reflected back normal to the surface. When this is considered, it should be clear that very flat surfaces are those best observed in the brightfield mode. Since IC's are relatively flat, brightfield is an excellent match. In cases where a rough surfaces is observed, illumination normal to the surface may not be desirable. When one observes the sides of a raised or depressed area, light entering from the normal direction will not be reflected back into the optical system but away from the optics and a dark area will be observed as shown in figure 3.

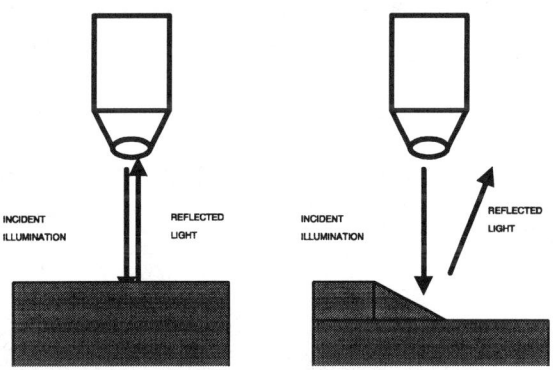

Fig. 3. Optical Microscopes in the brightfield mode provide good images of flat, orthogonal surfaces but little light is reflected back into the optics from a non-orthogonal surface.

In darkfield, illumination is from a slight angle, which can result in some light; being collected by the microscope from the sides of raised or depressed area as shown in figure 4. In the extreme case of a rough surface, little light is reflected back in the brightfield mode but darkfield may present a good image. This leads to a common error made by analysts. Dark optical images in brightfield mode are commonly misinterpreted as contamination with a dark or non-reflecting material. Darker areas in brightfield mode

can be an indication of a rougher surface as well. In general, brightfield is best applied to flat surfaces while darkfield is most applicable to rough surfaces. Since many semiconductor surfaces are very flat, brightfield microscopy tends to be more popular.

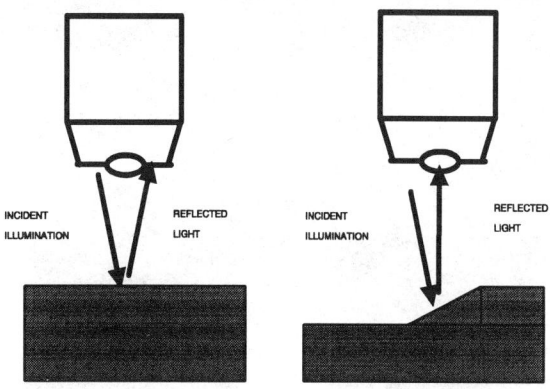

Fig. 4. In darkfield optical microscopy, more light is reflected back into the optical path from non-orthogonal surfaces than in a bright field mode. Because of this, darkfield can provide better images of rough surfaces.

Polarized Light Microscopy

The use of polarizers with light microscopes can provide added contrast in certain situation. The most common situation in failure analysis is use with liquid crystal. Polarizers filter out all of the light not polarized or oscillating in a single plane. The initial polarizing filter provides incident polarized light on the sample surface. A second polarizer, called an analyzer, is placed in the microscope optical path. The analyzer is rotated until the proper contrast is achieved.

Interference Contrast

Interference Contrast takes advantage of the refraction and polarization property of light. Interference contrast allows the analyst to detect subtle changes in surface topography. This is particularly useful in the analysis of silicon defects or observed stained silicon as shown in figure 5. Poor defined junction delineation can also be improved with interference contrast as shown in figure 6. An often-overlooked application is the similar use with cross sections to detect poorly defined edges.

Fig. 5. The bond pad shown is deprocessed to silicon and decorated with Schimmel etches. The decorated crystalline defects are observed in the interference contrast mode.

Fig. 6. Interference contrast image of device deprocessed to silicon and decorated. Resistor diffusions were not observable in other microscope modes.

Infrared Microscopy

Infrared microscopy is particular useful due to the transparency of silicon in the IR region. Since infrared light is not visible, the image cannot be directly viewed. The image is captured in some form of detector and converted to a visible image on a CRT. The most common failure analysis application is the viewing of devices from the backside. Material under the die including the package and die attach are removed mechanical and the backside is polished. Defects commonly observed include electrical overstress damage to junctions, bond intermetallic formation and metallization corrosion (single level metal with no barrier layer only) as well as other types of defects. (See example in figure 8.)

Fluorescence Microscopy

In fluorescence microscopy, the sample is exposed to ultraviolet light, which is absorbed by some materials. In some cases, the excited states produced decay to intermediate states with emission of visible light. This process is call fluorescence. Since a high density of UV light is required, mercury lamps are commonly used for fluorescence work. This light is focused on the sample of interest, visible light should be filtered out to reduce the background level of visible light. The visible light is then viewed through the optical system. The reflected UV light should be filtered since UV light can not be seen but cause damage to eyesight. Many organic materials emit green light. Silicon nitride emits an orange-red light. An example is shown in figure 7.

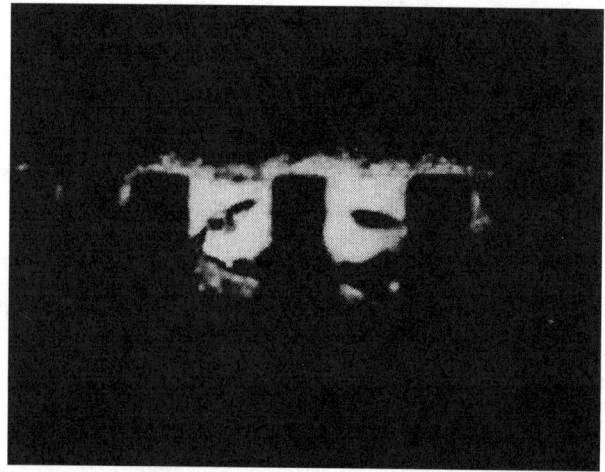

Fig. 7. Fluorescence image of package with non-halide flux contamination (abietic acid).

Fig. 8. IR microscope image of electrical overstress damage is viewed from the backside through silicon. Package materials have been polished away.

SUMMARY

These non-destructive evaluation techniques fill important niches in the failure analysis toolkit. While they are sometimes overlooked in discussion of failure analysis, they provide necessary complimentary information to more visible tools such as the field emission scanning electron microscope and transmission electron microscope.

Scanning Electron Microscopy (SEM) and Energy Dispersive X-Ray Analysis

John R. Devaney
Hi-Rel Laboratories Inc.
Spokane, Washington

Scanning electron microscopy (SEM) has been heralded as one of the most significant inspection and analytical tools to evolve since man first aided eyesight with the telescope and microscope. Its older, more familiar counterpart, the transmission electron microscope (TEM), has never achieved the widespread popular acknowledgment received by the SEM or "sem" as some call it. Rumors began to filter out of England, in the early part of the 60s, of a new, wonderful, magical, all-seeing microscope that could perform miracles! Early photos seemed to substantiate these claims. Marvelous photos of objects with great depth-of-field, good resolution, and no illumination problems appeared. In addition to these attributes, engineering reports talked of voltage mapping and displays depicting current flow in semiconductors!

Depending on one's point of view, the resulting marriage between the micro-electronics/semiconductor industry and the scanning electron microscope was either detrimental or beneficial . The new tool greatly aided the analyst-engineer in his efforts to examine and analyze devices and determine failure modes. It also made it very difficult to hide or ignore mistakes.

The SEM of today is basically third generation iteration of the laboratory models first sold in the mid 60s. These modern instruments are no longer bulky and cumbersome. They no longer require legions of mechanics and electronic technicians to keep them running. They are sleek and svelte.

Like most modern instruments, the actual operation of the SEM has been greatly simplified by complex electronics. Functions and adjustments once performed by the operator are now automatically performed by the internal electronics. Thus, what once required an operator with a full and complete working knowledge of the instrument, is now handled by an array of circuit boards and components. These advancements have made the SEM an instrument that many can operate, but few utilize to its full potential. For without the basic knowledge of the instrument and an understanding of electron beam interactions, the SEM is just an expensive camera. To the trained SEM technologist however, the instrument is much more.

When the electron beam strikes any sample, whether a semiconductor or ancient Roman coin, it produces a variety of signals. These signals exist whether or not the means are available to detect them with a particular instrument.

The basic signals created by electron beam-specimen interaction are as follows:

1. Emissive (Secondary Electron)

2. Backscattered (Primary Electron)

3. Absorbed

4. X-Rays

5. Auger Electrons

6. Beam Induced Hole-Electron Pair Generation (EBIC)

7. Voltage Field Enhanced (Voltage Contrast)

When electrons are injected into or onto a nonconductive surface the injected charge is accumulated and the specimen will "charge up". Thus under most conditions, the specimen should be somewhat conductive.

MICROSCOPE COMPARISONS

1. What are the operating parameters the SEM operator or user needs to be cognizant of to ensure maximum utilization of the instrument's potential?

2. What are the advantages of the SEM over other commonly available instruments?

3. What are the drawbacks or disadvantages of the SEM?

The typical application of the SEM in failure analysis is as "a microscope." Let's compare the SEM to the older, more common, optical and transmission electron microscopes.

Optical Microscope

The Optical Microscope, the mainstay of analysts for over 300 years, has several limiting features, especially for the microelectronic/semiconductor analyst. The useful magnification

Fig. 1　2kV, 50 degree tilt.

Fig. 2 3 kV.

Fig. 3 5 kV.

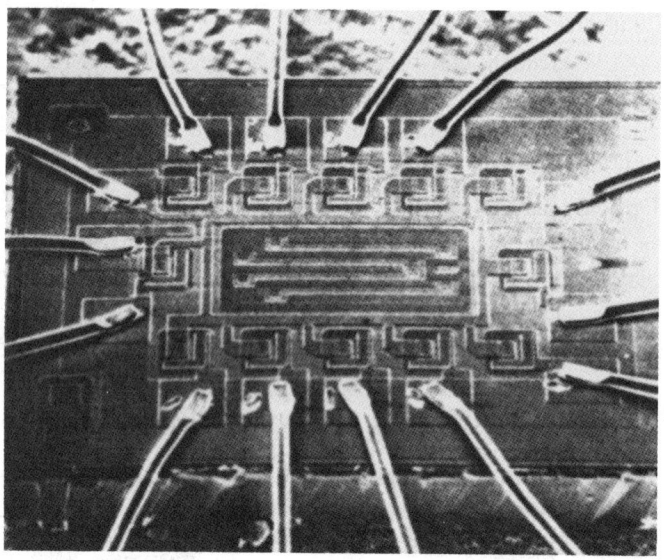

Fig. 4 10 kV.

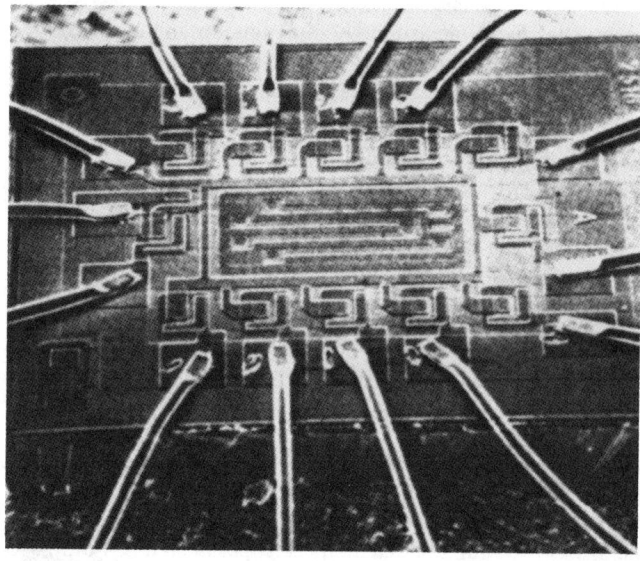

Fig. 5 20 kV.

range extends from very low, approximately 1×, to a useful high of 1,500×. If higher magnification and better resolution are needed, the microscopist can use an oil immersion lens. Even this only extends the range to 3000× and a useful resolution of 0.1 micron, or 1,000 Angstroms (Å). This has the shortcomings of contamination of the specimen with oil, a short working distance, and a sample with no vertical relief!

Transmission Electron Microscope (TEM)

It is surprising that the most important feature of the SEM is its great depth of field, or depth of focus. Although resolution (high magnification) is important, the long focal depth enables the microscopist to examine real, intact, 3-dimensional samples.

This is in sharp contrast to the TEM, which can only be applied directly to very small, thin sections of material or replicas of the sample to be examined. Sample preparation and replication is a difficult, time-consuming task if no artifacts are to be introduced.

Additionally, although the TEM has atomic level resolution (10-20 Å) and high magnification (300,000 -500,000×), it obviously has no apparent depth of field and the lower useful limit of its magnification range is 5000-l0,000×.

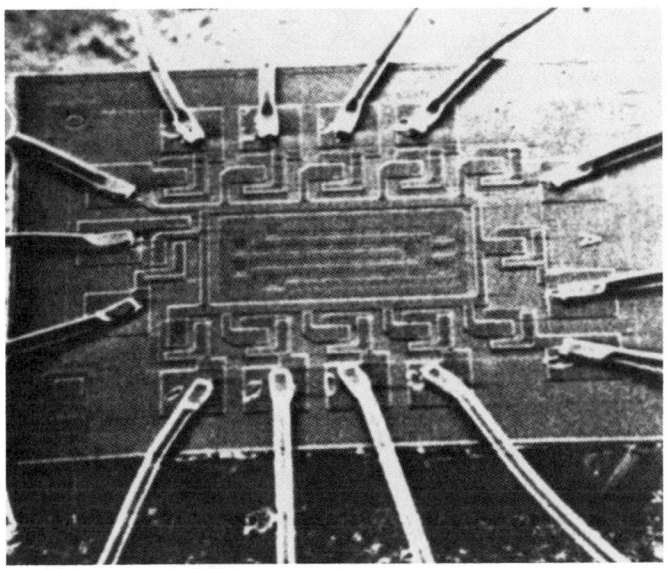

Fig. 6 30 kV

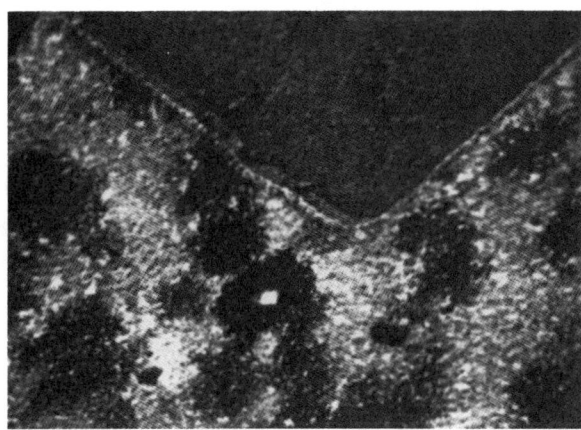

Fig. 7 2 kV, 45 degree tilt.

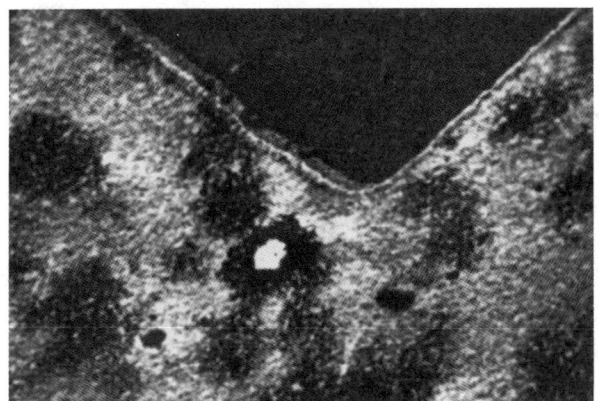

Fig. 8 3 kV, 45 degree tilt.

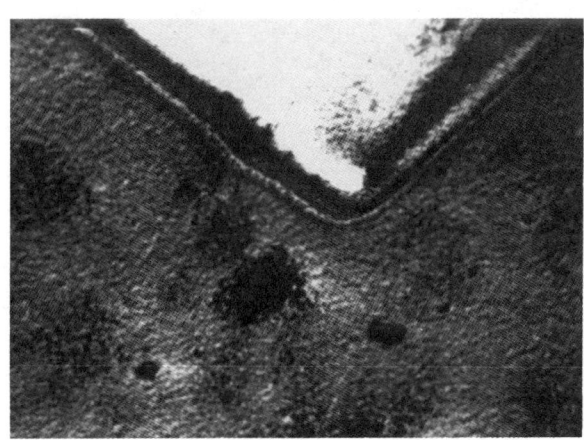

Fig. 9 5 kV, 45 degree tilt.

Scanning Electron Microscope (SEM)

The SEM, however, has a useful working range from 10× to 100,000×, a minimum attainable resolution of 20 Å, a depth of field (focus) 300 times that of an optical microscope, and good working distance. The point of focus or working distance from the final lens can be 15 to 20 mm. An obvious drawback, the specimen must withstand the rigors of vacuum!

In addition to these advantages there is one often overlooked feature which can save the working technician or engineer hours of frustrating effort.

The semiconductor or microelectronic component, for various reasons, is composed of many discrete physical elements. The typical IC, for instance, has smooth surfaced aluminum or gold wires. Gold wires have round shiny balls formed as a bond

on one end. The surface of the die is quite reflective as are many portions of the package, cavity surface, bond pads, etc.

Effective illumination of such a sample for examination or photographic documentation is almost impossible. This is especially so if the analyst attempts to examine the sample in some tilted condition.

With the SEM, due to the unique properties of signal generation and collection, many of these illumination difficulties are bypassed or overcome. For these reasons, the SEM still is very widely used as a low-power camera in the range of 10-20×.

The SEM is, however, a monochromatic microscope: THERE IS NO COLOR! Color enhancement may be accomplished, however, by special computer enhancement signal processing. The effect is purely artificial. With all these comments in mind, what then are the specimen and instrument parameters of interest?

277

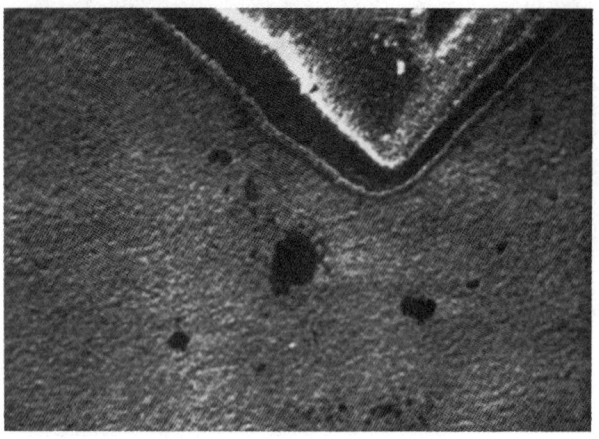

Fig. 10   10 kV, 45 degree tilt.

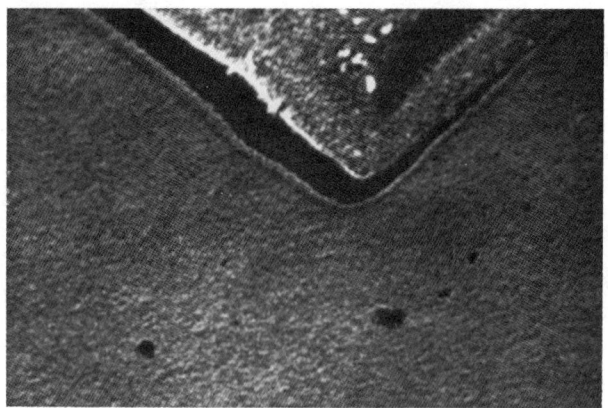

Fig. 11   20 kV, 45 degree tilt.

Fig. 12   20 kV, 40 degree tilt.

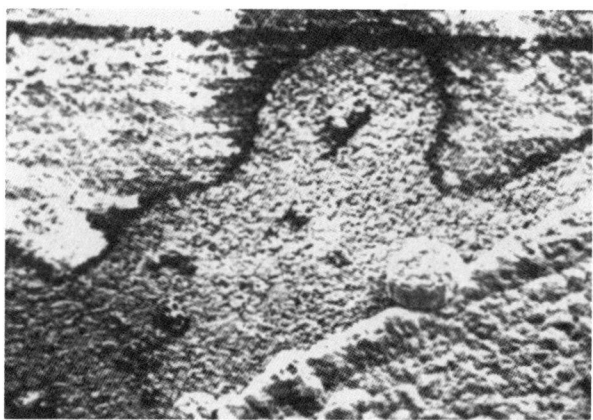

Fig. 13   10 kV, 40 degree tilt.

SPECIMEN AND INSTRUMENT PARAMETERS

The interaction of specimen and instrument parameter will be discussed in the context of the various imaging modes.

The Emissive Mode (Secondary Electron Imaging). With most modern instruments this is the standard imaging mode. The signal is comprised primarily (approximately 90%) of secondary electrons, although some backscattered electrons are collected and modify the image.

For the failure analyst, the primary consideration, in most applications, is to do nothing to alter or affect the sample. *A conducting coating is a major alteration!*

CAUTION: DO NOT COAT THE SPECIMEN, EXCEPT AS A LAST RESORT!

Examination of the sample surface directly results in significant and useful data only if properly interpreted. Surprisingly, the emissive image of an uncoated surface, especially one as complex as an IC, is parameter dependent. The parameters are instrument beam voltage and specimen tilt.

Beam Voltage. Early SEMs and even most in use today utilize a current-heated tungsten wire as a source of electrons, a "hot filament gun". Due to various electron-optic interactions, the electron beam is more sharply defined and smaller at high accelerating voltages (20-25 Kev) which results in increased point-to-point resolution. But a high accelerating voltage results in greater penetration of the sample surface by the beam. Thus, the various signals are generated from a greater volume of the sample. This penetration depth is directly proportional to the density/atomic number of the sample. For a material like aluminum, AT No. 13, or silicon, AT No. 14, this penetration depth at 25 kV can exceed 4 microns! If the beam voltage is reduced, the electron beam penetration is also reduced. This results in greatly improved contrast due to both an increase in secondary electron emission at the surface (point of interest) of the device and a decrease in background (subsurface) secondary emission. The latter is generated by backscattered electrons at considerable distances from the beam impact point. These backscattered electrons can also impinge upon other portions of the sample. If these areas are non-conducting, they can

Fig. 14 5 kV, 40 degree tilt.

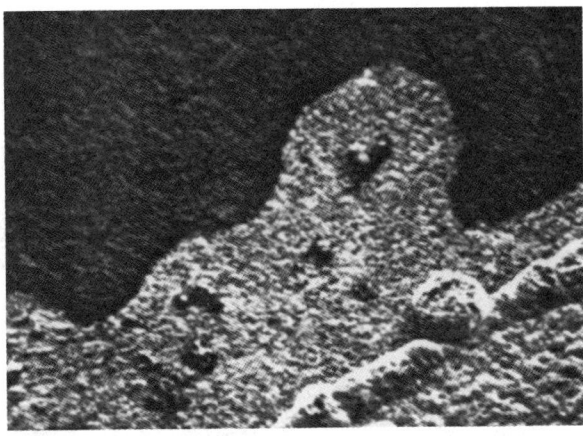

Fig. 15 3 kV, 40 degree tilt.

Fig. 16 2 kV, 40 degree tilt.

Fig. 17 3 kV, 50 degree tilt.

charge up and result in a drastic deterioration of the signal-to-noise ratio (in the area of observation) without the operator being cognizant of the effect. Thus, the selection of the beam acceleration voltage is dependent upon the specimen and the information desired.

Figures 1 thru 20 are examples of the difference of an image of a surface at low and high accelerating voltages. Figure 1 shows an area of a microcircuit examined at a beam voltage of 2Kev. Note the surface detail on the metal stripes. Figure 5 images the same area at a beam voltage of 20Kev. It is obvious that much of the finer surface detail is obscured. The loss of fine surface detail outweighs the apparent advantage of higher resolution. However, the use of higher beam voltages with greater penetration depths can result in information obtainable no other way. (See Fig. 19, 20).

Figure 7 shows the surface of a device covered with contamination (the image was obtained at a beam voltage of 2Kev). Since penetration and emission are shallow, only the contaminant is imaged. When the same area is examined at 20Kev, the greater depth of penetration and signal generation shown in Fig. 11 results. Im-

aging of the underlying surface and all traces of the contaminant are lost.

CAUTION: The analyst must constantly be aware of the fact that different beam voltages result in preferential charging and/or emission from different materials which radically alter the information content of the image (see Fig. 12 thru 16). Figures 1 thru 6 are photos of an IC surface. They illustrate the great diversity available by using different beam voltages. The specimen tilt is 50° in all six figures. The sample is not coated.

The series of micrographs in Fig. 7 thru 11 should be studied at the same time. They illustrate the advantages of using low beam voltages to examine a surface suspected of being contaminated with organic or silicone thin films. The sample is composed of thick film gold on a ceramic substrate. The dark spots in Fig. 7 are contaminant stains of varying thickness. Compare this to the apparently clean surface in Fig. 11.

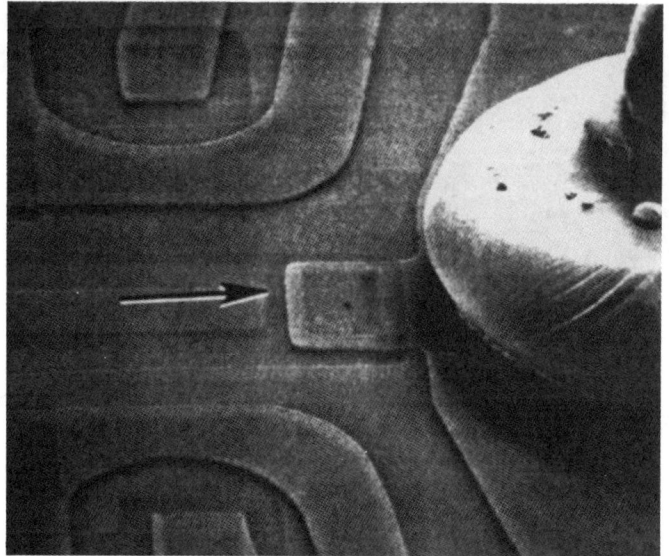

Fig. 18   30 kV, 50 degree tilt.

Fig. 19   3 kV, 60 degree tilt.

If surface contamination is suspected, always examine the sample at low beam voltages. High beam voltages will not detect surface films.

Figures 12 through 16 should be studied simultaneously. This sequence of SEM micrographs illustrates the dramatic change in photo information as the beam voltage is lowered. Most significant changes occur in the upper half of the image, which is bare ceramic. The region cutting diagonally across the lower right corner of the photo is a thick film of gold. The dark irregular shaped region in the photo center is a thin film chromium bleedout.

Figures 17 and 18 are photos which dramatically illustrate the loss of surface information on a low atomic number thin film, aluminum, at 30 kilovolts as compared to 3 kV.

Higher beam voltages do have some useful features. Figure 19 is an image of a badly contaminated IC surface. At 3 kV, the beam is charging the light element contaminant. In Fig. 20, at 20 kV, a significant portion of the beam penetrates the contaminant.

Tilt. Just as various beam voltages radically alter the emissive mode image, changes in specimen tilt also have a striking effect on the image. This is true not only for semiconductors, but many other thin film samples and samples where non-conductive films of varying thicknesses exist on the specimen surface.

The effect is due, in part, for the same reason that changes in beam voltage affect the image. As the angle between the beam and the normal to the specimen increase, the actual depth of penetration from the true surface decreases. The passage of the beam through various surface films at different angles results in varying penetration, emission, and charging rates.

Analogously, the sun illuminates and penetrates the same blanket of atmosphere (sky) at high noon and sunset, yet we see one as blue and the other as yellow or golden. The difference is angle! For specimens at very high tilt angles, more of the pri-

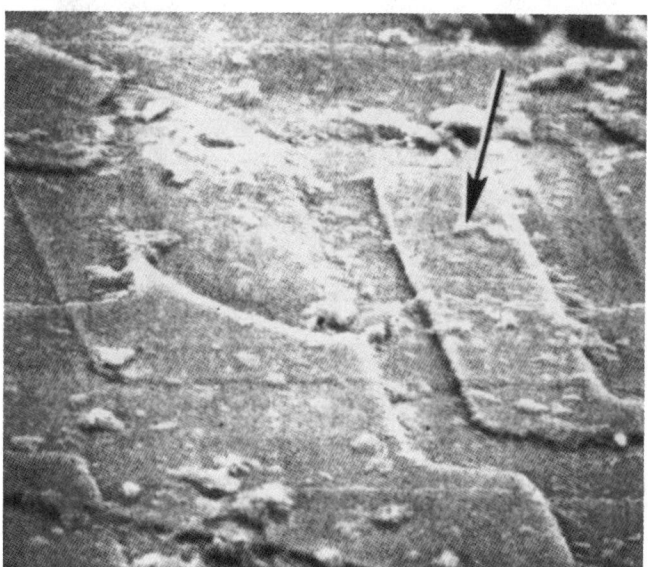

Fig. 20   20 kV, 60 degree tilt.

mary electron penetration volume is within signal escape depth of the surface. Thus more electrons are emitted. In the example of an IC, the analyst must recall that some of the enhanced contrast obtained by tilting the specimen is due to complex charging of thin oxides on the sample surface and preferential generation of the signal in the variously doped regions of the device. Figures 21 and 22 depict the effects on the image of merely tilting the specimen while keeping beam voltage fixed at 10 Kev and tilting from 0 to 75 degrees.

Another consequence of specimen tilt is illustrated in Fig. 23 and 24. Figure 23 shows a passivated microcircuit examined at

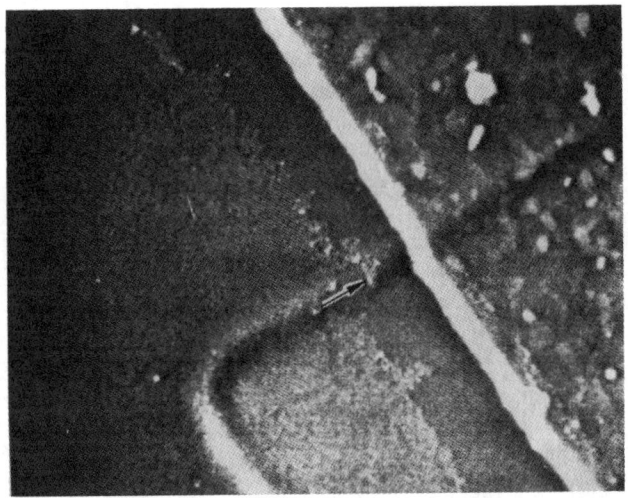

Fig. 21 10kV, 0 degree tilt.

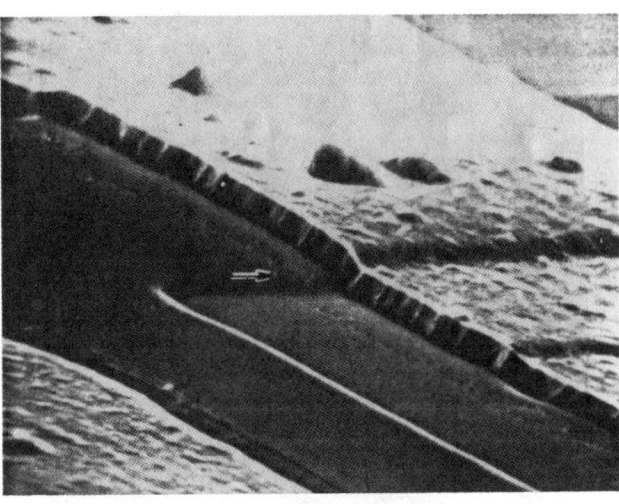

Fig. 22 Effects on image (from Fig. 21) by tilting the specimen while keeping beam voltage fixed at 10 kV, 75 degree tilt.

Fig. 23 5 kV, 0 degree tilt.

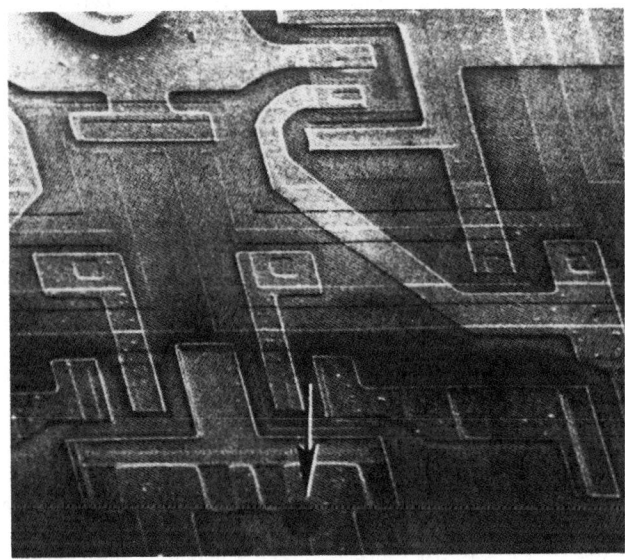

Fig. 24 5 kV, 40 degree tilt.

a zero degree tilt with no prior treatment to prevent charging. Charging typically equated with brightness (or enhanced emission) can also result in decreased emission (dark areas). Charging is thus manifested by very bright areas or spots with dark adjacent regions which results in loss of detail in both areas. Examination of the same area at the same beam voltage but a higher tilt angle (40 degrees) finds that the number of backscattered and secondary electrons has increased. This results in a reduction of residual electrons (charging) in the glass layers.

The resolution and surface detail in Fig. 24 is not spectacular, but a significant amount of information is obtained. Figures 23 and 24 are SEM micrographs of the same area at the same magnification, but two different tilt angles. Compare these two rather bland images to that of the same area at 50 degree tilt in Fig. 25.

Backscattered Mode—Primary Electrons (BSE). As with applications to samples other than semiconductors, the backscattered image is highly dependent on the constituents of the surface. More important for general applications is the fact that, unlike weak low-energy secondary electrons which can be collected out of holes and around corners, primary or backscattered electrons travel in relatively straight trajectories. Thus they are influenced very little by detector field voltages. Therefore, it is reasonable to expect them to be nearly unidirectional, and the images will have sharp shadows where the surface topography blocks the area imaged from the detector. Subtle features are often greatly enhanced by backscatter mode imaging.

Figure 26 illustrates a passivated microcircuit examined at a tilt angle of 60 degrees, in the emissive mode. This illustration

Fig. 25 Image of area shown in Fig. 23 and 24, except that angle of tilt has been increased to 50°.

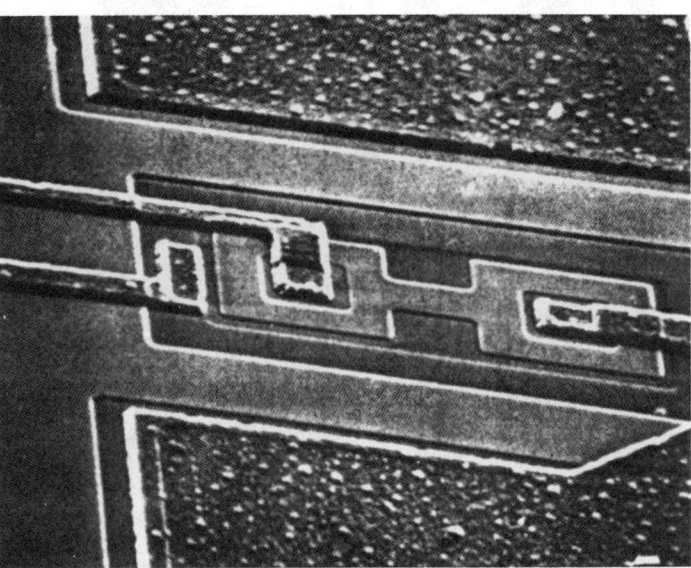

Fig. 26 Passivated microcircuit examined at tilt angle of 60 degrees in the emissive mode.

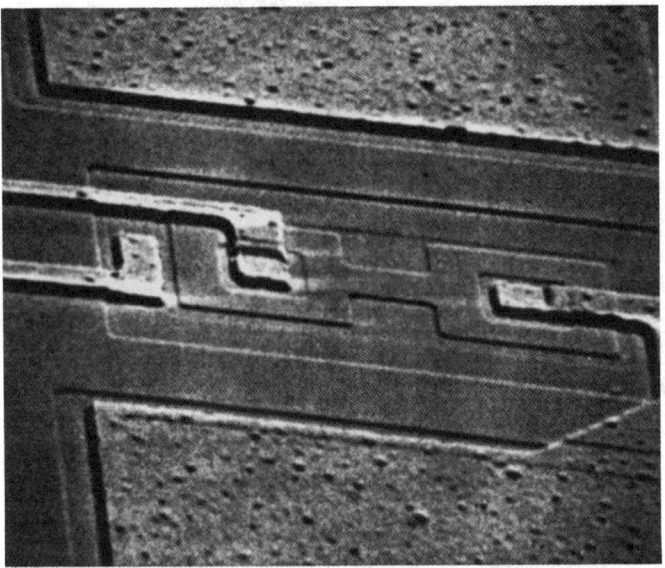

Fig. 27 Passivated microcircuit examined at tilt angle of 60 degrees but imaged only with backscattered electrons.

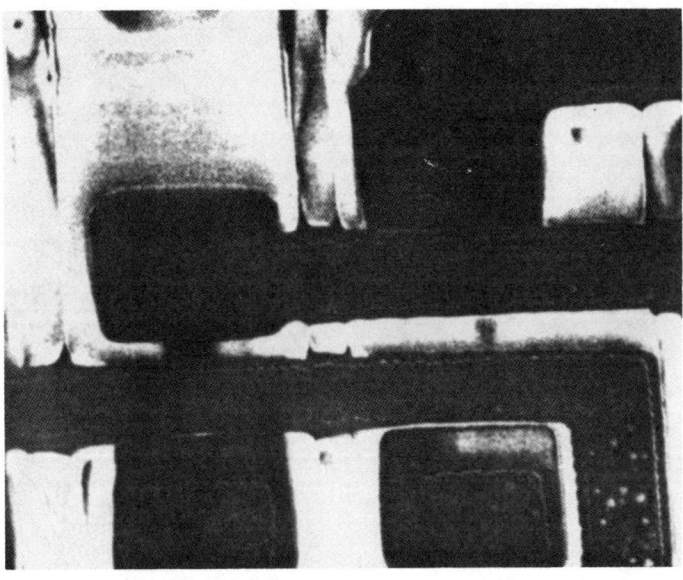

Fig. 28 Secondary emission

is followed by Fig. 27, which is a passivated microcircuit examined at a tilt angle of 60 degrees but imaged only with backscattered electrons. The two figures should be studied together.

Figures 28 and 29 show the radical differences between the same dual level passivated IC surface when imaged by secondary (Fig. 28) and backscattered electrons (Fig. 29). Preferential charging of the inter-layer glass has occurred in the secondary mode (brightness).

Voltage Field Enhanced. (See Fig. 30, 31, 32, and 33.) A special case of secondary electron imaging is "voltage contrast" imaging. This mode requires that connections be made to the device and the electrical biases be applied.

Voltage contrast is used primarily to pictorially display the electrical functions of devices. By careful adjustment of the beam accelerating voltage, aperture, beam sweep rate, and specimen electrical operation, voltage states and transitions can be detected on the device surface. This is accomplished by elec-

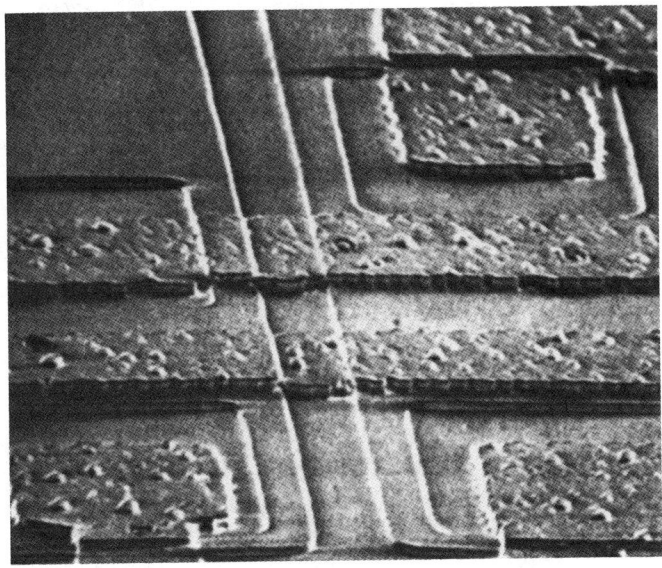

Fig. 29 Backscatter signal only, sample tilted 60 degrees.

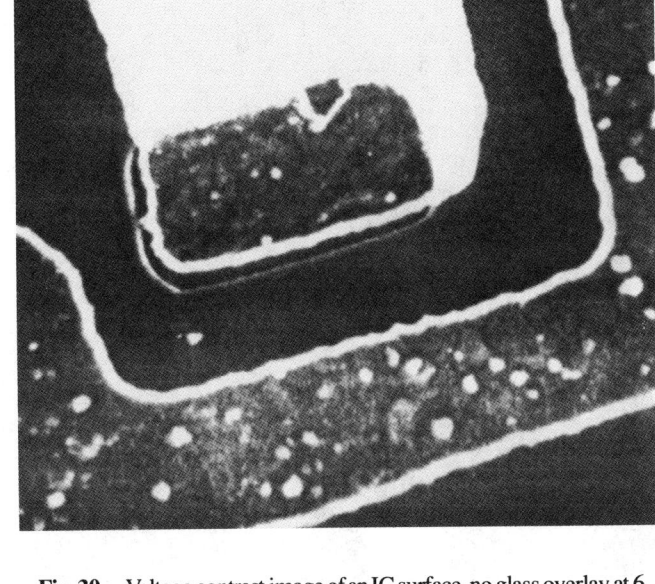

Fig. 30 Voltage contrast image of an IC surface, no glass overlay at 6 volts (black) and ground (bright metal contact). Open at step into contact window.

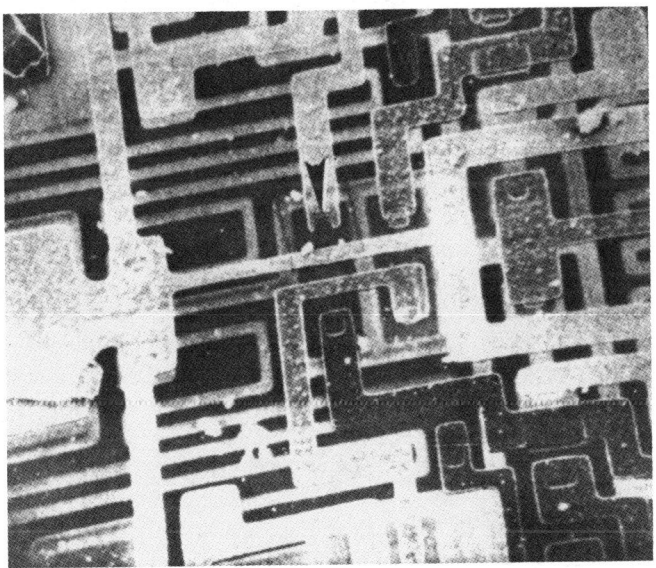

Fig. 31 Voltage contrast of a digital IC surface showing different voltage levels. Note the open at step in tunnel contact window.

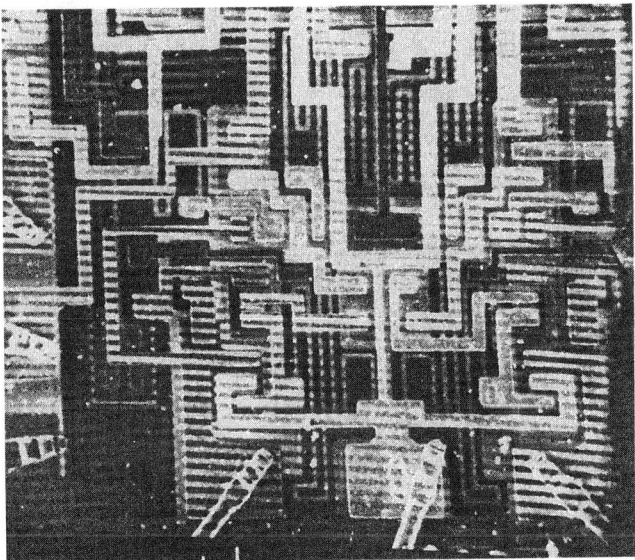

Fig. 32 Voltage contrast beat, or hetrodyne, pattern caused by pulsing the input voltages.

trically operating the device through external connections and adjusting the beam acceleration of the SEM so that secondary emission is affected by the charge states on the device. Positive areas will have a lower secondary emission efficiency than negative or ground potential areas. The more positive areas will thus appear dark in comparison to the more negative areas. By pulsing the device at a low repetition rate (approximately 10Hz), the effect is quite noticeable, and transitions from one state to another are detectable. Open circuits are very obvious since one side will be dark (positive) in comparison to the light side (ground).

Voltage contrast imaging can be performed on both single and multilayer interconnect devices with the glass passivation intact. The advantage of this mode of observation is that the glass does not need chemical treatment. The location of an open by VC, and verification of its presence, cannot be refuted because of an analytical procedure. Operation in this mode requires that the beam voltage be carefully selected for the layers of dielectric involved.

No special detector is required to achieve voltage contrast (differentials of 0.6v) on an ordinary semiconductor device. It is recommended that those unaccustomed to working with semi-conductors in the voltage contrast mode experiment with well-defined simple specimens where complications due to the specimen or package can be kept to a minimum.

The writer suggests that these experiments be initially performed on a transistor or integrated circuit which has no glass passivation over the metallization pattern. In addition, the package itself should be of the TO-5 or TO-18 "can" type, where the die sits on a conductive header with little or no glass or ceramic nearby. All efforts should be made to keep the specimen as far from the final lens as possible to prevent trapping of the signal electrons. *Under no conditions should the scanned beam be allowed to strike nearby glass or ceramic insulation. Charging of these will result in the loss of observable voltage contrast.* If possible, experiments should be carried out over the entire beam voltage range starting at 2 kV up to 30 kV, to determine which condition results in the maximum contrast. Scan rate and exposure of the sample will also alter visible contrast. In addition, higher beam voltages can result in alterations in the specimen itself.

Once the operator is fully familiar with one or two "standard" specimens, he should then begin to expand his experience by working with more complex devices and less ideal packages.

In addition to these very basic techniques, the literature abounds with excellent reports detailing various modifications to the instrument and signal collector which enhance observable contrast, and in some techniques quantifies it.

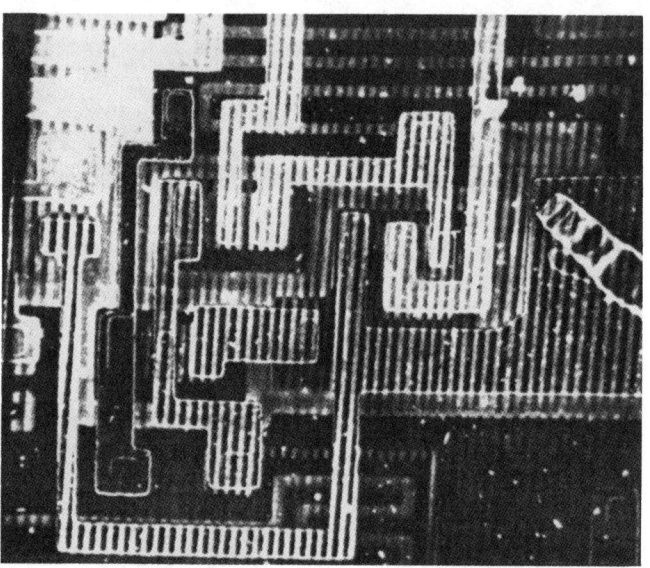

Fig. 33 Voltage contrast beat (or hetrodyne) pattern high magnification view of image in Fig. 32.

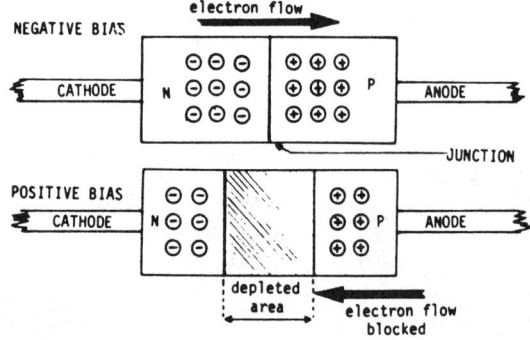

Fig. 34 Biasing a pn junction.

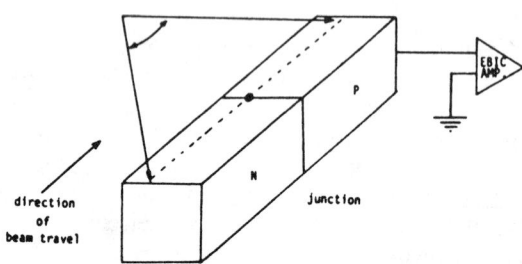

Fig. 35 Electron beam flow direction.

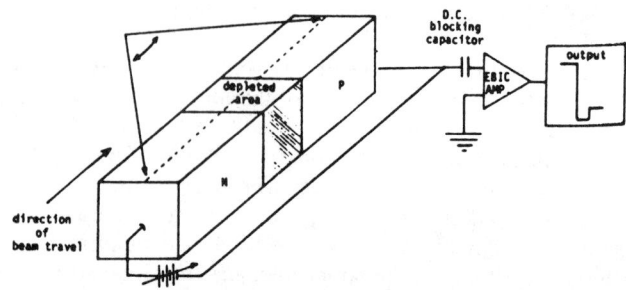

Fig. 36 Reaction of electron beam with depletion region.

Fig. 37 Micrograph of the specimen diode.

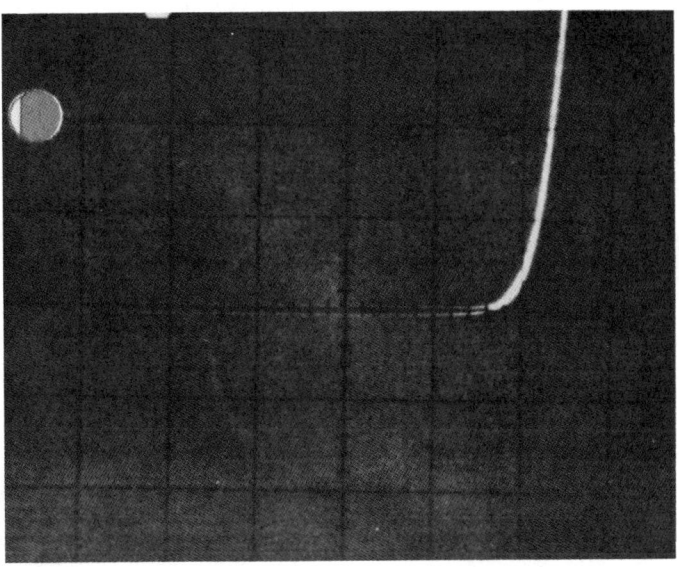

Fig. 38 Curve tracer photo of the forward breakdown characteristics

Other efforts have resulted in techniques which produce voltage contrast of a device in varying states of operation, such as alternating or pulse current mode, by strobing of the beam to the device. In fact, the creative electronics engineer is presented a basic tool with which he can experiment, and add signal processing instruments, and greatly increase the spectrum of data obtainable from the SEM.

EBIC utilizes the focused beam of an SEM to create hole-electron pairs in the examined specimens. The penetration depth of the beam is a function of the beam accelerating voltage. The induced current magnitude, on the other hand, is a function of electron beam current and the specimen.

The electrons generated flow in the specimen in a direction influenced by the depletion region and external circuitry. This induced current is supplied to an external high gain amplifier by attached wires. The amplifier and associated control circuit synchronize the signal with the x-y coordinates of the SEM's CRT display. The resultant image is manifested as light or dark regions, since the amplifier signal varies from positive to negative depending on the beam location on the specimen. (See Fig. 35.)

As a semiconductor surface is examined, the EBIC output signal is modulated by the various surface compositions. The current injection efficiency of the negatively charged beam is higher in N-doped areas than P-doped areas, and the output signal will be accordingly of higher amplitude.

The result of reverse biasing a P-N junction is that the depletion region reacts to electrical bias by acting as a barrier to current flow. The depletion region is an area depleted of free charges and thus, for all practical purposes, is incapable of supporting current flow. Theory contends that this depletion region expands with reverse bias (positive on N) and collapses with forward bias (positive on P).

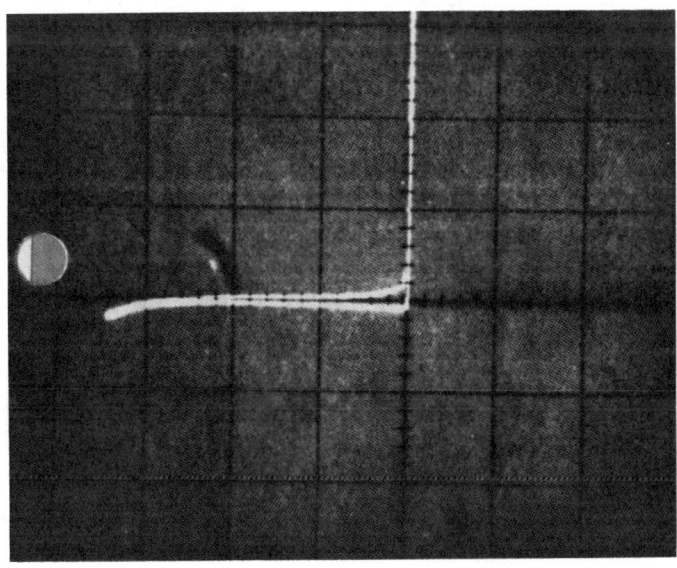

Fig. 39 Curve tracer photo of the forward and reverse breakdown characteristic of the diode. (1 mA/vert div, 200 V/hor div.)

The EBIC effect, when properly adjusted, can be utilized to examine the depletion region. The beam will react with the depletion region as though the area is a junction. Due to the last of free, available electrons, the current generation is much less in the depletion region. (See Fig. 36.)

Example

A 1N3209 power diode was selected as the experimental specimen. The choice of a power diode was made because of its rugged construction and relatively thick die. (See Fig. 37).

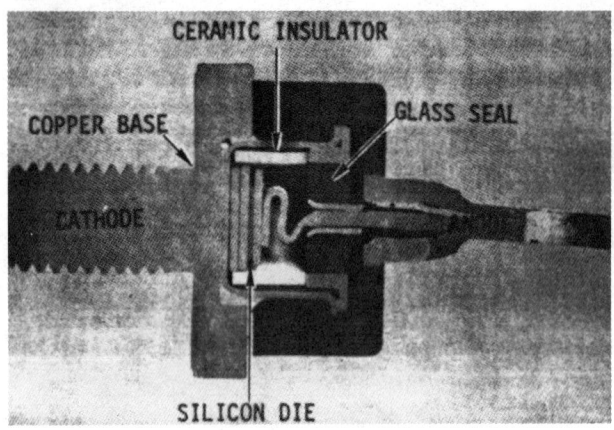

Fig. 40 Macrograph of the section of the diode.

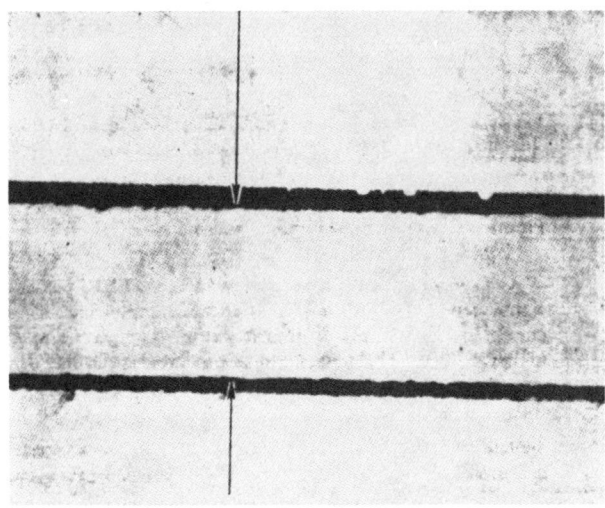

Fig. 41 Micrograph of the unetched die section.

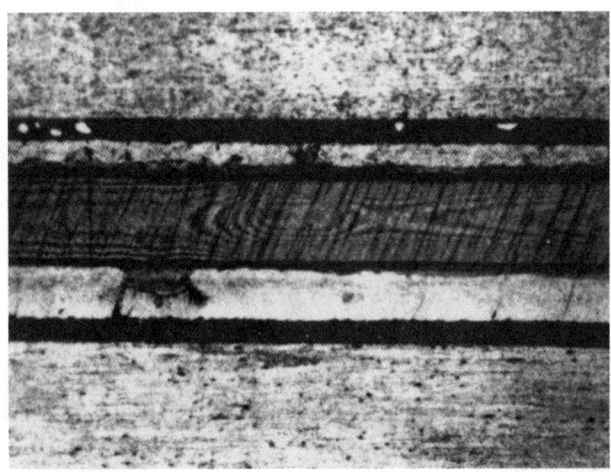

Fig. 42 Micrograph of the etched die section.

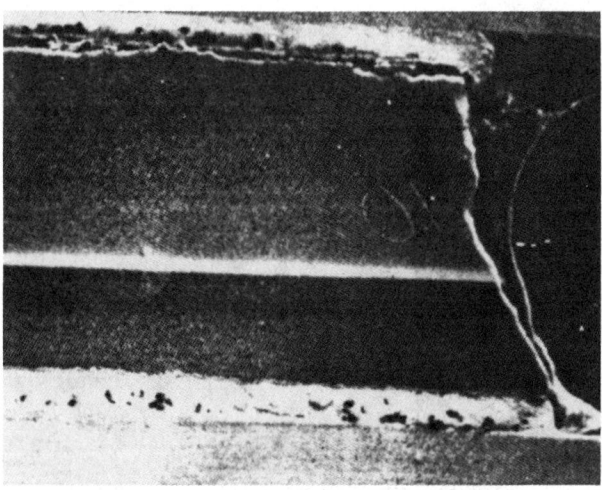

Fig. 43 SEM micrograph with superimposed EBIC image. O V bias.

The specimen was initially tested with a curve tracer and its operating characteristics recorded (see Fig. 38 and 39).

The device was then bisected with a low speed diamond wafering saw. (See Fig. 40 and 41). This step resulted in two experimental specimens.

The procedure continued with a fine polish of the sectioned surfaces to remove the sectioning-induced defects. Ultrasonic cleaning with acetone followed, to remove sectioning debris.

The final preparation step was a water rinse and high temperature (200 °C) bake. This allowed the polished silicon surfaces to form a protective oxide.

Curve tracer analysis revealed that both halves were still functional after the sectioning procedure. A slight degradation in reverse leakage was noted, but was not considered significant.

Specimen Examination. Microscopic examination of the sectioned die could not distinguish the junctions. One specimen half was then chemically etched to define the junction (Fig. 42). This sample was reserved for comparative purposes.

The unetched sample half was examined with the SEM. Again, without chemical etch, the junction was not detected. However, with EBIC analysis the junction area became quite visible. A definite correlation was noted between the EBIC display of the unetched sample and the optical results of the etched sample. This result demonstrated the ability of EBIC to accurately delineate junctions.

Under the influence of reverse bias, the EBIC image underwent a significant change. The junction area was replaced with the depletion region. With increasing reverse bias, the depletion region was observed to expand.

Figures 43 through 46 examine the specimen under several different reverse bias conditions.

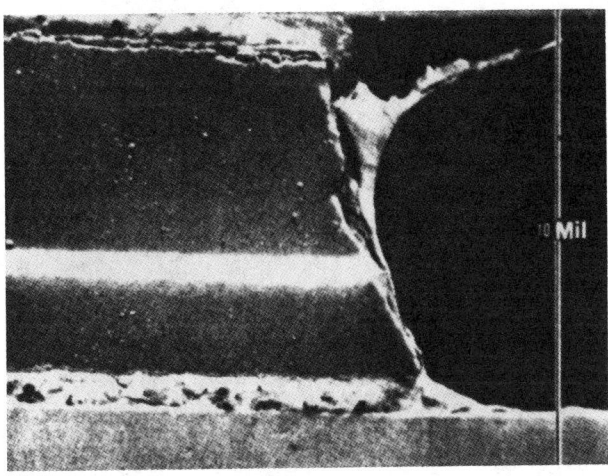

Fig. 44 SEM micrograph with superimposed EBIC image. 5 V reverse bias.

Fig. 45 SEM micrograph with superimposed EBIC image. 10 V reverse bias.

Fig. 46 SEM micrograph with superimposed EBIC image. 50 V reverse bias.

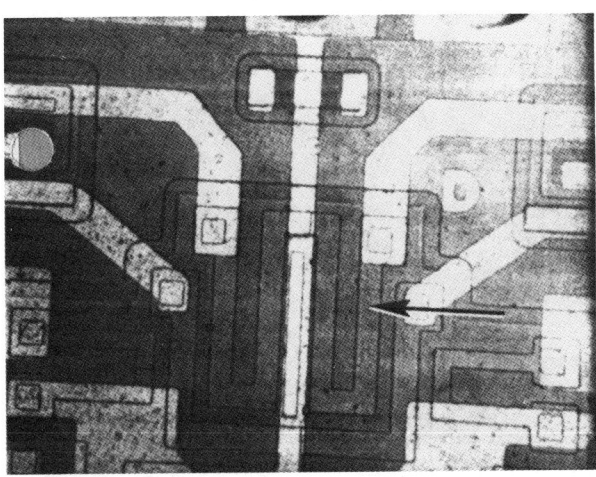

Fig. 47 Suspect resistor optical photo —device is a quad-NAND gate. Left and right sides are mirror images.

The depletion responded as anticipated, i.e., expanding with increasing reverse bias, contracting with decreasing reverse bias, and collapsing with forward bias. Once in stable conduction, the specimen would no longer react with the electron beam and no image response was observed.

The EBIC mode is relatively insensitive to charging artifacts. The package shape and materials are of minor consequence as compared to voltage contrast. A major drawback to the use of EBIC on complex devices is the proper interpretation of its data output.

The author has found the greatest success in using EBIC is in the comparison mode where EBIC maps of a good and (suspect) bad devices are obtained under identical conditions.

It should be remembered that EBIC mode imaging does not require that the device be biased with external power supplies. Any P-N junction will have an induced current flow when ex-

posed to a scanned electron beam. The only requirements are that any signals generated be collected and fed to the high gain amplifier. Because of the extremely high gain in the amplifier, long leads are undesirable since they act as antennas, feeding spurious signals to the amplifier.

Examples of suspect devices exposed to EBIC imaging are shown in the following figures:

1. Resistor, Fig. 47 & 48

2. Shift register, Fig. 49 & 50

3. Diode, Fig. 51

Artifacts Introduced by the Beam

Before the failure analyst can use any tool or instrument successfully and with a degree of confidence, he must be com-

287

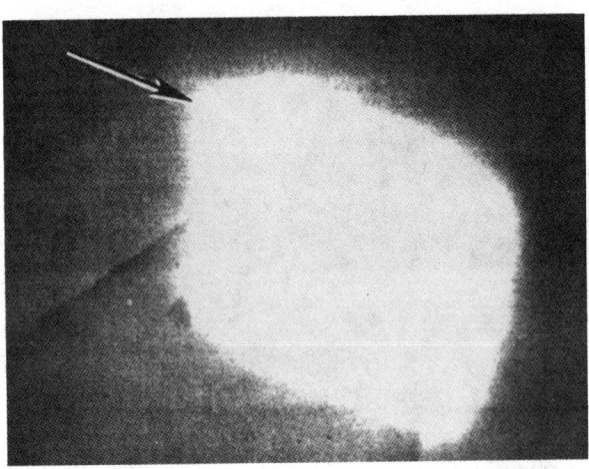

Fig. 48 Suspect resistor exposed to EBIC imaging. Identical electrical connections were made to each side. The arrow notes a resistor imaged by EBIC (bad side) not seen in the good side. Current induced resistor reached substrate thru a pinhole in the field oxide beneath that metal run.

Fig. 49 EBIC image of input protection diode with anomaly on junction perimeter.

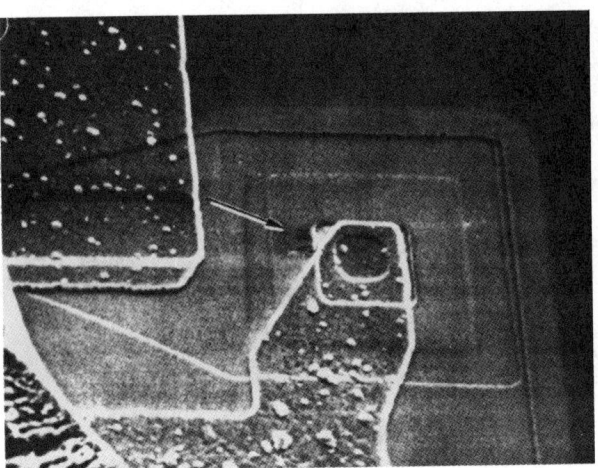

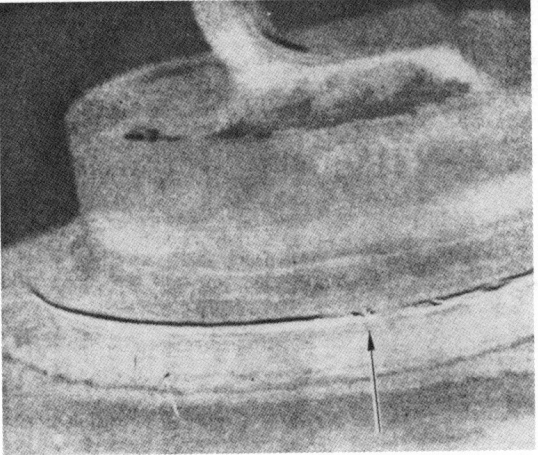

Fig. 50 Shift register EBIC image shows short across edge of protection diode diffusion, glass removed and surface etched to enhance damage.

Fig. 51 Suspect diode exposed to EBIC imaging—defect across junction is the leakage site.

pletely cognizant of any and all artifacts induced by the inspection/testing tool.

Thus he must be familiar with:

1. Artifacts induced by the beam

2. Specimen preparation

    a. Surface coating

    b. Chemical processing

3. Correlation with optical microscopy

In scanning electron microscopy, three of the most common artifacts introduced by the primary electron beam are electrostatic charging, vacuum pump oil polymerization, and electron beam damage. These artifacts present special problems for ex-

amining semiconductor devices because of the materials used in their construction and the fact that active electronic components are susceptible to radiation damage.

CHARGING

Electrostatic charging is due to the presence of a negative potential on an insulation portion or point of the specimen surface. Charging is caused by the collection of electrons on the surface of the insulator. The accumulation of electrons builds up a negative space charge which can deflect the incident beam, causing severe image distortion. The presence of the surface charge will also change secondary emission greatly. A charged area can cause the darkening of the image of a nearby un-

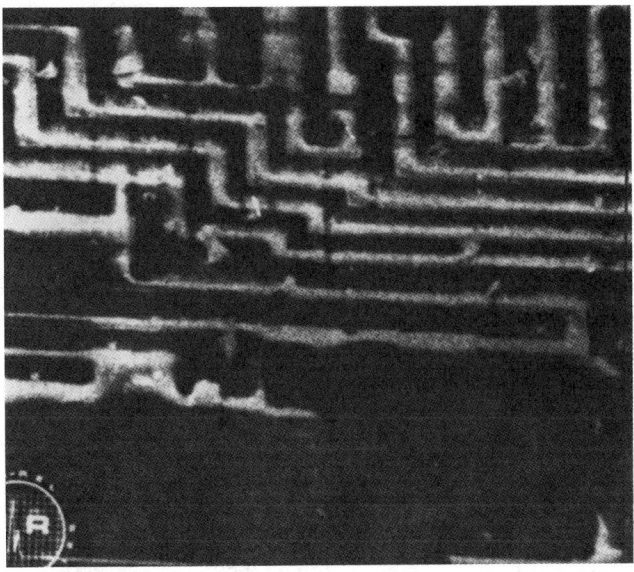

Fig. 52 Examination of passivated integrated circuit surfaces at 3 kV without a conductive overcoating results in extreme localized charging of the passivation layer.

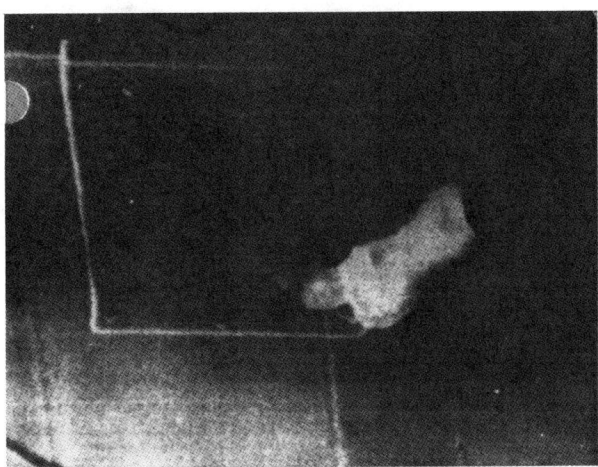

Fig. 53 Examination of an integrated circuit with an unattached non-conductive particle in the field of view results in charging of the particle and loss of information in the region around the particle.

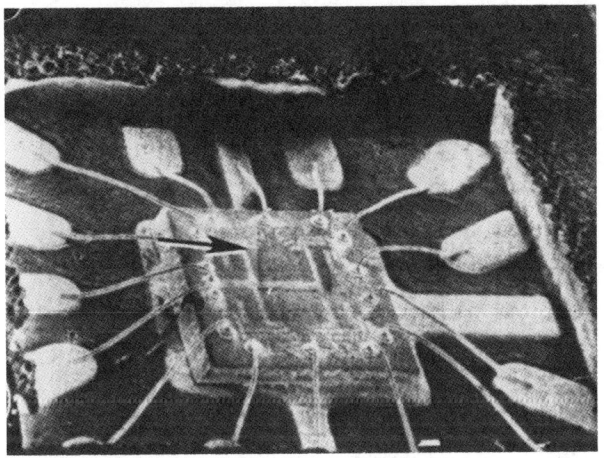

Fig. 54 Examination of a small region (dark square) of an integrated circuit die at a 10 kV beam voltage resulted in charging of a portion of the package out of the field of view by backscattered electrons (arrow).

Fig. 55 Initial image. Figures 55 and 56 show the image degradation due to the charge buildup on the package described in Fig. 54.

charged region by deflecting the emitted secondaries away from the detector. The charging effect is time-dependent, since a charged region may discharge by a breakdown to ground, and then the region can recharge.

Portions of semiconductor devices charge readily because of the insulating materials used in their fabrication. Glass, alumina, and oxide are examples of insulating materials used as substrates, packages, or passivating layers in semiconductor fabrication. These materials, along with nonconductive contaminant particles, are possible sites for charging.

In order to recognize how charging manifests itself, some examples are given. One example has already been seen in the discussion of tilt angle. Figure 52 shows glass passivation which has become charged between the metallization pattern. A single nonconducting particle whose image has become extremely bright is shown in Fig. 53. As a result of the high negative potential at its surface, the secondary electron collection has been reduced in the vicinity of the particle.

It is not necessary for an area of the specimen to be in the field of view for it to become charged! There can be a charge buildup where backscattered electrons strike a surface. Figure 54 shows a device which had been previously viewed with an accelerating voltage of 10kV at a higher magnification. The raster area during that exposure is shown by the darkened re-

Fig. 56 Degraded image.

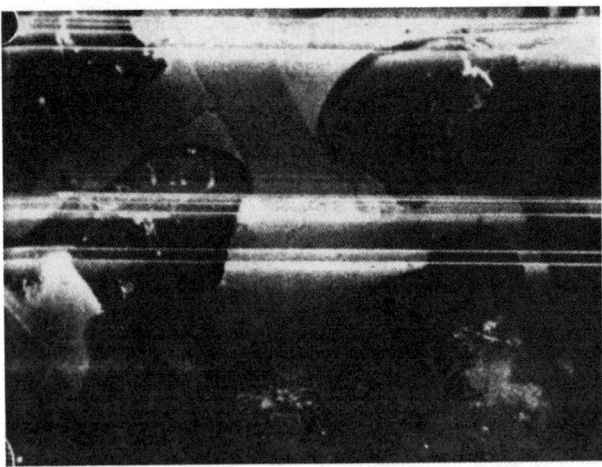

Fig. 57 Examination of a sample consisting of conducting and insulating areas without coating often results in severe image distortion due to charging and discharging in areas, which result in the bright lines in the photo.

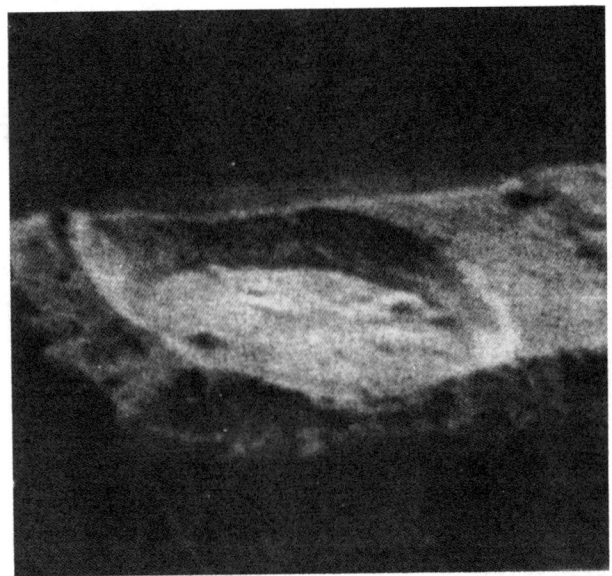

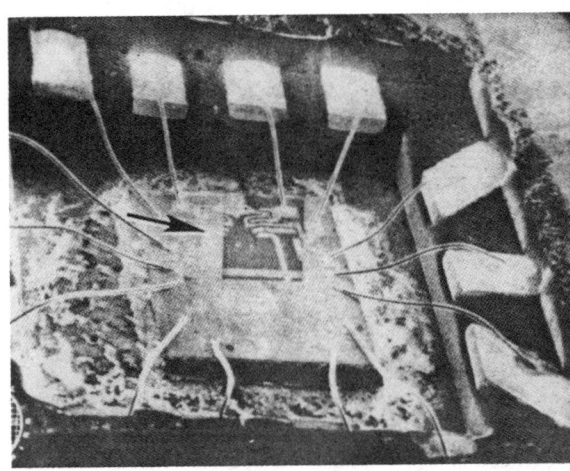

Fig. 59 Examination of the specimen from Fig. 54 at a beam voltage of 3 kV instead of 10 kV prevents charging of the package by energetic backscattered electrons.

Fig. 58 Charging often is not obvious and can result in subtle degradation of the image as in this case. The resolution is quite poor. It is not difficult to see that charging can render SEM images useless. Reducing the electron beam current will reduce, but not eliminate, charging.

gion on the die. The electrons scattered from this area formed a charge buildup on the ceramic package in the area that appears bright (indicated by the arrow). The darkened area around this bright region indicates the suppression of secondary electron collection in the vicinity of the charged region. The effect of this suppression can be seen more clearly in the next two figures which show two versions of the high magnification view.

Figures 55 and 56 show the image degradation due to the charge buildup on the package described in Fig. 54.

Charging is often seen as a distortion of the image in the photomicrograph. Figure 57 shows the characteristic bright areas of charging but in addition there are bright lines breaking the continuity of the image at the top and center. These are caused by unsynchronized discharges of the charged specimen surface during scanning.

Figure 58 is an example of a particularly deceptive charging symptom. There is poor point-to-point resolution, which no amount of manipulation of the SEM controls could significantly improve. In this case, there is a charge buildup in the residual glass left behind by poor etching. As a result, the primary beam cross section is distorted from its normal circular shape to

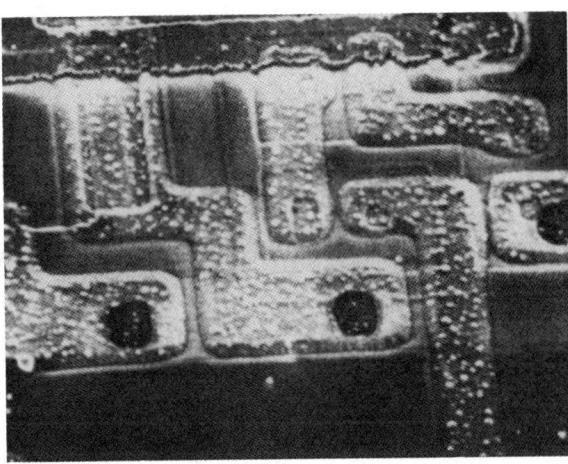

Fig. 60 Examination of a two-layer metallization system integrated circuit in an uncoated state results in preferential charging and signal collection. The dark areas are electrically isolated from the surrounding metallization, i.e., induced voltage contrast.

Fig. 61 Examination of the same sample with only backscattered signal results in the elimination of contrast due to charging effects. Considerable information has been lost, however.

that of an ellipse, which results in a loss of point-to-point resolution. This distortion is very similar to the uncorrectable instrumentation astigmatism caused by dirty apertures in the electron beam column.

It is not difficult to see that charging can render SEM images useless. Reducing the electron beam current will reduce but not eliminate charging.

A technique for reducing the effects of charging is illustrated in Fig. 59. The penetration of primary electrons into an insulating surface, and their storage on and beneath the surface, is the cause of charging. In Fig. 59 a microcircuit similar to the microcircuit in Fig. 54 is shown. However, the high magnification raster at the top of the die in Fig. 54 was irradiated using a high tilt angle and a 3 kV accelerating voltage instead of 10 kV, as in Fig. 54. The energy of the backscattered electrons is 3 keV, which is near the secondary electron crossover, unlike the 10 keV backscattered electrons which cause the charging seen in Fig. 54. In the section on specimen tilt, Fig. 7 demonstrates that specimen tilt could be used as a method to reduce electron penetration, and, therefore, charging. Using low accelerating voltage and specimen tilt is the most expedient means of preventing charging. It can be seen in Fig. 59 that another byproduct of these techniques is better contrast. The contrast in the dark square area in Fig. 59 is rich in comparison to that in Fig. 54.

Figure 60 shows a secondary electron image of a device with a charged passivation layer. Figure 61 shows a backscattered electron image of the same area. The characteristic dark areas adjacent to regions of charging is a result of the distortion in the paths of secondary electrons while higher energy backscattered electrons are not affected. Therefore, using backscattered electrons will eliminate contrast distortion due to charging. A treatment of electron beam interactions with metals and insulators and effects of specimen tilt and accelerating voltage has been given by Oatley.

Fig. 62 Examination of a clean integrated circuit surface at high magnification has resulted in polymerization of vacuum pump oil deposited on the surface in the SEM in the form of a dark square. This artifact can be removed by exposure to an O_2 plasma.

COATING OF SPECIMENS

One technique commonly used to prevent specimen charging, applying a conductive coating over the sample, has not been discussed. Because of the nature of semiconductor devices, such a coating should be a last resort and undertaken only after careful consideration of all phases of the examination to which the device will be subjected. Once coated, a device can no longer be operated, nor will use of voltage contrast and EBIC modes of SEM examination be possible. X-ray analysis can become more difficult after application of *any coating but carbon*.

Fig. 63 A microcircuit window which has been buried under large globs of aluminum from a poor evaporation process. This specimen is a total loss for examination purposes because of the excessive artifacts introduced.

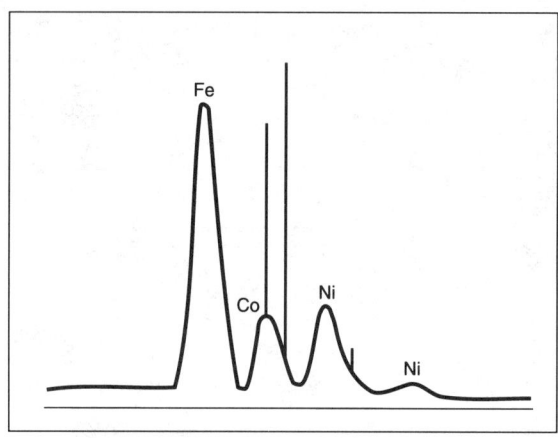

Fig. 64 Kovar, 20 kV. EDX spectrum of a Kovar particle at a beam voltage of 20 Kev. The Fe, Co, and Ni peaks are well defined with the Ni peak being higher than the low percentage Co peak.

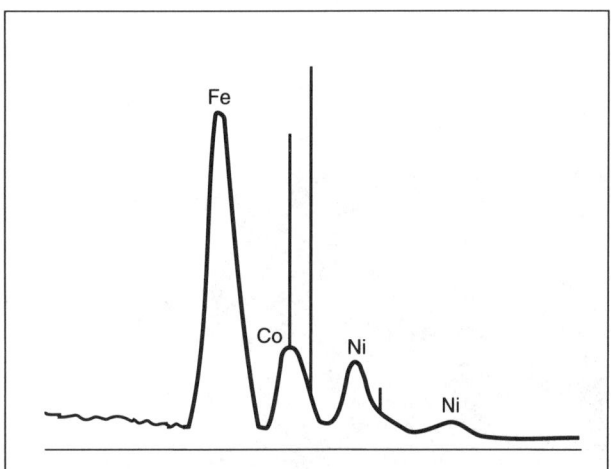

Fig. 65 Kovar, 10 kV. Same sample as 65 but at a 10 Kev beam voltage. All three elements are detected but the nickel peak is low due to poor excitation by the 10 Kev beam.

Scanning a semiconductor surface at high magnification can polymerize hydrocarbon monolayers deposited by an oil-pump vacuum system. This effect appears as a very dark square area as in Fig. 62. Such contamination is an unwanted artifact, and can be avoided by not making prolonged high magnification scans on a localized area. All focusing should be done outside, but adjacent to, the surface area to be photomicrographed. Alternatively, the hydrocarbon diffusion pump oil can be replaced with an oil that will not be polymerized by the primary beam. A good alternative is found in perfluorinated polyether oil. This oil fractures into molecules of low molecular weight which are swept away by the vacuum system. Another method is to employ a cold finger or trap to help reduce specimen contamination by capturing the oil contaminants on a cold surface. Pump oil contamination is not a problem with an SEM equipped with dry-pumped vacuum system.

The SEM primary electron beam damages the semiconductor device. Electron beam penetration into the device results in ionization processes which change the surface oxide properties and thus degrade the electrical parameters. Digital bipolar devices are susceptible to parameter changes during SEM examination; however, it is unlikely that it would cause them to cease functioning. MOS devices are readily damaged by exposure to an electron beam. Even a very short exposure can make them inoperative.

For devices for which optimum electrical performance is required after examination, the only way to eliminate electron beam damage is to not examine the device with an SEM. However, if an SEM examination must be performed, an accelerating voltage as low as possible, considering the resolution which is required, should be used. Also, the lowest practical beam current should be used and the specimen should be irradiated only as long as absolutely necessary. Appropriate alignment techniques are required so that time is not lost searching for the area of interest. If the examination does not include voltage contrast or EBIC modes, a bias should not be applied to the device junctions. It should be noted in subsequent use of the device that the electron beam could have made some irreversible changes in the device.

Semiconductor Specimen Preparation. One of the primary advantages of SEM over TEM and optical microscopy is the minimum amount of specimen preparation necessary. However, some preparation is required and if the study is a failure analysis, the microscopist must take care that the preparations do not obscure the cause of failure being studied. An examination with an optical microscope should be made of a semiconductor device immediately after delidding the package. The analyst should observe any contamination that may be present and which appears to be the result of device failure or poor fabrication. If any solder balls, fragments of broken passivation, or smeared metallization are noted, the sample surface should not be disturbed or valuable information could be lost. Photographic documentation after the optical examination is desirable.

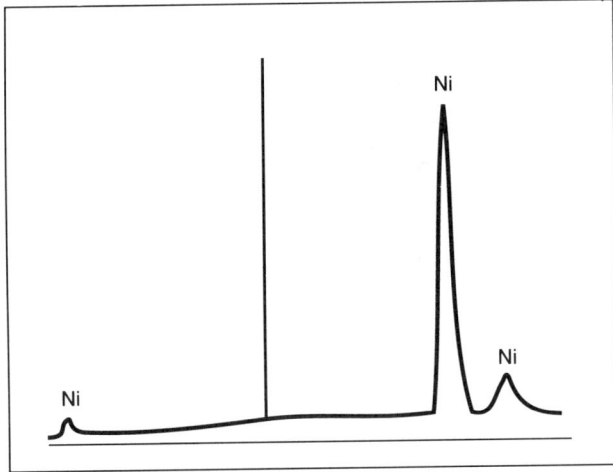

Fig. 66 Nickel, 20 kV. EDX spectrum of Nickel at a 20 Kev beam, both the low and high peaks are excited.

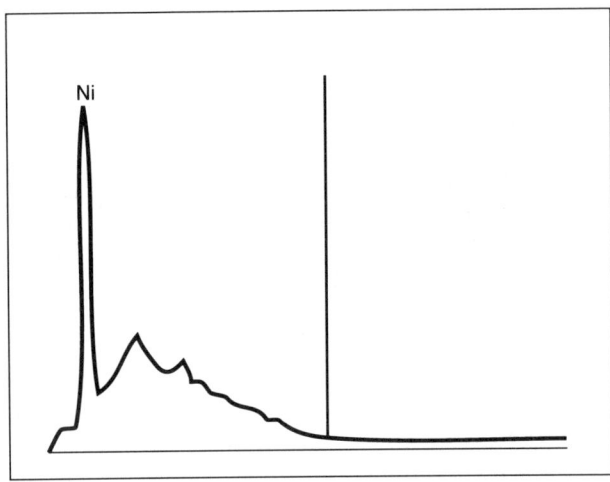

Fig. 67 Nickel, 5 kV. At a low, 5 Kev beam, only the low energy line of nickel is excited.

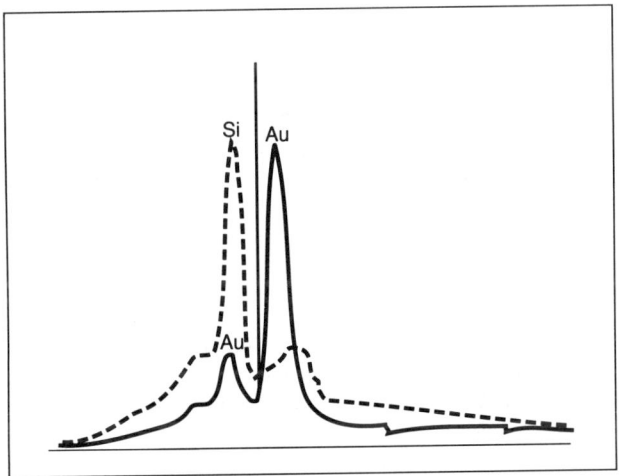

Fig. 68 Preform, 20 kV. EDX spectrum at 30 Kev of a Au-Si eutectic (bars) as compared to the same sample at 5 Kev (dots). Note the much higher silicon peak at 5 Kev, as compared to the gold peak.

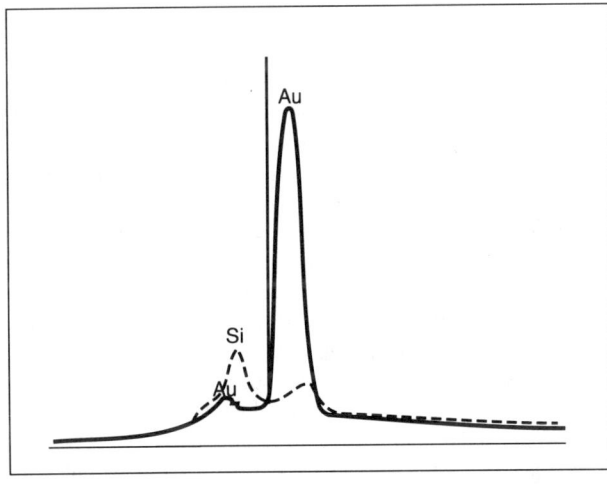

Fig. 69 Preform, 20 kV. EDX spectrum (bars) of gold at 20 Kev shows apparent silicon line at 1.65 Kev, which is actually an Au-M-Zeta line. Dot spectrum shows true silicon peak.

If the optical examination did not reveal any obvious device problems, the glass passivation layer should be removed from the semiconductor surface. Typically, the glass is removed by a wet acid or gas plasma etch. This is necessary because the glass does not replicate the underlying metallization stripes or oxide steps. Inadequate removal of the glass will leave a residue of glass on the metallization which would make topographical interpretation difficult. The residue also charges, which will normally manifest itself in an inability to focus, as illustrated in Fig. 58. The specimen should be firmly fixed to the specimen holder using an electrically conductive adhesive. Aquadag preparations and conductive silver paste can also be used. The holder should be firmly mounted in the stage and make electrical contact to it.

RECOMMENDED COATING PROCEDURES

Coating the specimen with a thin (5- to 20-nm thick) layer of carbon or metal may be used to enhance secondary emission and reduce charging. However, this deposited coating has potential drawbacks: its application is time consuming, there is die surface detail which is covered by the thin film coating, and the device is rendered inoperable. Despite these drawbacks, specimen coating is the method of suppressing charging and increasing the electrical emissivity of a specimen most often utilized. When applied with the specimen surface parallel to the rays of evaporating coating material, subtle changes in surface topography can be accentuated by shadowing.

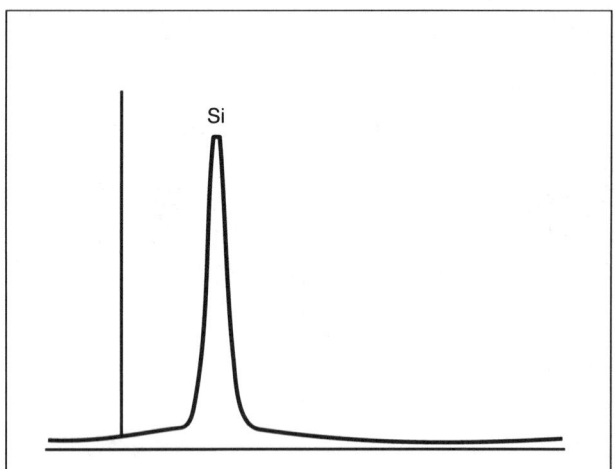

Fig. 70 Silicon, 5kV. EDX spectrum of a pure silicon sample excited with a 5 Kev beam.

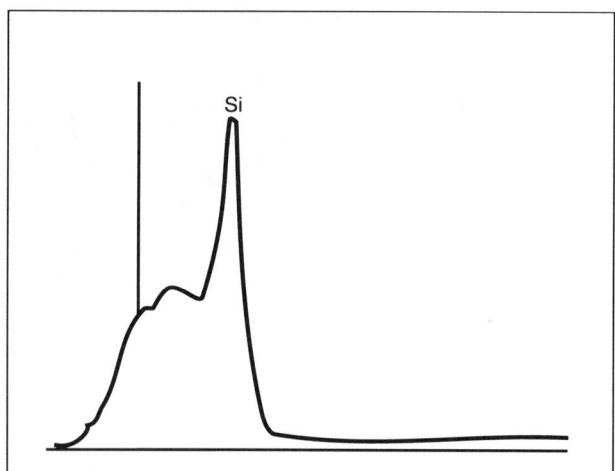

Fig. 71 Silicon, 2 kV. Same sample as Fig. 71 except at a beam voltage of 2 Kev. Silicon is still excited, but notice stray counts to its left.

Vacuum Evaporation

In order to reduce the amount of data which is obscured by specimen coating, certain guidelines should be observed. The selection of the evaporant material is the first consideration. Criteria for selection of a material include: electron emissivity, replication ability, oxidation rate, and melting point. Gold (Au) has excellent electron emissivity but, at high magnifications, can have an agglomerate (lumpy or spheroidal) appearance. Silver (Ag) has the best replicating ability, but in a few days it tarnishes and the specimen cannot be viewed. Gold-palladium (AuPd) alloy is a compromise between high electron emissivity and replicating ability, but the heat from its high melting point may alter the sample. Aluminum (Al) is a good alternative, having good replicating ability, a low melting point, and adequate electron emissivity. On a very irregular specimen surface, carbon is often used on a first layer to establish surface conductivity because its atoms deflect from surrounding surfaces and reapproach the semiconductor surface from many different angles. A second, very thin film of a noble metal is then applied to improve electron emissivity. For X-ray microanalysis, carbon can be used exclusively to establish conduction, while avoiding extraneous X-ray information.

The film should be the thinnest possible coating that will suppress charging. The best procedure is to deposit a very thin film of a noble metal (<10 nm, or 100 Å, thick) and then view the specimen in the SEM to see if the charging has been suppressed. This procedure can be repeated as often as necessary to satisfactorily suppress charging. Although very time consuming, this procedure will limit the amount of die surface data lost.

An evaporated thin film will not be uniform on a semiconductor die surface because of the irregular metallization and oxide steps. A gimbal mechanism may be used to spin the specimen through a complex movement exposing each die surface step for an equal amount of time to multiple evaporate sources, but it is still doubtful that the thin film will be truly uniform. There are two reasons for this. First, there are variations in the specimen-source distance while spinning and the thick-

ness of an evaporated thin film depends on the inverse square law. Also, there are changes during evaporation in the nominal angle of deposition.

To prevent the evaporant source heat from altering die surface topography during evaporation, three precautions should be taken. (1) The specimen should be kept at least 3 to 4 cm from the evaporant source(s). (2) A pre-evaporation shutter should be kept between source and specimen until the coating material has begun to evaporate. (3) A coating material with a low melting point, such as aluminum, should be used.

Sputter Coating

An alternative to vacuum evaporation as a specimen coating technique is sputter coating. In the evaporation technique, the coating material is heated to a sufficiently high temperature so that it evaporates, depositing a thin film on the die surface. In the sputter coating technique atoms from the coating material are ejected when bombarded by relatively high energy particles, usually argon ions. Some of the ejected atoms will land on the die surface to be coated. The particles must be of sufficiently high energy in order to overcome the binding energy of the atoms at the surface of the target. The fast heavy particles used to erode the target are usually derived from an ionized inert gas, and are produced by three main processes: ion-beam sputtering, radio-frequency sputtering, and direct current sputtering.

The main advantage of sputter coating is that a continuous thin film is deposited, even on parts of the die surface which are complex, and those surfaces which are not directly facing the rays of sputtered atoms. This occurs due to the fact that, under the high pressures used, the coating material atoms experience many collisions and approach the die surface from all directions. This effect is achieved without using a gimbal movement. Therefore, the sputter coating technique has good surface replication ability, and there is a reduced chance of surface damage by heating. Noble metals should be used because of their emissivity. *Aluminum is not adaptable to this technique because its surface oxide prevents sputtering.* The techniques advised for specimen coating should be rigorously followed. Thin film

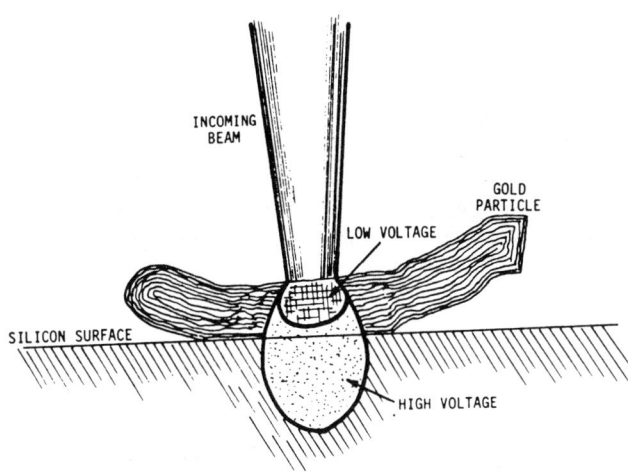

Fig. 72 Example of gold particle on silicon surface.

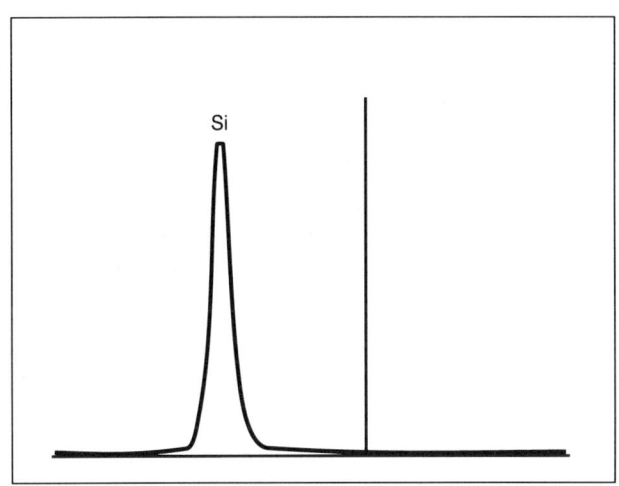

Fig. 73 20 Kev beam generated spectrum of a thin (20 nm) NiCr film on silicon dioxide. Only the silicon shows up.

evaporations which are performed improperly will ruin the possibility for SEM examination. Figure 63 pictures a microcircuit window which has been buried under large globs of aluminum from a poor evaporation process. This specimen is a total loss for examination purposes because of the excessive artifacts introduced.

SEM operation procedures must not become stereotyped for failure analysis applications if the maximum amount of information is to be obtained from the failed device. The analyst should recognize the need for SEM techniques, or imaging modes, which will yield the highest quality information. For failure analysis applications to be most productive, the SEM analyst must be fully aware of the various interactions between specimen and beam and be prepared to modify his operational modes to fit the requirements of the specimen.

In no other application does the operator need to be as informed about the subtle interactions between instrument and specimen as in semiconductor failure analysis.

Correlation with Optical Microscopy. Most often, initial phases of failure analysis include examination of the specimen by low (10×–20×), medium (100×–500×) and high power (1,000×–1,500×) optical microscopy. This is performed under both oblique and direct illumination. Unless the analyst is familiar with the SEM he can often be both misled and disappointed by the data as compared to the optical microscope. The two most common erroneous assumptions are: (1) since the SEM has better resolution it can "see" smaller defects, (2) it can image diffusion and oxide defects. Neither case is true, due to some (useful) properties of light. In the application to thin films and their defects, interference fringing and refraction often renders visible defects in the range of 3 to 8nm, which are still considerably smaller than the resolution limits of many SEMs.

Defects in the oxide layers which are obvious with optical microscopy, due to refraction or phase changes, can be quite invisible on the uncoated specimen due to oxide charging; then defects or features visible optically due to interference coloring will not be detectable in the SEM.

ENERGY DISPERSIVE X-RAY SPECTROSCOPY (EDXS)

The interaction between the energetic monochromatic electrons from an impinging electron beam and the electrons in the atoms of the specimen results in the generation of X-rays. Characteristic X-rays are always of a specific energy or wavelength and identify the elements in a specimen. The same principle is used to create the X-ray source for X-ray radiography, except the current of the electron beam is very high and the goal is the generation of a strong source of X-rays.

Based on this well-known effect, several analytical instruments have emerged to assist the failure analyst. The first, the electron microprobe, entered the field in about 1961-1962. This instrument utilized a stationary beam of electrons to excite X-rays from the sample area of interest. The X-rays were characterized by their specific wavelength with a crystal diffraction grating. The electron microprobe suffers from its selectivity and method of analysis. Various crystals are required to efficiently diffract different wavelengths of X-rays. In addition, a given crystal can diffract only one wavelength at a time. Therefore, ranging over all the wavelengths in the periodic chart is a very tedious and time consuming task. The second, much more common instrument, is an energy dispersive X-ray analyzer (EDXA) attachment to the SEM. This tool and its capabilities make use of the known binding energy in a doped wafer of silicon, which is used as a detector.

The generated X-rays impinge on the silicon detector surface. The silicon is doped with lithium. The penetration depth of the X-ray into the silicon is a direct function of the energy of the X-rays. Along the penetration track, interaction occurs between the X-rays and the silicon atoms, creating hole-electron pairs. The generation of each hole-electron pair requires a specific amount of energy. Thus a weak X-ray of a shallow penetration depth will generate a smaller current pulse than a more energetic X-ray of a longer penetration track. If the optical and physical appearances give no obvious clues to the particles' makeup, then with the energy dispersive analyzer (EDX) the analyst has to determine the particle compositions and speculate on a potential source for each.

Recommended EDX Procedure

1. The operator aligns the specimen in such a manner that a unobstructed path exists from analysis site to detector.

2. For a hybrid microcircuit, if uncoated, a low initial beam voltage should be used to locate the area(s) for analysis without charging up surrounding insulators.

3. The beam should be centered on the particle or area of interest and beam voltage increased to at least 20 or 25 Kilovolts. At 20-25 Kev, sufficient energy is available to produce X-ray emission for all elements up to lead (Pb), gold (Au), tungsten (W) and platinum (Pt) and/or bismuth (Bi). All these elements are atomic number 74 or heavier. Typically, a full spectrum count rate of 1000-1500 counts per second for 60-120 seconds will result in a spectrum of sufficient resolution to identify the major constituents of the particle.

4. If backscatter or penetration thru the particle to the underlying region is a concern, then a reduction of beam voltage may be feasible.

A beam voltage of 10 Kev is adequate to excite the chromium and iron K-α lines but not sufficiently for quantitative work. If, however, just detection and identification is required and only one of several typical elements exists of the Cr, Fe, Ni, Cu, Sn family, then the L-lines at approximately 1 Kev can be used. For confirmation of the tungsten, platinum, gold, lead group (W, Pt, Au, Pb,) then the M-series of lines between 1.5 and 2.5 Kev can be used at an incident beam voltage of 5Kev.

Caution

Due to the low energy spread at the L-series and M-series lines in the 1-3 Kev range, great difficulty is experienced in uniquely identifying the foregoing elements within a group.

For the four particles mentioned, then, the recommended conditions could be:

1. Kovar: Iron, Nickel, and Cobalt –20 Kilovolts, 1500 counts per second for 120 seconds —Spectrum must he carefully examined to spread the $K\alpha_1$ and $K\alpha_2$ lines because of the overlap between the three sets of lines.

2. Nickel: Nickel or Nickel and Phosphorus –20 Kilovolts for general spectrum (can be easily reduced to 10 Kilovolts) is also feasible since the $K\alpha$ and $K\beta$ phosphorus lines are easily excited and the Ni $L\alpha$ lines are also excited.

3. Eutectic: Gold and Silicon: 20 Kilovolts will excite all the M-series gold lines as well as the first three L-series lines. If gold is confirmed then great caution must be exercised because an M series (Zeta) lines exist at 1.65 Kev, which many mistake for silicon. If the sample is fairly small, then analysis at 5 Kev is recommended; this greatly reduces penetration, and the M-Zeta gold line does not overpower the silicon line at 1.74 Kev.

4. Silicon or Silicone: (Silicone contains Silicon). If the particle does not charge up, it is probably not silicone or SiO_2. Either

3 or 5 kV is more than adequate to confirm the existence of the silicon 1.74 Kev $K\alpha_1$ and α_2 lines.

The overlapping of peaks of different elements must be well recognized, understood and expected. Although newer EDX systems have a built-in data analyzer, they are not foolproof.

Efficient X-ray generation from specific elements requires use of an electron beam voltage at least twice as energetic as the X-ray line to be generated.

Example

1. Excitation of the iron $K\alpha$ line at 6.4 Kev energy requires an incoming beam voltage of a 12.8 Kev although 10 Kev is sufficient except for quantitative work.

2. Excitation of the gold $K\alpha$ line at 9.63 Kev requires at least an 18.8 Kev beam so most analysts use a 20, 25 or 30 Kev beam.

Caution

The presence of gold can be easily detected with a 5 Kev beam if the analyst uses the gold M (#4l) line at 2.12 Kev. Although not as intense, it is an effective approach.

Hydrocarbon or organic residue containing chlorine is a common foreign contaminant in the electronics industry. A beam voltage of 20 Kev is more than adequate to excite the 2.71 Kev chlorine line. Once the chlorine has been detected, an analysis with a 5 Kev beam may result in a much stronger chlorine peak, indicative of the fact that the chlorine is contained in a thin surface film.

X-ray detection suffers from the same restrictions as backscattered electron detection, i.e., shadowing. X-rays, similar to backscattered electrons, travel in an absolutely straight trajectory from the emission point, and cannot be focused or collected other than by the detector's exposed surface. *Although not nearly as sensitive to sample location and tilt as the crystal spectrometer, there are usually "dead spots", that is, combinations of height and tilt which result in poor X-ray detection.* The operator must constantly monitor the X-ray spectral count rate to ensure he is not imaging and analyzing such a "dead zone".

Light elements can result in an apparent "dead zone". Elements lower than atomic number 9 are not detected by a normal EDX system with a beryllium window. Thus an oxide, carbide or hydrocarbon cannot be analyzed by energy dispersive techniques. However, inferences can be drawn about a sample. If the analyst is sure he is not in a dead zone and the total count rate drops markedly when the beam traverses from a detectable matrix to the unknown, for instance, from 1500 cps to 50 cps, then the detector is not sensing the light element X-rays. Other knowledge about the specimen can assist the operator in postulating about the unknown as to whether it is oxide or hydrocarbon.

Several other artifacts are common enough to warrant mention:

1. High intensity backscattered electrons can generate X-rays from areas other than the area of interest. X-rays can also generate X-rays in the same manner.

2. Any film used to coat the specimen will alter the observed spectrum. Carbon will alter it the least. Aluminum, gold or

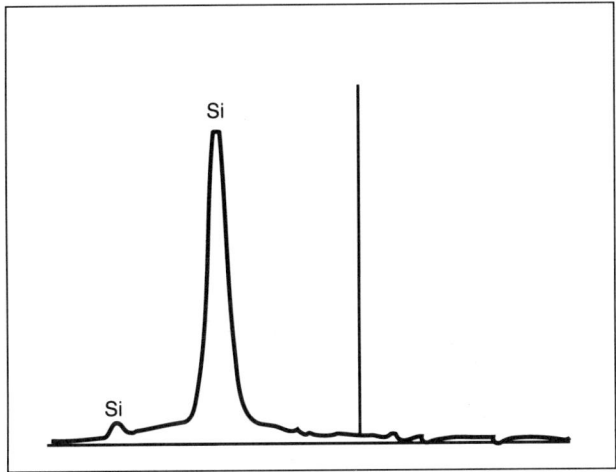

Fig. 74 10 Kev beam generated spectrum of sample from 74. Ni line at 1 Kev is starting to show up.

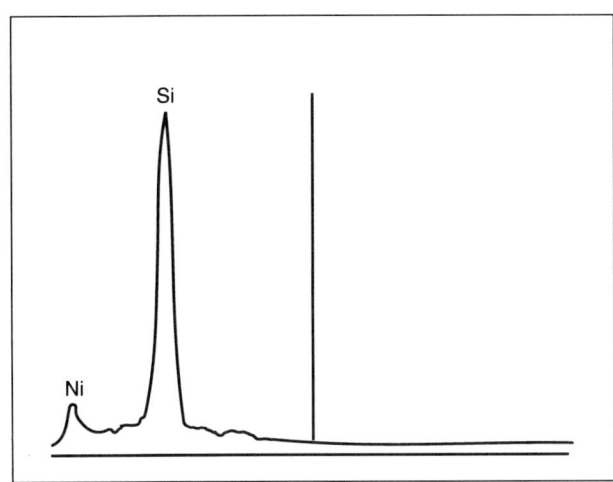

Fig. 75 At a beam voltage of 5 kEv, the low-order Ni line is becoming very obvious.

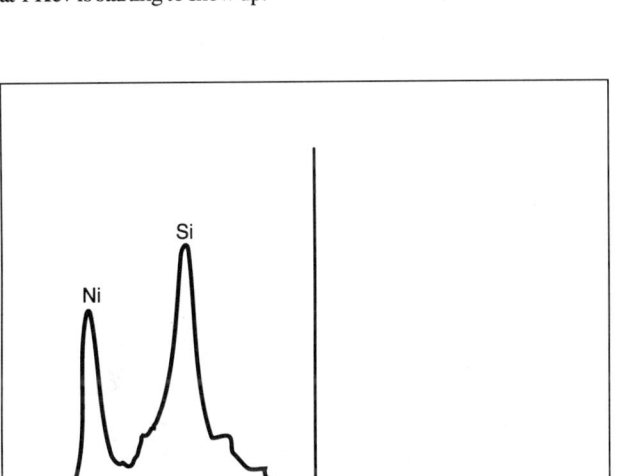

Fig. 76 At a beam voltage of 3 Kev, the Ni peak is almost as large as the silicon peak.

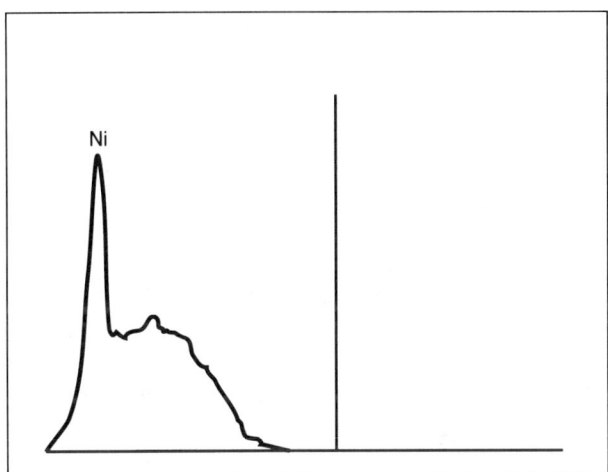

Fig. 77 At 2 Kev, the beam no longer penetrates the NiCr film to effectively excite the underlying silicon and only the Ni line is obvious.

gold-palladium should be used sparingly. Never coat with a film of the same material that is a subject of the analysis.

3. A common element in many spectra is iron. Under some conditions this is a false indication, which can be generated by the final lens pole piece.

Why, then, would the analyst want to use a low beam voltage for X-ray analysis? Penetration of the specimen is a direct function of beam voltage and sample density. Analysis of small particles or thin films is very dependent on operator experience and knowledge of the specimen. An example is illustrated in Fig. 73 thru 77 of thin film analysis.

As can be seen from Fig. 72, the gold particle on a silicon surface can easily be analyzed as containing silicon if a 20 Kev beam is used for analysis. Using the 5 Kev beam restricts penetration to the particle itself.

Because of the complexity of samples usually encountered in failure analysis, a thorough working knowledge of beam-specimen interactions is a necessity. Usually the author uses a 10 or 20 kilovolt beam to determine what elements are present. Then, the beam voltage is lowered to bracket the elements in the area of interest and so obtain an accurate spectra at the lowest possible beam voltage (Fig. 73 thru 77).

The relative intensities of spectral lines may vary significantly with changes not associated with excitation efficiency. This is especially the case with thin films consisting of relatively light elements.

As widespread and useful as the energy dispersive spectrometer (EDS) has become, it still is no panacea. Some problems, defects, films, chemicals, elements and compounds, are not readily or definitively analyzed by EDXA (EDXS).

EPM-EPMA-EMP. (See Figure 79). These are various acronyms for the same instrument, the electron probe microanalyzer (EPM).

EDX
EDXA
EDS
EDAX

Energy Dispersive X-ray Analysis or Spectroscopy

EXCITATION SOURCE IS AN ELECTRON BEAM SAMPLING DEPTH IS ~0.5-5μm

Attaches to any Electron Beam Instrument

Simple – Elemental – Z≠9⁻

Good spatial resolution

Semi-quantitative

Quantitative with Software

WDX
WDXS
EMP
EPMA

Wavelength Dispersive X-ray Analysis or Spectroscopy

Electron beam excitation source

Attachment to any instrument where Take-Off Angle of X-ray is known

More complex than EDS – Mechanically

Can be quantitative
DETECTS ALL ELEMENTS AT #5, BORON, AND HIGHER
DEPTH OF ANALYSIS 0.5-5.0μm

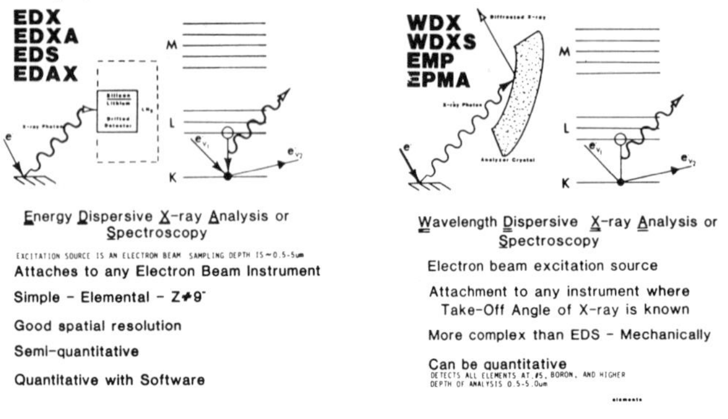

Fig. 78 (a) Energy dispersive X-ray analysis concept. (b) Wavelength-dispersive X-ray analysis concept.

Note: EDXA, EDXS, EDS, and EDX all may be used to indicate "energy-dispersive X-ray analysis." "EDAX" is a tradename for energy-dispersive X-ray detector product. Thus, instruments which only analyze thin surface films or have incredible sensitivity levels have been designed to extend our analytical capacities. The most common are:

AES - Auger (pronounced O-JAY) Electron Spectroscopy

ESCA - Electron Spectroscopy Chemical Analysis

SIMS - Secondary Ion Mass Spectroscopy

The following pages indicate the various acronyms each of these techniques is known by, and the fundamental physical mechanism by which the technique produces information about a sample. The final charts compare pertinent parameters for all the common analytical instruments.

For the majority of failure analysis, the energy dispersive X-ray spectrometer system is the basic working tool. Only after its capabilities have been exhausted, need the analyst resort to the use of more complex techniques.

EPM-EPMA-EMP. (See Fig. 78). These are various acronyms for the same instrument, the electron probe microanalyzer (EPM). The EPM is the oldest of the instrumental techniques. It was the forerunner of the SEM-EDXA system.

In review, constraints from the EPM are similar to the constraints for the SEM system. The EPM may use a light microscope or SEM image to view the specimen and may have slightly higher sensitivity for many of the elements. Sample size and shape are limited as compared to SEM.

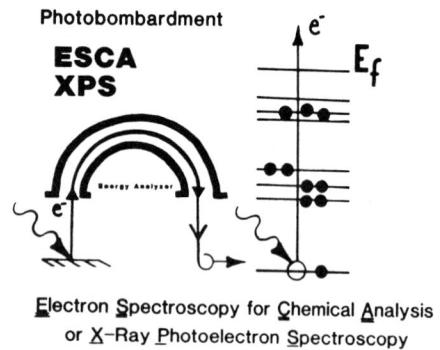

Photobombardment

ESCA
XPS

Electron Spectroscopy for Chemical Analysis or X-Ray Photoelectron Spectroscopy

Floods sample with X-rays and Valence

Band electrons are generated and detected
EXCITATION SOURCE IS AN X-RAY BEAM SAMPLING DEPTH IS ~30-50 Å

Fig. 79 Electron Spectroscopy for chemical analysis concept.

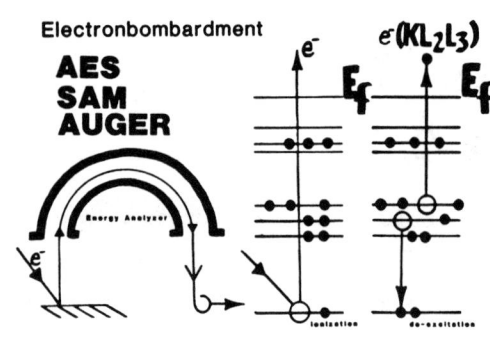

Electronbombardment

AES
SAM
AUGER

Auger Electron Spectroscopy

Electron beam excitation source high/clean vacuum

Sensitive to cleanliness of surface
DETECTS ALL ELEMENTS AT # 3, LITHIUM, AND HIGHER SAMPLING DEPTH ~30-50 Å

Fig. 80 Auger spectroscopy concept.

Results from EPM are often quantitative as compared to the EDXA. There is better light element detection down to boron and carbon. Determination of oxides and nitrides are a significant advantage.

ESCA. (See Fig. 79. Electron spectroscopy for chemical analysis provides valence state information of the material presentonthesurfaceofthepackage.Spatialresolutionispoor.

ESCA resolves an area as small as an integrated circuit die, no less. Usually no real visual observation is possible. Excellent resolution of the various carbon compounds are derived from ESCA. Since the excitation source is a soft X-ray beam, very shallow surface data is obtained. For this reason, ESCA is an excellent analytical tool for determining the molecular structure of polymer coatings and the identification of chemical states.

AES. Auger electron spectroscopy has an irradiation source that is a low voltage electron beam similar to that used in EDXA or EMP. The difference in the instrument techniques is in the detector. The cylindrical voltage analyzer/detector is sensitive to weak Auger electrons which are emitted from the first 15 to 30 Å of the surface.

Auger spectroscopy detects all elements from lithium upward and displays good spatial resolution. (See Fig. 81). If combined with a sputtering source the technique can provide shallow depth profiles.

SIMS. SIMS is an acronym for secondary ion mass spectrometry. The surface of a device to be analyzed is bombarded with ions which then cause secondary characteristic ions to be emitted (sputtered) from the surface. See Fig. 81, for a functional diagram. The sputtered ions are then fed into a mass spectrometer for identification.

The SIMS is the most sensitive of all techniques. In regard to minimum detectable limits, it is capable of detecting all elements and is a surface analytical tool. It is the only instrument capable of directly measuring dopant profiles in a semiconductor.

The drawbacks of SIMS are availability, expense, and poor spatial resolution.

References

Special note: Although the publication dates on some of these references are not current, they are classics in the field, and are applicable to the work discussed in this paper, as well as to current SEM tasks.

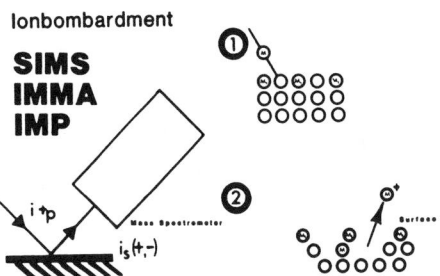

Fig. 81 Secondary ion mass spectroscopy concept.

1. *Proceedings of the Symposium on Scanning Electron Microscopy,* published by the IIT Research Institute, 1967 through 1980.

2. P.R. Thornton, *Scanning Electron Microsopy*, Chapman and Hall, Ltd. 1968.

3. O.C. Wells, A. Boyde, E. Lifshin, and A. Rezanowich, *Scanning Electron Microscopy,* McGraw-Hill, 1974.

4. J.W.S. Hearle, J.T. Sparrow, and P.M. Cross, T*he Use of the Scanning Electron Microscope,* Pergamon Press, 1972.

5. K.F. Heinrich, *Electron Beam Microanalysis,* Van Nostrand-Reinhold, 1981.

6. J.J. Hren, J.I. Goldstein, and D.C. Joy, *Introduction to Analytical Electron Microscopy,* Plenum Press, 1979.

7. M.L. Meny and R. Tixier, Microanalysis, *Scanning Electron Microscopy,* Les Editions des Physique, 1978.

8. B.L. Gabriel, *SEM: A User's Manual for Material Science,* published by American Society for Metals, Materials Park, Ohio, 1985.

Transmission Electron Microscopy: A Review and a Comparison with High Resolution Scanning Electron Microscopy

J.H. Rose, B. Miner, and C.A. Pelillo

INTRODUCTION

Transmission electron microscopes (TEM) have evolved over the past twenty-five years from awkward-to-operate instruments which were employed for microstructural observations to microprocessor-controlled instruments with various detectors attached, providing a range of chemical and structural information down to the atomic level. Today, metals, ceramics, semiconductors, and polymers are actively studied with TEM.

The latest generation of all-purpose analytical TEMs provide unique capabilities for materials structural and chemical analyses with exceptional spatial resolution. Microstructural imaging at magnifications up to 1,000,000 times, crystal lattice imaging below 3Å, and diffraction and chemical information from regions approaching 20Å across are nearly simultaneously possible. As device dimensions enter the submicrometer realm and with film thicknesses often much less than 1000Å thick, TEM is often required in process development and to assess the finest details of device structures. Direct observation in the TEM provides microstructural and chemical information unobtainable by other materials analysis techniques. The capabilities which distinguish TEM analysis can be summarized as follows:

1. Imaging internal structure with high resolution;

2. Observation of a class of features (crystalline defects) not readily observed by other techniques;

3. Chemical and phase analysis of features or films less than 100Å in size or thickness;

4. Atomic structure imaging.

This article reviews the instrument and types of information provided by TEM and is based in part on two previous ISTFA presentations. To accomplish this, the imaging and analytical features of the TEM will be outlined, followed by a brief description of the most common aspects of sample preparation. Next, the types of materials information provided by TEM are categorized. These considerations are then illustrated with a variety of examples demonstrating the value and place of TEM in the examination of VLSI-related thin films. Included in this last section is a comparison of TEM and high-resolution SEM for imaging device structures.

THE INSTRUMENT

Today's standard analytical TEM is a combined TEM and STEM (Scanning Transmission Electron Microscope). In the TEM mode of operation, magnetic lenses are used to image the sample in transmission in a manner analogous to a light microscope (Fig. 1). Lenses above the sample illuminate it with a defocused electron beam. Magnifying lenses below the sample focus the transmission image on a phosphor screen. Images are recorded by tilting the screen and exposing electron-sensitive film held in a chamber below.

IMAGE FORMATION : SEM vs TEM

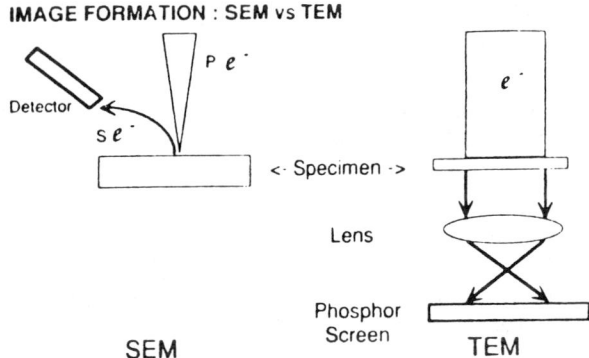

Fig. 1. A SEM forms an image by scanning a focused electron beam over a sample surface while modulating the intensity of a CRT beam in proportion with the signals of secondary or backscattered electron detectors. Imaging in the TEM is analogous to light optical microscopy with electrons replacing photons and magnetic lenses replacing glass lenses.

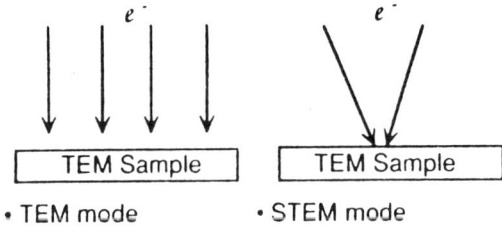

Fig. 2. Most TEMs have the ability to operate in a SEM-like mode (hence STEM mode—Scanning Transmission Electron Microscopy). In addition to the electron detectors found in a SEM, an additional detector is typically placed below the sample for transmission STEM images. Operation in STEM mode assists in probe placement and permits elemental mapping.

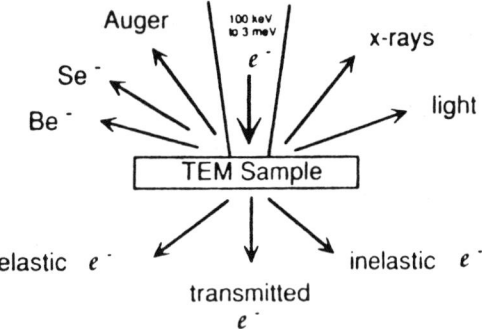

Fig. 3. While most electrons are transmitted through a typical TEM sample, various other scattered primary and ejected electrons and photons are emitted from the sample. The TEM makes use of transmitted and elastically scattered (diffracted) electrons for imaging and x-rays and inelastically scattered electrons from elemental analysis.

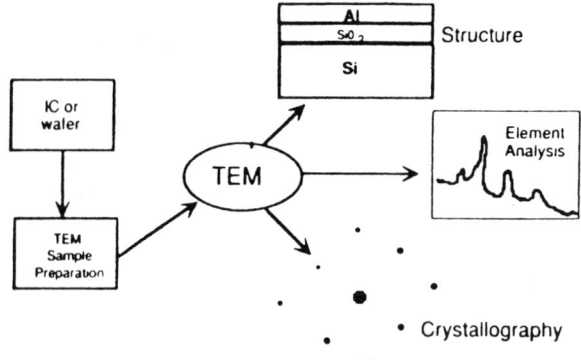

Fig. 4. After appropriate sample preparation, the TEM can quickly provide high spatial resolution microstructural, elemental, and diffraction data.

In the STEM mode of operation (Fig. 2), the microscope scans a focused electron beam over the sample, forming a video image with electrons collected by a secondary electron detector above the sample or a transmitted electron detector below the sample. Hence, it now has similarity to an SEM. This mode of operation allows placement of fine probes (about 20Å to 400Å in diameter) on regions of interest for diffraction or chemical analysis.

Chemical analyses can be performed during either mode of operation, the selected mode being a function of the size of the feature to be analyzed. STEM operation also permits production of elemental maps.

When the electron beam strikes the sample (commonly at 100 to 200 keV in TEMs), a variety of beam-specimen interactions occur which yield X-rays, light, Auger electrons, secondary and backscattered electrons, inelastically and elastically scattered electrons, and transmitted (unscattered) electrons (Fig. 3). This permits many potential applications of TEM. The following describes the types of image, diffraction, and spectroscopic data provided by TEM (Fig. 4).

Imaging

Imaging in the TEM makes use of transmitted and elastically scattered electrons, the latter being diffracted beams in the case of crystalline samples. This leads to three types of image formation, including mass thickness imaging.

Microstructural imaging (diffraction contrast imaging) employs the transmitted and at least one diffracted beam to observe the internal microstructure of a sample at magnifications typically up to a few hundred thousand times. A sample produces image contrast due to varying crystal orientation, presence of defect strain fields, or changing elemental make up. Phase morphology, grain boundaries, dislocations, and device structures may readily be observed in great detail (Fig. 5). Resolution is limited by the strain field surrounding defects, typically tens of Angstroms, though skilled microscopists employ the technique of weak beam imaging to reduce this to about ten Angstroms. This latter method revealed that disloca-

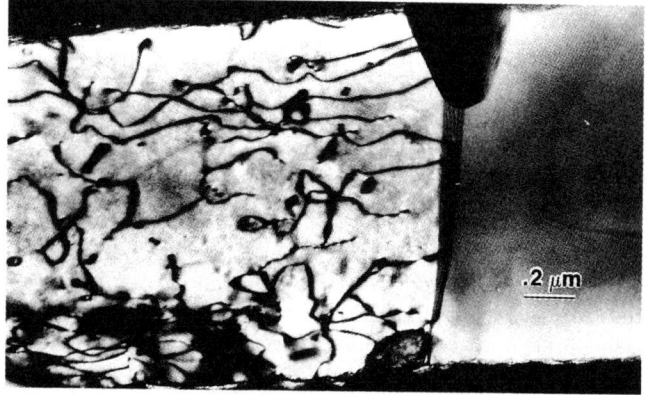

Fig. 5. Planar TEM image of aluminum—1% copper interconnect. Note the grain boundaries, dislocations, and copper rich precipitate at the grain boundary.

tions in semiconductors are typically dissociated into two partial dislocations bounding a stacking fault.

Lattice imaging (phase contrast imaging) is employed to obtain images of crystal atomic structure projected along low index directions (Fig. 6). This is typically accomplished in silicon by aligning a ⟨110⟩ crystal axis parallel to the electron beam. Interference between the transmitted and the {111} diffracted beams leads to periodicity in the image (phase contrast) corresponding to the planar spacings associated with the diffracted beams (3.13Å). Resolution below 2.5Å is available in all-purpose TEMs (below 2Å in dedicated high resolution TEMs).

This technique has been employed most heavily for study of semiconductors. Lattice imaging was first used to determine the atomic model for a crystalline defect in the late 70s (for a grain boundary in germanium) though it can routinely be employed for extremely precise internal magnification calibration, interfacial observation, and identification of defects in crystals.

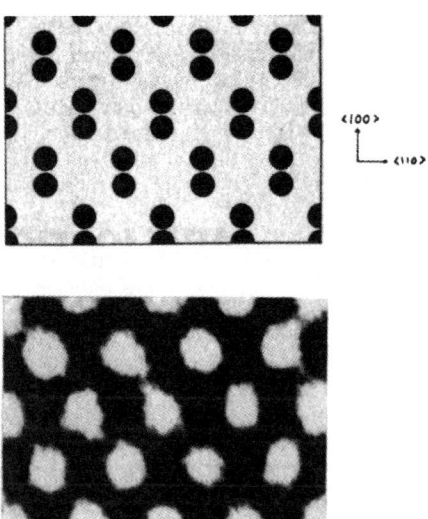

Fig. 6. Atomic model (top) and TEM lattice image (bottom) of ⟨110⟩ oriented silicon. The spacing of the "dumbbells" (1.35Å) is too small to resolve, leading to a single spot for each atomic column pair in the actual image.

Analytical Electron Microscopy

The modes of TEM operation which provide elemental, crystal structure, and phase information are collectively referred to as Analytical Electron Microscopy (AEM).[5] These techniques employ electron diffraction and spectrometers attached to the TEM. True AEMs have their designs optimized for sensitivity, small electron probe potential, and ease and flexibility of operation. For analytical studies with the best possible spatial resolution, AEMs are usually equipped for STEM operation. Here, probes down to about 20Å in diameter (5Å in dedicated STEMs) can be placed on areas of interest with the aid of the scanned image. STEM operation also permits elemental mapping and secondary and backscattered electron imaging.

Electron diffraction is readily available with a quick adjustment to the imaging lenses in the microscope (Fig. 7). This reveals crystallinity, grain orientation and texture, and phase identification. Since the electron beam can be focused to a probe well below 100Å in diameter, the technique of convergent beam diffraction, though outside the scope of this paper, provides additional crystal structure and electronic band information by analysis of details within the diffraction spots. Diffraction information can be obtained from very small regions, and requires nearly perfect crystals. This is achieved by converging the electron beam so that electrons strike the sample over a range of angles. See Ref. 6 for a review.

Elemental analyses are performed by utilizing inelastic events during which incident electrons lose discrete amounts of energy and sample elements give off characteristic X-rays. This is most commonly accomplished with Energy Dispersive X-ray Spectrometers (EDS) attached to the TEM. This technique provides sensitivity below 1 atomic %, and quantification of about

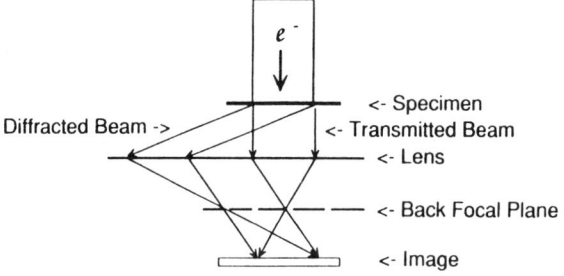

Fig. 7. By simply pushing a button on the TEM, one switches from image mode to diffraction mode. With a planar TEM sample, electron diffraction patterns can be obtained from films of any thickness. These provide phase a ndcrystal orientation information.

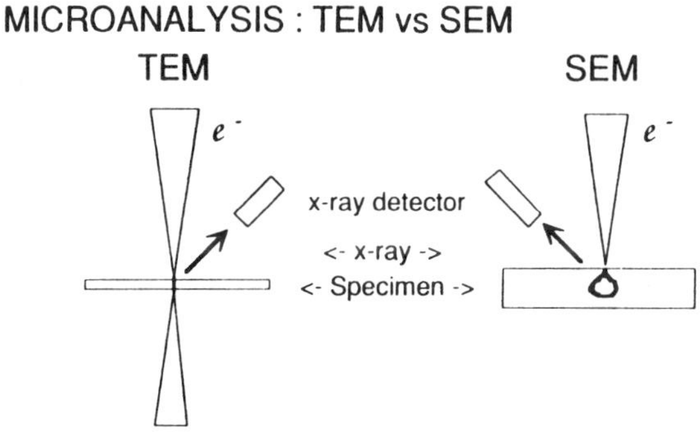

MICROANALYSIS : TEM vs SEM

Fig. 8. While TEM x-ray data spatial resolution is limited by the probe diameter, in the SEM it is limited by the scattering and absorption of electrons in a bulk sample. Hence, TEM resolution is about 1000× finer than SEM.

10%. Elements with atomic numbers down to boron can be detected with ultra-thin window or windowless detectors.

Electron energy loss spectrometry (EELS) can be performed with sensitivities similar to EDS for all light elements. EELS also has the potential of providing bonding information, though such application requires expertise with the technique.[7] The EELS spectrometer is attached beneath the electron column. The electron energy spectrum is obtained by raising the phosphor screen and allowing an image or diffraction spot to strike the spectrometer entrance aperture. EELS is practical only for extremely thin (< 500Å) samples.

Again, the fine probe available on the TEM permits chemical analysis of regions less than 100Å across. Since such observations are performed on a sample region typically less than a few hundred angstroms thick, the beam broadening limitations of SEM work are largely avoided (Fig. 8). In addition, quantification is easier and more reliable for such thin samples since bulk correction factors are unnecessary.

TEM SAMPLE PREPARATION

As in SEM and light microscopy there are two salient sample preparation geometries planar and cross-sectional (Fig. 9).

What distinguishes TEM is its requirement for electron transparent samples—typically one hundred to a few thousand angstroms thick, depending on material, type of observations, and accelerating voltage. This is a demanding requirement, however a number of methods have been developed over the years to obtain thin sections from virtually any material. These include electropolishing for metals, chemical polishing for semiconductors, ultramicrotomy for polymers and particles, and ion milling for ceramics and multi-component samples.

VLSI studies require techniques for examining thin films deposited on silicon wafers. Planar oriented samples provide a relatively large area of view of a thin-film normal to its surface (Fig. 10a) while cross-sections (Fig. 10b) permit a detailed analysis of structure versus depth.

Cross sections are extremely useful in VLSI studies given the high spatial resolution capability of the TEM. The various layers as well as the internal microstructure of the films (e.g., grain size and shape) can be observed in detail. Wafers with uniform thin films can be prepared easily, however small device features present the problem of obtaining an electron transparent section through the object of interest. This difficulty is sometimes remedied by the production of test patterns specifically designed to aid TEM sample preparation. A test pattern has a device-like structure which mimics all process aspects of a real IC. This structure is then repeated in the X direction and is unchanging in the Y-direction. Such test patterns can be included on actual wafers for TEM evaluation of production steps. Of course, test patterns cannot be electrically tested, nor serve for failure analysis of specific device sites.

Preparation of cross sections for TEM observation, though often routine, is still under active development. However a few essential steps are typical of most approaches. Wafer sections are sandwiched with epoxy, sliced, mechanically polished, and lastly ion milled.

In addition to production of samples of high quality, reliable and rapid generation of TEM samples is required for routine support of manufacturing and process development groups. Using the basic methods outlined above, workers in various industrial labs have modified techniques to suit their particular needs. Though TEM analysis can never be as rapid as SEM analysis, it can provide routine support for problem solving and development work that requires materials analysis of high spatial resolution.

CATEGORIES OF INFORMATION PROVIDED BY TEM

As described above, TEM provides a wealth of data via images, diffraction patterns, and spectroscopic analyses. Such data gives structural, crystallographic, phase, and chemical

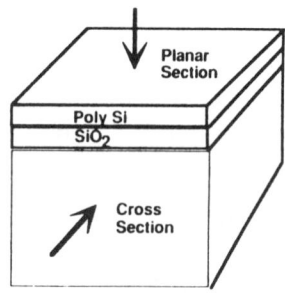

Fig. 9. For wafer or device observation, TEM samples are typically prepared for planar or cross sectional view of the surface films.

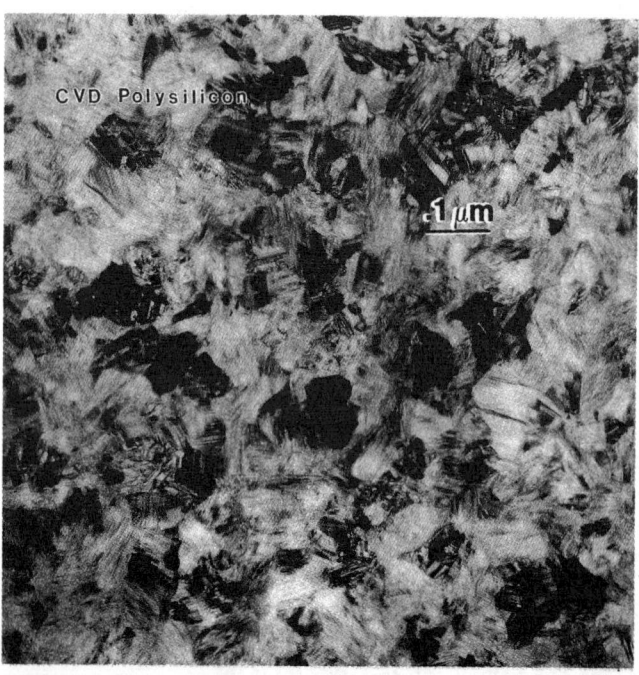

(a)

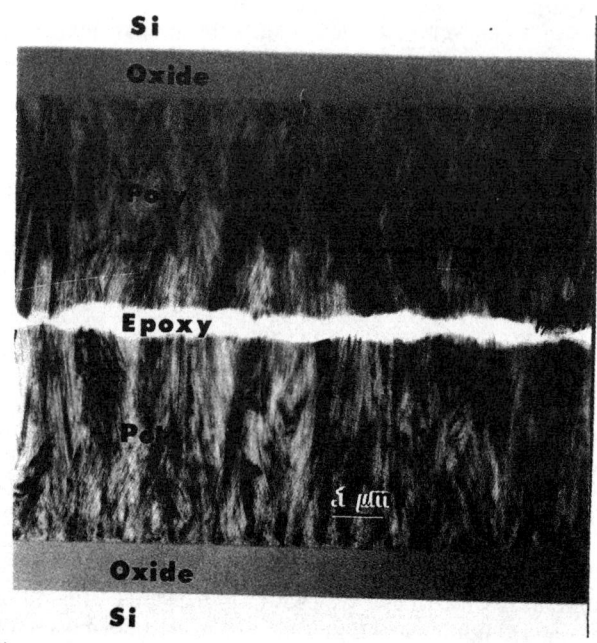

(b)

Fig. 10. TEM micrographs of a CVD polycrystalline silicon film deposited on oxide. a) planar section and b) cross section. Two wafer sections are face-to-face in the cross section.

TEM Data :

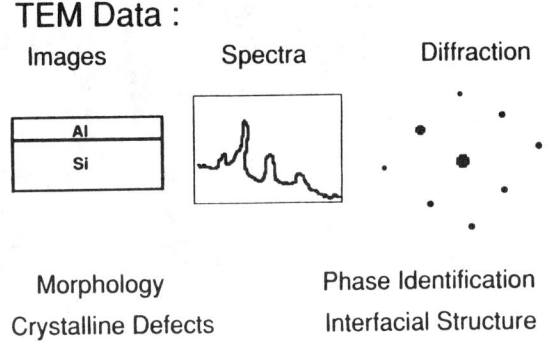

Fig. 11. Through the analysis of TEM data (images, spectra, and diffraction patterns) one obtains information about device morphology, crystalline defects, phase makeup, and the nature of thin film interfaces.

composition information (all with high spatial resolution). TEM provides materials analysis of such variety and detail as a consequence of its many modes of operation and its sample requirements.

For analyses of silicon-wafer-based thin-films, the salient types of materials information provided by TEM can be categorized into four areas, discussed in the following. This can be used for comparison to other materials analysis techniques and in determining when it is appropriate to seek TEM analysis.

Configuration. TEM imaging provides feature observation with the highest possible resolution. Device structures can be viewed in detail, film thicknesses can be measured with high precision, and grain structure can be evaluated. This area of application of the TEM is the most easy to interpret, but its use requires considerable expertise in crystallography. However, in device cross-section imaging applications, there exists much overlap in the information provided by TEM and SEM. Comparisons of the two methods for this application are instructive—one should be careful to select the most appropriate method for the problem at hand.

The new ultra-high-resolution SEMs, which use field emission filaments and in-lens imaging, combine the traditional strengths of SEMs with the higher resolution that has previously been attainable only with TEMs. Resolution of these new SEMs is ~ 1-10 Å, which is more than sufficient for many imaging tasks. They can be run at low kV, and excellent images can be obtained even from uncoated samples. There are several advantages of using the ultra-high-resolution SEM over using TEM; sample preparation time is much lower, field of view is much larger, depth of focus is large (i.e., device surface imaging), and recorded images are available quickly. On the other hand, TEM provides highest resolution of interfaces and, unlike SEM, does not require surface topography to distinguish different material regions at high resolution. However, one should not apply TEM to device imaging applications where SEM will suffice.

To understand which jobs must be performed on the TEM and which can be satisfactorily completed on the SEM, or the ultra-high-resolution SEM, the analyst must understand the different contrast mechanisms under which images are formed

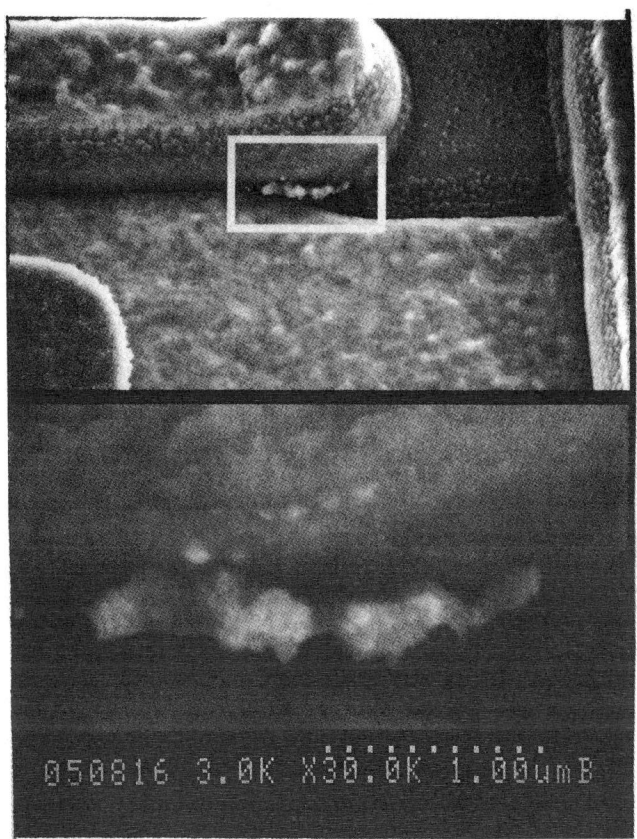

Fig. 12. SEM image of an isolated TiN stringer from a section of patterned wafer. The sample was uncoated and unstained. The image was recorded with a beam energy of 3.0 kV. The upper view of the stringer gives it relationship to the surrounding structures at a magnification of 30 kX, while the higher magnification view gives details of the stringers at a magnification of 150 kX.

and know what types of information are important. As described earlier, a TEM is NOT a SEM with higher magnification capabilities. Much information in a TEM image is fundamentally different from that in a SEM image. Some analytical problems require information that is uniquely available from SEM images, some problems require information that is uniquely available from TEM images, and some questions can be resolved with either technique; in the final case the SEM, with its quicker sample preparation time, should be used. The examples section of this article compares the use of TEM and SEM in this area.

Phase Identification. In addition to the configuration of the material components in a device, the identity of these components is often required. Thin film reactions and second phase precipitation are common. With chemical and diffraction analysis, the phase identity of a feature less than 100Å across can sometimes be determined. Such work on VLSI device cross-sections is impossible with an SEM.

Crystalline Defects. All crystalline materials typically possess point, line, and plane defects (e.g. grain boundaries, dislocations, and stacking faults). Such defects often have significant influence on device processing and performance.

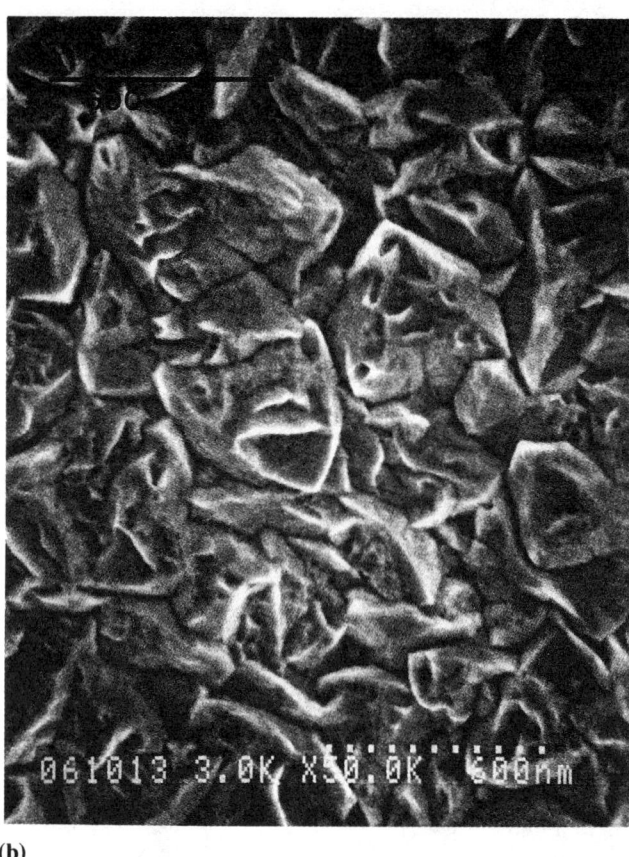

(a)
600 nm

(b)
061013 3.0K X50.0K 600nm

Fig. 13. (a) Plan-view TEM image of tungsten grains taken at 300 kV; magnification of 52 kX. (b) Plan-view SEM image of tungsten grains taken at 50 kX using a beam energy of 3.0 kV to optimize information from the surface.

For example, anomalous diffusion profiles can result from the presence of stacking faults while dislocations can serve to short a pn junction. Diagnosis of such problems is possible with TEM by direct imaging of these features. However, interpretation of the images of such features often requires the assistance of an experienced TEM scientist. This image analysis can be provided by the microscopist performing the TEM work.

Interfacial Structure. In this area, TEM can produce images with information not available by any other means. With lattice imaging performed on cross sections, the atomic level configuration of a semiconductor interface can be viewed and its influence on neighboring thin-film defects can be evaluated. Such information may prove vital in device analysis; device feature dimensions are now small enough that interfaces have a significant effect on electrical performance. Features will eventually be so small that interfacial effects will dominate.

ILLUSTRATIVE TEM ANALYSES

In illustrating the above categories, multiple use will be made of some case studies; usually, more than one type of information is required to evaluate a problem or provide process support. Most discussion will concentrate on configurational observations due to the value of imaging in device related studies. The focus in the following discussion will be on the data and its analysis rather than the technological significance of the structures.

Configuration. (Note: All ultra-high-resolution SEM images were recorded on the cold-cathode field emission in-lens Hitachi S900 SEM. TEM images were recorded on a JEOL 2000FX and a Philips CM30.) Use of TEM and SEM micrographs for device feature observation is common. (Ref. 8 catalogues many excellent TEM micrographs of device cross sections.) The following compares and contrasts TEM and SEM device imaging applications.

An SEM sample, even for in-lens, ultra-high-resolution SEMs, is larger than a TEM sample; more area is available for analysis. SEM analysis has always been used to search for abnormalities, defects, and residues. The higher resolution now available allows smaller defects to be imaged. Figure 12 is an image of a titanium nitride (TiN) stringer located in a particular position only on specific die. Several dozen RAM cells were viewed before any stringers were found. TEM is not suited to searches for such isolated defects. Part of the strength of SEM in device surface observations is due to its great depth of focus. The SEM is particularly well qualified for analyzing several layers of structure and the relationship of the layers to one another. In Fig. 12, the location of the stringer is also important.

The TEM is a transmission imaging instrument—a TEM image is inherently a two-dimensional projection of the sample structure. For planar orientation samples (see the tungsten film in Fig. 13a), there is no topographical or surface information in the image. Grains of tungsten appear as various shades of gray depending on their crystallographic orientation with respect to the electron beam direction. The black-appearing grains are oriented such that some zone axis is nearly parallel to the electron beam. If a particular orientation of grains is important, an-

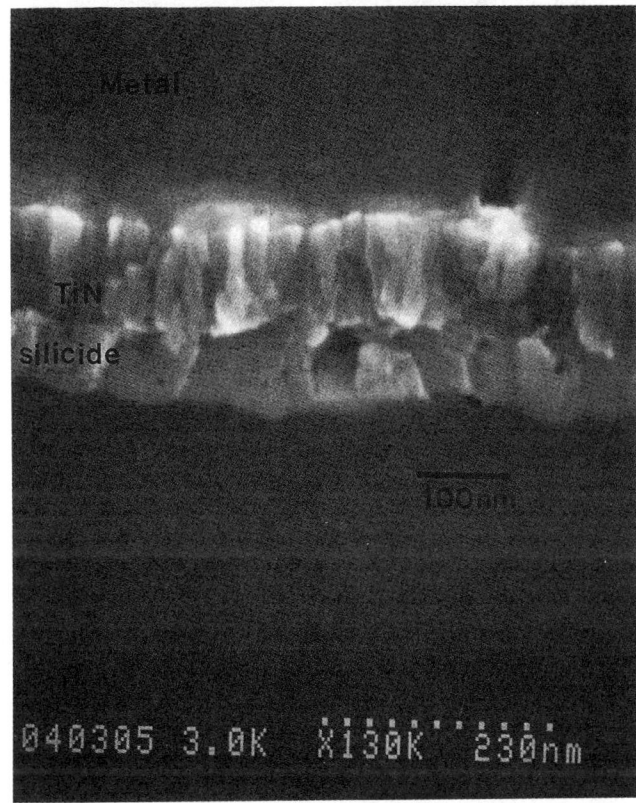

(a)

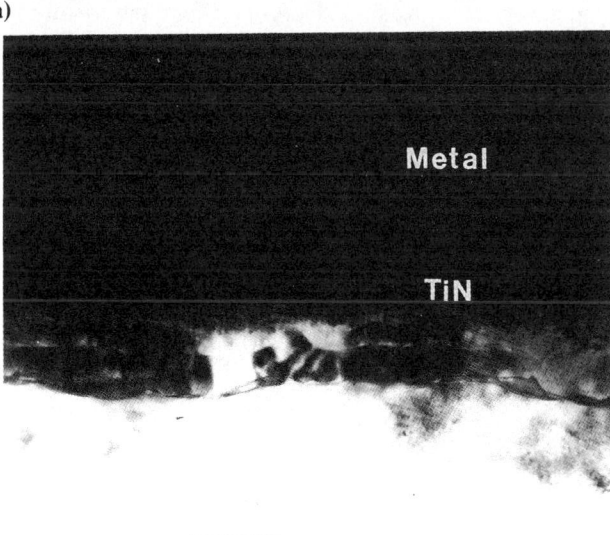

(b)

Fig. 14. (a) Cross-section SEM image of the metal layers. The sample was unstained and uncoated. Beam energy was 3.0 kV; magnification is 130 . (b) Cross- section TEM image of the same metal layers. Magnification of 120 kX.

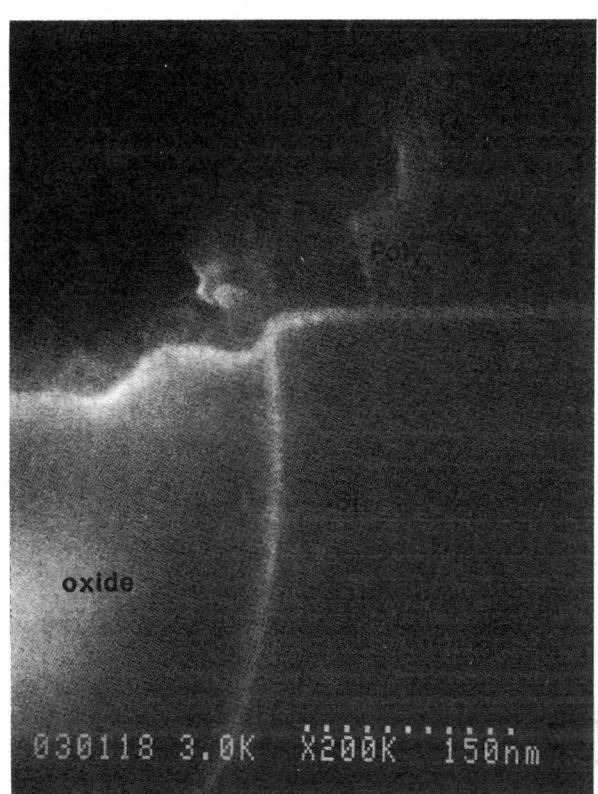

(a)

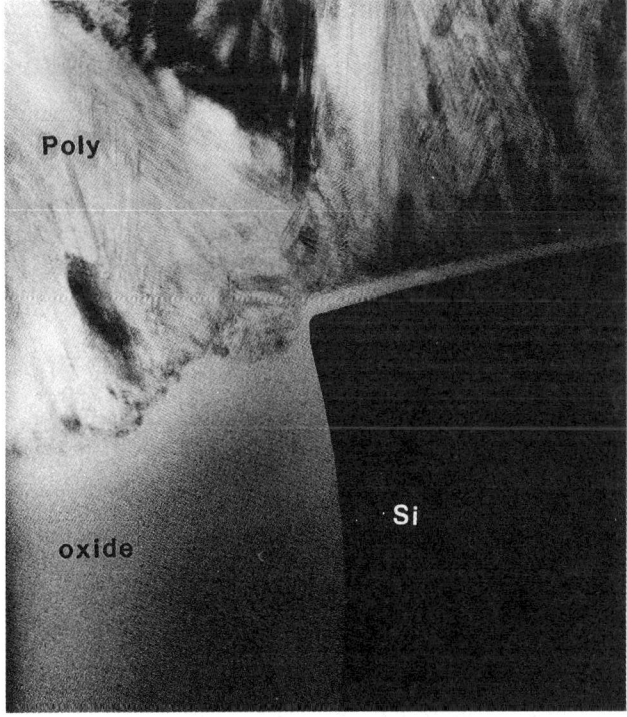

(b)

Fig. 15. (a) Cross-section SEM image of the Si corners at the edge of an oxide-filled trench. The image was recorded at 3.0 kV. The sample was uncoated. The oxide was highlighted using a 5 second dip into diluteHF. (b) TEM image of the same corners.

(a)

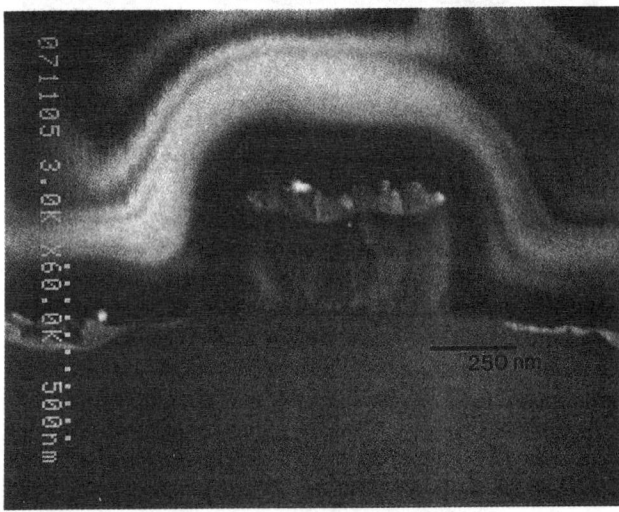

(b)

Fig. 16. (a) Cross-section TEM image of the gate and active areas of a transistor from a set of electrical test structures. Silicide covers the active area; end-of-range implant damage is visible beneath the silicide. The spacer is barely differentiated from the surrounding field oxide. The original magnification of this image was 10 kX. (b) Cross-section SEM image of a similar area. The sample was not capped, coated, or stained. Beam energy was 3.0 kV.

nular dark field TEM images, which show all the grains at a particular orientation, could be collected. In the SEM, the image formed by secondary electrons that are emitted from the surface of the sample has a three-dimensional appearance, as shown in Fig. 13b. The highest sample regions reflect more electrons into the detector (that is located above the sample) and appear brighter in the SEM image. This image of tungsten is from the same sample as the TEM image in Fig. 13a. Grain boundaries are also visible in this ultra-high-resolution SEM image of uncoated tungsten. However, there is no information available about the orientation of the tungsten grains and one can not be sure all grain boundaries are observed. If the surface roughness is important, the analysis is a SEM job. Grain size analysis could be done from either image, though TEM is more accurate (when properly done) and is required for fine grain sizes. Analysis of the relative orientation of grains is a TEM job as is any detailed study of grain boundaries.

Fig. 17. Planar sample of CoSi₂ film.

The ultra-high-resolution SEM image shown in Fig. 14a clearly shows the location of the silicon, titanium silicide (silicide), TiN, and metal layers. The roughness of the silicon/silicide interface can be measured. The contrast within the silicide layer relates to how the sample was cleaved, which can have some dependence on the crystallography within the layer. The entire field of view contains useful information. In the TEM image shown in Fig. 14b, the TiN layer is too thick to observe its internal structure. The TEM image is from the same sample and contains additional information. The silicide layer is composed of two separate layers of crystals; the contrast between the layers results from the differing orientation of the grains and the grain boundaries between the layers. The surface roughness of the silicon/silicide interface can be measured from either micrograph.

One of the advantages for TEM analysis of ICs is that they are formed on a large, perfect single crystal. Whether the substrate is GaAs or Si, the substrate is always present in cross-sections as a reference for calibrating magnification (if the TEM can resolve the lattice fringes), camera length for diffraction patterns, and for precise tilting of the sample so that the interface is observed exactly perpendicular to the electron beam. This orienting of the interface is completely independent of the orientation of the original cleave of the sample or the angle of the grinding or ion milling.

Figures 15a and b show the corners of an oxide trench. Figure 15a is the ultra-high-resolution SEM image, recorded at 200 kX. The sample is uncoated and recorded at 3 kV to minimize noise from secondary electrons created in the bulk of the sample. The shape of the trench corner is obvious; the gate oxide is clearly resolved. Figure 15b is a TEM image from the same sample recorded at the same magnification. The shape of the trench corner and the continuity of the oxide around the corner can clearly be seen in either image. Both TEM and ultra-high-resolution SEM provide information about the shape of the corner and the continuity of the oxide around the corner. If

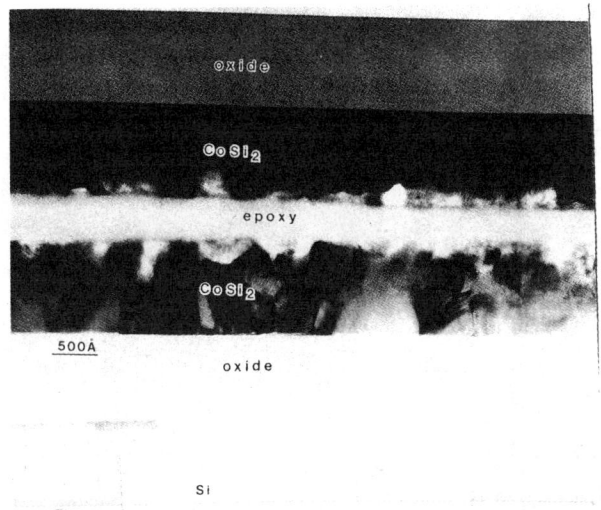

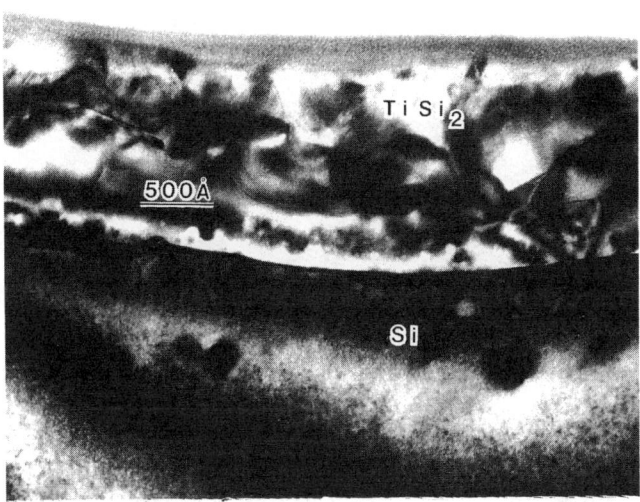

Fig. 18. Cross section of CoSi$_2$ sample in Fig. 17. Note voids near the surface of the film.

Fig. 19. Cross section of an annealed TiSi$_2$ film.

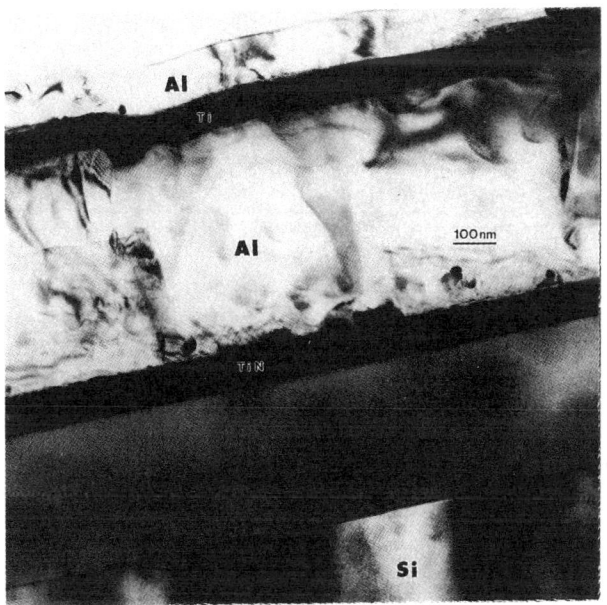

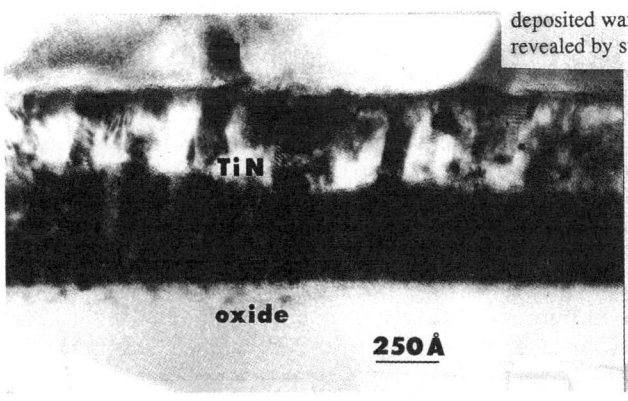

Fig. 21. High magnification view of a TiN layer showing columnar grain structure.

Fig. 20. Bright field image of a cross section of an as-deposited wafer. The detailed structure of the thin films is revealed by such samples.

variability of the corners is important, than the ultra-high-resolution SEM should be used. If an exact measurement of the gate oxide at particular points is important, the TEM must be used to ensure that the oxide is not shadowed, the interface is exactly parallel to the electron beam, and the silicon lattice plane can be used as an internal magnification calibration.

In a final comparison, Fig. 16a gives a low-magnification TEM image of a transistor. The poly gate is capped with silicide. There is also silicide over the active areas. The oxide spacer on either side of the gate is barely differentiated from the field oxide. The aluminum lines were broken during sample

preparation. A higher magnification view of a similar area is shown in Fig. 16b from the ultra-high-resolution SEM. This sample is uncoated, and so a halo effect from charging surrounds the poly gate. The silicide over the active area is obvious; the space is delineated from the field oxide. This sample was unstained and uncoated. If the appropriate stain were used, a qualitative dopant profile would be visible in the active area and differences between types of oxides would be more dramatic.

Other device features which are unexpected and invisible with other techniques can be found with TEM. In a planar sample prepared from a cobalt silicide thin-film, light-colored objects about 200Å across are found scattered through the film (Fig. 17). In a cross section prepared from the same wafer, these features are revealed to lie near the surface of the film (Fig. 18). These objects are voids, the identification being confirmed by observing the sample from different angles. Voids have been observed in a variety of silicides. The behavior of such voids

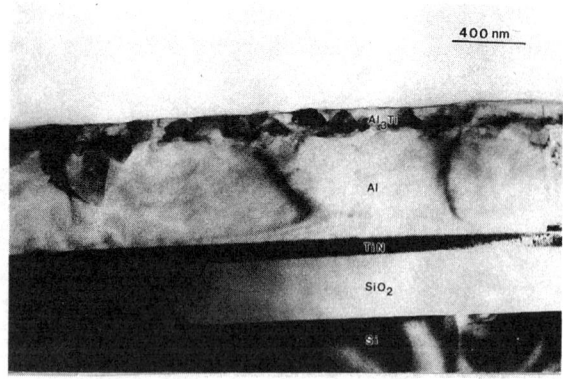

Fig. 22. Bright field image of a cross section of an annealed wafer. Aluminum has reacted with the titanium layer to form Al₃Ti.

must be monitored to make certain they do not trap chemicals on the wafer surface during processing steps.

Composition and Phase Identification. At times, the chemical identity of a device feature is unknown or must be confirmed. Conventional TEM imaging alone does not provide this information. Chemical composition can be obtained by use of the spectrometers described earlier.

As useful as chemical composition data is, at times it leaves ambiguities as to the phase identification of a given structural feature. Often, data is too inaccurate to distinguish between phases of related composition while a material of fixed composition can have a variety of stable or metastable crystal structures. This problem is solved with TEM by obtaining selected area or microdiffraction patterns. The crystal planar spacings extracted from the patterns (analogous to X-ray diffraction analysis) are compared to known spacings for the suspected phases. As an example, note the cross section of an annealed TiSi₂ film in Fig. 19. Prior to annealing, the film composition was that required for TiSi₂. However, diffraction analysis revealed a change in planar spacings subsequent to the annealing. The as-deposited film had the metastable C49 structure which transformed during heat treatment to the equilibrium C54 phase.[14] This transformation is significant since the C54 phase (formed above 800°C) has electrical properties superior to C49.

Figure 20 presents a multilayered interconnect scheme in as-deposited condition—a silicon wafer with oxide film followed by TiN, Al, Ti, and Al. The TiN film is smooth, continuous, and columnar grained (Fig. 21), while the Ti layer is continuous, although wavy. The Al has grown with an undulating surface, while the Ti has a uniform thickness and follows the undulations of the Al. Such undulations confuse analyses by thin-film techniques which average over relatively large sample areas, e.g., Auger Electron Spectroscopy (AES) or Rutherford Backscattering Spectroscopy (RBS). The non-flat interfaces blur the apparent location of the interface in depth profiles from such techniques. This could falsely be interpreted as a gradual, rather than abrupt, interface.

After annealing, the Ti layer reacts with neighboring Al (Figs. 22 and 23) to form a layer with a distinctly new structure.

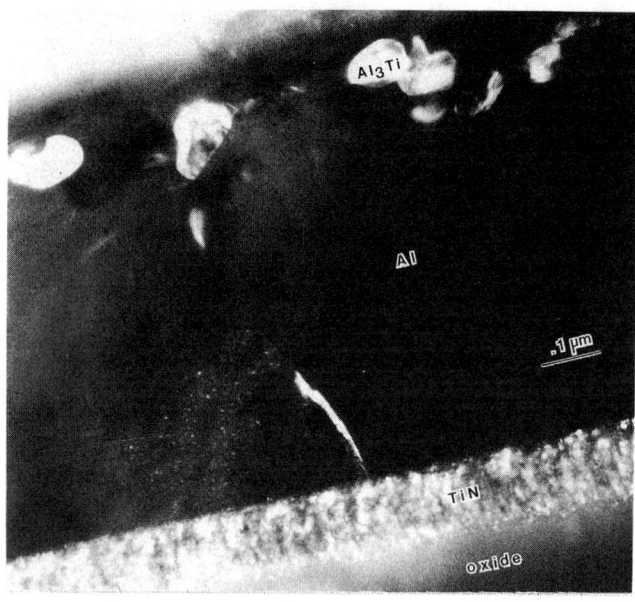

Fig. 23. Dark field image revealing grains of Al₃Ti.

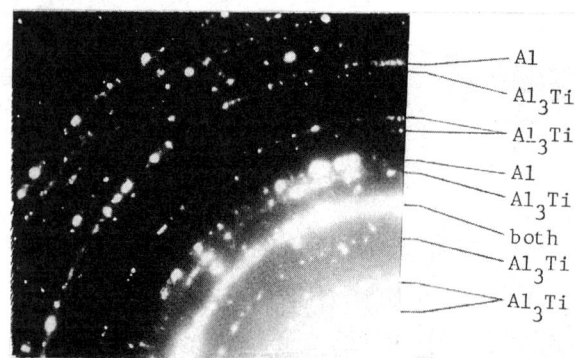

Fig. 24. Selected area diffraction pattern from annealed sample exhibits polycrystalline rings for Al and Al₃Ti.

The columnar grained Ti has been replaced by grains of irregular shape (most obvious in the dark field image, Fig. 22). This film appears to be discontinuous, though the shape and orientation of its grains make this uncertain. Further analysis of the composition of this new film requires observations beyond imaging. A number of phases are possible. However, since excess Al is available, Al₃Ti is expected if the reaction is allowed to attain an equilibrium state. This phase is desired since it is very stable against electromigration damage. The presence of this phase was confirmed by diffraction patterns taken from planar samples (Fig. 24). A ring pattern is obtained since many grains contribute to the diffraction pattern. Some Al film overlaps the Al₃Ti film so that rings are present for each material. Electron diffraction is very useful for planar spacing analysis in thin films due to the strong interaction of electrons with matter. Such detailed data is difficult to produce with X-ray diffraction techniques for very thin films.

The multilayered interconnect design discussed above is an attempt to prevent circuit opens caused by electromigration. This phenomenon causes voids to appear at grain boundary triple junctions and surface intersections. With two Al layers separated by an electromigration resistant Al_3Ti layer, current can bypass a void in one Al layer by utilizing the other layer for conduction. In the unlikely case of neighboring voids in both Al layers, the Al_3Ti serves as a current path for a short distance. TEM confirms the production of the desired structure and in the event of future failures would show why the failure occurred. Though TEM is more commonly applied to research and process development studies, failure analysis can be performed on completed devices. Figure 25 shows micrographs revealing the failure mechanism in a PROM device. Such devices are programmed by sending current through a fusible link, leading to the formation of insulating oxides in the formerly metallic link. In this case the link was composed of nichrome, NiCr. In this study, some fusible links still conducted after programming. SEM observations of the exterior of the device showed no differences between good and bad links. However, TEM comparisons of these fusible links revealed that metallic filaments were reforming across the fuse gap in bad samples, shorting the link.

Crystalline Defects. Defects typically encountered in device thin films include precipitates, dislocations, stacking faults, twin boundaries, and grain boundaries. These can all have direct or indirect effects on device performance by influencing the processing of the device or its electrical properties. Although grain size can be thought of as a structural feature, grain boundaries are in fact crystalline defects—the boundary between misoriented grains. These defects are invisible to the

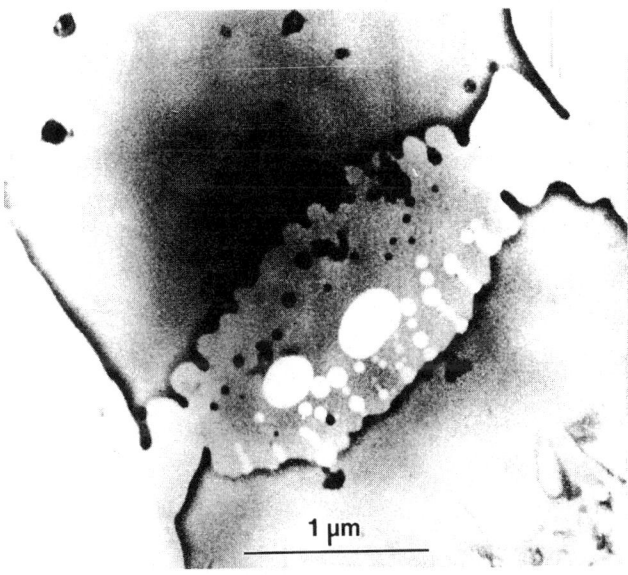

(a)

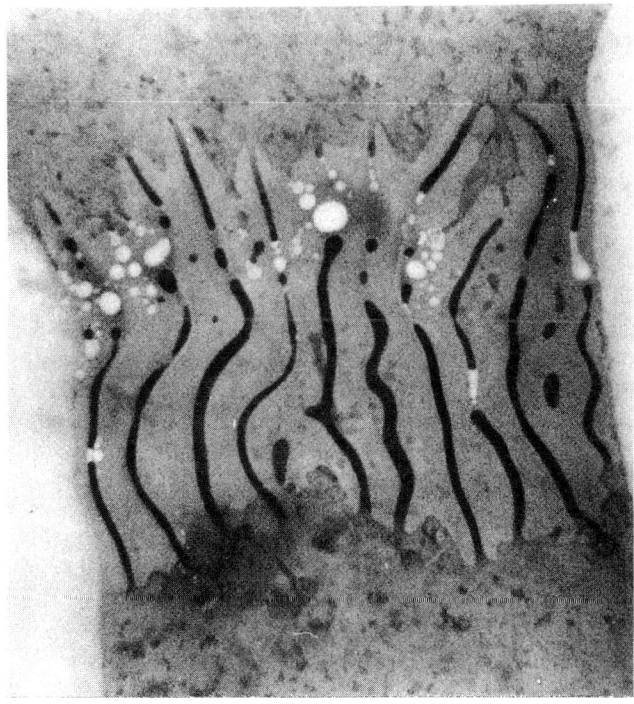

(b)

Fig. 25. Good (left) and bad (right) PROM NiCr fusible links. The bad link is shorted by metallic filaments, as determined by EDS analysis.

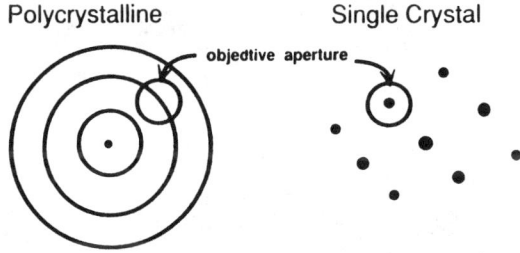

Fig. 26. By use of an aperture in the TEM, dark-field images can be formed by utilizing only a portion of the diffracted electrons. For single crystals, images are formed from a single diffracted beam while for polycrystalline samples a portion of a diffracted ring is used.

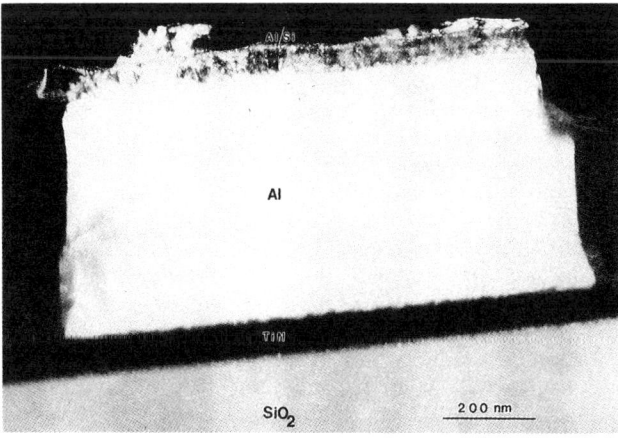

Fig. 27. Dark field image of the sample in Fig.20. The aluminum is very large grained.

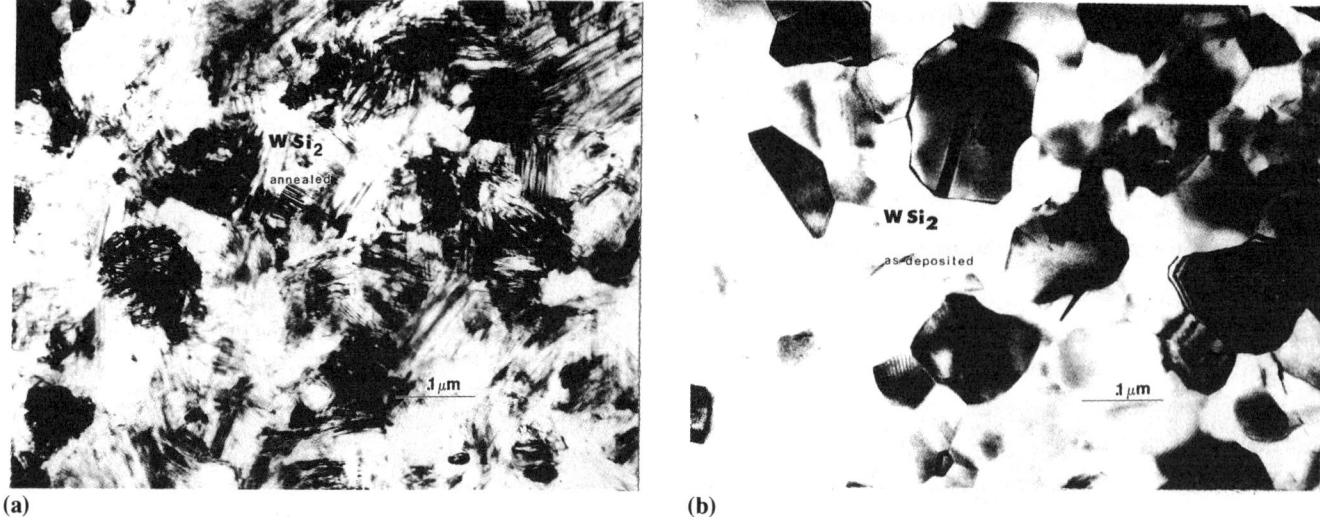

(a)   (b)

Fig. 28. (a) Planar sample of sputter deposited WSi₂ film. (b) WSi₂ film after annealing.

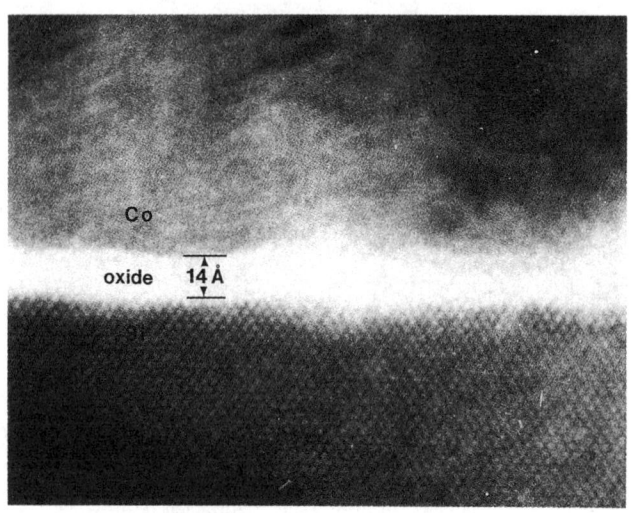

Fig. 29. Lattice image of silicon substrate with native oxide and Co film.

SEM, unless they intersect the sample surface AND have some surface topography at this intersection.

Dark-field images have many uses. In one application, by forming an image with diffracted electrons, high contrast is produced between grains of differing orientation or phase (Fig. 26). In Fig. 27, a large Al grain is seen. The Al grains span the full thickness of the film and are wider than they are thick. The boundaries between Al grains are perpendicular to the wafer surface. Also, see Fig. 23, discussed above.

Like grain boundaries, dislocations provide an enhanced diffusion path. An aluminum interconnect typically has a high dislocation density due to the large mechanical stresses encountered during the thermal cycles of IC processing (Fig. 5). Such dislocations may have a role in electromigration in narrow interconnects.[16] Returning to the transistor of Fig.16,

note in the TEM image that underneath the silicide, there is a row of dislocation loops that define the end-of-range of the original implant (not the junction depth). The depth of these crystalline defects is dependent on the dose, type of implant, and implant energy.

Sputtered films often contain a high fault density. Figure 28 shows micrographs of planar TEM samples of a tungsten silicide (WSi₂) film, as-deposited, and after annealing. A high density of stacking faults and twin boundaries (the striations in the grains of the as-deposited sample) are largely removed by annealing. The behavior of these defects is readily observed in the TEM. Note the same types of defects in the polycrystalline silicon film of Fig. 10.

Interfacial Structure. Though the atomic structure of an interface may appear to be an esoteric topic, dimensions employed in devices are approaching a size where scattering of electrons at interfaces will have a significant influence on the electrical performance of thin-films. Figure 29 is a lattice image of a cross sectioned silicon wafer with a deposited (Co) film. The native oxide layer had not been etched prior to deposition of the metal. The silicon is oriented for observation of a ⟨110⟩ projection; the spots in the image correspond to columns of atoms in the silicon. The cobalt film planar spacings were beyond the resolution of the TEM used for this micrograph while the oxide is amorphous, hence no periodic structure is seen in these layers. The boundary between the Co and oxide is observed because these materials are of different atomic number. This permits a thickness measurement of the oxide layer to an accuracy within a few angstroms.

The lattice imaging capability of the TEM is gradually extending from basic materials studies to become a routine problem-solving tool in device fabrication and failure analysis. Even with "conventional" TEM imaging, the ability to study the many critical device interfaces with great detail has proven of great use in device development and analysis.

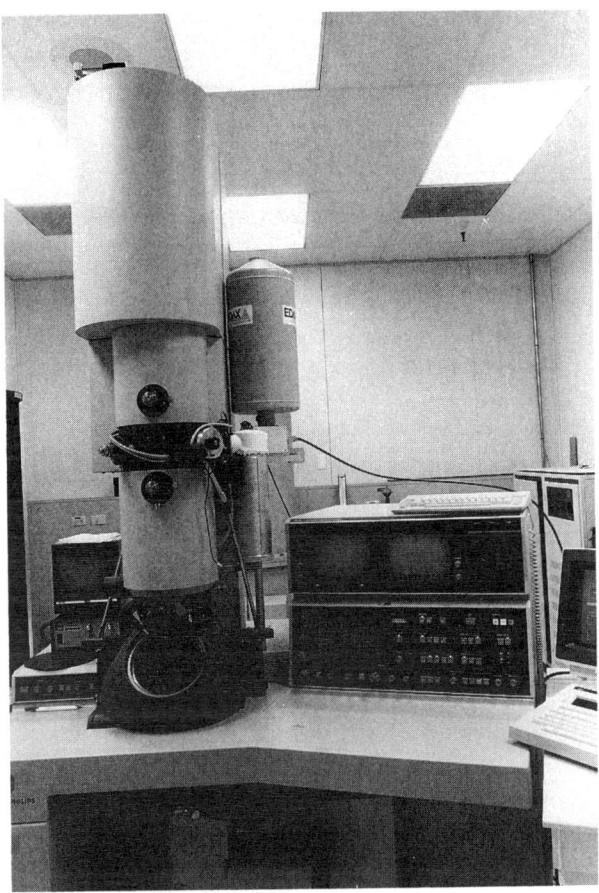

Fig. 30. View of a typical scanning transmission electron microscope system, consisting of major subsystem components from 3 manufacturers. The main TEM is in the left foreground, with the large 300 kV power supply in a pressurized tank filled with SF_6 gas immediately behind. The EDXA detector and Dewar are in the center of the photo, and the PEELS detector is in the kneehole of the desk. Notice that the operator looks through a thick leadglass window directly at the phosphor screen. With its 23 Å resolution, this large and expensive machine is a necessity in manufacturing sub-micron ICs, and in the failure analysis of these products.

SUMMARY

The TEM has a unique ability to observe internal structure and chemistry with resolution to a few Angstroms. Due to the extensive use of SEMs in device analysis, comparisons have been given for there instructive value. While the SEM is essentially a surface imaging instrument, the TEM observes internal microstructure by imaging with electrons which have traversed the sample. Internal structure not observable with SEM can be examined in great detail with TEM. Minimal beam spreading in thin samples permits chemical analyses from regions approaching the beam diameter—less than 100Å. Diffraction data can be obtained from crystalline films. Structural observations in the latest generation of field emission SEMs are approaching the resolution of TEMs. The purpose here has not been to determine which technique is better; the two are com-

plimentary and appropriate realms of application must be delineated. For problems where either technique is capable of providing the necessary information, SEM is preferred because of timeliness and lower costs.

The role of SEM is well established in the minds of the process developer, fabrication engineer, and failure analyst. Why is this not so for TEM? This is in part due to the specialized nature and less common availability of the technique, making it less familiar than the SEM to many engineers. TEM is also perceived as being so time consuming that it is impractical in manufacturing and failure analysis. In addition, full application and interpretation of TEM requires experience in materials science. However, these reasons are invalid for ignoring the technique, because many problems can only be solved through TEM study. Additionally, deeper materials understanding of many device related phenomena often only comes through studies which include TEM observations. The portion of problems in VLSI manufacturing which require TEM for solution are continually increasing. Manufacturers who do not pay sufficient concern to microstructural details will find themselves at a disadvantage to competitors with well developed TEM capabilities. Failure analysis applications of the TEM hinge on the ability to obtain thin sections from pre-specified device areas. Since the present smallest device dimension of interest is about 0.5μm, one requires the ability to cross section with a precision of about 500Å. This requirement applies to SEM cross sections as well! [17] A very few laboratories are presently performing such work. The recent introduction of Focused Ion Beam (FIB) methods for final thinning of TEM samples may prove a useful aid in such precise cross sectioning. The demands of failure analysis on a submicrometer scale will continue to be the primary driving force for advances in sample preparation methods.

In summary, the following are questions the engineer might ask himself/herself. Has the desired structure been produced? Have desirable or undesirable thin film reactions occurred? What defects have formed? Are there potentials for failure? What was the failure mechanism? When a device process development or manufacturing problem is encountered which would benefit from information obtained at high spatial resolution, TEM should be investigated as a potential problem solving tool. Armed with TEM analyses, the engineer can adjust design, process, and materials selection for improved performance and reliability.

REFERENCES

1. J. H. Rose, N. Riel, and R. Flutie, Proceedings of ISTFA 1987, pg. 95.

2. J. H. Rose, J. Kowalik, and N. Riel, Proceedings of ISTFA 1988, pg. 145.

3. J. W. Edington, "Practical Electron Microscopy in Materials Science", TechBooks, Herdon, VA (1976).

4. O. L. Krivanek, Chem. Scr. 14, 78-83, (1979).

5. D. B. Williams, Mat. Res. Soc. Symp. Proc., Vol 31, pgs. 11-22 (1983).

6. J. W. Steeds in "Introduction to Analytical Electron Microscopy", pgs. 387-422, J.J. Hren, J. L. Goldstein, and D. C. Joy (Eds.), Plenum Press, New York (1979).

7. David C. Joy in ref. 2, pgs. 223-244.

8. R. B. Marcus and T. T. Sheng, "Transmission Electron Microscopy of Silicon VLSI Circuits and Structures", John Wiley and Sons, Inc., New York (1983).

9. John C. Bravman, Ron M. Anderson, and Michael L. McDonald (Eds.), "Specimen Preparation for Transmission Electron Microscopy of Materials", Mat. Res. Soc. Symp. Proc., Vol 115 (1988).

10. Ron M. Anderson (Ed.), "Specimen Preparation for Transmission Electron Microscopy of Materials II", Mat. Res. Soc. Symp. Proc., Vol 199 (1990).

11. Richard Flutie, Mat. Res. Soc. Symp. Proc., 62, 105 (1986).

12. R. Pinizotto, F. Y. Clark, and M. L. Jarvis, Mat. Res. Soc. Symp. Proc., 62, 9 (1986).

13. A. E. Morgan, E. K. Broadbent, M. Delfino, B. Coulman, and D. K. Sadana, J. Electrochem. Soc., 134, 925 (1987).

14. Robert Byers and Robert Sinclair, J. Appl. Phys., 57, 5240 (1985).

15. D. Gardner, T. Michalka, K. Saraswat, T. Barbee, P. McVittie, and J. Meindl, IEEE J. of Solid State Circuits, 20, 94 (1985).

16. J. H. Rose, J. Lloyd, A. Shepela, and N. Riel, Proc. 49th Ann. Meeting EMSA, 820 (1991).

17. R. Anderson, Proc. 49th Ann. Meeting EMSA, 828 (1991).

18. S.J. Kirch, R. Anderson, S.J. Klepeis, Proc. 49th Ann. Meeting EMSA, 1108 (1991).

Transmission Electron Microscopy
for Failure Analysis of Integrated Circuits

Swaminathan Subramanian, Raghaw S. Rai, and Vidya S. Kaushik
Motorola
Austin, Texas

1. INTRODUCTION

A most important step in failure analysis is the physical and chemical characterization of the defect. The root cause of the failure is identified by physically inspecting a failure site identified by the electrical analysis. The physical inspection and chemical characterization can be performed by a scanning or a transmission electron microscope. In a scanning electron microscope (SEM), an electron beam is scanned over the surface of the die and an image is formed by recording the reflected (secondary or back-scattered) electrons. The contrast variations observed in the image are usually the result of variations in surface topography or the difference in atomic weight of elements on the scanned area. Hence, no information can be obtained regarding a defect that does not change the surface topography (e.g. crystal defects such as dislocations) or a defect that lies below the surface. In most situations, a decorative etch is required to change the topography of the defect for identification using a SEM. The decorative etch is a destructive process and hinders further chemical analysis. Even after a decorative etch, SEM cannot resolve most of the features and interfacial layers (dimensions of a few nanometers) fabricated in advanced technologies.

In a transmission electron microscope (TEM) a high energy (100 ~ 300 keV) electron beam is transmitted from the top through a thin section of a sample and the image is formed below the sample. Unlike a SEM image, the TEM image contains three-dimensional information from the thin section of the sample. The contrast variation in the image is a result of complex beam-specimen interactions that are unique to a TEM. The contrast is also sensitive to small variations in chemical, structural and topographical features of the sample. This property is frequently exploited to resolve the subtle effects of crystal defects and interfacial layers. In addition, the resolving power of the TEM is inherently better than the SEM because of the smaller wavelength (~ 0.0025 nm at 200 keV) of the high-energy electron beam.

Despite the above-mentioned advantages, the role of TEM in failure analysis has been limited because of the difficulties associated with site-specific sample preparation for characterizing the failure. Recently, the role of TEM in failure analysis have been driven by developments in specific-area TEM sample preparation using focused ion beam (FIB) instruments. In particular, the possibility of making precisely localized thin sections, from submicron regions, for TEM usage has advanced the failure analysis of ICs [1-22]. It may be pointed out that advanced dual-column instruments with high resolution drift-free ion and electron beam columns allow simultaneous imaging of the section using a electron beam while sectioning in the ion-beam mode.

In this paper basics of TEM analysis, imaging and interpretation, new developments in sample preparation procedures, and some specific applications of TEM in failure analyses are discussed.

2. TEM : BASICS

In a TEM, the image is formed by transmitting a high-energy (100 - 300 keV) electron beam through a thin section of the sample. As the electron beam is transmitted through the thin section, a variety of beam specimen interactions occur, which yield transmitted electrons, elastically and inelastically scattered electrons, X-ray photons and Auger electrons,. Most of the transmitted and elastically scattered electrons and some of the inelastically scattered electrons are used to form an image. X-ray photons and inelastically scattered electrons used for elemental analysis using energy dispersive spectrometer (EDS) and electron energy loss spectrometer [EELS] respectively. A schematic representation of the physical location of imaging plane, EDS and EELS in a transmission electron microscope is shown in Fig. 1.

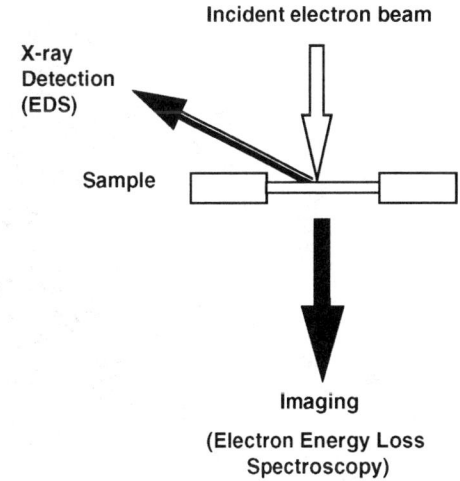

Fig. 1: A schematic representation of physical location of imaging plane, EDS and EELS in a transmission electron microscope.

In a crystalline sample, under parallel electron-beam illumination, the elastically scattered electron beam is split into one transmitted beam and several diffracted beams. An image of these beams can be easily recorded in a TEM by switching to diffraction mode and is known as a diffraction pattern. Single-crystal spot diffraction pattern recorded with the electron beam parallel to <110> direction of a cubic lattice and a polycrystalline ring pattern are shown in Figures 2 and 3 respectively. Each spot or ring corresponds to a set of crystallographic planes in the sample. The spacing of any spot or ring from the central (000) (or transmitted) spot is inversely proportional to the crystallographic interplanar or the 'd' spacing corresponding to the reflection. Spot and ring patterns are frequently used for identification of crystal structure of various phases in the sample. In order to form a TEM image from the diffraction pattern, the diffracted beams are selectively intercepted by an objective aperture located at the back-focal plane of the objective lens to produce desired contrast.

In semiconductor failure analysis, a TEM is usually used in the following diffraction conditions to provide different kinds of information about the sample:

Zone-axis imaging is the most commonly used condition for regular and lattice-imaging of cross-sectional samples in semiconductor devices. In zone-axis imaging the crystal is oriented, with respect to the electron beam, along a direction parallel to a low-index, i.e., high-symmetry crystal direction. Two common low-index crystal directions in silicon are [110] and [001]. Most semiconductor devices are fabricated on single crystal silicon wafer with the wafer normal parallel to [001] direction. In such a wafer the [110] direction is parallel to the wafer flat or notch, lies on the (111) plane (cleavage plane) and perpendicular [001] direction (perpendicular to surface of wafer). A cross-section image recorded with the electron beam oriented parallel to [110] direction of silicon substrate would ensure perpendicularity to the device features and eliminate overlapping of various features of devices on the projected image from the thin sample. For this reason most cross-sectional image of Si devices are recorded along the [110] zone axis. A schematic diagram of a TEM in zone-axis imaging mode is shown in Fig. 4. Conventional bright-field TEM images are recorded using the transmitted beam by intercepting the diffracted beams using a small objective aperture to improve contrast. The contrast observed

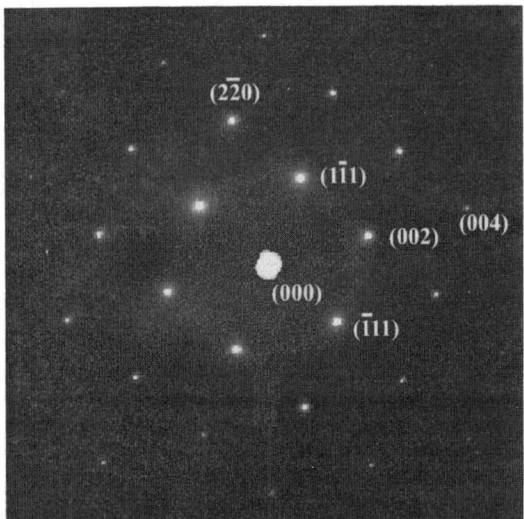

Fig. 2: A single crystal diffraction pattern recorded with electron beam oriented parallel to the <110> direction in silicon.

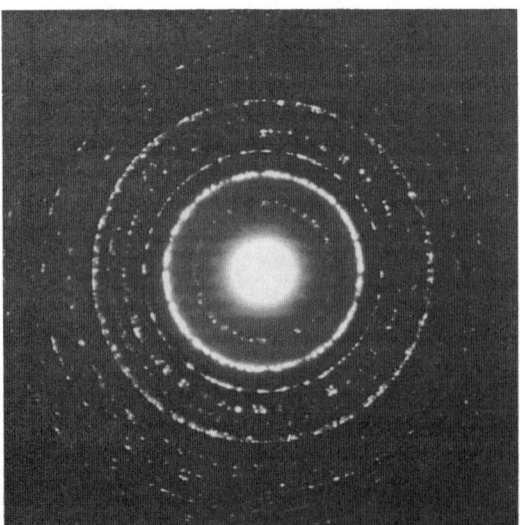

Fig. 3: A ring pattern record from a polycrystalline silicon sample.

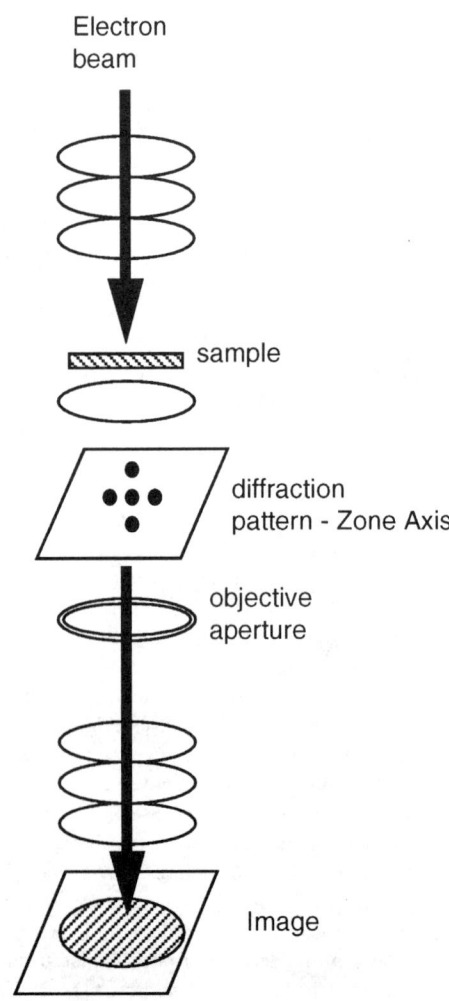

Fig. 4: A schematic representation of zone-axis imaging in a TEM.

in the image is a function (among other factors) of the size of the objective aperture. High-resolution phase contrast lattice images are recorded by selecting the transmitted and several diffracted beams using a large objective aperture. A detailed discussion of diffraction and phase contrast is presented in section 2.1.

Two-beam Bright Field and Dark-Field imaging is used to obtain crystallographic information about defects in crystalline samples. The silicon substrate of the TEM sample is oriented such that there are only two strong beams namely, the transmitted and any one diffracted beam in the diffraction pattern. A schematic representation of bright- and dark-field imaging in a TEM is shown in Fig. 5. In semiconductor devices, the silicon substrate is oriented off the [110] zone to achieve maximum intensity in the transmitted and one diffracted beam (usually (004)). The objective aperture is then used to select the transmitted or diffracted spot to form the bright-field or dark-field image respectively. A dark-field image formed this way is often referred to as a conventional dark-field image (CDF) to distinguish it from another type called weak-beam dark-field (WBDF). An example of the implant damage recorded under bright- and dark-field conditions is shown in Fig. 6.

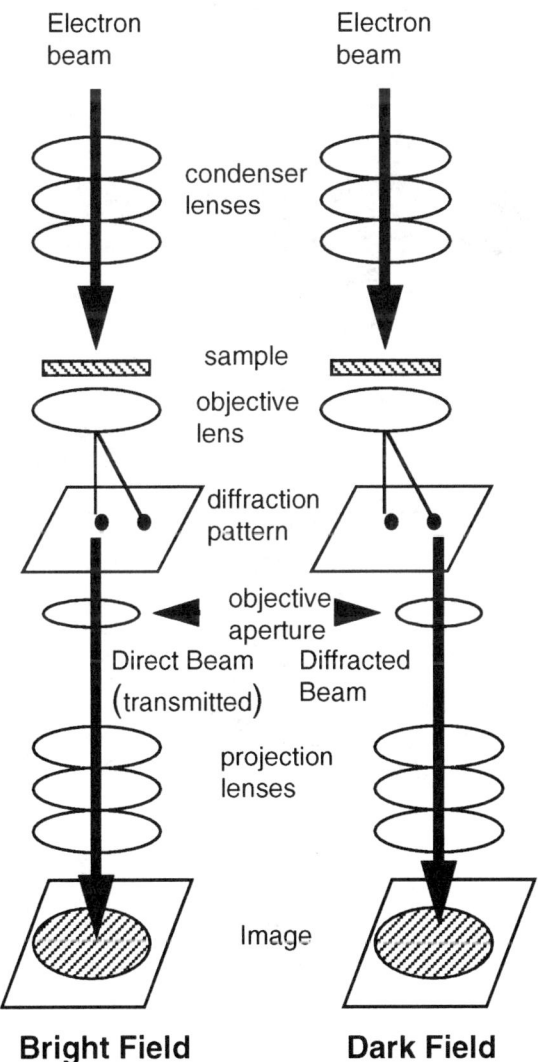

Fig. 5: A schematic representation of two beam bright-field and dark-field imaging in a TEM.

2.1 Image Contrast

The contrast in a TEM image is primarily formed by three different mechanisms, namely diffraction contrast, thickness-mass contrast, and phase contrast. The contribution from various contrast mechanisms can be optimized by an experienced TEM analyst to image the subtle effects of the defect. Ironically, for an untrained eye interpretation of the TEM image is complicated by multiple contrast mechanisms. Hence it is important to understand the principles and properties of various contrast mechanisms.

Diffraction Contrast: Diffraction contrast is obtained in practice by intercepting the diffraction pattern using an objective aperture in the back focal plane of the objective lens that allows only the transmitted beam to form the image. The image reveals variations in the intensity of the selected electron beam as it leaves the sample.

Diffraction contrast is extremely sensitive to the crystallographic orientation and thickness variations in the sample. In crystalline materials, the regions around crystal defect diffract differently compared to perfect regions due to elastic strains around defects causing the contrast variations in the image. Dislocations, stacking faults, and other crystal defects bend crystal planes and show up clearly in the image as a result of diffraction contrast (see Fig. 6).

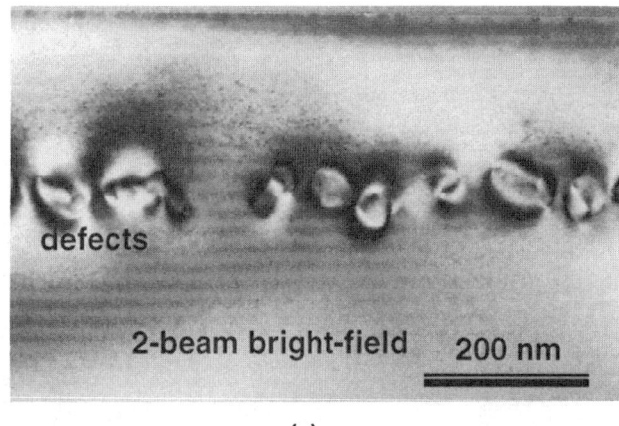

(a)

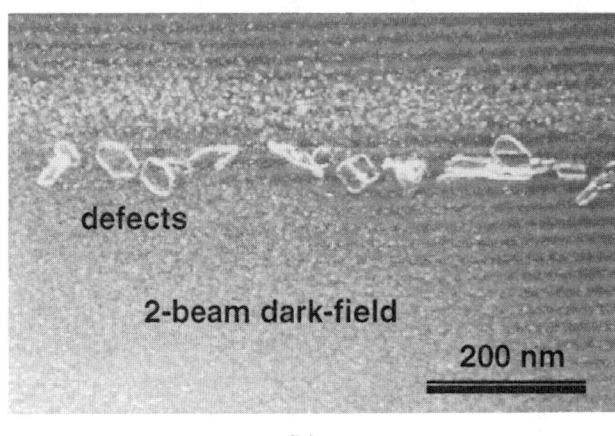

(b)

Fig. 6: Two beam (a) bright-field, and (b) dark-field diffraction contrast images of dislocation loops in silicon.

In polycrystalline materials such as poly-silicon or titanium nitride, diffraction contrast arises due to variations in orientation of adjacent grains relative to the electron beam. For semiconductor failure analysis diffraction contrast is useful to resolve microstructure of various components of the device such as poly-silicon layers and stringers, or interfacial features that may cause electrical opens or shorts.

One other very useful feature of the optics of the TEM is the combination of diffraction with imaging by the use of the selected area electron diffraction technique. This is possible because the focal lengths of electromagnetic lenses may be varied easily and rapidly. This selected area electron diffraction is of great value in enabling image contrast to be correlated with diffraction condition, especially for crystalline samples.

Thickness-Mass Contrast: Essentially all TEM samples have thickness variations. Electrons can be scattered, diffracted and absorbed to different extents with different sample thicknesses and lead to contrast variation known as the thickness contrast. In the case of mass contrast, the intensity variation is basically a map of the scattering power of the elements present in the specimen. Mass contrast electron microscopy can provide images with both atomic resolution and compositional sensitivity. The lighter elements absorb less and heavier elements absorb more electrons. The mass contrast is independent of the crystal structure of the device and useful to resolve multiple thin amorphous layers such as silicon-oxide, silicon-nitride and silicon-oxy-nitride.

an image will give a two dimensional periodic image, corresponding to the planar spacings associated with the diffracted beams, which may in some cases resemble the structure image of very thin crystal projected along its orientation. Resolution below 2.5 Å is available in most of current generation TEMs whereas resolution down to 2 Å is available in dedicated high-resolution TEMs. In silicon based integrated circuits, cross-sectional high resolution images are usually recorded with the electron beam parallel to <110> direction using diffraction spots from {111} planes. The periodic fringe spacing in such a image is equal to the planar spacing of {111} planes in silicon (3.14 Å). The fringe spacing is used as an internal magnification calibration to measure interfacial details near atomic level.

2.2 Elemental Analysis

The fine high intensity probe available in a TEM equipped with a field emission source permits elemental analysis with high spatial resolution (< 1 nm) from regions less than few hundred angstroms thickness. TEM based elemental analysis techniques utilize inelastic

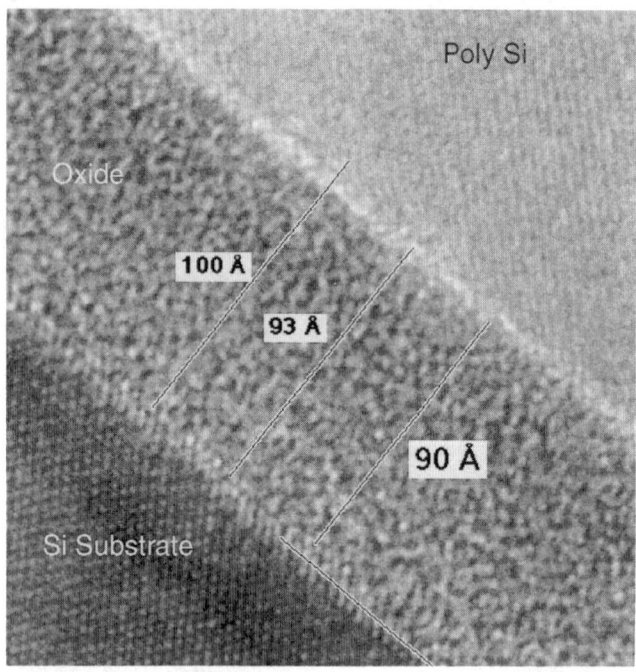

Fig. 7: High resolution phase contrast image of tunnel oxide. The best estimate of the tunnel oxide thickness is shown by the 93 Å marker.

Phase Contrast: Phase contrast is used to obtain high-resolution TEM images that can be used to precisely measure critical device dimensions, e.g., thickness of very thin tunnel/gate oxide (see Fig. 7). The high-resolution lattice images are formed as a result of interference between the transmitted and diffracted beams. If three or more, non-collinear strong diffracted beams with transmitted beam are symmetrically included in the objective aperture to form the image, two or more intersecting sets of parallel fringes will be observed. Such

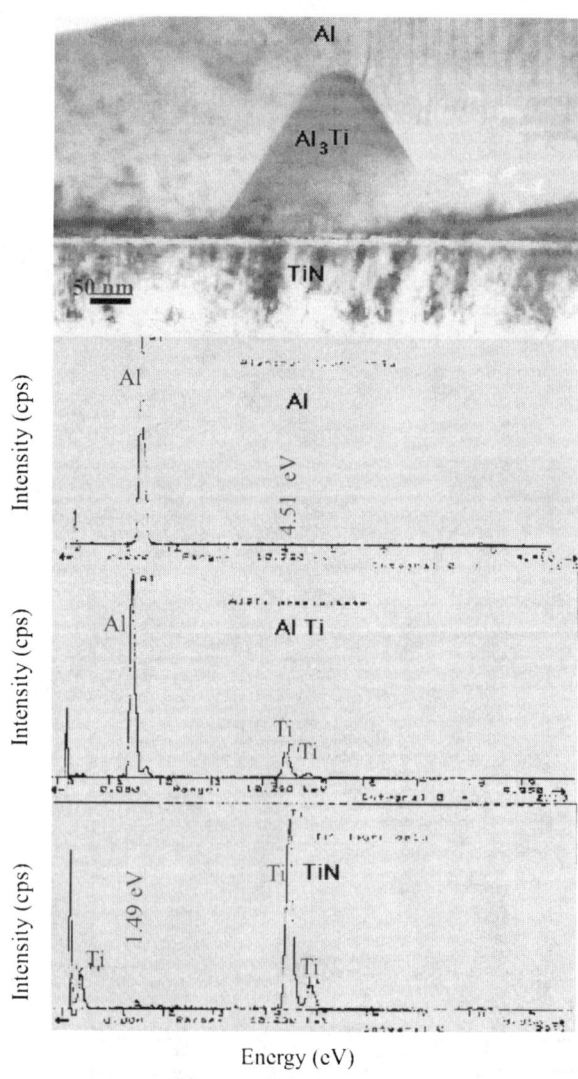

Fig. 8: Demonstration of EDS analysis at the Al-metal/TiN barrier interface. EDS spectrum from Al, Al3Ti and TiN are shown above.

318

scattering events that occur when the electron beam is transmitted through the sample. Two commonly used techniques are the energy dispersive spectroscopy (EDS) and electron energy loss spectroscopy (EELS). A schematic diagram of the implementation of EDS and EELS techniques in a TEM are shown in Fig. 1.

In EDS the characteristic X-rays emitted by various elements when electrons are incident on the sample are collected by a spectrometer. The characteristic X-rays generated by one element can be easily absorbed by a different element present in the same volume. Hence the sensitivity of the technique is a function of the element detected and the composition by volume in which the element is present. The detection limits for various elements decrease as a function of atomic weight. Using state-of-the-art ultra-thin or windowless detector elements, atomic weight down to boron can be detected provided if a reasonable quantity is present within a reasonable surrounding. An example of EDS analysis of Al_3Ti precipitates is shown in Fig. 8.

During the inelastic scattering process transmitted electrons lose an energy which is characteristic of the element. In electron energy loss spectrometry (EELS) a spectrometer placed underneath the TEM column is used to detect electrons with energy loss which is characteristic of the element detected. EELS offers better spatial resolution (1 nm) and light element detection capability than EDS. For best results extremely thin sample (~50 nm thickness) and a drift-free coherent electron source such as a field emission gun are recommended.

3. SAMPLE PREPARATION PROCEDURES

Sample preparation is the key to the success of failure analysis using a transmission electron microscope. In most situations, the sample preparation procedure is unique to the failure mechanism. A schematic diagram of the requirement of a TEM sample is shown in Fig. 9. In general, the sample consists of 3 mm diameter disc with a thickness less than 250 μm. The critical requirement of the sample is that it should be thinned to less than 250 nm at the failure site, to achieve reasonable electron transparency. Thinner samples (less than 100 nm) may be required for phase contrast high resolution imaging. In most cases, the area of the electron transparent region in a TEM sample is 25 μm X 25 μm. Hence it is very important to precisely isolate defect site to within *25 μm X 25 μm X .125 μm!*.

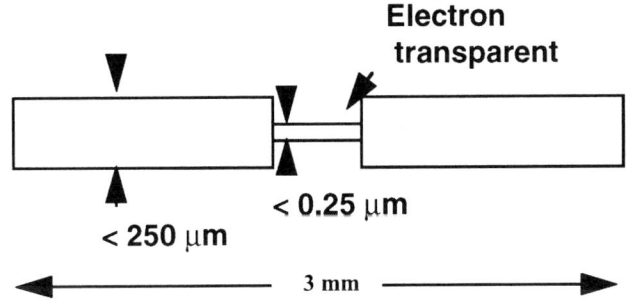

Fig. 9: A generic cross-sectional view of dimensional requirements for a TEM sample (not drawn to the scale).

A number of techniques [1-5] have been utilized in the past to prepare cross-sectional and plan-view TEM samples, the most common being that of conventional mechanical polishing and Ar-ion milling [6], used in the case of homogeneous large samples. The conventional cross-sectional TEM sample can be applied to ICs only if the region of interest (ROI) is repeated many times within the specimen and rarely used in failure analysis.

Recently, the focused ion beam tool (FIB) has been the dominant tool for selected area cross-sectional TEM (SAXTEM) and selected area planar TEM (SAPTEM) sample preparation. There are two approaches used to prepare specific SAXTEM samples. The first approach is a combination of mechanical polishing and final thinning to electron transparency using FIB milling. This approach was applied by Young *et al.* [11] in their planar and cross-sectional TEM specimen preparation using the FIB. Morris *et al.* [12], Kirk *et al.* [13], and Schraub and Rai [14] have used a similar approach of mechanical polishing and final thinning using FIB milling. Second approach is commonly known as 'lift-out' is given by Overwijk *et al.* [15], Herlinger *et al.* [16] and Giannuzi et al. [17] in their specific site TEM sample preparation. This method eliminates the requirement of mechanical polishing step and reduces the overall sample preparation time. In this approach the TEM specimen is actually separated with minimal damage to the substrate, transferred to a membrane coated grid and then to the TEM holder. A lift-out procedure for planar TEM sample preparation has been published by Stevie *et al.* [18].

Two methods of SAPTEM sample preparation have been developed by Anderson and Klepeis [19] and Subramanian *et al.* [20]. Anderson and Klepeis [19] used the tripod method to grind the sample Subramanian *et al.* [20] method utilized a back grinding step followed by a cross-section grinding step without the aid of a tripod. Both of the above methods used similar procedures to perform the FIB bulk cuts and polish. Subramanian *et al.* [20] demonstrated the use of a combination of wet chemical etch and FIB thinning to successfully address different failure mechanisms. A detailed description of commonly used TEM sample preparation procedures in failure analysis is presented in the following sections. These methods use a combination of procedures from the various publications referenced above.

3.1 General Area TEM techniques

The general area TEM techniques can be used in failure analysis in a case where the defect density is high. A schematic diagram for a general area cross-section TEM (XTEM) is shown in Fig. 10. Two pieces of the die (5 mm X 5 mm) are glued permanently face to face with the circuit in between. Two additional pieces of silicon are glued to the backsides for support. The stack is then ground from either side to 50 μm thickness. The sample is further thinned to 15 μm in the center by using a dimple grinder. A commercially available 3 mm diameter slotted Cu (can be Ni of Mo) ring is glued on to the specimen as is shown in Fig. 10(b). The sample is finally thinned to less than .25 μm using a Ar-ion beam as is shown in Fig. 10(c). The sample preparation procedure can be easily modified for general area planar analysis by changing the cross-sectional grinding and dimpling steps to backside grinding and dimpling.

3.2 Specific or Selected Area TEM techniques

The general area TEM technique is rarely used in failure analysis because the defect(s) causing the failures are usually limited to few

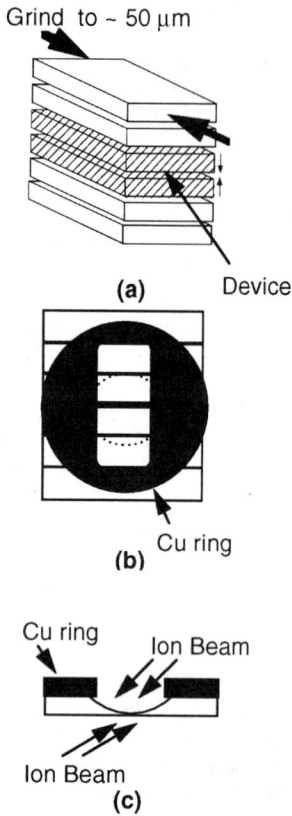

(a) Grind to ~ 50 μm

Device

(b) Cu ring

(c) Cu ring Ion Beam

Ion Beam

Fig. 10: A schematic representation of the general area sample preparation technique. (a) Two pieces of wafers are glued face-to-face and cross-sectionally ground to thickness less than 50 μm.(b) The thinned sample is further dimpled to about 15 μm and a slotted 3mm diameter copper ring is glued on to the sample for support. (c) The dimpled sample is milled using an Ar-ion mill to acheive electron tranparency.

transistors in a die. The sample has to be selectively thinned to less than .25 μm at the failing location to achieve electron transparency, hence the name selected area TEM techniques. The challenge in this procedure is to selectively thin the failure site without destroying the integrity of the sample. As noted previously, the complexity of thinning down the TEM sample, at the failure site, has been considerably reduced by recent developments in focused ion beam (FIB) instrument based

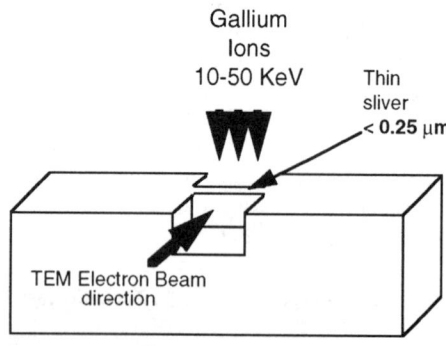

Gallium Ions 10-50 KeV

Thin sliver < 0.25 μm

TEM Electron Beam direction

Fig. 11: A schematic representation of focused ion beam milling technique for specific area TEM sample preparation.

techniques. A schematic representation of general principle of the FIB technique is shown in Fig. 11. In this technique, high energy focused gallium ions are used to ablate the sample leaving behind a thin, electron transparent sliver for TEM analysis. The FIB based sample preparation techniques usually requires complex preliminary steps before final FIB thinning of the samples. The preliminary procedures usually involve grinding to achieve less than 75 μm sample thickness and gluing a support ring for easy handling of the ground sample. The procedures vary depending on the geometry of the sample (i.e) cross-section or planar.

3.3.1 Specific area cross-section TEM (SAXTEM)

The goal of the SAXTEM procedure is to produce a thin cross-section , less than .25 μm, of a device encompassing the failing location. The technique is useful when the lateral location of the defect is precisely known and occurs in multiple layers.

A schematic representation of the pre-FIB sample preparation procedure for SAXTEM is shown in Fig. 12. In this procedure, the area of interest is permanently marked using a laser or a FIB. The sample is ground along a direction parallel to the final cross-section to a thickness of ~ 20 to 75 μm. The sample is usually glued to a glass

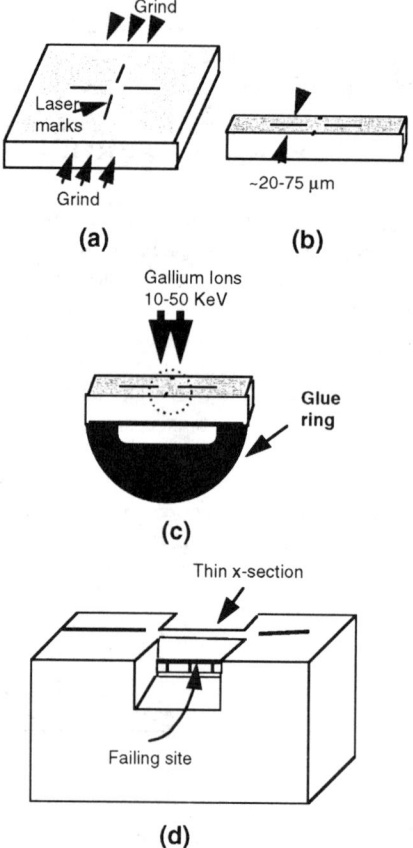

Grind

Laser marks

Grind

(a)

~20-75 μm

(b)

Gallium Ions 10-50 KeV

Glue ring

(c)

Thin x-section

Failing site

(d)

Fig. 12: A schematic representation of pre-FIB procedures for specific area cross-section TEM sample preparation (a) Area of interest is marked using a laser beam (b) The die is cross-sectionally ground to a thickness less than 75 micrometers. (c) A slotted 3 mm Cu semi-ring is glued on to the sample for support and area of interest is milled using the FIB. (d) A close-up view of the electron transparent region.

slide, using acetone soluble thermal plastic glue or 'jewellers wax' during the grinding for easy handling. The ground sample is removed from the glass slide and a thin slotted, one-half ring (3 mm diameter and 50 µm thickness) is glued permanently to the sample for easy handling. The sample is then loaded into the FIB instrument and thinned.

3.3.2 SAXTEM by 'LIFT-OUT' method

The pre-FIB thinning procedures described above usually are time consuming, expensive and require a lot of effort. A novel technique known as lift-out, plucking or pull-out method [15-17] eliminates all the pre-FIB sample preparation of SAXTEM. The total sample preparation time is cut down by more than 50%. In this technique a thin section at the area of interest is cut using the FIB and lifted out of the TEM sample. The thin section is then transferred to 3 mm diameter grid coated with formvar membrane. The lift-out technique is unpopular because it is very easy to lose the sample during the process. The success rate of the transfer can be easily improved to 100% by optimizing the factors that affect the lift-out process.

A version of the lift-out procedure is shown in Fig. 13. In this procedure, a small section with area of interest in the center, is cut using the FIB. The area of interest is thinned for electron transparency. The sample is isolated from the die by milling the bottom and the edges of the sample. A schematic representation of the lift-out sample preparation procedure for SAPTEM is shown in Fig. 13(a). A SEM picture of finished TEM sample is shown in the Fig. 13(b).

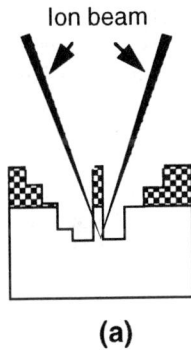

(a)

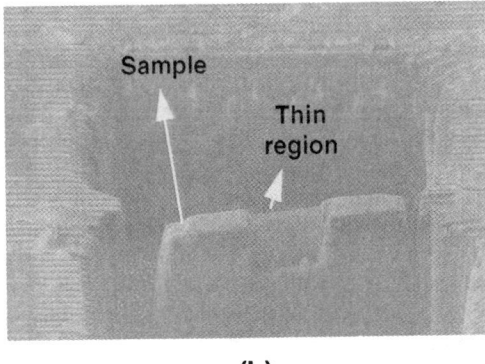

(b)

Fig. 13: The 'lift-out' sample preparation technique for specific area cross-section TEM. (a) A cross-sectional view of the milling geometry and sample cut-out procedure. Two stair-step cuts are made on either side of the area of interest. The sample is isolated from the die by cutting the region under and edges of the slice. (b) A scanning electron micrograph of completed TEM sample.

The most difficult part of the lift-out procedure involves transferring the finished TEM sample from the die to a grid. The common approach is to lift the sample out of the die using a glass needle with 'static charge' at the tip acting as glue. The sample is then transferred to the membrane grid. An optical image of the TEM sample on a grid is shown in Fig. 14.

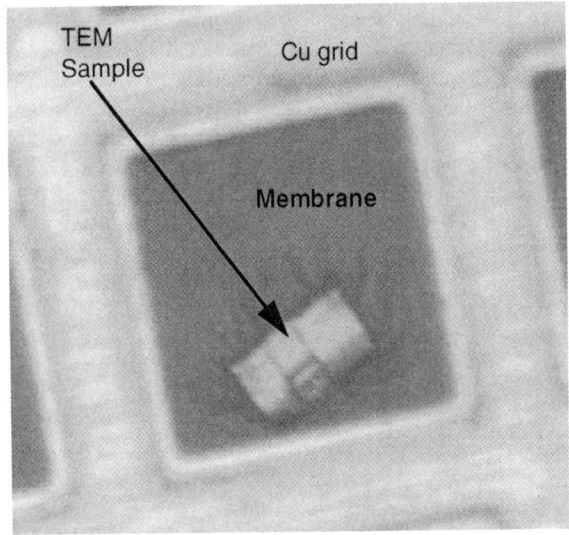

Fig. 14: A 'lift-out' TEM sample placed on a membrane-coated grid for TEM analysis.

The lift-out operation is performed under a long working distance optical microscope with at least 200X magnification. The needle is usually attached to a probe manipulator. Needle manipulators with hydraulic controls for precise movements are commercially available. A simple precision micrometer controlled microprobe needle handler may also do the job. The success of the transfer depends on the static charge at the tip of the needle, skill and patience of the operator. Various parameters that affect the lift-out process include the static charge at the tip of the needle, weight of the TEM sample in relation to the static charge and precision of the needle manipulator. The static charge is a factor of humidity level in the room, diameter of the tip of the needle and the composition of sample and the glass needle. If the strength of the static charge is too strong, the sample would exhibit a tendency to climb up the needle making the transfer to the TEM grid impossible. If the strength of the static charge is too weak, the sample would not stick to the needle. An optimum strength has to be established for successful lift-out and transfer of the sample. The strength of the static charge at the tip of the needle can be controlled by coating the needle and optimizing the tip diameter in relation to the weight of the sample.

3.3.3 Specific area planar TEM (SAPTEM)

The planar procedure offers unique geometric advantage over the cross-section method for failure analysis of problems that are limited to silicon or certain layers of the device. In the cross-sectional approach, a thin section (thickness less than .25 µm) of a device is available for failure analysis, whereas in the planar procedure a 20 µm² area of any layer (thickness less than .25 µm) of the device is available. Hence the planar approach can be used when there is an uncertainty in the lateral location of the defects provided if the defect is limited to a layer thickness of ~ .25 µm. Examples of such defects include dislocations at silicon-device interface, gate oxide defects and interfacial reactions.

Various steps involved in SAPTEM sample preparation are outlined in Fig. 15. Firstly the failed device is removed from the package and any required deprocessing and electrical characterization are performed to identify the failing site. It is advisable to leave as many layers as possible over the failing site to avoid introduction of artifacts during sample preparation. Laser marks are made from about 20 µm of the failing site with the aid of an optical microscope as shown in Fig. 15(a). A piece of the die, approximately 5 x 5 mm² dimension with the area of interest at the center is cleaved. This piece is mounted face down on to a 14 mm diameter transparent glass stub with clear thermal plastic material or 'jeweler's wax' as is shown in Fig. 15(b). The stub is then inserted into a disc grinder and ground from the back to ~ 30 µm die thickness. The end point of the grinding process (~ 30

µm die thickness) can be determined by using a backlit optical microscope and observing a dark brown color of the transmitted light. A precise (±5 µm) thickness measurement of the stub and the die prior to mounting and 'stub+die' after mounting will be useful for fast backside grinding process. Any commercially available 300, 600 and 1200 grid silicon carbide grinding paper can be used to perform the grinding in that sequence.

The second stage of SAPTEM sample preparation involves gluing one-half of a slotted (0.75 - 1.25 mm) 3 mm copper grid to the backside of the die. The purpose of the above is to provide support to the thinned die. The grid is glued on to the die with 'M-bond 610' thermal setting adhesive. The challenge in this step is to position the ring in the backside (blindly) of die such that the flat edge of the half-grid and the center of the slot are aligned with the laser mark. This is achieved by marking the approximate location of the laser mark on the glass stub using felt tipped pen. The location of the laser mark can be determined by using an optical microscope through the transparent glass stub. The gluing process is described in Fig. 15(c). The steps are as follows (i) copper grid is placed on a glass slide, (ii) glue is applied to the top of the ring, (iii) the ring is slid on the glass slide to a new dry location using a tweezers (iv) the glass slide is inverted, the ring is aligned to the marks made with the pen then transferred to the backside of the die. The glue is then cured and the 'die+ring' are removed from the stub using acetone to dissolve the 'jeweler's wax'.

The third stage involves a cross-section grind as is shown in Fig. 15(d). Firstly the thin die is glued face down (grid on the top) on to 1 mm thick glass slide using 'jeweler's wax' for support. The die and glass side are ground parallel to the flat edge of the grid to reach within 5 to 30 µm of the failing site. The lapping is performed using 15, 6 and 3 µm diamond impregnated paper in that sequence. The sample is removed from the glass slide using acetone and the extra material outside the half of the semi grid cleaved. The sample is ready for FIB milling.

The primary goal of the SAPTEM analysis is to be able to analyze any layer of the device at the failing site. This requires a precise alignment of the layer of interest, parallel to the ion beam. Any misalignment magnifies the 'waterfall effect' [19] resulting from different milling rate of different components of the device. In real life it is not always possible to achieve perfect alignment. The effect of misalignment and screen effect is minimized when the failing site is close to the top edge of the sample (5-8 µm). In a lot of cases it is difficult, risky and time consuming to grind within 5 to 8 µm of the failing site. Hence, in the fourth stage of SAPTEM sample preparation, a 25 µm wide slot is cut using focused ion beam to reach within about 5 to 8 µm from the failing site. In the present work the sample is glued flat with face-up on to metal coated wafer piece and a FIB cut is made using high ion beam current as shown in the Fig. 15(e). The sample is then removed from the wafer using acetone and mounted on to a holder for TEM sample preparation. The final planar bulk cuts and thinning are performed with the ion beam parallel to the device surface as is shown in Fig. 15(f). Detailed procedures involved in FIB thinning are outlined in the following section.

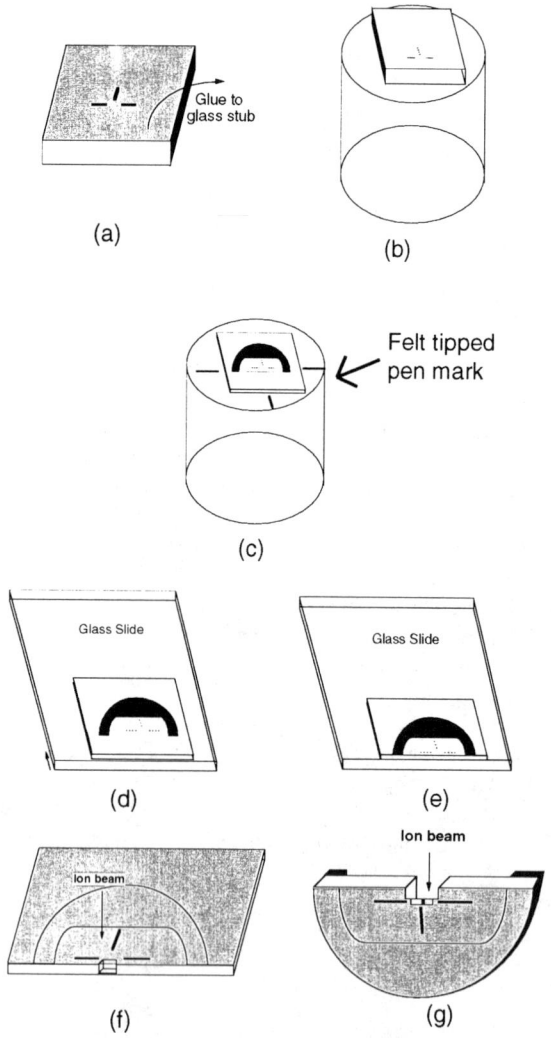

Fig. 15: A schematic representation of pre-FIB procedures for SAPTEM sample preparation (a) A 5 x 5 mm² piece of wafer with area of interest is marked using a laser beam. (b) Die glued face down on to a glass-stub and background. (c) Glue one-half Cu-ring slotted 3 mm. (d) Glue die on a glass-slide for x-section grind. (e) After x-section grind. (f) First FIB milling step, a slot is milled. (g) Second FIB milling step, SAPTEM sample is thinned.

3.3.4 FIB Thinning Procedures and Artifacts

A schematic representation of FIB thinning of TEM sample is shown in Fig. 11. High energy (30-50 keV) focussed gallium ions are used to ablate the sample leaving behind a thin electron-transparent sliver for TEM examination. The incident focused ion beam is rastered parallel to the surface final thin section. The sample is sequentially milled from edge to the final thin section in the middle, thus creating a window for TEM examination of the thin region. A high-current low-resolution ion beam is used for milling areas farthest from the sample followed by low-current high-resolution controlled milling of the final thin section. An advanced dual column FIB equipped with an electron column is useful for imaging the thin section as it is milled with the ion beam. Such an imaging capability is very useful in determining the final stopping point.

Usually about 1 to 2 μm of a protective layer of material, platinum (or tungsten), is deposited on top of the area of interest using a low current ion beam. The protective layer prevents sputtering of the top side of the specimen during FIB milling, prevents charging, and acts as a mask, effectively cutting off wide-spread wings for the beam. Since the protective platinum layer is amorphous, it plays a critical role by reducing the thickness variations and roughening of TEM specimens that may arise due to differences in sputter rate of various materials present in the specimen. If the platinum layer was polycrystalline, orientation dependent differences in sputter rate would cause considerable roughening of milled surfaces.

Normally the sample prepared using FIB milling are wedge shaped, i.e., thin at the top and thick at the bottom. Uniform sample thickness can be achieved with a combination of front side and back side milling at proper stage tilt using low beam current in the final phase of thinning of the sample [14,21]. Stage tilt during back side milling will depend on the sample geometry.

Surface damage occurs if the thin section is accidently scanned with ion beam at an angle more than few degrees parallel to the surface, or if the section is imaged using the ion beam. Such surface damage results in an amorphization of the surface of the sample that hinders high resolution imaging. In extreme cases, the long exposure to the ion beam at off-parallel axis to the surface of the thin section would result in anomalous defects/artifacts. Leslie et al. [22] have studied the effect of beam damage with varying tilt angles by recording selected area electron diffraction patterns from several Si samples prepared under the similar conditions with the final milling performed at different incident beam angle. Their results indicate that a specimen tilt angle of ±2 degrees from the vertical position provides optimum conditions for the sample being free from amorphization/phase transformation. Another common problem that induces artifact/amorphous layer on the surface is redeposition. Redeposition of an amorphous layer occurs if the region close to the thin sections are milled after the final area of interest is exposed. This problem can be easily prevented by avoiding any milling on the proximity after the thin region is exposed or by doing a cleanup milling.

4. SOME APPLICATIONS OF TEM IN FA

4.1 Defects in silicon

Diffraction contrast is the key to imaging crystal defects in silicon. Defects disturb the order in perfect lattice and the electrons are diffracted differently by the imperfect regions surrounding the defect. This property is frequently used to image the defects. The most frequently observed crystal defect in silicon is the implant induced dislocation loop. Implant damage is usually not limited to a specific failure site. Hence specific area TEM procedures are not required. Examples of dislocation induced by implant damage are demonstrated using the cross-sectional TEM micrographs in Fig. 5 and a planar TEM micrograph in Fig. 16. Another most common defect known as the spacer edge defect can also be seen in Fig. 16.

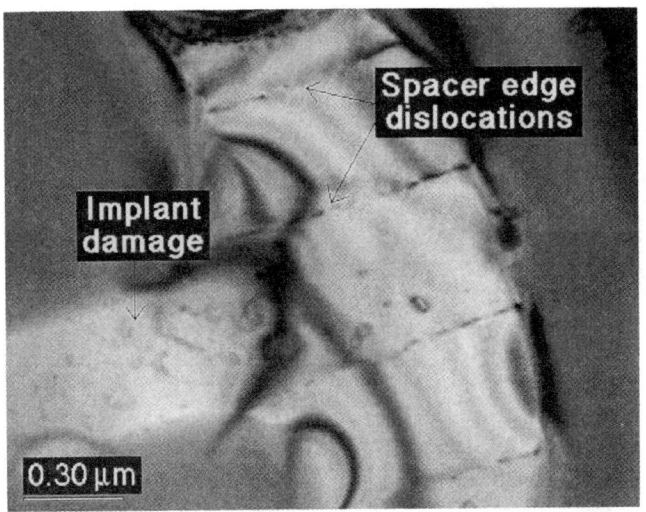

Fig. 16: A planar TEM micrograph containing implant induced dislocation loops and spacer edge defect. The defects are image with the aid of diffraction contrast.

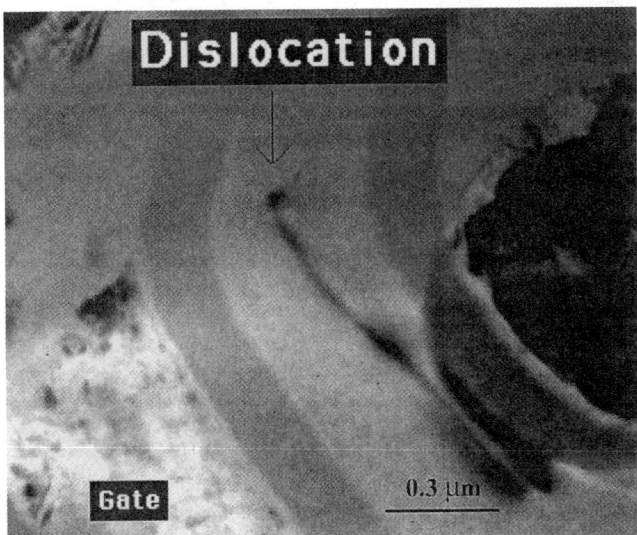

Fig. 17: A planar TEM micrograph containing unique dislocation orginating from the trench.

A classic example of a defect-induced failure in a silicon device involves dislocation-induced random single-bit failures in fast static random access memories. Electrical signature indicated low level leakage in one of the storage nodes of the failing bit. A dislocation, shown in Fig. 17, originating from the trench was observed in the silicon active region at the leaky storage node of the failing bit. The image is dominated by diffraction contrast at the defect location and polysilicon.

The drawback of FIB prepared TEM samples is the limited tilt available that poses problem to perform detailed characterization of dislocations. This limited tilting capability can be overcome using the SAPTEM sample preparation and it was possible to tilt the sample from a [100] orientation to [110] and [112] orientations. Fig.18(a,b) show contrast behavior of same dislocations imaged in two different diffraction vectors, g=022 and 004, under two beam imaging condition. Using a detailed diffraction contrast analyses, the Burgers vector of dislocations was found to be a/2<110>.

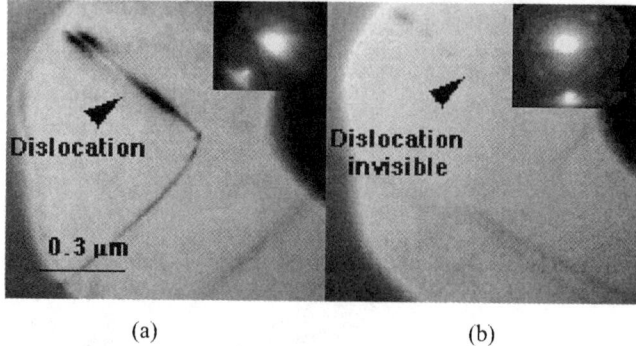

(a)   (b)

Fig. 18: Planar TEM micrographs of dislocations recorded under different two-beam conditions. (a) diffraction condition g = 022, shown as an inset, dislocation is visible, (b) diffraction condition g = 004, shown as an inset, dislocation is invisible.

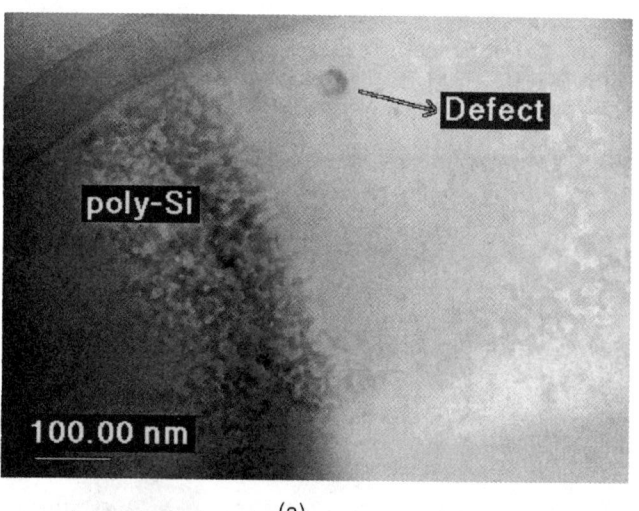

(a)

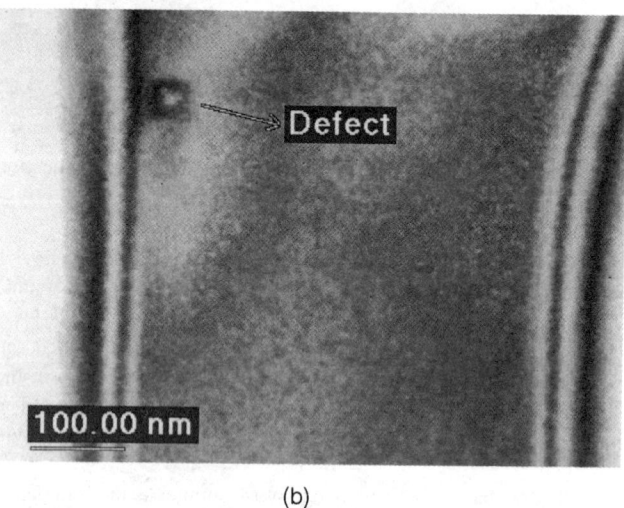

(b)

Fig. 19: A planar TEM micrograph of gate-oxide defects. (a) The SAPTEM sample was prepared by FIB thinning without chemical etching. (b) The SAPTEM sample was pre-thinned into the gate stack using the FIB. The sample was then subjected to a wet chemical etch followed by final backside thinning and polish using the FIB.

4.2 Gate oxide breakdown

Gate oxide defects are localized and difficult to detect. A combination of the large lateral area of analysis provided by SAPTEM and the strong diffraction contrast of the TEM can be used to observe the subtle effects (slight silicon damage) of gate oxide defects. In the examples demonstrated here, SAPTEM samples of random single bit failures in fast static random access memories were prepared by two approaches. In the first approach the TEM sample was thinned down to the gate oxide in the FIB. The defect shown in Fig. 19(a) was located in one of gates of the fast static random access memory cell. This procedure required very careful milling during the final stages of the thinning to stop within 70 Å of the gate. The residual poly can also be seen on the gate region. In the second approach the sample was thinned using FIB to about 3 μm thickness at the back side. In the front side the sample was milled up to the into the poly gate stack. The sample was then subjected to a surface clean in BOE followed by a silicon etch procedure. The sample was loaded in to a 15 x 15 x 5 mm teflon boat and transferred between various chemicals. The sample was then reloaded into the FIB and thinned from the substrate silicon side. The gate oxide damage was located in one of gates in the cell as is shown in Fig. 19(b).

4.3 Interfacial reaction

In the first example, a planar TEM study was conducted to investigate possible leakage paths in fast static random access memory cell. An interfacial reaction between titanium silicide and SiN that resulted in Ti+Si+N mixture was observed as is shown Fig. 20. The dark and bright polysilicon grains observed in the image is a result of various grains diffracting because of difference in the crystal orientation of those grains with respect to the ion beam. The amorphous silicon nitride spacer is sharper than the central amorphous silicon oxide because of the mass differences.

Fig. 20: A planar TEM micrograph showing the interfacial reaction between titanium silicide and titanium nitride.

In the second example a site specific cross-section TEM study was conducted to understand a timing issue. TEM analysis revealed a resistive layer that would caused the part to fail at tight timing. The diffraction/mass contrast and high magnification phase contrast TEM images of the resistive interface is shown in Fig. 21(a). The amorphous regions can be clearly identified in high resolution phase contrast image shown in Fig. 21(b).

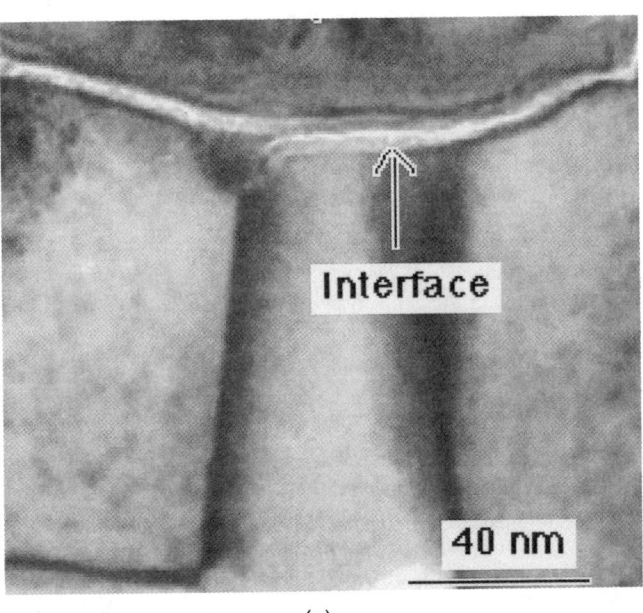

(a)

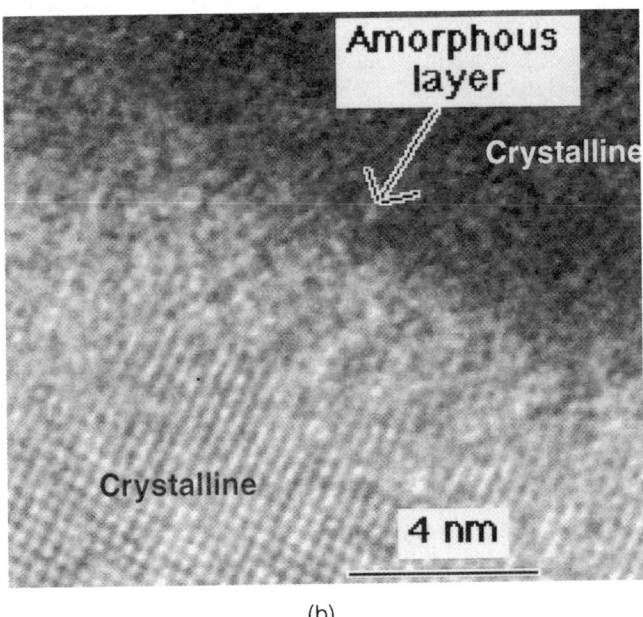

(b)

Fig. 21: (a) A low magnification cross-sectional view of a resistive interface imaged with the aid of diffraction contrast. (b) High resolution phase contrast image of the resistive amorphous layer.

4.4 Stringers and Protrusions

A cross-sectional TEM image of a polysilicon gate and the poly stringer is shown in Fig 22. The small but crystalline stringer is easily visible in a TEM image because of strong diffraction contrast produced

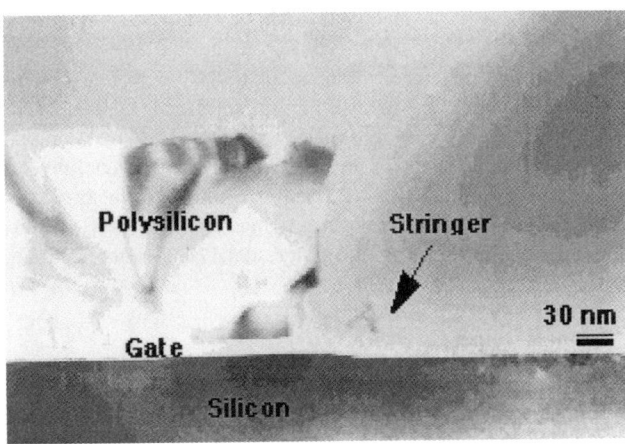

Fig. 22: A cross-sectional TEM image of a polysilicon stringer.

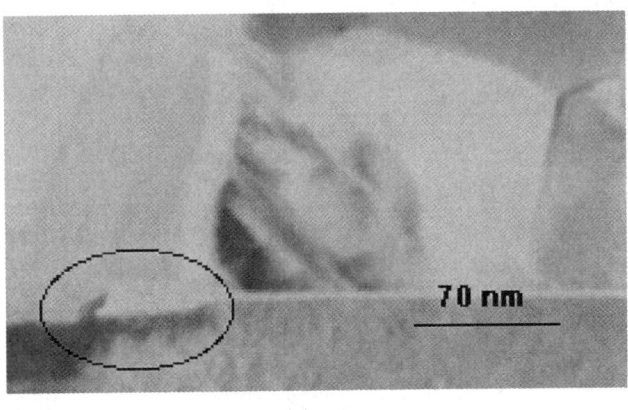

(a)

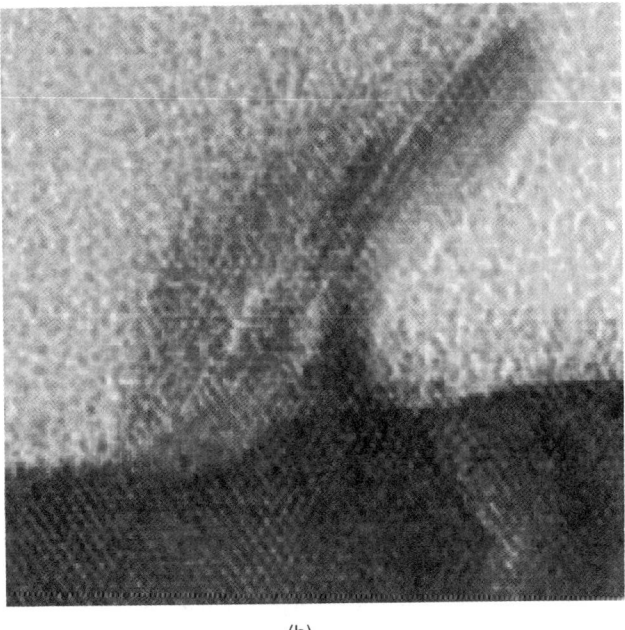

(b)

Fig. 23: (a) A low magnification cross-sectional view of a residue at the gate imaged with the aid of diffraction contrast. (b) High resolution phase contrast image of the crystalline residue embedded in the amorphous oxide.

325

by a crystalline polysilicon in a diffraction free amorphous background oxide.

High resolution lattice imaging is very useful in revealing presence of interfacial details at atomic resolution as well as local presence of different materials, e.g., poly stringer or any other residual material. One such example is shown in Fig. 23(a) where a residual layer is clearly visible in the bright-field image at one corner near salicide region. The details of residual layer at atomic resolution, crossed lattice fringes in salicide region as well as in residual layer, is shown in Fig. 23(b) .

5. SUMMARY

The role of TEM in the failure analysis of semiconductor ICs is described. The basics of TEM imaging techniques, origin of different image contrast mechanisms, their interpretation, and analytical techniques for composition analysis are discussed. New developments in the area of TEM sample preparation using FIB milling is reviewed. Site specific TEM sample preparation techniques, developed in the present work, suitable for failure analysis of ICs are described. Cause of the common problem that induces artifact/amorphous layer in the case of FIB prepared samples is discussed and suggestions for minimizing their effects are presented. Finally, application of TEM in failure analysis is presented using a number of examples from case studies of defects in silicon, gate oxide breakdown, interfacial reaction, and poly stringers.

ACKNOWLEDGMENTS

Authors would like to acknowledge fellow Motorolans (alphabetically) : Chuck Davin, Gloria Estrada, Stewart Rose, Phil Schani, Dave Sieloff, Vince Sooholtz, Pat Liston, Jamey Moss and Ed Widener for their support and contributions.

REFERENCES

1. R. Anderson, Proceedings of the 49th Annual Meeting of Electron Microscopy Society of America, edited by G.W. Bailey and E.L Hall, San Francisco Press, California, 1991, p828.
2. H. Oppolzer, Inst. Phys. Conf. Ser. No. 76, Sect. 11, 1985, p461.
3. S. J. Klepeis, J. P. Benedict, and R. M. Anderson, Mater. Res. Soc. Symp. Proc., Vol 115, 1987, p179.
4. J. Szot, D. Yong, A. Bourdilton, and K. E. Easterling, Phil Mag. Lett. Vol 55, 1987, p109.
5. R. M. Anderson and J. Benedict, Mater. Res. Soc. Symp. Proc. Vol 254, 1992, p141.
6. R. Alani, J. S. Jones, and P. R. Swann, Mater. Res. Soc. Symp. Proc. Vol 119, 1990, p85.
7. N. G. Chew and A. G. Cullis, Ultramicroscopy, Vol 23, 1987, p.175.
8. J. C. Bravman and R. Sinclair, J. Elect. Micro. Tech., Vol 1, 1984, p53.
9. C. W. T. Bulle-Lieuwma and P. G. Zalm, Surf. Interface Anal., Vol 10, 1987, p210.
10. J. Benedict, R. Anderson, and S. J. Klepis, Mater. Res. Soc. Symp. Proc., Vol 254, 1992, p121.
11. R. J. Young, E. C. G. Kirk, D. A. Williams, and H. Ahmed, Mater. Res. Soc. Symp. Proc. 199, 1990, p271.
12. S. Morris, S. Tatti, E.Black, N. Dickson, H. Mendez, B. Schwiesow, and R. Pyle, Proceedings of the 17th International Symposium for Testing and Failure Analysis, 1991, p417.
13. E. C. G. Kirk, D. A. Williams, and H. Ahmed, Inst Phys. Conf. Ser. No. 100 Section 7, 1989, p501.
14. D. M. Schraub and R. S. Rai, Progress in Crystal Growth and Characterization of Materials, Vol 36, 1998, p99
15. M. H. F. Overwijk, F. C. Van den Henvel, and C. W. T. Bulle- Lieuwma, J. Vac. Sci Technol. , Vol B11, 1993, p 2021.
16. L. R. Herlinger, S. Chevacharoenkul, D. C. Erwin, Proceedings of the 22nd International Symposium for Testing and Failure Analysis, 1996, p199.
17. L. A. Giannuzi, L. L. Drown, S. R. Brown, R. B. Irwin, and F. A. Stevie, Mater. Res. Soc. Symp. Proc., Vol 480, 1997, p19.
18. F. A. Stevie, R. B. Irwin, Tl L. Shofner, S R. Brown, J. L. Drown and L. A. Giannuzzi, Characterization and Metrology for ULSI Technology, AIP conference proceedings -449,1998, p868.
19. R. Anderson and S. J. Klepeis, Specimen Preparation for Transmission Electron Microscopy of Materials IV, Materials Research Society, Pittsburgh, Vol 480, 1997, p187.
20. S. Subramanian, P. Schani, E.Widener, J. Moss and V. Soorholtz, Proceedings of the 24th International symposium for Testing and Failure Analysis, 1998, p131.
21. TEM Sample Preparation-Application Note, FEI Company, May 1995.
22. A. J. Leslie, K. L. Pey, K. S. Sim, M.T. F. Beh, and G. P. Goh, Proc. 21st International symposium for Testing and failure Analysis, 1995, 353.

SELECTED REFERENCES

• M. M. Disko, C. C. Ahn and B. Fultz, Tranmission Electron Energy Loss Spectrometry in Materials Science, TMS, 1992.
• R.F. Egerton, Electron Energy Loss Spectroscopy in the Electron Microscope, Plenum Press, 1989.
• J.W. Edington and K.C. Thompson-Russell, Practical Electron Microscopy, Vol. 1-5: Monographs in Material Science, Macmillan Press, 1975.
• P.B. Hirsch, A. Howie, R.B. Nicholson, D.W. Pashley, M.J. Whelan, Electron Microscopy of Thin Crystals, Krieger, New York ,1977.
• D.C. Joy, A.D. Romig, and J.I. Goldstein, Principles of Analytical Electron Microscopy, Plenum Press, 1986.
• M.H. Loretto and R.E. Smallman, Defect Analysis in Electron Microscopy, Halsted Press, 1976.
• M.H. Loretto, Electron Beam Analysis of Materials, Chapman and Hall., 1984.
• G. Thomas, M.J. Goringe, Transmission Electron Microscopy of Materials, Wiley and Sons, New York, 1979.
• D. B Williams and C. Barry Carter, Transmission Electron Microscopy, Vol I-IV, Plenum press, 1996.

Atomic Force Microscopy:
Modes and Analytical Techniques
with the Scanning Probe Microscope

Yale Strausser
Digital Instruments
Santa Barbara, California

James Colvin
WSI
Fremont, California

1) Introduction

The Scanning Probe Microscope (SPM) has matured rapidly and is already a valuable tool for Failure Analysis and indispensable for FAB process issues. The SPM or more commonly known as the Atomic Force Microscope (AFM) can do much more than just image a surface as will be shown in each section. The theory of operation and various modes will be discussed. Examples from routine failure analysis will be presented.

2) The Scanning Probe Microscope Family

Scanning probe microscopes have been commercially available for just over 10 years now. In this time they have matured quite rapidly compared to other analytical instruments, such as scanning electron microscopes, surface analysis tools (Auger electron spectrometers, secondary ion mass spectrometers, etc.), and other frequently used instruments which have developed over the last 30 years. Scanning probe microscopes have not only improved the original scanning technologies (scanning tunneling microscopy and contact atomic force microscopy) but they have broadened their utility through the development of many new scanning technologies. In this article we will briefly review all of the scanning technologies which are currently helpful in IC failure analysis including some of the recently developed ones. New developments continue to come at a rapid pace so this article will not remain current for long.

During this first approximately ten years the applications of SPM's have also grown rapidly. Initially the applications were in fairly basic research areas while scientists began to become familiar with these new tools. Because, at least in part, of the ease of sample preparation and use of the SPM's the tool quickly found uses in more applied research and even production areas. They found uses in production process control in the data storage industry within the first five years of their availability. They were in use in analytical laboratories of the major semiconductor manufacturers within the first six to seven years of their availability. By now SPM's are in routine use in production control applications in data storage, semiconductor, polymer, contact lens, and other manufacturing facilities. These applications all focus on the measurement of some property, associated with a surface, at high spatial resolution.

Even though applications in the semiconductor wafer fabs were one of the first, and most successful industrial applications, the use of SPM's in IC failure analysis (ICFA) has not progressed very rapidly. The limitations have been in finding SPM technologies which measure surface related parameters which are valuable in ICFA, and in developing instrumentation which would allow the measurements to be made with minimal deprocessing of the failed, packaged device. Also the general familiarity of ICFA personnel to SPM's has not been as high as it is, for example, to SEM's. This is changing now with the development of new technologies such as scanning capacitance microscopy; with the availability of microscopes which allow access for measurements to be made on operating, packaged devices; and with broader exposure to semiconductor applications information, primarily from wafer process measurements.

In this section we will describe some of the newer scanning technologies which we feel are useful in failure analysis of integrated circuits. Atomic force microscopy (AFM) was improved dramatically over the original contact mode with the development of TappingMode AFM in 1993. An offshoot from the availability of TM AFM is phase contrast measurements in AFM. Lift Mode, introduced in 1994, enabled the separation of topography from magnetic or electric field measurements to make those measurements practical. Detailed surface electric potential measurements are also possible making high spatial resolution Kelvin probe maps of surface work functions and other potentials. High spatial resolution surface temperature measurements are available through scanning thermal microscopy (SThM)[1]. Optical probes or optical signal pickup can now be accomplished at dimensions below the traditional "diffraction limit" through the use of near field scanning optical microscopy (NSOM)[2]. And finally, carriers in semiconductors can be mapped out showing carrier type, concentration, junction locations[3], and even very local C-V measurements[4] using scanning capacitance microscopy (SCM).

In contact mode AFM a sharp tip mounted on the end of a long flexible cantilever is pressed against the sample surface with a known force (known from the curvature of the cantilever) and drug across the surface while keeping the force constant. The vertical movements required to maintain the force constant are the measurements of the surface topography. Contact AFM lacks the desired control of the applied force and produces a shear force from the dragging of the tip against the surface. TappingMode AFM (TM AFM) is an AC version of this technique[5]. Shown in figure 1 is a diagram of a TM AFM in which the tip/cantilever assembly is attached to the piezoelectric scanner and the sample surface is stationary. In this case the cantilever is oscillated so that the tip oscillates at an amplitude in the range from 5 to 100 nm before it is brought down to where it "taps" against the sample surface. When it is brought down and linked to the surface the oscillation amplitude reduces due to the tapping against the surface and the percentage reduction in the amplitude determines the force of the tip on the surface and is set by the operator and controlled by the feedback loop. The tip is then scanned across the surface while tapping against it and while maintaining a constant oscillation amplitude. The vertical motion required of the z piezoelectric crystal to maintain the constant oscillation amplitude is a precise measurement of the surface topography to .05 nm in all three axis. Compare this to the SEM, which is only capable of X-Y measurements since Z is relative topography only.

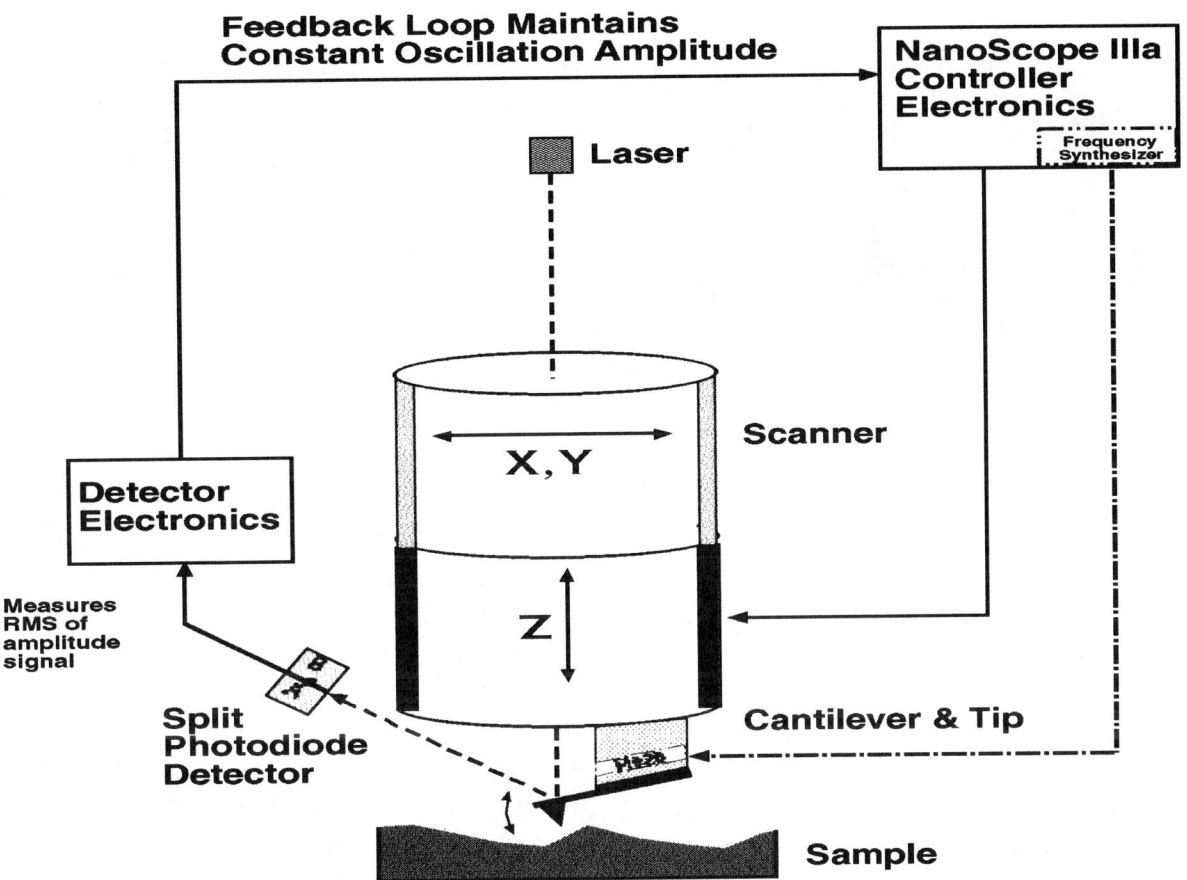

Feedback Loop Maintains Constant Oscillation Amplitude

NanoScope IIIa Controller Electronics

Frequency Synthesizer

Laser

Scanner

X, Y

Z

Detector Electronics

Measures RMS of amplitude signal

Split Photodiode Detector

Cantilever & Tip

Sample

Figure 1. Schematic diagram of TappingMode AFM.

Phase contrast measurements are an offshoot of the oscillating cantilever used in TM AFM[6]. The cantilever is driven in oscillation by an AC signal generator and a piezoelectric crystal. At the same time, the motion of the cantilever is monitored by the photodetector, which monitors the oscillation amplitude. We can compare the drive signal and the motion response signal and measure the phase shift between the two. When the cantilever bounces elastically off the surface in its tapping, there is a set phase shift. When the tip is pulled toward the surface by any of a variety of forces this extra pull required to get away from the surface causes an additional phase shift. Using the phase shift to produce contrast in an image shows a map of those locations on the sample where the tip/sample interaction is stronger than at others. An example of the use of phase imaging is shown in figure 2, which is a topography image (left) and phase image (right) taken simultaneously on the same area of a bond pad. This pad had been coated with a polyamide layer, which was to have been etched off. This bond pad was on a wafer, which showed poor bond pad adhesion. The phase image shows small spots of high contrast, which were later shown to be residual polyimide, which caused the poor bonding. This quick and easy measurement showed high sensitivity for the detection of the polyimide.

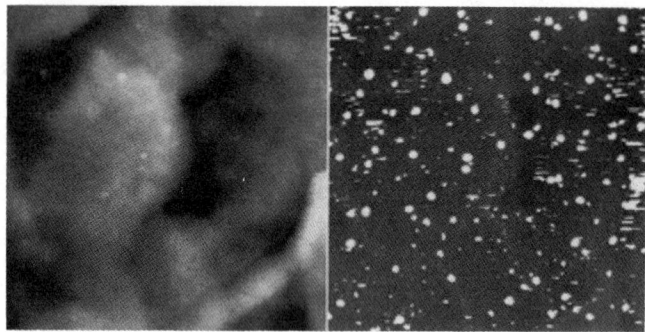 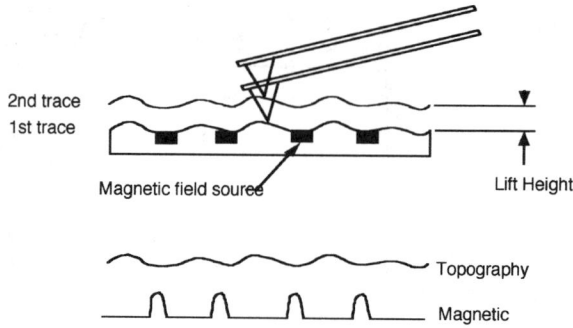

Figure 2. Topography (left) and phase contrast (right) images of an Al bond pad, which was on a wafer producing poor bond adhesion. The topography shows typical Al grains while the phase contrast shows polyimide residue.

Lift Mode is a technology that enables separate but concurrent measurements of surface topography and a field associated with the surface, through the use of interleaved scans. The principle is illustrated in figure 3. It works by making one scan line of the measurement area (left to right and then back, right to left) in TappingMode to measure the topography of the surface. Next the probe is lifted some height above the surface (typically 5 to 100 nm, chosen by the operator) and follows the same height pattern as it makes the next scan line. On this scan line it is lifted above the surface, oscillating, and sensing the field at this constant distance above the surface. This provides a field measurement at a constant height above the surface, without any interference from the tip striking the surface and thus gives a clean measurement of the electric or magnetic field. In this way both a topographical image and a field image are built up concurrently.

Figure 3. Schematic diagram of Lift Mode operation.

Magnetic field measurements require a tip which is either a small piece of magnetic material or a regular Si tip coated with a magnetic film. With the tip oscillating while it scans across and above the surface the vertical component of the magnetic field can be sensed by the force it exerts on the magnetic tip because this alters the resonant frequency and phase of the oscillating cantilever. The shift in either the phase or the frequency can be used to produce contrast in a map of the magnetic field associated with the sample surface[7].

Electric field measurements can be made in a similar way except that the tip only needs to be coated with a conductor, thus producing a capacitor with the tip as one plate and the region of the sample on which some charge is applied as the other plate. The tip scanning and oscillating above the surface again feels a force, which is dependent upon the vertical component of the electric field gradient at each pixel of the scan line. This causes a shift in the phase and frequency of the cantilever's oscillation, either of which can be used to produce contrast in a map of the field above the surface. As an example of an EFM image figure 4 shows an area of an operating device, with passivation intact. The right image is a TM AFM image of the surface

Figure 4. Electric force microscopy (left) and topography (right) images of an operating device. The potential distribution shown in the EFM image shows a transistor operating in saturation. This voltage distribution measurement was made through the passivation coating.

showing nothing out of place. The left image is an EFM image showing the potential distribution across this area and showing a transistor which is running in saturation because of a gate oxide short at another location on the device. The variations in intensity of the EFM signal are due in combination to variations in spacing between the tip and potential on the buried conductors.

Scanning thermal microscopy measurements require a tip that is constructed so as to measure the temperature or thermal conductivity of the surface, at high spatial resolution. There are currently many approaches being taken to this measurement. The seemingly obvious one is to produce a thermocouple at the tip of your probe. This is not straightforward. Some such probes have been made using Si technology by a couple of groups[8,9]. These have not yet been fully satisfactory. Others have made thermocouples and then attached a sharp diamond on the end to

act as the tip. Diamond is a good thermal conductor and is very hard so it works reasonably well. Still others have made thermocouples with one material (wire) going down the inside of a pipette, a junction formed just at the tip/opening of the pipette, and the other material running back out the outside of the pipette. These have had some success. A slightly different early approach was to use a Wollaston wire, which was etched to a point. Another approach in common use is to coat the end of a standard tip with a layer, which can act as a thermistor. A material is used which has a high temperature coefficient of resistivity. As it scans across the surface in contact AFM mode, encountering different temperatures, the resistivity of the film is monitored and used to produce a map of temperature over the sample surface. One of the early realizations to come out of SThM was that as thermal energy diffuses to a surface on which SThM can measure it, it also diffuses laterally. A point source of heat buried deep under the surface being measured shows considerable lateral extent in its signature on the surface. An example of this is seen in figure 5, which was taken on an operating device that had a gate oxide short. In the topography image, on the left, taken on top of the passivation, there is no indication of any problem. In the SThM image, on the right, the whiter area is the hot area. This shows where the gate oxide short is located but the heat has diffused outward to make a large area hot spot. In this instance the heat source is 2 to 3 microns below the surface which is being mapped.

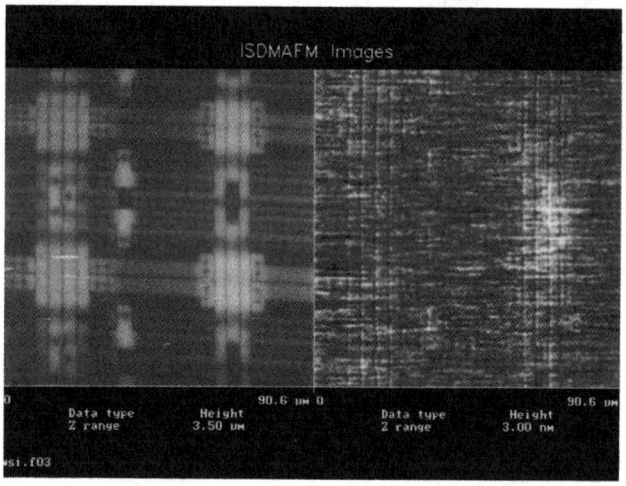

Figure 5. Topography (left) and temperature distribution (right) on an operating device. The temperature image points out the location of a gate oxide short, below 2 to 3 microns of material.

Tunneling mode AFM produces a topography map and a map of the voltage required to tunnel at a set current level[10]. The AFM feedback loop maintains a constant force on the tip while the tunneling current feedback loop adjusts bias voltage to maintain a constant tunneling current between tip and substrate. The current is typically set at around 1 pA.

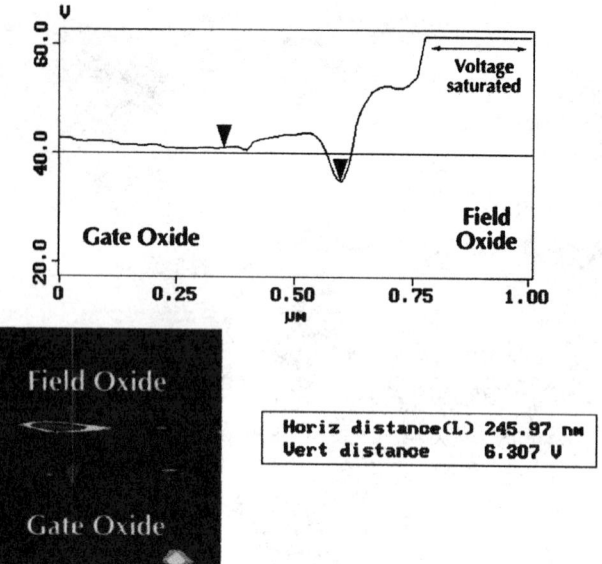

Figure 6. Gate oxide thinning at bird's beak measured by tunneling AFM at 1 pA.

Typical tunneling fields are reached at 10 V/nm of oxide thickness. Although this method is currently recognized as an inline measurement for the FAB, this technique has merit in FA for problems such as charge loss in EPROM or FLASH cells.

Scanning Capacitance Microscopy measures capacitance variations between a conductive tip and a semiconductor sample while scanning in contact mode. This technique produces a 2-D image of near surface variations in carrier concentration with a sensitivity range of 10^{15} to 10^{20} carriers/cm^3 and a lateral resolution of 10 to 20 nm. Sample preparation must be done properly for good results as outlined later in this paper. Figure 7 is a schematic of the Scanning Capacitance system.

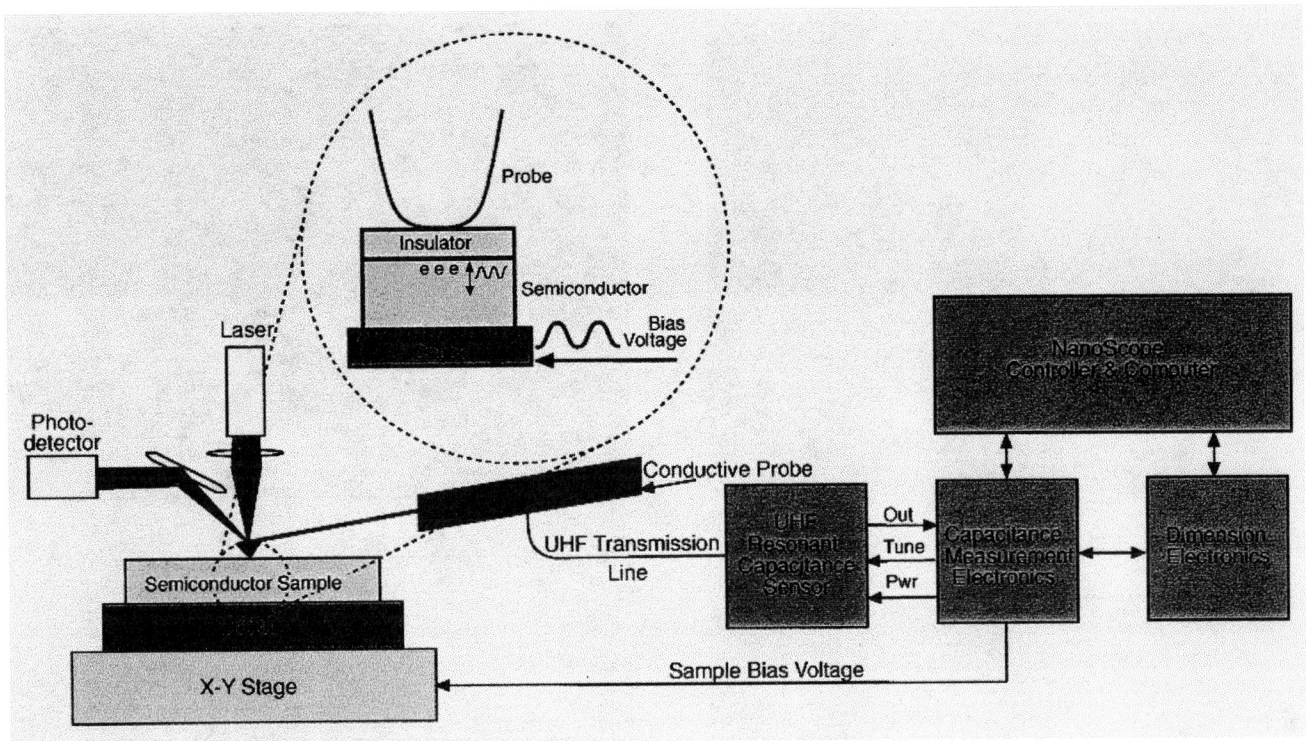

Figure 7. Schematic diagram of the Scanning Capacitance AFM.

3) Failure Analysis Case Histories

An EPROM process with a single bit charge loss (SBCL) problem was analyzed to identify why SBCL increased dramatically with increased implant dosages on a 1.2 um process. The affected cell was deprocessed to substrate and examined with the AFM and SEM respectively.[11] The residue which was quite difficult to see due to a lack of topography in figure 9 is easily seen in figures 8 and 10. The step height of the residue was measured with the AFM at 26.72 nm (figure 11) and believed to be a resist residue that interfered with subsequent implant steps. To confirm this hypothesis another SBCL failure was cross-sectioned through the failing cell and imaged after junction stain enhance with the AFM and SEM. Figure 12 is an AFM image of the cross-sectioned cell. Note the sloping substrate edge associated under the contaminant. This slope was due to a gradient in the implant due to the residue.

The sample preparation for the SEM differs from the AFM based on the amount of enhancement required. The SEM image of figure 13 is stain enhanced to allow the topography to be seen, however, the AFM image of figure 12 is clearly over-enhanced as evidenced by the tip convolution between the substrate and the defect. A guideline for AFM sample stain enhance is to back off on the etch time at least a factor of 10 from what is customary for the SEM.

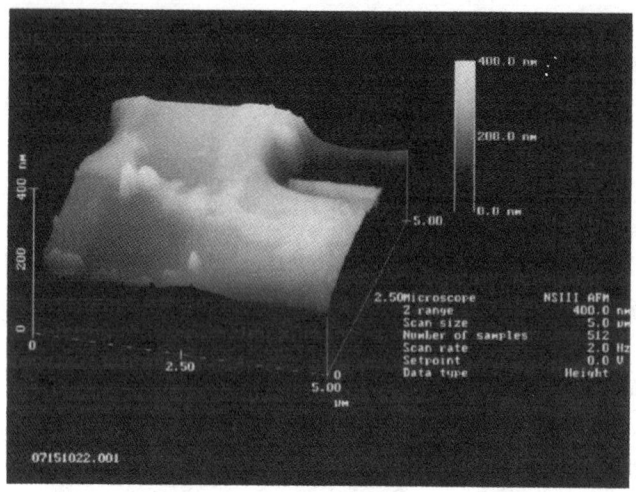

Figure 8. AFM image of residue on the substrate.

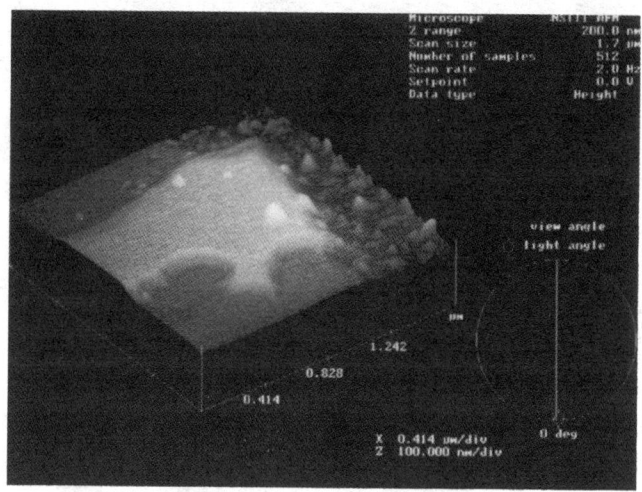

Figure 10. Increased magnification view of the residue.

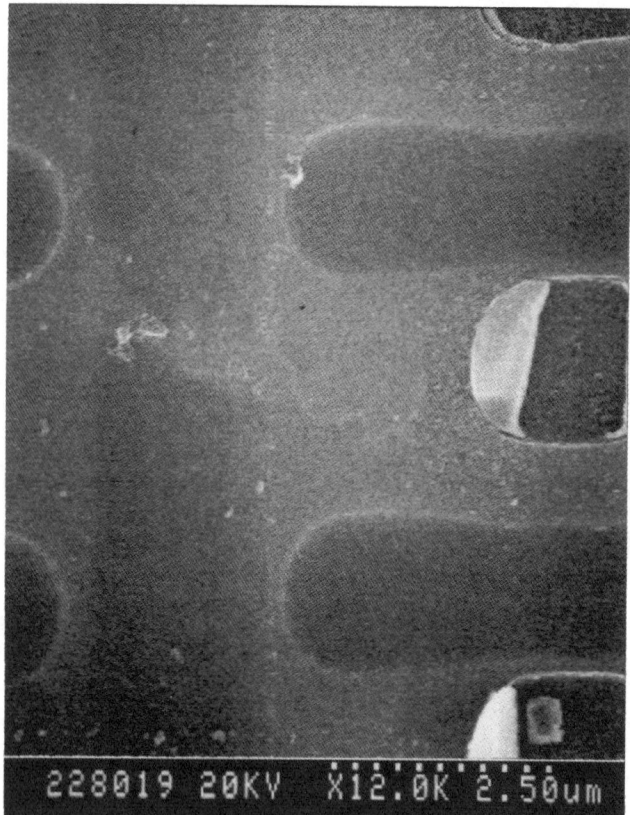

Figure 9. SEM image of the resist residue at substrate.

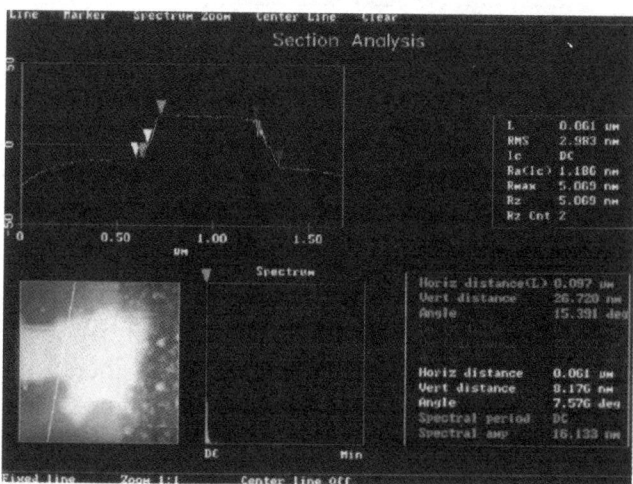

Figure 11. Section profile of the residue shows a total step height of 26.72 nm.

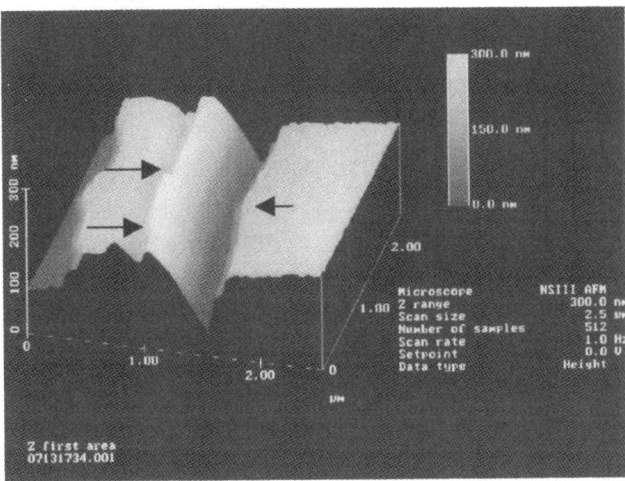

Figure 12. AFM image of cross-sectioned residue. The left arrows point to the residue. The right arrow points to the change in slope at the substrate / diffusion boundary.

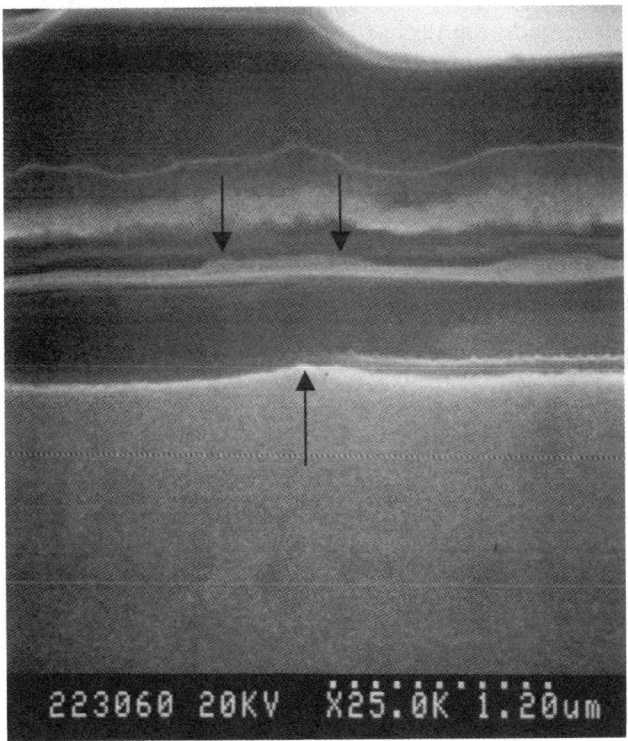

Figure 13. SEM image of cross-sectioned residue. The top arrows point to the reside. The bottom arrow points to the change in slope of the diffusion / substrate boundary.

Atomic Force Microscopy of LESD Damage

Latent electrostatic discharge (LESD) damage sites are normally identified and analyzed after the damaged gate oxide has been transformed from a low to a high level leakage. Failure analysis typically focuses only on the primary rupture site, neglecting the remaining LESD damage sites. Images from the Scanning Electron Microscope and the Atomic Force Microscope associated with the LESD damage sites are compared[12].

The sample was prepared using an ink planarization etchback method[13]. Note: Surface lapping the sample and using KOH etch will yield equivalent to superior results.

Figure 15 is an AFM view of the LESD rupture. Note the cone shaped hole. This is due to tip convolution as the edges of the tip enter the LESD hole. The edges of the hole were measured to be: $Y = .19$ μm, $X = .14$ μm. Notice that the etch undercut ring from figure 14 is not visible in the AFM image since the AFM reveals the true surface profile. In cases such as this, ultrasonic cleaning must be avoided due to the fragile nature of the gate oxide rupture. The AFM will image this oxide without damage in tapping mode.

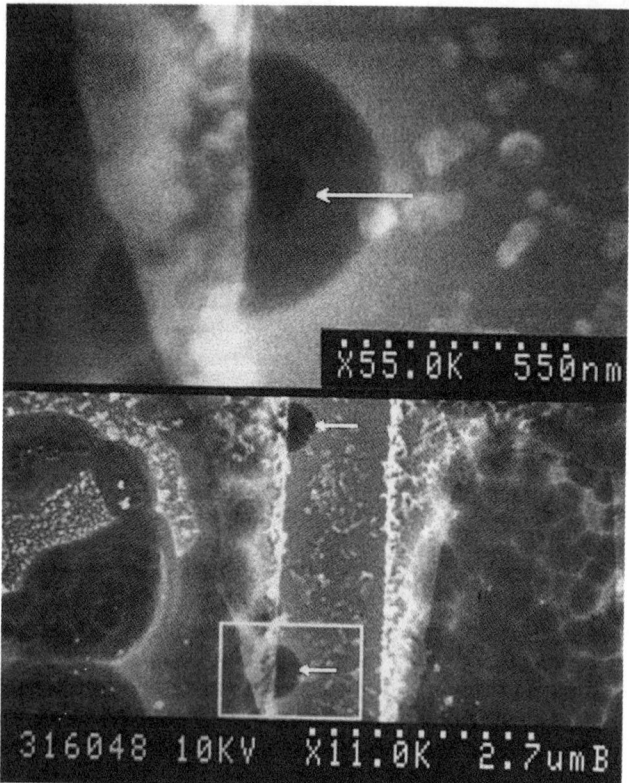

Figure 14 SEM high magnification split view of the LESD damage sites. Note the hole within a hole. The outside dark ring is due to plasma undercut that removed the underlying channel radially from the rupture. The size of the rupture ranges from .15 μm to .26 μm in the Y direction and from .12 μm to .19 μm in the X direction for various samples.

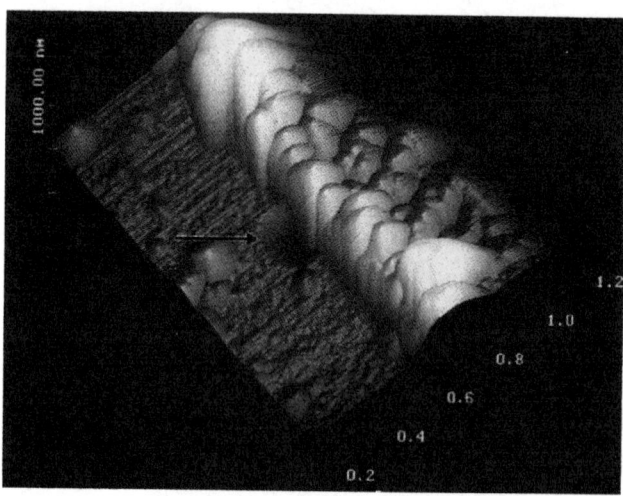

Figure 15 Atomic Force Microscope image of an LESD damage site. Compare the appearance of the oxide sidewall to the SEM image in figure 14.

Magnetic Force Microscopy

Magnetic force microscopy is used to map the local magnetic lines of flux associated with a scanning probe[14]. Sensitivities of around 1 mA dc and around 1 uA ac are attainable[15]. Comparable E-beam current mapping techniques are typically limited to greater than 100 mA of ac resolution[16]. A sufficient magnetic dipole moment is required at the tip in order to image the weak H field that is on the order of a few microteslas[17]. A coated tip for magnetic imaging has a weak magnetic domain and is suited to image intense magnetic domains such as recording media, whereas the converse is true with MFM current detection. In the following example, a rare earth magnet is mounted to an AFM cantilever (Figure 16) in order to provide the intense local magnetic domain[18]. The IC under test (Figure 17) has no permanent magnetic properties that will be disturbed with the tip. Figure 18 is an MFM image of 20 uA of ac current. The ac information is clearly visible even down to 2 uA.

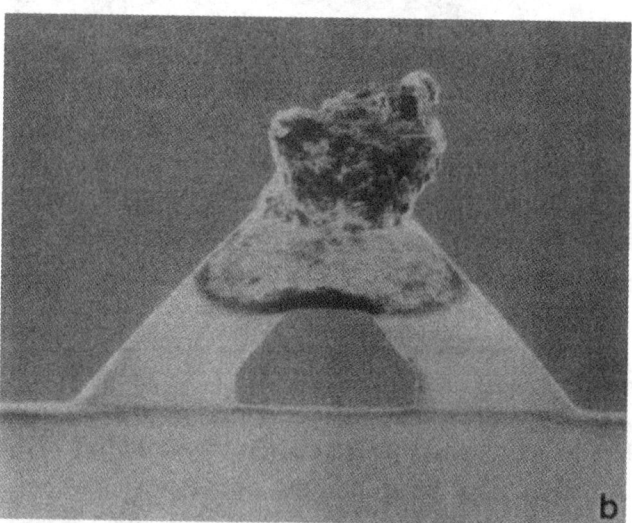

Figure 16 NdFeB magnet fragment attached to an AFM cantilever for Magnetic Force Microscopy.

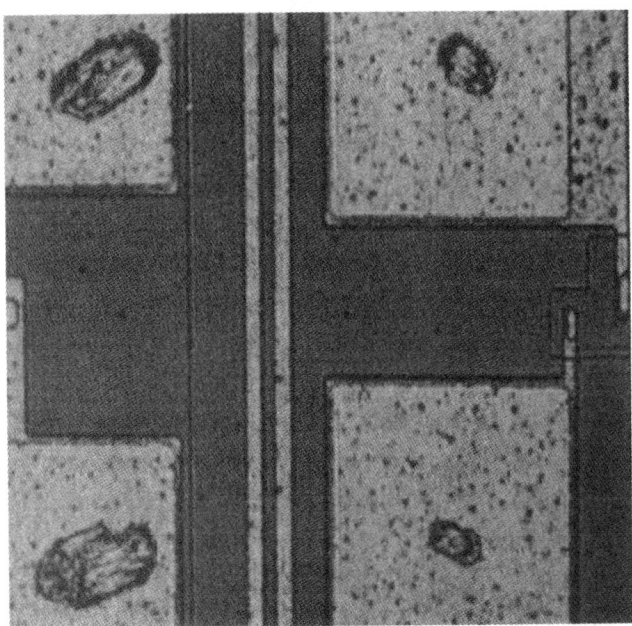

Figure 17 Optical photo of the 2 aluminum lines that were selected for imaging using magnetic force microscopy.

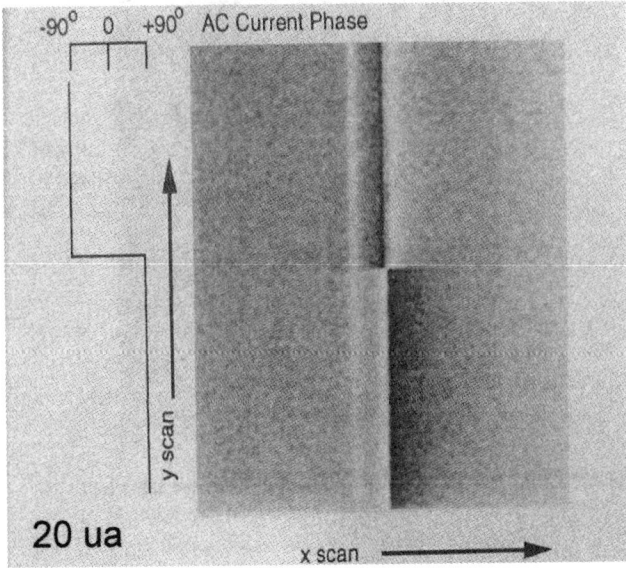

Figure 18 MFM image of 20 uA of ac current applied to the conductor on the right only.

Capacitive Probe Microscopy

A .8 um double metal process failed due to high IDDQ current values. The technology incorporates EPROM as well as standard 6 transistor SRAM cell design. Emission microscopy was used to locate the source of the problem as shown in figure 19. The location of the emission sites correspond to the field oxide isolation areas in the n-well area of the SRAM. All the cells were observed to photoemit in the field oxide area with varying intensity. A cross-section of a failing cell was made and stain enhanced with a combination of Wright and Dash etch. Figure 20 is an SEM image of the resulting cross section. Although the FOX isolation appears shallow, this data alone is insufficient due to the variance in traditional etch enhancement methods.

Scanning capacitive probe microscopy was chosen to image the surface of the SRAM in order to identify the integrity of the isolation. A cross section could also be done in the same manner, however, a topographical view of the dopants for the entire cell was desired. The die was stripped to substrate using 49% HF and cleaned with micro-organic soap and a swab before imaging in the AFM. Figure 21 is a view of the surface topography. Note that even though the height variation is around 500 um, capacitive probe techniques are quite effective.

Figure 22 is a capacitance image of a normal SRAM cell from a good lot, which shows a healthy isolation of P+ regions. Figure 23 is a capacitance image of the failing cell from the bad lot. Note the encroachment of the P+ active areas into the areas labeled FOX (field oxide). These areas correspond to the emission sites from figure 19. The time to obtain the Capacitance scan image was a matter of minutes since the problem is repetitive in the array.

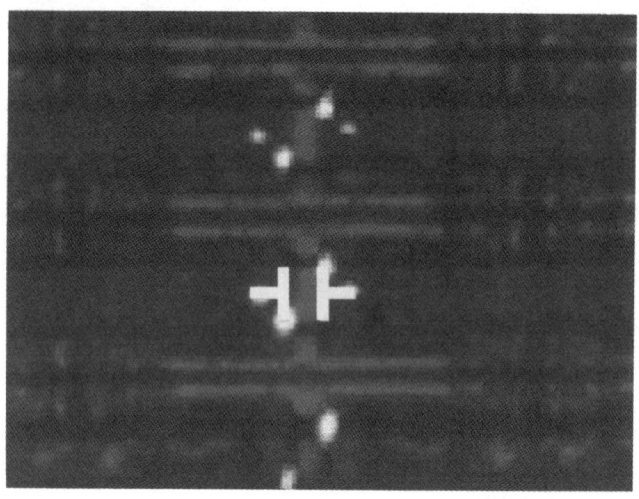

Figure 19 Photoemission image of the punchthrough associated with the SRAM array. Orientation of the field oxide boundary is shown on the cell in the center.

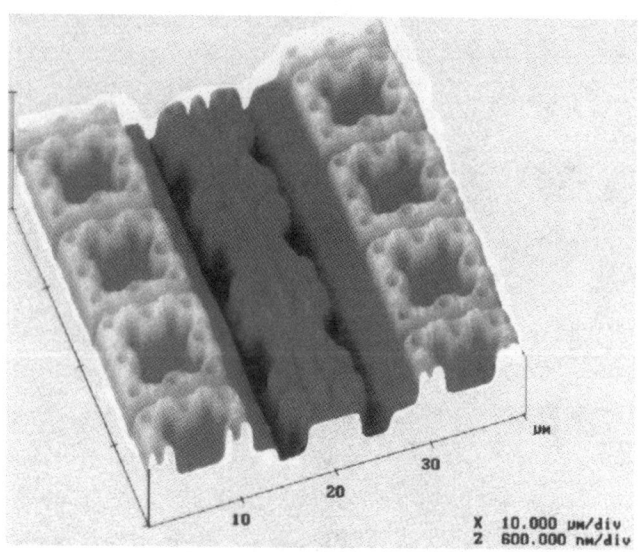

Figure 21 AFM image of the topography of the exposed substrate associated with the SRAM after HF etch.

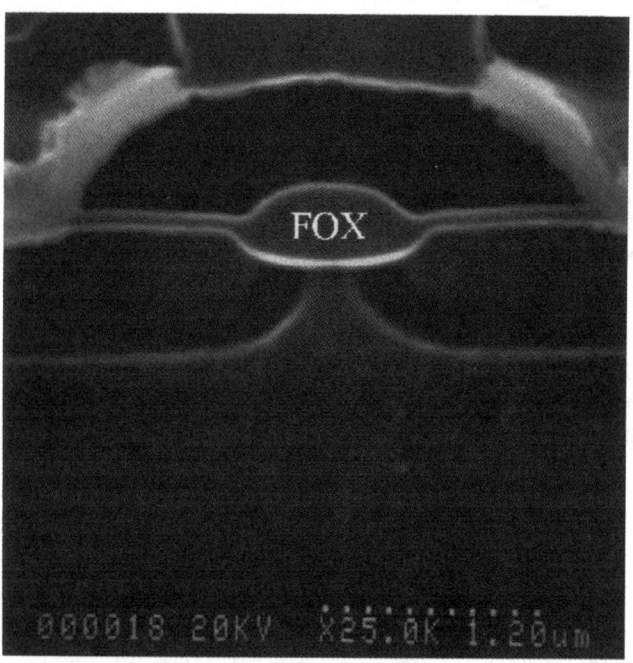

Figure 20 SEM cross section of a failing SRAM cell in the N-well area from a failing device. A comparison is required due to the inconsistencies of traditional etch enhancement methods.

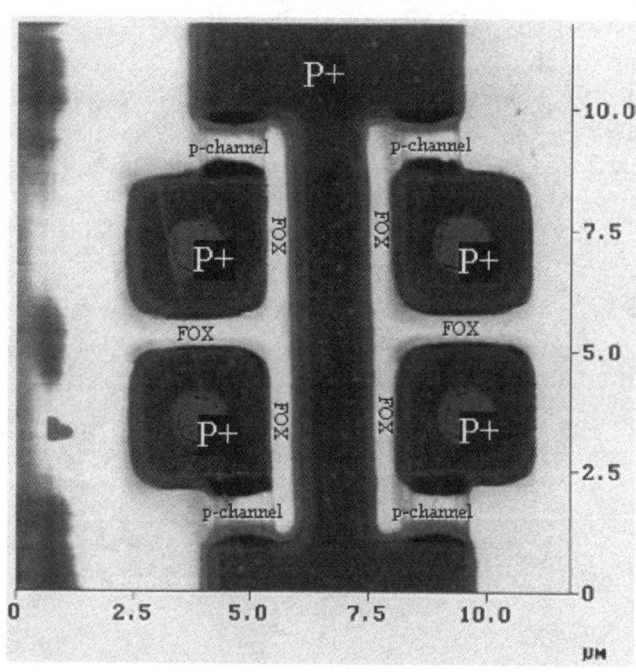

Figure 22 Capacitive probe image of an SRAM cell from a normal device in the N-well area. Note the field isolation (FOX) between the P+ areas is normal.

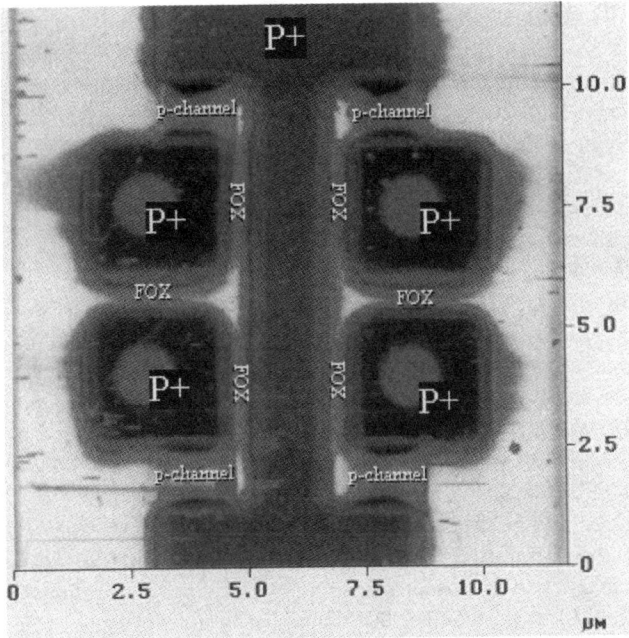

Figure 23 Capacitive probe image of a failing SRAM cell in the N-well area from a failing device. Note the lack of field isolation (FOX) between the P+ areas.

4) Sample Preparation Techniques

Sample preparation is straightforward and a few guidelines will result in repeatable clean results. For topographical images:

1. Use high selectivity etches such as KOH or CsOH to insure an endpoint on the point of interest. Other etches include 49% HF, or metal etchants. Etches which are not highly selective such as a poly etch containing nitric and HF will not leave crisp definition associated with silicon to oxide boundary and will, in many cases, remove too much gate oxide. Mechanical surface lapping techniques allow layers to be partially exposed and then removed using a selective etch. In order to expose the gate oxide of a failing transistor to examine the gate oxide integrity topographically, the following procedure is employed:

A. Surface polish the part using 1 micron diamond mylar lapping film or if planarity needs to be maintained use a napless cloth with colloidal silica such as chem-pol on an orbital polisher. Once the oxides over poly 2 are eliminated, use HCl to remove the silicide then heated KOH to remove the exposed poly surface. The etch will endpoint on the gate oxide with a 200:1 selectivity and remove any silicon radially under the defect. High oxide sidewalls will be seen which may interfere with imaging in the AFM due to tip convolution. The sample can at this point be stripped of oxide using HF and the substrate or poly 1 cells imaged as desired. Surface lapping eliminates etch undercut and artifacts common with traditional delayering and is useful for AFM as well as SEM imaging.

2. Preparation for capacitive probe requires that the surface be non-contaminated as is true with any surface science technique. Avoid finger oils or other contaminants as capacitive probe is capable of detecting into the attofarad level with a spatial resolution of 10-20 nm. Polished or lapped surfaces must be smooth in order to avoid local trapping of ionic contaminants. Several methods may be used as follows:

A. Use a glass wheel with water. This method will result in a differential surface height due to differing material hardness of the metal and silicide layers but will yield good SCM results.

B. Use diamond mylar films down to .1 um and finish with colloidal silica and a polishing cloth such as Final B to buff out the damage areas at the surface due to the diamond mylar film. Do not attempt SCM without the final polish or all that will be seen is the damage from the diamond. This method is preferred due to the improved planarity of the surface coupled with good SCM results.

5) Conclusions

The theory and practical application of the AFM has been presented with various modes of operation. Each year new modes of operation are added to the SPM which enhances its capability and allows us to routinely obtain data such as junction profiling and Fowler Nordheim tunneling maps of the surface with nanometer precision that just a few years ago would have been deemed nothing more than science fiction.

6) Additional Sources of Information

J.Ebel et al. "Cross-sectional Atomic Force Microscopy of Focused Ion Beam Milled Devices", International Reliability Physics Symposium Proceedings 1998. IEEE. pp. 157-162.

G.M. Fiege et al "Temperature Profiling with Highest Spatial and Temperature Resolution by Means of Scanning Thermal Microscopy (SThM)", ISTFA 1997 pp. 51-56.

R.M. Cramer et al "The use of Near-Field Scanning Optical Microscopy for Failure Analysis of ULSI Circuits", ISTFA 1996 pp. 19-26.

H. Yamashita, Y. Hata, "Grains Observation Using FIB Anisotropic Etch Followed by AFM Imaging", ISTFA 1996 pp. 89-94.

B. Ebersberger et al "Thickness Mapping of Thin Dielectrics with Emission Microscopy and Conductive Atomic Force Microscopy for Assessment of Dielectric Reliability", International Reliability Physics Symposium Proceedings 1996. IEEE. pp. 126-130

T. Hochwitz, et al "DRAM Failure Analysis with the Force-Based Scanning Kelvin Probe", International Reliability Physics Symposium Proceedings 1995. IEEE. pp. 217-222.

M. Masters et al "Qualitative Kelvin Probing for Diffusion Profiling", ISTFA 1995 pp. 9-14.

7) References

1.) A. Majumdar, Tutorial on scanning thermal microscopy, 1998 Tutorial Notes, International Reliability Physics Symposium, pp. 2b 1-16 (March 1998).

2.) L. Ghislain and V. Elings, Near-field scanning solid immersion microscope, Appl. Phys. Lett. **72**, 2779-2781 (1998).

3.) A. Erickson, D. Adderton, Y. Strausser, and R. Tench, Scanning capacitance microscopy for carrier profiling in semiconductors, Solid State Technology **40**, 125-133 (June 1997).

4.) P. Hansen, Y. Strausser, A. Erickson, E. Tarsa, P. Kozodoy, E. Brazel, J. Ibbotson, U. Mishra, V. Narayanamurti, S. DenBaars, and J. Speck, Scanning capacitance microscopy imaging of threading dislocations in GaN films grown on (0001) sapphire by metalorganic chemical vapor deposition, Appl. Phys. Lett. **72**, 2247-2249 (1998).

5.) Q. Zhong, D. Inniss, K. Kjoller, and V. Elings, Fractured polymer/silica fiber surface studied by tapping mode atomic force microscopy, Surf. Sci. Lett. **290**, L688-L692 (1993).

6.) S. Magonov, V. Elings, and M.-H. Whangbo, Phase imaging and stiffness in tapping-mode atomic force microscopy, Surf. Sci. Lett. **375**, L385-L391 (1997).

7.) K Babcock, M. Dugas, S. Manalis, and V. Elings, Magnetic force microscopy: recent advances and applications, Materials Research Society Symposium Proceedings **355**, 311-322 (1995).

8.) K. Luo, Z. Shi, J. Lai, and A. Majumder, Nanofabrication of sensors on cantilever probe tips for scanning multiprobe microscopy, Appl. Phys. Lett. **68**, 325-327 (1996).

9.) Yongxia Zhang, Yanwei Zhang, J. Blaser, T. S. Sriram, A. Enver, and R. B. Marcus, A thermal microprobe fabricated with wafer-stage processing, Rev. Sci. Instr. **69**, 2081-2084 (1998).

10.) A. Olbrich, "Nanoscale Electrical Characterization of Thin Oxides with Conducting Atomic Force Microscopy", International Reliability Physics Symposium Proceedings 1998. IEEE. pp. 163-168.

11.) Jim Colvin, "Advanced Methods for Imaging Gate Oxide Defects with the Atomic Force Microscope", Proceedings of the International Symposium for Testing and Failure Analysis, pp 271-276, 1992.

12.) Jim Colvin, "The Identification and Analysis of Latent ESD Damage on CMOS Input Gates", 1993 EOS/ESD Symposium Proceedings, EOS-15, pp 109-116.

13.) IBID, pp 113-115.

14.) Ann N. Campbell, Edward I. Cole Jr., Bruce A. Dodd, and Richard E. Anderson, "Internal Current Probing of Integrated Circuits Using Magnetic Force Microscopy", International Reliability Physics Symposium Proceedings 1993. IEEE. pp. 168-177.

15.) IBID, p. 168

16.) IBID, p. 168.

17.) IBID, p. 169.

18.) IBID, pp. 170-171.

Instrumental Analysis Techniques

Keenan Evans and Thomas A. Anderson
Motorola Inc.
Phoenix, Arizona

INTRODUCTION

In many instances it may be necessary to include detailed chemical characterization of a sample into the failure analysis report. The failure analyst may be required to gather this data himself or it may be necessary to send the sample to an outside laboratory for characterization by a variety of sophisticated instrumental analytical techniques. The following summary of ten common analytical methods is designed to provide both the analyst and the report recipient with a basic understanding of the various methods. The summaries are presented in outline form with a description of the Theory of Operation and Typical Applications provided for each of the techniques listed below.

A. Auger Electron Spectroscopy, AES

B. Scanning Electron Microscopy, SEM

C. Secondary Ion Mass Spectrometry, SIMS

D. Inductively Coupled Plasma-Atomic Emission Spectroscopy, ICP-AES

E. Residual Gas Analysis-Mass Spectrometry, RGA-MS

F. X-Ray Fluorescence, XRF

G. Gas Chromatography/Mass Spectrometry, GC/MS

H. Ion Chromatography, IC

I. High Pressure Liquid Chromatography, HPLC

J. Fourier Transform InfraRed Spectroscopy, FTIR

Instrumental Analytical Techniques

Auger Electron Spectroscopy, AES

Theory of Operation. Atomic core ionization of nuclides within the sample occurs as a result of interactions with the electron beam. As a result of this inner core ionization, an electron from a higher energy level (shell) within the ionized atom drops down in energy to fill the vacancy. The quantized energy difference between the two levels is then given up via either X-ray photon emission or via the radiationless Auger process yielding the *Auger electron*. (See Fig. 1.) Energy analysis of the emitted Auger electron provides elemental identification, and counting of the relative numbers of Auger electrons at the various energies provides semi-quantitative information. The technique is surface sensitive due to the limited escape depth of the Auger electrons, typically on the order of 3-30 Angstroms.

To provide information about subsurface composition and allow depth profiling capabilities, the surface is removed via in-situ ion milling (sputter etching) with an argon ion beam. Scanning or rastering of the electron probe beam also allows for elemental mapping capabilities. An ultra-high vacuum (10^{-10} Torr) in the analytical chamber is necessary in order to prevent re-adsorption of sputtered species or permanent gases onto the freshly sputtered sample surface.

Typical Applications. Auger analysis is used to characterize surface contaminants which may inhibit bondability and solderability, contribute to surface leakage or constitute visual rejects. It is also used for characterization of multi-layer structures to monitor the degree of interlayer blend or detect contaminants at various interfaces. Comparative analyses of "good" vs "bad" samples are common. The sample must be sized to a maximum of approximately 1 cm × 1 cm. Samples should not be coated. Due to charge build up at the surface from electron beam bombardment, analysis of thick insulators is limited. Auger electron spectroscopy should be considered a destructive analysis technique. Figures 2 & 3 illustrate a typical Auger survey spectrum and a depth profile.

Scanning Electron Microscopy, SEM

Theory of Operation. Under high vacuum (approximately 10^{-6} Torr), a 1 to 30 keV primary electron beam is rastered over the sample surface, creating secondary electrons. These are extracted from the sample and imaged to create a high resolution, high depth of field, secondary electron image at magnifications to 250k ×. (See Fig. 4.) Interactions of the sample atoms with the primary electron beam also result in inner core ionization of the atoms with the subsequent emission of quantized X-ray photon. Wavelength or energy analysis of the emitted X-rays provides qualitative elemental information and the relative intensity of these X-rays is used for quantitative purposes. In semiconductor materials, electron beam bombardment also creates electron-hole pairs, resulting in an electron beam induced current (EBIC), which can be imaged to correlate with

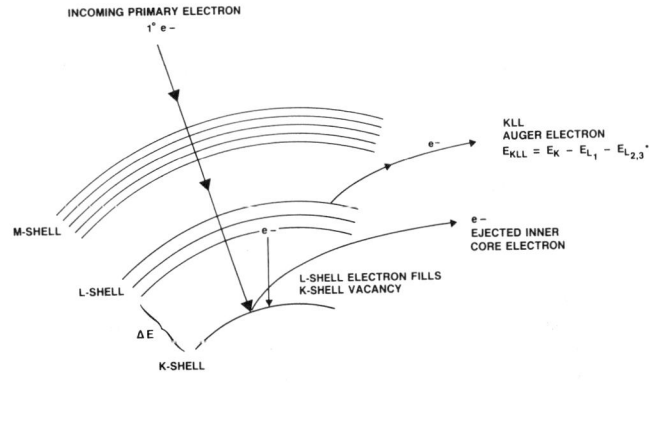

Fig. 1. Schematic representation of the Auger process.

AES PROFILE

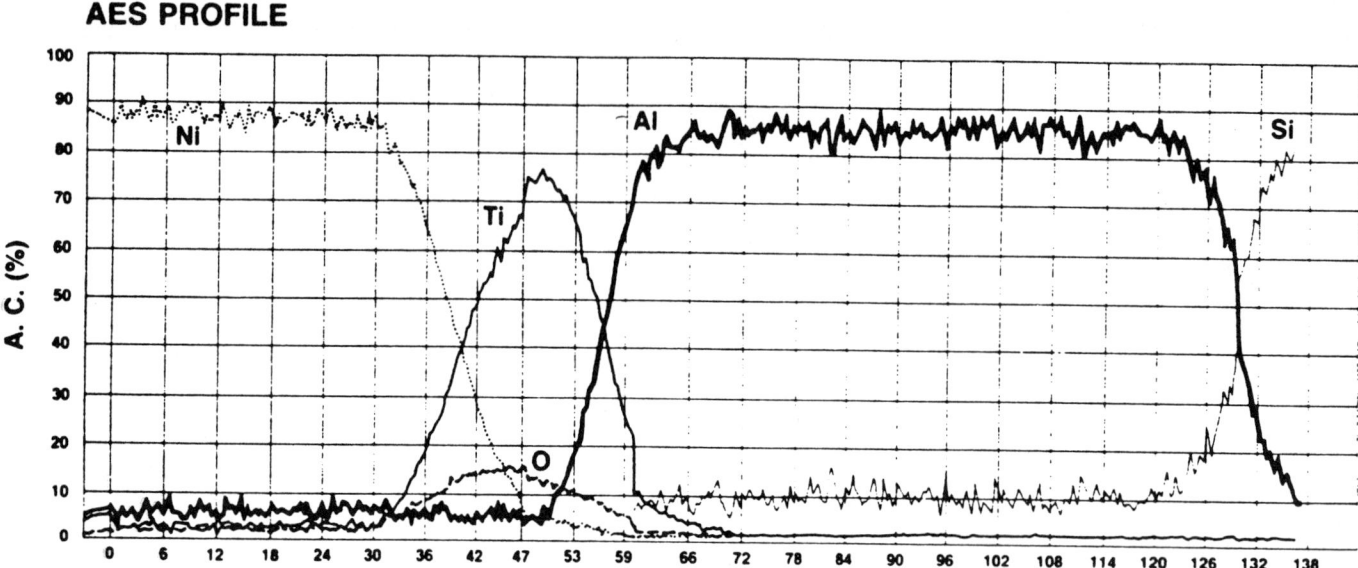

Fig. 2. Auger sputter depth profile through backmetal system of Si chip. Care should be used in the interpretation of A.C.% data since some quantification routines can result in false indication of elements at low AT%. Interface resolution can be improved by rotating the sample during ion milling. The determination of layer thickness from sputter time requires accurate sputter rates for each layer under identical experimental conditions. A surface profilometer may be used in conjunction with selective layer removal to measure true layer thickness.

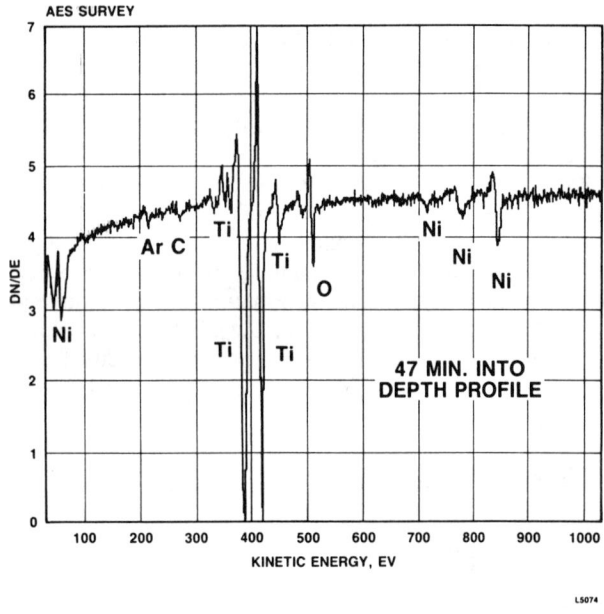

Fig. 3. Auger survey spectrum taken midway through a depth profile.

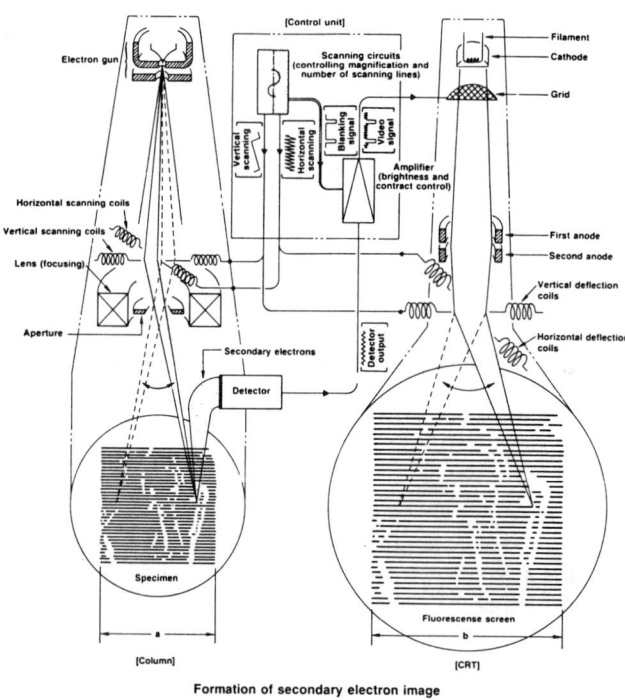

Formation of secondary electron image

Fig. 4. Schematic of a typical scanning electron microscope.

other sample features. Various stains and sample preparation techniques are utilized to enhance contrast between regions of interest within the sample. X-ray emission from the sample and primary electron beam penetration into the sample typically range from depths on the order of 0.5-2.0 μm, varying as a function of electron beam energy and sample density. Figure 5 gives range of penetration for typical electron beam energies in Si and GaAs, and Fig. 6 illustrates depths of quanta generation.

Typical Applications. SEMs are used in the characterization of microstructural surface topography, step coverage, grain size, oxide slope, construction parameters, sample composition, contamination, structural defects, bonding defects, intermetallic formation and degree of wire bond deformation. Digital X-ray maps of up to eight elements may be

344

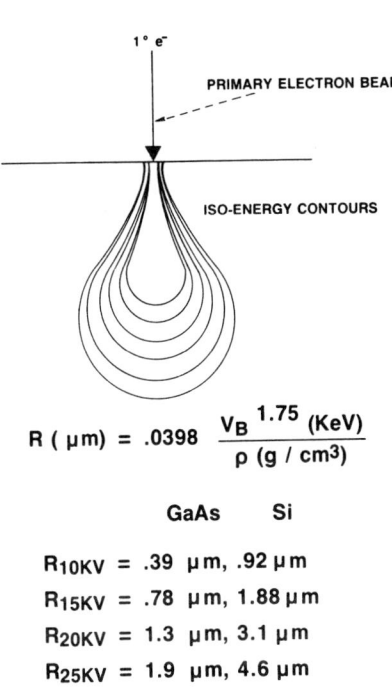

$$R\ (\mu m)\ =\ .0398\ \frac{V_B^{\ 1.75}\ (KeV)}{\rho\ (g\ /\ cm^3)}$$

	GaAs	Si
R_{10KV} =	.39 μm,	.92 μm
R_{15KV} =	.78 μm,	1.88 μm
R_{20KV} =	1.3 μm,	3.1 μm
R_{25KV} =	1.9 μm,	4.6 μm

Fig. 5. Electron beam ranges in Si and GaAs. (Drawing adapted from Goldstein 1, 2, 3).

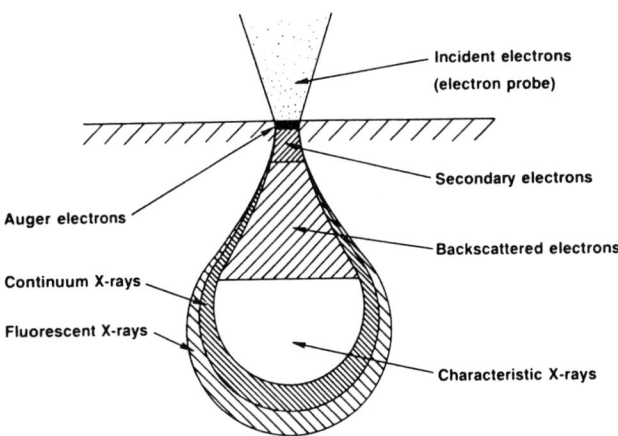

Fig. 6. Analytical depth of electron beam generated signals. (Drawing adapted from Goldstein 1, 2, 3).

generated simultaneously to document the geometrical distribution of elemental composition or contamination.

Secondary Ion Mass Spectrometry, SIMS

Theory of Operation. The basis of ion microprobe SIMS is the use of a focused primary ion beam to erode atoms from a selected region of a sample surface.

A portion of the eroded atoms undergo a charge exchange in the near surface environment, resulting in their conversion to negative or positive ions. These secondary ions are then extracted via an electrical potential and subsequently analyzed by a mass spectrometer. SIMS thus yields an analytical technique capable of elemental specificity, with detection limits in the range of ppm-ppb (atomic concentration). Figure 7 illustrates the ion beam-sample interactions and Fig. 8 displays a typical instrumental arrangement of the primary ion source and the mass analyzer. By comparison of relative ion yields in sample depth profiles to those of ion implanted reference samples in the same matrix, an approximate quantification is possible. The current density of the primary ion beam may be varied over several orders of magnitude to produce sample milling rates ranging from a few Å/min to several thousand Å/min, making the technique a powerful surface analysis, bulk analysis, thin film characterization and depth profiling tool. Minimum ion beam spot sizes are on the order of 2-5μm, thus limiting the lateral imaging resolution achievable with the primary ion sources to similar dimensions. However, SIMS is not useful for small areas. Data must be taken over much larger plastered areas in order to obtain any reasonable detection limit. Some instruments

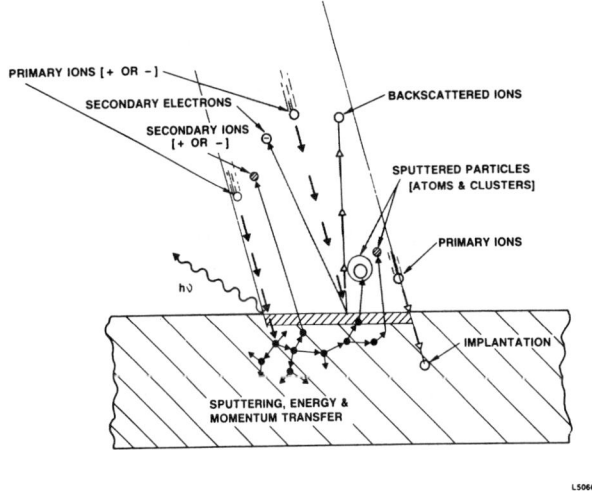

Fig. 7. Ion Beam / Sample interactions.

can analyze small areas for larger concentrations, such as in particle analysis, but dopant concentrations cannot be measured from an area of the same order of dimension as the spot site.

Typical Applications. SIMS is used to evaluate dopant profiling and trace contamination of surfaces, thin films, thick films, multilayer structures, and interfaces. It is used to measure relative levels of incorporated impurities or component elements as a function of processing parameters, correlation of trace ionic contaminant levels to electrical leakages, and mapping of impurities for correlation to device geometrical features. Samples must be compatible with the ultra-high vacuum

SIMS

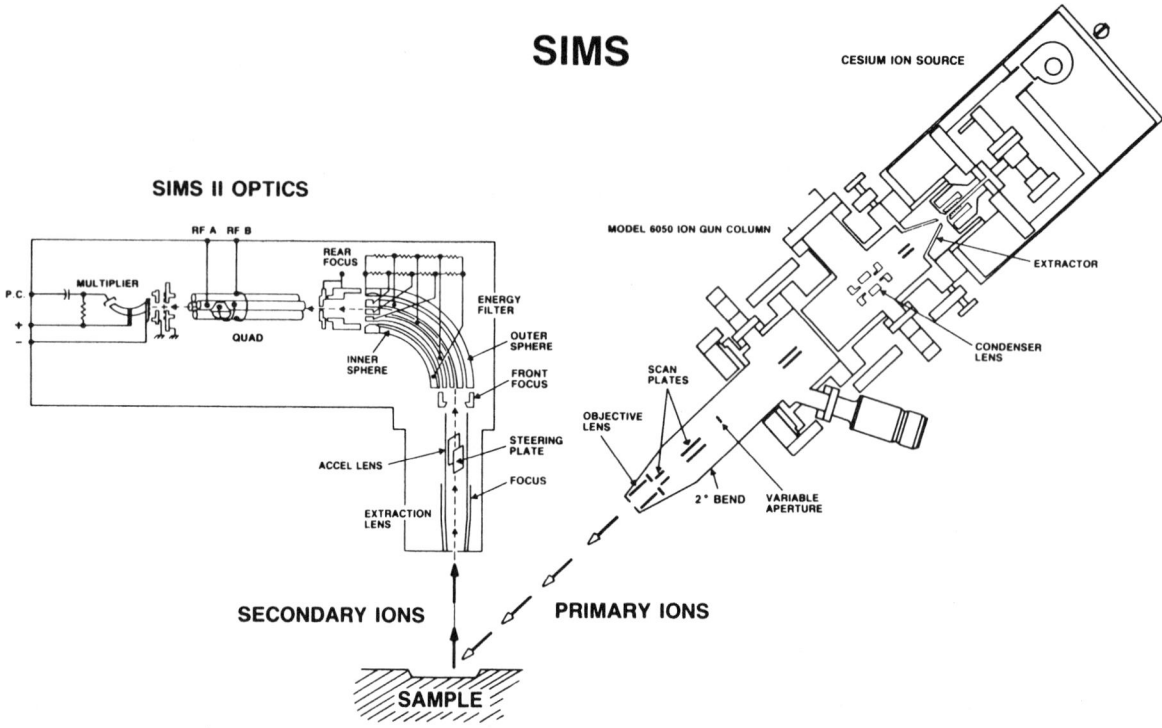

SIMS II OPTICS

Fig. 8. Typical arrangement of a quadruple ion microprobe.

(approximately 10^{-10} Torr) of the SIMS analytical chamber that is required to prevent reabsorption of sputtered species and permanent gases onto freshly ion milled sample surfaces.

Inductively Coupled Plasma-Atomic Emission Spectroscopy, ICP-AES

Theory of Operation. Three concentric quartz tubes are centered in the water cooled coils of an RF generator. An argon gas flow is introduced into the tubes and initially made electrically conductive by Tesla sparks. The magnetic field of the RF coil is axially oriented in the quartz tubes thus inducing an annular flow of electric current in the argon plasma, as shown in Fig. 9. This produces temperatures in the range of 5,000-10,000 Kelvin as a result of ohmic heating. Proper regulation of the argon gas flows allows a stable annular plasma to be maintained above the ends of the concentric tubes. A liquid sample is aspirated into the center tube, and it flows into the plasma. The complete plasma torch configuration is shown in Fig. 10. Solvent evaporates prior to reaching the actual plasma and atomization of the elements present in the sample solution occurs in the extremely high temperature of the plasma. Thermal excitation of valence electrons to higher allowed energy states takes place. When relaxation to the ground state occurs the energy difference between the excited state and the ground state is emitted as a quantized photon. The photon wavelengths emitted are commonly visible in the ultra-violet portion of the electromagnetic spectrum (UV-VIS). By separating the various emission wave-

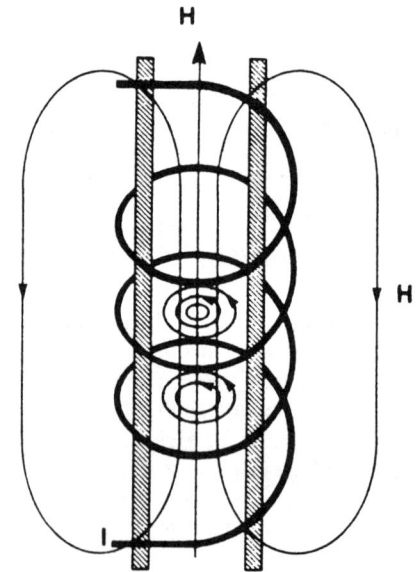

Fig. 9. Magnetic and electric field lines in a plasma torch.

lengths with a monochromator, qualitative information is gathered. By monitoring the relative intensity at a particular wavelength and comparing this intensity to that generated by introduction of a known external standard into the plasma, quantitation is achieved. Detection limits in solution are on the order of ppb for most elements. A typical instrumental arrangement is shown in Fig. 11.

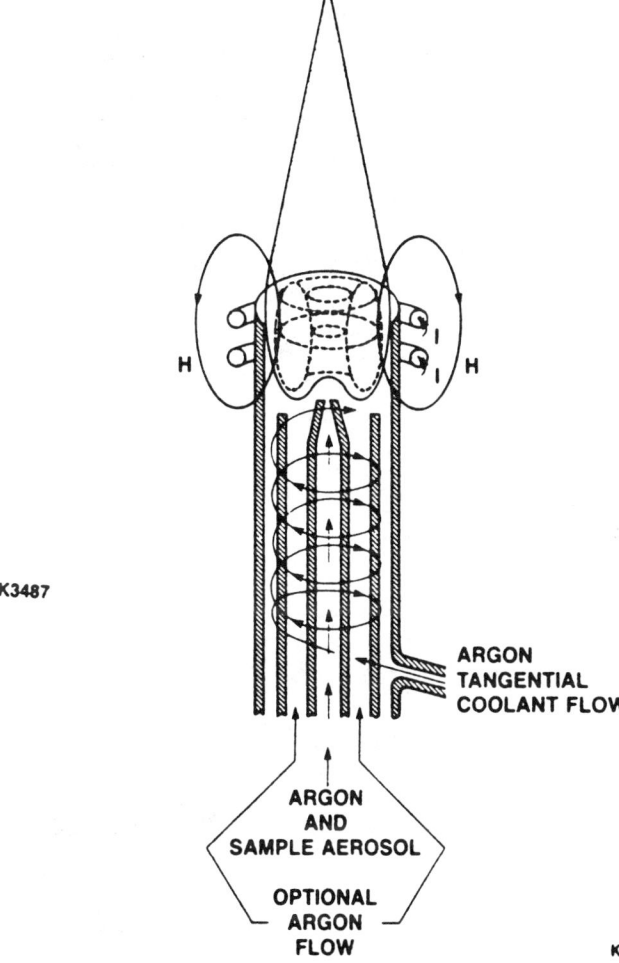

K3487

ARGON TANGENTIAL COOLANT FLOW

ARGON AND SAMPLE AEROSOL

OPTIONAL ARGON FLOW

K34

Fig. 10. Complete plasma configuration for an ICP-AES system.

Typical Applications. Trace metals in process solutions and waters, and concentration of major, minor and trace elements in alloys, thin films, solders, residues or other samples which can be digested in an appropriate solvent or mineral acid, are analyzed by this method.

Residual Gas Analysis-Mass Spectrometry, RGA-MS

Theory of Operation. Hermetically sealed packages are loaded into a sample chamber which is then evacuated to a base pressure of approximately 10^{-7} mbar. The samples are allowed to reach thermal equilibrium in the heated chamber. Then, a single sample is rotated into alignment with the package piercing assembly. A seal is made between the package lid and the vacuum system of the mass spectrometer. After an acceptable mass spectrometer background is reached, the package is pierced and the internal atmosphere of the hermetic device is sampled into the mass spectrometer for analysis. Up to 16 gas species which may have been trapped inside the package at sealing may be analyzed. Moisture levels are the primary concern. For quantitation of internal moisture levels, relative intensities of the mass

THE COMBINATION SYSTEM

A. ARGON GAS AT 12 TO 18 LITERS PER MIN.
B. SAMPLE, APPROX. 1ML/M
C. OPTIONAL TEFLON NEBULIZER SYSTEMS FOR HF SOLUTIONS AND SLURRIES
D. PLASMA 2.5 KW
E. THE FUNCTIONAL PATH PROCEEDS AS SHOWN ABOVE EITHER THROUGH THE SEQUENTIAL, THROUGH THE SIMULTANEOUS OR THROUGH BOTH SYSTEMS.

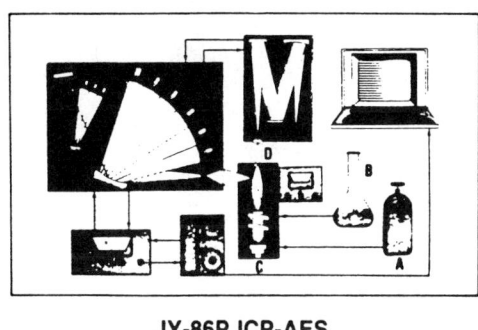

JY-86P ICP-AES COMBINATION SYSTEM

K3484

Fig. 11. Component layout for a dual detection ICP-AES system.

18 ion current resulting from the sample are compared to those of known concentrations of moisture which have been previously generated through an NBS traceable moisture hygrometer, moisture generator and four volume calibrator.

Typical Applications. RGA is used for process monitoring of internal package atmospheres to ensure minimum internal moisture levels. It is also used for tank gas analysis. RGA may be used as a measurement device for optimization of die coat processing parameters to minimize internal package moisture, oxygen and hydrocarbon levels.

X-ray Fluorescence, XRF

Theory of Operation. In an X-ray tube, electrons are generated from a cathode and accelerated at approximately 5-50 keV to impinge on the target anode thus generating broad band X-ray emission from the deceleration of the electrons (Bremsstrahlung) in the anode target material. Characteristic fluorescent X-rays also are generated from the target material due to atomic core ionization of atoms. The X-rays from the primary source are then directed on to the sample, causing atomic core ionization of the atoms. As a result of this inner core ionization, an electron from a higher energy level (shell) within the ionized nuclide drops down in energy to fill the vacancy. The quantized energy difference between the two levels is then given up via X-ray photon emission. Energy analysis of the secondary X-ray emission (or fluorescence) yields qualitative information about the elemental composition of the sample. Relative intensity of the X-ray fluorescence is used to quantify the amounts of the various species present in the sample. Detection limits are of the order of a few hundred ppm, and all elements above atomic number 11 (Na) may be detected with typical energy dispersive systems. Some EDXRF systems may also incorporate an optional secondary target into the system to allow sample excitation by the characteristic X-rays of any of the available secondary targets, such an arrangement is schematically depicted in Fig. 12.

347

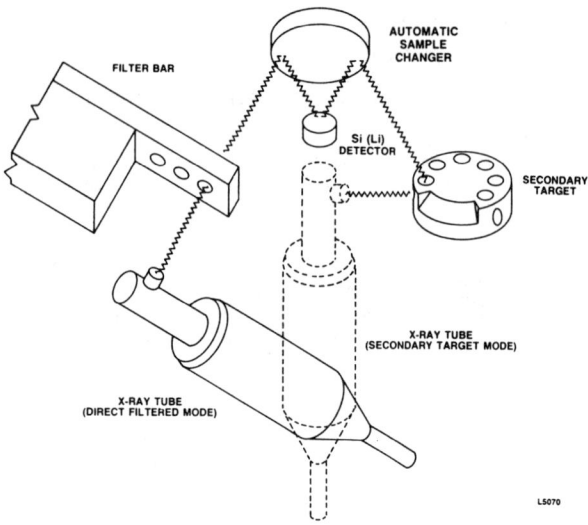

Fig. 12. Optional direct/secondary excitation in EDXRF.

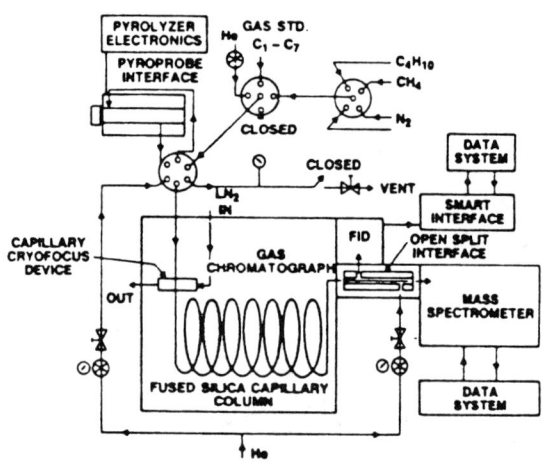

Fig. 13. Pyrolysis GC/MS Schematic. (See Reference 5)

Typical Applications. XRF is employed for qualitative elemental screening of unknown samples, often for subsequent characterization by other methods. Process control analyses of % Phosphorus in P-glass, thicknesses of multilayer backmetal systems, composition ratios of binary thin films, plating thicknesses, and solder bump thicknesses can all be performed with XRF.

Gas Chromatography-Mass Spectrometry, GC/MS

Theory of Operation. A solid, liquid, or gas sample is thermally vaporized, transported in a He carrier gas, and cryofocussed onto the head of a capillary chromatographic column. A capillary column GC/MS, equipped with a capillary pyroprobe and a parallel detection system is shown in Fig. 13. The various species in the sample matrix have different relative affinities for the coating of the capillary column and the carrier gas. These properties result in a partitioning and a time separation of the various sample components as they pass through the chromatographic column, and exit to the detectors. Upon exiting the capillary column, the carrier gas containing the separated sample components is split into the flame ionization and mass spectrometric detectors. As the samples pass through the hydrogen flame ionization detector, they are ionized and collected at a cathode, resulting in a detector current. The magnitude of this detector current is monitored to be used for quantitative purposes by comparison to the currents generated by an external standard of known concentration. The portion of carrier gas that is passed through the mass spectrometer is ionized via 70 eV electrons, resulting in ion fragments that are mass analyzed by the quadrupole mass spectrometer. The fragmentation patterns are characteristic of a particular molecular species, and are compared to those of known standards for qualitative identification of the various components of the sample matrix.

Typical Applications. GC-MS can be used for the identification of molding compounds by comparing the pyrolysis-chromatography-mass spectrometry fragmentation patterns or "fingerprints" to those of known compounds. Assessment of

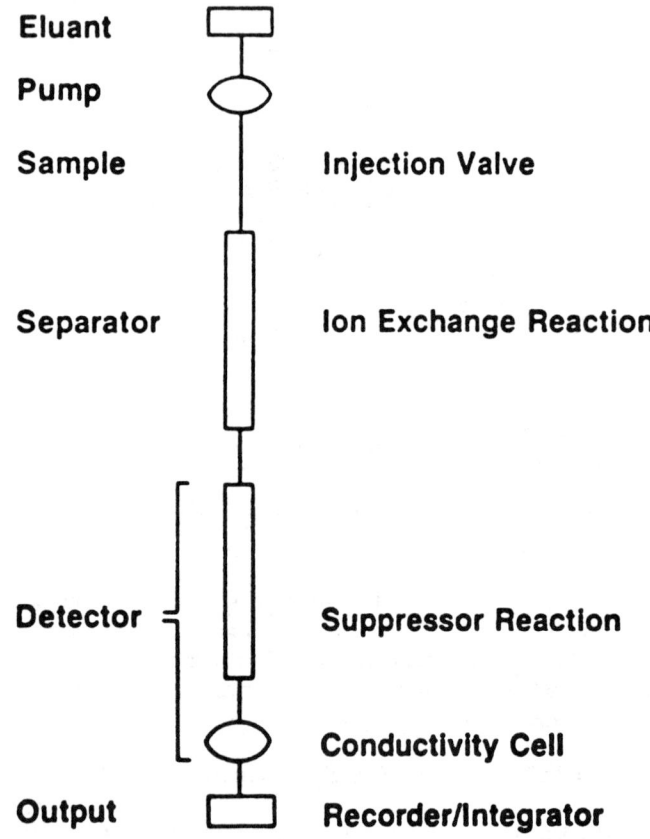

Eluant	
Pump	
Sample	**Injection Valve**
Separator	**Ion Exchange Reaction**
Detector	**Suppressor Reaction**
	Conductivity Cell
Output	**Recorder/Integrator**

Fig. 14. Components of an ion chromatograph.

"degree of cure" of die coat material by monitoring residual solvents and pyrolytic fragmentation patterns and comparing to "overcured" material can be performed. Optimization of curing process parameters, measurement of solvent contamination levels, polymer and co-polymer identification and relative ratios, and identification of organic contaminants on wafers and devices can be performed.

348

Ion Chromatography, IC

Theory of Operation. A known volume of aqueous sample solution is injected onto the head of a chromatography column packed with a pellicular ion exchange resin. Basic components of an ion chromatography system are illustrated in Fig. 14. A dilute acidic or basic eluant is pumped across the column and the various anions or cations of the sample are eluted off the column and passed through the system detectors. As the different ionic species have different relative affinities for the ion exchange resin, they spend more or less time in the mobile eluant phase depending on whether they have a lesser or greater relative affinity for the exchange resin. Thus, the ionic species exit the separation column at different times. The eluant carrying the various ionic species present in the sample then passes through a suppressor column or membrane where a second ion exchange process takes place, which results in the anions or cations of the eluant being exchanged for either OH^- or H^+, respectively. The separated ionic species of the sample are then passed through a flow-through conductivity detector in a background of de-ionized water. The relative magnitude of the resultant conductivity current is utilized for quantitative purposes and the elution time yields qualitative information. Detection limits in solution are typically in the ppm-ppb range and can be extended further with pre-concentration techniques. Ammonium, alkali, and alkaline earth cations can be determined in a single chromatographic run on the cation side, and all the common strong acid anions can be determined in a single run on the anion side.

Typical Applications. IC is better suited to process control analysis than to failure analysis. GC can measure trace ionic contamination in plant, process and rinse waters, major, minor and trace ions in plating baths and etch solutions, extractable ionic constituents of powders, encapsulation compounds, glasses, and devices via sample preparation by overnight reflux in ultra-pure de-ionized water.

High Pressure Liquid Chromatography, HPLC

Theory of Operation. A liquid sample solution or extract is injected in a known volume onto the head of the appropriate chromatographic column and transported under high pressure through the column by a liquid carrier solvent (or solvent mixture), termed the eluant. Due to various attractive forces or physical interactions with the coating or packing material of the chromatographic column, the various components of the injected sample will pass through the column at different rates, exiting at characteristic times for a particular species. The eluted species are then passed through either a refractive index detector (mg/L-μg/L) or an ultraviolet-visible photometric detector (μg/L-ng/L), or both. Qualitative and quantitative information for each of the sample components are determined through the use of external standards. In the gel permeation mode, components of the sample solution are separated on the basis of size by passing a column train of various pore sizes. The smaller (or lower molecular weight) components are able to penetrate into the smallest pores or voids of the column packing material and thus have a longer effective route through the

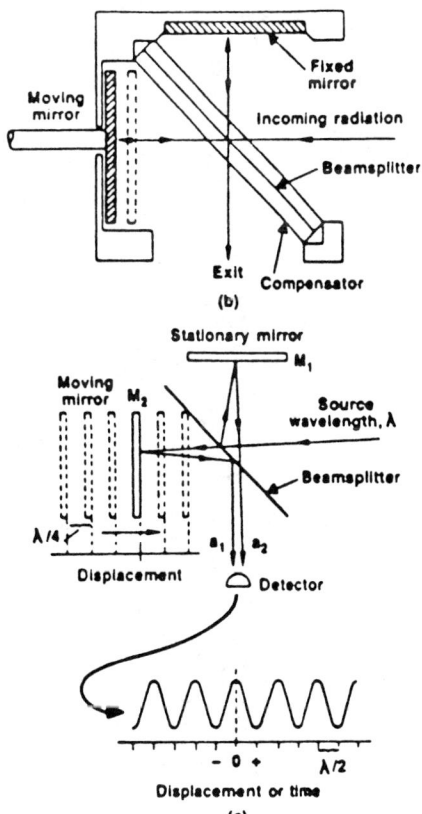

Fig. 15. Components of an FTIR system.

349

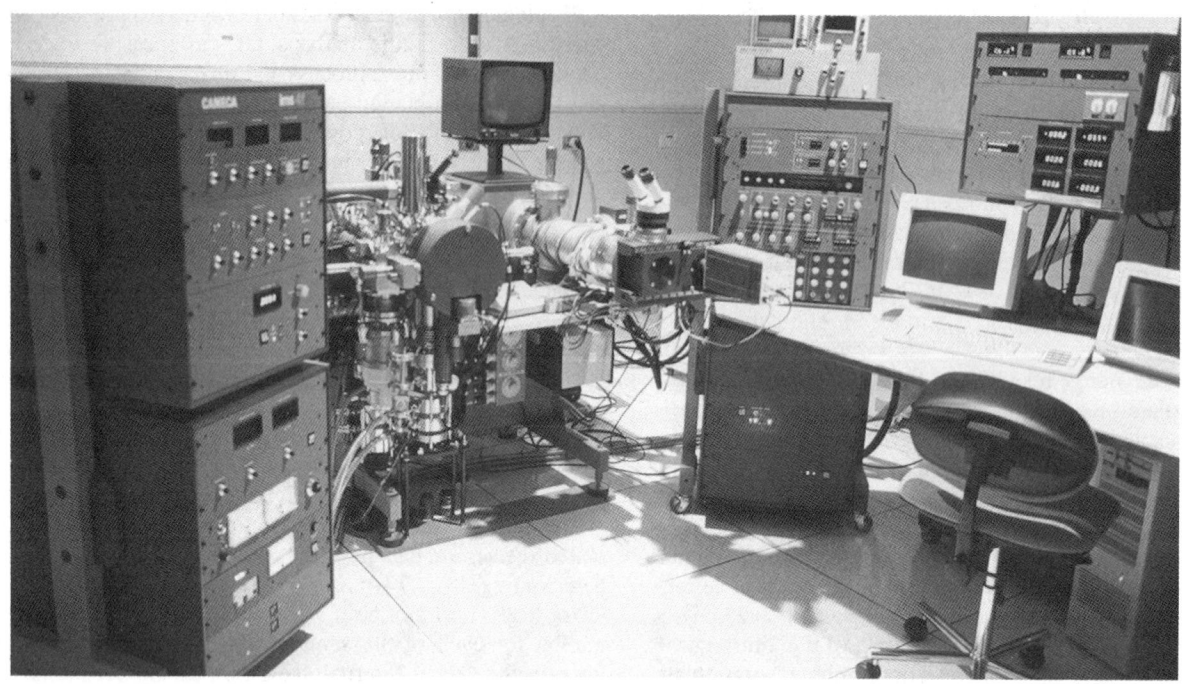

Fig. 16 Example of a SIMS system, which uses a magnetic sector to achieve a high mass resolution and a low energy spread for the identification of elements and compounds which have mass interferences, or overlaps. Examples are TiSi vs AsSi, and P vs SiH. Analysis time on this machine is expensive, and must be well justified. It costs over $1 million.

Fig. 17 Example of an ICP-MS sytsem, which uses a mass-spectrometer to analyze the mass of the atomic species from a plasma-volatilized sample. By contrast, an ICP-AES machine uses a diffraction grating to analyze the light emitted from the plasma flame containing the volatilized species. This machine is used primarily for analysis of the purity of process chemicals and deposited thin films.

column, exiting at later times than the larger (or higher molecular weight) components. Average molecular weight, and molecular weight distributions, are determined by the comparison of elution characteristics of the sample to those of a series of high purity polymer standards, usually polystyrene.

Typical Applications. HPLC is a process control tool suited for measurement of extractable organic components of encapsulation compounds and packaging materials. Component ratios of solvent mixtures and average molecular weight and molecular weight distributions of die coat materials can be evaluated.

Fourier Transform Infra-Red Spectroscopy, FTIR

Theory of Operation. The component atoms of polyatomic molecular groups are in constant motion with respect to each other, constantly changing between the molecular ground state and quantum mechanically allowed excited states due to thermal excitation. The twisting, bending, rotating and stretching motions of the atoms within a molecule occur at frequencies that are generally in the infra-red (IR) portion of the electromagnetic spectrum. (the wavelength, of this IR, lambda, is 0.7-500 μm). When infra-red light of energy coincident with the difference between the ground state and excited state of a particular molecular vibration is radiated onto a sample , the vibrations are stimulated, and the light of that particular wavelength is absorbed by the molecule (IR absorbance). By monitoring the absorbance of the various wavelengths or frequencies of an infra-red light as it is transmitted through or reflected from (IR reflectance) a sample, a characteristic infra-red spectrum can be obtained. In Fourier transform infra-red spectroscopy, the source is modulated with a Michelson interferometer. This results in a signal at the detector which has a distribution of frequencies determined by the speed of the moving mirror of the interferometer. The resulting interferogram in the amplitude-time domain is then transformed via the Fourier algorithm into the appropriate amplitude-wavelength domain, resulting in a characteristic infra-red spectrum for a particular compound. Typical instrumental components are depicted in Fig. 15.

Typical Applications. During failure analysis, FTIR can be used for the identification of extractable organic components from a device or wafer surface, and the identification of organic contaminants, including polymers, on surfaces of an appropriately reflective substrate. FTIR can identify unknown solid organic compounds, powders or residues and solvent or liquid organic sample identification.

Sample Requirements

Sample size, shape and physical state restrictions apply to all of the methods outlined above. For specific requirements of a particular technique it is best to consult with the appropriate analyst prior to sample submission. Generally, analyses performed to determine the reasons for a particular type of electrical device failure are best performed on a comparative basis. It is desirable to perform a side-by-side analysis of a "good" part and a "failed or rejected" part in order to distinguish the chemical, surface or physical sample features that differentiate them. As noted in the theory and applications sections describing the available instrumentation, many of the techniques are sensitive to elemental concentrations in the ppm to ppb range and/or to surface regimes on the order of 1-10 atomic layers. It is therefore imperative that samples to be analyzed are handled in an appropriate manner to avoid the introduction of extraneous contaminants prior to submission for analysis.

Special Note: Some of the equipment schematics in this article may be wholly, or in part, adapted from drawings contained in the operators manuals or descriptive brochures published by the manufacturers. These manufacturers include: Perkin-Elmer, Jobin-Yvon, JEOL, and Bio-Rad.

REFERENCE

1. *Metallography—A Practical Tool for Correlating the Structure and Properties of Materials*, by J.I. Goldstein, ASTM Special Technical Publication 55T, ASTM, 1974, pg 86.

2. *Scanning Electron Microscopy and X-ray Microanalysis*, by J.I. Goldstein, D.E. Newbury, P. Echlin, D.C. Joy, C. Fiori, and E. Lifshin, Plenum Press, New York, 1984.

3. *Monte Carlo Calculations on Electron Scattering in a Solid Target*, by K. Murata, T. Matsukawa, and R. Shimizu, Japan Journal of Applied Physics, 10, June, 1971, pp 678-686.

4. *Materials Characterization by Analytical Electron Microscopy*, by John E. Porter, Proceedings of ISTFA-84, pp 46-52 (see Fig. 2)

5. *Characterization of Molding Compounds Via Polymer Reconstruction Investigative Chromatopyrography*, by S.P. Rogers, K.L. Evans, and M.L. Parsons, ISTFA-84, pp 98-107 (see Fig. 2)

6. *Standard Practice for Reporting Data in Auger Electron Spectroscopy*, ASTM standard E 996-89, available from American Society for Testing and Materials, 1916 Race Street, Philadelphia, PA, 19103.

7. *Standard Practice for Reporting Sputter Depth Profile Data in Secondary Ion Mass Spectrometry (SIMS)*, ASTM Standard E-1162-87, available from American Society for Testing and Materials, 1916 Race Street, Philadelphia, PA, 19103.

8. *Problem-Solving Surface Analysis Techniques*, by Robert D. Cormia, Advanced Materials and Processes magazine, December, 1992, pp 16-23.

9. *Instrumental Methods of Analysis*, by H.H. Willard, L.L. Merritt, J.A. Dean, and F.A. Settle, D. Van Nostrand, 1981.

Microbeam Analysis and Semiconductor Reliability

Robert K. Lowry
Harris Semiconductor
Melbourne, Florida

1. INTRODUCTION

Microbeam analysis is an essential tool for reliability scientists and failure analysts. Electrical measurements are not always sufficient to fully understand performance of electronic devices. Indeed, every electronic malfunction has an underlying physicochemical cause. It is often necessary to supplement electrical parametric measurements with electron beam and/or ion beam imaging and analysis of the die or package under test, to fully understand causes of failure.

Microbeam analysis can be "micro" in two senses of the word. It can be micro in the z (height) dimension, in that ion and electron beams can give information about the uppermost few Angstroms of the sample. This can be true even with beams many millimeters in diameter. It can also be "micro" in the x,y dimension where finely-focused beams provide data from spatial areas fractions of a micron in diameter. Both types of "micro" beam analysis are important for electronic materials characterization. The former is important for broad areas of general film surface composition and/or cleanliness characterization. The latter is important for obtaining data from the closely-packed geometries of today's integrated circuits.

It is not the aim of this article to make surface scientists or microanalysis equipment experts out of failure analysis and reliability professionals. This article simply gives an overview of a few of the more common microanalysis techniques, and a potpourri of successful applications to reliability-oriented investigations.

The investigations discussed are chosen not just from failure analysis cases, but also from materials selection and assurance studies, studies to design and develop robust products and processes, and cooperative efforts to serve outside customers. These types of examples are included because reliability engineers

Table 1   Analytical Methods Overview

Analytical method	Incident particle, for sample excitation	Signal detected providing information about sample	Type of information (x,y) [z]	Elemental Sensitivity
Scanning electron microscopy (SEM)	electrons	secondary electrons emitted from sample atoms	visual image of topography	na
Energy dispersive and wavelength dispersive spectroscopy (EDS/WDS)	electrons	secondary and backscattered electrons and x-rays	chemical element identity, $(1\mu \times 1\mu)$ $[1\text{-}5\mu]$	0.1-1%
Auger electron spectroscopy (AES)	electrons	Auger electrons from near surface atoms	chemical element identity $(.1\mu \times .1\mu)$ [20A]	0.1-0.5%
X-ray photoelectron spectrometer (XPS)[1]	x-rays	photoelectrons from near surface sample atoms	chemical element identity & valence $(75\mu \times 75\mu)$ [20A]	0.1-0.5%
Secondary ion mass spectrometer (SIMS)	ions	secondary ions emitted from the sample	chemical element identity $(.2\mu \times .2\mu)$ [20A]	$\geq$ppm

[1] Also known as ESCA or Electron Spectroscopy for chemical analysis

Table 2   Applications of Microanalytical Methods

	SEM	EDS	AES	XPS	SIMS	TOPIC	DIE	PKG	IMPACT
1.					X	Qualify Product Handling Material	X	X	ESD; Ionic Contamination
2.	X	X	X			Soldering		X	Connection Reliability
3.			X			Seal Glass Contamination		X	Electrical Stability
4.			X		X	Die Cleaning	X	X	Molded Part Stability
5.	X	X				Human-Sourced Contamination	X	X	Corrosion; Ionic Contamination
6.	X		X			Bond Pad Quality	X		Wire Bond, Electrical Integrity
7.			X			Contact Resistance	X		Electrical Parameters
8.				X		Laser Processing	X		Electrical Parameters
9.					X	Device Ionic Contamination	X		Electrical Parameters

are not just failure analysts any more. They play an ever-increasing proactive role, often as part of TQM/continuous improvement activities, in design and analysis for failure prevention.

In these efforts, reliability scientists must often submit samples requiring microbeam analysis work to an in-house or an outside commercial analytical laboratory. The more background knowledge they have about strengths and limitations of various methods, the more fruitful will be the working relationship between themselves and analytical personnel, and the better the understanding they will have of the completed data.

Table 1 is a general overview of the analytical methods discussed in this article. The "type of information" column indicates information provided and the extent to which the analysis method is "micro", in terms of (x,y) and [z] dimensions from which the analysis information is obtained.

Table 2 lists the topics discussed in the following pages. The examples comprehend at least one application of the listed microanalytical methods. Examples were also chosen to represent both die level and package level types of reliability problems, characterized by applying the proper analytical method. Each application is discussed briefly in terms of the reliability hazard investigated and reasons for choosing the analytical method.

Table 4   Contaminant Effects

CONTAMINANTS	EFFECTS ON DIE	EFFECT ON PACKAGES
Positive Ionics e.g., Na+, K+, etc.	Electrical Anomalies: Inversion Drift Leakage Instability	Pin-to-pin Leakage
Negative Ionics e.g., CL⁻, F⁻, SO_4, etc.	Metallization Corrosion Bondability	Pin-to-pin Leakage Plating Defects Corrosion
Residuals, e.g., Hydrocarbons	Film Adhesion Contact Resistance Bondability	Solderability
Particulates	Lithographic Defects Metal Shorts	PIND Test Failures

Table 3   Selected Supplies for Handling Electronic Devices

Finger Cots	Sticks
Gloves	Foam
Wipers	Bags
Swabs	Staticides

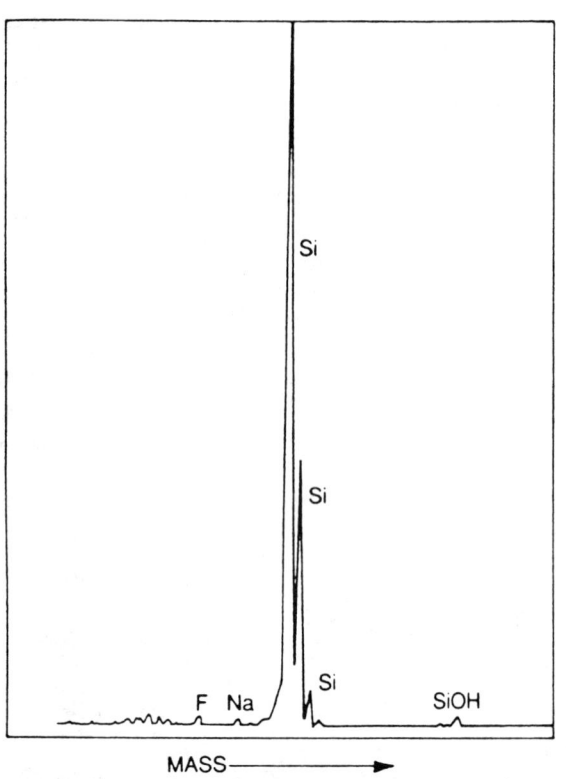

(a)

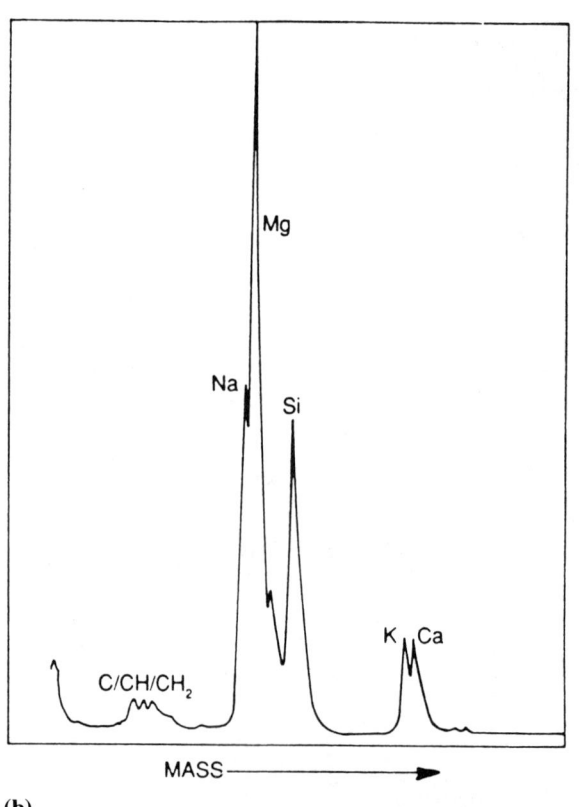

(b)

Fig. 1. (a) SIMS spectrum of untouched, deglazed silicon wafer surface. (b) Sample number 2. SIMS spectrum of deglazed silicon surface touched by finger cot.

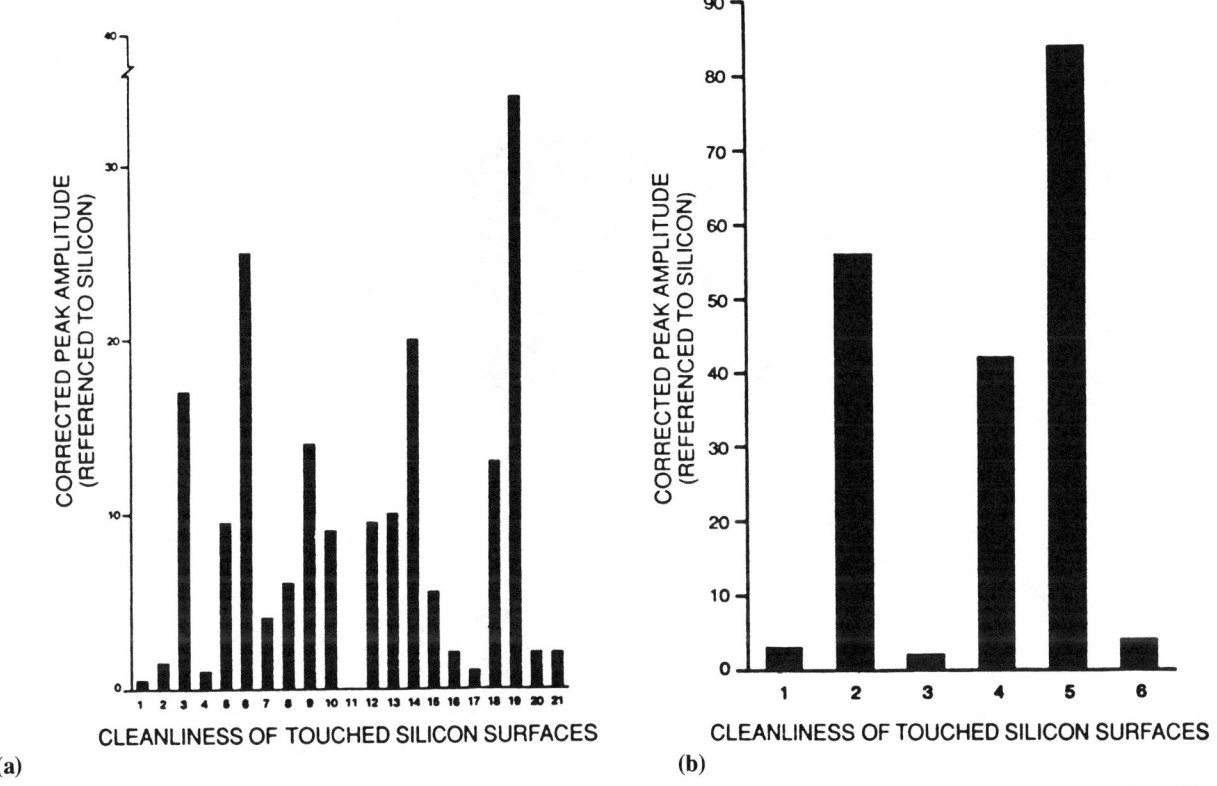

Fig. 2. (a) Relative cleanliness of clean silicon surfaces touched by finger cots. (b) Relative cleanliness of clean silicon surfaces touched by gloves.

One area of special concern regarding cleanliness of handling materials is that of controlling electrostatic discharge (ESD). Product handling materials are often treated with conductive coatings to help drain away electrical charge and keep it from building up on/in devices. Handling materials and supplies must offer the property of preventing static charge buildup which could induce ESD failure. These anti-static properties may be gained from chemical coatings. However, at the same time, they must not transfer chemical residues to surfaces which should remain clean, in order to avoid failure modes such as metal corrosion or electrical instability.

To ensure that product handling materials such as clean room gloves or sticks for shipping packaged IC's do not inadvertently contaminate products, it has been found appropriate to evaluate them for their surface contamination potential.

One of the ways to evaluate contaminating potential is to touch a clean "reference" surface with the item of interest and use surface analysis to measure the extent to which the touched surface was contaminated. This can identify types and amounts of contaminants transferred. In the materials evaluations described here, silicon wafers were used as reference surfaces. These reference surfaces were prepared by deglazing pieces of silicon wafers in 49% electronic grade hydrofluoric acid followed by ultrapure water rinse and blow dry with filtered ultrapure nitrogen. Immediately after deglazing, each clean

silicon substrate was touched with a sample of the material under test using nominal fingertip pressure for 2 seconds.

Touched, and therefore possibly contaminated, silicon pieces were placed in a sample mount, along with a similarly deglazed but untouched silicon piece for reference. The silicon surfaces were analyzed by Secondary Ion Mass Spectrometry (SIMS). SIMS spectra were obtained from the touched silicon surfaces using helium excitation at 4.5×10^{-5} Torr chamber pressure with an accelerating voltage of 2 KeV. Spectra covered the 3-100 amu range. Peak intensities from the clean reference surfaces were subtracted from those obtained from the touched surfaces. The result is a measure of amount and identity of transferred contaminants, permitting a judgment of relative cleanliness of the sample materials in terms of what they transfer to clean surfaces.

Figure 1 shows an example of the information that can be obtained this way. Figure 1a is the SIMS spectrum of a deglazed silicon surface. Only silicon appears in the spectrum, with barely discernable peaks for silicon hydroxide, fluoride (from the HF deglaze solution) and sodium, which appeared universally in all the spectra. Figure 1b is the SIMS spectrum of a deglazed silicon surface touched as described above while wearing one of Supplier B's finger cots (sample number 2). Magnesium now dominates on the surface, sodium is substantially increased above background, an enhancement of organic species is apparent, and potassium and calcium appear. Clearly

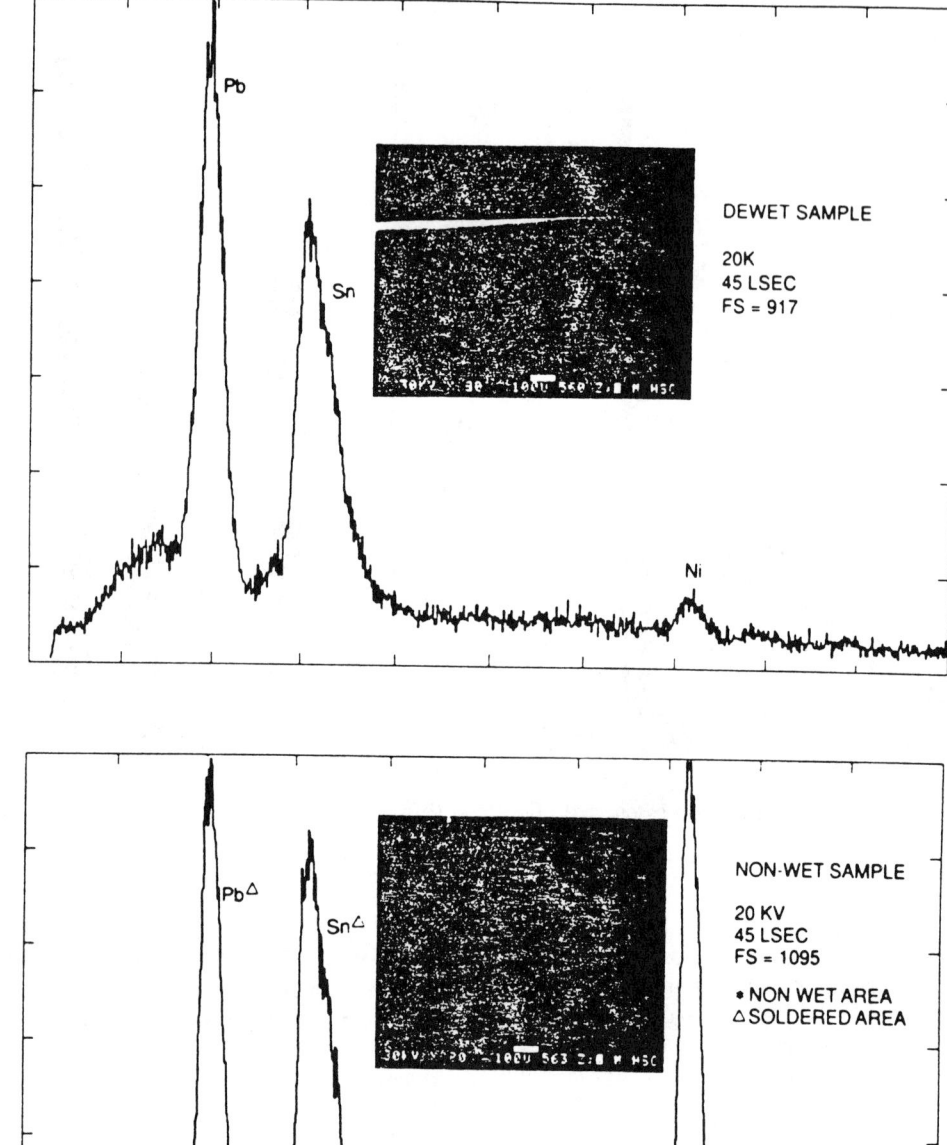

(a)

(b)

Fig. 3. (a) Photo and EDS spectrum of region of poor solder coverage on a package pin. (b) Photo and EDS spectrum of region of solder dewetting on a package pin.

Supplier B's finger cot transfers potentially dangerous ionic contaminant species to surfaces it contacts. The magnesium was likely sourced by talc, used to prevent sticking.

In controlled experimentation to select the cleanest finger cots and gloves, 6 samples of finger cots and 21 samples of gloves were subjected to these evaluations by SIMS. The peak

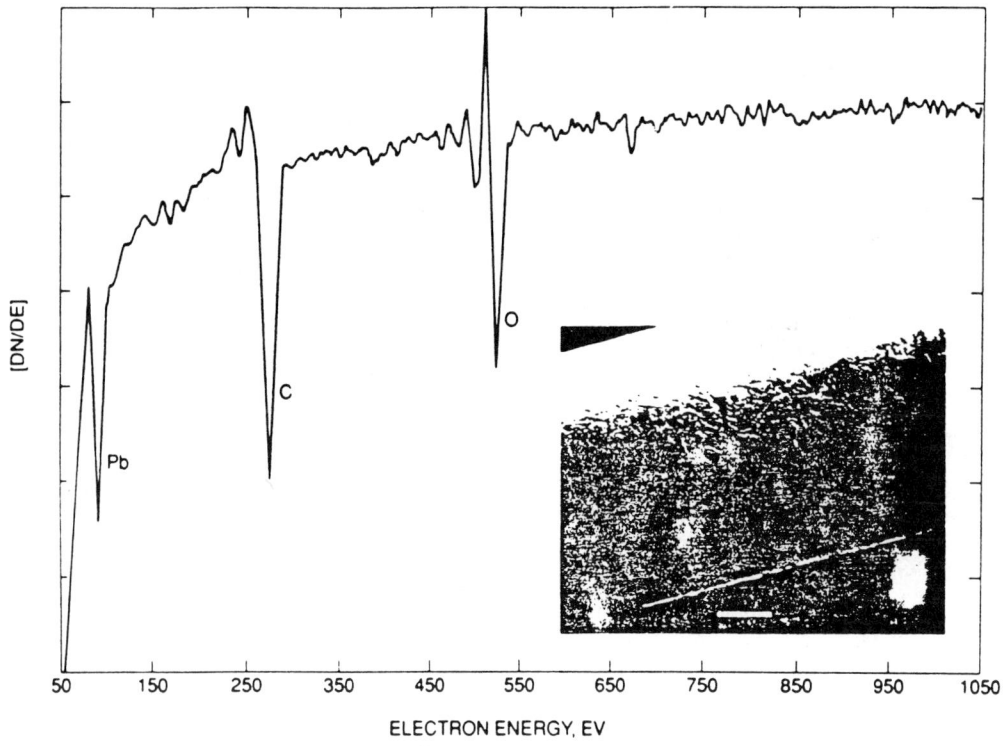

(a)

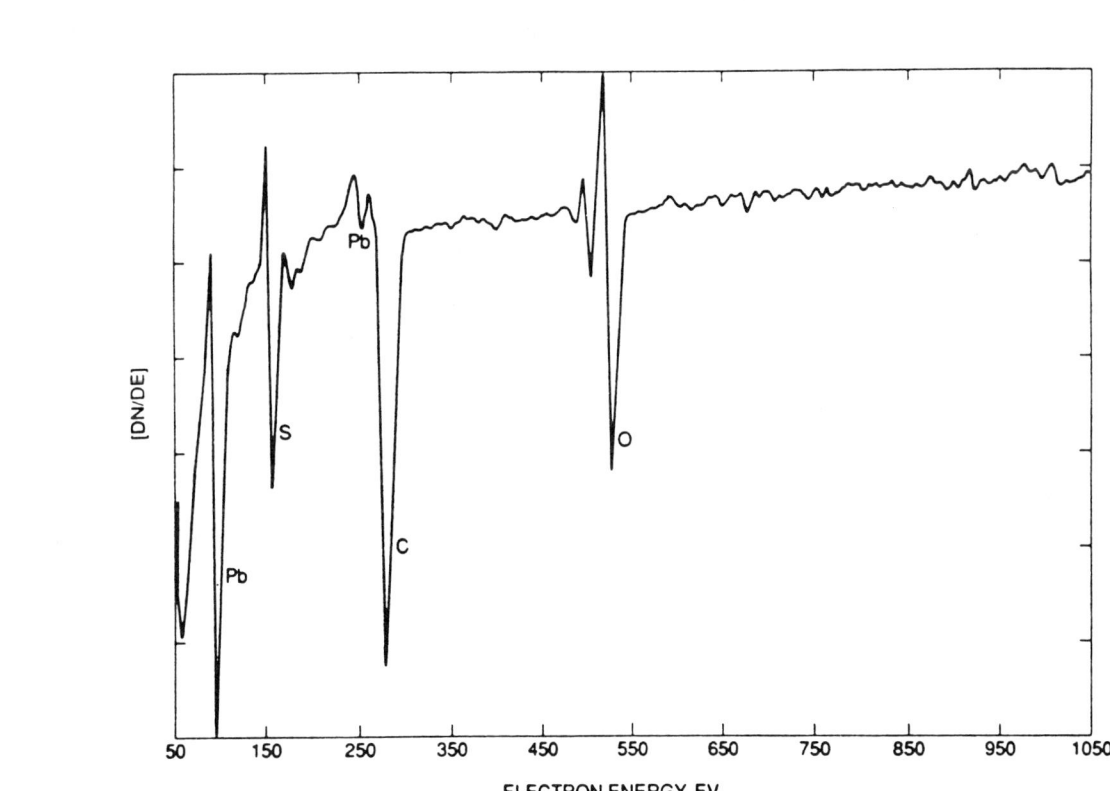

(b)

Fig. 4. (a) SEM photo of soldered region which appeared discolored under optical microscopy, and Auger survey spectrum of normal-appearing solder surface adjacent to anomalous region. (b) Auger spectrum obtained from surface of anomalous region shown in SEM photo in 4a.

amplitudes were summed and ratioed to the clean surface silicon peak intensity in order to display the relative surface cleanliness data in the bar graphs of Figure 2.

Based on these data no finger cots were qualified for use in IC assembly. Two kinds of ESD-protective gloves (Suppliers 1 & 4) were qualified for use.

While silicon was the reference surface for all transferable testing reported here, it is not necessarily the appropriate universal reference. For instance, plated metals (e.g. tin, gold, etc.) representative of package lead finishes may "receive" transferable contaminants differently than silicon. Test pieces should be chosen specifically for each situation being evaluated.

2. SOLDER WETTING

Soldering technologies for electronic connections have been a constant subject of study by reliability scientists and interconnection engineers for many years. Microbeam methods can be effectively used to evaluate soldering anomalies.

Determining the degree of solder wetting on surfaces can be done effectively using Energy Dispersive Spectroscopy (EDS) on a Scanning Electron Microscope (SEM). Photos in Fig. 3a and 3b show regions on package pins which exhibit varying degrees of poor solder coverage. Figure 3a shows a region where solder appears to be present, but with rather thin coverage. Figure 3b shows a region on a pin where there appears to be little or no solder coverage. The EDS point spectrum for the region in 3a shows a weak signal for nickel (from the pin's substrate metal). In contrast, the EDS point spectrum for the region in 3b exhibiting full solder coverage shows the same lead-tin response, while the EDS point spectrum from the center of the anomalous region shows a strong signal for nickel and no response for tin and lead. In these analyses, the relative sampling depth of EDS, which obtains signal from several microns of depth (compared to only a few tens of Angstroms for surface techniques like Auger), is used to advantage since this is really a "thick film" problem. The data characterizes very well the conditions of solder dewetting. In Fig. 3a, thin layers of solder remain, in contrast to the non-wetting condition in Fig. 3b, where no solder at all is present in the anomalous region.

Other types of anomalies include visual effects which suggest various contaminations on or in solder. Many of these, at first glance, could raise concern about the integrity of the solder coating. Figure 4 is a SEM micrograph of a local region of a soldered pin which appeared discolored under optical microscopy. The area is obviously locally roughened on the surface. Scanning Auger microprobe analysis (Auger electron spectroscopy, AES) was used to determine if the region contained significant levels of other chemical elements. The Auger spectra in Fig. 4a and 4b were taken off of, and directly on, the anomalous area respectively, after a brief 30 sec. argon ion sputter to remove surface-adsorbed species. The spectra show carbon impurity uniquely associated with the anomalous area. This carbon is present as a consequence of routine part handling and exposure to ambient conditions in the field. The roughened solder surface readily traps hydrocarbons from the surroundings; these hydrocarbons are surface-adsorbed species and are not actually an integral part of the bulk of the solder covering the pin.

The photographic and EDS data led to the conclusion that this anomalous region of the solder surface is a cosmetic aberration which contains no chemical impurities.

3. SEALING GLASS CONTAMINATION

Reliability scientists must often search long and hard for physical or chemical causes of electrical malfunctions. Parametric problems, where parts operate outside one or more of their expected electrical characteristics all or part of the time, are often the most difficult problem to associate with a physicochemical anomaly.

In this example involving sealing glass, microbeam analysis is used to identify contamination responsible for electrical leakage in operational amplifiers sealed within ceramic packages. Electrical leakage occurred on finished product sporadically. After much data accumulation, leakage was determined to be independent of all wafer processing functions. This fact, coupled with the observation that leakage seemed to be associated with particular pin-out configurations, led to an investigation of the packages. The problem was still sporadic, both between and within lots of the glass-sealed dual in-line packages. Eventually, careful Auger electron spectroscopy (AES) analyses of sealing glass between pins of samples exhibiting the problem vs glass between the same pair of pins on parts not exhibiting the problem were conducted. Figures 5a and 5b are examples of the Auger data which ultimately defined the problem. Devices with electrical leakage exhibited sulfur associated with the sealing glass; devices which were stable had no detectable sulfur on the surface of the sealing glass of their packages.

Sulfur was always found associated with the sealing glass of unstable device packages. It was never found on the sealing glass of stable devices. *The apparent leakage mechanism was the combination of chemisorbed moisture on the glass surface with the sulfur, forming a conductive chemical species which supported surface electrical conduction and therefore leakage between pins.* Through additional Auger analytical work conducted generally on the raw materials, it was found that package piece parts were received with sporadic occurrences of sulfur. All package lots had some fraction of packages whose sealing glass composition exhibited sulfur. Auger depth profiling showed that sulfur was not just a surface phenomenon. It persisted for thousands of *angstroms* into the glass. A program based on the analytical findings was initiated with the package supplier to eliminate the sulfur-contaminated sealing glasses.

This problem is an excellent example of the fact that many analytical investigations are a process and not a one-time event. The spectra in Fig. 5a and 5b could be declared definitive of the sulfur contamination problem only after months of studies on dozens of samples resulting from the ongoing, sporadic nature of the electrical leakage.

4. CHIP CLEANING

A reliability concern related to part and package qualification is the cleanliness of chips as they enter plastic molding assembly processing. Chip surfaces must be free of film contaminants to ensure intimate adhesion of mold compound. Additionally, they should be free of ionic residuals which could

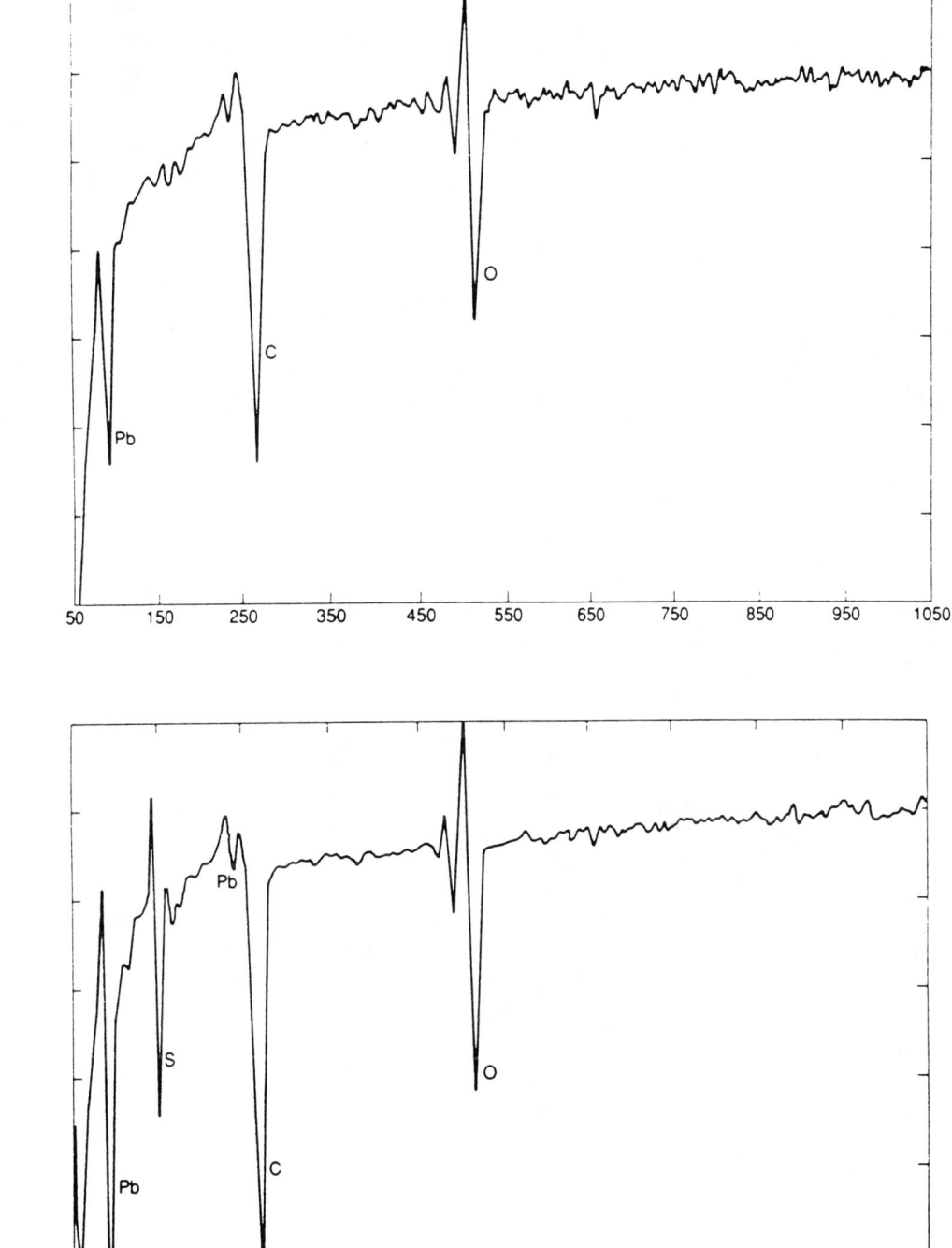

(a)

(b)

Fig. 5. (a) Auger spectrum of sealing glass between package leads of an electrically stable device. (b) Auger spectrum of sealing glass between package leads of electrically leaky device.

359

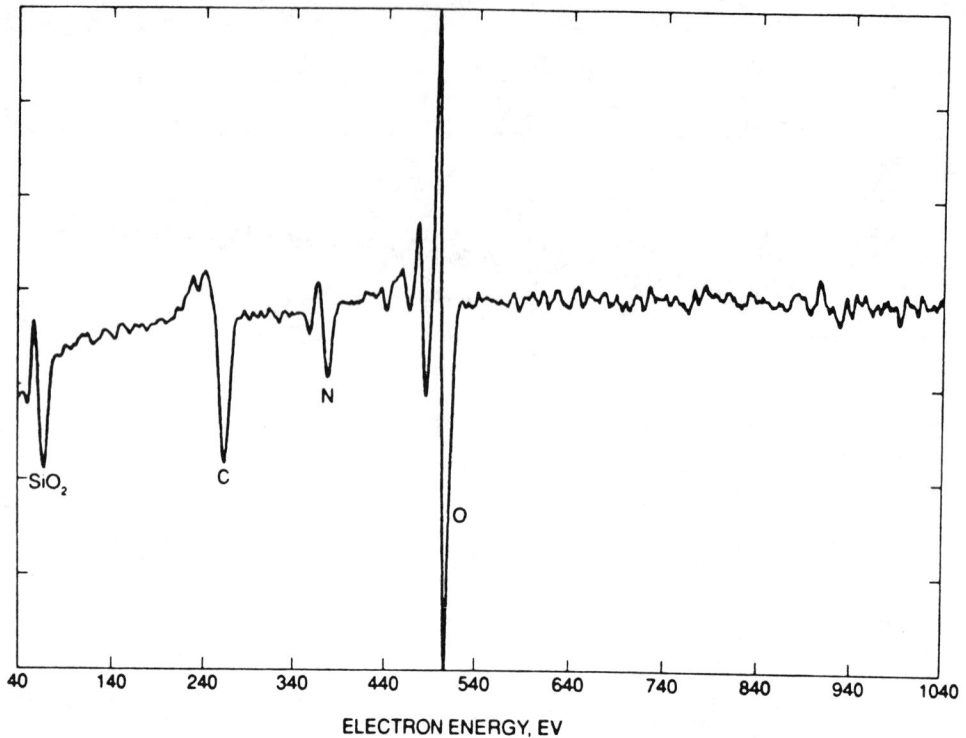

(a)

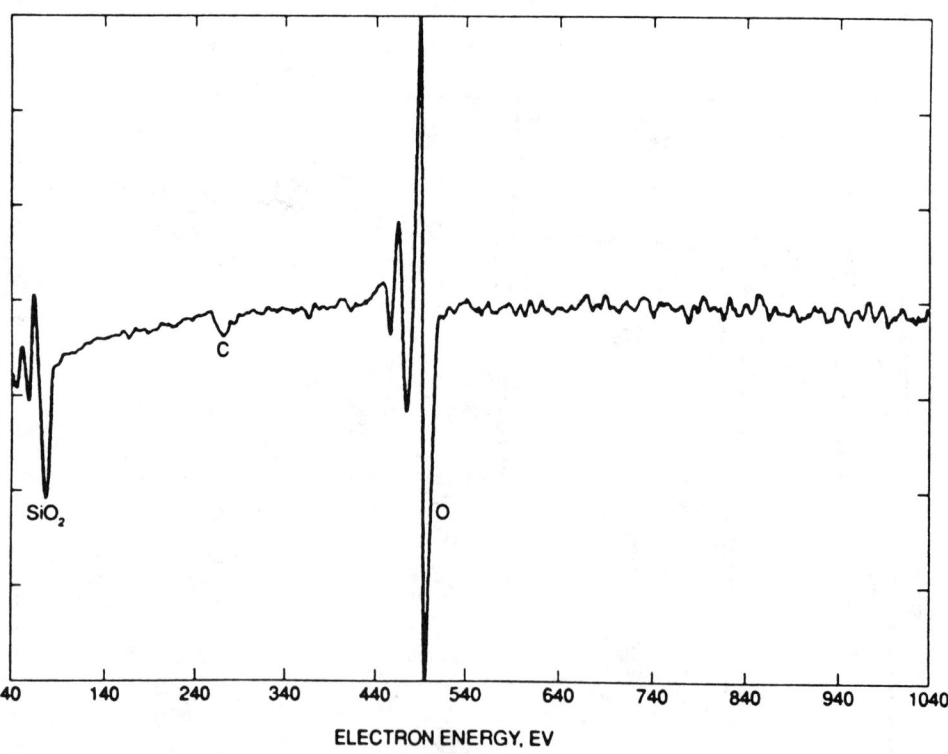

(b)

Fig. 6. (a) Auger spectrum of silicon nitride passivation subjected to UV/ozone cleaning. (b) Auger spectrum of silicon oxide passivation subjected to UV/ozone cleaning.

contribute to metal corrosion or to ionic-induced parameter changes. Molding compounds by themselves are a major potential contributor to such concerns, so chip surface cleanliness is imperative to ensure reliable device operation after plastic encapsulation.

Auger and SIMS analyses of chip surface cleanliness were conducted as part of cooperative technical support for a customer purchasing large numbers of die and assembling them in plastic packages. The customer was experiencing significant assembly yield loss with the product.

The die sales mix included parts that had surface protective passivation films of both silicon nitride and silicon dioxide. Figure 6a and 6b show Auger surface analysis spectra of a nitride passivated die and an oxide passivated die respectively on the as-received passivation surfaces. The spectra are identical with respect to surface cleanliness except for the high amounts of carbon on the surface of the nitride passivation. All product had been exposed to a pre-seal UV/ozone clean. Apparently, the UV/ozone process being used was a much less effective cleaner of nitride surfaces than of oxide surfaces.

Figures 7a and 7b are SIMS surface spectra of the nitride passivated and oxide passivated die. These spectra were taken to supplement the Auger spectra because of the better sensitivity of SIMS to low levels of ions. It was confirmed that elevated hydrocarbon levels were unique to nitride surfaces. Furthermore, SIMS analysis revealed that nitride surfaces were substantially more contaminated with sodium and potassium, and even exhibited the presence of titanium.

The cause of mobile ions was traced to a pre-UV/ozone chemical clean used exclusively for nitride passivated die. Using the data from these surface analyses, the customer was able to engineer more effective and more consistent die cleans prior to plastic encapsulation, which increased assembly yield to the high 90% range.

5. HUMAN CONTAMINATION

Despite spending huge sums for building and maintaining clean room production space, IC's are often inadequately protected from one of the most ubiquitous and sinister sources of dangerous contaminants: PEOPLE! Humans are abundant sources of organic, ionic, particulate, and moisture-laden materials that pose serious threats to both yield and reliability.

IC products are susceptible to human-sourced contamination from the time they enter the fab area as wafers until, as individual die, they have been sealed into a package. While cleanroom clothing is designed to reduce the spread of operator-sourced contamination, it can be less than adequate, especially in the facial area. Furthermore, post-wafer fabrication processes, such as probe, scribing, die storage/handling, and package seal, sometimes use clean room discipline less rigorous than that in wafer fabrication.

Human-sourced contamination is chemically reactive and easily capable of producing corrosive ionic solutions. It is also a source of solid particulates that can cause physical damage in hermetically sealed packages.

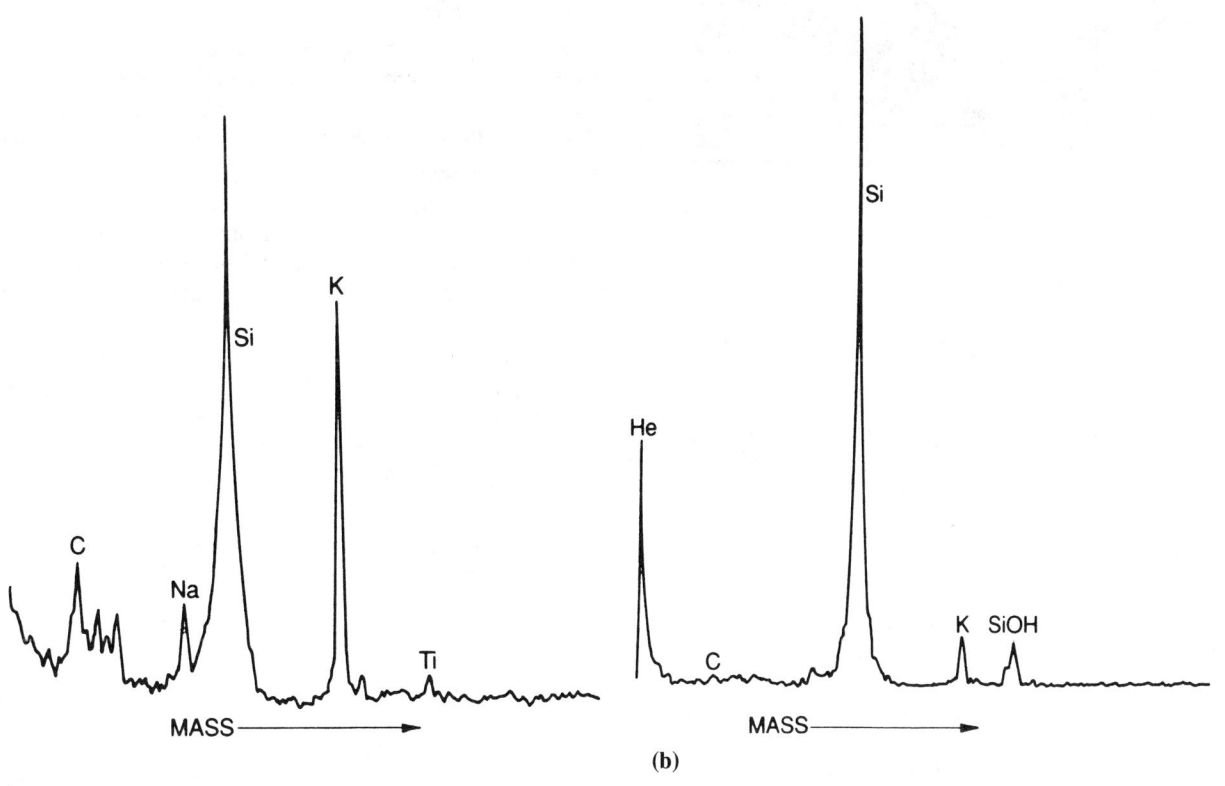

Fig. 7. (a) SIMS spectrum of silicon nitride passivation subjected to UV/ozone cleaning. (b) SIMS spectrum of silicon dioxide passivation subjected to UV/ozone cleaning.

361

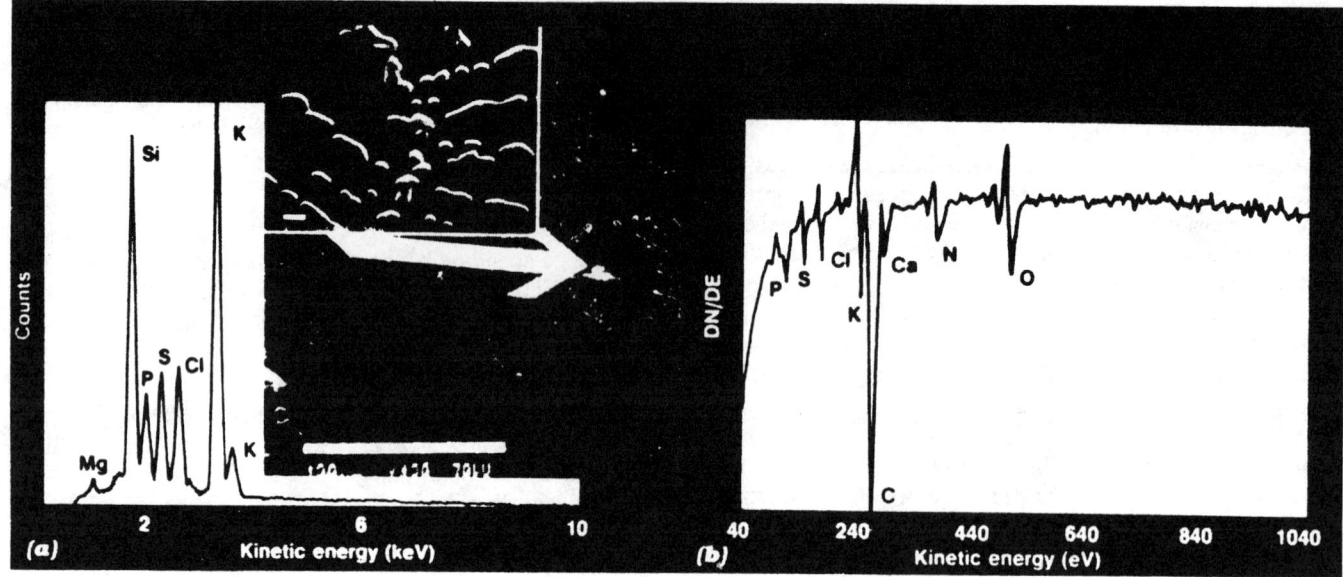

Fig. 8. Spittle contamination on a wafer, with a) the associated EDS spectrum, and b) Auger spectrum.

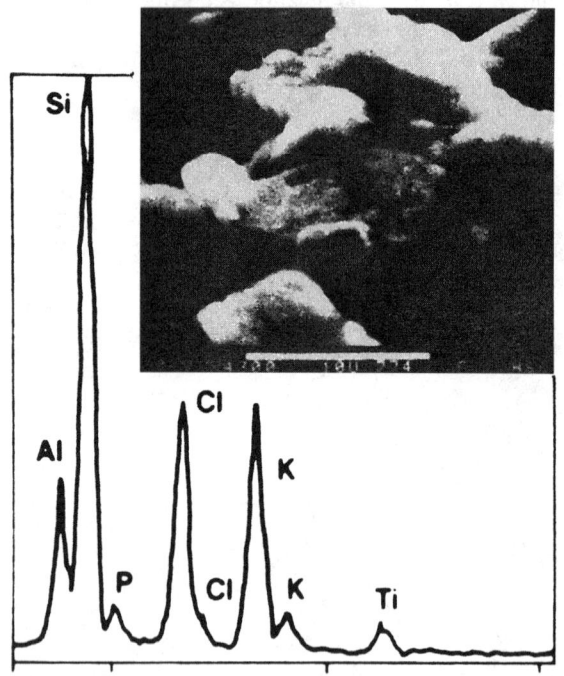

Fig. 9. Potassium chloride crystal, derived from human spittle, and its associated EDS spectrum.

On-die contamination is often first detected as "foreign material" by optical microscope inspection. When optical microscopy cannot plainly identify foreign material, it is submitted for microchemical analysis. This is usually done by **SEM/EDS**, where SEM is first used to locate and physically characterize the foreign material. EDS then provides chemical information. Characterization of three types of human-sourced contamina-

tion by SEM/EDS, spittle, perspiration, and cosmetics, is briefly described below.

Spittle

Figure 8 is a SEM photo of dried human spittle (airborne saliva droplets) on a silicon wafer. The inset in Fig. 8 is an enlargement of a small area of the patterned droplets.

The energy dispersive spectrum in Fig. 8a shows the elemental constituents to be potassium (K), chlorine (Cl), sulfur (S), phosphorus (P), magnesium (Mg), and traces of sodium (Na). The most noticeable feature is the potassium K-alpha line, indicating that a major constituent of spittle is potassium.

Complementary Auger data, shown in Fig. 8b, indicates that the surface of the dried spittle contains large amounts of carbon (C) and nitrogen (N). These elements are present in the liquid portion of the spittle. The potassium, chlorine, and sulphur are all readily visible in the Auger survey.

Another feature of spittle residue is the crystalline deposit shown by the arrow in Figure 8a. A close-up of these crystals, discovered on an actual circuit, is shown in Figure 9 along with the corresponding EDS spectrum. The spectrum shows the crystals to contain potassium (K) and chloride (Cl).

The two distinguishing features that uniquely define spittle residue are the high potassium concentration in the chemical spectrum and the presence of cubic potassium chloride crystals in the residual material.

Perspiration

Figure 10 shows a micrograph of dried perspiration on a silicon wafer. EDS analysis shows its elemental constituents to be sodium (Na), potassium (K), chlorine (Cl), and traces of sulfur (S) and aluminum (Al). Auger data in Fig. 10a and 10b support this and indicate that perspiration contains considerable carbon

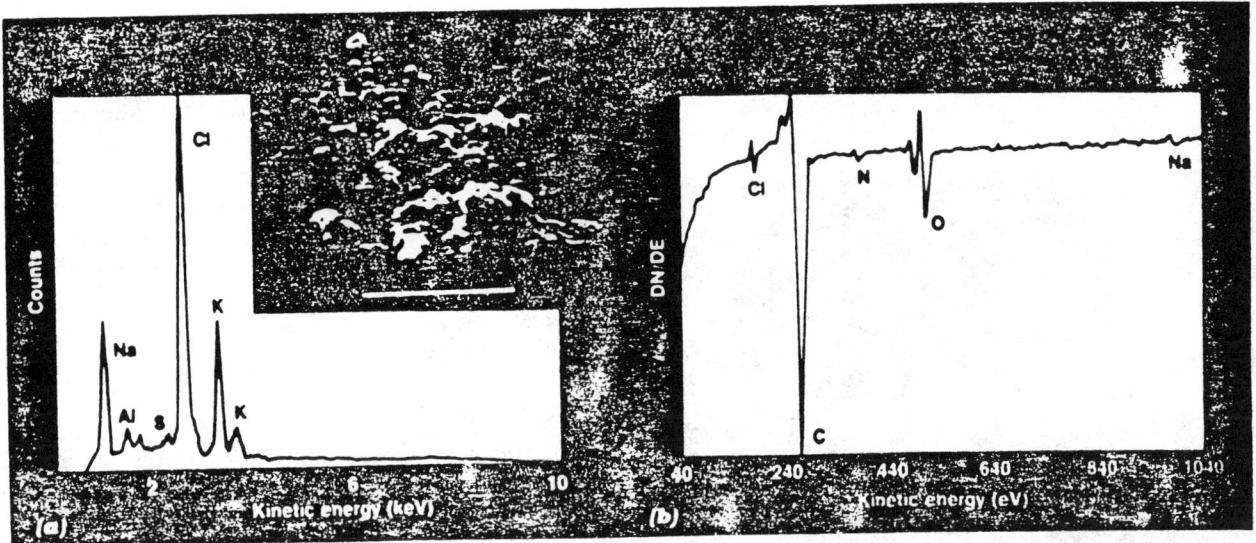

Fig. 10. Perspiration contamination on a wafer, with, a) the associated EDS spectrum, and, b) the Auger spectrum.

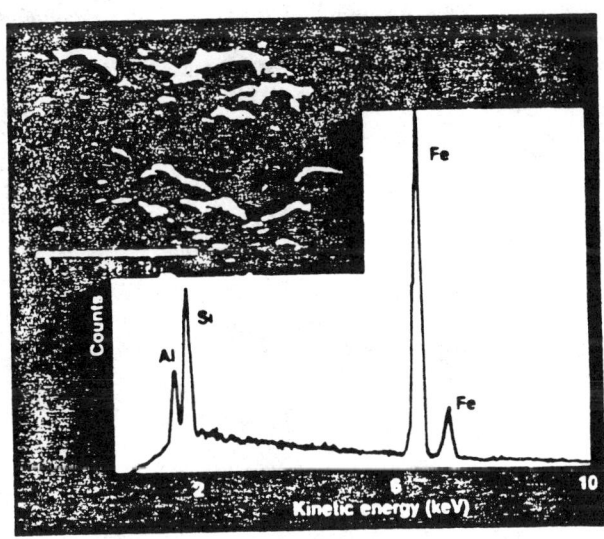

Fig. 11. Mascara deposit and associated EDS spectrum.

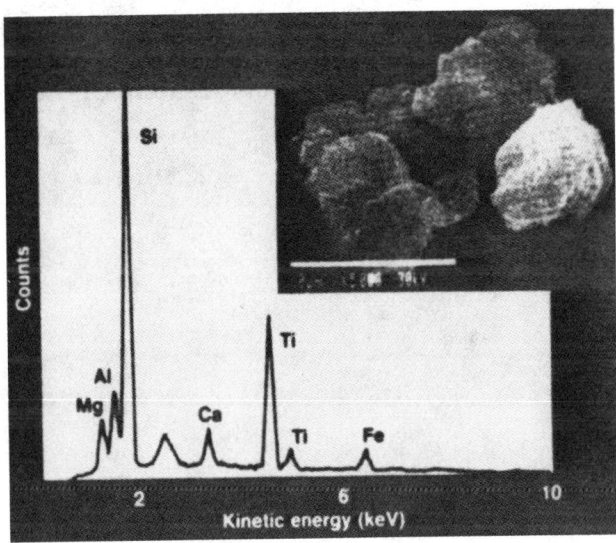

Fig. 12. Face powder deposit and associated EDS spectrum.

(C) and nitrogen (N) as well. *The main difference between perspiration and spittle is the sodium/potassium ratio. Spittle has a very low Na/K ratio when atomic sensitivity factors are considered, while perspiration has a large Na/K ratio. Perspiration also contains more nitrogen, due to release of amino acid derivatives from skin. Aluminum in perspiration is sourced by aluminum comounds used in antiperspirant formulations.*

Cosmetics

A third major type of human-sourced contamination is cosmetics. These can include mascara, facial powders, and fingernail polishes. For this study, samples of a popular mascara and a popular face powder were deposited onto a silicon wafer for analysis by SEM/EDS and Auger.

Figure 11 shows a SEM photo of dried mascara. EDS analysis of this area indicates that the mascara is composed of compounds containing the elements iron (Fe) and aluminum (Al). The prominent iron line is definitely an identifying feature. The iron is probably present as oxide, used as a pigment to give mascara its dark color. Significant quantities of carbon (C) were observed by Auger, but no other elements were identifiable. The carbon is a result of the solvent used to liquefy the mascara. Unlike the various types of body effluvia, no alkaline or halide-type ionic species such as sodium, potassium or chlorine are present in mascara studied.

Figure 12 shows a SEM photo of face powder deposited on a silicon wafer. EDS analysis shows titanium (Ti), iron (Fe), magnesium (Mg), aluminum (Al) and potassium (K). The most

Fig. 13 Photo of bond pad with mottled surface appearance.

obvious feature of the spectrum is the large titanium K-alpha line. The titanium, as well as the iron, magnesium, and aluminum, is probably present as an oxide and is the pigment in the powder. Other researchers have found that bismuth (Bi) and barium (Ba) can be present in cosmetics, and human contamination can generally contain: Al, Ca, Cl, Fe, K, Mg, Na, P, S.[1]

Microbeam characterizations obtained from purposely contaminated substrates, and from actual on-line occurrences of human contamination on product, have been instrumental in helping reliability scientists and production engineers optimize manufacturing and product handling procedures to greatly reduce contamination from human sources.

6. BOND PAD APPEARANCE

Process engineers and reliability engineers are often dismayed by the widely varying visual appearances of wire bonding pads on IC chips. Bond pads can certainly be a source of trouble. They expose a relatively large surface area of metal, slightly recessed from the actual top surface of the chip (covered everywhere else by the passivation film), and thus are a repository for sundry contaminants. Electrical probing mechanically (if not chemically) perturbs bond pads. In a finished device, bond pads contain metallurgical wire attachments which themselves can be sources of chemical and physical anomalies, or can be compromised by various types of contamination.

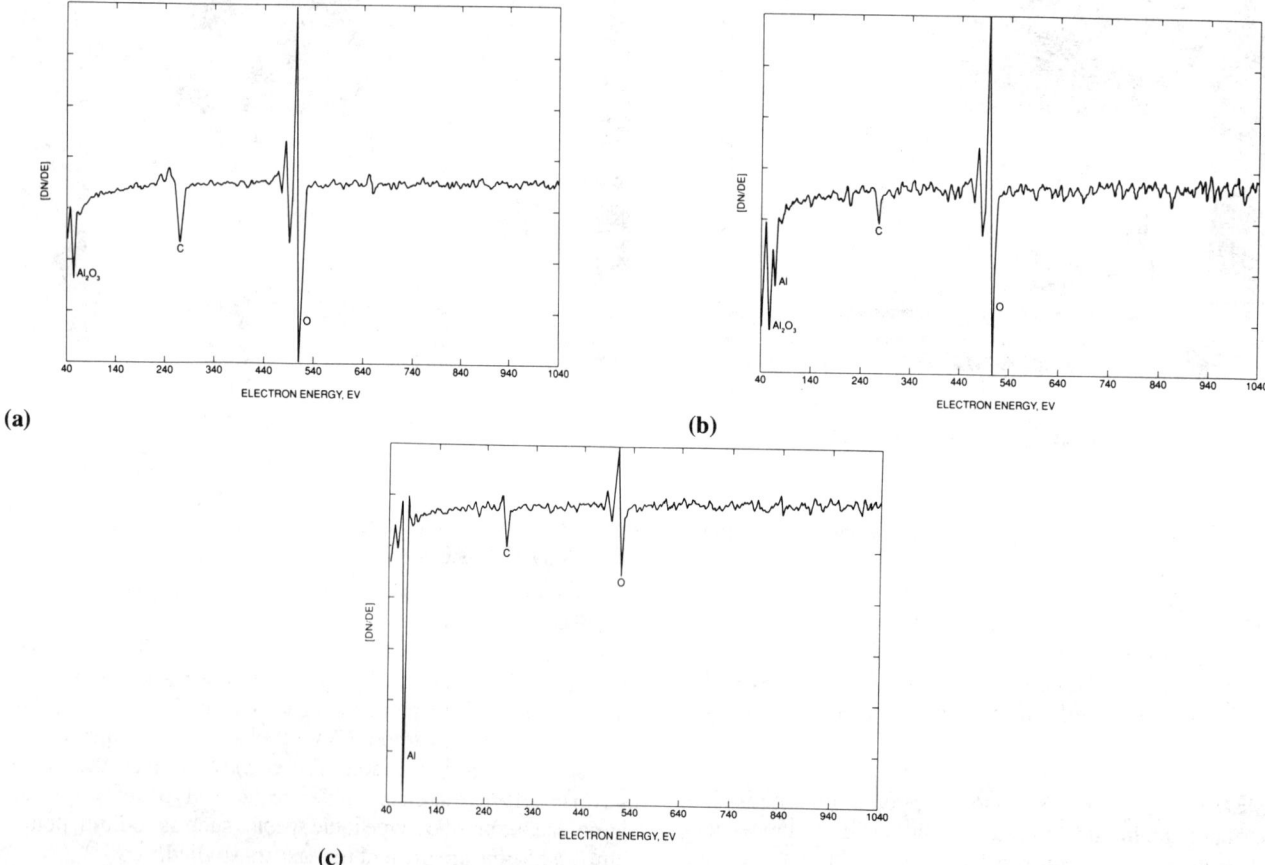

Fig. 14. (a) Auger survey spectrum of mottled bond pad surface as-received. (b) Auger survey spectrum of mottled bond pad after 0.5 minute Ar+ sputtering. (c) Auger survey spectrum of mottled bond pad after 1.0 minute Ar+ sputtering.

364

The general expectation is that bond pads will present the appearance of shiny, silvery aluminum. During routine visual inspections, any appearance to the contrary raises suspicions about the reliability of the device. However, some bond pad conditions, perceived by visual inspection to be problems, are actually shown by microbeam analysis to be purely cosmetic circumstances.

As an example, Fig. 13 shows a pad whose surface appears mottled and grainy. A sequence of Auger spectra on this pad is shown in Fig. 14a, b, and c for 0, 0.5, and 1.0 min. sputter times of the pad, respectively. These spectra show what is generally expected for an aluminum pad, DESPITE its appearance. The surface is aluminum oxide with a normal amount of native

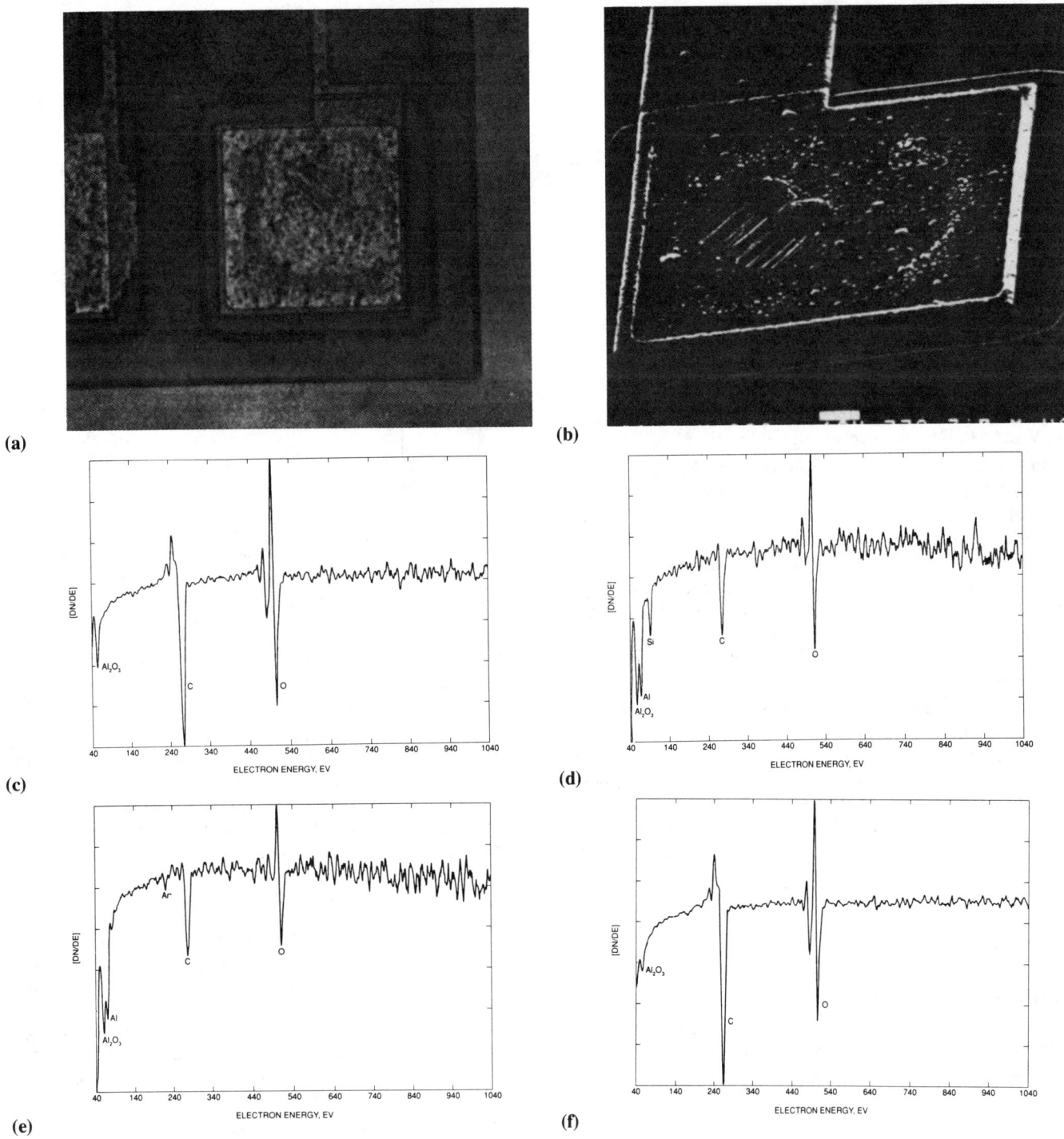

Fig. 15. (a) Optical photo of bond pad with ring anomaly. (b) SEM photo of bond pad in Figure 15a. (c) Auger survey spectrum on ring anomaly as-received. (d) Auger survey spectrum on ring anomaly after 0.5 minute Ar⁺ sputtering. (e) Auger survey spectrum on ring anomaly after 1.0 minute Ar⁺ sputtering. (f) Auger survey spectrum at center of ring anomaly as-received. **(continued)**

365

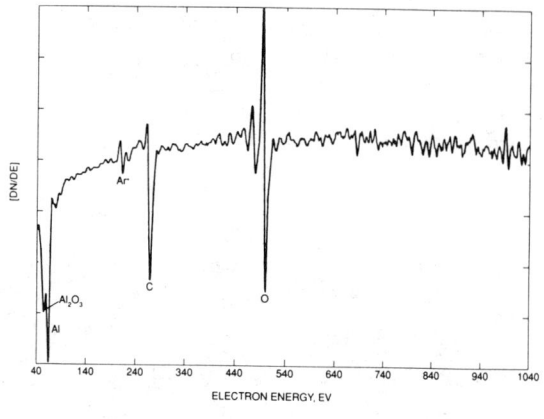

(g)

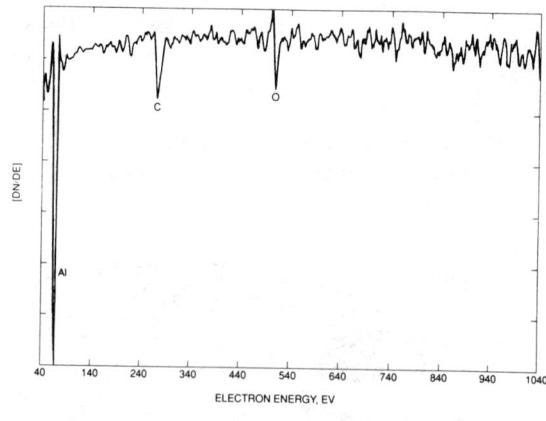

(h)

Fig. 15. (g) Auger survey spectrum at center of ring anomaly after 0.5 minute Ar$^+$ sputtering. (h) Auger survey spectrum at center of ring anomaly after 1.0 minute Ar$^+$ sputtering.

carbon (C) in Fig 14a; 0.5 min Ar$^+$ sputtering shows reduced carbon with a mixture of aluminum (Al) and aluminum oxide (Al$_2$O$_3$) in Fig. 14b; and 1 min. Ar$^+$ sputtering shows mostly elemental aluminum (Al) in Fig. 14c. This Auger study, differentiating between aluminum and its oxide as the pad surface is sputtered, takes advantage of the wide separation in Auger electron energies between elemental aluminum and its oxide, at 68eV and 51 eV, respectively.

Photos in Fig. 15a and 15b show another bond pad anomaly. This pad displays the expected shiny surface appearance, except for what appears to be a ring of material encircling the probe mark. This ring was characterized by Auger microanalysis, first at a point on the "ring", and then at a point in the center of the pad, which appears normal. The sequence of Auger profiles shows that the "ring" on the pad consists of aluminum enriched in aluminum oxide. This aluminum oxide is actually an area of pad material debris kicked up by the probe point and re-deposited around the circumference of the probed area.

The spectra in these two examples are as significant for what they *do not* show as for what they *do* show. No halide or positively charged ionic impurities are found with the pad anomalies. Thus, even though the pad surfaces appear contaminated, they are free of surface impurities which could lead to ion-induced failure events. Again, this is a common problem in IC assembly areas. It is not generally safe to assume that anomalous appearance of the metal can always be discounted. The point here is that the surface analysis can define the source of the problem and provide a scientific basis for an "accept" or "reject" decision.

7. CONTACT RESISTANCE

Reliability scientists often find that increased electrical resistance occurs where metal conduction layers make contact. When these specific contact points can be isolated, in-situ microbeam analysis, in the depth profile mode, can reveal the cause of elevated contact resistance.

Frequently, an interposed layer of carbon is found in the contact. Carbon can increase contact resistance so much that electrical parameters cannot be met. Figure 16 is a composite

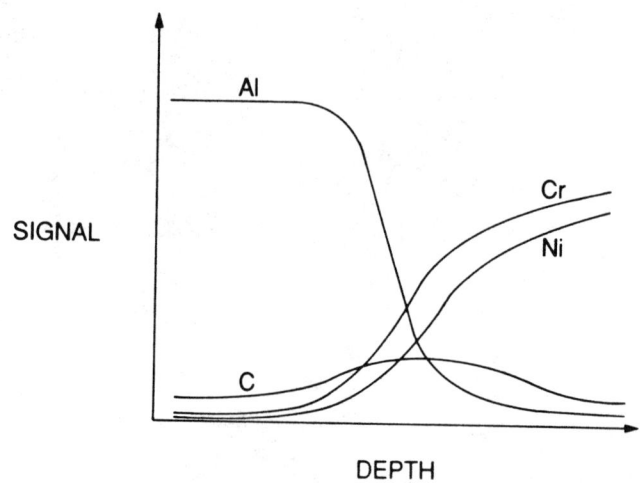

Fig. 16. Typical depth profile through a metal contact with high contact resistance.

depth profile ("averaged" from many similar types of analyses) showing the typical finding of carbon "sandwiched" between metal layers. Carbon is readily left behind by poor photoresist strips or other cleans that do not completely remove organic films. Contacts free of carbon never have high resistance.

8. THIN FILM LASER TRIMMING

Laser trimming is a process tool for in-situ fine-tuning of resistor films in devices. Achieving thin film resistors with reliable, stable electrical properties by in-situ lasing has been the subject of extensive microbeam analysis. In this example, understanding how to adjust resistance by laser alteration of the chemistry of as-deposited chromium silicide thin film was critical to obtaining devices of long term operational reliability.

Figure 17a presents X-ray Photoelectron Spectroscopy (XPS) spectra for silicon (2p) and chromium (2p) binding energies taken at the surface, 100 Å, and 200 Å deep, respectively, in a chromium silicide (CrSi) film as-deposited on thermal oxide, with no heat treatment. Figure 17b is the same set of infor-

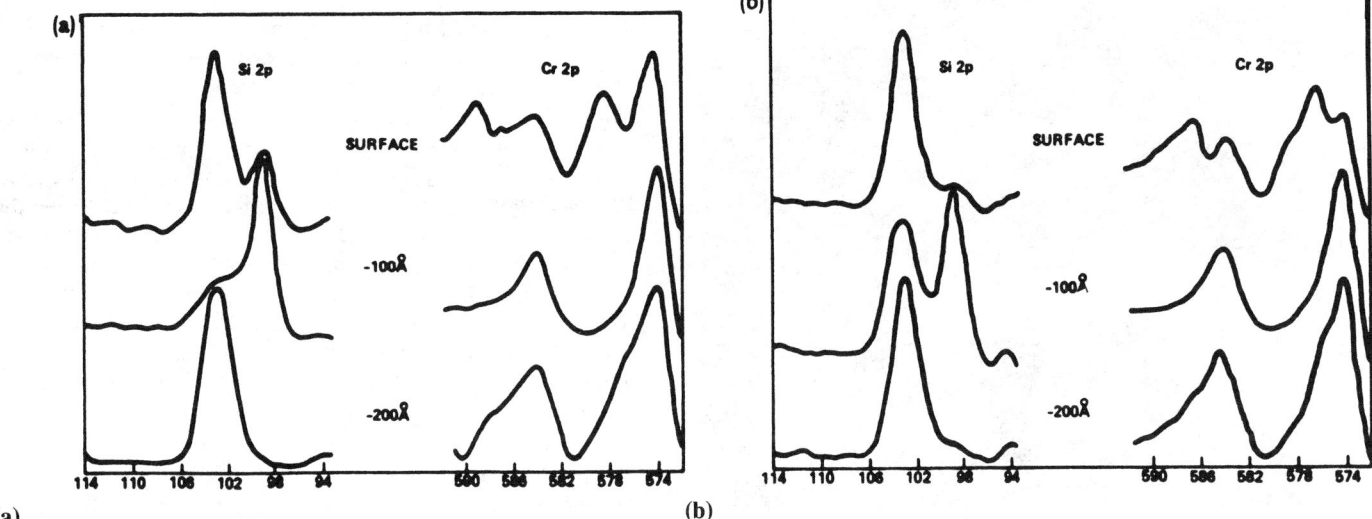

Fig. 17 (a) XPS spectra showing Si(2p) and Cr(2p) binding energies for an unbaked sample of CrSi deposited over silicon dioxide. (b) XPS spectra showing Si(2p) and Cr(2p) binding energies for a baked sample of CrSi deposited over silicon dioxide.

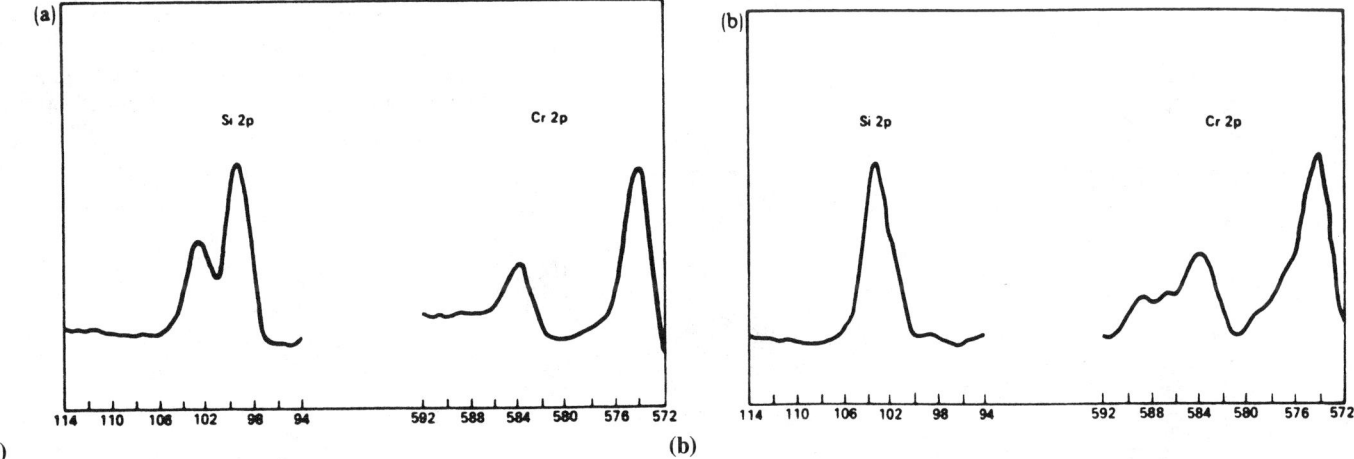

Fig. 18 (a) XPS spectra showing Si(2p) and Cr(2p) binding energies for a baked, passivated, and untrimmed sample of CrSi deposited over silicon dioxide. (b) XPS spectra showing Si(2p) and Cr(2p) binding energies for a baked, passivated, and trimmed sample of CrSi deposited over silicon dioxide.

mation for this sample after a stabilization bake. A layer composed of Cr_2O_3, $Cr(OH)_3$, and SiO_2, is detected on the surface of the unbaked sample. The baked sample shows a Cr_2O_3-SiO_2 layer free of $Cr(OH)_3$. SiO_2 is found in heavy concentrations on the surface, gradually decreasing to a minimum near the center of the film, and gradually increasing again towards the lower CrSi-SiO_2 interface.

XPS results of trimmed vs. untrimmed baked and passivated CrSi over SiO_2 are shown in Fig. 18a and 18b. These samples enabled gathering of XPS data on samples most closely resembling standard processing (i.e. baked, passivated, and trimmed). The control in this case is the baked, untrimmed sample, which represents processing up to the laser trim step. For these samples, the glass passivation was removed down to approximately 1000 Å with hydrofluoric acid. The remaining glass was removed down to the passivation-silicide interface by Auger profiling, monitoring closely the Cr and O signal to determine the actual location of the interface. The interface consisted of SiO_2, Cr_2O_3, and CrSi, as expected. The entire CrSi layer contained SiO_2, decreasing in concentration toward the center of the film, with Cr_2O_3 detected only at the upper and lower interfaces. The trimmed sample showed a diffuse layer composed of Cr in several oxidation states; Cr^0, Cr^{+3}, Cr^{+4}, Cr^{+6}, and SiO_2. These results indicate sufficient thermal energy in the laser beam to significantly alter the chemical makeup of the film. This XPS data, supporting a team effort by process and reliability groups, was an essential in a proactive process development initiative designed to build quality into the product.

9. A CONTAMINATED DEVICE STRUCTURE

Figure 19 is a photo of a test structure used as part of a yield/reliability evaluation for a process under development. It

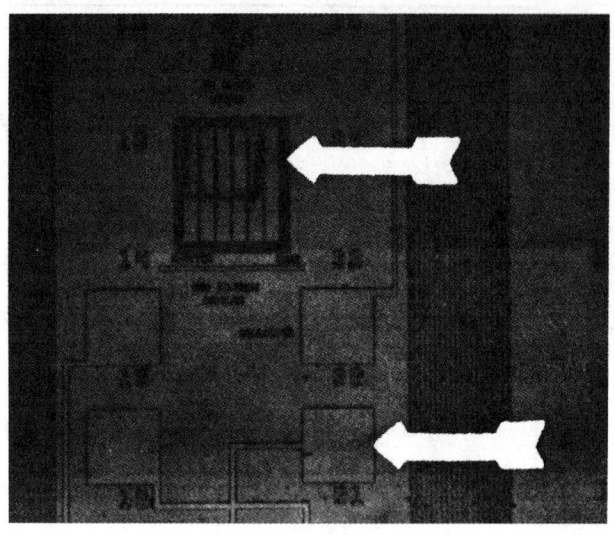

Fig. 19. Test structure. Small and large square features indicated by arrows to be depth profiled for elemental composition.

was desired to depth-profile through the two features marked by arrows in the photo. The larger feature (upper arrow) is about 166×200 while the smaller feature (lower arrow) is about 125μ on a side. These structures were analyzed by Secondary Ion Mass Spectrometry (SIMS) using O_2+ primary ions to sputter selectively through the regions of interest. The sputter crater in the upper feature is the black square favoring the upper left corner, and is $125 \times 125\mu$. Analytical information was obtained from an area within this sputter crater about 30μm on a side, providing very localized data about the structure. The lower feature was analyzed in exactly the same manner. Care was taken to ensure that the sputter crater's approximately $900\mu^2$ area, from which secondary ions were collected and analyzed, did not fall outside the boundaries of the sites of interest.

Figures 20a and b are the SIMS depth profiles for boron dopant and aluminum in these areas of interest. The upper (larger) structure is found to contain 3 to 4 times as much aluminum (as much as 10^{16} atoms of Al/cc) than the lower structure (in the N^+ layer, and near the pn junction). The presence of aluminum coincided with observed electrical anomalies. The in-situ depth profiling capability of the ion microprobe within small device features enabled the identification of impurity affecting device performance, allowing process corrections to be made. It is interesting to note, in this case, that the impurities

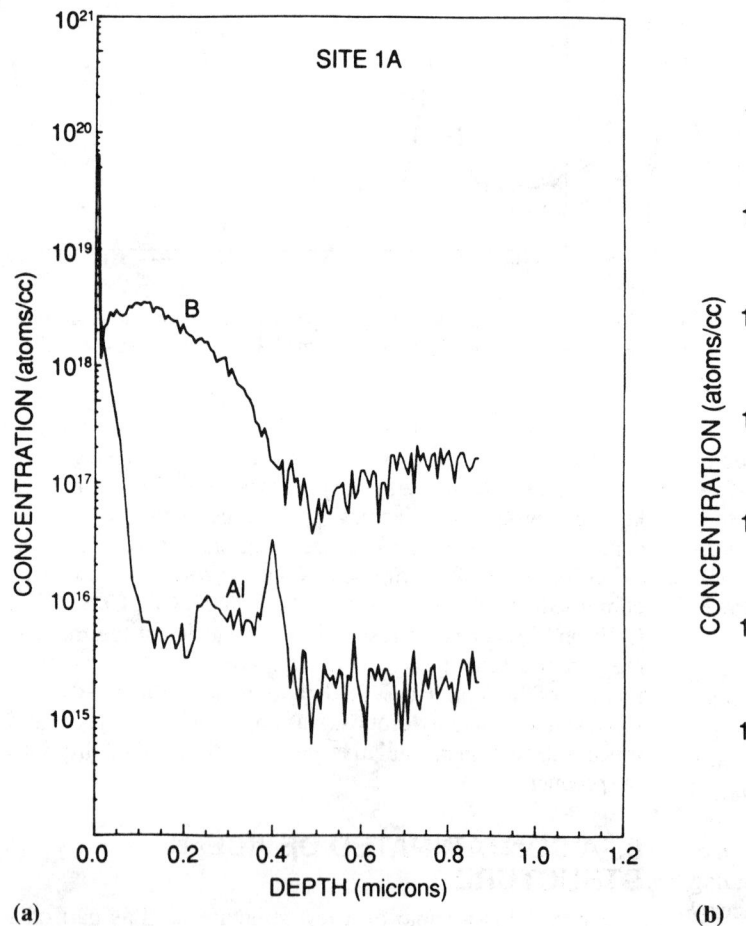

(a)

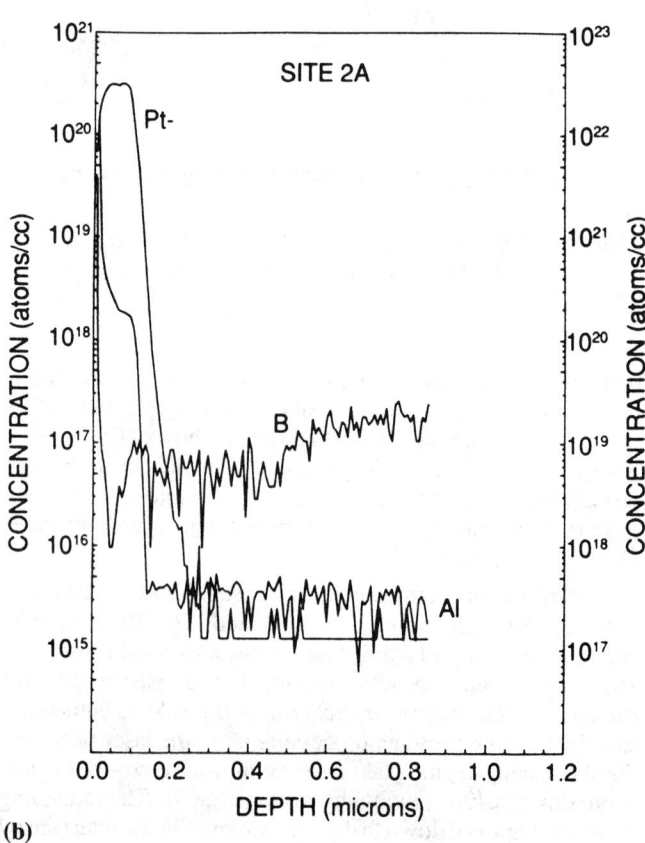

(b)

Fig. 20. (a) SIMS depth profile through small feature, (upper arrow). (b) SIMS depth profile through larger feature, (lower arrow).

responsible for the device problem was an element that is otherwise common to IC processing.

CONCLUSION

These are just a few examples of how microbeam analysis can characterize IC anomalies to help engineer high-quality processes and products as well as to understand failure mechanisms. The resourceful reliability scientist will be aware of and will use these and other microbeam techniques as supplementary tools to electrical methods in his quest for product excellence and customer satisfaction.

ACKNOWLEDGMENTS

The technical staff of the Beam Team in Harris Semiconductor's Analytical Services Laboratory contributed material for this tutorial.

REFERENCE

1. R.W. Thomas and D.W. Calabrese, *The Identification and Elimination of Human Contamination in the Manufacture of ICs*, ISTFA-85, pp 169-176.

Applications of FTIR Spectroscopy to Semiconductor Failure Analysis

Christopher D. Gondran and Ronald A. Carpio

Fourier Transform Infrared Spectroscopy (FTIR) is utilized extensively on a macroscopic as well as microscopic scale for both qualitative and quantitative analysis of inorganic and organic molecular species. The infrared spectrum is not only a unique fingerprint of the material being analyzed but it also provides information on the structural units present.

Common uses in the semiconductor industry include determination of the interstitial oxygen and substitutional carbon in processing wafers, epitaxial film thickness measurements, monitoring the phosphorus in phosphosilicate thin films and both phosphorus and boron in borophosphosilicate thin films, measurement of the hydrogen content in oxide and nitride thin films, and the identification of particles and residues. FTIR is utilized for incoming materials characterization, trace gas analysis, carrier concentration measurements, and the characterization of silicon surfaces after cleaning operations to list only a few of the lesser known applications. It is probable that the results of such FTIR measurements will be used at various times by the FA Engineer to help trace the source of device problems. So it is advisable to become at least acquainted with all these applications.

However, the most common use of FTIR in the FA Lab will for the identification of particles and residue contamination found on the partially or fully processed device, for the purpose of determining the origin of the contamination or to judge the probable effects of the contamination on device performance.

Thus, emphasis in this paper will be focused on the capabilities and limitations of FTIR for this particular application.

INTRODUCTION TO FTIR SPECTROSCOPY

Fourier transform infrared spectroscopy is a branch of vibrational spectroscopy. The sample is illustrated with infrared from a source such as a Glowbar® or Nernst Glower. Molecules will absorb radiation at characteristic frequencies in the infrared region of the electromagnetic spectrum whose wavelength extends from 2.5 microns to 50 microns or whose wavenumber (number of waves per centimeter) extends from 4000 cm^{-1} to 200^{-1}. *The wavenumber (cm^{-1}), which is used extensively in infrared spectroscopy instead of wavelength or frequency, is simply the inverse of the wavelength.*

$$\text{Wavenumber} = \frac{1}{\text{Wavelength}} \qquad \text{(Eq. 1)}$$

There are 3N-6 degrees of freedom for a non-linear molecule, where N is the number of atoms in the molecule. Those vibrations which occur with a change in dipole moment will absorb radiation at characteristic frequencies in the infrared region of the spectrum and will be observed in the infrared spectrum. Those that occur with a change in polarizability will be observed in the Raman Spectrum, which is thus complementary to infrared spectroscopy. In both cases, a spectrum is produced which is as unique as a fingerprint for the material being studied.

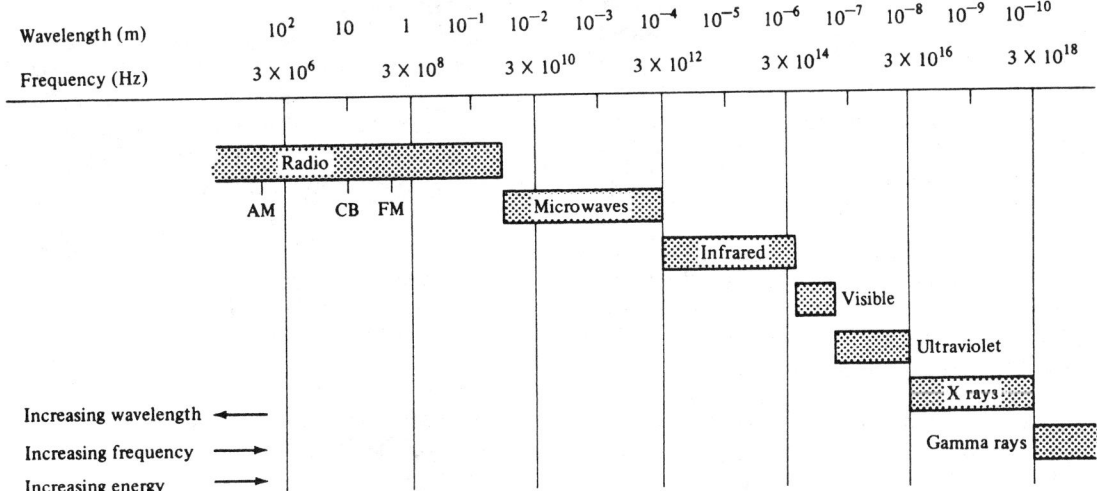

Fig. 1 The electromagnetic spectrum, showing the relative position of infrared between microwaves and visible light. (Adapted from Fundamentals of Organic Chemistry, by H. Richey, Prentice-Hall, 1983.)

Table 1 Commonly encountered bonds and their absorption regions.

Bond	Characteristic absorption
p=O	$1316\,cm^{-1}$
Si–O	$1080\,cm^{-1}$
B–O	$1370\,cm^{-1}$

In general, the FA Engineer can consult a reference text which tabulate the more common absorption frequencies for major vibrations of interest. In order to identify unknowns, most modern FTIR Systems can be equipped with reference libraries and a computer algorithm will allow the library to be rapidly searched and the best full spectral match or matches will be displayed. It should be noted, however, that no library currently on the market covers all possibilities which will be encountered in the course of semiconductor failure analysis work. This is especially true of plasma residues, which are unique. Moreover, in the case of a mixture of components, complexities are expected in interpretation. Nevertheless, it will always be possible to explain the more prominent vibrational bands which will provide clues as to the identity of the residue.

Table 1 provides an example of some commonly encountered boron, silicon, and phosphorus bonds and the absorption regions where the vibrational modes are expected. It should also be noted that frequency shifts of vibrational modes can be expected and are diagnostic of interactions as are changes in intensity and band shape. Moreover, in the solid state one must also consider phonons, or collective lattice motions, which can give rise to vibrational modes. In the case of polymer analyses, the infrared spectrum has bands which arise from absorption of energy by relatively localized parts of the chemical molecules in the sample. These parts are the functional groups contained in the polymer chain.

EXPERIMENTAL CONSIDERATIONS

Tremendous advancements have been made in the area of infrared spectroscopy in the past two decades due to the advent of (1) Fourier Transform Infrared Spectrometers, (2) computerized data processing procedures, (3) modern sampling techniques, and (4) the development of microscopic capabilities. These advancements are largely responsible for the widespread use of infrared spectroscopy in the semiconductor industry. The limitations of infrared spectroscopy can largely be attributed to the use of a black-body source and thermal detectors.

Almost all modern infrared spectrometers used in a semiconductor facility today are of the Fourier Transform Type. This type of instrument makes use in most cases of a Michelson Interferometer, and they appeared in the market in the 1960s. The FTIR concept was adopted largely to overcome the instrumental limitations of dispersive spectrometers based on gratings, and prior to the 1950s on prisms. These limitations included poor resolution, stray light, wavelength inaccuracies, and lack of sensitivity and speed.

To most effectively utilize FTIR, the FA Engineer should be acquainted with not only the specific instrument hardware and software but also sampling techniques used for obtaining the

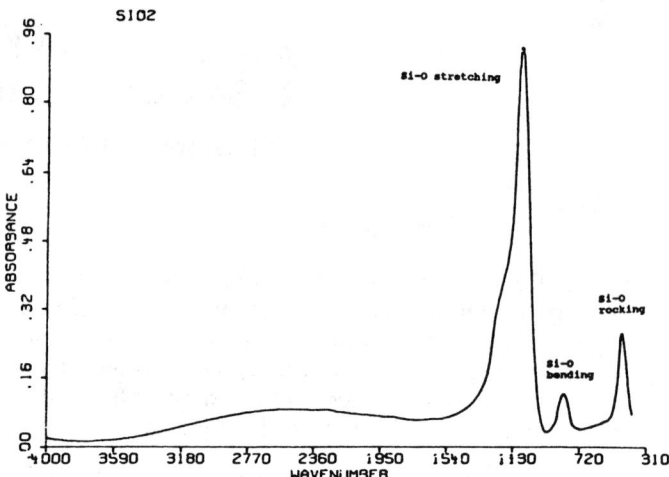

Fig. 2. Infrared spectrum of a silicon dioxide (SiO_2) thin film.

infrared spectrum. There are six techniques commonly used, which are transmission, external reflection, internal reflection, diffuse reflection, emission, and photoacoustic spectroscopy. Some of these modes are shown diagrammatically in Fig. 1. The transmission mode is the most popular while the emission mode is still in its infancy.

Transmission measurements should be utilized whenever possible, since the ease of obtaining and interpretation of the spectral information is simplified. Moreover, quantitative measurements are readily made from transmittance spectra by the application of Beer's Law, which is stated in the following equation:

$$A = eLC \qquad (Eq.\ 2)$$

where A= -log %T where %T= I/I_o and in turn I_o is the incident intensity and I is the transmitted intensity, e = molar extinction coefficient, L = the path length, and C = the concentration of the absorbing species

Generally, the incident beam is perpendicular to the sample surface, but this need not be the case. The key consideration in the use of the transmission mode of sampling is to ensure that the sample is thin enough, or diluted with a non-absorbing medium, to prevent total absorption of the beam. There are a variety of approaches for thinning or diluting the sample. One convenient approach is to dissolve the sample in a suitable volatile solvent, such as acetone, and to place the resulting solution on a silver chloride (AgCl) plate. The material of interest will be deposited on the support plate when the solvent evaporation is complete. Another potential complication that should be considered in the spectrum is the superposition of interference fringes upon the absorption spectrum. The separation between the fringes in terms of frequency f_2-f_1 is related to the thickness of the film or substrate having parallel faces, t, by the equation

$$f_2-f_1 = N/(2nt) \qquad (Eq.\ 3)$$

where N is the number of interference fringes between the two frequencies, and n is the refractive index of the medium. These

372

fringes are due to reflections at the interfaces where changes in refractive index are experienced. Such a fringe pattern will be observed for example if epitaxially grown silicon is present on top of bulk (Czorchralski-grown) silicon, and the doping concentrations are sufficiently different that there is a difference in refractive index. The equation above is also useful in determining the cell thickness.

Reflection techniques must often be used for studies of solids and thin films where it is inconvenient or not possible to prepare a sample for transmission work. This is often the case for FA work. Specular reflection is utilized when the surface is highly reflective, such as for metal surfaces, where the incident beam is reflected at the substrate interface. No beam penetration into the sample occurs, and the angle of reflection is equal to the angle of incidence. The bands will actually be derivative in shape. In cases where there is a thin absorbing layer on the surface of a more reflecting substrate, one must deal with the technique known as reflection absorption spectroscopy (1). When the angle of incidence is greater than 60 degrees, we are dealing with a special case of reflection absorption spectroscopy which is known as grazing angle reflection spectroscopy. The incident radiation is polarized parallel to the plane of incidence, and the angle of incidence is slightly less than 80 degrees. Extremely thin films on highly reflective surfaces can be studied using the grazing angle attachment, since the effective path length is much larger than the actual thickness. For example, an organic layer <10 nm thick can readily be detected on a metallic substrate using a grazing angle attachment. Diffuse reflectance is utilized for powders and rough surfaces, where the scattered radiation occurs in all directions.

Internal reflection, also known as attenuated total reflection or ATR, makes use of a crystal in contact with the sample. (2) It is desirable to select the crystal to ensure that there is some beam penetration into the sample at the crystal/sample interface. Intensification of the spectra will occur by multiple reflections. ATR is often useful for studies of liquids as well as solids. The contact between the sample and the internal reflection element is a critical factor in ATR. Thus, ATR is most useful for solid samples which are very flat and have a low degree of surface roughness. One of the benefits of ATR is that it makes use of multiple reflections which serve to maximize the band intensity, whereas in the case of external reflection a single reflection may be required to produce a usable spectrum. It is often possible to vary the angle of incidence between the IR beam and the ATR crystal and obtain depth profiling information. ATR spectra appear similar to transmission spectra, but the band intensities change significantly from one end of the spectrum to the other, which is due to the varying depth of penetration into the sample by the IR beam. In fact, the most often cited problem with reflection spectra is that they are generally distorted and hard to interpret. Reflection spectra of multi-film structure can be quite complex. Some of the techniques used in spectroscopic ellipsometry for simulation of spectra can be applied to infrared reflection spectra analysis.

There are a variety of accessories on the market which are utilized for determining spectra in the transmission or reflection modes. A current listing of suppliers can be found in the *Analytical Chemistry Lab Guide* and the *American Laboratory Buyers' Guide*. These accessories fit into the sample compart-

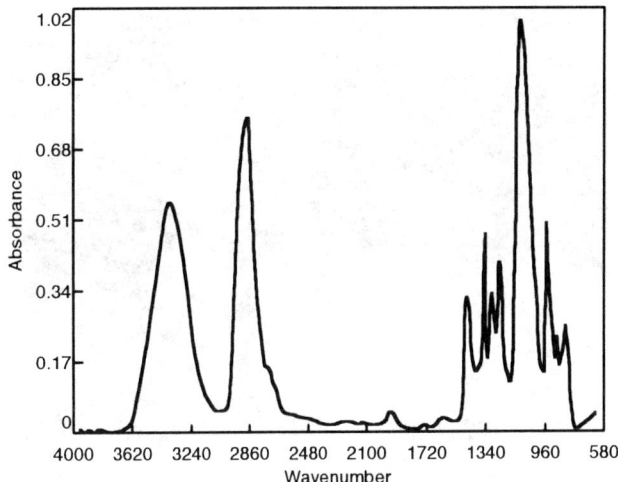

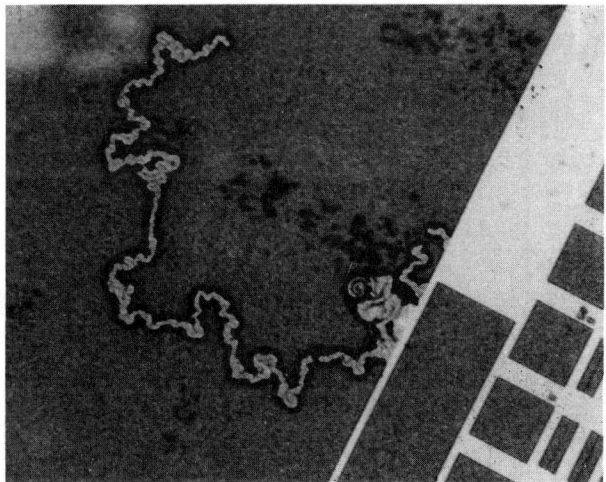

Fig. 3. Polymerized ethylene glycol on an integrated circuit, identified by infrared microanalysis.

ment of the spectrometer and generally require some alignment unless they are made specifically for the instrument that is being utilized. As a general rule, these accessories can be utilized if the sample is 200 microns or greater and at least 1 micron in thickness in size. Samples between approximately 200 microns and 20 microns, which is the diffraction limit, are handled with a special microscope attachment designed for transmissive and reflective modes of spectral acquisition. Special sampling accessories can be used in conjunction with an infrared microscope, and these accessories fit on the microscope stage. More information on microsampling is provided below.

Other general considerations which should be made in determining spectra is the selection of an appropriate reference. In some cases, the spectra of the sample is referenced against air. In other cases, such as the determination of a transmission spectrum borophosphosilicate glass (BPSG) thin films, it is desirable to use the wafer upon which the film was deposited or an identical wafer from the same lot as the reference. The spectrum of the reference will be mathematically subtracted from

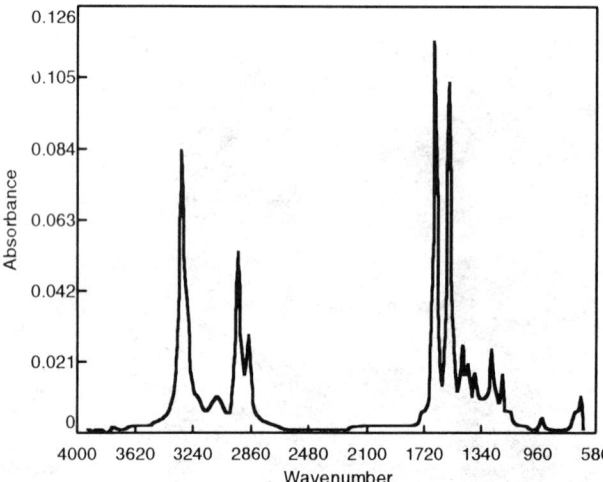

Fig. 4 Contaminant in photoresist identified as a polyamide (Nylon®) particle.

the composite spectrum leaving the spectrum of the BPSG film only. This will aid in interpretation and often in quantification. In the case of specular reflection being used for residue analysis on a bond pad, the reference chosen might be an area of the metal surface having no residue film.

Optimization of the signal-to-noise ratio can usually be achieved by increasing the number of scans. The number of scans generally must be changed by a power of 2 in FTIR, such as 526 or 1028. Of course, the time for the analysis will increase as the number of spectral scans is increased. The signal-to-noise ratio improvement generally goes up as $N^{0.5}$ where N is the number of scans. Other parameters which often must be optimized include sample purging, sample alignment, and selection of the interferometer data acquisition and Fourier transformation process itself.

The two most widely used **detectors** in the semiconductor industry are the liquid nitrogen cooled mercury-cadmium-telluride (mer-cad telluride, or MCT) photoconductivity detector and the triglycine sulfate (TGS) proelectric bolometer detector, which can be operated at ambient temperature. The MCT detector can be of the wide band or the narrow band type. Microsam-

pling generally requires a narrow band MCT detector, which makes it impossible to go to wavenumbers much below 600 cm^{-1}.

Modern data analysis systems afford the opportunity for spectral searches and comparisons as has already been noted. Software generally also exists for smoothing, spectral subtraction which is very useful for observing small differences, baseline correction, and other more advanced spectral analyses such as establishing derivatives or performing spectral deconvolution and bandshape analysis.

For more complete details on FTIR theory and instrumentation, the reader is referred to the more common references in this field. (3,4,5,6)

SPECIFICS OF MICROANALYSIS

In general, infrared microscopy has been widely used in a variety of fields besides semiconductor failure analysis, such as forensics and polymer science. It identifies or characterizes particles, fibers, and inclusions or imperfections which are embedded in a larger matrix and to perform spectral mapping. This technique allows spectra in many cases to be determined with less than a nanogram (10^{-9} gm) of material. Although the main discussion here is focused on the application of FTIR microscopy to particle analysis in the semiconductor industry. The reader who is interested in semiconductor applications of infrared microscopy to spectral mapping such parameters as epitaxial thickness, interstitial oxygen concentration, and boron and phosphorus oxide measurements is referred to the article by Krishnan.(7)

Particle Analysis

Proper isolation of the particles for study is key to the success of microanalysis. This should be accomplished in a manner which prevents the introduction of external contamination, which can lead to erroneous conclusions. This often means working within an area having laminar flow hoods. In cases where it is not practical to remove the particle or residue from the matrix or substrate, it will be necessary to use the reflection mode unless the substrate has a suitable transmittance. Typical examples of infrared microanalysis are demonstrated in Fig. 2 and 3.

The particles, if removed, must be placed on a support and then placed on the X-Y stage of the microscope. This support could be a silicon wafer itself or it might be a transmissive substrate such as an AgCl crystal or a potassium bromide (KBr) "salt plate". Another approach is the use of the Diamond Anvil Cell (available from High Pressure Diamond Optics, Arizona) in which the particles, fibers, residue, etc. is compressed under high pressure to a thin layer. In the use of this accessory, it is important to check for window parallelism before the application of pressure; good parellism is indicated by the appearance of a barely noticeable interference pattern. The nature of the sample will dictate not only the support but also whether the sample is studied in the transmissive or reflective mode.

The next step is to position the sample at a point where the IR beam will traverse during the analysis. To make this possible an IR microscope is designed with both an optical and a microspectroscopy system. Using familiar optical microscopic techniques, the desired particle or sample area is positioned and

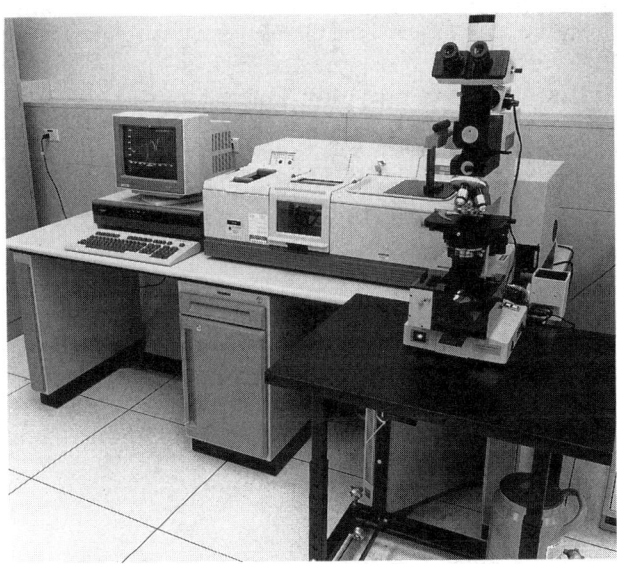

Fig. 5 An example FTIR system, with the main analyzer at center. The microscope in the foreground is equipped with both glass and Cassegrain optics. The system has a library of over 50,000 compounds, including organics and inorganics from commercial chemical suppliers, steroids, coatings, solvents, and flavors and fragrances.

brought into optical focus by using stage and focus adjustments. Once this is accomplished, the image is masked with a fixed or adjustable aperture. Most IR microscopes are equipped with variable apertures which can mask a sample of any shape. As the size of the aperture is decreased, the spectral noise level will increase.

The IR microscope and the optical system share a common intersection point at this remote image, so then it becomes a simple matter to simply covert the microscope from the optical viewing mode, which can either be transmissive or reflective, to the corresponding IR microspectroscopy mode. This conversion is most often achieved by rotating a knob or pulling a lever. To obtain a satisfactory spectrum, one must then optimize the acquisition parameters in a manner which is similar to that used for obtaining a quality spectrum for a macro-size sample.

In selecting an infrared microscope, the key parameters that must be considered are the numerical aperture, magnification, and the working distance. Since the numerical aperture is a measure of the collection efficiency, it is directly related to the sensitivity of the measurements. Cassegrain optical systems which are based on reflecting optics are utilized in such microscopes since ordinary lens material used as objectives in optical microscopes do not have acceptable transmittance properties for IR radiation.

COMPARISON OF FTIR WITH OTHER ANALYTICAL TECHNIQUES

It is very important for the FA Engineer to be able to select the appropriate analytical tool for the problem being investigated. This can best be accomplished by acquiring information on the sensitivity and capabilities of the wide array of analytical tools which are now available. In the case of particulate con-

tamination problems, the role of FTIR must be compared to a number of other readily available instrumental techniques. Chemical composition can often be established by an energy dispersive X-ray system (EDS) attached to a SEM. X-Ray fluorescence (XRF) is also useful if the contamination area is larger than 100 microns. Auger spectroscopy (AES) or scanning Auger spectrometry (SAM) can provide elemental identification with more sensitivity than an EDX system. Secondary Ion Mass Spectroscopy (SIMS) is even more sensitive than Auger Spectroscopy, but lacks the spatial resolution of Auger, especially if the Auger has field emission capability. Molecular level information can be sought with FTIR if the contaminants are greater than 20 microns in size. When the residual contaminants or particles approach or exceed 150 microns in size, electron spectroscopy for chemical analysis (ESCA) can also be utilized. Often useful data can be determined with particles as small as 10 microns with the FTIR.

If the particles are between 10 microns and 0.5 microns, the Raman Microscope is the clear choice. Moreover, the Raman spectral information will provide a valuable supplement to the infrared data for molecules of the particles which are larger than 10 microns. Besides improved size resolution, the Raman Microprobe has certain other advantages over FTIR which should be considered. The preferred method for determining Raman spectra is in the reflectance, or more appropriately the backscattering, mode in contrast to FTIR. No interpretative difficulties are encountered with Raman spectra which have been determined in the backscattering mode as is the case for FTIR. Less sample preparation is generally required for determining Raman spectra. Aqueous solutions present no problems for Raman spectroscopy but are difficult to analyze by for FTIR.

In general, the Raman spectra of ionic inorganic materials are sharp while the FTIR spectra are very broad. On the other hand, there are presently a number of advantages of the FTIR technique over the Raman technique. Raman Microprobes are not as readily available as FTIR Systems. The Raman instrumentation is also more expensive, and the need for high intensity laser excitation is accompanied by safety considerations. The Raman spectral data bases are less extensive than the FTIR data bases. There are also fewer individuals trained in the use of Raman spectroscopy. In the case of organic analysis, fluorescence problems can be encountered with the Raman technique. No comparable problems must be overcome in organic analysis using FTIR.

There are fewer instrument manufacturers which specialize in Raman spectroscopy in comparison to FTIR systems, and most of the Raman instruments are not domestic companies. In addition, it is generally necessary to construct a Raman System from components. Thus, at the present time and for sometime into the future the FTIR approach will be the method utilized for determining vibrational information which can be used for the identification of contamination sources in the semiconductor FA Laboratory.

SELECTED REFERENCES

1. Carter, R.O., III, C.A. Gierczak and R. A. Dickie, Applied Spectroscopy, 40, 649, 1986.

2. N. J. Harrick, Internal Reflection Spectroscopy Wiley Interscience, New York, 1967.

3. P.R. Griffiths, Chemical Infrared Fourier Transform Spectroscopy, Wiley, New York, 1975.

4. R. J. Bell, Introductory Fourier Transform Spectroscopy, Wiley, New York, 1975.

5. J.R. Ferraro and L. J. Basile (eds.), Fourier Transform Infrared Spectroscopy, Vol1, Academic Press, New York, 1982.

6. J. R. Ferraro and K. Krishnan, Practical Fourier Transform Infrared Spectroscopy: Industrial and Chemical Analysis, Academic Press, Inc., 1990.

7. K. Krishnan Characterization of Semiconductor Silicon Using the FT-IR Microsampling Techniques, in Infrared Microspectroscopy Theory and Applications, R.G. Messerschmidt and M.A. Harthcock, eds., Marcel Dekker, Inc., New York, New York, 1988.

Discrete/Passive Component Analysis

Failure Analysis
of Passive Components

Stan Silvus
Southwest Research Institute
San Antonio, Texas

INTRODUCTION

Why worry about passive-component failures? After all, passive components never fail; besides that, they are too cheap to be concerned about. Right?

WRONG! Failure of any component, passive or otherwise, on the production line translates directly into rework cost, increased scrap, and loss of efficiency. Worse yet, failure of any component within the product warranty results in even greater replacement expense, to say nothing about adverse effects on consumer confidence and, therefore, on future sales. Further, product liability and customer safety are (or should be in this era of frivolous lawsuits) of great concern. Clearly, though, a fire in the customer's facility caused by failure of a "cheap" passive component can result in consequent-damage and punitive assessments that far outweigh the cost of numerous failure analyses and corrective actions. The point is that failures of passive components cannot be economically or practically ignored. In some cases, a simple analysis will satisfactorily reveal the cause of the problem, but often, more in-depth analysis is required to establish the real cause of failure.

According to statistics compiled at a third-party electronic-equipment repair facility, field failures of components in electronic equipment are distributed as shown in Figure 1. Integrated circuits account for less than one-third of field failures. A significant reason for this is that integrated-circuit manufacturers do a lot of in-plant testing and analysis so that the product that reaches the customer has a low inherent failure rate; thus, a high percentage of integrated-circuit field failures are caused by some form of customer-induced overstress (electrical or mechanical). In contrast, many passive-component manufacturers do not test their final products at all or, at best, perform minimal testing. This is one reason why passive-devices have the greatest share of field failures; some of the subsequently presented case histories will emphasize this point.

TYPICAL PROBLEMS

Open circuits and short circuits are probably the most common passive-component failure modes, but more subtle embodiments of these extremes are also encountered; for example, open circuits or short circuits may be intermittent, or only device parameters may have changed. Mechanical damage is sometimes a failure mode.

Typical causes of these failure modes include electrical overstress, misapplication, improper handling, age (i.e., normal wear-out mechanisms), design or manufacturing defects, contamination, corrosion, metal migration or metal whiskers, hermeticity loss or fluid leakage, and fire or explosion. Sometimes, a simple external cause, such as misplaced or excessive solder, causes an apparent in-circuit malfunction that disappears when the "failed" component is removed from the board or equipment; in this case, out-of-circuit tests show that the component is fully functional.

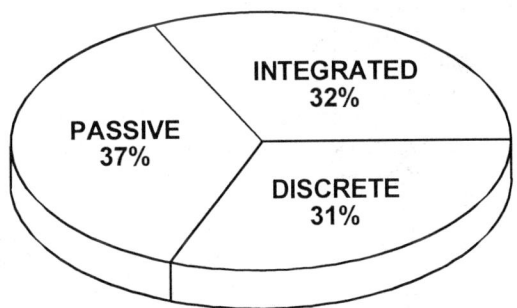

Figure 1. Distribution of electronic-equipment field failures among various types of components

PASSIVE-COMPONENT TYPES

This tutorial article will not attempt to address all possible classes and subclasses of passive components; the variety is just too great. However, selected important types of resistors, capacitors, wound components, electromechanical devices, and miscellaneous passive devices will be discussed.

Resistors

There are many types of resistors. In the early days of consumer electronics, carbon-rod resistors were common; however, these are now obsolete, so they will not be addressed further. The carbon-rod resistor was supplanted by the hot-molded composition type; even though composition resistors are no longer widely used in modern electronic equipment, they are frequently encountered in equipment that is only a few years old. Metal- or carbon-film resistors are currently the types of choice for axial-lead applications, and thick-film chip or thin-film-on-silicon resistors, both of which can be surface mounted, are widely used in compact electronic equipment. Finally, wirewound resistors are used in power applications, and variable resistors in various forms are still much in evidence.

Composition Resistor. A hot-molded composition resistor comprises a cylinder made of graphite and binders; the graphite-to-binder ratio sets resistivity of the material and, hence, terminal resistance of the resistor. Barbed ends of wire leads are imbedded in opposite ends of this cylinder, as apparent in Figure 2, and the assembly is molded in a thermosetting polymer, often a phenolic resin.

Figure 2. Longitudinal cross section through composition resistor showing internal structure

Composition resistors have a history of being moisture sensitive, so resistance change (i.e., a parametric change) is a common failure mode; the failure mechanism behind this parametric shift is absorption or desorption of moisture. Electrical overstress, principally overheating caused by dissipation of greater-than-rated power, is another important failure mechanism; this type of failure is usually accompanied by discoloration, swelling, and/or cracking of the molding compound. Finally, composition resistors are relatively fragile, so mechanical damage (e.g., a cracked or broken package) is another possible failure mechanism.

Film Resistor. A thin film of metal, carbon, or a metal oxide uniformly deposited on the external surface of a cylindrical ceramic core is the basis of a film resistor (Figure 3). After the film has been deposited, cup-shaped metal caps, to which wire leads have been welded, are pressed onto the ends of the ceramic rod. Then, a helical kerf is cut in the film using focused laser energy (formerly, this was done by abrasive-jet blasting) until the end-to-end resistance comes within the desired tolerance band. Following this operation, the resistor is conformally coated with or molded into a thermosetting polymer.

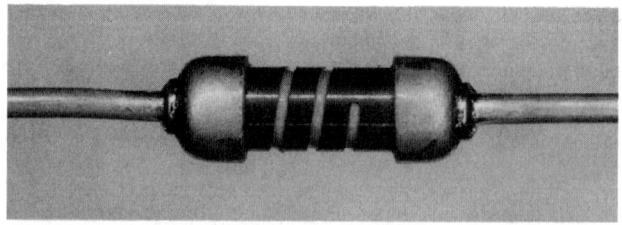

Figure 3. Decapsulated metal-film resistor

Resistances of most film resistors are fairly stable, even under adverse operating conditions, but open or short circuits and parametric changes are possible failure modes. The axial-lead film resistor and, especially, the thin film of such a resistor, is delicate, so mechanical damage (e.g., scraped or cracked film or a cracked or broken package) is a possible failure mechanism. Also, electrical overstress, accompanied by excessive heating, can cause swelling, cracking, and/or discoloration of the coating or package material. A failure mechanism unique to film resistors is intrusion of the coating

material or a corrosive material between an end cap and the film-coated ceramic core; when this occurs, the resistor usually opens.

Surface-Mount Chip Resistor. Numerous chip resistors are fabricated simultaneously on a thin ceramic substrate by a screen-deposition or photolithographic process. Typically, metallic terminations (pure silver or a silver-palladium alloy) are deposited and fired first; note that silver-palladium terminations are preferred because they are less susceptible to silver migration. Next, the resistive material, often ruthenium oxide, is deposited as a thick film, and this layer is covered by a fired overglaze. At this point in the process, the resistors may be probed and laser trimmed to final value. The last steps in fabrication include coating with a polymer, dicing the substrate to produce individual resistors, and applying solderable terminations to both ends of each device. Top and bottom views of typical chip resistors are presented in Figure 4. Terminations are usually solder; a thin barrier layer of nickel may be applied before the solder to prevent leaching of the fired termination. Thick-film resistor networks are fabricated by a similar process.

Figure 4. Top (left) and bottom (right) views of typical thick-film chip resistors

Like all resistors, chip resistors may fail open or shorted, or they may exhibit parametric changes. Similarly, they may be electrically overstressed, a failure mechanism that is usually accompanied by visible signs of overheating. Delamination between layers of a chip resistor may permit intrusion of solder flux, cleaning solvents, or water, and these foreign materials may cause corrosion or parametric changes.

Thin-Film-on-Silicon Resistor. Thin-film-on-silicon resistors and resistor networks are widely used in modern hybrid circuits and other very compact electronic products. These resistors, a typical example of which is illustrated in Figure 5, are manufactured using photolithographic techniques similar to those used in fabrication of semiconductor devices. These resistors are normally used in bare-die form, so they are not individually packaged.

Typically, the starting material is a surface-oxidized wafer of silicon. A thin film of metal is deposited on top of the field oxide, and the metallic layer is patterned to form numerous resistors or networks of resistors. Usually, an additional layer of metal is deposited in the bond-pad areas to improve wire-bonding properties. Normally, the resistors are not passivated. The last steps in fabrication include probing and laser trimming the individual resistors while they are still in wafer form. After trimming has been completed, the wafer is diced to yield the final product.

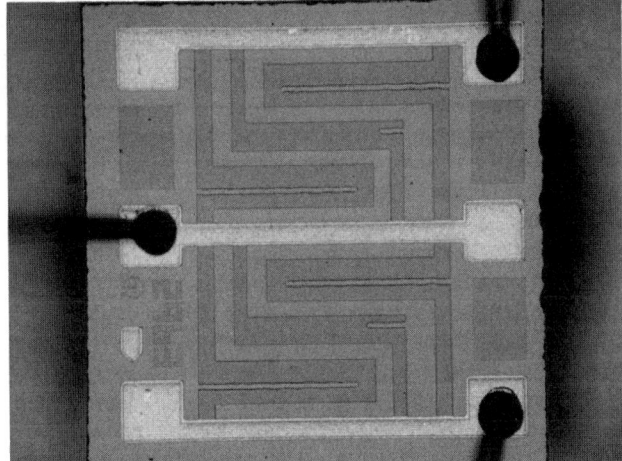

Figure 5. Thin-film-on-silicon resistor

Wirewound Resistor. A metal wire or ribbon wrapped around a ceramic-tube core is the major component of a wirewound resistor. The wire or ribbon is usually terminated by spot welding to metal caps slipped over the ends of the ceramic form or to circumferential metal bands crimped around the ceramic tube. The resulting assembly is conformally coated with a vitreous enamel or with ceramic. Sometimes, as shown in Figure 6, a longitudinal strip is masked while the conformal coating is being applied so that the resistor may be tapped or so that its resistance may be adjusted.

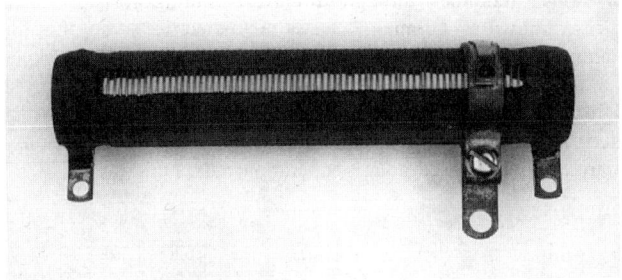

Figure 6. Adjustable wirewound resistor

The most common failure mode of a wirewound resistor is an open circuit, but parametric changes are possible. Electrical overstress, which usually causes fusing of the wire or ribbon, corrosion, which may sever the resistive material or establish a resistance-reducing path in parallel with the resistance element, and separation of the terminal weld are the most frequently observed wirewound-resistor failure mechanisms. In rare instances, poor process control causes nonuniform spacing between turns of the wire or ribbon; if sufficiently high voltage is impressed across such a resistor, arcing between turns may occur.

Capacitors

The earliest capacitor was the Leyden jar, which was invented circa 1745. Through the ensuing centuries, many improvements in design techniques, materials, and manufacturing processes have been developed. Despite these improvements, capacitors are still considered by many circuit designers to be weak spots in their products, and the trend is to minimize the quantity of capacitors included in a piece of electronic equipment. Nevertheless, in many cases, there is no satisfactory substitute for a capacitor.

Many types of capacitors exist; included are film, ceramic, mica, aluminum electrolytic, tantalum electrolytic, and variable types. Film capacitors are usually identified by the dielectric material, so there are, for example, polyester, polycarbonate, polypropylene, polytetrafluoroethylene, and polystyrene types; for purposes of this discussion, a paper dielectric is also considered to be a film. In some film capacitors the dielectric is a laminate (e.g., paper-polyester); properties of such capacitors may be tailored by selection of the dielectric materials in the laminate and their relative thicknesses.

Monolithic Ceramic Capacitor. In modern electronic equipment, the monolithic ceramic capacitor is probably the most widely used type. Such devices are manufactured by a screen-printing process in which one plate for each of many capacitors is printed on a sheet of green ceramic. A stack, comprising many layers of these printed sheets of green ceramic, is built, and an unprinted green-ceramic cover layer is applied to each side of the stack. The stack is then compacted in a hydraulic press (note that the ceramic is soft and tacky while it is in the green state), following which it is diced to form individual capacitors. The dice are then fired in a conveyor kiln. The final operations are application and firing of solderable terminations. Important construction features of a monolithic ceramic capacitor are visible in the cross section illustrated in Figure 7. However, the capacitor shown in Figure 7 has one additional unusual feature; that is, parts of the ceramic body and plates have been removed by abrasive-jet blasting to trim the capacitance to a specific value, and the eroded volume has been backfilled with a polymer.

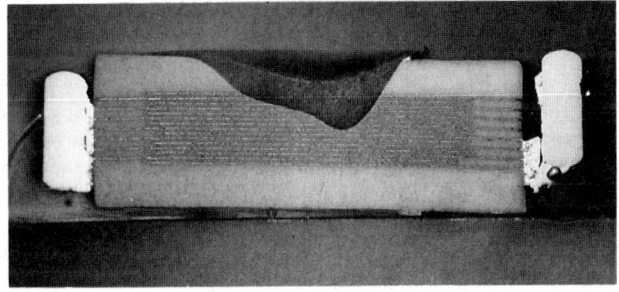

Figure 7. Cross section through trimmed monolithic ceramic capacitor

Formerly, plates in monolithic ceramic capacitors were pure silver, but now the plates are almost invariably a silver-palladium alloy that is less susceptible to silver migration. Terminations are usually silver, and if the capacitor is intended for surface-mount applications, terminations may be coated with thin nickel barrier layers and then solder.

Short circuits and capacitance changes are the most frequently encountered failure modes for monolithic ceramic capacitors. These failure symptoms may be caused by thermal or mechanical fracturing of the ceramic dielectric, internal delaminations or voids caused by too rapid heating during firing, or electrical overstress (usually, overvoltage). Also, silver migration is not totally out of the picture, so this failure mechanism must be considered during analysis.

Film Capacitor. In one type of film capacitor, two long, narrow strips of aluminum foil and two longer and slightly wider strips of film dielectric are stacked and rolled into a "jelly-roll" configuration. Often, plates in this type of capacitor are terminated by inserting wires between a dielectric layer and the corresponding plate. More modern capacitors incorporate two layers of dielectric film that have each been metalized on one side; these strips are stacked and rolled.

In more recent foil-and-film capacitors and in most metalized-film capacitors, "extended-foil" construction is used. One edge of one plate is flush with an edge of the dielectric, and the opposite edge of the other plate is flush with the opposite edge of the dielectric. Ends of the jelly roll are sprayed with metal (usually, zinc), and wire leads are soldered to the metallic termination. This configuration desirably decreases effective series resistance (ESR) and inductance of the capacitor. After the jelly roll has been terminated, the assembly may be packaged in a polymeric conformal coating or in a molded polymeric housing. Figure 8 shows a transverse cross section through a packaged "extended-foil" metalized-film capacitor.

Figure 8. Transverse cross section through packaged extended-foil metalized-film capacitor

The types of film capacitors discussed so far usually have wire leads, so they are intended for mounting on through-hole circuit boards or for point-to-point wiring. However, the modern trend toward surface mounting led to development of the stacked-film capacitor. This type of capacitor is fabricated by stacking sheets of dielectric film that have been metalized in a pattern of plates similar to the way that monolithic ceramic capacitors are made; however, the metalized dielectric-film sheets are held together by wax or another suitable adhesive. After the stack has been assembled and compacted, it is diced to yield individual devices, and solderable sprayed-metal terminations are applied to each device. The drawing presented in Figure 9 shows in cross section how a stacked-film capacitor is constructed.

Failed film capacitors may exhibit short circuits, open circuits, or parameter shifts. An open circuit is usually caused by separation of the lead wire from the termination. Generally, short circuits are caused by excessive voltage or by contamination (e.g., metallic inclusions in the jelly roll). Moisture intrusion, corrosion, and melting of the polymeric-film dielectric are possible causes of parametric changes. Metalized-film capacitors in which localized dielectric breakdown occurs may self-heal if energy dissipated during the short-circuit event is sufficient to evaporate the thin metallic film surrounding the failure site, but not great enough to cause an arc that melts or carbonizes the dielectric film.

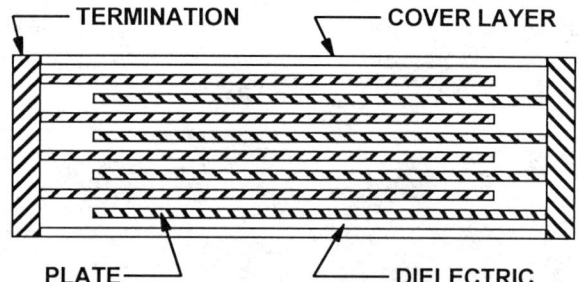

Figure 9. Cross-sectional drawing showing internal structure of stacked-film capacitor (definitely not to scale)

Aluminum Electrolytic Capacitor. Two etched aluminum-foil strips comprise the plates in an aluminum electrolytic capacitor. Etching greatly increases surface areas of the plates, and this is one of the reasons that high capacitance-voltage products are available in relatively small packages. Plates are terminated with transversely oriented aluminum ribbons that are crimped to the plate material. Two terminated plates and two or more slightly wider strips of porous paper are formed into a jelly-roll assembly. The paper separators are saturated with a liquid electrolyte, and the jelly roll is placed in an deep-drawn aluminum can that is plugged at the open end with a rubber seal that, in turn, is penetrated by the device terminals; it is important to note that this structure is not hermetically sealed. On all but the smallest aluminum electrolytic capacitors, a pressure-relief vent is provided; in many devices, this vent is a longitudinally oriented indentation in the can wall, but in some larger devices, there may be a rupture disk molded into the rubber seal plug. A typical aluminum electrolytic capacitor with the can removed is shown in Figure 10.

After the electrolyte-saturated jelly-roll assembly has been sealed in the can, a unipolar electrical potential is impressed across the device terminals through a current-limiting resistor. This process, known as "forming," causes the dielectric, a very thin film of aluminum oxide, to grow on the anode plate. Dielectric thickness is a function of working voltage, so the forming voltage is set slightly higher than the desired working-voltage rating to provide a safety margin. Note that an aluminum electrolytic capacitor is polarized, and reversed-polarity operation causes failure.

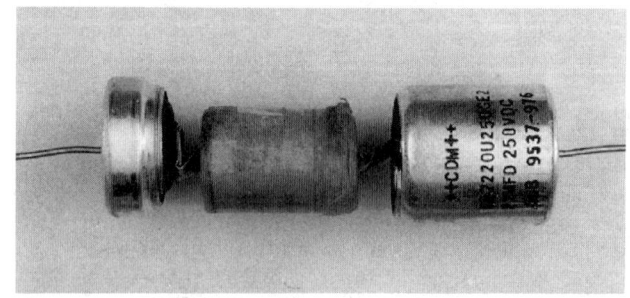

Figure 10. Decapsulated aluminum electrolytic capacitor

Short circuits, excessive leakage current, decreased capacitance, and increased effective series resistance (ESR) are common failure modes for aluminum electrolytic capacitors. Dielectric breakdown caused by overvoltage or metallic inclusions in the jelly roll are leading causes of short circuits.

For proper operation, it is necessary that the paper separators in an aluminum electrolytic capacitor be saturated with liquid electrolyte. The fact that the housing is not hermetically sealed gives rise to one of the most common failure mechanisms of this type of capacitor: namely, loss of electrolyte. As the electrolyte escapes from the capacitor housing, the separators dry out, and ESR increases. Because aluminum electrolytic capacitors are frequently used in power-supply ripple filters, they have significant alternating current flowing through them. Of course, this current flows through the ESR of the capacitor, and Joule (i.e., I^2R) heating occurs. As capacitor temperature rises as a result of this internal heating, both ESR and rate of electrolyte loss increase. An increase in ESR implies increased Joule heating, which, in turn, implies further increase in ESR. The result is a thermal runaway mechanism that ultimately leads to catastrophic failure of the capacitor (i.e., short circuit, rupture of the pressure-relief vent, and/or expulsion of flame). Thus, unlike many types of electronic components, the aluminum electrolytic capacitor has a definite wear-out mechanism.

Another mechanism that causes failure of numerous aluminum electrolytic capacitors is overcurrent. Each aluminum electrolytic capacitor has a frequency-dependent maximum-current limit that is usually published in the manufacturer's catalog. Many circuit designers ignore this limit for various reasons; however, lack of attention to this limit causes excessive internal heating and accelerated electrolyte loss that lead to the wear-out type of failure described in the previous paragraph; the difference is that failure occurs prematurely when the capacitor is subjected to excessive alternating current.

Finally, chlorinated solvents of the types that are sometimes used for cleaning circuit-board assemblies after soldering can penetrate the rubber seal of an aluminum electrolytic capacitor. In the environment that exists inside such a capacitor, chlorinated solvents decompose liberating hydrochloric acid that is highly corrosive to aluminum. This mechanism should be suspected when the interior parts of a failed aluminum electrolytic capacitor are corroded and the corrosion products contain chlorine. Seal ends of some aluminum electrolytic capacitors are coated with an epoxy resin in an attempt to prevent ingress of chlorinated solvents.

Tantalum Electrolytic Capacitor. The core of a dry tantalum electrolytic capacitor is a porous slug made of sintered tantalum particles; a tantalum anode wire is embedded in one end of the slug. After the slug has been sintered, it is anodized to produce the thin tantalum pentoxide dielectric film. Compacted around the outside of the anodized slug is the dry electrolyte, which comprises a mixture of graphite and manganese dioxide. A metallic cathode (usually silver, but sometimes copper) is sprayed on five surfaces of the dry-electrolyte-coated slug; of course, the cathode material should not cover the end of the slug that has the anode wire projecting from it. Solderable wire leads are attached; the anode lead is spot welded to the tantalum anode wire, and the cathode lead is soldered to the cathode material. Finally, the capacitor assembly is packaged; common package types include conformal coating with a polymer, molding in a thermosetting polymer, and sealing in a metal can. Figure 11 shows a decapsulated surface-mount tantalum electrolytic capacitor.

Particularly in surface-mount tantalum electrolytic capacitors, a short length of fusible wire may be included in series with the anode lead wire. This overcurrent-protection device helps to prevent a fire or other damage to the circuit board if the capacitor develops a short circuit in a high-energy-capacity circuit. Such a fuse is pointed out in the x-radiograph and cross-sectional view presented in Figures 12 and 13, respectively.

Typically, tantalum electrolytic capacitors fail with short circuits or open circuits. Open circuits are usually related to a blown fuse wire, which was probably caused by a short circuit.

Short circuits may be caused by overcurrent (see next paragraph), overvoltage, manufacturing defects, or reversed-polarity operation. Some short circuits in tantalum electrolytic capacitors will self-heal if energy dissipated at the short-circuit site is not too great. The healing mechanism is reoxidation of the tantalum, and this process occurs because interstices of the porous slug contain manganese dioxide, which is an oxidizing agent.

Figure 11. Decapsulated surface-mount tantalum electrolytic capacitor

Figure 12. X-radiograph of surface-mount tantalum electrolytic capacitor showing fuse

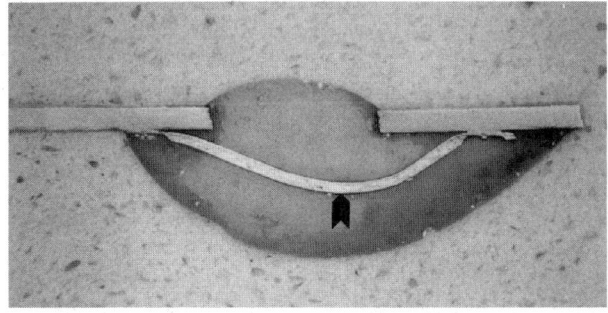

Figure 13. Cross section through fuse wire (arrow) in surface-mount tantalum electrolytic capacitor

Tantalum electrolytic capacitors are sensitive to surge currents. Hence, it is not good design practice to connect a tantalum electrolytic capacitor directly across a low-impedance voltage source. Instead, it is necessary to insure that there is sufficiently high impedance in the circuit to limit surge current through the capacitor (Reference 1).

Wound Components

There are many types of wound components, and within each type, there are numerous construction and package types. Because the variety is so great, wound components will be discussed generically.

Wound components include various types of transformers and inductors. Transformers may be relatively large components that handle power at frequencies in the 50-to-400-Hz range. Audio transformers are usually designed to have reasonably flat response over the audio-frequency range (i.e., 20 Hz to 20 kHz); such transformers also are designed to handle substantial power. However, many transformers are intended for use in radio-frequency amplifiers and other high-frequency applications. Such devices are usually small in size, and their construction features differ significantly from those of power and audio transformers.

Inductors usually have single windings, while transformers have multiple windings. Large laminated-iron-core inductors may be used as filter components in power supplies and in frequency-selective circuits. Smaller ferrite- or air-core inductors are widely used as decoupling devices and frequency-determining elements in radio-frequency circuits.

Windings for large power and audio transformers and large inductors often comprise numerous turns of insulated copper magnet wire wound on a polyamide bobbin or a cardboard form. When there are many layers in such windings, strips of paper may be incorporated between winding layers, but often, particularly in windings built on a bobbin, there is no interlayer insulation. In these large wound components, cores are usually made of laminations of electrical steel (i.e., silicon steel) or one of several special magnetic alloys.

Smaller transformers and inductors intended for higher-frequency applications usually have windings that are wound on polyamide bobbins. Such devices may have laminated cores or ferrite cores; in the latter case, the cup-core form is widely used. Toroidal cores with the windings wound directly around the core material are often used in filter inductors and higher-frequency power transformers.

Wound components may be packaged in a variety of ways, including no package at all. Strap-mounted open-frame construction is widely used for laminated-core wound components, but lead-mounted, surface-mounted, encapsulated, and hermetically sealed packages are also common.

Short circuits between adjacent turns within a single layer, short circuits between layers, and open circuits are the most frequently observed failure modes in wound components. Open circuits are usually caused by excessive current (i.e., fusing of the wire) or by mechanical abuse, such as overstress of a termination. Short circuits result from insulation failure, which may be caused by dielectric breakdown, overheating, or age. Dielectric breakdown may be caused by impressing an excessively high voltage across a winding; a high-voltage pulse may be caused by, among other things, inductive kickback when current through the winding is abruptly interrupted.

All types of electrical insulation, and especially the insulation used in wound components, is subject to aging, and aging is accelerated by elevated temperature. Wound components that must handle power get hot, so insulation aging is a built-in wear-out mechanism.

Corrosion of the wire can lead to either open-circuit or short-circuit failure. In fact, it is possible for a wound component to have turns that have been severed by corrosion, while continuity still exists through the winding because of an electrically conductive path through the corrosion products. In such cases, winding resistance and inductance are usually out of the specification range.

Defective solder joints, welds, or crimped connections in the current-carrying paths may cause open-circuit failures in wound components.

Electromechanical Devices

Rotating Machines. Motors, generators, and synchros are rotating machines that convert between electrical and mechanical energy. Such devices have windings that are similar in many respects to those in wound components, and, hence, these electromechanical devices have many of the same failure modes and mechanisms as wound components. Moreover, rotating components have bearings that are potential sources of failure, and some types of these devices have slip rings or commutators and brushes that provide additional opportunities for failure.

Switches. Switches may be of the rotary, toggle, pushbutton, or slide types. Common features of switches include some form of operating mechanism, which is subject to mechanical failure, and current-carrying contacts that are the primary causes of electrical failures. Contacts may wear mechanically, or they may be damaged by current flow or arcing. Usually, as contacts start to fail, contact resistance increases so that localized Joule heating increases; of course, elevated temperatures associated with such heating accelerate both contact damage and insulation deterioration (i.e., aging).

Relays. Relays are remotely actuated switches; hence, they have contacts that are subject to the same failure mechanisms as switch contacts. Further, relays have coils that are constructed like wound components and, therefore, have failure modes and mechanisms like those of wound components. Finally, relays have mechanisms of varying degrees of complexity, and, of course, mechanical failures may occur in relays.

Circuit Breakers. Circuit breakers are specialized switches that are designed to remain closed until an overcurrent event occurs. When excessive current is detected, contacts in the circuit breaker open, thus interrupting the current. Because the current must normally be interrupted under fault conditions, circuit-breaker contacts must be more rugged than those in a switch that has the same current-carrying rating.

The release or trip mechanism in a circuit breaker may be thermally and/or magnetically actuated. Generally, thermal circuit breakers are slow acting, so they are effective in detecting and interrupting long-term, moderate overcurrents. On the other hand, magnetic circuit-breaker actuators are fast (unless specifically retarded by a viscous damper or similar device), so they are useful for detecting and interrupting sudden short circuits. Of course, where both of these characteristics are desired simultaneously, both types of mechanisms may be included in the same circuit breaker.

Miscellaneous Components

The previous several paragraphs have been devoted to the mainstream electronic and electrical components; however, there are numerous other components that can cause electronic-equipment failures. Included are fuses, transducers and sensors, frequency-control crystals, connectors, lamps, circuit boards, and hardware.

Fuses. Most fuses include a fusible-alloy wire or strip that is connected in series with a current-carrying circuit. Alloy composition, fuse-element dimensions, and thermal characteristics of the structure are selected so that a predetermined value of current will produce enough Joule heating to melt the fusible link; when this occurs, the circuit opens. Of course, the blowing current of a fuse is normally set a little higher than the normal current that flows in the circuit to prevent nuisance blowing, while still providing protection against damaging overcurrent. Fuse bodies may be filled with a material that assists in quenching the arc that may occur when the fusible link opens; commonly used packing materials include sand, gypsum, and boric acid.

Some fuses incorporate two fusible systems that have different current-time characteristics; this design facilitates accommodation of short-duration inrush currents without nuisance blowing, while simultaneously providing protection against both short circuits and long-term moderate overloads.

Transducers and Sensors. Linear-variable differential transformers (LVDTs), accelerometers (piezoelectric or magnetic), resistance-temperature devices (RTDs), thermistors (i.e., special resistors that have high negative temperature coefficients of resistance), eddy-current probes, and quartz microbalances (variations of frequency-control crystals) are common examples of transducers and sensors. Each of these devices has unique construction features, failure modes, and failure mechanisms that are too numerous and varied for detailed discussion in this tutorial article.

Frequency-Control Crystals. A frequency-control crystal comprises a thin disk or small rod of a piezoelectric material, usually crystalline quartz. Thin-film metallic electrodes provide electrical connection to the crystal, which is supported, usually in a hermetically sealed housing, by metallic wires or ribbons that provide connections to package pins (Figure 14). Usually, the crystal is secured to the supports with small dabs of a metal-filled conductive polymer. Cracks in the crystal or in the attachments are common failure mechanisms.

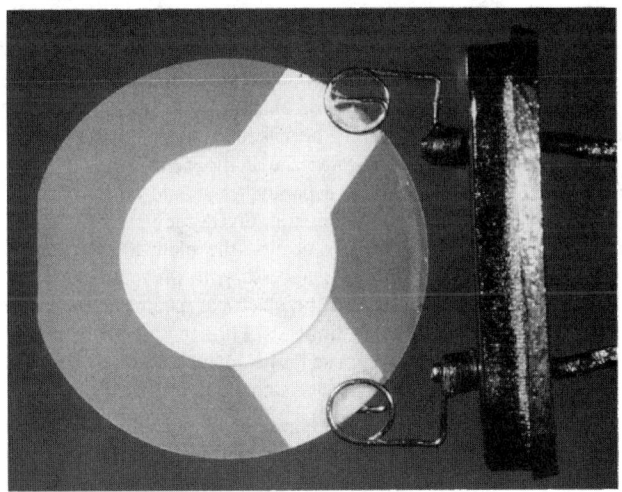

Figure 14. Depackaged frequency-control crystal

Connectors. Electrical connectors take many forms. The common feature is that mating electrical contacts are intended to be repeatedly engaged and separated. Contact-surface materials may be gold over nickel, silver, tin, solder, or, sometimes, bare copper or a copper alloy. Contamination, corrosion, and mechanical wear (e.g., relaxation of contact pressure and loss of rubbing action) are common causes of connector failure.

Lamps. Most lamps comprise a tungsten filament supported by metallic wires inside a sealed glass or quartz envelope. The atmosphere in the envelope may be a vacuum, an inert gas (e.g., argon, krypton, or nitrogen), or a halogen (e.g., iodine vapor). When sufficient current flows through the filament, it is heated to incandescence and, hence, it emits light. By far, the most common failure mode of an incandescent lamp is an open circuit, and the most common failure mechanism is wear out caused by evaporation of the filament material. Operation at slightly higher-than-rated voltage greatly accelerates filament evaporation and, hence, reduces lamp life; thus, when a lamp fails prematurely, overvoltage should be suspected.

Presence of a halogen in the lamp envelope slows, but does not prevent, filament evaporation; because of this phenomenon, filament temperature in a halogen-filled lamp is usually higher than in a more conventional incandescent lamp, so that the lamp emits more intense and whiter light. However, because of the higher operating temperature, lamp life may be short despite the evaporation-retarding characteristics of the halogen atmosphere inside the envelope.

Circuit Boards. Cracks in the metallic traces or separations of buried conductors from plated-through-hole walls are commonly encountered circuit-board failure mechanisms. Additionally, charring, often secondarily caused by failure of a board-mounted component, is another mechanism that damages circuit boards.

Hardware. As pointed out in one of the case histories to be discussed in a later section of this article, even normally benign hardware items (e.g., screws, nuts, insulated terminals, solder lugs, knobs, etc.) can be troublesome. Metal migration, metallic whiskers, dielectric breakdowns, and other failure mechanisms originating in hardware items can cause a piece of electronic equipment to fail.

ANALYSIS OPERATIONS AND THINGS TO LOOK FOR

Failure analysis of any component, passive or otherwise, must be a methodical and thorough process. The key to achieving these goals is planning. Before doing anything destructive or irreversible, the analyst should visually examine the failed component, obtain manufacturer's data sheets or catalog information, and prepare a written analysis plan. The plan may be in outline or bullet form, or it may be a detailed procedure document. Of course, it may be necessary to modify the plan as the analysis proceeds if observations indicate that the failure site can be better detected, exposed, or documented or that the failure mechanism can be more positively identified by using an alternative procedure. However, the point is that having a definite plan of action in mind at the outset of failure analysis is good engineering practice and scientific procedure.

In the paragraphs of this section, various failure-analysis operations are discussed in the order that they would normally be applied; note, however, that component-construction features, failure damage, and other factors may preclude use of some of the suggested analytical operations or may require that they be applied in a different order. Also, some of the operations may not be relevant to the type of component being analyzed; for example, a hermeticity test would not be an appropriate operation to apply to an unsealed component.

Preliminary Analysis

After the failed component has been examined by the analyst and a plan has been written, preliminary analysis should be initiated. If it is possible to do so, a known good part should be substituted into the circuit that failed, and the circuit should be checked for proper function. If the malfunction disappears, then it is likely that the suspect component was the cause of failure; on the other hand, if the circuit still does not work correctly, then the suspect component may not be bad, and it is probable that another component has failed. Of course, there is also a possibility that the suspect component AND another component have failed, and this possibility should be checked by additional troubleshooting of the malfunctioning circuit.

In some cases, it may be possible to place the suspect component in a working circuit to determine whether the malfunction follows the suspect component, but this procedure must be used with considerable caution. For example, if substituting the failed component in a working circuit would cause damage to that circuit or would cause additional damage to the suspect component, then this operation would not be appropriate. However, if a parametric change, for example, causes the circuit to malfunction, then substitution in a working circuit may provide valuable information about the effect of the failure on circuit operation.

In some cases, particularly if the failed component is inexpensive, it may be useful to duplicate the failure. Because the cause of failure may not be known early in the analytical process, it may be necessary to defer this step to a more appropriate time.

Relating failure symptoms to component parameters (e.g., increased ripple in a power-supply output may be caused by a decrease in capacitance of a filter capacitor) assists in troubleshooting. In addition, relating the component parameter to a component-construction feature (e.g., decreasing capacitance of an aluminum electrolytic filter capacitor may be caused by electrolyte loss) helps focus the analyst's attention on specific things to look for. However, as with all analytical techniques, this one must be applied with some degree of caution; that is, the analyst's focus must not be so narrow that other possibilities are overlooked or that a destructive operation that alters or disguises the real failure mechanism is performed.

Although difficult to obtain, information about conditions existing at the time of failure and history of the component (and others like it in the same or different application) can be a valuable asset to the failure analyst. Example points that may affect how the analysis is performed include, but are not limited to, the following:

- Was a thunderstorm in progress at the time of failure?
- Was there a power outage or glitch?
- Was someone probing the circuit or conducting a test on the equipment that contained the failed component?
- Was the equipment being subjected to an unusual operating condition (e.g., overvoltage, overload, vibration, moisture, higher-than-normal temperatures, etc.)?
- Had the equipment been subjected to contamination or a corrosive substance?

As noted earlier, it is a good idea to obtain the manufacturer's data sheets and/or catalogs relevant to the failed component. Often, these publications contain valuable information about specifications, operational limits, materials of construction, disassembly procedures, and design techniques that may facilitate determining whether the component had been misapplied or mishandled.

Whenever possible, it is very good practice to dissect a "good" component of the same type before performing any destructive operations on the failed device. This step provides the analyst with a vehicle for proving the selected disassembly procedure on a component that does not matter; it also provides valuable information about how the component was put together and how it works, and it facilitates determining in advance what materials were used in construction of the component.

Included under the heading "Preliminary Analysis" are such nondestructive operations as photography of the failed device in as-received condition and recording of markings on the device package. Additional operations that fall under this heading are x-radiography, hermeticity testing (if relevant), particle-impact-noise detection (again, if relevant), and electrical checks.

During visual inspection, such defects as cracks, physical damage, discoloration or charring, metal whiskers, carbon or metal tracks, corrosion, contamination, and excessive or misplaced solder should be noted. If contamination, corrosion, metal whiskers, or metal tracks are found, then scanning-electron micrography (SEM) and/or energy-dispersive x-ray (EDX) analysis should be performed before the observed materials are destroyed or altered by handling.

X-radiography is useful for determining internal structure of the component and for detecting deformed, misplaced, loose, or missing internal structures. Additionally, using x-radiography, it is often possible to identify open or shorted conductors, visualize inclusions and voids in encapsulating materials, and detect leakage paths through seals.

Confirmation of the failure mode is the most important function of electrical testing. Such failure modes as short circuits, open circuits, low leakage resistances, and deviations from specifications may be detected relatively easily in passive components with some simple instruments (e.g., a multimeter and an inductance-capacitance measurement instrument such as a self-balancing bridge). Note that capacitance-measuring instruments that depend on charging the capacitor with a constant current may give erroneous results if the capacitor has low leakage resistance.

If the failed component exhibits low or unstable leakage resistance or another failure mode that suggests the presence of moisture or corrosion, it should be baked at approximately 110°C for a couple of hours, and after the component cools to room temperature, the failing electrical parameter should be remeasured. If moisture or ionic contamination is involved, the failing parameter will recover. Of course, if this occurs, the planned disassembly procedure should be modified, if necessary, to prevent destruction of the contamination that caused the electrical parameter to deviate from specification tolerance limits. Further, as soon as the contamination has been exposed, its composition should be determined by a suitable analytical system (e.g., EDX).

If the failed component is hermetically sealed, it should be checked for leaks in the seals. In many cases involving passive components, a gross-leak test will suffice, but in others, both fine- and gross-leak tests must be performed in that order. The decision between these two alternatives should be based on sensitivity of the component to moisture contamination and size of the device package.

Particle-impact-noise detection (PIND) is a potentially useful method for detecting loose particles inside a device housing and for identifying rattling interior components. However, this technique may have limited usefulness in such components as relays that normally have internal moving parts.

Destructive Operations

Up to this point in the failure analysis, operations have been largely nondestructive. However, after all relevant nondestructive methods have been exhausted, it is time to become destructive with the ultimate goals of exposing the interior working parts of the device and revealing the failure site. The first destructive step is removing the device housing (i.e., decapsulation or disassembly, as appropriate to the device type). Housing removal may be accomplished using a chemical, thermal, mechanical, or combination process. As the housing is being removed, the analyst should be constantly alert to anything unusual; having previously disassembled a "good" sacrificial component provides the "norm" to which observations may be compared.

An integral part of the disassembly or decapsulation process is internal visual inspection. The analyst must look not only for the failure site, but also for unusual conditions that may be related to the failure (e.g., structural defects, corrosion, contamination, inclusions, voids, discoloration or charring, loose parts, metal migration, and metal whiskers). Additionally, the analyst must be alert to unusual, inappropriate, or incompatible materials. Foreign and out-of-place materials should be analyzed by an appropriate method.

Supplementary operations that assist in locating, identifying, and documenting the failure site and the cause of failure include microprobing (e.g., determining which two plates of a monolithic ceramic capacitor are short circuited), scanning-electron micrography (e.g., inspection of the failure site, corrosion sites, contamination, inclusions, metal migration, metal whiskers, etc.), energy-dispersive x-ray analysis (e.g., determination of material compositions, particularly contamination and corrosion products), cross sectioning (e.g., measuring depth of corrosion penetration or thicknesses of layers, exposure of hidden layers, etc.), and layer removal (e.g., exposure of underlying structures).

Reporting

Very often, the only contact that the customer has with the failure analyst is through the failure-analysis report. Hence, a poorly written or poorly illustrated report adversely affects the customer's view of the analyst's capabilities and competence. For this reason, if none other, failure-analysis reports should be planned and prepared with care so that the customer gets a correct (and, hopefully, good) impression of the analyst.

As a minimum, the failure-analysis report should contain an overall view or views of the failed component in as-received condition. Preferably, these views should show all surface-visible markings on the device package; however, if it is not feasible to show all the markings in photographs, then the markings should be listed in the textual body of the report. In addition, the report should include a low-magnification optical micrograph that shows the failure site and another readily identifiable feature (i.e., a "landmark"). This "road-map" photograph should be followed by a higher-magnification view of the failure site or, perhaps, two or three successively increasing magnification photographs that lead the reader's eye to the failure-site detail. Other views (e.g., scanning-electron micrographs, cross sections, etc.) may be included to support the conclusions.

In the written body of the report, the minimum elements include an introductory description of the device (e.g., manufacturer's name, part number, value, rating, date code, etc.), a description of the complaint (i.e., the failure mode), confirmation of the failure (i.e., electrical-test results), identification of the failure site with reference to the previously mentioned photographic documentation, and a discussion of the failure mechanism. The discussion of the failure mechanism may vary in detail depending on the level of knowledge of the customer (reader). Finally, the analyst(s) must be identified by name, and the report must be dated.

It is important that good grammatical construction and correct spelling be used in a failure-analysis report. Sloppiness in these items contributes to a poor report that raises questions in the customer's mind about competence of the failure analyst.

The Joint Electron Devices Engineering Council (JEDEC) has issued a standard (Reference 2) on format and content of failure-analysis reports relating to semiconductor devices. While not all parts of this standard are applicable to failure-analysis reports concerning passive components, use of relevant parts of this standard as guidelines will contribute greatly to completeness of report content and to uniformity of format, thereby making it easier for customers to understand, interpret, and compare failure-analysis reports from multiple sources.

CASE HISTORIES

Case histories are good vehicles for illustrating the types of failures that occur in actual practice and the methods employed in analyzing these failures. Accordingly, a number of passive-component case histories are discussed in the last part of this article.

Metal-Film Resistor

Figure 15. Bulged and cracked case (arrow) on failed metal-film resistor

The metal-film resistor shown in Figure 15 had a bulged, cracked, and charred spot on its cylindrical body; despite this damage, the device still had continuity, but its resistance was slightly out of tolerance. Circuit analysis showed that there was only one path through which a current high enough to cause the observed damage could flow, and this path included the collector-base diode of a transistor and a discrete diode connected between the base of the transistor and ground. Polarities were such that reverse current would have to flow through the collector-base diode of the transistor, but when this occurred, the base-to-ground diode was forward biased so that it offered little resistance to current flow. Electrical checks of the transistor revealed that it had a collector-to-base short circuit that was most likely precipitated by an accidental connection to a power-supply bus while someone was probing the circuit board; in fact, there were probe-tip marks in numerous solder joints on the board. Thus, failure of this resistor was secondary to failure of another component. In this unusual case, it was not necessary to perform destructive analysis of the resistor; instead, analytical emphasis was shifted to the transistor that was in series with the damaged resistor. It was concluded that the resistor had been subjected to electrical overstress that was caused by forced failure of another component in the circuit.

Dual-in-Line Thick-Film Hybrid Resistor Network

It was reported that some of the resistors in this network were open, and this failure mode was confirmed during preliminary analysis. Additionally, x-radiography showed that the pin-to-substrate joints associated with the open resistors were abnormal in appearance. Visual inspection of the decapsulated device showed that some of the pins were not soldered to the substrate metalization (Figure 16); only pressure connections existed. Apparently, during encapsulation, a thin film of the encapsulant had been forced between the pin fingers and the substrate metalization, thus causing the affected resistors to appear to be open. Destructive physical analysis of other samples from the same lot showed that they also had unsoldered pin-to-substrate-metalization joints.

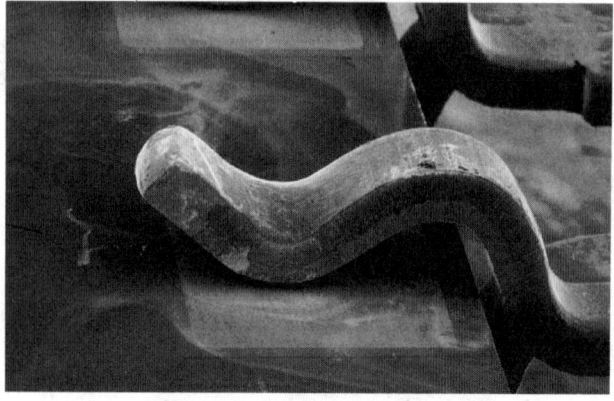

Figure 16.　Scanning-electron micrograph of unsoldered pin attachment on substrate of failed thick-film hybrid resistor network

The observed defects obviously occurred during fabrication of the device, so they were classified as manufacturing defects. Inadequate application of solder indicated poor process control, and the fact that failing devices were shipped to the customer suggested that the completed substrates had not been inspected prior to packaging and that the devices had not been tested after packaging.

Single-in-Line Thick-Film-Hybrid Resistor Network

During initial characterization of numerous thick-film-hybrid resistor networks, one unit failed. The usual preliminary analysis confirmed that one of the resistors in the network was open. Further, a crack in the ceramic substrate was visible in the x-radiograph. Decapsulation revealed that the failure site was a crack in the substrate (Figure 17). It was apparent that the crack passed through the narrowest neck of resistive material at the end of the

trim kerf, and this suggested that failure was caused by localized heating caused, in turn, by excessive current.

Analysis of the test circuit showed that when the 10-pin network was plugged into the 16-position socket such that network pin 1 entered socket pin 2, very high current would be forced through the resistor that failed. Numerous networks were intentionally plugged incorrectly into the test socket to duplicate the failure, and in all of these devices, failure occurred in the form of a substrate crack that passed through the narrowest region of the overstressed resistor. Accordingly, it was concluded that this failure was caused by a test-operator error.

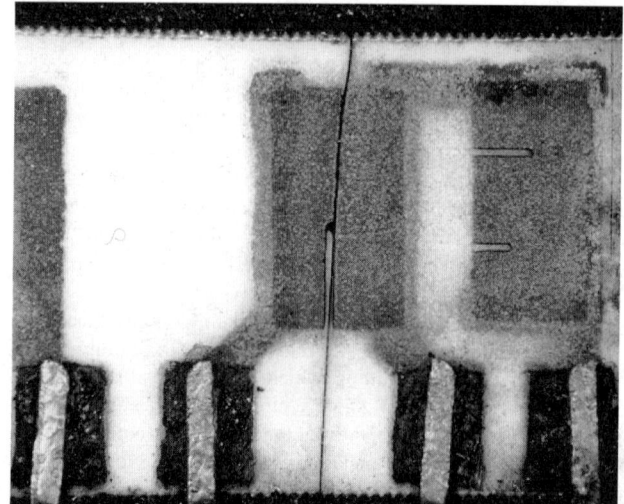

Figure 17.　Crack in substrate of failed single-in-line thick-film hybrid resistor network

Surface-Mount Thick-Film-Hybrid Resistor Network

A short circuit between two adjacent pins of a gull-wing surface-mount thick-film-hybrid resistor network was confirmed during preliminary analysis. Exterior visual inspection and x-radiography did not reveal anything unusual except for narrow openings between the bottoms of some of the pins and the plastic package material. After a stabilization bake, resistance between the failing pins returned to normal, thus indicating the possibility of ionic contamination. Inspection of the decapsulated substrate in the area between the failing pins revealed numerous metallic dendrites (refer to Figure 18) that were determined by energy-dispersive x-ray (EDX) analysis to be lead. Additionally, traces of sodium and chlorine were found in the same region of the substrate.

Review of the process used in manufacturing the circuit board on which the failing resistor network had been installed revealed that the device was mounted on the solder-wave side of the board; thus, it had been bathed in corrosive chloride-containing solder flux and then thermally shocked in molten solder. After soldering, the board had been washed in recirculated softened (i.e., sodium-containing) tap water before being rinsed in a distilled chlorofluorocarbon solvent; note that the chlorofluorocarbon solvent was a hydrophobic substance, so it did not remove any water that had entered crevices in the board-mounted components.

Fluorescent-dye-penetrant tests performed on similar resistor networks showed that liquids could enter the narrow openings below the package pins, previously observed during exterior visual inspection, and could wick across the top of the substrate (refer to the cross-sectional view presented in Figure 19). In the first wash after the board had been soldered, sodium from the softened water and chlorides from the solder flux (retained in ever-increasing concentration in the recirculated rinse water) entered the package through these openings, thus providing an electrolytic environment. When a potential difference was impressed between package pins, lead from the pin-to-substrate solder migrated across the surface of the substrate, thus forming the observed dendrites. However, the dendrites did not complete the short circuit between adjacent pins; instead, they greatly shortened the conduction path, but the short circuit occurred through the trapped liquid electrolyte.

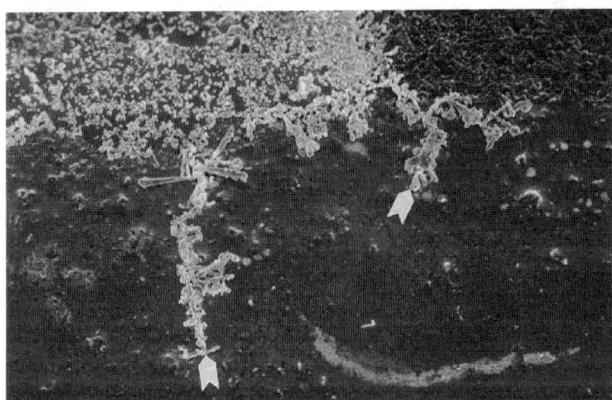

Figure 18. Scanning-electron micrograph showing metallic dendrites (arrows) on substrate of failed surface-mount thick-film hybrid resistor network

Figure 19. Cross section through pin entry into failed surface-mount thick-film hybrid resistor network showing separation between bottom of pin and encapsulant

It was concluded that the resistor network had been misapplied; that is, the package was not suitable for immersion in solder flux and molten solder. Substitution of an electrically similar, but differently packaged, resistor network solved the problem.

Thin-Film-on-Silicon Resistor

The thin-film-on-silicon resistor shown in Figure 20 was found during routine destructive physical analysis (DPA) of samples drawn from a lot of devices intended for installation in high-reliability, space-quality hybrid circuits. Short circuits between adjacent metal lines and thinned metal lines were causes for serious

concern about long-term reliability in the space environment. It was apparent that the devices had come from the manufacturer in the observed condition, so it was concluded that these devices had manufacturing defects. Probable causes for the observed defects included incomplete photoresist coverage, photolithographic-mask defects, and contamination (i.e., dirt) in the fabrication area.

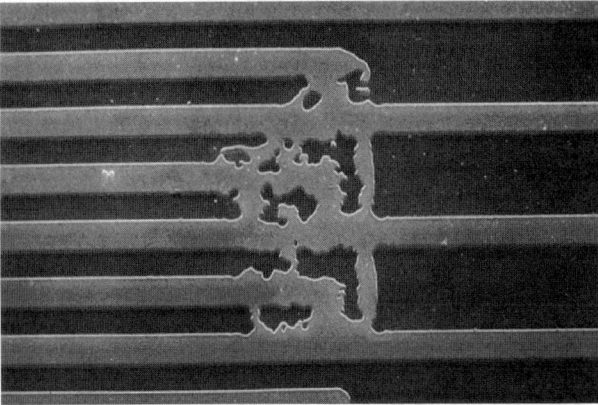

Figure 20. Scanning-electron micrograph of thin-film-on-silicon resistor showing short circuits between metal stripes and narrowed metal stripes

Monolithic Ceramic Capacitor - 1

A relatively large (i.e., 15 x 15 x 2.5 mm) 1.0-kV monolithic ceramic capacitor failed a high-voltage-withstand test. Preliminary analysis showed that surface-visible cracks existed on the unterminated sides of the device, and a dye penetrant was used to enhance visibility of these side-face cracks. These observations strongly suggested that delaminations and/or voids would be found inside the monolithic block. Accordingly, the capacitor was cross sectioned, and as expected there were some large delaminations that bridged between opposite-polarity plates (Figure 21).

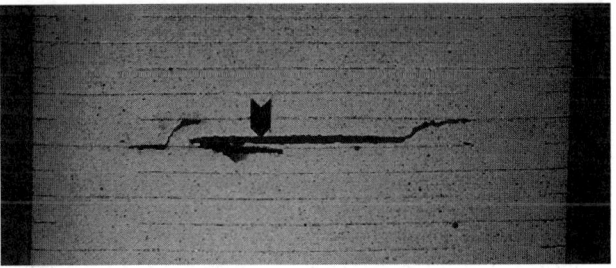

Figure 21. Bevel cross section through monolithic ceramic capacitor showing delaminations (arrow)

The green ceramic-dielectric material and the conductive ink that is screen printed on the ceramic both contain volatile constituents that must boil off during firing. In a relatively large device, such as the failed capacitor being discussed, the volatile products from the interior must diffuse to the surface so that they can escape. If temperature of the capacitor is increased too rapidly during firing, the volatile materials cannot diffuse to the surface fast enough, so they evaporate inside the device, thereby building internal pressure that delaminates the stacked structure. The conclusion of this analysis was that the device contained a manufacturing defect.

Monolithic Ceramic Capacitor - 2

Delamination is not necessarily confined to large monolithic ceramic capacitors. Figure 22 shows a cross section through a delamination in a small new, unused monolithic ceramic capacitor. Note that the ceramic dielectric layers and the plates on both sides of this delamination are deformed. This point is important because some monolithic-ceramic-capacitor manufacturers will argue that delaminations observed in cross sections are caused by the cross-sectioning process, and in some cases this is true. However, cross sectioning cannot cause the hard ceramic dielectric to bulge on either side of the delamination, so deformation around a delamination is indicative of a manufacturing defect (i.e., too steep a firing-temperature profile). Such was the case with the illustrated capacitor.

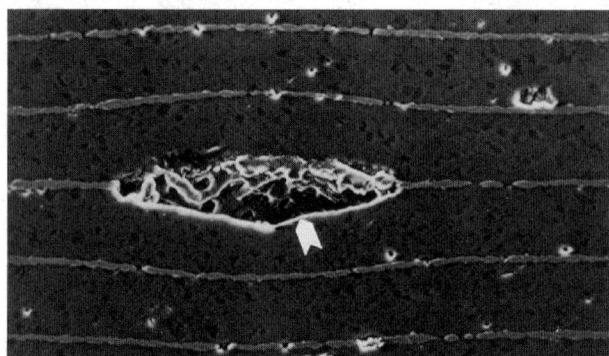

Figure 22. Scanning-electron micrograph of delamination (arrow) and adjacent plate and dielectric deformation in a monolithic ceramic capacitor

Monolithic Ceramic Capacitor - 3

A physically large 3-μF, 1.0-kV, 85°C monolithic capacitor was connected in parallel with three other similar capacitors, and this group was encapsulated in a plastic package. Two hundred of these 12-mF modules were connected in parallel to form a 2400-μF capacitor bank that was used in a down-hole well-logging-tool application in which operating temperature is typically 150°C. Further, voltage impressed across the capacitor bank was 600 V, which was 60% of the 85°C working-voltage rating. Thus, capacitors in the bank were being stressed beyond the normal operating limits, and both the user and the capacitor manufacturer were aware of this.

As one might expect, the failure rate was relatively high, and recognizing this, the component user provided a fusible link in series with each 12-μF package in the parallel combination to minimize the possibility of having to abort a logging run if a single capacitor short circuited. Further, the user had discussed reducing the failure rate with the capacitor manufacturer, but the manufacturer took the stance that the capacitor was being overstressed and that nothing could be done to improve reliability.

During a routine logging run, one of the 3-μF capacitors shorted, which was not an uncommon event; however, this time the fusible link on the circuit board did not open properly (another case history in itself). As a result, it was necessary to withdraw and repair the tool while the drilling rig and its crew stood idle and then perform another logging run; needless to say, this was a very expensive failure.

Preliminary analysis of the failed device provided only confirmation of the failure mode. Decapsulation revealed the failure site (shown in Figure 23), which was at the sharp corner of one of the capacitor plates. The sharp corner functioned as an electrical-stress concentrator, so this was the most logical location for the failure site. If the capacitor manufacturer had slightly rounded the corners of the plates (see Figure 24), electrical stress would have been more evenly distributed, and stress concentration would have been reduced. It is probable that this small design change would have decreased the failure rate, though not necessarily to zero.

Figure 23. Dielectric-breakdown site (arrow) at corner of plate in monolithic ceramic capacitor

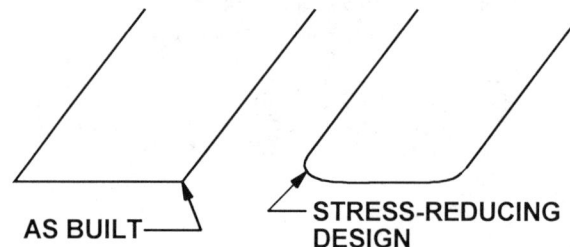

Figure 24. As-built square-corner plate (left) in failed monolithic ceramic capacitor and suggested rounded-corner electrical-stress-reducing plate design (right)

Conclusions drawn from this analysis were that the capacitor was being used beyond its capabilities and that plate design was not optimum, given the known harsh operating conditions. Hence, this failure had multiple causes: (1) electrical overstress and (2) plate-design inadequacy.

Rolled Film-and-Foil Capacitor

A rolled film-and-foil capacitor short circuited soon after being placed in service. Following preliminary analysis, during which the failure mode was confirmed, the jelly roll was carefully unrolled while terminal resistance was monitored. When the short circuit cleared, the plate-foil and dielectric-film layers at that location were separated. Metallic particles, shown in Figure 25, were found imbedded in one of the film-dielectric layers. EDX analysis of the particles, as well as other materials in the jelly-roll assembly, revealed that the particles had the same composition as the wire leads and that no other component of the jelly-roll assembly contained the elements found in the particles.

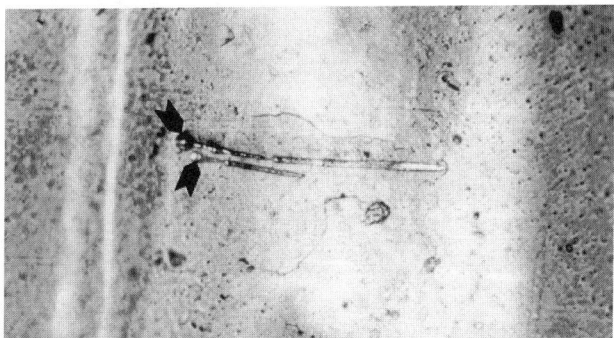

Figure 25. Metallic particles (arrows) imbedded in dielectric of rolled film-and-foil capacitor

In this capacitor, the wire leads had been spot welded to the aluminum-foil plates. A blowout had occurred while one of the spot welds was being made. When this happened, small droplets of molten wire-lead material were ejected from the weld, and these droplets landed on the foil surface and solidified; this type of metallic splatter can occur when the welder electrodes are worn, dirty, or improperly aligned. The particles were subsequently tightly rolled into the jelly-roll assembly where they remained until the capacitor had been through a few temperature cycles while under electrical stress. Eventually, at least one of the particles wore through the dielectric film, thus causing the short circuit. This failure was attributed to a manufacturing defect.

Stacked-Film Capacitor

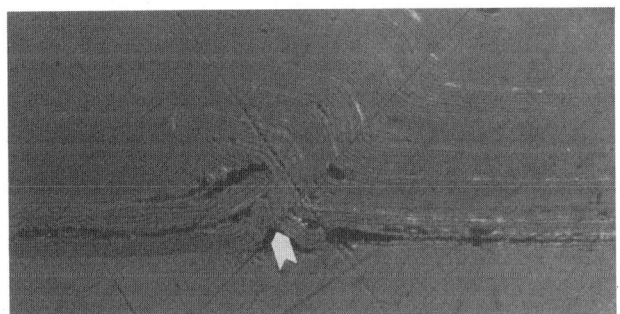

Figure 26. Wrinkled dielectric and plates (arrow) in stacked-film capacitor

During routine DPA of new and unused representative samples from a large lot of stacked-film surface-mount capacitors, it was observed that several of the dielectric-film layers and associated plates in one device were badly wrinkled (Figure 26). Such wrinkles concentrate dielectric stresses, thereby increasing the probability of dielectric breakdown at the affected location. Because these capacitors were intended for a high-reliability application, the entire lot was rejected. Actual cause of the wrinkles was not known, but it was apparent that wrinkling of the dielectric films and thin-film aluminum plates was a manufacturing defect.

Aluminum Electrolytic Capacitor - 1

The case of an aluminum electrolytic capacitor that had been used as a bypass capacitor in a 10-kW switching power supply ruptured at the built-in pressure-relief indentation. It was determined during preliminary analysis that the capacitor was also electrically open. The aluminum can that contained the capacitor was cut circumferentially just inside the seal, and the interior of the device was inspected. As apparent in Figure 27, one of the plate-termination ribbons had evaporated, and the jelly-roll assembly was badly charred.

Figure 27. Aluminum electrolytic capacitor with housing opened showing absence of termination ribbon (white arrow) and split pressure-relief indentation (black arrow)

Analysis of the circuit in which this capacitor was used revealed that the alternating-current (ac) component of current through this capacitor exceeded the maximum rated value by a factor of approximately four, which meant that internal heating was about sixteen times normal. It was clear from this analysis that the capacitor had been misapplied and, hence, electrically overstressed.

Aluminum Electrolytic Capacitor - 2

During DPA of representative samples from a lot of new, unused 250-V aluminum electrolytic capacitors, a dielectric-breakdown site was found in one unit (see Figure 28). This capacitor was not shorted (at least when checked with a low-voltage ohmmeter); in fact, its capacitance and dissipation factor were within the manufacturer's specified tolerance limits. Except for this obvious isolated defect, the capacitors were well made; however, because the capacitors were intended for a high-reliability application, it was necessary to reject the entire lot, and this rejection caused an expensive delay in the equipment production schedule.

Unrolling the jelly-roll assembly showed that ends of the plates had not been cut straight, but instead were cut in a sawtooth pattern similar to that produced by pinking shears. One of the sharp projections at the end of one plate had been bent out of the plate-foil plane, and this sharp point penetrated the paper separator and contacted the opposite-polarity plate. During forming, dielectric breakdown occurred at the high-electrical-stress point at the tip of this sharp projection. Energy stored in the capacitor dissipated in a highly localized area, thus blowing a hole in the side of the jelly-roll assembly; however, the current-limiting resistor in the forming circuit prevented greater damage.

Contact with the capacitor manufacturer revealed two interesting facts. First, capacitors that fail during forming are neither detected nor culled. Second, only capacitance and dissipation factor are measured after forming, and these measurements are made at low voltage. If the subject capacitor had been subjected to the full 250 V for which it was rated, it is highly likely that it would have failed and would have been rejected. Improper plate design (i.e., the jagged ends), a manufacturing defect (i.e., the bent point on the end of the plate), insufficient monitoring of the forming process, and inadequate final testing contributed to failure of this capacitor.

Figure 28. Dielectric-breakdown site (arrow) in jelly roll of aluminum electrolytic capacitor

Aluminum Electrolytic Capacitor - 3

Aluminum electrolytic capacitors can explode violently under certain conditions; this is the reason that pressure-relief vents are designed into all but the smallest of these devices. An example of the damaging effects of an aluminum electrolytic capacitor explosion is illustrated in Figure 29, which shows an epoxy-encapsulated power-supply module that has a piece blown out of it; note that the encapsulant in this module was rock hard, so it took considerable force to blow out this relatively large fragment.

Figure 29. Piece (black arrow) blown out of encapsulated power-supply module by explosion of aluminum electrolytic capacitor (white arrow)

There was not enough of the capacitor left for failure analysis (Figure 30), so it was necessary instead to resort to circuit simulation. SPICE analysis showed that the ac component of current flowing through this capacitor was about three times the manufacturer's specified current limit. Despite this overstress, the capacitor survived a few hundred hours of service with the module operating in a laboratory environment. Ultimately, however, the

electrolyte was driven out of the capacitor, and the explosion occurred. This failure was ascribed to electrical overstress caused by misapplication (i.e., excessive ac component of current).

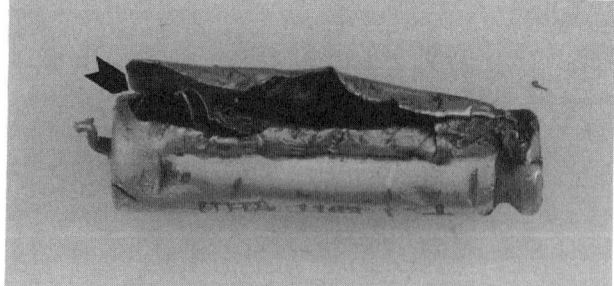

Figure 30. Remnants of aluminum electrolytic capacitor (arrow) that exploded inside encapsulated power-supply module

Aluminum Electrolytic Capacitor - 4

When the voltage regulator in a high-current power supply failed, an excessively high voltage was forced across the aluminum electrolytic filter capacitor; within a few seconds, the capacitor exploded. Pressure built up in the capacitor housing so fast that the rubber rupture disk in the end cap did not fail; instead, the entire end cap was blown off, and the jelly-roll assembly was extruded from the housing as shown in Figure 31. Fortunately for personnel in the vicinity of the capacitor when it exploded, wiring connected to the device terminals kept the end cap from hitting someone; however, there was concern that toxic or carcinogenic vapors had been released.

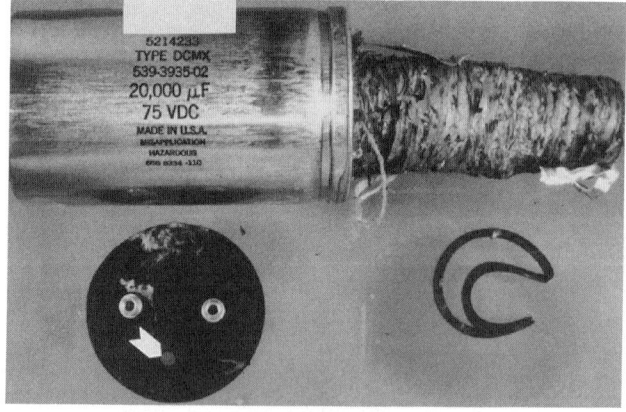

Figure 31. Exploded aluminum electrolytic capacitor (arrow points to intact pressure-relief rupture disk)

This failure was secondary to failure of the power-supply voltage regulator; however, the circuit designer could have prevented violent explosion by incorporating a "crow-bar" that would interrupt primary power in the event of voltage-regulator failure.

Tantalum Electrolytic Capacitor - 1

A new, unused molded-case tantalum electrolytic capacitor was found to be shorted during incoming inspection, and the failure mode was confirmed during preliminary analysis. This device was cross sectioned while terminal resistance was monitored. During cross sectioning, it was observed that the dry-electrolyte layer was not uniform and that it contained several voids and thin spots. After a few small increments of grinding, the failure site was exposed; as apparent in Figures 32 and 33, the cathode-lead solder had flowed through coincident openings in the cathode metalization and the dry-electrolyte layer into the porous anode slug. After this failure site had been photographically documented, grinding was continued; when the failure site had been ground away, the short circuit cleared, thus indicating that there was only one failure site. This failure was classified as a manufacturing defect.

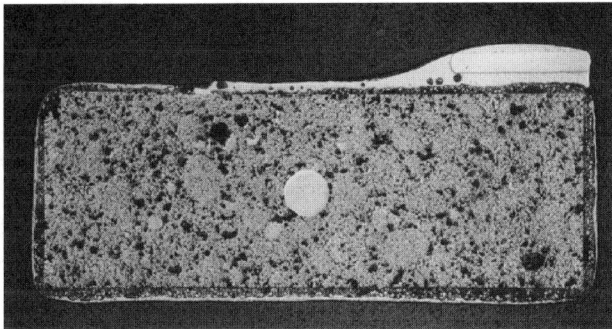

Figure 32. Overall view of transverse cross section through shorted tantalum electrolytic capacitor

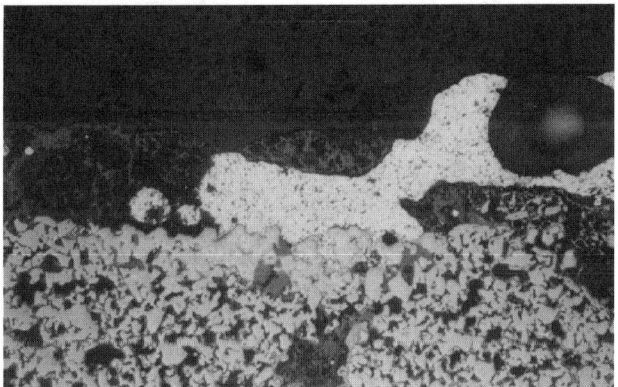

Figure 33. Detail cross-sectional view of short-circuited region of failed tantalum electrolytic capacitor

According to a representative of the manufacturer of this capacitor, no electrical testing is performed on the devices after they have been packaged, which means that short-circuited capacitors, such as the one discussed in this case history, are shipped as "good" units and that the user must screen incoming devices for failures. This is an interesting philosophy; usually, device manufacturers are equipped for performing very fast electrical characterizations of their products, while most customers are equipped for slow, probably incomplete, testing at best. Making the customer responsible for final testing permits a component manufacturer (not limited to capacitor manufacturers) to bid low prices to make the sale. However, in most cases, this is false economy for the customer, because the component manufacturer usually can test and screen devices much more efficiently, completely, quickly, and cost effectively than the customer.

Tantalum Electrolytic Capacitor - 2

Incoming inspection of another tantalum electrolytic capacitor, made by a different manufacturer, detected a short-circuited device. Failure analysis was performed using a procedure similar to that described in the previous case history. In the present device, however, it was found that the dry-electrolyte layer did not extend all of the way to the anode-wire end of the tantalum slug, but the cathode metalization did; a cross-sectional drawing comparing anode-wire ends of a normal tantalum electrolytic capacitor and the failed capacitor described in this paragraph is presented in Figure 34. Because the dry-electrolyte layer did not properly cover the slug in the failed device, there was a circumferential band near the anode-wire end in which the cathode metalization was in contact with the anode-slug surface. Absence in this band of the dry electrolyte with its oxidizing agent precluded self healing of the tantalum pentoxide dielectric, and dielectric breakdown eventually occurred. It appeared that the anode slug had not been dipped deeply enough into the electrolyte slurry during fabrication, so this failure was the result of a manufacturing defect.

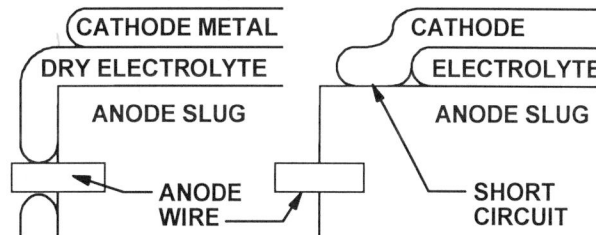

Figure 34. Anode-wire-end structure of normal tantalum electrolytic capacitor (left) and failed device (right)

400-Hz Power Transformer

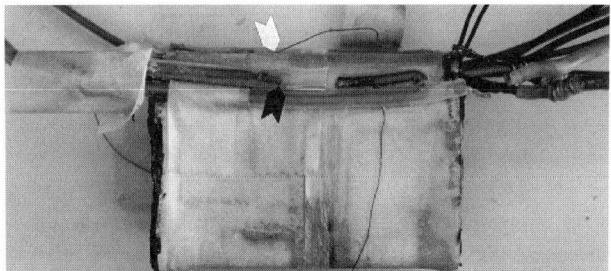

Figure 35. Center-tap lead termination in failed 400-Hz power transformer (white arrow points to severed end of one center-tapped winding pigtail; black arrow indicates absence of strain relief in lead wire)

Both halves of a center-tapped winding on a 400-Hz power transformer were reported as open, and this failure mode was confirmed during preliminary analysis. Further, detailed inspection of the device showed that the external lead wire connected to the center tap of the open winding could be moved in and out, but all other lead wires were firmly secured. When the transformer was disassembled and the winding buildup was unwound, it was found that both winding pigtails that had been soldered to the center-tap lead wire were broken. Additionally, it was found that all of the polytetrafluoroethylene-insulated lead wires entered the winding assembly through short lengths of polytetrafluoroethylene tubing so that the slick wire insulation was surrounded by slick tubing. This configuration offered little strain relief, so to compensate, the manufacturer had bent the inside end of each lead wire, except the center-tap lead, through 180° and had wrapped multiple layers of adhesive-backed tape over the bent ends of the wires. However, there was no such bend in the center-tap wire, and there was no adhesive-backed tape on this wire (Figure 35); as a result, the center-tap lead wire had essentially no strain relief. Failure was probably caused in the field when the center-tap lead wire was accidentally pulled. It appeared that this failure was related to a design or manufacturing defect.

Solenoid-Valve Coil

Numerous solenoid-valve coils in high-volume retail point-of-sale dispensing systems were failing open. Preliminary analysis included verification of the electrical failure mode. Additionally, x-radiography showed that the magnet wire was open near the terminals of some coils, but in most cases, x-radiography revealed nothing. However, when an open was visible in an x-radiograph, ends of the wires appeared ragged, and the usual ball associated with fusing in a conductor carrying direct current (dc) was absent.

Each coil was wound on a plastic bobbin, and the winding was covered with a molded-in-place polymeric jacket. Decapsulation was performed by a painstaking mechanical process that avoided damage to the magnet-wire insulation that would have been caused by chemical decapsulation. Because the failure mechanism turned out to be corrosion, selecting the mechanical method had an additional benefit; that is, a wet-chemical method would have destroyed the corrosion products.

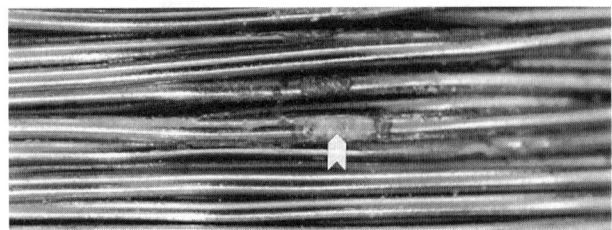

Figure 36.   Corroded conductor (arrow) in solenoid-valve coil

In all of the coils that were decapsulated, at least one turn had been severed by corrosion, and in some coils, numerous turns and even multiple layers of the winding were involved. Most of the failure sites were located near the soldered terminations, but some were farther down the sides of the coils, and others were at the ends opposite the terminals. At a typical failure site, such as the one shown in Figure 36, the magnet-wire insulation appeared yellow rather than the normal copper color; additionally, the insulation was slightly swollen, and small cracks in the insulation were observed. Further, there were several deposits of bluish-green corrosion products between nearby turns. EDX analysis of typical corrosion deposits revealed chlorine or bromine; chlorine was found in earlier-date-code coils, while bromine was found in recent-date-code coils. Contact with the coil manufacturer provided identification of the specific "no-clean" flux that was used in soldering the coil terminations, and subsequent contact with the flux manufacturer revealed that the flux contained halogens; however, the manufacturer would not say which halogens were used.

Clearly, the failure mechanism was corrosion of the copper winding conductors by a halogen-containing substance. While the solder flux was a highly likely candidate, other possible sources of corrosive contamination existed; for example, the dispensing equipment was cleaned periodically with a disinfectant, and some disinfectants contain chlorine and/or bromine. Armed with this knowledge, the equipment manufacturer undertook the task of working with its customers and the coil manufacturer to determine which material was the source of corrosive contamination. Thus, whether failure was caused by a manufacturing defect (i.e., inadequate cleaning of the coils after terminal soldering) or a design problem (i.e., inadequate sealing of the coil against externally applied disinfectants) was unknown.

Snap-Action Switch

One of the normally open contacts of a two-pole snap-action switch would not close when the actuator was depressed. Preliminary functional checks confirmed the failure mode, and x-radiography revealed that the over-center spring associated with the failing contact was not properly arched.

Disassembly and interior inspection of the switch (Figure 37) showed that the thermoplastic actuator, over-center spring, and knife-edge moving-contact pivot had been damaged by heat. Also, the hole in the actuator through which the over-center spring passed was enlarged (Figure 38) so that the spring could not be adequately deformed when the actuator was depressed, and the spring was no longer correctly arched.

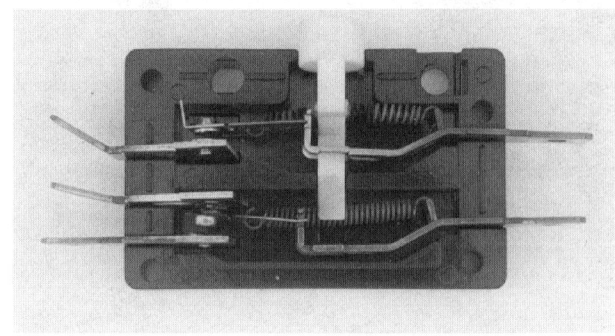

Figure 37. Overall view of failed snap-action switch with cover removed

The problem was that the primary path for the 12-A current that passed through the failed contact of the switch included the knife-edge pivot illustrated in Figure 39. Current-carrying area of this pivot was small, so eventually, it began arcing; arc damage increased resistance of the pivot, thereby diverting an increasing fraction of the load current through the over-center spring. Increased current in the spring produced abnormally high Joule heating, and the elevated temperature associated with this heating melted the thermoplastic actuator material so that the hole in the actuator was deformed by lateral force exerted by the spring. Ultimately, the spring sank deeply enough into the actuator material that the contact mechanism could no longer be forced over center.

Failure of this snap-action switch was the result of a wear-out mechanism. However, had the manufacturer provided an alternate low-resistance current path in parallel with the knife-edge pivot, life of this switch could have been prolonged.

Figure 38. Detail view of plastic actuator in failed snap-action switch showing heat-enlarged hole surrounding upper over-center spring

Figure 39. Detail view of knife-edge moving-contact pivot in failed snap-action switch

Relay

A direct-current (dc) control relay failed in a safety-related circuit in a nuclear power plant. Preliminary analysis showed that the 125-V dc coil was open and that the contacts and mechanism were in good condition. Additionally, a break in the fine-gauge coil wire and a small ball on one of the severed ends were visible near a coil-lead-wire termination during real-time x-radiography after the coil had been dismounted from the relay.

Using a spare coil for experimentation, it was determined that the coil could be disassembled easily without damaging the winding or its terminations. Disassembly of the failed coil revealed the failure site detected during real-time x-radiography (Figure 40). After this finish-end site had been photographically documented, the loose ends of the severed wire were microprobed, and continuity between each severed end and the corresponding coil lead wire was checked; it was found that the coil had another open.

Accordingly, disassembly of the coil was continued until the other (i.e., start-end) termination was uncovered. At this point, it became obvious that the failure site described in the previous paragraph was a secondary site; the primary failure site, shown in Figure 41, was caused by incorrect dress of the start-end pigtail of the coil. That is, the start-end pigtail was not entirely on the fluorocarbon-tape insulating pad provided for it. As a result, the start-end pigtail was in direct contact with the outer layer (i.e., finish end) of the coil winding so that the only insulation between the start and finish ends of the winding was two thicknesses of the thin insulation on the magnet wire.

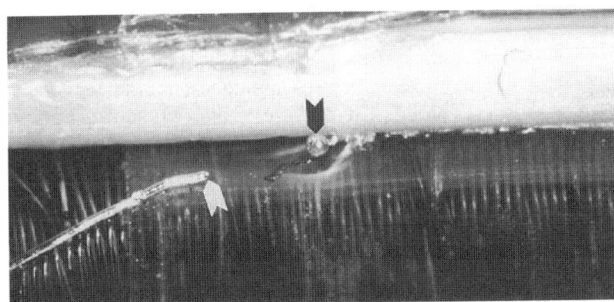

Figure 40. Secondary failure site at finish end of relay coil (black arrow points to ball on end of fused wire; white arrow indicates end of coil pigtail)

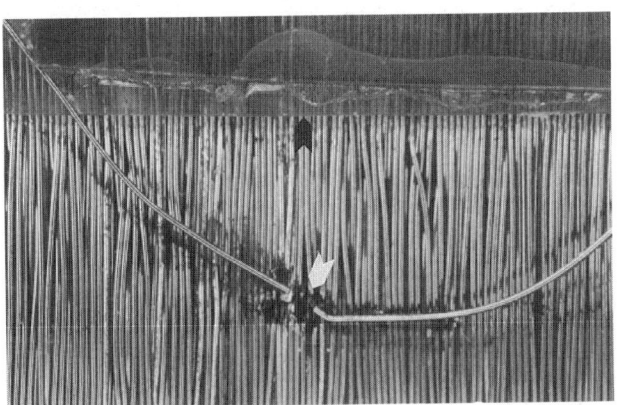

Figure 41. Primary failure site (white arrow) between start-end pigtail and finish-end layer of relay-coil (black arrow points to edge of start-end lead-termination pad)

While the normal 125-V coil voltage was not high enough to trigger dielectric breakdown of this insulation, the kilovolt-level inductive-kickback voltages that occurred each time current through the coil was interrupted were high enough. Thus, after a few actuations of the relay (in one reported case, a coil failed the first time the relay was energized), insulation between the start and finish ends of the coil would fail. Then, the next time that 125 V dc from a stiff source was applied to the coil, very high current flowed through the short circuit associated with the previously mentioned dielectric-breakdown site, and the fine-gauge coil wire fused near the finish-end termination.

Cause of this failure, which was classified as a manufacturing defect, could have been eliminated by proper routing of the start-end coil pigtail; that is, if the start-end pigtail had been kept entirely on its insulating pad, then insulation between start and finish ends of the coil would have been more than adequate to withstand service conditions. It is probable that coils containing improperly dressed start-end pigtails were produced by a single, inadequately trained and supervised assembly worker.

An interesting side note is that in coils having later date codes, the finish layer of the winding was fully overwrapped with insulating tape before the start-end termination pad was folded into place. Thus, even though the start-end pigtails in the more recent coils were also off their termination pads, there was adequate insulation between the start and finish ends of the coils. This added tape overwrap effectively eliminated the failure mechanism.

100-A, 3-Phase Molded-Case Circuit Breaker

After several years of satisfactory service, a 100-A, 3-phase molded-case circuit breaker began tripping each time the motor connected to it was started. Electrical tests showed that trip current of the circuit breaker had decreased, and replacement of the device eliminated the problem.

This circuit breaker had both thermal and magnetic trip mechanisms. The magnetic mechanism provided fast protection against short circuits and heavy overloads, while the thermal mechanism protected against sustained moderate overload conditions. A trip shaft that simultaneously operated contacts in all three phases was released from the "on" position by a force exerted by any one of the magnetic or thermal mechanisms; thus, when one of the current-sensing mechanisms detected an overload condition, all three contacts were opened at once.

Preliminary analysis did not reveal anything of significance. However, when the device was disassembled, corrosion was found in a metal-to-metal interface in the main current path of one phase (Figure 42). Heat generated at the corroded interface was conducted directly to the bimetallic-strip thermal-trip mechanism of that phase of the circuit breaker (Figure 43). Because of the extra heat generated in the corroded interface, the thermal-trip mechanism forced release of the trip shaft at lower-than-rated current.

Figure 42. Normal (left) and corroded (right) thermal-trip-mechanism feet removed from failed 100-A, 3-phase molded-case circuit breaker (arrows point to corrosion products)

EDX analysis of the corrosion deposits revealed potassium, but no corresponding anion; this suggested that the corrosion initiator was potassium hydroxide or an organic-acid salt of potassium. No source of such a substance was found in the circuit breaker; hence, it was speculated that plating on the corroded metal

part was deposited by a process inappropriate to the application or that the part had not been properly cleaned after plating. Hence, the failure was probably caused by a manufacturing defect.

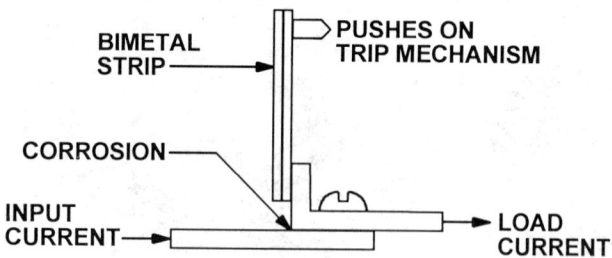

Figure 43. Principal components in main current path and thermal-trip actuator of failed circuit breaker

30-A Cartridge Fuse

During normal operation of a piece of electrical equipment, the subject fuse opened. Despite the fact that the fuse had opened, there was nothing wrong with the equipment, so a failure analysis was performed in the field. Preliminary analysis, which under the circumstances included only visual inspection and continuity checking, verified that the fuse was open and revealed that one of the brass end caps of the fuse was discolored. Disassembly of the fuse (Figure 44) showed that the element had not opened in one of the intended locations (i.e., at one of the punched holes or adjacent to the low-melting-temperature-alloy bead), but instead at the point where the fuse element was soldered to the discolored end cap. This observation suggested that the cause of failure was external to the fuse, so the fuse holder was visually inspected. As expected, the axial-pressure contact inside the fuse holder was corroded. With this corrosion in the current path, contact resistance was higher than normal, and localized Joule heating of the fuse end cap occurred; as a result, the fuse element melted just inside the end cap. Replacement of the fuse holder solved the problem.

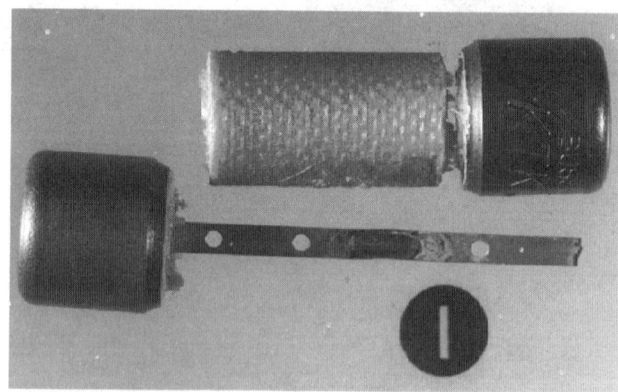

Figure 44. Disassembled components of failed cartridge fuse

Resistance-Temperature Detector (RTD)

An RTD is a very good temperature sensor; its resistance increases almost linearly with temperature, and deviations from linearity are well known, so they may be reliably and accurately compensated.

In the subject RTD, the sensitive element was a platinum-film resistor deposited and patterned on a ceramic substrate; a clear, colorless glaze protected the platinum resistor. Short lengths of platinum wire were spot welded to contact pads on the substrate and brazed to woven-glass-fiber-insulated stranded nickel lead wires; one of the brazed joints was insulated with heat-shrinkable tubing. These components, along with a silicone-impregnated woven-glass-fiber insulating sleeve that covered the resistance element and the lead-wire braze joints, were installed in a stainless-steel tube that was closed at one end, and the remaining volume of the tube was backfilled with granulated magnesium silicate. Finally, the open end of the tube through which the leads exited was filled with a high-temperature epoxy.

The failed RTD was the primary temperature-sensing device in a closed-loop temperature controller on a commercial oven. According to the RTD specifications, its maximum rated operating temperature was 450°C; however, in the oven application, it was normally subjected to a maximum temperature of 275°C, which should have been safe. Failure was detected when temperature of the oven went out of control and increased to the maximum value of which the oven heaters were capable. When this occurred, the oven was damaged, and its contents were destroyed.

Laboratory characterization of the RTD showed that at low temperatures, resistance tracked the known almost-linear curve quite well; however, as temperature approached the 275°C service temperature, RTD resistance fell below the predicted curve (Figure 45). Of course, in the closed-loop controller, less-than-expected resistance indicated that the oven temperature was low, so the controller called for additional heat, which caused the RTD resistance to fall farther below the curve. The result was thermal runaway.

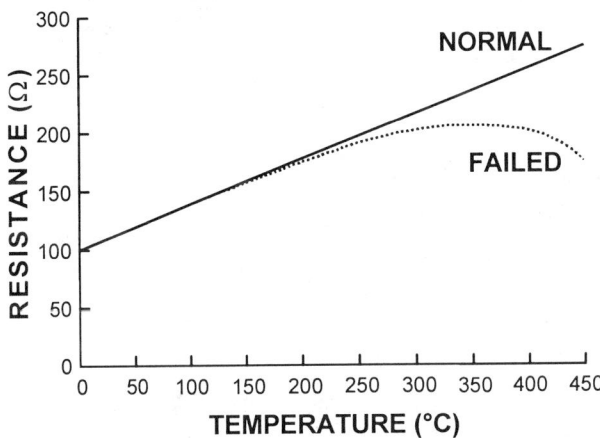

Figure 45. Resistance-versus-temperature characteristics of normal and failed RTDs

Disassembly of the failed RTD revealed that the aforementioned brazed joints were corroded and that all surfaces of interior components were coated with a gray contaminant film. EDX analysis showed that the film had high chlorine content. However, materials used in device construction were metals, glasses, ceramics, silicones, and high-temperature epoxies, all of which were suitable for the 450°C maximum rated temperature. The one material inside the device that was not suitable for high-temperature service was the heat-shrinkable tubing used to insulate one of the braze joints (Figure 46); EDX and Fourier-transform infrared (FTIR) analysis showed that the tubing was made of poly(vinyl chloride), a thermoplastic material that starts decomposing at approximately 105°C with evolution of hydrogen

chloride. The evolving hydrogen chloride combined with water vapor trapped inside the RTD housing to form hydrochloric acid.

Effectively, the hydrochloric-acid film on interior surfaces of the RTD formed a negative-temperature-coefficient resistor that was connected in parallel with the positive-temperature-coefficient platinum resistor. At low temperatures, resistance of the hydrochloric-acid path was very high compared with that of the platinum element, and hence, it had little or no effect on net resistance of the RTD. However, as temperature increased, decreasing resistance of the hydrochloric-acid path dominated the parallel combination so that net resistance of the RTD decreased.

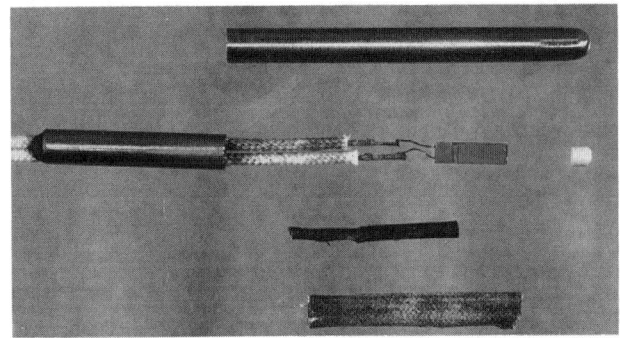

Figure 46. Components of disassembled failed RTD

If the RTD designer specified poly(vinyl chloride) heat-shrinkable tubing for use in the RTD, then this failure was caused by a design error. On the other hand, it is possible that the designer did not specifically call out any particular material for the braze-joint insulation and that production-line personnel, who did not understand the implications of using poly(vinyl chloride) in a high-temperature application, made the selection; in this case, the failure was caused by a manufacturing defect. Still another possibility is that the manufacturer designed the RTD for a maximum service temperature of, say, 100°C, but published an incorrect value in the product data sheet.

Hybrid Crystal Oscillator

During vibration testing of a piece of hardware intended for a space application, a hermetically sealed thick-film-hybrid crystal oscillator quit delivering an output. Preliminary analysis showed that the package was still hermetically sealed and that some interior component rattled when a particle-impact-noise (PIND) test was performed.

When the package cover was removed, the failure site was readily visible. The crystal, a disk of quartz mounted parallel to the plane of the ceramic substrate, was supported between two small vertically oriented springs, one end of each of which was soldered to the substrate (Figure 47). To prevent solder from wicking all the way up the spring, the manufacturer had bent the wire in each of the springs at approximately the half-way point of its length. Doing so introduced three types of damage: (1) the wire surface was nicked by the tweezers or pliers used to bend the wire; (2) the wire was plastically deformed, thus locally work hardening the material; and (3) solder wicked up the spring to the bend, thus establishing a fulcrum at the bend in the wire. The combined effect was that mechanical stress in the spring wire was concentrated at the bend where the wire material had been embrittled by work hardening and weakened by nicking. The result was that the spring wire fatigued and broke at the high-stress point (Figure 48). This failure was classified as a manufacturing defect.

Figure 47. Overall view of delidded hybrid crystal oscillator (arrow points to failed crystal-support spring)

Figure 48. Detail view of failed crystal-support spring (arrow) in hybrid crystal oscillator

Incandescent Lamp

An incandescent lamp prematurely quit emitting light. Preliminary analysis showed that the lamp was electrically open, but visual inspection of the interior structure through the transparent glass envelope revealed that the filament and its supports were intact. After careful removal of the base from the lamp, it was readily evident that the wire running from the center contact of the base to the seal of the glass envelope had corroded in two (Figure 49). EDX analysis of the corrosion products showed that the principal corrosive element was chlorine. Other materials in the lamp base were similarly analyzed, and chlorine was detected only in a residue of solder flux found inside the base. Review of the available history revealed that the lamp was used in a clean environment and that it had not been exposed to chlorine-containing materials while it was in service. Further, it was noted that the interior of the lamp base could not be cleaned after the center contact had been soldered. Thus, lamp failure was attributed to a manufacturing defect, namely, use of excessive or inappropriate solder flux.

Figure 49. Corroded wire inside base of failed incandescent lamp

Insulated Stand-Off Terminal

Even simple hardware items can cause failures in electronic equipment. During qualification of a piece of equipment intended for a long-term satellite application, a logic line was pulled to the low state, apparently by a short circuit to ground. Careful visual inspection of the circuitry revealed several long metal whiskers growing from the grounded tin-plated metal base of an insulated stand-off terminal (Figure 50). One of these whiskers was touching the bare wire lead of the pull-up resistor that normally held the failing logic line in the high state (Figure 51). After the whisker had been photographed, it was removed, and the circuit resumed normal operation. EDX analysis showed that the whisker was tin.

Whiskers can grow from tin-, cadmium-, and zinc-plated surfaces. In the case of tin, certain types of electroplating processes produce whisker-prone finishes. Potential for whisker growth can be reduced if the plating is thermally reflowed (e.g., in a hot-oil bath) after electrolytic deposition or if impurities (e.g., lead in tin) are included in the plating, but the best solution is to avoid these metals when electrically conducting components must be mounted very close to each other.

Figure 50. Scanning-electron micrograph of tin whiskers (arrows) growing from base of insulated stand-off terminal

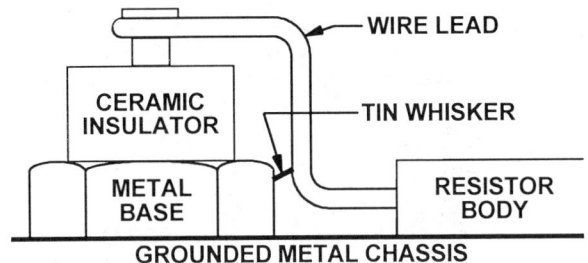

Figure 51. Spatial relationship among components involved in short circuit caused by tin whisker

CONCLUSIONS

Passive-component failures cannot be ignored. There are many consequences that may not be immediately apparent when such simple devices fail. For example, if production yield and efficiency are adversely affected, a consumer is injured, or property is damaged, then costs (real and intangible) can greatly exceed the cost of a failure analysis and ensuing corrective action. Thus, it is prudent to make frequent use of failure analysis as a quality- and safety-enhancement tool that helps insure product integrity.

REFERENCE MATERIAL

There is not much literature available on failure analysis of passive components. One good source is a paper by Johnson and Smith (Reference 3). Occasionally, a component manufacturer will publish a paper or brochure that contains information useful to the failure analyst (e.g., Reference 1), or a case history will be discussed in a technical or trade journal. Unfortunately, however, there is no consistent pattern for such information. If contacted by telephone, personnel associated with some component manufacturers will provide suggestions for opening packages and detecting certain types of failures; on the other hand, personnel working for other manufacturers will not admit that their products can fail for fear of lawsuits or adverse publicity.

A passive-component failure analyst must watch for, even search for, sources of information. Periodically searching the World-Wide Web and visiting home pages of some of the larger electronic-component manufacturers can yield useful information. Scanning various trade journals occasionally yields a useful article. Attending symposia and workshops, where it is possible to participate in discussions with other passive-component failure analysts, is one of the best ways of acquiring and supplementing one's knowledge. Listed after the references are other sources of information that are frequently useful to the passive-component failure analyst.

REFERENCES

1. J. Gill, "Surge in Solid Tantalum Capacitors," AVX Ltd., Tantalum Division, Paignton, England, Feb. 1995.

2. *Standard for Failure Analysis Report Format*, JESD38, Joint Electron Devices Engineering Council (JEDEC), Washington, DC, Dec. 13, 1995.

3. M. J. Johnson and S. E. Smith, "Failure Modes and Mechanisms of Non-Semiconductor Electronic Components,"

Microelectronic Failure Analysis Desk Reference, 3rd edition, T. W. Lee and S. Pabbisetty, editors, ASM International, Materials Park, OH, 1993, pp. 303-320.

OTHER SOURCES OF INFORMATION

- T. W. Lee and S. V. Pabbisetty, editors, *Microelectronic Failure Analysis Desk Reference*, ASM International, Materials Park, OH (Updated periodically).

- J. R. Devaney, G. L. Hill, and R. G. Seippel, *Failure Analysis Mechanisms, Techniques, and Photo Atlas, A Guide to the Performance and Understanding of Failure Analysis*, Failure Recognition & Training Services, Inc., Monrovia, CA, 1983.

- *Proceedings of the International Symposium for Testing and Failure Analysis* (ISTFA), especially the earlier volumes, ASM International, Materials Park, OH (Annual).

- *Proceedings of the Capacitor and Resistor Technology Symposium* (CARTS), The Component Technology Institute, Inc., Huntsville, AL (Annual).

- *Proceedings of the International Reliability Physics Symposium* (IRPS), especially the earlier volumes, The Institute of Electrical and Electronics Engineers, Inc., Piscataway, NJ (Annual).

- C. A. Harper, editor-in-chief, *Handbook of Components for Electronics*, McGraw-Hill Book Company, New York, NY, 1977.

- *Reference Data for Radio Engineers*, Howard W. Sams & Co., Inc., Indianapolis, IN (Updated periodically).

- M. L. Minges, Technical Chairman, *Electronic Materials Handbook, Volume 1, Packaging*, ASM International, Materials Park, OH, 1989.

- *Ceramic Chip Capacitor Handbook*, West-Cap Division, San Fernando Electric Manufacturing Co., San Fernando, CA, 1970.

- C. A. Hampel, *The Encyclopedia of the Chemical Elements*, Reinhold Book Corp., New York, NY, 1968.

- *Military Handbook 217*.

- *The Merck Index, An Encyclopedia of Chemicals, Drugs, and Biologicals*, Merck & Co., Rahway, NJ (Updated periodically).

- *Handbook of Chemistry and Physics*, The Chemical Rubber Co., Cleveland, OH (Updated periodically).

- Manufacturers' catalogs, data sheets, and handbooks.

- Manufacturers' WorldWide Web sites

Failure Modes and Failure Mechanisms

Failure Mechanisms
in Integrated Circuits

Seshu V. Pabbisetty, Dan Corum, P. Scott, and Kendall Scott Wills
Texas Instruments, Semiconductor Group
Stafford, Texas

INTRODUCTION

The two most important points, in the journey of the failure analyst, are the identification of the failure mode and failure mechanism. The failure mode is the observed electrical or visual symptoms that are anomalous to a component's advertised or implied behavior or appearance. Failure modes can range in extremity from catastrophic to slight degradation, and they are typically categorized as functional, parametric, or visual. Functional failures involve problems with the digital logic and the associated circuitry, which cause the device to behave differently than specified. Parametric failures involve analog outputs which are out of tolerance, e.g. I/O pin leakage and discontinuities. Visual reject failures involve devices which have been mismarked, corrosion occurrences, packaging issues, or other easily found defects.

The failure mechanism is the actual physical defect or condition that causes the failure mode to occur. Failure mechanisms can involve reliability, fabrication, design, test, packaging, and environmental or other use issues. For a given failure mechanism, its occurrence has a heavy dependence on the operational and environmental conditions under which the device has operated. A knowledge of the history of the device is critical to be able to effectively diagnose failures. [1].

Identification of the appropriate failure mechanism is nominally the endpoint of the failure analysis laboratory work. Once the failure mechanism has been identified, one could go further to identify the root cause of the failure. The root cause is the fundamental incident or condition that caused the component to fail and can usually be traced to pushing the limits of current fabrication technology, a machine problem, operator error, or sometimes a combination of all. Once the root cause is identified, actions can then be taken to eliminate future occurrences of such errors.

In this paper, we present an overview of common failure mechanisms illustrated with typical micrographs [2]. The main categories of failure mechanisms are chip related and assembly related failure mechanisms. Chip related failure mechanisms typically involve process issues such as particle defects, contamination, oxide defects, shorted metal, and diffusion or substrate defects. Assembly related failure mechanisms include problems which arise during the packaging of the device after processing. These can include chip and package cracks, protective over-coat (PO) damage, and bonding issues. The advent of the flip chip device technology has also ushered in a new breed of failure mechanisms to the world of failure analysis. Failures involving Flip Chip technology often result from the unique packaging structure of the device itself.

CHIP RELATED FAILURE MECHANISMS

In the section below, a few examples of typical chip related defects are reviewed: Particulates and Contamination, Gate Oxide Defects, Multi Level and Metal Interlayer Oxide Defects, Poly Layer Defects, Metal Layer Defects, Protective Overcoat (PO) Defects, Lithographic Defects, Open Contacts, Shorted Contacts, Laser Repair Defects, Silicide Extrusions, Filaments, and Substrate Defects.

Particle Defects/Contamination

Contamination related failures are typically categorized into three groups: ionic contamination, foreign contamination (particles), and process etch residue.

Ionic Contamination. Ionic contamination is a non-visual defect associated with the presence of ions (such as hydrogen or sodium) in sensitive areas within the device. Sodium is a common contaminate in the fabrication process due to its presence in the atmosphere, sweat, and breath, in the form of NaCl. Ionic fail mechanisms are typically bake recoverable and are usually seen following burn-in. Clearly, ionic contaminates can alter the electrical properties of the device and, in the case of sodium, will cause significant changes in the threshold voltage (V_t) and junction leakage. These types of defects subside with a high temperature bake with no bias as the mobile ions that collect at the PN junctions and gate oxide disperse.

Foreign Particles. Foreign particles can typically be traced to process equipment and personnel. Fig. 1 shows a DRAM device, which showed a column failure. Using the electrical signature, the failure was traced to gate oxide particles measuring approximately 0.4um, which created a leakage path at the p-channel pull-up transistor of the failing sense amplifier. The fail mechanism of this device was determined to be particulate contamination (possibly residual nitride) before gate oxidation. Similar examples of particulate contamination.are shown in Figures 2 and 3.

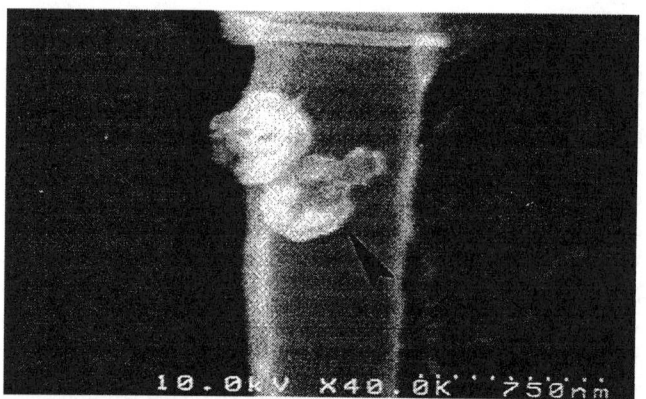

Fig. 1. Particulate gate oxide contamination.

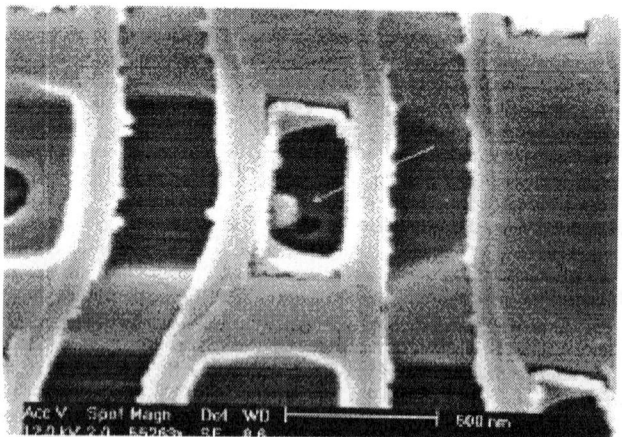

Fig. 2. Storage node contact particle

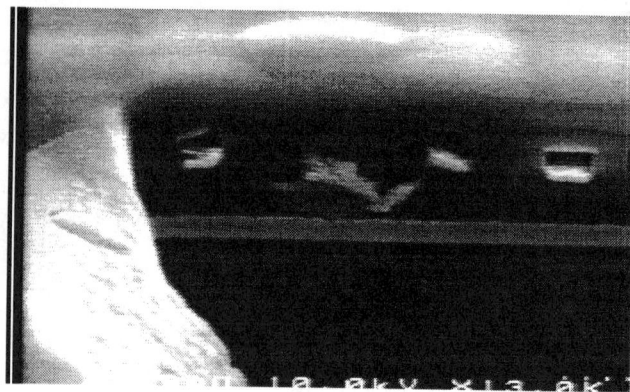

Fig. 3. Intermetal oxide particle shown after Focused Ion Beam cross sectioning.

Process Etch Residue. Process etch residue is usually associated with particles, but the contamination that blocks the etching process is removed during additional processing. In some cases, however, signs of the residue remain in tact. Fig. 4 shows residue contamination in the form of photoresist. This unit exhibited contamination blocking the guard-ring implantation in the periphery.

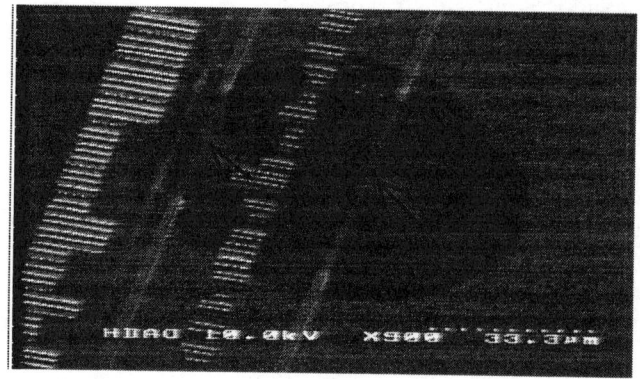

Fig. 4. Photoresist residue contamination.

Gate Oxide Defects

Pinholes. Many common failure mechanisms are related to the integrity of various dielectrics within the device. Defects in gate oxides can cause them to become leaky or shorted when a voltage is applied. Pinholes can be introduced during lithography steps or can arise from particulates into the area. They can also occur due to crystalline defects. With the complexity and size of today's devices, it is becoming evermore challenging to spot these kind of gate shorts visually. Even if a particulate does not produce a discernable hole, it can produce a region which has a lower breakdown strength than the rest of the oxide. Fig. 5 shows a DRAM device exhibiting a single bit logic failure. In this case, the gate oxide pinhole provided a leakage path from the word line to the cell bitline / cell capacitor. The root cause of the failure was determined to be dielectric failure in the cell transfer gate associated with contamination (most likely residual nitride) before gate oxidation.

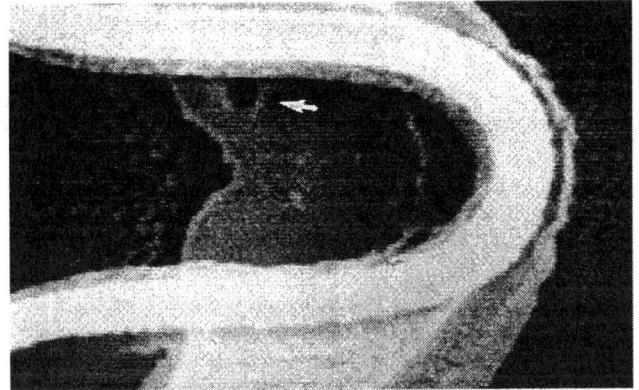

Fig. 5. Gate oxide pinhole pinhole providing a leakage path from the word line to the cell bitline / cell capacitor.

Figures 6 and 7 show other examples of gate oxide pinholes. In Fig. 6, the gate oxide pinhole provides a leakage path from Vpp to the passing word line in the row decoder causing this row to be "stuck on." The root cause was determined to be dielectric breakdown due to gate oxide thinning at the field oxide edge (birds beak).

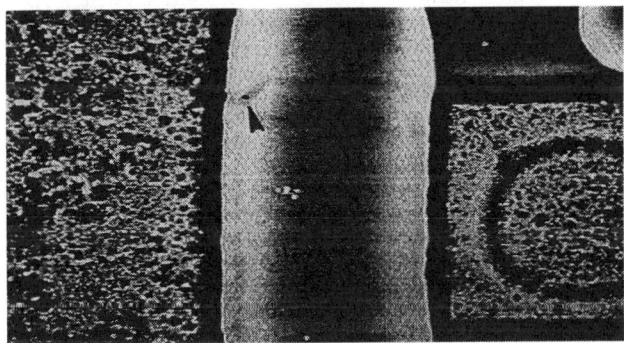

Fig. 6. Gate oxide pinhole providing a leakage path from Vpp to the passing word line in the row decoder.

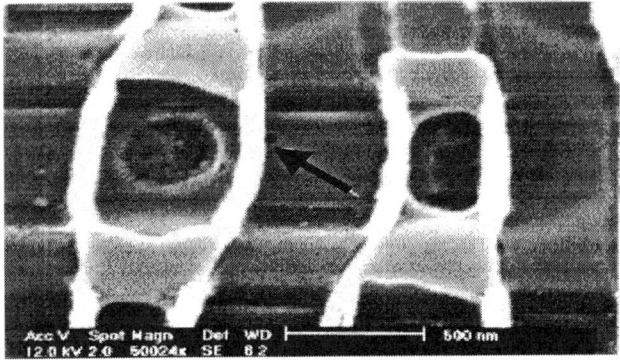

Fig. 7. Gate oxide pinhole in transfer gate.

Breakdown. Oxide pinholes can lead to extreme cases such as gate oxide breakdown. There are two types of breakdown which can exist. The first is due to EOS and ESD while the other is a time dependent breakdown, which can occur during operation within proper conditions. Clearly, EOS and ESD type failures are due to high electric fields which exceed the oxide breakdown field strength. Time dependent oxide breakdown can occur due to a number of reasons, including particulate contamination, ionic contamination, mechanical damage to the oxide, or uneven oxide growth. It has been found that oxide breakdown is not appreciably temperature dependent, but can be accelerated by voltage stressing. Figures 8-11 show examples of gate oxide breakdown.

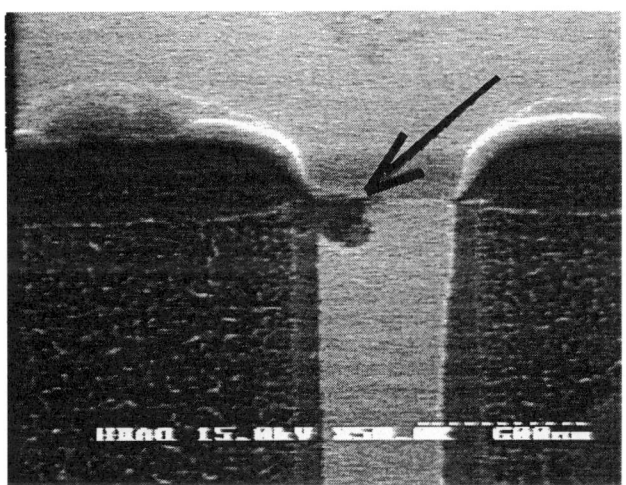

Fig. 8. Gate oxide breakdown due to EOS/ESD.

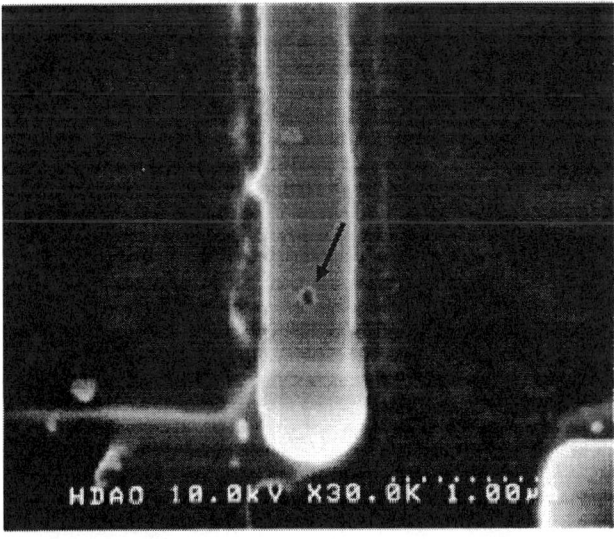

Fig. 9. EPROM device which showed multiple failing rows. After failsite isolation using emission microscopy, the failure was traced to a dielectric breakdown caused by electrical overstress.

405

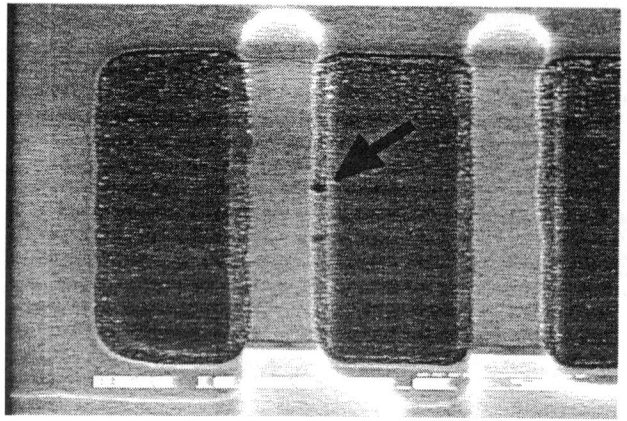

Fig. 10. Gate oxide breakdown caused by EOS/ESD.

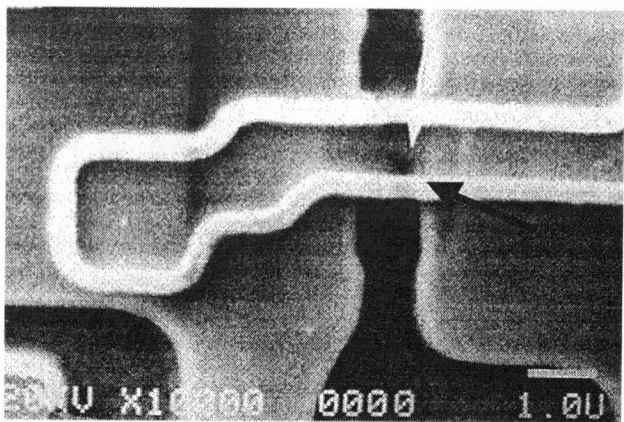

Fig. 11. DRAM cell which exhibited input leakage. After failsite isolation by emission microscopy, the defect was traced to oxide breakdown at the guard ring causing Vcc leakage. The root cause was a dielectric fail caused by the contamination of buried N+ oxide (most likely residual photoresist and nitride) before BN+ oxidation.

Etching Induced Damage. Other oxide related issues can also arise from processing. Fig. 12 shows an etch pit caused during an etching process.

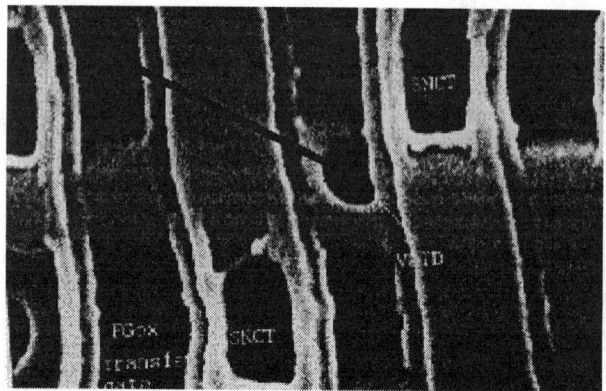

Fig. 12. Field oxide etch pit caused by stacked cell process.

Cracking. In Fig. 13, a DRAM device is shown which failed board level final tests. Failsite isolation by liquid crystal analysis traced the defect to side wall oxide crack which caused low BVDSS (breakdown voltage drain to source) at the field plate diode ESD structure. The root cause was determined to be poor oxide adhesion of side wall as a result of polymer residue left behind after poly gate etch.

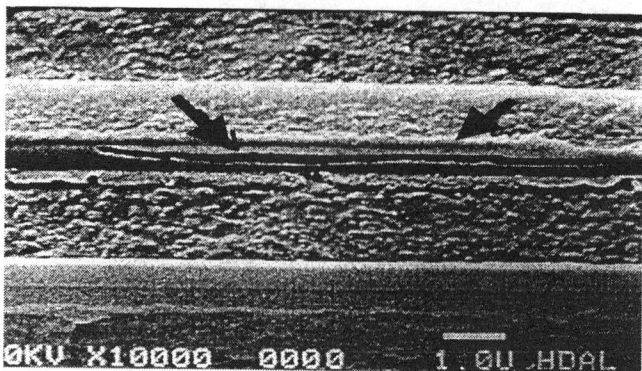

Fig. 13. Side wall oxide crack resulting in low drain to source breakdown voltage.

Multi Level and Metal Interlayer Oxide Defects

Gate oxides are not the only dielectrics which are susceptible to failures. Other oxide layers, such as multi level oxides (MLO) and metal interlayer oxides (MILO) are also candidates for defects.

Cracking. Fig. 14 shows a DRAM device which failed due to Icc power down and block failures. After liquid crystal analysis, the defect was discovered to be an interlayer oxide crack. The failure mechanism for this device was metal two to metal one leakage in the row decoder logic as a result of a metal inter layer oxide crack. The root cause was determined to be an abnormally thin MILO2 layer that was unable to compensate for the high tensile stress of the MILO1 layer.

Fig. 14. Metal inter-layer oxide crack causing metal two to metal one leakage.

Delamination. Delamination is also a concern in oxide layers. Fig. 15 shows a DRAM device which exhibited patterned array block failures. Visual analysis revealed that the gross functional failure was due to open metal lines associated with metal inter layer oxide delamination and corrosion. The root cause was determined to be moisture penetration and package stress which resulted in metal inter level oxide delamination. Poor adhesion between metal inter level oxide (MILO) and multi layer oxide (MLO) was also a contributing factor.

Fig. 15. Metal inter layer oxide delamination due to moisture penetration and package stress.

Particulates. Oxide layers are also susceptible to particulate contamination. In Fig. 16, a DRAM device is shown which exhibited shorts. After failsite isolation by liquid crystal analysis, the defect was traced to a particulate contaminate which can be seen at the metal interlayer oxide level after being cross sectioned using a Focused Ion Beam (FIB) workstation. The fail mechanism for this device was a metal interlayer oxide particle measuring approximately 0.5um which provided a leakage path between metal 3 and metal 2 in the device circuitry. The root cause for this mechanism was determined to be dielectric failure caused by particulate contamination prior to metal three lithography. Energy dispersive x-ray (EDX) analysis of the particle revealed calcium, potassium and chlorine.

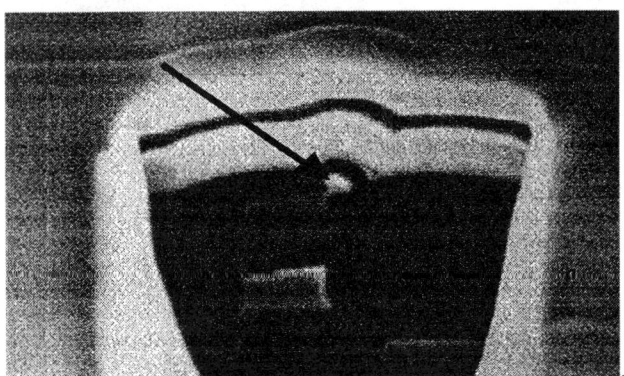

Fig. 16. Particulate contaminate in the metal interlayer oxide.

Poly Layer Defects

Another group of defects are associated with the polysilicon layer. These include mostly particulate contaminates. Figures 17 through 19 show examples of poly layer particulate contamination.

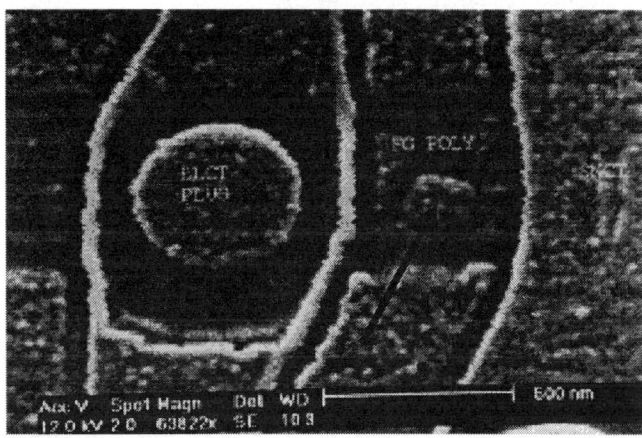

Fig. 17. DRAM device which was deprocessed to poly level using lapping techniques. The arrow indicates a poly particle which should not be there.

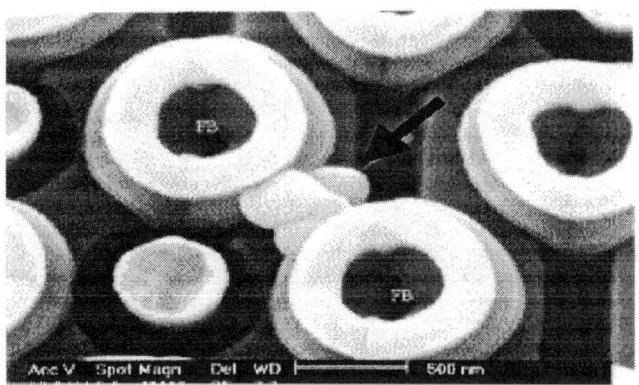

Fig. 18. Contamination of a storage cell (poly two level)

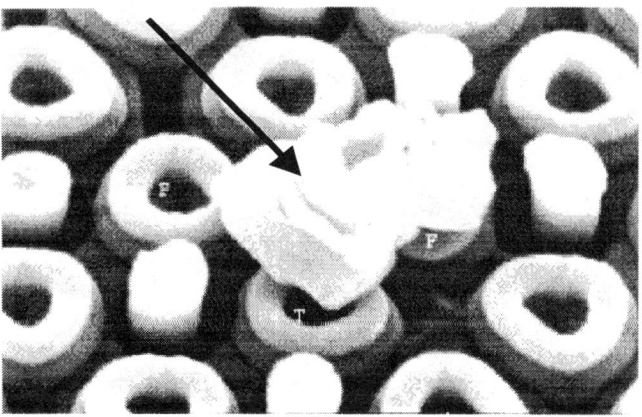

Fig. 19. Particle on top of poly plate

Metal Layer Defects

Typical defects occurring in metal layers are due to particulates and electromigration; however, other cases may be lithography, etch, or stress induced. (The stress can be mechanical, electrical, or thermal.)

Particulates. Figures 20 through 22 show examples of metal layer particulate contamination. Fig. 20 shows an example of metal particulate contamination at the metal one layer. Fig. 21 shows a an FIB cross section illustrating particulate contamination under the metal one layer resulting in degradation of the metal layer. Elemental analysis of the particle revealed the presence of titanium, which caused leakage and subsequent functional failure. Steep dielectric steps create metallization problems such as metal bridging and metal opens. Fig. 22 shows a particulate contamination at the pre- ONO layer.

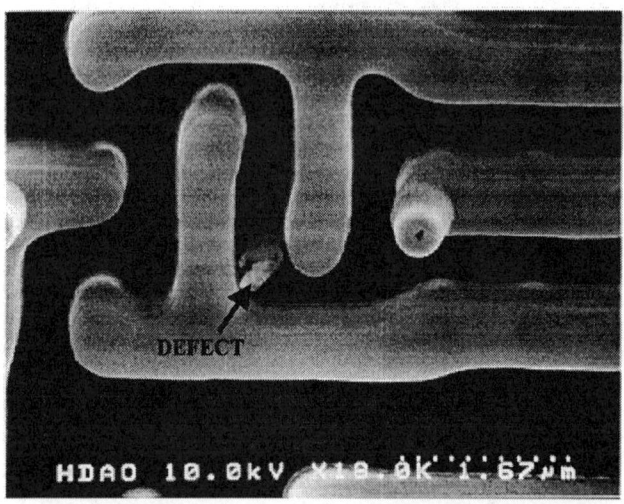

Fig. 20. Metal particulate contamination at metal one layer.

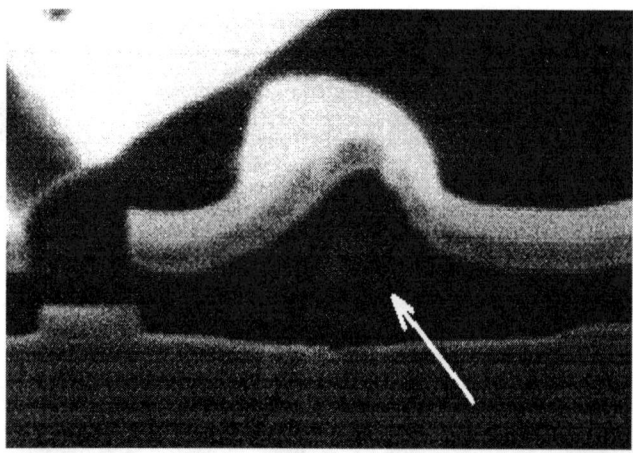

Fig. 21. FIB image revealing metal layer degradation due to particulate contamination.

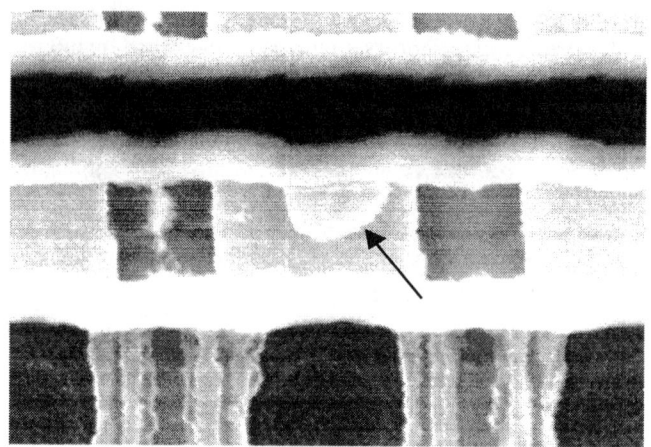

Fig. 22. Metal particulate contamination at pre-ONO layer.

Electromigration. Opens in metal lines due to electromigration occur as thin films are stressed at high current densities. This causes the metal atoms themselves to diffuse in the direction of electron travel and any flux divergence leads to a void or a pileup. Electromigration is a cumulative effect as stresses of short duration will eventually have the effect of a stress of long duration. Fig. 23 shows a DRAM device which failed a field functional array test. Visual analysis revealed an open metal in the clock circuitry as a result of electromigration. The root cause of the failure was determined to be electromigration of the Al:Si metalization due to extended high temperature / high current conditions.

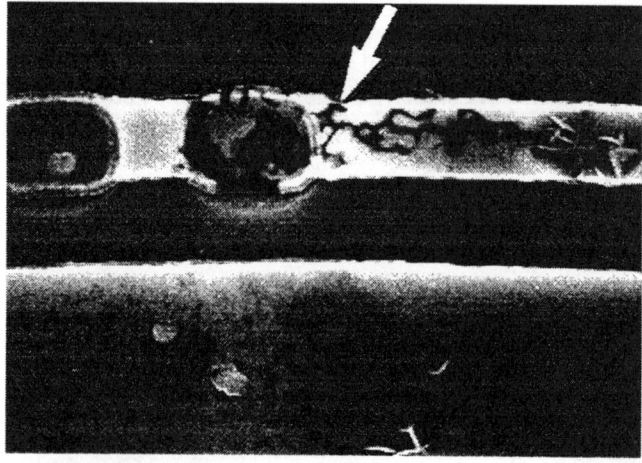

Fig. 23. Open metal due to electromigration

Stress Induced Failures. Fig. 24 shows a DRAM device which exhibited a partial column failure. In this case, the compressive stress of the nitride passivation was determined to be very high. The failure mechanism was an open metal bit line at the failing column due to metal voiding. The root cause was determined to be Al:Si metalization "creep" caused by high compressive stress of the overlying nitride passivation.

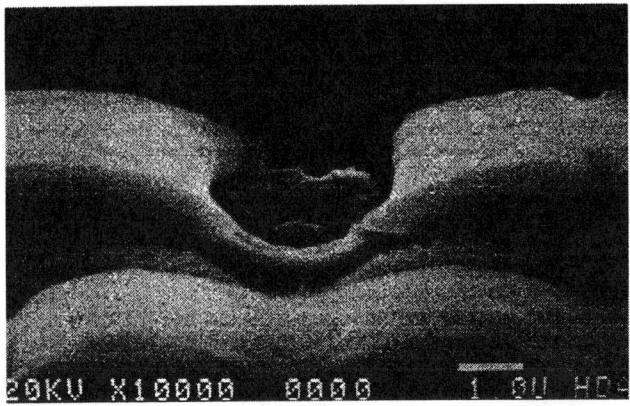

Fig. 24. Open metal at failing bit column due to metal voiding.

Protective Overcoat Related Defects

Breadloafing. PO contamination failures consist of those failures originating from particles or residue. Problems can also arise from PO coverage issues. Fig. 25 shows a poly level defect which was the result of a PO process. The image shows an EPROM device which exhibited a single bit data retention failure. After failsite isolation based on the electrical signature, the device was found to have a crack caused by poor PO coverage ("breadloafing"). The failure mechanism was determined to be a charge loss at the failing cell due to an MLO, poly 2, and ILO crack. The root cause was then determined to be that the MLO most probably originated at the PO layer as a result of PO breadloafing.

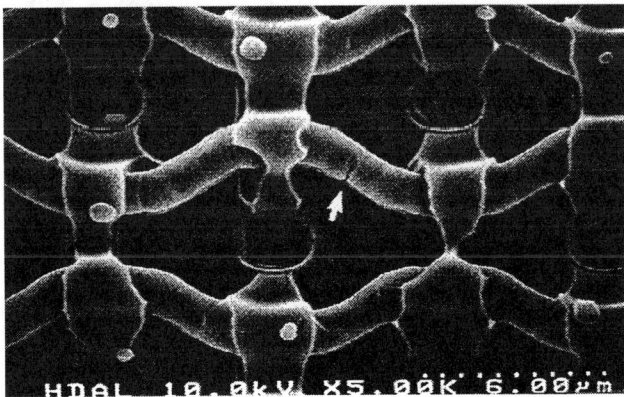

Fig. 25. Results of poor PO coverage seen at poly level.

Lithography Defects

Fig. 26 shows a DRAM device which exhibited moat to moat leakage in the internal logic circuitry Comparison of this unit to a control unit revealed a narrow moat-moat spacing. The control unit had a 1.1 um spacing whereas the failing one had a .50 um spacing. The root cause of this failure was determined to be improper moat sizing during patterning.

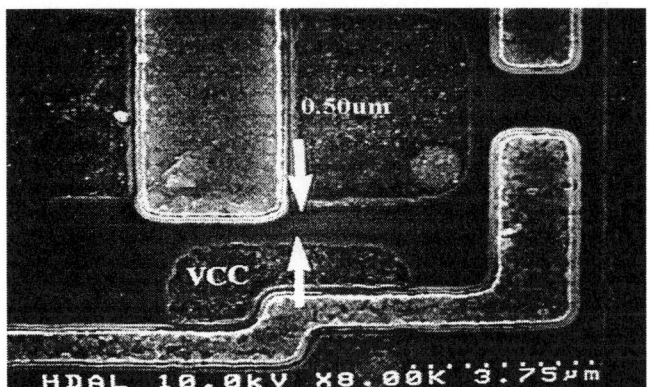

Fig. 26. Moat to moat leakage due to improper moat sizing.

Open Contacts

With the ever increasing aspect ratios required for multi level devices, proper contact formation has become an increasing challenge. Contacts may become open due to reasons ranging from the presence of silicon nodules, residual oxide, contamination, or incomplete etching.

Nodules. Silicon nodules occur when Al:Si metal is deposited into a contact via and is allowed to slowly cool. As the temperature drops, so does the solid solubility of the Si in the Al, and the result is the precipitation of Si onto the outside of the metal. This undoped silicon raises the resistance of the contact substantially. An example of silicon nodule resistance is seen in Figure 27.

Fig. 27. Silicon nodule causing open/resistive poly plug/metal one contact.

Contamination. Localized particulate contamination can also prevent complete contact etching and the proper filling of the via. Figures 28 and 29 show examples of incomplete metal deposition within the contact via. The root cause of these failures was determined to be incomplete metal step coverage as a result of contamination prior to metal deposition. Fig. 30 shows an example of incomplete etching during contact formation. In this case, the two step contact etch process only succeeded in performing one step at the failing contact. The root cause was most likely contamination related.

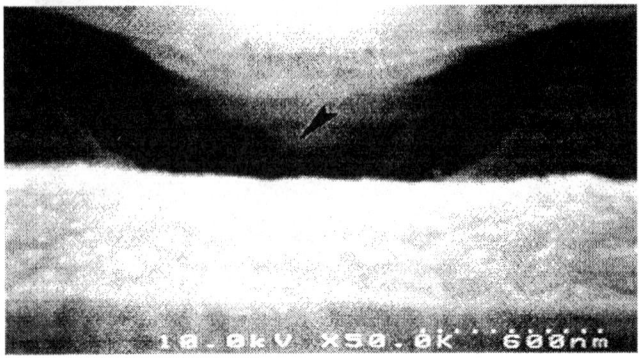

Fig. 28. Open / resistive bit line contact due to particulate contamination.

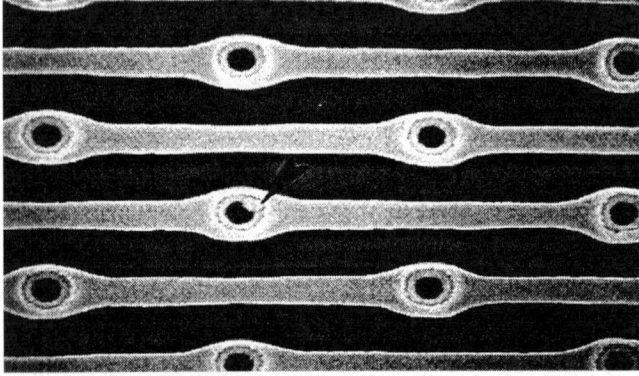

Fig. 29. Open/resistive bit line contact common to the failing bits as a result of particulate contamination.

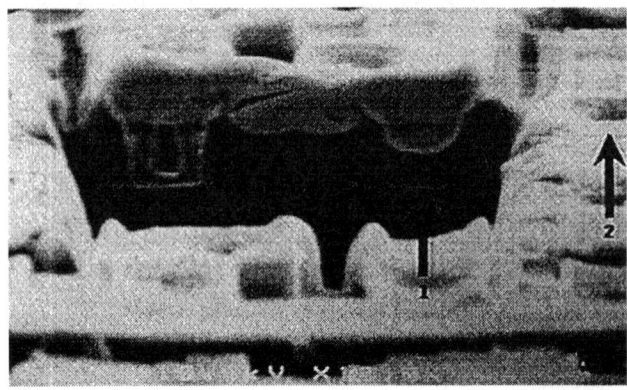

Fig. 30. FIB cross section of an SRAM device which revealed an incomplete oxide etch leading to an open metal to the moat contact at the failing RAM cell.

Stress Induced Failures. In many cases, a void in a metal trace can occur due to stress from an overlying passivation layer. This stress may be thermal mechanical, caused by the differing coefficients of expansions of the metal and the passivation. The compressive stress of the passivation can also be a factor. Fig. 31 shows a EPROM device which underwent failsite isolation by voltage contrast. It was determined that the row address line was disconnected from the nand gate control circuitry in row decoder as a result of open metal to poly contact. The root cause of the failure was determined to be an open contact caused by Al:Si voiding resulting from excessive compressive stress of the oxynitride passivation.

Fig. 31. Open contact due to compressive stress of overlying passivation.

Shorted Contacts

Shorted contacts are usually associated with electrical overstess (EOS) or electrostatic discharge damage (ESD). The example shown in Fig. 32 shows a blown contact due to spiking. Contact spiking can occur in Al:Si metal systems when the silicon content is low or when the sintering temperature is too high. Contact spikes can also occur during deprocessing if hot chucks are used above 450 degrees Celsius. Spiked contacts are similar to blown contacts but generally are characterized by tiny pits in the contact.

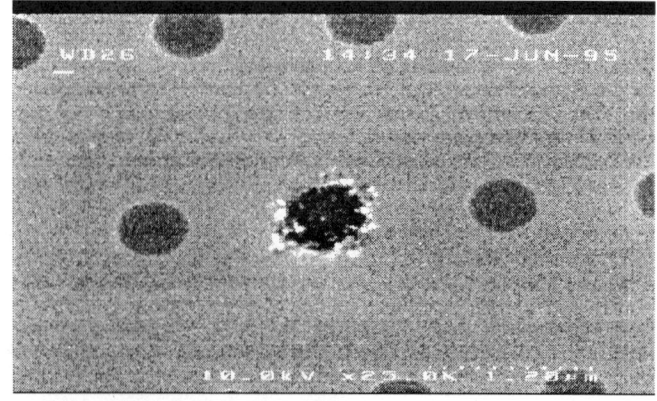

Fig. 32. Blown contact due to spiking.

Laser Repair Defects

Laser repair defects have become more common as more devices use redundancy in memory arrays. The laser fuses are usually polysilicon lines that are fused open with a pulse to set a specific memory address to be replaced by a redundant address. Fig. 33 shows a DRAM device which showed logic failures on half a row. After visual analysis, the defect was found to be a rresistive poly word line adjacent to laser repaired rows due to misaligned and overpowered laser repair.

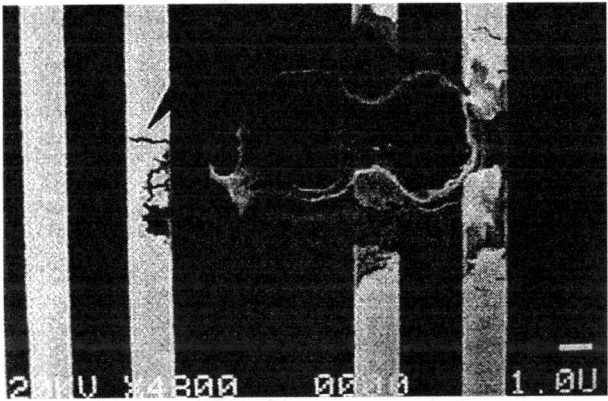

Fig. 33. Misaligned and overpowered laser repair.

Silicide Extrusions

Silicided polysilicon lines have become popular in reducing the poly interconnect resistance. One problem associated with this technology involves titanium silicide (TiSi2) and BPSG MLO. Fig. 34 shows a silicide extrusion was formed during the BPSG high temperature reflow and densification process. Although the defect appears to be silicide related, the root cause was primarily the MLO processing.

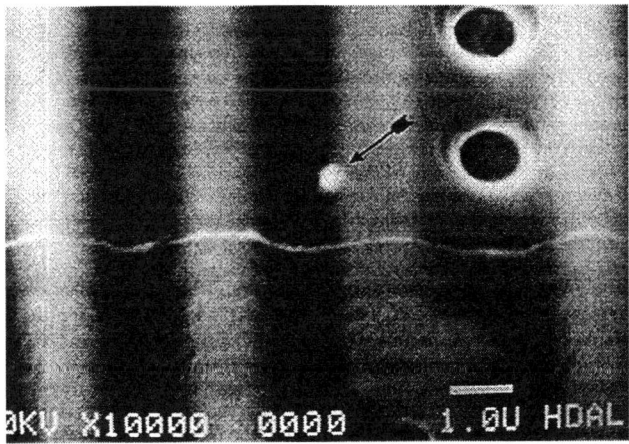

Fig. 34. Titanium silicide extruding from multilevel oxide layer, causing a short from the poly layer to the above metal line.

Filaments

Filament defects typically occur due to the incomplete etching of a metal layer which results in an inadvertent bridge between two conductors. Occurrences of filaments can also be caused by etch residue. The term has also come to include occurrences of poly melt filaments caused by EOS/ESD damage. Figures 35 through 37 show examples of metal and poly filaments.

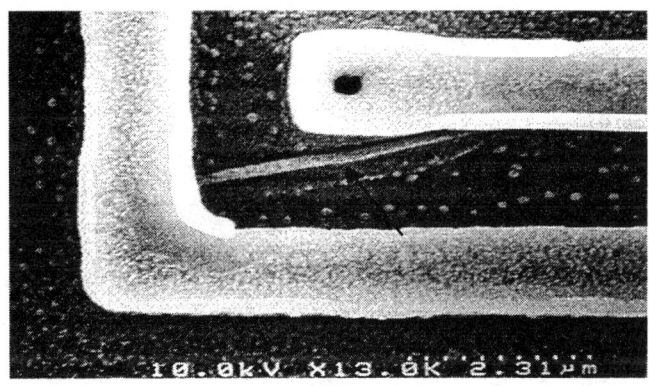

Fig. 35. Metal Filament which has broken off of the bottom metal line to short the two lines together.

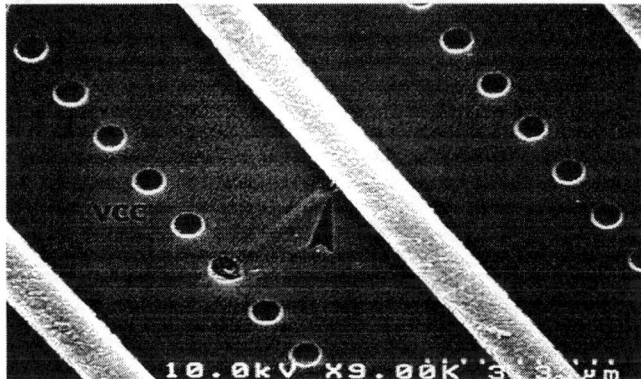

Fig. 36. Poly melt Filament resulting in a short.

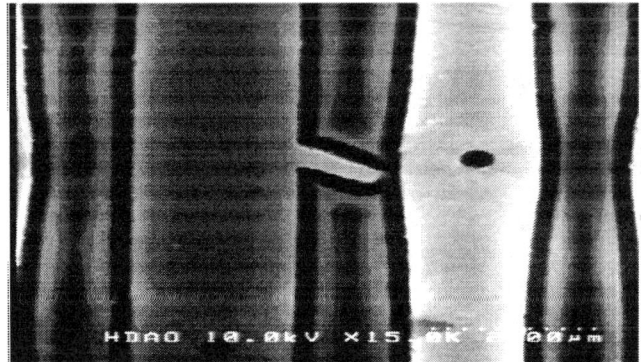

Fig. 37. Metal patterning defect causing a leakage/short to the adjacent failing column.

Substrate Defects

Substrate defects can be caused by oxygen precipitation during crystal growth, metal impurities in the silicon, or implant damage or stress induced during wafer processing. A dislocation can be present in the substrate without causing a failure. The dislocation must be electrically active to induce a failure. Most of the silicon dislocations which occur naturally are rendered inactive by H_2 sintering.

The example in Fig. 38 shows implant related dislocations at the gate region of P-channel transistors. Fig. 39 is a transmission electron microscope (TEM) cross section of a the failure, showing a loop dislocation in the silicon from source to drain under the channel region.

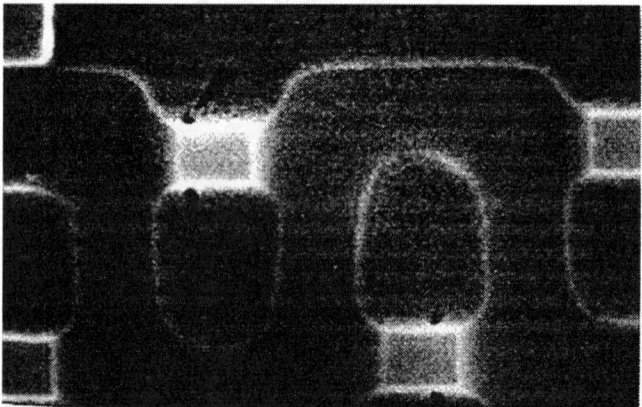

Fig. 38. Silicon dislocation.

Fig. 39. TEM cross section of Fig. 38.

ASSEMBLY RELATED FAILURE MECHANISMS

The major steps in IC assembly are wafer backgrind, saw, chip mount, wire bonding, molding, lead finishing, trim and form, and symbolization. The typical overall yield is in the upper 90's. Despite this high yield, the assembly process can still have many failure mechanisms under accelerated reliability testing. The accelerated testing is done by the application of high temperature, high humidity, high voltage, or a combination of these. In some cases, mechanical or themo-mechanical stresses are applied to screen for any structural flaws or design weaknesses. Typical failures are bond wire continuity, package cracks, chip cracks, corrosion, passivation cracks, interface delamination, leakage, and parametric failures.

The electrical signature of bond-related failures is typically continuity problems, opens followed by shorts or leakage. Some of the devices will show "input hi" or "input lo" level failures. Generally, these can be detectable if x-ray analysis is done to check for wire or bond related issues. In the case of intermittent contact, performing continuous testing with temperature ramps will typically bring out the failure. Bond strength problems due to low and high bond strengths are detectable by x-ray analysis when the x-ray is taken with a side view shot. They are indicators of potential yield and reliability problems.

Bond wire failures. Figure 40 gives a pictorial view of the failure mechanisms that are associated with the bonding process. Several possible failures can be observed in the diagram. First of all, one must consider the formation of intermetallics at the Al pad/Au wire interface. Some amount of alloying is necessary at this interface to form a good contact, but excessive alloying may cause bond failure. Kirkendall voiding may also be present, where voids within the metals have migrated until merging together at the bond interface [1].

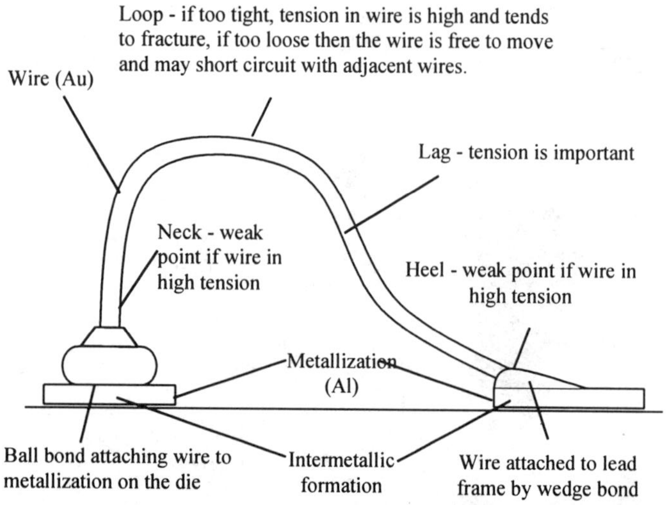

Loop - if too tight, tension in wire is high and tends to fracture, if too loose then the wire is free to move and may short circuit with adjacent wires.

Wire (Au)

Lag - tension is important

Neck - weak point if wire in high tension

Heel - weak point if wire in high tension

Metallization (Al)

Ball bond attaching wire to metallization on the die

Intermetallic formation

Wire attached to lead frame by wedge bond

Fig. 40. Reliability issues concerning bonding process [1].

Another parameter in the bonding process which may cause failure is the tension in the bond wire. If the tension is too low, the wire is said to have excessive lag and is susceptible to shorts with other bond wires. An example of this can be seen in Fig. 41, which shows an example of pin to pin leakage bond wires shorted to adjacent lead frame post. The root cause of this failure was determined to be a bonding machine tension feed failure. If the tension is too high, then the stress at the interfaces can lead to fracturing and opens. Examples of this can be seen in Figures 42 and 43.

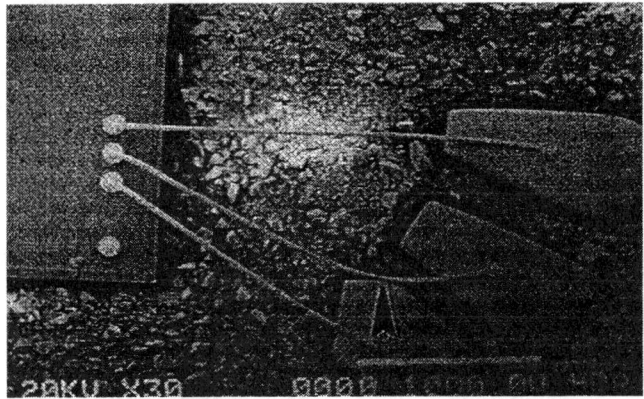

Fig. 41. Bond wires shorted to adjacent lead frame.

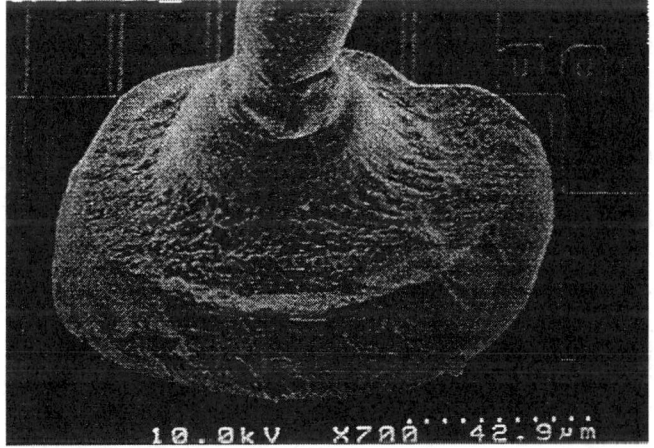

Fig. 42. Cracking at ball bond neck.

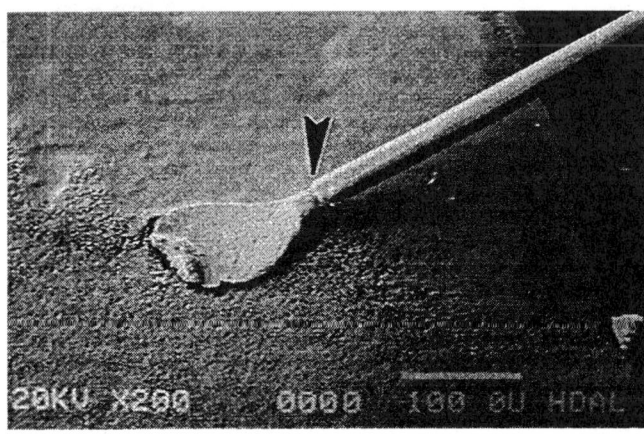

Fig. 43. Cracking at stitch bond.

There are other concerns with the bonding process as well. Moisture has been found to aggravate the metal migration process by the influence of electrolysis. Contamination by impurities such as carbon at the base of the bond can also cause voiding and lead to ball bond lifting. Contamination from chlorine and other impurities can lead to corrosion, as seen in Fig. 44. In this case, corrosion has caused a discontinuity in the bond pad aluminum. EDX analysis detected chlorine at the bond pad. Thus, the root cause of the failure was determined to be corrosion caused by moisture ingress and chlorine contamination.

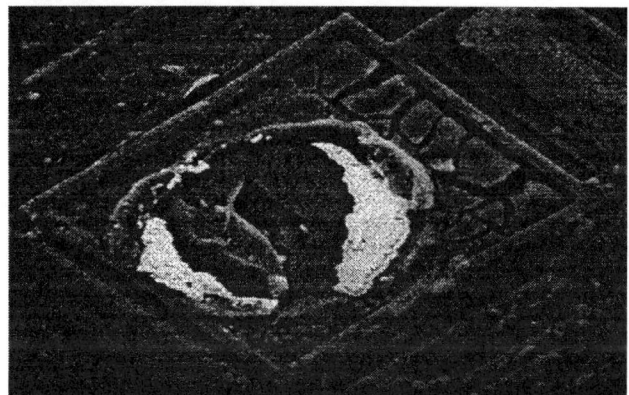

Fig. 44. Corrosion of bond pad due to chlorine presence.

Fractures within the bond wire itself can also occur. Fig. 45 shows a fractured bond wire at Vcc associated with thermally degraded mold compound. This failure was most likely caused by electrical overstress.

Fig. 45. Fractured bond wire due to EOS

Die Cracks. Fractures in silicon usually originate from the site of the flaw. This flaw can come from a backgrind operation, which leaves behind deep grooves. A high Rt, the maximum surface roughness, is usually indicative of this type of failure. Fractures can also result from saw operations, in which case, the crack origin can be traced to damage on the side of the chip. Other cracks may result from poker pin damage. In this case, one of the cracks should intersect the poker pin mark, which should be located near the center of the die. Fig. 46 shows a die exhibiting ICC leakage due to a die crack. The root cause of the failure was determined to be a cracked die caused by the lifter pin during the die mount process. Fig. 47 gives an overview of chip and package crack issues.

Fig. 46. Cracked die due to lifter pin damage.

Chip and Package Crack Issues	Causes	Distinguishing Feature
Cracked die and Package Cracks	Backgrind damage Saw Wafer mount poker pin Trim and form	Rough backside, High Rt Incomplete cut, Chip Pin mark, Crack thru mark Scuff marks, Imprint Delamination and Internal package crack
	Test handling	Scuff marks
	Surface mount	Package cracks at die edge
	Mold voids	Crack passes through void
	Molding and mold voids	Distortion of mount pad
	Mount epoxy voids	Epoxy viods
	Mount epoxy coverage	Imcomplete coverage
	Design/lay out weakness	PO at scribe
	Interaction with saw	Metal at scribe

Fig. 47. Chip and Package crack issues: their causes and their distinguishing physical features.

Filler Induced PO Damage

The electrical signature of failures due to PO damage for localized point defects in memory devices ranges from single bit cell failures to row and column failures. Recovery of the imprint of the mold is key if identification of filler as the cause is required. Locating PO defects can be done by the use of a metal etch or a Na ion drop test. Partial etching of the PO followed by inspection will also reveal crack patterns that are not obvious, or regions of stress through the crack

patterns. X-sections using metallographic techniques can be used to examine PO profiles.

Fig. 48 shows an example of mold compound filler induced PO damage. Analysis involving a mechanical decapsulation confirmed a protruding filler particle. Thus, the root cause was determined to be mold compound filler induced PO damage. Fig. 49 highlights some of the key failure mechanisms related to PO damage.

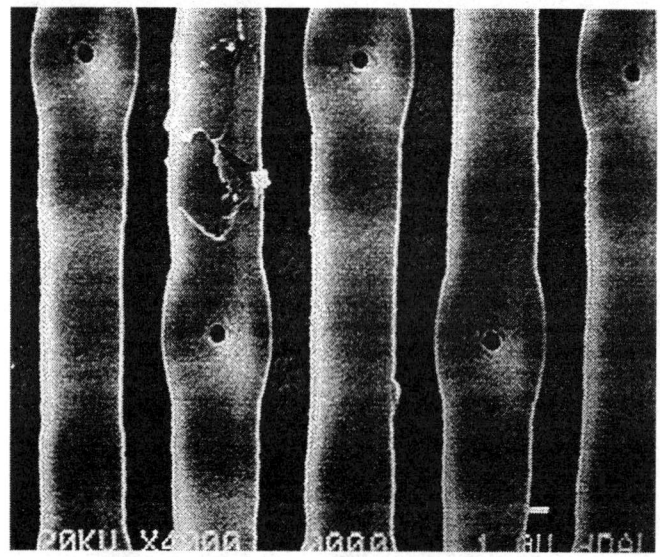

Fig. 48. Mold compound filler induced PO damage.

PO damage issues	Causes	Distinguishing feature
Localized PO defect	Mold compound filler and resin	Compressive fracture cracking at brittle levels
Linear fragment	Design/layout	Metal over poly, long metal runs
Linear scratch	Mold	Scratch lines along mold flow
Point defect	Bond wire	Touches PO during bond
Corrosion	PO stress/voids	X-section xtics, Sharp geometries, corners

Fig. 49. PO damage issues: their causes and their distinguishing physical features.

FLIP CHIP DEFECTS

One of the greatest concerns of the failure analyst at present is the ability to diagnose failures associated with flip chip style devices related to packaging. These failures may include solder bump defects or interfacial abnormalities between the solder bump/substrate and solder bump/die interfaces. An examination of the intermetallic composition of the metallurgy within the bump itself may be necessary to determine the failure mechanisms of some open bumps.

414

Proper package related fail mechanism identification for flip chip devices typically involves a die pull, die sheer or a traditional cross sectioning of the interfaces within the device. One of the more compelling reasons for performing the cross section is the ability to examine the metallurgical surfaces within the bump. Of particular interest is the consistency of the Sn:Pb bump metallurgy, including the Pb grain size within the bump. Fig. 50 shows an SEM cross sectional image of a bump failing continuity testing. From the cross section, two major anomalies were noted. The first anomaly was the void and possible interface dewetting between the bump and the die interface. The second anomaly was a large Pb band shown at the bottom of the bump in Fig. 50. The Pb band has a tendency to be located near the via, which is the path of the current flow. Due to large amounts of current flowing through the via, the location near it tends to be heated the most, leading to large Pb grain formation at this location. Subsequent curve trace analysis of the bump to the die revealed lack of continuity between the bump and the die, which corroborated our observations.

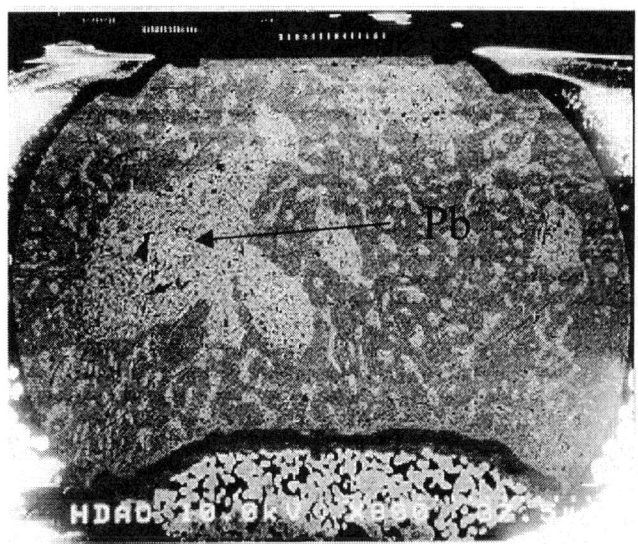

Fig. 51. Cross-sectional SEM micrograph of a bump after high temperature assembly, showing Sn and Pb grain sizes.

Fig. 52 shows the intermetallic formation of the bump shown in Fig. 51. The intermetallic formation is smooth and well defined indicating the diffusion of Sn into Ni.

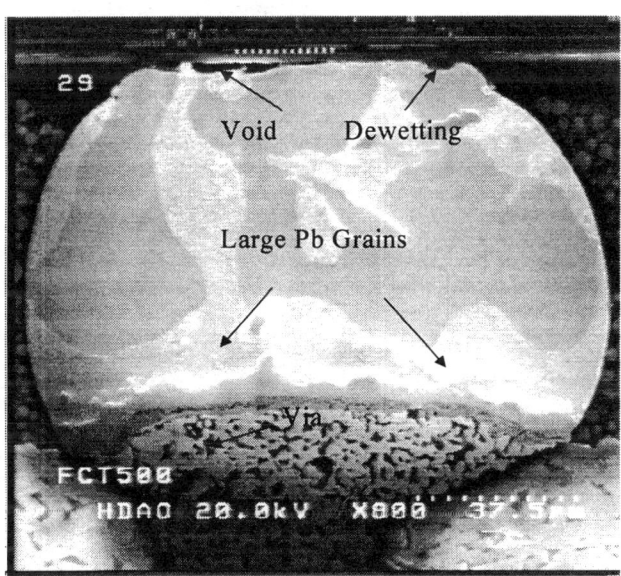

Fig. 50. SEM cross section image of a open bump which separated due to thermal and electromigration effects.

The bump structure shown in Fig. 50 uses a thin layer of Ni/V to separate the Sn of the eutectic (Sn:Pb, 63:37) solder bump from the Al of the top level metal of the CMOS device. If the Sn diffuses through the Ni and reaches the Al interface the bond will be broken. It has been found that this diffusion process is aggravated by the influence of high temperature and hence, the amount of diffusion is a good indicator of the age of the device.. The grain size of the Pb grains within the solder bumps are also good indicators of device age. Fig. 51 shows a cross section of a bump which has undergone tests involving high temperature conditions. The relatively large Pb grain sizes are noteworthy in this case.

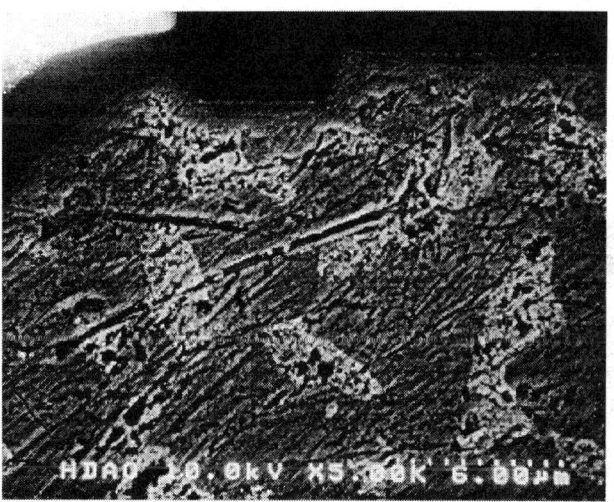

Fig. 52. SEM micrograph showing the intermetallic formation after high temperature assembly.

The extent of diffusion can also be determined by an EDX analysis of the bump , which reveals the amount of various elements at any point. Fig. 53 is an EDX linescan of the intermetallic formation seen in Fig. 52. As can be seen with respect to the reference line drawn across the linescans, Sn has already started to diffuse into the Ni during the assembly process. Any further heating of the bump will eventually cause failure.

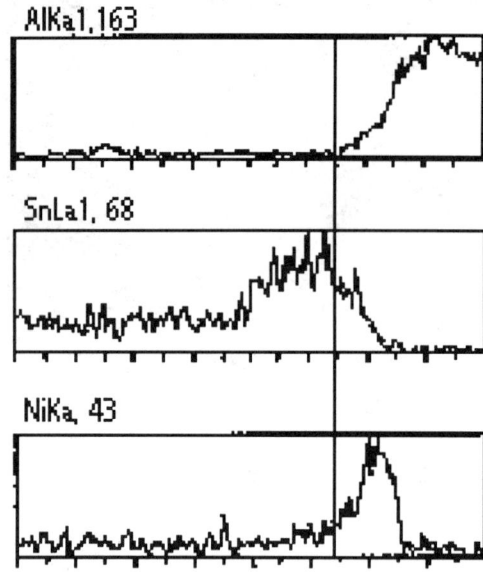

Fig. 53. EDX analysis showing the diffusion of Sn into the Ni, as indicated by the overlap of the elemental graphs

A complementary technique to cross sectioning is the die pull. In this case, the planes of the bump bond can be analyzed visually and by elemental analysis. When the surface of the bump from a die pull is spotted in texture this is an indication Sn has diffused through the Ni causing the bump to de-bond from the Al bond pad. The surface looks spotted in an SEM view as can be seen in Fig. 54.

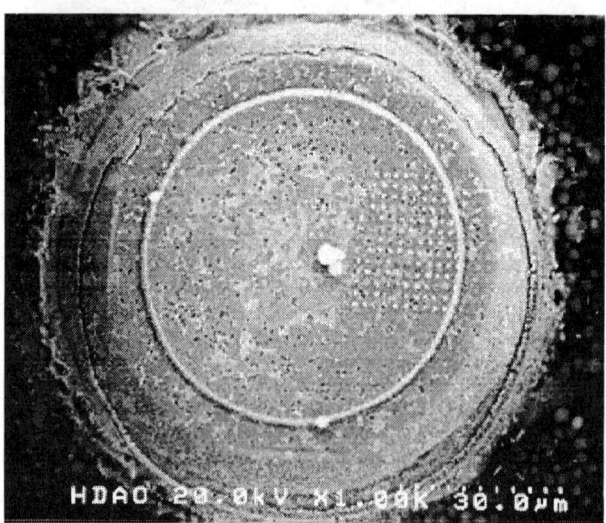

Fig. 54. SEM micrograph of a resistive bump. The bump de-bonded from the Al bond pad.

Fig. 55 shows a failure occuring due to the separation of the solder bump to metal five interface. The failure mechanism determined to be the diffusion of nickel into the

Sn:Pb solder causing de-bonding of the Under Bump Metal to metal five of the device.

Fig. 55. De-bonding of bump from metal five interface.

Another solder bump design containing a copper stud within the solder ball is shown in Fig. 56. After high temperature tests, it can be seen that the Pb grain sizes are very large, however, this is not a problem due to the influence of the Cu stud. As the temperature increases, Cu diffusion into the solder bump helps to stabilize the bump. The near infinite source of Cu in the stud prevents the Sn from making contact with the Al of the metal five in the die. The Cu stud does present a reliability issue as bump sizes reduce. Cu is rigid in comparison to the more malleable Sn:Pb solder. In smaller bumps, the Cu takes up a larger percentage of the height. The reduction in the amount of solder in the bump reduces the flexibility giving rise to stress related failures which will be noted later.

Fig. 56. SEM micrograph of a cross-sectional view of a bump after high temperature testing, showing the Sn and Pb grain sizes. Image used as reference to show alternate bump structure. No failure observed.

In Fig. 57 the result of a die pull of a bump subjected to high temperature operational life (HTOL) testing is presented. Inspection of the bumps after die pull indicated bump melting. The large region of Pb seen in the lower portion of the bump could have this appearance due to voiding in the bump as well as bump melting. However, the crystalline structure of the periphery of the bump (seen in the lower left corner of the image) is indicative of bump melting, which can only occur at temperatures exceeding 183C.

Fig. 58. Bump exhibiting signs of solder fatigue.

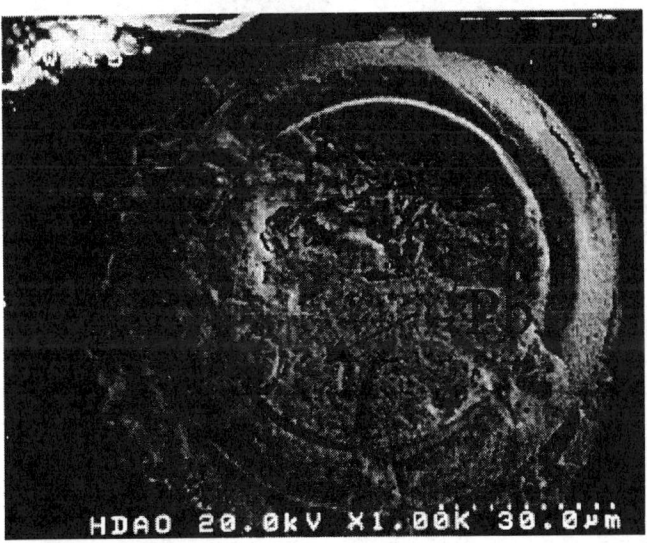

Fig. 57. SEM photograph of a failing bump, which was most likely melted during HTOL testing.

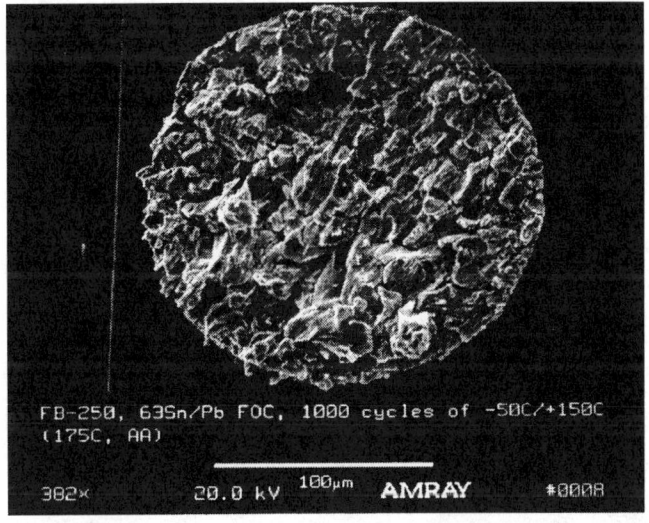

Fig. 59. Cross sectional view of bump in previous image.

There are a plethora of failure mechanisms associated with flip chip packages. Here, we will not discuss all the mechanisms in detail. However, as additional examples, we present Figures 58 through 66 to illustrate several other common failure mechanisms associated with flip chips. Figures 58 through 60 show examples of solder fatigue and creep. Fig. 61 shows a lack of supportive Ti:W layer which has allowed the "rocking" of the solder ball and the subsequent cracking of underlying structures in the die. Fig. 62 illustrates the results of electrothermal migration. The image shows the positive node of a bump in which voiding at the top edge of the bump has occurred due to the constant bombardment of electrons on that side. In this case, the Pb has migrated to the bottom edge under the influence of the electric field. Figures 62 through 66 show various other common failure mechanisms associated with flip chips.

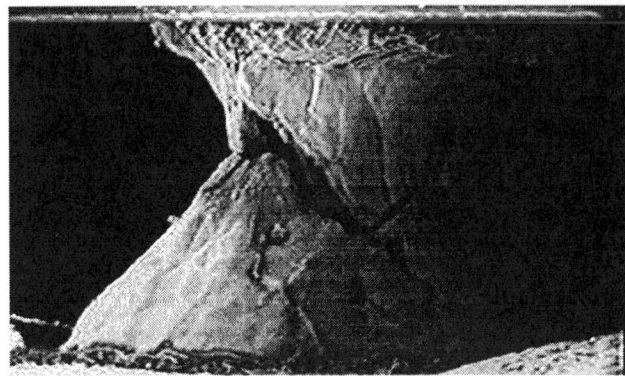

Fig. 60. Example of bump solder creep showing the two interfaces within the device separating from one another.

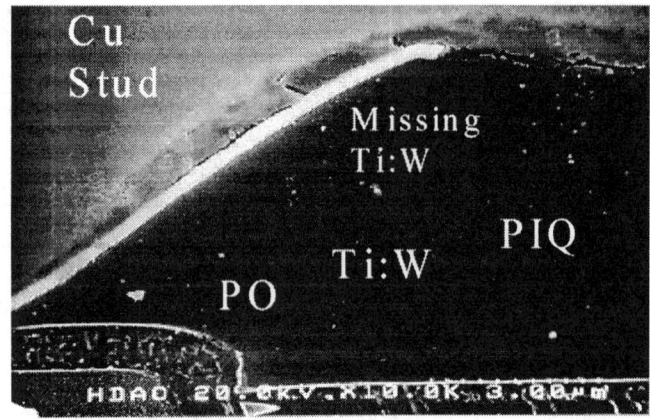

Fig. 61. Image shows lack of Ti:W metal near the cracked area which should extend out through it.

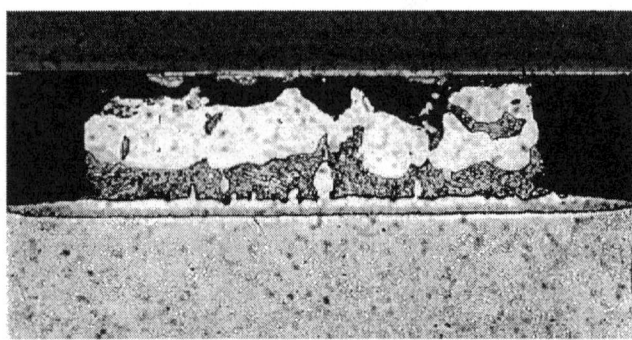

Fig. 62. Effects of the high electric field on the metallurgy within the bump. The image shows the positive node of a bump exhibiting signs of electro-thermal migration.

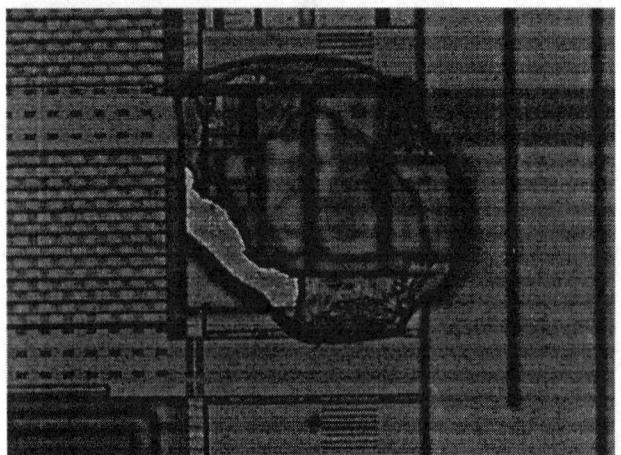

Fig. 63. Image showing a bond pad missing due to an oxide fracture. The fracture is caused by the thermal mismatch of the package to the die.

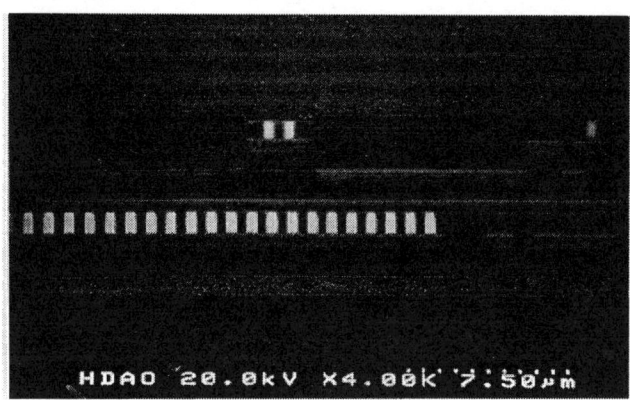

Fig. 64. ILO cracking due to thermo-mechanical package stress at the bump.

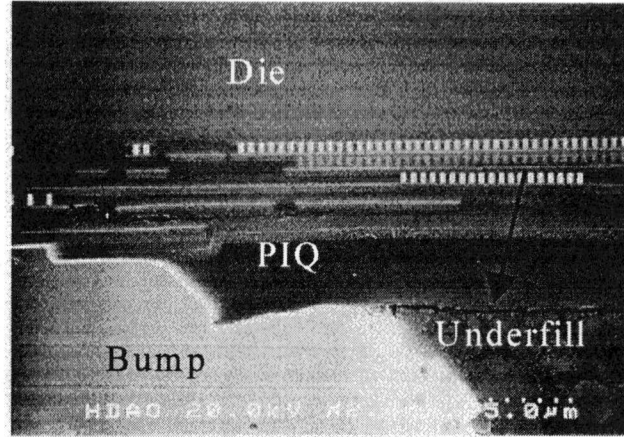

Fig. 65. SEM cross section image of underfill to PIQ delamination due to thermo-mechanical package stress.

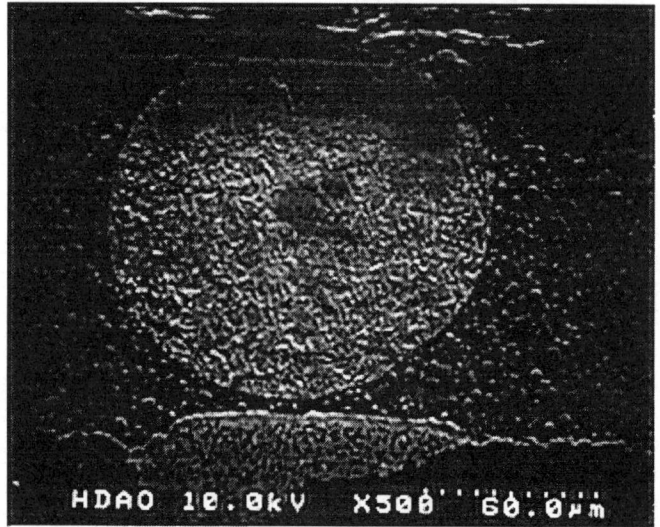

Fig. 66. Image showing a small bump due to the lack of solder plate.

CONCLUSIONS

The journey of the failure analyst is becoming ever more difficult due to the subtlety of the failure mechanisms present in today's devices. It is becoming increasingly important to have an detailed understanding of what the failure mechanism might be before actually finding it. The shrinking geometries of devices have made failure mechanisms out of subtle anomalies that were once irrelevant to the performance of the device. An increased knowledge of device properties on the micro scale has become necessary in order to identify these current failure mechanisms.

The evolution of device architectures has also created new failure mechanisms. One can trace the lineage of failure mechanism back to the most primitive devices. The evolution of the trench capacitor, the multi-level system, and the high aspect ratio contact are all technological advances which tested the abilities of the failure analyst and presented new suspects into the ever increasing file of integrated circuit failure mechanisms. Today, with the growing number of subtle defects which are virtually non- visual, the tools and techniques of the failure analyst are once again put to the test.

Packaging defects have also evolved over time. The mundane packaging issues of the past have given way to more complex issues such as filler damage. Today's packages involve a variety of different materials and some of the greatest problems have occurred due to mismatches in the expansion coefficients within the materials. Today, with the increase in usage of the Flip Chip packaging technology, an extensive knowledge of packaging defects is mandatory.

Inevitably, the list of failure mechanisms will see the addition of new mechanisms as package and device technologies continue to evolve. Thus, the failure analyst must constantly improve the tools and techniques to be able to succeed in the goal of producing needed information and solutions.

ACKNOWLEDGMENTS

The authors would like to acknowledge the entire HDAO Staff for their contributions to this paper. We would also like to thank David Lam and Wilson Tan of TI-Phillipines for images and technical information within the paper. S. Yeh and M. Varnau of Delco are also acknowledged for their images in the flip chip section of the paper.

REFERENCES

1. A. Amerasekhara and F. Najam, <u>Failure Mechanisms in Semiconductor Devices</u>, (2nd Edition), John Wiley and Sons, 1997.
2. D. Corum, S. Y. Kim, and K. S. Wills, "*Failures Mechanisms in Integrated Circuits*," <u>Microelectronic Desk Reference</u>, (1st Edition), pp. 278-300, 1993.

Differentiating between EOS and ESD Failures for ICs

Leo G. Henry
ORYX Instruments Corp

INTRODUCTION:

Distinguishing between EOS and ESD failures and differentiating the subtle differences between damage due to the several distinct ESD stress "Models" continue to challenge failure analysis capabilities as device dimensions shrink and critical defect sizes are reduced. Most of the ESD damage sites are not visible to optical microscopy and de-processing together with very high magnification examination using the SEM is most often necessary. However, the use of stress model simulators can most often replicate a *Failure Signature,* i.e., a unique die location and morphology associated with the ESD model [1, 2].

In the Semiconductor I/C industry, it has been well documented that the proportion of factory and customer field returns attributed to device damage resulting from Electrical Over-Stress (EOS) and Electro-Static Discharge (ESD) can amount to 40 to 50 % [3, 4]. These ESD events, which are the subset of EOS events, are associated with high voltages and can be replicated using one of several ESD *Failure Model Simulators.* It is to be noted that additional hard (functional) and soft (leakage) failures which also occur in the factory are normally detected by effective test programs. It is therefore still necessary to determine the probable cause of failure by Laboratory Simulation and Failure Analysis before effective corrective action can be accomplished.

This tutorial utilizes the results obtained from the evaluation performed on hundreds of devices over several years (mainly CMOS technologies- micron to Submicron). The study entailed EOS and ESD simulation using a variety of models, conducting detailed electrical and physical failure analysis and then comparing the results with documented analyses performed on customer field returns and factory failures. The immunity to EOS and/or ESD is indicated by parametric measurement and/or functional performance after exposure to the stress.

As a result of the differences in the EOS and ESD current stress magnitude and the associated time domain, we can determine the location, type, variation and magnitude of damage at the failure site. This is then used to establish *a* relationship between electrical *signature,* physical damage type and location. The actual physical FA procedure will not be described here as this topic is addressed in other sections of this desk reference [5].

Physics of Junction Failure: It has been shown that there is apparent similarity in damage resulting from ESD due to HBM or MM [6]. This is due to the common underlying physics for the thermal failure of P/N junctions. The dependence of failure power upon pulse length for high current square waveform pulses is illustrated in Figure 1 and described by the following one dimensional electro-thermal equation known as the Wunsch-Bell Model [6].

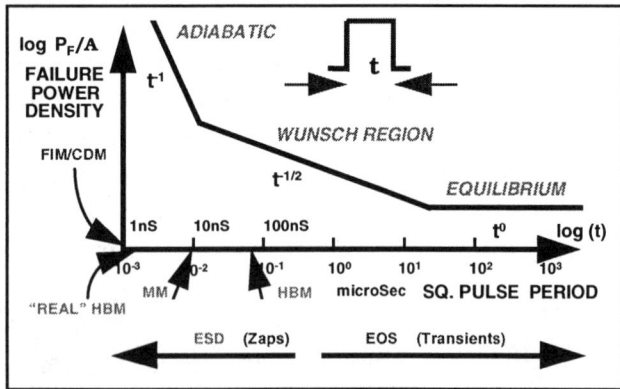

Fig. 1 - The Wunsch-Bell Model for Junction Failure.

$$P_F/A = K_1 \times t^{-1} + K_2 \times t^{-1/2} + K_3 \times t^0$$

Where P_F is the failure power, A is the junction area, t is the pulse period and K_1, K_2 and K_3 are proportionality constants.

This relationship describes the behavior of the instantaneous failure power density in a device plotted against the logarithm of the input pulse width (figure 1) and shows the characteristic "Wunsch" region which describes most ESD failures due to an external event, i.e., *discharge to a package pin from the human finger when a different*

(another) pin (or pins) is(are) grounded. Also shown is the "equilibrium" region for EOS events occurring for pulse widths greater than a few microseconds.

The time duration for typical EOS events is shown to be approximately one to 100 milliseconds, but for ESD events (from charged metallic objects and charged personnel) *to* the package, the duration will be 10 to 150 nano-seconds approximately. However, the failures due a Charged Package are below the 5-nsec pulse width region [7] and is at the fastest time domain region (< 500 psec rise time). Note that the region marked "real HBM" refers to the time domain below approximately one nano-sec, and relates to the electrical disturbances likely to cause system upset due to discharge from the human holding a metallic object [IEC   ].

EOS and ESD Failures:

While obvious EOS may be physically evidenced by cracked packages, carbonized plastic, burnt-out bond wires and massive visual damage to the metal on the die surface, the more subtle ESD failure is not visible at low magnification. However, in most instances, the observation of discoloration will provide a means of differentiating between the EOS and the ESD events [1, 2], since the EOS may result in a discoloration radius greater than 4 μ [May et al]. Thus EOS damage is visible at low magnification under an optical microscope, and the fail site will be surrounded by a region of discoloration; (note that the use of color Polaroid film facilitates the recording of any discoloration visible after de-capsulation). However, the relative absence of discoloration together with the need to de-process and to use high magnification (including the Scanning Electron Microscope) to locate the failure site indicates a type of ESD failure.

May and Guravage [9] have also shown that the region of apparent *discoloration* which *surrounds the damage site* due to EOS, visible at low magnification at the top glass to metal interface, has a radial extent that directly relates to the pulse length in the range of 100 nano-seconds to perhaps 30 milli-seconds. This occurs because the thermal transient has enough time to diffuse laterally away from the site of EOS damage across the die surface.

Dielectric Type Failure: *Dielectric failure* resulting from the *Electric Field* dependent Charged Device Model *(CDM)* is indicated in Figure 1 at the fastest end of the time domain. This type failure is due to *discharge from the package to ground via a single pin* with effective pulse duration of much less than 5 nano-seconds. The severity of the failure in this case will largely depend upon the package type used. Susceptibility is increased in SMT packages with very low lead inductance and with a large fraction of the lead-frame body capacitance upon the die paddle [10]

There are two charge modes for CDM [11]: the field induced mode (FIM) and the direct charged package mode (CPM). The distinction between these two electric field dependent failures is as follows:-

The FI mode type of failure occurs when a *single* pin is grounded while the package lead-frame is at elevated potential. Charged are induced on the lead frame. while the device is located within an electric field. Note that the failure occurs only after one of the pins has been grounded.

The CP mode type of failure also occurs when a *single* pin is grounded, but here induction on the lead frame is due to the *electrostatic charge already present on the package surface(s). The charged package may be due to* contact electrification or tribo-electric generation. Note again that the failure occurs only after one of the pins has been grounded.

The actual failure mode for both types is identical for the potential imposed upon the lead-frame, and normally results in a *unique failure signature far internal to the die consisting of gate oxide damage at the first Input buffer. This location is* beyond the HBM and MM protective structure at the bond pad. Most often the oxide failure is located beneath the poly gate in the gate oxide, or is located at the poly gate edge but still in the gate oxide.

In this tutorial, reference shall hereafter be made to both the FI mode and CP mode as CDM. The three major packaged device level ESD models were used; namely the Human Body Model (HBM), the Charged Device Model (CDM)and the so-called Machine Model (MM). Four EOS simulations were developed and applied: 1: *programming over-stress* (variation of the programming voltage applied to special high voltage pins); 2: forced I/O *reverse breakdown* (actually BV_{DSS}); 3A: *reverse insertion* for dual in-line packages (DIP) *and 3B: mis-orientation* for surface mount packages (example is PLCC); and 4: *transient latch-up [TLU]* using a capacitive discharge [12, 13]. TLU will not be discussed here as it is the subject of a more advanced tutorial.

For the ESD Simulation use was made of State-of-the-art commercial simulators, but for the EOS stresses, in-house apparatus was used as required. Stress procedures for ESD simulation conformed to those established by the ESD Association Standards: HBM-S-5.1 [14], MM-DS-5.2 [15], CDM-DS-5.3 [16], and where appropriate, the MIL STD-883 [17].

ESD Laboratory Simulations: The sample sizes for HBM ESD and MM ESD are as specified [{2N+1}xVxL] in the standards.

Here N is the number of independent power supplies, V is the number of voltage levels to be stressed and L is the number of different lots from which the samples are taken. This is typically three. For the HBM [sample size used was 60], the stress level ranged from +/- 500 Volts to +/- 6000 Volts. For MM [sample size used was 32], the stress ranged from +/-100 Volt to +/- 1000 Volt. For CDM only, the sample size (21) is simply VxL with a stress range from +/- 250 Volt to +/- 3000 Volt. All units were data-logged both pre and post test such that any leakage current equal or greater than 500 nano-Amps was "flagged" for analysis. Failing devices were analyzed using conventional electrical, chemical and electrochemical techniques. High magnification SEM examination had to be used to distinguish the physical damages.

HBM ESD Discharge: For the HBM simulation, the physical failures occurred mainly in the ESD protective structure at the input pads and within the "self protecting" output circuits, but may also occur in the core of the device if the failure is functional or Icc between the Vdd (or Vcc) and Vss pins. The failing devices can have a variety of electrical characteristics at different pins: resistive short-circuit, Icc failures, functional, leakage currents (figure 2) as low as 700 nano-Amps, and breakdown voltages down to 2-3 volts. These devices were all stressed to failure.

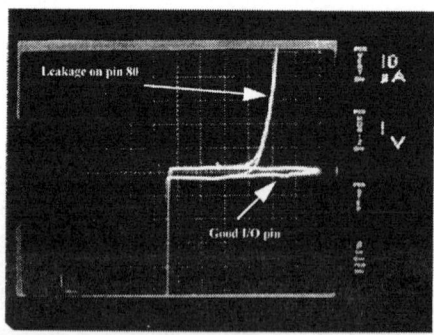

Figure 2: I-V characteristic curve comparing good (horizontal line) and leakage pins (curved vertical line).

The figure 3 below shows an optical photograph (<1000X) of the ESD failure. The failure site is revealed as a hot spot location (circled) revealed by liquid crystal. This is for an apparent electrical short circuit resulting from the I/O to V_{SS} pin combination subject to +/-3000 Volt HBM stress. Note that no physical damage is visible at this low magnification and there is no discoloration at the hot spot location. The physical damage is assumed to be below the surface at this stage. This figure shows the pad and the multiple fingered ESD protection structure connected also to the Vss bus line (left side looking at picture).

Note that for the I/O to Vcc pin combination stress, the ESD failure would occur (and did occur) on the other side (right side looking at the picture) of the ESD protection structure as this side is connected to the Vcc bus.

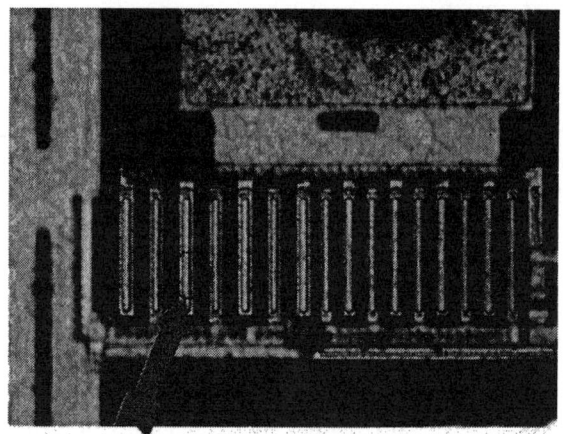

Fig. 3 - Hot spot location (circled) revealed by liquid crystal for an apparent short circuit in the ESD protective circuit (transistors) at the pad. There is No discoloration and No visible physical damage..

The ESD damage shown in figure 4 below is in the contact area and at the edge of one of the fingers. This SEM (20kX) photo represents the failure site from figure 3 above and is taken after de-processing down to the poly gate level. The physical failure then is below the surface, a distinction which will be used to distinguish between ESD and EOS failures.

We note also that there is was No discoloration at the ESD failure site, another distinction which will be used throughout this tutorial to differentiate the ESd failures from the EOS failures. Although th ESD failure is the result of I/O to Vss stress, the I/O to Vcc HBM stressing gives the same ESD failure signature.

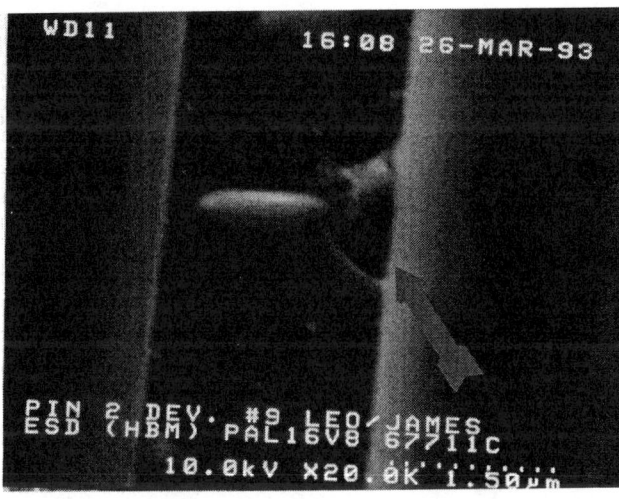

Fig. 4 – Subsurface ESD damage to the contact at the failure site in the ESD protection circuit. There is No discoloration.

Emission microscopy [5] is particularly useful to detect anomalous regions or damage, which results in the generation of an emission at wavelengths of 400nm – 1100nm. The bright spot in Figure 5 below taken after decapsulation only, indicates damage within the ESD protective structure. This is revealed by the EMMI for a leakage failure resulting from +/- 3000 volt HBM stress applied to an I/O to V_{CC} pin combination. At this the die surface, there is

no discoloration, no visible physical damage at the failure site.

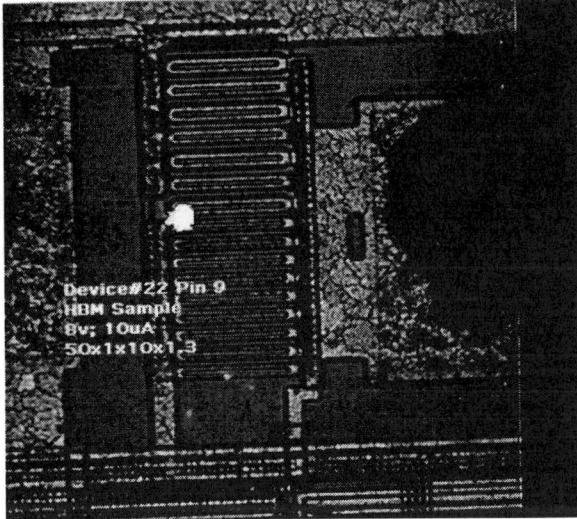

Fig. 5 - Site of leakage failure revealed by emission carriers that result in emission at optical wavelengths.

The SEM (mag. 7kX) photo shown in figure 6 below indicates contact damage after de-processing to the poly silicon level. Note the absence of discoloration, and the fact that the physical damage is revealed only after deprocessing.

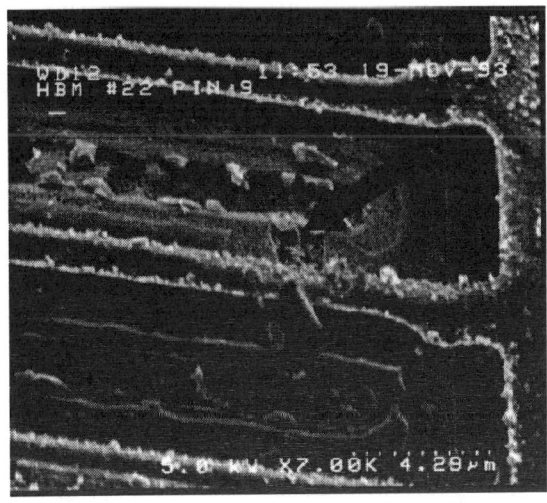

Fig. 6 - SEM of contact damage site seen in figure 5 after de-processing to the poly silicon level.

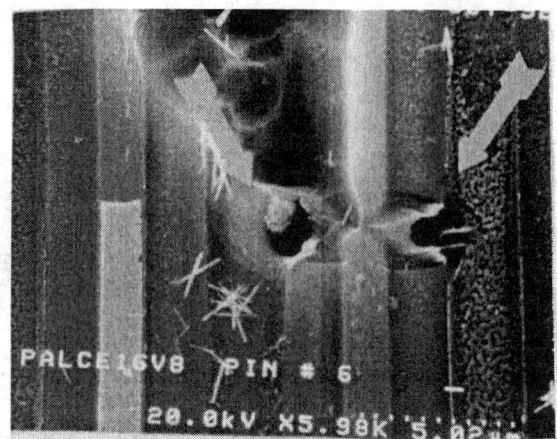

Figure 7. HBM. A field return showing arc-over failure between contacts in the ESD protection structure.

HBM Summary

In all cases no physical anomaly was observed after de-capsulation. SEM examination was required in conjunction with de-processing to establish physical failure site morphology and final location of the physical damage. These simulations were compared with similar damage on customer field failures (figure 7). The contact to contact damage shown in figure 7 is probably due to Icc failure failure after I/O to Vss or I/O to Vdd stressing.

MM ESD Discharge: The MM or "Japanese model" described by EIAJ was in fact intended as a worst case HBM[ ], but was found useful and developed in the US for failures which occurred due to the direct contact between the metallic leads of the device and the conducting metallic arm of robotic type equipment. Hence, the MM ESD simulations described here were conducted using a calibrated model compliant with ESD DS-5.2. Most failures occurred within the ESD protective structures (similar to HBM), but the physical damage revealed after de-processing to the poly silicon gate level was more severe as reported elsewhere [1].

Here also, the physical failures could also occur in the core of the device if the failure is functional or Icc between the Vdd (or Vcc) and Vss pins. The failing devices can also have a variety of electrical characteristics at different pins: resistive short circuit; Icc failures, functional, leakage currents and low breakdown voltages. Similar to HBM, no discoloration was observed after decapsulation, and the physical anomaly was observed after deprocessing to the polysilicon level. These devices were also all stressed to failure to thoroughly identify the physical damage.

Although the damages observed for MM were more severe than for HBM, similar degradation of electrical characteristics were found at failing pins, with leakage currents of 1.0 to 4.0 micro-amp and very low apparent breakdown voltages of 1-2 volts. Figure 8 shows typical leakage failure current induced by MM and Figure 9 illustrates the more severe damage where large deep pits occur at the contact/s suggesting high current parasitic bipolar action deep in the substrate.

Figure 8 MM: Severe damage at the contacts of the ESD protection structure.

The figure 9 below is a close up of a typical pitted contact where the hole is very deep compared to the relatively shallow holes found for HBM.

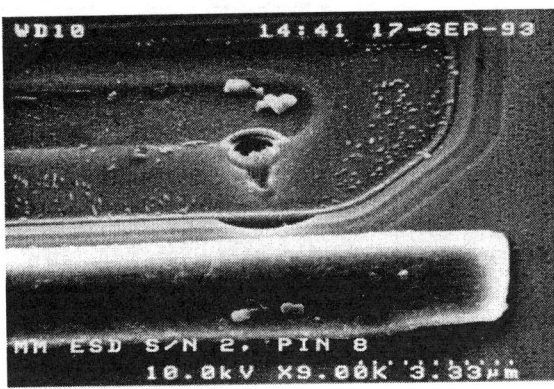

Figure 9. MM; magnified view of a deep pit in the contact.

This Failure Signature was characteristic of all MM induced damage irrespective of the stress magnitude used to cause failure. Corresponding signature has been reported [18] mainly by the auto industry.

MM Summary. Physical failure location is the same as for the HBM failures and also for the same pin combinations even though the failure thresholds are different.

Pin Combination Differences:

Failures can occur for all the pin combinations[19]: I/O to V_{SS}; I/O to $V_{DD)}$; I/O to I/O; Vss to Vdd and Vdd to Vss. It is established that ESD damage will occur at the core of the chip when these pin combinations are stressed. The HBM and MM stresses produced clear distinctions in the failure site for each pin combination stress mode when combined with the electrical characteristics. The Input/Output I-V electrical characteristics for leakage and Icc failures showed distinct differences between the various physical failure signatures. This enabled easy identification of the stress mode initiating the failure.

Current is known to flow through the internal circuitry (between Vss and Vdd) when I/O pins are stressed with respect to Vdd or Vss [20]. Physical damage below the die surface will occur to the NMOS pull-down output devices during I/O to Vss stress. This results in Source to Drain arcing or contact failure depending on the electrical characteristics. These two physical failure types were distinguishable because the source to drain arc-over was due to the Icc failures and the contact damage occurs for the pin leakages.

The I/O to Vcc failures occurred in the pull-up output devices with the same physical damage reflecting the distinction between leakages and the Icc failures. Note however that the I/O to I/O leakage failures produces physical damage in the contacts for both the pull-up and pull-down transistors. No Icc failures have been observed for I/O to I/O stressing.

Vss to Vdd (or Vdd to Vss) stress predominantly caused failures in the core logic of the device in the form of Icc failures which showed arc-over damage from source to drain (figure 9) at an internal location far from the ESD protection structure and far from the input buffer circuitry. The latter location is will be shown (later in this text) to be associated with CDM failures. For the devices analyzed here, the HBM and MM failure sites in the core were the parasitic lateral npn "structures" formed between the drain of the NMOS device connected to Vdd BUS and the source of another NMOS device connected to the Vss BUS.

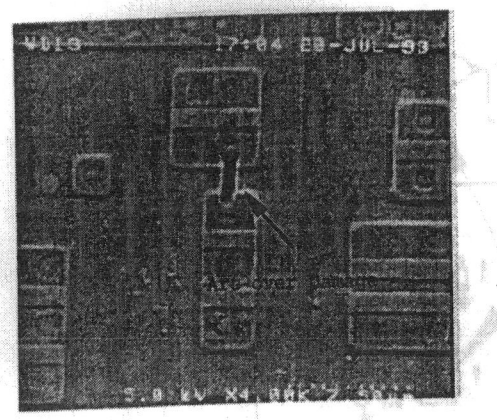

Figure 10. ESD Arc-over at the core of the device and occurring between Source and Drain for Vss to Vdd stress.

Note that a good power supply clamp would reduce the failures that would occur for the pin combinations I/O to V_{CC} , and to $V_{SS..}$ and also Vcc to Vss or Vss to Vcc.

427

Table for HBM and MM Pin Combination

Pin Combination	Failure type	Damage/location
1. I/O to Vss	Leakage	Contact/Junction in the pull-down transistor
2. I/O to Vss	Icc	Arc-over at pad to contact in the pull-down transistor
3. I/O to Vcc	Leakage	Contact/Junction – pull-up transistor
4. I/O to Vcc	Icc	Arc-over at pad to contact - Pull-up transistor.
5. I/O to I/O	leakage	Contact /Junction – pull-up or pull-down.
6. Vss to Vcc	Icc/func	Source to Drain arc-over in the core of the device.
7. Vcc to Vss	Icc/func	Drain to Source arc-over in the core of the device.

The above table is a summary of the electrical and physical morphological signatures associated with the different pin combination stress modes. Note for example that Leakage and Icc failures can occur for the pin combination of I/O vs Vcc but the physical signatures are different. Note also that the leakage for I/O vs Vss produce contact damage in the pull-down but occurs in the pull-up for I/O vs Vcc. Note finally, the physical failure differences between Vss to Vcc and Vcc to Vss.

CDM Discharge:
Exposure to CDM stress is carried out using a "dead-bug" socket-less configuration with air discharge. The setup results in charging by field induction as described by Renninger et al [21] and conforming to ESD DS-5.3. The CDM method is minimally influenced by parasitics since the stray capacitance of the socket is absent and series inductance to the RF ground plane is minimized because a relay is not used. This then allows for discrimination in the influence from the different package types (CDIP, PDIP, PLCC, LCC, PGA, BGA, TSOP and SSOP etc) with respect to the parasitics in the packages. This is quite different

from SDM (Socketed Device Model) where process is greatly influenced not only by the parasitics in sockets but also by the relay network etc of the simulator [22]. The effect of variation in air arc resistance is also minimized by controlling the local humidity and using multiple zaps (<5 per polarity) per pin.

frequently in The CDM failures occurred in the gate oxide at the first input buffer. This buffer circuit is located internally beyond the (HBM & MM) ESD protective structures (see figure 10). This is characteristic of the CDM failures. Note however that while the damage site occurs most the NMOS pull-down, the PMOS pull-up is not immune to damage.

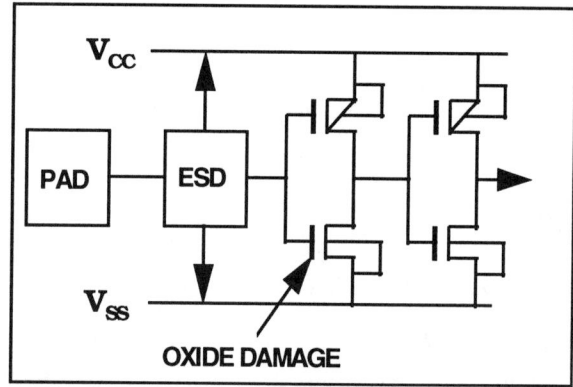

Fig. 10 - Location of CDM Damage in Input Buffer Circuit.

Failed pins can show enhanced leakage currents from 500nA to 100 micro-Amps, severely degraded breakdown voltage, but could still be functional when hand tested (taking full ESD preventative measures). Figure 11 is an optical photo (before CDM stressing) indicating the location of multiple data lines feeding into input buffers immediately below (see arrows in the figure).

428

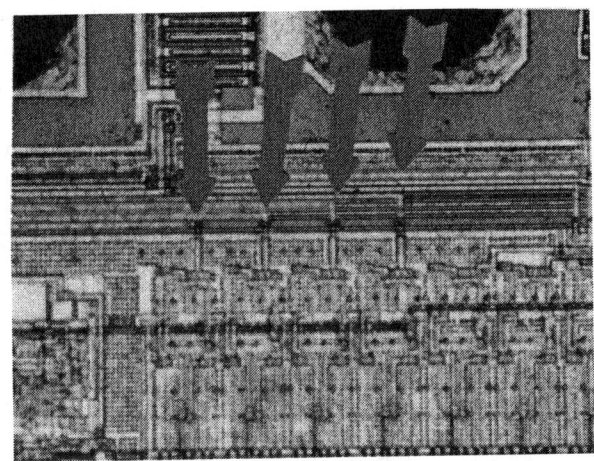

Figure 11. Optical photo showing location of multiple datalines feeding into buffers below.

For figure 12 at a lower magnification, a characteristic photoemission is seen at an input buffer after exposure to multiple discharges with the package leadframe charged to +/- 1000 volts with the CDM simulator. All CDM stressed input buffers showed emission at this same location for the same pin designation. No physical damage is visible and there is no discoloration.

Figure 12: Photoemission from beneath the metallization at the input buffer damaged by CDM stress. There is no visible damage.

After de-processing to the poly silicon gate level some perforation could just be discerned as shown in Figure 13 at the gate oxide edge and possibly in the gate oxide below. There is no discoloration, and as you can see, this subsurface damage was not visible in the photo of figure 12.

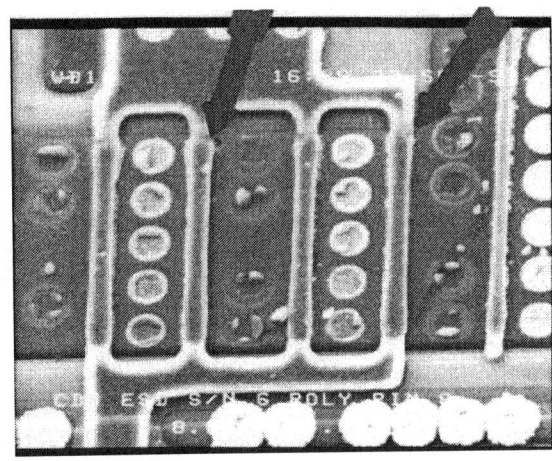

Figure 13. CDM. Perforations at gate oxide edge just visible after deprocessing to polysilicon level

After further stripping to the substrate pinholes are clearly indicated by pits in the silicon surface (figure 14) which is below the gate oxide. Most of the gate oxide has been stripped during the deprocessing.steps. this is clearly different from the HBM and MM failures which occurred in the contacts (see figures 6 & 7).

Figure 14: CDM. Pits in the silicon substrate showing clear evidence of gate oxide damage/rupture in the input buffer circuitry.

The next figure-15 shows a typical CDM failure corresponding to a customer field failure after de-processing to the poly silicon level. Again, there is no discoloration which is typical for ESD type failures.

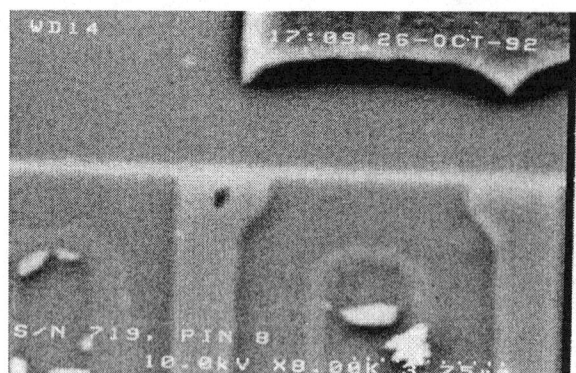

Figure 15 Typical customer return showing a single small hole in the gate oxide.

Reports of leakages as low as 100 nano-Amp for single zap events have been reported [23]. In our laboratory simulations the oxides show multiple perforations or are made up of more complex shape probably due to the use of the prescribed multiple zaps. The CDM ESD event for field or factory failure can be described by a *Fingerprint* consisting of enhanced leakage (at well below specification levels of 1-10 microamps). Such leakage is confirmed when the data-in line is laser cut just beyond the HBM protective structures at the bond pad. This confirms that the leakage isat the input bugtger circuitry.

CDM Summary

We note in summary that there is obvious difference between CDM ESD and the other two (HBM and MM) ESD models but the distinction between HBM and MM is not clear cut so simulation must be done. The CDM physical failure location is at the input buffer and in the gate oxide, where as both HBM and MM failures occur mostly in the contacts at the input protection structures. Neither do CDM failures exhibit arc-over type damages.

Is It ESD, EOS or ESD -induced EOS failures?

These failures can occur at anytime and is especially prevalent at burn-in. An EOS failure that cannot be directly duplicated using one of the "in-house" EOS models or by SLU or by TLU is prime suspect. However, should all EOS failures be checked as a possible ESD induced failure ? This may be a judgement call.

Here is some distinction. The suspect ESD induced EOS failure which should be clearly visible at low magnification is very carefully deprocessed with visual and or SEM examination at each step. If it is a pure EOS failure, the subsurface will be devoid of damage [24]. The extent of any physical damage beyond the metal layers will be at the dielectric layer below. The thermal nature of the very slow EOS pulse does not entend to the suburface. Recall that a pure ESD failure has no visible damage at the top surface and there is no discoloration [1, 24].

If it is an ESD induced failure, then further deprocessing will reveal damage down to the silicon level. The ESD damage would have already been present in the subsurface. Due to the very short pulse width (<< 10 nsec) and very fast risetime (< 500 psec) of the ESD pulse, the ESD event ruptures the oxide first. Additional stress can cause or induce an EOS failure in the layers above. SEM examination and some cross-section will reveal additional damage below the surface. This has been demonstrated and also published by several authors including [25].

EOS Simulations

A number of laboratory simulations can be devised and executed to try and replicate the major classes of EOS damage seen for customer returns and in factory and reliability rejects. Successful replication is described here for gross EOS damage evidenced by package distortion.
This allows ready distinction between incorrect insertion (for both PDIP and PLCC) and *latch-up related failures,* and for failures occurring most often on high voltage input programming pins.

Electrical Failures: These failures ranged from simple leakage to shorts (figure 16), low voltage breakdown, Icc and functional.

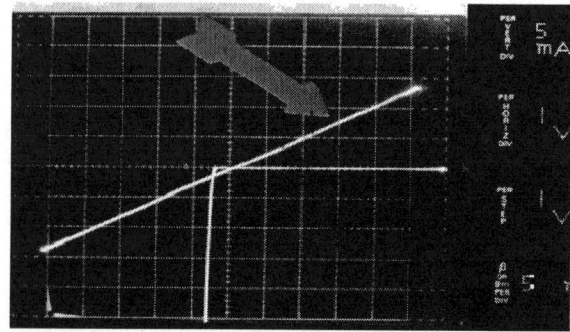

Figure 16. Degraded I-V curves showing sloped line for short failure (arrow) and good breakdown (horizontal line).

Physical Failures

"Burnt" Plastic: A class of failures showing damage to the package was characterized by "burnt" or carbonized plastic adhering either to the die surface (figure 17) or to the bond wires (figure 19), suggesting perhaps two different failure modes.

Bond wires: This is the first and simplest class of EOS returning from the field or from the factory, the test area or from the reliability lab. A curve tracer can be used to supply over-voltage or over-current to the pins resulting in EOS (figure18).

Incorrect Insertion: Use is made of dynamic burn-in boards to simulate possible mis-insertion. Here the device is powered up and the outputs are pulled high by a resistor to V_{CC}. This EOS simulation of reverse insertion (RVI) for PDIPs and of wrong orientation (WRO) for PLCCs revealed a *Failure Signature* of fused open V_{CC} power and/or V_{SS} ground bond wires with "Burnt" plastic around the damaged bond wires and/or on the die surface.

For PDIP reverse insertion (180 degrees only), I_{CC} failures occurred in times as short as 10 minutes at 25° C. However, in all cases the die was still functional when micro-probed at the bond pads and no metallization damage was observed. Under these conditions, this die functionality is a distinguishing feature and serves to differentiate between this regular and well-known EOS failure from the latch-up type failure. If probing the pad had resulted in an Icc current anomaly then further and a different simulation would have been necessary to confirm the likely hood of it being a latch-up type failure. Fig. 17 below illustrates the typical fused open Vcc (Vdd) or Vss bond wires with some carbonized lastic on the bond wires.

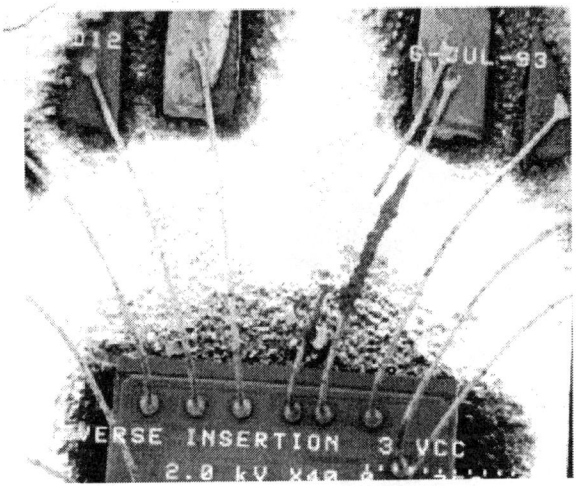

Fig17: Fused open Vcc bond wire from PDIP misinsertion (RVI).

The photo in figure 18 below illustrates the typical example of carbonized plastic adhering to the Vcc (Vdd) or Vss bond wires and to parts of the die surface. .

Figure 18: Carbonized plastic surrounding the Vcc dual bond wires and adhering to the die surface of a PDIP packaged device.

For the PLCC packaged device, incorrect insertion can result in misorientation or wrong orientation (WRO) by 90, 180 or 270°. For 180°, PLCC position, the *Failure Signature* was identical to the 180 degree position for the PDIP package, but no electrical failure occurred after 10 minutes at room temperature. All the PLCC units failed however after 24 hours at 125° C. Presumably the difference is explained by the difference in thermal impedance of the package/socket combination.

The Figure 19 below shows fused and opens V_{SS} bond wires for PLCC.

Figure 19: Fused open Vss bond wires for PLCC.

For the 90 and 270° WRO positions, none of the samples failed and negligible Icc current (< < one micro amp) was drawn from the supply. This negligible current may not normally be detected during burn-in and could lead to erroneous conclusions. Similar type failures (occurring at Burn-in at the customer site) had returned from the field. These devices had passed the JEDEC type EOS stress, called static latch-up (SLU). They passed/survived +/- 100 milli-amps at 125degreeC using the slow usec type square pulses. It is surmised that during burn-in, or while the devices are powered, they experienced a transient type pulse (nsec pulse width ranges).

The transient pulse simulations do show similar type failures in addition to internal core EOS failures (figure 20) which are not replicated using the incorrect insertion simulation. Note that this EOS failure occurred on the surface of the die, it is visible at low magnification and the physical failure shows discoloration. This is a clear distinction between EOS and ESD failures. However, Static and Transient Latchup simulation is beyond the scope of this tutorial and is detailed elsewhere [26].

Figure 20. Internal core EOS failure showing melted metal after transient (nsec pulse width) latch-up. Note the discolouration of the failed area (if photo is not black and white).

The next figure (21) represents a burn-in failure returned from the field. Here the carbonized plastic is adhering to the die surface, covering all of the Vcc or Vss bond wires and bonding pads. This carbonized material is very difficult if not impossible to remove, so the determination of the exact nature of failure is remote as the Icc (as measured on the bond pads) cannot easily be determined.

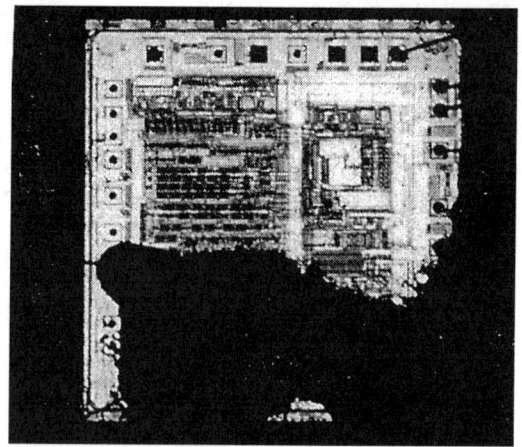

Figure 21: EOS. Heavy carbonization on bond wires for failure during actual burn-in.

Discolored Metallization:

The second class of EOS customer returns showed spiked and discolored metallization at the source and/or drain of the protective transistors adjacent to the bond pad. This is particularly prevalent for the HV programming pins suggesting over-voltage, which can be detected during the Routine Monitor Program in the factory or reliability lab.

Programming Over-Stress:

A Data I/O Unisite programmer can be used to investigate variation of programming parameters for both PDIP and PLCC packages. Varying the programming pulse width did not cause failure for these devices (but should not be ruled out). However, increasing the Vpp (above the nominal 9.0V) produced failures at between 11.5 and 15.0 Volt caused Pin failures for all units and the circuitry of adjacent pins was also damaged in some cases.

De-capsulation revealed that during simulation the dominant failure Signature was replicated, that is, spiked and discolored metallization (Figure 22).

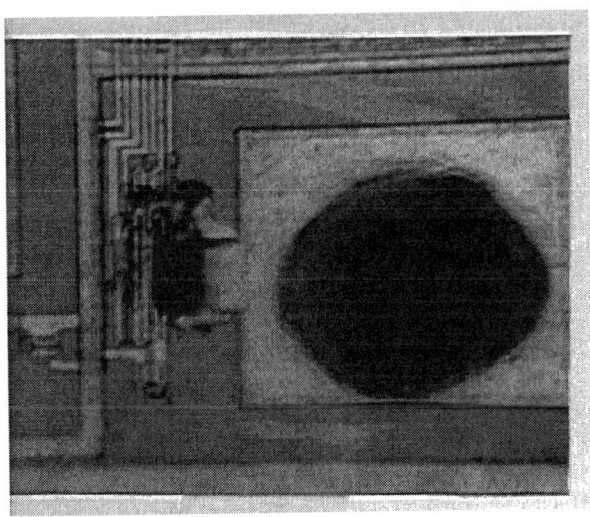

Figure 22: Simulated EOS damages with clearly visible discoloration (color photo required) on the transistor metal lines.

The figure 23 below shows a field return where the EOS damage is in the transistors and at the die surface. Discoloration is obvious (color photo necessary) and the failure is observed under low magnification. The figure 23 shows the EOS damage with clearly visible discoloration at the ends of the drain fingers of the input protection transistors adjacent to the bond pad. These EOS type failures have been the dominant type failures returning from the field.

Figure 23 show a field return where the damage is in the transistors and at the die surface.

Reverse Breakdown (BVDSS)

The input ESD protection structure shown in Figure 28 consists of two large NMOS transistors (Weff=45micron), one connected to VSS and the other to VCC with the gates connected to the substrate ground via a high value resistor in order to initiate bipolar action by dynamic gate to substrate voltage.

A standard curve tracer (Tektronix 575) and a manual power supply can both be used to apply an over-voltage to the input pad to force the input transistors into BVDSS mode breakdown, which occurred at 12-13 Volts. . Failure was immediately apparent; the electrical characteristics showed opens, gross leakage and shorts.

The characteristic spiking and discoloration visible after decapsulation is shown in the figure 29 for both terminated NMOS pull-up and pull-down transistors with no apparent damage to the adjacent pins.

It is apparent that the failure mode dominating these failures is due to over-voltage of sufficient magnitude that the protective transistors at the pad are damaged by being forced into BVDSS mode with no current compliance limit.

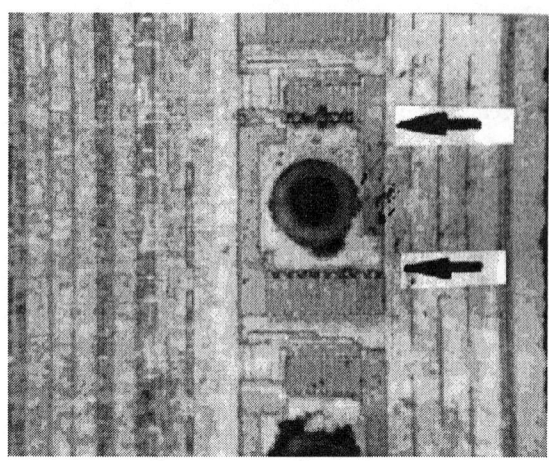

Fig.24- characteristic spiking and discoloration for both terminated NMOS pull-up and pull-down transistors.

Since this failure mode is absent as a return for product which are programmed in the factory, it is evident that many customer sites have excessive voltage on their programmers. This can be minimized by careful selection of a power line conditioner; routine calibration of voltage levels; verification of the correct programming algorithm; and by maintaining the condition of the programmer sockets. Routine periodic examination of the waveforms during programming using a wide band width oscilloscope to detect over-shoot will also prove beneficial.

Curve Tracing EOS

A simple simulation utilized the curve tracer to apply a high current or voltage to the input pin. At a voltage or current which is beyond the manufacturers rating of the device, the device failed. This is revealed by the melted metallization which is observable at low magnification and which shows discoloration. The EOS discoloration is a distinction and separates it from an ESD type failure.

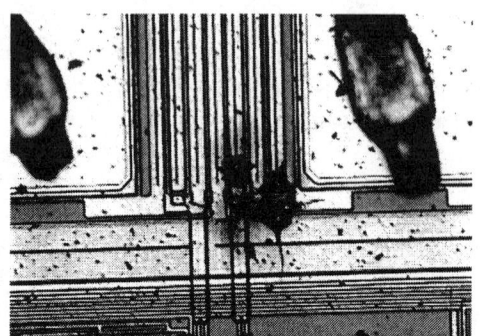

Figure. 25 EOS failure using the curve tracer to supply over voltage or over current

Discussion

Failure Signatures between HBM and MM show some variation in morphology but they are located in the same ESD protection structure. The failures for MM is more severe as is evidenced by the much lower (5-20x lower) voltage required for failure threshold. A clear distinction is achievable however after simulation and failure analysis because the input pulse required is quite different for each ESD event.

The physical failures for CDM occur at the input buffer circuit and are therefore quite different in location from that of HBM and MM. Even though both FIM and CPM type failures have exactly the same signature/fingerprint and are just subsets of the CDM event, it is necessary to distinguish between FIM and CPM as follows:- For Ceramic Packages:- FIM failures are the most likely due to the fact that tribocharging these ceramic surfaces are quite difficult, but possible. It is much easier to induce a charge on the metallic lead frame while the ceramic packaged device is in a field. This can occur quite easily when the devices are in an uncontrolled automatic piece of equipment.

For Plastic Packages:-CPM is most likely because of the propensity of the (insulating) plastic package to accumulate charge upon its surface from tribocharging or contact electrification, resulting in induced charges on the lead frame of the device.

However, a device in plastic packages can also fail due to FIM as a result of the package being in an electric field. Here, the charges will be induced on the leadframe. Recall that charges on insulators (plastic packages) are not mobile. It is seen then that the nature of the failure environment is (always) of utmost importance.

Summary

The various test "models" outlined here result in characteristic Failure Signatures that we have described. They are shown to replicate the morphology of actual factory and customer returns.

Clear correlation is shown for FIM and CPM, which also occurs in the field as well as in the factory. A sound low leakage screen in the test program will keep such damaged product from leaving the test floor and alerts staff to a possible problem.

Simulation then shows that some field returns are due to misinsertion with a failure Signature that includes; fused or open Vdd/Vss bond wires; burnt plastic on the damaged wires; but no metallization damage on the die. Transient Latch-Up simulation using capacitive discharge allows differentiation of a similar Failure Signature, but where burnt plastic adheres to the die in the region of the VCC/VSS bond pads. Damage is found to the metallization upon decapsulation and to the underlying structure upon further de-processing. Mis-insertion suggests the need for more care in customer handling methods, whereas the evidence for latch-up suggests the existence of power system faults of a transient nature that can be eliminated using suitable line conditioners and filters.

Conclusions

It is shown that a distinction can be made between EOS (discoloration) and ESD (no color) failures and between the characteristic failure Signatures produced by the various ESD models. This can then be used to correlate between actual factory and customer field failures. It is imperative then that the major EOS and ESD models be available to a Device Analysis facility since replication of Failure Signature is seen to be a powerful tool in determining probable cause of failure.

ACKNOWLEDGEMENTS

The author would like to extend special thanks to the authors of the original 1994 paper and to the rest of the engineering and technician staff. A lot has happened since then, but they were part of an original team trying to establish a clear distinction between ESD and EOS.

REFERENCES

1. LeoG. Henry, t.Raymond, > Mahanpour, I.H. Morgan, "EOS & ESD Lab. Simulations and Signature Analysis of a CMOS Prommable Logic Product". ISTFA, 1994. Los Angeles, CA., p117.
2. I.H. Morgan, Proceedings of the 3rd ESD Forum, Grainau, Germany, Dec. 1993. P27-33.
3. C.Cook, and S.Daniel, Proceedings EOS/ESD Symposium, Dallas, Texas, Sept., 1992, p 149-157.
4. B.L. Euzent,., T.J.Maloney and Donner II, J.C., Proceedings EOS/ESD Symposium, Las Vegas, Nevada, Sept., 1991, p 59-64.
5. Microelectronics Failure Analysis Desk Reference, 3rd Edition, 1996. Editors- T.N. Lee and S.W. Pabisetty.
6. D.C. Wunsch. and Bell, R.R,. IEEE Trans, Nucl. Sci, 1970.
7. Leo G. Henry, Hyatt H; Barth, J; Stevens, M; and Diep, T. Charged Device Model Metrology". Proc. 18th EOS/ESD Symposium, Orlando, Fl., Sept., 1996, p167-179. .
8. International Standard: IEC-1000-4-2. 1996. Electromagnetic Compatibility. Part 4. Testing and measurement. Section 2. ESD Testing. Note: Previously named- IEC-801-2.
9. J. T. May, and J.F. Guravage, Proceedings ISTFA, Los Angeles, CA, Oct., 1990, p 143-147.
10. L.R. Avery, EOS/ESD Symposium, Orlando, Florida, Sept., 1987, p 186-191; . and p 88-92.
11. P.R. Bossard., R.G.Chemelli, and B.A. Unger, Proc. EOS/ESD Symposium, San Diego, CA, Sept., 1980, p17-22.
12. T. Raymond, K.L. Chang, and LeoG.Henry AMD 3th Engineering Conference, Marriot Hotel, Santa Clara, CA, Feb., 1994
13. E.J. Chwastek,. Proceedings of EOS/ESDS Symposium, New Orleans, La Sep, 1989, p149.
14. ANSI/ESD Association Standard S-5.1, 1991.
15. ESD Assocn. Standard DS-5.2, 1994.
16. ESD Assocn Standard DS-5.3, 1994.
17. MIL-STD 883, Method 3015.7, March,1989
18. 18. A.Kelly, G.Servais, T.Diep, D. Lin, S. Twerefour, G. Shah, "A Comparison of ESD Models and Failure Signa for CMOS ICs." Proc. EOS/ESD Symposium, Phoenix, AZ, 1995. P175.
19. Leo G. Henry, Failure Signatures associated Pin combinations and the Core of the CMOS Device. AMD Internal Reliability report. , 1996.
20. A.Amerasekera, C. Duvvury. "ESd in Silicon Integrated Circuits". John Wiley, 1 London,1995.
21. R.G. Renninger, Jon, M.C., Lin, D.L., Diep,T.,and Welsher, T.L., Proceedings EOS/ESD Symposium, New Orleans, Louisiana, Sept.,89, p 59-71.

22. M.Chaine, L.Avery, K.Verhaege, LeoG. Henry, H. Geiser, M. Farris, T. Bodbeck, T. Meuse, K. Bock. "Investigation into Socketed CDM Tester Trans. Line Characteristics". Proc. EOS/ESD Symposium, Reno, Nevada, Oct.98.

23. H.A. Gieser, P.Egger, .M.R. Herrmann, J.C. Reiner, and A. Birolini, ESREF Proceedings, Bordeaux, France, 1993.

24. Leo G. Henry, & J.M. Majur, " Basic Physics Of color coded EOS Metallization Failures".

25. D.S.Kiefer, R.T. Milburn & K Rackley. "FA of EOS/ESD Damage to HCMOS Gate Arrays," Proc. Of the 15th ISTFA Conf, LA, CA p201, 1989.

26. Leo G. Henry "EOS" and ESD Laboratory Simulations for Failure Signature Analysis". Tutorial-R at EOS/ESD Symposium, Reno, Nevada. October, 1998.

ESD Damage Simulation and Failure Mechanisms

Thomas W. Lee
Motorola Inc.
Austin, Texas

BACKGROUND

Pareto diagrams of failure analysis data typically show that the most frequently occurring failure mechanisms in failure analysis (FA) are electrical overstress (EOS) and electrostatic discharge (ESD). Through questions posed by report recipients, it is known that the significance of the appearance, and in particular, the SIZE, of a meltthrough site in the determination of EOS versus ESD damage is generally not well understood. However, this understanding, through efforts shown here, is improving over the years. A common question asked of analysts who present their report recipients with evidence of electrical damage, is, "Was it EOS, or ESD?" Additionally, the frequency with which electrical damage is a conclusion in FA reports may cause all reports to be viewed with scepticism.

SEM photos of electrical damage sites are most helpful to customers. However, customers usually seek more specific information to help them in their search for the causes and solutions to ESD problems, once the problem is isolated by FA. Values such as approximate ESD voltage and polarity of the discharge, can help them more accurately identify the cause of the problem. Unfortunately, estimating these variables from the physical evidence found in damaged devices is inexact. Further, speculation, particularly when not identified as such, can lead to eventual loss of credibility; "Everything is electrical overstress". This is especially true when multiple or varied interpretations are possible, and the cause determination breaks down into contests of informed opinion. The following procedures can help alleviate this problem.

THE VALUE OF ESD SIMULATION IN F/A REPORTS

Experience has shown that customer confidence in reports is a strong contributer to the success of the problem-solving process. When accurate determination of the cause of electrical damage is important, the relatively small investment of time and effort necessary to simulate damage on a good part is rewarded by vastly more accurate interpretation. Report recipients will likely feel that investigative efforts on the problem are more likely to be targeted and productive.

A report in which simulations are used to form conclusions, should include side-by-side photos of real and simulated damage areas. Confidence in conclusions about electrical damage, when they are based on obvious similarity between the physical appearance of damage simulated on good parts, and damage found on returned parts, will be improved.

Over time, the collection of electrical damage simulation data in the FA laboratory will isolate device types having high sensitivity with respect to others. Externally generated or published ESD sensitivity data can be used as comparative refer-ences in reports. Over the past few years, an enormous amount of ESD sensitivity work has been performed and is available from the EOS/ESD Association, and Reliability Analysis Center (RAC), in particular, the VZAP 2 publication (Ref 1). Using these additional resources, the device being analyzed can be confidently placed on a relative sensitivity scale with respect to others in the industry. Figure 2 shows a typical example of this sensitivity scale.

Careful management of the magnitude of this effort is necessary to ensure that the demarcation between an FA lab which performs simulations as a part of the analytical process, and an electrical test lab, is clear. The two best indicators are level of effort and equipment set. The analyst time dedicated to the simulation of failures should be apportioned to only those reports where the extra work is justified by customer needs or criteria. The equipment set present in the FA lab should be limited to that specifically devoted to ESD or overstress testing, and should be carefully chosen so it does not duplicate equipment available in electrical test, electrical characterization, or reliability test facilities.

Objections are quite likely to be raised about the cost of sacrificing a good device for an ESD simulation, especially in complex ICs such as the example shown in Fig. 1. However, the confidence resulting from this work may be a good return on investment. Additionally, the device manufacturer has access to devices that are functional, and testable, but otherwise unsale-

Fig. 1 ESD threshold testing of complex ICs can be expensive. Thresholds can be low; from 4 kV in the R4000, 1 kV in drams and gate arrays, to as low as 500 V in some flash memories.

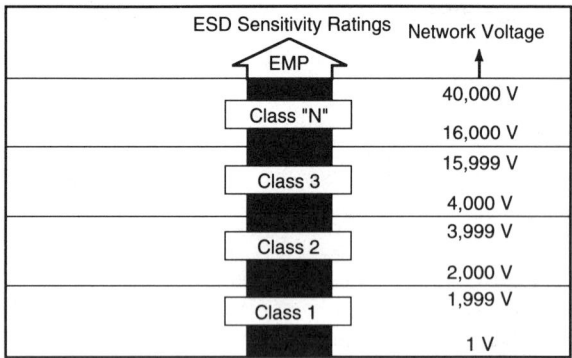

Fig. 2 ESD sensitivity rating scale, which applies to all device types.

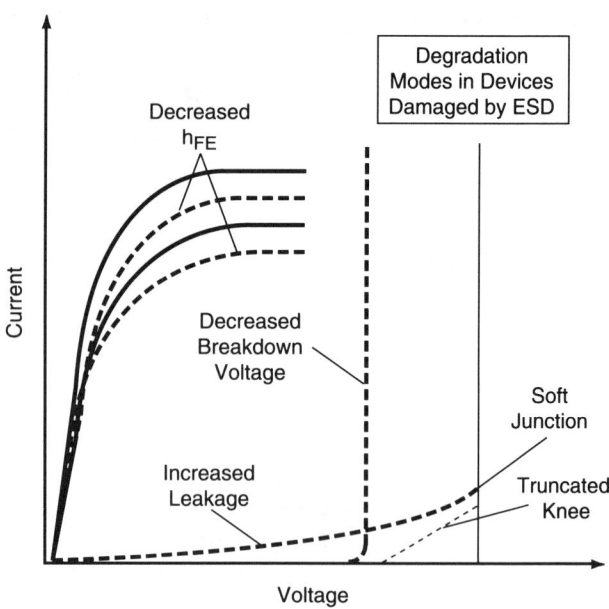

Fig. 3 Composite illustration showing some possible degradation characteristics in device junctions.

able. Examples are devices which have undergone various reliability tests and devices with cosmetic defects.

This article concentrates on the simulation of ESD damage in particular because ESD damage constitutes a significant portion of electrical damage found in FA labs. The equipment necessary to perform ESD simulations, when compared to production ATE, is relatively unobtrusive, inexpensive, and unique.

ESD TEST METHODS

Manual ESD testers may be applied to discrete device ESD measurements. The device can be monitored with the curve tracer and pulsed until it fails. The step-stress, or "threshold" method is recognized as valid by both MIL-STD-883A and MIL-STD-750C, and is well suited to this failure analysis task. This method finds the true failure voltage in a single device, rather than bracketing it by successive approximation with a large sample of devices. An example test circuit is shown in Fig. 4.

There are other important differences between the threshold and categorization methods. With the categorization method, the devices must be carried between the ESD tester and the production ATE. With the threshold method, the device junction reverse characteristics are continuously monitored for changes as pulsing is performed, at steadily increasing voltage levels. The most sensitive or vulnerable structure is exposed to testing. Rather than using ATE to simultaneously look for input pin degradation along with functional testing, threshold testing continuously monitors the device for *any significant degradation* in the characteristics, regardless of whether or not the value has moved out of specifications. At this subtle point, ESD degradation has already begun.

No attempt is made to pulse the device until it becomes a dead short, because the term "short" is not well defined. So decreased breakdown voltage, increased leakage, and truncated or soft knees, are interpreted as an indication of device failure. Figure 3 shows typical characteristics which can be monitored.

Devices are becoming more complex. Microprocessors,

ASIC, memories, gate arrays, and other types of complex logic circuitry may need to be evaluated by a lab. Hybrids, and other modules such as SIMMs may need evaluation. The ESD test must be performed at each pin, and there may be 10's to 100's of pins. Additionally, pin *combinations* need to be evaluated. This requires a larger and more complex ESD test system than for discretes. The minimum ESD tester size and cost for a lab is then a function of the complexity of the product analyzed. When the ESD damage site usually occurs within the input protection network (i.e., the network design is effective), degradation can again be monitored with the simple circuit and criteria shown in Fig. 4.

However, the ESD energy may bypass the protection network in some circuits and cause damage within the array itself before the network degrades, causing a parametric or functional failure. In this case, the detection of an ESD failure during simulation work may require return to the categorization method, with accompanying footwork for carrying the devices between the ESD tester and an ATE system. Which method to use is a choice which must be made by collaboration between the FA engineer and the customer. Certainly, the assumption that the input network design is effective in the prevention of transmission of destructive energies to the circuit interior is reasonable.

DISCHARGE NETWORKS

Pulsing is performed with a network specified by the requestor. The most common network is the human-body model (HBM), because it was standardized first, and is used in all government documents. The circuit diagram of this model appears in Fig. 4. However, as many as 40 other models are in use, and the component values differ. Most ESD testers have provision for plug-in interchangeability of

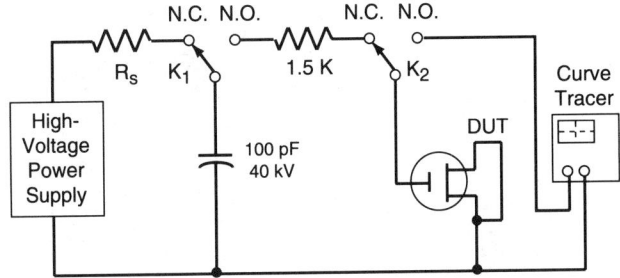

Fig. 4 Circuit diagram of an ESD step-stress, or threshold, test circuit using a curve tracer for failure indication.

E⁺B⁻	E⁻B⁺	C⁺B⁻	C⁻B⁺	C⁺E⁻	C⁻E⁺
220 µJ					
	3200 µJ				
		200 µJ			
			8450 µJ		
				2450 µJ	
					2450 µJ

Fig. 5 Table showing ESD energy necessary to damage 2N918 transistor junctions in all possible configurations.

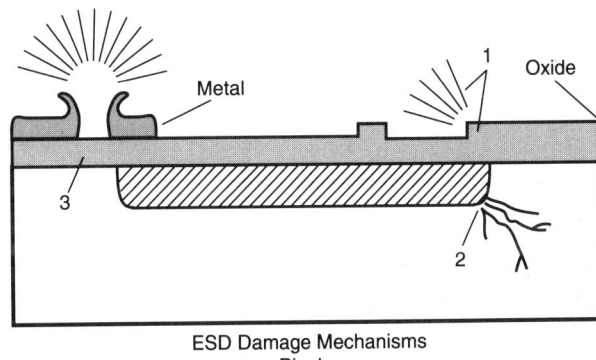

ESD Damage Mechanisms
–Bipolar–

1 Metal Transport Across Junction
2 Silicon Melting in Bulk
3 Fusing of Metal or Poly

Fig. 6 Cross-section drawing of ESD damage mechanisms in bipolar structures.

model networks. Below are listed example models which have been used:

Model Name	Cap, pF	Res, Ω	Notes
Charged device	various	various	
Energy quintupler	500	300	
ECMA	150	20	pushcart simulator
ECMA	150	1000	human-body
Field-induced	various	various	
Floating device	various	various	
IEC	150	150	
Machine 1	50	0	
Machine 2	200	0	Japanese model
MIL-STD-750;883	100	1500	human-body model
NEMA	100	1500	
Opt hum body	200	1500	
SAE	300	5000	human body
UL	250	1500	human body

In the data shown in this article, only the HBM was used. A drawing of the idealized output waveform of the HBM network is shown in the applicable military specifications (Ref 2). The most important parameter of this wave form is that it is *critically damped.*

With the threshold method, the voltage is manually adjusted in increasing increments until any change in the reverse characteristics of the junction or structure under test is indicated on the curve tracer. Again, this includes increased leakage, decreased breakdown voltage, soft knees, and truncated knees. Figure 3 is a composite drawing of the characteristics of a junction, showing some of the failure modes which degradation due to ESD exposure can produce. Notice that opens and shorts are also failure modes in discrete devices tested by the threshold method. This procedure is more stringent and accurate than simply detecting out-of-spec leakage values, but it cannot detect some parametric changes, such as hFE in discrete transistors, and functional failures in complex ICs.

It is clear that the voltage increment between tester pulses, or step size, is an important contributor to the accuracy of the tests. The resolution of the tester in this configuration is limited by the number of pulses applied at a given voltage increment. Using a very small voltage step would decrease error, but greatly increase test time, especially for devices with large threshold voltages. There is a direct tradeoff between test time and data accuracy. Furthermore, stress hardening, or even recovery of parameters, has been observed.

Previous work has well established that diode and transistor junctions are approximately an order of magnitude more sensitive to ESD energy when reverse-biased than when forward-biased. Figure 5 shows example data for the 2N918 bipolar transistor. If an evaluation on a sample of the device to be tested showed that the emitter-base junction was the most sensitive configuration, only the E-B, reverse-biased, case was pulsed and evaluated. The same consideration is true for the structures in IC ESD protection networks. Most of these networks consist of diodes or field-plate zeners to the rails, along with series poly resistors for current limiting. For a thorough review of input protection network schemes, refer to the book by Antinone et al. (Ref 4) This verifies the test procedure set forth in –883 and –1686, 50.1.2.

ESD SENSITIVITY CLASSIFICATION

As shown in Fig. 2, the ESD sensitivity of all devices, both active and passive, is classified into the following four categories:

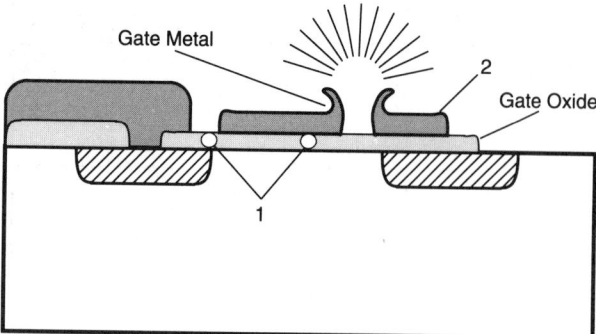

ESD Damage Mechanisms
–MOS–

1   Gate Oxide Rupture
2   Gate Metal Fusing

Fig. 7 Cross-section drawing showing some ESD damage mechanisms in MOS structures.

- Class 1—0 to 1,999 V
- Class 2—2,000 to 3,999 V
- Class 3—4,000 to 15,999 V
- Class N—16,000 V and above

These categories are well suited to indicate how carefully a device must be protected, and handled. The threshold test can quickly and accurately place a device into one of these categories. The actual threshold voltage found should also be reported to your customer. Depending on the needs of your customer, stating that a device "has an ESD damage threshold of 3.14 kV" may be more useful than stating that the device "is a class-2 part". The position of the device's ESD threshold in this classification system may be utilized as an indicator of which the ESD generation site in the factory was responsible for the damage (Ref 3). Determination of the satisfactory performance of the test circuit and threshold method can be made by comparison of the FA lab measurements with published data.

Through the performance of analyses and evaluations which required ESD data and damage simulations, a database containing about 400 discrete device types was assembled. The domain graph in Fig. 11 shows the results of ESD testing in this lab over a two-year period of time. It shows the relationships between device die size, device technology and ESD sensitivity. The motivation for showing this graph is to dispel the common perception that physically large devices have high resistance to ESD. Experience tells us that this is not true with large, leading-edge ICs. It is also not true of zener diodes (small die, high ESD damage threshold), and MOS power transistors (large die, low ESD damage threshold). Notice their relative positions on the graph. Examination of DOD-STD-1686 will show tables which indicate which of the discrete technologies are ESD sensitive in the range from 100 volts to 15 kV, and agreement is apparent between Fig. 11 and this publication. Examples of sensitive device technologies are: most MOS IC technologies, hot-carrier diodes, Schottky-barrier diodes, and small signal JFET, MOSFET transistors, and GaAs devices. The following discussions

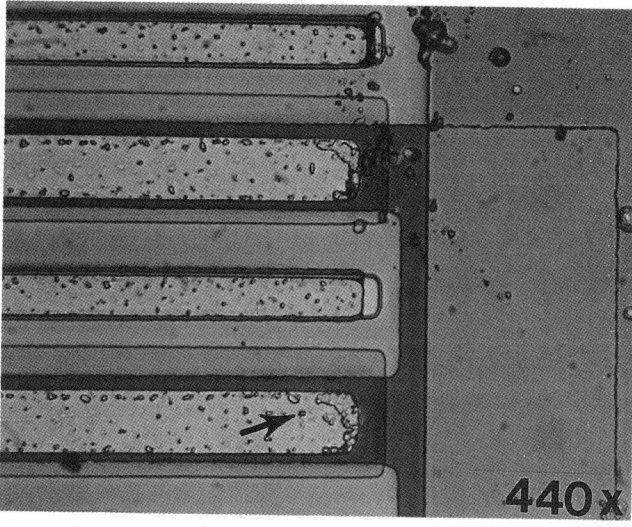

Fig. 8 Optical micrograph of bipolar structure with ESD melt-through in extremity of emitter finger. (440×)

partially explain the position of the various technologies on the domain graph. The testing and graph was originally extended to 40 kV because of military and automotive customer requests for specialized testing (Ref 5,6,7,8,9).

FAILURE MECHANISMS

The two drawings in Fig. 6 and 7 show idealized cross-sections of bipolar and MOS structures. Typical failure mechanisms are superimposed on these illustrations. In bipolar devices, a very common failure mechanism is the formation of a melt-through hole. The mechanism for the formation of this hole is explained in detail in the Thermomechanical Overstress section of this section, and in Dr. May's article.

Figures 8 and 9 show a typical ESD meltthrough hole. Note its small size with respect to EOS damage, which is frequently visible to the unaided eye. The energy of the pulse may be sufficient to transport metal across the junction and cause a short. This phenomenon has been termed the "silvery track", and it appears in area 1 of the bipolar drawing. The junction breakdown most likely occurs where there are small radii of curvature and electric field strength is most intense, such as at sharp edges. This is shown in area 2 of Fig. 6. Finally, any energy remaining in the external charged static source capacitance will cause joule heating and possible melting of metal runs, shown in area 3 of Fig. 6.

In MOS devices, puncture of the thin gate oxide is a very common mechanism. The gate oxide field intensity at normal operating voltages is already 10^4 to 10^6 V/cm, and overvoltages due to ESD rupture the film by exposing it to even greater field strengths (10^7 V/cm). This is shown in area 1 of the MOS drawing (Fig. 7). As in the bipolar case above, any energy remaining in the static source after forming the meltthrough hold may then damage the poly or metal runs, causing them to change resistance or melt and fuse open.

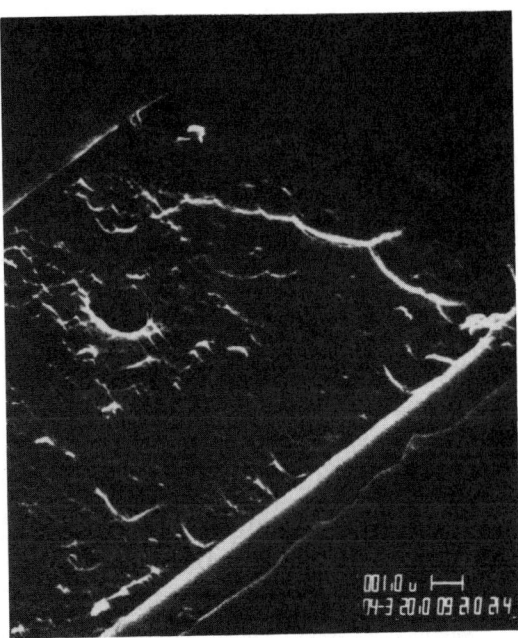

Fig. 9 SEM view of emitter shown in Fig. 8. (4,400×)

Figure 10 shows an example of ESD damage to a poly resistor in the protection network of a smart power IC. This damage was induced during device threshold testing, and the passivation has been chemically etched away. Notice that the ESD energy vaporized a trail of polysilicon from the centerline of the resistor, resulting in only a small increase in resistance. *This small increase was detectable during ESD threshold testing in the FA lab, but would have likely passed parametric testing.* Nevertheless, it has been proven that damage begins to occur at this level. This is endorsement for the FA threshold ESD test method.

CONCLUSIONS

1. For best ESD problem solving, show side-by-side field failure and simulated failure SEM photos.

2. The primary difference between ESD and EOS damage is the size of the damage feature. In ESD damage, a SEM must be used to view the damage area, and delineating etches must frequently be used. With EOS damage, the area can frequently be observed with the unaided eye, or in a low-power optical microscope. The amount of energy involved in the formation of ESD damage is also orders of magnitude smaller.

3. The factors which seem to influence ESD sensitivity are; small die size, small active area, high switching speed, low input capacitance, thin oxides, small feature size, and light doping.

4. It is difficult to determine if an EOS failure has been initiated by ESD, because the ensuing EOS energy will very likely be concentrated in the location of the ESD damage, and destroy it. You must look in the sample for ESD (lightly) damaged devices which for some reason, were not exposed to the EOS energy.

Fig. 10 SEM view of ESD damage to poly resistor. (902.8×)

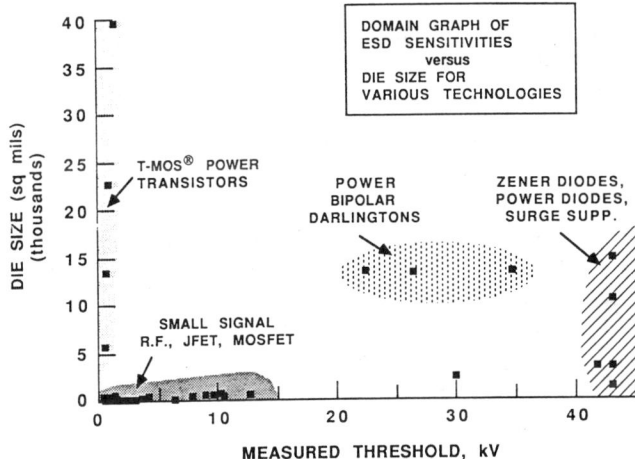

Fig. 11 Domain graph showing ESD sensitivities for various technologies vs die size.

Damage due to ESD can definitely result in EOS in some circuits. An example is the shorting of the gate and source of an MOS power transistor. The short may result in a voltage divider bias network between drain and source, internal to the device. If the signal source is not "stiff", that is low impedance, and the drain supply is not current-limited, the first turn-on of the circuit will place a device in a steady "on" state directly across the power supply.

CORRECTIVE ACTIONS FOR ESD

Customers alerted to the presence of an ESD problem through failure analysis can investigate their manufacturing and handling areas for sources of charge generation. Examples of high voltage ESD sources may include: CRT displays, plastic machine rails, carpeted offices, and certain types of clothing. The use of a hand-held static meter can locate voltage generation sites.

Corrective actions are widely available, and include proper conductive or dissipative material choices, humidity control, grounding, worker educational programs, detector networks, and charge neutralizing ion grids and blowers.

FRINGE BENEFITS

The sensitivity of devices needs to be known because ESD sensitivity could likely become a specification. Customers could begin to make purchasing decisions which are partially based on device ESD hardness, and the failure analysis ESD simulation activity could help in the evaluation process.

FA REPORTS

Use of simulations with the techniques outlined in this article can help you improve report effectiveness, particularly where suspected ESD damage is a major (expensive) problem, and conclusion of ESD damage is under question.

REFERENCES

1. *Electrostatic Discharge (ESD) Susceptibility of Electronic Devices (VZAP-2)*, by William K. Denson, Reliability Analysis Center, RADC/RAC, Griffiss AFB, New York 13441. (Presents tabulated data for I.C.s and discretes, along with data calculated with the Wunch-Bell model, and a preface discussing failure mechanisms and test methods).

2. *MIL-STD-883A*, method 3015.

3. *DOD-HDBK-1686.*

4. *Electrical Overstress Protection for Electronic Devices*, by Robert J. Antinone et al, Noyes Publications, 1986, pp 17-26, 75-80, 140-377.

5. *Economical ESD Testers for The Identification of Discrete Device Damage Thresholds*, by T. W. Lee, Evaluation Engineering Magazine, pp 80-99. (Details of the construction and application of testers for measuring ESD damage thresholds to 15 kV, and includes a nomograph for conversion of ESD test voltage to stored energy on page 98).

6. *Determination of the Threshold Failure Levels of Semiconductors Due To Pulse Voltages*, by D.C. Wunch and R.R. Bell, I.E.E.E. Transactions on Nuclear Science, Vol. NS-15, No. 6, 1968, pp 244-250. (A large study of the failure levels of semiconductors, presenting a power per unit area model for the prediction of thresholds)

7. *Electromagnetic Detection of Nuclear Explosions*, by Robert W. Cotterman, IEEE Transactions On Nuclear Science, pp 99-103. (Shows the general waveform of an NEMP event).

8. *Response of a System*, by Dr. Norman Rudie, Defense Science and Electronics Magazine, June, 1986, pg.31. (Includes a table showing the failure energy threshold for various discrete devices, including semiconductors).

9. *Electromagnetic Pulses: Potential Crippler*, by Eric J. Lerner, IEEE Spectrum, May, 1981, pp 41-46.

10. *Electrostatic Discharge (ESD) Protection Test Handbook*, published by the Keytek Instrument Corporation, Wilmington, Massachusetts, 01887, 1983, 1986.

11. *ESD Testing for ICs*, Second Edition, 1990, Keyter Instruments, Inc.

12. *Analysis of Electrical Overstress Failures*, by Jack S. Smith, in proceedings of 11th Annual Reliability Physics Symposium (IRPS), 1973, 1973, pp 105-107.

13. *Modeling of Electrical Overstress in Silicon Devices*, by N. Kusnezov and J.S. Smith, in proceedings of EOS/ESD Society, 1979, pp 133-139.

Thermomechanical Effects of EOS

Thomas W. Lee
Motorola Inc.
Austin, Texas

INTRODUCTION

Electrical overstress, or EOS, in a semiconductor device of any type usually takes the form of a hole or crater in the silicon die. Metallization is usually fused and alloyed with the underlying silicon. The device junctions may be totally shorted. Usually, failure analysis stops with this observation of catastrophic, and obvious damage. However, through research, it has been learned that the conditions responsible for this damage can be deciphered, to a degree, from the interpretation of the physical damage present. Its extent, its location, and the portions of the device which are involved all provide important clues. The article by Dr. Tom May preceding this one describes this method of interpretation by physical location thoroughly.

There is a second type of EOS, where the effects are not as obvious. The entire sample group can be studied for example devices in which damage is not total, as described in the above paragraph. In these devices, subtle damage or indications of degradation may reveal information about the conditions which resulted in the failure of the group. The analyst inspecting the surface of the device may find that the metallization has become darkened, and rough in texture. If the die was glassivated, a crack in the glass may allow the aluminum to extrude through the crack like toothpaste. X-ray shadowgraphs of the die bond solder alloy coverage show that the coverage is perfect. However, closer examination of the solder may show it to have enlarged grains, and it may also be extruded, oozing from beneath the die to some degree. This effect may be found in R.F. and microwave devices, power integrated circuits, and power transistors.

This phenomena is a life-limiting physical process in all devices made from silicon, silicon dioxide, aluminum, and soft solder, regardless of the company or country of origin. It is interchangeably identified as "thermal fatigue," "wearout," "reconstruction," or "restructuring." It can act alone to physically degrade a device, or it can be present as an indicator of abnormal conditions just prior to a failure. It is important to recognize and understand this wearout or failure mechanism.

The interpretation of the conditions responsible for this damage are not easily deciphered, but simulations can be somewhat more informative than in the case of catastrophic EOS. Accurate interpretation of the conditions resulting in such physical damage to semiconductors, and power transistors in particular, requires a brief review of their construction and principles of operation.

SEMICONDUCTOR DEVICE CONSTRUCTION

In our discussion, we proceed in a vertical section from the top of the die to the header. Figure 1 is a photograph of a typical power transistor, with its gross features identified. Figure 2 is a cross-sectional drawing of a hypothetical device, in which many typical metals and layers in common use have been su-

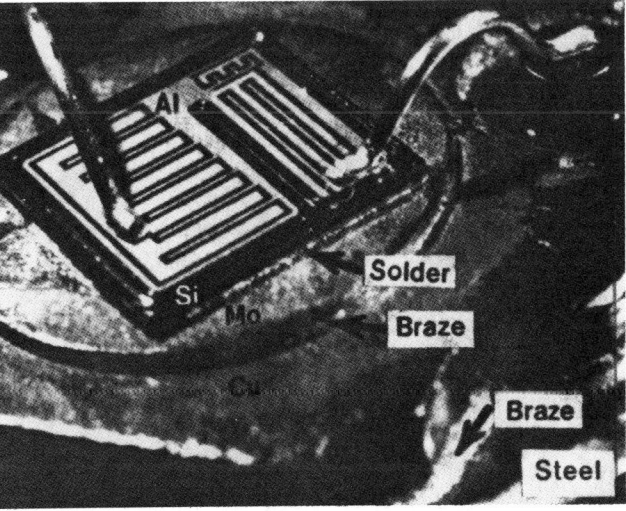

Fig. 1 Optical micrograph of the interior of a typical power transistor, with major structural features identified.

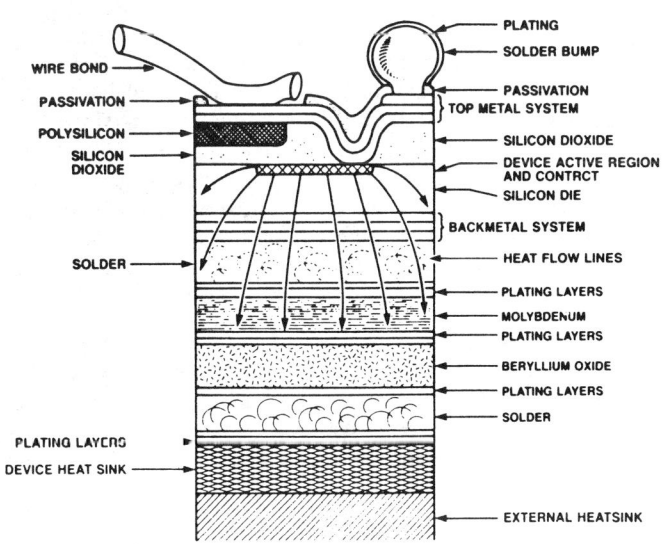

Fig. 2 Hypothetcial power device cross-section showing possible material layers (vertical dimensions not to scale).

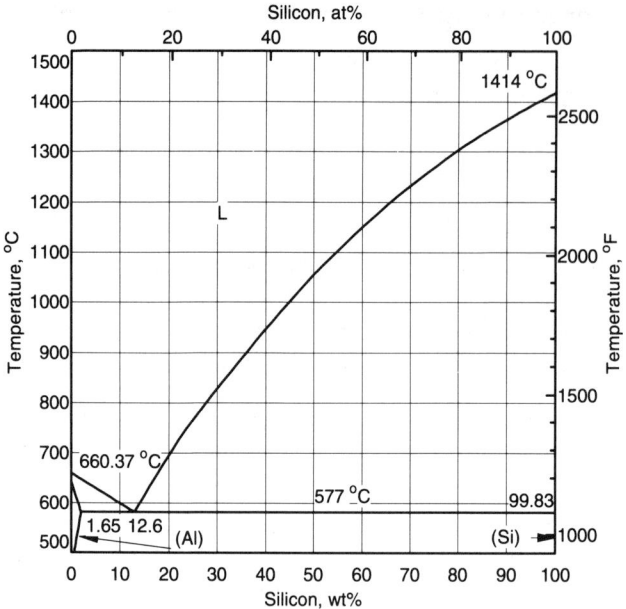

Fig. 3 Phase diagram for the aluminum silicon system. The area where the sintering of contacts takes place is marked by the arrow.

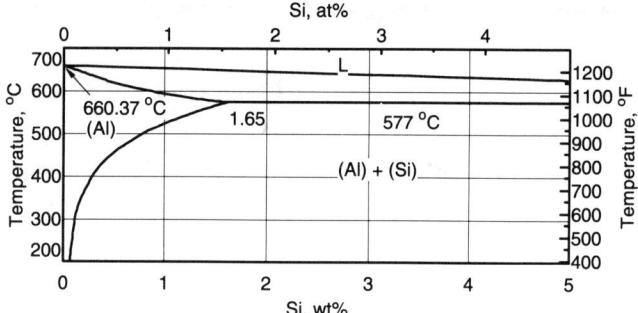

Fig. 4 Expanded area of the Al-Si diagram. Notice that alloying can take place below the eutectic temperature.

Silicide	Sintering temperature (°C)	Resultant resistivity (μΩ-cm)
Pd₂Si	400	30-35
CrSi₂	700	–600
PtSi	600-800	28-35
TiSi₂	900	13-18
VSi₂	900	13.3-66
CoSi₂	900	18-20
NiSi₂	900	50-118

Fig. 5 Resistivities and typical temperature of formation of some of the silicides.

perimposed into one device to illustrate the thermomechanical complexity of typical semiconductor devices.

Most devices have a thin deposited glass layer, or glassivation, over the surface to physically protect (passivate) it from damage, and to act as a shield against the penetration of contaminant ions. This deposited glass may be doped with phosphorus to neutralize the passage of contaminant ions to the device surface. It can be deposited by evaporation, RF or magnetron sputtering, or by chemical vapor deposition (CVD), in which a reaction which results in the formation of glass takes place in the vacuum chamber. Silicon nitride (Si_3N_4) can also be used.

The top surface of the die is occupied by the conductor runs, most commonly composed of aluminum (Al), which may also contain small percentages of silicon and/or copper. Al is a relatively easy metal to deposit. It can be evaporated or sputtered. Other metal schemes are also used, especially when a solderable surface is the end result. For example, some manufacturers evaporate, sputter, or plate the Al with a metal sequence ending with copper (Cu) or gold (Au) at the surface. A barrier metal, such as nickel (Ni), is usually placed between Cu and the Si.

Aluminum readily forms an ohmic contact with P-type silicon. All silicon (Si) surfaces quickly passivate themselves with a thin oxide skin only a few tens of Angstroms thick, which is an insulator. Aluminum on N-type silicon forms a Schottky barrier. Contact to the aluminum is made by "sintering" the Si and Al at a temperature well below the alloying temperature, 577 °C. This process is also called "densification" of the Al. The term "sintering" originally indicated the heating of a metal powder to a temperature just below its eutectic point, to bind the grains. It is an inaccurate description of the process of forming contacts in semiconductors because the melting point of Al

metal is not closely approached. Instead, alloying of a small percentage of aluminum with the silicon occurs at temperatures well below the eutectic line. The Si-Al phase diagram in Fig. 3 and 4 shows this area at the extreme left-hand end of the eutectic line. The thin oxide skin is removed, and the thin Al-Si alloyed region makes an ohmic contact. If heated to greater than 577 °C, Al would alloy with the Si to form a P-type semiconductor, and short the device.

In power devices, the thin films of Al are typically 10 to 20 kA thick. The emitter and base diffusions in bipolar devices are each about 0.1 mil thick. Because of their thinness, these films have significant resistances. The entire die, however, is much larger than these dimensions because a minimum thickness, usually 8 to 13 mils, is required so that the wafers can be handled without breaking. Additionally, a thick lightly-doped collector is required for high breakdown voltages. Figure 2 is a cross-sectional drawing of a hypothetical die, where the layer thicknesses are exaggerated. *It is important to remember that the die is generally 200 or more times thicker than the thin films on top that are the site of activity.* Aluminum wires, from 1 to 20 mils in diameter, are used for connection between the thin film conductors on the die surface die and the header pins, or posts. The aluminum on the die can be doped with copper to increase its current-carrying ability. The aluminum in the wires can be doped with Si, Mg, or Pt to improve its characteristics. Small gold wires are also used, particularly in small-signal transistors,

RF transistors, and ICs. The bonds can be thermocompression ball bonds, and stitch bonds, or ultrasonic wedge bonds. Other metal systems can be used over the aluminum, and may have a solderable top metal. One or more layers of barrier and adhesion metals may be used between the aluminum and the solderable top metal. Contact to the die can then be made with a clip, soldered wires, or in flip-chips as solder balls, also illustrated in Fig. 2. Wires may also function as inductors, in thin- and thick-film structures that are a part of the internal tuning networks in R.F. power devices and R.F. power modules.

Again refer to the drawing in Fig. 2. The silicon die must be fastened to a header for electrical contact to the backside of the die, and to remove heat. Epoxy may be used where ambient temperatures and power dissipation is low, but metal fastening systems are necessary for power-handling devices. The metals can be plated, sputtered, or evaporated onto the back of the die. It is helpful to think of Si as basically little different from glass in its physical properties. The metals in most solders do not adhere well to silicon at low temperatures. The backside of the die must be coated with a metal, or metal system, which is simultaneously solderable, and adherent to silicon. A single metal layer often cannot be made to satisfy both conditions. Backmetals vary from a single layer of a solderable metal such as nickel, to elaborate systems consisting of more than one metal layer and containing metal silicides, adhesion layers, and interphased layers. Figure 5 shows the properties of some of the metal silicides that could be present. At the formation temperature of many silicides, the dopants may "push", or continue to diffuse, depending on the dopant species, and alter or ruin the device. As a result, backmetals usually adhere by mechanical, not chemical, processes. Figure 6 is a chart showing the mechanical properties of some of the metals and materials which compose semiconductor devices. A variety of metals can be chosen for backing because of various desirable properties with respect to the aforementioned functions. For example, aluminum (Al) is used because its contact resistance is very low. Chromium (Cr) is used because it is tough and unreactive. Titanium (Ti) is used because it has a unit cell size close to that of silicon, making a strong bond. The resistivity of the metals themselves, the contacts, and the silicides are all important to this process.

A few metals will alloy with silicon at higher temperatures to form alloys. Gold, gold-tin, and gold-germanium eutectic (hard) solders were formerly used for many part types. They melt at temperatures above soft solders (about 300 °C), but far below temperatures which will move diffusions (about 950 °C). Their cost became prohibitive in the 1970s and 1980s, when gold prices rose to several hundred dollars per ounce. Now, these expensive solders are usually limited to R.F. and premium products. The lead-based soft solders were explored by almost every manufacturer. These lead alloy solders also contain, but are not limited to, antimony (Sb), bismuth (Bi), indium (In), silver (Ag), and tin (Sn). Antimony and Bi harden solder alloys. Silver prevents the scavenging of Ag from thin layers in die backmetal schemes when it is used. Indium, one of two metals known which wet glass, is used to promote adhesion and improve temperature cycling capability. Tin lowers the melting point. Some attention must be paid to the solder alloys which are available from the manufacturers. Understandably, lead-tin alloys are by far the most popular solders. Figure 6 also includes the properties of some of the metals used to make solder alloys.

Reexamining Fig. 2, a semiconductor device may be viewed as a virtual sandwich of materials and metals, and the particular materials found are chosen because of specific physical and chemical properties.

THERMOMECHANICS OF DEVICE OPERATION

The operation of all semiconductor devices generates heat. Various resistances present in the backing metal system, the silicon bulk, the epitaxial or doped layers, the contacts, thin surface layers, and metal connecting runs all generate IR drops when current is flowing. As feature sizes decrease, these problems become increasingly important.

With the exception of Bi, a temperature increase causes all the materials to expand slightly; cooling does the opposite. Any differential in expansion rate, in this case the difference in coefficient of expansion between the silicon and the aluminum, will result in the generation of stress, and strain within the material with the lowest yield strength. It will be the material to be deformed. The forces are enormous, a fact not well appreciated because of the smallness of the features and devices being discussed. A metal expands with a force equal to that required to compress it the same amount. This value is Young's Modulus, and it easily translates the tiny movements into large forces; tons/square inch (see Fig. 29). In Al, the force with which heat sinks are extruded is comparable to those within the aluminum top metal under some conditions.

Consider the case of aluminum deposited over oxide. Examining Fig. 6 shows that the aluminum has the highest coefficient of expansion of the two materials. This differential will generate large forces. The metallization will absorb the stress inter-

Thermomechanical Properties of Semiconductor Structural Materials		
Material	Coefficient of Linear Expansion (PPM °C⁻¹)	Thermal Conductivity (WATTS CM⁻² °C⁻¹)
Metals		
Aluminum (Al)	25	2.37
Chromium (Cr)	6.0	0.91
Copper (Cu)	16.6-18	3.98
Germanium (Ge)		1.5
Gold (Au)	14.2	3.15
Kovar	4.5-5.9	0.17-0.34
Lead (Pb)	29	0.346
Molybdenum (Mo)	5	1.4
Nickel (Ni)	13-14	0.899
Silicon (Si)	3	0.835
Silver (Ag)	17-19	4.27
Tin (Sn)	8.5	0.2
Tungsten (W)	4.5	1.78

Fig. 6 Table showing some of the physical properties of materials used in semiconductor packages.

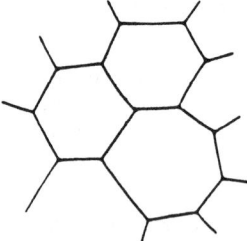

Fig. 7 Drawing of the crystal grains in an Al film, with triple points at intersections.

nally, and deform because there are differences in coefficients of expansion within the film itself. This occurs at its smallest structural discontinuities, the crystal grain boundaries. Figure 7 shows the hexagonal crystal grains in a metal film. Their size can vary tremendously, depending on the film composition and conditions of deposition. These individual crystals in the film expand at different rates along their three crystallographic axes. They are more or less evenly distributed, and randomly oriented. As they expand, the crystal grains in the film make collisions with each other where expansion rates have the highest vector sum in adjoining grains. These collisions raise the aluminum away from the device surface. The material pile-up is called a triple-point, or a hillock. Figure 8 is a SEM view of a large central hillock amid many smaller ones.

In glass passivated devices, this movement tends to be confined, but the metal may flow, and ooze, from beneath the edges, attesting to the tremendous forces involved. The glass may also crack, and the aluminum can penetrate, being extruded like toothpaste. These phenomena graphically verify the large forces developed. Comparatively, aluminum heat sinks are forcefully extruded through dies at 800 to 900 °F (425 to 480 °C), and pressures of 250 to 5500 tons/in^2.

DIE BONDS

Solder alloys, as established earlier, are composed of a mixture of grains or crystals of different metals and alloy phases, each with their own particular coefficient of expansion. Bismuth and Ga, which on occasion may occur in solder alloys, are the exceptions to this general rule; they actually expand upon cooling. Unfortunately, all the metals in alloys all expand at slightly different rates when heated; refer again to Fig. 6. As in the aluminum film, the material with the highest coefficient of expansion will move more than that with a low coefficient. The relative movement of the metals is tiny, usually a few ppm of length per degree. Thermal cycling causes the expansion and contraction of these grains, and the alloy slowly weakens with age.

At this point, with all these considerations, one may wonder how a semiconductor device works at all! In reality, the discussion of these conditions are no more consequential than, say, a tire manufacturer stating the tread life of a tire. We are all aware of the myriad mechanisms by which tires wear out. No one pushes a tire beyond its limits by, say, operating it at twice its rated load. The effects would be easily guessed. Similarly, in

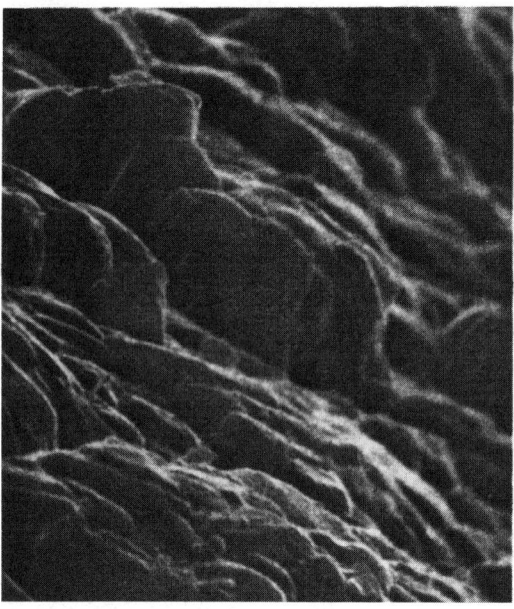

Fig. 8 SEM photo showing a group of pronounced hillocks in an Al film after stressing. (800×)

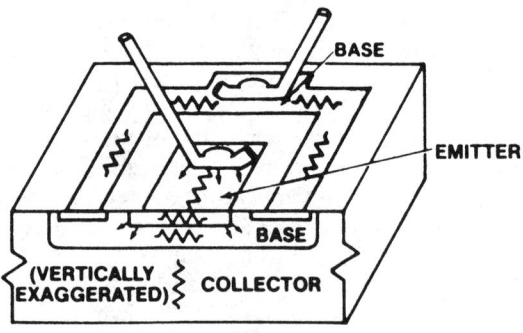

Fig. 9 Cross-sectional drawing of a transistor, with representative lumped elements representing layer resistances for equivalent circuits (Ebers-MOLL, etc.)

transistors, reliable operation is dependant upon conservative device design and rating. Examples of such parameters are maximum soldering temperature, RBSOA and FBSOA, Es/b, thermal cycling, etc. The specification sheets guide the circuit designer and ensure that the metal films are not stressed to failure within a period of time over which the device can be reasonably expected to be operated. Its use is obsoleted by technologically superior new products, not by catastrophic wear-out failure.

THERMOMECHANICAL OVERSTRESS

Refer to the drawing in Fig. 9. This drawing shows a hypothetical cross section of a device where the inherent resistances are lumped for illustration. Suppose the aluminum is 20 kA thick, and a finger is 20 mils wide and carries one amp, then the current density is 10^4 A/cm^2. The emitter finger can easily be 5

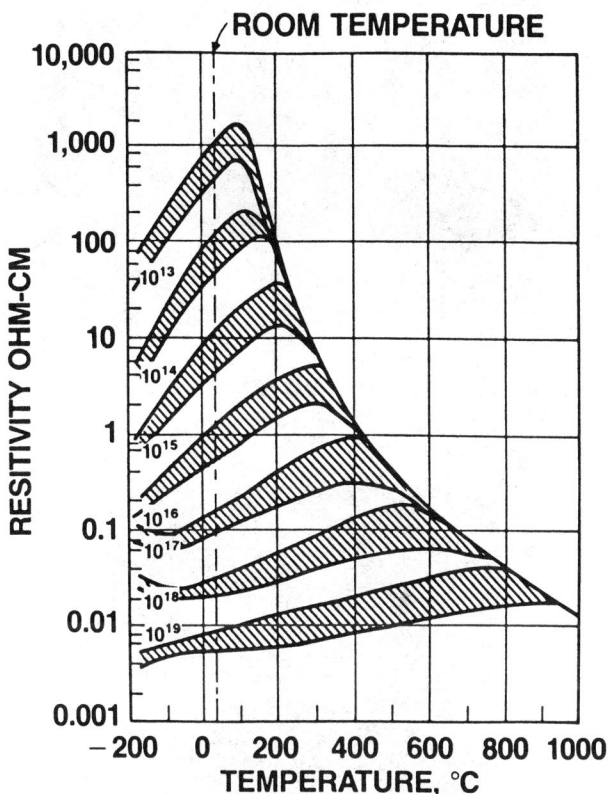

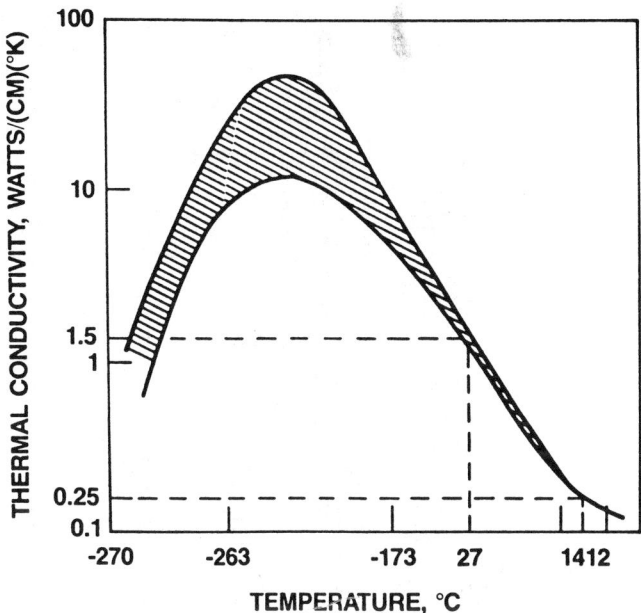

Fig. 10 Runyan's curves showing the very rapid drop in resistivity of doped Si with increasing temperature.

Fig. 11 Runyan's curves showing decreases in thermal conductivity of silicon with increasing temperature.

times as long as it is wide, or 5 squares long. At 1.4×10^{-2} ohms/square, this is 70 milliohms. In a five amp device, assuming current is distributed equally, there would be a 70 millivolt drop along the length of this finger. This is enough voltage to make a significant difference in conduction between the finger origin, and finger tip.

Similar considerations also apply to the emitter and base diffusions below the metal. A typical emitter doping level is about 10^{19} atoms/cc, and this will result in a resistivity of 10^{-2} ohm-cm in N or P-type material, comparing favorably with the value for the aluminum film. The more lightly-doped base will have drops and conductivity modulation, and these effects in both the emitter and base combine to make current distributions in the emitter nonuniform when the current is high, particularly when the device is operated at, or slightly beyond, its maximum ratings. These drops lead to resistive debiasing, one of the most important effects which can cause imbalance in the conduction of the device, and affects the sharing of current among its various structures. Conduction at the origin of the finger will be slightly greater than at its tip.

This process can be regenerative as the power applied to the device increases. Increasing temperature has two important effects in semiconductor devices: it decreases resistivity and increases thermal resistance of the silicon. Figure 10 shows the effect of temperature on the resistivity of silicon doped to a wide range of resistivities over the temperature range of –200

°C to 1000 °C. The resistivity decreases sharply in all cases. These are Runyan's curves. The carriers normally present are supplemented by thermally generated carriers, and the increased number of carriers will cause the transistor hFE to increase in the area that is hot.

The second effect is somewhat less pronounced. Figure 11 shows the second effect of increasing temperature in a semiconductor. The thermal conductivity drops. The plot is fuzzy because the results found by a number of researchers have been combined into one graph and the ranges are shaded-in. Note that silicon has a negative coefficient of thermal conductivity, with an increase in temperature, of only about one order at reasonable temperatures, while the change in resistivity can easily cover 2 orders. Combining the effects of increasing temperature in semiconductors, shown in the two figures, some conclusions involving thermomechanical behavior may be reached. Because of these two physical properties of silicon, an area of a transistor, usually close to the emitter wire bond, will tend to become hotter than its surroundings, and thermally isolated.

Heat travels in semiconductors by the phonon conduction process, a type of lattice vibration. During pulsed operation, the relatively low speed allows the heat to become confined in a small region of the device when power pulses are narrow.* A hot spot developed because heat is generated faster than it can be dissipated. At a critical point, carriers are thermally generated, and, the base injection no longer controls the collector current. This is thermal runaway. Depending on circuit conditions, the hot spot may also quickly develop temperatures at which the aluminum and silicon form an alloy, and the device is quickly shorted and destroyed, or "cratered."

*Note: the speed of sound in air is approximately 330 meters/sec. The speed of sound in silicon is 3,000 to 6,000 meters/second.

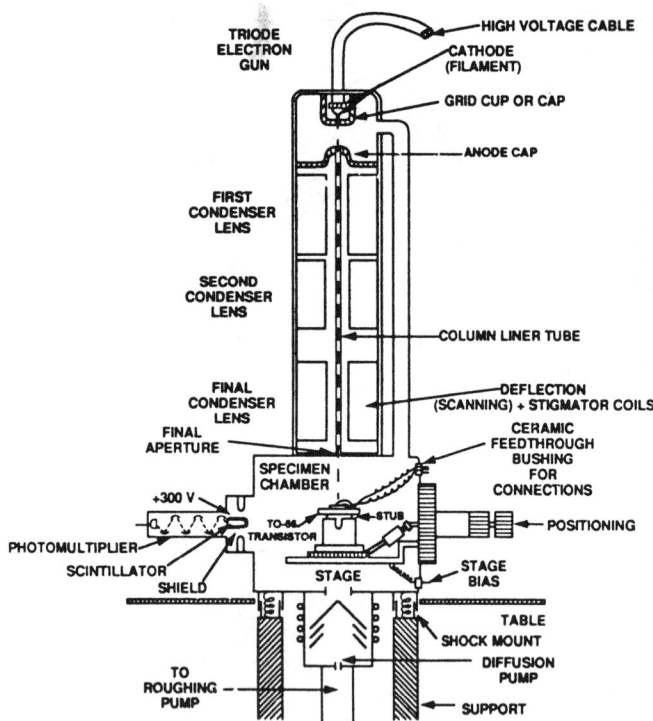

Fig. 12 Simplified cross-sectional view of typical scanning electron microscope. Note potential applied to shield in front of scintillator to attract electrons. This potential is variable in some SEMs.

It is this area of operation where device damage varying from slow degradation to instantaneous meltthrough can occur, and where the aging effects in the metal films and solder alloys are tremendously accelerated. This results in shortening of the device life, degraded characteristics, or catastropic failure. The aging of the metal film is caused by the same fundamental thermally driven mechanism described above, and the metal surface in the second breakdown spot will be microscopically identical in appearance to that found in a thermal or power-cycling failure.

Optical microscopy of the thermal cycling or power cycling failures generally shows that the metallization appears dark over some portion of the die surface, usually including some or all of the emitter region. Scanning electron microscopy (SEM) of the metallization at a magnification of a few hundred, and at a low angle, will show roughening of the metal surface. Inspection of the metallization at higher magnification will show the presence of numerous hillocks. If testing is continued to device destruction, the resistivity of the metal film will increase, and its mechanical integrity will be weakened until an open develops.

SIMULATIONS

In the past, experimental accelerated life test data was generated with the standard tests and fixtures, or with a special fixture which heated devices with their own power dissipation, then cooled them rapidly with a Freon spray. Repeat of these experiments with the immersion of devices in a Freon bath, accompanied by deliberate overstressing on a curve tracer, sug-

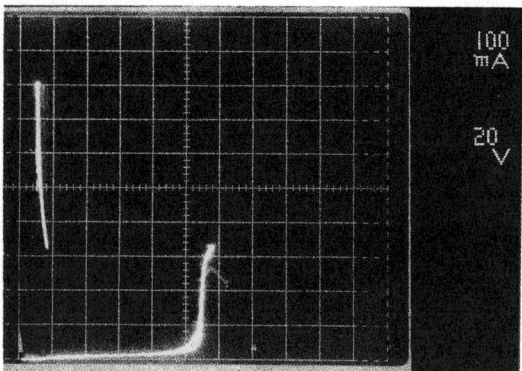

Fig. 13 Curve-tracer of display of V-I characteristic of a bipolar power transistor in second breakdown.

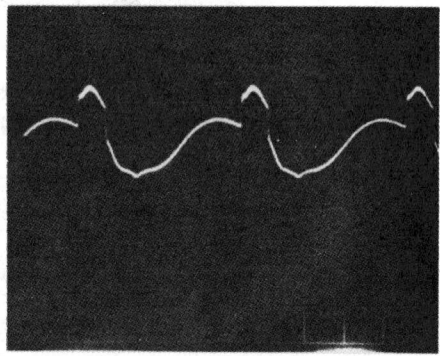

Fig. 14 Current waveform in device in Fig. 13. Pedestals are generated when device switches to low-voltage state.

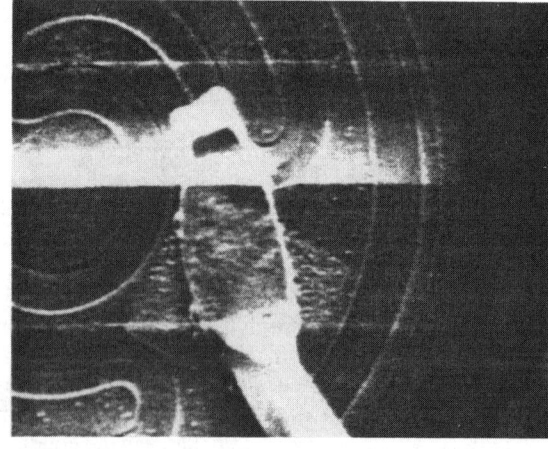

Fig. 15 SEM view of device when power is applied from curve tracer.

gested that the aging mechanisms described before could be greatly accelerated in time. The only difference between these two tests and standard power cycling, in terms of physical changes in the metallization, was the area of metallization invvolved This observation was responsible for the following experiments.

It was reasoned that the superior depth of focus of the SEM could enable a study of the effects of device degradation taking place under conditions of EOS in a very short period of time. Further, the entire event could be videotaped, and editing with time compression could be performed to adjust the real-time event length to be comfortable. Figure 12 is a cross-sectional view of a SEM for convenience in the following discussions.

Various small power devices were mounted to a specially-machined specimen mount with a large thermal mass. The leads ran through the chamber door connector. The collectors were operated at ground potential, and the emitters, in the case of NPN devices, were operated at high negative voltages with respect to ground. This was to ensure that the large collector flange was at ground potential. The devices were forced into second breakdown (Es/b) operation, as shown by the curve tracer Ic vs Vce plot in Fig. 13, and the Ic waveform shown in Fig. 14. They were not allowed to draw enough current to form a destructive meltthrough, and were manually maintained in this marginal overload condition throughout the experiment.

The initial SEM image was unusable, as it had bright lines through it, as shown in Fig. 15. It was reasoned that the beam was being deflected by the fields caused by the high emitter voltage. The backscattered electron image was examined. The energy of backscattered electrons is higher than that of secondary electrons, as shown in Fig. 16, and they are less easily deflected by stray fields. However, the image was still of unacceptable quality.

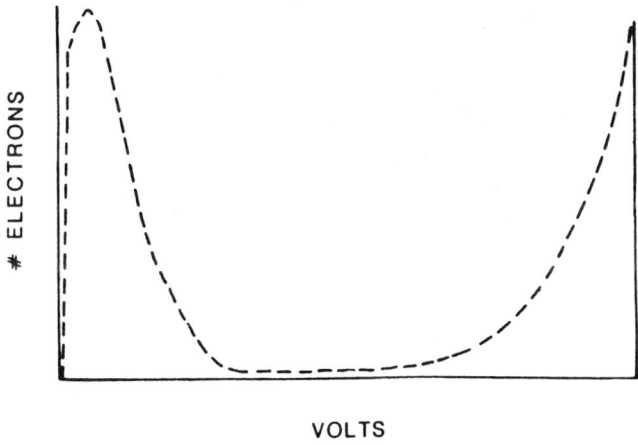

Fig. 16 Distribution of electron energies yielded from an incident beam in a SEM.

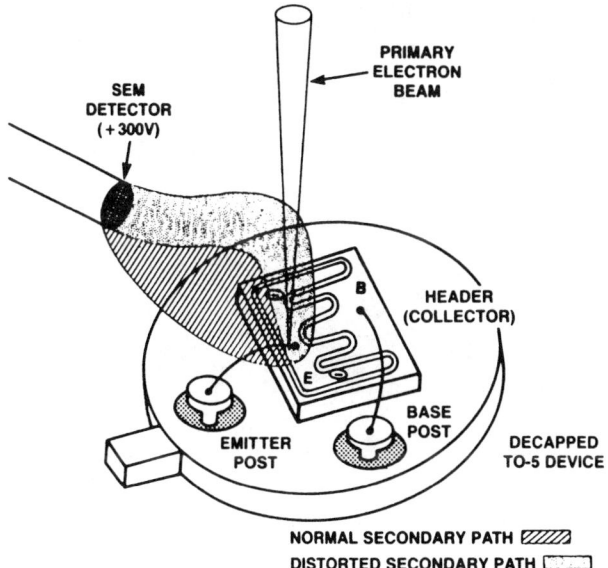

Fig. 17 Illustration of one effect of device bias on returning secondary and backscattered electrons in the SEM.

Fig. 18 View of device with Faraday shield in place.

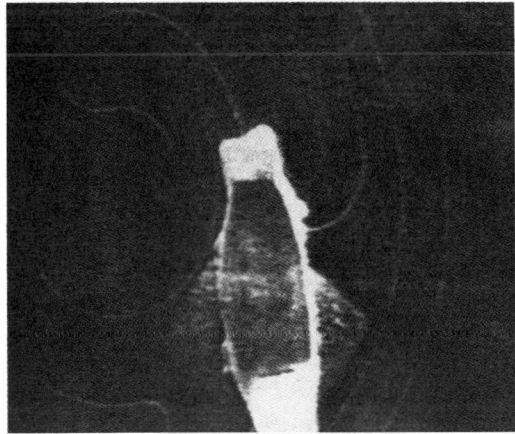

Fig. 19 SEM view of same device as Fig. 15, but with Faraday shield in place.

The interference problem was solved by constructing a Faraday Ring over the top of the device, whose operation is illustrated schematically in Fig. 17. The ring shielded the secondary electrons from the strong fringing fields from the device header. A photo of the ring appears in Fig. 18. The image was then usable with secondary emission, as shown in Fig. 19.

Operation of devices in the SEM with this improved image showed that the metallization could be quickly degraded by the mechanisms described earlier. Figure 20 shows a typical example. The operation of more samples of various device types, shown in the before-and-after Fig. 21-24, resulted in the appearance of the degradation effect in all cases. The SEM images of the degradation progress in all devices was videotaped, and the tape edited.

The metallization was stripped from the surface of a sample stressed by this method, and the surface studied. Figure 25 shows the darkened surface due to the nonreflectivity of the deformed metallization. Figure 26 shows the emitter contact region with the metallization removed, and the sintering pattern was found to be exaggerated in size and extent in the area where metallization reconstruction was most advanced, as shown in

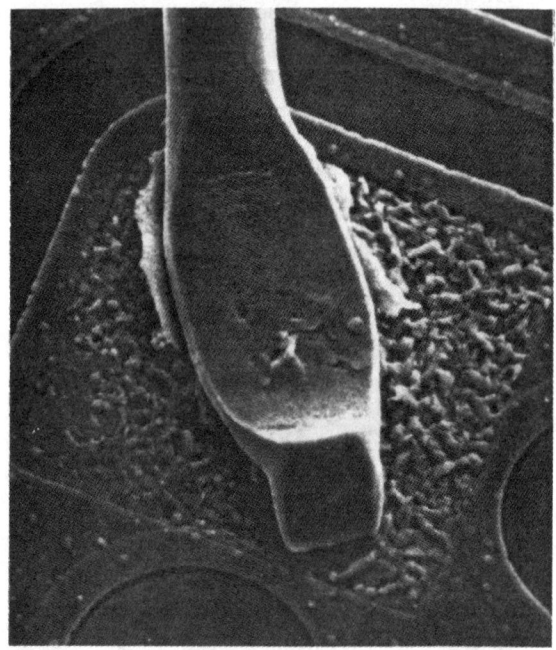

Fig. 20 Detailed SEM view of metallization in emitter pad window of device after stressing. (500×)

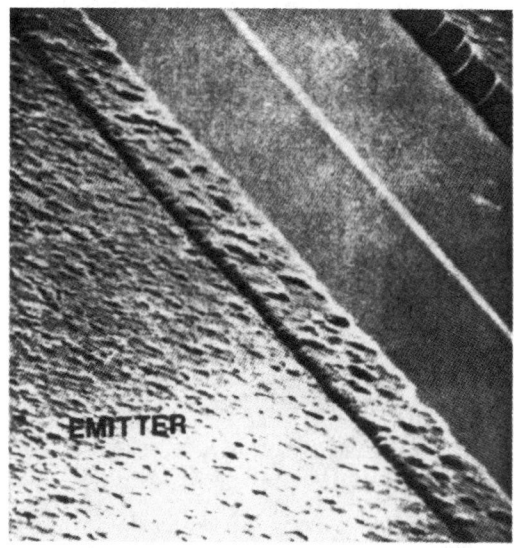

Fig. 21 Surface of power device at start of power cycling (800×)

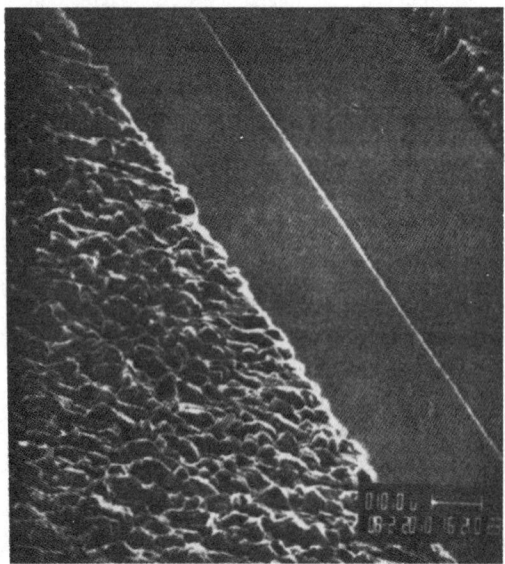

Fig. 22 Surface of power device after exposure to cycling. (800×) Compare with Fig. 21.

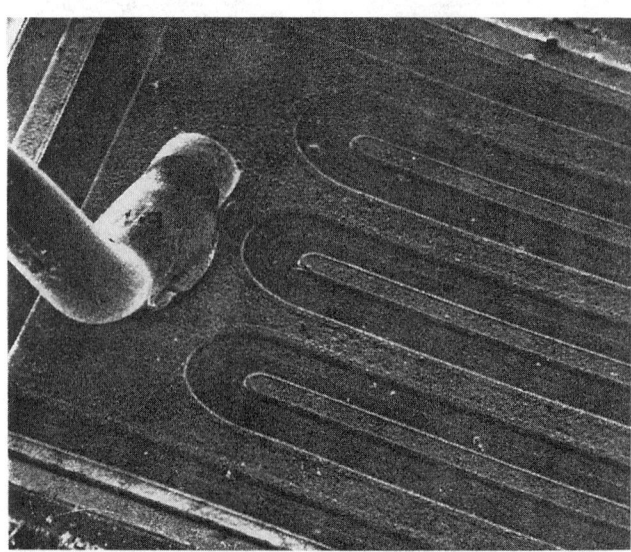

Fig. 23 Surface of device before powering in SEM. (2N5190) (75×)

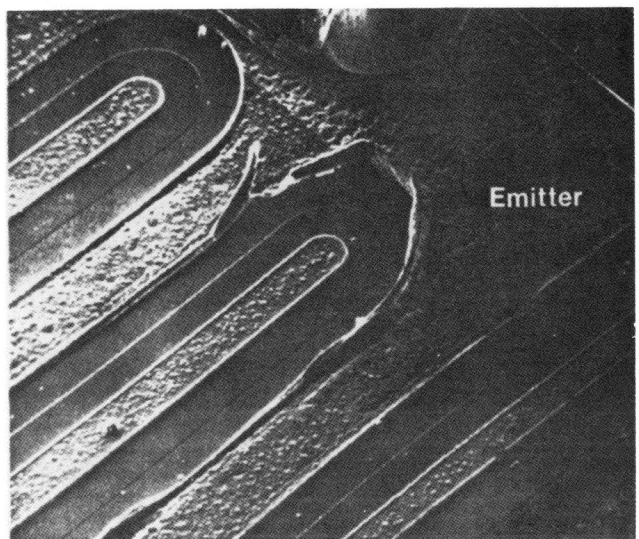

Fig. 24 Surface of power device after powering in SEM. (125×)

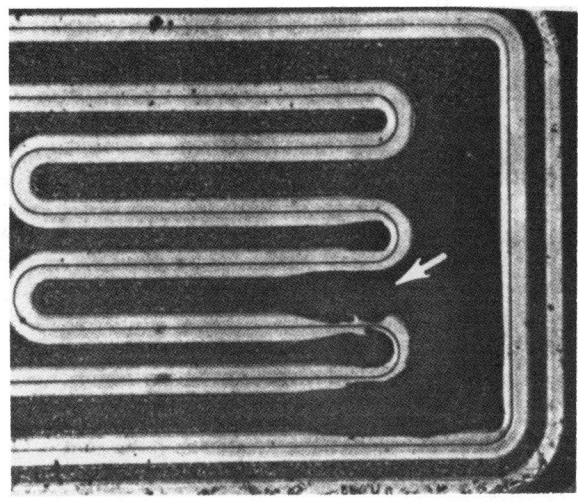

Fig. 25 Optical micrograph of die surface following overstress in SEM. (40×)

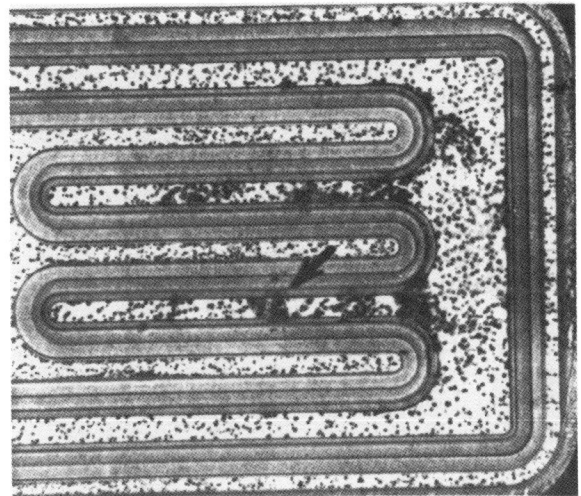

Fig. 26 Die in Fig. 25 following stripping of metallization (175×)

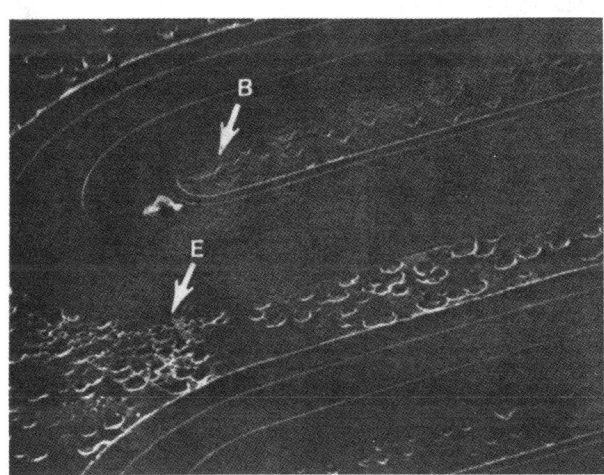

Fig. 27 SEM view of emitter region beneath most advanced metal reconstruction. (245×)

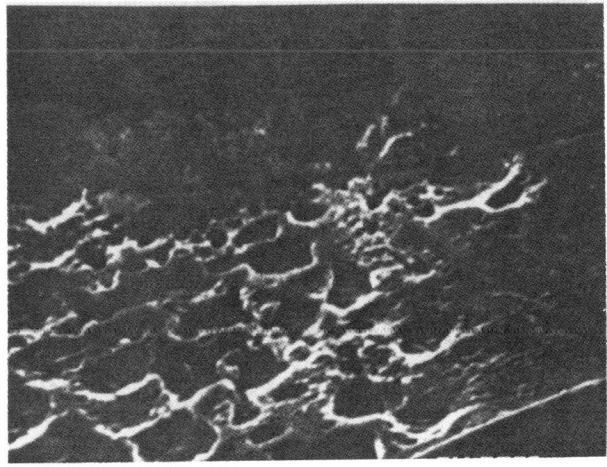

Fig. 28 SEM detail of area shown in view in Fig. 27, showing concentration and enlargement. (1050×)

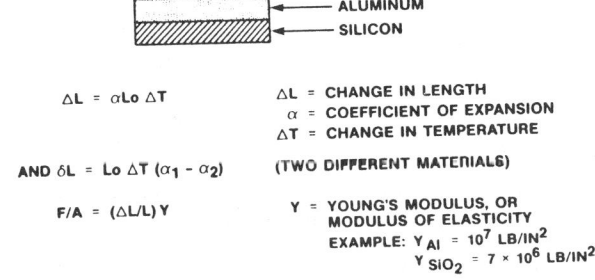

Fig. 29 Magnitude of forces in thin films.

451

Fig. 27 and 28. It was evident that the shallow alloying mechanism by which contacts are formed was continuing to operate in this test device, and the silicon was slowly being consumed dissolving into the aluminum emitter metallization Notice how quicky the effect took place, in only minutes.

The implied local instantaneous temperature was between 200° and 500°C.(Fig. 4) Althouth the measurement of the temperature of microscopic site with IR thermography is well refined, the measurement of the peak temperature at the silicon-aluminum interface (referring again to Fig. 27 and 28) is virtually impossible. Thus, the enlargement of the sintereing pattern is very important to this investigation. Keep in mind that simply maintaining a high temperature is not sufficient. It must be cycled with each pulse.

RESULTS

It was shown that the degradation mechanisms operating in a power transistor under conditions of marginal EOS are as predicted by theory. A hot spot forms, and all activity is concentrated in that area until a destructive meltthrough forms. Evidence of hot-spotting in a device can be positively identified by examining the metallization for areas in which hillocks are unusually numerous or large. Finally, the effects of aging of a semiconductor device due to this rapid thermal cycling could take place in a vacuum, because it could be done in the chamber of a SEM. The primary failure mechanism is an electrically powered thermomechanical effect. The effect of temperature cycling, whether a mechanical transfer of the parts from a high to low temperature environment; or from the alternate application and removal of current, is cumulative and indistinguishable. Power cycling will cause failure somewhat faster than thermal cycling because the devices carry current through the weakening metal film, and cycling can proceed at a much higher rate because physical transfer of the devices is not required.

The magnitude of stress generated is determined by the combined effects of maximum temperature, temperature change, rate of change, and coefficient of expansion of the metals involved. The die surfaces may have differing appearances, with the aluminum over the entire die roughening with exposure to thermal cycling, while power cycling generally confines the effect to the emitter region. As the power pulse narrows, the affected area on the emitter decreases in size. Again, heat travels in semiconductors by phonons, and the relatively low speed, approximately the speed of sound, allows the heat to become confined in a small region of the device when power pulses are narrow.

The effects of this wearout are not necessarily confined to the surface of the die. The heat travels through the silicon and is removed through the metal header. The metal grains in the solder alloy reconstruct, but the yield strength of the lead-based alloy and its components is less than that of the silicon die.Consequently, the solder alloy may expand and ooze from beneath the die. Figure 29 shows an example of this type of wearout mechanism. In some cases, this expansion may crack the die.

Refer again to Fig. 29. The difference in the CTE between two materials in contact with each other generates the force which drives the gradual tearing apart of the material with the lower yield strength. For example, the CTE of the alead in the die bond is about 29 PPM/°C, while the CTE of silicon is only about 3 PPM/°C. Thus, when heated, lead, tin, etc. expand more than the silicon, and the softer lead is stressed and deformed. Once again, this is a purely thermomechanical phenomenon. No moisture, air, solvents, or chemicals are necessary as participants. Early immersion boil experiments showed specifically that freon, methylen chloride, acetone, isopropyl alcohol, and methyle do not have any effect on this mechanism.

DEVICE CAPABILITY

The total number of thermal cycles that any semiconductor device can withstand is a complicated function of the following parameters:

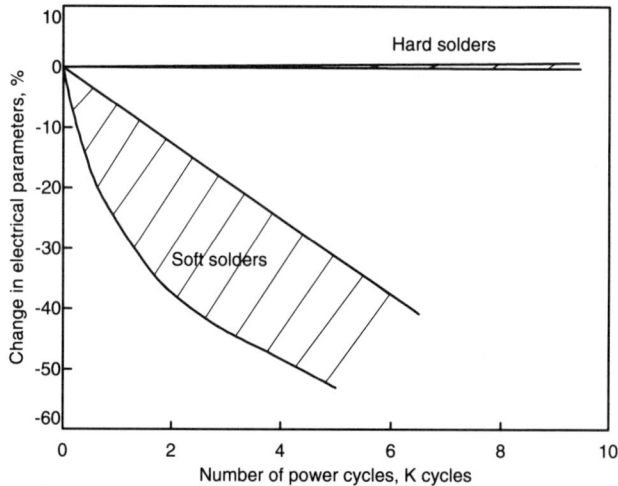

Fig. 30 Olsen and Berg's graph of the effect of power cycling on transistor parameters, excerpted from the IEEE Tran CHMT, June 1979, by permission.

Fig. 31 Die bond degraded as a result of excessive power cycling (80×).

452

- percent of maximum rated power being dissipated,
- pulse width,
- ambient temperature,
- materials used in device construction,
- delta T

Several published papers deal with the capability of devices under temperature cycling. With the foregoing discussions, we have established that that the effects of the thermal expansion mismatch, when combined with a large number of thermal cycles or large temperature excursions, or both, tear the device apart internally, both at the surface and at the die bond. All devices, regardless of company or country of origin, can be made to fail in this manner. These mechanical wearout phenomena are so universally prevalent in electronic devices that they also apply to, for example, electron tubes, capacitors, and transformers which are operated under pulsed conditions.

The solder deterioration effect has been studied in detail. In their paper on die bond alloys, Olsen and Berg show that tem-

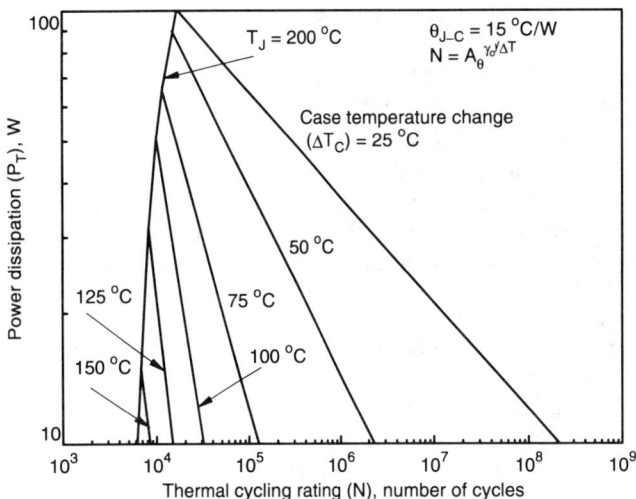

Fig. 32 Chart showing the effect of power cycling of transistor lifetime. Excerpted from the RCA High Reliability Devices Databook, SSD-207B, of 1974, Fig. 2-21, by permission.

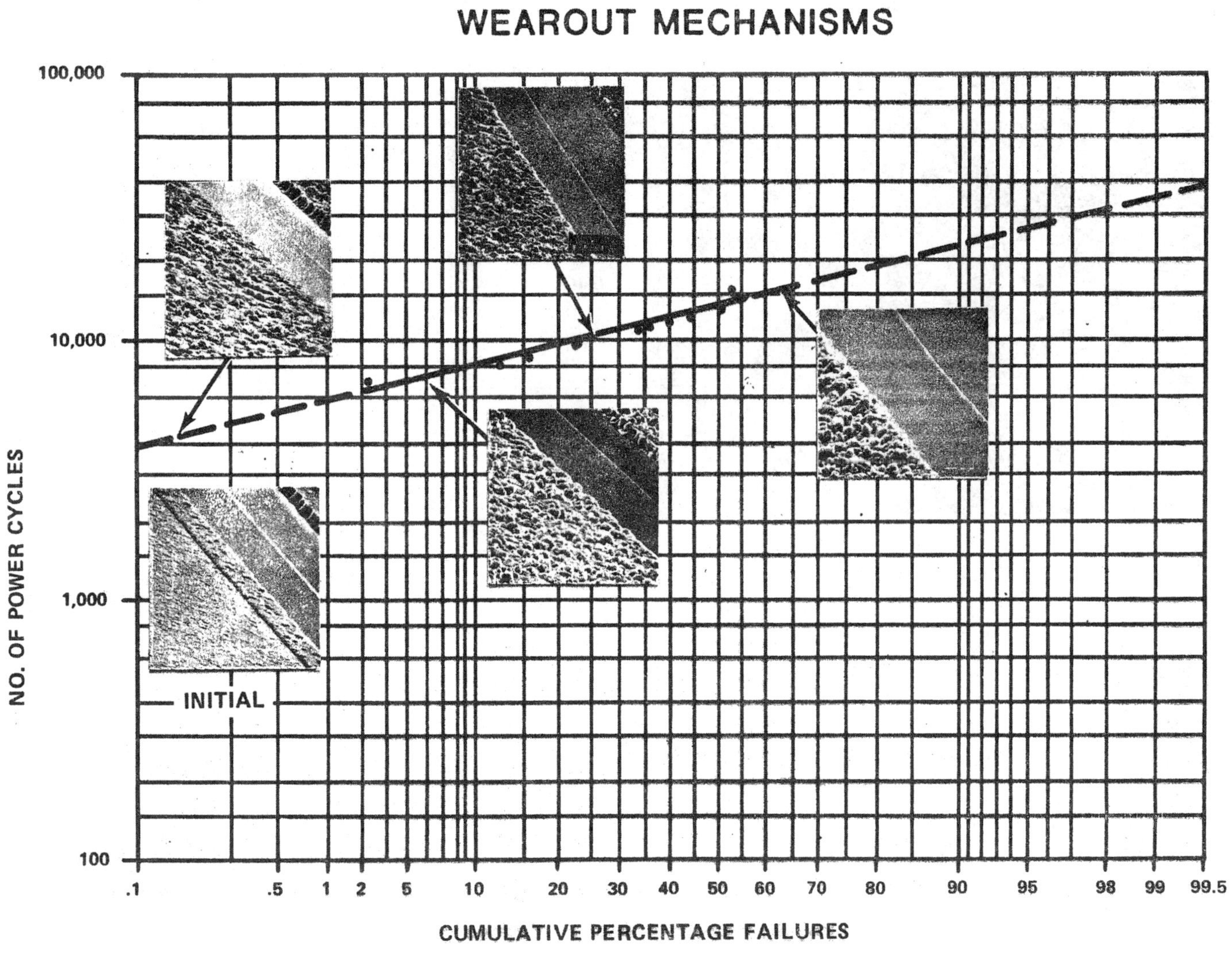

WEAROUT MECHANISMS

Fig. 33 Weibull chart of cycled devices with SEM photos of metallization. (600×)

453

perature cycling quickly reduces the capability of devices with soft solder die bonds, summarized in the adaptation of the graph shown in Fig. 30. The relative change in electrical parameters for transistors having hard and soft solder die bonds is dramatic

However, the lifetime of transistors with hard solder die bonds is not infinite, as Fig. 30 might imply. The stress generated by the silicon die-copper header mismatch, the driving force, moves to the die itself in hard-solder system. Because silicon is brittle and "notch-sensitive", electrically inactive mechanical imperfections can be propagated into cracks in the die. For this reason, a layer of metal having a CTE midway between silicon and copper, such as the molybdenum button shown in Fig. 1 and Fig. 2, is added. The stress is reduced, and resistance to cycling degradation improves. This increase in device lifetime is obtained at a slightly increased device piece-parts cost, but is amply rewarded by better performance.

Circuit Application Design for Reliability

The graph in Fig. 32 is excerpted from the RCA High Reliability Databook, SSD-207B of 1974. It shows the result of an extensive study with power devices in an actual circuit situation, a series-regulator type power supply. This plot clearly

Fig. 34 Typical field failure caused by excessive power and thermal cycling. (100×)

shows the degradation of electrical parameters in soft-soldered devices with increasing power dissipation and increasing number of cycles. Such data is precious and difficult to acquire. What it shows is that derating of components can help to extend their life expectancy beyond the product life cycle, eliminating the device as a reliability concern.

A Weibull plot is presented in Fig. 33, which shows cumulative percentage of failures of power transistors as a function of number of overstress power cycles, performed in a more conventional manner; a typical life-test. SEM photos of the metallization are superimposed on the graph. The gradual degradation of the metallization on the devices is evident from the increasingly roughened aluminum surface. These devices are being intentionally overstressed to failure, to evaluate their capability. They are not defective; they have been thermomechanically aggravated such that their end-of-life degradation mechanisms have been activated. If the test had been continued, all devices would have eventually failed.

There is no fundamental difference in the mechanisms operating in the good and failed devices when removed from any stage of this test. They are mechanically and electrically identical, with the exception of certain high-current or high-power tests, such as $V_{(SAT)}$, until the actual moment of catastrophic failure. The damage to the metallization in failures removed from test varied only in degree; the mechanism was always the same.

Catastropic failures are due to the weakening of the metallization beneath the emitter wire bonds, which progresses until the wire bonds lift and create an open. Unless prevented by careful and elaborate experimental control, this may be followed by arcing and a destructive meltthrough. As discussed earlier and shown in Fig. 31, the solder alloy may have expanded and weakened until the die power dissipation is lowered, and the device goes into destructive thermal runaway. A typical example of device which failed under these conditions in the field is shown in Fig. 34; notice the thermal involvement of the small base fingers.

Corrective Action

In all cases, the lifetime of any device, generically speaking, is extended by generous, but not extravagant, derating. Careful attention to data such as the foregoing, from reliability and product engineering departments, combined with attention to the published device thermal derating charts and close association with the customer and his application, can avoid overstress

Guide to the Interpretation of Physical Damage in Electrically-Damaged Devices

Physical Evidence	Most Probable Cause	Conclusion (Reported As)
Reconstructed metallization over entire die	FBSOA exceeded, device running too hot or excessive cycling	EOS./, FBSOA
Reconstructed metallization in characteristic area of emitter	RBSOA exceeded with any type of base drive, current-limited breakdown BVceo(sus) operation	EOS/RBSOA
Meltthrough hole in characteristic area of emitter	Es/b operation or repetitive high-voltage transient	EOS, second breakdown
Meltthrough hole in unusual area of emitter or base with edge arcovers	Non-repetitive transient in application	EOS, high-voltage transient
Bonding wire fused, die usually electrically OK	Forward-biased junction(s)	EOS, current pulse, reverse
Bonding wire fused, meltthrough hole in die, metal alloyed	Massive overstress	EOS, cause unknown
Tiny hole in oxide or silvery track visible in oxide between junctions	High-voltage, low energy	ESD event, test transient
Shorted junction or gate—no other physical damage is obvious	ESD event with damage located in silicon bulk	ESD event

Fig. 35 EOS Interpretation Guide

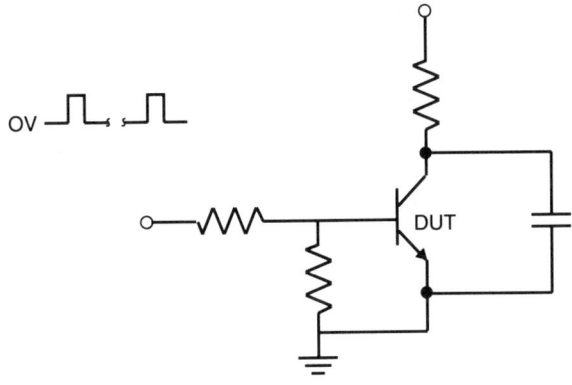

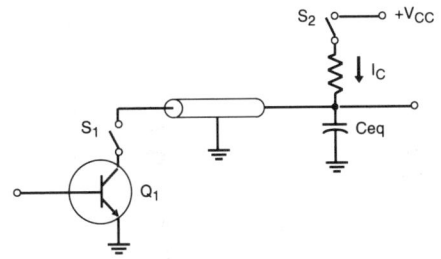

Fig. 36 Circuit showing capacitor discharge test.

Fig. 37 Circuit showing mechanism resulting in test failures at test.

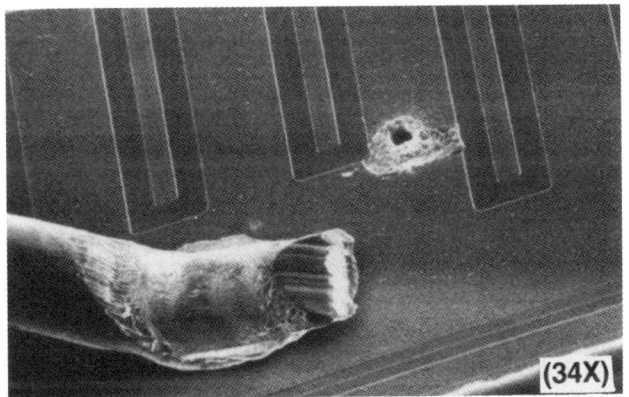

Fig. 38 Result of high-voltage transient on die. (34×)

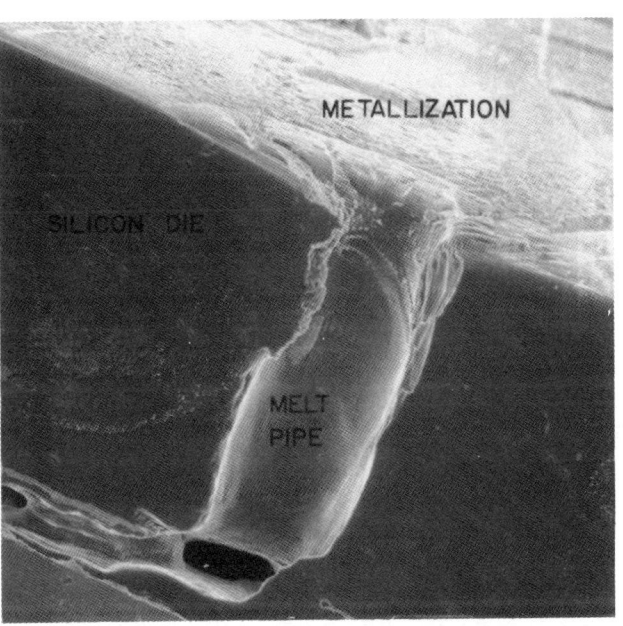

Fig. 39 Cross-section of a typical meltthrough in a die. (100×)

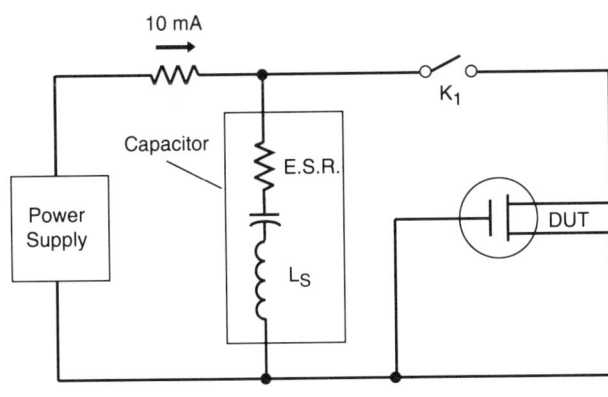

Note: K₁ Should be a
Large Mercury Contactor.

Fig. 40 Circuit for forward biased impulse testing.

Fig. 41 Typical forward bias impulse failure. (130×)

conditions and assure a reliable product application of the semiconductor power device.

INTERPRETATION OF CATASTROPIC OVERSTRESS PHYSICAL DAMAGE IN FA REPORTS

The deciphering of the conditions responsible for the failure of devices under these conditions is usually somewhat complicated. More than one of the forms of overstress presented in these discussions are usually present simultaneously. Careful

study of the circuit, device, and system in which failures occur can often produce better FA results than examining the device

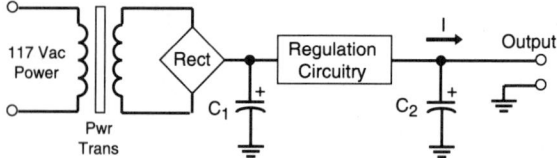

Fig. 42 Circuit arrangement which can result in impulse failures.

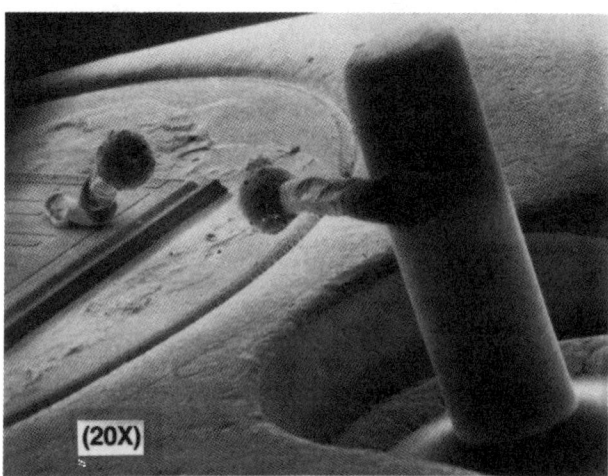

Fig. 43 Bonding wire melted due to forward bias impulse. (20×)

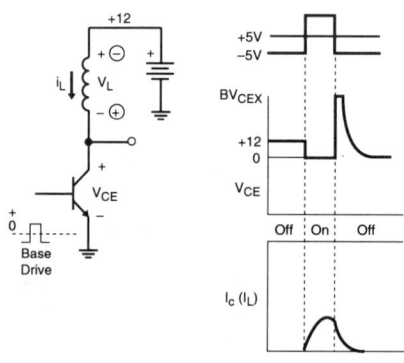

Fig. 44 Waveforms resulting from unsuppressed inductive tran-

Fig. 45 Damage to due resulting from condition in Fig. 44, when energy is below instantaneously destructive level. This damage was deliberately induced with a curve tracer. This is a simulation. Compare to Fig. 47.

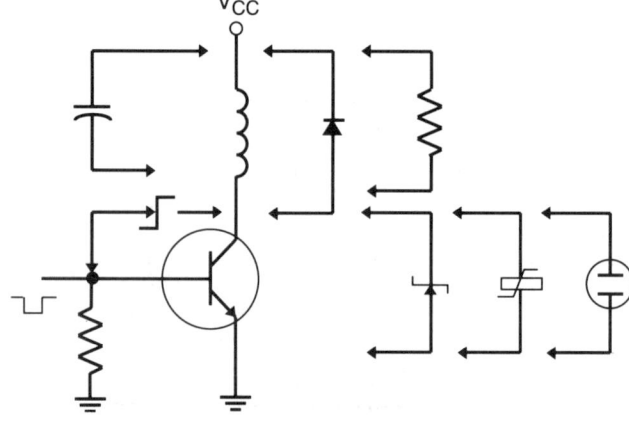

Fig. 46 Transient suppression possibilities for inductive switching.

Fig. 47 Actual field failure of a device from a drive for a 3-phase brushless motor in a loomfeeder. (20×) (compare to Fig. 45).

by itself. Simulations can also lead to data which can help pinpoint the problems.

The chart in Fig. 35 is intended to show the types of conditions under which some failures may be generated, and the following pages address some of the circuit and electrical conditions which can generate failures.

Figures 36 through 39 address the case where a capacitance is allowed to suddenly discharge into a device when the voltage is in excess of the C-E breakdown voltage, but the energy in the capacitor is small; of the order of 10s of millijoules. The result is usually a small meltthrough hole in the device, causing a short, with no other associated damage. If the external circuit impedance is low, and shunt current is available, the die can experience extensive damage (a crater), and the small meltthrough hole evidence may be lost.

Figures 40 through 43 show the result of the discharge of a capacitor in circuitry where the device junctions are instantaneously biased forward (in the case of the C-B junction), or reverse (in the case of the E-B junction) (reverse of normal). The result is usually a fused bond wire, and probing of the die probably shows it to be good to specifications. Devices plugged in backwards during test, and devices subjected to reversed battery or power supply faults, can exhibit this failure mechanism.

Figures 44 through 46 show a much more usual case, where the unsupressed (or unsnubbed) kickback from an inductive load is allowed to hammer the device into avalanche breakdown, but not sufficiently to cause second breakdown with its crowbarring of the power supply, and instantaneous catastrophic failure. Gradual degradation with **delayed** catastrophic failure will usually result. Examination of the unfailed, but aged, devices from the application will occasionally show the beginning signs of degradation in the metallization; **exactly** the same mechanism exercised in the tests producing the damage shown in Fig. 24. The "characteristic area" will vary in diameter with the pulse width, die design, and many other factors.

Figure 46 shows some of the many device protection schemes that may be used with a power switching device. The accidental omission of the protection device, oscillation in a snubber circuit, insufficient switching speed of the protection device, or a defective protection device can all result in failure in a protected application.

WAS IT CURRENT, OR VOLTAGE?

Probably one of the most often-asked questions following the pronouncement of EOS as the failure mechanism in a device, this question should be more easily answerable following the reading of this article. Figure 48 is an SOA chart with **typical**, not absolute, examples of transistor dice of various designs which have likely failed under the conditions indicated on the chart. This summary overview should help to decipher which of the device maximum ratings; current, voltage, power dissipation, temperature cycling, etc was likely exceeded when EOS failures cannot be explained by any device defect.

WAS IT EOS OR WAS IT ESD?

The most effective demarcation between EOS and ESD is the amount of energy which can be delivered to the device. The human body, and the human body model test, can only deliver microjoules to a few millijoules of energy. A power circuit, however, can deliver orders of magnitude more energy. Since it is the energy which causes the melting and vaporization of a volume of silicon to create a catastropic failure, ESD damage sites will be far smaller than EOS damage sites. Consequently, ESD damage must be found with a SEM, and EOS damage can be found with an optical microscope.

REFERENCES

1. *Protecting Computer Systems Against Power Transients,* by Francois Martzloff, IEEE Spectrum, April, 1990, pp 27-40.

2. *Surface Reconstruction of Aluminum Metallization - A New Potential Wearout Mechanism,* by E. Philofsky, K. Ravi, E. Hall, and J. Black, IEEE Proceedings of the Reliability Physics Symposium, 1971, pp 120-128.

3. *Thermal Fatigue in Power Transistors,* by G.A. Lang, B.J. Fehder, and W.D. Williams, IEEE Transactions on Electron Devices, Sept, 1970.

4. *Solder Fatigue Problems in Power Packages,* by James F. Burgess et.al, in the IEEE Transactions on Components, Hybrids, and Manufacturing Technology, Vol CHMT-7, No 7., December, 1984, pg 405.

5. *Properties of Die Bond Alloys Relating to Thermal Fatigue,* by Dennis R.Olsen and Howard M. Berg, IEEE Transactions on Components, Hybrids, and Manufacturing Technology, Vol CHMT-2, No.2, June 1979, pp 257-262.

6. *Quantitative Measurement of Thermal-Cycling Capability of Silicon Power Transistors,* by L. J. Gallace, Power Transistor Application Note AN-6163, RCA Solid State Division, Somerville, New Jersey.

7. *Protecting Computer Systems Against Power Transients,* by Francois Martzloff, IEEE Spectrum, April, 1990, pp 27-40.

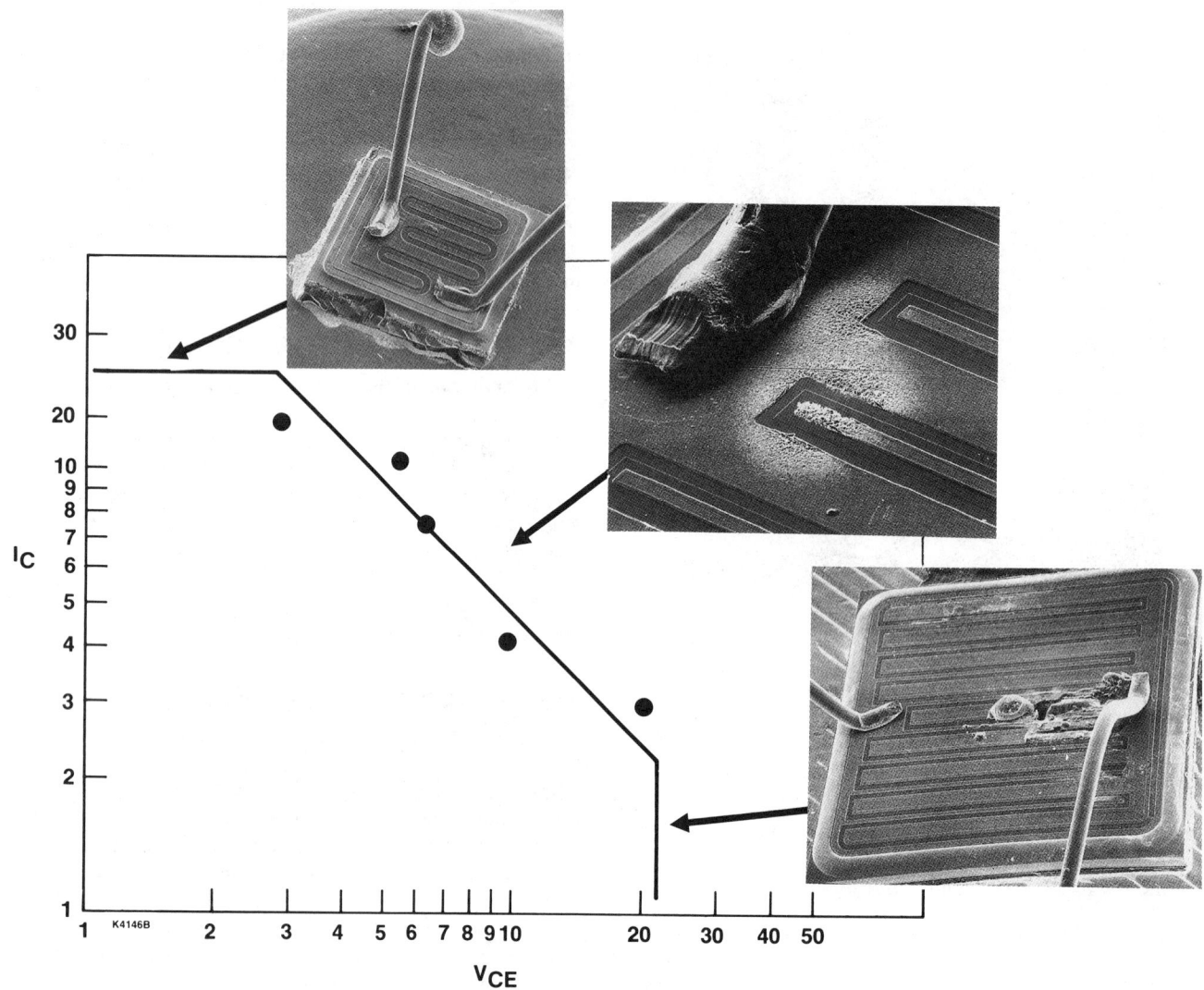

Fig. 48 SOA chart with typical power transistor damage. Appearance superimposed.

8. *Thermal Fatigue in Power Transistors*, by G.A. Lang, B.J. Fehder, and W.D. Williams, IEEE Transactions on Electron Devices, Sept, 1970.

9. *SEM Study of the Dynamics of Electrical Overstress (EOS) in Power Transistors*, by T.W. Lee, T.T. Guthrie, and G.P. Thome, Proceedings of ISTFA 1984, pp 221-226.

10. *Power Transistor Technology, and Safe Operating Area*, Chapters 1 and 2 in the Fairchild Power Data Book, 1976, pp 1-3 to 2-12. (Details the differences between double-diffused, single-diffused and epi-base transistor structures, design tradeoffs, and SOA in terms of I/sb and Es/b.)

11. *Limiting Phenomena in Power Transistors*, RCA Solid-State Devices Manual, (SC-16), 1975, pp 86-102.(Discusses base-width modulation, thermal debiasing, thermal cycle life, design tradeoffs, and variables included in the Es/b rating system.)

12. *Focus on Power Transistors and Thyristors,* by Louis Grossman, Electronic Design Magazine, #23, 8 November 1977, pp 52-60. (This article details the physical, electrical, and processing tradeoffs used by power device engineers in a non-technical manner and cau-

tions against the misinterpretation of published specifications, including Es/b.)

13. *Reverse-Bias Second Breakdown in Power Transistors,* by D. L. Blackburn and D. W. Berning, in Electrical Overstress/Electrostatic Discharge Symposium Proceedings, (EOS-I), RADC Reliability Analysis Center, 1979, pp 116-121. (Describes a test set which uses a high-speed clamp consisting of 16 sweep tubes connected in parallel to divert current from the device under test when it switches into second breakdown.)

14. *Transient Damage In Power Transistors* by T. W. Lee, Proceedings of ATFA-79, pp 130-135 (capacitor discharge simulation methods used to simulate field failure damage.)

15. *Transient Protection Devices*, by Chin-Lin Chen, in I.E.E.E. International Symposium on Electromagnetic Compatibility, 1975, pg 248.

16. *Electrical Overstress Failure Analysis in Microcircuits*, by J. S. Smith, International Reliability Physics Symposium, 16th Annual Proceedings, 1978, pgs 41-46.

17. *Handbook of Thermophysical Properties of Solid Materials*, Pergamon Press, 1961.

18. *Handbook of Material Science,* by Charles T. Lynch, CRC Press, 1974.

19. *Semiconductor Power Devices,* by Sorab K. Ghandi, Wiley, 1977. (Contains general discussions on the source, type, and effect of various defects on the charactristics of power devices in Chapter 6. This text is "required reading" for anyone associated with the manufacture of power devices.)

20. *Focus on Power Transistors and Thyristors,* by Morris Grossman, in Electronic Design Magazine, No. 23, Nov 8, 1977, pp 52-60.

21. *The Influence of Circuit and Device Parameters on the Switching Performance of Power Transistors,* by P.L. Hoover and K.S. Tarneja, Power Conversion International, pgs 10, 12, 14, 16, and 21.

22. *Basic Design Considerations for Power Transistors,* chapter 6 in the RCA Solid State Power Circuits Designer's Handbook, 1971, pp 80-112.

23. *Physical Basis for Power Transistor Ratings,* chapter 7 in the RCA Solid State Power Circuits Designer's Handbook, 1971, pp 113-149.

24. *Limiting Phenomena in Power Transistors and the Interpretation of EOS Damage,* by J. Thomas May (in this volume).

25. *The Application of Molybdenum Contacts for Improved Second Breakdown Performance,* by E.B. Hakim, Proceedings of IEEE, Vol 54, Jun 1966, pg 880.

26. *Second Breakdown in Power Transistors Due to Avalanche Injection,* by B.A. Beatty, Surinder Krishna, and Michael S. Adler, IEEE Transactions on Electron Devices, Vol ED-23, No 8, Aug 1976, pp 851-857.

27. *MIL-HDBK-217B,* U.S. Government Printing Office.

28. *On US Mil-Hdbk-217 and Reliability Prediction,* by Charles T. Leonard, IEEE Transactions on Reliability, Vol 37, No.5, December, 1988, pp 450-452.

29. *Reliability Prediction—Use It Wisely,* by Anthony J. Feduccia, IEEE Transactions on Reliability, Vol 37, No.5, December, 1988, pg 457.

30. *Determination of the Threshold Failure Levels of Semiconductor Diodes and Transistors Due to Pulse Voltages,* by D. C. Wunsch and R.R. Bell, in the I.E.E.E. Transactions on Nuclear Science, Vol. NS-l5, No. 6, pgs 244-247.

31. *Basic Integrated Circuit Engineering,* by Hamilton & Howard, McGraw-Hill, 1975, p. 245 and 246.

32. *Pulse Power Failure Modes in Semiconductors,* by D. M. Tasca, I.E.E.E. Transactions on Nuclear Science, Vol. NS-l7, No. 6, December, 1970.

33. *A Thermal Damage Model for Bipolar Semiconductors,* by G. A. Hjellen and T. J. Lange, I.E.E.E. Symposium on Electromagnetic Compatibility, 1977, p. 444.

34. *Second Breakdown in Power Transistors due to Avalanche Injection,* by Beatty, et.al., in the I.E.E.E. Transactions on Electron Devices, Vol. ED-23, No. 8, August, 1976, p. 852.

35. *Mesoplasma Breakdown in Silicon Junctions,* by A.C. English, Proceedings of IEEE, Mar 1963, pp 500, 501.

36. *Microplasmas in Silicon,* by D. J. Rose, 7. The Physical Review, Vol 105, No. 2, 15 Jan 1957, pp 413-147. (Reverse breakdown in silicon junctions was shown to consist of a microplasma of 5-600 Ångstroms dia, in which current was conducted in discrete pulses.)

37. *Thermal Instabilities and Hot Spots in Junction Transistors,* by R. M. Scarlett, W. Schockley, and R. H. Haitz, Physics of Failure in Electronics, Vol 1, 1963, pp 194-203. (A dated, but complete review of the physics of second breakdown. The experimental work used temperature-sensitive paints to find hot spot temperatures of 300 - 400 °C.)

38. *Avalanche Effects In Silicon P-N Junctions. I. Localized Photomultiplication Studies on Microplasmas,* by R. H. Ha I tz, A. Goetzberger, R. M. Scarlett, and W. Shockley, Journal of Applied Physics, Vo! 34, No 6, June 1963, pp 1581 - 1590. (Using a tiny light spot, carrier multiplication in microplasmas was shown to be as high as 106, microplasma diameter was shown to be a function of current, and microplasmas were found to have a dark core.)

39. *Second Breakdown and Crystallographic Defects in Transistors,* by H. A. Schafft, G. H. Schwuttke, and R. L. Ruggles, J. IEEE Transactions on Electron Devices, Vol ED-13, No 11, Nov 1966, pp 738-742. (The susceptibility to BVCEO second breakdown in 1500 transistors containing various defects was studied by means of x-ray diffraction microscopy) (The entire volume ED-13 is interesting reading.)

40. *Physical Investigation of the Mesoplasma in Silicon,* by A. C. English, IEEE Transactions on Electron Devices, Aug/Sep 1966, pp 662-667. (The properties of mesoplasmas were studied by observing melts in Zener diodes, and the mesoplasma properties are investigated with a computer program.)

41. *Second Breakdown—A Comprehensive Review,* by Harry A. Schafft, Proceedings of the IEEE, Vol 55, No 8, Aug 1967, pp 1272-1288. (A monumental summary of research on the subject from 1946 to 1967. Contains 134 references.)

42. *Stable Hot Spots and Second Breakdown In Power Transistors,* by P. L. Hower, D. L. Blackburn, F. F. Oettinger, and S. Rubin, PESC Record, 1967, pp 234-246. (Presents temperature contours of devices obtained with an infrared microadiometer.)

43. *Electrical Breakdown in Solids,* by N. Klein, a chapter in volume 26 of Advances in Electronics and Electron Physics, edited by L. Marton, Academic Press, 1968, pp 309-424. (A complete work covering the mechanisms of breakdown in semiconductors and insulators. Has 183 references.)

44. *Analysis of Requirements In Reliability Physics,* Alfred L. Tamburrino, Physics of Failure in Electronics (PFE) Vol 2, RADC, New York, 1964, pp 13, 16, 18.

45. *Silicon Semiconductor Technology,* by W.R. Runyan, McGraw-Hill, 1965. (This text book covers all phases of extraction of silicon and growth of crystals, doping diffusion, and optical, electrical and other physical properties.)

46. *High Current Transient Induced Shorts,* J. S. Smith, 9th Annual Proceedings Reliability Physics, 1971, pp 163-171.

47. *Basic Design Considerations for Power Transistors,* and *Physical Basis for Power Transistor Ratings,* chapters in RCA Power Circuits Designer's Handbook, Technical Series SP-52, 1971, pp 95-149. (Covers all concepts of power transistor design, multiple epitaxial (π-u) and multiple-emitter techniques. The second chapter discusses avalanche breakdown algebraically, RBSOA, FBSOA, and introduces forward bias capacitance-discharge as a useful test because it "approximates actual circuit conditions".)

48. *Non-Destructive Screening for Thermal Second Breakdown,* by Dante M. Tasca, et. al., IEEE Transactions on Nuclear Science, Vol NS-19, No 6, Dec, 1972, pp 57-67. (Among a large matrix of tests to screen switching diodes for second breakdown vulnerability, capacitors are placed in parallel with devices taken into second breakdown.)

49. *Reliability and Degradation,* edited by M. J. Howes and D. V. Morgan, Wiley, 1981. (An excellent study of degradation mechanisms in metal films; interfacial layers, interdiffusion, analytical methods, III-V devices, et cetera. Chapter 1, on metal films, has 457 references.)

50. *Safe Operating Area Information,* a section of the Motorola Power Device Handbook. Motorola Technical Information Center, 1982.

51. *Understanding Power Transistor Dynamic Behavior...and Power Transistor Safe Operating Area...,* by Warren Schultz, Motorola Application Notes AN-873 and -875 respectively, June and Sept 1982. (Discusses power transistor limitations in terms of dv/dt and Cob, concluding that snubbing networks are a cure, and shows that conditions resulting in power transistor failure may be subtle effects of the application conditions.)

52. *Atmospheric Electricity,* by B. F. J. Schonland, Meuthen & Co., London, 1953.

53. *Surge Voltages in Residential and Industrial Power Circuits,* by F. D. Martzloff and G. I. Hahn, in the I.E.E.E. Transactions on Power Apparatus & Systems, Vol. PAS-89 No. 6, July/August 1970, pg 65.

54. *Modeling of Electrical Overstress in Silicon Devices,* by N. Kusenov and J.S. Smith, in the Proceedings of the EOS/ESD Symposium, 1979, pp 133-139.

55. *Improved Cost Effectiveness and Product Reliability Through Solder Alloy Development,* by Dennis R. Olsen and Keith G. Spanger, Solid State Technology, Sept 1991, pp 121-126.

56. *RCA High Reliability Devices Databook,* SSD-207B, RCA Corporation, Sommerville, New Jersey, 1974.

57. *Thermal Cycling Rating System for Power Transistors, by Wally D. Williams,* RCA Application Note 4612, available from Harris Semiconductor Communications Department, Melbourne, Florida, 32902.

Advanced Techniques

Topics in Knowledge-Based
Failure Analysis

Christopher L. Henderson
Sandia National Laboratories
Albuquerque, New Mexico

ABSTRACT

Semiconductor failure analysis is a knowledge intensive discipline. Historically, failure analysts have relied on mentors (experts) to teach the discipline of failure analysis. Although printed material and training courses have helped to increase the availability of expert knowledge, there is still a considerable need for wider dissemination of failure analysis knowledge. This problem grows more critical as integrated circuit technology continues to advance. Knowledge-based systems can provide a unique ability to disseminate failure analysis information to the failure analyst.

A HISTORICAL PERSPECTIVE ON EXPERT SYSTEMS

Artificial Intelligence

Expert system research is a branch of the discipline of artificial intelligence. Artificial Intelligence in turn is a branch of computer science. Although you may not normally think that the study of artificial intelligence has produced much over its 30 years as an active field, it has produced several important products used extensively today. The windowing systems, made popular by Macintosh, then implemented by SUN, Microsoft, and the like, are the result of artificial intelligence work done by Alan Kay and his associates at Xerox, Palo Alto Research Center. Other important tools to emerge from the study of artificial intelligence are the LISP and PROLOG languages[1*], used not only for artificial intelligence applications, but also for drafting, graph search, and other engineering tasks.

Expert System Development

Expert systems began to appear in the 1960s, as researchers attempted to describe human reasoning in terms of computer program code. The earliest expert systems were developed in traditional programming languages such as FORTRAN. Traditional programming languages did not work well for expert systems, because expert systems are largely symbolic, that is word or character based. One of the earliest successful expert systems to be developed was MYCIN [1]. MYCIN was a system designed to diagnose and prescribe treatment for spinal meningitis and other bacterial infections of the blood. This system took some 20 man-years of effort to build. One of the earliest expert systems in the field of electrical engineering was a system called EL [2]. EL helped tutor students with the task of determining voltage levels and current values in an electrical circuit. One of the more famous expert systems in electrical engineering is XCON (also referred to as R1). XCON, developed jointly by Carnegie-Mellon University and Digital Equipment Corporation, configures VAX computer orders [3]. The majority of XCON was developed between 1980 and 1985. XCON is one of the most extensively used systems in industry today; consequently it has produced a wealth of information regarding the development, maintenance, and fielding of a large expert system.

XCON is a relatively large system, containing some 5000 rules, or pieces of information. It took approximately 10 man-years of effort to develop the system, and it takes approximately a man-year per year to maintain the software (add and test new rules as new computer hardware and software options become available for the VAX family of computers). XCON was written in a LISP like language called OPS. OPS (Original Production System language) is a high level language written in LISP to facilitate the development of expert systems. XCON did not put configuration order entry clerks

---

1.   *LISP is a computer language proposed by John McCarthy c. 1960 and developed by several universities and companies in the 1960's and 1970's. LISP is an acronym for LISt Processing. PROLOG is a computer language conceived by J. A. Robinson c. 1965. The majority of the development occurred between 1975 and 1979 at the University of Edinburgh. PROLOG is an acronym for PROgramming in LOGic.

out of business; rather it provided a means for the entry clerk to quickly check standard orders (XCON could configure a VAX computer in 90 seconds, whereas an order entry clerk took approximately 20 minutes). The clerk was then free to concentrate on more complex order configurations. XCON can successfully configure 95% of all VAX computer orders, although XCON has its drawbacks (large amount of maintenance and upkeep), overall, the system has been quite successful [4]. It stands as a target for which to aim in today's expert system development.

Knowledge-Based Systems for Failure Analysis

In the past several years, several companies have begun development of knowledge-based systems for failure analysis. The first such systems to appear were systems associated with the testing of ICs. [5,6]. Additionally, knowledge-based system development has occurred in conjunction with intelligent design validation tools. Companies such as IMAG/TIM3 Labs and Schlumberger Technologies have developed products that employ heuristics when diagnosing circuits [6,7,8]. A number of related knowledge-based systems have also been developed. These include reliability expert systems [9], optical defect inspection [10], and interactive fault correction systems [11]. More interesting to the overall subject of failure analysis, several systems have been developed to assist in the failure analysis of integrated circuits [12,13,14,15]. These four knowledge-based systems will be discussed in more detail later.

THE USES FOR A KNOWLEDGE-BASED FAILURE ANALYSIS SYSTEM

Capture Failure Analysis Expertise

The primary motivation in creating a knowledge-based system of any kind is to capture and retain expertise. Any manager who has had the misfortune of losing his or her most knowledgeable staff member knows the value of retaining expertise. Failure analysis is a discipline that has a long learning curve; it typically takes five years or more for an individual to become proficient in performing failure analysis on ICs. Much of this is due to the fact that the failure analyst must know something about a number of disciplines (chemistry, physics, electrical engineering, and mechanical engineering). To become proficient in device recognition alone can take many years. By and large, failure analysis has been a skill taught by companies, not by universities. These factors, coupled with a scarcity of failure analysts, makes the problem of retaining knowledge of failure analysis an important problem to address. If the expertise of the experienced analysts could be captured and retained effectively, the loss of an experienced analyst would be less catastrophic.

Train New Failure Analysts

In conjunction with retaining failure analysis expertise, there is also a problem associated with training new failure analysts. If the turnover rate is high in a failure analysis laboratory, then considerable time must be invested in training new failure analysts. This training can be costly in terms of courses and trainer time. Furthermore, the experienced analyst must watch the inexperienced analyst to ensure that he or she is doing the work correctly. This reduces the effectiveness of the more experienced analysts; they are not able to focus on the more difficult analyses. Either the backlog of failure analysis work grows, or the quality of the work suffers. If a knowledge-based failure analysis system could serve as a training tool, then the pressure on the experienced analysts could be lightened, freeing them to concentrate on the more difficult failure analysis work.

Centralize Failure Analysis Information

A third use of a knowledge-based failure analysis system is to centralize failure analysis information. Historically, this has been accomplished through the use of books. Books such as Rome Lab's *Failure Analysis Techniques* [16] and HiRel Lab's *Failure Analysis* [17] sought to centralize failure analysis knowledge and make it more accessible to the inexperienced analyst. A knowledge-based system can provide two main advantages that books on failure analysis cannot provide. First, the knowledge can be made available as part of each analysis if the system is interactive. Knowledge of a particular technique or procedure is useless if the analyst cannot recall or locate the information. Even experienced analysts cannot always remember particular techniques or recall where they read about them. A knowledge-based system can bring that information to the screen, making it available to the analyst. Second, the analyst may not know how to apply the information to the particular analysis being performed. A knowledge-based system can be constructed to describe the applicability of the technique to a particular analysis.

HOW FAILURE ANALYSIS KNOWLEDGE IS CODED

Embedded Heuristics

Most of the diagnostic software that is coupled with electron beam probing contains embedded heuristics. Embedded heuristics are "rules of thumb" that are coded into software to increase the knowledge of the diagnostic software. Most of the time these heuristics are coded in the same language as the diagnostic software, e.g. the heuristics employed in the Schlumberger tool IDA (written in C)

[7]. For example, in electron beam testing, it is preferable to examine top level metal whenever possible to measure waveforms. This heuristic can be coded into the diagnostic software, providing more accurate waveform measurement, hence better diagnosis.

Structured Query Language (SQL)

Structured Query Language (SQL) is a method developed for accessing data in a relational database. This method forms the backbone of most management information systems in use today. The SQL method is sufficiently powerful to provide "smart" access to a database. SGS-Thomson has developed a computer-aided reliability analysis system using this approach [9]. Texas Instruments successfully employs this technique to scan a large failure analysis database to retrieve similar failure analysis cases [14]. This allows the analyst to initially compare symptoms with already completed analyses to provide information on how to proceed with an analysis. An example from the Texas Instruments system is shown below that searches the database for all 1989 gold wire bond failures.

```
SELECT FAILMECH3, FAILMECH1, COUNT(*)
FROM JOB,UNIT,FAILMECH
WHERE JOB,JOBNUM=UNIT.JOBNUM
AND FAILMECH3=FAILCODE
AND STATUS='COMP'
AND DATECOMP BETWEEN '1989-01-01' AND '1989-12-31'
AND FAILMECH2 LIKE 'GOLD BOND FAIL.%'
GROUP BY FAILMECH3,FAILMECH
ORDER BY 3 DESC
```

Expert System Shells

In recent years, a number of expert system shells have become available on the market. The advantage of an expert system shell is that the developer does not need to know how to program in a low level artificial intelligence language such as LISP or PROLOG. A developer can quickly begin to write rules and not worry about writing an inference engine to process the rules. Today's expert system shells are powerful hybrid shells that allow a combination of objects and rules to create applications. The disadvantages of these shells are that flexibility and speed of operation are sacrificed for high level development. While speed is not a major concern for failure analysis applications, flexibility might be. A shell has to be carefully chosen to ensure that the desired control structure can be implemented. At least three companies have developed knowledge-based failure analysis systems using an expert system shell: Univ. of Arizona (CESM), Texas Instruments (Aion Development Software), and Sandia Labs (Level 5 Object).

Single Knowledge Base Several knowledge-based failure analysis systems use a single knowledge base. Because the amount of knowledge required to perform expert level failure analysis is very large, a single knowledge base system must have structure. The Texas Instruments system uses a single, highly structured, knowledge base (an Aion interface to a relational database). A highly structured knowledge base is necessary in order to provide a uniform method for retrieving information.

A second method for using a single knowledge base is to scope the problem narrowly. Several University of Arizona graduate students developed an expert system to perform automatic visual defect inspection on wafers [10]. The classification system uses 64 rules to distinguish six types of visual defects: voids, probe marks, scratches, bridging, cracks, and particles. This type of system has the advantage of incorporating highly specialized detail, but addresses only a portion of the subject of failure analysis.

Multiple Knowledge Bases/Blackboard Techniques
The University of Milan failure analysis assistant for linear integrated circuits [13] and the Sandia Labs Integrated Circuit Failure Analysis Expert System (ICFAX) [15] use a blackboard architecture with multiple knowledge bases. The blackboard architecture is a method for allowing independent knowledge bases to process information and post results asynchronously into a central global database known as a blackboard [18]. This allows failure analysis knowledge to be structured around different analytical techniques, providing modularity and structure. A blackboard system also reduces the "combinatorial explosion" problem associated with the number of possible paths that an analysis can take. The following code is an example of a rule used in a blackboard architecture:

```
RULE for e beam no netlist good die dfi fail scan die dbl step
IF (CCVC Results = "no anomalies" OR CCVC Results = "not
known")
AND (Light Emission Simple Setup Results = "no light emission
detected" OR Light
Emission Simple Setup Results = "not known")
AND e beam prober OF Voltage Contrast
AND ate connection to e beam prober OF Voltage Contrast
AND vector set available OF Voltage Contrast
AND NOT access to netlist OF Voltage Contrast
AND access to good die OF Voltage Contrast
AND NOT dfi located failure OF Voltage Contrast
AND scan die OF Voltage Contrast
AND type of anomaly found OF Voltage Contrast IS open at a
double step
THEN Dynamic Voltage Contrast Results := "open at a double
step"
AND CHAIN "icfax"
```

The "Dynamic Voltage Contrast Results" variable on the blackboard is modified as a result of rules firing in separate modules, in this case a module on voltage contrast testing.

Artificial Intelligence Languages

Artificial Intelligence languages have greater flexibility in terms of control and increased speed of execution when compared to expert system shells. Unfortunately, the development cycle is much longer because the inference engine (rule processor) must be written. Languages such as LISP and PROLOG are the most popular languages with which to develop. Several failure analysis systems have been developed using PROLOG. The first such system was an expert system to diagnose VLSI memories by IBM-France [5]. The developers choose PROLOG because of its flexibility and ability to interface to C language programs. A second system that uses PROLOG is PESTICIDE [6]. PESTICIDE is an expert system that interfaces with e-beam probing equipment to provide additional knowledge to aid in debugging complex VLSI circuits. The following piece of code is an example for several rules written in PROLOG:

```
rule((fails_continuity(continuity_test) :- shorted),80).
rule((fails_continuity(continuity_test) :- open),90).
rule((fails_functional(functional_test) :- fails_memory_test),90).
rule((fails_functional(functional_test) :- fails_logic_test),90).
rule((fails_timing(timing_test) :- fails_propagation_delay),70).
rule((fails_timing(timing_test) :- fails_max_frequency),80).
rule((fails_timing(timing_test) :- fails_rise_fall_time),90).
rule((fails_parametrics(parametric_test)                :-
fails_input_high_level_leakage),90).
rule((fails_parametrics(parametric_test)                :-
fails_input_low_level_leakage),90).
rule((fails_parametrics(parametric_test)                :-
fails_output_high_level_current),40).
rule((fails_parametrics(parametric_test)                :-
fails_output_low_level_current),40).
rule((fails_iddq(iddq_test) :- fails_iddq_single_vector),90).
rule((fails_iddq(iddq_test) :- fails_iddq_multiple_vectors),100).
```

Probably the most comprehensive knowledge-based failure analysis system developed from an artificial intelligence language is a system called IDA (Intelligent Diagnostic Assistant) [12]. IDA was developed by Cape Systems Inc. on a LISP-based machine (T. I. Explorer), to assist in the diagnosis of faulty hybrids. Although the system is quite flexible, it requires considerable rule entry from the purchaser.

Hypertext

Hypertext is a non-linear, computer-based method for viewing textual material. Markers in the material are coded in computer format that link two block of material together. For example, this could allow the end user to "click" on a word or phrase, and jump immediately to an expanded definition, or further description.

With the rise of the internet and CD-ROM applications, many knowledge providers are beginning to use hypertext as the preferred media for delivery of information. This is no exception in failure analysis as well. Already, many relevant conference proceedings for failure analysts are published on CD-ROM. These include: The International Symposium for Testing and Failure Analysis, The

International Reliability Physics Symposium, The Electrical Overstress/Electrostatic Discharge Symposium, The International Test Conference, and others.

One product that is currently available in CD-ROM format for failure analysts is FA Wizard from Knights Technology. This product is a comprehensive multimedia guide for failure analysis training and reference. The system uses the hypertext format, allowing users to quickly jump from topic to topic. The software makes extensive use of video to describe procedures and other "visual" information about failure analysis.

Probably the most important development in the use of hypertext is the World Wide Web. The World Wide Web is a collection of hypertext-linked documents that reside on numerous machines around the world. This document collection is large enough that subjects can be readily researched by the analyst from a networked computer. Search engines on the Internet such as Yahoo, Alta Vista, Lycos, and Excite, allow the analyst to search for particular subjects or terms. The main drawback of this type of research is that a single query can often yield tens of thousands of matches or "hits." The analyst must usually spend a good deal of time refining the search criteria in order to locate material of interest. Most material located on the World Wide Web is not of direct research or training use, but can be useful to the analyst.

Companies are beginning to use the World Wide Web format on their internal networks to store information related to failure analysis. Many large semiconductor manufacturers use the Web as a repository for basic training material, libraries of failure mechanisms, photographs of various defects, and completed failure analysis reports. This format can be an excellent way for a company to develop a corporate knowledge base in failure analysis.

One of the drawbacks of the internet is still the download speed of data. Even the fastest connections are not capable of downloading high quality video and photographs in a reasonable amount of time. This situation should improve quickly over the next few years as cable modems, digital subscriber lines (DSL) and satellite internet connections become prevalent and affordable.

A REVIEW OF KNOWLEDGE-BASED SYSTEMS FOR FAILURE ANALYSIS

The following tables describe some of the basic features of several knowledge-based systems for failure analysis. Although many systems have been developed as research projects, very few have made their way into commercial use. FDAL was used for a number of years within Texas Instruments. Of the systems listed below, only FA Wizard is currently available for commercial use. Figure 1 shows a screen view of the FA Wizard software..

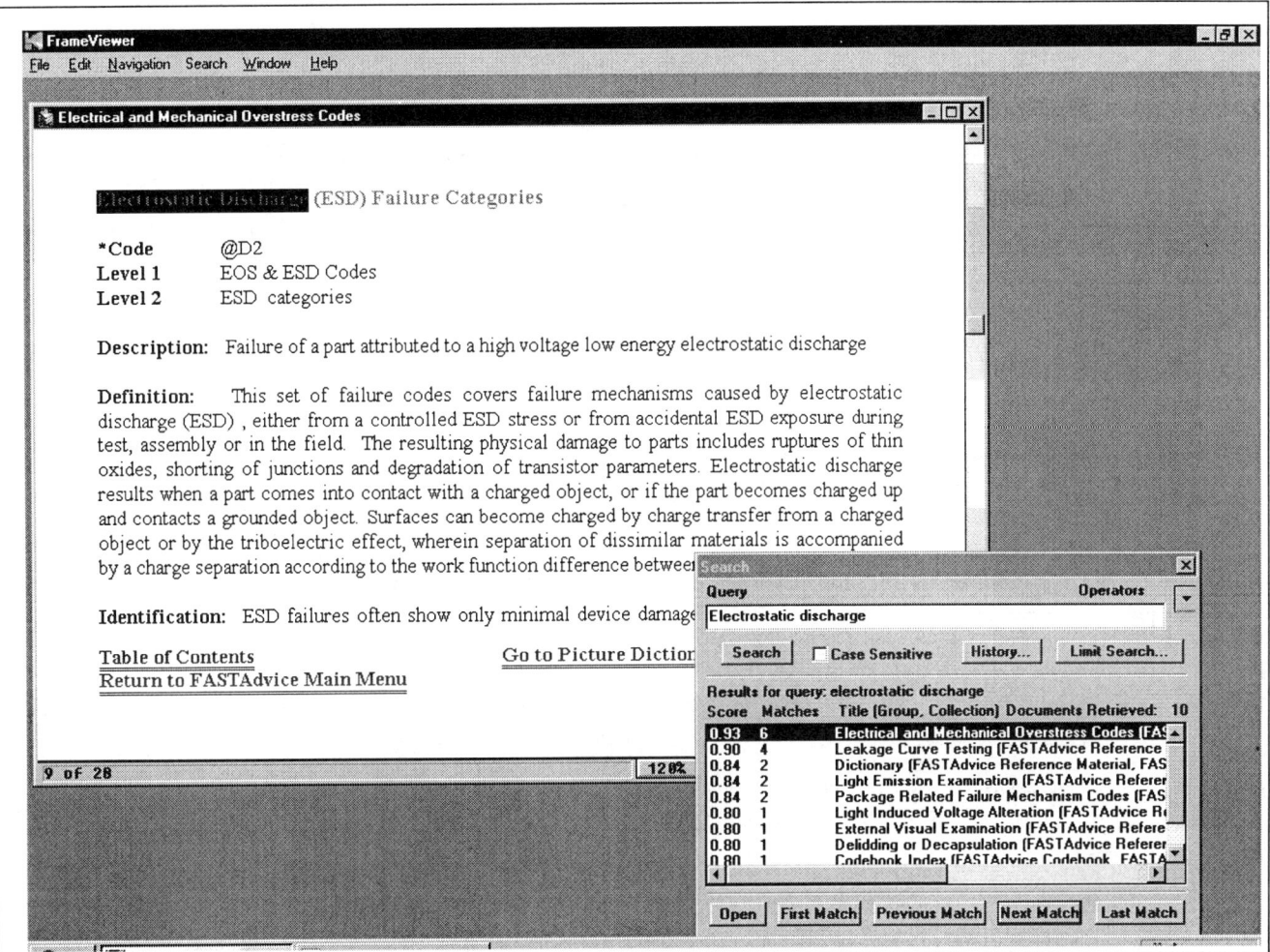

Figure 1: Screen shot of Knight's Technology's FA-Wizard software.

Figure 2: IBM France System

Developers	T. Viacroze, G. Fourquet, and M. Lequex
Computer Platform	IBM PC or compatible. Linked via RS232 to ATE testers
Software	Developed in PROLOG
Approx. Rule Count	120
Function	Software interfaces to ATE testers to analyze test data and diagnose the type of failure occurring on the memory device.

Figure 3: IMAG/TIM3 Labs' PESTICIDE

Developers	M. Marzouki, and B. Courtois
Computer Platform	SUN 3/160 workstation interfaced into e-beam equipment (2nd generation system developed on a BULL DPX-5000 workstation)
Software	Developed in PROLOG
Approx. Rule Count	N/A. No rules in the system. Knowledge maintained as specifications on device structure, behavior, and function.
Function	Software interfaces to e-beam probing system and CAD database to provide intelligent internal VLSI troubleshooting.

THE FUTURE OF INFORMATION SYSTEMS

Knowledge is Power

Frances Bacon coined the phrase *Nam et ipsa scientia potestas est,* which translated says "Knowledge is power." This

motivation, probably more than any other, will drive the future of knowledge-based systems. In the future, the companies that can acquire and utilize knowledge efficiently and correctly will be the companies that survive. This statement applies to the discipline of failure analysis as well. If we are to help our companies survive and

Figure 4: Allied Signal's FACES System

Developers	Tom Tetlow and Howard Dicken
Computer Platform	IBM Personal Computer
Software	Written in KAPPA PC
Approx. Rule Count	300
Function	Software guides the analyst through a safe sequence of activities for diagnosis of discrete transistor failures.

Figure 5: Texas Instruments' FDAL

Developers	IBM PC compatible computers linked to a main-frame computer
Computer Platform	IBM PC compatible computers linked to a main-frame computer
Software	Job submission program written in Turbo PAS-CAL, Database on mainframe in SQL format, AION Development System functions as the expert system software.
Approx. Rule Count	N/A. Access is via SQL commands. Database contains in excess of 10,000 records with 52 fields each relating device fabrication, assembly, etc.
Function	Software allows pareto analysis of database and suggests best course of analysis to pursue.

Figure 6: Sandia National Laboratories' ICFAX

Developers	Chris Henderson and Jerry Soden
Computer Platform	IBM PC compatible computers
Software	Level 5 Object Expert System Shell, Tanner Research (Layout Plots), CAD/CAM Group (Schematic Plots), and Microsoft Windows Help Compiler (Help Files).
Approx. Rule Count	8900
Function	Software provides comprehensive guidance through the failure analysis process for CMOS ICs (3 micron channel length, 1 level metal, 1 level polysilicon).

Figure 7: Knights Technology's FA Wizard

Developers	Chris Henderson
Computer Platform	IBM PC compatible computers
Software	FrameViewer (hypertext) and MPEG (for video).
Approx. Rule Count	N/A. Material is in hypertext format. Content is equivalent to approximately 2500 pages and 100 minutes of video
Function	Software provides computer-based training and knowledge preservation for failure analysis

and too many procedures to remember for the analyst to retain that information in his head. Knowledge-based systems for failure analysis provide a promising solution to this growing problem. With those thoughts in mind, the future is likely to bring several interesting developments in the area of knowledge-based systems.

More Comprehensive Systems

Most of the knowledge-based systems previously mentioned deal with specialized sets of problems within the discipline of failure analysis (ATE test, optical inspection, e-beam probing, etc.). While systems like these will continue to proliferate, scientists will develop more comprehensive systems dealing with the entire failure analysis process. There are still serious hurdles to overcome, however. The acquisition of failure analysis knowledge is a slow and tedious task. It may take many years to properly elucidate some aspects of failure analysis knowledge (turn "human" knowledge into computer code). The validation of complex, knowledge-based systems is extremely difficult. Once the knowledge is in computer code format, the code must operate correctly. The verification and testing of this code is not a trivial task.

Automatic Report Generation

One area where knowledge-based systems can aid the failure analyst is the area of report generation. Once the analyst has finished his or her work, he or she faces the mundane task of generating the failure analysis report. If a comprehensive knowledge-based system is used during the analysis, the significant steps and procedures are contained within the log of the session. This information can be used to automatically generate a failure analysis report. Such a report has the potential of being more accurate than the report generated from the analyst's memory.

Internet Training Environments

The internet will play an increasing role in the distribution

even prosper, we must be able to acquire and utilize failure analysis knowledge efficiently and correctly. The world of semiconductor technology is becoming too complex to ignore the issue of knowledge retention. There are simply too many technologies to analyze

and dissemination of knowledge. The internet model allows information to be updated rapidly at one location and then made immediately available to a large audience. In the near future, bandwidth on the internet will permit users to download video, high resolution photographs, and other forms of information with very little delay. This will permit training to occur quickly by computer. Most complex failure analysis equipment are now computer controlled. This will allow the user to be trained as they sit at the instrument. Large amounts of basic material on techniques will be downloaded from commercial sites. Company specific process information and data will be accessed from company servers. This information rich environment will allow failure analysis to proceed at an even quicker pace, allowing faster yield learning, improved customer response times, and more accurate analyses.

CONCLUSION

The history of artificial intelligence and expert system development sets the stage for the next generation of knowledge-based systems. The motivation and rationale for creating failure analysis knowledge-based systems is substantial. Several companies and universities have already demonstrated knowledge-based systems that are directly applicable to failure analysis. Although these systems are still quite simple, the stage is set for future development. As failure analysis knowledge increases, failure analysts must find new ways of capturing that knowledge and preserving it. Knowledge-based systems may provide the key to do so.

REFERENCES

1. B. G. Buchanan and E. H. Shortliffe, eds., *Rule-Based Expert Systems: The MYCIN Experiments of the Stanford Heuristic Programming Project,* Addison Wesley, Reading, MA., 1984.

2. G. J. Sussman, "Electrical Design: A Problem for Artificial Intelligence Research," Proc. of the 5th. Int. Joint Conf. on Artificial Intelligence, pp 894-900, 1977.

3. J. McDermott, *R1: A Rule Based Configurer of Computer Systems,* Technical Report, CarnegieMellon University, Dept. of Computer Science, 1980.

4. J. Bachant and J. McDermott, "Rl Revisited: Four Years in the Trenches," *The AI Magazine,* Fall 1984, pp.21-32.

5. T. Viacroze, G. Fourquet, and M. Lequex, "An E~pert System for Help to Fault Diagnosis on VLSI Memories," *Int. Symp. for Testing and Failure Analysis,* Nov. 1988, pp. 153-160. (IBM France, ATE test diagnosis)

6. M. Marzouki and B. Courtois, "Debugging Integrated Circuits: A.I. Can Help," *Proc. 1st European Test Conference,* pp. 184-191, April 1989. (IMAG~M3 Lab, E-beam coupled with A.I. to diagnose IC failures called PESTICIDE)

7. A. C. Noble, "A Diagnostic Assistant for Integrated Circuit Diagnosis," *Proc. 3rd European Conference on Electron and Optical Beam Testing,* pp. 78-85, Sept. 1991. (Schlumberger, Integrated Diagnostic Assistant)

8. M. Marzouki, and F. L. Vargas, "Using a Knowledge-Based System for Automatic Debugging: Case Study and Performance Analysis," *Proc. 3rd European Conference on Electron and Optical Beam Testing,* pp. 110-117, Sept. 1991. (IMAG/l~M3 Labs, reference to PESTICIDE)

9. P. Mauri, "Computer-Aided Analysis of Integrated Circuit Reliability," *Proc. NATO Advanced Research Workshop--Semiconductor Device Reliability,* June 1989, pp. 127-136 (SGS-Thomson, Integrated Circuit Reliability Expert System)

10. S. D. Chi, B. P. Zeigler, and T. G. Kim, "Using t'ne CESM Shell to Classify Wafer Defects from Visual Data," *Proc. S.P.I.E.--The International Society for Optical Engineering,* Nov. 1989, pp.66-77. (U. of Arizona, Optical defect inspection)

11. J. Krol, "ClRCOR--An Expert System for Fault Correction of Digital NMOS Circuits," *Proc. European Conference on Circuit Theory and Design,* Sept. 1989, pp. 674-676. (University of Gdansk Poland, design validation expert system called CIRCOR)

12. M. Kagan, J. Kudish, and A. Zelzion, "A New Method to Diagnose Chip-and-Wire Hybrids," *Hybrid Circuit Technology,* Vol. 6, No. 2, pp. 15-20, Feb. 1989. (Cape Systems Inc.,. Hybrid diagnostic expert system)

13. C. L. Henderson, "Computer-Based Training for Failure Analysis," Microelectronics and Reliability, Vol. 37, No. 10/11, pp. 1445-1448. (Knights Technology, FA Wizard)

14. G. Boella, P. Mussio, P. Mauri, and M. Picolli, "Design of an Automatic Assistant for Failure Analysis of Linear Integrated Circuits," *Proc. of the Tenth International Workshop--Expert Systems and TheirApplications,* pp.137-151, May 1990 (SGS Thomson/University of Milan, Linear integrated circuit failure analysis expert system)

15. L. M. Bellay, P. B. Ghate, and L. C. Wagner, "Computers in Failure Analysis," *Proc. Int. Symp. for Testing and Failure Analysis,* Nov. 1990, pp. 89-95. (Texas Instruments, Intelligent database system called FDAL)

16. C. L. Henderson and J. M. Soden, "ICFAX, An Integrated Circuit Failure Analysis Expert System," *Proc. 29th International Reliability Physics Symposium,* pp. 142-151, Apr. 1991. (Sandia National Laboratories, Integrated circuit failure analysis expert system)

17. E. Doyle Jr. and B. Morris eds., *Microelectronics Failure Analysis Techniques--A Procedural Guide,* IIT Research Institute, 1980.

18. J. R. Devaney, G. L. Hill, and R. G. Seippel, *Failure Analysis*

Mechanisms, Techniques, and Photo Atlas, Failure Recognition and Trauning Services, Inc., Monrovia, CA., 1986.

19 G. F. Luger and W. A. Stubblefield, *Artificial Intelligence and the Design of Expert Systems,* Benjamin/Cummings Publ. Co. Inc., pp. 561-563.

CAD Navigation Basics

Christopher L. Henderson
Sandia National Laboratories
Albuquerque, New Mexico

INTRODUCTION

Because of increases in integrated circuit (IC) complexity, it has become necessary to utilize computer-aided design drawings of integrated circuit layout features rather than hard-copy plots. Unlike printed material, the use of CAD Navigation for failure analysis (FA) requires an investment in computer resources and the development of a design process that facilitates the creation of the appropriate design files. The reasons for using CAD Navigation software are explained below. Next, an overview of the process used to create the design files is described. Finally, some of the uses for CAD navigation are explained.

REASONS FOR CAD NAVIGATION SOFTWARE

During the 1970's and early 1980's, IC designs were typically simple enough to be managed in paper format once they were generated. The features of interest could be seen on a layout plot generated on E4 (36 × 49 inch) paper. The schematics could typically be maintained in a document that was on the order of 50 pages or less. Hardware description languages (HDL) were still in their infancy and typically not required. Most importantly, the IC designers were familiar with the designs they created. Although the computer provided support for layout and routing of interconnect, the designer understood his or her design well enough to provide assistance to the FA engineer during debugging and failure analysis.

As designs became more complex, new strategies were needed to cope with the information required to design and lay out an IC. This resulted in several developments. One, IC designs became to complex to plot on paper. Designs of the complexity of the Intel 486 microprocessor (approximately 1.2 million transistors) took too long to plot. A plot of the 486 microprocessor showing enough detail for the FA engineer would take on the order of one month to plot and occupy approximately 200 square feet when assembled. And two, computer technology removed the design engineer from the details of the process. Design synthesis tools and silicon compilers replaced the basic layout and schematic capture tools. Hardware description languages such as VHDL, became common. Register-Transfer Languages (RTL) also became widespread. The computer-aided design tools were developed such that they could take RTL or VHDL code from the designer, convert it into a schematic, and realize the schematic in silicon. Design engineers simply specified various functions and behaviors in VHDL code, and the design software algorithms did the rest. This resulted in the design engineer not knowing the detailed design of the IC. He or she could not correlate the design information to physical features implemented in the silicon. This "lack of knowledge" on the part of the design engineers has required the development of new tools for the failure analyst.

CAD Navigation software provides the link necessary for the FA engineer to understand the correlation of layout features with design elements (see Fig. 1). The software takes design information and layout information and correlates it, linking the two pieces of information together on the computer. This enables the FA engineer to locate a design feature of interest and then locate the physical circuitry associated with that feature, or vice-versa. CAD Navigation software also provides the ability to zoom in on regions, pan through regions, show and hide various layers, and highlight interconnect (or nets) on the circuit.

SETUP PROCESS FOR CAD NAVIGATION

The setup process for CAD Navigation is not difficult; however, some planning is required to take full advantage of the software. The two major CAD Navigation software packages on the market for FA are from Knights Technology and Schlumberger Technologies. Both tools utilize a layout versus schematic (LVS) checker by Cadence called Dracula or Avanti's Hercules package. This software provides the ability to cross-link netlist and schematic information with layout information. The main files required to create the linked database of information are the layout file, the netlist file, the schematic file, and the technology file.

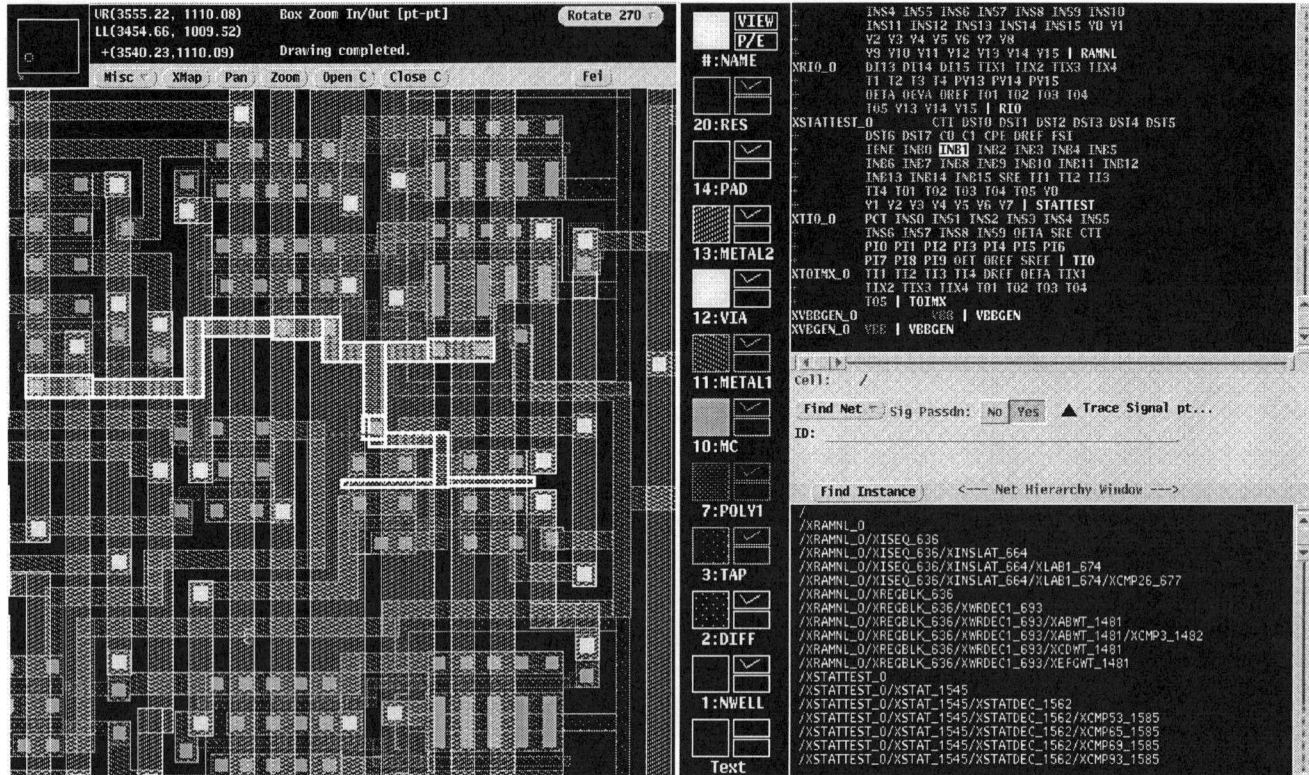

Figure 1 CAD Navigation Layout and Netlist Views. Note that highlighted node in netlist is also highlighted in layout. (Courtesy Knights Technology)

Layout File. This file contains the geometric representations or polygons for each mask level associated with the IC. This file is usually in a format called GDSII (pronounced "G-D-S-2", developed by Calma) or CIF (Caltech Intermediate Format). The GDSII format tends to be more widely used because it is more compact and can be compressed.

Netlist File. This file contains the connectivity information between transistors. There are several formats used for this file: SPICE, EDIF, ISS, TEGAS, SILOS, SDL, and LOGIS. The SPICE format is the most common format in use today. EDIF, ISS, and custom formats are also widely used.

Schematic File. This file contains the graphical representation of the logical elements of the design. This would include the shapes for gates such as AND, OR, INVERTOR, and larger functional blocks. The main format used in the commercial tools is EDIF 2.00. EDIF stands for Electronic Data Interchange Format and is typically annotated to include symbol data.

Technology File. This file is a flat text file that contains basic information used to define the design. It includes such terms as:

- Layer assignments
- Grid Spacing
- Top Cell Name

- Input File Names
- Definition Statements (to define transistors, vias, etc.)
- Output File Names

This file serves as a template for Dracula to process the design information.

One important aspect about this process is the need for the correct files and file formats. Many of the current design tool suites do not automatically archive the appropriate design files. The output of the design process from the design department's point of view is the files necessary to generate the mask set. Other files, such as the schematic and netlist files, are typically not needed after the design phase. It is crucial that these files be saved for CAD processing. It is also beneficial to save these files with their hierarchical formatting. This allows the FA engineer to "drill down" through the hierarchy, making the design much easier to understand and debug or troubleshoot. This underscores the need for system administration processes after design completion and before FA activities start.

Some design processes do not have the means to generate the appropriate file formats necessary for generating CAD Navigation files. It is important to communicate with your design departments about the need to generate netlist and schematic files for CAD Navigation. The generated files need to be compatible with the DRACULA LVS engine embedded in the CAD Navigation tools. A little bit of work

472

up front can save a lot of extra work and heartache in the long run.

The basic process for setting up CAD Navigation databases is shown in Fig. 2.

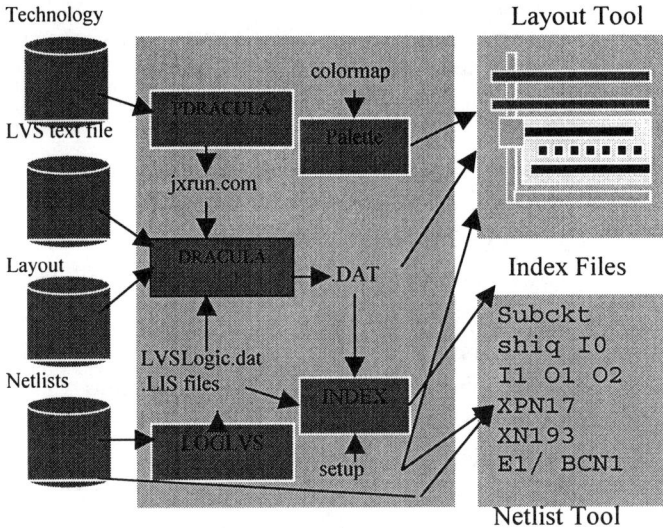

Figure 2 Process for creating CAD Navigation databases.

The four input files are first checked (to ensure the process will work correctly). This produces a file called jxrun.com; it is essentially a file in a format that the DRACULA LVS checker can utilize. The four file types are then run through the DRACULA processor. The output from the DRACULA processor is a series of data files that can be parsed and viewed in the layout viewer and netlist viewer. Because the information has to be run though the DRACULA LVS checker, a net (single interconnect) in the netlist window can be highlighted, causing the physical interconnect in the layout window to be highlighted (refer to Fig. 1).

Other Issues

There are a number of problems that can surface while trying to process a database for viewing[2]. These include:
- Incorrect top cell name
- Missing layout file
- Corrupted layout file
- Missing subcircuit
- Unrecognized keyword
- Duplicate subcircuit definition
- Incomplete derived layer definition
- Other missing files
- Problems with "connect" statement or "connect layer" sequence

In most cases, an error message will be generated that helps the user to understand what is occurring. A common source of errors is missing files, or missing declarations in the technology file. Be sure that all the necessary files are

present and that the appropriate declarations have been made. Another point that is commonly overlooked is that the top level cell name in the layout file can be a completely different name than the top level cell in the netlist file. It is important to note that some error messages do not stop the processing altogether. Many times, DRACULA will allow processing to continue, but the results may be different than anticipated (e.g. missing layers, layout not linked to netlist, etc.)

CAD processing can require up to twenty times the disk space the design files occupy[3]. This can result in overflow conditions on the disk drives in use, swap space defined, or even in the operating system itself. For large designs, 32-bit UNIX operating systems may not be able to handle the files because of the 2-Gbyte limit in file size. Some older computer platforms also have limited swap space sizes. There are several potential solutions to this class of problems. The first is to check the version of DRACULA being used. Some older versions of DRACULA will cause overflow problems when processing very large, flat memory arrays. The solution is to license DRACULA version 3 or newer, so that hierarchical processing can occur. This reduces the need for large amounts of swap space. If this doesn't work, it may be necessary to upgrade to a current computer architecture. Another solution may be to utilize a striped (concatenated) disk array such as SUN Microsystems' Online Disk Suite. Finally, it may be necessary to upgrade to a 64-bit version of UNIX. This removes many of the software limitations that occur because of 32-bit addressing space.

USES FOR CAD NAVIGATION SOFTWARE

Electron Beam Probing. CAD Navigation software was first developed for the application of fault localization with a scanning electron microscope. Richardson and Concina developed the concept of a fault isolation tool using voltage contrast effect produced in the scanning electron microscope to measure waveforms non-invasively on an IC[1]. They are also responsible for the initial development of CAD navigation software for design debug and failure analysis (see Fig. 3).

The tool used a workstation interface that combined the SEM image with CAD Navigation software. The two images could be locked together to allow simultaneous examination of the circuit under test with the design database. This tool became a necessity for design debug and failure analysis of complex ICs. The electron beam prober, as it is known, is still in wide use today.

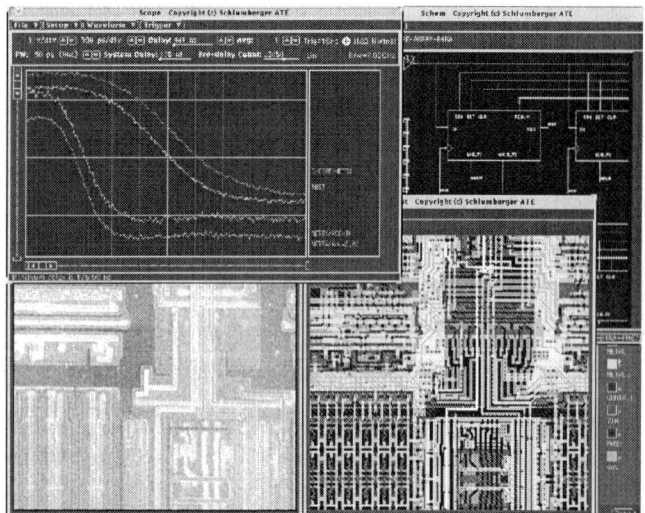

Figure 3. Screen shot showing the user interface on an electron beam prober. The four windows are (clockwise from upper left) acquired signal from a node, schematic, layout, and SEM image (Courtesy Schlumberger ATE).

Focused Ion Beam Activities. CAD Navigation software has also become an important tool for focused ion beam work. Because most IC technologies today utilize Chemical-Mechanical Polishing (CMP) technology, the surface of the IC exhibits very little topography. Even if the top level of metal is visible, underlying levels of metal are completely obscured. To allow modification of circuits, engineers developed the ability to overlay the CAD database with the focused ion beam image (see Fig. 4).

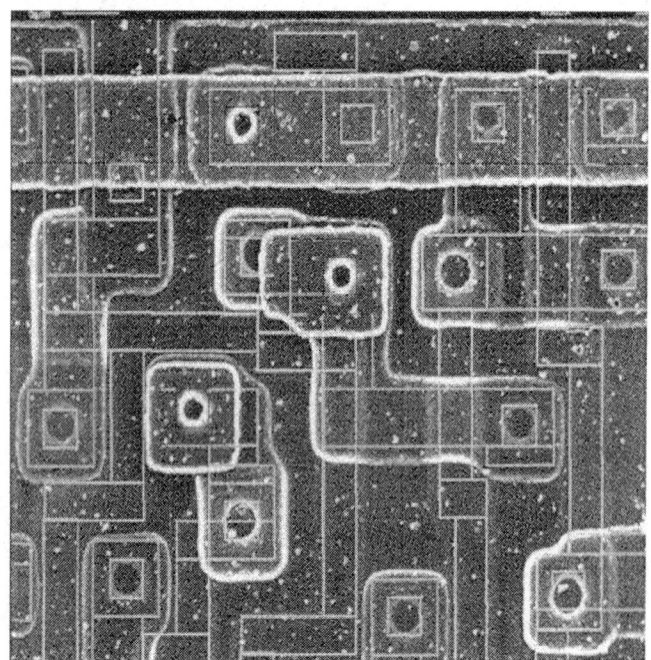

Figure 4. CAD overlay on a Focused Ion Beam image (Courtesy FEI Company).

This allowed more exact placement of the beam with respect to the feature of interest on the IC. On complex circuits with 5 or 6 levels of interconnect, CAD overlay is crucial for successful circuit editing.

Other Tools. To date, CAD Navigation has been incorporated into many failure analysis tools. A list of some of these tools is shown below:

- Analytical Probe Stations
- Atomic Force Microscopes (AFM)
- Electronic Design Automation (EDA) Verification & Test Software
- Emission Microscopes
- Energy Dispersive (EDS) X-Ray Microanalysis Systems
- In-Line Inspection Data Links
- Laser Cutters
- Laser Scanning Microscopes (LSM), IR Imaging
- Mechanical Stage Controllers
- Microchemical Laser Workstations
- Optical Beam Induced Current (OBIC) Instruments
- Optical Microscope Review Stations
- Scanning Electron Microscopes (SEM)

One advantage to having multiple tools with CAD Navigation software is that features of interest can be quickly located on the different tools. Coordinates of the defect or feature of interest can be transferred from one tool to the other with relative ease.

CONCLUSIONS

CAD Navigation software is becoming increasingly necessary to perform design debug and FA. Although CAD Navigation software is relatively easy to use, in practice it can be difficult to set up a procedure to generate the CAD Navigation databases. The biggest issues associated with implementing CAD Navigation software in an FA Laboratory are the design tools and design process flow. It is necessary to have the appropriate files and file formats for input into the CAD Navigation setup procedure. Because the setup procedure utilizes the LVS checker incorporated into DRACULA, the files must be compatible with this tool. The addition of netlists and/or schematics may require a different design process or modified procedures. The most important point is to work with the design department to develop an acceptable procedure so that the appropriate files are archived (in the appropriate format). Once the initial process is developed, subsequent designs can be set up for CAD Navigation in a straightforward manner.

ACKNOWLEDGEMENTS

The author would like to thank Daniel Barton, Richard Anderson, and Charles Hembree for their review and comments. Sandia National Laboratories is a multiprogram laboratory operated by Sandia Corporation, a Lockheed Martin Company, for the United States Department of Energy under Contract DE-AC04-94-AL85000.

REFERENCES

1. S. Concina and N. Richardson, Workstation Driven E-beam Prober, *Proc. Int. Test Conf.,* September 1987, pp 554-560.
2. Schlumberger Technologies IDS Application Note 103, September 1992
3. Schlumberger Technologies IDS Application Note 116, May 1995.
4. http://news.semiconductoronline.com/info-tech/ 19990126-2621.html

Signature Analysis, an Analytical Technique

Mike Pore
Advanced Micro Devices

INTRODUCTION

A failure analysis lab receives devices that have failed in some way. The lab then assesses the device's failure characteristics (the test data, morphology data, environmental data, etc.), and assigns the device to a failure category. We will call these failure categories "failure modes." While a failure mode categorization and other observable characteristics are usually referred to as a "signature," we will simply use "failure mode" in that more general sense here.

The failure analysis is (ideally) capable of analyzing each failed device (at great cost) to determine the mechanism causing (or more loosely, associated with) the failure.

The failure lab is unable, or unwilling, to analyze all of the devices that it receives, and wishes to implement a procedure of attributing a failure mechanism to some of the devices using only the failure mode categorization. This procedure for a given, failed device that has been categorized into a particular failure mode, will be:

(1) to observe the analysis history for devices from this failure mode,

(2) assess the risk of inferring the failure mechanism from this history, and

(3) if the risk is acceptable, make the inference; if not, analyze the device, thus adding to the history for this failure mode.

This procedure does not consider the similarities and differences among the failure modes (or categories of failure characteristics). Once a failure mode is determined, the analysis is for that mode only, and is not influenced by how similar modes are behaving. Nor are we assuming any relationship among the devices. That is, each device is an independent member of the failure mode category: the probability of the failure being from a particular mechanism is not changed by the mechanism attribution of the last (or the next) device categorized into that mode. Each failure mode is (treated as) unrelated to the others, and the devices in a failure mode are (treated as) identical with respect to the failure mode characteristics, but unrelated by other factors (such as proximity to other devices, order of discovery, etc.).

This paper will present how the history of devices can (and should) be characterized for (1) above, and how to characterize and assess the risk in (2). This is signature analysis.

The effect of a nonspecific failure mode definition

The concept of a failure mode is very loosely defined as a set of failure characteristics. The application of this concept to a set of failed devices (both in hand and those to come from a particular production line) can be quit subjective. It is a matter of judgement determining what failure characteristics to use in defining failure modes.

The attributes of using many failure characteristics to define a large number of failure modes are:

- the modes will be maximally pure with respect to failure mechanisms, thus inference will be easier with fewer analyses,

- there will be fewer devices in each mode, some modes with too few devices to build a history adequate enough for inference, and

- when all modes are pure, more analyses will be required.

The attributes of using only a few failure characteristics to define fewer failure modes are:

- the modes will be large (on average) in size yielding more opportunities to infer mechanisms (within the mode),

- the modes will be more mixed with respect to failure mechanisms yielding fewer opportunities for inference, or more analyses will be required to make riskier inferences, and

- when all modes are pure, fewer analyses will be required.

Both large, general failure modes and small, specific failure modes have their respective good points and weak points: we are torn in both directions. The most applicable solution is a compromise that balances the attributes of each approach for the particular product that this is being applied to. Two relatively pure failure modes that are suspected of being driven by the same failure mechanism might be combined into one larger mode. And a large, relatively mixed failure mode might be partitioned (using a failure characteristic not used hitherto) to form two relatively pure modes.

A general procedure

This technique for calculating the risk of inference, the signature analysis, is generic to any classification application where we are wishing to generalize about a category by "analyzing" (or classifying by a more costly procedure) units of any kind. This presentation is simply an application to the signature analysis problem.

Two Process Models

In some production processes, we have defined failure modes that will grow in size as more failures are discovered. And furthermore, we wish to draw conclusions about this failure mode that is (as of yet) of unknown size. This describes an "ongoing process," and will use a different (mathematical) model from the "finite population" model.

The finite population model is used when all the devices that will fail are observed and categorized, and the size of the failure mode under consideration is known. It must be emphasized that inferences in the finite population model are just for the unanalyzed devices that are currently categorized in the failure mode, and not for any future fails that might appear. For future fails, use the ongoing process model.

THE ONGOING PROCESS MODEL

The ongoing process model is the most general and widely used model, since the conclusions can readily be applied to future failures.

To characterize this model: we have N failures that have been categorized into the failure mode of interest. We have performed n failure analyses: analyses that determine the failure mechanism of the failure (the size of the failure mode, N, is irrelevant to the ongoing process model, but will be important in the finite population model; however, n is the important statistic here). The problem is, how do we use the distribution of the n analyzed devices to determine the risk of predicting that a future device (in this failure mode) will be associated with a particular failure mechanism?

More particularly, we will take the extreme case: suppose the failure analysis has determined that all n devices have the same failure mechanism. What is the risk in attributing future devices in this failure mode to the same failure mechanism without actually analyzing them?

In this section, we define terms and probability distributions, then in succeeding sections we will define our risk in terms of probabilities of correct classifications, determine how to solve for the number of required analyses, and finally we will present the statistical theory.

Definitions of Terms

For a particular failure mode (this analysis is repeated for each failure mode):

"n" denotes the number of devices analyzed with a particular failure mode; or the number proposed hypothetically in an inference calculation (as when solving for a required number).

"x" denotes the number of analyses (among the n) resulting in the primary failure mechanism: also denoted x_A (to associate x with failure mechanism A).

"θ" is the true, long-term proportion of devices in the failure mode under consideration that would result with the primary failure mechanism. It is always the case that $0 \leq \theta \leq 1$, and it is sometimes denoted θ_A (since it is the proportion for the particular failure mechanism A).

It follows from our assumption that there is no dependence (that we know of) between the devices: dependence such as devices next to each other in manufacturing or additional information, besides membership in the same failure mode, about the possible failure mechanism. Therefore x follows the binomial probability distribution (see [1] for an elaboration). This binomial distribution is denoted as

$$f(x|\theta) = \binom{n}{x} \theta^x (1-\theta)^{n-x} \quad \sim \quad B(n,\theta)$$

The binomial is actually a "family" of distributions that are indexed by n and θ: that is, there is a different binomial distribution for each (n, θ) pair.

Since θ, the proportion with a particular failure mechanism, is unknown to us, we will want to make probability statements about θ. We will model the unknown θ with the beta distribution, and denote it as

$$g(\theta) = \frac{\Gamma(\alpha + \beta)}{\Gamma(\alpha)\Gamma(\beta)} \theta^{\alpha-1} (1-\theta)^{\beta-1} \quad \sim \quad beta(\alpha, \beta)$$

α and β are positive parameters (constants that are > 0) that determine which distributions within the beta family that we are using. Γ is the gamma function: see any statistics text, such as [1] or [2] for a more precise definition.

Note: Notice that the binomial and beta families of distributions have the same form:

$$K_1 \theta^{K_2} (1-\theta)^{K_3}$$

for different "K_i" values in the two distribution families. To keep from confusing them, remember

(1) the beta is a distribution for θ, assuming α and β are known, and ranges from 0 to 1 continuously. It integrates to 1:

$$\int_0^1 g(\theta) d\theta = 1$$

(2) the binomial is a distribution for x, assuming n and θ are known, and ranges over the integers from 0 to n discretely. It sums to 1:

$$\sum_{x=0}^{n} f(x|\theta) = 1$$

The expression of risk

θ is the proportion that is of interest to us. For example, if we can be sure that θ_A is greater than or equal to .8 (or 80%), then we can infer that at least 80% of all future entries into this failure mode will have failure mechanism A. This would be a strong enough argument to stop analyzing failures from this mode (assuming the 80% is an acceptable criterion).

However, we cannot be "sure" what θ_A is. Even when ALL of the analyzed devices have failure

mechanism A, we still cannot be sure what θ_A is. Therefore, we will treat θ as a random variable, i.e., a proportion that we can make probability statements about. One such statement, for example, might be the probability that θ is greater than or equal to .8 is .9: P($\theta \geq$ .8) = .9. We will refer to this probability as being "90% confident that θ is at least 80%." (See the *Note for the Statistician* section below for a caveat on confidence interval statements.)

More generally, the expression that we are "A% confident that θ is at least B%" means

$$P(\theta \geq A/100) = B/100$$

And this will also be referred to as the A/B criterion. We have satisfied the 90/80 criterion if we have the data to support the statement that P($\theta \geq$ .8) = .9. (Notice that A/100 is simply the decimal equivalent of A%.) Do not confuse A%, which is an expression of uncertainty, with failure mechanism A. Since they are so different, the context of the use of A will make it difficult to confuse the two.

The calculation of probabilities

The probability distribution for θ is dependent on four things:

(1) which failure mode we are considering,

(2) which failure mechanism is being considered (we will call it A),

(3) how many devices from this failure mode have been analyzed (we will call it n), and

(4) how many of the n analyzed devices have failure mechanism A (we will call it x or x_A).

θ_A has a probability distribution from which we can calculate probability intervals. That probability distribution is

$$\theta_A \sim beta(1 + x_A, 1 + n - x_A).$$

Usually, we will drop the A subscript, and simply understand that we are referring to a particular failure mode and a particular failure mechanism:

$$\theta \sim beta(1 + x, 1 + n - x).$$

In figure 1, there is a typical beta distribution. The x-axis is θ. Since it is a continuous distribution, probabilities are integrals (areas under the curve). The probability that θ is greater than .8 is the area under the curve from .8 to 1.0. This is a one-sided confidence interval.

Two sided A% confidence intervals are any interval for θ where the area under the distribution and within the interval is equal to A/100. The minimum width A% confidence interval can be constructed (heuristically) by starting at the mode of the distribution (the high point on the distribution) and lowering a horizontal line, calculating the integral between the two points where the line meets the distribution (see figure 1). When the integral is equal to A/100, then that is the "high density" confidence interval, and it is also the minimum width confidence interval.

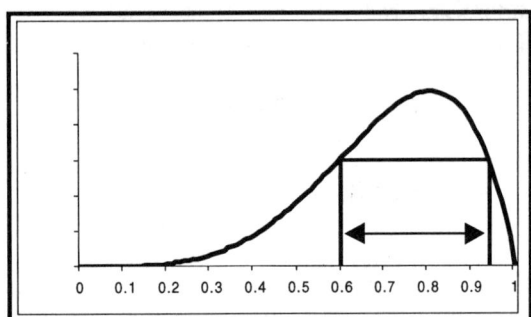

Figure 1: Confidence Interval Calculation

Although there are some applications in Failure Analysis for the two sided intervals, we will address only the use of the one sided interval here. One-sided intervals apply when the failure mode is relatively pure (one failure mechanism accounts for an overwhelming percentage of the devices in that failure mode).

When the failure mode is not relatively pure as, for example, when two failure mechanisms account for nearly all the devices in that failure mode, confidence intervals can be calculated for each of the mechanisms; however, it is unlikely that the analyst will want to make inferences about future fails. In fact, the analyst may wish to reexamine the failure mode with the objective of partitioning it into two modes that ARE relative pure (for inference purposes).

Beta distribution calculations by hand are impractical. Most desktop computer spreadsheet programs now have statistical functions programmed in, and this is the easiest source of accurate beta distribution calculations. Since Microsoft EXCEL is the most common spreadsheet, we will present examples of how to use it.

EXCEL allows the user to insert functions in any selected cell of the page. Simply select a cell, then from the top menu select (1) INSERT, (2) FUNCTION, (3) STATISTICAL, and (4) BETADIST. A menu will appear where the user will put in the "x value," alpha, beta, A, and B. If the user puts in .8, 1, 1, and leaves the A and B blank, the result will be .8 in the cell, and the entry in the cell will be seen at the top of the page as

BETADIST(.8,1,1).

This is the probability (or integral, or area) from 0 to .8 for θ. Hence the probability (for this alpha and beta) that θ is greater than .8 is 1- .8 = .2. The

BETADIST function has two additional input constants A and B (EXCEL's notation) that we left blank. The A and B define the interval that the distribution is defined over. The default is (0, 1), which is the interval that we will always use for proportions, hence they are left blank for signature analysis applications.

If you wish to find the "x value" such that the probability of $0 \leq \theta \leq x$ is a given value, say y, then use

BETAINV (y, alpha, beta, A, B).

This function is found in the same place and is implemented like BETADIST. For a y = .3 (we are seeking the 30th percentile), and if $\alpha = 5$, and $\beta = 2$, then

BETAINV (0.3, 5, 2) = .639642.

This says that the probability that θ is greater than or equal to .639642 is .7: or, we are 70% confident that θ is greater than .639642.

The information contained in the analysis is passed into the probabilities through the parameters, α and β, of the beta distribution. Before any analyses are performed, both n and x are 0: x = n = 0. Then the probability distribution for mechanism A is

beta $(1 + x, 1+ n - x)$ = beta (1, 1) = beta (α, β),

hence alpha = beta = 1 in the EXCEL BETADIST function. This beta (1, 1) is the Uniform distribution. The probability that $\theta \varepsilon$ (0, .8) is .8 (as the first example above verifies.

Suppose that we have now performed analyses on 5 devices and 4 were identified with mechanism A and one with mechanism B. Then the distribution that describes our uncertainty about mechanism A for another device is

beta $(1 + x, 1+ n - x)$ = beta (5, 2) = beta (α, β),

and from the second example above, we are 70% confident that θ is greater than ~.64.

Setting standards

The analyst will not wish to make inferences haphazardly, even if the risk is calculated and noted. The analyst can set acceptable risk levels in advance and continue to analyze all devices in a failure mode until the acceptable risk level is achieved. We will use EXCEL to calculate the required numbers of analyses in several examples.

Suppose that we have set the standard at the 90/90 criterion; that is, we wish to be 90% confident that θ is at least 90%. Let us further suppose, for this example, that each analysis identifies the same failure mechanism; that is, the failure mode is absolutely pure as far as our analyses have shown.

The question is: how many such analyses must be performed before the 90/90 criterion is met, and we can begin inferring the mechanism of future fails rather than analyzing them?

We will use EXCEL to derive the solution. First we check a relatively low number, say x = n = 5:

1 - BETADIST(.9,6,1) = .468559.

We use 1 – BETADIST (.9, 6, 1) for the following reasons:

(1) 1 – because we want the probability (or integral or area) in the right tail, that is, greater than x. The BETADIST statement gives the cumulative probability from the left, and we want its compliment.

(2) .9 because we want the θ greater than .9.

(3) 6 because $\alpha = 1 + x = 6$ when x = 5.

(4) 1 because $\beta = 1 + n - x = 1 + 5 - 5 = 1$.

The .468559 is not at least .9, therefore n = 5 analyses are not sufficient for the 90/90 criterion. So we increase the α to 7, and so on until the right side (.468559 for n = 5) first obtains .90. Figure 2 shows that the solution is where $\alpha = 22$ and therefore n = 21. The long decimal is presented in the table above to show the reader the actual EXCEL output.

x = n	alpha	beta	confidence
15	16	1	0.814697981
16	17	1	0.833228183
17	18	1	0.849905365
18	19	1	0.864914828
19	20	1	0.878423345
20	21	1	0.890581011
21	22	1	0.90152291
22	23	1	0.911370619
23	24	1	0.920233557

Figure 2: Obtaining 90/90 when x = n

This technique used BETADIST, but another approach would have been to use BETAINV. The BETADIST technique set x at .9 (x's are values on the θ axis) and solved for the confidence level, then varied the sample size until the confidence level was just greater than .9. The BETAINV approach sets the confidence level at .9 and solves for x, then varies the sample size until x is just greater than .9.

For a second example, suppose that the analyses of the n devices yields all but 1 with the same mechanism (call it A), and one with another mechanism (call it B). Then we use the same approach. This time we will start with x = 15, n - x = 1:

1 - BETADIST(.9,16,2) = .5497,

and continuing to increase alpha, we can generate Figure 3. Here we see that, when we are one-off from a pure mode, we obtain the 90/90 criterion at x = 36, n - x = 1, and n = 37.

x = n-1	alpha	beta	confidence
30	31	2	0.843576626
31	32	2	0.85578528
32	33	2	0.867116436
33	34	2	0.877623509
34	35	2	0.887358002
35	36	2	0.896369362
36	37	2	0.90470487
37	38	2	0.912409583
38	39	2	0.919526304

Figure 3: Obtaining 90/90 when x = n-1

For the third example, we have determined that analyzing n = 21 devices is too costly and we have decided to compromise to the 90/80 criterion; that is, we wish to be 90% confident that the mode is at least 80% pure. Again we start at a low sample size, say x = n = 5:

1 - BETADIST(.8,6,1) = .737856,

then we increase alpha until the term is greater than or equal to .9. Notice in figure 4, that this occurs at x = n = 10.

x = n-1	alpha	beta	confidence
5	6	1	0.737856
6	7	1	0.7902848
7	8	1	0.83222784
8	9	1	0.865782272
9	10	1	0.892625818
10	11	1	0.914100654
11	12	1	0.931280523
12	13	1	0.945024419
13	14	1	0.956019535

Figure 4: Obtaining 90/80 when x = n

These examples illustrate that the analyst can calculate "what if" scenarios to balance risk criteria and cost of analyses.

A visual understanding of the relationship between the A/B criterion and sample size can be obtained from figure 5. Also observe the interplay of A and B (in the A/B criterion) for given values of n, where x = n.

Figure 5 can also be used to determine the required sample size when the failure mode is anticipated to be pure, and in fact, the physical analyses bear that out. Observe that should we wish to satisfy the 90/70 criterion, we would go to the intersection of .9 on the y-axis and .7 on the x-axis, and note that the first curve up and to the right of

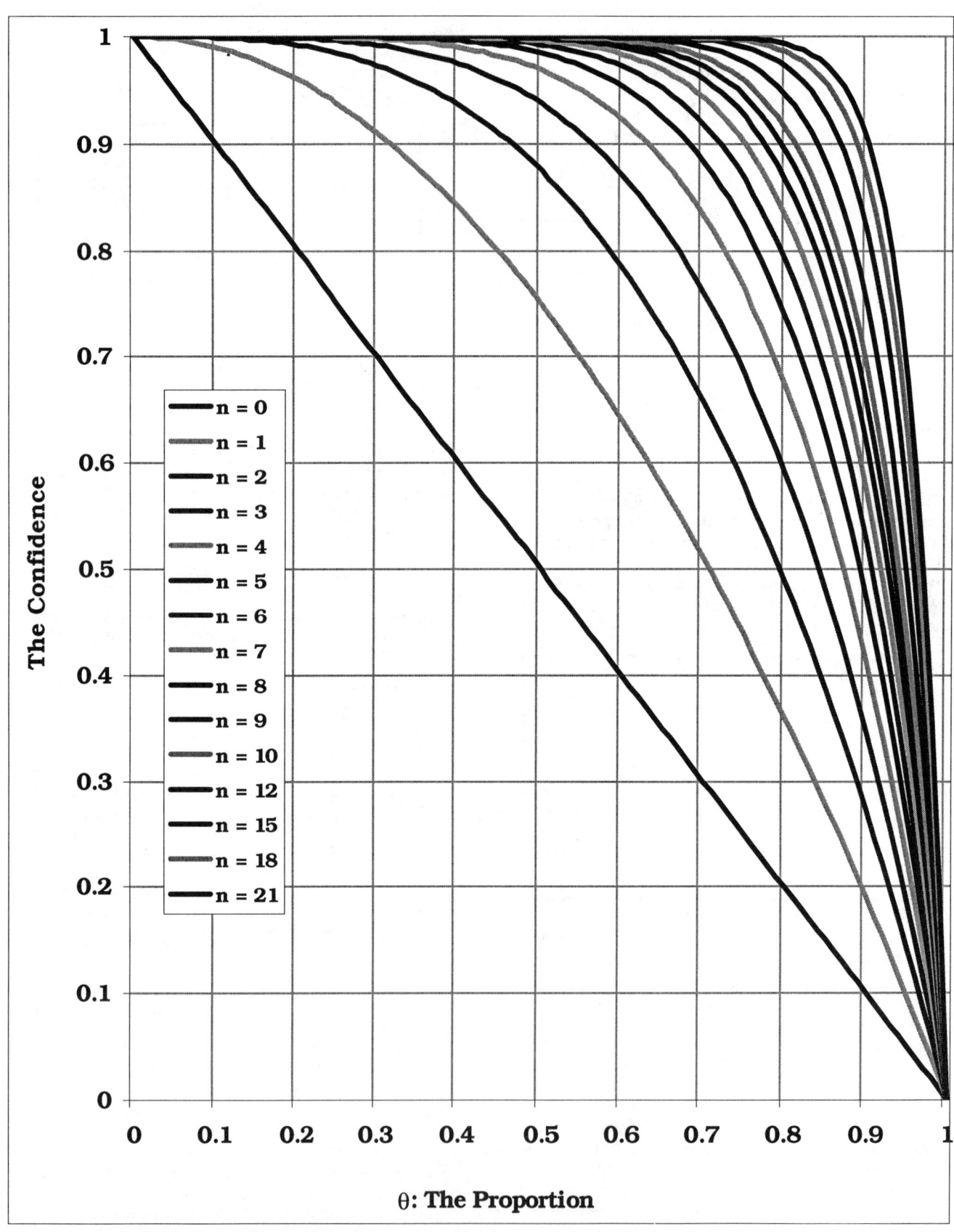

Figure 5: The A/B Criterion for the Ongoing Process

Letter notations	Number of devices with the signature of interest	Number of devices with the failure mechanism of interest
Population numbers	N (known)	M (unknown)
Sample numbers	n (known)	x (known)

Figure 6: Notation for the Finite Population Model

that intersection is the n = 6 curve. Hence we need n = x = 6 to satisfy the 90/70 criterion. In like fashion, we need n = x = 10 to satisfy the 90/80 criterion, and n = x = 21 to satisfy the 90/90 criterion.

THE FINITE POPULATION MODEL

The finite population uses another data point: the total number of devices that are in (and will ever be in) the failure mode. This failure mode size, we will denote as N, and we will let M denote the unknown true number of fails that are associated with mechanism A (the major mechanism for this mode): $0 \leq M \leq N$.

In order to extend the technique of the ongoing process model to this model, we need to define two additional distributions and shift our focus from making probability statements about θ, the proportion of mode purity, to making probability statements about M, the number of fails from the major mechanism among the N devices in the failure mode.

Figure 6 displays the notation for the finite population model. We will use a distribution that assumes that we know M and calculates probabilities about x. Then we will use that probability distribution to derive the probabilities about M when x is known. From the logic of the notation, we can state some ranges on the terms: $0 \leq x \leq n \leq N$ and $0 \leq x \leq M \leq N$.

When N, M, and n are known, then the probability that we will see x failure mechanism A's among the n devices analyzed is calculated using the hypergeometric distribution:

$$f(x \mid N, M, n) = \frac{\binom{M}{x}\binom{N-M}{n-x}}{\binom{N}{n}}.$$

We now use the hypergeometric distribution to calculate the probability that the total number of mechanism A's is M using the formula

$$g(M \mid N, n, x) = \frac{f(x \mid N, M, n)}{\sum_{M=x}^{N} f(x \mid N, M, n)}.$$

Since we are using the upper tail of the distribution, we will actually use a sum of these g(M|N, n, x) where M is large (close to N). For example, suppose that we have a lot of devices, N = 20 of which have failed the same electrical test and have been categorized into their own single failure mode. We have analyzed n = 5 of the devices, and found that all 5 failed due to the same mechanism: x = 5.

Using EXCEL to perform the work, figure 7 displays the results of the statistical analysis. The

M	f(x|N, M, n)	g(M|N, n, x)	cum sum
0			
1			
2			
3			
4			
5	0.000	0.000	1.000
6	0.000	0.000	1.000
7	0.001	0.000	1.000
8	0.004	0.001	0.999
9	0.008	0.002	0.998
10	0.016	0.005	0.996
11	0.030	0.009	0.991
12	0.051	0.015	0.983
13	0.083	0.024	0.968
14	0.129	0.037	0.945
15	0.194	0.055	0.908
16	0.282	0.080	0.852
17	0.399	0.114	0.772
18	0.553	0.158	0.658
19	0.750	0.214	0.500
20	1.000	0.286	0.286
sum	3.5	1	

Figure 7: Example of Finite Population Data Analysis

first column values are the possible values of M, the second column contains probabilities from the hypergeometric of

P(x=5 | N=20, n=5, M given in cloumn1),

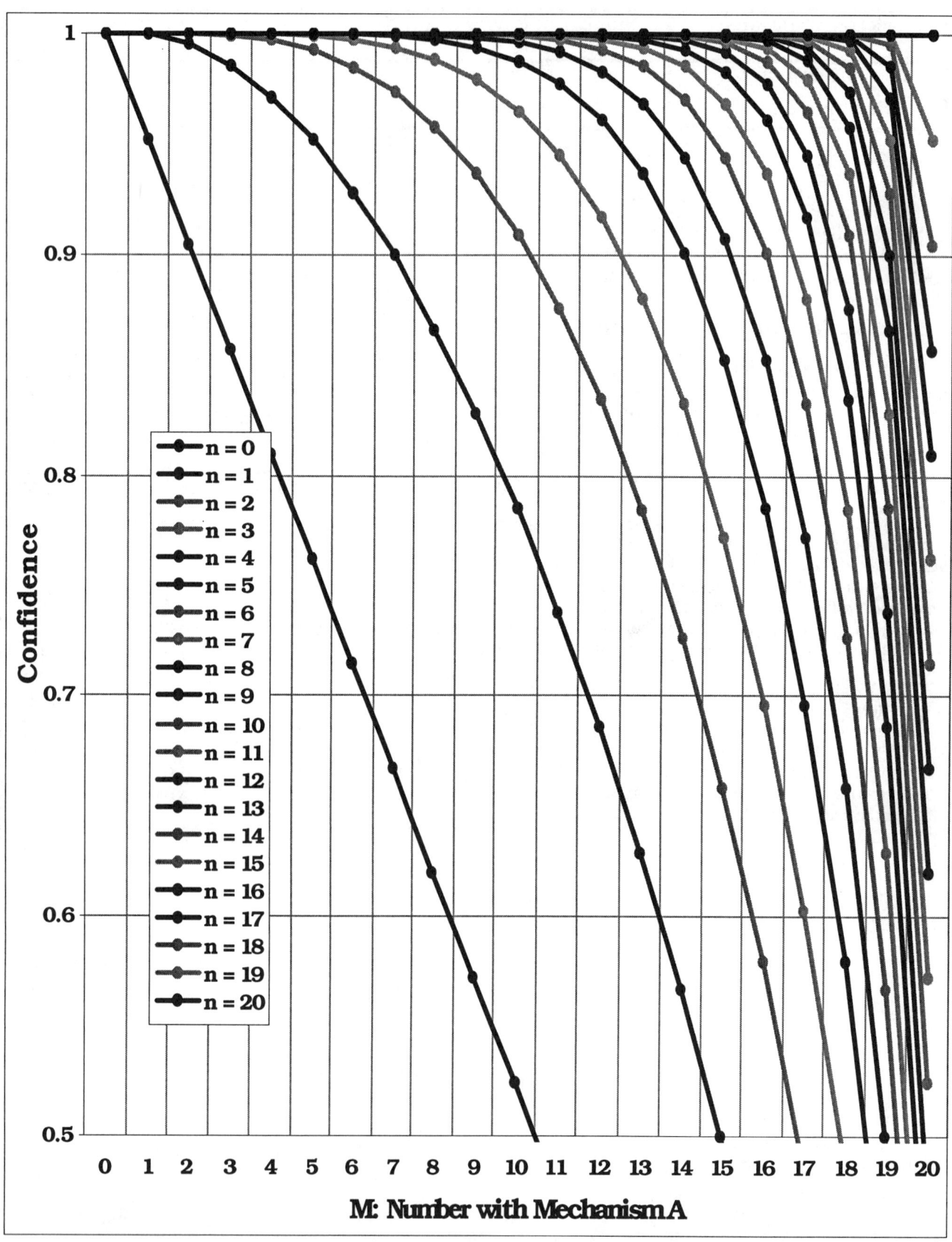

Figure 8: The A/B Criterion for the Finite Population Model, Special Case N = 20

484

the sum of these probabilities is given at the bottom of column 2, column 3 values are the column 2 values divided by the sum at the bottom of column 2, and the column 4 values are the sum of column 3 at and below the row of the entry.

The cum sum column is above .9 at M=15, so

$$P(M \geq 15) > .9,$$

Or, after analyzing n = 5 devices with the result that all x = 5 yielded the same mechanism, we have satisfied the 90/75 criterion: we are 90% confident that at least M = 15 (of the total N = 20, including the n = 5 that were analyzed) have the same failure mechanism.

Figure 8, like figure 5, illustrates the interplay of A and B, in the A/B criterion, for various values of n in the case that x = n, but only for the case N = 20.

THE STATISTICAL FOUNDATION

These techniques are based on the Bayesian approach to statistics. The assumed knowledge about the proportion θ in the ongoing process model, and M in the finite population model, is that no value is more likely than any other value; therefore, a "noninformative" distribution is used: in these two cases, the uniform distribution. This is reflected in the formula for the beta distribution when x = n = 0 is used:

$$\theta \sim beta (1 + x, 1 + n + x) = beta (1, 1) = uniform (0, 1).$$

This distribution for θ, that is used when there is no data, is referred to as the "prior" distribution, and generically denoted as $g(\theta)$.

The test data in the ongoing process model follows a binomial distribution:

$$x \sim f(x|\theta) = binomial (n, \theta).$$

This is called the "conditional" distribution since it is conditional on the unknown value of θ (and on n, but n is known).

What we are interested in, and all of this notation is building to, is the "posterior" distribution of θ; that is, the distribution of θ when the data, n and x, are taken into account. This posterior distribution is denoted $g(\theta|x)$. The g is the same as the prior distribution because the prior and posterior are both distributions for θ. The $g(\theta|x)$ is read as the distribution of θ given x, and is an "update" of $g(\theta)$ to incorporate the information from the data, x.

Bayes Theorem is used to derive $g(\theta|x)$ from $g(\theta)$ and $f(x|\theta)$:

$$g\left(\theta \,|\, x\right) = \frac{g(\theta)\, f\left(x|\theta\right)}{\int_0^1 g(\theta)\, f\left(x|\theta\right) d\theta}$$

We will not go through the derivation here, but it can be shown that when

1. $g(\theta)$ is a beta (α, β), and
2. $f(x|\theta)$ is a binomial (n, θ),

then $g(\theta|x)$ is a beta $(\alpha + x, \beta + n - x)$. In our application, $g(\theta)$ is our prior, $f(x|\theta)$ is our conditional distribution, and $g(\theta|x)$ is our posterior distribution. This can be found in many statistics texts, such as [1], but is left as an exercise for the student in others, such as [2]. In our application, α and β are both set equal to 1, reflecting no prior information about θ. Or more accurately, we assume a uniform prior distribution for θ, so we set $\alpha = \beta = 1$.

In the case of the finite population model, simplifying formulas for $g(\theta|x)$ when $g(\theta)$ is the uniform and $f(x|\theta)$ is the hypergeometric do not exist. But since these are discrete distributions with, at most, N points, we can perform complete enumeration using EXCEL.

Probability intervals calculated from the posterior distribution (which is what we are doing) are called credibility intervals and they play the same role that confidence intervals do in traditional statistical models. They are even sometimes called Bayesian confidence intervals. Actually, traditional confidence intervals are often misinterpreted as probability intervals in the same way that credibility intervals are correctly interpreted. However, the differences between confidence intervals and credibility intervals can be considered a subtle fine point in statistical theory that goes beyond the application of the practicing engineer. Hence, we have used "confidence interval" in the general sense to appeal to the more recognized term and facilitate understanding of the concept. Our confidence intervals, here, are the statisticians' Bayesian credibility intervals.

SOME ROOM FOR IMPROVEMENT

Is this the best we can ever do? ...analysis of 21 devices with same-mechanism results to satisfy the 90/90 criterion (for example)? No, there is a way to improve on these results.

In the Statistical Foundation section above, we stated that this analysis is based on the use of the uniform distribution for the prior. That is a reflection of the fact that we have no information about what the purity of the failure modes (which is the proportion θ) is likely to be.

However, since we suspect that by the way we are defining the failure modes, these modes will be relatively pure, and the proportions will be high (closer to 1 than to 0). In order to quantify this expectation of pure modes, we propose the use of an historical database. Maintaining records of the outcomes from analyses and organizing them to determine the empirical distributions of the different proportions could be used to model and derive a more informative prior distribution.

This more informative prior distribution could then be used to generate the probability intervals and justify decisions to classify devices with fewer failure analyses.

The development of such databases is no light undertaking, requiring a large data set of "like" devices. A statistician that understands the Bayesian approach will also be needed to model the updated prior.

SUMMARY

An analytical technique has been described for application to device failure signatures, which we have loosely termed failure modes; however the user may wish to include some auxiliary data that is not normally used in failure mode definitions to form signature categories. These failure modes (or signatures, if you prefer) and a number of physical failure analyses from some failure mode are used to ascertain the risk of inferring the failure mechanism of other devices in that failure mode.

Risk is expressed in terms of probability intervals of the type $P(\theta \geq B/100) = A/100$ and is denoted as the A/B criterion: A% confident that the failure mode is at least B% pure.

No special software is needed to perform the calculations. Examples are presented using the ubiquitous spreadsheet Microsoft EXCEL. Additionally, a set of graphs (figure 5) is presented that can be used as a reference to visually determine the risk level for a selected set of analyses results where the failure modes are empirically pure (i.e., x = n).

A distinction is made between the Ongoing Process Model, where devices are continuing to be categorized into the failure mode, and the Finite Population Model, where inference is made to the current failure mode members only. While the techniques and definitions are the same for the two models, the techniques are applied to different distributions. The Ongoing Process Model is the only one that allows for generalization to future devices in the failure mode.

This technique for risk assessment can be used to determine the number of devices to be physically analyzed to meet a particular criterion, or simply to assess the risk of inference at any point in the process of analyzing a sequence of devices.

ACKNOWLEDGEMENTS

This work was originally published in [3] and [4], and presented to the FA engineering world in [5].

The author would like to thank Glen Gilfeather, AMD, for posing the problem and championing the technique to the Sematech Product Analysis Forum. The author would also like to thank this Sematech Committee and its chair, Seshu Pabbisetty, for promoting these techniques to standardize their use in the semiconductor industry. It has led to wide industry acceptance and JEDEC consideration [6].

REFERENCES

1. J. Bernardo and A. Smith, *Bayesian Theory*, John Wiley & Sons, 1994, p 269-271.
2. J. Berger, *Statistical Decision Theory and Bayesian Analysis*, 2nd ed., Springer-Verlag, 1980, p 287.
3. M. Pore, "Sample Selection for Failure Analyses." American Statistical Association Proceedings of the Physical and Engineering Sciences Section, August 1995, p 206-211.
4. M. Pore, "Proportion Estimation in Signature Analysis." American Statistical Association Proceedings of the Physical and Engineering Sciences Section, August 1996, p 219-224.
5. M. Pore, G. Gilfeather, and L. Levy, "Risk Assessment in Signature Analysis." Conference Proceedings of the 22nd International Symposium for Testing and Failure Analysis, ASM International, November 1996, p177-182.
6. "Signature Analysis Guidelines." EIA/JEDEC Publication: JEP136. July 1999.

FLIP-Chip and "Backside" Techniques

Karoline Bernhard-Höfer
Infineon Technologies
München, Germany

Daniel L. Barton and Edward I. Cole Jr.
Sandia National Laboratories
Albuquerque, New Mexico

Abstract

State-of-the-art techniques for failure localization and design modification through bulk silicon are essential for multi-level metallization and new, flip-chip packaging methods. This tutorial reviews the transmission of light through silicon, sample preparation, and backside defect localization techniques that are both currently available and under development. The techniques covered include emission microscopy, scanning laser microscope based techniques (electro-optic techniques, LIVA and its derivatives), and other non-IR based tools (FIB, e-beam techniques, etc.).

1. Introduction

Quite often when performing failure analysis on a multi-layer device, the failing area of a circuit is physically located underneath upper layer metal lines such as power buses, high density routing signal lines, and bond pads. This makes electrical failure localization very difficult or even impossible using traditional front side failure analysis techniques (i.e. liquid crystal thermography, photoemission microscopy etc.). At the same time, new packaging techniques like Lead On Chip [1] and flip chip [2-6] enhance overall chip performance, but at the same time they can make conventional failure analysis techniques ineffective.

In Fig. 1, flip-chip technology is compared with face-up wire bonding which is still the most popular packaging method today. For wirebond packaging, the chip is typically mounted face up with the wire bonds that are connected to bond pads arranged along the periphery of the device. In contrast, flip chip is defined as mounting the chip on a substrate using a variety of interconnect materials and methods as long as the chip surface, i.e. the active circuit, is facing (oriented in the direction of) the substrate. Flip chips are cost effective and allow the realization of very slim and compact products. Due to the short interconnection length and the possibility of arranging the pads all over the chip surface the flip chip design enhances electrical performance and permits higher I/O densities at smaller chip size. In addition thermal dissipation is very efficient. These advancements present a challenge for failure analysts. Failure sites are simply no longer accessible from the front side of the die. Therefore the analysts has to gain access to the failure site from the backside of the die. In this case, however, the signal - usually light – has to penetrate through several hundred of microns of Si substrate. This tutorial deals with the issues of accessing the defects through the silicon, sample preparation, and the various techniques available for backside defect localization.

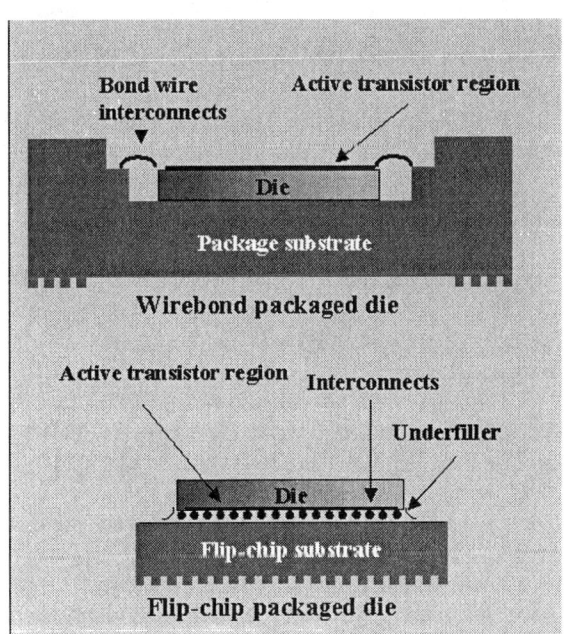

Fig. 1. Wirebond (a) versus flip chip packaging (b).

2. Basics

If light of intensity I_0 is incident on a piece of bulk Si, the light intensity is attenuated dependent on the thickness d_{Si}, wavelength λ, and doping concentration of the substrate [7]. The variation of the transmitted light intensity I_T is given by the so-called Lambert-Beer absorption law [7]:

$$T = \frac{I_T}{I_0} = \left(1 - R_{Si}\right)^2 \bullet \exp(-\alpha(\lambda)\,d_{Si}) \qquad (1)$$

with   T: light transmission

        R_{si}: reflection coefficient of Si

        $\alpha(\lambda)$: absorption coefficient of Si

The absorption coefficient α consists of two major contributions: the band-gap related absorption and the free carrier absorption [7]. Band-gap related absorption, which is also called interband absorption, is given by the band structure of Si (Fig. 2) [8].

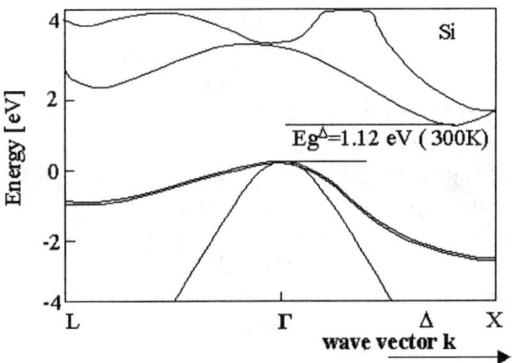

Fig. 2. Band Structure of Si [8].

An incident photon can only be absorbed if its energy can excite an electron from the valence band to the conduction band, i.e. if its energy is greater than or equal to the band-gap energy, E_{gap}, of Si (1.12eV corresponding to 1.1µm at 300K). In Fig. 3(a) the transmission of undoped Si is shown. Si acts as a low-pass filter absorbing all photons with energy greater than the band gap energy and being transparent for photons with energy less than E_{gap}. By contrast, free carrier (intraband) absorption induces a transition within the same band. The mechanism for free-carrier absorption is shown in Fig. 4 for an n-doped substrate. The absorbed photon looses its energy to an electron in the conduction band that is excited to higher energy within the same band. To conserve momentum a phonon has to be involved. Once excited to higher energy, the electron relaxes via phonon interaction to the conduction band minimum. Within the scope of a semi-classical model, the absorption coefficient increases linearly with the doping concentration, n,

of the substrate and quadratically to the light wavelength [9]:

$$\alpha \propto \frac{\lambda^2 n}{m^* \mu} \qquad (2)$$

with m^* the effective mass, μ the mobility, and n the doping concentration of the substrate.

In Fig. 3, the light transmission as a function of doping of the substrate is shown. The higher the doping concentration is, the lower the transmission due to free carrier absorption, especially in the infrared, i.e. transparent wavelength range. According to formula (1), light attenuation drops exponentially with decreasing substrate thickness. Thinning the substrate can minimize light attenuation above the band-gap.

Not only is the transmission dependent on the thickness and nature of the substrate but also the quality of the illuminated image taken from the backside of the die. To minimize diffuse light scattering (caused by scratches and irregularities at the surface of the die that deteriorate image quality) a mean surface roughness of less than 5 nm is recommended [10]. Even if the requirements to the surface quality are fulfilled, IR light has to be used for illumination due to the transmission characteristic of Si. As a result, the lateral resolution is reduced compared to standard frontside illumination. In addition, a lens failure, spherical aberration, deteriorates the illuminated image taken from the backside; especially for large numerical apertures (NA), i.e. for the 100x microscope objective. It can be shown that the lens failure increases with the numerical aperture of the objective lens and the thickness of the Si substrate. To ensure a good image quality even for large magnifications it is therefore recommended to thin the die.

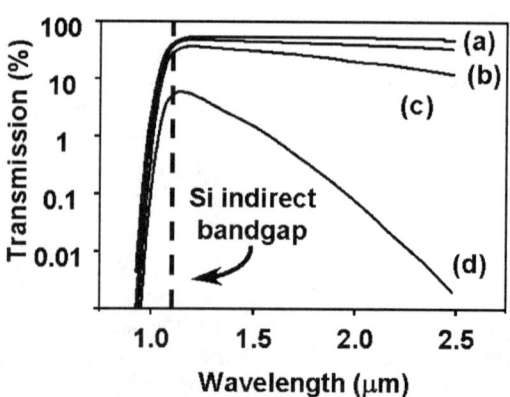

Fig. 3. IR transmission of 625 µm of p-doped Si with doping concentration $\times 10^{16}$ cm^{-3} of (a) 1.5, (b) 33, (c) 120, (d) 730.

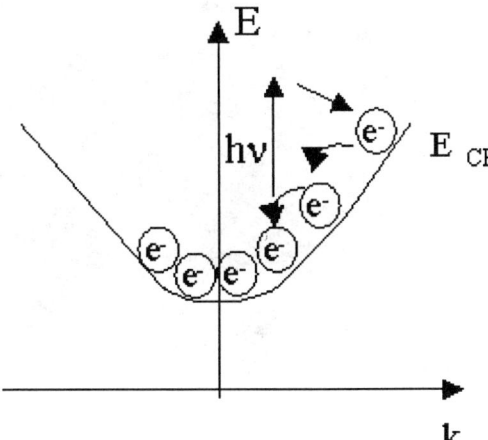

Fig. 4. Free-carrier absorption (n-doped semiconductor).

3. Backside preparation techniques

Backside preparation techniques can be roughly divided into two categories, i.e. global thinning techniques and local thinning techniques. Global thinning techniques such as CNC (Computer Numerically Controlled) milling, mechanical grinding as well as dry etching procedures permit to thin the entire die. By contrast, local thinning techniques such as FIB (Focused Ion Beam) and laser induced etching techniques permit to thin smaller areas but with higher lateral resolution. Thinning the entire die is required to save time whenever a local thinning technique is used. But it may also be required for emission based and optical probe based backside localization techniques whenever the signal strength or the quality of the illuminated image should be improved. The main operational areas of local thinning techniques are mechanical or e-beam probing and device modification from the backside.

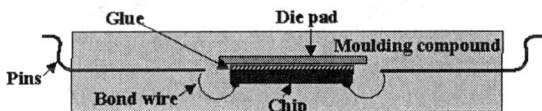

Fig. 5. Standard plastic package: to gain access to the chip the molding compound, the die pad and the glue have to be removed.

Backside sample preparation can be divided into two phases. The first phase involves gaining access to the chip. The second phase is modifying the backside surface of the die. Phase I is generally determined by package specifics. Flip-chip packages take advantage of the easy access to the backside of the die. Heat slugs attached to the die with thermally conductive epoxy can easily be detached for example by heating the package and removing the

heat slug with the help of a sharp metal object. The situation is more complex if you have to gain access to a chip mounted in a package (Fig. 5). To gain access to the chip first the molding compound, the die pad and the glue have to be removed. The process steps of phase I are very similar to those to partially decapsulate a device from the frontside.

The actual CNC milling of the Si die is usually done with a diamond milling head. This diamond head is run with a high-frequency spindle permitting up to 60000 rpm (Fig. 6). The device itself is fixed with wax on a sample holder. To cool the die and to remove the milled Si fragments, a water cooling system is used. With a setup like this, the Si die is usually thinned down to a remaining thickness of 100-150 µm. Thinning the die even more is not recommended as the die becomes mechanically unstable and is hard to handle. In addition, the heat management of the device would be disturbed which could cause a thermal collapse of the device during electrical operation. Depending on the geometry of the milling head used, the Si surface reveals graves and scratches which are typically in the order of several microns deep (Fig. 7a).

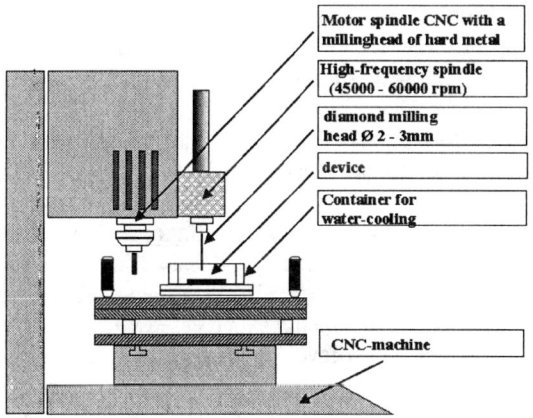

Fig. 6. CNC milling machine used to thin the Si die.

To achieve the required surface quality, several polishing steps are connected. Polishing is usually done with diamond paste. Dependent on the grain size used for the final polishing step, a mean surface roughness of 5 nm, or even better, can be obtained (Fig 7b).

The described technique is applicable for packaged devices and flip-chips and, with minor process modifications, even for a single chip on wafer level [11]. As it is a computer-controlled technique it provides highly reproducible and accurate results with a surface planarity of 20 µm or even better (milled area 1 mm^2).

A further global thinning technique is to mechanically grind and polish the die. To ensure a

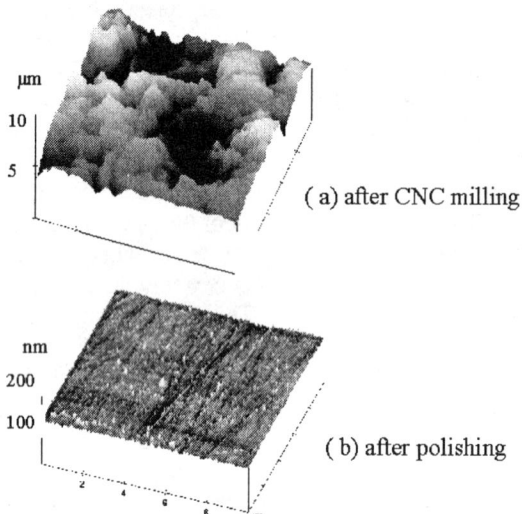

Fig. 7. Si surface roughness after CRC milling (a) and after polishing the milled surface (b).

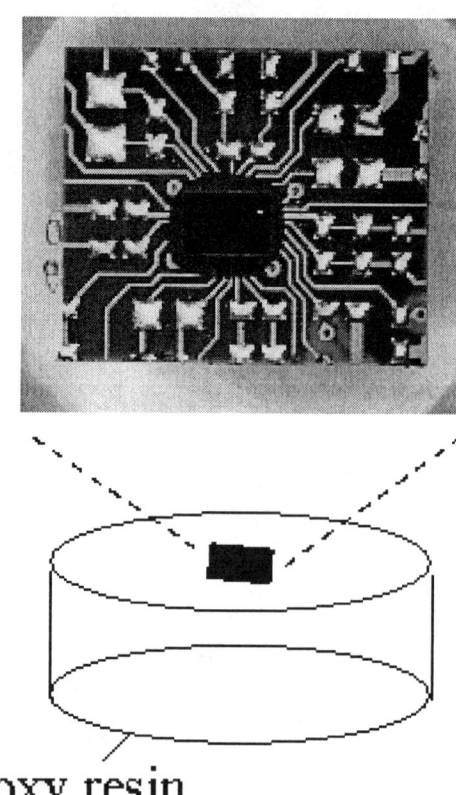

epoxy resin

Fig. 8. Encapsulated grinding to a homogeneous Si removal it is recommended to encapsulate the die.

homogeneous Si removal, it is recommended to encapsulate the die for example in epoxy resin (Fig. 8) or to place some wafer scraps around the die to be lapped. Grinding and polishing is usually done in several steps using a conventional lapping machine. For material removal, a coarse abrasive paper is used. To obtain the final mirror finish, polishing steps with diamond pastes of different grain sizes are performed. The benefits of mechanical grinding and polishing are very similar to those of CNC milling. Its use for packaged devices however, is limited as it permits only the thinning of the entire device inclusively package and leads.

For completeness, global thinning procedures based on dry etching processes should be mentioned. To achieve the required high etch rates, high plasma density machines are needed which can be realized for example through microwave excitation, in an inductive coupled plasma or an electron cyclotron resonance plasma [12]. Dry etching procedures using high plasma density systems however, are still under development and have not been routinely used for backside FA today.

To probe an IC electrically from the backside or to modify it from the backside of the die, FIB preparation procedures have been developed [13-17]. To gain access to the circuitry through the backside of the die, up to several hundred of microns of bulk Si have to be removed. Etching of Si is most effective using XeF_2 based gas chemistry [18]. XeF_2 molecules that are delivered from a gas nozzle are adsorbed on the Si substrate.

A surface reaction takes place and volatile reaction products are formed which are removed by the pumping system. This surface reaction is at least 10 times faster than pure milling. XeF_2 reacts spontaneously with Si; i.e. the reaction doesn't need to be activated by the ion beam. This spontaneous reaction however, can only take place if the surface layer is formed by pure Si. Even so, the Si removal rates achieved with a conventional FIB are still too low for backside applications. In addition, there is a tendency of the milled hole to become deeper on the side that is farthest away from the gas jet (Fig. 9a). This is due to the fact that the gas is injected in the process chamber under a certain angle. This effect is negligible in topside milling due to the relatively small depth of the milled trenches. It does become noticeable in backside milling since the holes milled are much deeper. Consequently new methods and equipment based on a high-pressure gas delivery system in combination with a high current density column have been developed which enable a high-speed Si etching process (Fig. 9b).

The commercially available high current density column provides beam currents up to 30 nA (compared to 6 nA in a conventional FIB). The nozzle configuration (either two gas nozzles or one nozzle with an opening for the ion beam) permits a very high local gas pressure (up to several Torr). The combination of both these components enhances Si etching up to a factor of 1000 compared to conventional FIB techniques [19]. In addition, nearly all gas delivered is instantaneously consumed. As a result etching takes place only where the beam is rastered and it is possible to mill a very precise pattern in the Si directly over the edit or probe area.

An infrared microscope, which is integrated into the FIB system, is an additional useful feature for backside FIB. As the backside of the chip is absolutely flat and opaque for the ion beam, the infrared microscope may be very useful for navigation and alignment purposes.

To do electrical/e-beam probing or device modification from the backside of the die with a high-speed FIB process some sample preparation is recommended (Fig. 10). To access the signal nodes it is not realistic to drill a small hole several square microns wide and several hundreds of microns deep through bulk Si down to the buried IC. Therefore the Si substrate is pre-thinned in 2 steps so that the final hole only needs to be drilled through a thin section of the substrate.

In the first step, the entire Si substrate is thinned down to a fraction of its original thickness (usually down to about 100 µm). This is done with the help of one of the global thinning techniques reviewed in the first part of this section. The removal of the majority of the bulk Si is essential to accelerate the preparation process.

Next, alignment points for CAD navigation have to be identified either with the help of the integrated infrared microscope or simply by estimating the locations of stepper alignment marks or other large non-critical geometries which are typically found at the corners of the device. After having carried out the alignment, it is possible to navigate precisely to buried geometries.

Next, a large trench is etched at the area of interest. This step is known as local thinning as Si is removed only above the selected nodes. This way the substrate rigidity is maintained, the electrical impact to the chip and the volume of the Si removed is minimized. For local thinning, the high speed FIB etching technique is used. Trench size is usually in the 100x100 µm^2 range; the remaining Si thickness is about 10 µm (Fig. 11).

During the final step, small holes are drilled in the bottom of the trench to expose the signal lines of interest. As the precise milling capability of the FIB is required for this task, this FIB process is very similar to standard frontside applications. However, compared to standard, frontside applications the substrate is covered with a thin silicon oxide film (Fig. 12). This oxide deposition step can be done in-situ with a FIB. By this thin oxide layer (thickness between 50-100 nm), undesired spontaneous etching of the Si substrate could be avoided during successive process steps based on XeF$_2$ gas chemistry. In addition, the oxide layer acts as an insulator between re-routed signal lines

and the Si substrate. Once the thin oxide layer has been deposited over the trench, the signal nodes of interest are exposed using XeF$_2$ assisted milling.

Next, the nodes that will be contacted are covered with oxide and then re-exposed. This way, shorting between the nodes and the bulk Si can be avoided when contact is made. Similar to standard frontside applications, the final step consists of metal deposition and cutting the original signal lines.

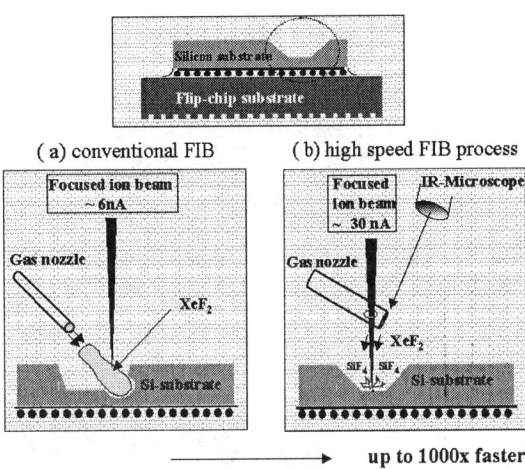

Fig. 9. Conventional FIB system versus high-speed FIB system.

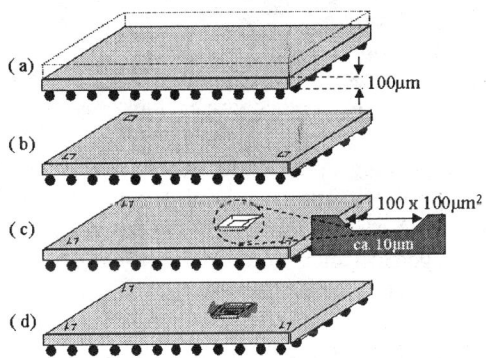

Fig. 10. Sample preparation with FIB: a) Global Si thinning, b) Identification of alignment points for C1D navigation, c) Local Si thinning with high-speed FIB etchings process, d) Precision probe milling with FIB.

Alternatively to the high speed FIB etching process for local Si thinning, a laser activated etching technique, the so-called laser microchemical (LMC) etching can be used [20 - 27]. This method is based on scanning a focused, high power Ar$^+$ laser beam over the backside of the Si die (Fig. 13). The substrate is in contact with a moderate pressure chlorine vapor that serves as the etchant. By the laser irradiation, a small volume (in the order of 1 μm³) of the Si substrate is brought just above its melting point. The surface of the molten zone reacts with the chlorine ambient to form volatile SiCl₄. Because of the high reactivity of liquefied Si, the etching rates are orders of magnitude faster than conventional chemical reactions. When the laser is scanned to the next address point, the thin unreacted liquid Si layer epitaxially re-grows to a single crystal substrate. This way, the surface quality of the etched area is very good and relatively insensitive to the surface quality before etching.

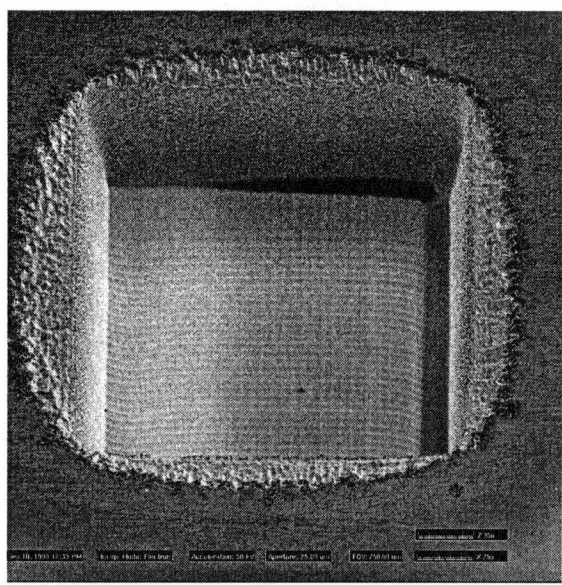

Fig. 11. High-speed FIB etched trench (100 x 100 μm², Al μm remaining Si thickness).

Depending on the raster sequence of the laser beam, various patterns (circular and rectangular geometries) can be etched into the substrate. This process permits etch rates up to 100000 μm³/sec, which is much faster than the fastest FIB method. By varying the gas ambient in the process chamber, the tool also permits a fast deposition of oxides and a variety of metals. Compared to a corresponding FIB process, these laser-activated processes are usually faster (up to 100x) and they are free from any process debris such as Ga⁺ contamination. A drawback however, of the LMC technique compared to FIB is lower lateral resolution. Due to non-linearity of the surface chemical reaction, the resolution can be better than the laser spot size but is limited to about 0.5 μm. This is significantly higher than FIB milling resolution. In practice, the chlorine-based Si etching process is used to locally thin the die to about 10 μm remaining Si thickness. As this laser-activated process is faster than the high-speed FIB process, typical trench dimensions are larger than for FIB preparation, i.e., an etched

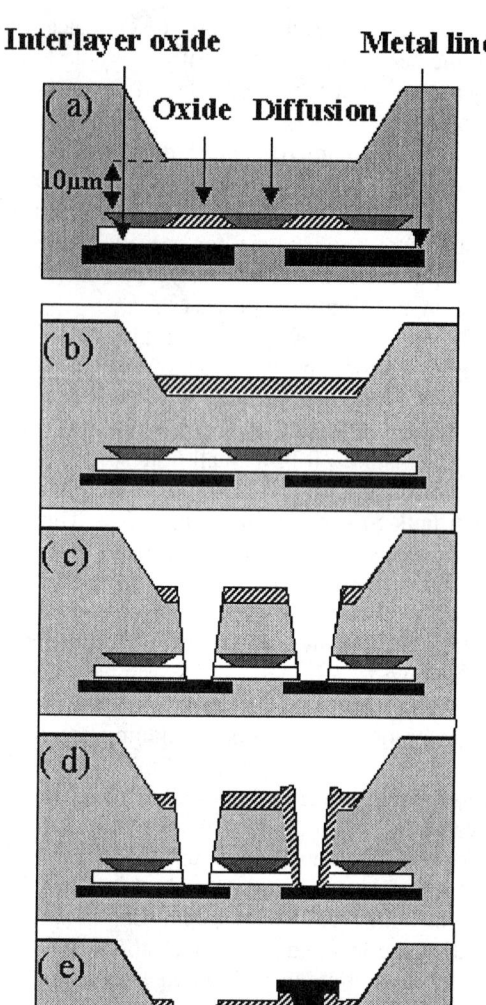

Fig. 12. Precision hole milling: a) schematics of the device structure, b) oxide deposition, c) XeF₂ assisted milling, d) SiO₂ deposition and re-opening of the holes, and e) metal deposition and cutting the old signal line.

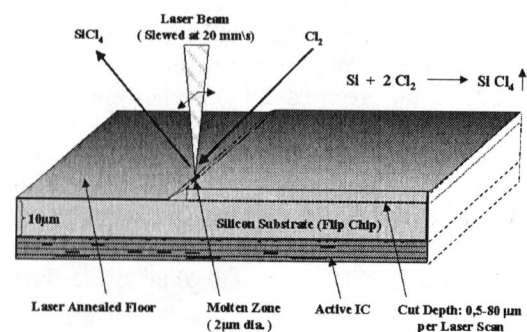

Fig. 13. Schematics of LMC etching, by scanning a laser beam across the backside of the die the Si substrate is removed.

492

window typically has dimensions of 200 to 500 μm² (Fig. 14). To optimize throughput, the laser-activated process can be used to deposit a dielectric for electrical insulation or longer interconnects with a length on the order of one millimeter. After these process steps, the die is loaded into a FIB where the local repairs at the base of the trenches are usually done because of the better lateral resolution of the FIB milling process.

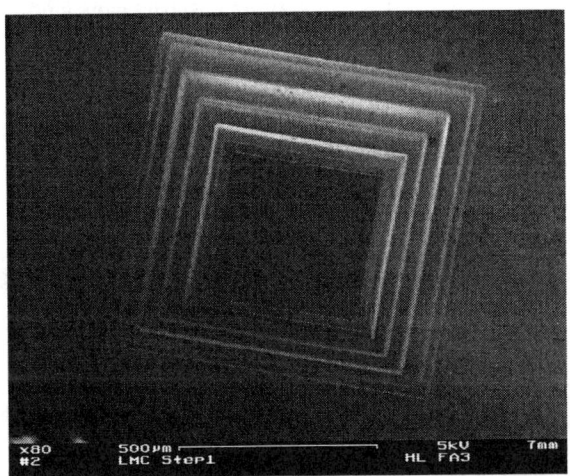

Fig. 14. LMC etched trench (typical depth up to 500 x 500 μm²).

Backside preparation techniques can roughly be divided into several categories. From the global thinning techniques, CNC milling, mechanical grinding and dry etching have been reviewed. These global Si thinning techniques serve as auxiliary methods to locally thin Si and are required for emission based and optical probe based backside FA techniques to increase signal strength. As a common feature of these techniques, large areas, i.e. the entire die, can be thinned. A remaining Si thickness of about 100 μm is recommended. At this thickness, the part still maintains adequate mechanical strength and can dissipate the required heat for operation. LMC etching and FIB are used to locally thin the die down to a remaining thickness of about 10 μm. These techniques are predominantly used for probing and device modification. Because of the limited lateral resolution precision, work has to be done with a FIB process analogue to standard frontside work.

4. Backside failure analysis techniques

Backside analysis techniques can be divided into two main categories: passive and active. Passive techniques measure the emissions from the sample and, aside from removing the packaging material, do not alter the sample. Photon emission

microscopy (PEM) and the recently developed PICA (picosecond integrated circuit analyzer) technique are passive methods that will be addressed. Active techniques use a probe to interact with the materials on the integrated circuit. The probe can vary from a physical microprobe to an electron beam to a photon beam. Active methods are more numerous than passive techniques, including traditional failure analysis techniques that have been modified for backside applications. Our focus will be on optically based, backside active techniques.

4.1. Photon (or light) emission microscopy

Photon (or light) emission microscopy (PEM) is a common failure analysis technique for frontside investigations of semiconductor devices [28, 29]. PEM is based on the detection of the weak electroluminescence radiation that is emitted from Si under numerous conditions in the visible and infrared wavelength range. The physical origins of light emission from *pn* junctions can be easily understood by studying recombination under forward and reverse bias conditions (Fig. 15, [30, 31]).

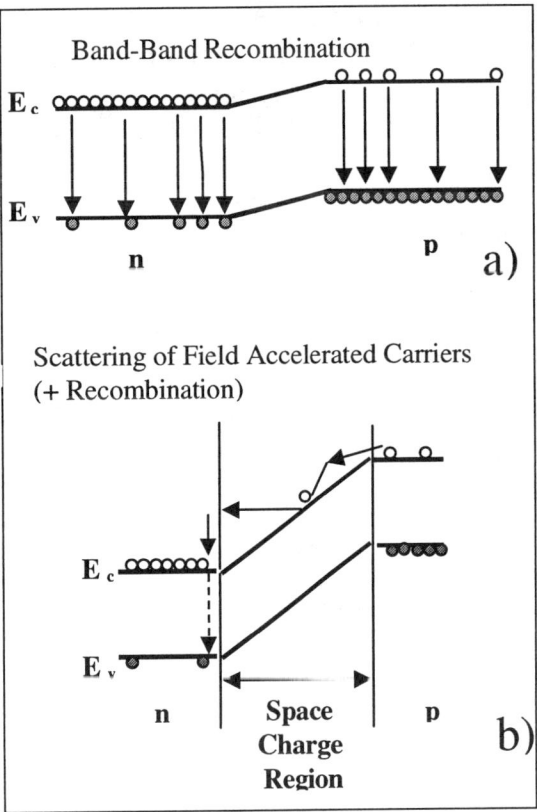

Fig. 15. Physical mechanism to light emission in a forward (a) and reverse (b) biased junction.

493

If the junction is forward biased, the light emission mechanism is similar to that of a light emitting diode (LED's) or semiconductor lasers. Under forward bias, the built-in potential is lowered and minority carriers are injected across the junction (holes are injected into the *n*-type side and electrons into the *p*-type side). Some of these injected carriers recombine radiatively with the corresponding majority carriers. Because of the low voltage bias conditions in the forward direction, there are only a few highly energetic carriers. Therefore, the light emission spectrum is very narrow with a maximum at the bandgap energy of Si at 1.1eV (Fig. 16). Under reverse bias, the built-in potential barrier and the electric field across the junction increase causing the light emission spectrum to broaden toward higher energies (shorter wavelengths).

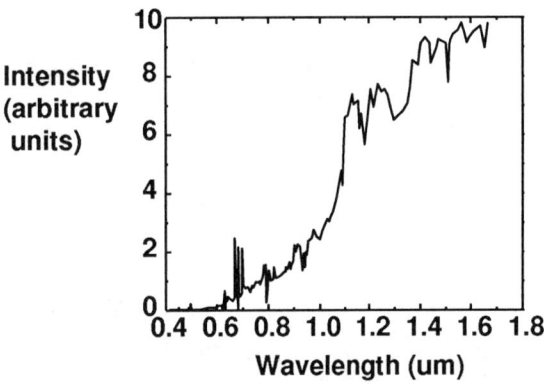

Fig. 16. Spectral content of light emitted from an *n*-channel MOSFET in saturation.

The current under reverse bias comes about from charge carriers that are thermally generated within a diffusion length of the junction, fall down the potential barrier, and are accelerated in the electric field of the space charge region. In the space charge region, the charge carriers loose their energy by various scattering events (scattering at phonons, crystal defects and charged Coulomb centers). The scattering at Coulomb centers (for example, ionized dopant atoms) goes according to classical electrodynamics hand in hand with the emission of light; the so-called Bremsstrahlung. Analogous to the continuum radiation of a x-ray tube, a broad spectrum from the visible to the near infrared wavelength range is emitted. From low temperature measurements, it is known that there are additional contributions due to band-to-band recombination and direct hole transitions between the different branches of the valence band, that are superimposed on the Bremsstrahlung. For FA, most of the relevant spectra are very similar to that of a reverse biased junction. As can be seen from the spectrum of an *n*-channel MOSFET in saturation

(Fig. 16), the spectrum extends from the visible to the near infrared wavelength range with at least about twice the intensity in the infrared than in the visible [32].

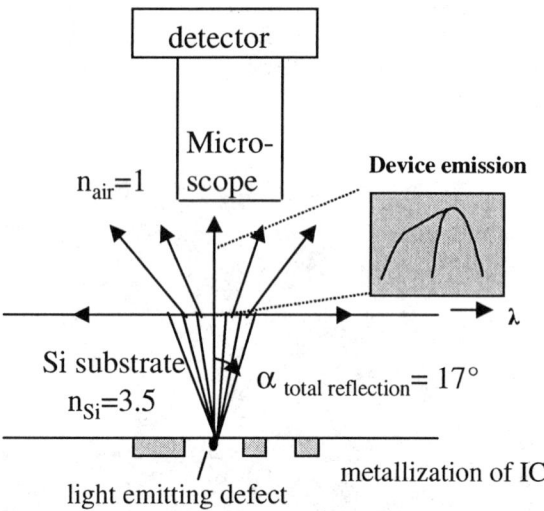

Fig. 17. PEM from the backside of the die.

For backside investigations, the photons which are emitted by a defect have to penetrate through the substrate and the backside of the die before they can be captured by the detector system (Fig. 17). Consequently, the emitted light propagates from the optically denser Si (refractive index $n_{Si} = 3.5$) to the optically thinner medium air (refractive index $n_{air} = 1$). Because of this, a large amount of the emitted light is lost due to total internal reflection and only a small cone with an angle of aperture of 34 degrees can escape from the surface. Beside these losses due to total reflection, the emitted light intensity is attenuated according to the processes described in section 1. As a consequence of the interband absorption of Si, only the fraction above about 900 nm is transmitted through the substrate from the total emitted spectrum. This infrared part of the emitted light intensity is further attenuated by an amount that depends on the doping concentration of the substrate.

From a historical perspective, light emission analysis of integrated circuits has relied on borrowed technology from military night vision applications. The benefit is that the technology exists and that performance (light amplification) is adequate for the needs of the application. The basic structure of an image intensifier is shown in Fig. 18. The drawback of image intensifiers is their spectral response and how well it matches the emission products from silicon devices. The spectral response of several commercially available image intensifiers is shown in Fig. 19. From Fig. 19, it is clear that intensifiers were designed to amplify visible light. As has been discussed already, a large portion of the emission from silicon devices is in the near infrared and for backside applications, the silicon is transparent only in the near infrared (and at longer wavelengths). This unfortunately limits the sensitivity of intensifiers to the light that escapes from the backside of an IC.

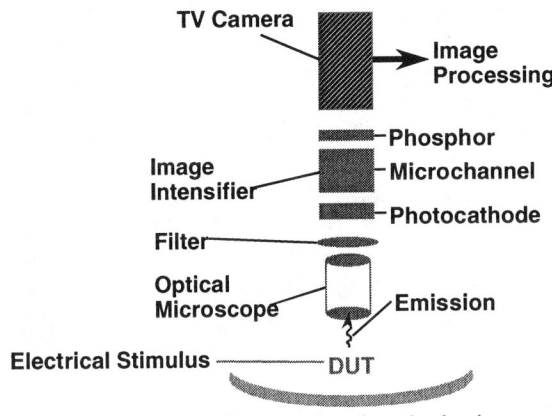

Fig. 18. Block diagram showing the basic components in an image intensifier.

The competition for intensified camera systems is cooled array cameras. While cooled array cameras do not offer any light amplification, they have extended spectral ranges and are extremely low noise.

The most cost effective and most readily available are cameras that use cooled silicon CCD arrays. These cameras benefit from the decades of process technology development for the high volume integrated circuit applications to reduce costs and increase performance. The spectral response of a typical silicon CCD array is shown in Fig. 20. The quantum efficiency curve in Fig. 20 shows that silicon arrays have some spectral response even at 1.1 μm. Usually arrays will have about 5% quantum efficiency at 1.1 μm but only if

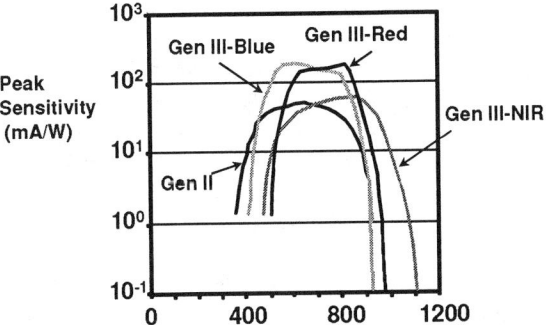

Fig. 19. Spectral response of several commercially available image intensifiers.

the camera is operating above –95 C (which is only a concern if the camera is liquid nitrogen cooled). The choice of cooling will depend on integration times needed for a typical analysis. Liquid/Peltier cooling is sufficient if integration times are less than a few minutes. For longer exposure times (up to many hours), liquid nitrogen cooling is preferred.

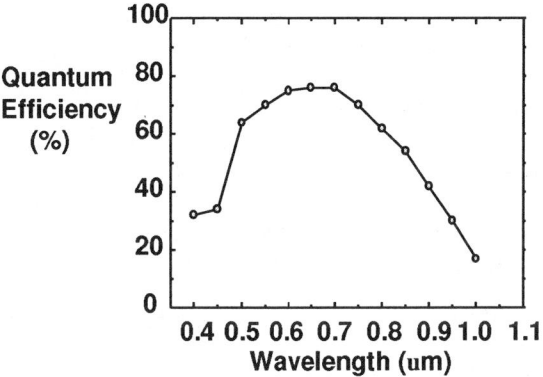

Fig. 20. Spectral response of a cooled Si CCD array.

Over the past few years, several non-Si imaging arrays have been developed that have performance similar to Si arrays but in the infrared. The technological breakthrough has been the use of a compound semiconductor sensing array coupled to a Si-based IC. This process allows the IR sensitive array to be biased and read with high performance Si IC electronics rather than the primitive circuits that can be fabricated on compound semiconductor substrates. The first such array to be used for failure analysis was the NICMOS-3 array made by Rockwell. This 256x256 pixel HgCdTe NIR array was developed to image in the 800 - 2500 nm band on the Hubble Space Telescope [33]. The quantum efficiency of this array is shown in Fig. 21. This camera used a cooled, anti-reflection coated ZnSe window to provide good transmission between 500 nm and 5000 nm as well as low thermal emission. The usual read noise of the NICMOS-3 array is 35 electrons but modifications using special read techniques can reduce this noise to less than 15 electrons. Because the temperature of the array is maintained near 77 K, the dark current is negligible for integration times from 100 milliseconds to greater than 100 seconds. Thus, it is possible to measure photon fluxes at the array of less than 1 photon per second. Thermal blackbody radiation from the sample starts to become noticeable beyond 1400 nm and limits the sensitivity to non-thermal emission. In order to eliminate thermal information, a cooled filter, that has a cutoff at 1400 nm and very low leakage at longer wavelengths must be placed in front of the array. PEM from the die backside is the application where infrared cameras significantly outperform CCD cameras.

Other arrays such as InSb and PtSi are now available for low noise NIR applications. InSb arrays are available in formats from 64 x 64 to 1024 x 1024. These arrays have similar quantum efficiencies to the HgCdTe arrays but offer a significant improvement in spectral range. InSb arrays have a spectral range from about 400 nm to 5000 nm. The increased response to wavelengths beyond 2500 nm significantly increases the difficulty to block thermal information from the sample from reaching the detector. Another issue with InSb is the need to cool the detector beyond liquid nitrogen temperatures. These arrays are designed to operate at 35 K necessitating the use of liquid helium. Alternatively, closed-cycle helium refrigerators can be used for cooling.

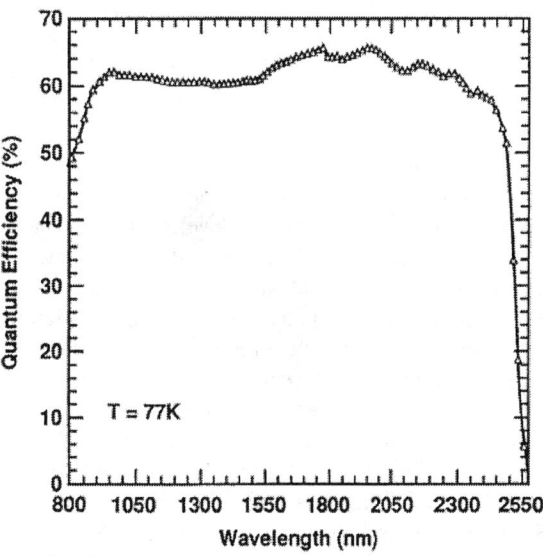

Fig. 21. Spectral response of a cooled HgCdTe CCD array.

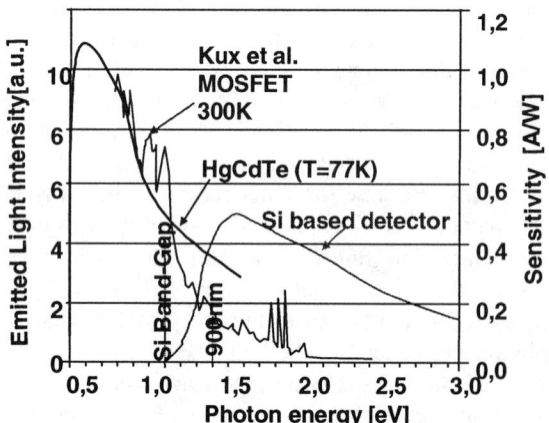

Fig. 22. A comparison of the spectral response of Si and HgCdTe arrays to the emission spectrum of an *n*-MOSFET in saturation.

Finally, Fig. 22 shows a combination of the spectral response of Si and HgCdTe array cameras along with the emission spectrum from an *n*-MOSFET in saturation. It is readily apparent from Fig. 22 that the infrared arrays are a better match to the emission spectrum from silicon-based devices.

Sample preparation for Backside PEM depends both on the doping level of the substrate and on the detector system used. In Fig. 23, a series of illuminated images taken from the backside with a Si CCD camera are shown. As can be seen in the case of a lightly doped substrate, thinning the device is not necessarily required. The only benefit of working with a thinned device is an improvement in the quality of the illuminated image at high magnifications (as described in section 1). However, for a heavily doped substrate, a remaining Si thickness of about 100 μm is required. For thicker substrates the image becomes faint and blurred. For thicknesses around 200 μm, structures on the front side of the die cannot be adequately resolved. In Fig. 24 for comparison, a series of backside illuminated images taken with a HgCdTe system are shown. Analogous to the results obtained with the Si CCD camera, the die does not need to be thinned down if the doping concentration is low. With increasing thickness, the optical quality deteriorates and for thicknesses above 400 μm, no features can be resolved. To summarize, compared with the use of a Si detector the requirements to thin the die are relaxed for a HgCdTe system. Even so, it is recommended that the die be thinned down to about 200 μm--at least in the case of a heavily doped substrate.

4.2. Picosecond imaging circuit analyzer (PICA)

PICA was developed at IBM and first presented by Kash and Tang [34]. The approach takes advantage of the light emission bursts that occur during switching of non-defective CMOS transistors. The light emission is strongest when the gate voltage is half the drain voltage; i.e. the midpoint of a digital logic transition as is shown in Fig. 25. Because "naturally" occurring optical information is used in PICA analysis, the method is non-invasive. PICA uses a gated, intensified detector to gather light intensity information as a function of time. Waveforms detecting logic state transitions are acquired stroboscopically by repeated cycling of the test device through test vectors while indexing the trigger signal for the detector through the clock cycle. Example waveforms are shown in Fig. 26.

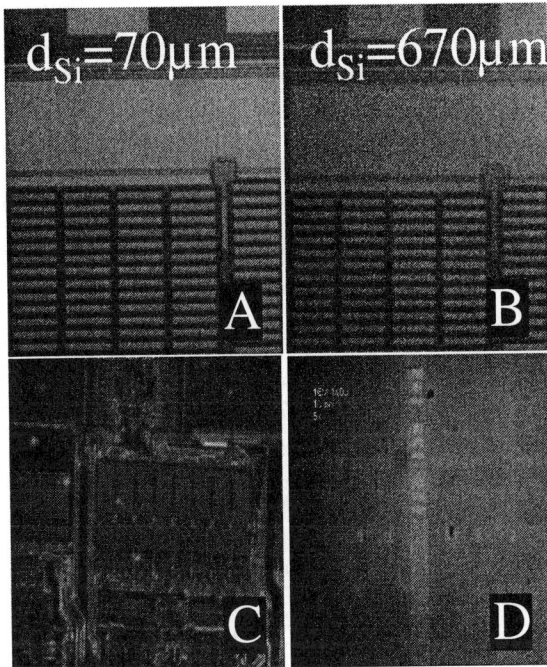

Fig. 23. Backside illuminated images taken with a Si CCD camera for a) 1×10^{15} cm⁻³, 70 μm thickness, b) 1×10^{15} cm⁻³, 670 μm thickness, c) 1×10^{19} cm⁻³, 70 μm thickness, b) 1×10^{19} cm⁻³, 670 μm thickness.

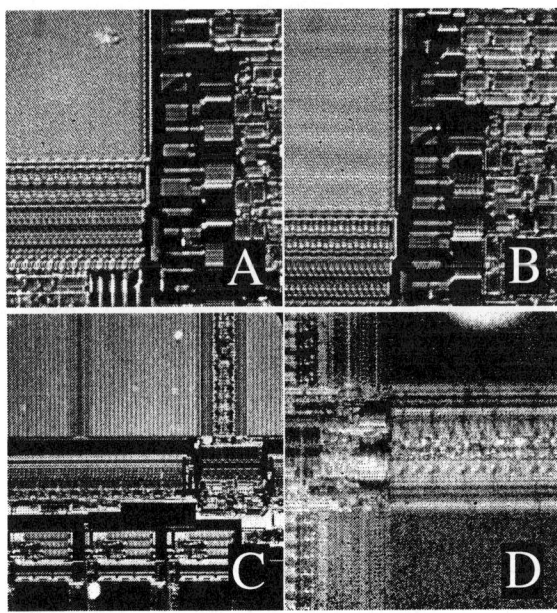

Fig. 24: Backside illuminated images taken with a HgCdTe camera for a) 1×10^{15} cm⁻³, 70 μm thickness, b) 1×10^{15} cm⁻³, 670 μm thickness, c) 1×10^{19} cm⁻³, 70 μm thickness, b) 1×10^{19} cm⁻³, 390 μm thickness.

A thermoelectrically cooled, microchannel plate, photomultiplier (MCP) with a position

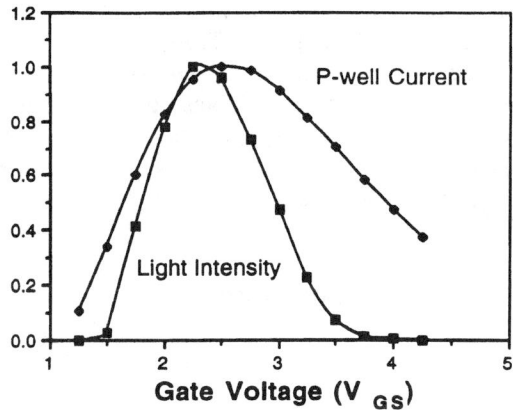

Fig. 25: Light emission intensity and its relation to p-well current.

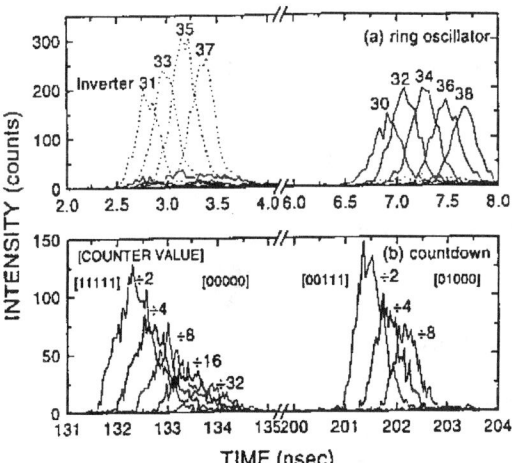

Fig. 26: Light intensity collected during various portions of a clock cycle.

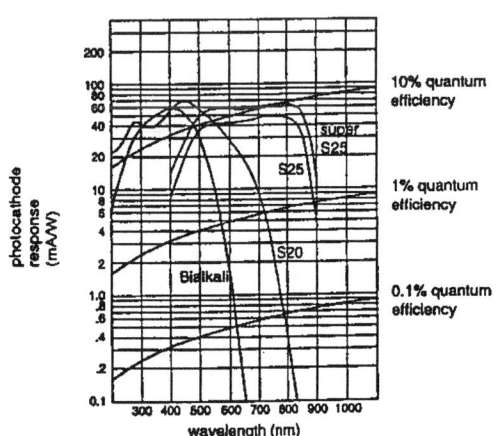

Fig. 27. Gated image intensifiers suitable for PICA and their spectral response [51].

sensitive anode is used to collect the light emission. The dark count per pixel of the PICA system has been reported as ~0.001/s, with a photon timing accuracy of 100 ps [34]. The present detector response is excellent for front side analysis, however the reported systems have marginal sensitivity for backside applications. The detector response of several gated MCP detectors are shown in Fig. 27.

4.3. Electro-optic effects in semiconductors

Scottish physicist John Kerr [43] first observed electro-optic effects in 1875. Kerr found that an isotropic, transparent substance becomes birefringent when placed in an electric field. When the optical axis corresponds to the direction of the applied electric field, the change in n (index of refraction) is given is given by:

$$\Delta n = \lambda_o K E^2 \qquad (3)$$

where K is the Kerr constant, E is the applied field, and λ_0 is the wavelength. This change in index leads to a polarization angle change that is dependent on the square of the applied voltage.

$$\Delta \varphi = \frac{2\pi K \ell V^2}{d^2} \qquad (4)$$

where d is the electrode spacing, l is the length of the Kerr cell, and φ is the polarization angle.
In 1893 German physicist Friedrich Pockels studied electro-optic effects and discovered that the birefringence is linearly proportional to electric field in certain crystals (those that lack center symmetry, such as KD*P and lithium niobate) [43]. In contrast to the Kerr effect, the Pockels effect gives a polarization rotation that is directly proportional to the applied voltage.

$$\Delta \varphi = \frac{2\pi n_0^3 r_{63} V}{\lambda_0} \qquad (5)$$

where r_{63} is the electro-optic constant. Fortunately, Si and other semiconductor materials exhibit the Pockels effect although the electro-optic constant has a much lower value in these materials than it does in KD*P for example.
Various combinations of the Kerr and Pockels effects have been used to measure potential changes from the backside of devices. The refractive index in Si is related to the electron and hole concentrations, which are related to the local applied electric fields.

$$\Delta n = \frac{n_0 q^2}{2\omega^2 \varepsilon} \left[\frac{b_e N_e^{1.05}}{m_{ce}^*} + \frac{b_h N_h^{0.80}}{m_{ch}^*} \right] \qquad (6)$$

where N_e and N_h are the electron and hole concentrations.

The basic idea is to detect changes in index through the backside of a device during operation. The Kerr and Pockels effects occur very rapidly, allowing GHz frequency measurements to be made. Normally a polarized reference laser beam is used and split into two beams with a Nomarski prism. One beam is reflected through the backside of the sample at the test point and the other beam to a reference site. A phase comparison between the reference and probe beam is performed. The phase difference is translated to a polarization change when the two beams are compared. The polarization modulation is then converted to a local voltage variation. An example experimental setup for measuring the electro-optic effect through an IC backside is shown in Fig. 28 [44 – 48]. This setup uses the direct electro-optic technique where the Si substrate is used as the sensor. Other arrangements have been used where an electro-optic crystal is placed in close proximity to an IC allowing the electric fields from the IC to interact with it [48 – 50]. This arrangement is called the indirect electro-optic technique. Both methods have been used for over a decade to measure waveforms on high frequency circuits made in GaAs mainly due to the lack of alternative measurement techniques.

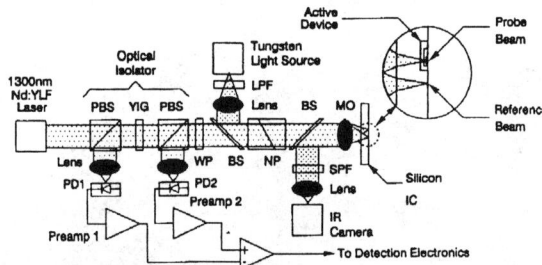

Fig. 28. Experimental setup for measuring electro-optic effects in silicon ICs.

4.4. Laser voltage probe (LVP)

The LVP produces voltage contrast images and waveforms similar to those produced by electron beam voltage contrast systems [35]. The early LVP work for Si was performed by Bloom and Heinrich at Stanford University. The LVP takes advantage of an electro-absorption phenomenon called the Franz-Keldysh effect. When the electric field exceeds 10^4 V/cm in Si, tunneling states are created that effectively reduce the band gap. Photons near the band gap wavelength that previously were transmitted are now absorbed. The change in absorption with applied electric field is used to measure changes in local potentials on the IC. The observed change in absorption with electric field is shown in Fig. 29. The maximum absorption change (shown in Fig. 30) occurs at 1065 nm, making the 1064 nm Nd:YAG laser line a convenient probe

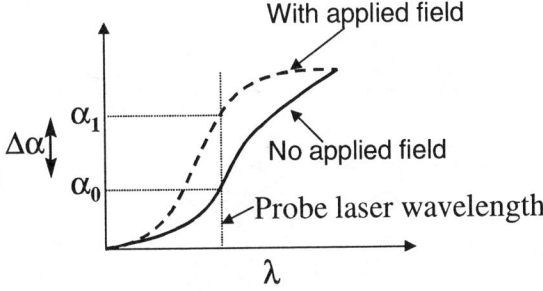

Fig. 29. LVP signal (electro-absorption) versus wavelength.

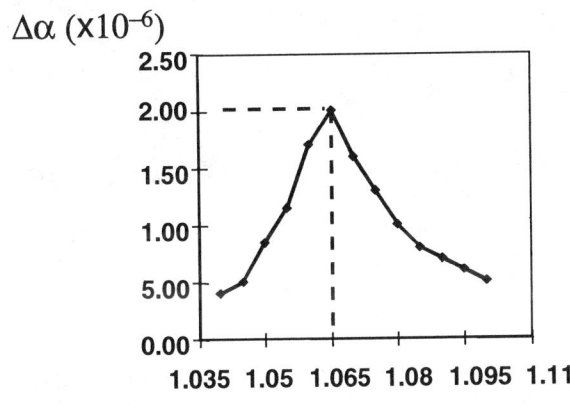

Fig. 30. Changes in F-K absorption with wavelength.

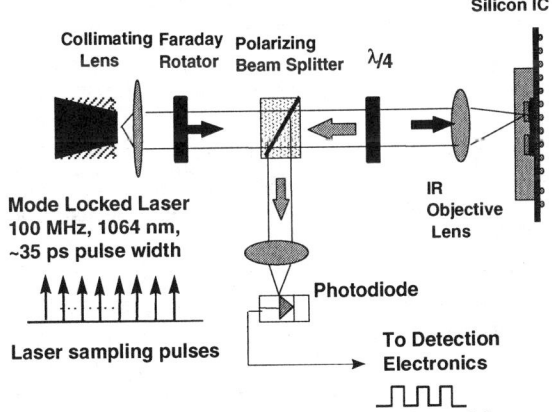

Fig. 31. Changes in F-K absorption with wavelength.

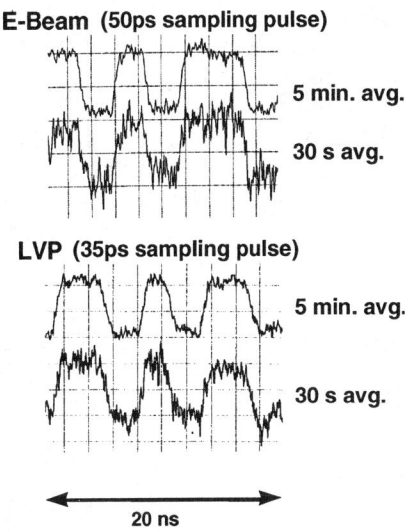

Fig. 32. E-beam and LVP waveform comparison.

source. The magnitude of the Franz-Keldysh effect is in the ppm range, so the system for detecting the change in absorption must be relatively sensitive. The use of a mode locked laser allows ~ 12 GHz bandwidth sampling. The use of an IR laser source limits the spot size to ~ 0.7 μm.

The experimental setup for LVP is shown in Fig. 31. A comparison of Fig. 31 to the electro-optic measurement setup in Fig. 28 shows several similarities with a noticeably simpler setup for LVP. Electro-optic techniques are further complicated by the need to locate a ground reference on the IC surface whereas LVP does not need any such reference.

A comparison of e-beam and LVP waveforms is shown in Fig. 32. Both methods use stroboscopic data acquisition. Note that the e-beam pulse width can be changed while the laser pulse width cannot. Improvements in IR laser technology will improve the LVP bandwidth.

4.5. Scanning optical microscopy techniques

Active photon probing takes advantage of the interactions of a scanned photon beam with an IC. In particular, for photon energies greater than the indirect band gap of silicon electron-hole pairs are generated in the semiconductor. Normally the electron-hole pairs will randomly recombine and there is essentially no net effect. However, when electron-hole pairs are generated near the interface between differently doped regions in an unbiased IC, the charge carriers are separated by the built-in potential between areas with different Fermi levels. Biasing an IC alters the Fermi levels and hence alters the magnitude of electron-hole pair separation or photocurrent. In contrast to passive photon probing techniques, the detector in active photon probing is the IC itself.

Future research on the interaction of the probe laser with circuit operation (such as occurs in LIVA and related techniques) will determine the non-invasiveness of the LVP approach.

The photon beam source for active probing is usually a scanning optical microscope (SOM). The basic SOM consists of a focused light spot that is scanned over the sample in a raster fashion. By using different laser wavelengths and intensities, variations in the amount of photocurrent can be obtained and backside photon probing can be performed.

4.5.1. LIVA (Light-Induced Voltage Alteration)

LIVA images are produced by monitoring the voltage changes of the constant current power supply as the optical beam from the SOM is scanned across an IC [36]. Voltage changes occur when the electron-hole pair recombination current increases or decreases the power demands of the IC. LIVA takes advantage of photon generated electron-hole pairs to yield information about IC defects and functionality.

In the "Voltage Alteration" mode, the IC acts as its own current-to-voltage amplifier, producing a much larger LIVA voltage signal than photocurrent signal with constant voltage biasing. This is in part due to the difference in "scale" for IC voltage (mV-V) and current (nA). Clearly the voltage signal is easier to measure.

Under identical illumination conditions, localized defects on ICs can generate LIVA signals 3 to 4 orders of magnitude greater than signals from non-defective ICs. This difference in LIVA signal depends upon the defect type, but two basic mechanisms can result in the large increase. First, the defect, because of its location in the IC amplifies the effects of normal photocurrents by altering the power demand of circuit elements connected to the defect region. Second, the defect region itself is a site of enhanced recombination compared to non-defective areas. Two types of defects illustrating the differences between these mechanisms are described below.

Junctions connected to open conductors amplify normal photocurrent effects to produce a larger LIVA signal (Fig. 33). As the photo-produced charge flows, the open line is unable to sustain its normal voltage, reducing the bias across the junction. This can change the saturation condition of the transistor directly associated with the open-circuited junction, changing the IC's power demands. The voltage of the open-circuited conductor will be the same as the p^+ diffusion. Therefore any other transistors connected to the open conductor may change their saturation condition, further amplifying the LIVA signal. When photon injection ceases, the junction voltage will slowly recover to its initial equilibrium voltage

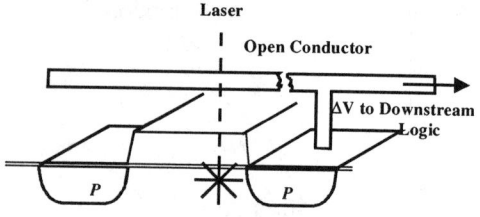

Fig. 33. Outline diagram showing how localized photon injection can affect open-circuited junctions and downstream logic.

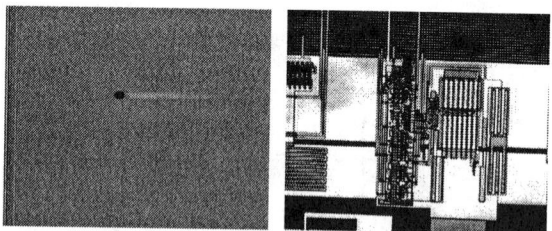

Fig. 34. Backside IR LIVA (left) and reflected IR (right) images of a microcontroller with open contacts.

which is determined by weak coupling of the open conductor to neighboring conductors and transistors, parasitic leakage conditions, and tunneling across the open [37]. Of course, if the IC logic state is such that there is no potential difference across the open-circuited junction, there will be no LIVA signal. The use of LIVA on an IC with open conductors is illustrated in Fig. 34. Generation of LIVA signals from backside IR illumination requires that the photon wavelength be long enough to penetrate through the silicon substrate but short enough (have enough energy) to produce electron-hole pairs in the junction regions. The 1064 nm line of a Nd:YAG laser serves both of these requirements. This example was performed from the backside using a 1064 nm, 1.2 W laser.

The other defect type with greatly enhanced LIVA signal is direct semiconductor damage such as overstress damage, crystal defects, and pinholes. Such semiconductor damage can cause a direct increase or decrease in recombination current. The changes in local Fermi levels caused by dopant redistribution and newly formed charge leakage paths will normally produce elevated I_{DDQ} with no illumination. Electron-hole pair generation and recombination due to illumination in the area of the defect will produce even greater amounts of "leakage current".

Logic State Mapping using LIVA can be acquired because photocurrent generation is dependent on the circuit bias. Fig. 35 illustrates how the logic states of transistors can be identified. Fig. 10b displays a LIVA difference image made from two images of the microprocessor in two different logic states. The field of view is the same as Fig. 35a. The difference image was produced by a simple subtraction of two LIVA images, with the resultant image showing only those transistors that changed logic state.

4.5.2 Localized Heating From Photon Beams

Thus far we have discussed the production of electron-hole pairs and the subsequent photocurrents in semiconductors through photon interactions. Another active photon probing approach uses the heat from a photon beam to effect IC functionality. Through localized heating shorted, resistive, and open interconnections on ICs can be localized. Two methods for detecting shorts and resistive conductors are Optical Beam Induced Resistive Change (OBIRCH) and its constant current analog, Thermally-Induced Voltage Alteration (TIVA). Open conductors can be localized using Seebeck Effect Imaging (SEI). All of the thermal probing techniques can be applied from the front and backside of an IC by use of the proper optical wavelengths.

To avoid photocurrent generation when performing localized thermal injection in these techniques, a laser wavelength with energy less than the silicon indirect bandgap is used. For the analysis described a 1340 nm laser is used.

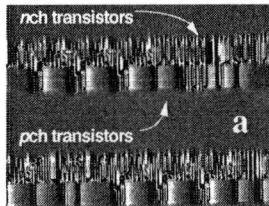 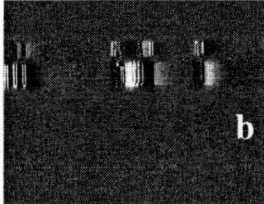

Fig. 35. Two images showing (a) a LIVA logic map of cell rows for one state, (b) a LIVA difference image between two different states. The 5 mW, HeNe laser was used for the images.

4.5.2.1 OBIRCH and TIVA imaging

Shorted conductors cause increased IC power consumption when the shorted conductors are at different electrical potentials, i.e. a short between V_{DD} and V_{SS}. The power consumption will depend upon the resistance of the short site and its location in the circuit. As a laser is scanned over an IC with a short circuit, laser heating changes the resistance of the short when it is illuminated, changing the IC power demand.

It has been found that thermally-induced power changes are usually greater for shorted signal lines than power busses. This results from signal line voltage fluctuations altering transistor gate voltages, producing the same amplification effect observed in LIVA. This resistance change with localized heating is the basis for the OBIRCH technique [38]. The change in IC power consumption is detected by an IC current change with constant voltage bias, yielding limited detection sensitivity in OBIRCH [38]. The same localized effect can be detected using a constant supply current biasing approach, which achieves greater sensitivity. This approach is known as constant current OBIRCH [39] or TIVA [40].

OBIRCH has also been used to localize defects with a high resistivity in conductors such as voids [39].

Recently it has been shown that the voltage signals produced in TIVA (and in SEI as well) have a non-linear response to increases in laser probe power [41]. An example of this non-linear response is shown in Fig. 36. The response results in a dramatic increase in sensitivity and decrease in image acquisition time. Fig. 37 is a backside TIVA example localizing a short site caused by a stainless steel particle on a 0.5 μm, 1Mb SRAM. The short was visible in earlier analyses using a 120 mW, 1.3 μm laser, and a 16-minute image acquisition time. Increasing the laser power by a factor of 5X decreased the acquisition time to 2 minutes. Moreover, the two shorted conductors that were not seen in earlier analyses are now clearly visible.

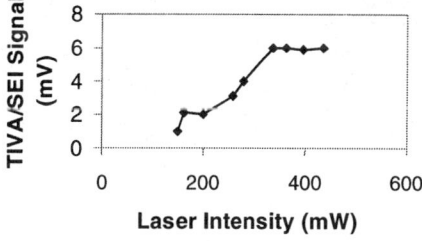

Fig. 36. Non-linear response of defects to laser intensity.

4.5.2.2 Seebeck Effect Imaging (SEI)

Thermal gradients in conductors generate electrical potential gradients with typical values on the order of μV/K [41]. This is known as thermoelectric power or the Seebeck Effect [41]. For IC analysis, the effect has been demonstrated as a means to localize voiding in metal test patterns [42]. If an IC conductor is electrically intact and has no shorts, the transistor or power bus electrically

501

driving the conductor readily compensates for the potential gradient produced by localized heating and essentially no signal is produced. However, if the conductor is electrically isolated from a driving transistor or power bus, the Seebeck Effect will change the potential of the conductor. This change in conductor potential will change the bias condition of transistors whose gates are connected to the electrically open conductor, changing the transistors' saturation condition and power dissipation. A Seebeck Effect Image (SEI) of the changing IC power demands displays the location of electrically floating conductors. The use of constant current, voltage change measurement is critical to the success of this technique because of the small voltage alteration which occurs, typically on the order of μVs. An example of a SEI image is shown in Fig. 38.

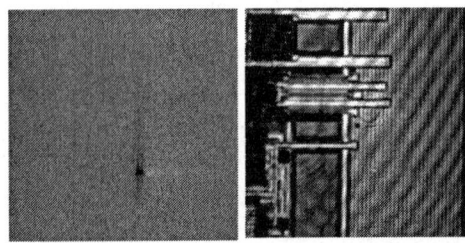

Fig. 37. Backside TIVA (left) and reflected (right) images of a shorted conductor.

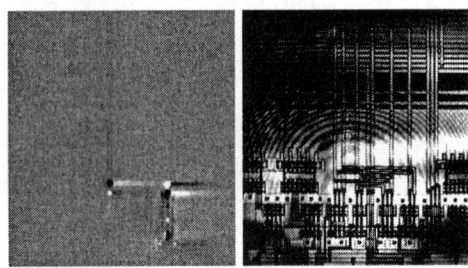

Fig. 38. Backside SEI (left) and reflected light (right) images of an open conductor.

5. Conclusions

The continued growth in the use of flip-chip packaging and multi-level metallization is driving the development of backside analysis tools and techniques. Backside sample preparation techniques and their effects have been discussed. The failure analysis tools reviewed primarily rely on infrared light for detection and stimulation of the sample under test. Non-optical tools were also discussed and can be very effective, but they are destructive and require extensive sample preparation. For the foreseeable future, backside failure analysis technique development and application will be at the forefront of advanced failure analysis interests.

Acknowledgements

The authors thank H. Millinger for CNC milling contributions and B. Ebersberger for AFM measurements, both with Infineon Technologies Munich. Sandia is a multiprogram laboratory operated by Sandia Corporation, a Lockheed Martin Company, for the United States Department of Energy under contract number DE-AC04-94AL85000.

References

[1] Lowrey T, Cloud G, Kelly D, Zagar P and Seyyedy M; 64Mb DRAM challenges. Semiconductor International. (1993) 49.

[2] DeHaven K and Dietz J. Controlled collapse chip connection (C4) – an enabling technology. IEEE. (1994) 1.

[3] Tummula R, Rymaszewski EJ. Microelectronics packaging handbook. Van Nostrand Reinhold, New York, ch. 6, 1989, pp 366-391.

[4] Kromann G, Huang W, Gerke D. A high-density C4/CBGA interconnect technology for a CMOS microprocessor. IEEE Electronics Component and Technology Conference. (1994) 1.

[5] Semmens JE and Adams T. Flip chip package failure mechanisms. Solid State Technology. (1998) 59.

[6] Lau J.J. Flip chip technologies. McGraw-Hill, New York, 1996.

[7] Pankove JL. Optical processes in semiconductors. Dover Publications, Inc. New York, 1971.

[8] Landolt-Börnstein. Zahlenwerte und funktionen aus naturwissenschaft und technik. Springer Verlag, Berlin, Chapter 17 Semiconductors, 1989.

[9] Soref RA and Bennett BR. Electrooptical effects in silicon. IEEE J. Quant. Electron. 23 (1987) 123.

[10] Weber G, diploma thesis, Infineon Technologies, Munich, 1998.

[11] Chiang CL and Hurley DF. Dynamics of backside wafer level microprobing. IRPS (1998) 137.

[12] Widmann D, Mader H, Friedrich H. Technologie hochintegrierter schaltungen. Springer Verlag Berlin, 1996.

[13] Livengood RH, Winer P, Rao VR. Application of advanced micromachining techniques for the characterization and debug of high performance microprocessors. Proc. of the Intern. Symp. on Electron, Ion and Photon Beams and Nanofabrication (1998)

[14] Lee R and Antoniou N. FIB micro-surgery on flip-chips from the backside. ISTFA. (1998) 455.

[15] Antoniou N. Focused ion beam systems keep pace. Back-end Supplement. (1998) 33.

[16] Ullmann PF, Talbot CG, Lee RA, Orjuela C, Nicholson R. A New robust backside flip-chip probing methodology. ISTFA. (1996) 381.

[17] Livengood RH and Rao VR. FIB techniques to debug flip-chip integrated circuits. Semicon. Intern. (1998) 111.

[18] Casey Jr. JD, Doyle AF, Lee RG and Stewart DK and Zimmermann H. Gas-assisted etching with focused ion beam technology. Microelectronic Engineering 24 (1994) 43.

[19] Micrion 9800 FlipChip Focused Ion Beam System, Data Sheet

[20] Ehrlich DJ, Tsao JY. Laser microfabrication – thin film processes and lithography. Academic Press, Inc., Boston, 1989.

[21] Bloomstein TM and Ehrlich DJ. Laser-chemical three-dimensional writing for microelectromechanics and application to standard-cell microfluidics. J. Vac. Sci. Technol. B 10 (1992) 2671.

[22] Bloomstein TM and Ehrlich DJ, " Stereo laser micromaching of silicon", Appl. Phys. Lett. 61. (1992) 708.

[23] Ehrlich DJ and Tsao JY. A review of laser-microchemical processing. J. Vac. Sci. Technol. B1 (1983). 969.

[24] Silvermann S, Laser microchemical technology enables real-time editing of first-run silicon. Solid State Technology (1996) 113.

[25] Treyz GV, Beach R and Osgood Jr. RM. Rapid direct writing of high-aspect ratio trenches in silicon: process physics. J. Vac. Sci. Technol. B 6, (1988) 37.

[26] Silverman S, Aucoin R, Mallatt J and Ehrlich D. Laser microchemical technology: new tools for flip-chip debug and failure analysis. ISTFA. (1997) 211.

[27] Silverman S, Aucoin R, Ehrlich D and Nill K. OBIC endpointing method for laser thinning of flip-chip circuits. ISTFA. (1998) 461.

[28] Hawkins CF, Soden JM, Cole Jr. EI, and Snyder ES. Use of light emission in failure analysis of CMOS ICs. ISTFA. (1990) 55-67.

[29] Soden JM and Cole Jr. EI. IRPS Tutorial. (1992) 4a1-4a.16.

[30] Pankove JI. Optical processes in semiconductors. Prentice-Hall, New Jersey 1971, Ch 6.

[31] Chynoweth AG and McKay KG, Phys. Rev., 102 (2), (1956) 369-376.

[32] Kux A, Lugli P, Ostermeir R, Koch F, and Deboy G. Mat. Res. Soc. Symp. Proc., 256. (1992) 223-226.

[33] Shivanandan K and Nyunt K. ISTFA. (1995) 69-71.

[34] Tsang JC and Kash JA. Appl. Phys. Lett., 70. (1997) 889-891 (1997).

[35] Paniccia M., Eiles T, Rao VRM, and Yee WM. Novel optical probing technique for flip chip packaged microprocessors. Int. Test Conf. (1998) 740-747.

[36] Cole Jr. EI, Soden JM, Rife JL, Barton DL, and Henderson CL. Novel failure analysis techniques using photon probing with a scanning optical microscope. IRPS. (1994) 388-398.

[37] Cole Jr. EI and Anderson RE, Rapid localization of IC open conductors using charge-induced voltage alteration (CIVA). IRPS. (1992) 288-298.

[38] Nikawa K and Inoue S. Various contrasts identifiable from the backside of a chip by 1.3 µm laser. ISTFA (1996) 387-392.

[39] Nikawa K and Inoue S. New capabilities of OBIRCH method for fault localization and defect detection. Proc. of Sixth Asian Test Symposium. (1997). 219-219.

[40] Cole Jr. EI, Tangyunyong P, and Barton DL. Backside localization of open and shorted IC interconnections. IRPS (1998) 129-136.

[41] Cole Jr. EI, Tangyunyong P, Benson DA, and Barton DL. TIVA and SEI developments for enhanced front and backside interconnection failure analysis. ESREF (1999).

[42] Koyama T, Mashiko Y, Sekine M, Koyama H., and Horie K, New non-bias optical beam induced current (NB-OBIC) technique for evaluation of Al interconnects. IRPS (1995) 228-233.

[43] E. Hecht and A. Zajac, Optics, Addison-Wesley, Reading, MA, Ch. 8, pp. 263 – 266, 1979.

[44] H. K. Heinrich, D. M. Bloom, and B. R. Hemenway, Noninvasive sheet charge density probe for integrated silicon devices, Appl. Phys. Lett. 48 (16), pp. 1066 – 1068, 21 April 1986.

[45] H. K. Heinrich, N. Pakdaman, J. L. Prince, D. S. Kent, L. M. Cropp, Picosecond backside optical detection of internal signals in flip-chip mounted silicon VLSI circuits, Proc. 3rd European Conf. on Electron and Optical Beam Testing of Electronic Devices, pp. 262 – 273, 1991.

[46] H. K. Heinrich, N. Pakdaman, D. S. Kend, and L. M. Cropp, Backside optical measurements of picosecond internal gate delays in a flip-chip packaged silicon VLSI circuit, IEEE Photonics Technology Letters, vol. 3, no. 7, pp. 673 – 675, July 1991.

[47] H. K. Heinrich, B. R. Hemenway, K. A. McGroddy, D. M. Bloom, Measurement of real-time digital signals in a silicon bipolar junction transistor using a noninvasive optical probe, Electronics Letters, vol. 22, no. 12, pp. 650 – 651, 5 June 1986.

[48] B. R. Hemenway, H. K. Heinrich, J. H. Goll, Z. Xu, and D. M. Bloom, Optical detection of charge modulation in silicon integrated circuits using a multimode laser-dioes probe, IEEE Electron Device Letters, vol. EDL-8, no. 8, pp. 344 – 346, Aug. 1987.

[49] W. Mertin, A. Leyk, F. Taenzier, T. Novak, G. David, D. Jager, and E. Kubalek, Characterization of a MMIC by direct and indirect electro-optic sampling and by network analyzer measurements, Microelectronic Engineering 24, pp. 377 – 384, 1994.

[50] G. Baur, R. Hoffman, G. Solkner, H. –J. Pfeiderer, High voltage resolution with a laser diodes based electro-optic measurement system, Microelectronic Engineering 24, pp. 393 – 400, 1994.

[51] Quantar Technology Mepsicron-II Series Single Photon Imaging Detector System Data Sheet, Quantar Technology Incorporated, 1997.

Non-Invasive Backside Failure Analysis of Integrated Circuits by Time-Dependent Light Emission: Picosecond Imaging Circuit Analysis

J.A. Kash, J.C. Tsang, and D.R. Knebel
IBM
Yorktown Heights, New York

D.P. Vallett
IBM Microelectronics Division
Essex Junction, Vermont

Abstract

A noninvasive backside probe of integrated circuits has been developed. This new probe can diagnose at-speed failures, stuck faults, and other defects. Because it is a highly parallel imaging technique, faults may be isolated which are difficult to locate by other methods. This optical technique has been named "PICA", for picosecond imaging circuit analysis. PICA relies on the fact that an FET in a CMOS circuit emits a picosecond pulse of light each time the logic gate changes state. The source of this emission is explained. The PICA technique, which combines optical imaging of the emission with picosecond time-resolution, is described. Because of the imaging, time-resolved emission data is acquired for many transistors in parallel. The use of the emission for failure analysis and AC characterization of integrated circuits is demonstrated. Because the emission can be detected from either the front or back side of the chip, it can be used for both front and back side analysis.

Introduction

With the submicron gate lengths typical of the FETs in modern integrated circuits, electric fields in the channel of an FET can readily exceed 10^5 volts/cm, particularly when the FET is in saturation and most of the source-drain voltage, V_{DS}, appears across a narrow region of the conduction channel near the drain. Because a typical mean free path for a carrier in the conducting channel of the FET is several tens of nanometers, maximum carrier energies in the region near the drain have been calculated to be greater than 1 eV[1]. Switching delays, rise times, and fall times are typically tens of picoseconds. Because the carrier-carrier thermalization time is much less (typically subpicosecond), the hot carriers in an FET are always in quasi-thermal equilibrium. The high energy part of the hot carrier distribution is described by a Boltzmann tail with temperatures that can be greater than 3000K. Although the matrix element responsible for the light emitting transition remains somewhat controversial[2], a small fraction of the hot carriers have been

shown to be able to give up this kinetic energy by creation of a photon[3]. Because the most energetic carriers have kinetic energies exceeding 1eV, an FET in saturation gives off measurable light in the near infrared and visible. An example of the spectrum of emission from a submicron nFET is shown in figure 1. Note that, as expected from the above discussion, the characteristic temperature of the emission increases as V_{DS} increases.

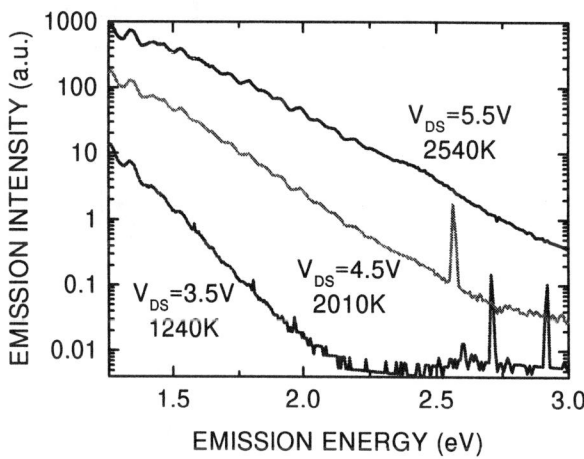

Figure 1. Optical emission spectrum from an nFET as a function of source drain voltage V_{DS}. For all these spectra, the gate and drain are connected together. Spectra are normalized for the system response. (The small oscillations and the sharp spikes are artifacts of the normalization.) The effective temperature of each spectrum is indicated.

This emission from the silicon FETs is present whenever the FET is in saturation. It is present in both nFETs and pFETs. However, the higher channel mobility of electrons in nFETs as compared to holes in a pFET gives a hotter electron temperature than hole temperature for FETs with comparable channel lengths[3]. As a result, the emission from an nFET is typically 10 to 50 times greater than from a pFET of the same dimensions. The emission is most significant when the FET is in saturation, since only under saturation conditions are the electric fields in the conducting channel concentrated in the drain region. In the linear region of transistor operation, the source-drain voltage V_{DS} generates an electric field which is uniform throughout the channel, and the resulting carrier distribution is much cooler. Measurements of the intensity of the emission as a function of V_{DS} and the gate voltage V_G show there is no emission outside of the device saturation region. The intensity is weak when the FET is conducting in the linear region, and increases rapidly as the transistor approaches saturation[4,5]. The intensity of the emission varies exponentially with V_{DS}.

The above discussion has considered optical emission from an FET under DC conditions. In a properly functioning CMOS gate, where a group of nFETs is in series with a group of pFETs, under DC conditions, no current flows. This absence of DC current, which gives CMOS circuits their characteristic low power consumption, arises because under steady state conditions, either the pFETs or the nFETs of any gate are turned off, so there is no conducting path between the power supply and ground. As illustrated in figure 2, however, when a CMOS circuit is switching from one logic level to another, there is a brief period of time (typically a few tens to hundreds of picoseconds in modern circuits), when both FETs are in saturation[6]. During this time current is flowing and light is emitted from the FETs. As a result, in a CMOS circuit, light comes from each FET whenever the associated circuit is switching. The fact that this switching-induced emission is coincident with the electrical switching allows its use as a diagnostic of circuit timing[7,8].

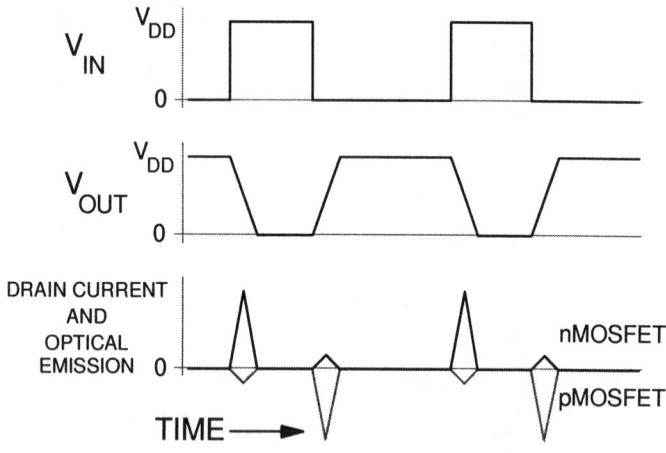

Figure 2. Schematic diagram of the input and output voltages of a CMOS inverter, and the nFET and pFET current associated with switching. For clarity, the nFET current is shown as positive, and the pFET current is shown as negative. The duration of the current pulses is comparable to the switching time of the circuit.

In figure 2, note that for a circuit attached to a real load, the nFET current is much larger when the inverter output is going from 1 to 0 than from 0 to 1, with the reverse being true for the pFET. The current through an inverter consists of two parts[6]. The first is a short circuit current going from the supply to ground which passes through both the n- and pFETs. However, the second involves the charge on the load. For the 0 to 1 transition at the inverter input, this charge passes through the nFET as the load is pulled down to ground. For the reverse transition (1 to 0 at the input), this charge goes from the supply to the load as the inverter output is pulled up to the supply voltage. This asymmetry in current also appears in the emission intensity, which allows determination of the actual logic level before and after switching is observed optically[8].

Technique

A distinct advantage of using this optical emission for circuit diagnostics and failure analysis is that it can be detected with an imaging photomultiplier tube. With this detector, an entire circuit or chip can be imaged and time-resolved with picosecond resolution simultaneously. This parallel acquisition of time-resolved optical emission has been named "PICA", for "picosecond imaging circuit analysis". The PICA apparatus uses a cooled microchannelplate photomultiplier with a position sensitive restive anode[9,10] to simultaneously image and time-resolve the optical emission from individual FETs in submicron CMOS circuits. Because the optical emission is quite weak, the system detects single photons. With the circuit operating repetitively, time-correlated photon counting[11] is used to time-resolve the emission. The time response of the measurement system limits the temporal width of the measured emission peaks to about 100 psec.

Our initial PICA measurements[7,8] have been made from the processed (front) side of chips. However, as the emission spectra of figure 1 make clear, a significant portion of the emission is at wavelengths which are at or near the 1.1 micron silicon bandgap. For such wavelengths, and depending somewhat on the doping level, silicon is partially or fully transparent. Thus, if a chip is suitably thinned and polished, the emission can be measured through the backside. Using our present detector, with usable response from the visible out to about 900 nm, we have been able to obtain PICA data from flipchip mounted normally operating ICs which have been thinned to roughly 100 microns and optically polished.

Results

Ring Oscillator - Frontside Analysis

To provide an example of failure analysis using PICA, a 47-inverter ring oscillator (effective gate length 0.6 microns) was modified by using a focused ion beam (FIB) tungsten deposition process. Approximately 3500 ohms of resistance was added between the supply voltage V_{DD} and the input to inverter 11, as indicated in figure 3. The resistance value was calculated and modeled so as to effect significant changes in the time delay, rise time, fall time, and low-level voltage of pulses propagating through the chain without causing the oscillation to cease.

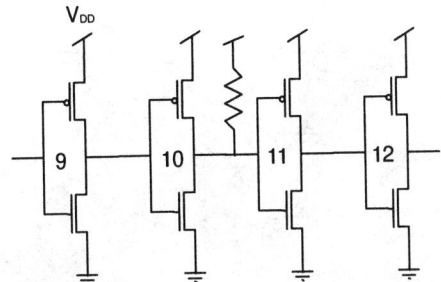

Figure 3. Schematic of a portion of the 47 inverter ring oscillator circuit, modified by adding a resistor with a focused ion beam.

This added resistor prevented the input to inverter 11 from going fully low. Thus, when this input was at logical 0, the nFET of inverter 11 was not fully off, resulting in leakage current through this nFET. The presence of this leakage current can clearly be observed in the time-integrated emission from the operating circuit, figure 4. Here, the nFET of inverter 11 is about 10 times brighter than the other nFETs in the ring. In addition, as will be shown below, the resistor changed the circuit delay through the inverter. In figure 4, the pFETs are not visible. One reason is that, as discussed above, pFET emission is generally more than 10 times weaker than nFET emission. In addition, the layout of this particular ring oscillator was such that the pFETs were largely covered by opaque metal lines.

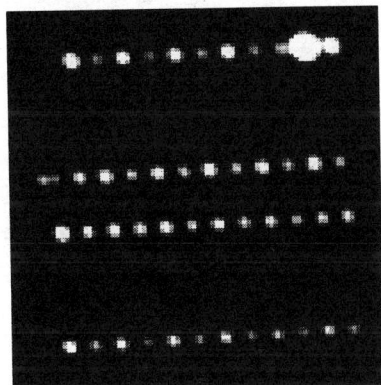

Figure 4. Time-integrated optical emission from the operating ring oscillator. Each emission spot corresponds to an inverter nFET. The brightest spot corresponds to inverter 11, with inverter 12 to its immediate right.

Figure 5 shows "optical waveforms" (i.e., light emission vs. time) from a series of nFETs including nFET 11. For the optical waveforms shown in figure 5, the input of each inverter is going from 0 to 1. While the other displayed waveforms consist simply of a short pulse of light coincident with the logic switch, the leakage current through nFET 11 is seen as light emission whenever the input to inverter 11 is 0. Only odd-numbered inverters are shown here. In the time interval displayed in figure 5, the even-numbered inverters are undergoing the reverse transition (i.e., 1 to 0 at the inverter input). As expected (see the discussion of figure 2), we measure a significantly weaker emission pulse which is not shown here.

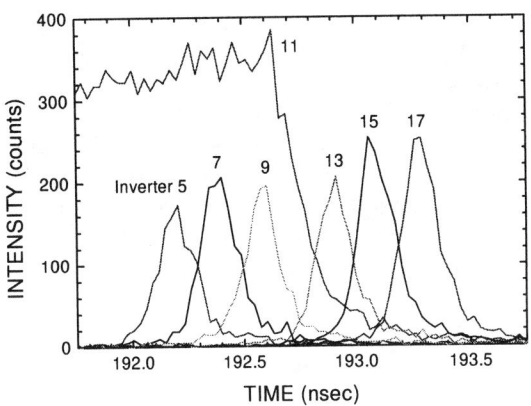

Figure 5. Time-resolved optical emission waveforms from several of the nFETs of the modified ring oscillator. For the odd-numbered waveforms shown, the inverter input is going from 0 to 1.

The period of the ring oscillator is about 9 nsec. So, about 4.5 nsec after the time shown in figure 4, these inverters are undergoing the reverse transition (i.e., 1 to 0 at the input), while their even-numbered neighbors have a 0 to 1 transition. The optical waveforms associated with these transitions are shown in figure 6. Here, the effects of the added resistor are less dramatic than in figure 5. However, the extra intensity of inverter 12 results from the slowed transition at the output of inverter 11 caused by the added resistor.

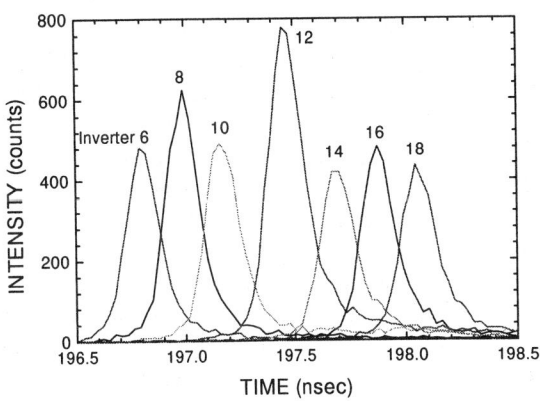

Figure 6. Same as figure 5, but one-half cycle of the ring oscillator later. Here, the even-numbered inverters have inputs going from 0 to 1.

In addition to looking at intensity, delays through each inverter can be determined from these optical waveforms. These measured delays are shown in table 1. For inverters which are not near the FIB resistor, stage to next-near-neighbor stage delays are all in the range 180 to 191 nsec. However, when the input to inverter 11 is at logical 0, we see in fig. 5 that nFET 11 is already conducting. As a result of this "pre-charging", the delay through the inverter is less. In fig. 5 it is difficult to directly measure delays associated with inverter 11 because the temporal shape of the emission is very different from the other inverters. Nonetheless, the 328 psec delay between inverters 9 and 13 is significantly less than, for example, the 382 psec delay between inverters 5 and 9, showing the effect of the pre-charging. Half a cycle of the ring later, when the opposite transitions are occuring, the resistor acts as an additional load when the output of inverter 10 is going low. This extra loading increases the circuit delay between inverters 10 and 12, and to a lesser extent, between inverters 12 and 14.

Between Inverters	Delay (psec)
5 and 7	191
7 and 9	191
9 and 13	328
13 and 15	182
15 and 17	191
6 and 8	182
8 and 10	182
10 and 12	294
12 and 14	234
14 and 16	180
16 and 18	180

Table 1. Relative delays between inverters, with data for odd-numbered inverters derived from the waveforms of figure 5, and the even-numbered inverters from figure 6.

Microprocessor - Backside Analysis

Figure 7 illustrates the ability of PICA to perform time-resolved imaging of single transistors through the silicon substrate. Here the sample is an IBM Power3 PowerPC microprocessor[12] which is built in IBM's CMOS6S2 0.25 μm, 2.5 volt, technology. Experiments were performed on a packaged Power3 microprocessor chip which was thinned on the backside to approximately 50 μm and polished to an optical finish to minimize signal attenuation of the infrared light emission through the substrate. The module was then socketed in a testboard and the clock circuitry was exercised. Figure 7 is a single frame of a movie showing the time-resolved emission from the phase-locked loop of the microprocessor. A layout of the circuit has been superimposed, which allows direct connection between the circuit schematic and the optical emission. By identifying each emission spot with an individual transistor, a detailed analysis of the circuit operation is possible.

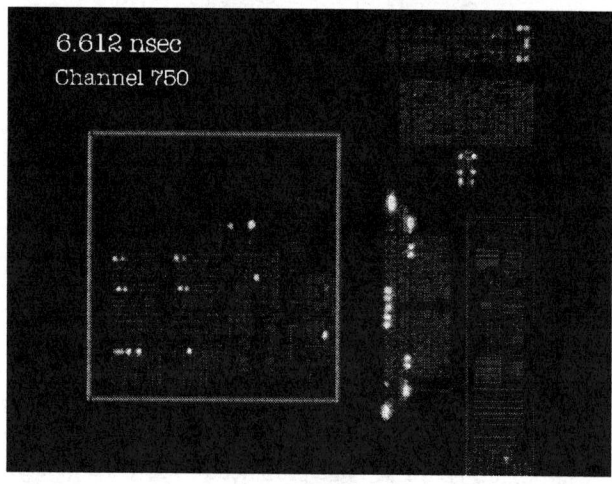

Figure 7. Single frame from a video of the optical emission from individual transistors in the phase-locked loop of a microprocessor. Each frame of the video corresponds to 28.5 psec. Superimposed on the bright spots of emission is a drawing of the circuit layout. The area on the left within the rectangle is physically distant from the rest of the circuit, but has been moved closer on the image for display.

An example is shown in figure 8. The two optical waveforms correspond to the system clock receiver and local clock delay circuits. A comparison of these waveforms shows that any offset between the system clock and local clock is less than one bit of the time digitizer, 28.5 psec .

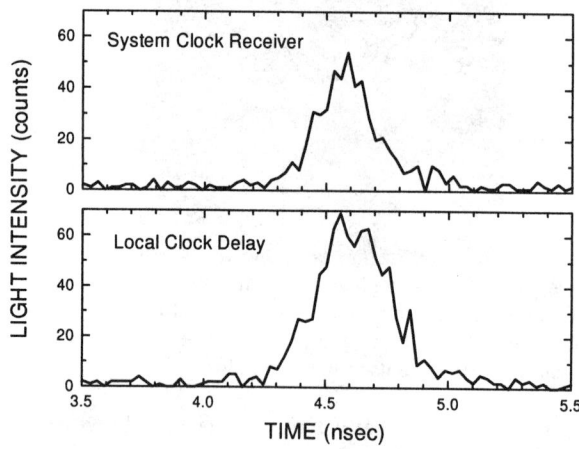

Figure 8. Optical waveforms from two of the transistors in the field of view of figure 7. The same data that is used to generate movie frames such as figure 7 can be used to derive these optical waveforms.

Microprocessor - Full Chip Analysis

The data shown in figures 7 and 8 were taken with a microscope objective so that individual transistors can be imaged. With a macro lens (approximately 1x magnification), all or most of a chip can be imaged onto the 25 mm diameter detector photocathode. With this large field of view, individual transistors cannot be resolved because of the 50 μm effective pixel size of the detector. Instead, complete circuit blocks are imaged.

An example is shown in figure 9, which is a series of time-resolved images of about two-thirds of the same Power3 microprocessor described above. Figure 9a. shows an emission image which corresponds to a time slice of 28.5 psec during which the central clock buffer is switching. Figure 9b shows an image 428 psec after Figure 9a, capturing the emission pulse of the two clock repeaters. Each repeater propagates the clock signal to eight sector buffers. Figure 9c shows a time slice 969 psec after Figure 9a, in which the emission from 12 of the 16 sector buffers are clearly observed. (The other 4 sector buffers are not in the field of view.) A continuous series of such can be assembled into a movie of the time resolved PICA emission. These movies help in the visualization of the time sequence of switching events. Alternately, an optical waveform can be derived for each emission spot to determine detailed switching activity of the corresponding circuits.

Conclusions

These time-resolved data show that critical performance characterization, debug, and failure analysis of timing problems and AC defects is possible with PICA. Analysis can be performed from the front side of the circuit. More importantly, the spectral characteristics of the emission also enable critical backside analysis of a flip-chip mounted logic chip.

Data are acquired for many transistors in parallel in a single image, allowing off-line analysis of any one transistor or group of transistors in a field of view. This feature creates an extremely productive analytical environment without the need for repetitive point-by-point probing. Further, with PICA there is no need to create backside 'holes', nor is there a requirement for encumbering the chip with designed-in test points or test structures. PICA is a purely passive test method that does not load circuits in operation.

Finally, FET emission intensity will increase as technology evolves. Supply voltage scales more slowly than channel length, increasing the electric field in the channel of the FET. The higher field creates more hot electrons, and hence more detectable emission. As a result, PICA should remain a viable backside analytical technique for some time.

Acknowledgments

The authors wish to thank P. J. Restle and S. E. Steen of the IBM Research Division, and H. F. Casal and E. Seewann of the IBM Server Development Group, for their contributions to this work. The authors also wish to acknowledge R. J. Evans, T. J. Hartswick, and P. S. Phoenix of the IBM Microelectronics Division for sample preparation and focused ion beam application.

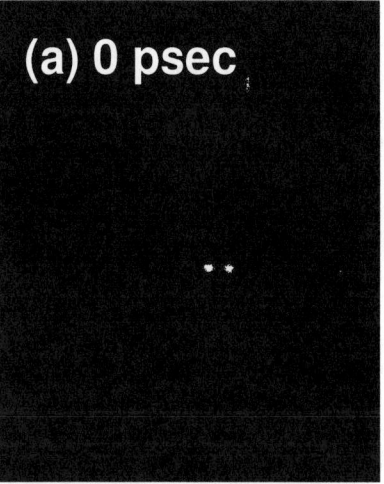

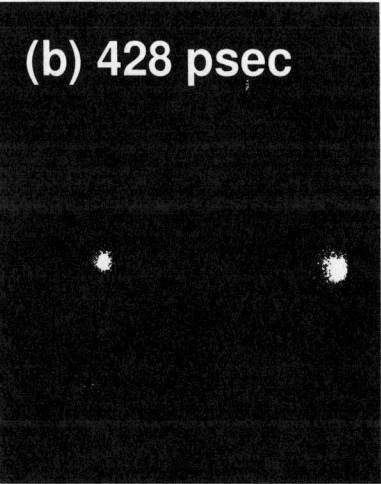

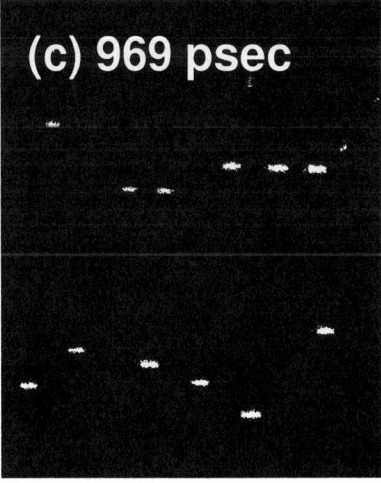

Figure 9. Time-resolved images from approximately two-thirds of the backside surface of a microprocessor. Figure 9a shows an emission image, chosen as t=0 psec, which corresponds to a time slice of 28.5 psec during which the central clock buffer is switching. Figure 9b shows an image 428 psec after figure 9a, where the repeaters are switching. Figure 9c shows a time slice 969 psec after figure 9a, where the sector buffers are switching.

References

1. M. V. Fischetti, S. E. Laux, and E. Crabbe, *J. Appl. Phys.*, 78, 1058-1087 (1995)

2. S. Villa, A. L. Lacaita, and A. Pacelli, *Phys. Rev. B*, 52, 10993-10999 (1995)

3. J. Bude, N. Sano, and A. Yoshii, *Phys. Rev. B,* 45, 5848-5856 (1992)

4. C. F. Hawkins, J. M. Soden, E. I. Cole, and E. S. Snyder, *Proc. International Symposium for Testing and Failure Analysis*, 1990,. 55-67

5. S. Tam and C. Hu, *IEEE Trans. Electron Devices*, 31, 1264-1273 (1984)

6. Neil H. E. Weste and Kamran Eshraghian, *Principles of CMOS VLSI Design, Second Edition*, Reading, Massachusetts: Addison-Wesley, 1993.

7. J. C. Tsang and J. A. Kash, *Appl. Phys. Lett.*, 70, 889-891 (1997)

8. J. A. Kash and J. C. Tsang, *IEEE Electron Device Letters,* 18, 330-332 (1997)

9. C. Firmani, E. Ruiz, C. W. Carlson, M. Lampton, and F. Paresce, *Rev. Sci. Instrum.* 53, 570-574, May, (1982)

10. J. C. Tsang, "Multichannel Detection and Raman Spectroscopy of Surface Layers and Interfaces", in *Light Scattering in Solids V*, eds. M. Cardona and G. Guntherodt, Berlin: Springer-Verlag, 1988), chapter 6, pp. 233-284.

11. S. Charbonneau, L. B. Allard, Jeff F. Young, G. Dyck and B. J. Kyle, *Rev. Sci. Instrum.*, 63, 5315-5319 (1992)

12. Peter Song, "Power3 to replace P2SC", *Microprocessor Report*, 11, 23-27 (1997)

Theory of Operation of High Resolution Liquid Metal Ion Source Focused Ion Beam Systems

Jon Orloff
University of Maryland
College Park, Maryland

INTRODUCTION

Focused ion beam (FIB) systems based on the liquid metal ion source (LMIS) are capable of high spatial resolution (to ~5 nm) while providing a high current density beam focused on a target (1-10 A cm^{-2}). These properties have been exploited widely in the past fifteen years or so, mainly by the semiconductor industry, where the ability to cross section samples and view them has been of tremendous value to failure analysis. The FIB can also be used for material deposition (from the gas phase) and can promote chemistry for selective removal of different materials from a surface. The reason FIBs have the properties they do is because they are based on high brightness ion sources - LMISs. These are field emission type ion sources with a very small (~ 5 nm) source size and a reasonably high intensity (~ 10 - 20 x 10^{-6} A sr^{-1}). Because of the small source size relatively simple optics can be used to achieve magnification in the range 0.1 - 1 with a working distance ~ 2-3 cm, so optical systems can be made quite compact.

This paper will review the operation of the LMIS and typical optics used in a FIB. The resolving power of the FIB used as a scanning ion microscope will be discussed and related to the beam size. Finally, a brief discussion of ion beam induced chemistry in the FIB will be given.

THE LIQUID METAL ION SOURCE

The heart of the high resolution FIB is the LMIS, because of its favorable optical properties. The LMIS usually consists of a blunt field emitter with an end radius ≈ 5 micrometers which is coated with the metal from which ions are to be created. Usually this metal is Ga, which is liquid at room temperature (M.P. = 29.8 C) and has an extremely low vapor pressure at room temperature (P ~ 10^{-15} torr). A photograph of an LMIS source is shown in Figure 1.

Taylor[1] showed that when a liquid is placed in a strong electric field it can assume a conical shape under the competing forces of electric field stress and surface tension. The equation expressing the equilibrium is:

$$\tfrac{1}{2}\,\varepsilon_0 F^2 = 2\,\gamma/r$$

where ε_0 is the permittivity of vacuum, F is the electric field strength, γ is the surface tension and r is the radius of curvature of the cone. If the liquid coated field emitter has a high voltage placed on it relative to a nearby electrode the liquid metal will assume a conical shape (this process can be helped by making the underlying substrate conical in shape[2]). Since the end of a perfect cone is a point the electric field rises to high values near the end as the metal cone takes shape when the field is applied.

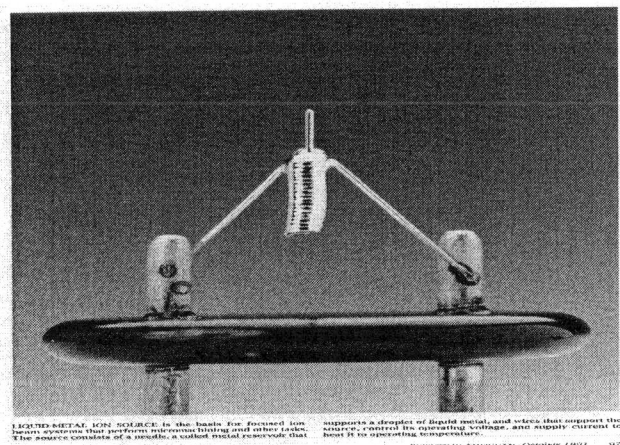

Figure 1. A gallium LMIS used in a commercial FIB system. The source consists of a W wire and spiral reservoir mounted on a metal/ceramic base. The diameter of the ceramic is about 1 cm.

When the field on the cone reaches a sufficiently small radius the field will be high enough to initiate field evaporation and the emission of ions commences. The process can be explained in terms of an atom evaporating from the metal surface and ionizing in the strong field as it moves away, as shown in Figures 2 and 3.

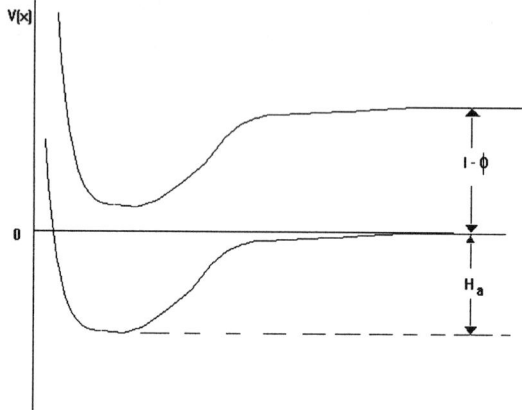

Figure 2. The energy levels for a neutral atom plus metal (lower curve) and an ion plus metal plus electron (upper curve). H_a is the energy needed for a neutral atom to evaporate from the metal, I is the ionization potential of a metal atom and φ is the metal work function.

Figure 2. shows the energy curves for a neutral atom and for an ion near a metal surface if there is no electric field. Figure 3. shows the situation when a strong field is applied. If the electric field is strong enough the energy curves for the neutral metal atom and the metal ion would cross near the metal surface. This is not permitted quantum mechanically - the two curves actually "repel" - and so at the position labeled X_c a transition can take place from {atom plus metal} to {ion plus metal plus electron}. The transition takes place with a certain probability (calculable from field ionization theory) that depends exponentially on the electric field and is reasonably large only for fields in the neighborhood of 10 V nm^{-1} or greater. Once ionization takes place the ions move rapidly from the emitter in the high electric field, following approximately radial paths.

In a classical field ionization source where gas molecules are ionized in the high field at the tip of a solid emitter with end radius ~ 0.1 micrometers, ray tracing shows the virtual source - the region from which ions appear to originate - to appear to be ~0.1 nm in diameter. Measurements on LMISs indicate a virtual source size ~ 50 nm. Calculations[3] on the actual physical source size indicate that it must be ~ 5 nm. The discrepancy is believed to be due to strong space charge effects in the beam that perturb the trajectories while the ions are close to (< 1 micrometer) the metal surface.

Ions created by field evaporation are replaced by a flow of atoms through the liquid cone to its apex, resulting in a negative pressure on the cone slightly distorting its shape. The complete problem of a LMIS taking into account the

flow of metal and the space charge of the ion beam was solved by Kingham and Swanson[4]. Their theory gives an explanation of source behavior that agrees reasonably well with experiment.

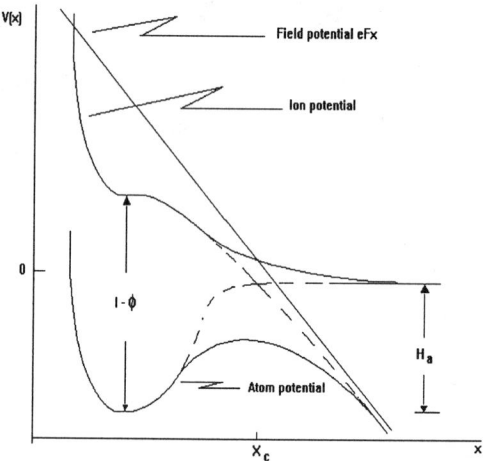

Figure 3. The energy levels for an atom and an ion near a metal surface when a high electric field F is applied. In the presence of a sufficiently high field the energy levels for metal plus atom and metal plus ion plus electron would cross and the atom can be ionized at the critical distance Xc.

From a practical point of view, the important aspects of the LMIS are its intensity, brightness, noise, energy distribution of ions, stability and lifetime. In this regard the LMIS is well suited for high-resolution applications, as shown in the table below. The high angular intensity

Angular Intensity A sr^{-1}	Brightness A sr^{-1} cm^{-2}	Noise	Lifetime	Energy Spread eV
10 x 10^{-6}	10^6	near shot noise	> 1000 microampere-hours	5 (at normal operating current)

Table 1. The important properties of the Ga LMIS for focused beam formation

means that currents in the nA range can be attained with modest beam apertures - a few milliradians - while the high brightness means that the optical source size is small so that a focused spot in the 10 - 100 nm range can be achieved with an optical column with only two lenses. The extremely low noise of the LMIS is due to space charge suppression of fluctuations in the emission process[5], while the relatively large energy spread arises from space charge interactions of the ions above the emitter.

The liquid metal reservoir in a standard LMIS contains 1 - 2 mm^3 of source material and since 1 mm^3 of Ga contains approximately 6 x 10^{19} atoms, at an emission current of 2 microamperes (12×10^{12} Ga$^+$ ions per second) a source will last for roughly 1000 hours. Therefore the ion source can be made highly compact.

Ion emission stability is very good. Typically, one sees drifts of the order of 1 microampere per hour, and these are easily correctable either by a servo mechanism or by hand. These are caused mainly by material sputtering from apertures in the ion gun back onto the liquid metal of the emitter. The reason for this seems to be that the back-sputtered material increases the flow impedance of the liquid metal along the substrate. Wagner[6] pointed out that the ion current is proportional to the emitter voltage divided by the impedance. A consequence of this is that if the current is measured as a function of current the slope of the I-V curve decreases with increasing flow impedance (see Figure 4 below). Experimentally we find that the slope decreases with source operating time and that, when the source is gently heated (a few hundred degrees C for Ga) the original I-V characteristic can be restored. If the emitter shank is observed by SEM after a number of heating cycles the shape will be seen to have changed considerably due to back-sputtered material that has precipitated out from the liquid.

It should be noted that if a LMIS is operated in a vacuum contaminated with organic vapors (such as pump oil) it will exhibit either drift or instability[7] as the organic material landing on the metal surface is cracked by returning secondary electrons generated when ions collide with nearby (~2 cm) apertures. Even in an uncontaminated vacuum the surface can oxidize. A general rule for operation is that the vacuum pressure should be kept below ~5×10^{-8} torr for reliable long-term operation.[7]

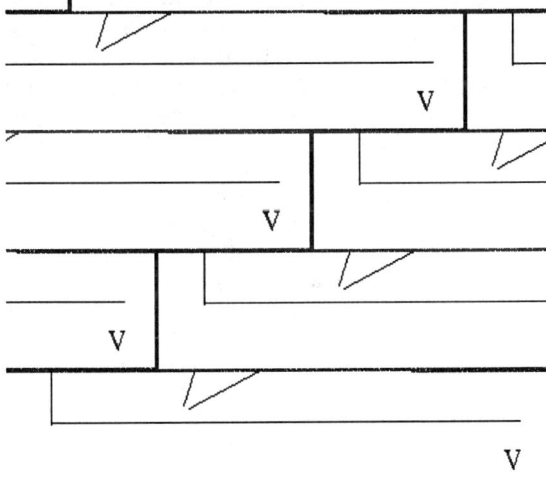

Figure 4. Schematic representation of the change in I-V characteristic of a LMIS as the impedance to flow of liquid metal on the emitter substrate increases: A = low impedance case,
B = high impedance case. Note that for the same operating voltage case A will provide higher current.

FIB OPTICAL SYSTEMS

Because the LMIS is almost a point source the optical system needed to produce a highly focused beam can be rather straightforward. It is found that the current density distribution in the focused spot of a system not limited by aberrations is approximately Gaussian (this corresponds to a system with a convergence angle of the beam less than about 1 mrad). To achieve a FWHM of 10 nm requires a demagnification of about 10X, which is readily achieved.

FIB systems use electrostatic optics almost exclusively except for mass filtering when alloy sources (e.g., AsPd$_2$) are used. The reason for this is the small charge to mass ration of ions relative to electrons. This is easily seen as follows: the force on a charged particle in an electrostatic element (lens or deflector) goes as qE where E is the electric field strength and q the charge. The acceleration a due to this force then goes as qE/M where M the mass. The deflection x of the particle is goes as at^2 where t is the time the particle spends in the lens or deflector. t in turn is proportional to the length L of the element divided by the particle speed v. v in turn goes as $(q\Phi/M)^{1/2}$, where Φ is the acceleration potential of the ion gun. Therefore the deflection x is proportional to qEL/Mv^2 ~ q/MEML/qΦ = EL/Φ. Thus, the deflection is independent of q/M. For magnetic systems x ~ $(q/M)^{1/2}$. For the case of Ga ($M_{Ga} \approx 10^5 M_e$) a magnetic lens with a given focal length for 30 keV electrons would have to be about 300 times stronger to focus Ga ions with the same focal length.

Considerable research has been done on electrostatic lenses in the 1970's and 1980's and lenses with quite good properties have been developed. Because it is not feasible to have either the object plane or the image plane of a lens in the electrostatic field (except in certain special cases such as mirrors) the focal lengths of electrostatic lenses have to be kept about an order of magnitude larger than for magnetic lenses. As a result the aberration coefficients are considerably larger for electrostatic lenses than for magnetic lenses. For a typical FIB system consisting of two lenses with 3 mm lens apertures and an overall optical column length of about 15 cm, one would find spherical aberration coefficients ~ 5000 mm and chromatic aberration coefficients ~ 100 mm. With an optical source size of 50 nm a two lens column is then capable of producing a beam with a FWHM as small as 5 nm at a working distance of a 1 - 2 cm with beam current in the pA range (current density on target ~ 10 A cm^{-2}). The usual arrangement for a FIB optical column is as shown in Figure 5.

Calculations show that the best optical performance is achieved if the beam is approximately collimated. The usual arrangement is to operate the first lens as a "zoom" lens - the voltage is variable with the first lens electrode held at the source extraction potential and the final electrode at the desired beam potential (relative to the source). The focal distance is determined by the central lens element. The second lens is operated as an einzel lens (first and third electrodes maintained at the same potential - ground).

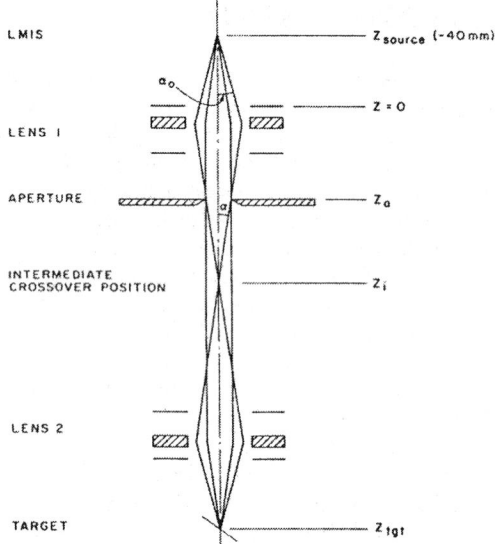

Figure 5. Schematic drawing of a two lens optical column for a FIB. If the crossover mode is used a single aperture suffices. If the beam is collimated, a variable aperture must be used to change beam current.

Beam positioning, stigmation and scanning are done with electrostatic multipole devices, generally quadrupoles for small alignment functions (a few tens of micrometers deflection) and octupoles or sextupoles for larger displacements and scanning. Numerous functions can be summed and applied to the multipole plates, so that the system can be kept compact.

Space charge effects can be severe in FIB systems. The problem is worse than the relatively benign defocus seen in electron beam systems that can be corrected by refocusing a lens. in a FIB system even at low currents radial broadening can spoil beam resolution. This is an effect due to random interactions of the ions and being stochastic in nature, it cannot be removed by focusing[8]. The problem depends on the current, mass and length of the optical system and inversely on the beam energy. Therefore, since the energy and beam current are usually fixed for any given application, it is necessary to minimize the length of the optical system as much as possible, and to aperture out unnecessary beam current as close to the ion source as possible, unlike many SEMs where the final beam aperture is located in or below the final lens polepiece. The problem of ion beam space charge dictates that the total ion current extracted from the LMIS - about 2 microamperes - be reduced at the first aperture to ~20 nA within about 2 cm of the source, at the first gun aperture. The beam defining aperture that selects current between about 20 nA and 1 pA is located below the first lens, so that through the majority of the distance of the optical column the beam contains no more current than is needed on the target.

IMAGING AND RESOLUTION

Perhaps the most important parameter for most FIB users is the imaging resolution their instrument is capable of. It turns out that, beyond a certain beam size the limit on imaging resolution is imposed by the sputter sensitivity of the target[10]. In order to resolve two points in an image there must be a contrast difference between them - the human eye can reliably detect a contrast difference of about 10%, for example. If the noise in an image prevents one from detecting a contrast difference at a particular level (e.g., 10%), and only a worse contrast difference can be reliably detected (e.g., 20%) then the resolution in the image will be worse[11]. Noise can be reduced by using longer imaging times - since the noise typically has a Poisson type distribution it will be reduced as the square root of the imaging time for a constant beam current. Unfortunately, the imaging time cannot be made arbitrarily long because of outside influences. In an e-beam instrument one has drift, possible specimen degradation, instabilities in power supplies, vibration from passing trucks etc. to contend with. In the case of ion beams the problem is much worse: there is a competition between collection of a signal from the specimen and its erosion by the ion beam. It comes down to the question of whether enough signal can be collected before the specimen disappears. The problem can be quantified and the results can be seen graphically as follows:

As can be seen in Figure 6, the contrast of an image decreases with increasing spatial frequency. Noise in the image place a limit on the maximum spatial frequency that can be detected. The inverse of this frequency can be taken as the resolution limit of the optical instrument (L_{min} in this case). If this idea is generalized[10] and applied to an ion beam then the resolution can be defined as the smallest object (volume$^{1/3}$) that can be detected at a given signal-to-noise ratio (S/N), as shown in Figure 7.

In a typical electron beam probe instrument resolution can have several meanings - it is the imaging resolution of the probe or, if an SEM is used for X-ray analysis, it may be the size of the volume in which X-rays are excited in the sample. The latter may be many times larger than the former, and the imaging resolution may be much larger than the probe size if secondary electrons are detected from the region over which the beam "blooms" or scatters on entering the sample.

The situation with ion beams is similar but somewhat more complicated. The probe size may be well defined, but interactions with the target allow one to define resolution differently depending on the application. For example, since secondary electrons generated by the ion beam have low energies (< 10 eV) the imaging resolution is more closely connected with the beam size. The extended tails on the current density distribution at higher currents may result in a usable beam size being considerably larger than the FWHM of the beam. This is especially true when the beam is used for deposition from a chemical in the gas phase, a process that is sensitive to low current densities. The region over which there is a measurable deposit may be many times larger than the nominal beam size. This problem can be analyzed by calculating the current density distribution of the beam and, in fact, such an analysis suggest ways to reduce the problem[9]. However, due to the limitations of space this will not be considered hers; the information can be found in the references given immediately above.

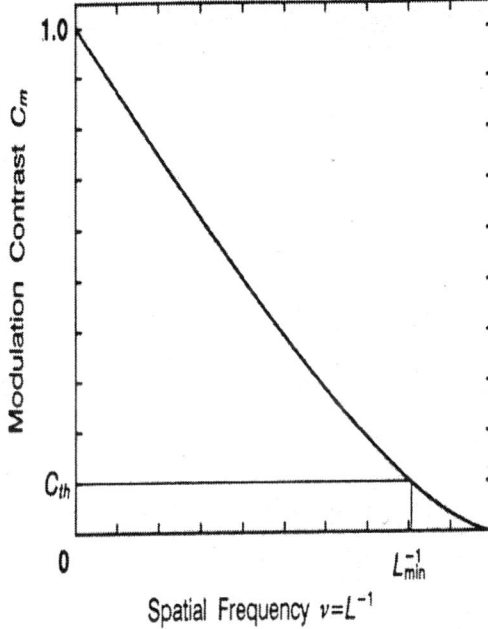

Figure 6. The contrast or modulation transfer function of an image made with an optical instrument (light or charged particle). C_{th} is the minimum contrast visible.

The present generation of FIB systems using Ga ion sources has demonstrated imaging resolution as good as 5 nm. An example is shown in Figure 8, where the contrast is shown as a function of resolution for a graphite specimen (graphite is a good target for measuring resolution, as it contains very fine detail and has low sputter sensitivity). It is unlikely that imaging resolution will improve much past this point, since it is so close to the theoretical limit. In order to achieve higher resolution it would be necessary to use lighter ions with a lower sputter yield. Since most applications of FIB depend on its ability to sputter materials, this would not be a happy choice, implying that an alloy source that can produce both heavy and light ions would be optimum. However, such a source would require the FIB to contain a mass filter which, because of its inherent aberrations, would spoil the resolution. A possible way out would be to use a gas ion source that could produce ions of H and Ar, depending only the gas valve opened. If a source of sufficiently high brightness can be developed this could extend the utility of the FIB. At this time the only way to achieve really high

resolution in a FIB is to add an electron column to the instrument.

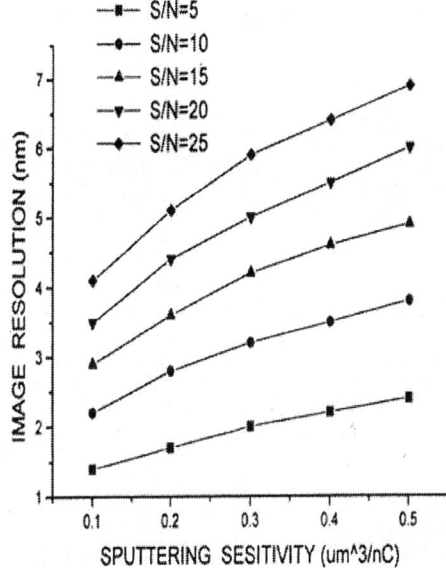

Figure 7. Imaging resolution for a variety of signal to noise ratios (S/N) as a function of sputter sensitivity.

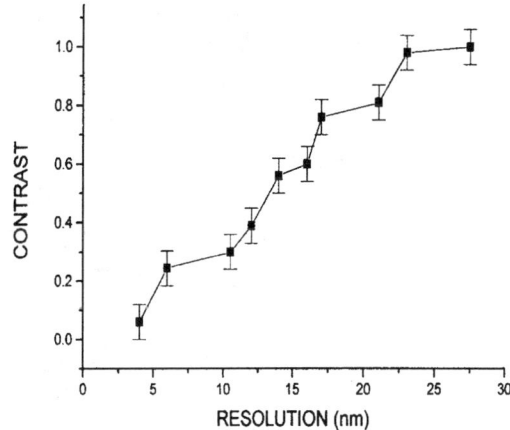

Figure 8. Contrast as a function of resolution (= inverse of spatial frequency) for a Ga LMIS FIB with a graphite target.

For more detailed discussion of many of these topics the reader is directed to references 12 and 13.

REFERENCES

1. G.I. Taylor, Proc. Royal Soc. **A280** (1964), 383

2. L.W. Swanson and J.-Z. Li, J. Vac. Sci. Tech. **B6** (1988), 1062

3. N.K. Kang and L.W. Swanson, Appl. Phys. **A30** (1983), 95

4. L.W. Swanson and D.R> Kingham, Appl. Phys. **A41** (1986), 223 and references therein

5. P.J. Pushpavati and A. van der Ziel, IEEE Trans. on Electron Devices **ED-12** (1965), 395

6. A. Wagner, Appl. Phys. Lett. **40** (1982), 440

7. P. Sudraud, J. Orloff and G. BenAssayag, J. de Physique, Coll. C7, supp. 11, **Tome 47** (1986), C7-381

8. G.H. Jansen, "Coulomb Interactions in Particle Beams", Acad. Press, San Diego, 1990

9. M. Sato and J. Orloff, J. Vac. Sci. Tech. **B9** (1991), 2602

10. J. Orloff, L.W. Swanson and M. Utlaut, J. Vac. Sci. Tech. **B14** (1996), 3759

11. M. Sato and J. Orloff, Ultramicroscopy **41** (1992), 181

12. J. Orloff, Rev. Sci. Inst. **64** (1993), 1105

13. "Handbook of Charged Particle Optics", J. Orloff, Ed., CRC Press, Boca Raton, 1997

Basic Technology and Practical Applications of Focused Ion Beam for the Laboratory Workplace

S.B. Herschbein, L.S. Fischer, and A.D. Shore
IBM Analytical Services
Hopewell Junction, New York

INTRODUCTION

Once a highly specialized tool only available to the most lavishly equipped labs, the focused ion beam (FIB) is now almost as common as the FE-SEM. And if you do not actually own one, the services are available at one of dozens of contract labs. Failure analysis and product support personnel are routinely called upon to use the FIB for a wide variety of complex tasks. No other tool in the arsenal offers the sheer versatility and multifunctional capability.

Basic FIB tasks can be divided into electrical and physical activity. Electrical includes functional chip repair, microprobing assistance, and voltage contrast. Physical is more widely varied, and includes several cross-sectioning techniques for defect determination, process monitoring, metrology & TEM preparation, materials and grain contrast imaging, material analysis via EDX or SIMS, plus the emerging field of micromachining. Details on these techniques and examples will be provided.

But to fully understand how to best configure and use these systems requires that we first review the basics. Column choices, imaging modes, chamber and stage designs, milling and chemically enhanced etching, deposition of insulators and conductors, automation, navigation aids, and accessory modules will all be discussed.

SYSTEM ARCHITECTURE

1, 2, or 3 Beams?

Common to all FIB systems is the primary ion column. The ion beam is generated from a liquid metal ion source. All commercially available FIB tools at this time use gallium, which becomes a liquid slightly above room ambient temperature. Positively charged gallium ions are drawn off of a field emitter point source and accelerated by a 25-50 kV potential. The emission is focused to submicron beam diameters with electrostatic lenses. In a companion paper to this text, Jon Orloff describes in graphic detail the full operation of the ion column, so it will not be repeated here.[1]

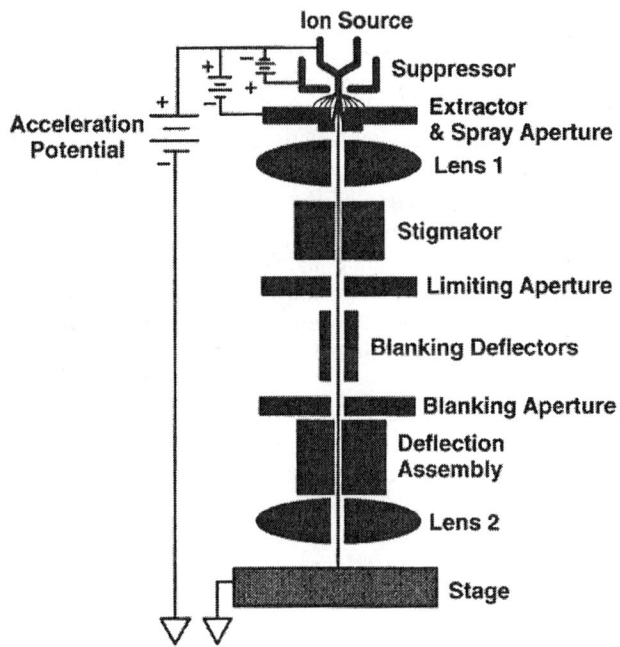

Figure 1. Basic ion column design.
Courtesy of Micrion Corp., used with permission

The primary ion beam is scanned across the target material in a pattern defined by the tool operator. Secondary atoms & molecules (neutrals), along with ions and electrons are ejected from the surface. The charged particles (ions and electrons) are drawn towards a biased grid and collected by one of several types of detectors. Their signal is amplified and displayed to provide a real-time image of the area.

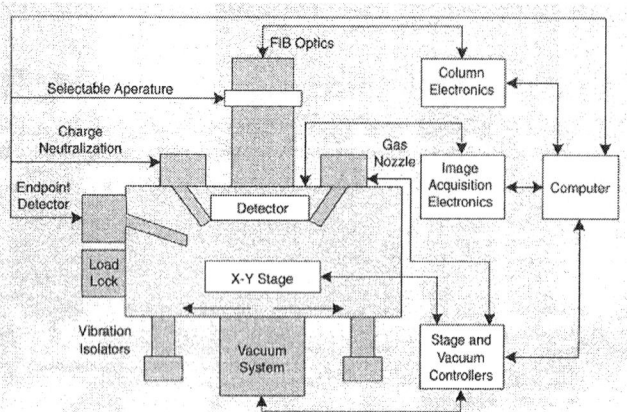

Figure 2. Block diagram of a single beam FIB system
Courtesy of Micrion Corp., used with permission

The most common FIB systems contain just an ion column. While they can provide decent imaging when operated in high resolution mode, the very act of imaging is destructive to the sample. Physical sputtering of the surface must be accomplished to liberate secondaries for capture. Excessive gallium implantation is another undesirable side effect of repeated imaging as it increases surface charging and the formation of an amorphous layer.

Dual beam tools (ion and electron) reduce the surface damage and gallium implantation. Supplementing the 'cutting' beam is an electron column for SEM imaging, usually placed at an angle of 45 to 60 degrees. While there may be some compromise in working distance or operating convenience, this configuration yields superior imaging with greatly reduced sample degradation. It also eliminates the need to tip the sample to view the cut edge, decreasing analysis time and positioning errors.

The term 'dual beam' could also mean combining an in-situ optical scope in a parallel, but off axis configuration with the primary ion beam. Secondary (ion or electron) derived imaging provides only surface topology information. An optical beam (especially IR) penetrates common semiconductor materials, showing buried lands and layers that the secondary detector cannot see. This greatly aids navigation in what might otherwise be a blind search. Focal point differences in the optical view can also be used to determine the amount of ion milling required to reach a certain depth.

A three-beam system combines the ion and electron columns, plus the optical scope, all in one tool. But while the current generation commercial offering is ideally suited for metrology and in-line applications, it may not serve general purpose laboratory use as effectively due to the non-tilting stage.

Detectors

Several types of ion and electron detectors are in use today. While preferences and placement vary by system manufacturer, the most often encountered detectors are the MCP, CDEM and Scintillator.[2,3]

MCP: The micro-channel plate is typically located around the column pole piece and is used to detect ions and electrons. It is a honeycomb structure of many small channel holes in a biased glass plate. Secondaries entering each small tube cause an electron cascade effect, yielding a high signal output.

CDEM: The continuous dynode electron multiplier is located off axis in the chamber. It is also used to detect both electrons and ions. Sometimes referred to as a 'particle funnel', it is essentially a single large area channel multiplier.

Scintillator: This type is an electron only detector. It may be used in an in-the-lens mode on the SEM column, or off-axis in the chamber. It can image in a conventional secondary mode, or backscatter.

As imaging by SEM columns is covered in other texts, we will concentrate on the image derived from the interaction with the ion beam, and the effect on the target material. The ion beam can be moved across the sample in a single direction raster or serpentine pattern. Users can usually vary parameters such as the spot size, beam current, pixel spacing/overlap, dwell time, and end of scan pattern dead time. On most systems, the range of beam currents and spot sizes is determined by a user selectable limiting aperture. The dose, or amount of ions striking the surface, is a function of beam current, duration and area scanned. The secondary yield, or number of ejected ions per primary ion, is also material dependent.[4]

All of these parameters interact to determine the amount of surface sputtered away by the gallium ions. Depending on the system, the work to be done may be inputted by the user as a selection of run time, dose, or predicted milling depth. Dose may be expressed as nanocoulombs/um^2, or as ions/um^2. Predicted mill depths are based on reference file estimates of known sputter rates for common materials.

The video image is formed one pixel at a time as the beam scans the surface. A portion of the liberated ions or electrons strikes the detector, depending on the detector bias. All other factors being equal, the sharpest, highest quality imaging is achieved with a small beam spot size (small aperture), the maximum number of pixels in a given area (high resolution), and adequate dose (higher beam current or longer dwell time). Shorter working distance (minimizing the space between the pole piece and sample) also improves image quality by increasing the detector collection efficiency, and by yielding a slightly sharper beam.

With beam parameters held constant, ion penetration into the sample will vary with the material, and within a given material, with grain orientation (lattice atoms in the beam path). This results in a variation in the secondary yield for a given area, and thus a wide range of image contrast. Differences can be seen when collecting electrons verses ions, so a comparison of both imaging modes may be useful to fully understand a given sample.

The high secondary electron yield compared to the relatively low ion count means that electron imaging provides the highest signal to noise ratio. But there are conditions in which electrons may be suppressed, such as when charge neutralization (to be covered later) is used. In this case imaging with ions, while providing a somewhat lower quality image, may be the only option.

EDX / SIMS: Materials identification is made possible when a variety of secondaries are available. Ion emission can be captured with the addition of a SIMS quadrupole detector. Dual beam tools can use the e-beam column to excite X-ray emission for capture by an EDX detector.

Chambers & Stages

Most FIB systems are still purchased by the semiconductor and related industries. As such, stage configurations and chamber sizes were created to primarily fit those needs. Chambers can be found in 50mm (2") for small samples and FA use, and wafer sizes of 150mm (6"), 200mm (8"), and 300mm (12"). The primary intended application usually dictates the system configuration.

In-Line Review & Metrology: Production systems use wafer transports for high volume work, and moderately precise X-Y stages. They may take location data from optical defect mapping tools located elsewhere in the FAB operation. Two & three beam machines are a real plus.

Head Trimming & Micromachining: These systems are more likely to use high precision encoded X-Y stages, augmented by pattern recognition routines for high accuracy, unattended operation. 'Handcrafted' and development work would benefit from a multi axis stage.

Chip Repair: Depending on the complexity of the product, circuit modification may be performed on general purpose platforms, or those with high precision encoded X-Y stages. The latter is especially true if guided by CAD or coordinate navigation, or working in a backside or other blind mode. A tilting stage degrades accuracy, but offers unique ability for cleanup in deep holes or unusual topology.

Cross-Sectioning: While dual-beam tools may decrease the need for tilting (SEM images directly), a multi axis stage still provides the most flexibility for ion imaging, face decorating, etc. TEM prep and other unattended operations may use pattern recognition as an aid.

Grain & Materials Analysis: This work requires a tilting stage to optimize ion channeling and angle to the detector.

Prober Modules & DUT Sockets: Electrical evaluation activity requires a purpose-built FIB endstation, or a general purpose system with a docking stage and electrical feedthroughs. While usually used flat, tilting can help give a perspective view for easier touch-down of probes.

Voltage Contrast: A socketed, precision stage with CAD navigation is a plus. Tilting is not usually required.

Photomask Repair: These specialized tools are likely to use high precision encoded X-Y stages and take defect location data from other mapping tools.

Milling, GAE, & Deposition

Conventional Sputtering: The 'straight' sputtering process involves the transfer of momentum from the primary ion beam to the target substrate material. The collisions follow the 'billiards ball' analogy, with the yield (number of removed ions per incident ion) dependent on the material. The amount of surface removed is again a function of the dose, beam parameters, and target material.

But as the yield for most common semiconductor materials is not substantially different, straight sputtering offers limited material selectivity. This becomes a problem, for example, when removing an oxide layer with the intent of stopping on an underlying conductor. Another limiting factor is redeposition of the sputtered material back into the region you are attempting to clear. Redeposition of oxide back onto the sidewalls limits the aspect ration, and can choke off a deep mill attempt in only a few microns. Redeposition of a conductive material can lead to unintentional shorts.

Gas Assisted Etching (GAE): The addition of a small amount of halogen, or halogen containing gas to the semiconductor milling process greatly improves the material selectivity and reduces redeposition. The addition of water has a similar effect on organic substances. Small nozzles within the chamber direct the gas to the target area. While it registers as a small increase in overall system vacuum base pressure, the localized pressure is raised to a greater extent. Some of the gas adsorbs to the material surface. Energy transferred from the primary beam leads to a dissociation of the gas molecule, resulting in localized plasma-like reactive species. In the presence of properly balanced beam parameters, the predominant mechanism becomes chemical etching, rather than physical sputtering. The gaseous chemical byproduct is volatile, and is easily removed by the vacuum system leaving little redeposition. [5]

Gas chemistries can be chosen that accelerate the removal of different materials commonly in use on today's IC products, resulting in a high degree of selectivity. But as of this writing, gases that yield acceptable results for copper films are still in development.

While the final delivery to the target is as a gas, precursor materials are consumed in infinitely small amounts, and may be stored in any phase. Evaporation of liquids or sublimation of solids work about as well as bottled gases, and are generally safer to handle.

Chemistry	State @ RTP	Enhanced Etch
XeF_2	Solid	Oxide, Si, Tungsten
Cl_2	Gas	Aluminum, Si
Br_2	Liquid	Aluminum, Si
H_2O	Liquid	Organics
Selected Salts	Solid	Organics
I_2	Solid	Metals / Universal

Figure 3. Common Precursors Chart

Endpoint Detection: Successful circuit repair is dependent on being able to drill through one layer and be able to stop before damaging the next. Determining when to stop a mill is one of the more difficult skills to be mastered.[6]

The most common indicator is a change in secondary yield that occurs at the material interface. But this can prove tricky when milling high aspect ratio holes due to the overall reduced secondary count.

Other methods include the use of an electrical detection method such as stage current monitoring, or possibly the addition of a SIMS detector to monitor material changes.

Ion Induced Deposition: FIB systems can deposit both conductive materials as well as insulator layers. Conductors may be in the form of new traces to rewire a circuit, or a protective/sacrificial layer to protect an underlying material from beam damage. Insulators are often used in circuit repair to prevent exposed or deposited conductors from unintentional leakage. As in GAE, nozzles within the chamber inject a gas near the surface.

Conductor Deposition: An organometallic gas containing tungsten or platinum (the two most common metals in use today) adsorbs to the sample surface. Beam parameters are adjusted to deliver total energy below the normal sputtering

threshold, but enough to dissociate the surface gas. Most of the byproducts are driven off. Repeated rastering in the presence of fresh gas results in the buildup of a metallic film. While the deposited metal contains some carbon, oxygen and gallium, the electrical characteristics are suitable for most digital circuit requirements. [7]

Platinum has a higher yield, and forms a thicker trace in less time. But tungsten has overall better electrical characteristics and seems to fill deep holes easier. Recent work also suggests that it may offer better resistance to electromigration in high current applications. [8]

FIB deposited conductors do not have perfect, square edged sidewalls. Overspray, a residual deposition tail alongside the line, tends to increase with higher beam currents. This 'drawing outside of the lines' is the result of a gaussian shaped beam energy pattern, raster retrace and overshoot, etc. Secondary scatter also contributes to overspray formation, and is particularly visible on adjacent topology sidewalls. When two lines are located nearby, the overspray must be removed to prevent leakage currents between them. A brief XeF_2 mill provides an effective cleanup.

Insulator Deposition: An SiO_2 layer can be formed in much the same manner as conductor deposition when a silicon containing gas precursor is delivered to the sample surface.[9] Most commonly used precursors are TEOS or a siloxane compound. Additional oxygen may be introduced through another nozzle to help balance stoichiometric requirements and speed the oxide growth, but the resulting oxide integrity and edge coverage may not be acceptable.[10]

Charge Damage and Control:

Gallium ions implanted by the scanning beam results in a strong positively charged localized region. Deflection of the beam (drift), transistor Vt shifts, and ESD type damage can occur if the charge is not bled off or otherwise neutralized.[11]

Charge control at the sample level can take the form of good mechanical grounding, conductive surface coatings, or intentionally drilled leakage paths. Maintaining a proper mechanical ground to insure electrical conduction requires that a variety of clamps, clips and specialized holders be available. Some additional sample preparation, such as physically scraping oxide off of a wafer edge can sometimes help.

System level charge control is becoming increasingly available on FIB tools. This involves the use of an electron source directed to the same region being scanned by the ion beam that effectively results in a net neutral condition on the surface.[12]

Automation & Navigation:

Like most laboratory work, the outcome of FIB tasks is still highly dependent on operator skill. But automation in the form of CAD navigation, coordinate maps and pattern recognition is beginning to make manufacturing-like operations possible.

The largest non-lab application of FIB is the production trimming of disk drive heads. These tools use automated transports combined with CAD navigation and pattern recognition to locate and mill with little operator intervention.

In the laboratory, routines are now available to permit largely unattended preparation of TEM samples. Sighting off of surface fiducial markers, the FIB can follow a prescribed milling pattern largely unattended.

Using CAD navigation packages such as Knight,™ the FIB stage can lock to and track the layout database, allowing blind repair of deeply buried circuit structures. Scanned photos or imported files of optical images can also be overlaid to the FIB image to reconcile the location of hidden features.

Accessory Modules:

FIB Chamber Prober: First introduced as an endstation adaptor on a FIB specifically designed for circuit debug, load-lock compatible probers are now available for general purpose tools. Probing by white light optics is becoming close to impossible in the sub-0.35um world. These high precision instruments supplement conventional stand-alone probe stations by making use of the great depth of field offered by secondary imaging. And as the FIB has the ability to cut and access buried nodes without moving between equipment, this application has the potential to greatly simplify failure analysis and characterization work.

FIB APPLICATIONS – Electrical

With the ability to selectively remove and deposit insulators and conductors, the FIB is the ideal tool for making electrical changes on finished integrated circuits.

Circuit Repair:

The act of repairing circuit design errors on finished product is sometime referred to as chip microsurgery, or 'cut and patch' technology. It gives circuit and system designers the opportunity to conceive and try out 'fixes' for common errors, without the expense and delay of a FAB run with a revised photomask set. Verification of a design fix can

sometimes prove to be extremely valuable when a primary fault obscures a secondary error that would have otherwise remained undetected. A day's worth of work can save a month of frustration.[13]

Most FIB tools are used to provide rapid turnaround time on a small volume of parts for confirmation purposes. But FIB can sometimes help product time-to-market by acting in a manufacturing adjunct capacity. By generating volumes of repaired chips, sometimes in quantities of hundreds of units, early user hardware is available to maintain aggressive card and system test schedules.[14]

The process begins with finding a suitable site on the physical layout to implement the change. The desire, for simplicity sake, is to work on metal lands as close to the surface as possible. Silicon IC's today can have as many as seven layers of interconnect metal, plus polycide transistor connections and gates. Filler metals are often placed in open areas to help even out RIE etching and chemical-mechanical polishing. The chance of finding a clear, unobstructed window down to a lower level line for a cut or connection is becoming increasingly rare. Designing a non-obvious electrically equivalent circuit, or rerouting a conflicting line, is often required.

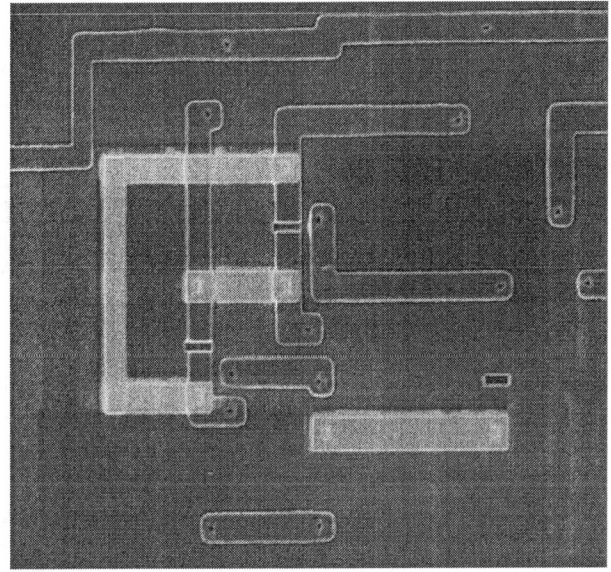

Figure 4. Basic chip edit of three straps and three cuts performed on a two level metal product.

Connections are made by drilling holes in the oxide until the desired trace is contacted, then backfilling the new via with deposited metal. The filling of deep holes requires some tricks, as metal 'growth' tends to occur on the top edges faster than at the bottom of the via. Strapping metal deposited on the top overlay glass connects the vias, completing the circuit. Isolation cuts are done by milling through the trace lines, leaving an open circuit. As described in the System Architecture section, high aspect ratio work requires the almost exclusive use of gas assisted etching. GAE is used to mill the holes in the oxide, cut traces at the bottom of holes, and cleanup metal deposition overspray.

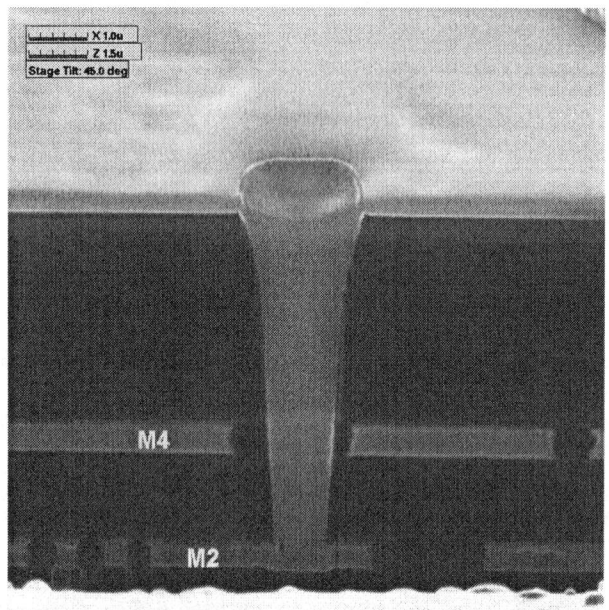

Figure 5. GAE via drill and tungsten fill to contact a deeply buried M2 line.

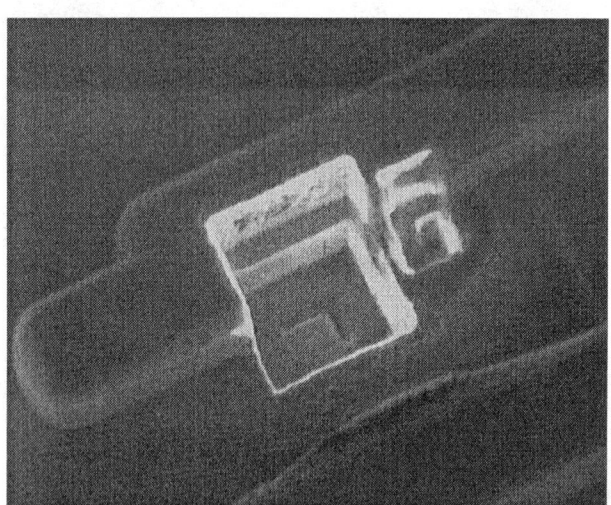

Figure 6. FIB repair at M2/M3 on a 6 level stack. Notches were taken out of M6 & M5 lines to reach repair site. Small deposited repair strap at bottom of large hole connects M3 to M2. Deep cut to right of main repair disconnects M2.

Sometimes the wire to receive the cut or connection is below another conductor. In the case of a wide buss in the way it may be possible to mill out a notch and work in the open area. Insulator deposition along the exposed sidewall of the buss will allow a FIB deposited conductor layer to run up and over while avoiding a short. When a narrow line blocks access it may be possible to remove it entirely and rerouted it around the work area.

When access from the topside becomes impossible due to packing density or exclusive use of flip-chip mounting, the only solution may be from the back, or well/tub side. Backside FIB repair begins with bulk or localized thinning of the Si substrate, from the standard backlapped die thickness of 15mils (375um) down to 1-2mils (25-50um). This can be done with mechanical polishing, or a chlorine based laser ablation tool. From there, a high XeF_2 concentration mill opens a wide hole down to within a few microns of the transistor implant region. This is usually done first at corner lock markers, then at the actual circuit site, as directed by the navigation aids. With FIB deposited oxide preventing leakage to the bulk silicon, small mills can now be made to contact the underside of source-drain regions, or through field isolation areas to access lower level conductors.[15]

Backside chip repair is the ultimate test of blind navigation and reference marker measurements. Even on tools equipped with an in-situ IR optical imaging system, the process is slow and difficult. The poor spatial resolution of IR makes it difficult to clearly see many of the small features of interest.

GAE gases compatible with the Aluminum/Silicon system have been in common use for several years. Unfortunately, the migration to copper metallurgy poses unique challenges. Straight milling is difficult due to very strong grain rate dependence. Halogen gases act in an uncontrolled manner, with spontaneous milling and the inability to reliably quench the reaction. Work continues to find a new chemistry set that is better suited to this interconnect technology.[16]

Electrical Probing Assistance:

This application is a logical extension of chip repair. It involves milling access holes to buried conductors for the purpose of making probing pads for mechanical or e-beam acquisition of electrical waveforms, current signatures, or simple curve tracing. Lines can also be cut to isolate nodes. But all of the problems of reaching lower level conductors apply here also, making backside FIB a possible solution.[17]

A dramatic example of assisted probe analysis is the thorough characterization of an SRAM 6T cell made possible by the FIB 'wagon wheel'. All of the nodes of interest reside within a 5um diameter, far too compact a region for conventional manipulator probing. In this particular case it was even too confined to drill all of the vias and connect them to pads without risking line leakage and cross-talk. The solution was to stack and cross lines, using FIB deposited insulator as a dielectric. A variety of standard sample preparation techniques were used to isolate the 6T nodes prior to the FIB work.

These techniques greatly aid circuit design verification and electrical failure analysis by again making microprobing possible. To further speed the process, the actual probing work currently done on a stand-alone station may eventually be replaced by a FIB chamber prober.

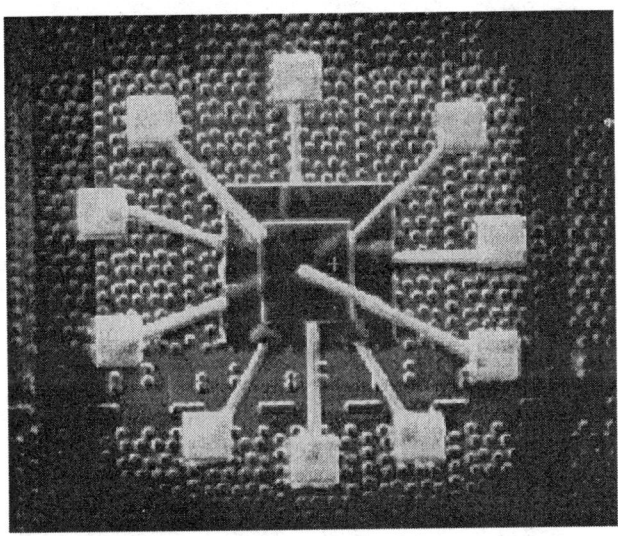

Figure 7. FIB generated 6T cell 'wagon wheel'

Voltage Contrast:

Voltage contrast circuit analysis was developed on e-beam systems, but is finding new acceptance on FIB tools. True VC requires a socketed stage for driving the device under test, but a passive form more akin to ebic is most often used. In electron imaging mode floating conductors go dark, grounded lines appear bright.[18]

The SEM version of these techniques is well suited to finding open circuits, but is generally more challenged when confronted with shorts. With no way to change the electrical circuit conditions real time, SEM VC requires a trip to a probe station laser or other modification tool. FIB solves this problem by being able to cut conductors and monitor the change to both halves of the split circuit. A divide-&-conquer approach that would otherwise require iterative trips is now possible.

FIB APPLICATIONS – Physical

With the ability to sputter away large quantities of material with pinpoint accuracy, the FIB is the ideal tool for defect investigation, structural analysis, process monitoring

and sample preparation. The addition of an EDX or SIMS detector to the FIB makes detailed materials analysis possible.

Cross Sectioning:

FIB cross sectioning is rapidly replacing conventional lapping and cleaving methods for failure analysis and process monitoring. Precision edge placement provides accuracy difficult to match by other means.[19] And multiple sections in different directions can be done in close proximity, not otherwise possible if you want to maintain the mechanical integrity of the sample.

A FIB section is done by sputtering away bulk material in front of the surface of interest, allowing the buried region of the sample to be viewed when tipped at a high angle. Two methods are commonly used to remove the bulk – staircase step mills and single progressive slopes.

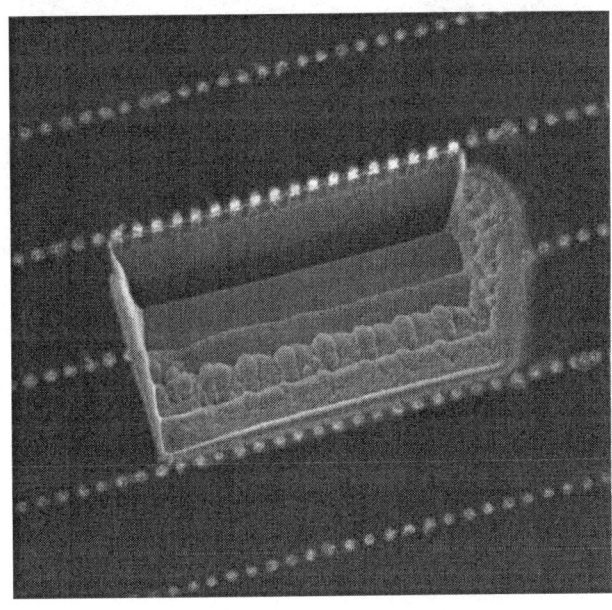

Figure 8. Low magnification view of a typical FIB cross-section. Stairsteps, sidewalls and polished 'face' are visible.

A staircase mill is a fairly simple pattern of successively deeper steps, cut with a high beam current and medium pixel spacing, to within a short distance of the intended site. This is a fairly course mill, designed for large area cleaning, but results in a moderate degree of redeposition onto the face. The final cut is done using 'fine polish' parameters (reduced current, tighter pixel overlap and longer dwell) that yields a smooth face ready for imaging.

A single progressive slope, or single raster mill makes use of tight parameters throughout. The beam only passes over the entire region a single time, unlike conventional mills that repeatedly scan the area hundreds or thousands of times. The spot dwell time is greatly extended with a high degree of overlap. The result is an ion cascade effect that multiplies the yield, giving a progressively deeper mill with each scan row. While giving a narrowed trench width due to greater sidewall redeposition, the face you want to image is fully polished and ready to view.

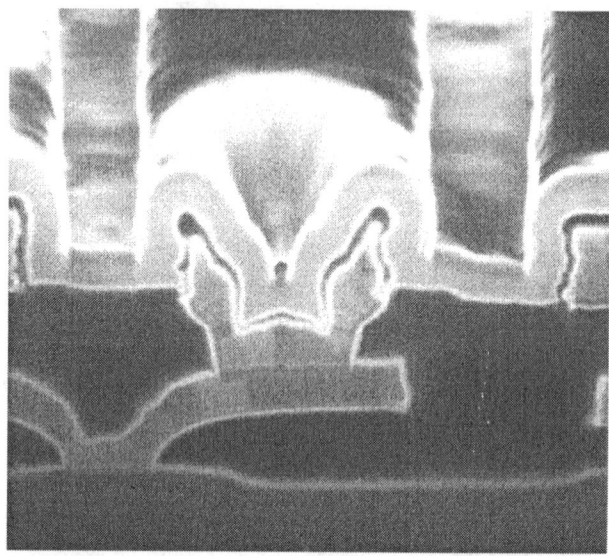

Figure 9. FIB generated and imaged cross-sectional view of metal contacts. Individual aluminum grains are visible, along with the metal-to-metal interface.

A dual beam tool simplifies the sectioning by allowing the simultaneous viewing by SEM while sputtering with the ion column. This is possible as the beam lines are at an angle of 45-60 degrees to each other. The advantage for the failure analyst of live image updates is never 'blowing through' the intended spot as the cut is not being done blindly. For routine process analysis, the advantage is speed – an immediate image without having to tip the sample.

TEM Sample Preparation:

This is an extension of cross sectioning, but fine FIB cuts may be needed from both sides. Successful TEM imaging requires that the sample be thinned to about 100nm to be transparent to a high kv electron beam. Traditional assisted TEM preparation only uses FIB for the final shaving of a sample that had been bulk polished by two sided mechanical methods.[20]

New techniques have been developed that allow TEM samples to be produced from any structure anywhere on the wafer. Carefully controlled FIB trench sections are cut from both sides, leaving a thin sliver barely held in place. A mechanical arm then removes the piece and transfers it to a TEM grid for imaging.[21] Automated software packages can perform the majority of the work unattended by relying on some level of pattern recognition to keep the beam properly placed.

The use of a dual beam FIB may prove to be an advantage for this work. Imaging sample face features only with the electron beam may reduce the formation of amorphous layers that can degrade TEM resolution.

Figure 10. FIB milled TEM section from a mechanically thinned sample.

Figure 11. FIB milled TEM section directly from bulk sample. *Courtesy of FEI Company, used with permission*

FAB Line Monitoring and Metrology:

Previously, understanding the nature of random defects found by optical mapping required mechanical sectioning and SEM imaging to comprehend. But as line costs increase, fabricators are becoming increasingly reluctant to break partially processed wafers with potentially good die.[22]

Accurate metrology requires knowing an exact point on a sidewall slope. But unless you can see the feature in cross-section, determination of the spot is only a guess.

Both of these applications can benefit from a FIB tool placed on or close to the process line. Sections can be done for defect review or metrology without sacrificing the wafer. Milled holes can be capped with oxide to reduce gallium migration, the wafers cleaned, and returned back to the line.

Micromachining & Head Trimming:

As the name implies, micromachining is the reshaping and finishing by FIB of micron scale products. Typical applications include the trimming of e-beam apertures, formation and sharpening of AFM and other tips, and more recently, the final shaping of recording heads.

Most applications generally require a substantial amount of operator involvement and the results tend to be 'one of a kind'. But with advanced polygon mill software, complex shaping of objects is beginning to become possible. Still, the problems associated with aspect ratio limitations and redeposition described in previous topics applies here also.

One success story is the production level trimming of magnetic recording heads. A number of specialized FIBs have been delivered to the disk drive industry for this use. These tools employ a precision stage and a high degree of automation, including pattern recognition, to trim long vertical edges off head strips, or even work on wafer level product.

Materials & Grain Imaging:

The visual contrast difference produced by ion channeling readily delineates material interfaces. This makes FIB imaging ideal for highlighting structural issues in semiconductor material, and makes it easy to differentiate dissimilar layers. The presence of impurities or voids becomes apparent, in either top-down, or cross-sectional analysis.

One of the most popular applications of contrast imaging is the determination of the grain structure of semiconductor metals. Each grain has a slightly different internal orientation and will appear a different shade of gray. The surface must initially be cleaned of any native oxide by a short duration mill before grains and boundaries can be clearly seen. To be sure that all grains are uniquely visible, two or more specific tilt angles are commonly used, and a composite image created. Automated grain software for measurement and statistical size distribution is available which makes comparison of films possible. With the ability to image large areas quickly in both planar and in section views, FIB grain analysis is a cost effective and convenient complement to TEM techniques.

Figure 12. FIB grain imaging of an aluminum film

Summary and Conclusions:

The Focused Ion Beam laboratory system is truly the wonder tool of the 1990's. With the ability to perform material removal through straight sputtering or chemically selective reactions, and deposit both conductors and insulators, all with pinpoint accuracy, the FIB can address a wide variety of complex tasks.

The FIB tool has been shown to be an excellent investment for any failure analysis, process or product/design support laboratory. No other single tool can fill so many different needs or return so much useful data.

ACKNOWLEDGEMENTS

The authors wish to thank Marsha Abramo and Loren Hahn of the IBM Analytical Lab in Essex Junction, Vermont, for their substantial contribution to the general body of FIB knowledge, and specifically to this manuscript. We would also like to thank Peter Carleson of the FEI Company, and Conrad Zagwyn of Micrion Corporation for the generous use of their original artwork and photos in the preparation of this text and the accompanying presentation slides.

REFERENCES:

1. J. Orloff, *Theory of Operation of High Resolution Liquid Metal Ion Source Focused Ion Beam Systems*, To be published: Microelectronics Desk Reference, ASM 1999

2. JEOL Training Material – New SEM system purchase operator classes, 1994

3. P. Carleson, FEI Corp. Private discussions on detectors and collection modes, 1999

4. Micrion System 9500 Operator's Reference Manual, 1996

5. M. Abramo, L. Hahn, & L. Moszkowicz, "Gas Assisted Etching: An Advanced Technique for FIB Device Modification", ASM Int. Symp. for Testing & Failure Analysis, p.439, 1995

6. S.T. Davies, B. Khamsehpour, End-point Detection Using Absorbed Current, Secondary Electron & Ion Signals During Milling of Multilayer Structures by FIB, J. Vac. Sci. Technology, Vol 11, No. 2, Mar/Apr 1993

7. J.P. Levin, P.G. Blauner, "Model for Focused Ion Beam Deposition", SPIE Vol. 1263 Electron Beam, X-Ray & Ion Beam Technology: Submicrometer Litho 1990

8. M. Zaragoza, "Reliability Test Results for Pt FIB Interconnect Structures", ASM Int Symp for Testing & Failure Analysis, 1999

9. H. Lomano, Y. Ogawa, T. Takigawa "Silicon Oxide Film Formation by FIB Assisted Deposition", Int. Symp. On MicroProcess Conference, pp 303-306, 1989

10. M. Abramo, E. Adams, M. Gibson, L. Hahn, A. Doyle "FIB Induced Insulator Deposition at Decreased Beam Current Density", IEEE Proc. Int. Rel Phys Symp pp 66-71, Apr 1997

11. A.N. Campbell, K.A. Peterson, D.M. Fleetwood, J.M. Soden, "Effects of Focused Ion Beam Irradiation on MOS Transistors" ", IEEE Proc. Int. Rel Phys Symp pp 72-81, Apr 1997

12. R.G. Lee, W.C. Monigle, "FIBs Probe and Fix Semiconductor Problems", Reprint – Test & Measurement World, May, 1998

13. L.S. Fischer, S.B. Herschbein, "FIB Assisted Circuit Debug of a Video Graphics Chip", ASM Int Symp for Testing & Failure Analysis, pp 37-39, 1996

14. M.T. Abramo, S.B. Herschbein, "FIB Techniques Provide Circuit Modification Capability for Rapid Design Debug", MicroNews, a publication of IBM Microelectronics, Vol 5, No. 1, 1999

15. R.H. Livengood, V.R. Rao, "FIB Techniques to Debug Flip-Chip Integrated Circuits", Reprint – Semiconductor International, March, 1998

16. P.F. Ulllman, C.G. Talbot, R.A. Lee, C.Orjuela, R. Nicholson, "A New Robust Backside Flip-Chip Probing Methodology", ASM Int Symp for Testing & Failure Analysis, 1996

17. S.B. Herschbein, L.S. Fischer, T.L. Kane, M.P. Tenney, A.D. Shore, "The Challenges of FIB Chip Repair & Debug Assistance in the 0.25um Copper Interconnect Millennium", ASM Int Symp for Testing & Failure Analysis, pp 127-130, 1998

18. A.N.Campbell, J.M. Soden, J.L.Rife, R.G.Lee, "Electrical Biasing and Voltage Contrast Imaging in a FIB System", ASM Int Symp for Testing & Failure Analysis, pp 35-41, 1995

19. R. Boylan, M. Ward, D. Tuggle, "FA of Micron Technology VLSI Using Focused Ion Beams", ASM Int Symp for Testing & Failure Analysis, pp 249-255, 1989

20. S. Morris, S. Tatti, E. Black, N. Dickson, H. Mendez, B. Schweisow, R. Pyle, "A Technique for Preparing TEM Cross Sections to a Specific Area Using the FIB", ASM Int Symp for Testing & Failure Analysis, pp 417-427, 1991

21. L.R. Herlinger, S. Chevacharoenkul, D.C. Erwin, "TEM Sample Preparation Using a FIB and a Probe Manipulator", ASM Int Symp for Testing & Failure Analysis, pp 199-205, 1996

22. A.B. Sato, S.L. Riley, "The Use of FIB in FA Support of In-Line Defect Inspection", ASM Int Symp for Testing & Failure Analysis, pp 419-424, 1993

Microsurgery Technology
for the Semiconductor Industry

Kendall Scott Wills and Seshu V. Pabbisetty
Texas Instruments
Stafford, Texas

Abstract

In this paper, we will analyze the tools and techniques, which allow the failure analyst to become a microsurgeon/micromachinist. Emphasis will be on areas such as localized deprocessing/delayering, specific area cross section analysis, and precise cut and paste operations on semiconductor devices. The pros and cons of the different tools will be discussed in conjunction with other key technologies. In this paper, two prevailing technologies will be considered - ion beam based and photon beam based. The more traditional method of ion beam milling will be covered first, following by promising techniques offered by photon beam based systems.

Focused Ion Beam Technology

Introduction - FIB

In the analytical area, Ion Beam Milling (IBM) was first used to polish samples for transmission electron microscopy (TEM). The beam was 1 – 2 cm in diameter and was composed of argon ions. Little thought was given to milling fine lines one at a time. It was in 1973 when a focused ion beam of 3.5 um diameter, 0.4 mA/sq.cm., was obtained using an implanter source at Hughes Research Laboratories. Other research activities followed which led to the introduction of the first commercial Focused Ion Beam (FIB) systems by Micrion and Seiko [1,2,3,4]. The following list gives a brief outline of the important events in the development of the current FIB process.

1974: Focused Ion Beams using cryogenic field ion emitter; 10K brighter. (Oregon Graduate Center, University of Chicago, Cornell University)

1975-1978 Development of Liquid Metal Ion Source (Cullam Laboratories and Oregon Graduate Center)

1979-1982 Gallium Focused Ion Beam; 0.1 um dia., 1.5 A/sq.cm. (Hughes Research and Oregon Graduate Center)

1985-1987 Micrion and Seiko introduced the first commercial FIB systems.

FIB milling was first used for mask repair with the greatest emphasis placed on the reduction of charge build up on the dielectric photo mask and making the repair opaque. These processes [2,3,4] facilitated not only conductor disconnects, but also conductor depositions. They also opened up new areas of applications, such as microcircuit analysis and structural failure analysis. End point detection allowed precise milling depth control.

In the evolution of FIB technology, semiconductor technology trends and market conditions have become the driving forces. These forces are outline in the following list. They are also understandably the same trends which drive the development of advanced failure analysis.

-Shrinking device geometries to submicron level
-Multiple level metal systems
-Increasing device complexity
-Ever-shortening time-to-market goals
-Requirement for better resolution of failures
-Increasing demand for better quality and reliability

Design proto-typing is becoming the most obvious direction for FIB technology for the failure analyst, who must find the fail site on the new VLSI designs. The FIB promises to speed their work and improve the resolution of the failure [2,3,5,6,7]. The direction of this paper is to provide the current failure analysis and process characterization/monitoring. Problems with the different techniques will be presented along with possible solution trends.

Principles

In principle, the focused ion beam (FIB) technology is closely related to that of the scanning electron microscope (SEM) and e-beam probing. Charged particles from a given source are accelerated and formed into a beam which is then focused to a small point. In FIB systems the charged particles are ions. For FIB systems used in integrated circuit repair, the charged particle is usually a liquid metal ion source (LMIS) made of gallium. The charged particles are accelerated off a point source by applying a strong electric field. Figure 1 details the reactions, which are present at the LMIS. Good source design must take into account the different reactions to prevent shorting in the column from evaporated liquid metal, while providing a stable source of ions [3,8,9].

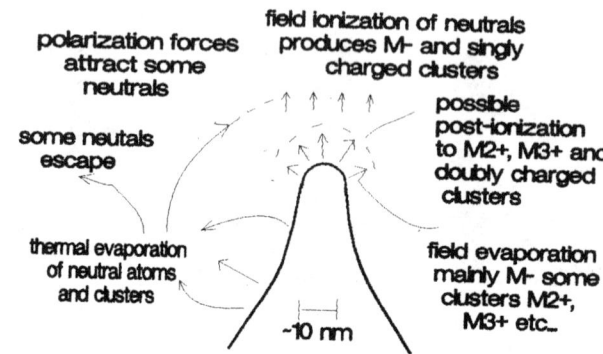

Figure 1. Schematic view of liquid metal ion source.

Electrostatic lenses are used to focus the beam to a size as small as 2 nm. having a beam energy in the range of 30 KeV. Figure 2 shows a diagrammatic view of a two lens ion column. The drawing shows the final deflection control below the final lens. This is not always the case. Some columns have an aperture after the final deflection control [1,3,10,11,12,13,14,15,16,17,18,19].

Ions are accelerated onto the surface of the sample where they dislodge or sputter ions and/or atoms from the surface. This removal of ions or atoms from the sample surface is known as sputtering or milling. The sputtering process generates several different types of particles including neutral atoms, secondary ions and secondary electrons. The collision of the primary ion with the target material can displace the target atom, generate phonons and the primary ion can become implanted in the target material. The ion beam interaction is shallow compared with the beam interaction of Scanning Electron Microscopes (SEM). The ion only affects the top 200 Angstroms, while the same energy electron would affect 4 microns or more of a silicon material. Figure 3 shows the material interactions [17,19, 20].

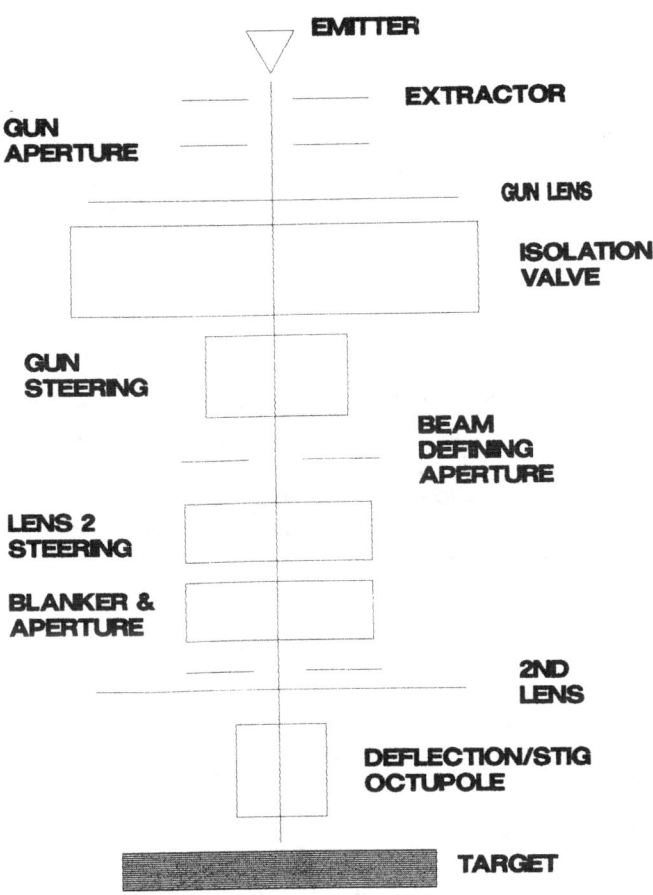

Figure 2. Two-Lens Ion column.

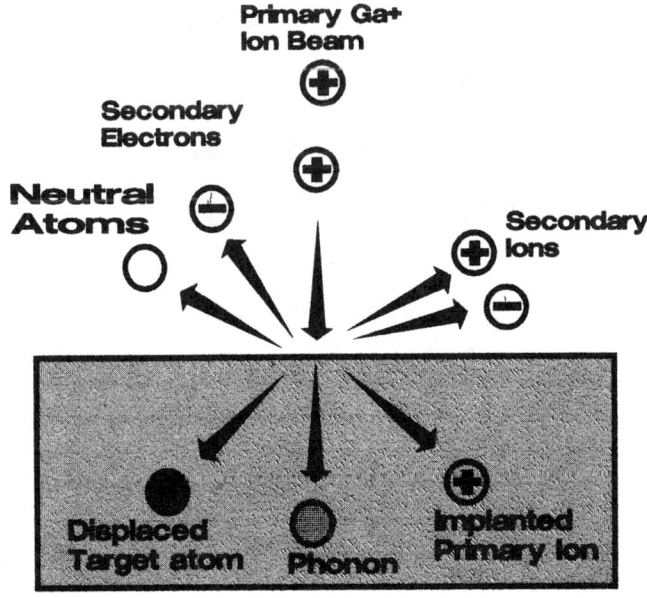

Figure 3. Specimen / Beam Interactions.

An image can then be formed either from the secondary electrons or the secondary ions by scanning the focal point of the primary ion beam across the sample. A detector, tuned to monitor electrons or ions, can give a signal that is used to produce an image of the sample on a monitor. There are several types of detectors [17,19,21,]. For electron imaging, a scintillator or channel electron multiplier (CEM) can be used. Some instruments use a microchannel plate (MCP) for electron imaging. A CEM can also be used to image secondary ions. FIB systems primarily see material differences. Therefore, the topological detail is not as clear as that typically seen in a SEM. Detector positioning is therefore extremely important to gain as much detail as possible from the secondary particles coming from the target material.

In the case of secondary ions, one could also get a chemically specific secondary ion mass spectrometry (SIMS) image by mass filtering the ions prior to detection and using only ions of a given mass when constructing the digitally scanned image [3,21]. Figure 4. shows a diagrammatic representation of a SIMS detection system.

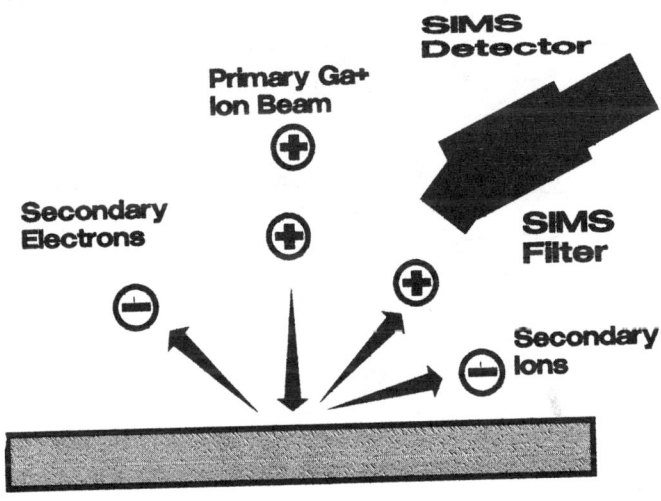

Figure 4. SIMS Configuration.

In FIB technology, the sputtering process of the primary ion beam is used to remove material in a localized area with excellent controllability and repeatability. By focusing on the area to be cut and rastering over the area, a cut of any desired shape can be made into the sample. Figure 5 shows a typical raster technique for scanning the ion beam over a target material [13,19,21,22].

Beam Raster

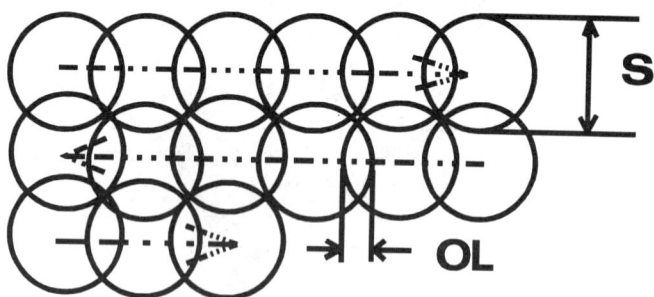

Figure 5. Beam Raster. OL is the overlap and S is the spot size.

By controlling the parameters shown in Figure 6, the raster can be changed to fit the requirements of the particular FIB process.

-Dwell time (Dt): The time period the beam resides at each point in the raster.

-Spot size (S): The beam diameter.

-Overlap (OL): The amount or percentage of beam which overlays an adjacent beam spot.

Figure 6. Beam parameters for controlling beam raster.

The total ion beam milling which takes place is related to the total "Dose" of ions which strike the surface. Figure 7 shows the beam interaction dose [13,15,19,21,22].

Dose = (J*T)/Q

-Current Density (J): The average current density.

-Exposure Time (T): The total pattern time or cumulative exposure frames.

-Elemental Charge (Q): The charge of a gallium ion ($1.6*10^{-19}$ Coulombs).

-Dose (ions/cm^2): The cumulative amount of gallium irradiating a unit area.

-For patterns with overlap and "f" is the number of frames.
$$T = {_0}^{f} Ltdf$$
$$= {_0}^{f} (D_t*I*W)/[S*(1-OL)]^2 df$$

-Metal line resistance (R): R = (•*L)/(Z*W)

where • is the metal resistivity for a given L_1,

L is the length of the metal line,
Z is the height of the metal line and
W is the width of the metal line [13,19,21,22].

Figure 7. Beam Interaction Dose.

The equations given for the beam interaction assume a perfectly round spot with uniform ion distribution across the entire surface. This is not quite true. The beam has a Gaussian shape, which means the ion density is the highest in the center of the beam and decreases toward the outside edge. When the beam current is increased the beam widens. Figure 8 illustrates the change in beam shape with increasing beam current. The unwanted ions around the center of the beam are called the beam tail. They are of lower ion density than desirable for FIB processing. Each ion column manufacturer tries to reduce the beam tail as much as possible. In most cases this is desirable. However, in some cases to be discussed later, the beam tail can be of some help.

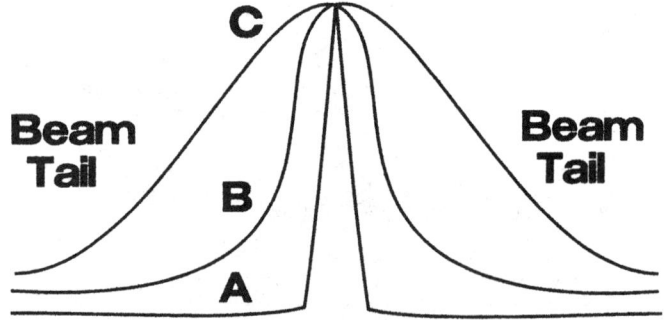

A= Low beam current
B= Medium beam current
C= High beam current

Figure 8. Ion Beam Shape.

Figure 9. Shows a typical mill in a semiconductor device. The mill is through passivation, top level metal and then an interlevel dielectric to reach an underlying metal level. In Figure 9 all the metal levels shown are aluminum, but they can be tungsten, titanium, molybdenum, copper or combinations of metals.

Notice the sloped walls shown in Figure 9. The cut starts out with a width W but ends up somewhat smaller. The cause of the taper is the beam shape and redeposition of milled material from the bottom of the hole. A better mill technique is required to make the holes with square walls.

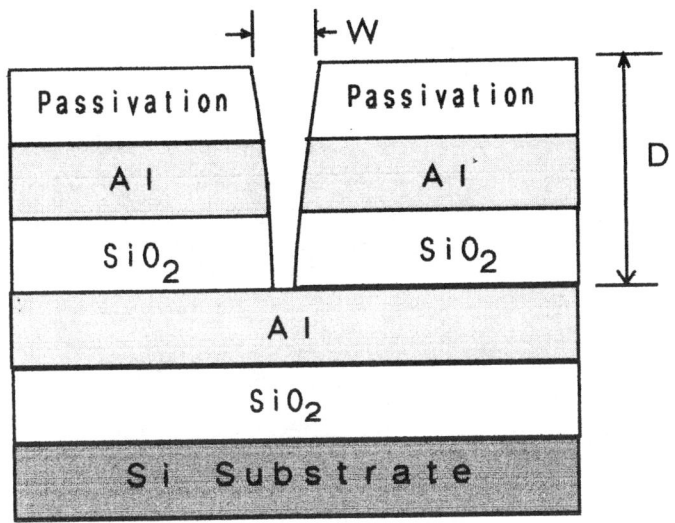

Figure 9. Typical FIB Mill.

In order to improve the milling process, an enhancing gas is required during the mill. The gas is introduced with an injection needle, which is placed near the surface of the sample, delivering the desired gases to the area being modified. Figure 10 shows an example of how a gas may be applied to the surface of a semiconductor device.

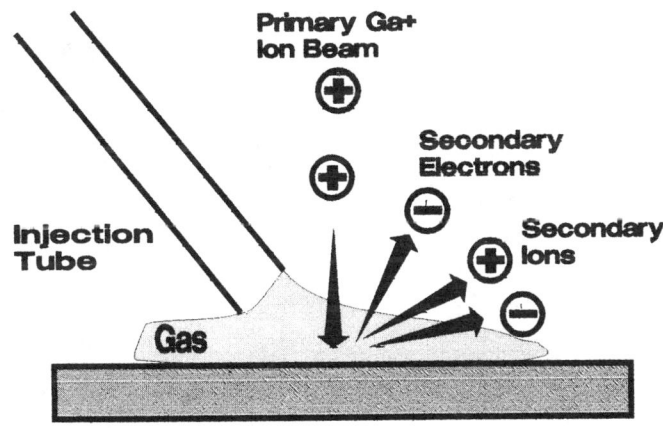

Figure 10. Gas Injection:

Etch / Deposition

As the energized ions from the sputtered surface are allowed to interact with the gas, the gas molecules split into their constituent parts. If the injected gas is an etchant, ion beam assisted chemical etching (IBASCE) takes place locally and converts the milled material into a gas that is pumped out of the vacuum chamber. This

process provides a significantly increased milling speed, as well as high aspect ratio cuts, when compared to the conventional milling process. Figure 11 shows the improvement gained with the enhanced etch gas [19,22,23,24,25].

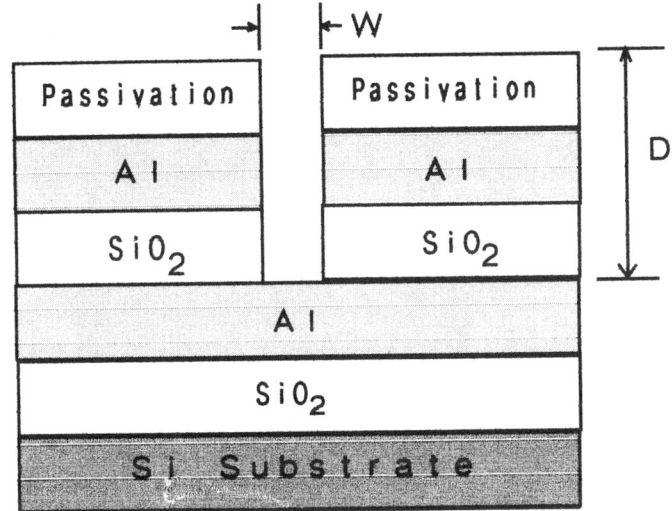

Figure 11. Gas Assisted Etching.

Another improvement is gained by use of an enhancing gas. The bottom of the milled area is flatter than without the enhancing gas. When a material is milled the primary ions from the beam tend to channel along crystal lines. As the region being milled becomes deeper, the roughness at the bottom of the region becomes more pronounced. In Auger milling this affect is the primary factor limiting the detection of thin materials in deep mills. With an enhancing gas, the ions react both along channel lines and chemically, reducing the tendency to create a rough bottom on the hole. Figure 12 shows the difference. "A" represents a mill without the enhancing gas and "B" represents the mill with an enhancing gas.

Enhancing gas also helps improve the aspect ratio of the holes, which can be milled. This improves the ability to connect to metals buried deep in the device. Figure 13 shows a comparison between a hole milled with and without an enhancing gas.[26]

Figure 12 gives a good idea of the problems associated with milling deep holes. Another problem is with end point detection [17,19,21,27]. There are several techniques for end point detection. First, the secondary ions can be monitored. This is an ideal technique in that the surface charges little from the loss of the ions. The difficulty with ion detection is the ion yield. There are nearly 10 times more electrons generated than ions. The electron signal is therefore stronger. Monitoring electrons is better than monitoring ions due to signal strength. The problem associated with electron monitoring is with surface charging. So many electrons are generated that the surface becomes depleted of charge when the electrons are pulled from the surface. Attempts to use a flood gun to control the charge masks the electron image and can only be used with ion monitoring. One way around the end point problem is to only look at the electrons, which come from the center of the mill. This improves the signal to noise ratio by removing from the endpoint the electrons generated by the edges of the mill where no real information regarding the end point exists.

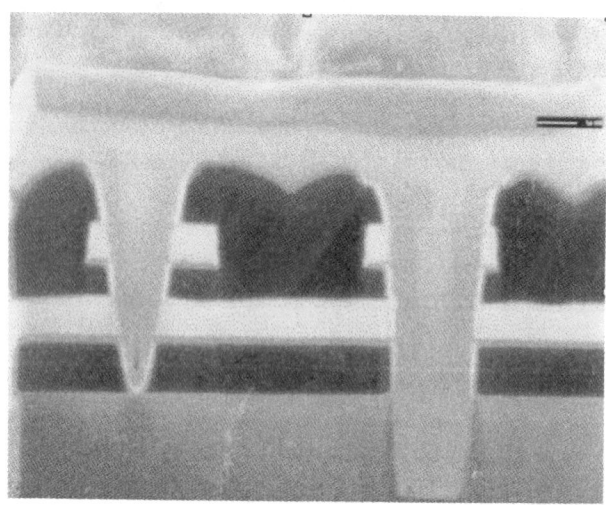

Figure 13. SEM image of a hole milled with and without enhancing gas.[26]

detection by creating a greater difference between one material and another as the mill process proceeds. The difference can be quite significant. For example, Metal 1 in a three layer device is almost impossible to see with a straight mill process. If I_2 is added as an enhancing gas, the endpoint becomes more distinct. If XeF_2 is used as the enhancing gas, the Metal 1 is distinctly visible at endpoint.

More complex endpoint techniques are available. SIMS can be used to monitor only the ions of a particular material. Even with this technique, deep holes present a problem.

Figure 14 gives a partial list of enhancing gases. The numbers in the columns are the rough improvements in milling times which can be expected by use of the particular gas / material combination [19,21,28,29,30].

If an organo-metallic gas is injected into the vacuum chamber instead of an etch gas, the metal atoms are deposited. This is a well-known process called ion beam assisted chemical vapor deposition (IBCVD). The most prevalent metals deposited today are tungsten and platinum, but gold, nickel and silver are among the many metals available with the IBCVD process [3,2,31,32].

It is the precision and repeatability of ion milling and metal-deposition processes, along with elemental analytical capability, that make the FIB technology so invaluable to the semiconductor industry for several key applications.

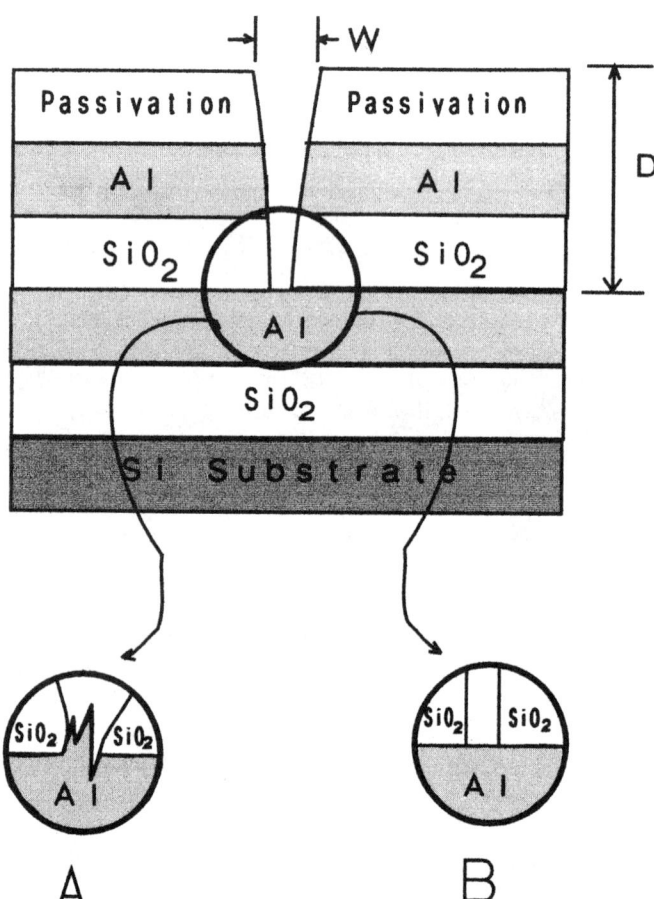

Figure 12. FIB Mill: A)-without enhancing gas. B)-with enhancing gas.

As the holes become deeper, the differences in materials are harder to detect. This is due to the reduction in electrons and ions, which can travel out of the deep hole. Enhancing gases help with end point

Material											
GAS	**Si**	**SiO$_2$**	**Si$_2$N$_4$**	**Al**	**W**	**Au**	**PMMA**	**GaAs**	**Ni**	**Polyimide**	**InP**
I$_2$	5-10	1	1-5	5-15			2	5-10	1-5		
XeF$_2$	>10	5-10	5-10	1	10		1-5			4-15	1
H$_2$O	<1	<1	<1	<1		1	>10			18	
Cl	>10	1-2	1-2	5-30	1	1	1-2	>10		1	11
Br	6-10	1	1	10-20	1					1	
CH$_3$OH							10				21
O$_2$							4				
H$_2$							1				
ICl	4-5	1	1	8-10	2-6		1			1	13

Figure 14. A partial list of enhancing gases. The numbers represent the potential improvement in mill rate expected with the gas / material combination.

Applications and Status

One could classify the applications into three broad categories as listed in Figure 15.

1. Microcircuit Analysis and Repair
   A) Microcircuit Analysis:
      -Probe-point access
      -Probe pad formation
      -Microcircuit isolation
   B) Repair or Reconfiguration:
      -Cuts or disconnects
      -Jumpers or interconnections
      -Capacitor/ Resistor trimming

2. Material Characterization and Failure Analysis:
   A) Material Characterization:
      -Grain boundary analysis w/ channeling contrast
      -Cross-sectional/ structural analysis
      -Elemental Analysis
   B) Failure Analysis:
      -Cross-sectioning and imaging
      -Sample preparation: SEM and TEM

3. Process Monitoring/Characterization:
   A) Off-line Monitoring (Post-)
      -Structural analysis
      -Yield loss analysis
      -New process characterization
   B) In-line Monitoring:
      -Grain size and orientation
      -Metal step coverage
      -Vertical CD measurements

Figure 15. FIB application areas.

To allow maximum utilization of the various capabilities of the technology, the FIB systems are now computer controlled with the ability to link to a Computer Aided Design (CAD) system.

1. Microcircuit analysis and repair

The primary use of FIB technology for device prototyping is to cut interconnects and paste in new interconnects. The cut and paste technique effectively changes the interconnect routing, thereby providing the designer with the capability to quickly modify the design to determine optimum performance conditions. Figure 16 shows two cuts and a jumper between the cut lines used to reroute the signal.

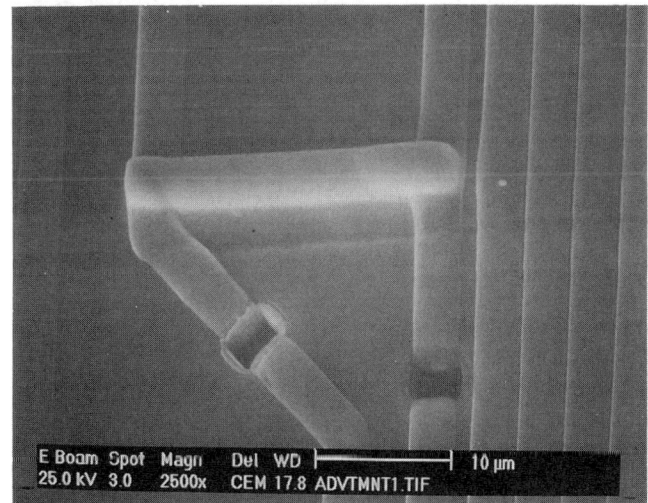

Figure 16. Two cuts and a Pt jumper to reroute the signal.

Figure 17 shows a cut "A" with a jumper "B". In this case the contacts at location "C" were considered to be suspect. Location "C" was cross-sectioned. This image shows how well multiple processes can be performed on a single sample.

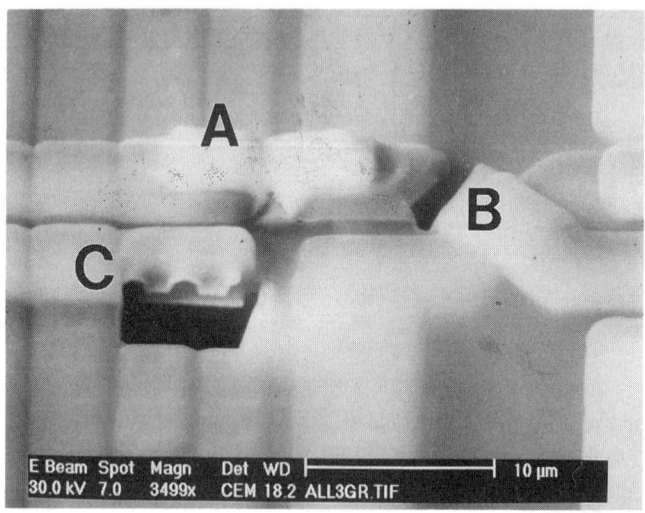

Figure 17. A)- Pt jumper, B)-Al line cut, C)-cross section of two contacts.

In the zeal to use the new FIB technology, sometimes analysts overlook some obvious problems. Figure 18 indicates that the pads just deposited will be shorted together if the metal between the pads is not cleaned up.

Consideration should be given to the fact that the ion beam impinging upon the sample has an energy of 25 KeV and could cause an undesirable change in the performance of the device. Theoretically, some of the electrons generated by the ion interaction should have an energy high enough to penetrate deep into the device structure where they generate a charge cloud of electron-hole pairs. Once the charge is trapped in the oxide or material it can upset the balance of the transistor.

Another problem with having a highly changed beam impinge upon a surface is surface charging. New semiconductor devices are moving to thin gate oxides, which can only withstand 10 to 20 Volts across them. Any trapped charge on the surface of the device, which can discharge into the semiconductor, will cause damage.

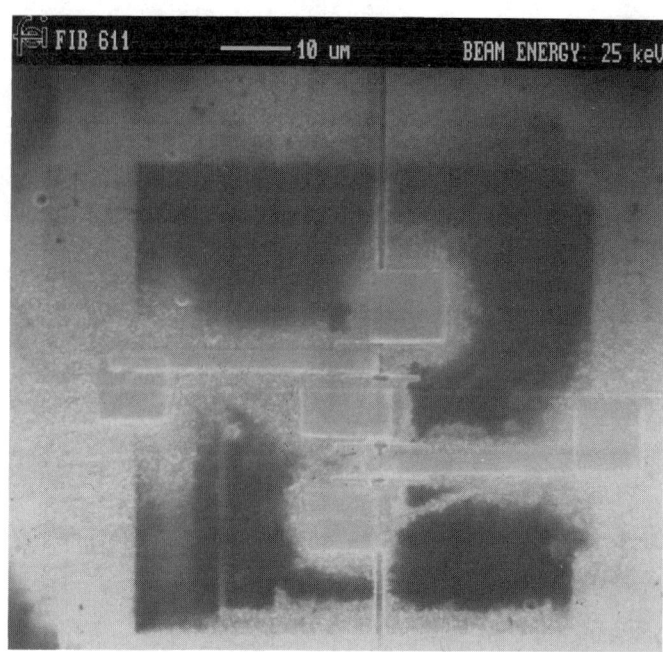

Figure 18. Metal pads deposited waiting for metal clean up to remove the excess material, which is shorting the pads.

Because the FIB technique is to be used to prototype functionally good devices, preventing inadvertent performance shifts is very important. In one test, transistor performance was measured for the test transistor and its neighboring transistors. The measured parameters were gate to substrate leakage and drain to substrate leakage. Each test FIB system then tried to cut the polysilicon interconnect to the transistor gate and tried to cut an interconnect to the drain of the transistor. After FIB milling was performed, each transistor was re-characterized. Figure 19 shows the results of the pre and post test for gate to substrate leakage. The label "A" is the post test measurement and "B" is the pre test measurement.

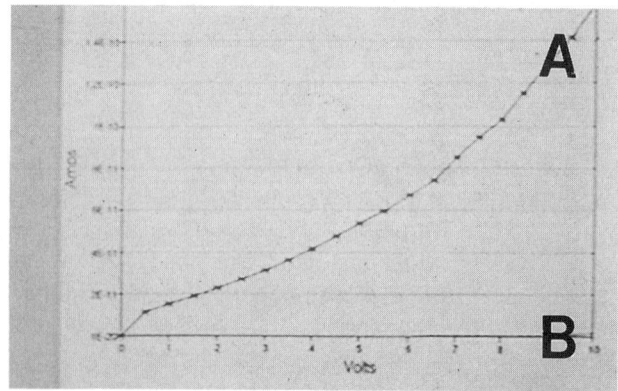

Figure 19. Leakage measurements of A)-Post FIB cut and B)-Pre FIB cut.

Figure 20 illustrates the change in drain to substrate leakage. The "post FIB" plot is along the X axis showing that there is a large change in leakage current. The label "A" is the pre test measurement and "B" is the post test measurement. None of the FIB systems could consistently prevent damage to the transistor being tested or the surrounding transistors. From a design engineer's point of view, the inadvertent change in performance is very undesirable.

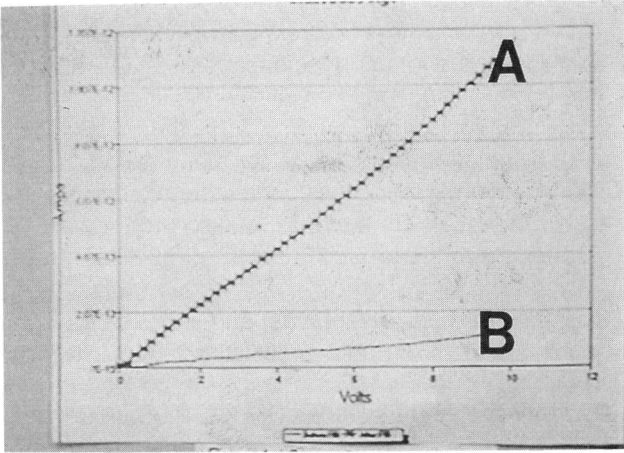

Figure 20. Drain to substrate A)-Pre FIB mill and B)-Post FIB mill.

Each FIB manufacturer tries to eliminate the problem of device damage in a different way. One vendor floods the device with electrons in an attempt to prevent surface charge build up [17,19,33]. Unconfirmed reports suggest that this technique can damage the adjacent transistors to the milled area while the transistor being milled is left unharmed. One company coats the device with Pt or C, which is an effective way to control surface charge, but brings about other problems. The Pt is hard to remove and the C can be "knock on implanted" into the top surface of the device. The carbon, when ashed to remove it from the surface after FIB process, can be ashed from under any depositions. The depositions are thereby removed with the carbon rendering the FIB job useless. Another company has established a second column on the FIB system that is used as a scanning electron microscope (SEM). This turns out to be an effective solution to charging and adds expanded imaging capability to the system. However, the addition of a SEM column can reduce surface charge but can alter device performance with new charging problems similar to the flood gun problem.

Control of surface charge is important during material removal. Some of the FIB companies suggest that the FIB operator strap metal lines to ground before any major mill operations are performed. Figure 21 shows that such a strategy may be required. The hole blown in the test structure is due to electrostatic discharge (ESD) across the surface of the device. The hole blown in the passivation did not occur at the location of the mill site. The blown area was several hundred microns down the line that was being milled. The cause of the ESD event was the charging of the metal line being worked on with respect to the surface and another metal line in the structure. The second metal line was floating.

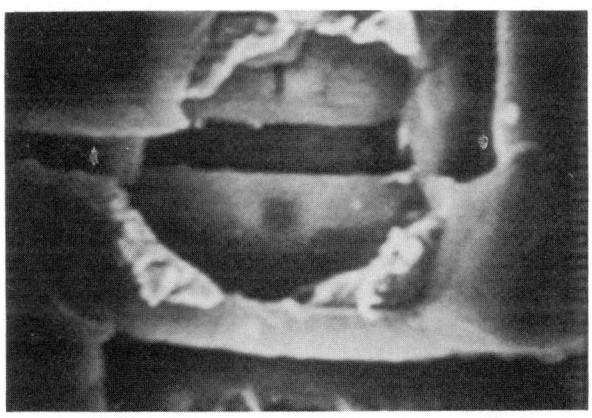

Figure 21. Hole blown in a test structure with at least one line grounded. The blown location did not occur at the mill site.

Some individuals have tried to shield the device from the ion beam with protective metal plates. Unconfirmed reports suggest the same problem as shown in Figure 21 can occur with a grounded plate over the surface of the device. Charge from an underlying metal line can gain enough electrical potential to discharge into the metal plate causing a hole in the device.

The flow of process gases has been shown to reduce the charging affects seen during FIB mill operations. A very powerful feature of FIB systems is the ability to deposit conductors. It is often desirable to create a connection between nodes of a device. A secondary electron image (SEI) of a deposited Pt jumper is shown in Figure 22. The connection is made by using FIB mill to expose the metal lines under the insulator, deposit metal in the openings and then connect the deposited metal by a metal line deposition.

At times the need arises to place bonds in the active circuit. Figure 23 shows a bond to a FIB deposited pad in the center of the active circuitry of an integrated circuit.

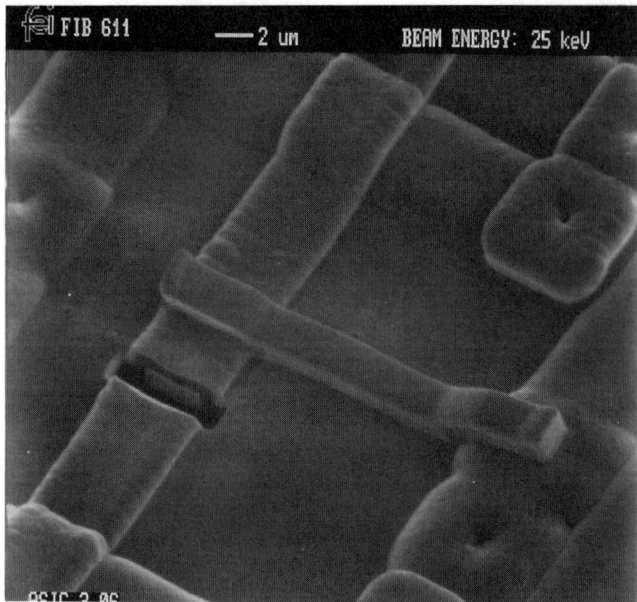

Figure 22. FIB jumper with an Al line cut.

The reason for the bond in the middle of an active circuit is the high resistance of the FIB deposited metal lines. There are ways around this problem. One is to deposit a thin line of metal and then electroplate the deposited line [34]. Another way is to laser deposit a metal line [35,36]. The laser deposition can be done from a liquid or a gas. The laser deposition is both clean and cost effective for long lines.

With years of usage of FIB in the IC industry, there are now a few design methodologies in place, some in public domain and some in private practice, to make the best use of FIB technology [3,4]

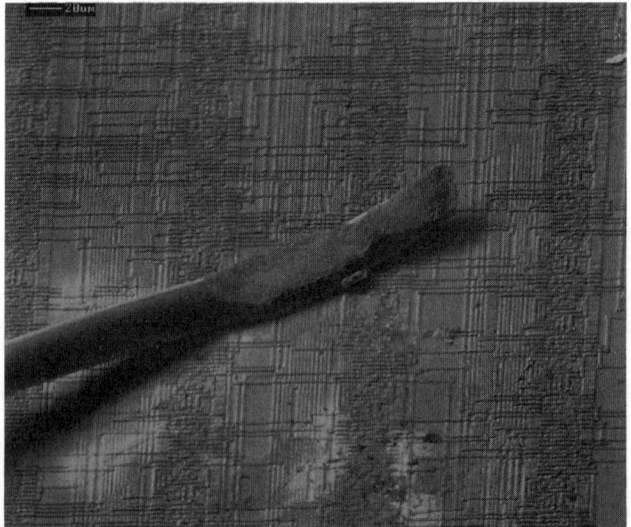

Figure 23. Gold bond to an FIB deposited pad in active circuitry of an integrated circuit.

Design Guidelines for FIB usage are given in Figure 24.

-Built-in signal access points: key electrical signals near device surface to minimize deep end-point detection with possible damage to the device.

-Strategically placed spare gates methodology in design. Do not put them all at the end of long rows. Long depositions are costly and unreliable.

-Spare long line interconnects placement: this saves time in depositing long line interconnects and decreases resistivity. One could use the metal lines already included, sometimes, for metalization density normalization. Precautions should be taken to avoid their inherent antenna effects. Make sure they are grounded and the ground is cut after FIB attachment.

-Designed-in metal layer/layout and windows for opening to look for photo emission.

-On multilevel metal designs with CMP or other extreme planerization: Add reference top level metal to use as overlay land marks.

Figure 24. Design guidelines for FIB usage.

2. Material characterization and failure analysis

- Grain boundary analysis

Both charging and ion channeling can make it difficult to image the surface of the sample. Ion channeling also causes non-uniform ion milling. As undesirable as ion channeling seems, it can be very helpful when trying to analyze grain structures. The FIB image in Figure 25 shows an Al:Si bond pad with its grain boundaries clearly delineated [3,37,38].

- Cross-sectional/structural analysis

The FIB system is useful in creating localized cross sections. Cross sections are made by milling a series of adjacent boxes with increasing depths creating a staircase effect as illustrated in Figure 26. Once the milling is complete, the sample can be tilted and the trace of interest can be viewed.

536

Figure 25. Al:Si bond pad with Al grain boundaries visible.

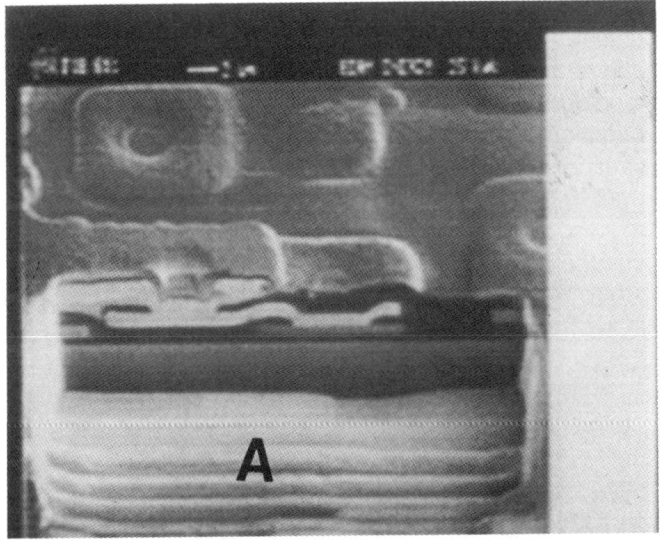

Figure 26. Stair step (A) entry to a cross section of a metal line and contact.

Perhaps a more practical application of grain boundary studies can be observed in the cross section of Figure 27, 28 and 29 [3,38,39,40]. Notice the grain detail in the aluminum lines. Figure 28 is noteworthy due to the versatility of the cross section in two directions. Figure 29 is a stand out as it is a cross section of a 0.35 micron line and contact.

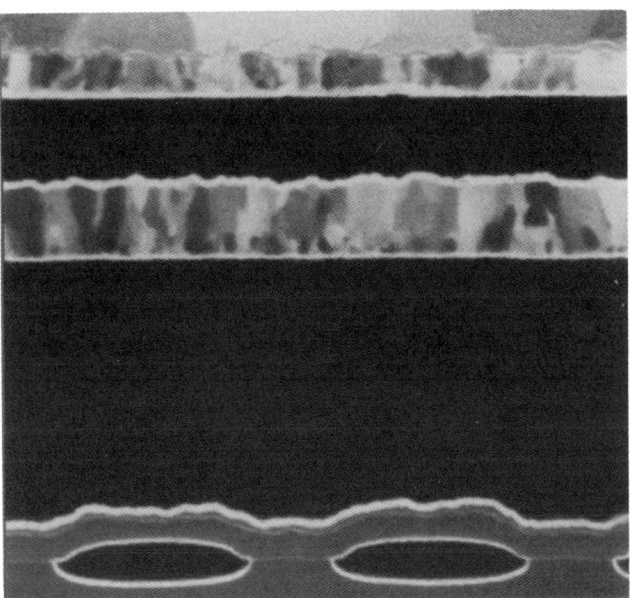

Figure 27. Aluminum line cross section showing the Aluminum crystal structure.

Figure 28. Cross section from two sides showing the Aluminum crystal structure and the versatility of the FIB process.

537

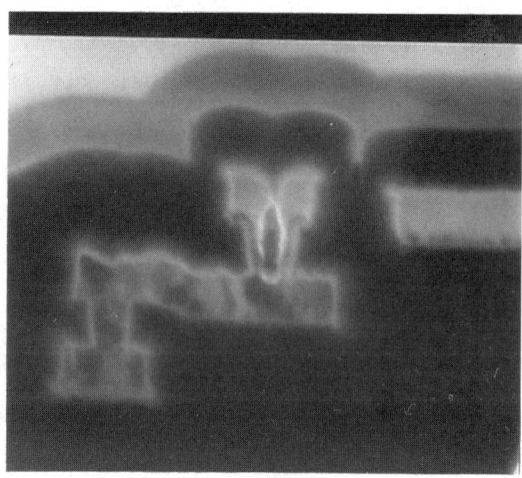

Figure 29. Cross section of a 0.35 micron line and contact.

Unfortunately, in the process of creating a cross section, metal lines are often isolated from ground and left floating. Figure 30 shows how these floating lines are difficult to observe in the FIB. The solution to this problem is to ground the floating lines as done in Figure 31. Having all the metal lines in the cross section at the same ground potential causes a uniform brightness of all the conductors during FIB imaging [3,38].

Figure 30. Dark spaces show metal lines which are not grounded due the cross section process.

Once all the conductors in a cross section are shorted to ground, voltage contrast reveals a well defined distinction between conductor and insulator. In some cases, this FIB cross section technique may provide more cross section information pertaining to conductor-insulator boundaries, than a SEM image of a cross section created with traditional methods.

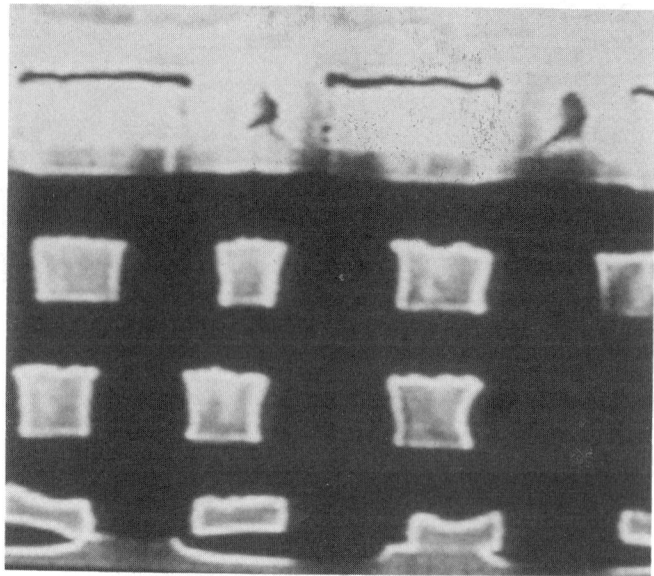

Figure 31. The dark spaces are now bright as all metal lines have been shorted together to ground.

The FIB cross section is very smooth. As stated earlier, the FIB image can only see material differences and the SEM is more adept at seeing topology. Several techniques have been developed to bring out the surface features of the FIB mill cross section. First, the cross section can be turned facing the FIB beam as shown in Figure 32. The beam is then allowed to mill the face of the cross section. Ion channeling will delineate some regions better than others leaving a distinct pattern of material as is shown in Figure 33.

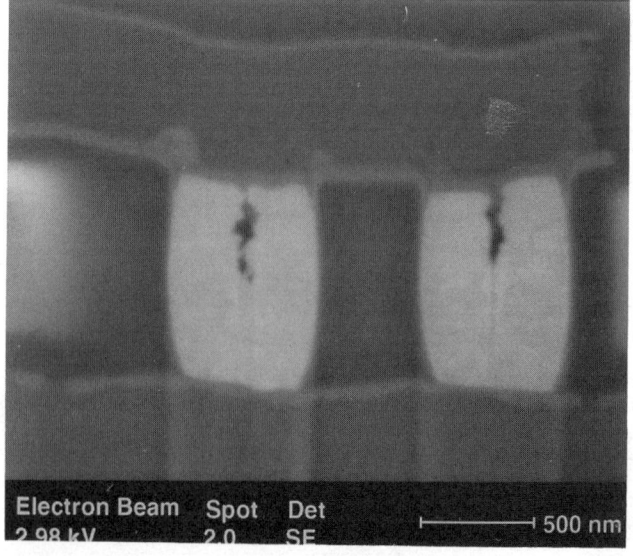

Figure 32. Cross section before FIB mill.

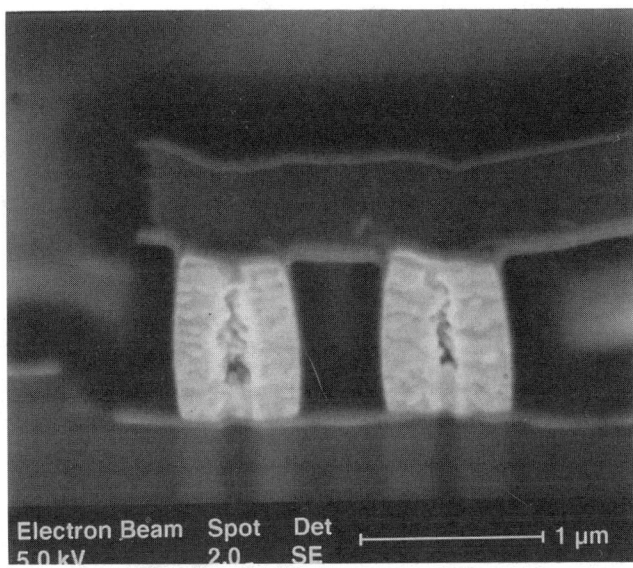

Figure 33. Cross section after FIB stain.

Using one of the enhanced gases can make further improvements in cross sectional viewing. Figure 34 shows the sample before the use of the enhanced gas. Figure 35 shows the sample after the use of the enhanced gas.

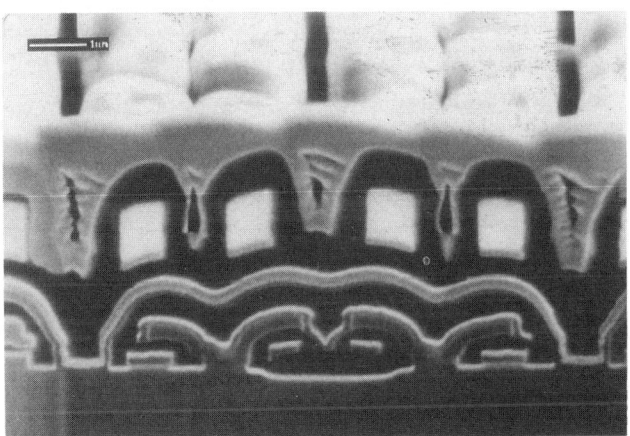

Figure.34 Cross section before FIB enhanced etch stain.

In some cases the use of an in-situ FIB stain is helpful. In other cases the failure analyst may need to use a wet chemical stain. Figure 38 shows the results of a 4.5% HF etch after FIB cross section milling. The mill was performed with I$_2$ as the enhancing gas to speed the substrate etching. The artifacts shown at the "?" marks are actually Al:I compounds, not real defects. The wet chemistry was used to meet government requirements on the method of cross sectional analysis.

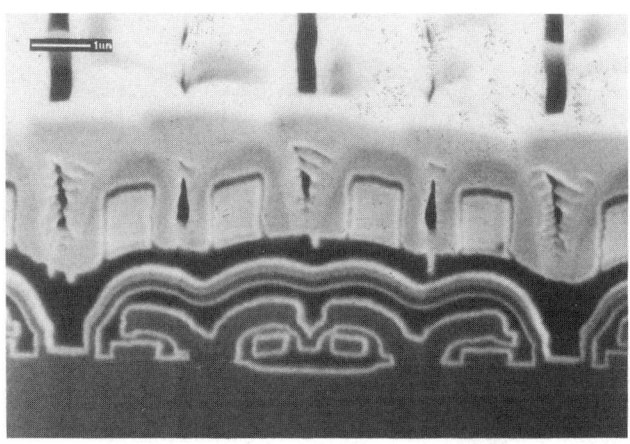

Figure 35 Cross section after FIB enhanced etch stain.

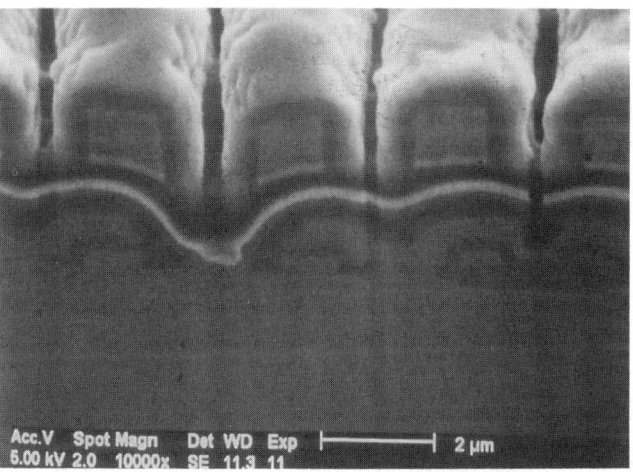

Figure.36.Cross section before FIB enhanced etch stain.

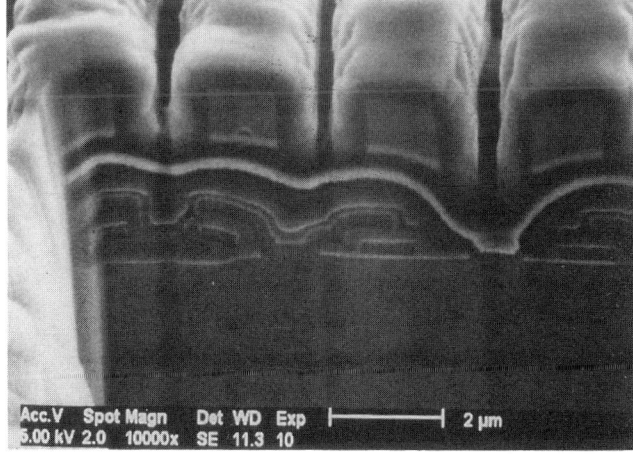

Figure 37 Cross section after FIB enhanced etch stain.

539

Figure 38. FIB cross section showing Al:I compounds at the "?" marks.

The FIB can be used to cross section more than the integrated circuit itself. As seen in figure 39, photoresist can be cross sectioned[26]. The use of water, oxygen and other mill enhancing chemicals make this possible.

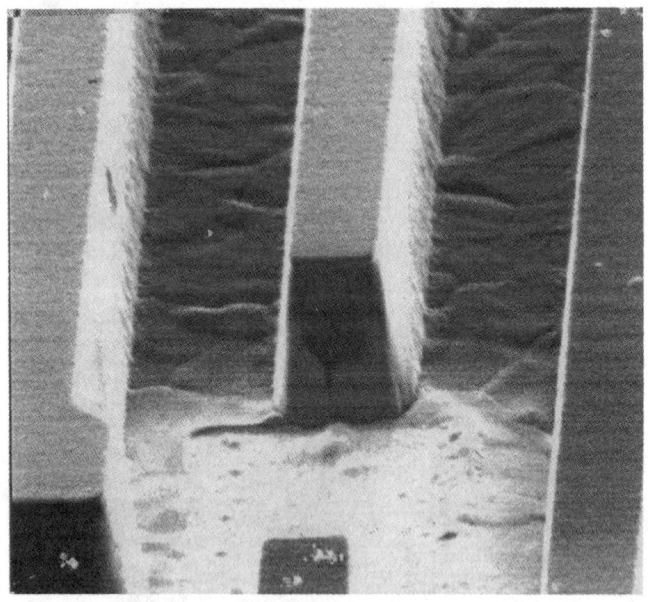

Figure 39. SEM image of photoresist cross sectioned with an FIB[26].

- TEM Analysis

Because TEM samples require such precise preparation, FIB systems are ideal for TEM sample preparation. Dr. Anderson of IBM has demonstrated

precise TEM lapping techniques, which are usable without a FIB. However, once the large scale polishing is performed by Dr. Anderson's technique on the TEM sample, the FIB can be used to make precise FIB cuts on either side of the point of interest leaving only a thin wall in the area of interest without the potential of damage to the sample inherent to mechanical methods [19,26, 41,42,43,44,45,46]. An example of a FIB prepared TEM sample is shown in Figure 40.

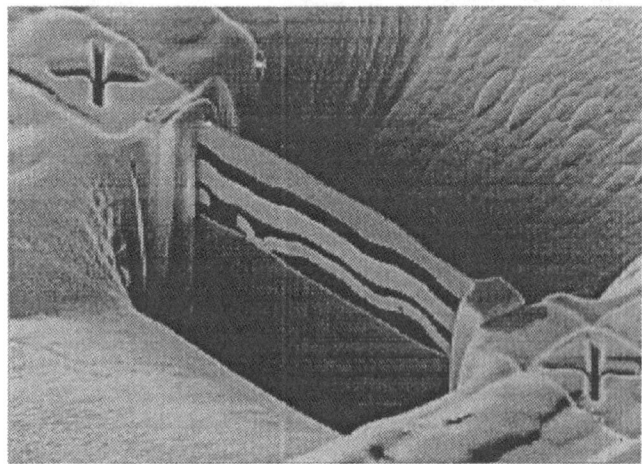

Figure 40. TEM sample prepared by FIB techniques.

To speed up the sectioning procedure several procedures are proposed. First, the sample can be cut by high power FIB milling to leave a wedge in the original sample. The wedge is then thinned in the center to the required thickness for TEM work. After the wedge is thinned, the wedge is trimmed and broken out. The wedge, once released from the bulk silicon, is picked up and placed upon a TEM grid.

A second proposal is to perform the bulk mill with a laser chemical technique instead of an FIB. A laser can be used to cut the major portion of the trench required for a cross section. A cross section for TEM work, such as the one shown in Figure 40, would take a long time if a laser were not used to pre-cut the trench before the FIB cross-sectioning.

- Elemental Analysis

FIB systems have the ability to effectively expose contaminated areas, so various tools can be used to answer questions as to the composition of these contaminates. There are tools that function well with the FIB system. Energy Dispersive X-ray (EDX) is available on dual beam FIB/E-Beam systems [16,17,19,47]. A typical Dual Beam system is shown in Figure 41.

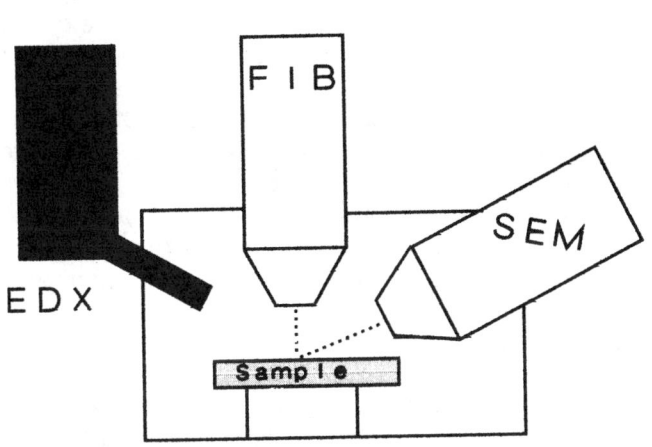

Figure 41. Typical Dual Beam FIB/SEM system with EDX capability.

SIMS, as discussed previously, can be used on standard FIB systems [19,21,45,48]. Both the EDX and the SIMS give valuable information on the material present in the semiconductor device. Figure 42 shows a FIB cross section of a particulate under a metal line. Figure 43 is a SIMS dot map showing the type of material in the particulate and the location of each material. [3,16]. Figure 44 shows a SEM picture of a cross section with an EDX spectrum transposed onto the screen. The EDX was performed on a dual beam system [3,16,37]. The advantage of SIMS over EDX is the resolution of the SIMS. The SIMS will only sputter material into the detector from the spot where the FIB hits the sample. The EDX signal, on the other hand, is dependent upon the material interaction of the SEM beam with the sample. As discussed earlier the SEM beam can penetrate deep into the sample. The EDX results are, therefor, more dependent upon the beam energy and the spread of the beam in the sample. The disadvantage of the SIMS is the fact that the sample must be sputtered. This means the SIMS technique is destructive.

Figure 45 shows a novel technique to improve the EDX spectra. The electrons from the primary electron beam in conventional EDX work penetrate deep into the substrate generating X-rays from portions of the material which may not be of interest. To alleviate the

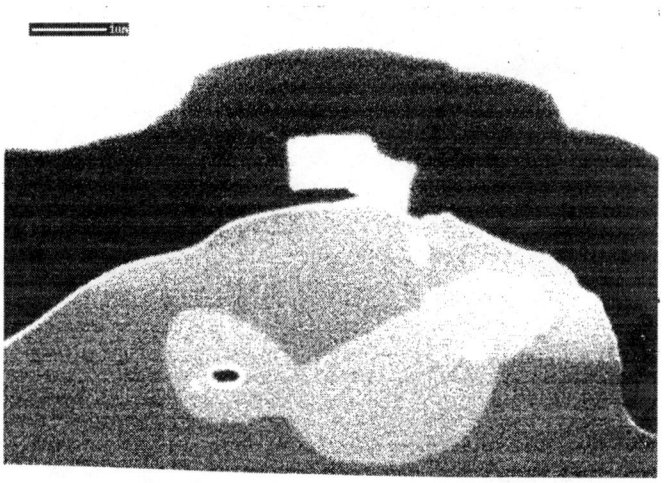

Figure 42. Top image is a cross section of a semiconductor device. The bottom image is the SIMS map of the elements which were found in the cross section of the top image[26].

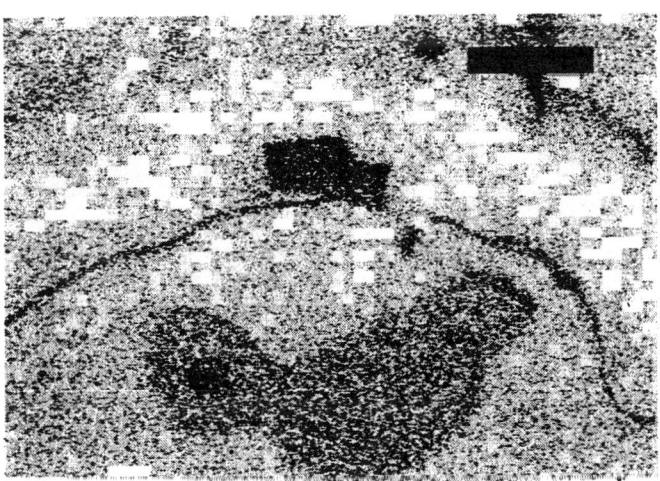

Figure 43. SIMS dot map of the region seen in figure 42. The image is normally in color where each color represents a different material. Here the difference in gray scale represents a different material[26].

problem, the unwanted material is removed from the back of the region of interest and then any X-rays reaching the EDX detector are generated by the penetration of the primary electron beam in the membrane. X-rays generated by the remaining back surface are generated at an angle where they can not reach the EDX detector.

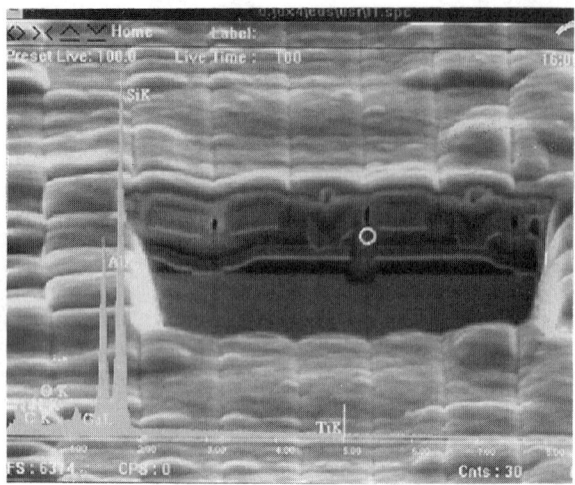

Figure 44. EDX spectra overlaying an image of the cross section from where the EDX spectra was taken. The circle marks the EDX location.

Figure 45 shows in the upper right corner material removed from the back of the region of interest. The upper left corner shows the EDX cross section and the lower left corner shows the EDX spectra produced [3,16,38].

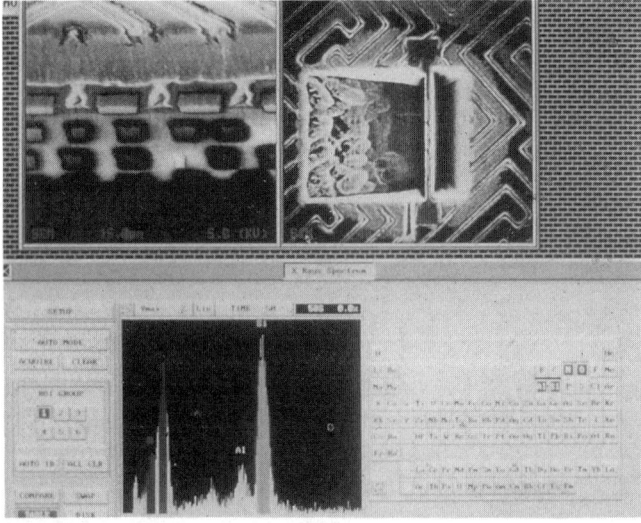

Figure 45. The upper right corner image shows material removed from the back of the region of interest. The upper left corner image shows the EDX cross section and the lower left corner shows the EDX spectra.

Multiple levels of metal in complex devices can cause navigation problems on FIB systems. To combat the dilemma, CAD navigation is sometimes used to assist

FIB operators in locating areas of interest and defining precise circuit modifications. Figure 46 show how several CAD features can be incorporated onto a single screen. Both schematics and layout can be observed on the same computer screen, as well as the image of the actual device. Hypothetical cross sections are also incorporated onto the screen to permit verification that correct location has been sectioned. CAD movements are linked to the stage allowing navigation to be performed from within the CAD program [3,39] One of the newer developments is the ability to overlay CAD information onto the FIB image providing the FIB operator with an accurate location of underlying metal layers. Some FIB vendors also allow the overlay of photographs with the sample being milled in the FIB. This is convenient when the CAD drawings are not available.

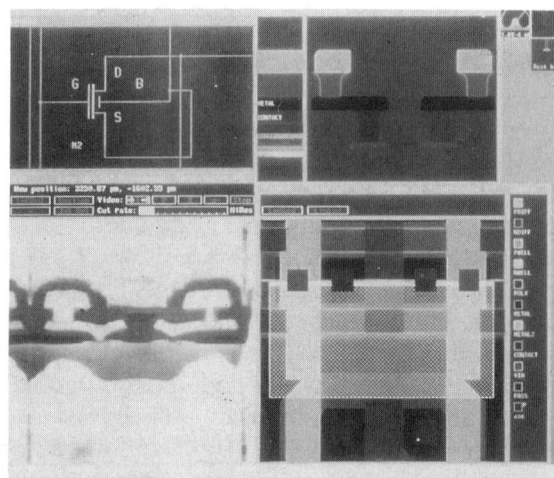

Figure 46. Workstation image of a cross section, lower left combines with schematic, layout and hypothetical cross section data.

The FIB system is ideal for localized cross sections that require a great deal of precision. Figure 47 shows a particle which needs to be cross-sectioned. Figure 48 shows a FIB cross section of a particle that was found during a process monitor [3,38].

Many times localized cross sections are also required on functional failures such as the cross section performed during failure analysis in Figure 49. Side (A) shows a malformed via that resulted in a functional failure. Side (B) shows a normal via in a

Figure 47. Particle found which needs cross sectioning to determine composition.

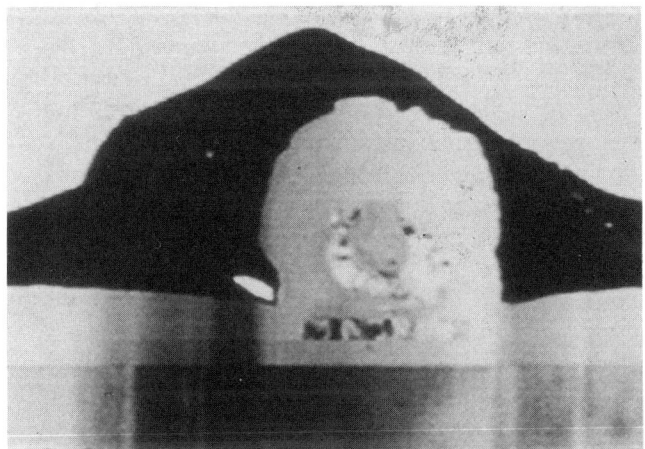

Figure 48. Cross section of particle ready for particle analysis.

similar location. Cross sections of this nature are very difficult, time consuming and have a low success rate when performed with traditional cross section methods.

Some practical issues.

- Channeling effects:

Although there are positive aspects of ion channeling, such as grain boundary delineation, the phenomena causes problems when trying to create uniform mill rates in multilevel metal layers especially with tungsten. Figure 50 shows how the channeling has caused damage to underlying oxide during the etch of a tungsten line. Fortunately, enhanced etch causes

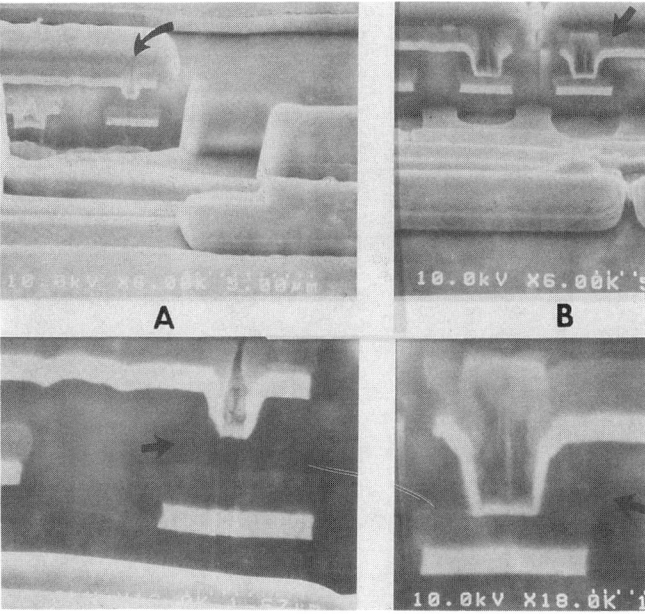

Figure 49. Side (A) shows a malformed via that resulted in a functional failure. Side (B) shows a normal via in a similar location.

a chemical reaction that reduces the effect of ion channeling on the underlying oxide. A similar cut, shown in Figure 51, was made on the tungsten line using the enhanced etch with minimal damage to the underlying oxide.

- Navigation problems:

Charging can sometimes cause the surface image to turn completely black, creating navigation problems. On some devices the top surface is completely flat, and there is no topography. All FIB vendors provide access to some form of CAD navigation tools. One company even has image processing which can help locate the exact spot to mill once the system has moved near the correct location. More advanced navigation techniques are on the way. Yet not much is said about the cost of CAD tools to help the FIB mill process. Nor, has anyone really proven expensive CAD tools are necessary.

CAD navigation is often an essential tool on complex devices for internal probing and voltage contrast testing. It is not necessarily a time efficient method of navigation on many common FIB jobs. Layout database translation is costly and time consuming. Many companies do not have the storage capacity to maintain a library of the primary database and the converted database for all the devices they have in production. The problem is compounded when there

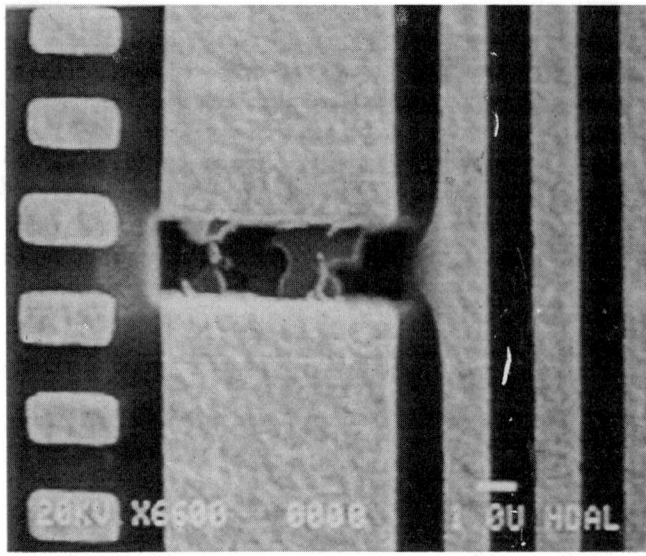

Figure 50. shows how the channeling has caused damage to underlying oxide during etch of a tungsten line.

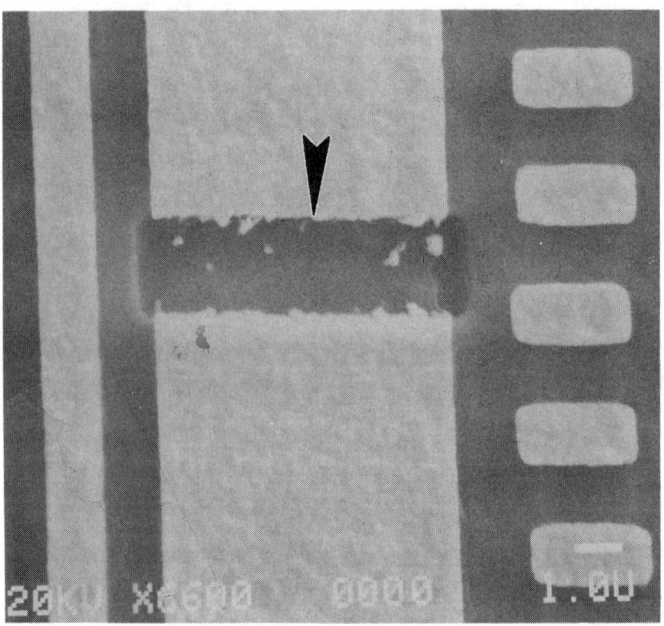

Figure 51. Etch on a tungsten line using enhanced etch with minimal damage to the underlying oxide.

may be several revisions of each device in the market at one time. Downloading of the database and graphical drawing can also be very time consuming and costly.

The FIB process has one other distinct difference from voltage contrast and internal probing. That is when the engineer goes to the FIB he knows exactly what he wants done. No trial and error testing is required. The selection of FIB locations has already been determined. Having a coordinate and a hard copy showing the point of interest from an external layout database is usually adequate information to identify the area of interest on the device once it is loaded into the FIB. In actual practice, all the CAD system is providing is a coordinate for the location to be milled. It just does the job with more style. For those who already have the CAD information at hand because of their interface to other equipment, which uses the same data, the CAD approach becomes addictive or even mandatory.

Planarized devices provide their own form of trouble. One company provides a wide field of view mode which allows the user to overlay the design layout with the top layer metal for lines as far away as 900 microns with .2 micron alignment accuracy [39]. This is adequate for most of today's devices. However, future devices will be come larger and the planarization process will hide more detail until not even this technique will work. One company proposes to resolve the issue by making marks on the top of the device with a laser or other mechanical means. The marks are photographed using a deep field of view microscope to include the underlying structure as it is aligned to the marks. A laser confocal microscope is perfect for this task. The optical image is then overlaid on the FIB system with the real time FIB image. The FIB/optical image comparison gives perfect alignment to structures, which would otherwise be totally lost to any other FIB technique. Even CAD systems can not make this claim. A metal level in a 4 metal system could be misaligned to an unrelated metal by as much as 0.5 microns. This means that alignment to the top-level metal may mean a miss in trying to hit the underlying metal if the alignment is not done optically. Figure 52 shows a layout image on the right side and the corresponding FIB image on the left.

Figure 53 shows the result of the removal of passivation over the line to be cut. Th e layout overlay verifies the correct location. The Figure 54 shows how the layout overlay can be used to verify the proper FIB location has been selected.

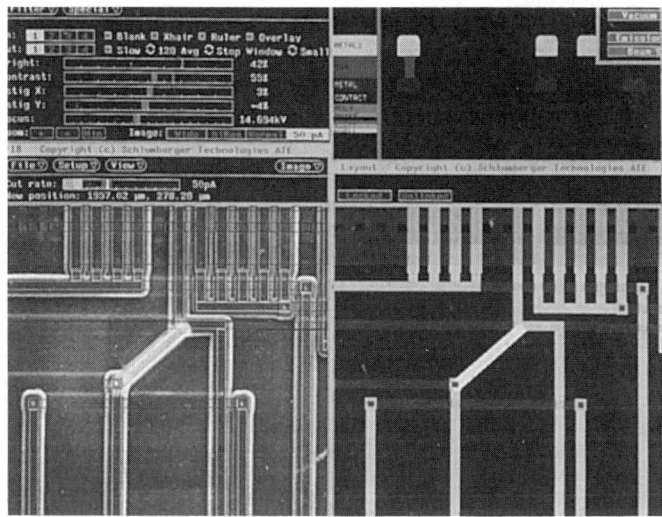

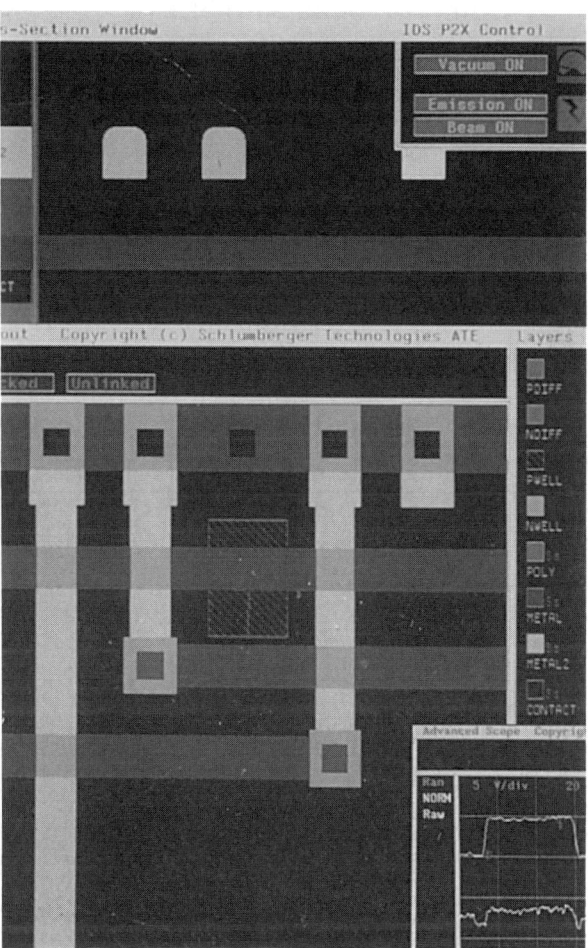

Figure 52. shows a layout image on the right side and the corresponding FIB image on the left.

Figure 54 shows on top a cross sectional image of the location to be FIB'ed. The bottom image in the layout is where a cut is to be made. By having the cross sectional view of the location, the best possible path can be selected for the cut. In some cases the engineer may see that the cut cannot be performed due to the close proximity of overlying metal.

Now that this paper reports CAD navigation is not necessary because the engineer already knows the locations to FIB, a new technique is being introduced into the FIB world, which changes the discussion completely. The addition of needle probes in the FIB chamber now allows the engineer to measure signals on lines that have been depassivated with the FIB technique. Figure 55 shows needles in an FIB system. They are positioned on 0.35 micron lines.

Figure 54 shows on top a cross sectional image of the location to be FIB'ed. The bottom image in the layout is where a cut is to be made.

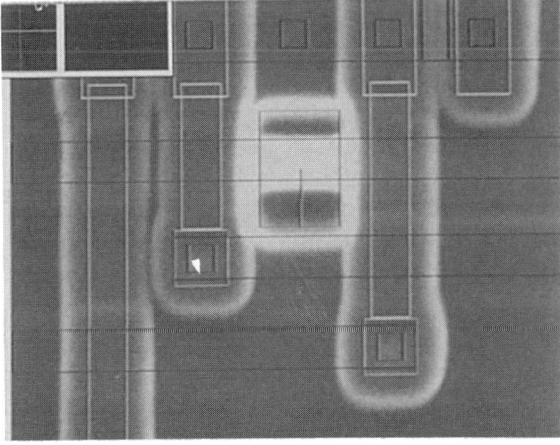

Figure 53 shows the result of the removal of passivation over the line to be cut.

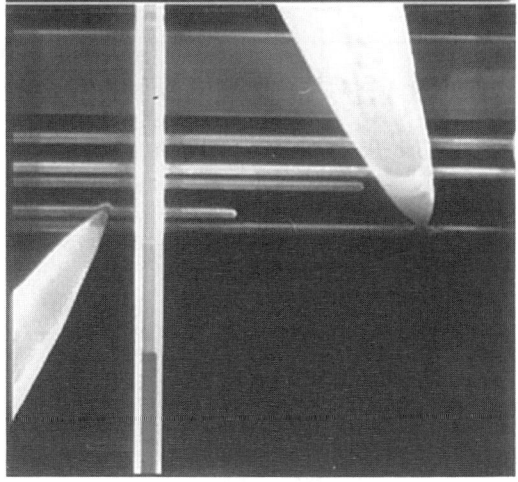

Figure 55. Needle probes in an FIB system probing 0.35 micron lines

Quality Control Issues with FIB

In some cases, the following figure will be a repeat of the previous text, however. Figure 56 highlights problems which lead directly into the next section on equipment selection.

Need for high current FIB for long depositions

Reason	Solution
Connect to circuits at great distance. 500pA+ currents	Electroless Cu plating of deposition, Carbon coat Use laser deposition instead of FIB.

VT shift due to FIB

Reason	Solution
Trapped charge during FIB	Carbon coat, H2 bake after FIB .5 hours at 400C

ESD damage before/after FIB

Reason	Solution
Improper handling	All parts handled on a grounded surface in an antistatic environment, Prevent device charging during FIB, Allow no floating lines in area of FIB mill (Floating lines charge independent of the rest of the circuit and can float to a voltage high enough to blow adjacent lines. Damage can occur even in areas away from the mill region.)

Wrong FIB performed

Reason	Solution
Operator error/Designer error	Limit interface personnel to FIB operator, Have each engineer do their own jobs to prevent communication errors, Set strict guide lines for communications of FIB work, Use three point coordinate alignment system and reference work by X,Y location on the die, Backup with design layout CAD system

Device charging during FIB

Reason	Solution
High quality oxide (dielectric) as part of the passivation own leakage currents.	Lower beam current (increases process time),Carbon coat, Flow an ionizing gas over device, Conductive coating other than carbon, Alternate technique to leak charge (Flood gun), Flow Pt during initial look at device to spread out charge area, Short device to ground and then ground surface to the ground lines.

Circuit sensitivity

Reason	Solution
Mirror circuits, floating lines, weak transistors.	Can not connect to or around a floating line, Must make modifications the same on mirror circuits, Carbon coat for weak transistors, Power device during FIB.

Can not see underlying layers:

Reason	Solution
Devices planarized, Detector	Use three point coordinate alignment system and reference work by X,Y location on the die, Backup with design layout or photograph overlay to image, Laser mark locations and use photographs of area to locate FIB repairs, CAD system overlay, Tilt sample or change detector position.

Poor contact to metal lines:

Reason	Solution
Small contacts, High power metal deposition into contact, High aspect ratio of contacts, oxidation of metal in contact.	Reduce power when Depositing into contact, Make large contacts, Change metal deposition process, improve system vacuum.

Cannot stop at Polysilicon level or Tungsten (Metal 1 of a 5 Metal process)

Reason	Solution
FIB mill is rough on bottom, emissivity of Polysilicon and W is the nearly the same as the surrounding material, Aspect ratio of contact is high.	Use EE, Find alternate Connection, Find alternate end point, use SIMS.

Limited FIB access to repair location/complex repair in small region

Reason	Solution
Design with submicron features, Overlying metal.	Laser marking to pinpoint location of repair, Ultra small spot size of FIB beam, Remove overlying material with Laser or FIB, Deposit dielectric in holes as needed to prevent shorting of different materials.

Figure 56. Quality control problems with some solutions.

Equipment Considerations

There are several important features that should be considered when trying to choose which FIB system and options to purchase. Some key features to consider when trying to distinguish between machines are given in Figure 57.

- Beam stability and control
- Image quality and resolution
- Stage repeatability
- Surface dimension repeatability
- Enhanced etch options
- Deposition options/characteristics
- Electrical performance
- Device integrity
- Narrowest cut
- Sample size accommodations (tilt and rotate limits)
- Ease of use
- Quality of CAD navigation
- Overlay accuracy/number of available overlay techniques

- Load/unload speed
- Ability to use multiple gasses quickly
- Ability to utilize modern FIB chemistries
- Device charging characteristics
- Anticipated use of FIB system
- Cost effectiveness/capitalization
  (In house or sub contract work)
- Device modeling capability
- Accessories (SIMS, FESEM, EDX)
- Electrical feed throughs
- Image capture techniques
- Training time
- Interface CAD/Electrical to other local equipment
- Ability to change gases quickly.

Figure 57. Equipment considerations.

Future Trends

The sudden growth, as seen recently in FIB usage, will certainly continue for some time. This can be anticipated from the fact that the number of new developments, techniques, applications, and FIB vendors has increased in the last several years. Also the detailed understanding of interaction of ion beams with the VLSI structures should continue to improve. Strategic direction for FIB technology is in the area of wafer fab support and eventually towards on-line or in-process applications inside the fab. Dual beam systems with SEM plus FIB imaging, as well as machining and metal deposition, are being specially designed and introduced for clean room environment. Special features such as fully digitized and computer controlled 8" wafer stages with precise tilt capability and precise positioning accuracy are becoming standard features. Advanced auto CAD navigation with built in analytical capabilities, such as EDX and SIMS, make the dual beam system an invaluable and essential tool for in-line process control, as well as for reducing and controlling the particle density through identification and elimination, especially when coupled with particle scanners.

In terms of navigation for planerized, multi-layer metal IC's, special layout overlay features for non-visible conductors are becoming available [3,39]. The overlay feature superimposes layout graphics onto an image of the physical circuit using the top level metal lines for alignment. Interactive alignment of graphic and physical metal lines allows greater precision in definition and position of FIB boxes without the circuit being altered.

The FIB, having a unique system architecture similar to that of an e-beam prober, can utilize voltage contrast to help identify underlying metal layers. Using this technique, the part is stimulated with an AC signal which creates a contrast in insulator and conductor on the FIB image. In addition to navigational enhancements, the voltage contrast phenomena can be used for end-point detection.

Because e-beam probers are sometimes utilized to confirm and analyze FIB fixes, it is convenient to have minimal load and alignment time between the FIB and e-beam system. In a multi-disciplinary mode, new systems are being introduced that eliminate device alignment by accurately repositioning device and fixture, as well as streamlining the vacuum system for fast pump down. In one case, the load module of the FIB has a provision for mechanical probes and was specially designed on a kinematics mount frame.

Having an interface between FIB and prober has cut probe point location times by a factor of 4 to 6 when compared with conventional FIB and probe setup. As a result, networked e-beam probers and FIB systems should become more common in the FA laboratory [3,39].

In the area of ion beam assisted chemical etching, a combination etch process called "Gas Assisted Etching" (GAE) is being introduced [3,17]. The etches are halogen containing molecules with the same principle of material removal, namely by the formation of volatile reaction products. For example, an aluminum line was cut with chlorine gas with little or no etch of underlying SiO2 (Selectivity: 10:1) thus eliminating the uneven SiO_2 surface caused by channeling effects of Ga+ ions.

Using a combination of gases, the effect of selective etch processing was demonstrated. For a 3 - level metal system, a hole was first made down to a large Al bus using XeF_2, a selective etch for SiO_2. Figure 58. Show the results of this step. The next step was to create a hole in the Al buses using the Al selective etch, Cl_2. Figure 59 shows the results of this step. In the final step, the metal levels 1and 2 were exposed using the selective XeF_2 to remove the interlevel oxide. Figure 60 shows the results of the final step. The total time required is less than 30 min[49].

Figure 58. Cut through oxide with XeF_2 to big metal line[49].

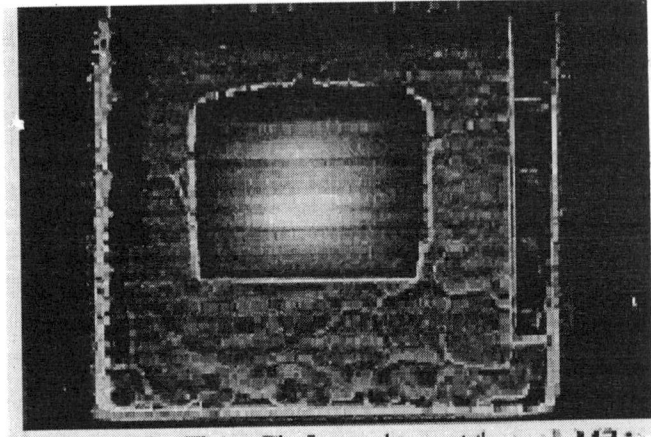

Figure 59. Cut through the metal line with Cl_2[49]

Quick changes in gases are becoming very important. Engineers want to deposit metal and move immediately to an etch location. If the metal deposition needle stays in place, excess metal is sprayed over the device potentially shorting modified areas of the device.

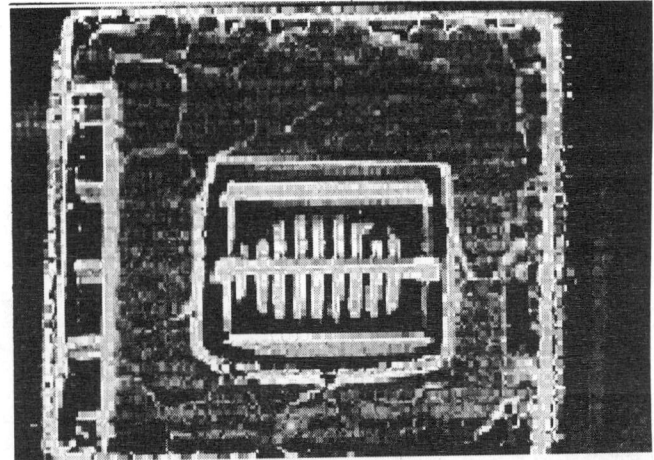

Figure 60. Cut through the interlevel oxide to reach the underlying metal lines [49].

Changing needles or moving needles in and out becomes cumbersome. Gas manifolds are required [39,40,51]. Currently most manifolds handle 4 gases but require some time after metal deposition before the device can be imaged, to allow the excess metal contained in the manifold needle to clear. Otherwise metal is deposited on the device as the device is imaged. Figure 62 is a graphical representation of a manifold system, which is optimized to reduce the time for gas changes. One company is offering a manifold system with 24 gases which reduces this problem [51].

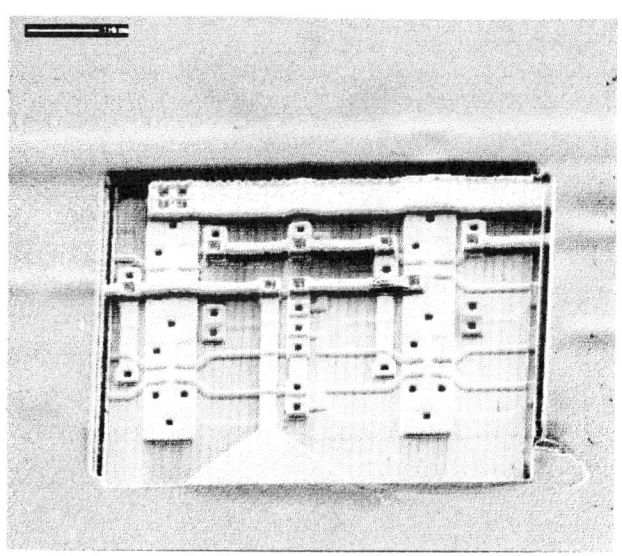

Figure 61 shows a two metal process where multiple layers of metal are exposed for probing by use of Xe_2 [26].

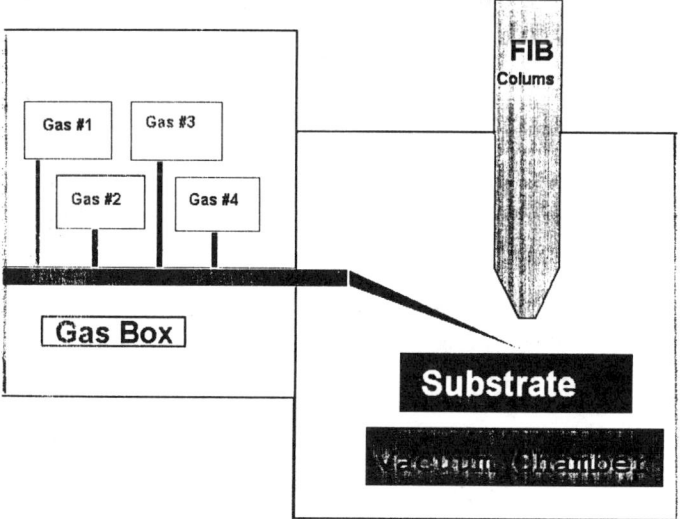

Figure 62. a graphical representation of a manifold system, which has four gas system.

Backside sample preparation is probably the most rapidly developing market segment. To meet the speed requirements of the next generation of devices, the die must be placed face down on the substrate. The die is electrically connected to the substrate with a metal solder bump. Due to the high number of interconnects and the nature of the connection, demounting the die is not practical. Even if the die could be dismounted the new designs are array bounded, which means there are bumps on all portions of the die. Because of these conditions only backside analysis can be performed.

The FIB vendors are gearing up for this problem. Fiducials, as shown in Figure 63, will be needed on the topside of the silicon which can be seen through the backside of the die [49]. The fiducials shown in figure 63 are seen through 500 microns of silicon by an infrared camera in the FIB chamber [49].

Once the device has been aligned to the fiducials, the FIB operator will move over the region of interest. There they will mill a deep trench to within 25 microns of the circuit to be studied. Figure 64 shows an example of a deep trench 200 X 200 X 200 microns [26]. The trench has been milled all the way to the oxide interface to permit viewing of the results, otherwise all we would see is the smooth bottom of the trench.

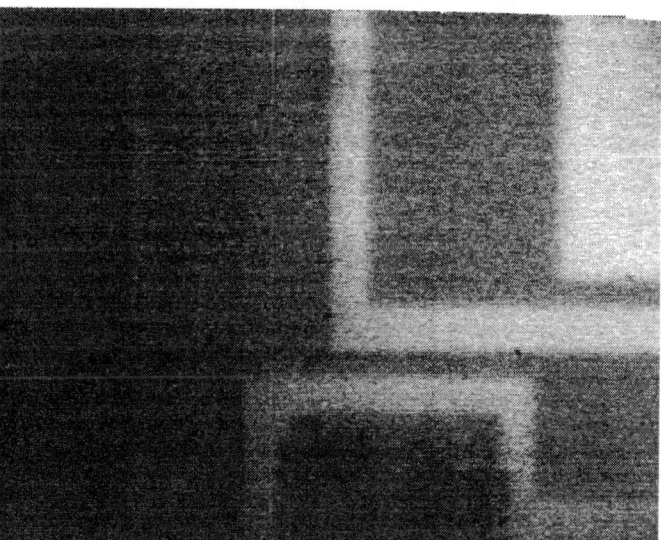

Figure 63. shows the fiducials through 500 microns of silicon. The fiducials can be any material which can be seen by the infrared camera [49].

Figure 64. shows an example of a deep trench 200 X 200 X 200 microns. The trench has been milled all the way to the oxide interface to permit viewing of the results, otherwise all we would see is the smooth bottom of the trench [26].

To align in the trench to the structures to be FIB'ed optical images must be made of the area. Figure 65 shows an optical infrared image of metal lines in the device to be modified [49]. A paper has been published where an FIB has been used to modify the circuit from the backside[52]. A laser could be used instead of an FIB to make the trench. The laser has the advantage of being faster in milling silicon. Then an FIB system could be used to make the interconnects through the silicon. The laser could then be used to make interconnects across the backside of the die to finish the cut and paste operation. Again the laser is faster at making depositions. The trade off with laser usage is resolution. The laser has less resolution and must therefor always need an FIB to perform the fine detailed work.

With the increased application of Scanning Tunneling Microscopy in VLSI technology and with narrower and steeper topological features, greater need is being felt for better probes. The unique capability of the FIB workstations to provide milling in a circular pattern makes it possible to sharpen tips for scanning probe instruments. Figure 66 shows the sharpened tungsten STM probe tip at high magnification - a unique custom micro-machining operation by FIB [3,8].

Figure 65. shows an infrared image of the metal in the die through the silicon of a trench which was milled by an FIB system. The image is taken in the FIB chamber [49].

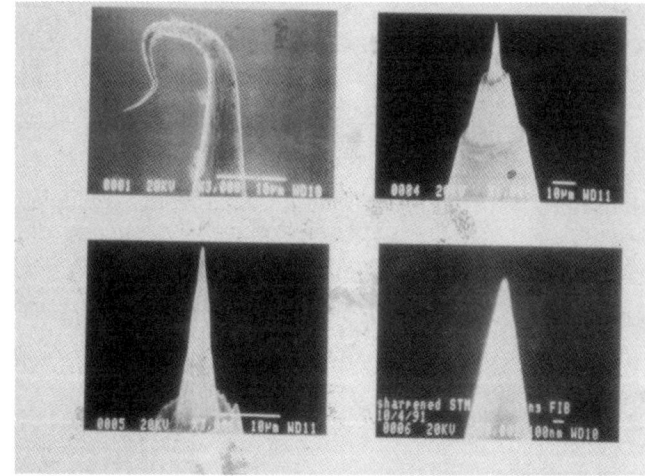

Figure 66. Top left image of bad AFM tip. Other images show the FIB sharpened tip.

As far as the future trend for ion sources and FIB columns is concerned, there is a significant on going effort to improve the beam resolution to less than 50 angstroms. One such most recent development is the new Gas Field Ionization Source (GFIS) technology [3,17]. The new GFIS operates a gas ambient, which supplies hydrogen, helium, neon, argon, or oxygen gases to be ionized and focused. Its brightness is greater than what is currently available from today's LMIS, thus allowing a focused beam system to approach the image resolution of a conventional SEM, while providing the unique features of material removal and deposition typical of the present generation of FIB equipment. In addition, the less massive ions from a GFIS may provide high resolution ion imaging with little or no noticeable sample damage. Many of the problems inherent in the use of today's LMIS, e.g. Gallium staining, implantation and associated material contamination, will no longer be a factor. However, the mass differences in ions will result in milling rate differences. Nevertheless, additional work is needed to overcome some of the limitations of the FIB repair technology. Figure 67 lists the problems.

-Accurate location of non-visible buried conductor

-Deep via contact hole limitations

-High resistivity nature of deposited metals

-Extensive time consumption for long line depositions

-Maintenance of electrical integrity of devices with complex geometries and multiple level metal systems.

Figure 67. FIB difficulties to overcome.

We are also clearly seeing the creation of refinement of several related techniques. In recent years, techniques such as laser ablation and other optically enhanced techniques have been rapidly developed. The combination of these technologies with emerging FIB will lead to a new generation of microsurgery techniques.

Also, a new horizon of applications of FIB technology is opening up in fields other than IC technology. For example, the additional benefits of the GFIS technology could open up applications for the use of a FIB system in many new areas, such as micro machining. Figure 68 shows a Fresnel lens and Figure 69 shows a cross section of the lens[26]. Mechanical structures such as bumps on flip chip devices can be analyzed. Figure 70 shows a bump cross section. Improvements in hard disk drives can be made through the machining of the write head as shown in Figure 71. Materials science and bio-medicine will benefit from the advancements in FIB technology. Figure 72. shows a cross-section of a brain cell obtained with a conventional LMIS FIB system [26].

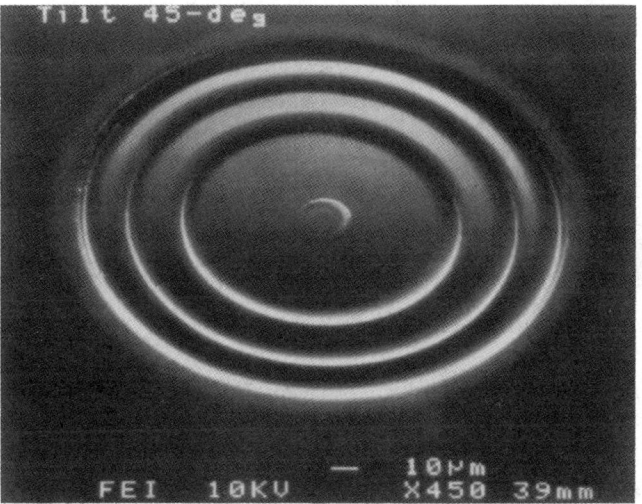

Figure 68. shows a Fresnel lens [26]

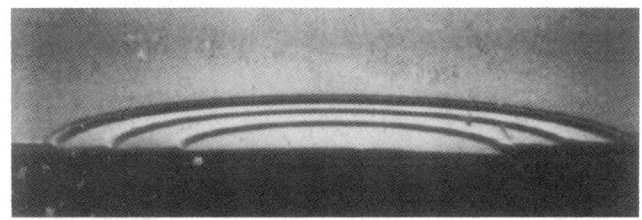

Figure 69. shows a cross section of the Fresnel lens [26].

Figure 70. shows a cross section of a C4 bump [26]

551

Figure 71. shows a hard disk drive where the write head has been altered by the FIB [26].

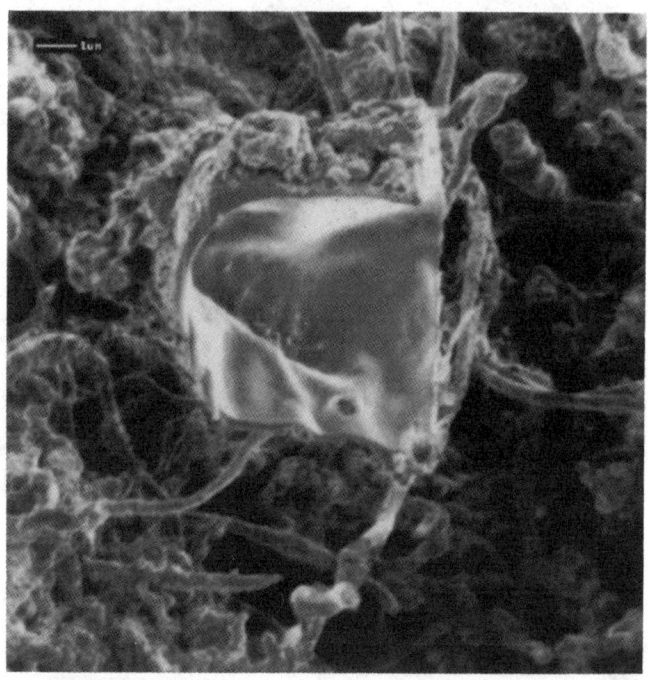

Figure 72. shows a cross-section of a brain cell obtained with a conventional LMIS FIB system [26].

PHOTONIC *TECHNOLOGY

*Name used by Photonics Spectra magazine.

Introduction

Photochemical reactions are not new. Our existence in the world today is dependent upon chlorophyll's conversion of light and CO_2 into food and O_2. As children we play with magnifying glasses to feel the heat from the sun. In failure analysis, photochemical reactions are used to help dissect the semiconductor device revealing the hidden problems inside. In this section of the paper, we will cover the principles behind the photochemical reactions. We will then describe the application of photochemical techniques to failure analysis.

Principles

Light is electromagnetic radiation. Light can interact with the materials being irradiated in 4 different ways as listed in Figure 73. The possible interactions are determined by the wave length of the light.

---

1. The light can cause the material to heat.

2. The light can cause the material to become disassociated and therefore reactive - bond breaking

3. The light can cause the material to change its reactive nature by selectively exciting chemical bonds: non - bond breaking

4. The light can cause atomic interactions, exciting the electronic structure of the material. The material may or may not become chemically reactive depending upon the loss of electrons.

Figure 73. Ways light reacts with a material.

To use any of the four types of interactions, the optical setup is basically the same. In Figure 74 a source of light is focused through an optical arrangement onto/into the material. The material may need to be in some form of chamber, if required, for the process being developed.

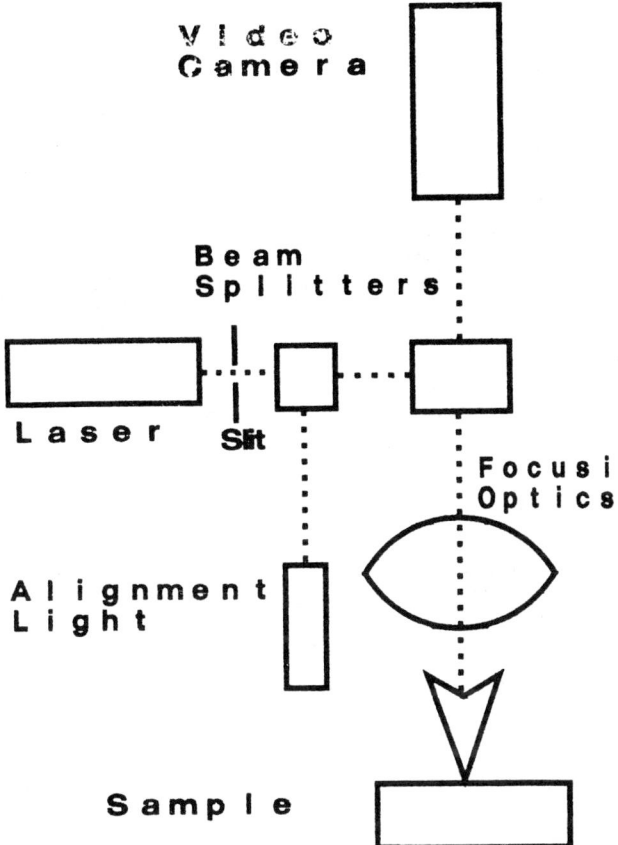

Video Camera

Beam Splitters

Laser **Slit**

Focusi Optics

Alignment Light

Sample

Figure 74. A typical laser excitation apparatus.

In Case 1, the light is focused onto/into the material to be processed. If the material is a solid, the light within the appropriate spectral range will be adsorbed, causing heating of the surface. Intense photo heating of the surface causes the material to melt. In the absence of reactive components, the material returns to a solid when the light is removed. This is the typical photo/laser anneal used to produce single crystal silicon from polysilicon. If the material being heated is placed in the presence of a reactive gas which is transparent to the wavelength of light being used, the molten surface pyrolyzes the gas causing a reaction. As an example, if silicon is heated in the presence of Cl_2, the silicon becomes a volatile compound $SiCl_x$. The gaseous $SiCl_x$ is swept away from the material surface and out of the process chamber leaving a void in the silicon where the reaction took place. The heating of the material can also be done in a liquid medium, with similar results, as if the heated material were in a gas environment.

Heating of the surface of a solid in the presence of a metal containing gas or liquid can cause the pyrolysis of the metal containing material, which will leave a deposition of the metal on the surface of the solid at the location of the heating.

If the surface of the material is heated rapidly to the vaporization point, the material can be ablated. This is the laser link blowing process used to repair memory devices.

Lasers commonly used to heat a semiconductor device are Ruby, Nd:YAG, CO_2, Ar+ and various Eximers.

In Case 2, the light interacts with molecular structures causing them to become disassociated or to break apart. This is most apparent in gas reactions. The light must have a wavelength short enough to be adsorbed by the compound. The compound is then broken into its respective atomic components, which are then ionized and become highly reactive. If the ionization takes place near the surface of a solid, the solid will be etched, if the ionized gas is reactive with the solid. If the material being ionized contains a metal and the ionization takes place near a solid, there may be deposition of the metal onto the surface of the solid. Since the solid does not control the location of the deposition, some over spray of the deposited metal takes place. To enhance the metal deposition process, additional gases may be introduced into the process to react with the unwanted by-products preventing them from slowing the reaction rate or becoming trapped in the deposition.

Lasers commonly used for decomposition have wavelengths in the blues or ultra violet. Some of the typical lasers are the frequency tripled Nd:YAG, Ar+ and Eximer lasers. The disassociation process is also easily controlled by the use of high intensity ultra violet light from a mercury light source.

In Case 3, the light causes the compound to become highly excited, but does not provide enough energy to disassociate the compound [53]. The compound becomes more reactive, but only to specific reactions involving the bond within the compound being excited. Such reactions are used for the separation of duterated methanol from hydrogenated methanol as part of the process of making heavy water. The process is quite selective resulting in high separation yields.

The lasers used to excite the molecules are long wavelength lasers. They couple with the harmonic vibration frequencies of the bond stretch in the molecules. For the carbon oxygen stretch frequency, a CO_2 laser is used. For other combinations a CO or other gas laser would be used. The important point of this process is to excite the stretch vibration within the molecule without breaking the bond. The selective excitation of the particular bond of interest forces the wanted reaction while preventing alternate reactions.

In Case 4, the light causes the electrons in the atom to become excited. If the energy of the photon is high enough, the electrons become ejected from the atom resulting in a highly charged particle. In a solid, the electron becomes free and in a metal can flow as current or leave the material altogether. Photo generation of electrons leaving the material is the principle of electron generation for femto second pulse generation for SEM voltage contrast. If the atom is in the gaseous state, the loss of the electron creates a highly charged particle, which in the presence of a solid, may etch the material. In the presence of a metal containing gas, the highly charged particle can participate in the break down of the metal compound causing a deposition. To enhance the quality of the metal deposition, additional gases may be added. The additional gases help speed the reaction and improve the metal quality by binding with the unwanted components of the reaction. This prevents them from participating further in the reaction.

Ultra violet lasers, Eximer lasers and frequency multiplied Nd:YAG lasers are used for this purpose. The reaction is non-specific in that all material in the process chamber is ionized to some extent by the light.

Applications and Status

Photo applications are generally thought of as additive or subtractive. However, as with Focused Ion Beam Mill, the resultant process can be analytical in nature. Figure 75 lists the broad categories for which photo excitation can be used by the failure analyst. Notice they are similar to the list for FIB. In most cases the laser can be used as a replacement for the FIB. The difference will be cost, spot size, process selectivity and type of damage to the device.

1. Microcircuit Analysis and Repair

A) Microcircuit Analysis:
        -Probe-point access
        -Probe pad formation
        -Microcircuit isolation
B) Repair or Reconfiguration:
        -Cuts or disconnects
        -Jumpers or interconnections
        -Capacitor or resistor trimming
        -Transistor characteristics
         modification

2. Material Characterization and Failure Analysis:

A) Material Characterization
        -Elemental analysis
B) Failure Analysis
        -Cross sectioning
        -Sample preparation for SEM
         and TEM analysis

3. Process Monitoring

A) Off-line monitoring (post)
        -Yield loss analysis
        -New process characterization
B) In-line Monitoring:
        -Yield loss analysis
        -New process characterization

Figure 75. List of uses for Laser microsurgery.

As with other techniques, the photo assisted processes can be computer controlled by Computer Aided Design (CAD) systems.

1. Microcircuit Analysis and Repair

The original use of lasers for analysis was to cut metal lines for circuit isolation. A xenon laser was used. The green light would couple with the aluminum to be ablated causing the aluminum to vaporize. Unfortunately, aluminum has a large delta between the melting point and the vaporization point. This, coupled with the large thermal conductivity of the aluminum meant the metal would splatter more than vaporize. Some companies actually took advantage of this phenomenon to connect one metal layer to another. They would open up the top metal layer, then use a relatively low power laser to splatter the underlying metal onto the upper metal shorting the two together.

Figures 76, 77, 78 and 79 show various forms of cuts using an ablation technique on metal lines. The metal splatter is common to all laser ablation processes regardless of the laser used [36].

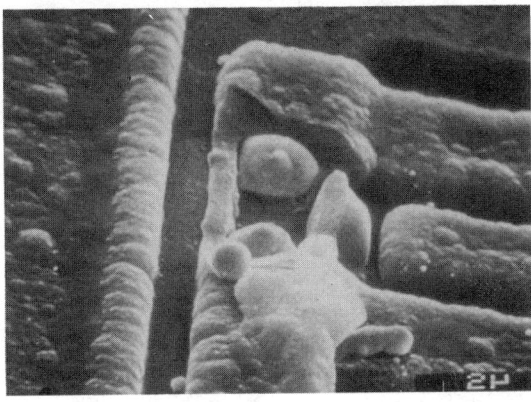

Figure 76. Gold metal line cut by laser ablation.

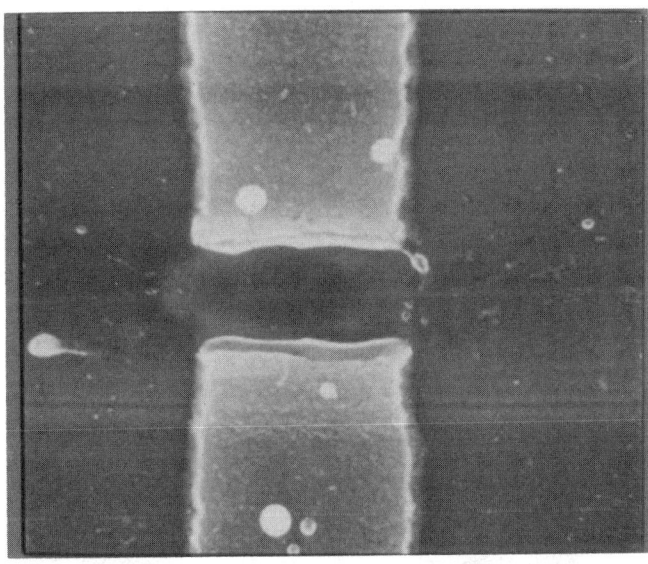

Figure 77. Silicon line cut by laser ablation.

Another company used the laser ablation process to remove the polysilicon lines. This process was used for dynamic random access memory (DRAM) [54] repair. The same process was used to isolate circuits. As a rule of thumb, empirical analysis found that for complete ablation of polysilicon lines, the spot size of a Nd:YAG laser at 1.06 um with a 35 nsec. pulse width had to be 4 times the width of the line being blown. For Al, the spot size had to be at least 7 times the line width and the pulse duration needed to be under 10 nsec.

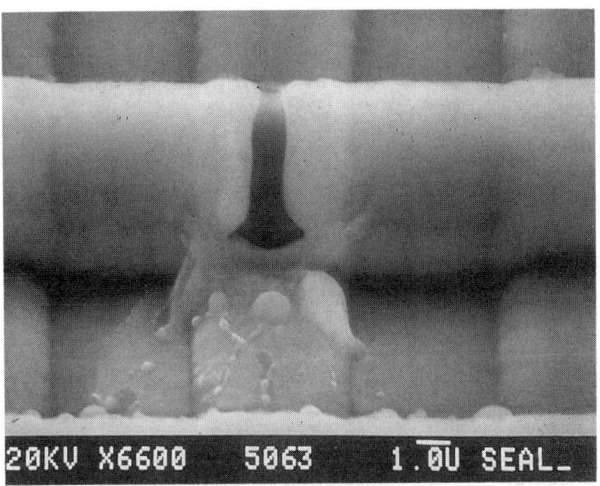

Figure 78. Metal line cut by laser ablation.

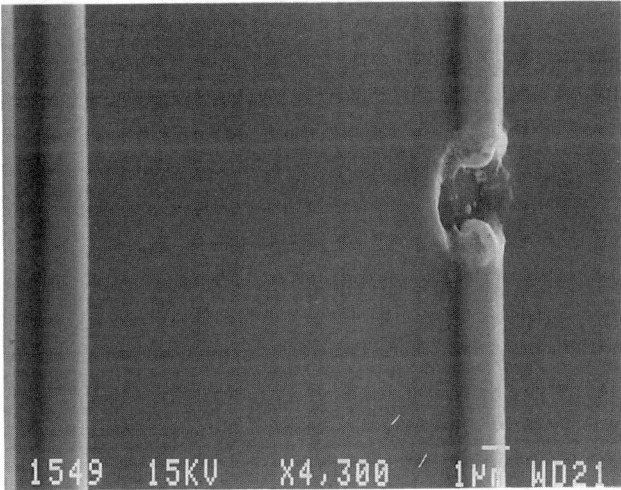

Figure 79. Metal line cut by laser ablation.

As laser prices dropped, more cost-effective lasers became available. The diode pumped Nd:YAG laser with frequency altering crystals provided light output at the primary wavelength of 1.06 um, and various higher frequency (shorter wavelengths): double, triple, quadruple and fifthtuple the primary frequency. The changes in laser technology provided the necessary low cost and reliability required to push the state of the art. Eximer lasers were developed with wavelengths in the deep UV region. All of these types of lasers provided high output powers with pulse widths short enough to vaporize the material to be removed without damage to underlying layers.

The real trick to using lasers is to select the correct wavelength. Both the frequency quadrupled Nd:YAG laser and the UV Eximer laser coupled with passivation very well to allow the laser to open up holes through which the metal line underneath could be probed. The material removal process could be controlled to permit only 1000 angstroms of material to be removed at a time. The new lasers also allowed the failure analyst to make shallow marks in the device surface for navigation aids in FIB repair where there is no topology or other reference marks for navigation. Even the wide field of view offered by some FIB companies is not large enough to locate on devices which have been Chemically Mechanically Polished (CMP) at all levels and the top level of metal only exists on the outer edge of a 2.5 X 2.5 cm die.

Figure 82. Shows the result of removing the overlying passivation, then cutting the underlying metal line using an ablation process [36].

The ablative process provided the failure analyst with access to the metal lines but was limited in its capability to allow for interconnection of metal lines. Metal carrying liquids were developed which could be applied to the surface of the device. Upon laser heating, the coating would pyrolize leaving a metal deposit on the surface of the substrate. The unreacted material would then be washed off. Figure 83 shows a metal line connecting two laser made contacts [36]. Figure 84 shows a laser ablated region opened to allow access to three metal lines [36]. The first line, A, was cut. Line B was cut then reconnected after test and line C was left untouched to show the process latitude.

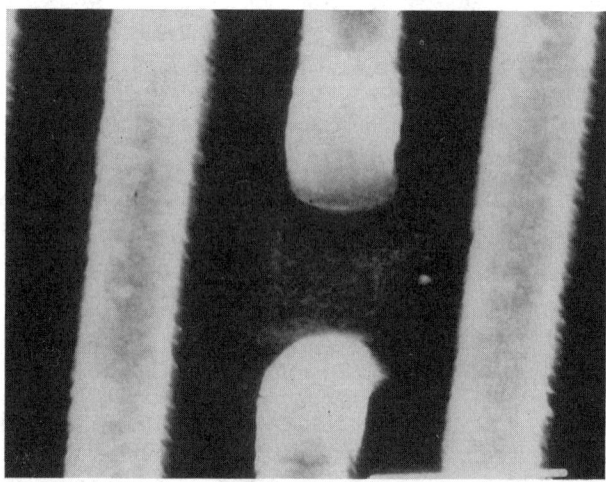

Figure 80. shows a line cut with an ablation process [36].

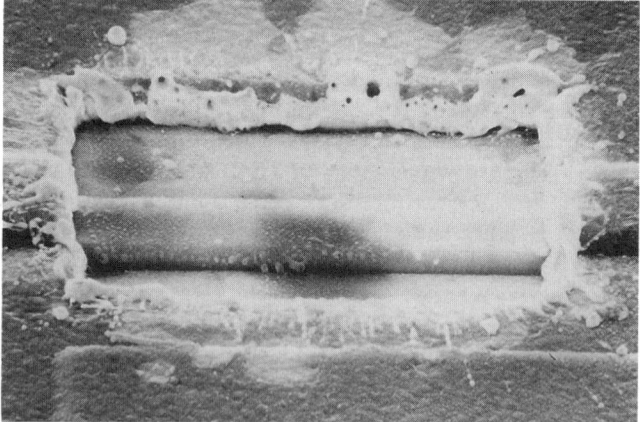

Figure 81. shows the passivation removed from over a metal line using an ablation process [36].

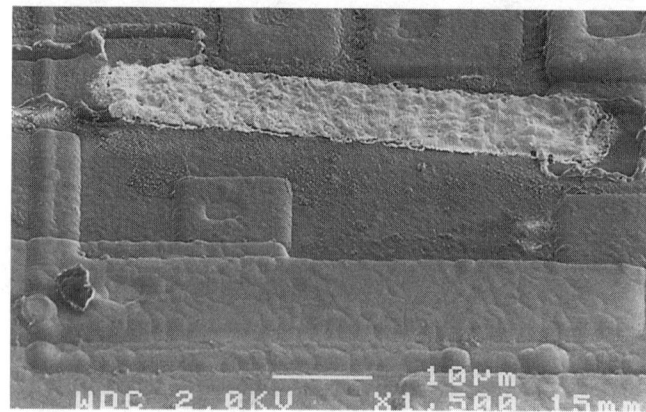

Figure 83. shows a metal line connecting two laser made contacts.

556

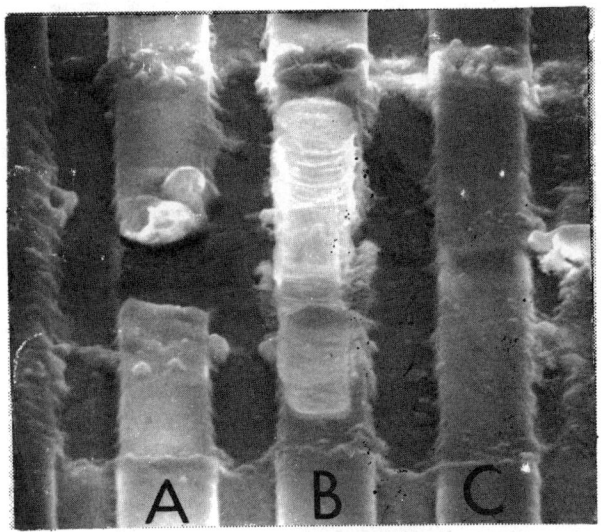

Figure 84. shows a laser ablated region opened to allow access to three metal lines [36]. The first line, A, was cut. Line B was cut then reconnected after test and line C was left untouched.

The interconnection would work on large structures but did not have the accuracy to connect fine structures. Figure 85 shows the use of the FIB system to connect the laser reacted material to the submicron contacts required for tight tolerance FIB work [36]. Where space allows, circuits can be drawn with the liquid process. Figure 86 shows a circuit with probe pads made from this process [36]. The advantage of this metal process is the speed of metal deposition and the ability to deposit low resistance long lines quickly. Many metals are capable of being deposited in this manner. Copper, gold, silver, platinum and nickel are some of them.

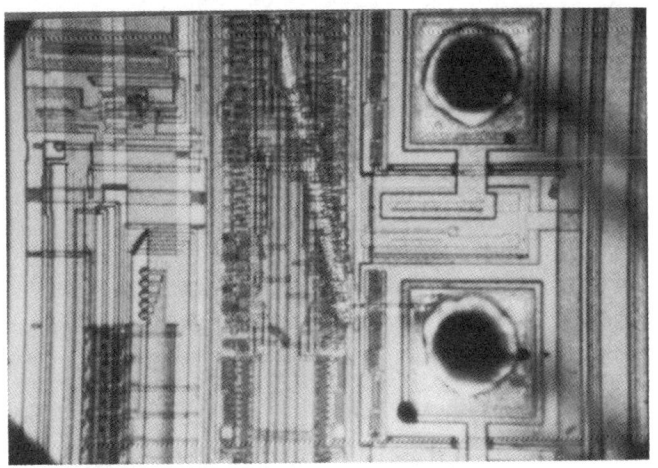

Figure 85. FIB connection to laser deposited metal from a liquid source.

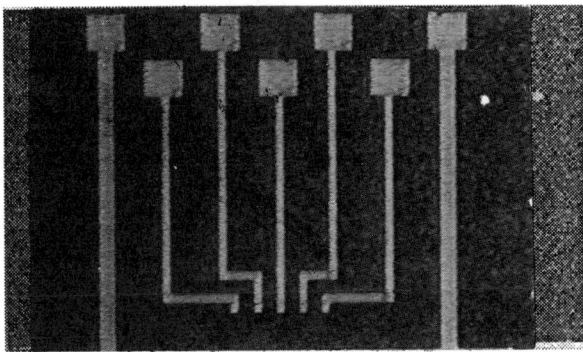

Figure 86. Laser deposited circuit with bond pads. Metal deposited from a liquid source.

Improvements in gas phase reactions are helping failure analysis. Like wet chemical laser processes, gas phase reactions are relatively clean. One company uses a chemically assisted process to form mechanical structures in silicon. Figure 87 shows a microelectromechanical system (MEMS) cell laser machined from silicon using a gas assisted laser process [35]. Figure 88 shows a pyramid etched using a laser gas assisted process [35]. Figure 89 shows a disk etched using a laser gas assisted process [35,55,56,57,58,59].

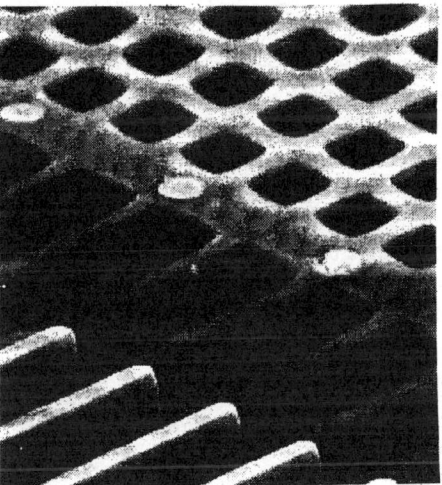

Figure 87. shows a MEMS cell laser machined from silicon using a gas assisted laser process.

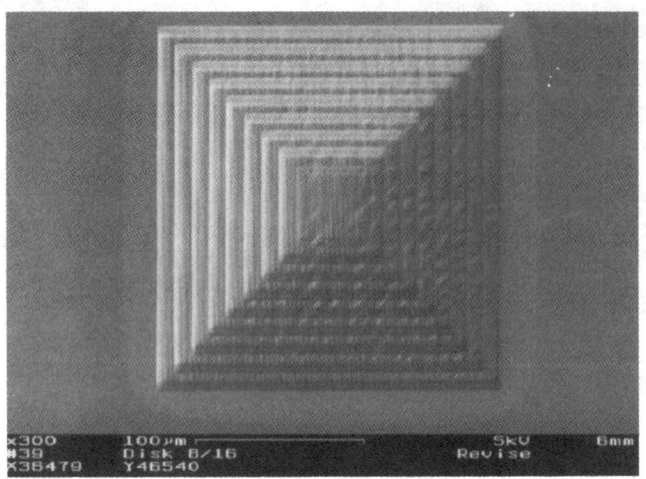

Figure 88. shows a pyramid etched using a laser gas assisted process.

The same process transfers over to the selective removal of passivation from active circuits as shown in Figure 90 [35]. The advantage of gas phase laser chemical reactions over ablation reactions is the reduction in power density, the ability to utilize natural etch stops and the lack of splattered material.

Gas phase reactions are available as well for metal deposition of interconnects on active devices. The major advantage of the gas phase reactions is the high quality of metal and the ease of depositing long lines. The only disadvantage is the inability to produce narrow
line structures. This is best left to the FIB system. Figure 91 shows a silicon line deposited by a laser gas assisted reaction [35]. Figure 92 shows an integrated circuit using laser deposited interconnects to modify circuit.

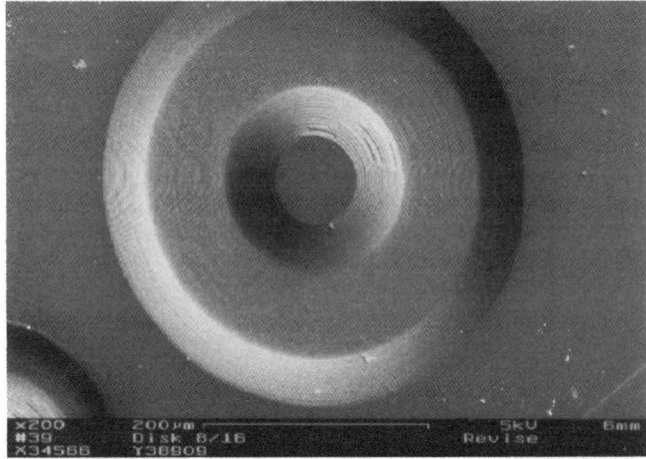

Figure 89. shows a disk etched using a laser gas assisted process.

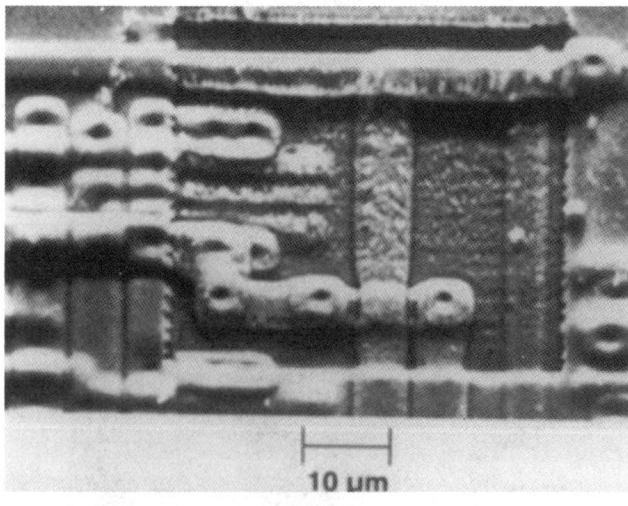

Figure 90. Laser gas assisted milling of passivation.

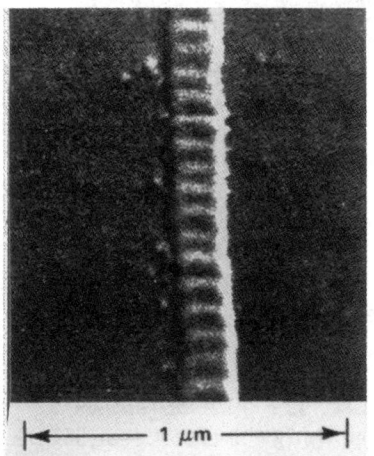

POLY-Si DEPOSITION FROM SiCl$_4$/H$_2$

Figure 91. Silicon line deposited by laser gas assisted reaction

Figure 92 shows an integrated circuit using laser deposited interconnects to modify circuit.

558

Figure 93 is a short list of substrates and etchants with the associated laser used [55,56,59].

Figure 94 lists the laser processes available for use with Eximers and Nd:YAG lasers [35].

Figure 95 is a short list of metals that have been deposited and the laser used [55,56].

Figure 96 is a list of the deposition processes available with Eximer and Nd:YAG lasers [35].

For active circuit repair and modification, the selection of laser wavelength and power is very important. If the laser energy coupled into the substrate is large enough to significantly heat the junction, then leakage can occur. Damage to the junction can occur with laser energies much lower than those required to show visible damage to the substrate. Because no visible damage can be seen at the laser site of a damaged junction, testing of the device must be conducted electrically to determine the laser energy which can be safely used for ablation, etching and deposition. Each laser wavelength couples differently. Each wavelength used will need to be tested independently. For a Nd:YAG laser of 1.06 um with a spot size of 4 um, a pulse width of 35 nsec and enough energy to blow a 1 um polysilicon link, the center of the laser heating must be at least 10 um from the junction.

Material	Etchant Laser	
Si	Br_2, Cl_2, HCl, XeF_2	Ar ion, CO_2
SiO_2	HF, Cl_2, CF_3Br	CO_2, Ar ion
Al	Cl_2, H_3PO_4:HNO_3:$K_2CR_2O_7$:H_2O	Ar ion,
C	H_2, UV lamp	
Mylar	Air, Ar ion	

Figure 93. A list of materials etched with laser gas assisted machining.

AVAILABLE LASER ETCHING PROCESSES
Nd:YAG and Eximer Laser Processes

Etched Material	Process	Minimum Linewidth	Maximum Thickness	Rate (um3/s)
Polyimide	Oxygen	2 um	None	>1000,000
Silicon Dioxide	Vacuum Evaporation	0.5 um	10 um	>100,000
Silicon Nitride	Vacuum Evaporation	0.5 um	10 um	>100,000
Metals	Vacuum Evaportaion	0.5 um	10 um	>100,000
Silicon	Halogen	0.2 um	10 um	>200,000

Figure 94 lists the laser processes available for use with Eximers and Nd:YAG lasers.

Material Deposited	Gas used	Laser
Si	SiH_4	Ar ion, CO_2
W	WF_6	Ar ion
Ni	$Ni(CO)_4$	Ar ion
Pt		Ar ion

Figure 95 is a short list of metals that have been deposited and the laser used.

AVAILABLE MICROCHEMICAL LASER DEPOSITION PROCESSES
Nd:YAG and Eximer laser processes

Deposited Interconnect	Material Properties	Typical Line Dimensions	Resistivity*	Write Rate**
Silicon	Low stress, Microcrystalline/ Polycrystalline	0.2 - 20 um wide 0.1 - 10 um thick	1mOhm-cm	1000 um/sec
Tungsten	Alpha-phase Low stress	2 - 2- um wide 0.1- 4 um thick	12-25uOhm-cm	200 um/sec
Nickel	Microcrystalline/ Polycrystalline	1 - 10 um wide 0.1 - 2 um thick	15 - 70 uOhm-cm	200 um/sec
Cobalt	Microcrystalline	1 - 25 um wide 0.1 - 2 um thick	13- 25 uOhm-cm	50. um/sec
Platinum	Microcrystalline	0.5 - 8 um wide	11 - 15 uOhm-cm	100 um/sec
Aluminum	Polycrystalline	0.5 - 10 um wide 0.1 - 4 um thick	4 uOhm - cm	200 um/sec
Copper	Microcrystalline	0.5 - 20um wide	7uOhm-cm	10 um/sec
Molybdenum	Microcrystalline	0.5 - 15 um wide 0.1 - 10um thick	24 uOhm-cm	100 um/sec
Gold	Polycrystalline	0.5 - 20um wide 0.1 - 10um thick	3uOhm-cm	10 um/sec

*Typical contact resistance 0.1 - 1 uOhm-cm
**Lines used for VLSI interconnect

Figure 96 is a list of the deposition processes available with Eximer and Nd:YAG lasers.

2. Material Characterization and Failure Analysis

Specialty lasers have paved the way for advanced materials characterization. In a laser ablation SIMS style system, a sample is placed under vacuum in the focal plane of the laser.. A high intensity light pulse is aimed at the suspect contamination and ablates a portion of the contamination from the surface. The material removed contains charged particles which can be swept into a secondary ion mass spectrometer. The laser ablation method of contamination removal has the capability of determining elemental composition as well as molecular formulation. The limiting factor in this technique is the size of the laser spot and the light adsorption properties of the contaminate.

Lasers can also aid in cross sectional studies of semiconductor circuits. The original technique was to expose photoresist in the area of interest, develop the resist and then expose the material to some form of etch. In this manner local isolation of the cross section could be performed without damage to the entire device. Later, laser ablation of the photoresist saved steps in the process and allowed for masking materials other than photoresist. Higher power lasers with short pulse widths permitted selective cross section without any mask material. The addition of etch enhancing gases to the cross section process permitted the side walls to be steeper and gave a smoother appearance.

The need for fast turn transmission electron microscope sample preparation lead to the use of lasers to pre mill the bulk material before FIB polish.

3. Process Monitoring/Characterization

Lasers are routinely used in the wafer fabrication process to measure particle density, size and location. On photo masks, particles left after cleanup can be removed with a short pulse of laser light to ablate the contamination from the surface. The same procedure can be applied to the wafer itself. In the case of laser ablation, there is no additional residue left from the process. With FIB processing, there is residual Ga left implanted into the surface of the device. To date, there has been no indication the Ga is a problem; however, more testing is necessary to determine the extent to which Ga may be left on the surface of the device.

Figure 97 Compares the use of FIB technology to that of laser technology [34]. The tradeoffs shown in the figure are time, natural etch stops and metal resistivity. The main factor left out is cost. The use of laser ablative systems is quite cost effective with the FIB system. The use of the liquid metal deposition is likewise very cost efficient. Gas assisted milling and deposition is still in the maturing stage. The efficiency is quite good but the cost factor is not as good as other laser techniques.

Future Trends

FIB systems are too expensive to be used for everyday types of jobs. The FIB process is time consuming and ultimately expensive. To the rescue comes laser processing. Depending upon the needs of the engineer an ablation technique, wet/dry chemical process or gas phase reaction may be used.

The laser process will extend beyond the range of the FIB process by adding capabilities not currently possible with ion milling. As an example, extremely large area milling using gas phase reactions can depassivate integrated circuits. For flip chip technology, where the engineer needs to see clearly into the workings of an integrated circuit from the backside, the laser process can smooth and thin the device to allow precise inspection of the junctions in the integrated circuit.

The laser will make its own optics for specialized milling applications. Perhaps laser processing will even trick the laws of physics, as is done with phase change projection printers, to etch and deposit materials using the diffraction of light.

The laser process is versatile enough that the same laser setup could conceivably mill holes, deposit dielectric, deposit interconnections, weld lids, inspect the device and stimulate the junctions for testing.

There will be a laser in the failure analysis laboratory. The question is not if, but when. How long can a company continue to use expensive equipment to perform jobs better suited to less costly techniques?

FIB vs LASER MICROSURGERY

COMPARISON OF LASER MICROCHEMICAL ASSIST TO FIB

Process	Focused Ion Beam	Laser
Metal Deposition Time: 1 mm long line	> 1 hour	3 Sec. - 1 min.
Resistance (uOhm-cm) 3 (Au)	200 - 500 (Pt)	15 -20 (Pt)
Purity 50% Metal	50% Carbon * (By Auger)	100 % metal
Sheet Resistance 0.2 um thick 4.0 um thick	>4 Ohms/Sq Impractical	0.4 Ohms/Sq (Pt) < 0.02 Ohms/Sq
Induced delay relative to lithographic interconnect	10 - 100X	None
Oxide/Polyimide Etch 100umX100um 3um deep	>3000,000 sec.	35 sec.
Silicon Etching 100 um X 100 um 20um deep	>10 hours	< 5 sec. 35nm mean surface roughness

*Newer FIB Pt deposition processes can exceed 88% metal and can be performed twice as fast as standard Pt depositions [55,56].

Figure 97. Compares the use of FIB technology to that of laser technology

Summary/Conclusions

Starting as a research tool, FIB technology has now become an essential technique for failure analysis making the almost impossible submicron device analysis not only feasible, but faster and more accurate. Its wide usage led to new terminology, such as FIBBED and FIBBING - a new set of words for Webster's Dictionary!! It is also making significant inroads as a post process monitoring tool to avoid design/mask changes that delay product introductions, where timing of samples to customers can make or kill a designing decision.

Because of the wide range of uses, a word of caution about proper selection of equipment with relevant features is appropriate. Complimentary usage of other techniques e.g. laser beam technology, will make the micro-surgery and analysis more cost effective. New laser technology with faster milling speeds and better deposition performance may surpass the FIB in capability for many of today's complex tasks.

Finally, we can only expect wider use of focused ion beam technology and laser technology in areas other than the semiconductor industry such as micro-machining and bio-medical areas, thus providing wider market base which should lead to FIB vendor stability and price reductions for the customer.

ACKNOWLEDGMENTS

The author would like to thank Preston Scott, Darwin Rusli, Wilson Tan, Marita Macy, Larry Reynolds, Olen Adams Jose Menendez, Nguyet Nguyen, and Huy Giang of Texas Instruments, K. N. Hooghan of Lucent Technologies and Dave Rogers of Optek for their assistance in preparation of the manuscript.

FIB EVALUATIONS

I would like to extend our sincere appreciation to Frank Bell of Motorola for making the evaluation of the Micrion 9000 system possible. Support from FEI, Micrion, Schlumberger and SEIKO in terms of their respective equipment evaluation and sharing of their results, is greatly appreciated. Also thanks is given to Nike Skvarta of Knight Labs, Cornell University for providing insight.

LASER EVALUATIONS

I would like to extend our sincere appreciation to Scott Silverman of Revise for the help in obtaining images of gas phase laser chemistry. Appreciation is extended to Rod Waters of Florod for his help in obtaining images of laser ablation and laser wet chemical techniques.

REFERENCES

1. J. Melngailis, Critical Review: "Focused Ion Beam Technology and Applications", J. Vac. Sci., March/April 1987, p. 469.

2. T. Lundquist, Schlumberger Corporation, 1601 Technology Dr., San Jose, CA 95110.

3. Seshu Pabbisetty, et al., "Focused Ion Beam Technology: Status and Trends", IPFA Singapore, Nov. 1993.

4. Floyd Pothoven, Florod Corporation, 17360 South Gramercy Pl., Gardena, CA 90247-5212.

5. S. J. Kirch, et al., Proc. ISTFA Symposium, P. 35, 1991.

6. J.A. Lange, et al., Proc. ISTFA Symposium, p. 397, 1991

7. P. Catinella, et al., Proc. ISTFA Symposium, 1993.

8. Courtesy of SEMATECH, U.S.A.

9. D. R. Kingham, et al., "Shape of a Liquid Metal Ion Source", Applied Physics A Solids and Surfaces, Accepted 11 January 1984.

10. J. Orloff, "Focused Ion Beams", Scientific American, October 1991, pp. 96-101.

11. R. Mackenzie, "Focused Ion Beam Technology: A Bibliography", 1990, Nanotechnology 1, pp. 163-201.

12. K. Gamo, "Focused Ion Beam Technology" Vacuum, Vol 42(1/2), 1991, pp. 89-93.

13. OGI Short Course on Focused Ion Beam Technology, September 24, 1991.

14. P. Prewett and G. Mair, Focused Ion Beams from Liquid Metal Ion Sources, Research Studies Press LTD., Copyright 1991.

15. S. Shuduri, " Precise Measurement of a Focused Ion Beam Profile", J. Electrochem. Soc., Vol. 134, no. 6, pp. 1536-1540.

16. FEI "FIB 600 Series Workstation Users Guide", FEI Corporation, 7451 N.E. Evergreen Pkwy, Hillsboro, OR 97124.

17. Micrion Focused Ion Beam Systems Applications Manual, Micrion Corporation, One Corporation Way, Peabody, MA 01960.

18. Jon Orloff, "An Optimized Two Lens Optical Column for Use with a Liquid Metal Ion Source", Elsevier Science Publishers B.V. (North-Holland), 1987, pp. 327 -332.

19. M. Thayer, "IC Failure Analysis: Focused Ion Beam ", IRPS Tutorial, pp. 4c.3 - 4c.16, 1996.

20. Z. Xu, Development of In situ Processing Techniques Using Focused Ion Beam, Masters Thesis, Osaka University, 1991.

21. J. Lindquist and J. Puretz, FEI Company, 7451 N. E. Evergreen Pkwy, Hillsboro, OR 97124.

22. G. A. Crow, Proc. ISTFA Symposium, p. 401, 1991.

23. D. Vechten, "Fundamentals of ion-beam-assisted deposition. 1. Model of process and reproducibility of film composition", J. Vac.Sci Technol. A, 8(2), 1990, pp. 821-830.

24. G. Hubler, "Fundamentals of ion-beam-assisted deposition, 2. Absolute calibration of ion and evaporant fluxes", J. Vac. Sci. Technol. A., 8(2), 1990, pp. 831-839.

25. A. Dubner, et al., "Summary Abstract: Ion-beam-induced deposition of gold by focused and broad-beam sources", J. Vac. Sci. Technol. B5(5), September/October 1987, pp. 1434-1435.

26. Courtesy of FEI Company, U.S.A.

27. Channel Electron Multipliers, Galileo Electro-Optics Corporation, Galileo Park, P.O. Box 550, Sturbridge, MA 01566.

28. T. Woodward, Materials Analytical Services, 2418 Blue Ridge Road, Suite 105, Raleigh, NC 27607.

29. C. Libby, Micrion Corporation, One Corporation Way, Peabody, MA 01960.

30. K. Nikawa, et al., Proc, IRPS Symposium, P. 43, 1989.

31. K. Gamo, et al., Japan J. Appl. Phys. vol. 23, p. 1293, 1984

32. J. Puretz, et al., "Focused ion beam deposition of Pt containing films", J. Vac. Sci. Technol B 10(6), Nov/Dec 1992, pp. 2695-2698.

33. J. Doherty, et al., "Focused Ion Beam Processing", United States Patent # 4,639,301, January 27, 1987.

34. K. Van Doorselaer, et al., " How to Prepare Golden Devices using Lesser Materials", ISTFA, pp. 405 - 414, 1993.

35. Scott Silverman, "Process Applications", Revise, Inc., 79 Second Avenue, Burlington, MA 01803-4479, Web page - http://www.revise.com/, 1996

36. C. G. Talbot, Tutorial Session #4, IDS Users Conference, Schlumberger Technologies, 1993.

37. William B. Thompson, et al. "Focused Ion Beam Process Monitoring", Micrion Corporation, One Corporation Way, Peabody, MA 01960.

38. Courtesy of SEIKO Instruments Inc., Japan.

39. Courtesy of Schlumberger Technologies, U.S.A.7

40. Courtesy of Micrion Corporation, U.S.A.

41. R. Young, et al., "Fabricatioin of Planar and Cross-Sectional TEM Specimens using a Focused Ion Beam", MRS Vol 199, 1990, pp. 205 - 216.

42. M. Thayer, "Enhanced Focused Ion Beam Milling Applications", ISTFA 1993, pp. 425-429.

43. S. Klepeis, et al., "A Grinding/Polishing Tool for TEM Sample Preparation", Mat. Res. Soc. Symp. Proc. Vol. 115, Dec. 3, 1987, pp. 179-184.

44. S. Morris, "A Technique for Preparing TEM Cross Sectioins to a Specific Area Using the FIB", 1991, ISTFA, pp. 417-427.

45. H. Lin, et al., "Ion beam profiling and end-point detection with microfocused secondary ion mass spectroscopy", J. Vac. Sci. Technol., A 8(1), Jan/Feb 1990, pp. 93-98.

46. S.J. Kirch, et al., Proc. EMSA., p. 1108, 1991.

47. D. Pramanik, et al., "Aluminum Film Analysis with the Focused Ion Beam Microscope", Solid State Technology, May 1990, pp. 77-80.

48. R. Levi-Setti, "Progress in High Resolution Scanning Ion Microscopy and Secondary Ion Mass Spectrometry Imaging Microananysis", (Scanning Electron Microscopy, AMF O'Hare, Chicago Il.), 1985, pp 535-551.

49. Nicholas Antoniou, Micrion Corporation, One Corporation Way, Peabody, MA 01960.

50. Schlumberger Technologies, Inc., "IDS Large Area Overlay User's Guide", May 1996

51. Kendall Scott Wills, et al., U.S. Patent pending on manifold construction. 1996.

52. Richard H. Livengood, et al., "FIB Techniques to Debug Flip-Chip Integrated Circuits, Semiconductor International", March 1998.

53. Kendall Scott Wills, et al., "Laser Induced Chemical Reactions of Ethyl Acetate and Ammonia at Various Frequencies", Master Thesis, Department of Physics, Southern Illinois University, August 1974.

54. Scott Wills, et al., "Characterization of laser blown molydisilicide links", Proceedings SPIE Laser Processing of Semiconductor Devices pp 82, - 86 Volume 385, 1983

55. T. M. Bloomstein, et al., "Stereo Laser Micromachining of Silicon", J. Vac. Sci. Technol. B 10, 2671 1992.

56. Bruce M. McWilliams, et al., "Wafer Scale Pantography: VI. Direct-Write Interconnection of VLSI Gate Arrays", Proceedings SPIE Laser Assisted Deposition, Etching, and Doping pp. 49 - 54 Volume 459, 1984

57. T. M. Bloomstein, et al., "Laser-Chemical Three-dimensional Writing for Microelectromechanics and Application to Standard-Cell Microfluidics", Appl. Phys. Lett. 61, 708 1992.

58. S. D. Allen, et al., "Optical and Thermal Effects in Laser Chemical Vapor Deposition", Proceedings SPIE Laser Assisted Deposition, Etching, and Doping pp. 42 - 48, Volume 459, 1984

59. J. R. Moulic, et al., "Laser microsurgery and fabrication of integrated circuits", Proceedings SPIE Laser Processing of Semiconductor Devices pp 87, - 92 Volume 385, 1983

60. N.G. Chew, et al., Appl. Phys. Letters vol. 44, p. 142, 1984.

61. R. Alani, et al., Proc. MRS Symposium Vol. 199 p. 85, 1990.

62. R. Young, et al., "Characteristics of gas-assisted focused ion beam etching", J. Vac. Sci. Technol. B 11(2), Mar/Apr 1993, pp. 234-241.

63. M. Vasile, et al., "Scanning Probe Tips Formed by Focused Ion Beams", Rev. Sci. Instrum. 62(9), September 1991, pp. 2167-2171.

64. R. Young, et al., "An applicatioin of foucused ion beam milling to studies on the internal morphology of small arthropods", FEI Copmpany, 7451 N.E. Evergreen Pkwy, Hillsboro, OR 97124.

65. R. DeFreez, et al., "Focussed Ion-Beam Micromachined Diode Laser Mirrors", SPIE Vol 1043, Laser Diode Technology and Applications, 1989, pp. 25-35.

66. L.L. Hsu et al., "The Study of Failure Mechanisms in IC Devices by Using Focused Ion Beam Technology", ISTFA Nov. 1991, pp 409 - 416.

67. J. G. Oellerin, D. P. et al., "Focused Ion Beam Machining of Si, GaAs, and InP", J. Vac.Sci. Technol. B3(6) Nov/Dec 1990, pp. 1945 - 1950.

68. F. A. Stevie, et al., "Applications of Focused Ion Beams in Microelectronics Production, Design and Development", Surface and Interface Analysis, Vol. 23, 61-68 1995.

69. Kendall Scott Wills, et al., U.S. Patent pending on FIB deposition processes. 1996.

Laboratory Operations and Management

Failure Analysis
Laboratory Management

Richard J. Ross
IBM Microelectronics
Essex Junction, Vermont

INTRODUCTION

Failure Analysis (FA) of electronic components is a highly technical activity, requiring highly skilled personnel, increasingly complex (and increasingly costly) equipment, and the development of increasingly sophisticated techniques and methods to discern the location, nature, and root-cause of defects causing non-conforming device operation. The devices to be analyzed range from discrete, passive components to ultra-high density integrated circuits (ULSI). In all cases, the operation and development of a Failure Analysis laboratory needs to be managed with an eye toward cost-effectiveness, customer satisfaction, and future challenges. The purpose of this article is to stimulate the reader to consider the various aspects of Failure Analysis laboratory operations and their respective business management requirements. While the general focus is primarily based on semiconductor integrated circuit (IC) operations—particularly those of IC suppliers—the general ideas and operations can be applied to a wide variety of other laboratory configurations and settings. References for further reading and/or examples of resource materials are also included.

OUTLINE

Generally speaking, FA operations and management requirements can be divided into the following key areas:

1. Staffing
2. Laboratory Organization
3. Laboratory Design & Operations
4. Strategic Development
5. Financial Management
6. Metrics & Measurements

STAFFING

The staffing function is a key, if not *the* key management function. No amount of complex and clever equipment, cost efficiencies, or fancy graphics-heavy reports can replace dedicated, motivated, skilled engineers and technicians. The staffing function can be divided into the areas of Recruitment, Retention, Training, and Skills Mix.

Recruitment

Recruitment of FA personnel is often a tricky business. A good analyst should have a working knowledge of device physics, circuit analysis, manufacturing processes, electrical characterization techniques, and chemistry. Further, (s)he must have strong oral and written communications skills, be a team player, and be able to manage the stress which comes from multiple tasks and multiple customers. Above all, analysts require a strong technical curiosity.

Where, then, does one go to recruit such people? The simple answer is that you *can't*. Engineers and technicians from good schools with degrees in electronics, chemistry, physics, or materials science can be found, but there are very few of these new graduates who will have the full set of requisite skills. Another source is to find experienced design, process, or characterization engineers/technicians from within the industry or within the company, but this leads to a fairly rapidly diminishing list of the same key prospects. The ultimate answer is to start with the newly minted engineer/technician as "raw material" and, by carefully applying time, patience, and mentoring, grow them into successful and competent analysts. This, then, leads us to…

Training

How do we take these new, eager people and turn them into failure analysts? There are few academic programs teaching the art of FA. Several workshops and seminars, such as those held in conjunction with the International Symposium for Test & Failure Analysis (ISTFA) and/or those given by a number of consultant firms exist, and these are very valuable, however, even these can only cover a broad overview of the skills necessary to become a full-fledged analyst. There are also self-paced training materials—books (such as this one) and multimedia reference works, which are excellent references, but again, can only provide broad guidance. The most prevalent and successful training methodology is on-the-job training (OJT) under the guidance of an experienced, highly skilled FA mentor. The model is similar to the "see one, do one, teach one" technique used in medical training. This requires, in addition to the mentor noted above, a clear training plan and well-documented procedures. Several independent estimates of the time required to transform an engineer/technician into a qualified failure analyst indicate that the process takes 18-24 months and may cost up to $250,000 including salary, benefits, materials, etc.

Further, this training need will continue, at less intensive levels, for the duration of the analyst's career. The industry, especially the IC industry, is moving at a rapid rate—so much so that each year, much of what we know and the tools we use become at best outmoded and inefficient, and at worst, completely obsolete. Tools and techniques exist today that were non-existent ten years ago (e.g. Focused Ion Beam, Fluorescent Thermal Imaging, Scanning/Atomic Force Microscopy). Simultaneously, materials and process complexities are being introduced which make current methods ineffective (e.g. Cu metallurgy, Flip-chip packaging, new dielectrics, GHz processors, sub-white-light optical pattern dimensions)—the need to grow and learn is endless. With this need and the time and cost involved, a major issue becomes…

Retention

It is an acknowledged fact that retaining skilled FA personnel continues to be a challenge across the industry. The average "lifetime" of a failure analyst is less than five years. Many young analysts get the impression that FA is not a "leading-edge" career path like design and development. Some feel that FA has low corporate visibility and that it is not good that the nature of the mission is dealing with "failure". I would make the argument that FA *is* a "leading-edge" career path. It is continuously varying, a challenge to the mind, an opportunity to master multiple disciplines, a way to keep up with the newest technology advances, and, we get to pay with some *really* cool toys! It all comes down to the fact that people need to be recognized for their contributions. The team needs to be built and management needs to insure that the FA team shares in the successes of the overall organization. Rotational assignments, certainly within the FA organization if it is a large one, are useful—even more so if rotation between FA and design, process engineering, and even marketing can be accommodated. Attendance at outside conferences and workshops are also key ways to reward and provide incentive to analysts—and to accomplish some training as well. Also, never forget the positive effect of just saying "thanks…"

Skills Mix

In a given setting, the "mix" of professional to technical personnel (engineers to technicians) is dependent on several factors. Product mix and complexity, task(s) to be performed, tool set automation, education/experience level, and proximity of design support all factor into the equation. In the IC world, typical engineer:technician ratios run form 1:2—1:4 in SRAM/DRAM to 3:1—1:1 for custom logic or microprocessors. The use of computer-aided diagnostics packages such as Level Sensitive Scan Design (LSSD) can reduce the logic ratios by up to a factor of three. Characterization FA operations in a manufacturing environment can typically run with several skilled operators working under the supervision of an FA engineer or technician to evaluate test structures or defect-density monitors. The key factor here is that one size does not fit all—the operation needs to be tailored to the task and setting.

LABORATORY ORGANIZATION

The organization aspect of FA Lab Management includes the reporting structure, the guiding management philosophy, the decision to operate as "generalists" or "specialists", and, especially significant in today's global economy, how to handle support of a world-wide business operating environment.

Reporting Structure

Who "owns" the FA operation depends in large part on the type and size of the overall business unit. In a large corporation, especially an IC supplier, the FA organization may report to corporate or division quality/reliability. An individual fabricator or design laboratory may have an FA operation linked to its local quality/reliability operation. Manufacturing and/or development functions may claim the rights to FA on their specific product sets. Or, the FA operation may be an independent entity (clearly, for commercial FA labs, this is the case). Again, there is no one "right" model, however, in actual practice, within the IC supplier industry, the most common reporting structures are to quality/reliability functions.

Management Philosophy

Many texts in management deal with the possible structures and philosophies of management of technical areas. The three most common FA management philosophies are the traditional functional organization, the matrix organization, and, a new and empowering option, the self-directed team. The traditional, *functional*, organization structure works best with less experienced personnel in a single task-oriented environment with fixed tools and established procedures. The *matrix* organization gives a mix of task and customer orientation, which allows the shifting of resources quickly to meet tactical needs, but can give rise to loyalty conflicts between the "owning" and the "using" organizations. The newest organizational paradigm, the self-directed team, gives the employees more control and empowerment over their assignments and procedures. It requires high levels of maturity, expertise, and trust—both among the employees and between management and employee. There is an up-front team training requirement to develop the team-building process and all factions must buy into the process from the start. When this concept is working, it provides the maximum in flexibility and, often, in productivity.

On the technical side, the organization of the lab can be segmented according to the various product types serviced (memory, logic, processor, discrete, passive,…), by technology (bipolar, CMOS, mixed,…), or by customer set (automotive, data processing, consumer, communications,…). It could also be divided by source of the work (qualification, reliability, returns,…) or by service type (test, electrical characterization, localization, deprocessing, inspection,…). Each of these methods has advantages and disadvantages and each may work in a given setting. The key challenge is the optimization of (usually very limited) resources to maintain operational flexibility and customer satisfaction.

Specialization versus Generalization

Analysts may be organized by task or by program supported. That is, they may be assigned one specific task or task set linked to key tools or they may be assigned the complete analysis task from start to finish. *Specialists* may be used to maximize critical expertise, to optimize key techniques or to optimize tool performance. They may become well-known experts in their specific field and may perform their specific task in a very efficient manner. The disadvantages to specialization include the risk of boredom or "burnout", the fact that the job has little variation or flexibility, and the risk of a tool or technique becoming obsolete and thereby the individual connected with it. *Generalists* have the advantage of very great variation in job tasks on a day-to-day basis. Flexibility is optimized as resources are more easily shifted from one job or product to another. The analyst feels a higher sense of ownership of the job from start to finish and gets a great feeling of accomplishment when it is completed. The disadvantages of generalization include the risk that there will be less than optimal results from a specific tool or technique, the potential lack of "experts" in one specific field, and an increase in training time as more operations need to be learned.

As is usual in FA, there is not a hard and fast rule as to which is the best mode of operation. Lab resources may dictate one model, as there may not be enough people to have single-procedure experts for all procedures. A good compromise is to use specialists for tools/processes where optimum performance is critical (e.g. Scanning Probe Microcroscopy, FIB Device Modification, Materials Analysis tools, etc) and generalists for broader applications (e.g. Deprocessing, Electrical Characterization, Inspection, etc). As always, the key is maintaining optimal flexibility and customer satisfaction.

World-Wide Operations

Today, all markets are becoming global. Personal computer revenue growth will increasingly come from non-US sources. The demand for wireless communications is worldwide, as the cost for developing countries to build huge landline infrastructures will be exorbitant as compared to the cost for wireless relay stations. Components for consumer, data processing, automotive, and other end items come from around the world, and the end item is itself shipped around the world. Alliances and partnerships are proliferating with many of these now multi-national in composition as the cost of development and fabrication facilities soars. All of these worldwide customers want FA answers and want them *now*. Within suppliers, there are often multiple fabricator locations around the world making the same products or exercising the same technologies—the ability to find and share information saves time and time is still money. Alternatively, some suppliers are defining specific fabricators as worldwide sources with all customers dependent on that single facility for key hardware. How, then, can FA labs respond to this global challenge?

Again, several possible operational models exist, each with its own advantages and disadvantages. For a fairly large company, a *geographic or regional model* can optimize the turn-around-time (TAT) from fail to root-cause analysis. The premise is that the failing device is transmitted from the customer to the nearest laboratory to provide the shortest analysis time. This model typically requires multiple, full capability labs—one in each general geography where the company operates or sells (e.g. North America, Europe, Pacific Rim)—with, if possible, common tools and procedures. Close communications and a worldwide, commonly accessible database among FA labs and design and fabrication facilities are essential to build the trust and "buy-in"

needed to accept FA results regardless of the location of analysis. This model works well when TAT is the overbearing issue. A second option is the *site of manufacture* model. In this model, the failing part is always returned to the location of manufacture for analysis. The premise is "the place that made it, owns it" and the fabrication site has the expertise required to link the specific product, process, and tooling which may have caused the defect to the fail mechanism. There is a TAT impact due to ship time from point of fail to fabrication location (customs, transit time, etc), but controversies caused by distance or differing tools sets are eliminated. The analysis and the corrective action are also now co-located. This model works well for worldwide single-sourced products as it requires only one set of process or product specific tooling or expertise.

LAB DESIGN & OPERATIONS

Lab Design

The design of a Failure Analysis lab revolves around the expected workload. Space requirements are driven by tools (number and footprint) and personnel (numbers). "Wet" space used in chemical deprocessing is typically the most expensive space (ventilation, drains, environmental concerns) as well as the most heavily utilized space. Safety concerns which, while primarily governed by national or local government regulations and laws, are also highly affected by specific corporate policies. The population of the lab is driven by the organization structure, the workload, cost constraints, TAT experience and requirements, and the tool set. The cost of operation can be divided into the costs of people and equipment to provide basic tool coverage and analytical support (*fixed cost*) and the costs driven by the quantity of work to be done and/or the TAT (*variable cost*). If, for, example, the average TAT for a particular product or product type is 2.0 days, than one analyst can, on average, complete 2.5 parts/week or about 125 parts/year. Therefore, for every 125 parts of a given type expected, one analyst would be required.

Lab Operations

Several models for day-by-day FA operations exist. They include:

- The Corporate Asset
- Cost (or Profit) Center
- "Open Campus"
- Dedicated Resource
- Tiered Support

Many of these models presume the lab exists in a corporate setting, usually a supplier of semiconductors, however several of them are applicable to independent laboratories and even university environments.

The *Corporate Asset* model implies that the FA lab is independent of any single/local manufacturing, development, or quality organization or product line. The costs are spread across all elements of the corporation as a "tax" as part of the cost of doing business and are not normally causally allocated. If the cost of the tooling is a major factor, and the product set is reasonably sized, this is a good model for medium-sized suppliers with multiple fabricators.

The *Cost (or Profit) Center* model implies again that the FA lab is an independent entity, but recovers its costs (or makes a profit) by causally charging its services to the requesting customer on an "as-used" basis. The lab is expected to be either self-funding (within a corporate environment) or to be a profit-making operation (corporate or independent lab) providing a return on investment over and above cost recovery. This model serves the purpose of accruing the cost of the FA operation to the customers in a causal fashion and allows the customers, in turn, to manage their costs by managing the submission of parts.

An interesting paradigm used in some companies is the *"Open Campus"* model. In this model, the FA organization consists of a small cadre of experts who provide the facilities, tools, and technique documentation/training to a large number of people from a variety of organizations who use the laboratory and perform their own analyses. The FA personnel do perform a few very complex analyses, but have the main function of maintaining the lab, providing training and certification, and developing or procuring new techniques/tools as needed. The lab costs, including personnel, are allocated across the using organizations as a percentage of lab time used. The effect of this is to minimize the size of the FA organization per se. Concerns include the potential loss of skills when an individual is an infrequent user, and the exposure to lack of careful tool use and resultant maintenance costs.

The *Dedicated Resource* model is normally employed by large companies, although it may be usable by independent contact laboratories or research organizations/universities working on specific contract items. The premise of this model is that resources are negotiated with and allocated to customers during the planning cycle, usually annually. Past history, planned volume (wafer starts, shipments, qualifications, etc), process/product complexity, etc. are normally used to project resource needs. Some flexibility is built in to allow for tactical changes, but the resource allocation is essentially like a service contract or insurance policy.

Tiered Support is not so much an operating model as it is a process to minimize TAT. The premise is that an analysis be performed at a facility as close in time and distance to the occurrence of failure. A hierarchy of laboratories is established, ranging from small, quick inspection facilities which may be at a sales office, through moderate-sized and equipped regional facilities, up to a full-function centralized total support facility. The paradigm is decentralized support and quick TAT so that relatively straightforward analyses (e.g. package mechanical damage) do not load down the main laboratory. This expected savings in time comes at some capital and staffing cost.

STRATEGIC DEVELOPMENT

As has been mentioned in many articles and publications, the pace of technology continues to accelerate. Whether or not Moore's law continues to hold true, new materials, product designs, processes, and customer quality requirements will drive the FA community to require more and more advance techniques and tools. The role of strategic development in FA is to insure the preparedness of the FA laboratory to analyze these advanced products and/or technologies. This process must be pro-active, as design schedules are tightening and time-to-market becomes as key competitive advantage. A strategy for identifying gaps in needs versus current capabilities, the prioritization of development dollars and resources, and project planning/scheduling/tracking are all key ingredients for strategic development teams. Another factor in the process is the increasing influence of external standards and expectations such as EIA Standard 671, CIQ-C, A and ISO-900x. Continuous process improvement in terms of TAT, accuracy, and financial efficiency also drive the need for strategic development activities. Unfortunately, given the too-frequent situation where an FA organization is understaffed and overloaded, the drive to get today's work out often conflicts with the need to address tomorrow's needs. The also too-frequent actuality is that tomorrow's needs go unaddressed.

A strategy for staffing and managing strategic development projects for FA can be established using standard project management techniques which have been published and taught for some time. The key techniques include Project Definition, Scheduling, Budgeting, Staffing, Supervising, Reporting, and Implementing. In addition to many books and management courses, several computer programs are available to aid in the process. In medium to large size FA organizations with mature populations, the matrix organization structure can be an excellent management method. Each employee is expected, as a part of his/her job performance, to spend some period of time (e.g. 20%) working on strategic development projects. Each project is assigned an owner and staffing is pulled from across the organization. Obviously, this requires agreement and championship from the management team and a realization that there may be some TAT impact to non-urgent analyses—customer involvement is also a key component.

In today's environment, complex problems—some of which push the fundamental laws of Physics (e.g. real-time optical inspection and/or mechanical probe placement of sub-0.25 micrometer features)—often surface. The cost and time of strategic development and the knowledge base/expertise of an individual FA organization or even an individual company may be insufficient. In addition to the "go it alone" approach which is traditional in organizations which have been typically highly competitive with one another, several alternative paradigms have emerged. The use, in larger corporations, of corporate resources such as Research & Development Labs, or, for small and large organizations, of independent labs, universities, and, increasingly, partnerships with equipment vendors can reduce the time and money requirements to resolve strategic issues. Consortia of companies, which have traditionally been bitter competitors (e.g. SEMATECH and JESSI), are also an increasing means of providing funding and impetus for the development of tools and techniques to meet common needs. One or all of these strategies may be applied simultaneously to maximize the effective resource brought to bear on difficult problems.

FINANCIAL MANAGEMENT

As the complexity of the problems continues to increase, the cost of analysis continues to increase. Personnel costs account for almost 66% of total FA lab costs and are driven by the number and skill mix of human resources, the cost of labor, and the associated overhead. Tooling (depreciation) accounts for the next largest percentage—about 20%. Once upon a time, FA tools were fairly inexpensive—the Scanning Electron Microscope (SEM) was probably the most expensive piece of equipment in a high-end FA laboratory and it cost under $100K. Today, new-generation SEMs cost upwards of $500K and some specialized tools (e.g. FIB, SIMS, TEM, High-Speed Electrical Characterization Testers, etc.) can cost in excess of $1.0M. Clearly, a million here and a million there, and pretty soon we're talking about real money. Financial management and cost-efficient operation are here to stay. Since FA is usually looked at as a service operation (and a necessary evil) as opposed to revenue-producing activities, cost constraints are often more severe in FA operations than in product development or manufacture. This is ironic, given that prompt, accurate FA can often result in improved yield (lower manufacturing cost), improved quality/reliability (improved customer satisfaction), and, indeed, often save the business for the company by providing quick and effective information for corrective action to customer concerns. In any event, prudence in financial matters is highly important and appropriate.

Given the high cost and high visibility of capital tooling, the need for care in tool evaluation and selection becomes ever more important. Decision points for tool purchases will vary with the size and scope of the organization. Key factors such as whether or not an existing tool can be upgraded to support the need, the cost of maintenance on older tools, the frequency at which the proposed operation is/will be needed, and personnel training costs will have significant influence on the decision process. If a process will be used infrequently, the best solution may be to purchase the service from an outside source. The drive to maximize the use of older tooling may seem sensible (tool is fully depreciated), but serviceability and increased maintenance may actually be costing more than the effective depreciation of a new piece of equipment.

Once the decision is made that a new piece of equipment is justified, how shall we decide which tool to buy? In some cases, where the market is essentially served by a dominant vendor, the choice is easy; however, this is not normally the case. Given variables like tool capability, cost (including parts, supplies, and service), vendor experience, user-friendliness, extendibility to future technologies, compatibility with existing equipment, delivery schedule, and service TAT, is there a cogent and methodic way to make an objective decision? There are, of course, several methodologies. One which provides a straightforward and logical approach is taught in management science courses as Multi-Criteria Decision Making (MCDM). The key elements of MCDM are the identification of critical decision elements, the assignment of weights to each element, the pair-wise comparison of each alternative solution to each decision element, and the aggregation of this data to a decision point. The decision elements may be sub-divided hierarchically as needed to establish the appropriate granularity of elements. The process can be done by hand for the case of three or fewer alternatives, but is best done using a spreadsheet or one of several commercially available software packages for larger numbers of alternatives and/or "what-if" analyses.

For techniques, the cost point comes down to the equivalent of the "make or buy" decision. Frequency, cost per use, service support, training, and proximity of service factor into this type of decision. If a university, vendor, or independent laboratory is sufficiently nearby, the frequency of need is low, and the cost to purchase favorable compares to the cost to develop or own, the service should be purchased. If TAT would significantly suffer, or the workload imposed upon the outside provider would overwhelm their capacity, the service should be developed in-house. A traditional payback analysis for short-term needs (18 months or less) is adequate, but for longer-tem needs, a financial analysis using an established Internal rate of return (IRR) is more appropriate.

Financial concerns can and will affect the general management philosophy and operation of the FA organization. Multi-shift operation may become cost-effective to improve utilization of high-priced assets. A vigorous program of preventive maintenance can reduce costly tool downtime. Cross-training of personnel and work schedule adjustments can optimize the efficiency of both human and capital resources. Reducing/eliminating duplication of effort among laboratories, control of expense items like chemicals, DP costs, travel, and supplies can also reduce the costs of operations. One approach toward the development of an understanding and methodology for operational cost control and efficiency which has been widely heralded in recent years is the concept of *Activity-Based Costing* (ABC). ABC involves the breaking down of all aspects of the operation into basic units or "activities". All the unitary costs of the operation are then assigned, using historical data for a specified reference period, to one or more of these activities. Each activity and associated cost is then inspected in detail to assess its relevance to the mission of the organization and its efficiency. The entire organization is invited to "brainstorm" ideas for better ways to perform the activity at reduced cost, or, even if the activity can be eliminated. As always, follow-up and implementation schedules are critical. Organizations employing ABC have reported overall cost efficiencies in the range of 15-30%, the reduction of "non-value-add time", improved use of capital and human assets, improved TAT, improved communications, a better understanding of the business process across the organization, and more thoughtful decision-making. Ideas can be as simple as the order quantity of supplies. For example, one organization saved 40% on chemical costs by purchasing high-use chemicals in larger unit sizes. Key ingredients for success of an ABC program include the commitment of senior management and the education of the personnel to obtain "buy-in". The program is not intended to be a reduction-in-force program and agreements and commitments must be fulfilled—it cannot become the "program du jour". Many books and articles exist on the process and concept of ABC and a growing number of independent consultants can be found to help tailor the process to an organization's specific culture and needs.

METRICS & MEASUREMENTS

Since FA is so important, and since we are all scientists and/or technical people, why should we measure anything? Just as the customer of an analysis can't resolve a problem until the root cause is established, the operational efficiency of the FA organization can't improve until we know where the root cause defects are. External to the FA organization, measurements of customer satisfaction can help us to improve service quality. Internal to the FA organization, we want to optimize the use of resources, control costs, justify needed capital or personnel additions, and assess the readiness for new technologies. Standards, such as customer requirements and/or EIA Standard 671 on TAT expectations for analysis of returns, or pending JEDEC/EIA standards on analysis report content and format may govern the operation's metrics. Corporate guidelines for expenses or other measures may be another source of standards. What is important is the decision of when to measure and when *not* to measure. A measurement which cannot or is not used is *worse* than no measurement at all, as it becomes a morale dissatisfier. What is it, then, that we really need to know? Key items for control and assessment of an FA operation include total TAT (queue time as well as process time), the rate (and trend) of unsuccessful or unresolved analyses, workload volume and source, and, most important, customer satisfaction. The measure of the components of TAT and unsuccessful analyses can serve to identify tooling pinch-points, obsolete tools, training deficiencies, overall staffing shortfalls, and customer communication issues. Measurements are least effective when used as a club and most effective when used as a lever. Often new tooling, increased resources, and/or enhanced training opportunities can be justified to higher technical or financial management with a cogent set of metrics—especially if the technical metrics are combined with financial data which show efficient use of existing resources.

SUMMARY

The management of a Failure Analysis laboratory requires a broad range of activities to optimize the efficiency of the operation. These have been described and some suggestions for approaches to the various activities have been outlined. The key item for consideration is that the pace of technology development continues to accelerate, and with it, the need for timely, cost-efficient Failure Analysis. As a colleague of mine, Gordon Babson, once said—"When all else fails, ask yourself: 'How would I do this if I were running the business out of my garage'". If it makes sense to the "Babson Garage Test", it's probably a good decision.

REFERENCE MATERIAL—FOR FURTHER READING

O'Guin, Michael C., The Complete Guide to Activity-Based Costing, Prentice-
      Hall, Englewood Cliffs, NJ, 1991

Krumwiede, Kip R., "ABC—Why It's Tried and Why It's Needed", Management
      Accounting, V79 N10, April 1998, pp. 32-38

Kerzner, Harold, Project Management—A Systems Approach to Planning,
      Scheduling, and Controlling, 2nd Edition, van Nostrand, New York,
      1984

Silverman, Melvin, The Technical Manager's Survival Book, McGraw-Hill, New
      York, 1984

Ross, Richard J., "The development of a Total Quality management System for a

Semiconductor Failure Analysis operation", Master's Thesis, National
      Technological University, 1991

Ranftl, Robert M., R&D Productivity, 2nd Edition, Hughes Aircraft Company Study
      Report, Los Angeles, 1978

Saaty, Thomas L., Decision-Making for Leaders, 2nd Edition, RWS Publications,
      Pittsburgh, PA, 1990

Kudva, S., et al, "The SEMATECH Failure Analysis Roadmap", Proceedings of
      21st International Symposium for test & Failure Analysis, ASM
      International, Materials Park, OH, 1995, pp. 1ff

Byham, William C., Zapp! The Lightning of Empowerment, Harmony Books,
      New York, 1991

Semiconductor Industry Association, "The National Technology Roadmap for
      Semiconductors", SIA, San Jose, CA, 1997

Frank, R. and Lee, T., "Business Aspects of failure Analysis", Proceedings of
      22nd International Symposium for Test & Failure Analysis, ASM
      International, Materials Park, OH, 1996, pp. 255ff

MATERIALS SOURCES (EXAMPLES)

MCDM:
      Expert Choice Software, Inc.
      4922 Ellsworth Avenue
      Pittsburgh, PA
      www.expertchoice.com

Project Management:
      Microsoft PROJECT for Windows
      Microsoft Corporation
      Redmond, WA
      www.microsoft.com

      CA-SuperProject
      Computer Associates International, Inc
      One Computer associates Plaza
      Islandia, NJ
      www.cai.com

Education/Training Sources and References

Christopher L. Henderson
Sandia National Labs
Albuquerque, New Mexico

Richard J. Ross
IBM Microelectronics Division
Essex Junction, Vermont

PURPOSE

The purpose of this section is to provide a sample listing of sources for further reference material. It is not intended to be all-inclusive, but to provide an idea of the types of additional information available to further explore the overall Failure Analysis community. Sources addressed include:

Formal courses (General and/or Specific)
Conferences
Books
Seminars/Workshops/Multimedia
Websites (Organizations/Vendors)

FORMAL COURSES

General FA

Hi-Rel Labs/DM Data Course
Technology Associates Failure Analysis Course
Integrated Circuit Engineering, Inc.
Lloyd Technology Associates
Sandia National Labs EQRC FA Short Course
SDG Analytic, Inc.

University/Technical School Courses

Arizona State University
National Technological University
National University of Singapore
San Joaquin Community College (CA)
University of Illinois
University of Maryland (Reliability/FA)
University of New Mexico

Tool/Technique Specific Courses

EDFAS Short Courses Website: www.edfas.org
SEM/TEM – Lehigh University (PA)
Surface Analysis Techniques – Charles Evans Labs
Accelerated Test/Reliability – Technology Associates
Reliability Institute – Univ. of Arizona

Vendor-Sponsored, Instrument-Specific Training

Emission Microscopy – Hypervision
Electron-Beam Testing – Schlumberger
Focused Ion Beam – FEI/Micrion
SEM – JEOL/LEO/Phillips/Hitachi
TEM Sample Preparation – Gatan
Energy-Dispersive X-Ray Analysis – Kevex/Noran

CONFERENCES

United States

ISTFA – November (Only conference *dedicated* to FA)
IRPS – usually April (Reliability, some FA)
ITC – usually Oct or Nov (Test, Defect Modeling)
EOS/ESD – usually Sept (ESD mainly, some FA)
Microscopy Society of America – spring
Electron/Ion/Photon Beam – usually summer or fall
IEEE Electron Devices Meeting – usually Dec
Materials Research Society – spring and fall
Int'l Microelectronic & Packaging Society -- Oct

Rest of World

ESREF – usually Oct (Reliability, some FA) -- Europe
IPFA – usually July (Reliability, some FA) – Singapore
(ALTERNATE YEARS—1999, 2001, etc)

BOOKS

General FA References

Electronic Failure Analysis Handbook; P. Martin; McGraw-Hill, 1999

Failure Analysis of Integrated Circuits: Tools and Techniques, L. C. Wagner editor, Kluwer, 1999

RADC Failure Analysis Techniques, Rome Air Defense Center, 1980

Physics of Semiconductor Failures, 1988

Failure Analysis, Mechanisms, and Photo Atlas, 1985

Tool/Technique Specific References

Transmission Electron Microscopy; Williams and Carter, 1996

Secondary Ion Mass Spectroscopy; Wilson, Stevie, and Magee, 1989

Practical Surface Analysis: Auger and X-Ray Photoelectron Spectroscopy; Briggs and Seah, 1996

Scanning Electron Microscopy and X-Ray Microanalysis; Goldstein, et al, 1992

Characterization of Integrated Circuit Packaging Materials; Moore and McKenna, 1993

Semiconductor Device and Material Characterization; Schroder, 1998

The Role of Microscopy in Semiconductor Failure Analysis; Richards and Footner, Oxford Univ. Press, 1992

Defect Modes/Mechanisms

Modeling of Electrical Overstress in Integrated Circuits; Diaz, Kang, and Duvvury, 1995

Latchup in CMOS Technology; Troutman, 1989

Electromigration and Electronic Device Degradation; Christou, 1993

Hot Carrier Effects in MOS Devices; Takeda, et al, 1995

Instabilities in MOS Devices, Davis, 1981

Semiconductor Physics and Design

Physics of Semiconductor Devices; Sze, 1981

Semiconductor Memories; Sharma, 1997

Bipolar and Analog Integrated Circuit Design; Grabene, 1984

Introduction to VLSI Silicon Devices; El-Kareh and Bombard, 1986

Physics and Technology of Semiconductor Devices; Grove, 1967

Principles of CMOS VLSI Design; Weste and Eshragian, 1993

SEMINARS/WORKSHOPS/MULTIMEDIA

EDFAS-sponsored videotapes/tutorials
ISTFA Seminars – Nov.
IRPS Workshops – Apr.
Sandia National Labs FASTAdvice
Semiconductor Picture Dictionary – DM Data
User Groups – ISTFA/IRPS/SEMATECH

SELECTED WEBSITES (examples only)

Organizations/Conferences/Universities/Magazines

EDFAS: www.edfas.org
ESREF (Conference): www.iae.dtu.dk/ESREF98
Int'l Reliability Physics Symposium: www.irps.org
Int'l Test Conference: www.itc.org
National University of Singapore/Center for
    Failure Analysis: www.cicfar.ee.nus.sg
SEMATECH: www.sematech.org/public/home.htm
Semiconductor International: www.semiconductor.net/
    semiconductor/lores/index.html
SIMS Webserver: www.simsworkshop.org/default.nclk

Analytical Labs (non-comprehensive listing)

Accurel Labs: www.accurel.com
Argonne National Laboratory Microscopy Site:
    www.amc.anl.gov
Charles Evans Labs: www.cea.com
Hi-Rel Labs: www.hrlabs.com
IBM Analytical Services: www.chips.ibm.com/services/
    asg/index.html
Integrated Circuit Engineering: www.ice-corp.com/ice/
    index.html
Rome Labs: www.rome.iitri.com/RAC
Sandia National Laboratoriies: www.sandia.gov/eqrc
Semiconductor Insights: www.semiconductor.com
Southwest Research Institute: www.swri.edu
Surface Science Labs: www.surface-science.com

Instrument Web Sites (non-comprehensive listing)

Test/Functional Characterization

Hewlett-Packard: www.tmo.hp.com
IMS: www.ims.com
MosAid: www.mosaid.com
Schlumberger Technologies: www.slb.com/ate

Electrical Characterization/Probing

Cascade Microtech/Alessi: www.cmicro.com
Hewlett-Packard: www.tmo.hp.com
Keithley Instruments: www.keithley.com
Lucas Signatone: www.signatone.com
MicroManipulator:
www.micromanipulator.com
Micron Force Instruments:
www.micronforce.com
(Karl) Suss: www.suss.com
Tektronix: www.tek.com
Temptronic: www.temptronic.com
Ultratest International: www.ultratest.com
Wentworth Labs:
www.wentworthlabs.com/co.htm

Chemicals/Fume Hoods

Alfa-Aesar: www.alfa.com
Cole-Parmer: www.coleparmer.com
Labconco: www.labconco.com
Kewaunee: www.kewaunee.com

Plasma/RIE/Evaporation

Balzers: www.bps.com
Denton: www.denton.com
Gatan: www.gatan.com
Leibold: www.leibold.com
March Instruments Corp: www.plasmod.com
Plasma-Therm: www.plasmatherm.com
Raines Technology: www.rainestech.com
STS/ElectroTech: www.stsystems.com
Trion Systems: www.triontech.com

Decapsulation/Sample Preparation

Allied Hi-Tech: www.alliedhightech.com
B&G International: www.bgintl.com
Buehler: www.buehlerltd.com
Sagitta: www.sagitta.co.il
SELA: www.sela.com

Package Inspection

Fein Focus: www.feinfocus.com
Nicolet: www.nicolet.com
Sonix: www.sonix.com
Sonoscan: www.sonoscan.com

Microscopes (Optical)

A-Zoom: www.a-zoom.com
Bausch & Lomb: www.bausch.com
Leica: www.leica.com
Mitutoyo: www.mitutoyo.com
Nikon: www.nikonusa.com
Noran: www.noran.com
Olympus: www.olympus.com
(Carl) Zeiss: www.zeiss.com

Electron Microscopes/EDX Systems

JEOL: www.jeol.com
Hitachi Instruments: www.hitachi.co.jp
Kevex: www.kevex.com
Leo Electron Microscpy, Ltd:
www.leo-em.co.uk
Noran: www.noran.com
R. J. Lee Instruments: www.rjleeinst.com

PhotoEmission Microscopes

Hypervision: www.hypervisioninc.com
Hamamatsu: www.hamamatsu.com
QFI: www.quantumfocus.com

E-Beam /Laser Probers

Schlumberger Technologies: www.slb.com/ate

Scanning Probe Microscopes

Burleigh Instrument: www.burleigh.com
Digital Instruments Corp.: www.di.com
Molecular Imaging: www.molec.com
Technical Instruments: www.techinst.com
Zygo: www.zygo.com

Focused Ion Beam Tools

FEI Corp.: www.fei.com
Micrion Corp.: www.micrion.com
Schlumberger Technologies: www.slb.com/ate
Seiko: www.seiko-usa.com

CAD Navigation Software/Hardware

Kensington Labs: www.kensingtonlabs.com
Knights Technology: www.knights.com
Raith GmbH: www.raith.de
Schlumberger Technologies: www.slb.com/ate

Lasers

New Wave: www.new-wave.com
Revise: www.revise.com

Magnetic Imaging Systems

Digital Instruments Corp.: www.di.com
Neocera: www.neocera.com

DISCLAIMER

The above listings are not intended to be all-inclusive, nor to represent any ranking or preference for any individuals, organizations, laboratories, and/or vendors listed. The lists are merely intended to be a starting point for further investigation into the opportunities for education, tools, and services available in the field. Neither the authors nor EDFAS have any affiliation nor interest in the above organizations or companies, other than those already disclosed, nor do they herein endorse or recommend any particular organization, tool, or company.

Afterword

The New Millennium:
Challenges and Opportunities

Christian Boit
Infineon Technologies AG

Abstract

The question is: Can Failure Analysis keep pace with the advances in semiconductor technology? The answer is yes from a technical viewpoint, if some new recipes for new materials and packages can be established, but the realization will require a revolution in failure analysis (F/A) in terms of methods, skills and tools.

From a business point of view, the answer is only yes if F/A can offer adequate results for reasonable costs in order to meet the cost reduction requirements of semiconductor products.

With all the innovations in F/A necessary to meet semiconductor development requirements, F/A becomes a relevant cost factor in IC business. F/A road maps are discussed for their implications on investment, required skills of the analyst, lab organization and time to result.

This presentation concludes that the resulting cost explosion in F/A cannot be compensated by any conceivable measures to enhance F/A productivity. This holds even under the assumption that a rising number of today's F/A problems will be solved by modern testing techniques.

Several strategies to reduce F/A costs while keeping techniques state of the art are presented. The most important approach is to integrate F/A in semiconductor product and technology development and introduce it as part of their projects. Only this way F/A can deliver custom-taylored solutions that are optimized in productivity and time to result.

1. Introduction

Readiness of F/A for new technologies, as laid out in development roadmaps, results in the implementation of new methods, tools, routines and consequences for lab organization if obsolete techniques go out of business.

Together with the requirement of ultra-short cycle times, this means an explosion of investment, skills, and training, in other words cost. The expense of F/A now becomes a serious factor in semiconductor business. F/A is not a playground for some unworldly scientists anymore (what critics of F/A often suggested in the past). The predicted increase of method development, investment and of analyst hours/result, means a multiplication of F/A budgets. Cost in F/A starts to really matter. F/A management needs to face the fact that analysis productivity will be a key issue for product cost reduction.

This is not a new idea. W. Maly pointed out in ISTFA 97 (ref. 2) that F/A becomes such a substantial cost factor in yield learning that testing must be empowered to do the F/A job as well (which he called rapid F/A). This may work to a large extent for yield learning, but despite all the testing-based rapid F/A that is in sight there still remain additional requirements to an F/A.

This presentation highlights the productivity implications that hide behind F/A roadmaps. It is not a complete productivity analysis, just a step into this direction.

2. Correlation of IC Technology and F/A Roadmaps

Technology development and reverse engineering or F/A use a separate set of techniques. In consequence, the SIA technology roadmap (fig. 1, ref. 1) and an F/A roadmap do not have necessarily much in common. In fact, if the Sematech F/A roadmap (fig. 2, ref. 3) is taken as a reference here, for more than 50% of the 17 F/A topics that are listed in fig. 2 there is no change specified between 1995 and 2007. The complete section of defect tracing does not seem to undergo significant development in that period of time.

The reason for this lack of correlation is that technical

Key failure analysis drivers extracted from the SIA roadmap					
Intro to mfg. Date	1995	1998	2001	2004	2007
Primary roadmap drivers					
Minimum feature size, µm	0.35	0.25	0.18	0.12	0.1
Memory size (DRAM), MB	64M	256M	1G	4G	16G
Gates per chip	800K	2M	5M	10M	20M
Interconnect levels	4 to 5	5	5 to 6	6	6 to 7
Max power – hi -perf., W	15	30	40	40-120	40-200
Max power – portable, W	4				
Min supply voltage, V	2.2		1.5		
Number of I/O-pins	750	1500	2000	3500	5000
Speed– off chip, MHz	100	175	250	350	500
Speed– on chip, MHz	200	350	500	700	1000
Chip size, Logic, mm sq.	400	600	800	1000	1750
Chip size, DRAM, mm sq.	200	370	500	750	1000
Wafer diameter, mm	200		300		400
Additional drivers					
Package processor (leading)	Flip-chip				
Package memory (leading)	DIP	Cube?/Others			
Gate ox thickness T-eff, nm	7-12	4-6	4-5		< 4
Metallization	Al		Al, Cu		
LD-oxide	Air, polyimide, low dielectric				
Design – database	Flat		Hierarchical		
Design – model	Behavioral		Symbolic		M-block
In-lab logic testing	Low cost logic testers + IDDQ, BIST				

Fig. 1: SIA Roadmap (from ref. 1)

F/A-Step Localize/Characterize Faults					
Capability	1995	1998	2001	2004	2007
Intelligent fault localization	Software traceback				
		Fault simulation s/w f. combination logic			
	Increasing level of designed –in diagnosability (DMA, Scan, BIST etc.)				
Node-measurement	Mechanical probers				
	Electron beam based tools				
		Scanning Probe testers, backside msmt.			
Thermography/ mapping	Frontside heat sensing				
		Hi resolution backside thermal mapping			
Light sensing/ mapping	Visible emission micr.				
		Infra red emission microscopy			
Micro-surgery	Focused ion beam tools				

F/A-Step Defect-Tracing/Sample Preparation					
Capability	1995	1998	2001	2004	2007
Delayering	Reactive ion etch/plasma based etch tools				
	Wet etch techniques				
	Mechanical polishing				
Delid/Demount	Mechanical				
	Wet/dry etch				
	Die removal jigs and tools				
X-Sectioning	Focused ion beam tools				
	Mechanical cross sectioning				
Sample prep for analysis tools	Plasma coaters, focused ion beam tools				
Assembly integrity Verification	X-ray radiography				
	Acoustic microscopy (Delamination only)				
Positioning on analysis site	Laser markers, CAD-based navigation				

F/A-Step – Physical/Chemical Characterization					
Capability	1995	1998	2001	2004	2007
Thin film delamination	Acoustic microscopy				
Optical imaging	Optical microscopy				
	Confocal microscopy				
			IR optical laser scanning scopes		
Surface imaging	Scanning electron microscopy				
		Scanning probe techniques			
Elemental analysis	X-ray analysis				
		Auger electron spectroscopy			
Structural characterization	Transmission electron microscopy				
Bulk analysis	Chem. lab				

Fig. 2: Sematech F/A Roadmap (from ref. 2)

revolutions in F/A come in two flavours: one is driven by the semiconductor development requirements, like analysis from the chip backside, the other is driven by the progress of analytical tools, independent of the F/A demand by semiconductor development. The onset of professional use of FIB or more recently Scanning Probe Techniques, two revolutionary techniques, would have occurred as soon as the tools hit the market.

This independence of F/A development from semiconductor technology development explains why it is tough to justify the investment of a new machine. Often a hard criterion which new analysis tools are required for the next technology generation is not easy to define.

3. Innovation: Readiness of F/A for New Technologies

In the Sematech roadmap, the F/A techniques are listed without exact recourse to the failure analysis drivers extracted from the SIA roadmap.

In order to discuss the consequences of semiconductor technology development on the F/A process, let's go back to the SIA roadmap, highlight major F/A innovation drivers and correlate them with technical action items.

In general, the closer today's IC technology operates to the limits of scientific laws the more specific rather than general are the solutions that R&D offers, and in turn the degree F/A can profit from such solutions is decreasing. Accordingly, new generations of F/A tools will be increasingly specific. This means, more tools are required to cover a general F/A approach.

Feature size starts to have a strong impact on the application range of optical microscopy:

With SEM established as an integral part of microelectronics F/A by the mid-eighties, optical microscopes have remained a key for rough identification and preparation process control as long as the polysilicon and metal lines or at least their pitches could be resolved. These days, we need to introduce UV, confocal or immersion techniques to push optics to our demands or find workarounds like low-vacuum SEM, but main-road technology developments like phase-shift or other lithography tricks will not find their ways into F/A.

Inline physical failure analysis on productive wafers based on FIB needs to replace wafer-sacrificing techniques for technology monitoring with growing wafer size.

CAD Navigation and S/W network on all stages in F/A is a condition sine qua non feature for F/A of multi-level metallization (MLM) and complicated delayering and localization jobs.

Planarization as required for MLM makes circuit modification with FIB very complex and time consuming for small feature sizes due to the orientation problems of ion beam imaging on surfaces that lack topology (fig. 3).

New Packages like BGA, CSP and Flip Chip require new recipes for local decapsulation, from top or backside, if a decapsulation is possible at all.

Some of the packages may refrain from local decapsulation access at all which is still a big question mark for F/A of the future. However, all these new packages were already out of the range of F/A if not a new localization approach had conquered the world of semiconductor F/A:.

Failure localization from the chip backside would not have been emphasized as a development need without semiconductor technology development of flip chip packages. Now, with a set of methods for that task at hand, it almost seems like this has always been the localization technique of our desire: Free optical access to the active device level and the lower metallization levels - very clear detection of the failure origin (fig. 4).

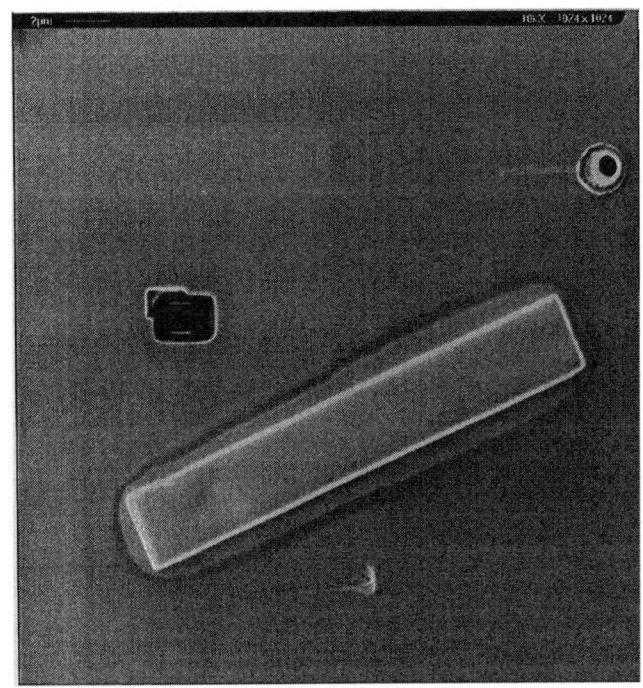

Fig. 3: FIB image for circuit modification of chip from previous technology with topology (lower left) and of chip from recent technology with CMP (above) (from ref. 4)

High Frequencies will become more and more manageable for F/A localization by concentrating on optical methods from the chip backside that are able to handle a much higher bandwidth than E-Beam Probing does.

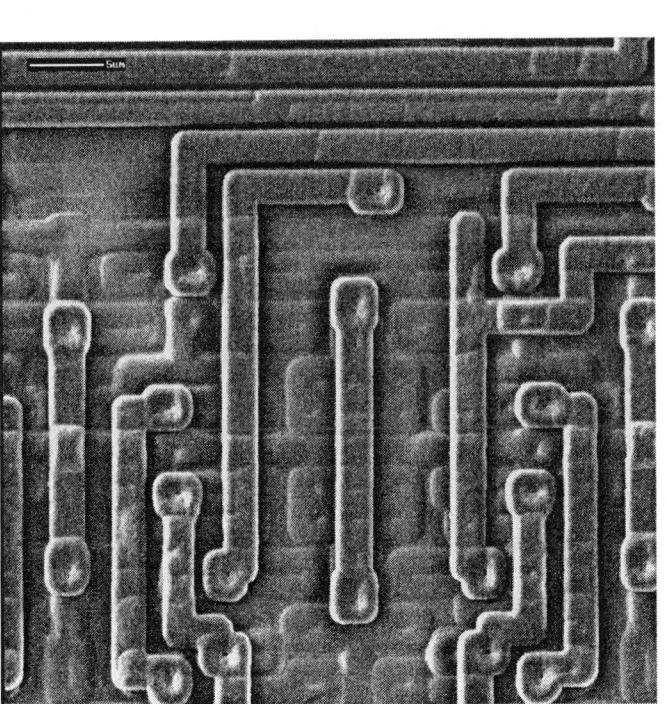

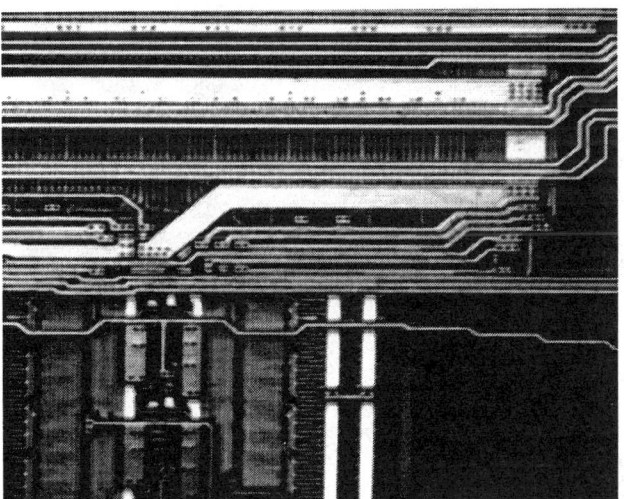

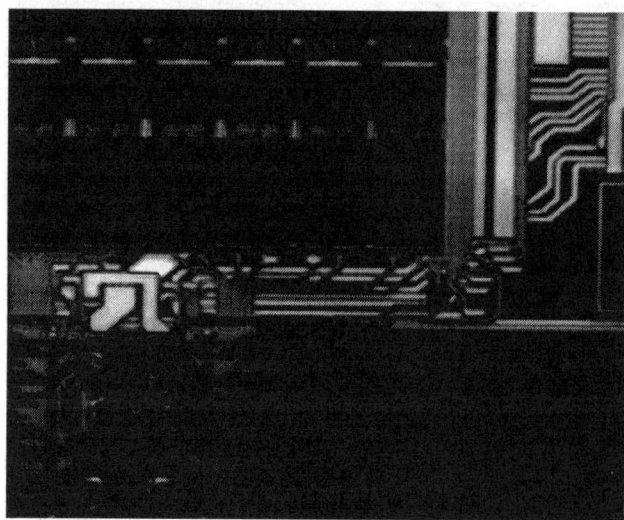

Fig. 4: Optical micrograph (100x) of DRAM
top: frontside; bottom: backside (from ref. 5)

Low Supply Voltage will shift Photoemission signals to the IR-regime. Tools for analysis from chip backside also need to go in that direction.
New Materials like Copper, di- and ferroelectrics require new recipes but probably not many new tools in F/A.

4. Complexity: Techniques that go out of Business – Challenges for lab organization

On the other end of the roadmap (fig. 2), 6 out of 33 techniques are listed to go out of business for F/A on leading edge IC technologies until 2001: Software trace back, mechanical probers, frontside thermography, frontside emission microscopy, x ray material analysis and optical microscopy. This seems to be a small number. However, they belong to the most general techniques used in F/A. When they will be replaced by tools for more specific applications the toolset will grow substantially.

In the past, new techniques in F/A usually were established as add-ons to the existing process flow in the lab. It took some time of the most experienced and skillful engineers to start practise, develop routines for the tools and integrate the method in the F/A process flow. Workload volume ramp up on new methods used to be slow.

Now, for small feature sizes or analysis from chip backside, new techniques must be ready for volume workload from the beginning and, with the established methods to become obsolete, the F/A staff needs to gain flexibility and skills to manage the new analysis process flow, which is also accompanied by a trend to a larger portion of engineers.

This is the bulk of enhanced F/A complexity. It is not the introduction of complex methods alone, it is the shift of the major workload to complex processes.

5. Productivity: Innovation as a Cost Factor

In summary, F/A roadmaps show the change of the techniques in order to fulfill the technical demands of technology development. However, there is another goal of semiconductor development that seems to point in the opposite direction: cost reduction. For this reason, it is

F/A Support for Wafer Manufacturing

FA-requirement	Semiconductor Feature size	topic Wafer size	New materials
Key features	Lateral Resolution	Inline F/A on productive wafer	Copper Dielectrics Ferroelectrics
High end tools	Sophisticated microscopy, SEM,FIB	Dual Beam FIB inline	RIE w/ high plasma density
Obsolete tools	Optical microscopy	Cleave	
Skills	All imaging tools	Dual Beam FIB, contamin.	New recipes
Time to result	No difference	x0.5 feedback to fab engineer x2 F/A work	No difference
Cost	Invest UV/LSM Training x3 in 1st year x1.5 next year	Invest DB FIB Training x2 in 1st year Saving: wafers productive	x1.3 in 1st year

FA-requirement	Semiconductor Planarization	topic Defect monitoring	Yield improvement
Key Features	Roughness, Resolution	Inline inspect. on productive wafer	Automated F/A
High end tools	AFM and Derivatives	DB FIB, TOFSIMS	Rapid Testing
Obsolete tools			
Skills		DB FIB, contamination	
Time to result		x0.5 feedback to fab engineer x2 FA work	x0.2
Cost	Invest AFM $ 0.3 M Training 0.3 MY	Invest DB FIB $ 2 M Training x2 in 1st year Saving: wafers productive	5 engineer MY experience in generation of Rapid Testing Routines

Fig. 5: Cost of F/A innovation

important to investigate the cost of an F/A that is ready for the technological challenges. Only when the F/A roadmap complies with the cost reduction milestones, it can really keep pace with IC technology development.

To achieve that, here is in addition to the road maps a different approach:

For the main technology challenges highlighted in section 3 the F/A requirement is listed with all the consequences in investment, techniques out of business and skills. It allows a rough estimation of expenses and time to result that an adjustment of F/A to the respective technology would imply. The factors in the time to result and cost rows of fig. 5 and the following estimates are only rough approximations to give just a taste of the order of magnitude F/A is going to deal with.

F/A Support for Product Development

FA-requirement	Semiconductor topic New packages	MLM	Low Voltage
Key features	Partial decapsul. Backside-localization. CAD navigation New SAM approach	Backside localis. (see new packages), Parallel polish	EMI detection
High end tools	Backside EMI w/ IR-detector Laser absorption LSM, PICA LIVA, TIVA OBIRCH etc	FIB w/ CAD for rewiring failing circuitry after delayering	EMI w/ IR-detector
Obsolete tools	E Beam Prober, Direct Probing		EMI of visible light
Skills	Prep. tailored to product, In depth know how of localis. + Si/light interaction	In depth know how of localization and circuit modification	Same
Time to result	x3 if F/A lab serves large product variety	x2 of regular analysis	Same
Cost	Invest from EM-upgrade $ 15 k to $ 2 M (Laser absorption) 3-4 skilled engrs. 1-2 years x3 of topside analysis	1-2 skilled engineers, not for operators much FIB workload, x1.5 of regular analysis	Invest $ 0.5 M

FA-requirement	Semiconductor topic High frequency	Planarization	New Materials
Key features	Localization up to 1GHz	Circuit modification	Circuit modification
High end tools	Laser absorption PICA etc	FIB w/ CAD-navigation	New recipe
Obsolete tools	E Beam Prober	FIB w/o CAD	?
Skills	Experienced engineers	Experience w/ alignment	?
Time to result	x2, complex methods	x5 at start, x2 after 1 year	?
Cost	Invest $2 M/ tool 1 MY to routine	x2-3 workload	?

If all the innovation items are going to be accomplished within the timeframe the SIA roadmap suggests there is an invest of necessary tools of ca. $5 million for a wafer manufacturing F/A and ca. $10 million for a product supporting F/A until 2001. The time to result per analysis will rise by about x2-3,

A shift of the employees to engineers in the order of 20% and time for method development, evaluation and lab maintenance mean a budget explosion for F/A in 2001 of about 100%. Even taking into account productivity gains in F/A due to improved routines etc. of 20% per year, complexity of analysis will overcompensate the cost reduction potential of F/A itself.

6. Strategies for Cost Reduction

One effect of semiconductor technology development is cost reduction. This does not hold for failure analysis. The technological progress in semiconductor industry causes in turn a cost explosion in F/A if not a number of measures are taken against this. Also the Sematech roadmap postulates: Breakthroughs are needed for cost efficiency. With productivity now established as a key issue in F/A, the main cost factors should be inspected:

- **Investment** will go up by estimated 30% minimum: To justify the increased depreciation the uptime rate is a key from the beginning. In order to load the tools to capacity shared investments or F/A groups with a large range of customers contribute to a competitive F/A.

- **Complexity** requires experienced and skillful analysts that develop specific process flows for each technology, in long term personnel strategies this means a shift to a larger portion of engineers. Here, the most important productivity gain is a minimized time to result. Semiconductor companies that use external labs need to exchange layout data for CAD navigation. Otherwise, time to result and solution rate would be increasingly poor.

- **Lab organization** - The increased number of high end tools requires a joint lab for a broad set of customers to optimize productivity in terms of load of tools to capacity, maintenance and lab organization.

- **Staffing** becomes an even more essential asset of an F/A lab. All efforts need to be taken to keep a core team of skillful analysts - on the other side, with the focus on the motivation of engineers, F/A should be a well respected joint in the chain of semiconductor business in order to establish F/A as integral part of career planning.

- **Job handling** - tight interaction with the F/A customer helps the failure analyst focus on the issues: update of priority, reduction of the number of jobs by evaluation of failure analogies, analysis depth depending on likeliness of corrective actions etc. are checked in continuous control loops.

- **Electrical signature analysis** and other programs to get the most F/A information out of electrical testing need to be pushed to their limits: the remaining F/A jobs are time and cost consuming enough.

The F/A requirements of microelectronics business are a dramatic increase in performance for minimum cost, due to decreasing revenue per development cycle. For this purpose, the complete F/A logistics need a revision. F/A requires a new role in semiconductor industry.

Concentration of the F/A activities minimizes cost for investment and lab organization. All the other topics can only be reached with one strategy: full integration of F/A in IC development projects from early on. F/A will not survive as a separate service to semiconductor business but needs to cooperate with development projects in order to identify possible analysis showstoppers from the beginning, work jointly on solutions, workarounds and method developments, and have some productivity ramp up before the workload of the new analysis routines starts.

A full integration is established when:
- no technology project is released without evaluation of F/A risks
- no package is released without evaluation of F/A requirements on them
- no library is released that does not contain design features and rules for analysis.

These two approaches, concentration of the F/A groups and integration in semiconductor development projects, must be combined in F/A labs that will be competitive in future semiconductor business.

7. Conclusion: The Key for an F/A Ready for the Future at a Reasonable Price is Integration of F/A in Microelectronics Process Flow

Can F/A keep pace with IC technology development? Yes, the selected topics show it can be done, although new recipes are still pending for some new materials and local decapsulation of some packages. But, opposite to technology development, innovation in F/A does not mean cost reduction. The productivity gain produced by rapid failure analysis via testing and rationalization of analysis flow will be more than compensated by investment requirements, enhanced complexity and process steps per analysis. A growing portion of the F/A work needs to be performed by experienced engineers.

Key for successful F/A is not only the art of analysis but more and more the art of cost containment. At upcoming technologies and products, this goal can only be achieved with concentration of F/A in big units and simultaneously full integration of F/A into microelectronics R&D projects.

Acknowledgement

I want to thank all my colleagues in the world-wide F/A community for the continuous exchange of improvements and visions; my management that always focused me on productivity issues and helps integrating F/A in the product process flow; our happy and unhappy customers at Infineon Technologies for their feedback and all the skilled failure analysts at Infineon Failure Analysis who create our innovation.

Many thanks go to Wiltrud Golombek for the processing of words and figures and to Bernd Ebersberger for critical revision of the manuscript.

References

[1] The National Technology Roadmap for Semiconductors – *Semiconductor Industry Association, 1994*

[2] Testing-Based Failure Analysis; W.Maly; Proc. 23rd *ISTFA 1997*, p.3

[3] Sematech Failure Analysis Roadmap; *Future Fab International 1997*, p.237

[4] Layout Overlay Techniques to Improve Failure Analysis; C. Burmer, S.Görlich, S.Pauthner; Proc.24th *ISTFA 1998*, p.365

[5] Flipchip and "Backside" Techniques; K. Bernhard-Höfer and D. Barton; Tutorial at 24th *ISTFA 1998*, to be published in Microelectronic Failure Analysis Desk Reference 1999, ASM International

Appendixes

A Review of Wet Etch Formulas for Silicon Semiconductor Failure Analysis

Thomas W. Lee
Motorola Inc.
Austin, Texas

Abstract

WET ETCHING is an important part of the failure analysis of semiconductor devices. Analysis requires etches for the removal, delineation by decoration or differential etching, and study of defects in layers of various materials.

Each lab usually has a collection of favored etch recipes. Some of these etches are available premixed from the fab chemical supply. Some of these etches may be unique, or even proprietary, to your company.

Additionally, the lab etch recipe list will usually contain a variety of classical "named etches". These recipes, such as Dash Etch, have persisted over time. Although well-reported in the literature, lab lists may not accurately represent these recipes, or contain complete and accurate instructions for their use. Time seems to have erased the understanding of the purpose of additives such as iodine, in some of these formulas.

To identify the best etches and techniques for a failure analysis operations, a targeted literature review of articles and patents was undertaken. It was a surprise to find that much of the work was quite old, and originally done with germanium. Later some of these etches were modified for silicon. Much of this work is still applicable today.

Two main etch types were found. One is concerned with the thinning and chemical polishing of silicon. The other type is concerned with identifying defects in silicon.

Many of the named etches were found to consist of variations in a specific acid system. The acid system has been well characterized with ternary diagrams and 3-D surfaces. The named etches were plotted on this diagram. The original formulas and applications of the named etches were traced to assure accuracy, so that the results claimed by the original authors, may be reproduced in today's lab.

The purpose of this paper is to share the condensed information obtained during this literature search. Graphical data has been corrected for modern dimensions. Selectivities have been located and discussed. The contents of more than 25 named etches were spreadsheeted.

It was concluded that the best approach to delineation is a two-step etch, using uncomplicated and well-characterized standard formulas. The first step uses a decoration or differential etch technique to define the junctions. Formulations for effective decoration etches were found to be surprisingly simple. The second step uses a selective etch to define the various interconnections and dielectric layers. Chromium compounds can be completely eliminated from these formulas, to meet environmental concerns.

This work, originally consisting of 30 pages with 106 references, has been condensed to conform with the formatting requirements of this publication.

INTEREST in the subject of this paper grew from a conversation in a lab, during which an analyst related that the name "Dash Etch" was derived from the manner in which it was used. That is, the etch solution was quickly "dashed onto" the sample with a cotton swab.

In actuality, the etch is named for its developer, W.C. Dash[1,2,3,4]. Dash Etch is an old formulation, originally developed for germanium devices[31] in the 1950's, then modified for silicon.

Every lab has a list of etches and their components. When mixing some Dash Etch according to your lab's etch list, are you mixing the Ge formula, the Si formula, or an "adjusted" version ?

These considerations suggest that accurate information on etches may fade with time, and a study of them was in order. Reviewing the original work can help eliminate confusion and errors.

ETCH RATE data is easy to find. Device production processing requires the maximization of etch rates. The etch rate for thermal oxide is the usual standard, since its composition is fairly predictable.

Etch selectivity is the property of a chemical etch that both process and failure analysis engineers strive to characterize and understand. Selectivity arises from etch rate differentials. If we are to effectively decorate or delineate device structures in cross-section, we need to identify the most selective etches.

Etch selectivity data is difficult to find. This is especially true if seeking comparative etch rates for a broad spectrum of differing materials. Deviations in doping concentration, composition, or temperature, will cause the etch rate for a material to be different from the thermal oxide rate. When performing an analysis, these numbers are generally not easily available.

If we were to perform an experiment to find the best and most selective etches for all common silicon semiconductor materials, the matrix would be unreasonably large, for even the simplest of semiconductor device structures. This translates to time-consuming and expensive work. It follows that selectivity data is very valuable. Without reliable selectivity numbers, the analyst will be forced to guess the etch times, risking the ruin of samples through an inappropriate etchant choice.

Review of the original characterizations of etches can supply numbers piecemeal. Some examples are included in this paper. When constructing etch selectivity data from these comparative rates, we have to be careful to be knowledgeable of the factors which influence the rates we would compare. Such research efforts can help eliminate uncertainties, and speed the etch selection process for particular labs, devices, and material combinations. Defect identification, where necessary, may be more accurate. By eliminating uncertainties due to poor etch selectivity, measurement of critical dimensions (C-D's); junction depths, gate length, oxide thicknesses, and linewidths, may all be more accurate.

Silicon Etching Systems

HYDROFLUORIC ACID, HF, will dissolve glass, or silicon dioxide, SiO_2. However, HF does not dissolve silicon. Silicon spontaneously forms a layer of oxide on its surface a few molecules thick, known as *nascent oxide*. This oxide will dissolve in HF, but the reaction will then stop. The surface of the silicon must be further oxidized to SiO_2 before the reaction can start again.

NITRIC ACID is the most common **oxidizer**. Other oxidizers used in etches are hydrogen peroxide, chromic acid, potassium dichromate $(K_2Cr_2O_7)$ and potassium permaganate, $(KMnO_4)$. If only an oxidizing agent is placed on bare silicon, an SiO_2 layer quickly forms. However, the resulting oxide layer is insoluble in the oxidizer solution. The oxide layer is thick enough to prevent diffusion of the oxidizer to the silicon surface, protecting it, and the oxidizing reaction will automatically stop. This oxide layer is a type of **passivation**.

When both an oxidizing agent and HF are present in the etch solution, a two-step reaction takes place. The oxide is dissolved by HF. In this process of changing SiO_2 into a soluble anion, HF is known as a **complexing agent**. After the SiO_2 is dissolved away, the freshly exposed silicon surface is re-oxidized by the oxidizer. This process will continue until the acids are depleted, or the silicon is entirely dissolved.

Robbins and Schwartz[5], and Turner[30] suggest the following reaction:

$$3Si + 18 HF + 4 HNO_3 = 3H_2SiF_6 + 4NO + 8H_2O$$

ETCH RATE CONTROL is achieved through the use of a **buffer**, or **diluent**, or **moderating agent**[7]. These terms all seem to be used interchangeably, but inaccurately, for this etch component.

DILUENTS are etch components which control the reaction rate by reducing the concentrations of kinetically important species. Robbins and Schwartz consistently to refer to acetic acid and water as diluents. Note that **HAc** and **HOAc** are not chemical formulae, but acronyms for acetic acid. Glycerin and ethylene glycol are additional examples of diluents.

BUFFERS are used to maintain a constant concentration of a species in solution, and to control pH. The fluoride ion can quickly become depleted in a dilute HF etch. Adding NH_4F to the HF, buffers the etch by making additional F ions available for reaction. This maintains a more constant etch rate for the bath. This is very important in processing, because the bath has to be drained and refilled less often. Spiking of the bath to restore its proper concentration, becomes unnecessary. *A dilute HF etch is not a buffered etch.*

Also note that **BHF** is not a chemical formula, but an acronym for "buffered HF". BHF and BOE, an acronym for buffered oxide etch, are the same thing.

The term "motoring agent" was coined by Mills[32,35] and professional collaborators. However, most authors in the etch literature, along with standard chemical dictionaries, prefer to use the term **moderating agent** interchangeably with "diluent". A typical example is Table One in Wang's paper[25].

The HF-HNO₃-H₂0 / HC₂H₃O₂ system

Robbins and Schwartz work stands out in the literature, as they have very thoroughly characterized the etch rates of this acid system for silicon.

They were investigating some peculiar behaviors in this system. The first was to gain an understanding of the purpose of *"acetic acid, bromine, and heavy metal salts"* added to the named etches, which were already in existence at the time.

Secondly, Robbins and Schwartz[5,6] were also motivated by the peculiar and complex manner in which "white etch" behaves. The solution gives off red fumes during etching. The reaction rate tends to accelerate with time. The addition of water beyond a certain critical amount, or stirring the solution, will suddenly stop the reaction. If no reaction is taking place, "seeding" or "catalyzing" the solution with a small piece of wafer having a rough surface, will start the reaction on the smooth wafer.

Robbins and Schwartz' first paper contains very useful graphs showing the etch rate of the system HF-HNO₃-H₂0 versus HNO₃ concentration and added water. They also show the effects of crystal orientation, temperature, etc.[7] Fig 1 is a typical example.

They comment, "...differently oriented surfaces, as well as *p*-(type) and *n*-type silicon of different resistivities, *all appeared to etch at the same rate , whereas large differentials in the etch rate of these surfaces are generally observed in other etching systems".*[7]

Increasing the temperature of an HF-HNO₃ etch from 10°C to 50°C roughly doubles the etch rate, independant of the acid ratios used.

Figure 1. In their paper, Robbins and Schwartz show that the etch rate of silicon at any HF:HNO₃ ratio, increases linearly with temperature.[7]

Their fourth paper[8] , the result of more research, recognized that this acid system can accentuate (like defect etches) or ignore (like chem polishes)

the crystal structure of the silicon, *depending on the etch composition..*

Robbins and Schwartz present a ternary diagram for the system HF-HNO₃-H₂0 / HC₂H₃O₂, shown in Fig 2. Constant etch rate contours are plotted, and the units have been converted from mils/min to uM/sec. This diagram allows us to plot the named etches. We may then compare their composition and behavior with that found by Robbins and Schwartz for etch mixtures with no additives. Note that the ternary diagram suggests that use of acetic acid as a diluent allows more control over the etch rate at low concentrations. Acetic acid etches are not stable at elevated temperatures, however.

We may conclude some important things from this plot. In general, fast etches don't show defects, and slow ones do. That is, the chem polishes are near the HF-HNO₃ axis of Fig 2, and the defect etches are nearer to the diluent corner of Fig 2.

Diluent choice affects the etch rates. Note that H₂0 and HC₂H₃O₂ -diluted etch rates are plotted separately. Note that CP-4, Dash, Hg etch, and 10-1-7 use CH₃COOH diluent. Also note that 100 to 1, 50 to 1 Stain, and WAg etch use H₂O as the diluent. The white etch family and 20:1 stain have no diluent at all.

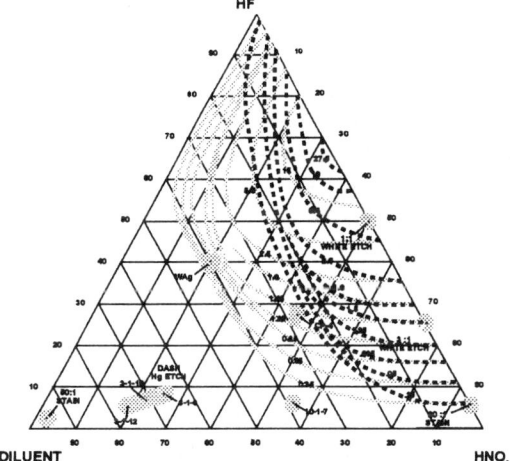

Figure 2. Ternary diagram, with some of the more popular chem-polish and delineation etches superimposed. Adapted from Robbins and Schwartz[7].

In their fourth paper, Robbins and Schwartz summarize the acid system's behavior, shown in Fig 3. It is disappointing to have never seen this graph in an FA lab etchant collection. This work stands as the ideal model for other etch rate characterizations.

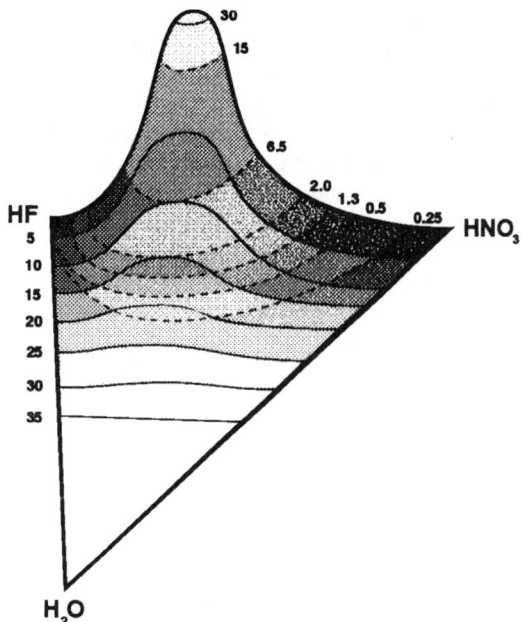

HF 5 10 15 20 25 30 35

HNO₃

H₂O

Figure 3. Etch rate surface as a function of etch solution composition in the 48% HF-70% HNO₃ system. Scale has been converted from mils/min to uM/sec. (Adapted from Robbins and Schwartz[8]).

The sobering fact is, that this plot is for *silicon etching only*. If you would like results comparable to Fig 3 for, say BPSG, you would have to repeat Robbins and Schwartz' work at many acid compositions, and for many glass compositions and temperatures. The result would be more useful than the usual piecemeal data, to which Robbins and Schwartz sarcastically refer. However, such a project is a major research effort, requiring a significant allocation of funding and time.

These facts help us realize the utility and economy of gleaning selectivity data from the literature, wherever it happens to occur.

Oxide etching with HF-HNO₃-H₂O mixtures is very dependent upon the doping of the oxide. Notice in figure 4 that the etch rate of the phosphorus-doped oxide is orders of magnitude faster than with undoped oxide or thermal oxide. The rate for Si in Fig 2 is very small. This difference will result in selectivity of the etch to silicon and various types of oxide.

Consider the data collected by Beadle et al[19] from the work of Pliskin[29], for etch rates of various materials in "p-etch" (plotted in Fig 2), consisting of 15 parts HF, 10 parts HNO₃, and 300 parts H₂O. The rates for 15:1 HF are also given:

Material	P-etch rate	15:1 rate
thermal oxide	1-2 A/sec	160 A/sec
14% B-doped oxide	3 A/sec	
pyrolitic oxide	13 A/sec	
34% B-doped oxide	50 A/sec	
TEOS oxide densified		260 A/sec
P-doped SiO₂	500 A/sec	
B-doped thermal ox		1000 A/sec
TEOS oxide		2000 A/sec
P-doped TEOS oxide		10000 A/sec
P-doped thermal ox		20000 A/sec

$$SiO_2 + P_2O_5 = 1.8 \text{ A/sec}$$
$$SiO_2 = 570 \text{ A/sec}$$

Figure 4. Illustration of the dramatic difference in etch rates of doped and undoped oxides with "p-etch". Similar selectivity occurs with BOE. After the much-quoted work of Pliskin and Knall.[29]

The Named Etches

ETCHES WITH ADDITIVES based on the HF-HNO₃-H₂O / HC₂H₃O₂ system, abound in the literature.

Many of the better-known etches are discussed in detail in the following paragraphs. Etches containing additives may have been named for their proponents or inventors; Sirtl is an example. They may have been named for their function; pinhole etch is an example. They may have been named to honor the company in which they were developed; RCA Clean and WAg Etch are examples.

It should become apparent that there is considerable similarity among many of the named etch formulas. For example, in the spreadsheet, note the large quantity of named etches which are based on the HF-HNO₃-H₂0 / HC₂H₃O₂ acid system. Consequently, the named etches have been plotted on the ternary diagram in Fig 2 to illustrate that their behavior is largely influenced by their position in the HF-HNO₃-H₂0 / HC₂H₃O₂ continuum.

Many chemicals are used in the named silicon etches. Robbins and Schwartz conclude in their fourth paper[8], that the addition of Br in CP-4 etch, a chemical polish; and the copper in Purdue etch[34], a decoration etch, was "pointless". Both of these statements turn out to be true, *but only for silicon*. Heidenrich explains the reason in his patent.[41]

Finally, Robbins and Schwartz[8] emphasize their point unmistakably by stating: *"...The diversity of circumstances under which the etching is performed, coupled with the infinite variability of the etching formulas, has created a fertile field for the sorcerer and alchemist to practice their arts.*

It was inevitable that some excellent etching procedures should have been discovered in the course of time. However, these procedures were applicable only to the attainment of limited objectives with specific devices".

At some point or other, many of us have wondered about this. However, as we shall discuss, the judgment was somewhat severe.

NUMBERED ETCHES in the HF-HNO$_3$-H$_2$0 system are commonly encountered in FA labs. Although they are said to follow some kind of standard, none seems to have been written down. The convention varies among persons, labs, and papers. This is a source of confusion and error. Some insist that we list the ingredients of etches in the order; HNO$_3$-HF-H$_2$O/CH$_3$COOH. The formula for Dash Etch then would become "3-1-8" or "3-1-12".

However, Robbins and Schwartz list their etch ingredients in the order, "HF-HNO$_3$- H$_2$O" consistently throughout their four-paper series. Dash himself also follows this convention; that is, quasi-alphabetical. *So Dash would have called his own formula "1-3-12".*

Analysts who are mixing etches from a lab formula collection should be sure they understand the convention in use by whoever transcribed, the formulas into numbered recipes. *Remember that each of these etch formulas represent only one point* on the HF-HNO$_3$-H$_2$O ternary diagram., Fig 2.

Ge or Si, or both, may have been the semiconductor material of concern at the time the named etch was developed. The formula may or may not have been altered for use with silicon. Be sure that you have the Si-version, not the Ge-version, by looking up the original work.

One may think that the addition of Br$_2$, I$_2$, Hg, and other salts to the old Ge etch formulas, may be to speed up the etch when it is used for silicon. The opposite is true. Turner[30] reminds us that Robbins and Schwartz thoroughly characterized the etch rates of Si; and Cretella and Gatos[12] characterized the etch rates of Ge, both in HNO$_3$-HF-H$_2$O/CH$_3$COOH solutions. He presents their data on a single graph, showing that *Si etches about an order of magnitude* **faster** than Ge.

Detailed Etch Descriptions

CP-4 is a chemical polish etch, originally developed for germanium by R.D. Heidenreich, and covered by U.S. Patent 2,619,414.[41] The etch consists of 15 ml HF, 25 ml HNO$_3$, 15 ml CH$_3$COOH, and 0.3 gm Br$_2$ dissolved in the acetic acid. The Br$_2$ is present to speed up the etch, but it has a disadvantage solved with CP-4a. The composition of this etch varies in the literature. Vogel et al[14] used CP-4 etch for Ge in their paper in 1953. Their formulation consisted of 5 parts HNO$_3$, 3 parts HF, 3 parts CH$_3$COOH, and 0.1 part Br$_2$.

When CP-4 was used on silicon, it was modified to 3 parts HF, 5 parts HNO$_3$, and 3 parts CH$_3$COOH, and it would chemically polish (remove mechanical damage from) silicon in 2 to 3 minutes.

It was used to polish and etch on {100} and {111} surfaces, and to reveal sharp p-n junctions, grain boundaries, and twin boundaries. CP-4 was used for 1.5 minutes to etch, and 2 or more minutes for polishing. CP-4 etches Si at approximately 35 uM/min (350,000A/min) and thermal oxide at about 2750 A/min.

CP-4a has also been called CP-6 and CP-8. CP-4a is approximately the same formulation as CP-4, with the Br$_2$ omitted. One formula for CP-4a consists of 3 parts HF, 5 parts HNO$_3$, and 3 parts CH$_3$COOH.

This was done to eliminate the adsorption of Br$_2$ on the device surfaces. Bromine is a contaminant in many processes, adversely affecting contacts, carrier lifetime, and leakage *in silicon*. CP-4a etch reveals twins, p-n junctions, grain boundaries, and dislocations in 1.5 to 3 minutes. It has been used warm, up to 70°C.

Dash etch[1,2,3,4,31] was first used for etching out dislocations on Ge after precipitation of lithium. This Dash etch consisted of 2 parts HF, 4 parts HNO$_0$, and 15 parts CH$_3$COOH. It was used for 20 sec to 1 minute to reveal dislocations.

When etching silicon, Dash modified his etch formula[2,3] to 1 part HF, 3 parts HNO$_3$, and 8 to 12 parts CH$_3$COOH, and 1 to 16 *hours* were required to develop dislocation pits.

CP-I (Iodine etch) was developed as a chemical polish and etch for {100} and {111} surfaces on Ge by Wang[25] in 1958. He references patents, one by Ellis and one by Wolsky, both in 1956. His formula was 5 parts HF, 10 parts HNO$_3$, 11 parts CH$_3$COOH, and 30 mg I$_2$ crystals dissolved in the acetic acid. It is was applied for about 4 minutes. Holmes shows CP-I as being specifically for germanium

Mills and Sponheimer[32] used the iodine formula, and found excellent results with cross-sections having, "junctions, implants, oxides (thermal and deposited), nitrides, aluminum..., tungsten...., and polysilicon."

Mercury Etch was first described by Faust[18] and utilized by Vogel, et al[26,27] in their study of edge and screw dislocation etch puts in silicon. The paper is interesting, because Vogel actually heated and bent silicon crystals to affirm the defects identified by the etch.
When working with Ge in 1953, Vogel used "5 parts HNO_3, 3 parts HF, 3 parts CH_3COOH, and 1/10 part Br2."

In his 1956 paper on dislocation etch pits in Si, he chose a "polish-etch" also based on the HF-HNO_3 system. The formula consisted of "3 parts HF, 5 parts HNO_3, 3 parts CH_3COOH, and 2 parts of 3% $Hg(NO_3)_2$ solution."

Salier Etch is often misspelled "Sailor's Etch", both in lab etch formula collections, and in the literature. It is very similar to SD-1 etch. Salier Etch consists of 300 ml HNO_3, 600 ml HF, 2 ml Br_2, and 24 gm $Cu(NO_3)_2$, diluted 10:1 with water. The large volumes probably indicate it was used in processing or evaluating wafers.

Schimmel Etch[9,10] was originally intended to remove thin films of stain on wafers, not to decorate defects. Schimmel Etch consists of 2 parts HF and 1 part 0.038 molar potassium permaganate, $KMnO_4$. Refer to the discussion on stains later in this paper. Landgren's patent[39] 2,847,287 of 1956 describes an etch consisting of 3% potassium permanganate, $KMnO_4$, as the oxidizer, and 97% HF. Schimmel's etch[9], documented 21 years later, contains the same ingredients, but different quantities.

SD-1 etch appears to be another derivative of CP-4. SD-1 contains roughly the same ingredients as Salier's etch; 25 ml HF, 18 ml HNO_3, 5 ml CH_3COOH, 0.1 gm Br_2, 10 ml H_2O, and 1 gm $Cu(NO_3)_2$. This etch reveals all types of dislocations in 2 to 4 minutes, and copper is plated out for decoration. The use of both acetic acid and water as diluents, would make little sense, from Myers and Robbins' viewpoint.

This is one of the complex etches in which *the advantages of several separate etches are all combined into one*. M.W. Jenkins' Wright Etch is another example.

Sopori Etch[15] is a defect etch consisting of a mixture of 36 parts 49% HF, 20 parts CH_3COOH, and x parts of HNO_3, where x is varied from 1 to 2.

Superoxol Etch has also been called No.2 etch. It consists of 1 part H_2O_2, 1 part HF, and 3 parts H_2O. It typically developed etch figures *on germanium* in 1 to 3 minutes.

WAg Etch is an acronym standing for "Westinghouse silver etch". WAg etch consists of 40 ml HF, 20 ml HNO_3, 40 ml H_2O, and 2.0 gm $AgNO_3$. WAg etch was used by Wynne and Goldberg[28] in 1953 as a preferential defect etch *for germanium*.

White etch is HF:HNO_3, usually in the ratio of 3 parts HNO_3 to 1 part HF; but 4:1 and 50%-50% are listed as the composition by some companies. White etch is more generally, HNO_3-HF undiluted with water, which places the composition along the HNO_3-HF axis of Figure 2. This will yield maximum concentration, and maximum etch rate. In addition to dissolving silicon, white etch will also attack chromium, tantalum, titanium, and tungsten. It will dissolve most silicides,[37] and most other objects with which it comes into contact. Operators have great respect for it, sometimes referring to it as "death etch".

Other Silicon Etch Systems

Water-Amine-Pyrocatechol System

Finne and Klein[11] characterized the water-amine-pyrocatechol (or catechol) system with respect to water content and temperature, and pointed out that the system is highly selective to silicon in the presence of SiO_2. Additionally, if metals such as Ag, Au, Cu, or Ta are used as a mask for etching, that the HF-HNO_3 system cannot be used.

HF-Glycol System

The HF-glycol etches were never named, but they were patented. Deckert[12] characterized HF/glycol system in the etching of CVD Si_3N_4 over thermal SiO_2. It was known that HF would attack Si_3N_4 faster than SiO_2 if the temperature was raised above 100°C. *The purpose of adding the glycerol was to raise the boiling point of the solution; to prevent the evaporation of water from the etch solution.* Her paper contains graphical data on the etch rates vs HF concentration for various temperatures, in A/min. Ethylene glycol is commonly used in the etch, but propylene glycol has a slightly higher boiling point and also works.

Silicon Defect Etches Using Chromium Compounds

The development of these etches was originally not intended for failure analysis; it was intended for quality control in the manufacturing of silicon crystals for wafers by the Czochralski[23] method.

A slice (or wafer) from the ingot was/is polished, etched, and then inspected by optical microscopy with Nomarski[24] interference contrast. Since the boule takes about a week to grow, the decision to throw away these crystals has serious economic consequences for a company. These etches had to be highly reliable indicators of defect type and density, without the formation of spurious etch artifacts, which confused this task. This demanded research, and the various formulas emerged from that work.

The oxidizing agent in most of these formulas is chromic acid. Another oxidizer used in these formulas is potassium dichromate ($K_2Cr_2O_7$).

Again, the etches discussed below are formulated specifically to accentuate defects. *Improperly used, they can generate artifacts* which accentuate the normal crystal structure.

ASTM F47-88[20] is a standard defect etch which describes how to test silicon ingots (or boules, slugs, crystals, rods) for imperfections. It contains succinct and clear definitions for the various types of crystal imperfections revealed by preferential etches.

Secco Etch[13] is also called Dow Etch because its inventor was working for Dow Chemical at the time. It is also occasionally called D'Aragona Etch. Secco was not pleased with the long etch times required by Dash Etch, or its relative insensitivity to crystal orientation. Although the speed of Sirtl Etch was satisfactory, Secco wanted better performance on (100) material.

Secco tuned his etch for good performance on 100 material; high etch rate, strong definition of (100) feature, and low tendency to form etch artifacts which can complicate etched wafer defect counts and interpretation. Good sensitivity to (111) came with the deal.

Secco Etch produces elliptical dislocation etch pits on both {100} and {111} silicon. This etch uses potassium dichromate ($K_2Cr_2O_7$) instead of CrO_3 as the oxidizer. The formula specifies; one part by volume 0.15 molar with two parts HF. Secco etch is slow-acting, requiring about 20 minutes to produce decoration in material with

resistivities from 1 to 10,000 Ohm-cm. Ultrasonic agitation reduces this time to 5 minutes.

Sirtl Etch[14] resulted from what was apparently the first investigation of the system $HF:H_2CrO_3$. Sirtl Etch is effective in the delineation of defects such as dislocations and stacking faults. The mix specified in the paper was 46 gms CrO_3 in 100 gms of 40% HF solution. It was reported to be stable with use, and useful over a wide range of concentrations.

Futagami[27] found Sirtl etch superior to Secco and Wright etches in distinguishing oxygen-induced stacking faults. He found that the morphology of stacking faults decorated with Sirtl etch compared favorably with TEM images. He states that the etching rates of Sirtl, Secco, and Wright etches are about 3.0, 1.5, and 1.0 um/min in *p*-type material with resistivities of 40-60 Ohm-cm.

The Wright Etch[16] is a defect etch which is an improved version of Sirtl Etch. The Wright etch consists of 30 ml 5 molal chromic acid, 60 ml HF, 30 ml HNO_3, 60 ml glacial acetic acid, 60 ml water, and 2 gms copper sulfate or copper nitrate. The Wright etch decorates defects on both {100} and {111} planes in both *n*-type and *p*-type material, over a wide range of resistivities. Margaret's is the only paper on defect etches, which presents graphical etch rate data.

Yang Etch[17] was the result of his experimentation with the composition of Secco and Sirtl Etches. Yang basically had the same problem with other named etches as Secco; they not work well on (100) material.

The performance of Secco and Sirtl Etches was investigated. Yang found that the etching of dislocations was sensitive to the ratio of CrO_3 to HF, and his optimum formula was 1:1 - 1.5 M CrO_3:HF. Yang studied his etch figures with the TEM.

Alkaline Etch Systems

Cesium hydroxide, CsOH, will etch silicon from 200 to 8000 times faster[30] than undoped SiO_2. The maximum etch rate occurs at a concentration of 50 to 60%. By contrast, KOH will etch silicon only about 200 times faster than SiO_2. These selectivity numbers are very good.

Cyanide etch, or **Ferricyanide Etch**, was used by Billig[22] to delineate defects in both silicon and germanium crystal boules. This etch has an unusual alkaline formulation. The samples were immersed in a boiling solution of 8gm $K_3Fe(CN)_6$, 12 gm KOH, and 100 ml distilled water, for 1 to 5 minutes. The etch shows grain boundaries, and etch pits on {111}. This solution is also called "tungsten etch".

NaOH and KOH seem to be used with about equal formulations and results. Solutions with concentrations of 1 to 30%, used at 50 to 100°C will develop etch figures in 1 to 5 minutes.

RCA Clean[18] is a formulation originally developed for cleaning electron tube parts. It is now widely used as a semiconductor process cleaning solution. RCA Clean is actually a slow alkaline silicon etch utilizing the weak base ammonium hydroxide, NH_4OH. The formula also includes water and hydrogen peroxide, H_2O_2. Ratios of 5:1:1 and 7:2:1 are used.

Syetems Based on $NH_4HF:HF$

Buffered Oxide Etch (BOE) searching in the Inspec[R] database produced a suprisingly anemic response; only 35 hits.

This glass etch consists of a mixture of ammonium fluoride, NH_4F, and HF in varying ratios, from 4:1 to 50:1.

Mix	Etch rate, A/sec
5:1	1000 to 1200
10:1	500 to 600
25:1	200 to 300
50:1	100 to 200

Figure 5. Approximate thermal oxide etch rates for BOE of varying concentrations.[21]

BOE is used to prepare the silicon surface for hot processing, and to etch openings in SiO_2 layers.

The NH_4F is the buffer, supplying an abundant source of F ions. This keeps the fluoride concentration high, so that the solution does not become depleted, with resultant changes in the etch rate. For this reason, BOE etches more uniformly with time than dilute HF. This is important for etch baths in processing.

Also, the NH_4F buffer performs pH control. It is a weak base, and it raises the pH of HF-containing etch solution. Raising the pH prevents chemical attack of the photoresist materials and aluminum metallization by the HF.

Note that these graphs always use thermal oxide as the standard, since doped oxides, such as phos-glass or BPSG, can have etch rates

orders of magnitude faster than thermal oxide. From this fact comes the delineation ability of BOE for cross-sections.

Fortunately, one chemical manufacturer has carefully characterized BOE. The data appears in Fig 5; the etch rate depends on the ratio. Ratios greater than 4:1 will result in crystals of NH_4F dropping out of solution.

BOE is also known by other names, applied to specific concentrations or applications. Examples are Bell's, COE (common oxide etch), silicon-oxide (S/O), and tantalum stain. That is, BHF, BOE, COE, and S/O are all about the same thing.

BOE will etch 8% phosphorus-doped CVD oxide 5 to 10 times faster than thermal oxide.

Figure 6. BOE is important in failure analysis because the differential in its etch reaction rate between layers of varying composition, results in steps in the cross-section, which can be measured.

Decoration Etches

Terminology causes great confusion and disagreement in the literature, as the words "stain" and "decorate" are used interchangeably. In actuality, the two processes are very different. Both will **delineate** device structures. Clarification in terms can result from working through the mechanisms inherent to these techniques.

Delineation can occur through a differential in the etch rates of two or more dissimilar materials exposed in the cross-section. This etch rate differential translates into a height difference on the specimen.

When examined under Nomarski[24] interference contrast, colors will appear. This interference phenomenon translates small height differentials into color differences, so that they may be easily observed. The Nomarski interference colors are not due to interference in a film on the specimen, they are deliberately created within the microscope's optical system. However, interference colors within the microscope and interference colors from a thin film on the specimen, *can also be present simultaneously.*

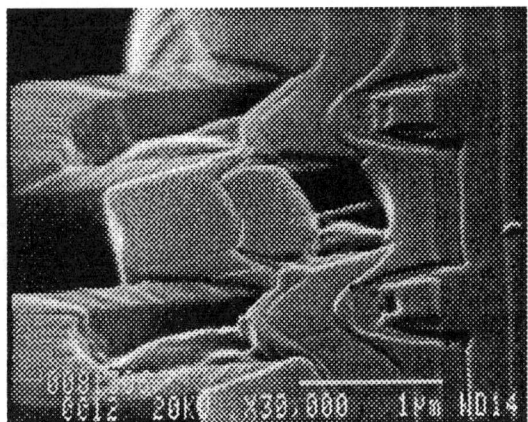

Figure 7. Low-angle SEM view of a cross section etched with BOE. The height differences result from etch rate differentials. They are what you see as different colors in an optical scope using Nomarski interference contrast. This is DELINEATION. (30 kx)

Decoration is preferential electroplating on one side of the junction. Decoration occurs by transition metal electroplating, usually copper from a solution containing copper sulfate or copper nitrate.

The acid forms an electrochemical cell with the variously doped regions as the electrodes. The cell voltage can cause selective plating.

Wu et al[42] and Subrahmanyan[43] utilized a bright light. Under strong illumination, a junction creates a photovoltage of about 0.5 V. Copper plates out onto the negatively-biased regions of the sample. *It is vital to note that this photovoltage can overcome the electrochemical acid cell voltage, and actually reverse the type of silicon which is decorated.*

Decoration with Pinhole Etch

Pinholes in oxide layers can be exposed by etching the silicon layer beneath. This results in a void in the silicon beneath the pinhole. The void will be visible through the oxide, and its large size will reveal the location of the pinhole.

Stains for silicon junctions were originally *undesirable deposits* on wafers; processing anomalies. Schimmel and Elkind[9,10] used ion-scattering spectroscopy (ISS) and secondary-ion mass spectroscopy (SIMS) to study the mechanism of formation and composition. We can utilize this information to deliberately produce stains on silicon for FA purposes.

They showed that a stain is actually a thin film about 20 to 100A thick, formed on the silicon surface during exposure to the etch solution. The

film was determined to be silicon suboxide ($SiO_{<2}$) or silicon monoxide (SiO).

The stain film is commonly brown or violet, and it is *insoluble in HF.* Only a bright light and an acid etch solution are required to stain one side of a junction. In their study, a 4:1:2 HNO_3-HF-CH_3COOH chemical polish solution was being used in process. *They fiddled with the etch composition, and found that HNO_3 to HF ratios of 1:4 to 1:250 would **all** form stains.*

Figure 8. This polished bipolar transistor cross-section was stained with only HF-HNO_3 and a bright light. No other additives were used. The silicon is not etched. This is DECORATION. (500x)

Etch Stability With Time

Many etches are not stable with time, or at elevated temperature. Some of the reasons are discussed below.

Peroxide based etches are not stable with time. Kern[18] has done a thorough job of characterizing his RCA Clean (etch) with respect to time. The results are a good benchmark for all H_2O_2 - containing etches. Although acidic solutions are stable for about an order of magnitude longer time than basic solutions, his graph shows that 70% or less of the original H_2O_2 remains after only 48 hours of storage at room temperature. There is no doubt that etches containing H_2O_2 should always be mixed fresh. These etches go flat about as quickly as an open can of carbonated soda.

Acetic acid diluted etches are generally not stable at elevated temperatures. Acetic acid has a high vapor pressure and is easily volatilized and lost. Its sharp odor even at room temperature is partially due to this fact.

BOE will grow crystals of NH_4F as the water evaporates from the solution, particularly when the ratio of NH_4F to HF is low.[21]

Reaction with their containers causes instability in some etches. For example, the storing of Sirtl Etch in Nalgene[R] labware is not recommended. Over time, the plastic bottle will turn green and become brittle. It can break in

597

ETCH NAME	Acetic acid, CH_3COOH	Acetone or toluene	Ammonium fluoride, NH_4F	Ammonium hydroxide, NH_4OH	Bromine, Br	Cesium hydroxide, $CsOH$	Choline	Chromic oxide, CrO_3	Copper nitrate, $Cu(NO_3)_2$	Copper sulfate, $CuSO_4$	Ethylenediamine	Hydrofluoric acid, HF	Hydrogen peroxide, H_2O_2	Iodine, I	Isopropanol, IPA	Mercuric nitrate, $Hg(NO_3)_2$	Nitric acid, HNO_3	Periodic acid, H_5IO_6	Phosphoric acid, H_3PO_4	Potassium dichromate, $K_2Cr_2O_7$	Potassium ferrocyanide, $K_3Fe(CN)_6$	Potassium hydroxide, KOH	Potassium iodide, KI	Potassium permaganate, $KMnO_4$	Pyrocatechol	Silver nitrate, $AgNO_3$	Tetramethyl ammonium hydroxide (TMAH)	Water, H_2O
ASTM	X											X					X											
CP-5	X											X					X											
Dash Etch	X											X					X											
Planar etch	X											X					X											
Polysilicon	X											X					X											X
Sopori Etch	X											X					X											X
Bell's #1												X					X											
Silicon etch												X					X											
White Etch												X					X											
Pinhole Etch												X					X											X
CP-4, 4a	X				X							X					X											
Salier's Etch					X				X			X					X											
Thomic etch	X							X				X					X											X
Wright Etch	X							X	X			X					X											X
Seiter Etch								X				X																X
Sirtl Etch								X				X																X
Yang Etch								X				X																X
Purdue etch									X			X					X											X
Copper decoration										X		X					X											X
Iodine etch	X											X		X			X											
Russian Etch	X											X		X			X									X		X
Mercury etch	X											X				X	X											
Sponheimer-Mills		X										X						X						X				X
Secco (Dow) Etch												X								X								X
Schimmel Etch												X													X			X
WAg												X					X									X		X
Bell's #2			X									X																
Buffered oxide etch, BOE			X									X																
Common oxide etch (COE)			X									X																
S/O (silicon-oxide) Etch	X		X									X																
RCA Clean				X									X															
Silicon etch					X																							X
Silicon etch						X																						
Silicon etch											X														X			X
Superoxyl Etch												X	X															X
Silicon etch															X							X						X
Schell Etch																		X										X
Nitride etch																			X									
Cyanide Etch																				X	X							X
Pinhole etch																					X							
Silicon etch																										X	X	X

598

your hands. The etch solution will become contaminated with bits of reacted plastic.

Vogel claimed that his mercury etch was better after it had been stored in the bottle for 6 weeks. No particulars were ever given.

Etch Recipe Spreadsheet

As mentioned previously, inquiry throughout the processing and analytical organizations within most semiconductor manufacturers for etch formulas usually produces a portfolio consisting of a mixture of formal papers and references, process recipes, typed lists of favored recipes, handwritten notes, and verbal instructions ("Use This" etch). Formulations also change over time, change with the particular user, and change with the materials. Further, the etch compositions, or mixing instructions are generally not given; lists frequently consist only of the names and quantities of the etch constituents. Listing all the names and recipes in the collection would produce an excessively large and confusing spreadsheet.

Again, Robbins and Schwartz[8] showed that the $HF-HNO_3$ etch solution compositions are a continuum; there are no magic ratios. *The ratios stated in the profusion of custom recipes, are generally those which have been optimized for the best etch rate and selectivity for a given device, with its manufacturer's specialized process for a narrow set of materials and dimensions.*

The strategy for limiting the extent of the spreadsheet was to list only the formulas traceable to the original developers of the named etches. Even these may become complicated. For example, the original etch developers usually found more than one favored etch composition; usually specialized for the materials with which their companies manufactured products. The work of Dash is an example. Instructions for mixing the etches may be more complex than can be conveniently reduced to spreadsheet format; some of the named etch solutions start with another named etch. Some start with a "stock solution". The formulas for the named etches are listed in variable and even mixed units, such as parts, grams, ml, cc's, mg, molar, molal, etc. The times, temperatures, and methods of application may be highly variable.

For complete instructions for mixing and applying the etches, the reader is referred to the specifics discussed under the major headings for the named etches.

The purpose of the spreadsheet, is to uncover the basis of the most familiar named etches, and show the universal applicability of the work of Schwartz and Robbins to failure analysis. The frequent hits in the HF, HNO_3, H_2O, and CH_3COOH columns, drive this vision.

Conclusions

1. The work of Robbins and Schwartz is an outstanding example of what a good characterization looks like. It is also a premier example of research project management applicable to FA; the project spanned many years and a job change. Information transfer in a case such as this, can help you avoid asking your people to periodically recharacterize what's already been characterized.

2. The majority of silicon etch formulas occur in the system $HF-HNO_3-H_2O/CH_3COOH$. There are no magic ratios of ingredients. Many of the most widely used formulas are variations based on relatively few fundamental constituents. There are no etches which will decorate and stain all layers of everything simultaneously, regardless of how many different chemicals you mix together.

3. Copper decoration is electroplating. Shining a light on the junction turns it into a photovoltaic (solar) cell, which powers the electrochemical plating reaction.

4. Junction staining is easily accomplished by shining a bright light on a junction covered with an $HF-HNO_3$ mixture which is rich in HNO_3.

5. Use etches with long-demonstrated and well-characterized high selectivities. Global selectivity numbers are rare and valuable. The best initial leverage of your research dollars is obtained with a review of appropriate papers for selectivity numbers specific to your needs.

6. Avoid the use of formulas which contain explosives, carcinogens, and intense poisons unless it is absolutely necessary. The defect delineation etches utilizing chromic acid, appear to have no particular advantage, for layer or junction delineation in FA.[36] Elimination of Cr compounds from the lab can greatly reduce specialized handling mandated by environmental considerations.

7. When mixing a named etch, obtain a copy of the research by the originator the etch, to assure correct composition, method of use, and anticipated results.

Acknowledgements

Thanks to Jim Jordan (Motorola hi-performance embedded systems) for his suggestion of this project, and Harvey Alford (consultant, TI retired), for kick-starting the literature search. The presentations at TI by Stacy Witkowski were valuable.

Gene Thome's experiment with bright light delineation, Figure 8, is an outstanding example of the results which can be achieved with this method. Leigh Myering's paper[38] is an excellent example of well-organized etch research.

I appreciate the expert review and comments from Jim Baylis, Richard Chapman, Barbara Denton, Larry Foreman, Cotton Hance, Mary Harrison, Carol Keate, Warren Lee, Sally Macrae, Ron Pyle, and Rose Ring.

References

NOTES:
The original collection of 103 references has been foreshortened to the list below, to adjust the length of the paper to conform with ASM publication standards.
* Texas law prohibits the use of the title "engineer" by those employed as engineers, unless the individual is a registered engineer in the State of Texas.

1. W. C. Dash, Bull. Amer. Phys. Soc., vol. 30(11), 1955.

2. W. C. Dash, "Copper precipitation on dislocations in silicon," Journal of Applied Physics, vol. 27(10), pp. 1193-1195, Oct. 1956.

3. W. C. Dash, "Evidence of dislocation jogs in deformed silicon," Journal of Applied Physics, vol. 29(4), p. 705, Apr. 1958.

4. W. C. Dash, "Gold-induced climb of dislocations in silicon," Journal of Applied Physics, vol. 31(12), pp. 2275-2283, 1960.

5. H. Robbins and B. Schwartz, "Chemical etching of silicon I. The system HF, HNO3, and H2O," Journal of the Electrochemical Society, vol. 106(6), pp. 505-508, Jun. 1959.

6. H. Robbins and B. Schwartz, "Chemical etching of silicon II. The system HF, HNO3, H2O, and HC2H3O2," Journal of the Electrochemical Society, vol. 107(2), p. 108-111, 1960.

7. H. Robbins and B. Schwartz, "Chemical etching of silicon III. A temperature study in the acid system," Journal of the Electrochemical Society, vol. 108(4), pp. 365-372, Apr. 1961.

8. H. Robbins and B. Schwartz, "Chemical etching of silicon IV. Etching technology," Journal of the Electrochemical Society, vol. 123(12), pp. 1903-1909, Dec. 1976.

9. D. G. Schimmel and M. J. Elkind, "An examination of the chemical staining of silicon," Journal of the Electrochemical Society, vol. 125(1), pp. 152-155, Jan. 1978.

10. D. G. Schimmel, "Defect etch for <100> silicon ingot evaluation," Journal of the Electrochemical Society, vol. 126(3), pp. 479-483, 1979.

11. R. M. Finne and D. L. Klein, "A water-amine-complexing agent system for etching silicon," Journal of the Electrochemical Society, vol. 114(9), pp. 965-979, Sep. 1967.

12. C. A. Deckert, "Pattern etching of CVD Si_3N_4/SiO_2 composites in HF/glycerol mixtures," Journal of the Electrochemical Society, vol. 127(11), pp. 2433-2438, Nov. 1980.

13. F. Secco d'Aragona, "Dislocation etch for (100) planes in silicon," Journal of the Electrochemical Society, vol. 119(7), pp. 948-951, Jul. 1972 and Phys. Status Solidi (a), vol. 7, p. 577, 1971.

14. Von Erhard Sirtl and Annemarie Adler, "ChromsÑure-FluasNure als spezifisches System zur érzgrubenentwicklung auf Silizium (Chromic acid-hydrofluoric acid as specific reagents for the development of etching pits in silicon)," Zeitschrift fur Metallkunde, vol. 52, pp. 529-531, 1961.

15. B. L. Sopori, "A new defect etch for polycrystalline silicon," Journal of the Electrochemical Society, vol. 131(3), pp. 667-672, Mar. 1984.

16. Margaret Wright Jenkins, "A new preferential etch for defects in silicon crystals," Journal of the Electrochemical Society, vol. 124(5), pp. 757-762, 1977.

17. K.H. Yang, "An etch for delineation of defects in silicon," Journal of the Electrochemical Society, vol. 131(5), pp. 1140-1145, 1984.

18. Werner Kern and David. A. Puotinen, "Cleaning solutions based on hydrogen peroxide for use in silicon semiconductor technology," RCA Review, vol. 31, pp. 187-206, Jun. 1970.

19. W. E. Beadle, J. C. C. Tsai and R. D. Plummer, eds., Quick Reference Manual for Silicon Integrated Circuit Technology, New York: Wiley, 1985.

20. "ASTM F47-88: Standard test method for crystallographic perfection of silicon by preferential etch techniques," in Annual Book of ASTM Standards, vol. 10.05, Philadelphia, PA: ASTM, 1988.

21. Olin Hunt Process Chemical Data Sheets, Olin Hunt Specialty Products, West Paterson, NJ, 1990.

22. E. Billig, "Some defects in crystals grown from the melt. I. Defects caused by thermal stresses," Proceedings of the Royal Society of London, Series A, vol. 235, pp. 37-54, 1956.

23. J. Czochralski, "Ein neues Verfahren zur Messung der Kristallisationsgeschwindigkeit der Metalle", Zeitschrift fur Physikalische Chemie, vol. 92, pp. 219-221, 1917.

24. M.G. Nomarski, "Microinterferometre differentiel a ondes polarisees", Le J de Physique et le Radium, 16, 59, 1955, pp. 9-16.

25. P. Wang, "Etching of germanium and silicon," The Sylvania Technologist, vol. 11(2), pp. 50-58, Apr. 1958.

26. F. L. Vogel, Jr. and L. C. Lovell, "Dislocation etch pits in silicon crystals," Journal of Applied Physics, vol. 27(12), pp. 1413-1415, Dec. 1956.

27. F. L. Vogel, W. G. Pfann, H. E. Corey, and E.E. Thomas, "Observations of dislocations in lineage boundaries in germanium (letter)," Physical Review, vol. 90, pp.489-490, 1953.

28. R. H. Wynne and C. Goldberg, "Preferential etch for use in optical determination of germanium crystal orientation," Transactions of AIME, Journal of Metals, vol. 5, p.436, Mar. 1953.

29. W. A. Pliskin and R. P. Gnall, "Evidence for oxidation growth at the oxide-silicon interface from controlled etch studies," Journal of the Electrochemical Society, vol. 111(7), pp. 872-873, Jul. 1964.

30. D. R. Turner, "On the mechanism of chemically etching germanium and silicon," Journal of the Electrochemical Society, vol. 107(10), pp. 810-816, Oct. 1960.

31. W. W. Tyler and W. C. Dash, "Dislocation arrays in germanium," Journal of Applied Physics, vol. 28(11), pp. 1221-1224, Nov. 1957.

32. T. Mills and E. Sponheimer, "Precision VLSI cross-sectioning and staining," Proceedings of the International Reliability Physics Society (IRPS), 1982, pp 214-220.

33. I. F. Nicolau, "Junction delineation and dislocation revealing in silicon by the HIO4-HF-H2O system," Solid State Electronics, vol. 12, pp. 446-448, 1968.

34. MIT Radiation Lab Series, Volume 15, "Crystal Rectifiers", McGraw-Hill, 1948, pg 369.

35. T. Mills, "Staining cheat sheet: fine-tuning of staining formulations for ICs made simple by the use of a reference staining poster," Tutorial Notes, International Reliability Physics Symposium, 2 Apr. 1984, pp. 2-9.1 to 2-10.7. Also private communication.

36. T. C. Chandler, "MEMC etch - a chromium trioxide-free etchant for delineating dislocations and slip in silicon," Journal of the Electrochemical Society, vol. 137(3), pp. 944-948, Mar. 1990.

37. S. P. Murarka, Silicides for VLSI Applications, New York: Academic Press, 1983.

38. Leigh Meyering, Chemical Delineation of Silicon and Gallium Arsenide Devices, Motorola internal unpublished paper, 1995.

39. Clarence R. Landgren, U.S. Patent 2,847,287, 1956.

40. Somner P. Wolsky, U.S. Patent 2,734,806, 1954.

41. Robert D. Heidenreich, "Surface treatment of germanium circuit elements," U.S. Patent 2,619,414, 1952.

42. C. P. Wu, E. C. Douglas, C. W. Mueller, and R. Williams, "Techniques for lapping and staining ion-implanted layers," Journal of the Electrochemical Society, vol. 126(11), pp. 1982-1987, Nov. 1979.

43. R. Subrahmanyan, H. Z. Massoud, and R. B. Fair, "Experimental characterization of two-dimensional dopant profiles in silicon using chemical staining," Applied Physics Letters, vol. 52(25), pp. 2145-2147, Jun. 20, 1988.

GENERAL

Cornelius A. Johnson, ed., Metallographic Principles and Procedures, St. Joseph, MO: Leco Corporation.

Gunter Petzow, Metallographic Etching, Materials Park, OH: American Society for Metals, 1976, pp. 18-23, 68,69.

"Chapter 7: Etching", in Struers Handbook, Westlake, OH: Struers, Inc., 1976.

Failure Analysis Terms and Definitions

Christopher L. Henderson
Sandia National Laboratories
Albuquerque, New Mexico

Why do we need to define terms and definitions?

Confusion among terms and acronyms

Long-term problem that interferes with effective communication between analysts and requestors and within the failure-analysis community

Although there are myriad sources of terms & definitions addressing various technical subjects comprising the multidisciplinary failure-analysis field, certain terms, expressions, acronyms, etc. are prevalent in and peculiar to this profession

To determine which terms and definitions are focused directly on failure analysis

To provide a common language (foundation) that improves communication within the failure-analysis community and between failure analysts and the electronics industry by providing and promoting standardization of descriptive technical language

Acknowledgments

ASTM definitions have been reprinted, with permission, from the Annual Book of ASTM Standards, copyright American Society for Testing and Materials, 1916 Race Street, Philadelphia, PA, 19103–1187.

ASME definitions have been reprinted with permission from the American Society of Mechanical Engineers, 345 East 47th Street, New York, NY, 10017–2392.

EIA definitions have been reprinted with permission from the Electronic Industries Association, 2500 Wilson Boulevard, Arlington, VA, 22201–2392.

IEEE definitions have been reprinted from IEEE Std 100–1992 IEEE Standard Dictionary of Electrical and Electronics Terms, copyright 1993 by the Institute of Electrical and Electronics Engineers, Inc. The IEEE disclaims any responsibility or liability resulting from the placement and use in this publication. Information is reprinted with the permission of IEEE.

National Technology Roadmap for Semiconductors definitions are reprinted with permission from Semiconductor Industry Association (SIA), 4300 Stevens Creek Boulevard, Suite 271, San Jose, CA, 95129.

SEMI definitions are reprinted by SEMATECH with permission from Semiconductor Equipment and Materials International. Copyright Semiconductor Equipment and Materials International, 805 E. Middlefield Road, Mountain View, CA, 94043.

SEMATECH definitions are reprinted from the SEMATECH Official Dictionary Version 5.0, copyright 1995 by SEMATECH, 2706 Montopolis Drive, Austin TX, 78741.

acceptor n : in a semiconductor, an impurity in a semiconductor that accepts electrons excited from the valence band, leading to hole conduction. [SEMI M1-94 and ASTM F1241] Also see hole.

access time n : a time interval that is characteristic of a storage device and is essentially a measure of the time required to communicate with that device. [IEEE]

accumulation condition n : the region of the capacitance-voltage (C-V) curve for which a 5-V increment toward a more negative voltage for p-type material, or toward a more positive voltage for n-type material, results in a change of less than 1% in the maximum capacitance, Cmax . [ASTM F1241]

active area n : the region of thin oxide on a die or wafer in which transistors and other circuits reside. [SEMATECH]

active devices n : semiconductor devices that have active function, such as integrated circuits and transistors. [SEMI G35-87] Contrast passive devices.

adhesion, resist edge n : the ability of the edge of an image in a developed resist coating to adhere to its substrate under applied physical or chemical stress. [ASTM F127-84]

adhesive stringer n : on a photolithographic pellicle, any detectable protrusion from the edge of the adhesive. [SEMI P5-94]

aeolotropic : see anisotropic.

AES (Auger-electron spectrometry)/AES (Atomic-emission spectroscopy)/SAM (Scanning Auger microprobe): see Auger electron spectroscopy.

AFM : see atomic force microscopy.

alignment n 1 : the accuracy of the relative position of an image on a reticle with reference to an existing image on a substrate. [SEMATECH] 2 : a procedure in which a wafer is correctly positioned relative to a reticle. [SEMATECH] 3 : the mechanical positioning of reference points on a wafer or flat panel display substrate (also called alignment marks or alignment targets) to the corresponding points on the reticle or reticles. The measure of alignment is the overlay at the positions on the wafer or substrate where the alignment marks are placed. [Adapted from SEMI P18-92 and D8-94] Also see direct alignment and indirect alignment.

alloy n 1 : a composite of two or more elements, of which at least one is metal. [SEMATECH] 2 : a thermal cycle in which two or more discrete layers (of which at least one is metal) react to allow good electrical contacts. [SEMATECH]

aluminized area n : in a cerdip or cerpack semiconductor package, the leadframe area coated with aluminum to provide a surface suitable for wire bonding. The maximum area is defined by the inside dimension of the cap or ceramic ring. In some cases, the die attach area is also coated if a full leadframe is used. The coating may be vacuum deposited or bonded. [SEMATECH]

aluminized width n : in a semiconductor package, the width of the area coated with a protective layer of aluminum. This area covers most of the top formed width. [SEMATECH] Also see package, bond finger, top formed width, and aluminized area.

aluminum (Al) n : a metal used to interconnect the devices on a wafer and to interconnect external devices or components. [SEMATECH]

ambient temperature (TA) 1 : the temperature of the surrounding medium, such as air or liquid, that comes into contact with the device or apparatus. [SEMATECH] 2 : the temperature of the specified, surrounding medium (such as air, nitrogen, or a liquid) that comes into contact with a semiconductor device being tested for thermal resistance. [SEMI G38-87]

ammonium fluoride (NH4 F) : a white crystalline salt used to buffer hydrofluoric acid etches that dissolve silicon dioxide but not silicon. An example of such an etch is the buffered oxide etch. [SEMATECH] Also see pinhole.

ammonium hydroxide (NH4 OH) : a weak base formed when ammonia is dissolved in water. [SEMATECH]

amorphous silicon : silicon with no discernible crystalline structure. [SEMATECH] Contrast polycrystalline silicon.

analog adj : A signal in an electronic circuit that takes on a continuous range of values rather than only a few discrete values; a circuit or system that processes analog signals. [1994 National Technology Roadmap for Semiconductors] Contrast discrete.

angle-resolved scattering (ARS) n : technique that measures light scattered from particles as a function of angle; used to characterize particles. [SEMATECH]

angstrom (Å) n : unit of linear measure equal to one ten billionths of a meter (10 -10 m). (The diameter of a human hair is approximately 750,000 Å.) The preferred SI unit is nanometers. 10 Å = 1 nm. [SEMATECH]

anion n : an ion that is negatively charged. [SEMATECH]

anisotropic adj : exhibiting different physical properties in different directions. NOTE—In semiconductor technology, the different directions are defined by the crystallographic planes. [SEMI M1-94 and ASTM F1241] Also called nonisotropic and aeolotropic. Also see anisotropic etch.

anisotropic etch n : a selective etch that exhibits an accelerated etch rate along specific crystallographic planes. NOTE—Anisotropic etches are used to determine crystal orientation, to expose crystal defects, and to facilitate dielectric component isolation. [SEMI M1-94 and ASTM F1241] Also called preferential etch. Also see anisotropic.

anneal n : a high-temperature operation that relieves stress in silicon, activates ion-implanted dopants, reduces structural defects and stress, and reduces interface charge at the silicon-silicon dioxide interface. [SEMATECH]

anomaly : see defect.

antireflective coating (ARC) n : a layer of dielectric material deposited on a wafer before resist to minimize reflections during resist exposure. [SEMATECH]

antimony (Sb) n : a brittle, tin-white, metallic chemical element of crystalline structure. Antimony is used as an n-type dopant in silicon, often for the buried layer. [SEMATECH]

ARC : see antireflective coating.

architecture n : of a computer system, a defined structure based on a set of design principles. The definition of the structure includes its components, their functions, and their relationships and interactions. [SEMATECH]

area contamination n : foreign matter on localized portions of a wafer or substrate surface. [SEMI M3-88]

arsenic (As) n : a highly poisonous chemical element, which is brittle and steel gray in color. Arsenic is often used as an n-type dopant for buried layer predisposition. [SEMATECH] Also see n-type.

artifact n 1 : a physical standard against which a parameter is measured; for example, a test wafer used for testing parametric drift in a machine. [SEMATECH] Also called standardreference material. 2 : a superficial or unessential attribute of a process or characteristic under examination; for example, a piece of lint on a lens that appears through a microscope to be a defect on a die. [SEMATECH] 3 : in surface characterization, any contribution to an image from other than true surface morphology. Examples include contamination, vibration, electronic noise, and instrument imperfections. [SEMATECH]

ash v : to apply heat to a material until the material has been reduced to a mineral residue. [SEMATECH]

asher n : a machine used to remove resist from substrates. [SEMATECH]

ashing n : the operation of removing resist from a substrate by oxidation; a reaction of resist with oxygen to remove the resist from the substrate. [SEMATECH]

aspect ratio n 1 : in etch, the depth-to-width ratio of an opening on a wafer. [SEMATECH] 2 : in feature profile, the ratio of height to width of a feature. [SEMATECH]

atomic force microscopy (AFM) n : a microscopy technique based on profilometry using an atomically sharp probe that provides three-dimensional highly magnified images. During AFM, the probe scans across a sample surface. The changes in force between the sample and the probe tip cause a deflection of the probe tip that is monitored and used to form the magnified image. [SEMATECH]

atomic percent n : in electron spectroscopy for chemical analysis (ESCA) of plastic surface composition, the number of atoms of a particular element present in every hundred atoms within the ESCA detection volume. [SEMATECH]

ATPG : see automatic test pattern generation.

at-speed test n : any test performed on an integrated circuit that tests the device at its normal operating clock frequency. [1994 National Technology Roadmap for Semiconductors]

Auger electron spectroscopy (AES) n : the energy analysis of Auger electrons produced when an excited atom relaxes by a radiationless process after ionization by a high-energy electron, ion, or X-ray beam. [SEMATECH]

Auger process n : the radiationless relaxation of an atom involving a vacancy in an inner electron shell. An electron is emitted, which is referred to as an Auger electron. [ASTM E673-90]

autodoping n : in the manufacture of silicon epitaxial wafers, the incorporation of dopant originating from the substrate into the epitaxial layer. [SEMI M1-94 and ASTM F1241] Also called self-doping. Also see doping and substrate.

automatic test pattern generation (ATPG) n : the automatic development of vectors which, when applied to an integrated circuit, permit faults to be detected in the performance of the integrated circuit. [1994 National Technology Roadmap for Semiconductors]

back-end of line (BEOL) n : process steps from contact through completion of the wafer prior to electrical test. Also called back end. [SEMATECH]

backgrind n : an operation using an abrasive on the back side of a substrate to achieve the necessary thinness for scribing, cutting, and packaging of die. [SEMATECH]

back oxide n : a layer of silicon dioxide formed on the back of a wafer during oxidation. [SEMATECH]

backside : see back surface.

back surface n : of a semiconductor wafer, the exposed surface opposite to that on which active semiconductor devices have been or will be fabricated. [ASTM F1241] Also called backside.

bake n : in wafer manufacturing, a process step in which a wafer is heated in order to harden resist, remove moisture, or cure a film deposited on the wafer. [SEMATECH]

ball-grid array (BGA) n : an integrated circuit surface mount package with an area array of solder balls that are attached to the bottom side of a substrate with routing layers. The die is attached to the substrate using die and wire bonding or flip-chip interconnection. [SEMATECH] Also called land-grid array, pad-grid array, or pad-array carrier.

bar : see die, crossbar, and bar end.

bare die n : individual, unpackaged silicon integrated circuits. [1994 National Technology Roadmap for Semiconductors]

barrier n : a physical layer designed to prevent intermixing of the layers above and below the barrier layer; for example, titanium-tungsten and titanium-nitride layers. [SEMATECH]

barrier layer : see depletion layer.

base n 1 : in semiconductor manufacturing chemicals, a substance that dissociates in water to liberate hydroxyl ions, accepts a proton, has an unshared pair of electrons, or reacts with acid to form a salt. A base has a pH greater than 7 and turns litmus paper blue. [SEMATECH] 2 : in facilities and safety, a corrosive material with the chemical reaction characteristic of an electron donor. [SEMI S4-92] 3 : in quartz and high temperature carriers, the material at the bottom of a wafer carrier on which the wafer carrier rests when placed on a flat surface. [SEMI E2-93] 4 : of a cerdip or cerpack package, the bottom ceramic portion. A leadframe, a window frame, and the cap are attached to the base—generally with devitrifying solder glass—during package/device manufacture. [SEMI G1-85] Also see cap and window frame.

behavioral n : a level of logic design that involves describing a system at a level of abstraction that does not involve detailed circuit elements, but instead expresses the circuit functionality linguistically or as equations. [1994 National Technology Roadmap for Semiconductors]

BEOL : see back-end of line.

BGA : see ball grid array.

biCMOS design n : the combination of bipolar and complementary metal oxide semiconductor design and processing principles on a single wafer or substrate. [SEMATECH]

bimetal mask : see mask, bimetal.

binding energy n : the value obtained by subtracting the instrumentally measured kinetic energy of an electron from the energy of the incident photon, corrected for an instrument work function. [SEMATECH]

bipolar adj : a semiconductor device fabrication technology that produces transistors which use both holes and electrons as charge carriers. [SEMI M1-94 and ASTM F1241]

bird's beak n : a structural feature produced as a result of the lifting of the edges of the nitride layer during subsequent oxidation. [SEMATECH]

BIST : see built-in self test.

blister ceramic n : an enclosed, localized separation within or between the layers of a ceramic package that does not expose an underlying layer of ceramic or metallization. [SEMI G61-94] Also called bubble ceramic.

blister metal n : in packaging, an enclosed, localized separation of a metallization layer from its base material (such as ceramic or another metal layer) that does not expose the underlying layer. [SEMI G8-94] Also called bubble metal, blister metallization, and bubble metallization. Also see package.

bonding pads n : relatively large metal areas on a die used for electrical contact with a package or probe pins. [SEMATECH]

boundary scan n : a scan path that allows the input/output pads of an integrated circuit to be both controlled and observed. [1994 National Technology Roadmap for Semiconductors]

bridge n 1 : a defect in which two adjacent areas connect because of misprocessing such as poor lithography, particle contamination, underdevelop, or etch problems. [SEMATECH] Also called short. 2 : software that allows access to, and combination of, data from incompatible databases. [SEMATECH]

bridging fault n : a fault modeled as a short-circuit between two nets on a die. [1994 National Technology Roadmap for Semiconductors]

brightfield illumination n (transmission electron microscopy) : the illumination of an object so that it appears on a bright background. [ASTM E7-93]

buffered hydrofluoric acid n : an extremely hazardous corrosive used to etch silicon dioxide from a wafer. This acid has a 20- to 30-minute reaction delay after contact with skin or eyes. [SEMATECH]

built-in self test (BIST) : any of the methods of testing an integrated circuit (IC) that uses special circuits designed into the IC. This circuitry then performs test functions on the IC and signals whether the parts of the IC covered by the BIST circuits are working properly. [1994 National Technology Roadmap for Semiconductors]

buried contact n : a conductive region between two less conductive regions. [SEMATECH]

buried layer n 1 : a conductive layer between two less conductive films; for example, a localized n+ region in a p-type wafer that reduces the npn collector series resistance for integrated circuit transistors fabricated in an n-type epitaxial layer deposited on the p-type wafer. [SEMATECH] 2 : in epitaxial silicon wafers, a diffused region in a substrate that is, or is intended to be, covered with an epitaxial layer. [SEMI M18-94 and ASTM F1241] Also called subdiffused layer and diffusion under film.

burn-in n : the process of exercising an integrated circuit at elevated voltage and temperature. This process accelerates failure normally seen as "infant mortality" in a chip. The resultant tested product is of high quality. [1994 National Technology Roadmap for Semiconductors] Also see infant mortality.

C4 (controlled collapse chip connect) : see flip chip.

cap deposition : see passivation.

carrier n 1 : an entity capable of carrying electric charge through a solid; for example, mobile holes and condition electrons in semiconductors. [SEMI M1-94 and ASTM F1241] Also called charge carrier. Also see majority carrier and minority carrier. 2 : slang for wafer carrier. [SEMATECH]

cavity-down packages n : in cofired ceramic packages, packages on which the die surface faces the mounting board. [SEMI G61-94]

cavity-up packages n : in cofired ceramic packages, packages on which the die surface faces away from the mounting board. [SEMI G61-94]

cerdip : abbreviation for ceramic dual inline package. See dual inline package.

cerpack : abbreviation for ceramic package.

channel : the portion of a MOS integrated circuit that allows current to flow when biased by the overlying gate region [Sandia Labs].

chemical-mechanical polish (CMP) n : a process for the removal of surface material from a wafer. The process uses chemical and mechanical actions to achieve a mirror-like surface for subsequent processing. [SEMI M1-94 and ASTM F1241] Also called chem-mech polish.

chemical vapor deposition (CVD) n : in semiconductor technology, a process in which a controlled chemical reaction produces a thin surface film. [SEMI M1-94 and ASTM F1241] Contrast physical vapor deposition.

chem-mech polish : See chemical-mechanical polish.

chip n 1 : in semiconductor wafers, a region where material has been unintentionally removed from the surface or edge of the wafer. [ASTM F1241] Contrast indent. 2 : see die. 3 : in packaging, a region of material missing from a component; for example, ceramic from a package or solder from a preform. The region does not progress completely through the component and is formed after the component is manufactured. The chip size is given by its length, width, and depth from a projection of the design plan-form. [SEMI G61-94] Also called chip-out. Contrast pit. 4 : in flat panel display substrates, a region of material missing from the edge of the glass substrate, which is sometimes caused by breakage or handling. [SEMI D9-94]

chip carrier (CC) n : a small footprint semiconductor package generally with terminals on all four sides. The package may be manufactured by cofired ceramic or multilayer printed circuit board technologies. [SEMATECH] Also see castellation and ceramic chip carrier.

chip-out : see chip.

circuit n : the combination of a number of connected electrical elements or parts to accomplish a desired function. [SEMATECH]

circuit design n : techniques used to connect active (transistors) and passive (resistors, capacitors, and inductors) elements in a manner to perform a function (that is, logic, analog). [1994 National Technology Roadmap for Semiconductors]

circuit geometries n : the relative shapes and sizes of features on a die. [SEMATECH]

CMOS : see complementary metal oxide semiconductor.

CMP : see chemical-mechanical polish.

comet n : on a substrate, a buildup of resist shaped like a comet and generated by a defect. [SEMI P3-90] Also called motorboat.

complementary metal oxide semiconductor (CMOS) n : a fabrication process that incorporates p-channel and n-channel MOS transistors within the same silicon substrate. [SEMATECH]

component n 1 : an individual electronic part, such as a device, diode, or capacitor that is fabricated in a metal oxide semiconductor or bipolar process. [SEMATECH] 2 : an individual piece or a complete assembly of individual pieces, including industrial products that are manufactured as independent units, capable of being joined with other pieces or components. The typical components referred to by the specification are valves, fittings, regulators, gauges, instrument sensors, a single length of tubing, several pieces of tubing welded together, tubing welded to fittings, and the like. [SEMI F1-90] 3 : the fundamental parts of an object, its entities, or relationships. [SEMATECH] 4 : the hardware and software that work in sets (functional entities) to perform the operation(s). [SEMATECH]

conchoidal fracture n : a fracture having smooth convexities and concavities like a clamshell. [SEMATECH] Also see chip.

conductor n : a substance through which electricity can readily flow. Contrast insulator. [SEMATECH]

confocal n : refers to a microscope design with superior abilities to image submicron features on a wafer. [1994 National Technology Roadmap for Semiconductors]

contact n : in an oxide layer, an opening that allows electrical connection between metal and silicon layers. [SEMATECH] Also see window and via.

contamination n 1 : the presence of particles, chemicals, and other undesirable substances, such as on or in a process tool, in a process liquid, or in a cleanroom environment. [SEMATECH] Also see area contamination and particulate contamination. 2 : three-dimensional foreign material adhering to a package (plastic or ceramic) or leadframe, or parent material displaced from its normal location and similarly adhered. Adherence means that the particle cannot be removed by an air or nitrogen blast at 20 psi. [SEMATECH] Also see foreign material and stain.

controlled collapse chip connect (C4) : see flip chip

correlation n 1: a relation existing between phenomena or things or between mathematical or statistical variables which tend to vary, be associated, or occur together in a way not expected on the basis of chance alone. [Webster's Dictionary]

crack n 1 : on semiconductor wafers, a cleavage or fracture that extends to the surface and may or may not pass through the entire thickness of the wafer. [ASTM F1241] 2 : of a semiconductor package or solder preform, a cleavage or fracture that extends to the surface. The crack may or may not pass through the entire thickness of the package or preform. [SEMI G61-94] 3 : in flat panel display substrates, a fissure located at the sheet edge or central area. [SEMI D9-94]

crater n : on the surface of a slice or wafer, an individually distinguishable bowl-shaped cavity. A crater is visible when viewed under diffused illumination. [SEMATECH]

cratering n : on a slice or wafer, a surface texture of irregular closed ridges with smooth central regions. [ASTM F1241]

crescents n : structures with parallel major axes, attributed to substrate defects either above or below the surface plane of silicon substrates after epitaxial deposition. [ASTM F1241] Also see fishtails.

critical area n : the area in which the center of a defect must occur to cause a failure or fault. [SEMATECH] Also see fault and fault probability.

critical dimension (CD) n : the width of a patterned line or the distance between two lines, monitored to maintain device performance consistency; that dimension of a specified geometry that must be within design tolerances. [ASTM F127-84] Also see linewidth.

crosstalk n : the undesirable addition of one signal to another in a circuit usually caused by coupling through parasitic elements. An example would be inductive or capacitive coupling between adjacent conductors. [1994 National Technology Roadmap for Semiconductors]

crossunder n : on a die, the point at which a conductor crosses under a second conductor without making electrical contact. [SEMATECH]

crow's foot n : on a semiconductor wafer, intersecting cracks in a pattern resembling a "crow's foot" Y on {111} surfaces and a cross "+" on {100} surfaces. [ASTM F1241]

crystal n : a solid composed of atoms, ions, or molecules arranged in a pattern that is periodic in three dimensions. [ASTM F1241]

crystal defect n : departure from the regular arrangement of atoms in the ideal crystal lattice. [ASTM F1241] Also see crystal lattice and damage.

crystal indices : see Miller indices. Also see crystallographic notation.

crystal lattice n : in a crystal, the three-dimensional and repeating pattern of atoms. [SEMATECH]

crystallographic notation n : a symbolism based on Miller indices used to label planes and directions in a crystal as follows: (111) plane [111] direction {111} family of planes <111> family of directions
[SEMI M1-94 and ASTM F1241]

crystal originated particle (COP) n : a surface depression that is formed during soft alkaline chemical treatment of silicon wafer surfaces that contain crystal defects at or close to the wafer surface and that scatters light similarly to a very small particle. [ASTM F1241] Also called surface micro defect.

CTE : see coefficient of thermal expansion.

CVD : see chemical vapor deposition.

cycle time n : (1) the length of time required for a wafer to complete a specified process or set of processes. [SEMATECH] (2) the length of time required to complete a failure analysis job from receipt in the failure analysis lab to the time results (written or verbal) are communicated back to the immediate requestor. [Sandia Labs] Also see equipment cycle, minimum theoretical cycle time, and theoretical cycle time.

damascene n : an integrated circuit process by which a metal conductor pattern is embedded in a dielectric film on the silicon substrate. The result is a planar interconnection layer. The creation of a damascene structure most often involves chemical mechanical polishing of a nonplanar surface resulting from multiple process steps. A damascene trench is a filled trench. [1994 National Technology Roadmap for Semiconductors]

damage n 1 : of a single-crystal silicon specimen, a defect of the crystal lattice in the form of irreversible deformation that results from mechanical surface treatments such as sawing, lapping, grinding, sandblasting, and shot peening at room temperature without subsequent heat treatments. [ASTM F1241] Also see crystal lattice. 2 : any yield or reliability detractors other than those related to design, process specification violations, or particles. [SEMATECH]

DC test : A sequence of direct current (DC) measurements performed on integrated circuit pads to determine probe contact, leakage currents, voltage levels on input and output, power supply currents, etc. [1994 National Technology Roadmap for Semiconductors]

deep level impurity n : a chemical element that, when introduced into a semiconductor, has an energy level (or levels) that lies on the midrange of the forbidden energy gap, between the energy levels of the dopant impurity species. [ASTM F1241]

defect n : for silicon crystals, a chemical or structural irregularity that degrades the ideal silicon crystal structure or the thin films built over the silicon wafer. 2 : a pit, tear, groove, inclusion, grain boundary, or other surface feature that is either characteristic of the material or a result of its processing and that is not a result of the sample preparation. [SEMATECH] Also called anomaly.

defect density n : the number of imperfections per unit area, where imperfections are specified by type and dimension. [ASTM F127-84] Also see defect.

defect level n : the number of die in parts-per-million that are shipped to customers and that are defective even though the test program declares them to be good. [1994 National Technology Roadmap for Semiconductors]

defect, photomask n : any flaw or imperfection in the opaque coating or functional pattern that will reproduce itself in a resist film to such a degree that it is pernicious to the proper functioning of the microelectronic device being fabricated. [SEMI P2-86]

delamination n : in a cofired ceramic package, chip carrier, dual inline, pin grid array, etc., the separation of one ceramic layer from another. [SEMI G61-94] Also see package.

delay fault n : a fault that has the effect of causing a signal to appear late in arriving at a destination. [1994 National Technology Roadmap for Semiconductors]

design for test (DFT) n : design of logic circuits to facilitate electrical testing. [SEMATECH]

destructive physical analysis n 1: the examination and testing of components to ensure proper operation and behavior. [Sandia Labs]

device n : a specific kind of electronic component (such as an MOS transistor, resistor, diode, or capacitor) on a die. The diode and transistor are referred to as active devices; the capacitor and resistor, as passive devices. [SEMATECH]

dew point n : the temperature at which liquid first condenses when vapor is cooled. [SEMI C3-94]

DFT : see design for test.

die n (sing or pl) : a small piece of silicon wafer, bounded by adjacent scribe lines in the horizontal and vertical directions, that contains the complete device being manufactured. [SEMATECH] Also called chip and microchip. Obsolete: bar, slice.

die attach area n : the nominal area designated for die attaching to the package or leadframe. [SEMI G22-86] Contrast effective die attach area and die attach pad.

die attach pad n : the nominal area designated for die attaching to the package or leadframe.. Die attach pad is usually applied to leadframes. The term die attach area is usually applied to ceramic packages. [SEMATECH] Also see package and die.

die attach surface n : in a ceramic semiconductor package, a dimensional outline designated for die attach. [SEMI G33-90] Also see package and die.

die bonding (D/B) : an assembly technique that bonds the back side of an integrated circuit die to a substrate, header, or leadframe. [SEMATECH]

dielectric n 1 : a nonconductive material; an insulator. Examples are silicon dioxide and silicon nitride. [SEMATECH] 2 : a material applied to the surface of a ceramic or preformed plastic package to provide functions such as electrical insulation, passivation of underlying metallization, and limitations to solder flow. [SEMI G33-90]

dielectric isolation (DI) n : a nonconductive barrier layer grown or deposited between two adjacent regions on a die to prevent electrical contact between the regions. [SEMATECH] Also see isolation.

diffusion n : a high-temperature process in which desired chemicals (dopants) on a wafer are redistributed within the silicon to form a device component. [SEMATECH]

dimple n : on a semiconductor wafer, a shallow depression in a wafer surface with a concave, spheroidal shape and gently sloping sides. NOTE—Dimples are macroscopic features that are visible to the unaided eye under proper lighting conditions. [ASTM F1241]

DIP : see dual inline package.

dislocation n : a line imperfection in a crystal that either forms the boundary between slipped and nonslipped areas of a crystal or that is characterized by a closure failure of the Burger's circuit. [ASTM F1241] Also called line defect. Also see slip.

dopant n : in silicon technology, a chemical element incorporated in trace amounts in a semiconductor crystal or epitaxial layer to establish its conductivity type and resistivity. [Adapted from SEMI M9-90 and M8-84] Also see conductivity type, n-type, and p-type.

dopant density n : in an uncompensated extrinsic semiconductor, the number of dopant impurity atoms per unit volume, usually given in atoms/cm 3 , although the SI unit is atoms/m 3 . Symbols: ND for donor impurities and NA for acceptor impurities. [ASTM F1241]

doping n : the addition of impurities to a semiconductor to control the electrical resistivity. [SEMI M1-94 and ASTM F1241]

drain n : one of the three major parts of a complementary metal oxide semiconductor transistor. [SEMATECH]

edge crown n : an increase of epitaxial layer thickness around the periphery of the wafer arising from differences in deposition rate. [SEMATECH]

electrostatic discharge (ESD) n 1 : a sudden electric current flow, such as between a human body and a metal oxide semiconductor semiconductor, with potential damage to the component. [SEMATECH] 2 : the transfer of electrostatic charge between bodies at different electrostatic potentials. [SEMI E33-94]

energy-dispersive X-ray spectrometer n : a detector used to determine which elements are present in a sample by analyzing X-ray fluorescence for energy levels that are characteristic of each element. [SEMATECH]

epitaxial layer n : in semiconductor technology, a layer of a single crystal semiconducting material grown on a host substrate which determines its orientation. [SEMI M2-94 and ASTM F1241]

epitaxy (epi) n : a silicon crystal layer grown on top of a silicon wafer that exhibits the same crystal structure orientation as the substrate wafer with a dissimilar doping type or concentration or both. Examples are p/p+, n/n+, n/p, and n/n. [SEMATECH] Also see epitaxial layer.

ESD : see electrostatic discharge.

etch 1 n : a category of lithographic processes that remove material from selected areas of a die. Examples are nitride etch and oxide etch. [SEMATECH] 2 : in the manufacture of silicon wafers, a solution, a mixture of solutions, or a mixture of gases that attacks the surfaces of a film or substrate, removing material either selectively or nonselectively. [SEMI M1-94 and ASTM F1241] Also see anisotropic etch, preferential etch, dry plasma etch, reactive ion etch, and wet chemical etch.

etchant n : an acid or base (in either liquid or gaseous state) used to remove unprotected areas of a wafer layer. Examples are potassium hydroxide, buffered oxide etch, and sulfur hexafluoride. [SEMATECH]

etch pit n : a pit, resulting from preferential etching, localized on the surface of a wafer at a crystal defect or stressed region. [ASTM F1241]

eutectic n : alloy or solution with components distributed in the proportions necessary to minimize the melting point. [SEMATECH] Also see azeotrope.

excessive leakage n : in the testing of semiconductors, current that is above the specified limit for the particular test being conducted. [Sandia Labs]

failure n : in failure analysis, an event where the semiconductor component does not function according to its intended use or specifications. [Sandia Labs]

failure mechanism n : in failure analysis, a fundamental process or defect responsible for a failure. [SEMATECH]

failure mode n : in failure analysis, the electrical symptoms by which a failure is observed to occur. Failure mode types include a catastrophic failure that is both sudden and complete and degraded failure that is gradual, partial, or both, as well as intermittent failures. [Sandia Labs]

failure mode and effects analysis (FMEA) n : an analytically derived identification of the conceivable semiconductor failure modes and the potential adverse effects of those modes on the system and mission. [SEMATECH]

fault n 1 : an accidental condition that causes a functional unit to fail to perform its required function. [SEMATECH] 2 : a defect-causing out-of-spec operation of an integrated circuit. [SEMATECH] Also see exception condition and defect.

fault coverage n : the percentage of a particular fault type that a test vector set will detect when applied to a chip. [1994 National Technology Roadmap for Semiconductors]

fault dictionary n : a list of faults that a test vector will detect in a failing circuit, or a list of all such faults for each vector in a vector set. [1994 National Technology Roadmap for Semiconductors]

fault model n : a model of the behavior of defective circuitry in an integrated circuit. Physical defects result in improper behavior in a circuit which must be modeled in order for test patterns to be designed to properly detect them. Examples include stuck-at model, timing model, and bridging model. [1994 National Technology Roadmap for Semiconductors]

FET : see field-effect transistor.

FIB : see focussed ion beam

field-effect transistor (FET) n : a transistor consisting of a source, gate, and drain, the action of which depends on the flow of majority carriers past the gate from the source to the drain. The flow is controlled by the transverse electric field under the gate. [SEMATECH]

fishtails n : structures, attributed to substrate defects, either above or below the surface plane after epitaxial deposition; the "tails" are aligned in a particular crystallographic direction. [ASTM F1241] Also see crescents.

fissure : see crack.

flake n : material missing from one but not the other side of a semiconductor wafer. [SEMI M10-89]

flake chip : see chip and peripheral chip.

flaking : see peeling.

flip-chip n : a leadless, monolithic structure that contains an integrated circuit designed to electrically and mechanically interconnect to a hybrid circuit. Connection is made to bump contacts covered with a conductive bonding agent on the face of the hybrid. [SEMATECH] Also called controlled collapse chip connect or C4

fluorescence n : the emission of light as the result of, and only during, the absorption of radiation of shorter wavelengths. [IEEE]

Fluorescent Microthermographic Imaging n : a failure analysis technique that uses a temperature dependent fluorescent compound and an optical pumpiing source to image temperature changes on a semiconductor device with near optical spatial resolution. [Sandia Labs]

FMEA : see failure mode and effects analysis.

FMI : see Fluorescent Microthermographic Imaging.

focussed ion beam (FIB) n : an imaging tool that can be used to deposit or etch materials on wafers. A focussed ion beam is often used in the etch mode to selectively cleave structures for failure analysis. It is also used in photomask repair for removing or adding material, as necessary, to make the photomask defect free. [SEMATECH]

Front end of line (FEOL) n 1: in semiconductor processing technology, all processes from wafer start through final contact window processing [SEMATECH].

FTIR : see Fourier transform infrared spectroscopy.

functional pattern : see pattern, functional.

functional probe n : the electronic testing of die on a wafer to determine conformance to specifications. [SEMATECH]

functional test n : one or more tests to determine whether a circuit's logic behavior is correct. [1994 National Technology Roadmap for Semiconductors]

gate n : an electrode that regulates the flow of current in a metal oxide semiconductor transistor. [SEMATECH]

gate electrode n : the electrode of a metal oxide semiconductor field effect transistor (MOSFET); it controls the flow of electrical current between the source and the drain. [SEMATECH]

gate oxide n : a thin, high-quality silicon dioxide film that separates the gate electrode of a metal oxide semiconductor transistor from the electrically conducting channel in the silicon. [SEMATECH]

glass n : a deposited film of silicon dioxide with additives to adjust coefficient of thermal expansion, color, conductivity, and melting point, generally doped with boron or phosphorus or both. [SEMATECH] Also see silicon dioxide.

groove n : in a semiconductor wafer, a shallow scratch with rounded edges that is usually the remnant of a scratch not completely removed by polishing. [SEMI M1-94 and ASTM F1241]

growth hillock : see pyramid.

hermetic seal n : a coat applied in the final stage of thermal processing to seal the ceramic package and to protect the device from the external environment. [SEMATECH]

hillock n : a defect caused by stress that raises portions of a metal (such as aluminum) film above the surface of the film. Localized stress within the metal film may elevate portions of the film through the adjacent dielectric layer, resulting in a metal extrusion and a short to the next metal layer. [SEMATECH] Also see pyramid.

hole n 1 : of a semiconductor, a mobile vacancy in the electronic valence structure that acts like a positive electron charge with positive mass; the majority carrier in p-type material. [SEMI M1-94 and ASTM F1241] 2 : in plastic and metal wafer carriers, the area through which a pin from another wafer carrier can enter for the transfer of wafers. [SEMI E1-86] Also see wafer carrier.

hot carriers n : those carriers, which may be either electrons or holes, that have been accelerated by the large traverse electric field between the source and the drain regions of a metal oxide semiconductor field-effect transistor (MOSFET). They can jeopardize the reliability of a semiconductor device when these carriers are scattered (that is, deflected) by phonons, ionized donors or acceptors, or other carriers. The scattering phenomenon can manifest itself as substrate current, gate current, or trapped charges. [SEMATECH] Also see trapped charges.

IC : see integrated circuit.

IDDQ : abbreviation for direct drain quiescent current. See static current test.

impact test n : in component testing, a test performed to determine particle contribution as a result of mechanical shock to the component. [SEMATECH] Also called particle impact noise detection or PIND

implant : see ion implantation.

impurity n : a chemical or element added to silicon to change the electrical properties of the material. [SEMATECH] Also see dopant, ion implantation.

inclusion n : discrete second phases (oxides, sulfides, carbides, intermetallic compounds) that are distributed in a metal matrix. [SEMATECH]

indent n : on a semiconductor wafer, an edge defect that extends from the front surface to the back surface. [ASTM F1241] Contrast chip.

insulator n : a substance that will not conduct electricity; for example, silicon dioxide and silicon nitride. [SEMATECH] Contrast conductor.

integrated circuit (IC) n 1 : two or more interconnected circuit elements on a single die. [SEMATECH] 2 : a fabrication technology that combines most of the components of a circuit on a single-crystal silicon wafer. [SEMI Materials, Vol. 3, Definitions for Semiconductor Materials]

interference contrast microscope n : a microscope that reveals surface details of an object in which there is no appreciable absorption by using the interference between two beams of light. [Adapted from ASTM F1241] Also called Nomarski Interference Contrast

interlevel dielectrics n : an insulating film between two conductive film layers, as between poly and aluminum or between layers of aluminum. [SEMATECH]

interstitial n : in a crystalline solid, an atom that is not located on a lattice site. [SEMATECH]

intrinsic semiconductor n : a semiconductor in which the density of electrons and holes is approximately equal. [SEMATECH] Contrast extrinsic semiconductor.

ion implantation (I 2 , II) n : a high-energy process that injects an ionized species such as boron, phosphorus, arsenic, or other ions into a semiconductor substrate. [SEMATECH]

I/O pins n : connections to an integrated circuit through which input and/or output (I/O) signals pass. [1994 National Technology Roadmap for Semiconductors]

isolation n : an electrical separation of regions of silicon on a wafer; for example, boron diffusion to isolate a transistor. [SEMATECH] Also see dielectric isolation.

junction spiking n : the penetration of a junction by aluminum, which occurs when silicon near the junction dissolves in aluminum and migrates along the interconnect lines. Aluminum then replaces silicon at the junction. [SEMATECH]

Kirkendall void n : voids induced in a diffusion couple between two metals that have different interdiffusion coefficients. [SEMATECH]

large scale integration (LSI) n : the placement of between 100 and 1000 active devices on a single die. [SEMATECH]

laser-scattering light event n : a signal pulse that exceeds a preset threshold, generated by the interaction of a laser beam with a localized light scatterer (LLS) at a wafer surface as sensed by a detector. [ASTM F1241] Also see haze.

layout n 1 : the physical geometry of a circuit or die. [1994 National Technology Roadmap for Semiconductors] 2 : the process of creating the physical geometry of a circuit or die. [1994 National Technology Roadmap for Semiconductors] 3 : see composite drawing.

LDD : see lightly doped drain.

life test n : in semiconductor reliability, a test designed to operate the semiconductor until it fails by elevating both temperature and voltage to accelerate the aging process. [Sandia Labs]

lightly doped drain (LDD) n : a metal-oxide semiconductor (MOS) device design in which the drain doping is reduced to improve breakdown voltage. [SEMATECH]

line defect : see dislocation.

LSI : see large scale integration.

metallization void n : the absence of a clad, evaporated, plated, or screen printed metal layer or braze from a designated area. [SEMI G58-94] Also called metal void.

metal void : see metallization void.

MFM-CCI : see magnetic current imaging.

microchip : see die.

moon crater n : on a semiconductor wafer, surface texture that results when a wafer floats during the initial stages of chemical polishing in a rotating cup etcher. [ASTM F1241]

motorboat : see comet.

mottled adj : pertaining to the existence on a wafer of material in a window that prevents the window from being properly opened. [SEMATECH]

mound n : on a semiconductor wafer, an irregularly shaped projection on a semiconductor wafer surface with one or more irregularly developed facets. [ASTM F1241] Contrast pyramid. Also see haze.

mouse nip n : a semicircular intrusion into a straight edge of a film or etched pattern on a wafer or reticle. [SEMATECH] Also called mouse bite.

nick : see chip.

notch n 1 : an unexpected intrusion or reduction of linewidth in patterned geometries. May also be a V-shaped intrusion into the perimeter of a wafer. The intrusion is used to align the wafer during process. [SEMATECH] 2 : on a semiconductor wafer, an intentionally fabricated indent of specified shape and dimensions oriented such that the diameter passing through the center of the notch is parallel with a specified low index crystal direction. [SEMI M1-94 and ASTM F1241]

oil canning n : in metal lid/preform assembly, lid concavity after sealing. [SEMI G53-92]

overcoat : see passivation.

oxide defect n : an area of missing oxide on the back surface of back-sealed wafers discernible to the unaided eye. [ASTM F1241]

oxide etch n : an etch process in which unprotected areas of the oxide layer are eroded by use of a chemical to expose the underlying layer. [SEMATECH]

parametric test n : wafer-level testing of discrete devices such as transistors and resistors. [SEMATECH]

parasitics n : unwanted circuit components (for example, capacitors or resistors) present in a design. [1994 National Technology Roadmap for Semiconductors]

particle n 1 : a minute quantity of solid or liquid matter. [SEMATECH] Also see dirt. 2 : in the manufacture of photolithographic pellicles, material that can be distinguished from the film, whether on the film surface or embedded in the film. [SEMI P5-94] 3 : the replating step in which a catalytic material, often a palladium or gold compound, is absorbed on a surface to act as sites for initial stages of deposition. [ASTM B374-93]

particulate 1 n : discrete particle of dirt or other material. [ASTM F1241] Also see dirt. 2 n (dust) : discrete particle of material that can usually be removed by (nonetching) cleaning. [SEMI M10-89] 3 adj : describes material in small, discrete pieces; anything that is not a fiber and has an aspect ratio of less than 3 to 1. Examples are dusts, fumes, smokes, mists, and fogs. [SEMATECH]

particulate contamination n : on a semiconductor wafer, a particle or particles on the surface of the wafer. [ASTM F1241]

passivation n : deposition of a scratch-resistant material, such as silicon nitride and/or silicon dioxide, to prevent deterioration of electronic properties caused by water, ions, and other external contaminants. The final deposition layer in processing. [SEMATECH] Also called overcoat and cap deposition.

peeling n : any separation of a plated, vacuum deposited, or clad metal layer from the base metal of a leadframe, lead, pin, heatsink, or seal ring, from an underplate, or from a refractory metal on a ceramic package. Peeling exposes the underlying material. [SEMI G61-94] Also called flaking. Contrast blister metal.

peripheral chip n 1 : crystallographic damage along the circumference of a wafer. [SEMATECH] 2 : on a wafer surface, shallow crater formed in the periphery of the specimen through conchoidal fracture and resultant spalling. [ASTM F1241] Also called flake chip or surface chip.

pinhole n 1 : minute defect or void in a film, mask, or resist, usually the result of contaminants. [SEMATECH] 2 : a small opening that extends through a covering, such as a resist coating or an oxide layer on a wafer. [SEMI P2-86]

pit n 1 : in a wafer surface, a depression in a wafer surface that has steeply sloped sides which meet the surface in a distinguishable manner, in contrast to the rounded sides of a dimple. [ASTM F1241] Also see slip and dislocation. 2 : in semiconductor packages, plastic or ceramic, or in the leadframes, a shallow depression or crater. The bottom of the depression must be visible in order for the term to apply. A pit is formed during the component manufacture. [SEMI G61-94] Contrast chip. 3 : in flat panel display substrates, a small indentation on the glass substrate surface. [SEMI D9-94]

point defect n : a localized crystal defect such as a lattice vacancy, interstitial atom, or substitutional impurity. [ASTM F1241] Contrast with localized light scatterer.

poly : see polycrystalline silicon.

polycrystalline adj : describes a form of semiconductor material made up of randomly oriented crystallites and containing large-angle grain boundaries, twin boundaries, or both. [SEMI M10-89 and ASTM F1241] Contrast single crystal. Also see amorphous silicon.

polycrystalline silicon (poly) n 1 : a nonporous form of silicon made up of randomly oriented crystallites or domains, including glassy or amorphous silicon layers. [ASTM F399-88] 2 : silicon formed by chemical vapor deposition from a silicon source gas or other methods and having a structure that contains large-angle grain boundaries, twin boundaries, or both. [SEMI M16-89] Also called poly and polysilicon. Contrast amorphous silicon and single crystal.

polysilicon (poly) : see polycrystalline silicon.

precipitate n 1 : within a silicon lattice, a region of silicon oxide frequently manifested as an etch pit. [ASTM F1241] Also see crystal lattice and pit. 2 : in a gallium arsenide wafer, a localized concentration of dopant that is insoluble. Precipitate is formed during crystal growth and during any process in which the temperature is sufficient to provide the necessary impurity mobility. [SEMI M10-89]

process-induced defect (PID) n : defect(s) added to the wafer as a result of a processing step. The PID wafer undergoes the same process sequence as a product wafer. PID wafer data is a closer approximation of actual process defect contributions than particles per wafer pass (PWP) wafer data. [SEMATECH]

pyramid n : a structure displaying |111| facets that appears on surfaces after epitaxial growth. DISCUSSION—A pyramid originates at the interface of the substrate and the epi layer and is due to various imperfections at the beginning of epi growth. [ASTM F1241] Also called growth hillock. Also see hillock and mound.

registration n 1 : the accuracy of the relative position of all functional patterns on any reticle with the corresponding patterns of any other reticle of a given device series when the reticles are properly superimposed. [ASTM F127-84] 2 : a vector quantity defined at every point on the wafer. It is the difference, R, between the vector position, P1 , of a substrate geometry and the vector position of the corresponding point, P0 , in a reference grid. [SEMATECH] 3 : in the overlay capabilities of wafer steppers, a vector quantity defined at every point on the wafer. It is the difference, R, between the vector position, P1 , of a substrate geometry, and the vector position of the corresponding point, P0 , in a reference grid. [SEMI P18-92]

residue n : any undesirable material that remains on a substrate after any process step. [ASTM F127-84 and SEMI P3-90]

root cause n 1: in failure analysis, the fundamental incident or condition that initially caused the failure to occur. [Sandia Labs]

saucer pits : see shallow etch pits.

saw-blade defect n 1 : on semiconductor wafers, a roughened area visible after polishing with a pattern characteristic of the saw blade travel. [ASTM F1241] Also see saw marks. 2 : a depression in the wafer surface made by the blade, which may not be visible before polishing. [SEMI M10-89]

saw exit chip n : in gallium arsenide technology, an edge fragment on a wafer broken off at the point at which the saw completed its cut of the wafer. A saw exit chip is typically straight or arc shaped, not irregular, and sometimes can be confused with the orientation flats. [SEMI M10-89] Contrast saw exit mark.

saw exit mark n : in silicon technology, a ragged edge at the periphery of a wafer consisting of numerous adjacent small adjoining edge chips resulting from saw blade exit. [ASTM F1241] Also see saw marks, saw exit chip.

saw-kerf : see scribe line.

saw marks n : on a wafer, surface irregularities in the form of a series of alternating ridges and depressions in arcs, the radii of which are the same as those of the saw blade used for slicing. [ASTM F1241] Also see saw exit mark.

scanning electron microscope (SEM) n : a device that displays an electronically scanned image of a die or wafer for examination on a screen or for transfer onto photographic film; displays a higher magnification than an optical microscope. [SEMATECH]

scanning tunneling microscope (STM) n : an instrument for producing surface images with atomic scale lateral resolution, in which a fine probe tip is raster scanned over the surface and the resulting tunneling current is monitored. [SEMATECH]

scratch n : on semiconductor wafers, a shallow groove or cut below the established plane of the surface, with a length to width ratio greater than 5:1. [ASTM F1241] Also see macroscratch, microscratch.

scum n : resist residue located in a window or along the foot of patterned geometry. [SEMATECH]

scumming n : residual resist located in areas that should have been cleaned in the develop operation. [SEMATECH]

SEM : see scanning electron microscope.

semiconductor n : an element that has an electrical resistivity in the range between conductors (such as aluminum) and insulators (such as silicon dioxide). Integrated circuits are typically fabricated in semiconductor materials such as silicon, germanium, or gallium arsenide. [SEMATECH]

shallow etch pits n : on a wafer, etch pits that are small and shallow in depth under high magnification greater than 200X. [ASTM F1241] Also called saucer pits. Also see haze.

short : see bridge.

silicon (Si) n : a brownish crystalline semimetal used to make the majority of semiconductor wafers. [SEMATECH]

silicon dioxide (SiO2) n : a passivation layer thermally grown or deposited on wafers. It is resistant to high temperatures. Oxygen or water vapor is used to grow silicon dioxide at temperatures above 900°C. Silicon dioxide is used as a masking layer as well as an insulator. [SEMATECH] Also called quartz. Also see glass.

silicon nitride (Si3 N4) (abbr. SiN) n : a passivation layer chemically deposited on a wafer at temperatures of between 600°C and 900°C to protect the wafer from contamination. Silicon nitride is also used as a masking layer and as an insulator. [SEMATECH]

silicon on insulator (SOI) n : a novel substrate for high-performance, low-power, and radiation-hard CMOS applications that offers process simplification, improved scalability, latch-up free and soft-error free operation, improved subthreshold slope, and drastic reduction in parasitic capacitances. At this writing, there are two manufacturing-oriented techniques to build SOI: SIMOX and bonded. Also see SIMOX and bonded. [SEMATECH]

slice : see wafer.

slip n : in semiconductor wafers, a process of plastic deformation in which one part of a crystal undergoes a shear displacement relative to another in a manner that preserves the crystallinity of each part of the material. DISCUSSION—After preferential etching, slip lines are evidenced by a pattern of one or more parallel straight lines of dislocation etch pits that do not necessarily touch each other. On |111| surface, group of lines are inclined at 60° to each other; on |100| surfaces, they are inclined at 90° to each other. [SEMI M10-89 and ASTM F1241] Also see pit.

slip line n : a step occurring at the intersection of a slip plane with the surface. [ASTM F1241]

slip plane n : the crystallographic plane on which the dislocations forming the slip move. [ASTM F1241]

small scale integration (SSI) n : the placement of between 2 and 10 active devices on a single die. [SEMATECH] Also see die.

smudge n : dense local area of contamination usually caused by handling or fingerprints. [SEMI M1-94 and ASTM F1241] Also see dirt.

snowball n : on a semiconductor wafer, a track with the appearance under magnification of a snowball rolled through snow. [ASTM F1241]

SOI : see silicon on insulator.

SOS : see silicon on sapphire.

source n : one of the three major components of a CMOS transistor. [SEMATECH]

spike n 1 : in an epitaxial wafer surface, a tall, thin dendrite or crystalline filament that often occurs at the center or recess. [ASTM F1241] 2 : an extreme structure that has a large ratio of height-to-base width and no apparent relation to epitaxial film thickness. [SEMATECH] Also see pyramid and mound.

SSI : see small scale integration.

stacking fault n : in a crystal, a two-dimensional defect caused by a deviation from the normal stacking sequence of atoms. [ASTM F1241]

stain n 1 : a solution applied to a cross-sectioned silicon device to reveal the location of various structures. [SEMATECH] 2 : contaminant in the form of streaks that are chemical in nature and cannot be removed except through further lapping or polishing. Examples are "white" stains that are seen after chemical etching as white or brown streaks. [SEMI Materials, Vol. 3, Definitions for Semiconductor Materials] 3 : a two-dimensional, contaminating foreign substance on a component surface. [SEMATECH] Also see contamination and foreign material. 4 : in flat panel display substrates, any erosion of the surface; generally cloudy in appearance, it sometimes exhibits apparent color. [SEMI D9-94] 5 : area contamination that is chemical in nature and cannot be removed except through further lapping or polishing. [ASTM F1241]

step coverage n : the ratio of thickness of film along the walls of a step to the thickness of the film at the bottom of a step. Good step coverage reduces electromigration and high-resistance pathways. [SEMATECH]

STM : see scanning tunneling microscope.

stuck-at fault n : a fault in a manufactured circuit causing an electrical node to be stuck at a logical value of 1 or a logic value of 0, independent of the input to the circuit. [1994 National Technology Roadmap for Semiconductors]

substrate n : in the manufacture of semiconductors, a wafer that is the basis for subsequent processing operations in the fabrication of semiconductor devices or circuits. [ASTM F1241]

surface chip : see peripheral chip.

surface defects n 1 : in the manufacture of silicon on sapphire (SOS) epitaxial silicon wafers, mechanical imperfections, SiO2 residual dust, and other imperfections visible on the wafer surface. Some examples of surface defects are: dimples, pits, particulates, spots, scratches, smears, hillocks, and polycrystalline regions. [SEMI M4-88] 2 : in flat panel display substrates, a marking, tearing or single line abrasion on the glass surface. [SEMI D9-94]

tester pattern generation (TPG) n : the generation of a program that runs on an integrated circuit hardware tester (integrated circuit tester). The purpose of this program is to permit test vectors to be applied to the pins of the integrated circuit, and measurements made to determine the performance of the integrated circuit. [1994 National Technology Roadmap for Semiconductors] Also called tester program generation.

test pattern : see pattern, test.

test techniques n : any methods used for the expressed purpose of testing integrated circuits. Examples include built-in self test (BIST), automatic test pattern generator (ATPG), static current test (IDDQ), and boundary scan. [1994 National Technology Roadmap for Semiconductors]

test vectors n : sequences of signals applied to the pins of an integrated circuit to determine whether the integrated circuit is performing as it was designed. [1994 National Technology Roadmap for Semiconductors]

total reflection X-ray fluorescence (TXRF) n : an analytical method usually used to characterize the level of metallic (and nonmetallic elemental) surface contamination. In TXRF, an X-ray beam excites fluorescence from the contamination that is present on a silicon surface. Since the beam is incident at grazing angles, it totally reflects from the surface, thus maximizing the signal. [SEMATECH]

trapped charges n : charges trapped either in the gate oxide or, in the case of a lightly doped drain (LDD) metal-oxide semiconductor field-effect transistor (MOSFET), in the spacer region. Trapped charges in the gate or the spacer lead to threshold voltage shift or to transconductance degradation, respectively. [SEMATECH]

TXRF : see total reflection X-ray fluorescence.

undercutting n : the lateral etching into a substrate under a resistant coating, as at the edge of a resist image. [ASTM F127-84]

unencapsulated thermal test chip n : an unpackaged, specially designed silicon die with standard test junctions that, after mounting into a package, may be used to thermally characterize that package. This technique is useful in determining the difference between various vendors' packages and package designs. [SEMATECH]

via n 1: a connection between two conducting layers above the silicon surface that is created by a different material or deposition step. [Sandia Labs]

void 1 : see dielectric void. 2 : see glass void. 3 : see metallization void.

wafer n : in semiconductor technology, a thin slice with parallel faces cut from a semiconductor crystal. [ASTM F1241] Also called a slice. Also see substrate.

well n : a localized n-type region on a p-type wafer or a p-type region on an n-type wafer. [SEMATECH]

X-ray fluorescence n 1 : the property of atoms to absorb X rays and emit light of characteristic wavelengths. [SEMATECH] 2 : a material diagnostic technique that determines the surface concentration of contaminants. [SEMATECH]

Failure Analysis Acronym List

Christopher L. Henderson
Sandia National Laboratories
Albuquerque, New Mexico

AA	atomic emission spectroscopy	OBIC	optical beam-induced current
ACS	American Chemical Society	OBRICH	optical beam-induced resistance change
AES	atomic emission spectroscopy	ppb	parts per billion
AE	atomic emission	ppm	parts per million
BC	bias contrast	ppt	parts per trillion
CIVA	charge-induced voltage alteration	PEELS	parallel electron energy-loss spectrometry
CVD	chemical vapor deposition	PEM	photon emission microscopy
DTA	differential thermal analysis	PDVC	phase-dependent voltage contrast
EDS	energy-dispersive spectrometry	PIND	particle-impact noise detection
EDXA	energy-dispersive x-ray analysis	quad	quadrupole mass spectrometer
EBIC	electron beam-induced current	RGA	residual gas analysis
EELS	electron energy-loss spectroscopy	RIE	reactive ion etching
EMP	electron probe microanalysis	RBS	Rutherford backscattering spectroscopy
EOS	electrical overstress	RBS	Robinson backscattering spectroscopy
ESCA	electron spectroscopy for chemical analysis	SAM	scanning Auger microprobe
ESD	electrostatic discharge	SAM	scanning acoustic microscopy
FIB	focused ion beam	SEI	Seeback effect imaging
FMI	fluorescent microthermographic imaging	SEM	scanning electron microscope
FTIR	Fourier transform infrared spectroscopy	SIMS	secondary ion mass spectroscopy
GAE	gas assisted etching	SLAM	scanning laser acoustic microscopy
GC	gas chromatography	SOM	scanning optical microscopy, scanning optical microscope
GC/MS	gas chromatography - mass spectroscopy	SRP	spreading resistance profiling
GPC	gas permeation chromatography	STEM	scanning transmission electron microscopy
HPLC	high-pressure liquid chromatography	STM	scanning tunneling microscope
IC	ion chromatography	TEM	transmission electron microscopy
ILD	interlevel dielectric	TDDB	time-dependent dielectric breakdown
IMMA	ion microprobe mass analysis	TDS	total dissolved solids
ICP/AES	inductively-coupled plasma - atomic emission spectroscopy	TGA	thermo-gravimetric analysis
		TIVA	thermally-induced voltage alteration
ICP/MS	inductively-coupled plasma - mass spectroscopy	TXRF	total reflection x-ray fluorescence
IR	infrared, infrared microscopy, infrared thermography	UV/VIS	ultraviolet - visible
ISS	ion scattering spectroscopy	VC	voltage contrast
LC	liquid chromatography	WDXA	wavelength-dispersive x-ray analysis
LE	light emission	WDS	wavelength-dispersive spectrometry
LEM	laser emission microprobe	XPS	x-ray photoelectron spectroscopy
LIVA	light-induced voltage alteration	XRD	x-ray diffraction
NMR	nuclear magnetic resonance	XRF	x-ray fluorescence
NAA	neutron activation analysis		

Applicable Industry Standards and References

Richard J. Ross
IBM Microelectronics Division
Essex Junction, Vermont

TECHNOLOGY DIRECTION

The National Technology Roadmap for Semiconductors: Technology Needs, 1997 Edition. Semiconductor Industry Association, 1997. www.semichips.org

FAILURE ANALYSIS TRENDS

The SEMATECH Failure Analysis Roadmap, Sematech Product Analysis Forum; Proceedings of the 21st International Symposium for Test and Failure Analysis, 1995; ASM International, 1995. www.sematech.org

INTERNATIONAL QUALITY STANDARDS

ISO 9000/1/2/3 and ISO 14000; International Standards Organization. www.iso.ch

TERMS AND DEFINITIONS

SEMATECH Official Dictionary, Rev. 5.0; www.sematech.org

INDUSTRY STANDARDS AND PUBLICATIONS

EIA/JEDEC Subcommittee 14.6 (Quality & Reliability of Solid State Devices—Failure Analysis) www.eia.org/www.jedec.org

Standards

Component Problem Analysis and Corrective Action Requirements; EIA Standard EIA-671, November 1996

Standard for Failure Analysis Report; JEDEC Standard JESD 38

Publications

Failure Mechanisms and Models for Silicon Semiconductor Devices; EIA/JEDEC Publication EIA/JEP 122, February 1996

Guidelines for Preparing Customer-Supplied Background Information Relating to a Semiconductor Device Failure Analysis; EIA/JEDEC Publication EIA/JEC 134, September 1998

Signature Analysis; EIA/JEDEC Publication EIA/JEC 136, July 1999

ISTFA Subject Index

Christopher L. Henderson
Sandia National Laboratories
Albuquerque, New Mexico

The subject index is a compilation of various topics and techniques within failure analysis that have been addressed in previous proceedings of the International Symposium for Testing and Failure Analysis (ISTFA). A few key papers from other conferences describing techniques and topics on failure analysis have also been included. Each topic word is followed by a list of papers in which it is addressed. The title of the paper is followed by the year (two-digit format), a colon, and the page number on which the paper begins.

Topic Words

accelerated testing
acoustic microscopy
alloying defects
alpha particles
analog circuits
analytical electron microscopy (AEM)
analytical pyrolysis
analysis, surface
atomic force microscope (AFM)
Auger energy spectroscopy (AES)
automated testing
backmetal
backside etching
batteries, NI-Cd
beam leads
birefringence
bolometers
bond pads
bonding wire, aluminum
bum-in testing
capacitance testing
capacitors
capacitors, ceramic
capacitors, metallized film
capacitors, tantalum
capacitors, trench
cathode ray tubes (CRT)
cathodoluminescence
ceramic resonators
charge-coupled devices (CCDs)

chemicals, purity of
CIVA
CMOS (complimentary metal oxide-silicon)
coax cable
cold-start effects
construction analysis
contacts (vias)
contamination
cordwood
corrosion
crystal oscillators
data retention failure
decapping (delidding)
decapsulation techniques
defects, crystallographic
defects, layout-dependent
defects, polysilicon
defects, surface, oxide
delamination (metals)
delamination (plastics)
delineation (decoration, staining)
die bonds
die coating
die cracking
dielectric breakdown
DPA (destructive physical analysis)
DRAM dynamic random access memory
drift
e-beam testing
EBIC--electron beam induced current
EDXA, EDS energy dispersive X-ray spectroscopy
EGA (evolved gas analysis)
electrical testing and characterization
electromechanical devices
electromigration
electroplating
emission microscopy
encapsulant material
EOS (electrical overstress)
EPROM (erasable programmable read only memory)

epoxy
equipping the FA lab
ESCA (electron spectroscopy for chemical analysis)
ESD (electrostatic discharge)
excimer lasers
expert systems
failure analysis methodology
failure mechanisms
fault detection, location, isolation
fault detection theory
ferroelectric liquid crystals
field crystallization
field failures, FFRP
filler particles
fits
flip-chip bonds
fluorescent die testing
fluorescent microthermal imaging (FMI)
focused Ion beam (FIB)
FTIR (Fourier transform infrared spectroscopy)
fuses
gallium arsenide (GaAs)
gas analysis
gate arrays
gate oxide
glass deposition
glass voiding
gold metallization
HAST (highly accelerated stress testing)
heat pumps
HEMTs
hillocks (Al)
holography
hybrid circuits
ICP-AES (inductively coupled plasma-atomic emission spectroscopy)
IDDQ (quiescent power supply current)
Image analyzers
Image processing
IMMA-ion microprobe mass analyzer
IMPATT diodes
Inclusions
Indium phosphide (InP)
Inductance testing
Infrared microscopy
Infrared thermal imaging
interconnects (contacts)
internet (World Wide Web)
ionic contamination
ion beams
ion migration
IR detectors
isolation
isolation, photoresist
isolation, probe
ISS (ion scattering spectroscopy)
lasers

laser diodes
laser microscopy
latch-up
lead crystals
LED (light-emitting diode)
leads
leak testing
leakage current
LIMS (laser ionization mass spectrometry)
liquid crystal techniques
LIVA
LMMA, LAMMA (laser microprobe mass analysis)
low temperature testing
LSM (laser scanning microscope. confocal LSM, etc)
magnetic decoration
magnetic media
marking
memories
Microelectromechanical (MEMS) Devices
MESFET (metal semiconductor field effect transistor)
metallization
metallization, stress relief In
metallography (cross-sectioning)
microanalysis-overviews
microprocessors
Micro-Raman spectroscopy
microsectioning
microwave devices
MLCC (multilayer ceramic capacitors)
MMIC (monolithic microwave integrated circuit)
moisture resistance
moisture sensors
MOS capacitor
MOSFET (metal oxide semiconductor field effect transistor)
nodules, in silicon
noise
optical beam-induced current (OBIC)
optical microscopy
optoelectronics
oxide defects
oxide trapped charge
packages, IC, ceramic
particles
passivation cracking
pattern recognition
Peltier devices
phos-glass PSG (phosphosilicate glass)
photoacoustic spectroscopy
photoemission
photoresist
photoresist; masking of specimens
PIND (particle Impact noise detection)
pinholes
planar cell capacitor
plasma-enhanced chemical vapor deposition (PECVD)
plasma etching and reactive Ion etching (RIE)
plating

accelerated testing

Voltage Measurements on Semiconductor Devices Using Auger Spectroscopy 83:57

Failure Analysis of Electroplated Nickel on Ceramic IC Packages Using Electron Beam Techniques 83:182

Microcircuit Failure Analysis Using a Scanning Auger Multiprobe 83:53

Sputter Ion Depth Profiling Limitations of Microelectronic Materials 82:8

Scanning Auger Analysis of IV Semiconductor Hetemepitaxial Interfaces 78:165

Development of Failure Analysis Techniques Including Auger Spectroscopy for CMOS/SOS LSI Devices 78:162

Analysis of Bonds and Interfaces with Auger Electron Spectroscopy 77:246

Surface Analysis by Auger Spectroscopy and ESCA 76:9

automated testing

ATE Failure Isolation Methodologies for Failure Analysis, Design Debug and Yield Enhancement 98:235

Testing-Based Failure Analysis: A Critical Component of the SIA Roadmap Vision 97:3

Test and Failure Analysis Implications of a Novel Inter-Bit Dependency in a Non-Volatile Memory 97:25

Experimental Figures for the Defect Coverage of IDDQ Vectors 97:9

A Structured Approach for Failure Analysis of a 256k BiCMOS SRAM 89:167

GRAPHATT/GSI-A Paperless Approach to VLSI RAM Failure Analysis with Automatic Wafer Stager Control from Graphic Bit-Fail Map Displays 89:37

Automating Failure Analysis Pin-to-Pin Leakage Measurements 88:227

An Automatic LCD Defect Inspection System Using an Area Image Sensor 88:189

Die Attach Screen on A.T.E. 84:147

Automated Electro-Optical Testing for High Resolution and High Performance Analog VLSI Imagers 83:257

Software Model for Parallel Test Systems 83:250

Testing of a Microprocessor with Pseudo-Random Vectors 83:247

Problems Encountered during Automated Testing of Microelectronic Devices 82:140

An Automated Device Characterizer Adds the Edge to Successful Failure Analysis 82:162

backmetal

Characterization of Multi-Layer Backmetal Systems via Multi-Technique Instrumental Analysis 87:137

Factors Resulting in Bond Delamination of Au Backmetal Si Die Eutectically Bonded to Ni/Au 83:263

backside etching

Laser Microchemical Technology: New Tools for Flip-Chip Debug and Failure Analysis 97:211

Failure Analysis of DRAM Storage Node Trench Capacitors for 0.35-Micron and Follow-On Technologies Using the Focused Ion Beam for Electrical and Physical Analysis 96:401

Practical Applications of Backside Silicon Etching 95:263

Back-Etch: An Effective Tool for Characterization and Failure Analysis of MESFET Devices 88:235

Characterization of Polyimide Defects Utilizing a Backside Etch/ Delamination Technique 86:223

Backside Etching Techniques 85:147

batteries, NI-Cd

Failure Analysis of Ni-Cd Battery Cell for Space Application 80:148

beam leads

Process Controls to Increase the Effectiveness of Beam Lead Devices in High Rel Applications 77:251

birefringence

Use of Birefringent Patterns to Determine the Strain induced in Silicon Chips by Bonding 77:36

bolometers

Reliability of Thermistor Bolometers 77:295

bond pads

Characterizing Integrated Circuit Bond Pads 92:165

bonding wire, aluminum

Failure Analysis of Bonding Wires in Power Transistor Modules 91:237

Ultrasonic Aluminum Wire Bonding and Lead-Indiurn Soldering to Gold Alloy Thick Film Conductors-Performance and Failure Mechanism 77:227

Behavior of 1 Mil Aluminum Wire at High Temperature for Hybrid Microcircuits 77:94

bum-in testing

VLSI Design Considerations for Bum-In 91:377

'FITS'-An Analytical Method of Choosing Burnin Processes 82:53

capacitance testing

The Use of Infrared Imaging and Measurements of Inductance and Capacitance for Location of Faults in Multi-Layer Ceramic Circuit Boards 92:201

capacitors

Capacitor End-Cap Dissolution Testing by Solder Reflow 93:222

Capacitors, Thermal Rating/Derating (AC-DC Operation) 80:96

capacitors, ceramic

Acoustic Visualization of Stressed Ceramic Chip Capacitors 82:215

MLCS Outperform Metallized-Film Capacitors Under Actual Operating Conditions 82:209

Identification of Ceramic Capacitor Shorts by Voltage Contrast in Scanning Electron Microscope 82:203

Extended Electrical Characterization of Ceramic Capacitors Failing Under Low-voltage Conditions 82:194

New Tools in the Failure Analysis of Multilayer Ceramic Capacitors 81:269

Solder Coating of Ceramic Capacitors; Wettabililty Problems81JU

Low Voltage Failures of Monolithic Ceramic Capacitors and Their Screening Method 81:105

Low-Voltage Failure Mechanisms for Ceramic Capacitors 81: 101

Innovative Screening for Ceramic Capacitors to Remove Failure Mechanisms 80:237

Comparison of Screening Techniques for Ceramic Capacitors 80:230

Low Voltage Screening of Ceramic Capacitors from Leakage Failures 80:225

624

Failure Analysis of a Resistance Temperature Detector (RTD) 89:395

Degradation Mechanisms of GaAs MESFET Devices in High Humidity Conditions 89:141

Failure of an Embedded Bond Wire in a CERDIP: Reliability Implications 89:133

Pin-Hole Failure Analysis of Plastic-Based MagnetoOptical Disk 89:223

Corrosion Mechanism Analysis of Salt Spray Test and S02 Test on Gold Plated Connector Contact 88:205

Highly Accelerated Stress Test (HAST) on VLSI Plastic Components 88:167

Evaluation of NiCd Cell Terminals' Durability for Alkaline 87:233

Deterioration of ZnO/SiO2 Diode Packages in High Humidity 87:207

Moisture-Resistance Test Using Unsaturated Pressure Cooker Equipment 86:189

Testing and Analysis of Photovoltaic Modules for Electrochemical Corrosion 86:39

Corrosion Mechanism of Thin Film Disk Magnetic Recording Structures 86:35

Analysis of Fractures in Half-Size Crystal Can Relay Moveable Contact Arms 84: 196

A New Accelerated Test for Moisture Resistance on IC's 84:108

Fourier Transform Infrared Spectroscopy (FTIR): Its Application to Failure Analysis 84:53

Microanalytical Characterization of Semiconductor Device Contamination Using the Laser Microprobe Mass Analysis (LAMMA) Technique 83:63

Stress Corrosion Cracking of Dual In Line Package Leads 82:346

Search for a Test Simulating Indoor Corrosion of Electrical Contacts 82:115

crystal oscillators

Test Techniques Useful in Identifying Sources of Anomalous Behavior in Quartz Crystal Filters and Oscillators 77:178

data retention failure

Filler Particle induced Data Retention Failures in Plastic Encapsulated EPROM's 92:391

decapping (delidding)

Opening Techniques for IC Ceramic Packages 83:100

decapsulation techniques

Selective Wet Etch of Silicon Nitride Passivation Layers 98:429

Mechanical/Plasma Decapsulation Method and Thermal Finite Element Analysis Provide Explanation for SMB Zener Failures 98:353

Non-Destructive Chemical Decapsulation Techniques for TBGA Package Devices 98:297

Semiconductor Device Decapsulation Using a CNC Milling machine 91:77

Plastic Mold Opener That Uses Fuming Nitric Acid as Dissolving Liquid 88:137

Chemical Decapsulation Revisited 87:113

Expedited Failure Analysis of GaAs FET Transistors, Si 111APATT Diodes and Schottky Diodes Using X-ray, SEM and EBIC Techniques 87:53

Failure Analysis Applications of Plasma, 87:35

Failure Analysis of Metallization Corrosion, ASM Conference on Electronic Packaging, 1987, p 275.

Techniques and New Etch Block Design to Enhance the Jet Etch Decapsulation 85:134

An Improved Decapsulation Technique for Plastic Encapsulated Opto Coupler Devices 85:114

Micro-Surgery as Used in Epoxy Laminate Failure Analysis 84:130

Mechanical Decap Method for Plastic Devices 84:95

Opening Techniques for IC Ceramic Packages 83:190

Decapsulation of Silicone-Epoxy Copolymer Packages 82:73

Three Decapsulation Methods for Epoxy Novolac Type Packages, IRPS-1980, p 107.

Improved Technique for Decapsulation of Epoxy Packaged Semiconductor Devices and Microcircuits, by B. Wensink, Solid State Technology, 1979, p 107.

Tough Analysis Problems That Have a Solution 78:124

Nondestructive Decapsulation of Epoxy B/Novolac Packages 78:90

defects, crystallographic

An Application of Breakthrough Failure Analysis Techniques in Eliminating Silicon Dislocation Problem in Sub-Micron CMOS Devices 97:165

A New Chemical Method of Wright Etch in the Delineation of Stacking Faults and Crystalline Defects in Fabrication Silicon Wafer Substrate 97:69

Crystallographic Imperfections and Their Effect on Micro-Electronic Performance 76:52

defects, layout-dependent

Characterization of Layout-Dependent Defects on the 16-Mb DRAM 92:117

defects, polysilicon

Novel Deprocessing Techniques to Analyze Gate Level Defects 95:275

Characterization of a Unique FSRAM Sub-Half Micron Gate Polysilicon Defect 94:341

Case History: Failure Analysis of a 16K ROM with a Polysilicon Gate Defect 93:191

Failure Analysis and Failure Mechanisms of Triple Polysilicon VLSI Devices 86:201

Failure Due to Pinholes in Polysilicon Conductors on CMOS Memory Devices 77:60

defects, surface, oxide

Advanced Methods for Imaging Gate Oxide Defects With the Atomic Force Microscope 92:267

Failure Analysis Techniques in Thin Gate Oxide Defects 92:181

Failure Analysis of Double Polysilicon Thick Interlevel Oxide Failures 82:104

Failure Analysis Technique for Examination of SNOS Type Defects 82:20

Silicon Dioxide Defect Location and Analysis on VLSI DPS Structures 81:216

Analysis of Surface Defects on Hybrid Microcircuits 79:153

delamination (metals)

Factors Resulting in Bond Delamination of Au Backmetal Si Die Eutectically Bonded to Ni/Au 83:263

delamination (plastics)

RAM Cell Defect Localization and Analysis on VLSI Structures 82:13

Development of the Method to Determine the Critical Energy for the Soft Error in Dynamic RAM 79:191

drift
Advanced Technique for Analysis of Drift Failures in PN Junction Devices 78:69

e-beam testing
Automatic Fault Tracing Using an E-Beam Tester with Reference to a Good Sample 97:243

Carbon Coating for Electron Beam Testing and Focus Ion Beam Reconfiguration 96:333

Next Generation E-Beam Probe Station 94:135

Electron Beam Probing for Multichip Modules 94:465

FACE: An Approach for Automatic IC Diagnosis by E-Beam Testing 92:15

New Developments in the Failure Analysis of Passivated CMOS ICs by Electron Beam Testing 91:389

Electron Beam Testing for Verification of Voltage Distribution on VLSI Circuits 82:149

EBIC--electron beam induced current
Microprobing and EBIC for VLSI Technology 92:43

EBIC and TEM Analyses of Piped Multi-Emitter Transistors 92:55

CMOS Characterization Using EBIC: Applications and Performance 91:199

Electron Beam Analysis of Laser Diodes and LEDs Using EBIC 89:87

EBIC Observation of Subsurface Damage 88:225

Failure Analysis of Inversion Mechanisms: Beyond Bake Recoverable 88:217

Expedited Failure Analysis of GaAs FET Transistors, Si IMPATT Diodes and Schottky Diodes Using X-ray, SEM and EBIC Techniques 87:53

Quantitative Measurement of Metallization Integrity Using EBIC 78:175

Investigation of Gate Shorts in CMOS Devices by Electron Beam Induced Current and Internal SEM Probes 81:5

Application of Scanning Electron Microscope, EBIC Mode to Semiconductor Evaluation and Failure Analysis 79:136

EDXA, EDS energy dispersive X-ray spectroscopy
Quantitative Analysis of Thin Layers Using Microanalytical Techniques 86:5

Analysis of On-Line Organic Microcontaminants in Semiconductor Assembly Plants 86:1

Failure Analysis of ESD Damage in MOSFET Power Devices 86:83

Materials Characterization by Analytical Electron Microscopy 84:46

Failure Analysis of Electroplated Nickel on Ceramic IC Packages Using Electron Beam Techniques 83:182

Reliability of SEM/EDX Analysis of P in SiO2 82:23

Non-Destructive Method to Approximate SiO2 Passivation Thickness on Fabricated Semiconductor Devices Using WDXA for Oxygen 82:6

Comparison of Minimum Detection limits using Wavelength and Energy Dispersive Spectrometers 77:49

EDXRF Energy Dispersive X-ray Fluorescence Characterization of Multi-Layer Backmetal Systems via Multi-Technique Instrumental Analysis 87:137

EGA (evolved gas analysis)
Analytical Pyrolysis and Evolved Gas Analysis of Electronic Polymers 86: 11

electrical testing and characterization
Customer and Manufacturer Analytical Tool for Solving Testhole Issues 94:273

Failure Analysis of CMOS Devices with Anomalous IDD Currents 91:381

Automated Digital Integrated Circuit Tester (ADICT) 89:297

Failure Analysis of Complex and High Pin Count Devices Using Computer Aided Electrical Characterization 89:261

Failure Mode Analysis Methodology on High Density CMOS SRAMS 89:389

Locating High Resistance Shorts in CMOS Circuits by Analyzing Supply Current Measurement Vectors 89:231

Application of Quiescent Supply Current Signature Analysis to Failure Analysis of Integrated Circuits 89:27

Nodal Waveform Analysis of Recoverable Gate Array Functional Failures 88:41

Analysis of CMOS Microprocessor Failures Using a New Electrical Signature Analysis Technique 87:1

Failure Analysis of VLSI Logic Functional Fails 86:179

A Dual Trace, Multi-Pin Microcircuit Test Adaptor for Standard Curve Tracers 86:99

A Diagnostic Technique for CMOS ICs Based on Changes in the Quiescent Power Supply Current Caused by Toggling the Inputs 86:185

Data Retention Test Method of EPROMS 85:104

Analysis of Failures in Large Arrays 85:85

Signature Analysis Technique for Defect Characterization of CMOS Static RAM Single Cell Fails 84:141

High Speed Approach to Die Bond Integrity Testing 83:254

Software Model for Parallel Test Systems 83:250

Testing of a Microprocessor with Pseudo-Random Vectors 83:247

Microprocessor Test Experience 83:244

Novel VLSI SRAM Failure Analysis Technique 83:40

Fault Isolation in LSI Circuits Using Logic Simulator Driven Computer Guided Diagnosis 83:1

Laser Die Probing for Complex CMOS 82:178

Electron Irradiation Tests on Small Signal and Power GaAs FETs 82:168

An Automated Device Characterizer Adds the Edge to Successful Failure Analysis 82:162

Internal Node Testing by Tester Aided Voltage Contrast 82:156

Electron Beam Testing for Verification of Voltage Distribution on VLSI Circuits 82:149

Problems Encountered During Automated Testing of Microelectronic Devices 82:140

An Electrical Test to Detect Mechanical Damage in Junction Transistors 81:39

Results of Testing and Programming of Fusible Link PROMs 81:50

Simulation of Stuck-Open Faults in CMOS Integrated Circuits 81:53

Electrical Testing for Failure Analysis of Memory Devices-The Role of Memory Checker 81:57

Test Pattern Generation for Large Digital Hybrid Microcircuits 80:262

EPROM (erasable programmable read only memory)

epoxy

equipping the FA lab

ESCA (electron spectroscopy for chemical analysis)

ESD (electrostatic discharge)

excimer lasers

expert systems

failure analysis methodology

failure mechanisms

fault detection, location, isolation

A Technique for Preparing TEM Cross Sections to a Specific Area Using the FIB 91:417

Imaging Dielectric Breakdown in MOS Transistors Using a Combination Transmission Electron Microscope and Focused Ion Beam 91:429

Integration of a Focused Ion Beam System in a Failure Analysis Environment 91:85

Advanced Micromachining Techniques for Failure Analysis 91:35

Failure Analysis of Micron Technology VLSI Using Focused Ion Beams 89:249

A New VLSI Diagnosis Technique: Focused Ion Beam Assisted Multi- Level Circuit Probing IRPS-1987 p 111.

FTIR (Fourier transform infrared spectroscopy)

Identifying Plastic Encapsulant Materials by Pyrolysis Infrared Spectophotometry 98:399

Application of FTIR Microspectroscopy in Contamination Analysis 92:175

An Analysis of Power Transistors with a Polyimide Die Coat 87:213

Analysis of On-Line Organic Microcontaminants in Semiconductor Assembly Plants 88:1

Analysis of On-line Organic Microcontaminants in Semiconductor Assembly Plants, 86:1

Microscopic FTIR for Failure Analysis 85:177

Fourier Transform Infrared Spectroscopy: Its Applications to Failure Analysis 84:53

FTIR Microspectrophotometry-Advances and Applications 84:43

fuses

Failure Analysis Techniques of TiW Fuse Failures on PLDs for Improved Programming Yield 91:271

Reliability of Polysilicon Fuses in Plastic Molded Devices 84:112

Study to Establish Limits for Reliable Fuse Application in Transient Environment 82:111

Incremental Failure Mechanism in Subminiature Hermetically Sealed Fuses 81:26

gallium arsenide (GaAs)

Interpretation of Life Test Results on DHBC InGaAsP/InP Laser Diodes 92:253

Storage Tests and Failure Analyses on AlGaAs/GaAs HEMTs 92:73

High Reliability of Low-Noise Self-Aligned Gate GaAs MESFET with Noise Figure of 0.85db at 12GHz 91:191

Reliability Evaluation on GaAs MMIC Components 89:335

Reliability Evaluation of GaAs MESFETs and ICs 89:329

Degradation Mechanism Due to Hot Electron Trapping in High Density CMOS DRAM 88:89

The Effects of the Passivation Film on the Reliability of High Power GaAs MESFETS 83:302

A Well Breakdown Phenomenon in High Density RAM 93:270

Reliability Study of (AlGa)As Laser Diodes for Long Life Space Applications 82:25

Reliability and Failure Mechanisms of GaAs-FETs 81:69

Microwave Power GaAS FET Failure Mechanisms 80:93

Voids in Aluminum Gate Power GaAsFET.% Under Microwave Testing 78:155

A Novel Electron Beam Technique for Studying GaAsFET%77:114

gas analysis

Analysis of Organic Gases Generated Inside a Magnetic Disk Drive 91:519

gate arrays

Failure Analysis of ESD/EOS Damage to HCMOS Gate Arrays 89:201

Photoemission Testing for EOS/ESD Failures in VLSI Devices: Advantages and Limitations 89:183

Nodal Waveform Analysis of Recoverable Gate Array Functional Failures 88:41

Fault Isolation in LSI Circuits Using Logic Simulator Driven Computer Guided Diagnosis 83:1

gate oxide

Detection of Gate Oxide Defects Using Electrochemical Wet Etching in KOH:H$_2$O 97:279

Case Study: Unique Stress Induced Gate Oxide Defects in a CMOS Analog/Digital Device Revealed by Backside Silicon Removal 96:169

Novel Deprocessing Techniques to Analyze Gate Level Defects 95:275

Advanced Methods for Imaging Gate Oxide Defects With the Atomic Force Microscope 92:267

Failure Analysis Techniques in Thin Gate Oxide Defects 92:181

A New Technique to Rapidly Identify Gate Oxide Leakage in Field Effect Semiconductors 90:331

Gate Oxide Reliability and Defect Analysis of a High Performance CMOS Technology 89:109

The Role of Failure Analysis in Problem Solving: Gate Oxide Integrity Problems 87:11

Determination of Particulate Contamination in Gate Oxide 85:48

glass deposition

TEOS Glass Deposition by PECVD, A New Failure Analysis Technique 92:141

glass voiding

Novel Microscopic Techniques in the Failure Analysis of Trimmed Ceramic Capacitors 86:131

gold metallization

Reliability investigation on Gold Metallized RF Transistors 83:180

HAST (highly accelerated stress testing)

Highly Accelerated Stress Testing on VLSI Plastic Components 88:167

heat pumps

Thermoelectric Heat Pump Failures 91:167

HEMTs

Storage Tests and Failure Analyses on AlGaAs/GaAs HEMTs 92:73

hillocks (Al)

Submicron Precision Die Cross-Sectioning 89:243

Die Attach/Die Shear and Metal Migration 83:317

Application of Infrared Microscopy in the Design and Production of High Power Semiconductor Devices 82:188

Application of Infrared Microscopy in the Design and Production of High Power Semiconductor Devices 82:188

Electron Irradiation Tests on Small Signal and Power GaAs FETs 82:168

IC Thermal Response to Transient Current Overloads 77:28

Short Pulse Thermal Measurements Using IR for Reliability Studies 77:24

Noncontacting Infrared Measurements of Operating Temperatures in Semiconducting Devices 77:21

interconnects (contacts)

Silicon Inclusion Failures in Aluminum Interconnects 83:30

internet (World Wide Web)

Realistic Database for Semiconductor Devices Analysis 98:247

Making the Most of the Internet for Failure Analysis 98:323

ionic contamination

Failure Isolation of Mobile Ions Using Secondary Ion Mass Spectroscopy 97:273

Model for Ionic Contamination Induced Failure of a Semiconductor 92:237

ion beams

VLSI Failure Analysis and Characterization Applications of Photon and Ion Beams 91:1

ion migration

Possible Causes for Ion-Migration in the Space-Use Electronic Equipments 92:93

IR detectors

Solder Seal Failure Analysis of an IR Detector 82:359

isolation

Fault Isolation in LSI Circuits Using Logic Simulator Driven Computer Guided Diagnosis 83:1

Isolation of Polysilicon for Failure Analysis 82:101

Failure Analysis Techniques Used to Physically Isolate LSI Circuits 79:177

Low-Cost LSI Device Failure isolation Equipment 76:92

isolation, photoresist

Application of image Reversal Photoresist for Failure Analysis 87:17

isolation, probe

Methods of Trace Cutting for Diagnostic Probing of Semiconductor Devices 84:136

A Laser Cutter for Failure Analysis 83:69

Microprobing IRPS-1980, p 117.

Failure Analysis Techniques Used to Physically Isolate LSI Circuits 79:177

ISS (ion scattering spectroscopy)

Applications of ISS/SIMS for Surface Analysis 76:13

lasers

Advanced Micromachining Techniques for Failure Analysis 91:35

Failure Analysis of Laser Optical Components 83:178

A Laser Cutter for Failure Analysis 83:69

laser diodes

Interpretation of Sudden Failures in Pump Laser Diodes 97:189

Interpretation of Life Test Results on DHBC In GaAsP/InP Laser Diodes 92:253

laser microscopy

Application of Laser Scanning Microscope to Analyze Forward Voltage Snapback of Compound Semiconductors 97:185

Latchup Testing of Integrated Circuits 89:285

Laser Scanner For Solar Cell Evaluation and Failure Analysis 78:16

Holographic Technique for Failure Evaluation of Piece Pail Leads Due to Vibration 76:75

latch-up

Bilateral Latch-Up Failure in CMOS ESD Protection Circuits 92:223

Advanced Photoemission Technique for Distinguishing Latch-up from Logic Failures on CMOS Devices 91:335

Latchup Testing of Integrated Circuits 89:285

lead crystals

Electrical Failures Caused by Lead Crystal Growth on Isolation Glass-A New Version of an Old Problem 92:397

LED (light-emitting diode)

Automated Long Term, Uninterrupted, LED Test Complex 78:59

leads

Failure Analysis of Electronic Component Leads 79:28

A Holographic Technique for Failure Evaluation of Piece Part Leads Due to vibration 76:75

leak testing

Failure Mechanisms for Seals on Large Packages 88:181

Helium Leak Test for Small Components 86:195

Zyglo® Penetrant Testing of Plastic Package Integrity 85:126

A Novel Technique for Evaluating a Gap between Lead and Resin for Plastic Encapsulated Devices 85:120

Fluorescent Penetration Testing for Failure Analysis on Plastic Encapsulated Integrated Circuits 82:125

leakage current

A Technique to locate leakage current on Semiconductor Devices 77:90

LIMS (laser ionization mass spectrometry)

Recent Advances in the Use of Laser Ionization Mass Spectrometry (LINIS) for Failure Analysis 84:22

liquid crystal techniques

Thermal and Optical Enhancements to Liquid Crystal Hot Spot Detection Methods 97:57

A Novel, Proactive Failure Analysis Technique Using Liquid Crystals 92:219

VLSI Chip Testing Method Using Ferroelectric Liquid Crystals 92:107

Advanced Liquid Crystal for Improved Hot Spot Detection Sensitivity 92:341

Micro-Raman spectroscopy

Micro-Raman Spectroscopy Evaluation of the Local Mechanical Stress in Shallow Tench Isolation CMOS Structures: Correlation with Defect Generation and Diode Leakage 98:11

microsectioning

A Method of Cross Sectioning Polyimide Passivated Semiconductors 98:387

Cross Sectioning with a Pivoting Sample Block 97:79

Evaluation of the Planar Grinding Stage to Optimize Electrical Component Polishing 92:363

Microstructural Preparation of Microelectronic Components and Devices Having Ceramic Substrates 91:483

Microstructural Examination for Failures in Microelectronic Components 91:471

New Cross Sectioning Techniques Using the Reactive Ion Etcher 91:88

Precise Polishing Techniques for Technology Development and Failure Analysis 91:55

Failure Analysis of Micron Technology VLSI Using Focused Ion Beams (FIB) 89:249

Failure Analysis of DRAMs 88:31

A Microfinishing Technique for Semiconductor Failure Analysis 88:161

Single Stain Delineation of Semiconductor Devices 86:95

Analysis of VLSI Metallization Microstructure by High Resolution Mechanical Cross Sectioning and Auger Analysis 85:98

Improved Microsectioning Materials andTheirApplications83:93

A Nonchemical Method of Delineating Junctions in Selected Semiconductor Devices 83:88

Novel Techniques for Analyzing Device High via Resistance 83:44

Non-encapsulated Microsectioning as a Construction and Failure Analysis Tool IRPS-1982, p 221.

Precision Diffusion Delineation in Silicon Microcircuits 81:1

An Improved Scribe and Break Method to Obtain a Semiconductor Device Cross-Section 79:143

microwave devices

Defect Analysis of MIC Substrates 83:218

Failure Mechanisms in Microwave Solid State Devices 80:193

Microwave Power GaAS FET Failure Mechanisms 80:83

MLCC (multilayer ceramic capacitors)

Advanced Failure Analysis Techniques for Multilayer Ceramic Chip Capacitors 87:25

Novel Microscopic Techniques in the Failure Analysis of Trimmed Ceramic Capacitors 86:131

Nondestructive Inspection of Encapsulated Capacitors: Flaw Growth Detection of Parts Submitted to 85% Humidity, 85 IC 95:193

Improved Microsectioning Materials and Their Applications 83:93

MLCs Outperform Metallized-Film Capacitors under Actual Operating Conditions 82;209

Extended Electrical Characterization of Ceramic Capacitors Failing under Low-Voltage Conditions 82:194

Identification of Ceramic Capacitor Shorts by Voltage Contrast in Scanning Electron Microscope 82:203

Acoustic Visualization of Stressed Ceramic Chip Capacitors 82:215

MMIC (monolithic microwave integrated circuit)

Reliability Evaluation on GaAs MMIC Components 89:335

moisture resistance

Acceleration Factor at a Pressure Cooker Test for the Surface Mount Device (SMD) 87:283

Moisture-Resistance Test Using Unsaturated Pressure Cooker Equipment 86:189

A Novel Method of Evaluating Moisture Resistance of Soldered Plastic Encapsulated LSI by a New

Ultrasonic Inspection System 86:173

Humidity Resistance of (SOP-ICs) Small Outline Package ICs 86:235

A Novel Technique for Evaluating a Gap between Lead and Resin for Plastic Encapsulated Devices 85:120

Fluorescent Penetration Testing for Failure Analysis on Plastic Encapsulated Integrated Circuits 82:125

A New Accelerated Humidity Test for Plastic Encapsulated ICs 82:65

moisture sensors

Recent Evaluations of Al2O3 in Situ Moisture Sensors for Industrial Electronic Hermetic Components Application 83:300

MOS capacitor

Investigation of Trapped Oxide Charge in MOS Devices 86:219

MOSFET (metal oxide semiconductor field effect transistor)

ESD-Induced Damage in Vertical Power MOSFETs 88:111

Simulation and Characterization of EOS and ESD Damage in MOSFET Transistors and Integrated Circuits 88:95

Failure Analysis of ESD Damage in MOSFET Power Devices 86:83

Determination of Particulate Contamination in Gate Oxide 95:48

Observed Physical Effects and Failure Analysis of EOS/ESD on MOS Devices 84:205

nodules, in silicon

Silicon Precipitate Nodule-Induced Failures of MOSFETs 91:161

noise

Noise Analysis & Potential Failure Mechanisms in Hybrid Microcircuits and Components 77:64

optical beam-induced current (OBIC)

High Spatial Resolution OBIRCH and OBIC Effects Realized by Near-Field Optical Probe in the Analysis of High Resistance 200 nm Wide TiSi Line 98:25

OBIC Endpointing Method for Laser Thinning of Flip-Chip Circuits 98:461

Novel Failure Analysis Technique "Light Induced State Transition (LIST)" Method Using an OBIC System 97:159

Failure Analysis from Back Side of Die 96:393

Various Contrasts Identifiable from the Backside of a Chip by 1.3 µm Laser Beam Scanning and Current Change Imaging 96:387

Verification and Improvement of the Optical Beam Induced Resistance Change (OBIRCH) Method 94:11

Optical Beam Induced Current Testing of Power Up Failures to Determine Root Cause 93:297

Optical Beam Induced Current (OBIC) Imaging with a Scanning Laser Acoustic Microscope (SLAM) 93:311

Novel BIC Observation Method for Detecting Defects in Al Stripes Under Current Stressing 93:303

Logic Failure Analysis of CMOS Using a Laser Probe IRPS-1984, p 69

optical microscopy
Microlenses 85:165

Innovative Dark Field Optical Techniques for VLSI Evaluation 83:8

Application of the Optical Scanner to Semiconductor Failure Analysis 78:13

Optical Techniques for Semiconductor Material and Circuit inspection 76:31

Optical Microscopic Methods for Accentuating Surface Contrast 76:18

optoelectronics
Typical Failure Mode and Mechanism of Opto-Electronic Devices 92:433

Evaluation of Optical Fibers for Space Use 87:275

An Improved Decapsulation Technique for Plastic Encapsulated Opto Coupler Devices 85:114

Facet Degradation in High Power AlGaAs Laser Diodes from -20°C and 50°C Life Tests 85:31

Failure Analysis of Laser Optical Components 83:178

Screening for Opto-Isolator Failure Modes to Improve Long Life Performance 77:294

Characterization and Analysis of Certain Types of Defects in Channel Electron Multiplier Arrays (CEMAs) 77:280

oxide defects
Gate Oxide Reliability and Defect Analysis of a High Performance CMOS Technology 89:109

Application of Quiescent Supply Current Signature Analysis to Failure Analysis of Integrated Circuits 89:27

Leakage Detection Techniques: A Comparative Study 89:5

The Role of Failure Analysis in Problem Solving: Gate Oxide Integrity Problems 87:11

Bubbles Induced under Thermally Grown SiO_2 by the Interaction of Phosphorus and Moisture in an Oxygen Plasma 86:121

Failure Analysis and Failure Mechanisms of Triple Polysilicon VLSI Devices 86:201

Determination of Particulate Contamination in Gate Oxide 85:48

Failure Analysis of Double Polysilicon Thick Interlevel Oxide Failures 82:104

oxide trapped charge
Recent Failure Analysis Applications of Secondary Ion Mass Spectrometry, SIMS 89:45

Failure Analysis of Inversion Mechanisms: Beyond Bake Recoverable 88:217

Investigation of Trapped Oxide Charge in MOS Devices 86:219

packages, IC, ceramic
Failure Analysis of Electroplated Nickel on Ceramic IC Packages Using Electron Beam Techniques 93:182

Factors Influencing Microcircuit Package Reliability 78:108

particles
Circuit Failure Due to Fine Mode Particulate Air Pollution 92:329

Techniques for Analysis of Small Particles 78:151

Coverseal Particle Detection Using X-Rays 77:88

Techniques in Post PIND Test Examination of Particle Contamination in Semiconductor Devices 77:56

passivation cracking
Stress Analysis for Passivation and Interlevel-Insulation Film Cracks in Multilayer Aluminum Structures for Plastic-Packaged LSI 87:75

Stress Analysis of Passivation Film Crack for Plastic Molded LSI Caused by Thermal Stress 83:275

pattern recognition
GRAPHPATT/GSI-A Paperless Approach to VLSI RAM Failure Analysis with Automatic Wafer Stager Control from Graphic Bit-Fail Maps 89:37

Pattern Recognition Analysis: An Aid to Component Problem Solving 85:138

Memory Array Failure Pattern Recognition 78:52

Peltier devices
Thermoelectric Heat Pump Failures 91:167

phos-glass PSG (phosphosilicate glass)
Comparative Study of Various Methods for Measuring Phosphorus Concentration in Phosphosilicate Glass 83:174

Controlled Selective Etching of Silicon Dioxide and Phossil Glass 82:80

Reliability of SEM/EDX Analysis of P in SiO_2 82:23

Geometry Differences In PSG Films After Etching with HF Allows the Determination of Phosphorous Concentration 78:137

photoacoustic spectroscopy
Photoacoustic Spectroscopy as a Tool for the Nondestructive Evaluation of Materials and Devices 86:25

photoemission
Advanced Photoemission Technique for Distinguishing Latch-up from Logic Failures on CMOS Devices 91:335

photoresist
A Study for insoluble Layer Formation of Chemically Amplified Positive Resist Using Time-of Flight Secondary Ion Mass Spectrometer 92:295

photoresist; masking of specimens
Micro-Control of Photoresist Deposition for Failure Analysis of Microelectronic Circuits 92:335

PIND (particle Impact noise detection)
Methods of Particle Retrieval from Metal Packages 80:217

PIND Testing and Its Problems 80:203

Comparative PIND Tests of Hybrid Microcircuits 80:200

Loose-Particle Detection in Micro-electronic Devices-A Review of an 1155 Task 79:1

Development of the Method to Detect Foreign Materials In IC Packages 79:20

PIND's Role as a Failure Analysis Tool 79:23

The Effectivity of PIND Testing 79:27

Techniques in Post PIND Test Examination of Particle Contamination in Semiconductor Devices 77:56

Electrochemical Voiding of Silicon-Chromium Thin Film
  Resistors: A Case History 91:157

RF transistors
Reliability Investigation on Gold Metallized RF Transistors 83:180

RGA (residual gas analysis)
Failure Mechanisms for Seals on Large Packages 88:181

An Analysis of Power Transistors with a Polyimide Die Coat
  97:213

optimization of Polyimide Die Coat Cure Cycle via Chemical
  Analysis and Electrical Characterization 87:41

RIE (reactive Ion etching)
Reactive Ion Etch Process Optimization for Failure Analysis Applications 92:33

New Cross Sectioning Techniques Using the Reactive Ion Etcher
  91:88

ROM (read only memory)
Filler Particle induced Data Retention Failures in Plastic Encapsulated EPROM's 92:391

Analysis of Failures in Large Arrays 85:85

SAM (scanning Auger multiprobe)
Microcircuit Failure Analysis Using a Scanning Auger Multiprobe
  83:53

SAM (scanning acoustic microscopy)
Scanning Acoustic Microscopy: Reliability in Processing High
  Power Semiconductor Devices 91:119

sample preparation
Techniques to Remove the C4 Die from a Ceramic Package 98:303

High Yield and High Throughput TEM Sample Preparation Using
  Focused Ion Beam Automation 98:329

A Study of Effects of Backside Thinning on Integrated Circuits
  Using a Precision Diamond Wheel Apparatus 98:441

Cold Sample Preparation Technique for Soft Solder Die Attach
  Examination 92:135

Sample Preparation Techniques for Optical and Electron Beam
  Analysis 76:62

Schottky diodes
Progressive Deterioration of Glass-to-Metal Seals in Glass Body
  Diode Packages 89:361

Expedited Failure Analysis of GaAs FET Transistors, Si IMPATT
  Diodes and Schottky Diodes Using X-ray, SEM and EBIC
  Techniques 87:53

An Automated Device Characterizer Adds the Edge to Successful
  Failure Analysis 82:162

SCR (silicon controlled rectifier)
The Interpretation of EOS Damage in Power SCRs 85:42

scratch test
Adhesion and Cohesion Testing of Hard Coatings 86:87

screening and testing
Screening Techniques for Detecting Latent Defects in Adhesive
  Bonds Associated with Microwave Hybrid Packages 88:173

Screening Design System Generated VLSI Logic Chips for Performance 84:157

Method of Estimating the Benefits of FRU Bum-in 83:290

A New Screening Method for Metal-Nitride Oxide Silicon (MNOS)
  Non-Volatile Memory IC 83:285

Substrate Voltage Bump Test for Dynamic RAM Devices 83:281

Failure Mechanism and Assurance Technique 79:105

Results of a Screening Program for Distributor-Procured Microcircuits 79:111

Screening and Testing Techniques for GaAs power MESFETs
  79:114

Study of Reliability of GaAs MESFETs 79:123

ATE, the Device Specification and Part Failure 78:63

Automated Test Design for Analog Circuits 77:17

Limitations and Attributes of Microprocessor Testing Techniques
  77:13

SEM (scanning electron microscope)
SEM Equipment Capabilities Evaluated for Sub-Half Micron Semiconductor Applications 97:329

A New Fast Signal Processing Devoted to Real Time Differential
  Imaging in Scanning Electron Microscopy 91:126

Utilization of High Resolution SEM in the Failure analysis of Bipolar Chips 91:7

Multibeam Wafer Inspection SEM 81:22

SEM Stroboscopic Techniques-Their Application to Failure
  Analysis of LSIs 80:9

SEM Techniques for the Isolation of Failures in Memory Circuits
  80:1

A Micromechanical Probe in the SEM 79:145

SEM X-Section Investigation of IC Structures 78:160

Methods to Augment SEM Metallization Inspection 78:170

Applications of the Scanning Electron Microscope in the Development of Microtechnology 77:323

Computer Controlled SEM for Image and Microtopography
  Analysis 77:108

sensors, In situ
Recent Evaluations of Al, 0, In Situ Moisture Sensors for Industrial
  Electronic Hermetic Components Application 83:300

shear test (die)
Characterization of Die Attach Integrity Using Destructive and
  Nondestructive Techniques 88:77

Factors Resulting in Bond Delamination of Au Backmetal Si Die
  Eutectically Bonded to Ni/Au 83:263

shear test (wire bond)
Thermalsonic Ball Bond Evaluation by a Bond Pluck Test 84:237

shift register
Fault Isolation of Dynamic Shift Registers Using Unorthodox Electronics or SEM Voltage Contrast 77:111

sichrome (SICr)
Electrochemical Voiding of Silicon-Chromium Thin Film
  Resistors: A Case History 91:157

SIDP (sputter ion depth profiling)

639

Sputter Ion Depth Profiling Limitations of Microelectronic
Materials 82:8

signature analysis

Signature Analysis: Statistical Models and Their Application to FA
96:183

Signature Analysis of Package Delamination Using Scanning
Acoustic Microscope 96:287

Risk Assessment in Signature Analysis 96:177

A Signature Analysis Method for IC Failure Analysis 96:189

Latch-Up Signature Analysis Technique for Plastic Dual-In-Line
Package (PDIP) Devices Using Acoustic Microscopy 95:325

EOS and ESD Laboratory Simulations and Signature Analysis of a
CMOS Programmable Logic Product 94:117

Application of Quiescent Supply CUrrent Signature Analysis to
Failure Analysis of Integrated Circuits 89:27

Analysis of CMOS Microprocessor Failures Using a New Electrical
Signature Analysis Technique 87:1

Signature Analysis Technique for Defect Characterization of
CMOS Static RAM Single Cell Fails 84:141

silicides

Transmission Electron Microscopy (TEM) in Silicon Device Technology 87:83

Auger Analysis of Refractory Metal Silicides 84:12

MeV Ion Beam Analysis for Quality Control 82:2

silicon

Silicon Precipitate Nodule-Induced Failures of MOSFETs 91:161

silicon nitride

Low Temperature Silicon Nitride protection of plastic Encapsulated
Integrated Circuits 79:185

silicon-germanium

Analysis of Epitaxial SiGe Structures via SIMS 91:17

silver migration

Silver Migration in Lead Borosilicate Glass on Soda Lime Glass
Substrate 80:153

SIMS (secondary Ion mass spectrometry)

Failure Isolation of Mobile Ions Using Secondary Ion Mass Spectroscopy 97:273

A Study for Insoluble Layer Formation of Chemically Amplified
Positive Resist Using Time-of-Flight Secondary Ion Mass
Spectrometer 92:295

Analysis of Epitaxial SiGe Structures via SIMS 91:17

Recent Failure Analysis Applications of Secondary Ion Mass Spectrometry, SIMS 89:45

Surface Analysis of Contamination in Thin Film Coatings 87:155

Role of Surface Chemistry in Packaging Failures 86:27

Modem Analytical Usage of Secondary Ion Mass Spectrometry
84:1

Investigation of Integrated Circuit Failures by Liquid Metal (SIMS)
Imaging and Multitechnique Surface Analysis 84:3

Microanalytical. Characterization of Semiconductor Device
Contamination Using Laser Microprobe Mass Analysis
Technique 83:63

Direct Lateral and In-Depth Distributional Analysis for Ionic Contaminants in Semiconductor Devices Using Secondary ion Mass
Spectrometry 79:148

Evaluation of Pt/PZT/Pt Capacitors Using SIMS 98:185

single bit failures

Single Bit Failure Analysis: Once Straight Forward is Now Challenging-A Case Study of an Advanced Microprocessor SRAM
Failure 92:79

single-event upset (upset)

Measurement of Single Event Upsets in the Laboratory 83:137

Single Event Upset Mechanisms and Predictions 83:128

SLAM (scanning laser acoustic microscope)

Scanning Laser Acoustic Microscopic (SLAM) Study of Die Attach
Integrity 85:202

Nondestructive Evaluation of Ceramic Chip Capacitors by Means
of Scanning Laser Acoustic Microscope (SLAM) 81:259

Scanning Laser Acoustic Microscope Applied to Failure Analysis
78:25

Software

The Logic Mapper 98:319

Layout Overlay Techniques to Improve Failure Analysis 98:365

solar cells

Diagnostic Test and Numerical Simulation of Solar Cells Subjected
to High Reverse Current 89:323

Influence of Cracks on GaAs Solar Cells 87:59

Testing and Analysis of Photovoltaic Modules for Electrochemical
Corrosion 86:39

A Portable, X-Y Translating Infrared Microscope for Remote
inspection of Photovoltaic Solar Arrays 80:21

solder, indium

Ultrasonic Aluminum Wire Bonding and Lead Indium Soldering to
Gold Alloy Thick Film Conductors-Performance and Failure
Mechanism 77:227

solder failure (non die-bond)

The Identification of Thermal Fatigue Testing Method of Soldered
Joints for Space Use 97:91

Investigation of Solder Joint Failure 84:201

Capacitor End-Cap Dissolution Testing by Solder Reflow 83:222

Solder Evaluation Test 83:232

Thermal Cycling Induced Cracks in PWB Solder Connections
93:204

Solder Coating of Ceramic Capacitors; Wettabililty Problems
81:111

solder heat testing

An Advanced Evaluation Method of Soldering Heat Resistance for
Ultra Thin Plastic Encapsulated LSIs 91:213

solderability

Reduced Solderability Effectiveness Due to Base Metal Oxide Formation 91:103

Solderability Problem with Solder Coated Package Leads 87:201

Investigation of Problems Resulting from Surface Mounted Device
Board Attachment 85:56

Comparative Analytical Techniques for the Determination of Ph in Solder Plated Leads 84:247

solder joints
Cold Sample Preparation Technique for Soft Solder Die Attach Examination 92:135

Correlation of Accelerated Test Failure Modes with Solder Joint Integrity and Performance 81:84

SOS (silicon-on-sapphire)
Case Study of ESD Damage and Long-Term Implications on SOS RAMs 92:213

space-level parts
Parts Technology Support to European Space Programs 78:83

specimen preparation
Evaluation of the Planar Grinding Stage to Optimize Electrical Component Polishing 92:363

spin-on glass (SOG)
Spin-On-Glass (SOG) Contamination Causing Single Via Failure 97:287

Applications of Spin on Glass to Precision Metallography of Semiconductors 92:157

SRAM (static random access memory)
Voltage Contrast Application on 1M SRAM Single Bit Failure Analysis 97:339

A Microprobing Technique for DC Characterization of Defects and Instabilities in Static RAMs 92:229

Single Bit Failure Analysis: Once Straight Forward is Now Challenging-A Case Study of an Advanced Microprocessor SRAM Failure 92:79

A Structured Approach for Failure Analysis of a 256K BiCMOS SRAM 89:167

Anisotropic Etching for Failure Analysis Applications 99:161

Gate Oxide Reliability and Defect Analysis of a High Performance CMOS Technology 89:109

Failure Mode Analysis Methodology on High Density CMOS SRAMS 89:389

Application of Quiescent Supply Current Signature Analysis to Failure Analysis of Integrated Circuits 99:27

Screening for Faults in High Density SRAMs 88:195

Failure Analysis by Dynamic Voltage Contrast Development of a Semi-Automatic System 87:67

Analysis of CMOS Microprocessor Failures Using a New Electrical Signature Analysis Technique 87:1

A Signature Analysis Technique for Defect Characterization of CMOS Static RAM Single Cell Fails 84:141

Investigation of Bipolar Microcircuit Cold Start Problem and its Impact on Reliability 84:85

A Novel VLSI SRAM Failure Analysis Technique 83:40

Reliability Evaluation of CMOS RAMs 82:133

step coverage
Analytical Techniques and Procedures to Successfully Analyze Metal Step Coverage on Semiconductors 84:124

STM (scanning tunneling microscope)

Applications of Scanning Kelvin Probe Force Microscope (SKPFM) for Failure Analysis 92:9

strain
Use of Birefringent Patterns to Determine the Strain Induced in Silicon Chips by Bonding 77:36

stress induced voiding
Stress Voids, Creep Voids, and Silicon Nodules in Custom LSI Circuits: A Case History 89:355

Causes, Failure Analysis, and Accelerated Testing of Creep Voids in Aluminum Interconnect on Integrated Circuits 87:225

surface mount devices (SOT, SOIC)
Identification of Package Defects in Plastic-Packaged Surface Mount IC's by Scanning Acoustic Microscopy 89:61

Acceleration Factor at a Pressure Cooker Test for the Surface-Mount-Device 97:293

Failure Analysis of Surface Mounted Interconnection 86:73

A Novel Method of Evaluating Moisture Resistance of Soldered Plastic Encapsulated LSI by a New Ultrasonic Inspection System 86:173

Investigation of Problems Resulting from Surface Mounted Device Board Attachment 85:56

Capacitor End-Cap Dissolution Testing by Solder Reflow 83:222

TAB (tape automated bonding)
Acoustic Microscopy: A Nondestructive Tool for Bond Evaluation on Tab Interconnections and Die Attach 84:243

tantalum capacitors
Reliability Evaluation and Failure Diagnosis of Tantalum Capacitors Using Computer-Aided Frequency Analysis and Design of Experiments 92:189

Life Estimation for Slug Tantalum Capacitor 86:213

TEM (transmission electron microscopy)
Transmission Electron Microscopy (TEM) Specimen Preparation Technique using Focused Ion Beam (FIB): Application to Material Characterization of Chemical Vapor Deposition of Tungsten (W) and Tungsten Silicides (WSI_x) 97.237

BiCMOS Die Sort Yield Improvement from Isolation of a Localized Defect Mechanism and Precision TEM Cross Section 97:321

EBIC and TEM Analyses of Piped Multi-Emitter Transistors 92:55

Specific Area TEM Analysis of ESD-Induced Silicon Melt Filaments on Output Transistors 92:27

Sample Preparation Methods for TEM Analysis of Semiconductor Devices 91:41

Comparison of the Quality of Foreign and Domestic Integrated Circuits by Transmission Electron Microscopy 89:75

A Technique for Producing Precise Large Area XTEM Samples of VLSI Devices 89:1

TEM Studies of VLSI Metallization 88:145

Identification of Nodules in Dual-Layer Aluminum Copper Metallization on a Programmable Read Only Memory (PROM) 87:251

TEM Sample Preparation Methods to Inspect Integrated Circuit Structures 87:103

Applications of Transmission Electron Microscopy in LSI Process Development 87:95

Transmission Electron Microscopy in Silicon Device Technology 87:83

IC Process Flow Analysis Using Transmission Electron Microscopy 87:91

Backside Etching Techniques 85:147

Sample Preparation for Ion Milling Sections of Semiconductor Structures for TEM Inspection 84:119

Transmission Electron Microscope Techniques Applied to Microelectronics and Failure Analysis 78:30

temperature sensors

Failure Analysis of a Resistance Temperature Detector (RTD) 89:385

The Failure Analysis of a Bimetallic Thermostat 89:369

Failure of Platinum Resistance Temperature Sensors 86:53

TEOS (tetraethylorthosilicate)

TEOS Glass Deposition by PECVD, A New Failure Analysis Technique 92:141

testing

ReTAB for Test and Failure Analysis 91:221

Test Pattern Generation for Large Digital Hybrid Microcircuits 80:262

Hybrid Die Acceptance Testing 80:250

Automated Test Program Generation and Fault Isolation for Boards and ICs 78:185

Limitations and Attributes of Microprocessor Testing Techniques 77:13

testing, high temperature

Electric Field Equivalent-High Temperature Test (EFE-HTT) for Nonpackaged Dice 78:39

test chips

A Microprocessor Test Experience 93:244

TGA (thermogravimetric analysis)

An Analysis of Power Transistors with a Polyimide Die Coat 87:213

thermal balancer

Outgassing Method Using Thermal Balancer 84:118

thermal wave imaging

Failure Analysis of HgCd Infrared Detector Arrays 84:170

Nondestructive Evaluation with Thermal Waves 83:80

Thermal-Wave Microscopy 82:92

thermoelectric devices

Thermoelectric Heat Pump Failures 91:167

thick films

The Effect of Edge Preparation on the Failure of Thin Film Network Alumina Components 91:507

Techniques for the Evaluation of Thick Film Materials 78:34

thin film resistors

A Safe-Operating Area (SOA) Concept for Pulse-Operated Thin Film Resistors 83:294

threshold shift

Degradation Mechanism Due to Hot Electron Trapping in High Density CMOS DRAM 88:89

tin whiskers

Tin Whisker Induced Failure in Vacuum 92:407

titanium-tungsten (TiW)

Failure Analysis Techniques of TiW Fuse Failures on PLDs for Improved programing Yield 91:271

traveling-wave tubes (TWTs)

Failure Mode Analysis and Life Time Estimation of Traveling Wave Tubes 80:196

tunneling current microscopy

Case History: Wafer-Level Failure Analysis of Functional ICs with Elevated IDDQ Caused by Resistive Gate Oxide Shorts 91:138

ultrasonic Inspection

A New Ultrasonic Technique for the Inspection of Semiconductor Package Diebonds; 76:86

vapor bubble test

Thermal Analysis of Microelectronic Devices Using a Hot Spot Detection System Based Upon Vapor Bubble Technology 82:43

vibration effects

Screening Techniques for Detecting Latent Defects in Adhesive Bonds Associated with Microwave Hybrid Packages 88:173

voltage contrast

Passive Voltage Contrast Technique for Rapid In-Line Characterization and Failure Isolation During Development of Deep-Submicron ASIC CMOS Technology 98:221

Voltage Contrast Application on 1M SRAM Single Bit Failure Analysis 97:339

Failure Analysis of Vendor-Produced CMOS Modules 89:99

Practical Applications of Dynamic Fault Imaging 89:269

Voltage Contrast Imaging of Passivated Devices 88:9

Dynamic Voltage Contrast SEM Failure Analysis of a 32K ROM 88:15

Electron Beam Testing of VLSI Circuits 88:21

Overview of Voltage Contrast Techniques and Applications 88:1

Failure Analysis by Dynamic Voltage Contrast Development of a Semi-Automatic System 87:67

Electron-Beam Testing and Its Application to VLSI Technology, SPIE Conference on Characterization of Very High Speed Semiconductor Devices and ICs: Critical Review of Technology, March, 1987.

A Practical VLSI Characterization and Failure Analysis System for the IC User, IRPS-1986, p 87.

Fault Contrast: A New Voltage Contrast VLSI Diagnosis Technique, IRPS-1986, p 109.

E-Beam Testing: Image & Signal Processing for the Failure Analysis of VLSI Components 85:89

LSI Failure Analysis Using an Electron-Beam Tester Directly Combined to an LSI Tester 85:78

Digital Processing and Color Coding of Voltage Contrast Images 85:63

Observation of Latch-Up Phenomena in CMOS ICs by Means of Digital Differential Voltage Contrast 84:265

Fundamentals of Electron Beam Testing of Integrated Circuits, Scanning, Vol 5, 1983, p 103

Electron Beam Testing Methods and Applications, Scanning, Vol 5, 1983, p 14

Internal Node Testing by Tester Aided Voltage Contrast 82:156

Identification of Ceramic Capacitor Shorts by Voltage Contrast in Scanning Electron Microscope 82:203

Electron Beam Testing for Verification of Voltage Distribution on VLSI Circuits 82:149

Automated Contactless SEM Testing for VLSI Development and Failure Analysis, IRPS-1982, p 163

Failure Analysis and Fault isolation of a I Kbit Schottky RAM by SEM Voltage Contrast 80:131

The Practical Implementation of Voltage Contrast as a Diagnostic Tool, IRPS-1983, P 167.

Application of SEM EBIC Mode to Semiconductor Evaluation and Failure Analysis, 79:136

Fault Isolation of Dynamic Shift Registers Using Unorthodox Electronics or SEM Voltage Contrast 77:111

WDXA (wavelength-dispersive x-ray analysis)

Non-Destructive Method to Approximate SiO_2 Passivation Thickness on Fabricated Semiconductor Devices Using Wavelength Dispersive X-Ray Analysis (WDXA) for Oxygen 82:6

Comparison of Minimum Detection limits using Wavelength and Energy Dispersive Spectrometers 77-49

welds

Weld Reliability Data on Plastic Encapsulated Solid-state Devices 80:111

Failure Analysis of EB Welded Metal Hybrid Packages 76:80

whiskers

Elimination of Whisker Growth on Tin Plated Electrodes 97:305

Degradation Mechanisms of GaAs MESFET Devices in High Humidity Conditions 89:141

Tin Whiskers on Flat Pack Lead Plating between Solder Dip and Sealing Glass 85:16

Crystalline Growth on RF Tuning Screws 85:10

Comparative Analytical Techniques for the Determination of Pb in Solder Plated Leads 84:247

wire bond

Case History of Epoxy Contaminated Wire Bond Failures on Space Shuttle Hybrids 92:61

Wire Bonding-A Closer Look 91:525

Failure of an Embedded Bond Wire in a CERDIP: Reliability Implications 89:133

Thermal Fatigue Failures of Large Scale Package Type Power Transistor Modules 89:309

Au-Al Bond Degradation Mechanism and the Analysis with Infrared Microscope 89:15

Recognition and Morphology of Ultrasonic Vibration-Induced Damage in Gold Wire Bonds 86:227

High-Temperature Long-Term Reliability Evaluation of Plastic Encapsulated LSI 85:129

Thermalsonic Ball Bond Evaluation by a Bond Pluck Test 84:237

Investigation of the Effect of Thallium on Gold/Aluminum Wire Bond Reliability 84:227

wire fracture

Failure Analysis of Copper-Nickel Alloy Wire Fracture in Electrical Relay Components 87:259

wire fusing current

Electromigration in Aluminum Bond Wires Subjected to Current Pulsing 91:183

The DC Fusing Current and Safe Operating Current of Microelectronic Bonding Wires 89:121

wire pull

A Preview of 100% Prestress Wire Bond Testing (Nondestructive and Pull Techniques Hazards and Economics) 78:106

x-ray diffraction

Failure Analysis of Shorts in Color TV Tubes 86:163

x-ray radiography

Transmitted Microfocus X-Ray Techniques 92:279

Non-Destructive Failure Analysis of ICs Using Scanning Acoustic Tomography (SCAT) and High Resolution X-Ray Microscopy (HR)CM 89:69

Characterization of Die Attach Integrity Using Destructive and Nondestructive techniques 88:77

A High Speed X-Ray Topographic Camera For Wafer Defect Analysis 78:9

Analysis of Electronic components by Micro X-Radiography 77:217

Applications of Stereo X-Radiography to Failure Analysis 77:105

Techniques for Improvement of Radiography for Electronic Parts 77:104

XPS (x-ray photoelectron spectroscopy)

Surface Analysis of Contamination in Thin Film Coatings 87:155

Non-Destructive Multitechnique Surface Analysis The Role of XPS and ISS 84:40

Using ESCA in Failure Analysis: Some Recent Applications, 84:35

Zener diodes

Degradation of Zener Diodes Caused by Alloying Irregularities 86:145